Handbook of Numerical Analysis

General Editors:

P.G. Ciarlet

Analyse Numérique, Tour 55–65
Université Pierre et Marie Curie
4 Place Jussieu
75005 PARIS, France

J.L. Lions

Collège de France
Place Marcelin Berthelot
75005 PARIS, France

ELSEVIER

Amsterdam • Lausanne • New York • Oxford • Shannon • Singapore • Tokyo

Solution of Equations in $\mathbb{R}^n$ (Part 3)

Techniques of Scientific Computing (Part 3)

Volume VII

Solution of Equations in $\mathbb{R}^n$ (Part 3)

Techniques of Scientific Computing (Part 3)

N·H

2000

ELSEVIER

Amsterdam • Lausanne • New York • Oxford • Shannon • Singapore • Tokyo

ELSEVIER SCIENCE B.V.
Sara Burgerhartstraat 25
P.O. Box 211, 1000 AE Amsterdam, The Netherlands

For information on published and forthcoming volumes URL = http://www.elsevier.nl/locate/hna

Library of Congress Catalog Card Number: 89-23314

ISBN: 0 444 50350-1

This book is printed on acid-free paper.

Printed in The Netherlands

General Preface

During the past decades, giant needs for ever more sophisticated mathematical models and increasingly complex and extensive computer simulations have arisen. In this fashion, two indissociable activities, *mathematical modeling* and *computer simulation*, have gained a major status in all aspects of science, technology, and industry.

In order that these two sciences be established on the safest possible grounds, mathematical rigor is indispensable. For this reason, two companion sciences, *Numerical Analysis* and *Scientific Software*, have emerged as essential steps for validating the mathematical models and the computer simulations that are based on them.

Numerical Analysis is here understood as the part of *Mathematics* that describes and analyzes all the numerical schemes that are used on computers; its objective consists in obtaining a clear, precise, and faithful, representation of all the "information" contained in a mathematical model; as such, it is the natural extension of more classical tools, such as analytic solutions, special transforms, functional analysis, as well as stability and asymptotic analysis.

The various volumes comprising the *Handbook of Numerical Analysis* will thoroughly cover all the major aspects of Numerical Analysis, by presenting accessible and in-depth surveys, which include the most recent trends.

More precisely, the Handbook will cover the *basic methods of Numerical Analysis*, gathered under the following general headings:

– Solution of Equations in $\mathbb{R}^n$,
– Finite Difference Methods,
– Finite Element Methods,
– Techniques of Scientific Computing,
– Optimization Theory and Systems Science.

It will also cover the *numerical solution of actual problems of contemporary interest in Applied Mathematics*, gathered under the following general headings:

– Numerical Methods for Fluids,
– Numerical Methods for Solids,
– Specific Applications.

"Specific Applications" include: Meteorology, Seismology, Petroleum Mechanics, Celestial Mechanics, etc.

Each heading is covered by several *articles*, each of which being devoted to a specialized, but to some extent "independent", topic. Each article contains a thorough description and a mathematical analysis of the various methods in actual use, whose practical performances may be illustrated by significant numerical examples.

Since the Handbook is basically expository in nature, only the most basic results are usually proved in detail, while less important, or technical, results may be only stated or commented upon (in which case specific references for their proofs are systematically provided). In the same spirit, only a "selective" bibliography is appended whenever the roughest counts indicate that the reference list of an article should comprise several thousand items if it were to be exhaustive.

Volumes are numbered by capital Roman numerals (as Vol. I, Vol. II, etc.), according to their *chronological appearance*.

Since all the articles pertaining to a given *heading* may not be simultaneously available at a given time, a given heading usually appears in more than one volume; for instance, if articles devoted to the heading "Solution of Equations in $\mathbb{R}^n$" appear in Volumes I and III, these volumes will include "Solution of Equations in $\mathbb{R}^n$ (Part 1)" and "Solution of Equations in $\mathbb{R}^n$ (Part 2)" in their respective titles. Naturally, all the headings dealt with within a given volume appear in its title; for instance, the complete title of Volume I is "Finite Difference Methods (Part 1) — Solution of Equations in $\mathbb{R}^n$ (Part 1)".

Each article is subdivided into *sections*, which are numbered consecutively throughout the article by *Arabic numerals*, as Section 1, Section 2, ..., Section 14, etc. Within a given section, *formulas, theorems, remarks, and figures*, have their own independent numberings; for instance, with Section 14, formulas are numbered consecutively as (14.1), (14.2), etc., theorems are numbered consecutively as Theorem 14.1, Theorem 14.2, etc. For the sake of clarity, the article is also subdivided into *chapters*, numbered consecutively throughout the article by *capital Roman numerals*; for instance, Chapter I comprises Sections 1 to 9, Chapter II comprises Sections 10 to 16, etc.

P.G. CIARLET
J.L. LIONS
May 1989

Contents of Volume VII

Contents of the Handbook

VOLUME VII

SOLUTION OF EQUATIONS IN $\mathbb{R}^n$ (PART 3)

TECHNIQUES OF SCIENTIFIC COMPUTING (PART 3)

Solution of Equations in $\mathbb{R}^n$ (Part 3)

Gaussian Elimination for the Solution of Linear Systems of Equations

Gérard Meurant

Commissariat à l'Energie Atomique
CEA/DIF
BP12 91680 Bruyeres-le-Chatel, France

HANDBOOK OF NUMERICAL ANALYSIS, VOL. VII
Solution of Equations in $\mathbb{R}^n$ (Part 3)
Techniques of Scientific Computing (Part 3)
edited by P.G. Ciarlet and J.L. Lions

Enseigner la vie sans la vivre était le crime de l'innocence la plus détestable

Albert Cossery
Mendiants et orgueilleux, Paris 1955

This approximately translates into:
Teaching life without living it was the crime of the most hateful innocence

Contents

Foreword

I have tried to do my best and to provide a review of the main stream of research on Gaussian elimination that was done during the last thirty years. I hope I have been fair to everybody. Anyway there are probably some mistakes and some topics or some people could have been forgotten and I apologize for that.

Notice that some of the given references are not cited in the text. We give them for completeness. Some interested readers could like to learn more on some particular points and may want to have a look at these papers.

The method of successive elimination of the unknowns for linear systems is quite old. It is stated in IFRAH [1994] that in the second century the Chinese were already solving linear systems. Many works and papers have been published since then. From STEWART [1991], we know that Lagrange was already using what we call Gaussian elimination before Gauss. However, this method is so well known under the name of Gaussian elimination that we choose to continue the tradition.

This work is dedicated to the fond memory of James H. Wilkinson (1919–1986). I had the privilege to know him quite well in 1982 when he was a professor in Stanford. He was certainly a great numerical analyst that has inspired many of us, particularly for his work about round off error analysis for Gaussian elimination but, more importantly, he was a very nice human being.

For an outline of his career and his work, see the interesting paper by Beresford Parlett in NASH [1990].

James H. Wilkinson, Stanford, 1982

This work is also dedicated to the memory of my father Georges Meurant (1921–1994).

CHAPTER I

Numerical Solution of General Linear Systems

1. Introduction to Gaussian elimination

The problem we are concerned with is obtaining the numerical solution of a linear system

$$Ax = b \tag{1.1}$$

on a computer, where A is a square nonsingular matrix (i.e. $\det(A) \neq 0$) of order n, b is a given vector and x is the vector we are looking for. The entries of A and the elements of b and x can be real or complex numbers denoted respectively by $a_{i,j}$, b_i and x_i, $i, j = 1, \dots, n$.

Of course, the solution x of Eq. (1.1) is given by

$$x = A^{-1}b,$$

where A^{-1} denotes the inverse of A. Unfortunately, in most cases, A^{-1} is not explicitly known, except for some special problems and/or for small values of n. But, as we all know, the solution can be expressed by Cramer's formulae (see, e.g., GANTMACHER [1959]):

$$x_i = \frac{1}{\det(A)} \begin{vmatrix} a_{1,1} & \cdots & a_{1,i-1} & b_1 & a_{1,i+1} & \cdots & a_{1,n} \\ a_{2,1} & \cdots & a_{2,i-1} & b_2 & a_{2,i+1} & \cdots & a_{2,n} \\ \vdots & \cdots & \vdots & \vdots & \vdots & \cdots & \vdots \\ a_{n,1} & \cdots & a_{n,i-1} & b_n & a_{n,i+1} & \cdots & a_{n,n} \end{vmatrix}, \quad i = 1, \dots, n. \tag{1.2}$$

The computation of the solution x by Eq. (1.2) requires the evaluation of $n + 1$ determinants of order n. Clearly, this implies that this method will require more than $(n + 1)!$ operations (multiplications and additions) to deliver the solution. This is far too much even for small values of n. It gives already more than $4\,10^7$ operations for $n = 10$. Today, it is well known and recognized that there are much better methods than Cramer's rule which is almost never used for solving linear systems.

There are two main classes of algorithms to obtain a solution to Eq. (1.1): iterative methods and direct methods. Iterative methods define a sequence of approximations that are expected to be closer and closer to the true solution in some given norm, stopping the iterations by using some predefined criterion, obtaining a vector which is only an approximation of the solution. Direct methods try to compute the solution doing some combinations and modifications of the equations and after a finite number of floating point operations. Of course, since computer floating point operations are only done with a certain precision, the computed solution is generally different from the exact solution even with a direct method.

The most used direct methods for general matrices belong to a class collectively known as Gaussian elimination. There are many variations around the same basic idea and we will describe some of them in the next chapters.

The basis of the method is easily explained on a small example and then, it can be extended to any value of n. The main idea is to successively eliminate the unknowns. Consider the following set of linear equations,

$$\begin{aligned} x_1 + x_2 &= 2, \\ 4x_1 + 5x_2 + 3x_3 &= 12, \\ 4x_1 + 6x_2 + 7x_3 &= 17, \end{aligned} \tag{1.3}$$

whose unique solution is $x_1 = x_2 = x_3 = 1$. Eqs. (1.3) can be written in matrix form as

$$Ax = \begin{pmatrix} 1 & 1 & 0 \\ 4 & 5 & 3 \\ 4 & 6 & 7 \end{pmatrix} x = \begin{pmatrix} 2 \\ 12 \\ 17 \end{pmatrix}. \tag{1.4}$$

The determinant of A is equal to 1. Therefore, as we said, there is a unique solution to Eq. (1.3). The first equation of Eq. (1.3) is used to express x_1 as a function of x_2,

$$x_1 = 2 - x_2. \tag{1.5}$$

Then, using Eq. (1.3), x_1 is eliminated in the second and third equations giving a new (reduced) system of order 2, involving only x_2 and x_3,

$$\begin{aligned} x_2 + 3x_3 &= 4, \\ 2x_2 + 7x_3 &= 9. \end{aligned} \tag{1.6}$$

In the second step, x_2 is expressed as a function of x_3 using the first equation of Eq. (1.6),

$$x_2 = 4 - 3x_3. \tag{1.7}$$

Using Eq. (1.7) in the last equation of (1.6), we get a linear system involving only x_3 whose solution is immediately given,

$$x_3 = 1.$$

Knowing the value of x_3, we can compute x_2 with Eq. (1.7) which gives

$$x_2 = 1.$$

Finally, Eq. (1.5) gives

$$x_3 = 1.$$

The previous elementary elimination method can be cast in a matrix framework. Eliminating x_1 from the second equation of Eq. (1.3) amounts to do a linear combination of the first two equations. This is obtained by (left) multiplying A by the matrix

$$E_{2,1} = \begin{pmatrix} 1 & 0 & 0 \\ -4 & 1 & 0 \\ 0 & 0 & 1 \end{pmatrix}.$$

Multiplying the system of Eq. (1.4) by $E_{2,1}$ replaces the second equation by the second equation minus four times the first one, leaving the two others invariant,

$$E_{2,1}A = \begin{pmatrix} 1 & 1 & 0 \\ 0 & 1 & 3 \\ 4 & 6 & 7 \end{pmatrix}.$$

Elimination of x_1 from the last equation of Eq. (1.3) is obtained by (left) multiplying by

$$E_{3,1} = \begin{pmatrix} 1 & 0 & 0 \\ 0 & 1 & 0 \\ -4 & 0 & 1 \end{pmatrix}$$

and we obtain

$$E_{3,1}(E_{2,1}A) = \begin{pmatrix} 1 & 1 & 0 \\ 0 & 1 & 3 \\ 0 & 2 & 7 \end{pmatrix}.$$

Then, all the elements of the first column of A below the diagonal element have been reduced to 0. Remark that the subsystem that corresponds to the second and third rows and columns of this matrix is precisely the same as in Eq. (1.6).

Elimination of x_2 from the third equation is obtained by (left) multiplying by

$$E_{3,2} = \begin{pmatrix} 1 & 0 & 0 \\ 0 & 1 & 0 \\ 0 & -2 & 1 \end{pmatrix}$$

and

$$E_{3,2}(E_{3,1}E_{2,1}A) = \begin{pmatrix} 1 & 1 & 0 \\ 0 & 1 & 3 \\ 0 & 0 & 1 \end{pmatrix}.$$

Therefore, when A is successively (left) multiplied by $E_{2,1}$, $E_{3,1}$ and $E_{3,2}$, it is reduced to an upper triangular form. Notice that $E_{3,1}$ and $E_{2,1}$ commute, since we have

$$\begin{pmatrix} 1 & 0 & 0 \\ 0 & 1 & 0 \\ \beta & 0 & 1 \end{pmatrix}\begin{pmatrix} 1 & 0 & 0 \\ \alpha & 1 & 0 \\ 0 & 0 & 1 \end{pmatrix} = \begin{pmatrix} 1 & 0 & 0 \\ \alpha & 1 & 0 \\ 0 & 0 & 1 \end{pmatrix}\begin{pmatrix} 1 & 0 & 0 \\ 0 & 1 & 0 \\ \beta & 0 & 1 \end{pmatrix} = \begin{pmatrix} 1 & 0 & 0 \\ \alpha & 1 & 0 \\ \beta & 0 & 1 \end{pmatrix}.$$

Remark that the first column of the product is the "superposition" of the first columns of $E_{3,1}$ and $E_{2,1}$. Matrices $E_{2,1}$, $E_{3,1}$ and $E_{3,2}$ are nonsingular as their determinants are equal to 1. Therefore, the linear system of Eq. (1.4) can be transformed into the equivalent system,

$$E_{3,2}E_{3,1}E_{2,1}Ax = E_{3,2}E_{3,1}E_{2,1}b, \tag{1.8}$$

where

$$b = \begin{pmatrix} 2 \\ 12 \\ 17 \end{pmatrix}.$$

Let

$$L^{-1} = E_{3,2}E_{3,1}E_{2,1} = \begin{pmatrix} 1 & 0 & 0 \\ -4 & 1 & 0 \\ 4 & -2 & 1 \end{pmatrix}.$$

Equation (1.8) reduces to

$$Ux = \begin{pmatrix} 1 & 1 & 0 \\ 0 & 1 & 3 \\ 0 & 0 & 1 \end{pmatrix} x = L^{-1}b = \begin{pmatrix} 2 \\ 4 \\ 1 \end{pmatrix}. \tag{1.9}$$

Since the matrix U of Eq. (1.9) is upper triangular, this system which is equivalent to Eq. (1.4), is easily solved starting with the last equation and moving backwards. This process is usually called backward (or back) substitution.

Besides solving the linear system, we have also seen that

$$L^{-1}A = U.$$

Therefore,

$$A = LU.$$

L^{-1} being lower triangular, its inverse L is also lower triangular. The matrix A has been factored into the product of a lower and an upper triangular matrices. The matrix L is easily obtained from L^{-1}. First, we verify that if

$$E_{3,1}E_{2,1} = \begin{pmatrix} 1 & 0 & 0 \\ \alpha & 1 & 0 \\ \beta & 0 & 1 \end{pmatrix},$$

then

$$E_{2,1}^{-1}E_{3,1}^{-1} = \begin{pmatrix} 1 & 0 & 0 \\ -\alpha & 1 & 0 \\ -\beta & 0 & 1 \end{pmatrix}.$$

Moreover, if

$$E_{3,2} = \begin{pmatrix} 1 & 0 & 0 \\ 0 & 1 & 0 \\ 0 & \gamma & 1 \end{pmatrix},$$

then

$$E_{3,2}^{-1} = \begin{pmatrix} 1 & 0 & 0 \\ 0 & 1 & 0 \\ 0 & -\gamma & 1 \end{pmatrix},$$

and, hence

$$L = E_{2,1}^{-1}E_{3,1}^{-1}E_{3,2}^{-1} = \begin{pmatrix} 1 & 0 & 0 \\ -\alpha & 1 & 0 \\ -\beta & 0 & 1 \end{pmatrix}\begin{pmatrix} 1 & 0 & 0 \\ 0 & 1 & 0 \\ 0 & -\gamma & 1 \end{pmatrix} = \begin{pmatrix} 1 & 0 & 0 \\ -\alpha & 1 & 0 \\ -\beta & -\gamma & 1 \end{pmatrix}.$$

L is obtained straightforwardly from the elementary matrices $E_{i,j}$ by simply changing the signs of the nonzero off diagonal elements.

Of course, not all systems are as simple as the previous example. Remark that:

(1) in this example, we were able to work only with integers. In real life problems, we have to work with the computer representations of real (or complex) numbers. We will look later on at the perturbation and stability problems related to this fact. This sometimes can make the computed solution to greatly differ from the exact solution.

(2) not all systems have an LU factorization, even if they are nonsingular. Consider, e.g., a system with the following matrix,

$$A' = \begin{pmatrix} 1 & 1 & 0 \\ 4 & 4 & 3 \\ 4 & 6 & 7 \end{pmatrix},$$

which differs only from A by the $(2, 2)$ coefficient. If we use the same right-hand side as in Eq. (1.4), the first reduced system is

$$0x_2 + 3x_3 = 4,$$

$$2x_2 + 7x_3 = 9.$$

The coefficient of x_2 in the first equation is 0. So, we cannot use this equation to express x_2 as a function of the other unknowns. Of course, the solution is to use another equation and this corresponds to applying a permutation to the matrix A'. In this simple example, we directly get $x_3 = 4/3$ and then, we compute x_2 using the last equation.

As we shall see, the general result for all nonsingular matrices is that there exists a permutation matrix P such that,

$$PA = LU,$$

where L is lower triangular and U is upper triangular.

2. Examples of problems

An important source of problems which require the solution of linear systems is the numerical solution of elliptic and parabolic partial differential equations (PDEs). These equations (or systems of equations) modeling some physical phenomena are discretized with different kind of methods like finite differences, finite elements or spectral methods. Much in favor today are the finite element methods. However, finite differences or finite volume methods are still in use, particularly in computational fluid dynamics. All these discretization schemes lead to some large linear (and sometimes nonlinear) systems of equations. Generally, the matrices involved in these systems contain many zero entries. They are called sparse systems.

Of course, PDEs are not the only source of sparse matrices. Many examples are given in the book by DUFF, ERISMAN and REID [1986] like, e.g., power networks, problems in chemistry, etc....

However, solving large dense (with no or a few zero entries) linear systems still occur. An example is given by boundary integral methods. A problem that is of great importance today is solving the Maxwell equations for computing Radar Cross Sections giving the response of some objects to incoming radar waves. This problem is posed in an unbounded domain surrounding the object to study. Therefore, one way to get back to a bounded region is to write down an integral equation on the surface of the object which is discretized with finite elements and gives rise to a large dense linear system. Spectral methods also need to solve dense linear systems. Other examples of dense systems are given by EDELMAN [1993].

It is generally thought that solving linear systems is one of the most important operations in scientific computing since most physical problems lead to linear systems. At least, this is probably where most of the computer time is spent for scientific computing.

3. The general form of Gaussian elimination

In this section, we generalize the method we have introduced on a small example in the first Section of this chapter. First, we describe the method without permutations, exhibiting the necessary and sufficient conditions for a matrix to have an LU factorization. Then, we will introduce permutations to handle the general case.

Gaussian elimination without permutations

Let

$$A = \begin{pmatrix} a_{1,1} & \dots & a_{1,n} \\ \vdots & \vdots & \vdots \\ a_{n,1} & \dots & a_{n,n} \end{pmatrix},$$

and b be given. The problem to be solved is the linear system

$$Ax = b.$$

The first step of the algorithm is the elimination of x_1 into the equations 2 to n. This is done through $n - 1$ steps. Suppose that $a_{1,1} \neq 0$, $a_{1,1}$ is called the first pivot. To eliminate x_1 from the second equation, we (left) multiply A by

$$E_{2,1} = \begin{pmatrix} 1 & & & & \\ -\frac{a_{2,1}}{a_{1,1}} & 1 & & & \\ 0 & 0 & 1 & & \\ \vdots & \vdots & \ddots & \ddots & \\ 0 & \dots & \dots & 0 & 1 \end{pmatrix}.$$

More generally, to eliminate x_1 from the ith equation, we (left) multiply by

$$E_{i,1} = \begin{pmatrix} 1 & & & & & & & \\ 0 & 1 & & & & & & \\ \vdots & & \ddots & & & & & \\ 0 & & & \ddots & & & & \\ -\frac{a_{i,1}}{a_{1,1}} & 0 & \dots & 0 & 1 & & & \\ 0 & & & & & \ddots & & \\ \vdots & & & & & & \ddots & \\ 0 & \dots & & & & \dots & 0 & 1 \end{pmatrix},$$

the nonzero terms of the first column being in positions $(1, 1)$ and $(i, 1)$.

LEMMA 3.1. *Let $j > i$, then*

$$E_{i,1}E_{j,1} = E_{j,1}E_{i,1} = \begin{pmatrix} 1 & & & & & & & & & & & \\ 0 & 1 & & & & & & & & & & \\ \vdots & & \ddots & & & & & & & & & \\ 0 & & & \ddots & & & & & & & & \\ -\frac{a_{i,1}}{a_{1,1}} & & & & \ddots & & & & & & & \\ 0 & & & & & \ddots & & & & & & \\ \vdots & & & & & & \ddots & & & & & \\ 0 & & & & & & & \ddots & & & & \\ -\frac{a_{j,1}}{a_{1,1}} & & & & & & & & \ddots & & & \\ 0 & & & & & & & & & \ddots & & \\ \vdots & & & & & & & & & & \ddots & \\ 0 & & & & & & & & & & & 1 \end{pmatrix}$$

PROOF. Straightforward matrix multiply. □

Let $L_1 = E_{n,1}E_{n-1,1}\cdots E_{2,1}$ and $A_2 = L_1 A$. We denote the elements of A_2 by $a_{i,j}^{(2)}$. A_2 has the following structure,

$$A_2 = \begin{pmatrix} a_{1,1} & x & \dots & x \\ 0 & x & \dots & x \\ \vdots & \vdots & & \vdots \\ 0 & x & \dots & x \end{pmatrix},$$

where the x's correspond to (possibly) nonzero elements that are defined in the following lemma.

LEMMA 3.2.

$$a_{i,j}^{(2)} = a_{i,j} - \frac{a_{i,1}a_{1,j}}{a_{1,1}}, \quad 2 \leqslant i \leqslant n,\ 1 \leqslant j \leqslant n,$$

$$a_{1,j}^{(2)} = a_{1,j}, \quad 1 \leqslant j \leqslant n.$$

PROOF. This is just the result of the multiplication by L_1. □

Now, we describe the kth step of the algorithm. Suppose we have zeroed the elements below the diagonal in the $k-1$ first columns and let A_k be the matrix that has been obtained,

$$A_k = \begin{pmatrix} a_{1,1}^{(k)} & \cdots & \cdots & \cdots & \cdots & a_{1,n}^{(k)} \\ & \ddots & & & & \vdots \\ & & a_{k,k}^{(k)} & \cdots & \cdots & a_{k,n}^{(k)} \\ & & a_{k+1,k}^{(k)} & \cdots & \cdots & a_{k+1,n}^{(k)} \\ & & \vdots & \vdots & \vdots & \vdots \\ & & a_{n,k}^{(k)} & \cdots & \cdots & a_{n,n}^{(k)} \end{pmatrix}$$

and $a_{k,k}^{(k)} \neq 0$, $a_{k,k}^{(k)}$ is called the kth pivot.

For further references, we define a quantity g_A which is called the growth factor,

$$g_A = \frac{\max_{i,j,k} |a_{i,j}^{(k)}|}{\|A\|_\infty}.$$

Sometimes, g_A is defined as

$$g_A = \frac{\max_{i,j,k} |a_{i,j}^{(k)}|}{\max_{i,j} |a_{i,j}|}.$$

Let

$$E_{i,k} = \begin{pmatrix} 1 & & & & & & & \\ & \ddots & & & & & & \\ & & 1 & & & & & \\ & & 0 & 1 & & & & \\ & & \vdots & & \ddots & & & \\ & & 0 & & & \ddots & & \\ & & -\frac{a_{i,k}^{(k)}}{a_{k,k}^{(k)}} & & & & \ddots & \\ & & 0 & & & & & \ddots \\ & & \vdots & & & & & & \ddots \\ & & 0 & & & & & & & 1 \end{pmatrix}, \quad i > k,$$

where the element $a_{i,k}^{(k)}/a_{k,k}^{(k)}$ is in row i and column k. The nondiagonal elements that are not explicitly given are 0. Let $L_k = E_{n,k}E_{n-1,k}\cdots E_{k+1,k}$ and $A_{k+1} = L_k A_k$.

LEMMA 3.3.

$$L_k = \begin{pmatrix} 1 & & & & & \\ & \ddots & & & & \\ & & 1 & & & \\ & & -\frac{a_{k+1,k}^{(k)}}{a_{k,k}^{(k)}} & 1 & & \\ & & \vdots & & \ddots & \\ & & -\frac{a_{n,k}^{(k)}}{a_{k,k}^{(k)}} & & & 1 \end{pmatrix}$$

and A_{k+1} has the following structure

$$A_{k+1} = \begin{pmatrix} a_{1,1}^{(k+1)} & \cdots & \cdots & \cdots & \cdots & a_{1,n}^{(k+1)} \\ & \ddots & & & & \vdots \\ & & a_{k,k}^{(k+1)} & \cdots & \cdots & a_{k,n}^{(k+1)} \\ & & 0 & a_{k+1,k+1}^{(k+1)} & \cdots & a_{k+1,n}^{(k+1)} \\ & & \vdots & \vdots & & \vdots \\ & & 0 & a_{n,k+1}^{(k+1)} & \cdots & a_{n,n}^{(k+1)} \end{pmatrix}$$

PROOF. Straightforward. □

As for the first column, the elements of the jth column are given by similar expressions.

LEMMA 3.4.

$$a_{i,j}^{(k+1)} = a_{i,j}^{(k)} - \frac{a_{i,k}^{(k)} a_{k,j}^{(k)}}{a_{k,k}^{(k)}}, \quad k+1 \leqslant i \leqslant n,\ k \leqslant j \leqslant n,$$

$$a_{i,j}^{(k+1)} = a_{i,j}^{(k)}, \quad 1 \leqslant i \leqslant k,\ 1 \leqslant j \leqslant n$$
$$\text{and } k+1 \leqslant i \leqslant n,\ 1 \leqslant j \leqslant k-1.$$

PROOF. This is the formula we obviously get when multiplying by L_k. □

LEMMA 3.5. *L_k is nonsingular and*

$$L_k^{-1} = \begin{pmatrix} 1 & & & & & \\ & \ddots & & & & \\ & & 1 & & & \\ & & \frac{a_{k+1,k}^{(k)}}{a_{k,k}^{(k)}} & 1 & & \\ & & \vdots & & \ddots & \\ & & \frac{a_{n,k}^{(k)}}{a_{k,k}^{(k)}} & & & 1 \end{pmatrix}.$$

PROOF. Let

$$l_k = \begin{pmatrix} 0 \\ \vdots \\ 0 \\ \frac{a_{k+1,k}^{(k)}}{a_{k,k}^{(k)}} \\ \vdots \\ \frac{a_{n,k}^{(k)}}{a_{k,k}^{(k)}} \end{pmatrix}.$$

Clearly $L_k = (I - l_k e_k^{\mathrm{T}})$ where $e_k = (0 \quad \dots \quad 0 \quad 1 \quad 0 \quad \dots \quad 0)^{\mathrm{T}}$ with the 1 in the kth position. Then, we verify that

$$L_k^{-1} = I + l_k e_k^{\mathrm{T}},$$

because,

$$\begin{aligned} L_k^{-1} L_k &= \bigl(I + l_k e_k^{\mathrm{T}}\bigr)\bigl(I - l_k e_k^{\mathrm{T}}\bigr), \\ &= I - l_k e_k^{\mathrm{T}} l_k e_k^{\mathrm{T}}, \end{aligned}$$

and $e_k^{\mathrm{T}} l_k = 0$. □

The previous algorithm is summarized in the following lemma.

LEMMA 3.6. *If for all k, $1 \leqslant k \leqslant n-1$, $a_{k,k}^{(k)} \neq 0$, then there exists a factorization*

$$A = LU,$$

where L is lower triangular and has a unit diagonal and U is upper triangular.

PROOF. The elimination process goes as follows,

$$\begin{aligned} A_1 &= A, \\ A_2 &= L_1 A_1, \\ &\vdots \\ A_n &= L_{n-1} A_{n-1}. \end{aligned}$$

A_n is upper triangular and is therefore denoted by U. Hence,

$$L_{n-1} L_{n-2} \cdots L_1 A = U.$$

The matrices L_i, $1 \leqslant i \leqslant n-1$, are nonsingular. Then,

$$A = \left(L_1^{-1} \cdots L_{n-1}^{-1}\right) U.$$

The product of unit lower triangular matrices is unit lower triangular. Hence,

$$L = L_1^{-1} \cdots L_{n-1}^{-1}$$

is a lower triangular matrix with a unit diagonal. Moreover, it is easy to see that

$$L = \begin{pmatrix} 1 & & & & & \\ \frac{a_{2,1}^{(1)}}{a_{1,1}^{(1)}} & \ddots & & & & \\ \vdots & \ddots & 1 & & & \\ \vdots & & \frac{a_{k+1,k}^{(k)}}{a_{k,k}^{(k)}} & 1 & & \\ \vdots & & \vdots & \ddots & \ddots & \\ \frac{a_{n,1}^{(1)}}{a_{1,1}^{(1)}} & \cdots & \frac{a_{n,k}^{(k)}}{a_{k,k}^{(k)}} & \cdots & \frac{a_{n,n-1}^{(n-1)}}{a_{n-1,n-1}^{(n-1)}} & 1 \end{pmatrix}.$$

Since $L_i^{-1} L_{i+1}^{-1} = I + l_i e_i^{\mathrm{T}} + l_{i+1} e_{i+1}^{\mathrm{T}}$, we have

$$L = L_1^{-1} \cdots L_{n-1}^{-1} = I + l_1 e_1^{\mathrm{T}} + \cdots + l_{n-1} e_{n-1}^{\mathrm{T}}. \qquad \square$$

LEMMA 3.7. *If the factorization $A = LU$ exists, it is unique.*

PROOF. Suppose there exist two such factorizations, $A = L_1 U_1 = L_2 U_2$, then

$$L_2^{-1} L_1 = U_2 U_1^{-1}.$$

The matrix on the left-hand side is lower triangular with a unit diagonal and the matrix on the right-hand side is upper triangular. Therefore,

$$L_2^{-1} L_1 = U_2 U_1^{-1} = I$$

and the decomposition is unique. □

Now, we derive the conditions under which there exists an LU factorization.

THEOREM 3.1. *A nonsingular matrix A has a unique LU factorization if and only if all the principal minors of A are nonzero. That is*

$$A\begin{pmatrix} 1 & 2 & \dots & k \\ 1 & 2 & \dots & k \end{pmatrix} \neq 0, \quad k = 1, \dots, n,$$

where

$$A\begin{pmatrix} i_1 & i_2 & \dots & i_p \\ k_1 & k_2 & \dots & k_p \end{pmatrix} = \begin{vmatrix} a_{i_1,k_1} & a_{i_1,k_2} & \dots & a_{i_1,k_p} \\ a_{i_2,k_1} & a_{i_2,k_2} & \dots & a_{i_2,k_p} \\ \vdots & \vdots & & \vdots \\ a_{i_p,k_1} & a_{i_p,k_2} & \dots & a_{i_p,k_p} \end{vmatrix}.$$

Moreover,

$$a_{k,k}^{(k)} = \frac{A\begin{pmatrix} 1 & 2 & \dots & k \\ 1 & 2 & \dots & k \end{pmatrix}}{A\begin{pmatrix} 1 & 2 & \dots & k-1 \\ 1 & 2 & \dots & k-1 \end{pmatrix}}$$

and

$$a_{k,j}^{(k)} = \frac{A\begin{pmatrix} 1 & 2 & \dots & k-1 & k \\ 1 & 2 & \dots & k-1 & j \end{pmatrix}}{A\begin{pmatrix} 1 & 2 & \dots & k-1 \\ 1 & 2 & \dots & k-1 \end{pmatrix}}, \qquad j > k.$$

PROOF. Suppose that there exists an LU factorization. We know from the proof of Lemma 3.6 that

$$A_{k+1} = L_k \cdots L_1 A.$$

Therefore,

$$A = L_1^{-1} \dots L_k^{-1} A_{k+1},$$

and we have also

$$A = LU.$$

We can write these matrices in block form:

$$A = \begin{pmatrix} A_{1,1} & A_{1,2} \\ A_{2,1} & A_{2,2} \end{pmatrix}, \qquad L = \begin{pmatrix} L_{1,1} & 0 \\ L_{2,1} & L_{2,2} \end{pmatrix}, \qquad U = \begin{pmatrix} U_{1,1} & U_{1,2} \\ 0 & U_{2,2} \end{pmatrix},$$

and from Lemma 3.6,

$$L_1^{-1} \dots L_k^{-1} = \begin{pmatrix} L_{1,1} & 0 \\ L_{2,1} & I \end{pmatrix}, \qquad A_{k+1} = \begin{pmatrix} U_{1,1} & U_{1,2} \\ 0 & W_{2,2} \end{pmatrix},$$

where all the matrices in position $(1, 1)$ are square of order k. By identification, we have

$$\begin{aligned} A_{1,1} &= L_{1,1}U_{1,1}, \\ A_{2,2} &= L_{2,1}U_{1,2} + L_{2,2}U_{2,2}, \\ A_{2,2} &= L_{2,1}U_{1,2} + W_{2,2}. \end{aligned}$$

Therefore, $L_{1,1}U_{1,1}$ is the LU factorization of the leading principal submatrix of order k of A and $L_{2,2}U_{2,2}$ is the factorization of the matrix $W_{2,2}$ in the bottom right-hand corner of A_{k+1} since we have $W_{2,2} = L_{2,2}U_{2,2}$.

Notice that $\det(A) = \det(A_{k+1})$. We have $\det(L_{1,1}) = 1$ and $\det(A_{1,1}) = \det(U_{1,1})$. $U_{1,1}$ being upper triangular, its determinant is equal to the product of the diagonal elements. Therefore, for all k,

$$\det(A_{1,1}) = a_{1,1}^{(1)} \cdots a_{k,k}^{(k)}.$$

This shows that the principal minors are nonzero and the first formula. Now, we proceed in the same way as in WILKINSON [1965]. We have

$$\begin{pmatrix} A_{1,1} \\ A_{2,2} \end{pmatrix} = \begin{pmatrix} L_{1,1} \\ L_{2,1} \end{pmatrix} U_{1,1}.$$

Let $A_{1,1}^i$ denotes the matrix formed by the first $k-1$ rows and the ith row of the first k columns of A and let $L_{1,1}^i$ be defined in a similar way, then

$$A_{1,1}^i = L_{1,1}^i U_{1,1}.$$

It is easy to see that $L_{1,1}^i$ is triangular and

$$\det(L_{1,1}^i) = l_{i,k} = \frac{a_{i,k}^{(k)}}{a_{k,k}^{(k)}}.$$

Therefore, as $\det(A^i_{1,1}) = l_{i,k} \det(U_{1,1})$,

$$l_{i,k} = \frac{\det(A^i_{1,1})}{\det(A_{1,1})} = \frac{A\begin{pmatrix} 1 & 2 & \cdots & k-1 & i \\ 1 & 2 & \cdots & k-1 & k \end{pmatrix}}{A\begin{pmatrix} 1 & 2 & \cdots & k \\ 1 & 2 & \cdots & k \end{pmatrix}}.$$

Similarly, we have

$$(A_{1,1} \quad A_{1,2}) = L_{1,1}\,(U_{1,1} \quad U_{1,2}).$$

This leads to

$$u_{k,i} = \frac{A\begin{pmatrix} 1 & 2 & \cdots & k-1 & k \\ 1 & 2 & \cdots & k-1 & i \end{pmatrix}}{A\begin{pmatrix} 1 & 2 & \cdots & k-1 \\ 1 & 2 & \cdots & k-1 \end{pmatrix}},$$

for the elements of U. The converse of the proof is easily derived by induction. □

Gaussian elimination with permutations (partial pivoting)

We now allow for possible zero pivots at each step. If the first pivot $a_{1,1}$ is zero, we permute the first row with a row p such that $a_{p,1} \neq 0$. This is always possible since $\det(A) \neq 0$. This is done by left multiplication of A by a permutation matrix P_1. P_1 is equal to the identity matrix except that rows 1 and p have been exchanged,

$$P_1 = \begin{pmatrix} 0 & 0 & \dots & 0 & 1 & 0 & \dots & 0 \\ 0 & 1 & 0 & \dots & 0 & \dots & & \vdots \\ \vdots & \ddots & \ddots & \ddots & \vdots & & & \\ 0 & \dots & 0 & 1 & 0 & \dots & & \\ 1 & 0 & \dots & 0 & 0 & 0 & \dots & \\ 0 & \dots & & \dots & 0 & 1 & \ddots & \vdots \\ \vdots & & & & \vdots & \ddots & \ddots & 0 \\ 0 & \dots & & \dots & 0 & \dots & 0 & 1 \end{pmatrix}.$$

Notice that $P_1^{-1} = P_1$. On the permuted matrix, the algorithm is the same as without permutations. We construct L_1 such that $A_2 = L_1 P_1 A$.

Let us describe the kth step. The main difference with what we saw before is that the pivot can possibly be zero. If this is the case, it is possible to find a row p such that $a^{(k)}_{p,k} \neq 0$. The reason for this being that $\det(A_k) = \det(A) \neq 0$ (as seen from the proof of Theorem 3.1) and the determinant $\det(A_k)$ is equal to the product of the first $k-1$ (nonzero) pivots and the determinant of the matrix in the right-hand bottom corner.

Therefore, this matrix is nonsingular. In fact, we choose the nonzero element which has the maximum modulus. This strategy of choosing the pivot in the kth column is called partial pivoting.

Then, we multiply A_k by the corresponding permutation matrix P_k and apply the elimination algorithm,

$$A_{k+1} = L_k P_k A_k.$$

So, finally, we have

$$U = L_{n-1} P_{n-1} \cdots L_2 P_2 L_1 P_1 A.$$

It may seem that we have lost the good properties of the Gaussian algorithm as permutation matrices come in, even if some of them are equal to the identity matrix. However, we have the following result.

LEMMA 3.8. *Let P_p be a permutation matrix representing the permutation between indices p and $q > p$ then, $\forall k < p$*

$$L_k P_p = P_p L'_k,$$

where L'_k is deduced from L_k by the permutation of entries in rows p and q in column k.

PROOF. Recall that $P_p^{-1} = P_p$ and $L_k = I - l_k e_k^{\mathrm{T}}$,

$$L'_k = P_p L_k P_p = P_p\big(I - l_k e_k^{\mathrm{T}}\big) P_p = I - P_p l_k e_k^{\mathrm{T}} P_p.$$

Since $p > k$, $P_p e^k = e^k$, therefore

$$L'_k = I - l'_k e_k^{\mathrm{T}},$$

where $l'_k = P_p l_k$. Notice that the same is true for L_k^{-1} since $L_k^{-1} = I + l_k e_k^{\mathrm{T}}$. □

Now, we have the main result of this section.

THEOREM 3.2. *Let A be a nonsingular matrix. Then, there exists a permutation matrix P such that*

$$PA = LU,$$

where L is lower triangular with a unit diagonal and U is upper triangular.

PROOF. We have seen that

$$A = P_1 L_1^{-1} P_2 \cdots P_{n-1} L_{n-1}^{-1} U.$$

Then, we have from Lemma 3.8,

$$A = P_1 P_2 \cdots P_{n-1} \left(L_1''\right)^{-1} \cdots \left(L_{n-1}''\right)^{-1} U,$$

where

$$\left(L_k''\right)^{-1} = P_{n-1} \cdots P_{k+1} L_k^{-1} P_{k+1} \cdots P_{n-1},$$

corresponding to a permutation of the coefficients of column k. □

We notice that, by definition, in the factorization of Theorem 3.2, we have $|l_{i,j}| \leqslant 1$ since we have chosen the pivot as the element of maximum modulus.

Once the factorization has been obtained and the permutation matrix P is known (usually stored as a vector of indices as row permutations are not explicitly done during the factorization), the linear system

$$Ax = b,$$

is transformed into

$$PAx = LUx = Pb,$$

and is solved as

$$\begin{aligned} Ly &= Pb, \\ Ux &= y, \end{aligned}$$

by two triangular solves.

Gaussian elimination with other pivoting strategies

Pivoting strategies different from partial pivoting can be used. We can search for the pivot not only in the lower part of the kth column but in all the current submatrix, the pivot being the element that realize $\max_{i,j} |a_{i,j}^{(k)}|$. This is called complete pivoting. Then, we have to introduce not only row permutations but also column permutations. This is obtained by multiplying from the right by a permutation matrix. Finally, we obtain two permutation matrices P and Q such that

$$PAQ = LU.$$

The solution of the linear system is obtained through

$$\begin{aligned} Ly &= Pb, \\ Uz &= y, \\ x &= Q^{\mathrm{T}} z. \end{aligned}$$

We will see later that complete pivoting have some advantages regarding stability. However, the cost of finding the pivot is much larger than for partial pivoting.

Another strategy called rook's pivoting has been introduced by POOLE and NEAL [1992]. At the kth step, the algorithm is the following. Let

$$r_1 = \min\{r \mid |a_{r,k}^{(k)}| \geqslant |a_{i,k}^{(k)}|,\ k \leqslant i \leqslant n\}$$

and

$$c_1 = \min\{c \mid |a_{r_1,c}^{(k)}| \geqslant |a_{r_1,j}^{(k)}|,\ k \leqslant j \leqslant n\}.$$

If $c_1 = k$, then $a_{r_1,k}^{(k)}$ is the selected pivot. If $c_1 \neq k$, column c_1 is searched for the entry with maximum modulus. Let

$$r_2 = \min\{r \mid |a_{r,c_1}^{(k)}| \geqslant |a_{i,c_1}^{(k)}|,\ k \leqslant i \leqslant n\}$$

and

$$c_2 = \min\{c \mid |a_{r_2,c}^{(k)}| \geqslant |a_{r_2,j}^{(k)}|,\ k \leqslant j \leqslant n\}$$

and so on.

Therefore, rook's pivoting searches for coefficients of maximum modulus in rows, then columns and then, rows and columns until an entry $a_{r,c}^{(k)}$ verifies $|a_{r,c}^{(k)}| \geqslant |a_{i,c}^{(k)}|,\ k \leqslant i \leqslant n$, and $|a_{r,c}^{(k)}| \geqslant |a_{r,j}^{(k)}|,\ k \leqslant j \leqslant n$. Numerical experiments using this strategy are given in NEAL and POOLE [1992].

Operation counts

We are interested in computing (approximately) the number of floating point operations that must be done to obtain the LU factorization of A.

For computing the kth column of L, we need 1 division by the pivot and $n-k$ multiplications. To compute the updated matrix A_{k+1}, we need (after having computed the multipliers $-a_{i,k}^{(k)}/a_{k,k}^{(k)}$ which are the elements of L) $(n-k)^2$ additions and the same number of multiplications. To get the total number of operations, we sum these numbers from 1 to $n-1$

$$\sum_{k=1}^{n-1}(n-k) = n(n-1) - \tfrac{1}{2}n(n-1) = \tfrac{1}{2}n(n-1),$$

$$\begin{aligned}\sum_{k=1}^{n-1}(n-k)^2 &= n^2\sum_{k=1}^{n-1}1 - 2n\sum_{k=1}^{n-1}k + \sum_{k=1}^{n-1}k^2 \\ &= n^2(n-1) - n^2(n-1) + \frac{(n-1)^3}{3} + \frac{(n-1)^2}{2} + \frac{n-1}{6} \\ &= \tfrac{1}{3}n(n-1)\big(n-\tfrac{1}{2}\big).\end{aligned}$$

THEOREM 3.3. *To obtain the factorization*

$$PA = LU$$

of Theorem 3.1, *we need*

$$\frac{2n^3}{3} - \frac{n^2}{2} - \frac{n}{6}$$

floating point operations (multiplies and adds) and $n-1$ *divisions. The solutions of the triangular systems to obtain the solution* x *give* $n(n-1)$ *floating point operations for* L *and* $n(n-1)+n$ *for* U.

4. Gaussian elimination for symmetric systems

The factorization of symmetric matrices is an important special case that we are going to consider in more details. Let us specialize the algorithm of Section 3 to the symmetric case. We are looking for a factorization

$$A = LDL^{\mathrm{T}},$$

where L is lower triangular with a unit diagonal and D is diagonal. There are several possibilities to obtain this factorization. We are going to study three different algorithms that will lead to several ways of programming the factorization.

The outer product algorithm

The first method to construct the LDL^{T} factorization is in the same way as we have seen for general systems, columns by columns. Suppose $a_{1,1} \neq 0$, let

$$L_1 = \begin{pmatrix} 1 & 0 \\ l_1 & I \end{pmatrix}, \qquad D_1 = \begin{pmatrix} a_{1,1} & 0 \\ 0 & A_2 \end{pmatrix},$$

and

$$A = \begin{pmatrix} a_{1,1} & a_1^{\mathrm{T}} \\ a_1 & B_1 \end{pmatrix} = L_1 D_1 L_1^{\mathrm{T}}.$$

By identification, we obtain

$$l_1 = \frac{a_1}{a_{1,1}},$$

$$A_2 = B_1 - \frac{1}{a_{1,1}} a_1 a_1^{\mathrm{T}} = B_1 - a_{1,1} l_1 l_1^{\mathrm{T}}.$$

Notice that A_2 is a symmetric matrix. If we suppose that the $(1,1)$ element of A_2 is nonzero, we can reiterate this and we write

$$\begin{aligned} A_2 &= \begin{pmatrix} a_{2,2}^{(2)} & a_2^{\mathrm{T}} \\ a_2 & B_2 \end{pmatrix} = \begin{pmatrix} 1 & 0 \\ l_2 & I \end{pmatrix} \begin{pmatrix} a_{2,2}^{(2)} & 0 \\ 0 & A_3 \end{pmatrix} \begin{pmatrix} 1 & l_2^{\mathrm{T}} \\ 0 & I \end{pmatrix}, \\ l_2 &= \frac{a_2}{a_{2,2}^{(2)}}, \\ A_3 &= B_2 - \frac{1}{a_{2,2}^{(2)}} a_2 a_2^{\mathrm{T}} = B_2 - a_{2,2}^{(2)} l_2 l_2^{\mathrm{T}}. \end{aligned}$$

We remark that if we denote,

$$L_2 = \begin{pmatrix} 1 & 0 & 0 \\ 0 & 1 & 0 \\ 0 & l_2 & I \end{pmatrix},$$

then,

$$D_1 = \begin{pmatrix} a_{1,1} & 0 \\ 0 & A_2 \end{pmatrix} = L_2 \begin{pmatrix} a_{1,1} & 0 & 0 \\ 0 & a_{2,2}^{(2)} & 0 \\ 0 & 0 & A_3 \end{pmatrix} L_2^{\mathrm{T}} = L_2 D_2 L_2^{\mathrm{T}}.$$

Therefore, we have

$$A = L_1 L_2 D_2 L_2^{\mathrm{T}} L_1^{\mathrm{T}}.$$

We notice that

$$L_1 L_2 = \begin{pmatrix} 1 & 0 \\ l_1 & \begin{pmatrix} 1 & 0 \\ l_2 & I \end{pmatrix} \end{pmatrix}.$$

If all the pivots are nonzero, we can go on and at the last step, we have

$$A = A_1 = L_1 L_2 \cdots L_{n-1} D L_{n-1}^{\mathrm{T}} \cdots L_1^{\mathrm{T}} = L D L^{\mathrm{T}},$$

where L is unit lower triangular and D is diagonal.

There is a variant of this algorithm where a decomposition

$$A = \bar{L} \bar{D}^{-1} \bar{L}^{\mathrm{T}}$$

is obtained with $\bar{L}$ being lower triangular, $\bar{D}$ diagonal and $\operatorname{diag}(\bar{L}) = \operatorname{diag}(\bar{D})$. We can obtain this variant from the first algorithm by writing

$$A = L D L^{\mathrm{T}} = (LD) D^{-1} \big(D L^{\mathrm{T}}\big),$$

and $\bar{D} = D$, $\bar{L} = LD$.

In the previous algorithm, the matrix L is obtained column by column. This method is called the outer product form of the algorithm since, at each step, an outer product aa^{T} is involved.

The bordering algorithm

We partition the matrix A in a different way as

$$A = \begin{pmatrix} C_n & a_n \\ a_n^{\mathrm{T}} & a_{n,n} \end{pmatrix},$$

with obvious notations. Suppose that C_n has already been factored as

$$C_n = L_{n-1} D_{n-1} L_{n-1}^{\mathrm{T}},$$

with L_{n-1} unit lower triangular and D_{n-1} diagonal. We can write,

$$A = \begin{pmatrix} L_{n-1} & 0 \\ l_n^{\mathrm{T}} & 1 \end{pmatrix} \begin{pmatrix} D_{n-1} & 0 \\ 0 & d_{n,n} \end{pmatrix} \begin{pmatrix} L_{n-1}^{\mathrm{T}} & l_n \\ 0 & 1 \end{pmatrix}.$$

By identification,

$$\begin{aligned} l_n &= D_{n-1}^{-1} L_{n-1}^{-1} a_n, \\ d_{n,n} &= a_{n,n} - l_n^{\mathrm{T}} D_{n-1} l_n. \end{aligned}$$

Therefore, by induction we can start with the decomposition of the 1×1 matrix $a_{1,1}$, keeping on adding rows and obtaining at each step the factorization of an enlarged matrix. The only operation we have to perform at each step is solving a triangular system. To be able to proceed to the next step, we need the diagonal entries of D_n (i.e. $d_{n,n}$) to be nonzero. For obvious reasons, this method is called the bordering form of the algorithm.

The inner product algorithm

Another way to compute the factorization is simply to write down the formulas for the matrix product

$$A = LDL^{\mathrm{T}}.$$

Suppose $i \geqslant j$, we have

$$a_{i,j} = \sum_{k=1}^{j} l_{i,k} l_{j,k} d_{k,k}.$$

If we set $i = j$ in this formula as $l_{i,i} = 1$, we obtain

$$d_{j,j} = a_{j,j} - \sum_{k=1}^{j-1} (l_{j,k})^2 d_{k,k},$$

and for $i > j$,

$$l_{i,j} = \frac{1}{d_{j,j}} \left(a_{i,j} - \sum_{k=1}^{j-1} l_{i,k} l_{j,k} d_{k,k} \right).$$

Since, locally, we have to consider the product of the transpose of a vector times a vector, this method is called the inner (or scalar) product form of the algorithm.

Considering the number of floating point operations, these three variants all need about $1/2$ of the number of operations for the general algorithm, i.e. about $n^3/6$ multiplies and the same number of adds.

Programming the factorization algorithms

We are considering the different ways of coding the three algorithms that we have described for general (dense) symmetric matrices. Then, we will discuss the consequences of these different implementations, depending on the computer architecture.

The codes are written in Matlab-like language, although for clarity, we do not always use the most compact and efficient Matlab constructs.

We consider first the outer product algorithm. In this variant the matrix L is constructed column by column. At step k, the column k is constructed by multiplying by the inverse of the pivot and then, the columns at the right of column k are modified using the values of the entries of column k. This is summarized in Fig. 4.1.

The modification of columns $k + 1$ to n can be done by rows or by columns and this leads to the two codes given below.

For clarity, we store the matrix D in a vector denoted by d and L in a separate matrix although in practice, it can be stored in the lower triangular part of A (if A is not to

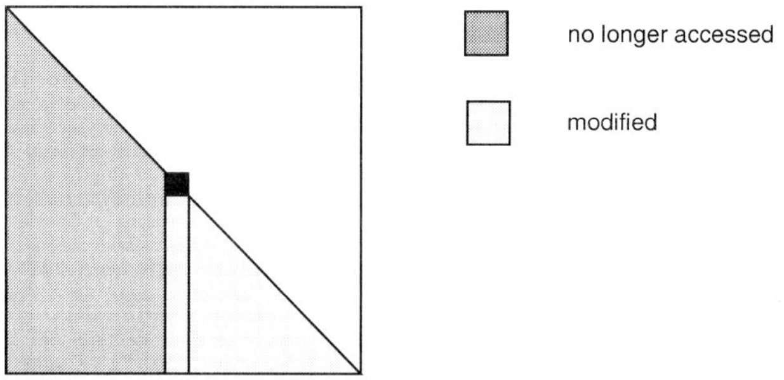

FIG. 4.1. The outer product algorithm data layout.

be saved). `temp` is a temporary vector whose use can sometimes be avoided. We use it mainly for the clarity of the presentation. Notice that the codings are different from the ones in DONGARRA, GUSTAVSON and KARP [1984]. They are arranged such that in the main loop, `i` is a row index, `j` is a column index and `k` can be both.

The strictly lower triangular part of L is initialized to that of A by

```
function [l]=init(a)
[m,n]=size(a)
for i=1:n
  for j=1:i-1
    l(i,j)=a(i,j)
  end
end
```

OUTER PRODUCT kij ALGORITHM

```
function [l,d]=kij(a)
[m,n]=size(a)
d=zeros(n,1)
temp=zeros(n,1)
d(1)=a(1,1)
l(1,1)=1
for k=1:n-1
  dki=1./d(k)
  for i=k+1:n
    temp(i)=l(i,k)*dki
  end
  for i=k+1:n
    for j=k+1:i
      l(i,j)=l(i,j)-temp(i)*l(j,k)
    end
  end
  for i=k+1:n
    l(i,k)=temp(i)
  end
  d(k+1)=l(k+1,k+1)
  l(k+1,k+1)=1
end
```

To reflect the way the three loops are nested, this algorithm is called the kij form. We can get rid of the temporary vector `temp` by using the upper part of the matrix `l`, leading to the following code. However, we think the coding is clearer using `temp`.

```
function [l,d]=kijbis(a)
[m,n]=size(a)
d=zeros(n,1)
temp=zeros(n,1)
```

```
l=init(a)
d(1)=a(1,1)
l(1,1)=1
for k=1:n-1
  dki=1/d(k)
  for i=k+1:n
    l(k,i)=l(i,k)
    l(i,k)=l(i,k)*dki
  end
  for i=k+1:n
    for j=k+1:i
      l(i,j)=l(i,j)-l(i,k)*l(k,j)
    end
  end
  for i=k+1:n
    l(k,i)=0.
  end
  d(k+1)=l(k+1,k+1)
  l(k+1,k+1)=1;
end
```

Modifying by rows, we get

OUTER PRODUCT kji ALGORITHM

```
function [l,d]=kji(a)
[m,n]=size(a)
d=zeros(n,1)
temp=zeros(n,1)
l=init(a)
d(1)=a(1,1)
l(1,1)=1
for k=1:n-1
  dki=1/d(k)
  for i=k+1:n
    temp(i)=l(i,k)*dki
  end
  for j=k+1:n
    for i=j:n
      l(i,j)=l(i,j)-temp(i)*l(j,k)
    end
  end
  for i=k+1:n
    l(i,k)=temp(i)
  end
  d(k+1)=l(k+1,k+1)
```

```
    l(k+1,k+1)=1
  end
```

Now, we turn ourselves to the bordering algorithm whose data accesses are summarized in Fig. 4.2.

For each row i, we have to solve a triangular system. There are two algorithms to do this. One is column oriented, the other is row oriented.

BORDERING ijk ALGORITHM

```
function [l,d]=ijk(a)
[m,n]=size(a)
d=zeros(n,1)
temp=zeros(n,1)
l=init(a)
d(1)=a(1,1)
l(1,1)=1.
for i=2:n
  for k=1:i
    temp(k)=a(i,k)
  end
  for j=1:i
    if j  ~= i
      l(i,j)=temp(j)/d(j)
    end
    for k=j+1:i
      temp(k)=temp(k)-l(k,j)*temp(j)
    end
  end
  d(i)=temp(i)
  l(i,i)=1
end
```

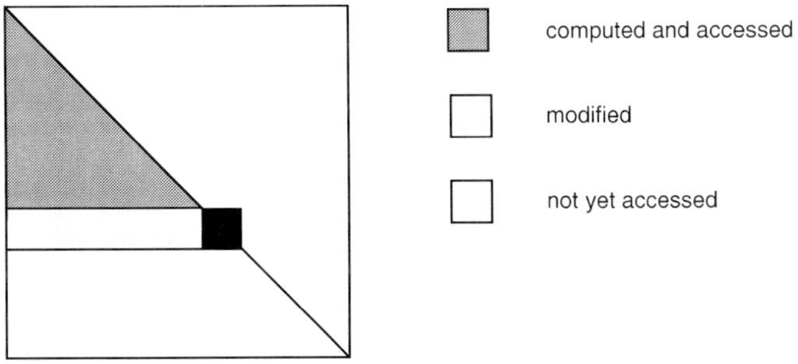

FIG. 4.2. The bordering algorithm data layout.

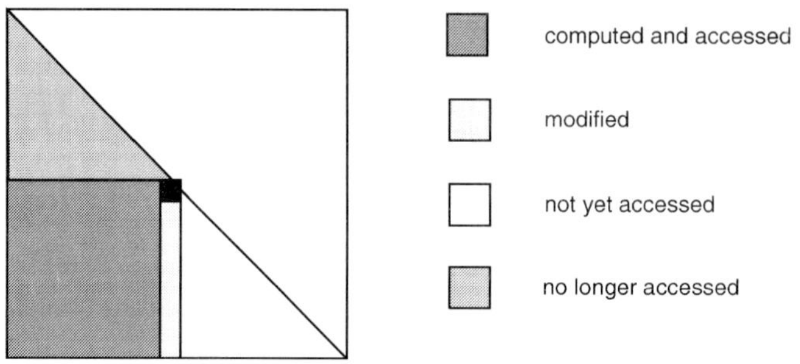

FIG. 4.3. The inner product algorithm data layout.

Notice there are too many divisions in the previous coding since the divisions are in a k loop. They can be avoided by storing the inverses of d as they are computed.

BORDERING *ikj* ALGORITHM

```
function [l,d]=ikj(a)
[m,n]=size(a)
d=zeros(n,1)
temp=zeros(n,1)
l=init(a)
d(1)=a(1,1)
l(1,1)=1
for i=2:n
  for k=1:i
    temp(k)=a(i,k)
  end
  for k=1:i
    for j=1:k-1
      temp(k)=temp(k)-temp(j)*l(k,j)
    end
    if k  ~= i
      l(i,k)=temp(k)/d(k)
    else
      d(i)=temp(i)
      l(i,i)=1
    end
  end
end
```

Finally, we consider the inner product algorithm. This algorithm is schematically depicted in Fig. 4.3.

INNER PRODUCT *jik* ALGORITHM

```
function [l,d]=jik(a)
```

```
[m,n]=size(a)
l=init(a)
for j=1:n
  for k=1:j-1
    l(j,k)=l(j,k)/d(k)
  end
  d(j)=a(j,j)
  for k=1:j-1
    d(j)=d(j)-l(j,k) ^ 2*d(k)
  end
  for i=j+1:n
    for k=1:j-1
      l(i,j)=l(i,j)-l(i,k)*l(j,k)
    end
  end
  l(j,j)=1.
end
```

In the computation of $a_{i,j} - \sum_{k=1}^{j-1} l_{i,k} l_{j,k} d_{k,k}$, one can compute $a_{i,j} - l_{i,k} l_{j,k} d_{k,k}$ for a fixed value of k looping on i provided that the divide by $d_{j,j}$ is done afterwards. Then, we obtain the following algorithm.

INNER PRODUCT jki ALGORITHM

```
function [l,d]=jki(a)
[m,n]=size(a)
l=init(a)
for j=1:n
  for k=1:j-1
    l(j,k)=l(j,k)/d(k)
  end
  d(j)=a(j,j)
  for k=1:j-1
    d(j)=d(j)-l(j,k) ^ 2*d(k)
  end
  for k=1:j-1
    for i=j+1:n
      l(i,j)=l(i,j)-l(i,k)*l(j,k)
    end
  end
  l(j,j)=1.
end
```

We have obtained six different ways to code the LDL^{T} factorization of a symmetric matrix A. The same can be done for the LU factorization of a nonsymmetric matrix. Of course, we are interested in knowing what is the best implementation, that is the one giving the smallest computing time. Unfortunately, this is dependent on the computer architecture and also on the languages that are used for coding and executing the algorithm. It also depends on the data structure that is chosen for storing L, as for today computers the performance depends mainly on the way the data is accessed in the computer memory.

Consider first L to be stored in a two-dimensional array or in the lower triangular part of A. In the Fortran language, which is up to now the most widely used for scientific computing, two-dimensional arrays are stored by columns, that is, consecutive elements in a column of an array have consecutive memory addresses. Therefore, to avoid memory access conflicts, it is much better to use algorithms that access data by columns. This could be different for other languages. For instance, in C two-dimensional arrays are stored by rows. Moreover, in computers with data caches, it pays to do operations on data in consecutive memory locations. This increases the cache hit ratio as the data is moved into the cache by blocks of consecutive addresses.

The data access is by columns for algorithms kji and jki, by rows for ikj and jik and by rows and columns for kij and ijk. This favors algorithms kji and jki.

Form kji accesses the data by columns and the basic operation involved is a so called SAXPY (for single precision a times x plus y), i.e.

$$y = y + \alpha x,$$

where x and y are vectors (columns of L) and α is scalar. Notice that the vector y has to be stored after it is computed. This particular form was used in the famous LINPACK package (BUNCH, DONGARRA, MOLER and STEWART [1979]).

Form jki also accesses that data by columns and the basic operation is also a SAXPY. However, the same column (j) is successively accessed many times. This is called a generalized SAXPY or GAXPY. This can sometimes be exploited when coding in assembly language (one can keep the data in registers). These algorithms are analyzed for a vector computer in DONGARRA, GUSTAVSON and KARP [1984], their notations being slightly different. On this type of vector architectures, the GAXPY jki form is generally the best one.

If L is not stored in the lower triangular part of A, it is better to store it in a one-dimensional array of dimension $n(n-1)/2$. Consecutive elements can be chosen by rows or columns. If consecutive elements are chosen by rows, it is better to use algorithms ikj and jik since the data accesses are going to be in consecutive addresses. kji and jki forms will be chosen if the data is stored by columns.

As an example, we consider the form jki, when the matrix is stored by columns. The matrix L is stored in a one-dimensional array ll. To simplify a little bit the coding, we explicitly store the ones on the diagonal. Therefore, we have,

$$l(i, j) = ll(k),$$

with $k = n(j-1) + \frac{j-j^2}{2} + i$.

```
function [ll,d]=lljki(a)
[m,n]=size(a)
```

```
d=zeros(n,1)
ll=initll(a)
for j=1:n
  nj=n*(j-1)++(j-j ^ 2)/2
  for k=1:j-1
    nk=n*(k-1)+(k-k ^ 2)/2
    ll(nk+j)=ll(nk+j)/d(k)
  end
  d(j)=a(j,j)
  for k=1:j-1
    nk=n*(k-1)+(k-k ^ 2)/2
    d(j)=d(j)-ll(nk+j) ^ 2*d(k)
  end
  for k=1:j-1
    nk=n*(k-1)+(k-k ^ 2)/2
    for i=j+1:n
      ll(nj+i)=ll(nj+i)-ll(nk+i)*ll(nk+j)
    end
  end
  ll(nj+j)=1.
end
```

So far, we have supposed it was not necessary to do any pivoting for a symmetric system. We will study some particular cases where it can be shown that pivoting is not needed, at least to be able to run the algorithm to completion.

Positive definite systems

Let A be symmetric and positive definite. We are looking for an LDL^{T} factorization. The fundamental result is the following.

LEMMA 4.1. *Let A be a symmetric positive definite matrix partitioned as*

$$A = \begin{pmatrix} A_{1,1} & A_{2,1}^{\mathrm{T}} \\ A_{2,1} & A_{2,2} \end{pmatrix},$$

where the blocks $A_{1,1}$ and $A_{2,2}$ are square. Then,

$$S_{2,2} = A_{2,2} - A_{2,1} A_{1,1}^{-1} A_{2,1}^{\mathrm{T}}$$

is symmetric and positive definite. $S_{2,2}$ is called the Schur complement (of $A_{2,2}$ in A).

PROOF. This can be proved in many different ways. Perhaps, the simplest one is to consider A^{-1} and to compute the bottom right-hand block of A^{-1}. Let

$$\begin{pmatrix} A_{1,1} & A_{2,1}^{\mathrm{T}} \\ A_{2,1} & A_{2,2} \end{pmatrix} \begin{pmatrix} x_1 \\ x_2 \end{pmatrix} = \begin{pmatrix} b_1 \\ b_2 \end{pmatrix}.$$

Then,

$$\begin{aligned} A_{1,1}x_1 + A_{2,1}^{\mathrm{T}}x_2 &= b_1 \\ \Rightarrow x_1 &= A_{1,1}^{-1}\left(b_1 - A_{2,1}^{\mathrm{T}}x_2\right). \end{aligned}$$

Therefore,

$$\left(A_{2,2} - A_{2,1}A_{1,1}^{-1}A_{2,1}^{\mathrm{T}}\right)x_2 = b_2 - A_{2,1}A_{1,1}^{-1}b_1.$$

This means that the inverse of A can be written as

$$A^{-1} = \begin{pmatrix} X & Y \\ Z & S_{2,2} \end{pmatrix}.$$

A being positive definite, the diagonal blocks of A and A^{-1} are also positive definite, so $S_{2,2}$ is positive definite. □

If we look at the outer product algorithm of Section 4, we see that in the first step, $A_2 = B_1 - (1/a_{1,1})a_1 a_1^{\mathrm{T}}$ is a Schur complement, the matrix A being partitioned as

$$A = \begin{pmatrix} a_{1,1} & a_1^{\mathrm{T}} \\ a_1 & B_1 \end{pmatrix}.$$

Therefore, if A is positive definite, A_2 is also positive definite and the next pivot is nonzero. The process can be continued until the last step. All the square matrices involved are positive definite and so the diagonal elements are strictly positive. All the pivots are nonzero and the algorithm can go through without any pivoting. This is summarized in the following result.

THEOREM 4.1. *A matrix A has a factorization $A = LDL^{\mathrm{T}}$, where L is a unit lower triangular matrix and D is a diagonal matrix with* $\operatorname{diag}(D) > 0$, *if and only if A is symmetric and positive definite.*

PROOF. Lemma 4.1 and the discussion afterwards have shown that if A is positive definite, it can be factored as LDL^{T}. Reciprocally, if $A = LDL^{\mathrm{T}}$, then of course A is symmetric and if $x \neq 0$, then

$$x^{\mathrm{T}}Ax = x^{\mathrm{T}}LDL^{\mathrm{T}}x = y^{\mathrm{T}}Dy,$$

where $y = L^{\mathrm{T}}x \neq 0$. Notice that

$$y^{\mathrm{T}}Dy = \sum_{i=1}^{n} d_{i,i}y_i^2 > 0,$$

as the diagonal elements of D are strictly positive. □

In the factorization of Theorem 4.1, as the diagonal elements of D are strictly positive, we can introduce a diagonal matrix S such that $s_{i,i} = \sqrt{d_{i,i}}$, $i = 1, \ldots, n$. Therefore, $S^2 = D$ and let $\bar{L} = LS$. Then,

$$A = LDL^{\mathrm{T}} = LSSL^{\mathrm{T}} = \bar{L}\bar{L}^{\mathrm{T}}.$$

Strictly speaking this is what is called the Cholesky factorization of A. However, this form of factorization is not much used today as the computation involves square roots. On modern computers, square roots are much slower than multiplies and adds and must be avoided whenever possible. Factorizations like LDL^{T} have sometimes been called root free Cholesky. Nowadays, the generic name Cholesky factorization is often used for any LDL^{T} factorization.

An interesting property of positive definite matrices is that there is no growth of the entries of the reduced matrices during the factorization.

THEOREM 4.2. *Let A be symmetric positive definite. Consider the matrices D_i, $i = 1, \ldots, n$ of the outer product algorithm, then,*

$$\max_k \Bigl(\max_{i,j} \bigl|(D_k)_{i,j}\bigr|\Bigr) \leqslant \max_{i,j} |a_{i,j}| = \max_i (a_{i,i}).$$

We first prove a well known but useful lemma.

LEMMA 4.2. *Let A be a symmetric positive definite matrix. Then,*

$$\max_{i,j} |a_{i,j}| = \max_i (a_{i,i}).$$

PROOF. It is obvious that the diagonal entries of A are positive. Suppose there is a couple i_0, j_0 such that $|a_{i_0,j_0}| > |a_{i,j}|$, $i_0 \neq j_0$, $\forall i, j$ different from i_0, j_0. There are two cases.

(i) Suppose $a_{i_0,j_0} > 0$, then let

$$x = (0 \ldots, 0, 1, 0, \ldots, 0, -1, 0, \ldots, 0)^{\mathrm{T}},$$

the 1 being in position i_0 and the -1 in position j_0. We have,

$$x^{\mathrm{T}}Ax = a_{i_0,i_0} + a_{j_0,j_0} - 2a_{i_0,j_0} < 0.$$

Therefore, A is not positive definite which is a contradiction;

(ii) suppose $a_{i_0,j_0} < 0$, we choose

$$x = (0 \ldots, 0, 1, 0, \ldots, 0, 1, 0, \ldots, 0)^{\mathrm{T}},$$

and get

$$x^{\mathrm{T}}Ax = a_{i_0,i_0} + a_{j_0,j_0} + 2a_{i_0,j_0} < 0,$$

which, again, is a contradiction. □

PROOF OF THEOREM 4.2. By Lemma 4.1, we already know that the matrices D_k are positive definite. Therefore, by Lemma 4.2, it is enough to consider the diagonal to find the maximum of the absolute values of the entries. It also suffices to consider the first step as the proof is the same for the other steps.

As the diagonal entries are positive, we have

$$\operatorname{diag}(A_2) \leqslant \operatorname{diag}(B_1).$$

Therefore, either $\max_i (a_{i,i}) = a_{1,1}$ and then, $\max_i (D_1)_{i,i} = a_{1,1}$ or the maximum is on the diagonal of B_1 and then,

$$\max_i (D_1)_{i,i} = \max\Bigl(a_{1,1}, \max_i \bigl[\operatorname{diag}(A_2)_{i,i}\bigr]\Bigr),$$

with $\max_i[\operatorname{diag}(A_2)_{i,i}] \leqslant \max_i[\operatorname{diag}(B_1)_{i,i}]$. In both cases,

$$\max_i (D_1)_{i,i} \leqslant \max_{i,j} |a_{i,j}|. \qquad \square$$

Indefinite systems

When factorizing an indefinite matrix, there can be some problems as the following example from GOLUB and VAN LOAN [1989] shows if ε is small,

$$\begin{pmatrix} \varepsilon & 1 \\ 1 & 0 \end{pmatrix} = \begin{pmatrix} 1 & 0 \\ 1/\varepsilon & 1 \end{pmatrix} \begin{pmatrix} \varepsilon & 0 \\ 0 & -1/\varepsilon \end{pmatrix} \begin{pmatrix} 1 & 1/\varepsilon \\ 0 & 1 \end{pmatrix},$$

The term $1/\varepsilon$ can be very large and the factorization can be spoiled. One can use pivoting to avoid this. However, if symmetry is to be preserved, pivoting must be done from the diagonal and does not always solve the problems. Moreover, zero pivots can be the only alternative sometimes, as, e.g., in the following example

$$A = \begin{pmatrix} 0 & 1 & 1 \\ 1 & 0 & 1 \\ 1 & 1 & 0 \end{pmatrix}.$$

One possibility to overcome these difficulties is to use a method due to AASEN [1971]. Here, a factorization

$$PAP^{\mathrm{T}} = LTL^{\mathrm{T}},$$

where L is unit lower triangular and T is tridiagonal is obtained. This is not really Gaussian elimination, therefore, we will not give more details on this. Another idea, introduced in BUNCH and PARLETT [1971] and further developed in BUNCH and KAUFMAN [1977] is to used diagonal pivoting with either 1×1 or 2×2 pivots. Suppose

$$P_1 A P_1^{\mathrm{T}} = \begin{pmatrix} A_{1,1} & A_{1,2} \\ A_{1,2}^{\mathrm{T}} & A_{2,2} \end{pmatrix},$$

where $A_{1,1}$ is of order s with $s = 1$ or 2 and P_1 is a permutation matrix. Then, this matrix can be factored as

$$P_1 A P_1^{\mathrm{T}} = \begin{pmatrix} I_s & 0 \\ A_{1,2}^{\mathrm{T}} A_{1,1}^{-1} & I_{n-s} \end{pmatrix} \begin{pmatrix} A_{1,1} & 0 \\ 0 & A_{2,2} - A_{1,2}^{\mathrm{T}} A_{1,1}^{-1} A_{1,2} \end{pmatrix} \begin{pmatrix} I_s & A_{1,1}^{-1} A_{1,2} \\ 0 & I_{n-s} \end{pmatrix}.$$

The algorithm can go through provided that $A_{1,1}$ is nonsingular. It can be proved that A being nonsingular, it is always possible to find a nonzero pivot ($s = 1$) or a nonsingular 2×2 block ($s = 2$). A strategy has been devised in BUNCH and KAUFMANN [1977], see also GOLUB and VAN LOAN [1989] to find the block pivots.

5. Gaussian elimination for H-matrices

There are other types of matrices (not necessarily symmetric) than positive definite matrices for which there is no necessity to use pivoting (at least to obtain a factorization without permutations). Let us first introduce a few definitions.

- A matrix A is reducible if and only if there exists a permutation matrix P such that

$$P^{-1} A P = \begin{pmatrix} D_1 & 0 \\ F & D_2 \end{pmatrix},$$

  D_1 and D_2 being square matrices. A matrix that is not reducible is said to be irreducible.
- – A is (row) diagonally dominant if

$$|a_{i,i}| \geqslant \sum_{j=1,\ j \neq i}^{n} |a_{i,j}|, \quad \forall i.$$

  – A is (row) strictly diagonally dominant if

$$|a_{i,i}| > \sum_{j=1, j \neq i}^{n} |a_{i,j}|, \quad \forall i.$$

  – A is irreducibly (row) diagonally dominant if, (a) A is irreducible, (b) A is (row) diagonally dominant and (c) there exists an i_0 such that

$$|a_{i_0,i_0}| > \sum_{j=1, j \neq i_0}^{n} |a_{i_0,j}|.$$

Notice that similar definitions hold for column oriented diagonal dominance. Matrices which are symmetric and strictly diagonally dominant or irreducibly diagonally dominant are positive definite. Their cases are covered in Section 4.

- A is an M-matrix if and only if $a_{i,j} \leqslant 0$ for $i \neq j$ and $A^{-1} \geqslant 0$.

If A is a symmetric M-matrix then A is positive definite. Once again, this is covered in Section 4.

Let A be a matrix, define $M(A)$ as the matrix having entries $m_{i,j}$ such that

$$m_{i,i} = |a_{i,i}|, \qquad m_{i,j} = -|a_{i,j}|, \quad \forall i, j, \ i \neq j.$$

- A is an H-matrix if and only if $M(A)$ is an M-matrix.

The previous definitions for diagonally dominant matrices can be slightly generalized. This will lead to a characterization of H-matrices.

- A matrix A is generalized (row) diagonally dominant if there exists a vector d with $d_i > 0, \ \forall i$ such that

$$|a_{i,i}|d_i \geqslant \sum_{j=1, j \neq i}^{n} |a_{i,j}|d_j, \quad \forall i.$$

  A is generalized (row) strictly diagonally dominant if

$$|a_{i,i}|d_i > \sum_{j=1, j \neq i}^{n} |a_{i,j}|d_j, \quad \forall i.$$

  Clearly, this means that if we denote

$$D = \begin{pmatrix} d_1 & & & \\ & d_2 & & \\ & & \ddots & \\ & & & d_n \end{pmatrix},$$

  then, AD is (row) diagonally dominant or (row) strictly diagonally dominant. The same is true of $D^{-1}AD$.

We have the following results that we state without proof, see BERMAN and PLEMMONS [1979].

THEOREM 5.1. *A is an M-matrix if and only if* $a_{i,j} \leqslant 0, \forall i \neq j$ *and A is generalized (row or column) strictly diagonally dominant.*

THEOREM 5.2. *A is an H-matrix if and only if A is generalized (row or column) strictly diagonally dominant.*

It is obvious that strictly diagonally dominant, irreducibly diagonally dominant and M-matrices are H-matrices. For these types of matrices, the important fact is that at each step of Gaussian elimination, the property in question is maintained, that is $A_k, \forall k$ possesses the same property as A.

Let us first consider A being diagonally dominant.

THEOREM 5.3. *If A is (row or column) diagonally dominant, then*

$$A = LU,$$

where L is unit lower triangular and U is upper triangular.

PROOF. Suppose A is (row) diagonally dominant. Then, $a_{1,1} \neq 0$, otherwise all the elements in the first row are 0 and A is singular. We shall prove that A_2 is also (row) diagonally dominant and then, the proof will go on by induction. The case of the first row is already handled. Now, we have

$$a_{i,j}^{(2)} = a_{i,j} - \frac{a_{i,1}a_{1,j}}{a_{1,1}}, \quad 2 \leqslant i \leqslant n,\ 2 \leqslant j \leqslant n, \qquad a_{i,1}^{(2)} = 0, \quad 2 \leqslant i \leqslant n,$$

$$\sum_{j,\ j\neq i} |a_{i,j}^{(2)}| = \sum_{j,\ j\neq i,\ j\neq 1} |a_{i,j}^{(2)}| \leqslant \sum_{j,\ j\neq i,\ j\neq 1} |a_{i,j}| + \left|\frac{a_{i,1}}{a_{1,1}}\right| \sum_{j,\ j\neq i,\ j\neq 1} |a_{1,j}|.$$

But,

$$|a_{1,1}| \geqslant \sum_{j,\ j\neq i,\ j\neq 1} |a_{1,j}| + |a_{1,i}|.$$

Therefore,

$$\begin{aligned}
\sum_{j,\ j\neq i} |a_{i,j}^{(2)}| &\leqslant \sum_{j,\ j\neq i,\ j\neq 1} |a_{i,j}| + \left|\frac{a_{i,1}}{a_{1,1}}\right| \big(|a_{1,1}| - |a_{1,i}|\big) \\
&\leqslant \sum_{j,\ j\neq i} |a_{i,j}| - \frac{|a_{i,1}a_{1,i}|}{|a_{1,1}|} \\
&\leqslant |a_{i,i}| - \frac{|a_{i,1}a_{1,i}|}{|a_{1,1}|} \\
&\leqslant \left| a_{i,i} - \frac{a_{i,1}a_{1,i}}{a_{1,1}} \right| = |a_{i,i}^{(2)}|.
\end{aligned}$$

This shows that all the pivots are nonzero and the computations can go through. If A is column diagonally dominant, the same proof can be done with A^{T}.

From the above discussion, it is obvious that $|l_{i,j}| \leqslant 1$. □

We then consider M-matrices. The following result has been proved by FIEDLER and PTÀK [1962].

THEOREM 5.4. *If A is an M-matrix, then*

$$A = LU,$$

where L is unit lower triangular and U is upper triangular.

PROOF. The proof can be found for instance in FIEDLER'S book [1986] or in BERMANN and PLEMMONS [1979]. □

This can be proved through the following lemma.

LEMMA 5.1. *Let A be an M-matrix written in block form as*

$$A = \begin{pmatrix} B & F \\ E & C \end{pmatrix},$$

where B and C are square matrices, then the Schur complement $S = C - EB^{-1}F$ is an M-matrix.

PROOF. It is obvious that the principal submatrices of an M-matrix are M-matrices. Therefore, B is an M-matrix and $B^{-1} > 0$. Since, by definition, the entries of E and F are nonpositive, the entries of $EB^{-1}F$ are nonnegative. Therefore, the nondiagonal entries of S are nonpositive.

Now, since A is an M-matrix, we know there is a diagonal matrix D with strictly positive diagonal entries such that AD is (row) strictly diagonally dominant.

Let

$$D = \begin{pmatrix} D_1 & 0 \\ 0 & D_2 \end{pmatrix},$$

AD being (row) strictly diagonally dominant means that if $e = (1 \ \dots \ 1)^T$, then $ADe > 0$. But,

$$AD = \begin{pmatrix} BD_1 & FD_2 \\ ED_1 & CD_2 \end{pmatrix}$$

and let $e = \binom{e_1}{e_2}$. The Schur complement of AD is (row) strictly diagonally dominant (CONCUS, GOLUB and MEURANT [1985]). This means that,

$$0 < \left[CD_2 - ED_1(BD_1)^{-1}FD_2\right]e_2 = SD_2e_2.$$

This shows that S is (row) generalized strictly diagonally dominant. Hence, S is an M-matrix. □

PROOF OF THEOREM 5.4. We apply readily Lemma 5.1 that shows that starting from an M-matrix, the reduced matrices that are obtained at each step are M-matrices. Moreover, L and U are M-matrices. □

We now consider H-matrices. Let B be an M-matrix, we define

$$\Omega_B = \{A \mid B \leqslant M(A)\}.$$

That is

$$|a_{i,i}| \geqslant b_{i,i}, \quad 1 \leqslant i \leqslant n,$$
$$|a_{i,j}| \leqslant |b_{i,j}|, \quad i \neq j,\ 1 \leqslant i, j \leqslant n.$$

This means that A is at least as diagonally dominant as B.

LEMMA 5.2. *Let B be an M-matrix. Each $A \in \Omega_B$ is (row) generalized strictly diagonally dominant.*

PROOF. There exists a diagonal matrix D (with $\operatorname{diag}(D) > 0$) such that BD is strictly diagonally dominant and let $A \in \Omega_B$, we have

$$BD \leqslant M(A)D = M(AD).$$

Therefore,

$$0 < BDe \leqslant M(AD)e,$$

which implies that AD is (row) strictly diagonally dominant. □

THEOREM 5.5. *Let B be an M-matrix. For each $A \in \Omega_B$,*

$$A = LU,$$

where L is unit lower triangular and U is upper triangular. In particular, for every H-matrix, there exists an LU factorization.

PROOF. We have seen in the proof of Lemma 5.2 that AD is (row) strictly diagonally dominant. Then, by Theorem 5.3, there exist $\bar{L}$ and $\bar{U}$, lower and upper triangular matrices such that

$$AD = \bar{L}\bar{U}.$$

We have,

$$A = \bar{L}\bar{U}D^{-1},$$

and the result follows. □

In FUNDERLIC, NEUMANN and PLEMMONS [1982], it is proved that if $A = LU$ and $B = L'U'$, then

$$|l_{i,j}| \leqslant |l'_{i,j}|, \quad 1 \leqslant i, j \leqslant n,$$
$$|u_{i,j}| \leqslant |u'_{i,j}|, \quad i \neq j,\ 1 \leqslant i, j \leqslant n,$$
$$|u_{i,i}| \geqslant |u'_{i,i'}|, \quad 1 \leqslant i \leqslant n,$$

and let

$$\beta_D = \frac{\max_i(D_{i,i})}{\min_i(D_{i,i})}.$$

Then,

$$|l_{i,j}| \leqslant \beta_D,$$
$$|u_{i,j}| \leqslant 2\beta_D \max_i |a_{i,i}|.$$

This gives

$$g_A \leqslant 2\beta_D,$$

and for an M-matrix, FUNDERLIC, NEUMANN and PLEMMONS [1982] proved that $g_A \leqslant \beta_D$. The proofs of these inequalities are easily obtained by induction.

The proof that an H-matrix possesses an LU factorization can also be established by showing, as in BERMANN and PLEMMONS [1979], that all the leading principal minors are nonsingular. Similar results for the case where A is singular have been studied in VARGA and CAI [1981], FUNDERLIC and PLEMMONS [1981] and FUNDERLIC, NEUMANN and PLEMMONS [1982].

For the case of H-matrices, we have seen that the growth factor is bounded

$$g_A \leqslant 2\beta_D.$$

It has been shown also that any symmetric permutation of A has an LU factorization. However, even if g_A is bounded, it can be large. Consider, e.g., the following example (AHAC, BUONI and OLESKI [1988]),

$$A_x = \begin{pmatrix} 2 & 0 & -x \\ -x & x & -1 \\ 0 & -1 & x \end{pmatrix}, \quad x > 0.$$

The matrix A_x is an M-matrix if $x > \sqrt{2}$ and

$$A_x = \begin{pmatrix} 1 & 0 & 0 \\ -\frac{x}{2} & 1 & 0 \\ 0 & -\frac{1}{x} & 1 \end{pmatrix} \begin{pmatrix} 2 & 0 & -x \\ 0 & x & -\frac{x^2}{2} - 1 \\ 0 & 0 & \frac{x}{2} - \frac{1}{x} \end{pmatrix}.$$

If x is large, the growth factor is large (in fact $O(x)$). We will see in the next sections that this is bad for the stability of the algorithm. This can be avoided by using some form of symmetric pivoting. For M-matrices, AHAC and OLESKI [1986] chose (in the reduced matrix) the column which has the largest column sum.

In the example, we choose the second column. The permuted matrix is

$$A' = \begin{pmatrix} x & -1 & -x \\ -1 & x & 0 \\ -x & 0 & 2 \end{pmatrix},$$

and

$$A' = \begin{pmatrix} 1 & 0 & 0 \\ -\frac{1}{x} & 1 & 0 \\ -1 & \frac{x}{1-x^2} & 1 \end{pmatrix} \begin{pmatrix} x & -1 & -x \\ 0 & x - \frac{1}{x} & 1 \\ 0 & 0 & 2 - x + \frac{x}{x^2-1} \end{pmatrix}.$$

The growth factor of A' is bounded independently of x. In AHAC and OLESKI [1986], it is proved that $g_A \leqslant n$ for M-matrices. This result is extended to H-matrices in AHAC, BUONI and OLESKI [1988].

6. Block methods

Block methods are obtained by partitioning the matrix A into blocks (submatrices). Consider, e.g., a 3×3 block partitioning. Then, A is written as

$$A = \begin{pmatrix} A_{1,1} & A_{1,2} & A_{1,3} \\ A_{2,1} & A_{2,2} & A_{2,3} \\ A_{3,1} & A_{3,2} & A_{3,3} \end{pmatrix}.$$

The matrices $A_{i,i}$ are square of order n_i, $1 \leqslant n_i \leqslant n$. Then, a block LU factorization can be obtained. There are several ways to do that. One is the following,

$$A = \begin{pmatrix} I & & \\ L_{2,1} & I & \\ L_{3,1} & L_{3,2} & I \end{pmatrix} \begin{pmatrix} U_{1,1} & U_{1,2} & U_{1,3} \\ & U_{2,2} & U_{2,3} \\ & & U_{3,3} \end{pmatrix}.$$

The factorization proceeds in the same way as for the standard (point) factorization. Therefore, to understand the algorithms, it is enough to look at a 2×2 case,

$$A = \begin{pmatrix} A_{1,1} & A_{1,2} \\ A_{2,1} & A_{2,2} \end{pmatrix} = \begin{pmatrix} I & 0 \\ L_{2,1} & I \end{pmatrix} \begin{pmatrix} U_{1,1} & U_{1,2} \\ 0 & S \end{pmatrix}.$$

Then,

$$U_{1,1} = A_{1,1}.$$

$L_{2,1}$ is obtained by solving $L_{2,1}A_{1,1} = A_{2,1}$ and finally

$$S = A_{2,2} - A_{2,1}A_{1,1}^{-1}A_{1,2} = A_{2,2} - L_{2,1}A_{1,2}.$$

Then, the algorithm is repeated on S. Here, no pivoting takes place. The stability is investigated in DEMMEL, HIGHAM and SCHREIBER [1995]. Block LU factorization (without pivoting) is unstable in general, although it has been found to be stable for matrices that are block diagonally dominant by columns, i.e.

$$\|A_{j,j}^{-1}\|^{-1} \geqslant \sum_{i \neq j} \|A_{i,j}\|.$$

7. Particular systems

Tridiagonal matrices

Tridiagonal matrices are particularly easy to handle. Let T be a symmetric tridiagonal matrix,

$$T = \begin{pmatrix} a_1 & -b_2 & & & \\ -b_2 & a_2 & -b_3 & & \\ & \ddots & \ddots & \ddots & \\ & & -b_{n-1} & a_{n-1} & -b_n \\ & & & -b_n & a_n \end{pmatrix}.$$

We will also look at a particular case of a tridiagonal Toeplitz matrix,

$$T_a = \begin{pmatrix} a & -1 & & & \\ -1 & a & -1 & & \\ & \ddots & \ddots & \ddots & \\ & & -1 & a & -1 \\ & & & -1 & a \end{pmatrix}.$$

The Cholesky factorization of T is easily obtained,

$$T = LD_L^{-1}L^{\mathrm{T}},$$

$$L = \begin{pmatrix} \delta_1 & & & & \\ -b_2 & \delta_2 & & & \\ & \ddots & \ddots & & \\ & & -b_{n-1} & \delta_{n-1} & \\ & & & -b_n & \delta_n \end{pmatrix}, \quad D_L = \begin{pmatrix} \delta_1 & & & & \\ & \delta_2 & & & \\ & & \ddots & & \\ & & & \delta_{n-1} & \\ & & & & \delta_n \end{pmatrix}.$$

By inspection, we have

$$\delta_1 = a_1, \qquad \delta_i = a_i - \frac{b_i^2}{\delta_{i-1}}, \quad i = 2, \ldots, n.$$

This involves only $n-1$ additions, multiplications and divisions. Extensions are easily obtained to nonsymmetric tridiagonal matrices as long as pivoting is not needed. If $T = T_a$, then we have

$$\delta_1 = a, \qquad \delta_i = a - \frac{1}{\delta_{i-1}}, \quad i = 2, \ldots, n,$$

and the explicit solution of this recurrence is known, see, e.g., MEURANT [1992],

$$\delta_i = \frac{r_+^{i+1} - r_-^{i+1}}{r_+^i - r_-^i},$$

where

$$r_\pm = \frac{a \pm \sqrt{a^2 - 4}}{2}$$

are the two solutions of the quadratic equation $r^2 - ar + 1 = 0$ if $a \neq 2$. If $a = 2$, then $\delta_i = (i+1)/i$. Of course, there are more efficient methods to deal with Toeplitz matrices, see GOLUB and VAN LOAN [1989].

The Cholesky factorization of tridiagonal matrices has been used by MEURANT [1992] to characterize the inverse of such matrices.

Block tridiagonal matrices

The previous method is easily extended to block tridiagonal symmetric matrices. Let

$$A = \begin{pmatrix} D_1 & -A_2^{\mathrm{T}} & & & \\ -A_2 & D_2 & -A_3^{\mathrm{T}} & & \\ & \ddots & \ddots & \ddots & \\ & & -A_{n-1} & D_{n-1} & -A_n^{\mathrm{T}} \\ & & & -A_n & D_n \end{pmatrix},$$

each block being of order n. Denote by L the block lower triangular part of A then, if such a factorization exists we have,

$$A = (\Delta + L)\Delta^{-1}(\Delta + L^{\mathrm{T}}),$$

where Δ is a block diagonal matrix whose diagonal blocks are denoted by Δ_i. By inspection, we have

$$\Delta_1 = D_1, \qquad \Delta_i = D_i - A_i(\Delta_{i-1})^{-1}A_i^{\mathrm{T}}, \quad i = 2, \ldots, n.$$

Therefore, obtaining this block factorization involves only solving (small) linear systems with matrices Δ_i and several right-hand sides. Notice that whatever are the structures of the matrices D_i, the Δ_is are dense matrices.

Again, this block factorization can be used (MEURANT [1992]) to characterize the inverse of block tridiagonal matrices.

CHAPTER II

Error Analysis

8. Round off error analysis

The algorithms that we have seen so far are supposed to give the LU factorization of a given matrix with real or complex coefficients after $n-1$ steps (and possibly some permutations). These algorithms are programmed using some languages like Fortran, C or Matlab and the codes that are produced are run on different computers which are ranging from PCs to supercomputers.

Unfortunately, all the parts in a computer are finite. In particular, registers and memory words designed to store the data and the intermediate and final results have a finite length or capacity and cannot store all real numbers. Moreover, when computations are performed on a computer, each arithmetic operation $(+, -, *, /)$ is generally affected by round off errors.

The subject of round off error analysis is to try to understand what are the effects of these limitations on the result of solving a problem, in our case, using Gaussian elimination. Before going into these problems, we must define the floating point arithmetic model we are using.

Floating point arithmetic model

Here, we follow the expositions of FORSYTHE and MOLER [1967] and GOLUB and VAN LOAN [1989]. The numbers that can be represented in the computer are a (finite) subset of the real line and are denoted by F. This set is characterized by four integers: the base β, the number t of base–β digits in the fractional part (also called the mantissa) and the exponent range $[e_L, e_U]$. Then, (normalized) numbers in F consists of all real numbers f of the form

$$f = \pm .d_1 d_2 \cdots d_t \, \beta^e, \quad 0 \leqslant d_i < \beta,\ d_1 \neq 0,\ e_L \leqslant e \leqslant e_U,$$

where e is an integer, to which we add zero and a representation for results whose absolute value will be smaller (resp. larger) than the smallest (resp. largest) absolute value of a nonzero number in F.

For a nonzero $f \in F$, we have

$$m = \beta^{e_L - 1} \leqslant |f| \leqslant M = \beta^{e_U}\left(1 - \beta^{-t}\right).$$

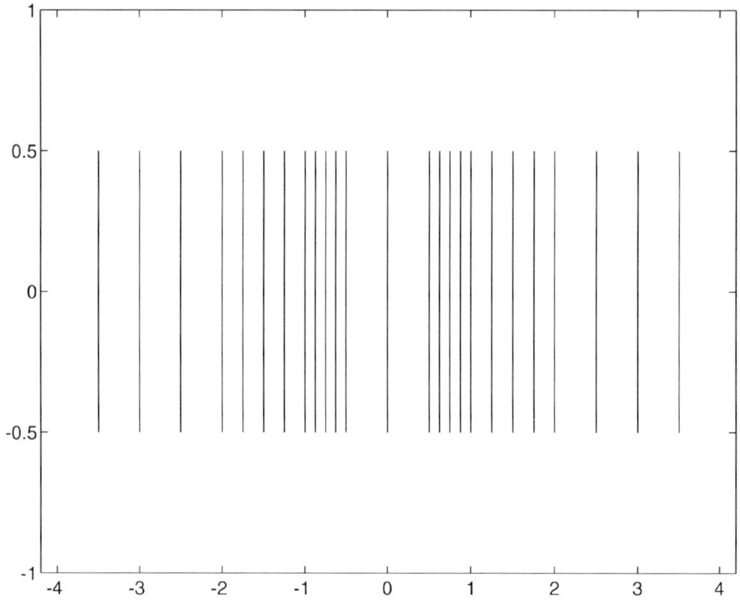

FIG. 8.1. Elements of F for $\beta = 2, t = 3, e_L = 0, e_U = 2$.

On the real number line, the elements of F are not equally spaced (see Fig. 8.1 that shows the elements of F for $\beta = 2, t = 3, e_L = 0, e_U = 2$).

A real number x that we would like to represent is approximated by a number $fl(x)$ that can be defined as an operator from

$$G = \{x \in R,\ m \leqslant |x| \leqslant M\} \cup \{0\},$$

into F, by

$$fl(x) = \begin{cases} \text{nearest } x_{\mathrm{R}} \in F \text{ to } x \text{ if rounded arithmetic is used,} \\ \text{nearest } x_{\mathrm{C}} \in F \text{ s.t. } |x_{\mathrm{C}}| \leqslant |x| \text{ if chopped arithmetic is used.} \end{cases}$$

Today, most computers use $\beta = 2$, although there had been in the past computers with $\beta = 8$ or 16. In the examples, we will use $\beta = 10$ as this is more familiar to most people. Typical values of the other parameters for some (1996) computers are given in Table 8.1.

For more details on floating point arithmetic particularly IEEE formats, see GOLBERG [1991]. The following results are proved in FORSYTHE and MOLER [1967].

THEOREM 8.1. *If $x \in G$, then*

$$x_{\mathrm{R}} = x(1 + \varepsilon_{\mathrm{R}}), \quad |\varepsilon_{\mathrm{R}}| \leqslant \tfrac{1}{2}\beta^{1-t},$$

$$x_{\mathrm{C}} = x(1 + \varepsilon_{\mathrm{C}}), \quad |\varepsilon_{\mathrm{C}}| \leqslant \beta^{1-t}.$$

TABLE 8.1
Examples of floating point formats.

	t	e_L	e_U
CRAY single	48	−8192	8191
IEEE single	24	−125	128
IEEE double	53	−1021	1024

Next, we have to define the operations on elements of F. Let $\bullet$ be one of the four operations $+, -, *, /$ and $\odot$ its implementation on a computer. We say that a computer arithmetic is correct if

$$x, y \in F, \quad fl(x \bullet y) = x \odot y.$$

Not all arithmetics are correct. The one used on CRAYs Y–MP and C90 is not correct. The IEEE norm defines a correct arithmetic.

We define the unit round off u as,

$$u = \begin{cases} u_{\mathrm{R}} = \frac{1}{2}\beta^{1-t} & \text{for rounded arithmetic,} \\ u_{\mathrm{C}} = \beta^{1-t} & \text{for chopped arithmetic.} \end{cases}$$

For IEEE single precision arithmetic, the unit round off (rounded arithmetic) is $u_{\mathrm{R}} = 5.9605\ 10^{-8}$ and $u_{\mathrm{R}} = 1.1102\ 10^{-16}$ for double precision. We have,

$$fl(x) = x(1+\varepsilon), \quad |\varepsilon| \leqslant u.$$

We have the following result,

$$fl(x \bullet y) = (x \bullet y)(1+\varepsilon), \quad |\varepsilon| \leqslant u, \tag{8.1}$$

and then

$$\frac{|fl(x \bullet y) - (x \bullet y)|}{|x \bullet y|} \leqslant u.$$

This shows there is a small relative error associated with each operation (provided t is large enough). Note that (8.1) is not verified by CRAY arithmetic because of the lack of guard digits. Relaxed assumptions have to be made. Fortunately however, most of the results carry over to this case.

From this last result we can construct some other error bounds. We use the following lemma from FORSYTHE and MOLER [1967].

LEMMA 8.1. *If* $0 \leqslant u < 1$ *and* $n \in N$,

$$1 - nu \leqslant (1-u)^n.$$

If $0 \leqslant nu \leqslant 0.01$,

$$(1+u)^n \leqslant 1 + 1.01nu.$$

The first statement comes from Taylor's formula and the second one is because we have

$$e^x \leqslant 1 + 1.01x \quad \text{for } 0 \leqslant x \leqslant 0.01. \qquad \square$$

Notice that the largest n that satisfies the inequality $0 \leqslant nu \leqslant 0.01$ is $n \approx 167772$ for IEEE single precision arithmetic and $n \approx 910^{13}$ for double precision.

Then, we can show

THEOREM 8.2. *Let* $w = x - yz$ *and* $e = fl(w) - w$. *If* $nu \leqslant 0.01$, *then*

$$|e| \leqslant 3.02\, u \max\big(|x|, |w|\big).$$

PROOF. $fl(w)$ is defined as $fl(x - fl(yz))$. But, $fl(yz) = yz(1+\varepsilon_1)$, $|\varepsilon_1| \leqslant u$ and

$$\begin{aligned} fl(w) &= \big(x - yz(1+\varepsilon_1)\big)(1+\varepsilon_2), \quad |\varepsilon_2| \leqslant u, \\ &= x(1+\varepsilon_2) - yz(1+\varepsilon_1)(1+\varepsilon_2). \end{aligned}$$

This can be written with the help of Lemma 8.1,

$$\begin{aligned} fl(w) &= x(1+\theta_2 u) - yz(1+2.02\,\theta_1 u), \quad |\theta_i| < 1, \\ &= x - yz + u(x\theta_2 - 2.02\,yz\theta_1). \end{aligned}$$

Therefore,

$$|e| \leqslant u\big(|x| + 2.02\,|w - x|\big),$$

From which the results follows. $\square$

The last result can also be written as

$$|e| \leqslant u\big(2.02|yz| + |x|\big).$$

We have also,

$$|e| \leqslant u\big(2|yz| + |x|\big) + \mathrm{O}\big(u^2\big).$$

The values of the constants involved in these inequalities like 2.02 or 3.02 are not really important, therefore, most of the time in the statements of the results, we will replace them by a generic constant C or C_i. However, for the sake of completeness, we will indicate their possible values.

Now, we are able to analyze the effect of round off errors in Gaussian elimination. There are several different ways to approach this problem. The one that is more popular today is called inverse or backward error analysis. It has been introduced by W. Givens and developed, in particular for Gaussian elimination, by WILKINSON [1965]. In this type of analysis the rounding errors are related to the data. The result of floating point operations is interpreted as the exact result of ordinary arithmetic on some perturbed data and some bounds are derived for the perturbations. This allows to be able to use ordinary arithmetic.

Examples of difficulties

Let us give a few small examples showing some of the problems that can happen with Gaussian elimination and finite precision computations.

We suppose $\beta = 10$, $t = 3$ (we do not care about limits on the exponent range) and, for the sake of simplicity, chopped arithmetic, the operations being done on normalized numbers. Although these assumptions are not realistic, the examples will give us an idea of what could happen on real problems. Consider the following linear system (Example 8.1),

$$3x_1 + 3x_2 = 6,$$
$$x_1 + \delta x_2 = \delta + 1,$$

where δ is a given real number ($\in F$). The exact solution is $x_1 = x_2 = 1$.

Choosing 3 as the pivot and noticing that the computed value of the multiplier $1/3$ is $0.333\ 10^0$, we obtain for the second equation

$$\left(\delta - 0.999\ 10^0\right)x_2 = \delta + 1 - 6 \times \left(0.333\ 10^0\right) = \delta - 0.990\ 10^0.$$

Therefore, if $\delta = 0.990$, then $x_2 = 0$ and $x_1 = 0.199\ 10^1$!

One can argue that this system is close to being singular, but this is not so as

$$\det(A) = 3 \times 0.99 - 3 = -0.03.$$

Notice that even though the solution is wrong, the residual $b - Ax$ is small being $(0.03\ 0)^{\mathrm{T}}$. Remark also that the pivot is not small and that, in this case, if we consider the permuted system (Example 8.2),

$$\begin{aligned} x_1 + 0.99x_2 &= 1.99, \\ 3x_1 + 3x_2 &= 6, \end{aligned}$$

then, the reduced system is

$$(3 - 3 \times 0.99)x_2 = 6 - 3 \times 1.99,$$
$$0.03x_2 = 0.03$$

giving $x_2 = 1$ and then $x_1 = 1$.

Notice that the first system is just obtained by using partial pivoting on the second one. This gives an example where it does not pay to use partial pivoting. If one considers a more general 2×2 system (Example 8.3),

$$\begin{aligned} \alpha x_1 + \beta x_2 &= \alpha + \beta, \\ \gamma x_1 + \delta x_2 &= \gamma + \delta. \end{aligned}$$

Then, it can be easily shown that if $fl(x \bullet y) = (x \bullet y)(1 + \varepsilon)$, the computed value of x_2 is

$$1 + \varepsilon\left(1 - \frac{2\alpha\gamma}{\alpha\delta - \beta\gamma}\right) + \mathrm{O}(\varepsilon^2).$$

Therefore (as it is well known) it is not necessary to have a small determinant $\alpha\delta - \beta\gamma$ to have a large error. This can happen also in this case if the product $\alpha\gamma$ is large.

Consider, e.g., the system (Example 8.4),

$$\begin{aligned} 30x_1 + x_2 &= 31, \\ 10x_1 + x_2 &= 11. \end{aligned}$$

The determinant is 20. The computed value of $10/30$ is 0.333, therefore the reduced equation is

$$\begin{aligned} 0.667x_2 &= 11 - 0.310\ 10^2 \times 0.333\ 10^0 \\ &= 11 - 0.103\ 10^2 \\ &= (0.110 - 0.103)10^2 = 0.700\ 10^0 \end{aligned}$$

and $x_2 = 0.104\ 10^1$.

This is a 4% error in the second component and

$$\begin{aligned} x_1 &= \left(0.310\ 10^2 - 0.104\ 10^1\right) \times 0.333\ 10^{-1} \\ &= (0.310 - 0.010)\ 10^2 \times 0.333\ 10^{-1} \\ &= 0.3\ 10^2 \times 0.333\ 10^{-1} \\ &= 0.990\ 10^0 \end{aligned}$$

a 1% error. However, small pivots may also lead to large errors. Consider (Example 8.5),

$$\begin{aligned} 0.3\ 10^{-3}x_1 + x_2 &= 1, \\ x_1 + x_2 &= 2. \end{aligned}$$

Then, the computed value of x_2 is 1 and x_1 is 0. The "exact" solution is $x_1 = 1.0003$ and $x_2 = 0.9997$.

Partial pivoting has been invented to prevent this to happen. If we permute the equations

$$\begin{aligned} x_1 + x_2 &= 2, \\ 0.3\ 10^{-3} + x_2 &= 1, \end{aligned}$$

then, $x_2(1 - 0.3\ 10^{-3}) = 1 - 0.6\ 10^{-3} = 1$, giving $x_2 = 1$ and the first equation gives also $x_1 = 1$.

Errors in the LU factorization

Of course, the rounding errors depend on the order in which the operations are done. Therefore, all the variants and implementations of Gaussian elimination are different in this respect.

Let us analyze the standard algorithm that has been described in Section 3. Notice that permutations have no effects on rounding errors (they are just recorded by pointers). For convenience in this section we will denote by the same notations as before the computed quantities as there will be no ambiguities.

Let us first analyze the kth step.

THEOREM 8.3.

$$L_k A_k = A_{k+1} + E_k,$$

where

$$\left|(E_k)_{i,j}\right| \leqslant Cu \max\left(\left|a_{i,j}^{(k+1)}\right|, \left|a_{i,j}^{(k)}\right|\right),$$

with $C = 3.02$.

PROOF. The multipliers (e.g. the elements of L) that we denote by $l_{i,k}$ are

$$l_{i,k} = fl\left(\frac{a_{i,k}^{(k)}}{a_{k,k}^{(k)}}\right), \quad i \geqslant k+1,$$

$$a_{i,j}^{(k+1)} = \begin{cases} 0, & i \geqslant k+1,\ j = k, \\ fl\left(a_{i,j}^{(k)} - l_{i,k} a_{k,j}^{(k)}\right), & i \geqslant k+1,\ j \geqslant k+1, \\ a_{i,j}^{(k)}, & \text{otherwise.} \end{cases}$$

Let us first consider the multipliers, $i \geqslant k+1$,

$$l_{i,k} = \frac{a_{i,k}^{(k)}}{a_{k,k}^{(k)}}(1 + \varepsilon_{i,k}), \quad |\varepsilon_{i,k}| \leqslant u.$$

This translates into

$$a_{i,k}^{(k)} - l_{i,k}a_{k,k}^{(k)} + a_{i,k}^{(k)}\varepsilon_{i,k} = 0.$$

If we denote by $e_{i,k}^{(k)}$, the elements of E_k, this shows that

$$e_{i,k}^{(k)} = a_{i,k}^{(k)}\varepsilon_{i,k}, \quad i \geqslant k+1.$$

For $i \geqslant k+1$ and $j \geqslant k+1$, we have

$$a_{i,j}^{(k)} = fl\big(a_{i,j}^{(k)} - fl\big(l_{i,k}a_{k,j}^{(k)}\big)\big).$$

From Theorem 8.2, we have

$$\big|e_{i,j}^{(k)}\big| \leqslant Cu \max\big(\big|a_{i,j}^{(k+1)}\big|, \big|a_{i,j}^{(k)}\big|\big).$$

As this bound is certainly also true for $e_{i,k}^{(k)}$, we have

$$\big|e_{i,j}^{(k)}\big| \leqslant Cu \max\big(\big|a_{i,j}^{(k+1)}\big|, \big|a_{i,j}^{(k)}\big|\big), \quad i \geqslant k+1, \ j \geqslant k,$$

and the other entries of E_k are zero. □

With the notations of Lemma 3.5, we have seen that L_k has the form

$$L_k = I - l_k e_k^{\mathrm{T}}.$$

LEMMA 8.2. *If B_k is a matrix whose first k rows are zero, then*

$$L_i B_k = B_k, \quad i \leqslant k.$$

Similarly, $(L_i)^{-1} B_k = B_k$.

PROOF.

$$L_i B_k = \big(I - l_i e_i^{\mathrm{T}}\big) B_k = B_k - l_i e_i^{\mathrm{T}} B_k = B_k,$$

as $e_i^{\mathrm{T}} B_k = 0$. □

THEOREM 8.4. *Let F be defined as*

$$F = F_1 + \cdots + F_{n-1},$$

where

$$(F_k)_{i,j} = \begin{cases} 1, & i \geqslant k+1, \ j \geqslant k, \\ 0, & \textit{otherwise}, \end{cases}$$

i.e.,

$$F = \begin{pmatrix} 0 & 0 & 0 & \dots & 0 \\ 1 & 1 & 1 & \dots & 1 \\ 1 & 2 & 2 & \dots & 2 \\ \vdots & \vdots & \vdots & & \vdots \\ 1 & 2 & 3 & \dots & n-1 \end{pmatrix}.$$

Then,

$$A = LU + E,$$

with

$$|E| \leqslant Cu \max_{k,i,j} |a_{i,j}^{(k)}| F,$$

with $C = 3.02$.

PROOF. We have

$$L_k A_k = A_{k+1} + E_k \Longrightarrow A_k = L_k^{-1} A_{k+1} + E_k.$$

By Lemma 8.2,

$$(L_1)^{-1} \cdots (L_{k-1})^{-1} A_k = (L_1)^{-1} \cdots (L_k)^{-1} A_{k+1} + E_k.$$

We sum these equalities for $k = 1$ to $n - 1$. Most of the terms cancel and we obtain,

$$A = (L_1)^{-1} \cdots (L_{k-1})^{-1} A_n + E_1 + \cdots + E_{n-1}.$$

Therefore,

$$A = LU + E,$$

with

$$E = E_1 + \cdots + E_{n-1}.$$

Our next task is to bound the elements of E. It is easy to see that

$$|E| \leqslant Cu \max_{k,i,j} |a_{i,j}^{(k)}| F. \qquad \square$$

Theorem 8.4 means that the matrices L and U are the exact factors of the factorization of a perturbed matrix $A - E$. Since the elements of F are bounded at most by $n - 1$, $|E|$ being small depends on $\max_{i,j,k} |a_{i,j}^{(k)}|$.

The previous bound can be written in terms of the growth factor g_A defined in Chapter I,

$$|E| \leqslant Cu\, g_A \|A\|_\infty F.$$

It can be shown easily that,

$$\|E\|_\infty \leqslant Cu\, g_A\, n^2 \|A\|_\infty.$$

Errors in the triangular solves

It is a little easier to bound the error arising when solving a triangular system. Let us consider

$$Ly = b.$$

Then, on the computer we get,

$$\begin{aligned} y_1 &= fl\left(\frac{b_1}{l_{1,1}}\right), \\ y_i &= fl\left(\frac{b_i - l_{i,1}y_1 - \cdots - l_{i,i-1}y_{i-1}}{l_{i,i}}\right), \quad i = 2, \ldots, n. \end{aligned}$$

The analysis is done in FORSYTHE and MOLER [1967] by the same technique as in the last section. One can see also HIGHAM [1989a] where the following result is proved.

THEOREM 8.5. *We have,*

$$(L + K)y = b,$$

with

$$|K| \leqslant Cu \begin{pmatrix} |l_{1,1}| & & & & \\ |l_{2,1}| & 2|l_{2,2}| & & & \\ 2|l_{3,1}| & 2|l_{3,2}| & 2|l_{3,3}| & & \\ \vdots & \vdots & \vdots & \ddots & \\ (n-1)|l_{n,1}| & (n-1)|l_{n,2}| & (n-2)|l_{n,3}| & \ldots & 2|l_{n,n}| \end{pmatrix},$$

with $C = 1.01$ *and*

$$\begin{aligned} &|K| \leqslant Cnu|L|, \\ &\|K\|_\infty \leqslant \frac{n(n+1)}{2} Cu \max_{i,j} |l_{i,j}|. \end{aligned}$$

Errors in the solution of a linear system

When the inexact factors are used to obtain the computed solution of the linear system, what we get is

$$(L+K)y = b,$$
$$(U+G)x = y,$$

with

$$|K| \leqslant Cnu|L|,$$
$$|G| \leqslant Cnu|U|.$$

THEOREM 8.6. *The computed solution x satisfies*

$$(A+H)x = b,$$

with

$$|H| \leqslant C_1 u \max_{k,i,j} \left|a_{i,j}^{(k)}\right| F + C_2 nu|L||U| + \mathrm{O}\left(u^2\right),$$

with $C_1 = 3.02$ *and* $C_2 = 2.02$.

PROOF.

$$(L+K)(U+G)x = (LU + KU + LG + KG)x = b$$

and

$$LU = A - E.$$

Then, if we denote by H the following matrix

$$H = KU + LG + KG - E,$$

we have $(A+H)x = b$ and

$$|H| \leqslant |E| + |K||U| + |L||G| + |K||G|,$$
$$|H| \leqslant C_1 u \max_{k,i,j} \left|a_{i,j}^{(k)}\right| F + C_2 nu|L||U| + C_3 n^2 u^2 |L||U|,$$
$$\|H\|_\infty \leqslant \left(C_3 n^3 + C_4 n^2\right) u g_A \|A\|_\infty. \qquad \square$$

The role of pivoting

Pivoting techniques are not usually only used to insure that the pivots are nonzero but, also to reduce the growth factor. If partial pivoting is used, then

$$|l_{i,j}| \leqslant 1.$$

Therefore, $\|L\|_\infty \leqslant n$ and

$$\left|a_{i,j}^{(k+1)}\right| \leqslant \left|a_{i,j}^{(k)}\right| + \left|a_{k,j}^{(k)}\right| + \varepsilon \max\left(\left|a_{i,j}^{k+1)}\right|, \left|a_{i,j}^{(k)}\right|\right).$$

Let $\rho_k = \max_{i,j} |a_{i,j}^{(k)}|$, then

$$\rho_{k+1} \leqslant 2(1+\varepsilon)\rho_k.$$

Hence, by recurrence

$$\rho_n \leqslant 2^{n-1}(1+\varepsilon)^{n-1}\rho_1.$$

The term $(1+\varepsilon)^{n-1}$ is insignificant as ε is small. In fact, WILKINSON [1965] exhibited some contrived examples where an exponential growth is actually obtained. A well known example is

$$A = \begin{pmatrix} 1 & 0 & \dots & \dots & 0 & 1 \\ -1 & 1 & 0 & \dots & 0 & 1 \\ -1 & -1 & \ddots & & \vdots & \vdots \\ \vdots & \vdots & \ddots & \ddots & 0 & 1 \\ -1 & -1 & \dots & -1 & 1 & 1 \\ -1 & -1 & \dots & -1 & -1 & 1 \end{pmatrix}.$$

The bound is also attained for all $B = DAD$ with $D = \mathrm{diag}(1, -1, \dots, (-1)^{n+1})$. Although there exist such examples, in most practical cases the growth factor is small and Gaussian elimination with partial pivoting is a safe algorithm. However, there are cases where large growth can occur and the algorithm must be used cautiously.

WRIGHT [1993] has exhibited problems arising from solving two point boundary value problems with multiple shooting methods for which an exponential growth is observed. Due to the structure of the problem, no pivoting occurs when using partial pivoting. Moreover, not only the growth factor is large but also, large errors are measured.

FOSTER [1994] has also given practical examples where g_A grows exponentially. These examples come from Volterra integral equations. Again, no row interchange is needed by partial pivoting. Analytic asymptotic value of the growth factor is obtained as well as lower bounds involving the coefficients of the equation. Examples are given in FOSTER [1994] from population dynamics.

The average case stability of Gaussian elimination with partial pivoting has been studied in TREFETHEN and SCHREIBER [1990]. Random matrices of order $\leqslant 1024$ have been looked at. The average growth factor was approximately $n^{3/2}$ for partial pivoting and $n^{1/2}$ for complete pivoting.

These examples show that users of partial pivoting must carefully analyze the results.

Another possible choice is complete pivoting where the pivot is searched in $i, j > k$. Then, WILKINSON [1965] showed that the growth factor is bounded, as

$$|a_{i,j}^{(k)}| \leqslant k^{1/2}\big(2 \cdot 3^{1/2} \cdots k^{1/(k-1)}\big)^{1/2} \max_{i,j} |a_{i,j}|.$$

It was conjectured that in this case $g_A \leqslant n$.

CRYER [1968] proved that this is true for $n \leqslant 4$. However, the conjecture has been shown to be false for $n > 4$ if rounding error is allowed by GOULD [1991]. EDELMAN and OHLROCH [1991] modified the Gould counterexample to show that the conjecture is also false in exact arithmetic.

HIGHAM and HIGHAM [1989] have exhibited some matrices of practical interest which have a growth factor at least $n/2$ for complete pivoting. They proved the following result.

THEOREM 8.7. *Let* $\alpha = \max_{i,j} |a_{i,j}|$, $\beta = \max_{i,j} |(A^{-1})_{i,j}|$, $\theta = \frac{1}{\alpha\beta}$.
Then, $\theta \leqslant n$ *and for any permutation matrices* P *and* Q *such that* PAQ *has an LU factorization, the growth factor* g *without pivoting on* PAQ *satisfies*

$$g \geqslant \theta.$$

PROOF.

$$\big|(U^{-1})_{n,n}\big| = \big|e_n^{\mathrm{T}} U^{-1} e_n\big| = \big|e_n^{\mathrm{T}} U L^{-1} e_n\big| = \big|e_n^{\mathrm{T}} Q^{\mathrm{T}} A^{-1} P^{\mathrm{T}} e_n\big|,$$

as $L^{-1} e_n = e_n$. Therefore,

$$\big|(U_{n,n})^{-1}\big| = \big|(U^{-1})_{n,n}\big| = \big|(A^{-1})_{i,j}\big| \leqslant \beta,$$

for some (i, j). Clearly,

$$\max_{k,i,j} \big|a_{i,j}^{(k)}\big| \geqslant |U_{n,n}| \geqslant \frac{1}{\beta}. \qquad \square$$

With no pivoting at all, $(U^{-1})_{n,n} = (A^{-1})_{n,n}$, therefore it is easy to construct examples with a large growth factor by building matrices whose inverses have a small (n, n) entry. Notice that a large growth factor is a property of the matrix itself and not of the algorithm.

Perturbation analysis

So far, we have been concerned with the consequences of running the Gaussian elimination algorithm but we should also look at the consequences of perturbations on the data. When the matrix A or the right-hand side b are obtained (from scratch or from some other computations) some errors could also be introduced.

There are different ways to measure these perturbations (see CHATELIN and FRAYSSE [1993]) and a lot of literature on this topic. One of the oldest method has been introduced by A. Turing in 1949 and then developed by WILKINSON [1965]. It is called normwise analysis.

Normwise error analysis

To the data (matrix A and right-hand side b), we associate the computed solution y:

$$(A, b) \to y.$$

We have to choose how to measure distances in both spaces. Let x and y be such that

$$\begin{aligned} Ax &= b, \\ (A + \Delta A)y &= b + \Delta b. \end{aligned}$$

ΔA and Δb will be chosen such that

$$\|\Delta A\| \leqslant \alpha\omega, \qquad \|\Delta b\| \leqslant \beta\omega,$$

ω is given, α (resp. β) will be 0 or $\|A\|$ (resp. $\|b\|$) depending on whether A, or b, or both are perturbed. ω defines the normwise relative perturbation. Following GOLUB and VAN LOAN [1989], we have

LEMMA 8.3. *If* $\xi = \alpha\omega\|A^{-1}\| < 1$, *then*

$$\frac{\|y\|}{\|x\|} \leqslant \frac{1}{1-\xi}\left(1 + \frac{\xi\beta}{\alpha\|x\|}\right).$$

PROOF. The first question is to know if $A + \Delta A$ is nonsingular. But, we notice that

$$A + \Delta A = A\big(I + A^{-1}\Delta A\big)$$

and, with the hypothesis

$$\big\|A^{-1}\Delta A\big\| \leqslant \alpha\omega\big\|A^{-1}\big\| = \xi < 1.$$

Then, $A + \Delta A$ is nonsingular by Lemma 2.3.3 of GOLUB and VAN LOAN [1989]. We have

$$\big(I + A^{-1}\Delta A\big)y = A^{-1}(b + \Delta b) = x + A^{-1}\Delta b.$$

Taking norms,

$$\|y\| \leqslant \frac{1}{1-\alpha\omega\|A^{-1}\|}\left(\|x\| + \beta\omega\|A^{-1}\|\right).$$

But,

$$\omega \leqslant \frac{1}{\alpha\|A^{-1}\|} \Longrightarrow \|y\| \leqslant \frac{1}{1-\alpha\omega\|A^{-1}\|}\left(\|x\| + \xi\frac{\beta}{\alpha}\right).$$

Therefore,

$$\frac{\|y\|}{\|x\|} \leqslant \frac{1}{1-\alpha\omega\|A^{-1}\|}\left(1 + \xi\frac{\beta}{\alpha\|x\|}\right).$$

If $\alpha = \|A\|$ and $\beta = \|b\|$, then

$$\frac{\|y\|}{\|x\|} \leqslant \frac{1+\xi}{1-\xi}. \qquad \square$$

THEOREM 8.8. *Under the conditions of Lemma* 8.3,

$$\frac{\|x-y\|}{\|x\|} \leqslant \omega\left(\|A^{-1}\|\frac{\alpha\|x\|+\beta}{\|x\|}\right)\left(\frac{1}{1-\xi}\right).$$

PROOF. There are different ways to prove this result. For instance,

$$\begin{aligned}
y - x &= A^{-1}\Delta b - A^{-1}\Delta A y,\\
\|y-x\| &\leqslant \beta\omega\|A^{-1}\| + \alpha\omega\|A^{-1}\|\,\|y\|,\\
\frac{\|y-x\|}{\|x\|} &\leqslant \|A^{-1}\|\left(\frac{\beta\omega}{\|x\|} + \frac{\alpha\omega}{1-\alpha\omega\|A^{-1}\|}\left(1+\frac{\xi\beta}{\alpha\|x\|}\right)\right)\\
&\leqslant \frac{\|A^{-1}\|}{\|x\|}(\alpha\|x\|+\beta)\left(\frac{\omega}{1-\alpha\omega\|A^{-1}\|}\,\frac{\alpha\|x\|+\beta\xi}{\alpha\|x\|+\beta} + \frac{\beta\omega}{\alpha\|x\|+\beta}\right)\\
&\leqslant \omega\|A^{-1}\|\frac{\alpha\|x\|+\beta}{\|x\|}\left(\frac{1}{1-\xi}\,\frac{\alpha\|x\|+\beta\xi}{\alpha\|x\|+\beta} + \frac{\beta}{\alpha\|x\|+\beta}\right)\\
&\leqslant \omega\|A^{-1}\|\frac{\alpha\|x\|+\beta}{\|x\|}\left(\frac{1}{1-\xi}\right).
\end{aligned}$$

If $\alpha = \|A\|$ and $\beta = \|b\|$, then

$$\alpha + \frac{\beta}{\|x\|} = \|A\|\left(1 + \frac{\|b\|}{\|A\|\;\|x\|}\right) \leqslant 2\|A\|,$$

and

$$\frac{\|y-x\|}{\|x\|} \leqslant 2\omega\|A\|\|A^{-1}\|\left(\frac{1}{1-\xi}\right). \qquad \square$$

Let us define

$$K_{\mathrm{T}}(A,b) = \|A^{-1}\|\frac{\alpha\|x\|+\beta}{\|x\|}.$$

This is the (normwise) condition number of the problem that "measures" the sensitivity of the solution of $Ax = b$ to perturbations. The subscript T refers to Turing. Notice that if there are only perturbations in A, then $\beta = 0$ and $K_{\mathrm{T}}(A) = \|A^{-1}\|\,\|A\|$.

Now, let η_{T} be defined on the set of all possible perturbations such that $(A+\Delta A)y = b+\Delta b$, where y is the computed solution (see CHATELIN and FRAYSSE [1993]),

$$\eta_{\mathrm{T}} = \inf\{\omega \mid \omega \geqslant 0,\ \|\Delta A\| \leqslant \omega\alpha,\ \|\Delta b\| \leqslant \omega\beta,\ (A+\Delta A)y = b+\Delta b\}.$$

THEOREM 8.9. *Let $r = b - Ay$ be the residual,*

$$\eta_{\mathrm{T}} = \frac{\|r\|}{\alpha\|y\|+\beta}.$$

PROOF. As $(A+\Delta A)y = b+\Delta b$, we have

$$\Delta Ay - \Delta b = r.$$

Therefore,

$$\|r\| \leqslant \omega(\alpha\|y\|+\beta),$$

which implies that for all cases

$$\omega \geqslant \frac{\|r\|}{\alpha\|y\|+\beta}.$$

Now, we only have to exhibit a case for which equality occurs. Consider $\alpha = 0$ (e.g., $\Delta A = 0$) and $\Delta b = \omega b$. Then,

$$Ay = (1+\omega)b \quad \text{or} \quad \omega b = -r.$$

This gives

$$\omega = \frac{\|r\|}{\|b\|},$$

which is the equality we were looking for. $\square$

η_T is sometimes called the (normwise) backward error. It measures the minimal distance to a perturbed problem that is solved exactly by the computed solution. We have seen that (approximately), the forward error ($\|y - x\|/\|x\|$) is the condition number times the backward error. Unfortunately, most of the time, the bounds that we have just established are too pessimistic and do not reflect the reality of the computation.

A topic that has been widely studied (see, e.g., CLINE, MOLER, STEWART and WILKINSON [1979]) is finding estimates for the normwise condition number. This is much less interesting today as people are mainly considering other ways to measure sensitivity to perturbations.

Beginning in the sixties, notably in the work of F.L. Bauer, a new style of perturbation analysis has been developed considering componentwise perturbations. This has later been studied by SKEEL [1979] and is much in favor today. This perturbation analysis tries to assess the consequences of having perturbations on individual elements of A or b.

Componentwise error analysis
We consider perturbations ΔA and Δb such that

$$|\Delta A| \leqslant \omega E, \qquad |\Delta b| \leqslant \omega f,$$

and

$$(A + \Delta A)y = b + \Delta b,$$

where E and f are given matrix and vector of nonnegative entries and for a matrix B, $|B|$ denotes the matrix whose entries are the absolute values or modulus of the entries of B. Clearly, if $E_{i,j} = 0$ (resp. $f_i = 0$), then $(\Delta A)_{i,j} = 0$ (resp. $(\Delta b)_i = 0$), e.g., the corresponding entry is not perturbed.

Common choices for E and f are $(|A|, |b|)$, $(0, |b|)$, $(|A|, 0)$. This allows also special choices for sparse matrices taking into account the sparsity structure. We choose to measure distances with the $\|\bullet\|_\infty$ norm. The analysis is almost the same as for the normwise case.

THEOREM 8.10. *If* $\omega \| \, |A^{-1}|E\|_\infty < 1$,

$$\frac{\|y - x\|_\infty}{\|x\|_\infty} \leqslant \omega \frac{\| \, |A^{-1}|(E|x| + f)\|_\infty}{\|x\|_\infty} \frac{1}{1 - \omega \| \, |A^{-1}|E\|_\infty}.$$

PROOF. We have

$$y - x = A^{-1}\Delta b - A^{-1}\Delta A(y - x) - A^{-1}\Delta A x.$$

Taking absolute values,

$$|y - x| \leqslant \omega |A^{-1}| \big(f + E|x|\big) + \omega |A^{-1}| E |y - x|.$$

By taking norms, we obtain the result. □

The (componentwise) condition number of the problem is defined as

$$K_{\mathrm{BS}}(A,b)=\frac{\|\,|A^{-1}|(E|x|+f)\|_\infty}{\|x\|_\infty}.$$

The subscript BS refers to Bauer and Skeel.

Other condition numbers can be exhibited depending on the metric that is chosen, see CHATELIN and FRAYSSE [1993]. Remark that if $f=0$ and $E=|A|$ then,

$$K_{\mathrm{BS}}\leqslant \big\||A^{-1}||A|\big\|.$$

Let

$$\eta_{\mathrm{BS}}=\inf\big\{\omega\mid\omega\geqslant 0,\ |\Delta A|\leqslant\omega E,\ |\Delta b|\leqslant\omega f,\ (A+\Delta A)y=b+\Delta b\big\},$$

be the componentwise backward error. OETTLI and PRAGER [1964] proved the following result.

THEOREM 8.11.

$$\eta_{\mathrm{BS}}=\max_i\frac{|(b-Ay)_i|}{(E|y|+f)_i}.$$

An algorithm is said to be backward stable when the backward error is of the order of the machine precision u. Gaussian elimination with partial pivoting is both normwise and componentwise backward unstable. As we have said, there are examples where η_{T} or η_{BS} are large compared to machine precision.

Despite this fact, the method can be used safely on most practical examples. We will also see later that there are some remedies to this backward instability.

Componentwise condition numbers

CHANDRASEKARAN and IPSEN [1995] analyzed the errors in components of the solution of a linear system when the right-hand side is perturbed.

THEOREM 8.12. *Let*

$$\begin{aligned}Ax&=b,\\ Ay&=b+\Delta b\end{aligned}$$

and c_i^{T}, $1\leqslant i\leqslant n$, *be the rows of* A^{-1}. *Moreover, let* β_i *be the angle between* c_i *and* b, ψ_i *be the angle between* c_i *and* Δb *and* $\varepsilon_b=\|\Delta b\|/\|b\|$, *then (when* $x_i\neq 0$*)*

$$\frac{y_i-x_i}{x_i}=\frac{\|\Delta b\|\cos\psi_i}{\|b\|\cos\beta_i}=\frac{\|b\|}{\|A\|\,\|x\|}\frac{\|x\|}{x_i}\|A\|\|c_i\|\varepsilon_b\cos\psi_i.$$

PROOF. We have

$$y = A^{-1}(b + \Delta b).$$

Then,

$$y - x = A^{-1}\Delta b.$$

Therefore,

$$y_i - x_i = c_i^{\mathrm{T}}\Delta b = \|\Delta b\|\,\|c_i\|\cos\psi_i,$$

and

$$x_i = c_i^{\mathrm{T}} b = \|c_i\|\,\|b\|\cos\beta_i,$$

from which the result follows. □

Perturbing the matrix, we get the following result.

THEOREM 8.13. *Let*

$$\begin{aligned} Ax &= b, \\ (A + \Delta A)y &= b. \end{aligned}$$

Moreover, let ψ_i *be the angle between* c_i *and* $\Delta A\, y$ *and* $\varepsilon_A = \frac{\|\Delta A\, y\|}{\|A\|\,\|y\|}$, *then (when* $x_i \neq 0$)

$$\frac{y_i - x_i}{x_i} = -\frac{1}{\cos\beta_i}\frac{\|\Delta A y\|}{\|b\|}\cos\psi_i = -\frac{\|y\|}{x_i}\|A\|\,\|c_i\|\varepsilon_A\cos\psi_i.$$

PROOF. We have

$$y - x = -A^{-1}\Delta A y.$$

Therefore,

$$y_i - x_i = -c_i^{\mathrm{T}}\Delta A y = -\|c_i\|\,\|\Delta A y\|\cos\psi_i,$$

from which the result follows. □

These results lead CHANDRASEKARAN and IPSEN [1995] to define the componentwise conditions numbers as

$$\frac{\|y\|}{|x_i|}, \quad \|A\|\,\|c_i\|.$$

A geometric interpretation of these choices is given in CHANDRASEKARAN and IPSEN [1995]. Relating these results to componentwise perturbation analysis, we have the following result.

THEOREM 8.14. *If we have* $|\Delta b| \leqslant \omega |b|$, *then,*

$$\frac{|y_i - x_i|}{|x_i|} \leqslant \omega \frac{|c_i^{\mathrm{T}}||b|}{|c_i^{\mathrm{T}} b|}.$$

If $|\Delta A| \leqslant \omega |A|$, *then*

$$\frac{|y_i - x_i|}{|x_i|} \leqslant \omega \frac{|c_i^{\mathrm{T}}||A||y|}{|x_i|}.$$

PROOF. Straightforward. □

A posteriori errors bounds

Suppose that the matrix A is symmetric and positive definite and we have an approximate solution x_0. Then, we know that the error e satisfies

$$Ae = r_0 = b - Ax_0,$$

and therefore

$$\|e\|^2 = \left(A^{-1} r_0, A^{-1} r_0\right) = \left(A^{-2} r_0, r_0\right).$$

Since A is symmetric, it can be written as

$$A = Q \Lambda Q^{\mathrm{T}},$$

where Q is an orthonormal matrix whose columns are the normalized eigenvectors of A and Λ is a diagonal matrix whose diagonal elements are the eigenvalues λ_i,

$$\lambda_1 \leqslant \lambda_2 \leqslant \cdots \leqslant \lambda_n.$$

Then, if u and v are two vectors and f is a given function,

$$\begin{aligned} \left(u, f(A)v\right) &= \left(u, Qf(\Lambda)Q^{\mathrm{T}} v\right) \\ &= \left(\alpha, f(\lambda)\beta\right) \\ &= \sum_{i=1}^{n} f(\lambda_i)\alpha_i \beta_i. \end{aligned}$$

This last sum can be considered as a Riemann–Stieltjes integral

$$I[f] = \big(u, f(A)v\big) = \int_{\lambda_1}^{\lambda_n} f(\lambda)\,\mathrm{d}\alpha(\lambda),$$

where the measure is piecewise constant and defined by

$$\alpha(\lambda) = \begin{cases} 0 & \text{if } \lambda \leqslant \lambda_1, \\ \sum_{j=1}^{i} \alpha_i \beta_i & \text{if } \lambda_i < \lambda \leqslant \lambda_{i+1}, \\ \sum_{j=1}^{n} \alpha_i \beta_i & \text{if } \lambda_n < \lambda. \end{cases}$$

The case we are interested in is $u = v = r_0$ and $f(x) = 1/x^2$. This was considered in DAHLQUIST, EISENSTAT and GOLUB [1972]. Numerical methods for obtaining bounds on the integral (and therefore on the error) are given in GOLUB and MEURANT [1994]. These methods use quadrature formulas (Gauss, Gauss–Radau and Gauss–Lobatto) to approximate the integral. These formulas rely on orthogonal polynomials (for the given measure) which are computed by running a few iterations of the Lanczos process. Therefore, these computations involve only matrix $\times$ vector products and operations on tridiagonal matrices generated by the Lanczos algorithm. As long as only a few iterations are needed, this involves only $\mathrm{O}(n^2)$ operations. These methods give bounds on the elements of the inverse of A or for the norm of the error e.

Back to the examples

It is interesting to look at these condition numbers and backward errors for some of the examples we have defined before. Remember that we have chosen $\beta = 10$ and $t = 3$.

In Example 8.1, we have

$$A = \begin{pmatrix} 3 & 3 \\ 1 & 0.99 \end{pmatrix},$$

the exact solution is $x = (1\ 1)^{\mathrm{T}}$ and the computed solution is $y = (1.99\,0)^{\mathrm{T}}$. The (exact) inverse of A is

$$A^{-1} = 10^2 \begin{pmatrix} -0.33 & 1 \\ 1/3 & -1 \end{pmatrix}.$$

Then, $K_{\mathrm{T}}(A) = 666$ and $\eta_{\mathrm{T}} = 0.0033$. The product of these two quantities is 2.19.

Looking at componentwise analysis, we have $K_{\mathrm{BS}} = 399$, much smaller than K_{T} and $\eta_{\mathrm{BS}} = 0.005$. Then, the product of these two quantities is about 2.

Notice that we have

$$\frac{\|y - x\|}{\|x\|} = 0.995, \qquad \frac{\|y - x\|_\infty}{\|x\|_\infty} = 1,$$

the bounds being only off by a factor of 2. Componentwise, we have

$$\left[\frac{y_i - x_i}{x_i}\right]_{i=1,2} = \begin{pmatrix} 0.989 \\ -1 \end{pmatrix}.$$

Notice that $|c_i^{\mathrm{T}}|$ is 105.3 for $i = 1$ and 105.4 for $i = 2$. Finally, we remark that the two lines corresponding to the two equations are almost the same explaining the problems we got computing the intersection.

Now, we consider Example 8.2 where

$$A = \begin{pmatrix} 30 & 1 \\ 10 & 1 \end{pmatrix},$$

the exact solution is $x = (1\,1)^{\mathrm{T}}$ and the computed solution is $y = (0.99\,1.04)^{\mathrm{T}}$. The (exact) inverse of A is

$$A^{-1} = \begin{pmatrix} 0.05 & -0.05 \\ -0.5 & 1.5 \end{pmatrix}.$$

$K_{\mathrm{T}}(A) = 50.08$ and $\eta_{\mathrm{T}} = 0.00587$. The product of the condition number and the backward error is 0.294.

For componentwise analysis, $K_{\mathrm{BS}} = 31.999$ and $\eta_{\mathrm{BS}} = 0.0084$. Then, the product of these two quantities is about 0.27.

Looking at relative errors, we have

$$\frac{\|y - x\|}{\|x\|} = 0.0291, \qquad \frac{\|y - x\|_\infty}{\|x\|_\infty} = 0.04.$$

Notice this is much smaller than the bounds involving the condition numbers. Componentwise, we have

$$\left[\frac{y_i - x_i}{x_i}\right]_{i=1,2} = \begin{pmatrix} -0.01 \\ 0.04 \end{pmatrix}.$$

Notice that $|c_i^{\mathrm{T}}|$ is 0.0707 for $i = 1$ and 1.5811 for $i = 2$. This indicates that the error should be larger on the second component, which is what we observe.

Let us look at Example 8.5,

$$A = \begin{pmatrix} 3\,10^{-2} & 1 \\ 1 & 1 \end{pmatrix},$$

the exact solution is $x = (1.0003 \;\; 0.9997)^{\mathrm{T}}$ and the computed solution is $y = (0\,1)^{\mathrm{T}}$. The (exact) inverse of A being

$$A^{-1} = \begin{pmatrix} -1.003 & 1.003 \\ 1.003 & -0.003 \end{pmatrix}.$$

$K_T(A) = 2.63$ and $\eta_T = 0.618$. The product of the condition number and the backward error is 1.6237.

For componentwise analysis, $K_{BS} = 3$ and $\eta_{BS} = 1$. Then, the product $K_{BS}\eta_{BS}$ is 3.

For relative errors, we have

$$\frac{\|y-x\|}{\|x\|} = 0.709, \qquad \frac{\|y-x\|_\infty}{\|x\|_\infty} = 1.$$

Componentwise, we have

$$\left[\frac{y_i - x_i}{x_i}\right]_{i=1,2} = \begin{pmatrix} -1 \\ 0.003 \end{pmatrix}.$$

We note that $|c_i^T|$ is 1.418 for $i = 1$ and 1.003 for $i = 2$, indicating that we have a larger error in the first component.

Scaling

Scaling is a transformation of the linear system to be solved trying to give a better behaved system before using Gaussian elimination. Let D_1 and D_2 be two nonsingular diagonal matrices. The system $Ax = b$ is transformed into

$$A'y = (D_1 A D_2)y = D_1 b,$$

and the solution x is recovered as $x = D_2 y$. Notice that left multiplication by D_1 is a row scaling and right multiplication by D_2 is a column scaling. The entry $a_{i,j}$ of A is changed into $d_i^1 d_j^2 a_{i,j}$ where d_i^l, $l = 1, 2$ are the diagonal entries of D_l.

Notice that if one uses Gaussian elimination with partial pivoting to solve the scaled system, row scaling influences the choice of the pivot. Classical strategies can be found in CURTIS and REID [1972]. Other proposals were done by HAGER [1984]. A common strategy for row scaling is to divide the entries of a row by the infinity norm of the row. SKEEL [1979] showed that a good scaling matrix is choosing the diagonal elements of D_1 as $d_i = (|A||y|)_i$ where y is the computed solution. Of course, this is impractical as the solution y depends on the scaling. However, if an approximation c of the solution is known, then A could be scaled by $(|A||c|)_i$.

Today, no scaling strategy has been shown to give consistently better results than not using any scaling at all although many different strategies have been proposed over the years. If the diagonal elements of D_1 and D_2 are (as it has been suggested) powers of the machine base then, if there are no overflows or underflows and no pivoting is used, the computed solution is the same as without scaling.

It has been argued that the real role of scaling is to alter the pivoting sequence. This can result in a better or worst solution. The rule of thumb given in POOLE and NEAL [1992] is that when scaling can lead to a system which is more diagonally dominant, it could be useful and otherwise it should be avoided.

Further remarks: nearness to singularity

When perturbing only the matrix entries, the normwise condition number is

$$K_I(A) = \|A\| \|A^{-1}\|,$$

the index I referring to the fact that this is also the condition number for matrix inversion. The singular value decomposition of A can be written, see GOLUB and VAN LOAN [1989],

$$A = U \Sigma V^{\mathrm{T}},$$

where U and V are n by n orthogonal matrices and

$$\Sigma = \operatorname{diag}(\sigma_1, \ldots, \sigma_n), \quad \sigma_1 \geqslant \cdots \geqslant \sigma_n \geqslant 0.$$

σ_n is the distance to the nearest singular matrix, if we measure distances in the l_2 norm or in the Frobenius norm. If we use the l_2 norm and if $\sigma_n > 0$, it is easy to see that

$$K_I(A) = \frac{\sigma_1}{\sigma_n}.$$

If we normalize A such that $\|A\|_2 = 1$, then $K_I(A) = 1/\sigma_n$. The condition number is the inverse of the distance to the nearest singular matrix. In DEMMEL [1992c], J. Demmel studied the possible extension of these results to componentwise analysis. We briefly state these results informally. If we only perturb A, the condition number is

$$K(A, E) = \| |A^{-1}| E \|.$$

Let $\omega(A, E)$ be the smallest ω satisfying $|\Delta A| \leqslant \omega E$ and such that $A + \Delta A$ is singular. It has been proved (ROHN [1990]), that

$$\omega(A, E) = \frac{1}{\max_{S_1, S_2} \rho(S_1 A^{-1} S_2 E)},$$

where S_1 and S_2 are diagonal matrices with ± 1 on the diagonal and ρ is the spectral radius (max of modulus of the eigenvalues). If $E_{i,j} = 1$ and $\Omega = \{(x, y) \mid x_i = \pm 1,\ y_i = \pm 1\}$, then

$$\omega(A, E) = \frac{1}{\max_{(x,y) \in \Omega} |x^{\mathrm{T}} A^{-1} y|}.$$

Computing $\omega(A, E)$ is an NP-complete problem. There are no such results as for the normwise case relating the condition number to nearness to singularity. However, DEMMEL [1992c] proved the following bounds,

$$\omega(A, E) \geqslant \frac{1}{K(A, E)}$$

and

$$\frac{1}{\max_{i,j}(|A^{-1}_{i,j}|E_{i,j})} \geqslant \omega(A, E).$$

9. Iterative refinement

We have seen in the round off error analysis that the computed solution satisfies

$$(A + H)y = b,$$

with

$$\|H\|_\infty \leqslant u\, C\|A\|_\infty,$$

if the growth factor is bounded. Let $r = b - Ay$ be the residual, then

$$\|r\|_\infty \leqslant \|H\|_\infty \|y\|_\infty \leqslant uC\|A\|_\infty \|y\|_\infty.$$

Hence, if C is not too large, Gaussian elimination produces a small residual. But, small residuals do not always imply high accuracy in the solution. Let $e = x - y$, then

$$Ae = b - Ay = r.$$

Therefore, one natural idea is to solve

$$Ae = r.$$

This will produce a computed solution $\tilde{e}$, satisfying

$$(A + H)\tilde{e} = r$$

and we set $\tilde{x} = y + \tilde{e}$ as the new approximation to the solution. If we like, we can iterate this process. This algorithm is known as iterative refinement (or iterative improvement).

Here, the main question is to know to which precision the residual r has to be computed since we are not able to get the exact answer. If we have $\tilde{r} = fl(b - Ay)$ and $(A + H)\tilde{e} = \tilde{r}$, SKEEL [1979] has shown that computing the residual with the same precision as the computations is enough to make Gaussian elimination with partial pivoting backward stable. This is stated in the following theorem.

THEOREM 9.1. *If $u|A|\,|A^{-1}|$ is small enough, one step of iterative refinement with single precision residual computation is componentwise backward stable.*

PROOF. We just give a rough informal sketch of the proof, see SKEEL [1980] for details. All the constants below are functions of n only. The computed solution y verifies $(A+H)y=b$ and

$$|b-Ay| \leqslant uC\,|A|\,|x|,$$

with $|Hy| = \mathrm{O}(u)$. More specifically

$$|Hy| \leqslant Cu\,|A|\,|x| + \mathrm{O}\big(u^2\big).$$

The residual $\tilde{r} = fl(b-Ay) = Hy + z$ is such that

$$|z| \leqslant Cu|A|\,|x| + \mathrm{O}\big(u^2\big).$$

When solving for the error, we have

$$(A+H)\tilde{e} = \tilde{r}.$$

Clearly, $|H\tilde{e}| = \mathrm{O}(u^2)$. Finally

$$\begin{aligned} b - A\tilde{x} &= b - Ay - A\tilde{e} \\ &= b - Ay + H\tilde{e} - Hy - z. \end{aligned}$$

Therefore,

$$|b - A\tilde{x}| \leqslant Cu\,|A|\,|x| + \mathrm{O}\big(u^2\big).$$

We also have,

$$|A|\,|\tilde{x}| = |A|\,|x| + \mathrm{O}(u).$$

These two statements imply the componentwise backward stability. □

However, one step of iterative refinement only gives a small backward error. It does not guarantee a better accuracy. If this is wanted, the residual must be computed in double precision.

10. Geometric analysis

In POOLE and NEAL [1991] and NEAL and POOLE [1992], was studied the geometry of Gaussian elimination. The solution of a linear system of order n is viewed as finding the intersection of n hyperplanes.

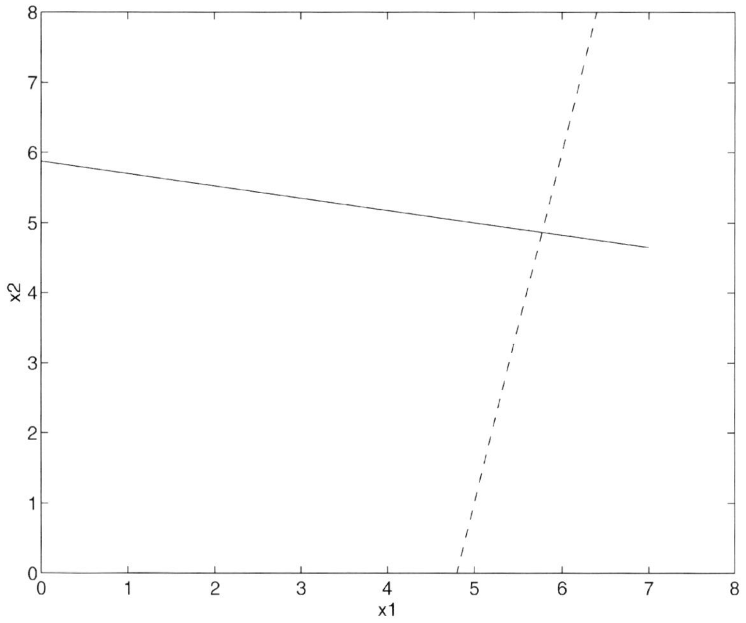

FIG. 10.1.

Consider the following simple example, using three decimal digits arithmetic, close to an example given in POOLE and NEAL [1991],

$$H_1\colon\ 5x_1 - x_2 = 24,$$
$$H_2\colon\ 14x_1 + 80x_2 = 470$$

whose exact solution using 3-digits arithmetic is $(5.77, 4.86)$. The geometric interpretation of this system is shown on Fig. 10.1.

Notice that H_1 and H_2 are nearly orthogonal. If we use partial pivoting, this system is transformed into

$$\begin{aligned} H_2\colon\ 14x_1 + 80x_2 &= 470, \\ H_1'\colon\ 29.5x_2 &= 143. \end{aligned}$$

This gives $x_2 = 4.85$

H_1' is parallel to the x_1 axis but with an approximate ordinate value. As H_2 is also nearly parallel to the x_1 axis, the error on x_2 is amplified and we find $x_1 = 5.85$.

Generalizing this example, an hyperplane H_i (in the upper triangular system obtained by the forward sweep) is said to be poorly oriented with respect to the x_i-axis if there exists an entry $u_{i,j}$, $i \neq j$, such that $|u_{i,j}| \geqslant |u_{i,i}|$. In this case, the backward sweep may lead to large errors.

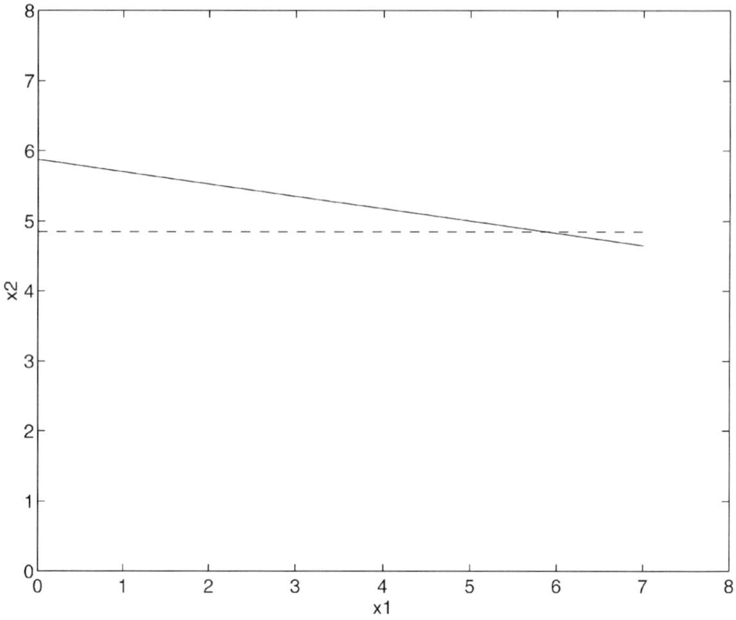

FIG. 10.2.

Even if there are no rounding errors in the forward sweep, everything can be spoiled by the backward sweep. Consider the following bidiagonal matrix,

$$B = \begin{pmatrix} 1 & -\alpha & & & \\ & 1 & -\alpha & & \\ & & \ddots & \ddots & \\ & & & 1 & -\alpha \\ & & & & 1 \end{pmatrix}.$$

If we solve the system

$$B\varepsilon = (0 \dots 0\ \beta)^{\mathrm{T}},$$

then, we have

$$\begin{aligned} \varepsilon_n &= \beta, \\ \varepsilon_i &= \alpha^{n-i}\varepsilon_n. \end{aligned}$$

Consequently, if $\alpha > 1$, the value of ε_n is amplified by this backward sweep even if the computations are done in exact arithmetic.

Therefore, in Gaussian elimination and under certain circumstances, if there is an error in the last component (coming for the forward sweep), it may be amplified by the backward sweep.

In POOLE and NEAL [1991], it is stated that a good pivoting strategy must address potential problems in both phases of Gaussian elimination. Notice that partial pivoting address only problems in the first phase.

The pivoting strategy should be such that during the first phase, the arrangement of equations gives the ith hyperplane as nearly orthogonal as possible to the x_i-axis. Partial pivoting gives only a good hyperplane orientation in L. Complete pivoting addresses also the orientation problem in U. So, generally, there are less problems in the backward sweep with complete pivoting.

However, examples can be found (POOLE and NEAL [1991]) where partial pivoting is better than complete pivoting. It is stated in POOLE and NEAL [1991] that although the geometric analysis does not favor the use of partial pivoting, empirical evidence shows that for "most" linear systems partial pivoting gives an acceptable computed solution. This is attributed to the fact that in most systems hyperplanes are not nearly parallel to any axis at all and that computer precision is sufficiently good to postpone instability that could occur. Rook's pivoting was designed to address these problems.

In NEAL and POOLE [1992], the concept of error multipliers was introduced to look at the amplification of errors in the backward sweep. Let H_k be the upper Hessenberg submatrix of U of order k such that $H_k = [u_{i,j}]$, $n-k \leqslant i \leqslant n-1$, $n-k+1 \leqslant j \leqslant n$, and ε be the difference of the computed and exact solutions of $Ux = b$. Then,

$$\varepsilon_{n-k} = \frac{(-1)^k \det H_k}{\prod_{i=n-k}^{n-1} u_{i,i}} \varepsilon_n.$$

The multiplicative factor here is called the error multiplier (EM). It is clear that if some of the EMs are large then, a large amplification of errors can occur. When EMs are less or equal than 1, the backward sweep is generally without problems.

Interesting examples are shown in NEAL and POOLE [1992], particularly Example 2.5 where a 5×5 system is given for which there is no rounding error using IEEE single precision arithmetic during the forward sweep. However, very large errors arise in the backward sweep coming from large EMs. Using IEEE double precision on this system, a good solution is obtained.

The examples in NEAL and POOLE [1992] show that, usually, partial pivoting produces an upper triangular system whose exact solution is very close to the solution of the original system. Partial pivoting also tends to produce hyperplanes that are well oriented with respect to each other.

From these studies, it is also clear that contrived examples exhibiting almost any behaviour (with partial pivoting, complete pivoting or rook's pivoting) can be constructed.

CHAPTER III

Vector and Parallel Algorithms for General Systems

11. Introduction

In this chapter, we consider the implementation of Gaussian elimination on vector and parallel computers.

It is always difficult to write about such a topic in a book or a review like this one, as the computer architectures evolve so fast that the risk is to be outdated almost the next day. Since the first years after World War II and the advent of the first real computers (with stored programs), there has been a tremendous progress both in the ease of use of scientific computers and in their floating point performances.

When I started learning numerical analysis in 1970, top scientific computers like the IBM 360-91 were running at about 1–2 Mflops (that is 10^6 floating point operations per second). Although there were already some forms of parallelism hidden in the computer architectures (particularly for the I/Os), they were mainly sequential machines. This culminated in the well known Control Data CDC 7600 that was running at about 5–10 Mflops on typical codes. Then, came the era of vector computers which has not ended yet in 1998. The first one commercially available (1976) was the CRAY 1 which had a peak speed of 160 Mflops and an average speed of 10 to 30 Mflops on real production codes.

These machines introduced a new kind of implementation problem. Roughly speaking, on sequential computers the most efficient algorithm was the one with good numerical properties that had the smallest total number of floating point operations. This is not true anymore on a vector machine. There is a lot of difference between the scalar and vector speed processing, typically a factor of 10. To run fast an algorithm has to be expressed (if possible) in terms of vectors. Moreover, the longer is the vector length, the faster is the speed of execution (up to an asymptotic value). The memory traffic issue is also very important to get good performances. Usually the fastest vector computers are the ones with the largest memory bandwith and the speed of algorithms depend also on the ratio of floating point operations to memory references.

The vector computers have now evolved into machines with several vector processors sharing a common large memory. Today (1998), these machines run at a few (1–100) Gflops (10^9 floating point operations per second).

Recently (end of the 80's, beginning of the 90's), a new kind of scientific computers have appeared on the commercial market. These machines are parallel computers with distributed memory. Although this is a quite old idea (the parallel computer Illiac IV was built in 1972), only recently had these machines attained a level of reliability good enough to allow their use in the industry. Right now, they deliver merely about the same performance as the top of the line parallel vector supercomputers. However, it has been shown on specific examples that they can reach Teraflops (10^{12} floating point operations per second). Now, some people are even seriously considering building Petaflops computers (10^{15} floating point operations per second) in about 20–25 years from now (1998). Of course, to fully exploit these architectures, we need to have parallel algorithms and suitable programming models.

It is remarkable that through all these changes the analysis that has been developed in the 60's by WILKINSON [1965] is still relevant. However, it should be noticed that together with the evolution of computers and particularly with huge shared or distributed memories, larger and larger problems are solved (see EDELMAN [1993]) and this can possibly raise some new computational problems concerning the accuracy of computations.

There is such a lot of different architectures available today, all with their details, that it is almost impossible to have a unified treatment of the implementation problems of Gaussian elimination on these computers. Therefore, we will concentrate on what we feel is the most important ways to construct reliable and fast software, that is basic algorithms. We will study this for parallel distributed architectures without looking too much into the details of implementation or performance issues.

12. BLAS routines

Software reuse and portability are issues that are almost as important as performances. In the 70's, LAWSON, HANSON, KINCAID and KROGH [1979] described a set of basic routines commonly called the BLAS (Basic Linear Algebra Subprograms) for linear algebra problems. They are an aid to clarity and portability and also performance, as these routines can be implemented as efficiently as possible by each manufacturer, possibly in assembly language.

It turns out that the basic frequently occurring operations are most often the same in all linear algebra problems, so it was useful to develop a standard interface for these kernels. The set of operations described in 1979 is now referred as Level 1 BLAS or BLAS1. These routines are mainly concerned with vector operations like

$$\begin{aligned}
y &= \alpha x + y \quad \text{(Saxpy)},\\
\alpha &= x^{\mathrm{T}} y \quad \text{(Sdot)},\\
x &= \alpha x \quad \text{(Sscal)},\\
x &= y \quad \text{(Scopy)}\\
\alpha &= \|x\|_2 \quad \text{(Snrm2)},
\end{aligned}$$

where x and y are vectors and α is a scalar, the S standing for single precision. For a complete list, see ANDERSON, BAI, BISCHOF, DEMMEL, DONGARRA, DU CROZ, GREENBAUM, HAMMARLING, MCKENNEY, OSTROUCHOV and SORENSEN [1992].

The well known linear algebra package LINPACK (for linear systems solves and least square problems) was written using BLAS1 and published in 1979 (BUNCH, DONGARRA, MOLER and STEWART [1979]). BLAS1 involves $O(n)$ floating point operations on $O(n)$ data items, n being the length of the vectors.

At the time where the BLAS1 appeared and LINPACK was completed, the vector computers appeared on the market. One can think this was fine since the BLAS1 defined vector operations. Unfortunately, this is not completely true and the BLAS1 and LINPACK had poor performance on vector machines. The reason for that is the value of the ratio of floating point operations to data loads and stores. It is too low to keep the processor busy all the time, in which case the performance could be smaller (sometimes by a large amount) than the peak theoretical speed. In BLAS1, there are very few possibilities of data reuse in vector registers or memory caches.

An additional set of routines called the Level 2 BLAS (BLAS2) was then designed, (DONGARRA, DU CROZ, HAMMARLING and HANSON [1988]). They are based on matrix × vector operations. Examples are

$$
\begin{aligned}
y &= \alpha Ax + \beta y, \\
x &= Tx, \\
A &= \alpha xy^{\mathrm{T}} + A, \\
x &= T^{-1}x,
\end{aligned}
$$

where α, β are scalars, x, y are vectors, A and T are matrices, T being triangular.

Most algorithms of linear algebra can be coded using Level 2 BLAS including Gaussian elimination. Level 2 BLAS involves $O(n^2)$ floating point operations on $O(n^2)$ data items. Therefore, the ratio of Level 1 BLAS is not improved. But, e.g., in the first kernel above, data can be kept in the vector registers, improving the computational speed.

To improve on this point and to increase the data locality, a level 3 BLAS (BLAS3) has been proposed in 1990, see DONGARRA, DU CROZ, HAMMARLING and DUFF [1990], that defines matrix × matrix operations. Examples are

$$
\begin{aligned}
C &= \alpha AB + \beta C, \\
C &= \alpha AA^{\mathrm{T}} + \beta C, \\
C &= \alpha AB^{\mathrm{T}} + \alpha BA^{\mathrm{T}} + \beta C, \\
B &= \alpha TB, \\
B &= \alpha T^{-1}B,
\end{aligned}
$$

α, β are scalars, A, B, C, T are matrices, T being triangular.

There, we have $O(n^3)$ floating point operations on $O(n^2)$ data items helping to improve the data reuse. Level 3 BLAS shows very good performances on vector supercomputers and computers with a memory hierarchy. Performance close to the peak speed is frequently obtained.

13. LAPACK (and follow ons)

At the end of the 80's, at the same time the Level 2 BLAS and Level 3 BLAS have appeared, a new software project has been developed. Its goal was to supersede both LINPACK and EISPACK (a well known package for eigenvalues computations) and also to obtain better performances. The computers targeted were parallel vector supercomputers with shared memory. Another goal was to improve the quality and accuracy of the algorithms, particularly for eigenvalue computations.

The first version of LAPACK (for Linear Algebra PACKage) has appeared officially in 1992, see ANDERSON, BAI, BISCHOF, DEMMEL, DONGARRA, DU CROZ, GREENBAUM, HAMMARLING, MCKENNEY, OSTROUCHOV and SORENSEN [1992] and the second one in 1994. The strategy of LAPACK to obtain portable codes that are also efficient is to construct the software as much as possible using calls to the BLAS. The BLAS 2 and 3 can achieve near peak performance on the targeted architectures. Moreover, it allows also to exploit parallelism in a transparent way.

Partitioned block forms of the LU (or other) factorizations (where point algorithms are used in which operations have been grouped together) are used in LAPACK in order to use Level 2 and 3 BLAS. Some routines also exist in LAPACK that return bounds on the componentwise backward error.

LAPACK has been extended in an almost transparent way to distributed memory parallel computers, see the ScaLAPACK library (CHOI, DONGARRA, POZZO and WALKER [1994]).

14. Triangular systems solvers on distributed memory computers

Introduction

Although only $O(n^2)$ operations are required for the forward and backward solves compared to the $O(n^3)$ operations of the LU factorization, the solution of triangular systems is an interesting challenge on parallel computers. Moreover, quite often, several systems with different right-hand sides have to be solved and then, the cost of triangular solutions can be as large as the one for factorization.

In considering an algorithm for a triangular solve on a distributed memory parallel computer, two main issues have to be studied. The first one is to find some parallelism in a process which, at first sight, seem mostly sequential. The second issue is finding good data distributions in the local memories associated with each processor to minimize memory transfers.

Many parallel algorithms have been devised for solving triangular systems, see HELLER [1978]. However, many of them assumed having $O(n^3)$ processors available and are not of practical use on present machines that have between a hundred and a thousand processors. Here, we mainly follow HEATH and ROMINE [1988] and EISENSTAT, HEATH, HENKEL and ROMINE [1988] and then, we describe some alternatives.

Let us start with serial algorithms and let

$$Lx = b,$$

to be solved, where L is lower triangular. There are basically two ways to compute the solution. The first one is the classical way, in which components of x are computed exactly one after another,

```
for i=1:n
  for j=1:i-1
    b(i)=b(i)-l(i,j)*x(j)
  end
  x(i)=b(i)/l(i,i)
end
```

This algorithm is called the scalar product algorithm as the main operation is computing the scalar product of the ith row of L (except the diagonal element) with the vector of the components already computed.

The second algorithm uses the fact that after a component is computed, the right-hand side can be modified at once,

```
for j=1:n
  x(j)=b(j)/l(j,j)
  for i=j+1:n
    b(i)=b(i)-l(i,j)*x(j)
  end
end
```

We called this algorithm the Saxpy algorithm since the main loop is clearly a Saxpy operation, $x(j)$ being the scalar. Notice that this corresponds to swapping the two loops of the algorithm.

We will first examine parallel algorithms where the data is distributed either by rows or by columns. Then, the important issue is the mapping of rows and columns to processors' memories. We define this mapping by `map(j)`: the number (or address) of the processor to which row (or column) `j` is mapped.

The most commonly used mapping is wrapping. We suppose $n \gg p$ where p is the number of processors. For simplifying purposes, suppose that p divides n exactly. The simplest way of defining a wrap mapping is the following,

$$\begin{pmatrix} \texttt{j:} & 1 & 2 & 3 & \dots & p & p+1 & \dots & 2p & 2p+1 & \dots & n \\ \texttt{map(j):} & 1 & 2 & 3 & \dots & p & 1 & \dots & p & 1 & \dots & p \end{pmatrix}$$

The advantage of this mapping is its simplicity. However, a potential problem is that each processor does not receive the same number of matrix entries, particularly if only the lower triangular part of L is stored. Suppose we distribute the (nonzero) elements of columns. Processor 1 will have

$$\sum_{i=0}^{n/p-1} n - ip = \sum_{i=1}^{n/p} n - (i-1)p = (n+p)\frac{n}{p} - \frac{p}{2}\frac{n}{p}\left(\frac{n}{p}+1\right) = \frac{n^2}{2p} + \frac{n}{2},$$

elements while processor p will receive

$$\sum_{i=1}^{n/p} n - ip + 1 = (n+1)\frac{n}{p} - \frac{n}{2}\left(\frac{n}{p}+1\right) = \frac{n^2}{2p} + \frac{n}{p} - \frac{n}{2}.$$

This can cause some load imbalance. The problem can be fixed (in this simple case) by reflecting the mapping in the following way:

$$\begin{pmatrix} \texttt{j:} & 1 & 2 & 3 & \dots & p & p+1 & \dots & 2p & 2p+1 & \dots & n \\ \texttt{map(j):} & 1 & 2 & 3 & \dots & p & p & \dots & 1 & 1 & \dots & \dots \end{pmatrix}$$

One can easily check that in two consecutive sets of indices, each processor has the same number of elements. Therefore, if n/p is even, each processor receives the same number of entries.

These mappings can be generalized by considering blocks of consecutive rows or columns instead of individual rows or columns. Doing so decreases communication time but increases load imbalance. Methods using these mappings are called panels methods in Rothberg's Ph.D. thesis (ROTHBERG [1993]).

Fan out and fan in algorithms

These two algorithms seek the parallelism in the inner loops of the Saxpy and the scalar product algorithms. We consider first the Saxpy algorithm. Clearly, the components of b can be computed in parallel. Each component b_i is going to be computed by one particular processor. To be able to achieve this in parallel, b_i must be in the memory of the processor computing this entry. Therefore, the data must be distributed by rows.

Suppose that each processor has a set { `myrows` } containing the indices of rows the memory of the processor is storing. As soon as the x_j component of the solution is computed, it must be broadcasted to all other processors. This is done in a fan-out operation. `Fan-out(x,proc)` means that processor `proc` sends `x` (located in its memory) to all other processors. The algorithm (the code running on one processor) is the following (HEATH and ROMINE [1988])

```
for j=1:n
  if j ∈ { myrows }
    x(j)=b(j)/l(j,j)
    fan-out(x(j),map(j))
  end
  for i ∈ { myrows }
    b(i)=b(i)-l(i,j)*x(j)
  end
end
```

This code is not exactly the one that will really be used on a parallel machine as it uses a global indexing scheme. In a real code indices local to each processor may have to be used, depending on the programming model. Nevertheless, this code helps understanding what is going on.

The implementation of the fan-out (broadcast) operation depends on the computer architecture, particularly the topology of the communication network. It provides the necessary synchronization as one processor sends data and all the others wait to receive it. One problem is that only one word (x_j) is sent at a time. Usually, sending a message of l words costs

$$t = t_0 + \tau l,$$

t_0 is the start-up time (or latency). The efficiency depends on the value of t_0 relatively to τ and l. For sending only one word, the cost is essentially the start-up time. Therefore, this algorithm can be efficient only on computers with a small latency.

Now, we look at the scalar product algorithm. Parallelism is found in the inner loop, computing the scalar product. To be able to do so, the data has to be distributed by columns. Then, each processor can compute the `l(i,j)*x(j)` term for `j` in its column index set { `mycolumns` }. These partial contributions must be added to the ones of the other processors. This is done in the fan-in operation: `Fan-in(x,proc)` means that processor `proc` receives the sum of all the `x`'s over all processors. Again, the implementation of this operation depends on the computer architecture. One naive way to do it, it that each processor sends its contribution to processor 0 that does the summation and then, broadcast the result to all processors. However, depending on the architecture, there are much more efficient ways to implement this operation. Generally, one can obtain an $\log_2 n$ computational time.

Wavefront algorithms

Fan-in and fan-out algorithms are seeking parallelism in the inner loop. The algorithms in this section look for parallelism in the outer loop. However, here some data dependencies have to be respected, e.g., x_{i-1} has to be computed before x_i.

Consider the Saxpy algorithm with data distributed by columns this time. Informally, the algorithm is the following. Let y be an n-vector, processor `map(1)` computes x_1 and starts computing the updates $y_i = l_{i,1}x_1$. After computing σ such components ($1 \leqslant \sigma \leqslant n-1$) that we called a segment, processor `map(1)` sends them to `map(2)` and resumes computing the next σ update components. As soon as the data is received, processor `map(2)` computes x_2 and the updates with x_2 in the first segment. When it is completed, it is send to processor `map(3)`, etc....

Having completed the update for a segment, `map(j)` sends it to `map(j+1)`. After a start-up phase, all processors are working concurrently. A smaller segment size increases parallelism particularly at the beginning, but increases also the number of messages required. A value of $\sigma = n-1$ gives a purely serial algorithm.

The same idea can be applied to the scalar product algorithm with the data distributed by rows.

Cyclic algorithms and variations

The algorithms in the two previous sections can be used with any mapping of rows or columns although this choice will have an impact on performance. In this section, following HEATH and ROMINE [1988], we study an algorithm that has been introduced by LI and COLEMAN [1989] to exploit the wrap mapping and to reduce the communication volume.

Consider the Saxpy algorithm with the columns distributed using the wrap mapping on p processors. A segment of size $p-1$ passes from processor to processor and accumulates all the necessary updates. At step j, processor `map(j)` receives the segment from processor `map(j-1)` and uses its first element to compute x_j. It deletes the first element, updates the other components and appends a new element to the segment. The processor `map(j)` sends the new segment to processor `map(j+1)` and starts computing update components involving x_j that will be used when the segment will return to `map(j)` according to the wrap mapping.

```
for i=1:n
  t=0
  for j ∈ { mycolumns } & j<i
    t=t+l(i,j)*x(j)
  end
  s=fan-in(t,map(i))
  if i ∈ { mycolumns }
    x(i)=(b(i)-s)/l(i,i)
  end
end
```

These algorithms have several shortcomings. The main one is that they do not exploit all the parallelism that can be found in the problem. For example, take the Saxpy algorithm. After $x(j)$ has been received, each processor updates all of its b components. However, after b_{j+1} has been updated, processor `map(j+1)` can immediately compute x_{j+1} and start sending it to everybody else before updating the other components of b.

Similar modifications can be made to the scalar product algorithm. However, complete asynchronism is difficult to implement. The data dependencies of the classical forward sweep are illustrated in Fig. 14.1 on a 5×5 example where an arrow indicates that the computation where it started from must be completed before the computation where it is pointing to can take place. It illustrates the data flow of the algorithm.

As stated in HEATH and ROMINE [1988], the efficiency of the previous algorithm depends on how much computation can be done in the time between successive appearances of the segment in a processor.

Numerical experiments in HEATH and ROMINE [1988] show that for a given problem size, when increasing the number of processors, there is a point where the computer time increases rather than decreases.

Cyclic algorithms have been improved in this respect in EISENSTAT, HEATH, HENKEL and ROMINE [1988]. One obvious modification is to use ideas from the wavefront algorithm and break the segment into several subsegments. A subsegment is broadcasted to the next processor as soon as possible. In this modification, there are more

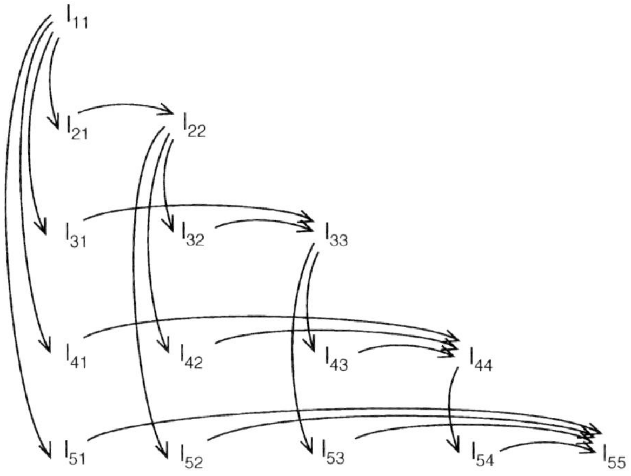

FIG. 14.1.

messages to be sent but the communication pattern is still a ring-like one. This is called a pipelined cyclic algorithm.

Another modification is to send the subsegments not anymore to `map(j+1)`, but directly to `map(k)` where `k` is the index of the first element in the subsegment. Of course, additional synchronizations need to be done to be sure that processor `map(k)` has received all the necessary information before proceeding to compute x_k. Another variant is not using equally sized subsegments but instead having subsegments of size $2, 4, \ldots, p/2$ and one of size 1 sent to the predecessor in the ring.

Numerical experiments comparing these different algorithms are given in EISENSTAT, HEALTH, HENKEL and ROMINE [1988].

Other algorithms

The algorithms in the previous sections are parallel implementations and variations of sequential algorithms. They have the same number of floating point operations, although they are processed in a different order.

However, one can imagine other ways of computing the solution of $Lx = b$. Some are based on computing the inverse of L by various means. The following algorithm is due to SAMEH and BRENT [1977].

Suppose for simplicity that $l_{i,i} = 1$. Then, it is easy to see that

$$L = \prod_{i=1}^{n-1} N_i^{-1},$$

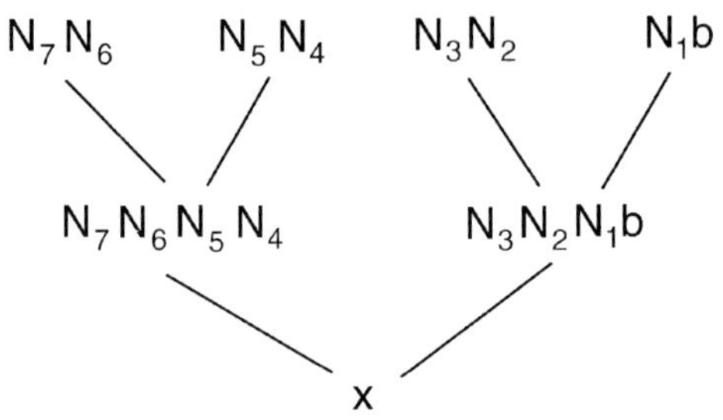

FIG. 14.2.

where $N_i^{-1} = I + l_i e_i^{\mathrm{T}}$ with

$$l_i = \begin{pmatrix} 0 \\ 0 \\ l_{i,i+1} \\ \vdots \\ l_{n,i} \end{pmatrix}.$$

By Lemma 3.5, we know that

$$N_i = I - l_i e_i^{\mathrm{T}}.$$

Therefore,

$$x = L^{-1} b = \prod_{i=n-1}^{1} N_i b.$$

The product on the right-hand side can be computed by a logarithmic algorithm. Suppose $n = 8$, then

$$x = N_7 N_6 N_5 N_4 N_3 N_2 N_1 b.$$

This can be written as

$$x = (N_7 N_6)(N_5 N_4)(N_3 N_2)(N_1 b).$$

The products within each parenthesis can be formed in parallel, the results combined again two by two, etc. . . .

This algorithm requires $\log_2 n$ stages. However, the total number of floating point operations is about $n^3/10$, that is much larger than the usual $\mathrm{O}(n^2)$ operations of the serial algorithm. For this algorithm to be usable, a very large number of processors is required.

Another way of writing the matrix is

$$L = I - \bar{L},$$

where $\bar{L}$ is strictly lower triangular and therefore $\bar{L}^n = 0$. Then,

$$\begin{aligned} x &= (I - \bar{L})^{-1} b \\ &= \big(I + \bar{L} + \bar{L}^2 + \cdots + \bar{L}^{n-1}\big) b \\ &= \big(I + \bar{L}^{2n-1}\big)\big(I + \bar{L}^{2n-2}\big) \cdots (I + \bar{L}) b. \end{aligned}$$

Again, this can be evaluated in $\log_2 n$ steps.

Finally, another algorithm is obtained by partitioning the matrix L as

$$L = \begin{pmatrix} L_{1,1} & 0 \\ L_{2,1} & L_{2,2} \end{pmatrix}.$$

Then,

$$L^{-1} = \begin{pmatrix} L_{1,1}^{-1} & 0 \\ -L_{2,2}^{-1} L_{2,1} L_{1,1}^{-1} & L_{2,2}^{-1} \end{pmatrix}.$$

The process can be reiterated in parallel for $L_{1,1}$ and $L_{2,2}$ and so on. Then, if $x = (x_1, x_2)^{\mathrm{T}}$ and $b = (b_1, b_2)^{\mathrm{T}}$,

$$\begin{aligned} x_1 &= L_{1,1}^{-1} b_1, \\ x_2 &= L_{2,2}^{-1} b_2 + L_{2,2}^{-1} L_{2,1} L_{1,1}^{-1} b_1. \end{aligned}$$

15. LU factorization on distributed memory computers

A very large number of papers have been written over the years on the parallelization of the LU factorization algorithms. Actually, it is easier to exhibit parallelism in the LU factorization than in the triangular solves. It can be easily seen that in the variants of LU factorization we have studied, there is a phase where some elements are modified and that these modifications are independent of each other, giving rise to some parallelism. A problem that we had not to face with triangular solves is pivoting and we will see that it can cause some troubles.

There are a lot of variants that can be thought of for the parallel LU factorization. Here, we first follow the exposition of GEIST and ROMINE [1988].

As for the triangular solves, we consider the matrix to be distributed by rows or columns. Notice that, for a better efficiency, this distribution has to be consistent with the one used for the triangular solve. Otherwise, some data redistribution has to be done between the two phases. We use partial pivoting on the kij form of the factorization.

Row storage scheme

Suppose first that the matrix is stored by rows. A possible algorithm is the following.

```
for k=1:n-1
  find pivot row r
  if r ∈ { myrows }
    broadcast pivot row
  else
    receive pivot row
  end
  for i>k & i ∈ { myrows }
    m(i,k)=a(i,k)/a(k,k)
    for j=k+1:n-1
      a(i,j)=a(i,j)-m(i,k)*a(k,j)
    end
  end
end
```

Analyzing the communications, we see that since the kth column is scattered amongst processors, communication must take place between processors to find which row is the pivot row. The way it is done depends on the computer architecture. Note that in the previous algorithm – which is denoted by RSRP (Row Storage Row Pivoting) in GEIST and ROMINE [1988] – no explicit exchange of rows takes place and this can cause some load imbalance depending on the pivot choices. CHU and GEORGE [1987] proposed to use explicit exchange of rows with a wrap mapping. Obviously, this will produce a communication overhead and the question is to know if this is offset by an improved load balancing. Examples in GEIST and ROMINE [1988] demonstrate that, in some cases, it is beneficial to use explicit row exchanges. Geist and Romine show how to decrease the number of communications by relaxing the requirement that the final distribution of rows be a wrap mapping. They ask that rows kp through $(k+1)p-1$ lie in distinct processors for each k. A processor that contains one of these pivot rows cannot have another one and otherwise must exchange rows with a processor that does not contain one. Different strategies for choosing the processor to exchange the row with are given in GEIST and ROMINE [1988].

Column storage scheme

This algorithm is denoted by CSRP (Column Storage Row Pivoting) in GEIST and ROMINE [1988]. Here, updating the matrix is done by columns (kji form) instead of being done by rows in RSRP. However, the main differences are when computing the multipliers and searching for the pivot row. As the coefficient matrix is stored by columns, the computation of the multipliers is done serially by the processor owning the pivot column. This, of course, reduces the parallel efficiency of the factorization. On the other hand, finding the pivot row is done without communication by only one processor. Moreover, partial pivoting has no incidence on load balancing as the mapping of columns to processors is not changed by pivoting.

```
for k=1:n-1
  if k ∈ { mycolumns }
    find pivot row r
    for i=k+1:n
      m(i,k)=a(i,k)/a(k,k)
    end
    broadcast m and pivot index
  else
    receive m and pivot index
  end
  for j> k & j ∈ { mycolumns }
    for i=k+1:n
      a(i,j)=a(i,j)-m(i,k)*a(k,j)
    end
  end
end
```

Numerical experiments in GEIST and ROMINE [1988] show that the results of CSRP as defined before are worst than those of RSRP. This is essentially due to the serial part of the algorithm.

A way to improve this algorithm is to use some form of pipelining. For instance, the processor containing column k can compute some multipliers and send them allowing some of the other processors to start their computations. Using these modifications CSRP is competitive with RSRP.

Block storage scheme

In DONGARRA and WALKER [1993], it is proposed to use a block cyclic data distribution and block partitioned versions of LU factorization for distributed memory parallel computers. In the approach of Dongarra and Walker, all the parallelism is to be find at the level of the BLAS routines, implying that the top layers of the source code of the parallel version look very similar to the ones of LAPACK.

Independent data distributions are used for rows and columns. An object m (a piece of row or column) is mapped to a couple (p, i), p being the processor number and i the location in the local memory of this processor. By using wrapping as before, we have

$$m \to \left(m \bmod p, \lfloor m/p \rfloor\right).$$

Blocking consists of assigning contiguous entries to processors by blocks,

$$m \to \left(\lfloor m/L \rfloor, m \bmod L\right), \quad L = \lceil m/p \rceil.$$

The block cyclic distribution is a combination of both. Blocks of consecutive data are distributed by wrapping,

$$m \to (q, b, i),$$

q is the processor number, b the block number in processor q and i the index in block b. If there are r data objects in a block, then

$$m \to \left(\left\lfloor \frac{m \bmod T}{r} \right\rfloor, \left\lfloor \frac{m}{T} \right\rfloor, m \bmod r \right), \qquad T = rp.$$

A slight generalization of this is given in DONGARRA and WALKER [1993] by adding offsets to the processor number.

To distribute the matrix, independent block cyclic distributions are applied for the rows and columns. The processors are supposed to be (logically) arranged in a two-dimensional mesh and referred by couples (q_1, q_2). For general data distributions, communications are required for the pivot search and the computation of the multipliers. The communications to be done are broadcast to all processors and broadcast to all processors in the same row (or column) in the 2D mesh of processors.

Of course, the logical arrangement of processors has to be mapped to the physical layout. This is architecture dependent. In DONGARRA and WALKER [1993], experiments are reported on the Intel Delta computer.

Dense block-oriented factorizations are also studied in Rothberg's Ph.D. thesis (ROTHBERG [1993]). In his approach, if there are p processors arranged in a 2D mesh, block (i, j) of the matrix is mapped to processor $i \bmod \sqrt{p},\ j \bmod \sqrt{p}$. Then, two variants are studied; the first one is a destination–computes approach where all updates for a block are computed on the processor that owns the destination block. The second one is a source–computes approach where updates are computed by a processor that owns one of the source blocks. It is shown that the first approach gives better results. Details can be found in ROTHBERG [1993].

CHAPTER IV

Gaussian Elimination for Sparse Linear Systems

16. Introduction

An area that has seen a rapid development since the end of the 60's is research on sparse matrices. So far, in the previous chapters, we have addressed several properties of the matrices we considered like symmetry or positive definiteness, but we did not care if some of the matrix entries were zero or not.

Now, we are going to look at this possibility in details. A sparse matrix is one with many zero entries. However, it is difficult to give a precise definition of what a sparse matrix is and to say how many zeroes must be there or even what percentage of zeroes we should have.

We will see later on that special techniques are used to store sparse matrices and that special algorithms are going to be defined in order to minimize the storage and the number of operations during Gaussian elimination. Therefore, a definition that has sometimes been given is that a matrix is sparse when it pays (either in computer storage or in computer time) to use these special sparse techniques as opposed to the more traditional dense (or general) algorithms that we described before.

Nevertheless, exploiting sparsity allows solving very large problems with even millions of unknowns by 1996.

There are several good books about sparse linear systems. Let us mention those of GEORGE and LIU [1981] for symmetric positive definite systems and DUFF, ERISMAN and REID [1986] for more general systems.

A drawback of these sparse techniques is that they are quite complex (actually more complex than dense algorithms) and sometimes difficult to optimize. Therefore, it is usually not feasible for the average user to write a sparse code from scratch. Fortunately, there exist some good packages available containing well tuned codes like the Harwell Library, DUFF and REID [1993], or SPARSPAK, GEORGE and LIU [1981], to mention just a few.

In this part about sparse matrices, we are going first to introduce some definitions and to stress some basic facts. We will look at the case of symmetric and particularly positive definite matrices as it is a little easier to introduce some of the techniques in this framework. Then, we will move on to the general case of nonsymmetric sparse matrices.

17. Basic storage schemes and fill-in

Storage schemes

Dealing with sparse matrices, our aim is to be able to avoid storing the zero entries of the sparse matrix A and to avoid doing operations on these zeroes. The most natural way to do this is to store only the nonzero entries $a_{i,j}$ of A, together with the row and column indices i and j. Therefore, if nz is the number of nonzeroes of A, the storage needed is nz floating point numbers and 2 nz integers. In most modern computers, integers use the same number of bits as floating point numbers. In that case, the total storage is 3 nz words.

However, this mode of storage (which is sometimes called the coordinate scheme) is not very convenient for Gaussian elimination as we have seen that most variants of the method require easy accesses to rows and/or columns of the matrix.

One common way to store a sparse matrix which is more suited to Gaussian elimination is to hold the nonzeroes of each row (resp. column) as a packed sparse vector `AA`, together with the column (resp. row) index of each element in a vector `JA`. These two vectors have a length of nz words. A third vector `IA` of integers of length $n + 1$ is needed to point to the beginning of each row in `AA` and `JA`. Let us look at this storing scheme on a small example.

Let

$$A = \begin{pmatrix} a_1 & 0 & 0 & a_2 \\ a_3 & a_4 & a_5 & 0 \\ 0 & a_6 & a_7 & 0 \\ a_8 & 0 & 0 & a_9 \end{pmatrix},$$

the stored quantities are

	1	2	3	4	5	6	7	8	9
`AA`	a_1	a_2	a_3	a_4	a_5	a_6	a_7	a_8	a_9
`JA`	1	4	1	2	3	2	3	1	4
`IA`	1	3	6	8	10				

Notice that `IA(n+1)=nz+1`. This allows to compute the length of row i by `IA(i+1)-IA(i)`. Equivalently, the lengths of rows can be stored in place of `IA`.

Sometimes, something special is done for the diagonal elements. As often, they are all nonzero, they are stored in a special vector of length n or the diagonal entry of each row could be stored in the first (or last) position of the row. If the matrix is symmetric only the lower or upper part is stored.

Another storage scheme which is much used is linked lists. Here, each nonzero element (together with the column index) has a pointer `IPA` to the location of the next element in the row. To be able to add or delete entries easily in the list, it is sometimes handy to have also a second pointer to the location of the previous entry. Finally, we must know the beginning of the list for each row in a vector `IA`. Going back to our small example, we have the following storage when using only forward links. Notice the elements could be stored in any order,

	1	2	3	4	5	6	7	8	9
AA	a_3	a_2	a_9	a_1	a_4	a_8	a_6	a_5	a_7
JA	1	4	4	1	2	1	2	3	3
IPA	5	0	0	2	8	3	9	0	0
IA	4	1	7	6					

Notice that a value of zero in `IPA`(•) indicates the end of the list for the given row.

Later on, we will see other schemes for storing sparse matrices that are more dedicated to some particular algorithms. Some other storage schemes exist that are more suited to iterative methods, specially those which need only matrix vector products. However, whatever the choice is for the storage scheme, it must be able to deal with the phenomenon of fill-in.

The fill-in phenomenon

Remember that at the kth step of the algorithm, we have to compute

$$a_{i,j}^{(k+1)} = a_{i,j}^{(k)} - \frac{a_{i,k}^{(k)} a_{k,j}^{(k)}}{a_{k,k}^{(k)}}.$$

Therefore, even if $a_{i,j}^{(k)} = 0$, $a_{i,j}^{(k+1)}$ can be nonzero if $a_{i,k}^{(k)} \neq 0$ and $a_{k,j}^{(k)} \neq 0$. Nonzero entries in the L and U factors in positions (i, j) for which $a_{i,j} = 0$ are called the fill-ins.

Consider the small example below where the matrix is symmetric and the x's stand for the nonzero entries,

$$A = \begin{pmatrix} x & x & 0 & x & 0 \\ x & x & x & 0 & 0 \\ 0 & x & x & 0 & x \\ x & 0 & 0 & x & 0 \\ 0 & 0 & x & 0 & x \end{pmatrix}.$$

We look at the different steps of Gaussian elimination where fill-ins are denoted by •,

$$A_2 = \begin{pmatrix} x & x & 0 & x & 0 \\ 0 & x & x & \bullet & 0 \\ 0 & x & x & 0 & x \\ 0 & \bullet & 0 & x & 0 \\ 0 & 0 & x & 0 & x \end{pmatrix},$$

$$A_3 = \begin{pmatrix} x & x & 0 & x & 0 \\ 0 & x & x & \bullet & 0 \\ 0 & 0 & x & \bullet & x \\ 0 & 0 & \bullet & x & 0 \\ 0 & 0 & x & 0 & x \end{pmatrix}.$$

Notice that the fill-in in position (4, 3) has been created by the fill-in in position (4, 2) at the previous step,

$$A_4 = \begin{pmatrix} x & x & 0 & x & 0 \\ 0 & x & x & \bullet & 0 \\ 0 & 0 & x & \bullet & x \\ 0 & 0 & 0 & x & \bullet \\ 0 & 0 & 0 & \bullet & x \end{pmatrix},$$

$$A_5 = \begin{pmatrix} x & x & 0 & x & 0 \\ 0 & x & x & \bullet & 0 \\ 0 & 0 & x & \bullet & x \\ 0 & 0 & 0 & x & \bullet \\ 0 & 0 & 0 & 0 & x \end{pmatrix},$$

and

$$L = \begin{pmatrix} x & & & & \\ x & x & & & \\ 0 & x & x & & \\ x & \bullet & \bullet & x & \\ 0 & 0 & x & \bullet & x \end{pmatrix}.$$

In this example, three elements which were initially zero in the lower triangular part of A are nonzero in L.

The aim of sparse Gaussian elimination is to avoid doing operations on zero entries and therefore to try to minimize the number of fill-ins. This will have the effect of both minimizing the needed storage and the number of floating point operations.

It is clear that the number of fill-ins depends on the way the pivots are chosen if pivoting is allowed. As the following well known example proves, there can be large differences in the number of fill-ins with different pivoting strategies.

Consider

$$A = \begin{pmatrix} x & x & x & x \\ x & x & & \\ x & & x & \\ x & & & x \end{pmatrix}.$$

Then,

$$L = \begin{pmatrix} x & & & \\ x & x & & \\ x & \bullet & x & \\ x & \bullet & \bullet & x \end{pmatrix},$$

i.e., all the zero entries fill. If we define a permutation matrix P such that the first element is numbered last, then

$$PAP^{\mathrm{T}} = \begin{pmatrix} x & & & x \\ & x & & x \\ & & x & x \\ x & x & x & x \end{pmatrix}.$$

In this case, there is no fill-in at all. This is called a perfect elimination.

The way the fill-in problem is handled depends on the properties of the matrix. If the matrix is nonsymmetric (and without any special properties), we have seen that we generally need to pivot to have an acceptable numerical accuracy. If in addition the matrix is sparse, we now have another requirement which is to minimize the fill-in. Therefore, these two goals have to be dealt with at the same time. Moreover, this implies that the data structure for the LU factors cannot be set up before the numerical factorization as the pivot rows and therefore the potential fill-in are only known when performing the numerical factorization.

If the matrix is symmetric and, e.g., positive definite, we do not need to pivot for numerical stability. Therefore, we have the freedom to choose symmetric permutations only to minimize the fill-in. Moreover, the number and indices of fill-ins can be determined before doing the numerical factorization since this depends only on the structure of the matrix. Hence, everything can be handled within a static data structure built in a preprocessing phase called the symbolic factorization.

Finding an ordering that minimizes the fill-in is an NP complete problem, see YANNAKAKIS [1981]. Hence, we will have to rely on heuristics that can be computed quickly.

18. Definitions and graph theory

Basic definitions

It is well known that a graph can be associated with every matrix. For a general nonsymmetric square matrix A of order n, we associate a directed graph (or digraph). A digraph is a couple $G = (X, E)$ where X is a set of nodes (or vertices) and E is a set of directed edges. For a given matrix A of order n, there are n nodes and there is a directed edge from i to j if $a_{i,j} \neq 0$. Usually, self loops corresponding to $a_{i,i} \neq 0$ are not included.

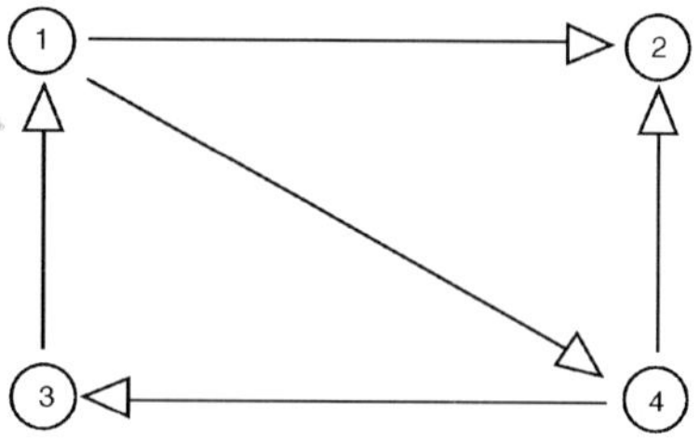

FIG. 18.1.

FIG. 18.2.

Let

$$A = \begin{pmatrix} x & x & 0 & x \\ 0 & x & 0 & 0 \\ x & 0 & x & 0 \\ 0 & x & x & x \end{pmatrix}.$$

Then, the associated digraph is given in Fig. 18.1.

Graphs are much more used in problems involving symmetric matrices. If $a_{i,j} \neq 0$, then $a_{j,i} \neq 0$. Therefore, we can look at undirected graphs and drop the arrows on the edges.

Let

$$A = \begin{pmatrix} x & x & x & 0 \\ x & x & 0 & x \\ x & 0 & x & x \\ 0 & x & x & x \end{pmatrix},$$

then, the graph of A is shown on Fig. 18.2.

Let us introduce a few definitions. Let $G = (X, E)$ be a (undirected) graph. We denote the nodes of the graph by x_i or sometimes i.

- $G' = (X', E')$ is a subgraph of G if $X' \subset X$ and $E' \subset E$.
- Two nodes x and y of G are adjacent if $\{x, y\} \in E$. The adjacency set of a node y is defined as

$$\text{Adj}(y) = \{x \in X \mid x \text{ is adjacent to } y\}.$$

If $Y \subset X$, then

$$\mathrm{Adj}(Y) = \{x \in X \mid x \in \mathrm{Adj}(y), x \notin Y, y \in Y\}.$$

- The degree of a node x of G is the number of its adjacent nodes in G,

$$\deg(x) = |\mathrm{Adj}(x)|.$$

- Let x and $y \in X$. A path of length l from x to y is a set of nodes $\{v_1, v_2, \ldots, v_{l+1}\}$ such that $x = v_1$, $y = v_{l+1}$ and $\{v_i, v_{i+1}\} \in E$, $1 \leqslant i \leqslant l$. A path $\{v_0, v_1, \ldots, v_l, v_0\}$ is a (simple) cycle of length $l+1$.

A graph is connected if for every $x, y \in X$, there exists a path from x to y. This corresponds to the matrix being irreducible.

A chord of a path is any edge joining two nonconsecutive vertices in the path. A graph is chordal if every cycle of length greater than three has a chord (BLAIR and PEYTON [1993]).

- An important kind of graphs is when there is no closed paths. A particular node is labeled as the root. Then, there is a path from any node to the root. Such a (connected) graph is called a tree. If it is not connected, we have a set of trees called a forest.
- Let $Y \subset X$, the section graph $G(Y)$ is a subgraph $(Y, E(Y))$ with

$$E(Y) = \big\{\{x, y\} \in E \mid x \in Y,\ y \in Y\big\}.$$

- A set $Y \subset X$ is a separator for G (a connected graph) if $G(X/Y)$ has two or more connected components.
- The distance $d(x, y)$ between two nodes x and y of G is the length of the shortest path between x and y.

The eccentricity of a node $e(x)$ is

$$e(x) = \max\big\{d(x, y) \mid y \in X\big\}.$$

The diameter δ of G is

$$\delta(G) = \max\big\{e(x) \mid x \in X\big\}.$$

A node x is peripheral if $e(x) = \delta(G)$.

- A clique is a subset of nodes such that they are all pairwise connected.
- A level structure of a graph G is a partition $\mathcal{L} = \{L_0, L_1, \ldots, L_l\}$ of X such that

$$\begin{aligned}
\mathrm{Adj}(L_i) &\subset L_{i-1} \cup L_{i+1}, \quad i = 1, \ldots, l-1, \\
\mathrm{Adj}(L_0) &\subset L_1, \\
\mathrm{Adj}(L_l) &\subset L_{l-1}.
\end{aligned}$$

Note that each L_i is a separator for G. For each node $x \in X$, a level structure $\mathcal{L}(x)$ can be defined as

$$\mathcal{L}(x) = \{L_0(x), \dots, L_{e(x)}(x)\},$$
$$L_0(x) = \{x\},$$
$$L_i(x) = \operatorname{Adj}\left(\bigcup_{k=0}^{i-1} L_k(x)\right), \quad 1 \leqslant i \leqslant e(x),$$

where $e(x)$ is the eccentricity of x. The width of a level structure $\mathcal{L}(x)$ is

$$w(x) = \max\{|L_i(x)|,\ 0 \leqslant i \leqslant e(x)\}.$$

Characterization of the fill-in for a symmetric structure

Now, we turn to looking at the interpretation of Gaussian elimination in terms of graphs. This was first studied in PARTER [1961], see also ROSE [1970]. We consider a sequence of graphs $G^{(i)}$, $G^{(1)} = G$ corresponding to the different steps of the elimination on a matrix with a symmetric pattern.

THEOREM 18.1. *$G^{(i+1)}$ is obtained from $G^{(i)}$ by removing the node x_i from the graph as well as all its incident edges and adding edges such that all the remaining neighbors of x_i in $G^{(i)}$ are pairwise connected.*

PROOF. It is enough to look at the first step, eliminating the node x_1 (or the corresponding unknown in the linear system). Then,

$$a_{i,j}^{(2)} = a_{i,j} - \frac{a_{i,1}a_{1,j}}{a_{1,1}}.$$

$a_{i,j}^{(2)}$ is nonzero if either $a_{i,j} \neq 0$ or, $a_{i,j} = 0$ and $a_{i,1}$ and $a_{1,j}$ are nonzero. The last possibility translates into x_i and x_j being neighbors of x_1 in the graph. When x_1 is eliminated, they will be connected by an edge representing the new element $a_{i,j}^{(2)} \neq 0$. This occurs for all the neighbors of x_1. We did not consider zeroes that arise by cancellation. In this way, $G^{(2)}$ is obtained corresponding to the submatrix obtained from $A^{(2)}$, by deleting the first row and column. □

Taking the graph $G(A)$ of A and adding all the edges that are created in all the $G^{(i)}$s during the elimination, we obtain G_F ($F = L + L^{\mathrm{T}}$), which is called the filled graph

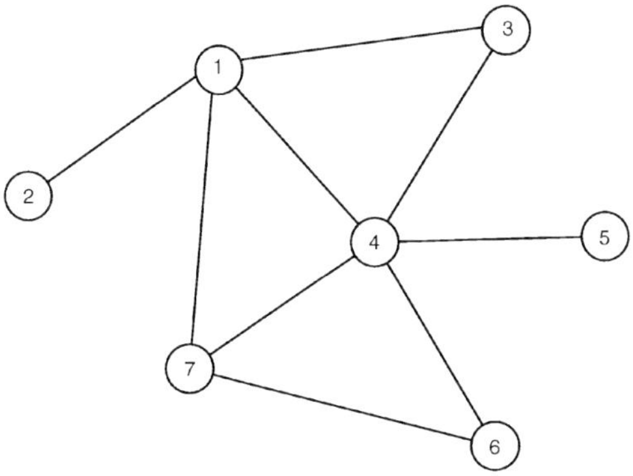

FIG. 18.3.

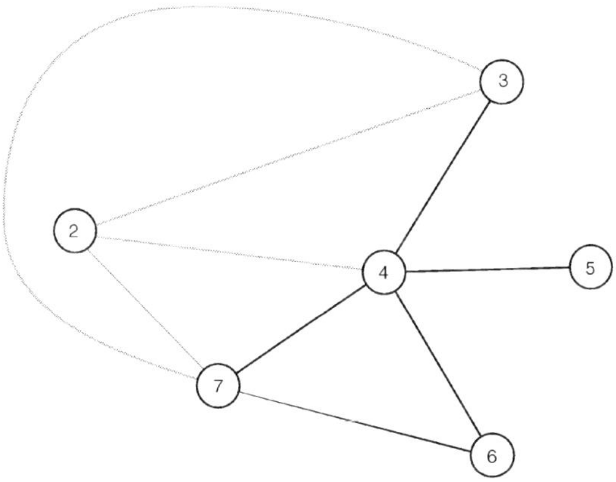

FIG. 18.4.

$G_F = (X, E^F)$. An example of elimination graphs is given below. Let

$$A = \begin{pmatrix} x & x & x & x & & & x \\ x & x & & & & & \\ x & & x & x & & & \\ x & & x & x & x & x & x \\ & & & x & x & & \\ & & & x & & x & x \\ x & & & x & & x & x \end{pmatrix},$$

then, Fig. 18.3 displays the graph $G(A)$. The graph $G^{(2)}$ is given in Fig. 18.4.

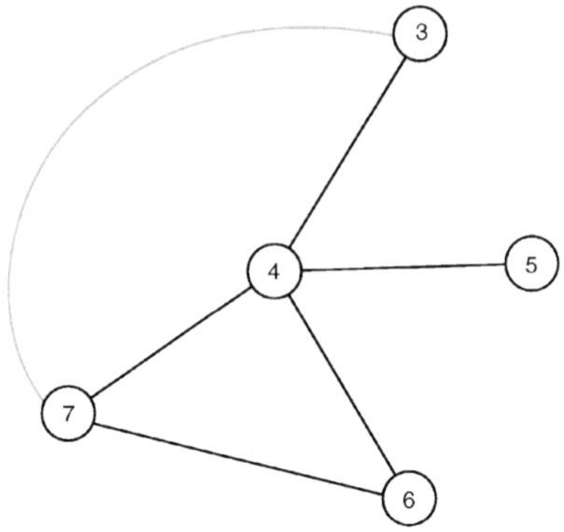

FIG. 18.5.

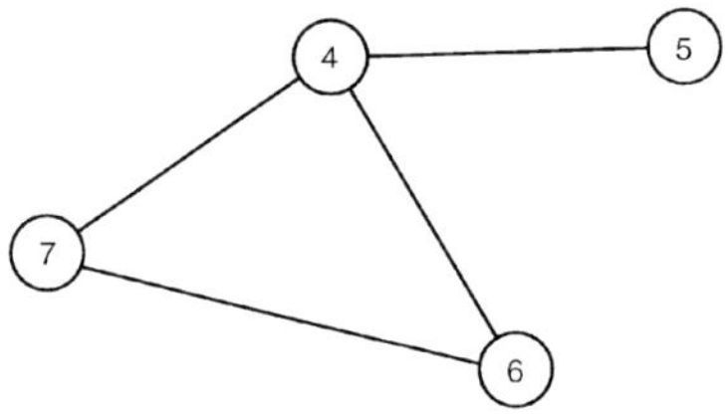

FIG. 18.6.

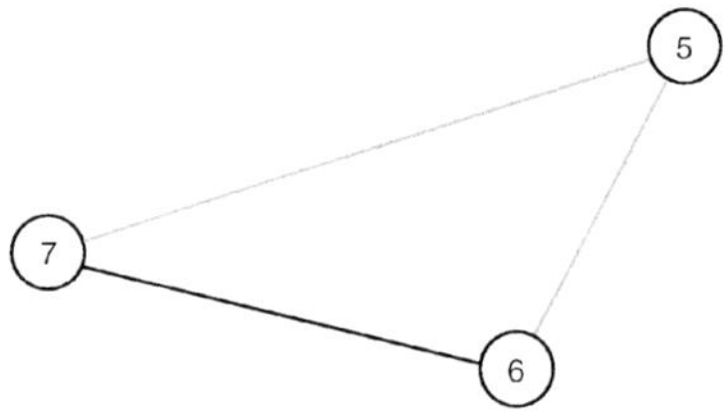

FIG. 18.7.

Here the edges corresponding to fill-ins are denoted by grey lines. The graph $G^{(2)}$ corresponds to the matrix,

$$A^{(2)} = \begin{pmatrix} x & x & x & x & & & x \\ x & x & \bullet & \bullet & & & \bullet \\ x & \bullet & x & x & & & \bullet \\ x & \bullet & x & x & x & x & x \\ & & & x & x & & \\ & & & x & & x & x \\ x & \bullet & \bullet & x & & x & x \end{pmatrix}.$$

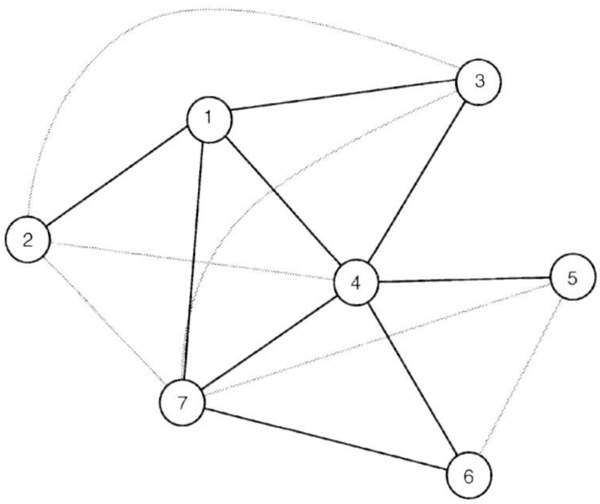

FIG. 18.8.

Then, the elimination of x_2 does not cause any fill-in since all its neighbors form already a clique. $G^{(3)}$ is given in Fig. 18.5 and $A^{(3)} \equiv A^{(2)}$, $\equiv$ meaning that the two matrices have the same structure. The next step is to eliminate x_3. Here, again, there is no fill-in as x_4 and x_7 are already connected. $G^{(4)}$ is displayed on Fig. 18.6 and $A^{(4)} \equiv A^{(3)}$. Elimination of x_4 connects x_5, x_6 and x_7. $G^{(5)}$ is shown on Fig. 18.7.

$$A^{(5)} = \begin{pmatrix} x & x & x & x & & & x \\ x & x & \bullet & \bullet & & & \bullet \\ x & \bullet & x & x & & & \bullet \\ x & \bullet & x & x & x & x & x \\ & & & x & x & \bullet & \bullet \\ & & & x & \bullet & x & x \\ x & \bullet & \bullet & x & \bullet & x & x \end{pmatrix}.$$

$G^{(5)}$ is a clique, the corresponding 3×3 submatrix is dense thus there will be no other fill-in before the end of the elimination and $A^{(7)} \equiv A^{(5)}$. The filled graph is given in Fig. 18.8.

In total there are six fill-ins in the elimination. An interesting point is to remark that the (position of) the fill-in entry created between x_3 and x_7 by the elimination of x_1 would have been created anyway by the elimination of x_2.

The number of fill-ins depends of the order of elimination, that is on the numbering of the vertices in the graph (or of the unknowns in the linear system) or otherwise said on the pivoting sequence.

For instance, in our example, at the beginning we can eliminate x_2 and x_5 as they have only one neighbor and their elimination do not create any fill-in. This gives the graph in Fig. 18.9.

Now, x_3 and x_6 can be eliminated without fill-in as they are already forming a clique with their neighbors. After eliminating x_3 and x_6, the situation is shown on Fig. 18.10.

Again, this is a clique and the nodes can be eliminated in any order without fill-in.

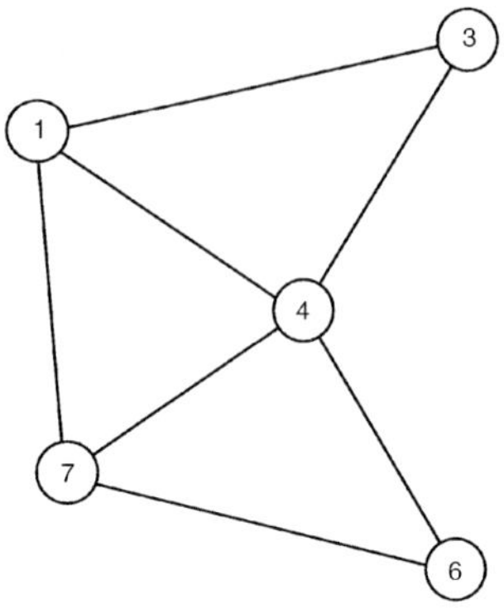

FIG. 18.9.

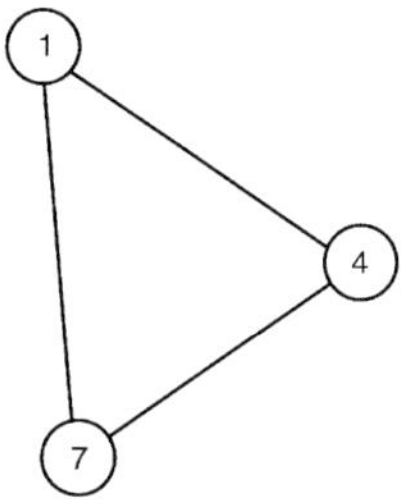

FIG. 18.10.

Therefore, we have just seen that for this example, there are several numberings that lead to a perfect elimination without any fill-in. For instance,

$$2, \quad 5, \quad 3, \quad 6, \quad 1, \quad 4, \quad 7.$$

The permuted matrix is

$$A' = PAP^{\mathrm{T}} = \begin{pmatrix} x & & & & x & & \\ & x & & & & x & \\ & & x & & x & x & \\ & & & x & & x & x \\ x & & x & & x & x & x \\ & x & x & x & x & x & x \\ & & & x & x & x & x \end{pmatrix}.$$

Of course, this is not a general situation. One thing that we must realize is that the more fill-ins we create in the early stages, the more fill-ins we will get later on. Thus, an heuristic rule is that it is likely to be beneficial to start by eliminating nodes that do not create much fill-in. These are the nodes with a small number of neighbors or nodes in cliques.

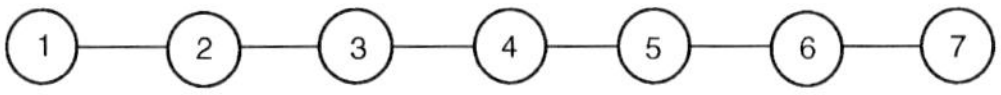

FIG. 18.11.

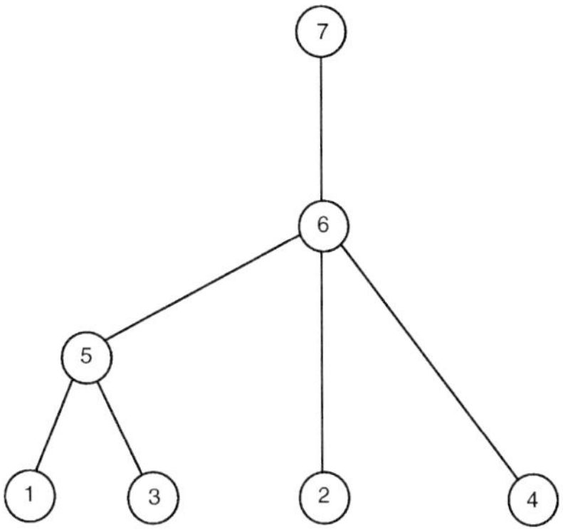

FIG. 18.12.

Let us now introduce a few more definitions.

- The elimination tree of A, symmetric matrix of order n, is a graph with n nodes such that the node p is the parent of node j, if and only if

$$p = \min\{i \mid i > j,\ l_{i,j} \neq 0\},$$

where L is the Cholesky factor of A. The elimination tree of A will be denoted by $T(A)$ or simply T if the context makes it clear that we refer to A.

Clearly, p is the index of the first nonzero element in column j of L. For the previous example without renumbering, the elimination tree $T(A)$ is simply a chain shown on Fig. 18.11.

$T(A')$ is given in Fig. 18.12 (renumbering the unknowns according to P).

This exhibits the fact that x'_1, x'_2, x'_3, x'_4 (corresponding to x_2, x_5, x_3, x_6 in the initial ordering) can be eliminated in any order (or even in parallel) as there are no dependencies between these variables.

It is of course interesting to know if there will be some fill-in between two nodes of $G(A)$ before doing the elimination. This has been studied by George some time ago, see GEORGE and LIU [1981] for a detailed exposition. It can be understood intuitively in the following way.

We have seen that there can be a fill-in between x_j and x_k only if at some step m, they are not already connected together and both neighbors of x_m, $m < j$, $m < k$. Therefore, either they were already neighbors of x_m in G or they were put in this situation by the elimination of other nodes x_l, $l < m$.

Recursively, we see that at some stage, x_j was a neighbor of one of these nodes and the same for x_k with another of these nodes. This means that in G, there is at least one

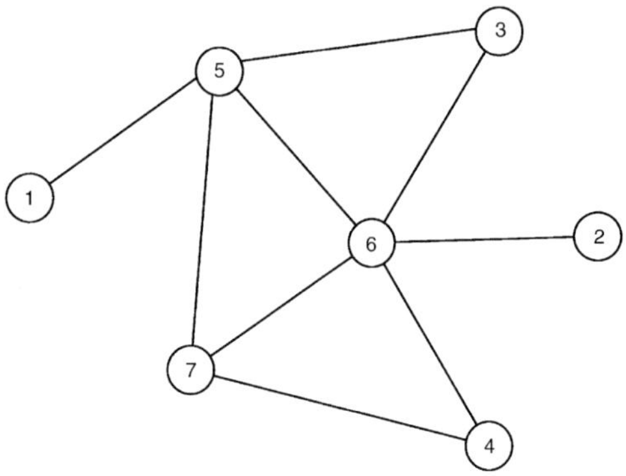

FIG. 18.13.

path between x_j and x_k and that all nodes on this path have numbers smaller than j and k. If there is no such path, there won't be a fill-in between x_j and x_k.

In our example and in the initial ordering, we see that x_2 and x_3 are linked by a path of length 2 through x_1. The same is true for x_2 and x_4 and also x_2 and x_7. No other node can be reached from x_2 in this way as nodes x_3, x_4 or x_7 are on the paths to the other nodes and they have numbers greater than 2. There will be a fill-in between x_3 and x_7 as they can be reached through x_1.

For the permuted matrix A', we have the graph of Fig. 18.13.

It can be seen that on any path from one node to another one which is not a neighbor, there is always a node with a larger number. This discussion can be formalized in the following way.

- Let $S \subset X$ and $x \in X$, $x \notin S$, x is said to be reachable from $y \notin S$ through S if there exists a path $(y, v_1, \ldots, v_k, x)$ from y to x in G such that $v_i \in S$, $i = 1, \ldots, k$. We define

$$\mathrm{Reach}(y, S) = \{x \mid x \notin S, x \text{ is reachable from } y \text{ through } S\}.$$

THEOREM 18.2 (George). *Let $k > j$, there will be a fill-in between x_j and x_k if and only if*

$$x_k \in \mathrm{Reach}\big(x_j, \{x_1, \ldots, x_{j-1}\}\big).$$

We first prove a lemma due to PARTER [1961].

LEMMA 18.1. *$\{x_i, x_j\} \in E^F$ if and only if $\{x_i, x_j\} \in E$ or $\{x_i, x_k\} \in E^F$ and $\{x_k, x_j\} \in E^F$ for some $k < \min\{i, j\}$.*

PROOF. If $\{x_i, x_k\} \in E^F$ (filled graph) and $\{x_k, x_j\} \in E^F$ for some $k < \min\{i, j\}$, then the elimination of x_k will create a fill-in between x_i and x_j. Therefore, $\{x_i, x_j\} \in E^F$.

Conversely, if $\{x_i, x_j\} \in E^F$ and $\{x_i, x_j\} \notin E$, then we have seen in the previous discussion that at some stage, x_i and x_j must be neighbors of a node, say x_k, that is going to be eliminated before x_i and x_j. Thus, $k < \min\{i, j\}$. □

PROOF OF THEOREM 18.2 (see George (GEORGE and LIU [1981]), ROSE, TARJAN and LUEKER [1976]). Suppose $x_k \in \text{Reach}(x_j, \{x_1, \dots, x_{j-1}\})$. There exists a path $\{x_j, v_1, \dots, v_l, x_k\} \in G$ with $v_i \in \{x_1, \dots, x_{j-1}\}$, $1 \leqslant i \leqslant l$. If $l = 0$ or $l = 1$, the result follows from Lemma 18.1. If $l > 1$, it is easy to show that $\{x_k, x_j\} \in E^F$ by induction.

Conversely, we assume $\{x_i, x_j\} \in E^F$, $j < k$. The proof goes by induction on j. For $j = 1$, $\{x_1, x_k\} \in E^F$ implies $\{x_1, x_k\} \in E$ since there is no fill-in with the first node. Moreover, the set $\{x_1, \dots, x_{j-1}\}$ is empty.

Suppose the result is true up to $j - 1$. By Lemma 18.1, there exists some $l \leqslant j - 1$ such that $\{x_j, x_l\} \in E^F$ and $\{x_l, x_k\} \in E^F$. By the assumption, there exists a path between x_j and x_l and another one from x_l to x_k. Clearly, this implies that there is a path from x_j to x_k whose nodes have numbers $\leqslant l \leqslant j - 1$. □

George (GEORGE and LIU [1981]) demonstrated that reachable sets can be implemented efficiently by using the notion of quotient graph.

- Let $\mathcal{P}$ be a partition of X,

$$\mathcal{P} = \{X_1, X_2, \dots, X_p\}, \quad \bigcup_{k=1}^{p} X_k = X, \; X_i \cap X_j = \emptyset \text{ if } i \neq j.$$

  The quotient graph of G is $(\mathcal{P}, \mathcal{E})$, where $\{X_i, X_j\} \in \mathcal{E}$ if and only if

$$\text{Adj}(X_i) \cap X_j \neq \emptyset, \quad i \neq j.$$

  It is denoted by $G/\mathcal{P}$.

Let $S \subset X$, following George (GEORGE and LIU [1981]), we define a partitioning of X by

$$\mathcal{C}(S) = \bigl\{C \subset S \mid G(C) \text{ is a connected component in the subgraph } G(S)\bigr\},$$
$$\bar{\mathcal{C}}(S) = \{y \mid y \in X - S\} \cup \mathcal{C}(S).$$

$G/\bar{\mathcal{C}}(S)$ can be viewed as the graph obtained by amalgamating nodes in connected sets in S.

Let $S_i = \{x_1, x_2, \dots, x_i\}$, $1 \leqslant i \leqslant n$. S_i induces a partitioning $\bar{\mathcal{C}}(S_i)$ and a quotient graph $\mathcal{G}_i = G/\bar{\mathcal{C}}(S_i)$. George proved the following result.

THEOREM 18.3. $\forall y \in X - S_i$,

$$\text{Reach}_G(y, S_i) = \text{Reach}_{\mathcal{G}_i}\bigl(y, \mathcal{C}(S_i)\bigr).$$

The advantages of the quotient graph are:

(1) elimination graphs are easily obtained from quotient graphs

(2) the quotient graph can be implemented in place. It requires no more space than the original graph structure.

Usually, the quotient graph structure is implemented via linked lists, see GEORGE and LIU [1981].

Characterization of the fill-in through elimination trees

In LIU [1990] the use of elimination trees in sparse factorization was reviewed. We will follow Liu's exposition to explain how elimination trees can be used to describe the fill-in.

Remark first that if the graph of a matrix A is a tree, then there exists a permutation P such that a perfect elimination is possible on PAP^{T}. A topological ordering of a rooted tree is defined as a numbering that orders children nodes before their parents.

The elimination tree $T(A)$ of a matrix A has, as we have seen, the same nodes as G and is a spanning tree of the filled graph G_F. Notice that $T(A)$ and $T(F)$ are identical.

We denote by $T[x]$, the subtree of $T(A)$ rooted at node x. $y \in T[x]$ is a descendant of x and x is an ancestor of y. From the definition of $T(A)$, we have that if x_i is a proper ancestor of x_j in $T(A)$, then $i > j$.

THEOREM 18.4 (Liu). *For $i > j$, the numerical values of columns i of L ($L_{*,i}$) depend on column j of L ($L_{*,j}$) if and only if $l_{i,j} \neq 0$.*

PROOF. This is obvious from what we have seen before. □

In the filled graph G_F, we can use a directed edge from x_j to x_i to indicate that column i depends on column j. The result is a digraph of the Cholesky factor. From this digraph, LIU [1990] derived the transitive reduction: if there is a directed path greater than one from x_j to x_i and a directed edge from x_j to x_i, this last edge is removed. Removing all these redundant edges gives the transitive reduction.

The transitive reduction of the filled graph generates the elimination tree structure (LIU [1990]). This is a way to compute the elimination tree from G_F. Algorithms for determining the elimination tree structure are given in LIU [1990].

Moreover,

THEOREM 18.5 (SCHREIBER [1982]). *If $i > j$ and $l_{i,j} \neq 0$, then x_i is an ancestor of x_j in $T(A)$.*

As a consequence, if $x_s \in T[x_i]$ and $x_t \in T[x_j]$ a disjoint subtree from $T[x_i]$, then $l_{s,t} = 0$. Now, we have a result related to Theorem 18.2.

THEOREM 18.6. *Let $i > j$, $l_{i,j} \neq 0$ if and only if there exists a path*

$$x_i, v_1, \ldots, v_k, x_j, \quad v_i \in X,$$

in $G(A)$ *such that* $\{v_1, \ldots, v_k\} \subseteq T[x_j]$.

The row structure of L is characterized in the following result.

THEOREM 18.7 (LIU [1990]). *Let* $i > j, l_{i,j} \neq 0$ *if and only if* x_j *is an ancestor of some* x_k *in* $T(A)$ *such that* $a_{i,k} \neq 0$.

The column structure can be characterized as well.

THEOREM 18.8 (LIU [1990]). *The structure of column* j *of* L *is given by*

$$\mathrm{Adj}_G\big(T[x_j]\big) \cup \{x_j\} = \{x_i \mid l_{i,j} \neq 0,\ i \geqslant j\}.$$

PROOF. This is a simple consequence of Theorem 18.7. □

Elimination trees are the basis for efficient implementations of symbolic factorization (LIU [1990]).

19. Band and envelope numbering schemes for symmetric matrices

Definitions

Most of the first attempts to exploit sparsity were considering band or envelope storage schemes and were trying to minimize the storage using these schemes. This can be explained after a few definitions.

- $f_i(A) = \min\{i \mid a_{i,j} \neq 0\}$.
  $f_i(A)$ is the index of the column with the first nonzero element of row i.
- $\beta_i(A) = i - f_i(A)$ is the bandwith of row i. The bandwith of A is

$$\beta(A) = \max_i\big\{\beta_i(A),\ 1 \leqslant i \leqslant n\big\}$$

  and

$$\mathrm{band}(A) = \big\{(i,j) \mid 0 < i - j \leqslant \beta(A),\ i \geqslant j\big\}.$$

These notions have led to ideas for storing the matrices A and L. If $\beta_i(A)$ is almost constant as a function of i, then it makes sense to store the entries corresponding to all the indices in band(A). However, most of the time this is not the case as they are a few rows with a larger bandwith than the other ones and too much storage is wasted by the band scheme. Then, one can use the variable band or envelope storage scheme JENNINGS [1977].

- $\mathrm{Env}(A) = \{(i,j) \mid 0 < i - j \leqslant \beta_i(A),\ i \geqslant j\}$ is the envelope of A. The profile of A, denoted by $P_r(A)$ is given by

$$P_r(A) = \big|\mathrm{Env}(A)\big| = \sum_{i=1}^{n} \beta_i(A).$$

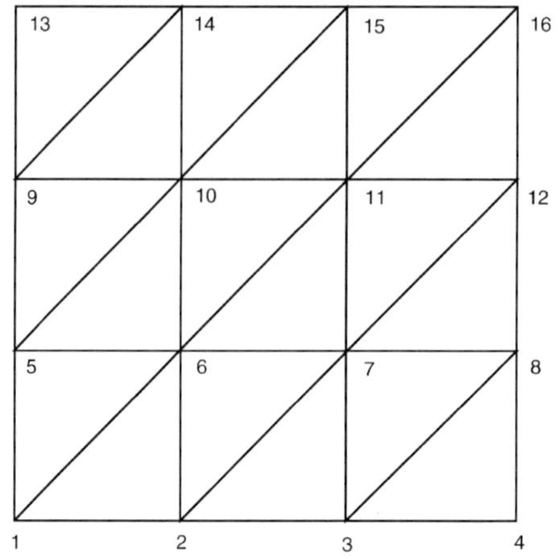

FIG. 19.1.

It is clear that a simple storage scheme is obtained by storing for each row all the elements of the envelope in a vector. We only need another vector of integers to point to the start of each row. The invention of this storage scheme was motivated by the following result.

THEOREM 19.1. *Let* $\mathrm{Fill}(A) = \{(i, j) \mid i > j,\ a_{i,j} = 0,\ l_{i,j} \neq 0\}$ *be the index set of the fill-ins,*

$$\mathrm{Fill}(A) \subset \mathrm{Env}(A).$$

PROOF. This is a consequence of Theorem 18.2 as it is clear from that result that there cannot be any fill-in from a node x_i to a node x_j whose number is smaller than the smallest number of the neighbors of x_i. All the paths going from x_i to x_j will have a node with a number larger than x_j. □

Giving these simple storage schemes, it was natural to try to devise ordering schemes that minimize the bandwith or the profile of the matrix. This is an NP-complete problem.

The Cuthill–McKee and reverse Cuthill–McKee orderings

The Cuthill–McKee algorithm is a local minimization algorithm to reduce the profile of A.

It is clear that if we number the nodes sequentially and, at some stage, we would like to minimize $\beta_i(A)$, then we must number immediately all the nonnumbered nodes in $\mathrm{Adj}(x_i)$. Then, the algorithm is the following

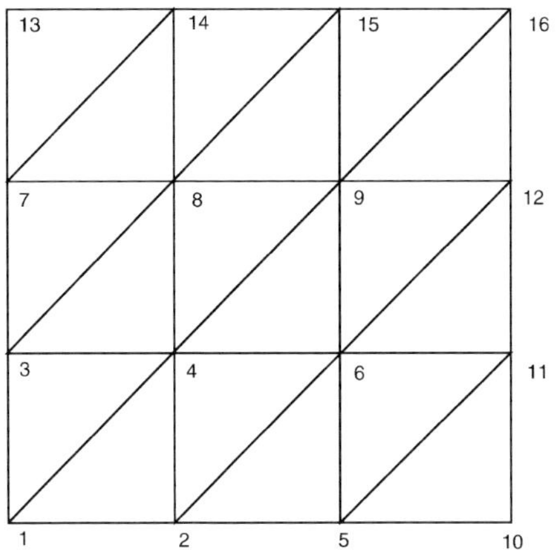

FIG. 19.2.

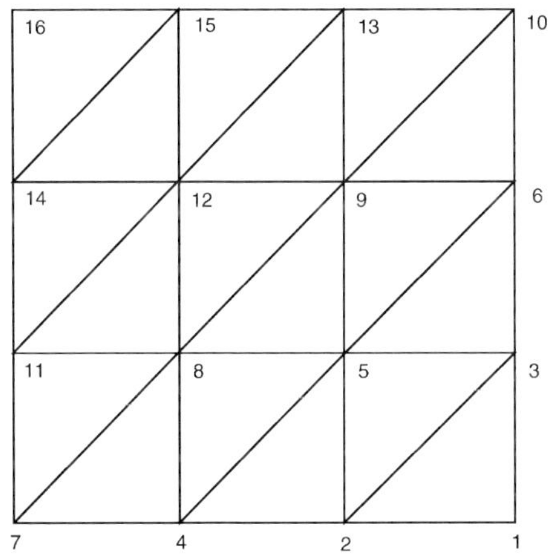

FIG. 19.3.

ALGORITHM CM.

(1) we choose a starting node,

(2) for $i = 1, \ldots, n-1$ we number all the (nonnumbered) neighbors of x_i in $G(A)$ in increasing order of degree,

(3) we update the degrees of the remaining nodes.

The profile resulting from this numbering is quite sensitive to the choice of the starting node. Consider the initial graph shown on Fig. 19.1.

If we choose 1 as a starting node, we obtain Fig. 19.2. There are 38 fill-ins in L.

If node 4 of the initial ordering is chosen as a starting node the ordering of Fig. 19.3 is obtained. With this ordering, there are 14 fill-ins in L.

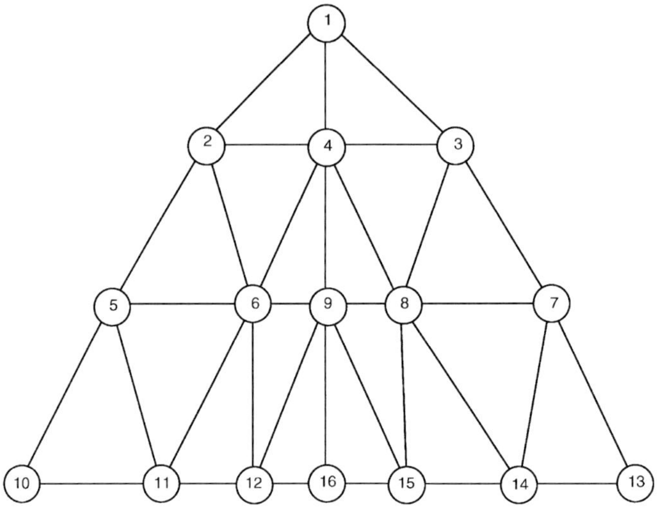

FIG. 19.4.

In the first ordering, the maximum difference of node numbers between neighbors is 7. With the second one, the maximum difference is 4.

It is interesting to look at the level structure $\mathcal{L}(1)$ for both orderings. The level structure of the first ordering is given on Fig. 19.4 using the numbers of the initial graph. The height is 4 and the width 7.

The level structure of the second ordering is given in Fig. 19.5. The height is 7 and the width is 4. Clearly, the second choice is better. The higher is the structure, the narrower it is. Starting nodes which give rise to narrow level structures are clearly better choices.

A good choice will be to choose as a starting node, a peripheral node, that is one whose eccentricity equals the diameter of the graph. Peripheral nodes are not that easy to find quickly. Therefore, people have devised heuristics to find "pseudo-peripheral" nodes, i.e. nodes whose eccentricities are close to the diameter of the graph. Such an algorithm was proposed by GIBBS, POOLE and STOCKMEYER [1976].

ALGORITHM GPS.

(1) choose a starting node r,

(2) build the level structure $\mathcal{L}(r)$

$$\mathcal{L}(r) = \{L_0(r), \ldots, L_{e(r)}(r)\},$$

(3) sort the nodes $x \in L_{e(r)}(r)$ in increasing degree order,

(4) for all nodes $x \in L_{e(r)}(r)$ in increasing degree order build $\mathcal{L}(x)$. If the height of $\mathcal{L}(x)$ is greater than the height of $\mathcal{L}(r)$, choose x as a starting node ($r = x$) and go to (2).

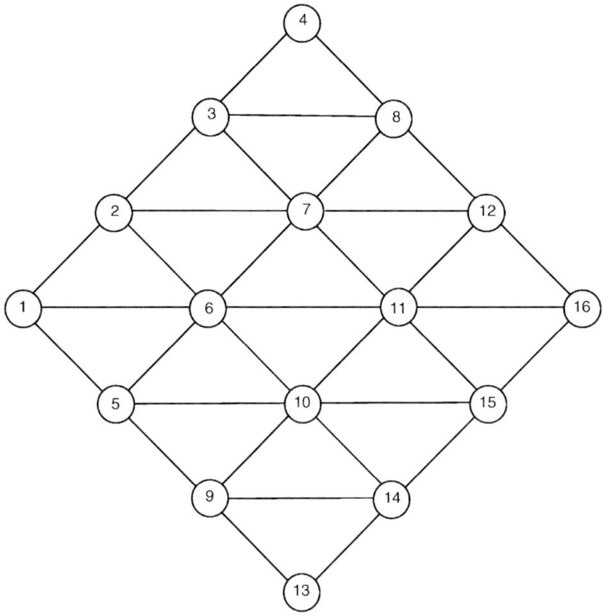

FIG. 19.5.

We are sure that this algorithm will converge as eccentricities are bounded by the diameter of the graph. However, it can be very costly.

GEORGE and LIU [1979a] tried to shorten the computing time by eliminating structures with large width as soon as possible. Step (4) of the algorithm is modified as

(4′) let $w(x)$ be the width of $\mathcal{L}(x)$. For all $x \in L_{e(r)}(r)$ in order of increasing degree, we build

$$\mathcal{L}(x) = \{L_0(x), \dots, L_{e(r)}(x)\}.$$

At each level i if $|L_i(x)| > w(r)$, we drop the current node and we pick another one x. If $w(x) \leqslant w(r)$ and $e(x) > e(r)$, we choose x as a starting node ($r = x$) and go to (2).

GEORGE and LIU [1981] also proposed to use the following simple algorithm,

(1) choose a starting node r,

(2) build $\mathcal{L}(r)$,

(3) choose a node x of minimum degree in $L_{e(r)}(r)$,

(4) build $\mathcal{L}(x)$. If $e(x) > e(r)$, choose x as a starting node and go to (2).

As an ordering scheme, George (GEORGE and LIU [1981]) proposed to reverse the Cuthill–McKee ordering.

ALGORITHM REVERSE CUTHILL–MCKEE (RCM).

(1) find a pseudo–peripheral starting node,

(2) generate the CM ordering,

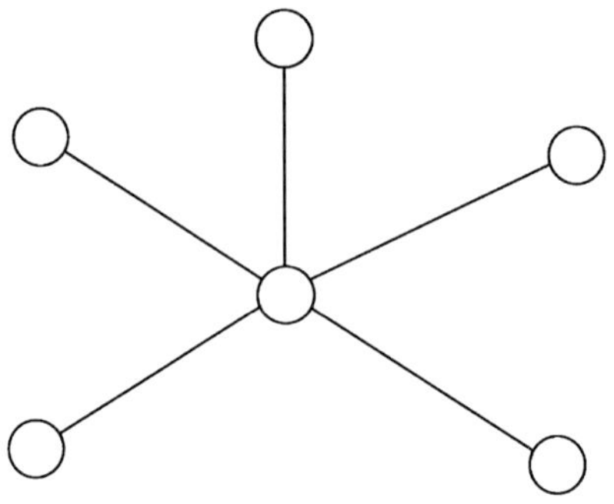

FIG. 19.6.

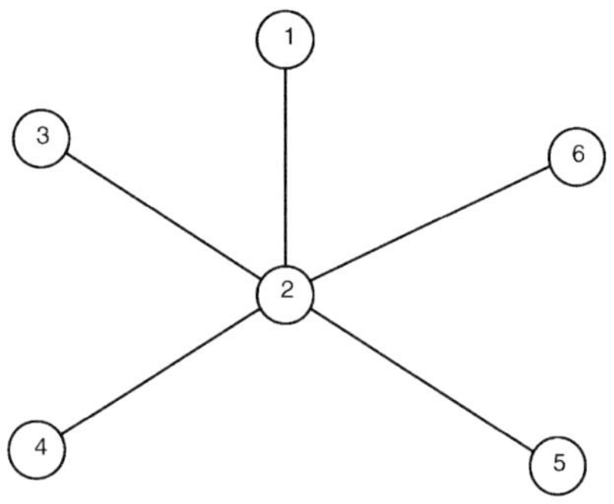

FIG. 19.7.

(3) reverse the numbering. Let $x_1, \dots, x_n$ be the CM ordering, then the RCM ordering $\{y_i\}$ is given by $y_i = x_{n+i-1}$, $i = 1, \dots, n$.

Let us look at a small example. We would like to number the graph of Fig. 19.6.

With the Cuthill–McKee algorithm, we find the graph of Fig. 19.7 with a Cholesky factor

$$L = \begin{pmatrix} x & & & & & \\ x & x & & & & \\ & x & x & & & \\ & x & \bullet & x & & \\ & x & \bullet & \bullet & x & \\ & x & \bullet & \bullet & \bullet & x \end{pmatrix}.$$

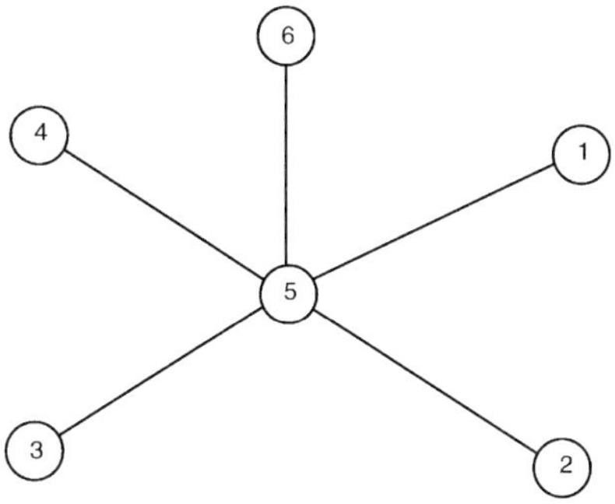

FIG. 19.8.

We have 6 fill-ins with this ordering. If we reverse the ordering, we obtain Fig. 19.8.

$$L = \begin{pmatrix} x & & & & & \\ & x & & & & \\ & & x & & & \\ & & & x & & \\ x & x & x & x & x & \\ & & & & x & x \end{pmatrix}.$$

There is no fill-in at all. We are going to show that in terms of number of fill-ins, RCM is always as good as CM. So, there is no reason to use CM.

THEOREM 19.2. *Let A be an irreducible matrix and A_{CM} be the matrix corresponding to reordering (the graph of) A by the Cuthill–McKee scheme. Then,*

$$\forall i,\, j,\ i \leqslant j,\ f_i \leqslant f_j.$$

Moreover, $f_i < i$ if $i > 1$.

PROOF. Suppose that the conclusion does not hold. Then, there exists a column k and rows $p, l, m,\ p < l < m$, such that

$$f_p \leqslant k,$$
$$f_l > k,$$
$$f_m \leqslant k.$$

This means that

$$a_{p,k} \neq 0 \Rightarrow x_p \in Adj(x_k),$$
$$a_{m,k} \neq 0 \Rightarrow x_m \in Adj(x_k),$$
$$a_{l,k} = 0 \Rightarrow x_l \notin Adj(x_k).$$

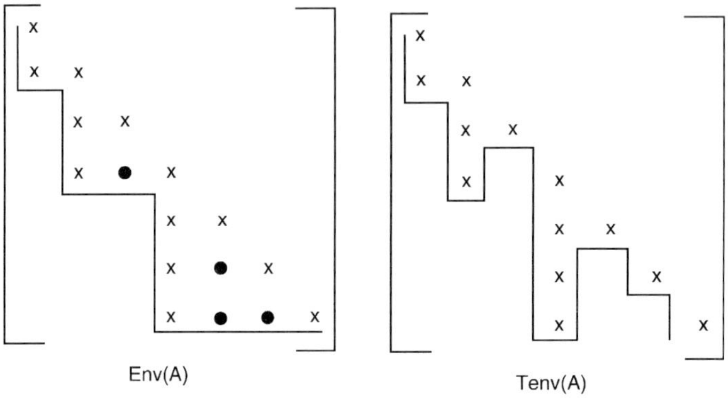

FIG. 19.9.

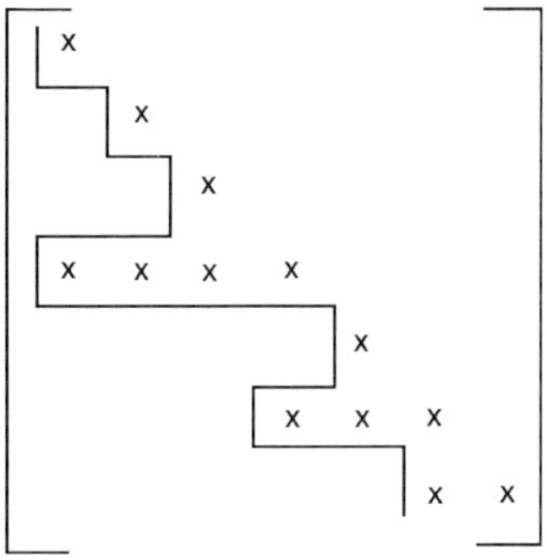

FIG. 19.10.

But this is impossible as the Cuthill–McKee algorithm has numbered successively all nodes in $\mathrm{Adj}(x_k)$. □

We introduce a new definition.

• $\mathrm{Tenv}(A) = \{(i, j) \mid j \leqslant i,\ \exists k \geqslant i,\ a_{k,j} \neq 0\}$.

$\mathrm{Tenv}(A)$ is the "transpose envelope" of A. Let us consider an example on Fig. 19.9. If we use the reverse Cuthill–McKee algorithm, we have to reverse the ordering. Thus, if, e.g., we have the previous matrix, we obtain Fig. 19.10. The rows of A_{RCM} are the columns of A_{CM}.

LEMMA 19.1.

$$\left|\mathrm{Env}(A_{\mathrm{RCM}})\right| = \left|\mathrm{Tenv}(A_{\mathrm{CM}})\right|.$$

PROOF. Straightforward. □

THEOREM 19.3.

$$\mathrm{Tenv}(A_{\mathrm{CM}}) \subseteq \mathrm{Env}(A_{\mathrm{CM}}).$$

PROOF. Looking at matrix pictures, the result is obvious. However, let us formalize the proof. Suppose we have $(i, j) \in \mathrm{Tenv}(A_{\mathrm{CM}})$ and $(i, j) \notin \mathrm{Env}(A_{\mathrm{CM}})$. Then, $\exists k \geqslant i$ such that $a_{k,j} \neq 0$. Either

(1) $a_{i,j} \neq 0 \Rightarrow (i, j) \in \mathrm{Env}(A_{\mathrm{CM}})$,

(2) $a_{i,j} = 0$. If $(i, j) \notin \mathrm{Env}(A_{\mathrm{CM}}) \Rightarrow \forall l \leqslant j,\ a_{i,l} = 0 \Rightarrow f_i > j$.

On the other hand, we have $f_k \leqslant j$. This implies $f_k < f_i$ which is impossible as $k \geqslant i$ by Theorem 19.2. □

Obviously, we have

$$\left|\mathrm{Env}(A_{\mathrm{RCM}})\right| \leqslant \left|\mathrm{Env}(A_{\mathrm{CM}})\right|$$

and George proved the following result,

LEMMA 19.2. *If* $\forall i > 1,\ f_i < i$, *the envelope* $\mathrm{Env}(A)$ *fills completely.*

This implies

$$\left|\mathrm{Fill}(A_{\mathrm{RCM}})\right| \leqslant \left|\mathrm{Fill}(A_{\mathrm{CM}})\right|.$$

There are cases for which equality holds. Notice that the previous results are true for every ordering such that $k \geqslant i \Rightarrow f_k \geqslant f_i$. If we look back at the example given after Theorem 18.1, as a function of the initial ordering, we have if we choose 2 as the starting node (notice that 2 is a peripheral node),

	1	2	3	4	5	6	7
CM	2	1	3	4	6	7	5
RCM	6	7	5	4	2	1	3

With the initial ordering we get six fill-ins. With CM we get three fills and zero with RCM.

It has been shown that RCM can be implemented to run in $O(|E|)$ time. For a regular $N \times N$ grid and P_1 triangular finite elements, the storage for RCM varies as $O(N^3)$ ($\approx 0.7N^3$), i.e., $O(n^{3/2})$.

Several other algorithms have been proposed to reduce the profile of a symmetric matrix, for example the King algorithm, see GEORGE and LIU [1981].

20. The minimum degree ordering for symmetric matrices

This ordering scheme was introduced in TINNEY and WALKER [1967]. It is one of the ordering schemes that are most often used today. It is a local minimization algorithm.

The minimum degree ordering works with the elimination graphs $G^{(i)} = (X_i, E_i)$. The ith step of the algorithm is

ALGORITHM MD.

(1) in $G^{(i)}$, find a node x_i such that

$$\deg(x_i) = \min_{y \in X_i} \{\deg(y)\},$$

(2) form $G^{(i+1)}$ by eliminating x_i,

(3) if $i+1 < n$ go to (1) with $i \leftarrow i+1$.

So, to use this algorithm we must have a way to represent the elimination graphs and to transform them. Of course, the degree of some nodes change after the deletion of edges incident to x_i and the addition of new edges.

The biggest problem arising with the minimum degree algorithm is that quite often, there may be several nodes of minimum degree. This must be resolved using a tie breaking strategy. Unfortunately, the final number of fill-ins is quite sensitive to the tie breaking strategy, see GEORGE and LIU [1989].

As being a local minimization algorithm, the minimum degree does not always give a minimum fill-in ordering. There are cases, like trees, for which it gives no fill-in at all, but there are examples for which it generates fill-in that is more than a constant time greater than the minimum fill-in, see BERMAN and SCHNITGER [1990]).

If we look again at the example after Theorem 18.1, we get the following ordering,

	1	2	3	4	5	6	7
MD	4	1	3	5	2	6	7

and there is no fill-in.

During the years many improvements have been suggested to the basic algorithm, mainly to shorten the computer time needed to run the algorithm rather than to improve the ordering. A summary of these results can be found in GEORGE and LIU [1989]. The main points are the following.

Mass elimination

When x_i is eliminated, often there are nodes in $\mathrm{Adj}_{G^{(i)}}(x_i)$ that can be eliminated immediately. This is because, when x_i is eliminated, only the degrees of nodes in $\mathrm{Adj}_{G^{(i)}}(x_i)$ change and some of them can be $\deg(x_i) - 1$. For instance, if a node in a clique is eliminated, then the degree of all the other nodes in the clique decreases by 1. Therefore, all these nodes can be eliminated at once, before the degrees are updated. This leads to the concept of indistinguishable nodes.

• Two nodes u and v are indistinguishable in G if

$$\mathrm{Adj}_G(u) \cup \{u\} = \mathrm{Adj}_G(v) \cup \{v\}.$$

GEORGE and LIU [1989] proved the following result.

THEOREM 20.1. *Let* $z \in \mathrm{Adj}_{G^{(i)}}(x_i)$, *then* $\deg_{G^{(i+1)}}(z) = \deg_{G^{(i)}}(x_i) - 1$ *if and only if*

$$\mathrm{Adj}_{G^{(i)}}(z) \cup \{z\} = \mathrm{Adj}_{G^{(i)}}(x_i) \cup \{x_i\}.$$

By merging indistinguishable nodes together, we need only to update the degrees of the representatives of these nodes.

Incomplete degree update

- v is said to be outmatched by u in G if

$$\mathrm{Adj}_G(u) \cup \{u\} \subseteq \mathrm{Adj}_G(v) \cup \{v\}.$$

THEOREM 20.2 (GEORGE and LIU [1989]). *If v is outmatched by u in $G^{(i)}$, it is also outmatched by u in $G^{(i+1)}$.*

If v becomes outmatched by u, it is not necessary to update the degree of v until u is eliminated.

Multiple elimination

This slight variation of the basic scheme LIU [1985] proposed is when x_i has been chosen, to select a node with the same degree than x_i in $G^{(i)}/(\mathrm{Adj}_{G^{(i)}}(x_i) \cup \{x_i\})$. This process is repeated until there are no nodes of the same degree and then the degrees are updated.

Therefore, at each step, an independent set of minimum degree nodes is selected. Notice that the ordering that is produced is not the same as for the basic algorithm. However, it is generally as good as the "true" ordering.

Early stop

In many implementations of Gaussian elimination for sparse matrices, a switch is done to dense matrices when the percentage of nonzeroes in the remaining matrix is large enough. Of course, the ordering can be stopped at that stage since later stages are meaningless.

Tie breaking

An issue that is really important is the choice of a tie breaking strategy. Unfortunately, not much is known about how to decide which nodes to choose at a given stage. Some experiments (GEORGE and LIU [1989]) show that there can be large differences in the number of nonzeroes and factorization times when several random tie breakers are chosen.

Most often, the initial ordering determines the way ties are broken. It has been suggested to use another ordering scheme like the Reverse Cuthill–McKee algorithm before running the minimum degree.

Approximate minimum degree

In AMESTOY, DAVIS and DUFF [1994] it was proposed to use some bounds on the degree of nodes instead of the real degree. This allows a faster update of the information when nodes are eliminated. Techniques based on the quotient graph are used to obtain these bounds. The quality of the orderings that are obtained are comparable to the ones from the genuine minimum degree algorithm although the algorithm is much faster, see the performances in AMESTOY, DAVIS and DUFF [1994].

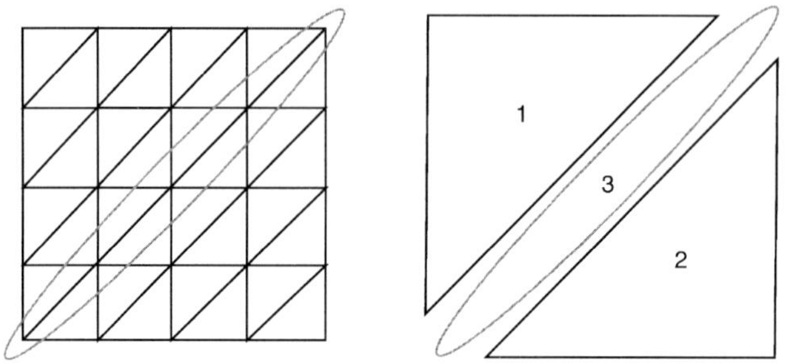

FIG. 21.1.

21. The nested dissection ordering for symmetric matrices

Introduction

This algorithm has been introduced by GEORGE [1973] for finite-element problems and then generalized to general sparse matrices. It is very close to an old idea used in Mechanics called substructuring and also to what is now called domain decomposition.

The idea is based on Theorem 18.2 that essentially says that there won't be any fill-in between x_i and x_j if, on every path from x_i to x_j in G, there is a node with a number greater than x_i and x_j.

Consider the graph of Fig. 21.1 (arising, e.g., from a finite-element matrix) and its partitioning given in the right-hand side.

The main diagonal separates the graph into three pieces. The diagonal 3 is called a separator. From Theorem 18.2, it is clear that, if we first number the nodes in part 1, then the nodes in part 2 and finally the nodes of the separator 3, there won't be any fill-in between sets of nodes 1 and 2. With this ordering and obvious notations, the matrix has the following block structure

$$A = \begin{pmatrix} A_1 & 0 & A_{3,1}^{\mathrm{T}} \\ 0 & A_2 & A_{3,2}^{\mathrm{T}} \\ A_{3,1} & A_{3,2} & A_3 \end{pmatrix}.$$

It is obvious that the Cholesky factor L has the following block structure

$$L = \begin{pmatrix} L_1 & & \\ 0 & L_2 & \\ L_{3,1} & L_{3,2} & L_3 \end{pmatrix},$$

this means that blocks A_1 and A_2 can be factored independently.

The basis of the nested dissection algorithm is to apply this idea recursively to sets 1 and 2. If we look at a mesh graph like the one in Fig. 21.2, there are two basic ways to partition it. The first one is to partition the graph into vertical (or horizontal) stripes, see Fig. 21.3.

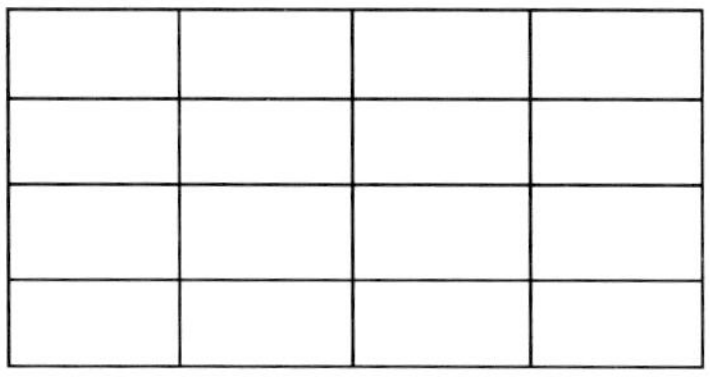

FIG. 21.2.

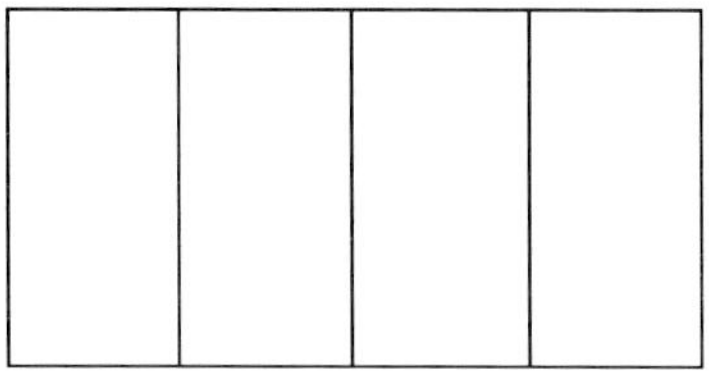

FIG. 21.3.

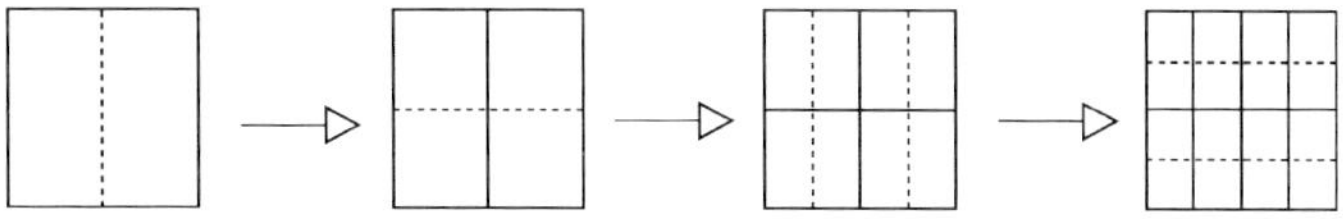

FIG. 21.4.

This is called one-way dissection. The other way is to alternate between vertical and horizontal partitioning, see Fig. 21.4. This is called nested dissection. Of course, one-way dissection is a little simpler to implement.

One-way dissection

GEORGE [1980] considered a mesh graph consisting of an m by l grid. It is partitioned by σ vertical grid lines.

George showed that the required storage for storing the LU factors using the one-way dissection ordering is (if $m \leqslant l$)

$$S(\sigma) = \frac{ml^2}{\sigma} + \frac{3\sigma m^2}{2}.$$

This is approximately minimized (as a function of σ) by

$$\sigma = l\left(\frac{2}{3m}\right)^{1/2},$$

giving $S_{\text{opt}} = \sqrt{6}\, m^{3/2} l + \mathrm{O}(ml)$. For comparison, numbering the graph by columns would yield a storage of $m^2 l + \mathrm{O}(ml)$.

The operation count for the factorization is approximately

$$\theta = \frac{ml^3}{2\sigma^2} + \frac{7\sigma m^3}{6} + \frac{2m^2l^2}{\sigma}.$$

This is approximately minimized by

$$\sigma = l\left(\frac{12}{7m}\right)^{1/2},$$

yielding $\theta_{\text{opt}} = (28/3)^{1/2} m^{5/2} l + \mathrm{O}(m^2 l)$.

One-way dissection can be somehow generalized to general sparse matrices by using level structures of the graph, see GEORGE and LIU [1978b].

Nested dissection of a mesh graph

Consider a square mesh. A partition function Π is defined for integers i from 0 to N $(=2^l)$ as

$$\begin{aligned} \Pi(0) &= 1, \\ \Pi(N) &= 1, \\ \Pi(i) &= p+1, \quad \text{if } i = 2^p(2q+1). \end{aligned}$$

For example, for $N = 16$, we obtain

i	0	1	2	3	4	5	6	7	8	9	10	11	12	13	14	15	16
$\Pi(i)$	1	1	2	1	3	1	2	1	4	1	2	1	3	1	2	1	1

For $k = 1, \ldots, l$ we define sets P_k of mesh nodes (i, j) as

$$P_k = \{(i, j) \mid \max(\Pi(i), \Pi(j)) = k\}.$$

For a 17×17 mesh we have Fig. 21.5 where the numbers denote the set P_k to which the nodes belong and the lines separate these sets.

Nodes in P_1 are numbered first, then nodes in P_2, etc... up to nodes in P_l. GEORGE [1973] has shown that

$$\begin{aligned} S &= \mathrm{O}(N^2 \log_2 N), \\ \theta &= \mathrm{O}(N^3). \end{aligned}$$

DUFF, ERISMAN and REID [1976] generalized the partition function when $N \neq 2^l$ as

(1) l is defined as the smallest integer such that $2^l \geqslant N$,

(2) $\tilde{\Pi}(0) = \tilde{\Pi}(N) = 1$,

(3) $\tilde{\Pi}(i) = 0$, $i = 1, \ldots, N-1$,

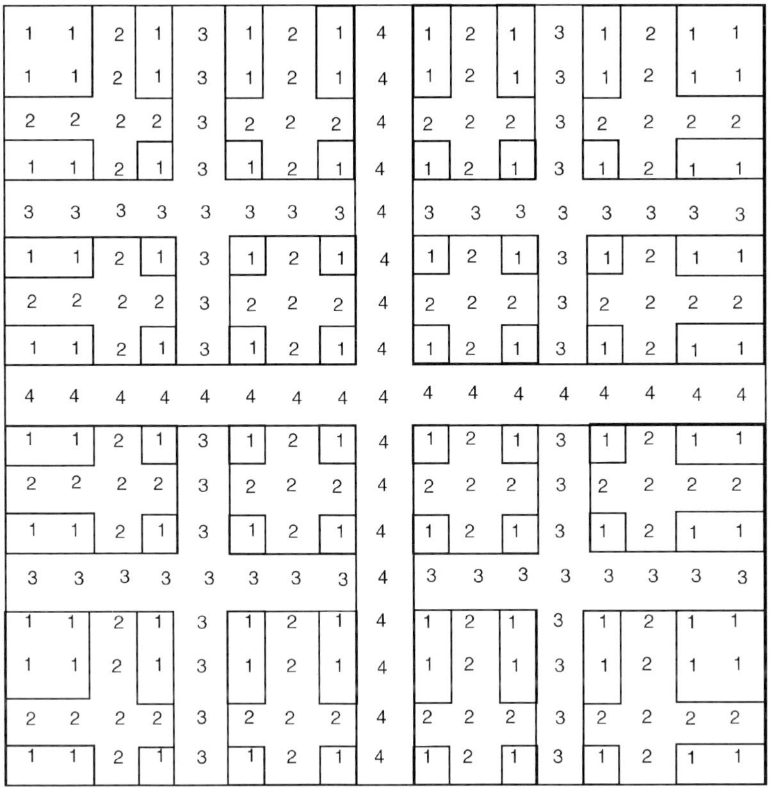

FIG. 21.5.

(4) For $m = l, l-1, \ldots, 1$ we look for the mid point x of groups of adjacent nodes such that $\tilde{\Pi}(i) = 0$ and we set $\tilde{\Pi}(x) = m$.

EXAMPLE. $N = 11 \Rightarrow l = 4$,

Step	0	1	2	3	4	5	6	7	8	9	10	11
1	1					4						1
2	1			3		4			3			1
3	1		2	3	2	4	2		3	2		1
4	1	1	2	3	2	4	2	1	3	2	1	1

The partition function is applied to each side of the rectangular mesh.

It must be noticed that the storage schemes that we have described at the beginning of this chapter are not well suited for nested dissection orderings. In nested dissection for mesh problems, there is a natural block structure that arises. Each block corresponds to subsets of each P_k. Diagonal blocks are stored by rows in a one-dimensional array together with an integer pointer that gives the position of the diagonal element.

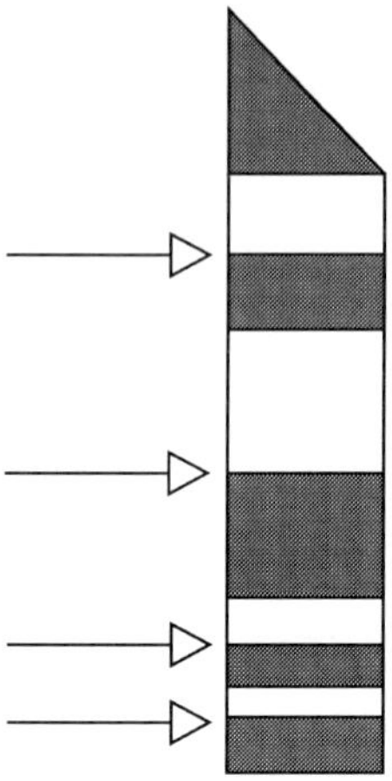

FIG. 21.6.

Nondiagonal blocks are stored in a one-dimensional array. It is necessary to know the beginning of each block, see Fig. 21.6.

For an $N \times N$ 2D grid, the storage for nested dissection is smaller than the one for RCM for $N > 37$.

Nested dissection for a general matrix

For a general sparse matrix, we have to work directly on the graph of the matrix and the problem is to find a small separator that partitions the graph in two or more components of almost an equal number of nodes. There are many ways to handle this problem.

Geometric (or greedy) algorithms can be used, see CIARLET and LAMOUR [1994], FARHAT [1988] or spectral bisection techniques, SIMON [1991] relying on computation of the smallest nonzero eigenvalue of the Laplacian matrix of the graph. The Laplacian $\mathcal{L}$ of the graph is constructed as follows: for the row i $\mathcal{L}_{i,j} = -1$ if node j is a neighbor of node i in the graph and the diagonal term is minus the sum of the other entries. From the corresponding eigenvector a partition is obtained by considering the components of the eigenvector larger or smaller than the median value.

Earlier, GEORGE and LIU [1981] proposed an algorithm based on the level structure. Consider the graph $G = (X, E)$,

(1) let $\mathcal{P} = \emptyset$, $R = X$, $N = |X|$,

(2) if $R = \emptyset$ we stop. Otherwise, we choose a pseudo peripheral node y,

(3) we build the level structure $\mathcal{L}(y) = \{L_0, \dots, L_l\}$ and let $j = (l+1)/2$,

(4) find a separator $S \subseteq L_j$,

$$S = \{y \in L_j \mid \mathrm{Adj}(y) \cap L_{j+1} \neq \emptyset\},$$

(5) nodes in S are numbered from $N - |S| + 1$ to N using, e.g., RCM. Set $N = N - |S|$,

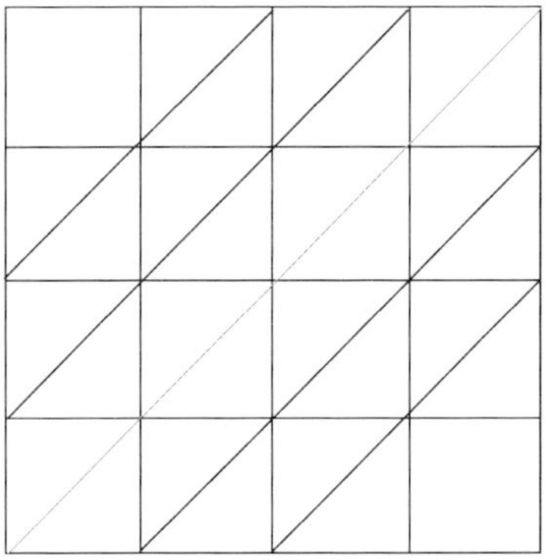

FIG. 21.7.

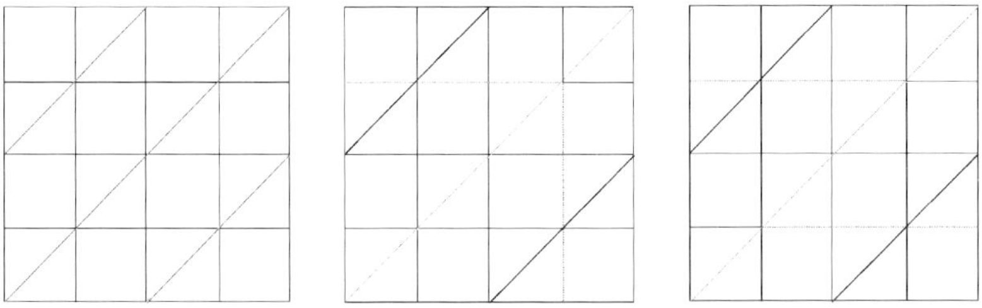

FIG. 21.8.

(6) set $R = R - S$, $\mathcal{P} = \mathcal{P} \cup \{S\}$. Go to (2).

In this algorithm, there can be some ties and the result depends on how they are broken. For instance, we consider the graph of Fig. 21.7.

The first separator is the diagonal. Then, we have several different choices given in Fig. 21.8, giving rise to different orderings.

Considering again the same small example, we can choose the set $\{1, 4\}$ as a separator and the following ordering,

	1	2	3	4	5	6	7
ND	6	3	2	7	1	4	5

With this ordering, there is no fill-in.

The storage scheme used for mesh problems has to be generalized to cope with the ordering scheme, see GEORGE and LIU [1981].

General theorems have been proved using graph theory about the existence of good separators, see LIPTON, ROSE and TARJAN [1979], ROMAN [1985], CHARRIER and ROMAN [1989]. Without going into too much details, let us summarize the kind of results that have been established.

• Let $\mathcal{S}$ be a family of graphs. There is an $f(n)$-separator theorem for $\mathcal{S}$ if there exists α and β positive real numbers such that $\forall G \in \mathcal{S}$, the nodes of G can be partitioned into three subsets A, B, C with the following properties

- there is no edges between nodes of A and nodes of B,
- $\mathrm{card}(A) \leqslant \alpha n$, $\mathrm{card}(B) \leqslant \alpha n$,
- $\mathrm{card}(C) \leqslant \beta f(n)$,

where $\mathrm{card}(A)$ is the number of elements in the subset A. C is called a separator of G.

LIPTON and TARJAN [1980] proved that $f(n) = \sqrt{n}$ for planar graphs and 2D finite-element graphs.

The elements of the separator C are numbered last and the same partition algorithm is applied recursively to A and B. It has been shown in ROMAN [1985] that

THEOREM 21.1. *Let G be a graph satisfying the previous definition with $f(n) = \sqrt{n}$ with n nodes and m edges,*

- *the time to construct the partitions is* $\mathrm{O}(n + m)$,
- *the number of nonzeroes in L is* $\mathrm{O}(n \log_2 n)$,
- *the number of floating point operations for the factorization is* $\mathrm{O}(n\sqrt{n})$.

Therefore, we have the same results as for a mesh grid. These results can be slightly generalized (ROMAN [1985]).

22. The multifrontal method

Introduction

The multifrontal method has been introduced by DUFF and REID [1984] as a generalization of the frontal method developed by IRONS [1970] for finite-element problems.

The essence of the frontal method was that in finite-element problems the two phases, assembly of the matrix (from integral computations) and factorization of the matrix can be mixed together. However, a variable can be eliminated only when it has been fully assembled.

The main goal of the multifrontal method is to be able to use dense matrix technology for sparse matrices. A possible drawback of the method is that technical details are quite complex and many refinements are necessary to make the method efficient. A nice exposition of the principles of the method has been given in LIU [1992]. We will follow his lines.

We consider a symmetric matrix A. The basis of the method is the block outer product Cholesky factorization,

$$A = \begin{pmatrix} D & B^{\mathrm{T}} \\ B & C \end{pmatrix} = \begin{pmatrix} L_D & 0 \\ BL_D^{-\mathrm{T}} & I \end{pmatrix} \begin{pmatrix} I & 0 \\ 0 & C - BD^{-1}B^{\mathrm{T}} \end{pmatrix} \begin{pmatrix} L_D^{\mathrm{T}} & L_D^{-1}B^{\mathrm{T}} \\ 0 & I \end{pmatrix},$$

$$D = L_D L_D^{\mathrm{T}}.$$

The Schur complement $C - BD^{-1}B^{\mathrm{T}}$ is the next matrix to be factored. Let D be of order $j-1$,

$$BD^{-1}B^{\mathrm{T}} = (BL_D^{-\mathrm{T}})(L_D^{-1}B^{\mathrm{T}}) = \sum_{k=1}^{j-1} \begin{pmatrix} l_{j,k} \\ \vdots \\ l_{n,k} \end{pmatrix} (l_{j,k} \quad \dots \quad l_{n,k}),$$

$l_{i,k}$ being the elements of the Cholesky factors.

The multifrontal method uses the definition of an elimination tree. We have the following result that we state without proof,

THEOREM 22.1. *If node k is a descendant of node j in $T(A)$, then the nonzero structure of $(l_{j,k}, \dots, l_{n,k})^{\mathrm{T}}$ is contained in the structure of $(l_{j,j}, \dots, l_{n,j})^{\mathrm{T}}$. If $l_{j,k} \neq 0$, $k < j$, node k is a descendant of node j in $T(A)$.*

As in LIU [1992], let us consider a small example,

$$A = \begin{array}{c} \\ 1 \\ 2 \\ 3 \\ 4 \\ 5 \\ 6 \end{array} \begin{array}{c} \begin{array}{cccccc} 1 & 2 & 3 & 4 & 5 & 6 \end{array} \\ \begin{pmatrix} x & & & x & x & x \\ & x & & & x & \\ & & x & & x & x \\ x & & & x & & x \\ x & x & x & & x & x \\ x & & x & x & x & x \end{pmatrix} \end{array}.$$

The graph of the matrix A is given on Fig. 22.1.

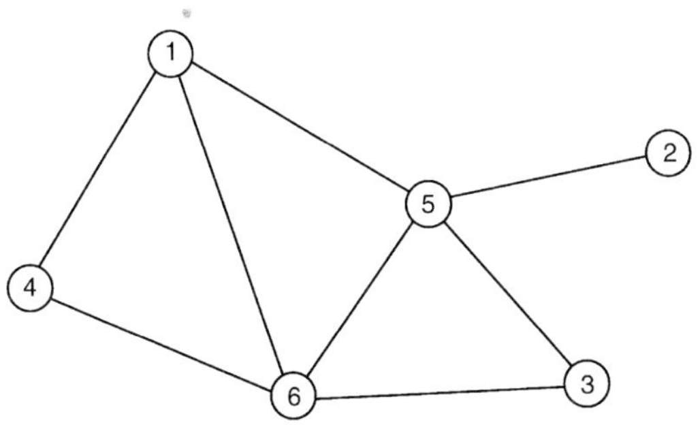

FIG. 22.1.

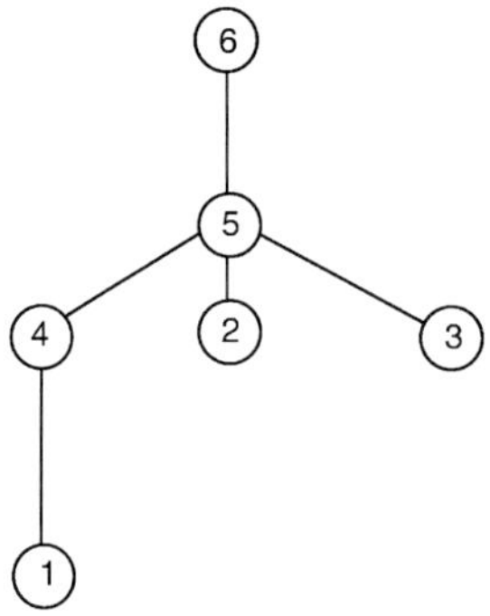

FIG. 22.2.

Doing Gaussian elimination, we get

$$L = \begin{array}{c|cccccc} & 1 & 2 & 3 & 4 & 5 & 6 \\ \hline 1 & x & & & & & \\ 2 & & x & & & & \\ 3 & & & x & & & \\ 4 & x & & & x & & \\ 5 & x & x & x & \bullet & x & x \\ 6 & x & & x & x & x & x \end{array}.$$

The elimination tree $T(A)$ is shown on Fig. 22.2.

It is clear that we can eliminate $1, 2$ and 3 independently. If we consider 1, we can restrict ourselves to the following matrix (rows and columns where there are nonzeroes in the first row and first column,

$$F_1 = \begin{array}{c|cccc} & 1 & 4 & 5 & 6 \\ \hline 1 & a_{1,1} & a_{1,4} & a_{1,5} & a_{1,6} \\ 4 & a_{4,1} & & & \\ 5 & a_{5,1} & & & \\ 6 & a_{6,1} & & & \end{array}.$$

Eliminating 1 will create contributions in a reduced matrix $\bar{U}_4$,

$$\bar{U}_4 = \begin{array}{c|ccc} & 4 & 5 & 6 \\ \hline 4 & x & \bullet & x \\ 5 & \bullet & x & x \\ 6 & x & x & x \end{array},$$

where, as before, the • represents a fill-in. In parallel, we can eliminate 2, defining

$$F_2 = \begin{matrix} & 2 & 5 \\ 2 & a_{2,2} & a_{2,5} \\ 5 & a_{5,2} & \end{matrix}.$$

Elimination of 2 will create a contribution to the (5, 5) term,

$$\bar{U}_5^2 = \begin{matrix} & 5 \\ 5 & (x) \end{matrix}.$$

We can also eliminate 3,

$$F_3 = \begin{matrix} & 3 & 5 & 6 \\ 3 & a_{3,3} & a_{3,5} & a_{3,6} \\ 5 & a_{5,3} & & \\ 6 & a_{6,3} & & \end{matrix}.$$

Elimination of 3 will create contributions,

$$\bar{U}_5^3 = \begin{matrix} & 5 & 6 \\ 5 & x & x \\ 6 & x & x \end{matrix}.$$

Then, we eliminate 4. For this, we have to consider the matrix resulting from the elimination of 1, i.e.

$$F_4 = \begin{matrix} & 4 & 5 & 6 \\ 4 & a_{4,4} & 0 & a_{4,6} \\ 5 & 0 & & \\ 6 & a_{6,4} & & \end{matrix} + \bar{U}_4.$$

Elimination of 4 creates contributions,

$$\bar{U}_5^4 = \begin{matrix} & 5 & 6 \\ 5 & x & x \\ 6 & x & x \end{matrix}.$$

Now, before eliminating node 5, we must sum the contributions from the original matrix and what we get from the eliminations of nodes 2, 3 and 4. To do this, we must extend $\bar{U}_5^2$ to the proper set of indices, i.e. 5, 6. We do this as in LIU [1992] by considering an operator that we denote by $\circ$.

For two matrices A and B, $A \circ B$ takes as the set of indices of the result, the union of the sets of indices of A and B and whenever they coincide, the result is the sum of the entries. Let

$$\bar{U}_5 = \bar{U}_5^2 \circ \bar{U}_5^3 \circ \bar{U}_5^4,$$

$$F_5 = \begin{pmatrix} a_{5,5} & a_{5,6} \\ a_{6,5} & 0 \end{pmatrix} + \bar{U}_5.$$

Elimination of 5 gives a matrix of order 1 that is added to $a_{6,6}$ to give the last term of the factorization.

On this example, we have seen that all the elimination steps can be carried out by working on small dense matrices of different orders, extending and summing these matrices by looking at the elimination tree.

Description of the method

Following LIU [1992], let us formalize the process of the multifrontal method. Giving the elimination tree $T(A)$, we define the subtree update matrix for column j as

$$\bar{U}_j = -\sum_{k \in T[j]-\{j\}} \begin{pmatrix} l_{j,k} \\ l_{i_1,k} \\ \vdots \\ l_{i_r,k} \end{pmatrix} \begin{pmatrix} l_{j,k} & l_{i_1,k} & \dots & l_{i_r,k} \end{pmatrix},$$

where, $i_0 = j, i_1, \dots, i_r$ are the row indices of the nonzeroes in column j of L, $(L_{*,j})$.

$\bar{U}_j$ contains the contributions for the columns preceding j which are proper descendants of j in the tree. The frontal matrix F_j is defined as

$$F_j = \begin{pmatrix} a_{j,j} & a_{j,i_1} & \dots & a_{j,i_r} \\ a_{i_1,j} & & & \\ \vdots & & & \\ a_{i_r,j} & & & \end{pmatrix} + \bar{U}_j.$$

The first column of F_j contains all the nonzero updates entries to column j. Then, we perform one step of elimination on F_j,

$$F_j = \begin{pmatrix} l_{j,j} & 0 \\ l_{i_1,j} & \\ \vdots & I \\ l_{i_r,j} & \end{pmatrix} \begin{pmatrix} 1 & 0 \\ 0 & U_j \end{pmatrix} \begin{pmatrix} l_{j,j} & l_{i_1,j} & \dots & l_{i_r,j} \\ 0 & & I & \end{pmatrix}.$$

U_j is called the update matrix. It is a dense matrix. It is proved in LIU [1992] that

$$U_j = -\sum_{k \in T[j]} \begin{pmatrix} l_{i_1,k} \\ \vdots \\ l_{i_r,k} \end{pmatrix} \begin{pmatrix} l_{i_1,k} & \dots & l_{i_r,k} \end{pmatrix}.$$

If $c_1, \dots, c_s$ are the children of j in $T(A)$, then

$$F_j = \begin{pmatrix} a_{j,j} & a_{j,i_1} & \dots & a_{j,i_r} \\ a_{i_1,j} & & & \\ \vdots & & & \\ a_{i_r,j} & & & \end{pmatrix} \circ U_{c_1} \circ \dots \circ U_{c_s}.$$

Then, the multifrontal method is defined as

```
for j=1:n
```
(1) Form the update matrix $U_{c_1} \circ \dots \circ U_{c_s}$,
(2) Form the frontal matrix F_j,
(3) Factorize F_j,
```
end
```

A lot of issues have to be considered to derive an efficient multifrontal matrix code. Let us consider what the main problems are.

(1) Storage of the frontal and update matrices.

Update matrices must be stored and easily retrieved when they are needed in the algorithm to contribute to a frontal matrix. A nice way to do this is to use a topological ordering of $T(A)$ and to number the nodes in every subtree consecutively (this is called a postordering). Then, the update matrices can be stored in a stack using a last-in first-out algorithm. Using this scheme, the update matrices appear at the top of the stack in the order they are needed.

To manage the storage working space, Duff proposed to use a buddy system. In a buddy system, each block of storage has a buddy with which it can be combined to form a larger block. In a binary buddy system, the sizes of the block are $c2^i$. Each block keeps some associated information: a flag to indicate if the block is free or not and the logarithm of the size of the block. Free blocks are linked through a doubly linked list. There is also a free list for each block size.

When a working area of memory of size m is needed, the system allocates a block of list i, where $c2^i \geqslant m$. If there is no block on the ith free list, a level $i+1$ block must be split in two. A part is used to serve the request, the other part is put on the ith free list. The algorithm is applied recursively.

When a block is deallocated, the system checks if the block's buddy is free and of the correct size. If the answer is positive, the two blocks are combined and put on the $(i+1)$rst free list.

Liu proposed a postordering of $T(A)$ that minimizes the working storage. Generally, reduction in the working storage can be obtained by tree restructuring. In particular, tree rotations can be used, see LIU [1988].

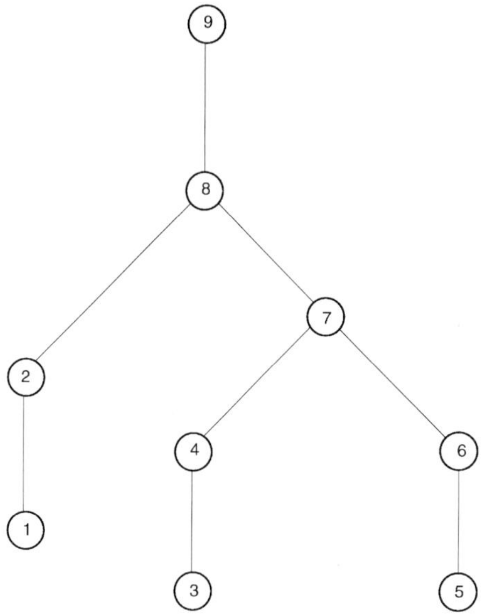

FIG. 22.3.

(2) Supernode methods.

Some nodes in $T(A)$ can be grouped together and treated as a single computational unit. Formally (LIU [1992]), a supernode is a set of contiguous nodes

$$\{j, j+1, \ldots, j+s\},$$

such that

$$\mathrm{Adj}_G\big(T[j]\big) = \{j+1, \ldots, j+s\} \cup \mathrm{Adj}_G\big(T[j+s]\big).$$

Consider the tree of Fig. 22.3, corresponding to the graph of Fig. 22.4. The supernodal elimination tree is given in Fig. 22.5. As an example, consider

$$\mathrm{Adj}_G\big(T[7]\big) = \{8, 9\} = \{8, 9\} \cup \mathrm{Adj}_G\big(T[9]\big) = \{8, 9\} \cup \emptyset.$$

Therefore, $\{7, 8, 9\}$ is a supernode.

Then, all the nodes in a supernode are eliminated in the same step of the algorithm. The advantages of using supernodes are that the supernodal tree is smaller and the frontal matrices have a larger order giving better performances on modern computers.

(3) Dense techniques.

An advantage of the multifrontal method is to allow to use dense matrix techniques in the sparse case. Particularly, dense algorithms based on the use of Level 3 BLAS can

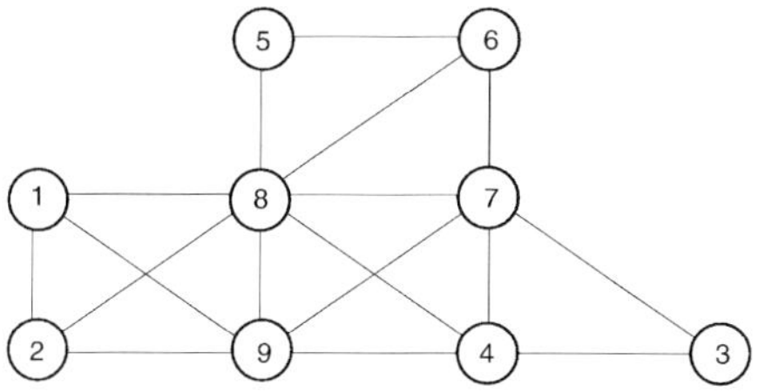

FIG. 22.4.

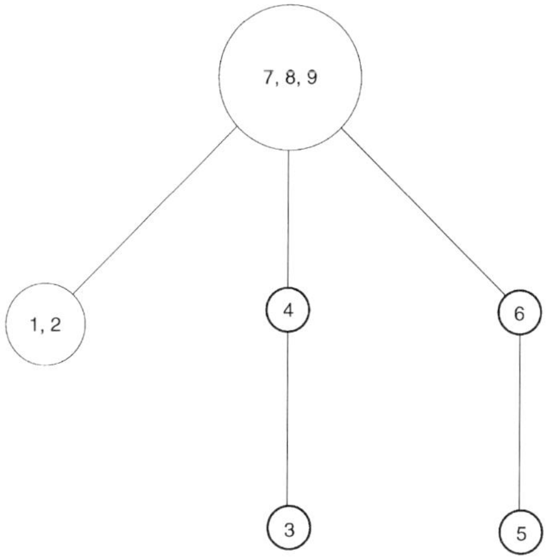

FIG. 22.5.

be used when factoring the frontal matrices. Furthermore, there is much less indirect addressing.

(4) Node amalgamation.

There are some performance advantages to have large supernodes. Therefore, it has been suggested (DUFF and REID [1984]) to amalgamate some nodes, treating some zeroes as nonzeroes to obtain larger supernodes. For a detailed study of this point, see AMESTOY [1990].

Software packages are available that implement the multifrontal method. See various programs in the Harwell Scientific Library, or the MUPS code developed by Amestoy and Duff.

23. Nonsymmetric sparse matrices

Introduction

There is an additional difficulty in Gaussian elimination for nonsymmetric sparse problems, namely the need of pivoting to improve numerical stability. When dealing with sparse symmetric positive definite systems, the ordering of the unknowns can be chosen only for the purpose of maintaining sparsity as much as possible during the elimination. This is not true anymore for nonsymmetric problems.

There are still a lot of active research going on to try to make the algorithms for nonsymmetric systems as efficient and robust as those we have seen in the previous sections. We are going to briefly review these efforts.

The Markowitz criterion

If we choose the pivots as it is done for dense systems (e.g., partial pivoting), there is no room for preserving sparsity. Therefore, for sparse matrices, we have to relax the constraints for choosing the pivot. The usual strategy is to consider candidate pivots satisfying the inequality,

$$\left|a_{i,j}^{(k)}\right| \geqslant u \max_{l} \left|a_{l,j}^{(k)}\right|,$$

where u is a user defined parameter such that $0 < u \leqslant 1$. This will limit the overall growth as

$$\max_{i} \left|a_{i,j}^{(k)}\right| \leqslant \left(1 + \frac{1}{u}\right)^{p_j} \max_{i} |a_{i,j}|,$$

where p_j is the number of off diagonal entries in column j of U, see DUFF, ERISMAN and REID [1986].

From these candidates, one is selected that minimizes

$$\left(r_i^{(k)} - 1\right)\left(c_j^{(k)} - 1\right),$$

where $r_i^{(k)}$ is the number of nonzero entries in row i of the remaining $(n-k) \times (n-k)$ matrix in A_k. Similarly, $c_j^{(k)}$ is the number of nonzeroes in column j. This criterion is due to MARKOWITZ [1957].

This method modifies the least entries in the remaining submatrix. Notice that if A is symmetric, this is exactly the minimum degree algorithm that was introduced historically after the Markowitz criterion. Many variations of the Markowitz criterion have been studied. For a summary, see DUFF, ERISMAN and REID [1986]. However, most of these other methods are generally not as efficient as the Markowitz criterion.

One possibility is to choose the entry (which is not too small) that introduces the least amount of fill-in at step k. Unfortunately, this is much more expensive than the Markowitz criterion. Moreover, having a local minimum fill-in does not always gives a globally optimal fill-in count. There are even some examples where the Markowitz criterion does a better job at globally reducing the fill-in.

As for the minimum degree algorithm, the tie breaking strategy is quite important for the result of the Markowitz algorithm. Details of the implementation of the Markowitz algorithm are discussed in DUFF, ERISMAN and REID [1986]. A switch to dense matrix techniques is done when the nonzero density is large enough.

Elimination structures for nonsymmetric matrices

GILBERT and LIU [1993] have introduced some definitions to characterize the structures of the triangular factors of a nonsymmetric matrix (without pivoting).

This starts by considering a triangular matrix L and its directed graph $G(L)$ which is acyclic (there is no directed cycles). An acyclic directed graph is called a dag. Let $w=(w_1,\dots,w_n)^{\mathrm{T}}$, then we define

$$\mathrm{Struct}(w)=\{i\in\{1,\dots,n\}\mid w_i\neq 0\}.$$

THEOREM 23.1. *If $Lx=b$ then* Struct(x) *is given by the set of vertices reachable from vertices of* Struct(b) *in the dag $G(L^{\mathrm{T}})$.*

An economical way to represent the information contained in a dag G is to consider its transitive reduction G^0. Then, in Theorem 23.1, we can replace $G(L^{\mathrm{T}})$ by $G^0(L^{\mathrm{T}})$. The transitive closure G^* of a directed graph G is a graph that has an edge (u,v) whenever G has a directed path from u to v.

Let A be factored as $A=LU$ without pivoting. $G^0(L)$ and $G^0(U)$ are called the lower and upper elimination dags (edags) of A. For a symmetric matrix, $G^0(L)$ and $G^0(U)$ are both equal to the elimination tree.

If B and C are two matrices with nonzero diagonal elements then $G(B)+G(C)$ is the union of the graphs of B and C, i.e. the graph whose edge set is the union of those of $G(B)$ and $G(C)$. $G(B)\bullet G(C)$ is the graph with an edge (i,j) if (i,j) is an edge of $G(B)$ or (i,j) is an edge of $G(C)$ or there is a k such that (i,k) is an edge of $G(B)$ and (k,j) is an edge of $G(C)$.

GILBERT and LIU [1993] proved the following result.

THEOREM 23.2. *If $A=LU$ and there is a path in $G(A)$ from i to j, then there exists a k, $1\leqslant k\leqslant n$, such that $G^0(U)$ has a path from i to k and $G^0(L)$ has a path from k to j. That is*

$$G^*(A)\subseteq G^{0^*}(U)\bullet G^{0^*}(L).$$

If there is no cancellation in the factorization $A=LU$, then

$$G(L)\bullet G(U)=G(L)+G(U).$$

From these results, the row and column structures of L and U can be derived.

THEOREM 23.3. *If $l_{i,j}\neq 0$, then there exists a path from i to j in $G^0(L)$.*

Let $i>j$, $l_{i,j}\neq 0$ if and only if there exists $k\leqslant j$ such that $a_{i,k}\neq 0$ and there is a directed path in $G^0(U)$ from k to j.

$$\mathrm{Struct}(L_{*,j})$$
$$=\mathrm{Struct}(A_{*,j})\cup\bigcup\{\mathrm{Struct}(L_{*,k})\mid k<j,\ u_{k,j}\neq 0\}-\{1,\dots,j-1\}.$$

In the last statement, $k < j, u_{k,j} \neq 0$ can be replaced by (k, j) is an edge of $G^0(U)$. Similarly the structure of U can be characterized.

From these results, an algorithm can be derived for the symbolic fill computation when there is no pivoting, see GILBERT and LIU [1993]. When pivoting is required for stability, edags can also be useful, this time to symbolically compute the fill at each stage of Gaussian elimination.

The multifrontal method

The multifrontal algorithm we have studied for symmetric matrices previously has been applied to nonsymmetric matrices by DUFF and REID [1984]. The idea was to consider the sparsity pattern of $A + A^{\mathrm{T}}$ to construct the elimination tree. Numerical pivoting takes place within the frontal matrices. This works well if the pattern of A is nearly symmetric. However, the results are poor if the pattern of A is far from being symmetric.

Recently Davis and Duff have introduced an extension of the multifrontal algorithm to nonsymmetric matrices in a series of papers: DAVIS and DUFF [1993], DAVIS [1992, 1993], HADFIELD and DAVIS [1992, 1994a, 1994b]. This results in a package called UMFPACK, see DAVIS [1993].

In the nonsymmetric case, the frontal matrices are rectangular. The elimination tree is replaced by a directed acyclic graph (DAG) and update matrices are no longer assembled by a single parent.

We briefly summarize the work of DAVIS and DUFF [1993]. A frontal matrix F_k is a dense $r_i^{(k)} \times c_j^{(k)}$ submatrix that corresponds to the choice of $a_{i,j}^{(k)}$ as a pivot. The columns in F_k are defined by the set of indices $\mathcal{U}_k$ and the rows by the set $\mathcal{L}_k$. As in the symmetric case, amalgamation can take place if some other pivots have the same row and column patterns. Let g_k be the number of such pivots. Then

$$\mathcal{L}_k = \mathcal{L}'_k \cup \mathcal{L}''_k, \qquad \mathcal{U}_k = \mathcal{U}'_k \cup \mathcal{U}''_k,$$

where $\mathcal{L}'_k$ and $\mathcal{U}'_k$ correspond to the g_k pivots. We write

$$F_k = \begin{pmatrix} E_k & B_k \\ C_k & D_k \end{pmatrix},$$

where E_k is of order g_k. A numerical factorization $E_k = L'_k U'_k$ is computed as well as the block column L''_k of L and the block row U''_k of U and the Schur complement

$$D'_k = D_k - L''_k U''_k.$$

In the factorization any entry of E_k can be selected as a pivot. A bipartite graph $\mathcal{A}^k = (\mathcal{A}^k_\nu, \mathcal{A}^k_\varepsilon)$ represents the active submatrix,

$$\mathcal{A}^k_\nu = \mathcal{A}^k_{\mathrm{R}} \cup \mathcal{A}^k_{\mathrm{C}},$$

where $\mathcal{A}_{\mathrm{R}}^{k}$ (resp. $\mathcal{A}_{\mathrm{C}}^{k}$) is the index set of the rows (resp. columns) and

$$\mathcal{A}_{\varepsilon}^{k} = \left\{(i, j) \mid i \in \mathcal{A}_{\mathrm{R}}^{k},\ j \in \mathcal{A}_{\mathrm{C}}^{k},\ a_{i,j}^{(k)} \neq 0\right\}.$$

The factorized frontal matrices are described by a directed acyclic graph

$$\begin{aligned}
G^{k} &= \left(\nu^{k}, \varepsilon^{k}\right),\\
\nu^{k} &= \{t \mid F_{t} \text{ is a frontal matrix created before step } k\},\\
\nu^{k} &\subseteq \{1, \ldots, k-1\},\\
\varepsilon^{k} &= \varepsilon_{L}^{k} \cup \varepsilon_{U}^{k},\\
\varepsilon_{L}^{k} &= \Big\{(i, j) \mid i < j < k,\ i \in \nu^{k},\ j \in \nu^{k},\ \text{one or more entire rows of } F_{i}^{j}\\
&\qquad\qquad \text{are assembled in } F_{j}\Big\},\\
\varepsilon_{U}^{k} &= \Big\{(i, j) \mid i < j < k,\ i \in \nu^{k},\ j \in \nu^{k},\ \text{one or more entire columns of } F_{i}^{j}\\
&\qquad\qquad \text{are assembled in } F_{j}\Big\},
\end{aligned}$$

F_{i}^{j} is a submatrix of F_{i} formed by rows and columns that are nonpivotal and not yet assembled into subsequent frontal matrices before step j.

Edges in ε^{k} are called inactive edges. Node i is an LU child of node j if F_{i}^{j} is assembled into F_{i}, i.e. $(i, j) \in \varepsilon_{L}^{k} \cap \varepsilon_{U}^{k}$. If $(i, j) \in \varepsilon_{L}^{k}$, i is an L-child of j. If $(i, j) \in \varepsilon_{U}^{k}$, i is an U-child of j.

An active edge connects a frontal matrix in G with a row or column in $\mathcal{A}$ for which it has an unassembled contribution. They can be partitioned into active L-edges and active U-edges.

The actual algorithm is quite complex. Therefore, we refer the reader to DAVIS and DUFF [1993] and we consider just the beginning of a small example.

Before the start, the DAG G^{1} is empty and $\mathcal{A}^{1}$ is the bipartite graph of the original matrix. A pivot is chosen (based on the Markowitz strategy) and permuted to the first row and column. Consecutive pivots with the same row and column patterns are also included in the first frontal matrix F_{1} which is then factorized.

The first node in G refers to F_{1}. $\mathcal{L}_{1}$ (resp. $\mathcal{U}_{1}$) is given by the set of edges in $\mathcal{A}^{1}$ incident to columns (resp. rows) nodes 1 through g_{1}. Pivot row and column nodes 1 through g_{1} and edges in $\mathcal{A}^{1}$ incident to these nodes are removed from $\mathcal{A}^{1}$. $\mathcal{L}_{1}''$ (resp. $\mathcal{U}_{1}''$) defines new active L (resp. U) edges.

The example taken from DAVIS and DUFF [1993] is the following,

$$A = \begin{pmatrix}
a_{1} & & & x & x & & \\
x & a_{2} & x & & x & & x\\
x & x & a_{3} & & & & x\\
x & & & a_{4} & x & & \\
 & x & x & & x & x & \\
 & & & & & x & \\
x & x & & & x & & x
\end{pmatrix}.$$

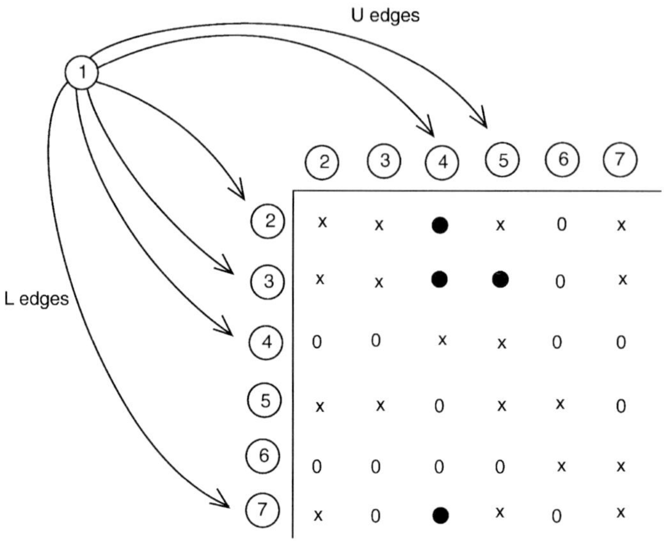

FIG. 23.1.

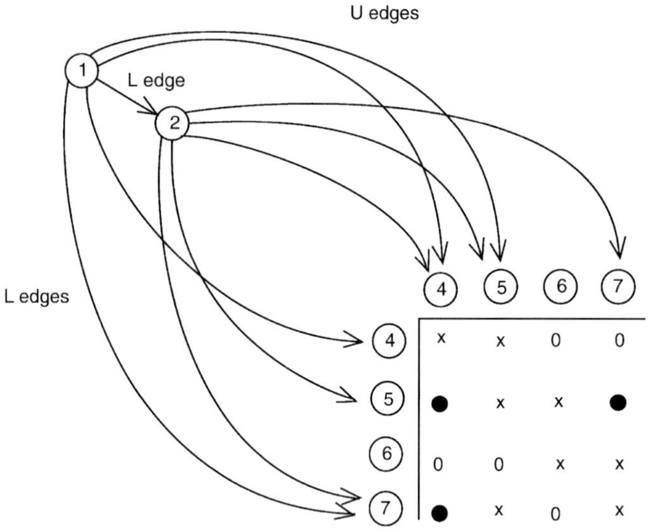

FIG. 23.2.

Node 1 is an L-child of node 2. Two pivots are considered in the second step as rows and columns 2 and 3 have the same pattern in L and U. The graphs are given in Fig. 23.2.

Then, node 4 is eliminated, $u_{2,4}$ and $u_{3,4}$ are nonzero as well as $l_{4,1}$ and $l_{1,4}$. This gives Fig. 23.3. and the algorithm goes on....

Some edge reduction operations are applied during the factorization, see DAVIS and

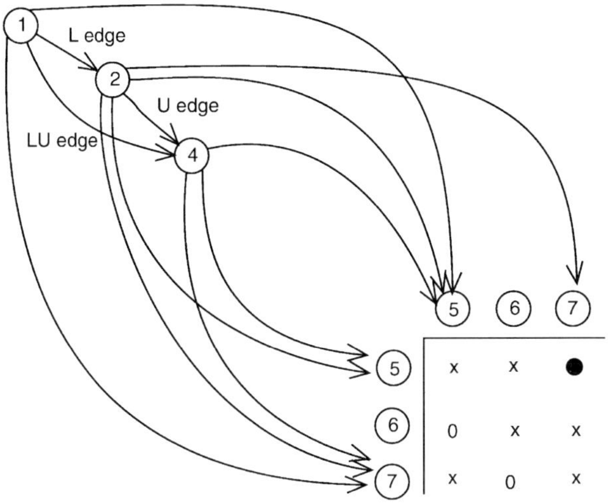

FIG. 23.3.

DUFF [1993]. With no additional edge reductions, a frontal matrix persists until the latest possible step of the elimination when it is completely assembled into other frontal matrices. The edge reduction discussed in Davis and Duff tries to assemble these contributions as soon as possible.

Some node amalgamation can also be introduced to give larger frontal matrices at the price of having a little more fill-in. Additionally, to speed up the pivot search, only upper and lower bounds of the degree of each row and column are computed. Moreover, only the first few columns with minimum upper bounds are scanned.

The DAG (the edge set) is constructed for being reused if some matrix with the same pattern is to be factored. Numerical experiments and comparisons with other methods are given in DAVIS and DUFF [1993].

SuperLU

In 1995, J. Demmel, S. Eisenstat, J. Gilbert, X. Li and J.W. Liu described and made available a code named SuperLU that combines several interesting techniques.

(1) The first thing is to extend supernodes to nonsymmetric matrices. Several possibilities were studied but the one retained in SuperLU is the following. A supernode is a range of columns in L with a full triangular diagonal block and the same structure below the diagonal block. This will allow a supernode to column update using BLAS2 routines. All the updates to a column from a supernode are done together.

In SuperLU (written in C), the matrix is stored by columns as well as L and U but the diagonal blocks in the supernodes are stored in L (as if they were full). This allows to treat supernodes as two-dimensional arrays.

As many supernodes are not large enough, some columns are merged into artificial supernodes even if they don't have the same structure. Also several consecutive columns can be factored at the same time (this set is called a panel).

(2) The column elimination tree is defined as the elimination tree of $A^{\mathrm{T}}A$. The column elimination tree T gives the dependencies among columns in the LU factorization. In particular, if $l_{i,j} \neq 0$ then, i is an ancestor of j in T, if $u_{i,j} \neq 0$ then, j is an ancestor of i in T.

Before factoring the matrix, the columns are permuted according to a postordering of the column elimination tree.

(3) Before each supernode-panel update, a symbolic factorization is done using a depth first traversal of the graph (this is in fact done on a reduced graph).

Careful analysis and experiments were done by Demmel et al. showing that this code is one of the most efficient one for nonsymmetric matrices.

24. Numerical stability for sparse matrices

The study of componentwise error analysis for sparse systems has been considered in ARIOLI, DEMMEL and DUFF [1989]. In this paper, they show how to compute estimates of the backward error. The perturbation f of the right-hand side is chosen in an a posteriori way and is not equal to $|b|$ to keep the perturbation on A sparse and the iterative refinement algorithm convergent (see ARIOLI, DEMMEL and DUFF [1989]).

Let $w = |A||y| + |b|$, y being the computed solution. A threshold τ_i is chosen for each w_i such that if $w_i > \tau_i$, then $f_i = |b_i|$. Otherwise if $w_i \leqslant \tau_i$, f_i is chosen larger. The value of τ_i suggested in ARIOLI, DEMMEL and DUFF [1989] is $\tau_i = 1000nu(\|A_{i,*}\|_\infty \|y\|_\infty + |b_i|)$ where $A_{i,*}$ is the ith row of A.

Let $f^{(2)}$ be the components of f for which $w_i \leqslant \tau_i$, then $f^{(2)}$ is defined as $f^{(2)} = \|b\|_\infty e$ where e is the column vector of all ones. With this choice, we can compute an estimate of the backward error

$$\frac{|b - Ay|_i}{(|A||y| + f)_i}.$$

Remember that the condition number is

$$K_{\mathrm{BS}}(A, b) = \frac{\| |A^{-1}|(E|x| + f)\|_\infty}{\|x\|_\infty}.$$

But we may just want to estimate $\||A^{-1}||A|\|_\infty = \||A^{-1}||A|e\|_\infty$. This can be estimated by an algorithm due to HAGER [1984] that uses multiplications by the matrix and its transpose. This is obtained by forward and backward solves using the LU factorization of A.

Numerical experiments in ARIOLI, DEMMEL and DUFF [1989] using iterative refinement show that it is possible to guarantee solutions of sparse linear systems that are exact solutions of a nearby system with a matrix of the same structure. Estimates of the condition number and of the backward error can be obtained easily using the previous strategies giving estimates of the error.

CHAPTER V

Parallel Algorithms for Sparse Matrices

25. Introduction

As for dense matrices, it is important to be able to solve efficiently sparse linear systems on parallel architectures. From one side, it is easier to consider sparse matrices rather than dense ones as, in the sparse case, there is more natural parallelism. Data dependencies are weaker in the sparse case since, in the factorization process, some columns are independent of each other. On the other side, it is more difficult to obtain significant performances since the granularity of independent tasks is often quite small and indirect addressing could lead to a lack of data locality.

Nevertheless, a lot of research is going on to obtain efficient algorithms. As for dense matrices, we will only consider algorithms for distributed memory architectures. A good review has been done in HEATH, NG and PEYTON [1991].

26. Symmetric positive definite systems

As we have seen, for symmetric positive definite matrices, Gaussian elimination proceeds in three phases: ordering, symbolic factorization and numerical factorization. The problem we are faced with is to have parallel implementations of these three phases. Therefore, things are complicated as, e.g., for the first phase, not only we have to find an ordering that reduces the fill-in and gives a good degree of parallelism during the solution phase, but also ideally, we need to be able to compute this ordering in parallel.

Orderings

As we have seen, the most commonly used heuristic algorithm to reduce fill-in is the minimum degree or one of its variants. Although there is not much theory about it (except negative results), it works quite well in practice.

Unfortunately, the basic minimum degree algorithm is quite sequential by nature. Modifications have to be made to generate parallelism and to be able to run the algorithm on parallel computers. Several attempts have been made or are under consideration by now.

One idea is to look for multiple elimination of independent nodes of minimum degree, see LIU [1985]. Another idea is to relax a bit the constraint of finding a node of minimum degree and just to look at nodes whose degree are within a constant factor of the minimum degree.

However, the implementation of the algorithm on sequential computers is now really efficient but quite complex and it is not that easy to port it to a parallel computer.

An ordering that is much more promising for parallelism is the nested dissection algorithm. As it is a divide and conquer algorithm, it is well suited to parallel computers. However, as we have seen, the algorithm is not easy to implement for general sparse matrices. For this, we need an efficient algorithm for partitioning a graph satisfying certain constraints on the separators. Research is underway to reach these goals (SIMON [1991], CIARLET and LAMOUR [1994], FARHAT [1988]), even though these algorithms are not parallel themselves. Some research is under way in this field. Combinations of the minimum degree and nested dissection have been proposed by LIU [1989b]. Therefore, the problem of finding in parallel a good ordering for parallel computation cannot be considered to be completely solved by 1996.

The current approach of the ordering problem is to separate the two (conflicting) goals that we have. First of all, an ordering for (approximately) minimizing the fill-in is chosen. Then, the ordering is modified by restructuring the elimination tree in order to introduce parallelism.

This approach has been described by JESS and KEES [1982]. The method starts by looking at PAP^{T}, where the permutation P was chosen to preserve sparsity. Then, the natural ordering is a perfect elimination one for $F = L + L^{\mathrm{T}}$. Now, the goal is to find a permutation matrix Q that gives also a perfect elimination but with more parallelism.

A node in G_F whose adjacency set is a clique is called simplicial. Such a node can be eliminated without causing any fill-in. Two nodes are independent if they are not adjacent in G_F. The Jess and Kees algorithm is the following,

- Until all nodes are eliminated, choose a maximum set of independent simplicial nodes, number them consecutively and eliminate these nodes.

It has been shown, LIU [1988], that the Jess and Kees method gives an ordering that has the shortest elimination tree over all orderings that do a perfect elimination of F.

The problem now is to implement this algorithm. This question was not really addressed by Jess and Kees. A proposal using clique trees was described in LEWIS, PEYTON and POTHEN [1989]. We will return to this method later. Another implementation was proposed in LIU and MIRZAIAN [1989].

Heuristically, it can be seen that larger elimination trees (having more leaf nodes) introduce more parallelism. The number of nodes being fixed, larger trees mean shorter trees. Therefore, it seems that finding an ordering that gives a shorter tree would increase the level of parallelism.

LIU [1988] has proposed to use tree rotations to reach this goal. The purpose of this algorithm is to find a reordering by working on the structure of PAP^{T}, namely the elimination tree.

A node x in a tree $T(B)$ is eligible for rotation if

$$\mathrm{Adj}_{G(B)}\big(T[x]\big) \neq \mathrm{Anc}(x),$$

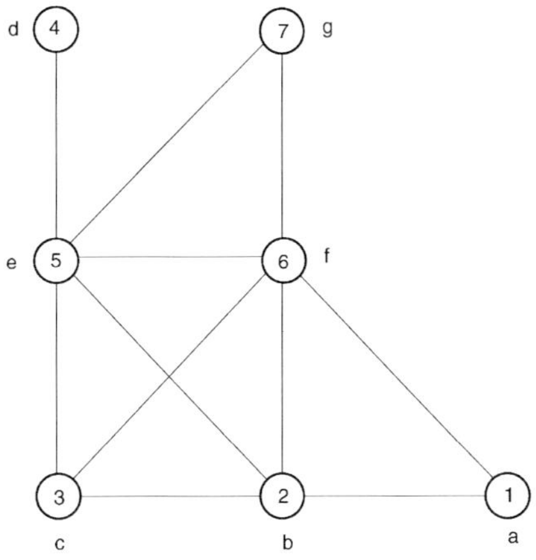

FIG. 26.1.

where $\mathrm{Anc}(x)$ is the set of ancestors of x in T and

$$\mathrm{Adj}_{G(B)}\big(T[v]\big) = \mathrm{Anc}(v), \quad \forall v \text{ ancestor of } x.$$

A tree rotation at x is a reordering of $G(B)$ such that the nodes in $\mathrm{Adj}_{G(B)}(T[x])$ are labeled last while keeping the relative order of the nodes.

Consider a small example similar to the one in LIU [1988], see Figs. 26.1 and 26.2.

$$\begin{aligned}\mathrm{Adj}\big(T[c]\big) &= \{e, f\} \neq \{e, f, g\} = \mathrm{Anc}(c),\\ \mathrm{Adj}\big(T[d]\big) &= \{e\} \neq \{e, f, g\} = \mathrm{Anc}(d).\end{aligned}$$

Thus, c and d are eligible for tree rotations. A rotation at c gives what is seen in Figs. 26.3 and 26.4. A rotation at d gives Figs. 26.5 and 26.6.

Let $h_{\mathrm{T}}(v)$ be the height of $T[v]$ and $\bar{h}_{\mathrm{T}}$ be defined as -1 if every subtree of T intersects $T[v]$ or otherwise,

$$\bar{h}_{\mathrm{T}} = \max\big\{h_{\mathrm{T}}(w) \mid T[w] \cap T[v] = \emptyset\big\}.$$

In the original tree,

$$\begin{aligned}h_{\mathrm{T}}(a) &= 0,\ h_{\mathrm{T}}(b) = 1,\ h_{\mathrm{T}}(c) = 2,\ h_{\mathrm{T}}(d) = 0,\ h_{\mathrm{T}}(e) = 3,\ \text{etc}\ldots,\\ \bar{h}_{\mathrm{T}}(a) &= 0,\ \bar{h}_{\mathrm{T}}(b) = 0,\ \bar{h}_{\mathrm{T}}(c) = 0,\ \bar{h}_{\mathrm{T}}(d) = 2,\ \bar{h}_{\mathrm{T}}(e) = -1,\ \text{etc}\ldots.\end{aligned}$$

The algorithm proposed by LIU [1988] is the following:

- If x is an eligible node with $\bar{h}_{\mathrm{T}}(x) < h_T(x)$, then apply a tree rotation at x relabeling the nodes.

In the above example, only c can be chosen by the algorithm. Results are proven in LIU [1988] that supports this choice. The implementation details and experimental

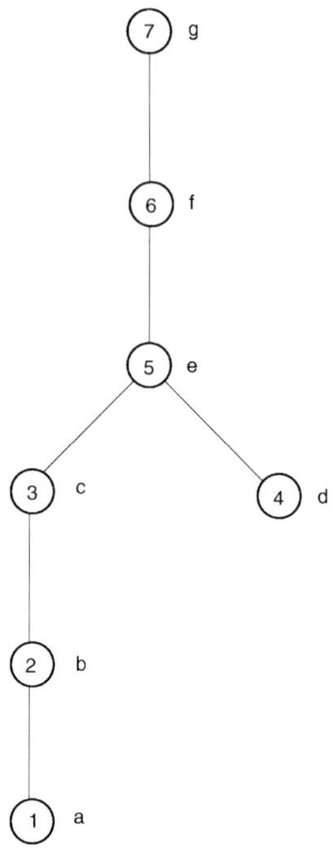

FIG. 26.2.

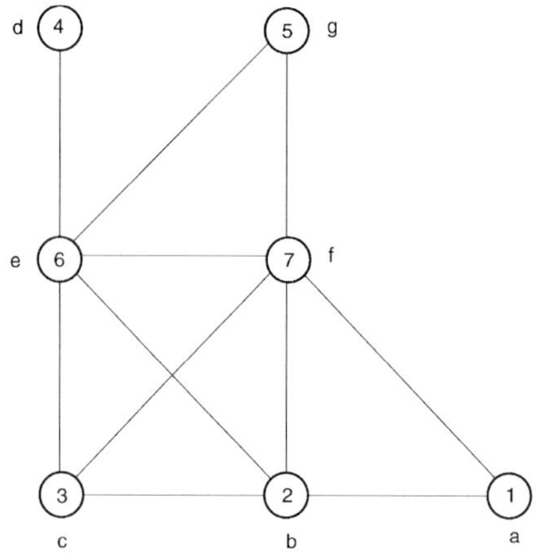

FIG. 26.3.

results are given in LIU [1988]. However, tree rotations do not always give a tree of minimum height.

More than the height of the elimination tree, it will be better to minimize the parallel completion time. This was defined by Liu as:

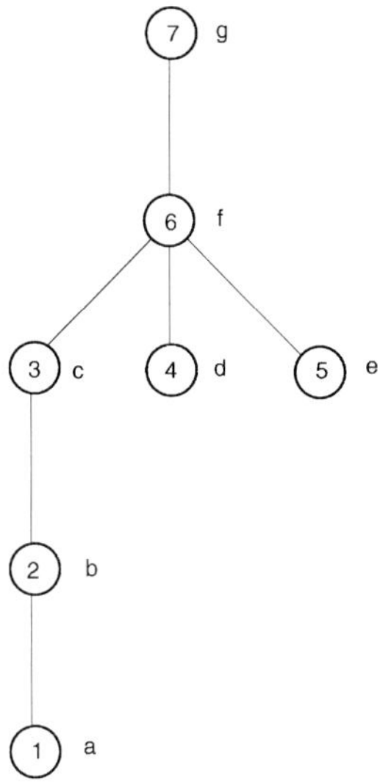

FIG. 26.4.

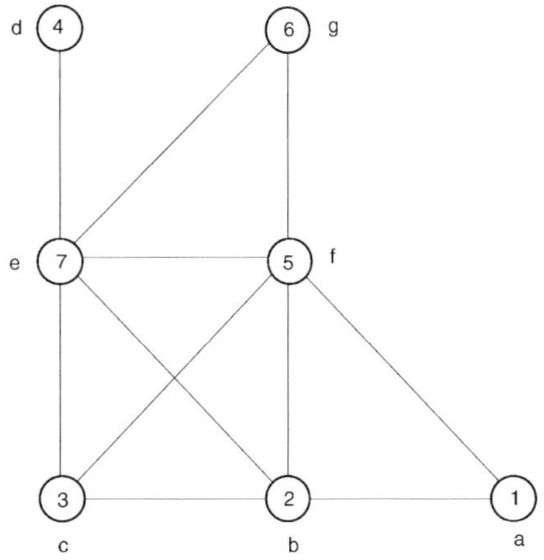

FIG. 26.5.

- Let time[v] be the execution time for the node v in T (e.g., a constant times the number of operations).
  Then,

$$\text{level}[v] = \begin{cases} \text{time}[v] & \text{if } v \text{ is the root,} \\ \text{time}[v] + \text{level[parent of } v] & \text{otherwise.} \end{cases}$$

  The parallel completion time is $\max_{v \in T} \text{level}[v]$.

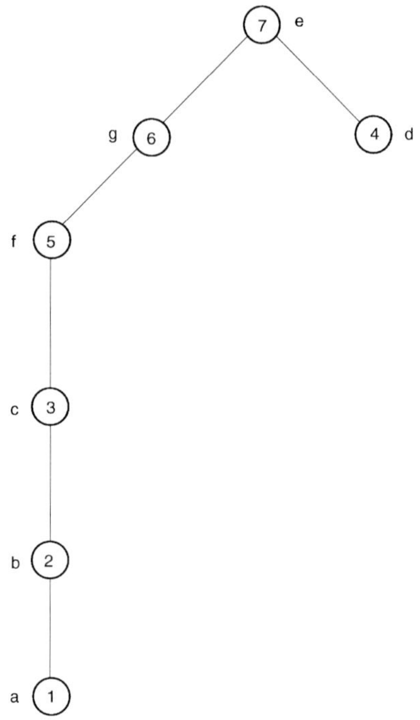

FIG. 26.6.

LIU and MIRZAIAN [1989] proposed an implementation of the Jees and Kees algorithm. In their method, the cost of detecting simplicial nodes is $O(n\nu(F))$ where $\nu(F)$ is the number of off diagonal elements in $L + L^{\mathrm{T}}$. The tree rotations heuristic is faster than that.

In LEWIS, PEYTON and POTHEN [1989], an implementation of the Jees and Kees algorithm was proposed using clique trees. They gave a characterization of simplicial nodes,

THEOREM 26.1. *A node is simplicial if and only if it is contained in only one maximal clique.*

PROOF. See LEWIS, PEYTON and POTHEN [1989]. □

Let $\mathrm{Madj}(v) = \{u \mid u \in \mathrm{Adj}(v),\ u > v\}$, then

THEOREM 26.2. *If K is a maximal clique in G_F and v is the lowest numbered node in K, then*

$$K = \{v\} \cup \mathrm{Madj}(v).$$

PROOF. See LEWIS, PEYTON and POTHEN [1989]. □

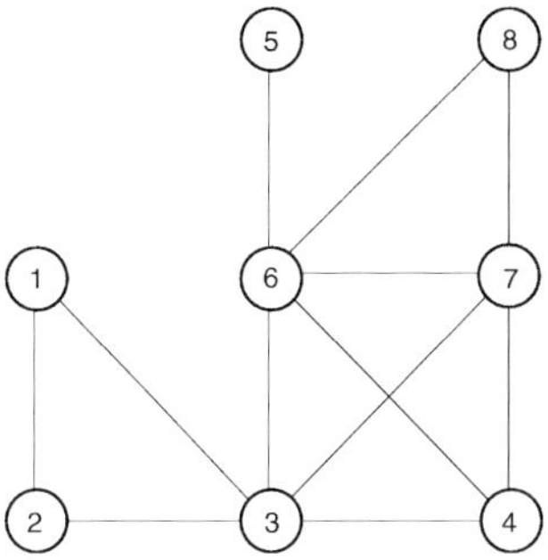

FIG. 26.7.

The clique $K(v)$ is represented by such a v which is called a representative node. Lewis, Peyton and Pothen proposed a way to find the representative nodes.

THEOREM 26.3. *A node v is not representative for any maximal clique if and only if there exists a representative node z, $z < v$, such that*

$$\{v\} \cup \mathrm{Madj}(v) \subset \{z\} \cup \mathrm{Madj}(z).$$

PROOF. See LEWIS, PEYTON and POTHEN [1989]. □

An algorithm based on these results can be constructed to find the representative nodes, hence the maximal cliques. Considering the details of the algorithm will take us too far as it is rather technical. However, during the marking procedure nodes are partitioned into sets $\mathrm{new}(K(v))$, $\mathrm{anc}(K(v))$.

$\mathrm{New}(K(v))$ consists of v together with the nodes marked as nonrepresentative while considering $K(v)$. $\mathrm{Anc}(K(v))$ is the ancestor set of $K(v)$. Rather than looking at the details, let us consider the example given in LEWIS, PEYTON and POTHEN [1989] which comes from Liu, see Fig. 26.7.

$$K(1) = \{1, 2, 3\},\ K(3) = \{3, 4, 6, 7\},\ K(5) = \{5, 6\},\ K(6) = \{6, 7, 8\},$$

$$\mathrm{new}\,K(1) = \{1, 2\},\ \mathrm{new}\,K(3) = \{3, 4\},\ \mathrm{new}\,K(5) = \{5\},\ \mathrm{new}\,K(6) = \{6, 7, 8\},$$

$$\mathrm{anc}\,K(1) = \{3\},\ \mathrm{anc}\,K(3) = \{6, 7\},\ \mathrm{anc}\,K(5) = \{6\},\ \mathrm{new}\,K(6) = \emptyset.$$

The number of nodes of the clique tree is the number of maximal cliques and

$$\mathrm{parent}(K) = \{K' \mid \mathrm{first\ anc}(K) \in \mathrm{new}(K')\},$$

where first $\mathrm{anc}(K)$ is the smallest element of $\mathrm{anc}(K)$,

$$\mathrm{parent}[\mathrm{new}\,K(1)] = \mathrm{new}\,K(3),$$

$$\text{parent}\big[\text{new } K(3)\big] = \text{new } K(6),$$
$$\text{parent}\big[\text{new } K(5)\big] = \text{new } K(6),$$

and the clique tree is the one in Fig. 26.8.

LEWIS, PEYTON and POTHEN [1989] described how to update the clique tree during elimination using the Jees and Kees algorithm.

Let $R(K)$ be defined as

$$R(K) = \begin{cases} K, & \text{if } K \text{ contains no simplicial nodes,} \\ K - \{u\}, & \text{where } u \text{ is the eliminated simplicial node of } K. \end{cases}$$

Given the clique tree, when a simplicial node u contained in K is eliminated, the new clique tree is constructed in the following way,

- if $R(K)$ is maximal, make no changes
- otherwise,
  - if K is a leaf, delete K,
  - if K is an ancestor, let C be a child with the largest ancestor set
    - assign all nodes in new(K) to new(C), removing them from anc(C),
    - attach C as a child of parent(K) with first anc(C) = first anc(K),
    - attach children of K other than C as children of C,
    - delete K.

Proofs are given in LEWIS, PEYTON and POTHEN [1989]. Experimental results in Lewis, Peyton and Pothen show that this implementation is about as fast as the tree rotation heuristic and faster than the LIU and MIRZAIAN [1989] Jees and Kees implementation.

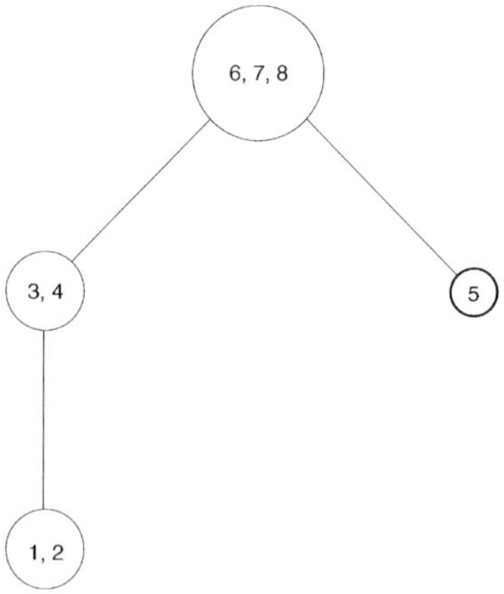

FIG. 26.8.

Symbolic factorization

There is not much parallelism to be found in the symbolic factorization phase. The natural structure for increasing the level of parallelism is the elimination tree. But, then we are faced with a bootstrapping problem as we have also to compute the elimination tree in parallel.

GEORGE, HEATH, LIU and NG [1988] have proposed a column oriented parallel symbolic factorization algorithm. Experimental results showed only modest speed ups. Further research is needed in this area.

Numerical factorization

The first algorithms that were studied were column oriented. Traditionally, the two main operations of Gaussian elimination are denoted:

- cmod(j, k): modification of column j by column k, $k < j$,
- cdiv(j): division of column j by a scalar (the pivot).

In the fan-out and fan-in algorithms, data distribution is such that columns are assigned to processors. As in the dense case, column k is stored on processor $p = \text{map}(k)$. Leaf nodes of the elimination tree are independent of each other and can be processed first. Let mycols(p) be the set of columns owned by processor p and

$$\text{procs}(L_{*,k}) = \left\{\text{map}(j) | j \in \text{Struct}(L_{*,k})\right\}.$$

The fan-out algorithm is the following,

```
For j ∈ mycols(p)
  if j is a leaf node
    cdiv(j)
    send L_{*,j} to p' ∈ procs(L_{*,j})
    mycols(p) = mycols(p)-{ j}
  end
  while mycols(p) ≠ ∅
    receive column L_{*,k}
    for j ∈ Struct(L_{*,k}) ∩ mycols(p)
      cmod(j,k)
      if all cmods are done for column j
        cdiv(j)
        send L_{*,j} to p' ∈ procs(L_{*,j})
        mycols(p) = mycols(p)-{ j}
      end
    end
  end
end
```

When columns are sent to other processors, it is also necessary to send their structures to be able to complete the cmods operations. There are too many communications in this

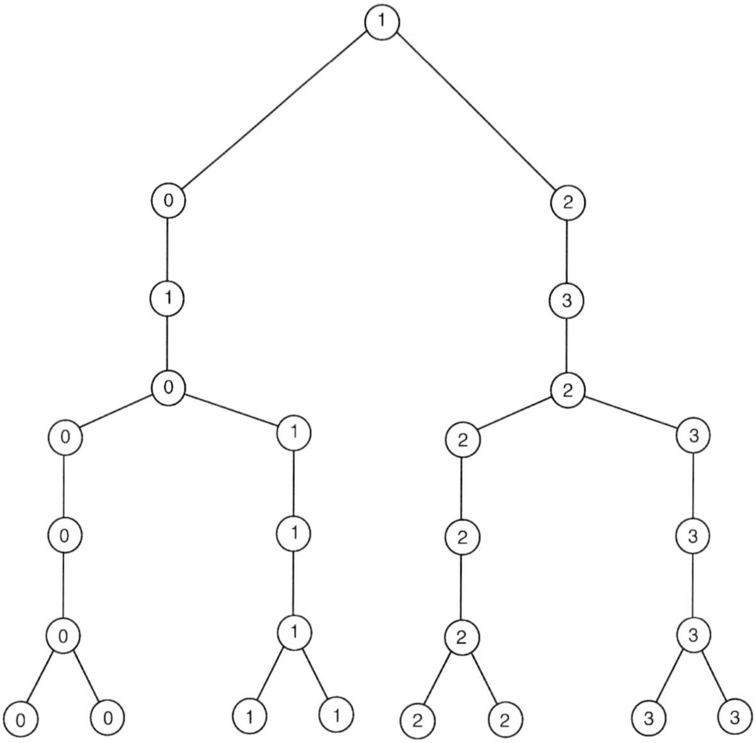

FIG. 26.9.

algorithm. This has been improved in the fan-in algorithm (ASHCRAFT, EISENSTAT and LIU [1990]),

- Processing column j, processor p computes the modification $u(j,k)$ for $k \in \text{mycols}(p) \cap \text{Struct}(L_{j,*})$. If p does not own column j, it sends $u(j,k)$ to processor $map(j)$. If p owns column j, it receives and processes the aggregated modifications and then completes the $\text{cdiv}(j)$ operation.

When processing column j, if there are no modifications arrived yet for this column, processor p can proceed to compute some columns $i > j$ by aggregating modifications already received or receiving updates. In practice, this look ahead is limited to column i in the same supernode as column j.

An important issue in these column oriented algorithms is the mapping of columns to the processors. Currently, most implementations use a static mapping of computational tasks to processors. This can lead to load balancing problems. In the fan-out or fan-in algorithms, the assignment of columns to processors is guided by the elimination tree. The goals are good load balancing and few processor communications.

The first implementations were based on wrap mapping of the levels of the elimination tree starting from the bottom up. This gives good load balancing properties but too many communications.

Another technique that has been much used is the subtree to subcube mapping. This was specifically designed for hypercube architectures but can be easily generalized to other distributed memory architectures.

This is illustrated on the following example, see Fig. 26.9, distributing the tree on 4 processors.

In KARYPIS and KUMAR [1994], a subforest to subcube mapping is introduced which seems to improve on the subtree to subcube mapping.

Another algorithm that is in favor is the multifrontal algorithm. We have already seen that there is a natural parallelism in the early (bottom) stages of the multifrontal method. Dense frontal matrices are assigned to one processor. The problem is that when moving to the root of the tree, there is less and less parallelism of this kind. However, frontal matrices are getting larger and larger and dense techniques used to handle these matrices can be distributed on several processors (using level 3 BLAS primitives).

In KARYPIS and KUMAR [1994], the multifrontal method is implemented using the subforest to subcube mapping and very good efficiencies are obtained. In this scheme, many subtrees of the elimination tree are assigned to each subcube. They are chosen in order to balance the work. Algorithms are given in KARYPIS and KUMAR [1994] to obtain this partitioning.

All these mappings are based on column distribution of the matrix to the processors. ROTHBERG and GUPTA [1994] proposed to use a block oriented approach of sparse Gaussian elimination. These partitionings have good communication performances.

27. Nonsymmetric systems

As for dense systems, the problem here is complicated by the issue of pivoting. There are not many efficient implementations of Gaussian elimination on parallel architectures for general sparse matrices.

The best algorithms are based on the multifrontal method, either the methods based on the structure of $A + A^{\mathrm{T}}$ or the truly nonsymmetric ones.

The nonsymmetric version of the multifrontal method has been studied by HADFIELD and DAVIS [1994b] for parallel computing. As for the symmetric case, different levels of parallelism must be used. Experimental results are given in HADFIELD and DAVIS [1994b].

DAVIS and YEW [1990] have proposed another nondeterministic algorithm called D2. This algorithm is based on selecting an independent set of m pivots whose updates can be computed in parallel

$$S = \big\{a_{i,j}^{(k-1)} \mid a_{i,j}^{(k-1)} = a_{j,i}^{(k-1)} = 0, \text{ for } k \leqslant j \leqslant m+k-1,\ k \leqslant i \leqslant m+k-1, \\ \text{and } i \neq j\big\}.$$

Then, a sparse rank-m update is done. This algorithm gives good results on matrices that have a very nonsymmetric pattern.

References

AASEN, J.O. (1971), On the reduction of a symmetric matrix to tridiagonal form, *BIT* **11**, 233–242.

AHAC, A.A., J.J. BUONI and D.D. OLESKY (1988), Stable LU factorization of H-matrices, *Linear Algebra Appl.* **99**, 97–110.

AHAC, A.A. and D.D. OLESKY (1986), A stable method for the LU factorization of M-matrices, *SIAM J. Alg. Disc. Meth.* **7** (3), 368–378.

ALAGHBAND, G. (1989), Parallel pivoting combined with parallel reduction and fill-in control, *Parallel Comput.* **11**, 201–221.

ALVARADO, F.L. and R. SCHREIBER (1993), Optimal parallel solution of sparse triangular systems, *SIAM J. Sci. Statist. Comput.* **14** (2), 446–460.

AMESTOY, P.R. (1990), Factorisation de grandes matrices creuses nonsymétriques basée sur une méthode multifrontale dans un environnement multiprocesseur, Ph.D. thesis, CERFACS report TH/PA/91/2.

AMESTOY, P., D.A. DAVIS and I.S. DUFF (1994), An approximate minimum degree ordering algorithm, Report TR-94-039, University of Florida.

AMESTOY, P.R. and I.S. DUFF (1989), Vectorization of a multiprocessor multifrontal code, *Internat. J. Supercomputer Appl.* **3** (3), 41–59.

AMODO, P., L. BRUGNANO and T. POLITI (1993), Parallel factorizations for tridiagonal matrices, *SIAM J. Numer. Anal.* **30** (3), 813–823.

ANAND, I.M. (1980), Numerical stability of nested dissection orderings, *Math. Comp.* **35** (152), 1235–1249.

ANDERSON, E., Z. BAI, C. BISCHOF, J. DEMMEL, J. DONGARRA, J. DU CROZ, A. GREENBAUM, S. HAMMARLING, A. MCKENNEY, S. OSTROUCHOV and D. SORENSEN (1992), *LAPACK User's Guide* (SIAM).

ANDERSON, E. and Y. SAAD (1989), Solving sparse triangular linear systems on parallel computers, *Internat. J. High Speed Comput.* **1** (1), 73–95.

ANDERSON, D.V., A.R. FRY, R. GRUBER and A. ROY (1987), Gigaflop speed algorithm for the direct solution of large block tridiagonal systems in 3D physics applications, Report UCRL-96034, Lawrence Livermore Nat. Lab.

ANGELACCIO, M. and M. CLAJANNI (1994a), The row/column pivoting strategy on multicomputers, *Parallel Comput.* **20** (2), 197–214.

ANGELACCIO, M. and M. CLAJANNI (1994b), Subcube matrix decomposition, a unifying view for LU factorization on multicomputers, *Parallel Comput.* **20** (2), 257–270.

ARIOLI, M., J.W. DEMMEL and I.S. DUFF (1989), Solving sparse linear systems with backward error, *SIAM J. Matrix Anal. Appl.* **10**, 165–190.

ASHCRAFT, C. (1991), A taxonomy of distributed dense LU factorization methods, Report ECA-TR-161, Boeing Computer Services.

ASHCRAFT, C., S.C. EISENSTAT and J.W.-H. LIU (1990), A fan-in algorithm for distributed sparse numerical factorization, *SIAM J. Sci. Statist. Comput.* **11** (3), 593–599.

ASHCRAFT, C. and R.G. GRIMES (1989), The influence of relaxed supernode partitions on the multifrontal method, *ACM Trans. Math. Soft.* **15**, 291–309.

ASHCRAFT, C., R.G. GRIMES, J.G. LEWIS, B.W. PEYTON and H.D. SIMON (1987), Progress in sparse matrix methods for large linear systems on vector supercomputers, *Internat. J. Supercomputer Appl.* **1** (4), 10–30.

AXELSSON, O. and V.A. BARKER (1984), *Finite Element Solution of Boundary Value Problems* (Academic Press).

BABUŠKA, I. and H.C. ELMAN (1989), Some aspects of parallel implementation of the finite element method on message passing architectures, *J. Comput. Appl. Math.* **27**, 157–187.

BANK, R.E. and D.J. ROSE (1990), On the complexity of sparse Gaussian elimination via bordering, *SIAM J. Sci. Statist. Comput.* **11** (1), 145–160.

BANK, R.E. and R.K. SMITH (1987), General sparse elimination requires no permanent integer storage, *SIAM J. Sci. Statist. Comput.* **8** (4), 574–584.

BARLOW, J.L. (1986), A note on monitoring the stability of triangular decomposition of sparse matrices, *SIAM J. Sci. Statist. Comput.* **7** (1), 166–168.

BARNARD, S.T., A. POTHEN and H.D. SIMON (1993), A spectral algorithm for envelope reduction of sparse matrices, Tech. Rept. CS-93-49, Univ. of Waterloo.

BARWELL, V. and A. GEORGE (1976), A comparison of algorithms for solving symmetric indefinite systems of linear equation, *ACM Trans. Math. Soft.* **2** (3), 242–251.

BAUER, F.L. (1974), Computational graphs and rounding error, *SIAM J. Numer. Anal.* **11**, 87–96.

BENNETT, J.M. (1965), Triangular factors of modified matrices, *Numer. Math.* **7**, 217–221.

BERMAN, A. and R.J. PLEMMONS (1979), *Nonnegative Matrices in the Mathematical Sciences* (Academic Press).

BERMAN, P. and G. SCHNITGER (1990), On the performance of the minimum degree ordering for Gaussian elimination, *SIAM J. Matrix Anal. Appl.* **11** (1), 83–89.

BISCHOF, C.H. (1990), Incremental condition estimation, *SIAM J. Matrix Anal. Appl.* **11**, 312–322.

BISCHOF, C.H., J.G. LEWIS and D.J. PIERCE (1990), Incremental condition estimation for sparse matrices, *SIAM J. Matrix Anal. Appl.* **11** (4), 644–662.

BLAIR, J.R. and B. PEYTON (1993), An introduction to chordal graphs and clique trees, in: A. George, J.R. Gilbert and J.W.-H. Liu, eds., *Graph Theory and Sparse Matrix Computation* (Springer-Verlag) 1–29.

BODEWIG, E. (1959), *Matrix Calculus*, 2nd ed. (North-Holland).

BOISVERT, R.F. (1991), Algorithms for special tridiagonal systems, *SIAM J. Sci. Statist. Comput.* **12** (2), 423–442.

BOTHE, Z. (1975), Bounds for rounding errors in the Gaussian elimination for band systems, *J. Inst. Math. Appl.* **16**, 133–142.

BRAYTON, R.K., F.G. GUSTAVSON and R.A. WILLOUGHBY (1970), Some results on sparse matrices, *Math. Comp.* **24** (112), 937–954.

BROYDEN, C.G. (1973), Some condition number bounds for the Gaussian elimination process, *J. Inst. Math. Appl.* **12**, 273–286.

BUNCH, J.R. (1971), Analysis of the diagonal pivoting method, *SIAM J. Numer. Anal.* **8** (4), 656–680.

BUNCH, J.R. (1973), Complexity of sparse elimination, in: J.F. Traub, ed., *Complexity of Sequential and Parallel Numerical Algorithms* (Academic Press) 197–220.

BUNCH, J.R. (1974a), Partial pivoting strategies for symmetric matrices, *SIAM J. Numer. Anal.* **11** (3), 521–528.

BUNCH, J.R. (1974b), Analysis of sparse elimination, *SIAM J. Numer. Anal.* **11** (5), 847–873.

BUNCH, J.R. (1987), The weak and strong stability of algorithms in numerical linear algebra, *Linear Algebra Appl.* **88**, 49–66.

BUNCH, J.R., J.W. DEMMEL and C. VAN LOAN (1989), The strong stability of algorithms for solving symmetric linear systems, *SIAM J. Matrix Anal. Appl.* **10** (4), 494–499.

BUNCH, J.R., J.J. DONGARRA, C. MOLER and G.W. STEWART (1979), *Linpack User's Guide* (SIAM).

BUNCH, J.R. and J.E. HOPCROFT (1974), Triangular factorization and inversion by fast matrix multiplication, *Math. Comp.* **28** (125), 231–236.

BUNCH, J.R. and L. KAUFMAN (1977), Some stable methods for calculating inertia and solving symmetric linear systems, *Math. Comp.* **31**, 162–179.

BUNCH, J.R. and B.N. PARLETT (1971), Direct methods for solving symmetric indefinite systems of linear equations, *SIAM J. Numer. Anal.* **8** (4), 639–655.

BUNCH, J.R. and D.J. ROSE, eds. (1976), *Sparse Matrices Computations* (Academic Press).

BURGESS, I.W. and P.K.F. LAI (1986), A new node renumbering algorithm for bandwith reduction, *Internat. J. Numer. Methods. Engrg.* **23**, 1693–1704.

CARNEVALI, P., G. RADICATI, Y. ROBERT and P. SGUAZZERO (1987), Efficient Fortran implementation of the Gaussian elimination and Householder reduction algorithms on the IBM 3090 vector multiprocessor, Report ICE-0012 IBM European Center for Scientific and Engineering Computing.

CARTER, R. (1991), Y–MP floating point and Cholesky factorization, *Internat. J. High Speed Comput.* **3** (3), 215–222.

CHAN, T. (1984), On the existence and computation of LU factorizations with small pivots, *Math. Comp.* **42** (166), 535–547.

CHAN, T.F. and D.E. FOULSER (1988), Effectively well conditioned linear systems, *SIAM J. Sci. Statist. Comput.* **9**, 963–969.

CHAN, T. and D.C. RESASCO (1986), Generalized deflated block elimination, *SIAM J. Numer. Anal.* **23** (5), 913–924.

CHAN, W.M. and A. GEORGE (1980), A linear time implementation of the reverse Cuthill–McKee algorithm, *BIT* **20**, 8–14.

CHANDRASEKARAN, S. and I. IPSEN (1995), On the sensitivity of solution components in linear systems of equations, *SIAM J. Matrix Anal. Appl.* **16** (1), 93–112.

CHARRIER, P. and J. ROMAN (1989), Algorithmique et calculs de complexité pour un solveur de type dissection emboîtée, *Numer. Math.* **55**, 463–476.

CHARRIER, P. and J. ROMAN (1991), Analysis of refined partitions for a parallel implementation of nested dissection, Report LABRI 91-38, Université de Bordeaux I.

CHATELIN, F. and V. FRAYSSE (1993), Qualitative computing, elements of a theory of finite precision computation, CERFACS Rept.

CHEN, S.C., D.J. KUCK and A.H. SAMEH (1978), Practical parallel band triangular system solvers, *ACM Trans. Math. Soft.* **1** (3), 270–277.

CHOI, J., J. DONGARRA, R. POZO and D.W. WALKER (1994), ScaLAPACK: A scalable linear algebra library for distributed memory concurrent computers, in: *Proceedings of Fourth Symposium on the Frontiers of Massively Parallel Computation (McLean, Virginia)* (IEEE Computer Society Press).

CHU, E. and A. GEORGE (1987), Gaussian elimination with partial pivoting and load balancing on a multiprocessor, *Parallel Comput.* **5**, 65–74.

CIARLET, P. JR. and F. LAMOUR (1994), On the validity of a front oriented approach to partitioning large sparse graphs with a connectivity constraint, Report CAM 94-37, UCLA, submitted to Numerical Algorithms.

CLEARY, A.J. (1990), Parallelism and fill-in in the Cholesky factorization of reordered banded matrices, Report SAND90-2757, Sandia Nat. Lab., Albuquerque.

CLINE, A.K., A.R. CONN and C.F. VAN LOAN (1982), Generalizing the LINPACK condition estimator, *Lecture Notes in Math.* **909** (Springer) 73–83.

CLINE, A.K., C.B. MOLER, G.W. STEWART and J.H. WILKINSON (1979), An estimate of the condition number of a matrix, *SIAM J. Numer. Anal.* **16**, 368–375.

CONCUS, P., G.H. GOLUB and G. MEURANT (1985), Block preconditioning for the conjugate gradient method, *SIAM J. Sci. Statist. Comput.* **6** (1), 220–252.

COSNARD, M., Y. ROBERT, P. QUINTON and M. TCHUENTE, eds. (1986), *Parallel Algorithms and Architectures* (North-Holland).

COSNARD, M., Y. ROBERT and D. TRYSTRAM (1986), Résolution parallèle de systèmes linéaires denses par diagonalisation, EDF, Bulletin DER, série C no 2, 67–88.

CRYER, C.W. (1968), Pivot growth in Gaussian elimination, *Numer. Math.* **12**, 335–345.

CSANKY, L. (1976), Fast parallel matrix inversion algorithms, *SIAM J. Comput.* **5** (4), 618–623.

CURTIS, A.R. and J.K. REID (1971), The solution of large sparse unsymmetric systems of linear equations, *J. Inst. Math. Appl.* **8**, 344–353.

CURTIS, A.R. and J.K. REID (1972), On the automatic scaling of matrices for Gaussian elimination, *J. Inst. Math. Appl.* **10**, 118–124.

CUTHILL, E. (1972), Several strategies for reducing the bandwith of matrices, in: D.J. Rose and R.A. Willoughby, eds., *Sparse Matrices and Their Applications* (Plenum Press).

DAHLQUIST, G. and Å. BJÖRCK (1974), *Numerical Methods* (Prentice-Hall).

DAHLQUIST, G., S.C. EISENSTAT and G.H. GOLUB (1972), Bounds for the error of linear systems of equations using the theory of moments, *J. Math. Anal. Appl.* **37** (1), 151–166.

DAVE, A.K. and I.S. DUFF (1993), Sparse matrix calculations on the Cray 2, *Parallel Comput.* **5**, 55–64.

DAVIS, T.A. (1992), Performance of an unsymmetric pattern multifrontal method for sparse LU factorization, Report TR-92-014, University of Florida.

DAVIS, T.A. (1993), Users' guide for the unsymmetric pattern multifrontal package, Report TR-93-020, University of Florida.

DAVIS, T.A. (1994), A combined unifrontal/multifrontal method for unsymmetric sparse matrices, Report TR-94-005, University of Florida.

DAVIS, T.A. and I.S. DUFF (1993), An unsymmetric pattern multifrontal method for sparse LU factorization, Report TR-93-018, University of Florida.

DAVIS, T.A. and P.C. YEW (1990), A nondeterministic parallel algorithm for unsymmetric sparse LU factorization, *SIAM J. Matrix Anal. Appl.* **11**, 383–402.

DAYDE, M.J. and I.S. DUFF (1989), Use of Level 3 BLAS in LU factorization on the Cray 2, the ETA-10P and the IBM 3090/VF, *Internat. J. Supercomputer Appl.* **3** (2), 40–70.

DAYDE, M.J., I.S. DUFF and A. PETITET (1993), A parallel block implementation of level 3 BLAS for MIMD vector processors, Report RAL-93-037, Rutherford Appleton Lab.

DEMMEL, J.W. (1984), Underflow and the reliability of numerical software, *SIAM J. Sci. Statist. Comput.* **5**, 887–919.

DEMMEL, J.W. (1987), On condition numbers and the distance to the nearest ill-posed problem, *Numer. Math.* **51** (3), 251–289.

DEMMEL, J.W. (1989), On floating point errors in Cholesky, Lapack Working Note 14, University of Tennessee.

DEMMEL, J.W. (1992a), Open problems in numerical linear algebra, Report CS-92-164, Univ. of Tennessee.

DEMMEL, J.W. (1992b), Trading off parallelism and numerical stability, Report CS-92-179, Univ. of Tennessee.

DEMMEL, J.W. (1992c), The componentwise distance to the nearest singular matrix, *SIAM J. Matrix Anal. Appl.* **13** (1), 10–19.

DEMMEL, J.W., M.T. HEATH and H.A. VAN DER VORST (1993), Parallel numerical linear algebra, *Acta Numerica*, 111–197.

DEMMEL, J.W., N.J. HIGHAM and R.S. SCHREIBER (1995), Stability of block LU factorization, *Numer. Lin. Alg. Appl.* **2** (2).

DESPREZ, F., B. TOURANCHEAU and J.J. DONGARRA (1994), Performance complexity of LU factorization with efficient pipelining and overlap on a multiprocessor, Report Univ. of Tennessee.

DONGARRA, J. and J.W. DEMMEL (1991), LAPACK; a portable high-performance numerical library for linear algebra, *Supercomputer* **46**, 33–38.

DONGARRA, J., J. DU CROZ, S. HAMMARLING and I.S. DUFF (1990), A set of level 3 basic linear algebra subprograms, *ACM Trans. Math. Soft.* **16** (1), 1–17.

DONGARRA, J., J. DU CROZ, S. HAMMARLING and R. HANSON (1988), An extended set of Fortran basic linear algebra subprograms, *ACM Trans. Math. Soft.* **14** (1), 1–17.

DONGARRA, J., I.S. DUFF, P. GAFFNEY and S. MCKEE, eds. (1989), *Vector and Parallel Computing, Issues in Applied Research and Development* (Ellis Horwood).

DONGARRA, J., I.S. DUFF, D.C. SORENSEN and H.A. VAN DER VORST (1991), *Solving Linear Systems on Vector and Shared Memory Computers* (SIAM).

DONGARRA, J., F.G. GUSTAVSON and A. KARP (1984), Implementing linear algebra algorithms for dense matrices on a vector pipeline machine, *SIAM Rev.* **26**, 91–112.

DONGARRA, J. and T. HEWITT (1986), Implementing dense linear algebra algorithms using multitasking on the Cray X-MP-4 (or approaching the Gigaflop), *SIAM J. Sci. Statist. Comput.* **7** (1), 347–305.

DONGARRA, J. and A.H. SAMEH (1984), On some parallel banded system solvers, *Parallel Comput.* **1**, 223–235.

DONGARRA, J. and D. WALKER (1993), The design of linear algebra libraries for high performance computers, Report Univ. of Tennessee.

DRMAC, Z., M. OMLADIC and K. VESELIC (1994), On the perturbation of the Cholesky factorization, *SIAM J. Matrix Anal. Appl.* **15** (4), 1319–1332.

DU CROZ, J.J. and N.J. HIGHAM (1992), Stability of methods for matrix inversion, *J. Inst. Math. Appl.* **12**, 1–19.

DUFF, I.S. (1974), On the number of nonzeros added when Gaussian elimination is performed on sparse random matrices, *Math. Comput.* **28** (125), 219–230.

DUFF, I.S. (1977), A survey of sparse matrix research, *Proc. of the IEEE* **65** (4), 500–535.

DUFF, I.S., ed. (1981a), *Sparse Matrices and Their Use* (Academic Press).

DUFF, I.S. (1981b), MA32, a package for solving sparse unsymmetric systems using the frontal method, Report AERE R10079, Harwell Lab.

DUFF, I.S. (1983), The solution of sparse linear equations on the Cray 1, Report CSS125, Harwell Lab.

DUFF, I.S. (1984a), Design features of a frontal code for solving sparse unsymmetric linear systems out-of-core, *SIAM J. Sci. Statist. Comput.* **5**, 270–280.

DUFF, I.S. (1984b), Direct methods for solving sparse systems of linear equations, *SIAM J. Sci. Statist. Comput.* **5** (3), 605–619.

DUFF, I.S. (1984c), Data structures, algorithms and software for sparse matrices, Report CSS 158, Harwell Lab.

DUFF, I.S. (1986a), Parallel implementation of multifrontal schemes, *Parallel Comput.* **3**, 193–204.

DUFF, I.S. (1986b), The use of vector and parallel computers in the solution of large sparse linear equations, Report AERE R12393, Harwell Lab.

DUFF, I.S. (1986c), The influence of vector and parallel processors on numerical analysis, Report AERE R12329, Harwell Lab.

DUFF, I.S. (1988), Multiprocessing a sparse matrix code on the Alliant FX/8, Report CSS 210, Harwell Lab.

DUFF, I.S. (1993), The solution of augmented systems, Report RAL-93-084, Rutherford Appleton Lab.

DUFF, I.S., A.M. ERISMAN and J.K. REID (1976), On George's nested dissection method, *SIAM J. Numer. Anal.* **13** (5), 686–695.

DUFF, I.S., A.M. ERISMAN and J.K. REID (1986), *Direct Methods for Sparse Matrices* (Oxford Univ. Press).

DUFF, I.S. and S.L. JOHNSON (1989), Node ordering and concurrency in structurally symmetric sparse problems, in: G. Carey, ed., *Parallel Supercomputing: Methods, Algorithms and Applications*.

DUFF, I.S., N.I. GOULD, M. LESCRENIER and J.K. REID (1987), The multifrontal method in a parallel environment, Report CSS 211, Harwell Lab.

DUFF, I.S., N.I. GOULD, J.K. REID, J.A. SCOTT and K. TURNER (1990), The factorization of sparse symmetric indefinite matrices, Report RAL-90-084, Rutherford Appleton Lab.

DUFF, I.S., J. LAMINIE, A. LICHNEVSKY and F. THOMASSET (1987), An experiment with arithmetic precision in linear algebra computations, *Internat. J. Numer. Methods Fluids* **7**, 1077–1092.

DUFF, I.S. and J.K. REID (1974), A comparison of sparsity orderings for obtaining a pivotal sequence in Gaussian elimination, *J. Inst. Math. Appl.* **14**, 281–291.

DUFF, I.S. and J.K. REID (1979), Some design features of a sparse matrix code, *ACM Trans. Math. Soft.* **5** (1), 18–35.

DUFF, I.S. and J.K. REID (1983), The multifrontal solution of indefinite sparse symmetric linear systems, *ACM Trans. Math. Soft.* **9**, 302–325.

DUFF, I.S. and J.K. REID (1984), The multifrontal solution of unsymmetric sets of linear systems, *SIAM J. Sci. Statist. Comput.* **5** (3), 633–641.

DUFF, I.S. and J.K. REID (1993), MA48, a Fortran code for direct solution of sparse unsymmetric linear systems of equations, RAL-93-072, Rutherford Appleton Lab.

DUFF, I.S. and J.K. REID (1995), MA47, a Fortran code for direct solution of indefinite sparse symmetric linear systems, RAL-95-001, Rutherford Appleton Lab.

DUFF, I.S., J.K. REID, N. MUNSKGAARD and B. NIELSEN (1979), Direct solution of sets of linear equations whose matrix is sparse, symmetric and indefinite, *J. Inst. Math. Appl.* **23**, 235–250.

DUFF, I.S. and J.A. SCOTT (1993), MA42 – A new frontal code for solving sparse unsymmetric systems, Report RAL-93-064, Rutherford Appleton Lab.

DUFF, I.S. and J.A. SCOTT (1994), The use of multiple fronts in Gaussian elimination, Report RAL-94-040, Rutherford Appleton Lab.

DUFF, I.S. and G.W. STEWART, eds. (1979), *Sparse Matrix Proceedings 1978* (SIAM).

ELDEN, L. and G. SVENSSON (1991), Matrix computations on an SIMD parallel computer, Report LiTH-MAT-R-1990-19, Linköping University.

EDELMAN, A. (1993), Large dense numerical linear algebra in 1993, the parallel computing influence, *Internat. J. Supercomputer Appl.* **7** (2), 113–128.

EDELMAN, A. and M. OHLROCH (1991), Editor's note, *SIAM J. Matrix Anal. Appl.* **12**.

EISENSTAT, S.C., M.T. HEATH, C.S. HENKEL and C.H. ROMINE (1988), Modified cyclic algorithms for solving triangular systems on distributed memory multiprocessors, *SIAM J. Sci. Statist. Comput.* **9** (3), 589–600.

EISENTAT, S.C., M.H. SCHULTZ and A.H. SHERMAN (1975a), Efficient implementation of sparse symmetric Gaussian elimination, in: R. Vichnevetsky, ed., *Advances in Computer Methods for Partial Differential Equations*, 33–39.

EISENTAT, S.C., M.H. SCHULTZ and A.H. SHERMAN (1975b), Application of sparse matrix methods to partial differential equations, in: R. Vichnevetsky, ed., *Advances in Computer Methods for Partial Differential Equations*, 40–45.

EISENTAT, S.C., M.H. SCHULTZ and A.H. SHERMAN (1981), Algorithms and data structures for sparse symmetric Gaussian elimination, *SIAM J. Sci. Statist. Comput.* **2** (2), 225–237.

ERISMAN, A.M., R.G. GRIMES, J.G. LEWIS, W.G. POOLE AND H.D. SIMON (1987), Evaluation of orderings for unsymmetric sparse matrices, *SIAM J. Sci. Statist. Comput.* **8** (4), 600–624.

EVANS, D.J., ed. (1985) *Sparsity and Its Applications* (Cambridge Univ. Press).

FADDEEV, D.K. and V.O. FADDEEVA (1963), *Computational Methods of Linear Algebra* (W.H. Freeman and company).

FARHAT, C. (1988), A simple and efficient automatic FEM domain decomposer, *Computers and Structures* **28** (5), 579–602.

FARHAT, C. (1990), Redesigning the skyline solver for parallel/vector supercomputers, *Internat. J. High Speed Comput.* **2** (3), 223–238.

FIEDLER, M. (1986), *Special Matrices and Their Applications in Numerical Mathematics* (Martinus Nijhoff).

FIEDLER, M. and V. PTAK (1962), On matrices with non-positive off-diagonal elements and positive principal minors, *Czechoslovak Math. J.* **12**, 123–128.

FONG, K. and T.L. JORDAN (1977), Some linear algebraic algorithms and their performance on CRAY-1, Report LA-6774, Los Alamos Scientific Laboratory.

FORSYTHE, G.E. (1953), Tentative classification of methods and bibliography on solving systems of linear equations, *Nat. Bur. Stand. Appl. Math.* **29**, 1–28.

FORSYTHE, G.E. and C.B. MOLER (1967), *Computer Solution of Linear Algebraic Systems* (Prentice-Hall).

FOSTER, L.V. (1994), Gaussian elimination with partial pivoting can fail in practice, *SIAM J. Matrix Anal. Appl.* **15** (4), 1354–1362.

FOX, L., D. HUSKEY and J.H. WILKINSON (1948), Notes on the solution of algebraic linear simultaneous equations, *Quart. J. Appl. Math.* **1**, 149–173.

FUNDERLIC, R.E., M. NEUMANN and R.J. PLEMMONS (1982), LU decompositions of generalized diagonally dominant matrices, *Numer. Math.* **40**, 57–69.

FUNDERLIC, R.E. and R.J. PLEMMONS (1981), LU decomposition of M-matrices by elimination without pivoting, *Linear Algebra Appl.* **41**, 41–99.

GALLIVAN, K.A., M.T. HEATH, E. NG, J.M. ORTEGA, B.W. PEYTON, R.J. PLEMMONS, C.H. ROMINE, A.H. SAMEH and R.C. VOIGT (1990), *Parallel Algorithms for Matrix Computations* (SIAM).

GALLIVAN, K.A., R.J. PLEMMONS and A.H. SAMEH (1990), Parallel algorithms for dense linear algebra computations, *SIAM Rev.* **32** (1), 54–135.

GANTMACHER, F.R. (1959), *Matrix Theory*, Vol. I (Chelsea).

GEIST, G.A. (1985), Efficient parallel LU factorization with pivoting on a hypercube multiprocessor, Report ORNL-6211, Oak Ridge Nat. Lab.

GEIST, G.A. and E. NG (1989), Task scheduling for parallel sparse Cholesky factorization, *Internat. J. Parallel Programming* **18** (4), 291–314.

GEIST, G.A. and C.H. ROMINE (1988), LU factorization algorithms on distributed memory multiprocessor architectures, *SIAM J. Sci. Statist. Comput.* **9** (4), 639–649.

GEORGE, A. (1973), Nested dissection of a regular finite element mesh, *SIAM J. Numer. Anal.* **10** (2), 345–363.

GEORGE, A. (1974), On block elimination for sparse linear systems, *SIAM J. Numer. Anal.* **11** (3), 585–603.

GEORGE, A. (1977), Numerical experiments using dissection methods to solve n by n grid problems, *SIAM J. Numer. Anal.* **14** (2), 161–179.

GEORGE, A. (1980), An automatic one way dissection algorithm for irregular finite element problems, *SIAM J. Numer. Anal.* **17** (6), 740–751.

GEORGE, A., J.R. GILBERT and J.W.-H. LIU, eds. (1993), *Graph Theory and Sparse Matrix Computation* (Springer-Verlag).

GEORGE, A., M. HEATH and J.W.-H. LIU (1986), Parallel Cholesky factorization on a shared memory multiprocessor, *Linear Algebra Appl.* **77**, 165–187.

GEORGE, A., M.T. HEATH, J.W.-H. LIU and E. NG (1988), Sparse Cholesky factorization on a local memory multiprocessor, *SIAM J. Sci. Statist. Comput.* **9** (2), 327–340.

GEORGE, A. and J.W.-H. LIU (1975), A note on fill for sparse matrices, *SIAM J. Numer. Anal.* **12** (3), 452–455.

GEORGE, A. and J.W.-H. LIU (1978a), Algorithms for matrix partitioning and the numerical solution of finite element systems, *SIAM J. Numer. Anal.* **15** (2), 297–327.

GEORGE, A. and J.W.-H. LIU (1978b), An automatic nested dissection algorithm for irregular finite element problems, *SIAM J. Numer. Anal.* **15** (5), 1053–1069.

GEORGE, A. and J.W.-H. LIU (1979a), An implementation of a pseudoperipheral node finder, *ACM Trans. Math. Soft.* **5** (3), 284–295.

GEORGE, A. and J.W.-H. LIU (1979b), The design of a user interface for a sparse matrix package, *ACM Trans. Math. Soft.* **5** (2), 139–162.

GEORGE, A. and J.W.-H. LIU (1980a), A minimal storage implementation of the minimum degree algorithm, *SIAM J. Numer. Anal.* **17** (2), 282–299.

GEORGE, A. and J.W.-H. LIU (1980b), An optimal algorithm for symbolic factorization of symmetric matrices, *SIAM J. Comput.* **9** (3), 583–593.

GEORGE, A. and J.W.-H. LIU (1980c), A fast implementation of the minimum degree algorithm using quotient graphs, *ACM Trans. Math. Soft.* **6** (3), 337–358.

GEORGE, A. and J.W.-H. LIU (1981), *Computer Solution of Large Sparse Positive Definite Systems* (Prentice-Hall).

GEORGE, A. and J.W.-H. LIU (1989), The evolution of the minimum degree ordering algorithm, *SIAM Rev.* **31** (1), 1–19.

GEORGE, A., J.W.-H. LIU and E. NG (1988), A data structure for sparse QR and LU factorizations, *SIAM J. Sci. Statist. Comput.* **9** (1), 100–121.

GEORGE, A., J.W.-H. LIU and E. NG (1989), Communication results for parallel sparse Cholesky factorization on a hypercube, *Parallel Comput.* **10** (3), 287–298.

GEORGE, A. and D.R. MCINTYRE (1978), On the application of the minimum degree algorithm to finite element systems, *SIAM J. Numer. Anal.* **15** (1), 90–112.

GEORGE, A. and E. NG (1985), An implementation of Gaussian elimination with partial pivoting for sparse systems, *SIAM J. Sci. Statist. Comput.* **6** (2), 390–409.

GEORGE, A. and E. NG (1987), Symbolic factorization for sparse Gaussian elimination with partial pivoting, *SIAM J. Sci. Statist. Comput.* **8** (6), 877–898.

GEORGE, A. and E. NG (1988), On the complexity of sparse QR and LU factorization for finite element matrices, *SIAM J. Sci. Statist. Comput.* **9** (5), 849–861.

GEORGE, A., W.G. POOLE and R.G. VOIGT (1978), Incomplete nested dissection for solving n by n grid problems, *SIAM J. Numer. Anal.* **15** (4), 662–673.

GEORGE, A. and H. RASHWAN (1980), On symbolic factorization of partitioned sparse symmetric matrices, *Linear Algebra Appl.* **34**, 145–157.

GIBBS, N.E., W.G. POOLE and P.K. STOCKMEYER (1976), An algorithm for reducing the bandwith and profile of a sparse matrix, *SIAM J. Numer. Anal.* **13** (2), 236–250.

GILBERT, J.R. (1994), Predicting structure in sparse matrix computations, *SIAM J. Matrix Anal. Appl.* **15** (1), 62–79.

GILBERT, J.R. and H. HAFSTEINSSON (1990), Parallel symbolic factorization of sparse linear systems, *Parallel Comput.* **14**, 151–162.

GILBERT, J.R. and J.W.-H. LIU (1993), Elimination structures for unsymmetric sparse LU factors, *SIAM J. Matrix Anal. Appl.* **14** (2), 334–352.

GILBERT, J.R., E.G. NG and B.W. PEYTON (1994), An efficient algorithm to compute row and column counts for sparse Cholesky factorization, *SIAM J. Matrix Anal. Appl.* **15** (4), 1075–1091.

GILBERT, J.R. and T. PEIERLS (1988), Sparse partial pivoting in time proportional to arithmetic operations, *SIAM J. Sci. Statist. Comput.* **9** (5), 862–874.

GILBERT, J.R. and R. SCHREIBER (1992), Highly parallel sparse Cholesky factorization, *SIAM J. Sci. Statist. Comput.* **13** (5), 1151–1172.

GILBERT, J.R. and E. ZMIJEWSKI (1990), A parallel graph partitioning algorithm for a message passing multiprocessor, *Internat. J. Parallel Programming* **16**, 427–449.

GILL, P.E., G.H. GOLUB, W. MURRAY and M.A. SAUNDERS (1974), Methods for modifying matrix factorizations, *Math. Comp.* **28** (126), 505–535.

GOLBERG, D. (1991), What every computer scientist should know about floating point arithmetic, *ACM Computing Surveys* **23** (1), 5–48.

GOLDSTINE, H.H. (1977), *A History of Numerical Analysis from the 16th Through the 19th Century* (Springer).

GOLUB, G.H., ed. (1984), *Studies in Numerical Analysis*, Vol. 24 (Amer. Math. Soc.).

GOLUB, G.H. and G.A. MEURANT (1983), *Résolution Numérique des Grands Systèmes Linéaires* (Eyrolles).

GOLUB, G.H. and G.A. MEURANT (1994), Matrices, moments and quadrature, in: D.F. Griffiths and G.A. Watson, eds., *Numerical Analysis 1994*, Pitman Research Notes in Math. Series 103 (Longman Scientific and Technical) 105–156.

GOLUB, G.H. and J.M. ORTEGA (1992), *Scientific Computing and Differential Equations, an Introduction to Numerical Methods* (Academic Press).

GOLUB, G.H. and C. VAN LOAN (1979), Unsymmetric positive definite linear systems, *Linear Algebra Appl.* **28**, 85–97.

GOLUB, G.H. and C. VAN LOAN (1989), *Matrix Computations*, 2nd ed. (Johns Hopkins Univ. Press).

GOULD, N. (1991), On growth in Gaussian elimination with complete pivoting, *SIAM J. Matrix Anal. Appl.* **12** (2), 354–361.

GRIFFITHS, D.F. and G.A. WATSON (1989), *Numerical Analysis 1989*, Pitman Research Notes in Math. Series 228 (Longman Scientific and Technical).

GRIFFITHS, D.F. and G.A. WATSON (1994), *Numerical Analysis 1994*, Pitman Research Notes in Math. Series 103 (Longman Scientific and Technical).

GRIMES, R.G., D.J. PIERCE and H.D. SIMON (1990), A new algorithm for finding a pseudoperipheral node in a graph, *SIAM J. Matrix Anal. Appl.* **11** (2), 323–334.

HADFIELD, S.M. and T.A. DAVIS (1992), Analysis of potential parallel implementations of the unsymmetric pattern multifrontal method for sparse LU factorization, Report TR-92-017, University of Florida.

HADFIELD, S.M. and T.A. DAVIS (1994a), A parallel unsymmetric pattern multifrontal method, Report TR-94-28, University of Florida.

HADFIELD, S.M. and T.A. DAVIS (1994b), Potential and achievable parallelism in the unsymmetric pattern LU factorization method for sparse matrices, Report TR-94-006, University of Florida.

HADFIELD, S.M. and T.A. DAVIS (1994c), Potential and achievable parallelism in the unsymmetric pattern LU factorization, Report TR-94-027, University of Florida.

HAGER, W.W. (1984), Condition estimates, *SIAM J. Sci. Statist. Comput.* **5** (2), 311–316.

HARARY, F. (1971), Sparse matrices and graph theory, in: J.K. Reid, ed., *Large Sparse Sets of Linear Equations* (Academic Press) 139–150.

HARROD, W.J. (1986), LU decompositions of tridiagonal irreducible H-matrices, *SIAM J. Alg. Disc. Meth.* **7** (2), 180–187.

HEATH, M.T., E. NG and B.W. PEYTON (1991), Parallel algorithms for sparse linear systems, *SIAM Rev.* **33** (3), 420–460.

HEATH, M.T. and P. RAGHAVAN (1993a), Distributed solution of sparse linear systems, Report UT CS-93-201, University of Tennessee.

HEATH, M.T. and P. RAGHAVAN (1993b), A cartesian parallel nested dissection algorithm, *SIAM J. Matrix Anal. Appl.* **16** (1), 235–253.

HEATH, M.T. and C.H. ROMINE (1988), Parallel solution of triangular systems on distributed memory multiprocessors, *SIAM J. Sci. Statist. Comput.* **9** (3), 558–588.

HEGLAND, M. (1990), On the parallel solution of tridiagonal systems by wraparound partitioning and incomplete LU factorization, Report 90-14, IPS, ETH.

HELLER, D. (1978), A survey of parallel algorithms in numerical linear algebra, *SIAM Rev.* **20** (4), 740–777.

HENDRICKSON, B.A. and D.E. WOMBLE (1994), The torus–wrap mapping for dense matrix calculations on massively parallel computers, *SIAM J. Sci. Statist. Comput.* **15** (5), 1201–1226.

HENRICI, P. (1964), *Elements of Numerical Analysis* (John Wiley).

HIGHAM, N.J. (1986), Efficient algorithms for computing the condition number of a tridiagonal matrix, *SIAM J. Sci. Statist. Comput.* **7** (1), 150–165.

HIGHAM, N.J. (1987), A survey of condition number estimation for triangular matrices, *SIAM Rev.* **29** (4), 575–596.

HIGHAM, N.J. (1989a), The accuracy of solutions to triangular systems, *SIAM J. Numer. Anal.* **26** (5), 1252–1265.

HIGHAM, N.J. (1989b), How accurate is Gaussian elimination?, in: D.F. Griffiths and G.A. Watson, eds., *Numerical Analysis 1989*, Pitman Research Notes in Math. Series 228 (Longman Scientific and Technical).

HIGHAM, N.J. (1990), Bounding the error in Gaussian elimination for tridiagonal systems, *SIAM J. Matrix Anal. Appl.* **11** (4), 521–530.

HIGHAM, N.J. and D.J. HIGHAM (1989), Large growth factors in Gaussian elimination with pivoting, *SIAM J. Matrix Anal. Appl.* **10** (2), 155–164.

HIGHAM, N.J. and D.J. HIGHAM (1992), Backward error and condition of structured linear systems, *SIAM J. Matrix Anal. Appl.* **13** (1), 162–175.

HIGHAM, N.J. and A. POTHEN (1994), Stability of the partitioned inverse method for parallel solution of sparse triangular systems, *SIAM J. Sci. Statist. Comput.* **15** (1), 139–148.

HOFFMAN, A.J., M.S. MARTIN and D.J. ROSE (1973), Complexity bounds for regular finite difference and finite element grids, *SIAM J. Numer. Anal.* **10** (2), 364–369.

HOOD, P. (1976), Frontal solution program for unsymmetric matrices, *Internat. J. Numer. Methods Engrg.* **10**, 379–400.

HOUSEHOLDER, A.S. (1953), *Principles of Numerical Analysis* (McGraw-Hill).

HULBERT, L. and E. ZMIJEWSKI (1991), Limiting communication in parallel sparse Cholesky factorization, *SIAM J. Sci. Statist. Comput.* **12** (5), 1184–1197.

IFRAH, G. (1994), *Histoire Universelle des Chiffres* (Robert Laffont).

IPSEN, I.C., Y. SAAD and M.H. SCHULTZ (1986), Complexity of dense linear system solution on a multiprocessor ring, *Linear Algebra Appl.* **77**, 205–239.

IRONS, B.M. (1970), A frontal solution program for finite element analysis, *Internat. J. Numer. Methods Engrg.* **2**, 5–32.

JANKOWSKI, M. and H. WOZNIAKOWSKI (1977), Iterative refinement implies numerical stability, *BIT* **17**, 303–311.

JENNINGS, A. (1977), *Matrix Computation for Engineers and Scientists* (John Wiley).

JENNINGS, A. and A.D. TUFF (1971), A direct method for the solution of large sparse symmetric simultaneous equations, in: J.K. Reid, ed., *Sparse Sets of Linear Equations* (Academic Press) 97–104.

JESS, J. and H. KEES (1982), A data structure for parallel LU decomposition, *IEEE Trans. Comput.* **C-31**, 231–239.

JOHNSSON, L. (1984), Odd–even cyclic reduction on ensemble architectures and the solution of tridiagonal systems of equations, Report CSD/RR-339, Yale University.

JOHNSON, L. (1987), Solving tridiagonal systems on ensemble architectures, *SIAM J. Sci. Statist. Comput.* **8** (3), 354–392.

JOHNSON, L. and K.K. MATHUR (1990), Data structures and algorithms for the finite element method on a data parallel supercomputer, *Internat. J. Numer. Methods Engrg.* **29**, 881–908.

JONES, M.T. and M.L. PATRICK (1994), Factoring symmetric indefinite matrices on high-performance architectures, *SIAM J. Matrix Anal. Appl.* **15** (1), 273–283.

JORDAN, T.L. (1974), Gaussian elimination for dense systems on STAR and a new parallel algorithm for diagonally dominant tridiagonal systems, Report LA-5803, Los Alamos Scientific Laboratory.

KAHANER, D., C.B. MOLER and S. NASH (1988), *Numerical Methods and Software* (Prentice-Hall).

KARYPIS, G. and V. KUMAR (1994), A high performance sparse Cholesky factorization algorithm for scalable parallel computers, Report 94-41, University of Minnesota.

KOWALIK, J.S., ed. (1984), *High Speed Computation*, NATO ASI Series, Vol. 7 (Springer).

LASCAUX, P. and R. THEODOR (1993), *Analyse Numérique Matricielle Appliquée à l'Art de l'Ingénieur*, tome 1, 2nd ed. (Masson).

LAWSON, C.L., R.J. HANSON, D.R. KINCAID and F.T. KROGH (1979), Basic linear algebra subprograms for Fortran usage, *ACM Trans. Math. Soft.* **5** (3), 308–323.

LEUZE, M.R. (1989), Independent set orderings for parallel matrix factorization by Gaussian elimination, *Parallel Comput.* **10**, 177–191.

LEWIS, J.G. and R.G. GRIMES (1981), Condition number estimation for sparse matrices, *SIAM J. Sci. Statist. Comput.* **2**, 384–388.

LEWIS, J.G., B.W. PEYTON and A. POTHEN (1989), A fast algorithm for reordering sparse matrices for parallel factorization, *SIAM J. Sci. Statist. Comput.* **10** (6), 1146–1173.

LEWIS, J.G. and H.D. SIMON (1988), The impact of hardware gather/scatter on sparse Gaussian elimination, *SIAM J. Sci. Statist. Comput.* **9** (2), 304–311.

LI, G. and T.F. COLEMAN (1989), A new method for solving triangular systems on distributed memory message passing multiprocessors, *SIAM J. Sci. Statist. Comput.* **10** (2), 382–396.

LICHTENSTEIN, W. and L. JOHNSON (1993), Block–cyclic dense linear algebra, *SIAM J. Sci. Statist. Comput.* **14** (6), 1259–1288.

LIPTON, R.J., D.J. ROSE and R.E. TARJAN (1979), Generalized nested dissection, *SIAM J. Numer. Anal.* **16** (2), 346–358.

LIPTON, R.J. and R.E. TARJAN (1980), Applications of a planar separator theorem, *SIAM J. Comput.* **9** (3), 615–627.

LIU, J.W.-H. (1985), Modification of the minimum degree algorithm by multiple elimination, *ACM Trans. Math. Soft.* **11**, 141–153.

LIU, J.W.-H. (1987a), An adaptive general sparse out-of-core Cholesky factorization scheme, *SIAM J. Sci. Statist. Comput.* **9** (4), 585–599.

LIU, J.W.-H. (1987b), A note on sparse factorization in a paging environment, *SIAM J. Sci. Statist. Comput.* **8** (6), 1085–1088.

LIU, J.W.-H. (1988), Equivalent sparse matrix reordering by elimination tree rotations, *SIAM J. Sci. Statist. Comput.* **9** (3), 424–444.

LIU, J.W.-H. (1989a), Reordering sparse matrices for parallel elimination, *Parallel Comput.* **11**, 73–91.

LIU, J.W.-H. (1989b), The minimum degree ordering with constraints, *SIAM J. Sci. Statist. Comput.* **10** (6), 1136–1145.

LIU, J.W.-H. (1990), The role of elimination trees in sparse factorization, *SIAM J. Matrix Anal. Appl.* **11**, 134–172.

LIU, J.W.-H. (1992), The multifrontal method for sparse matrix solution: theory and practice, *SIAM Rev.* **34** (1), 82–109.

LIU, J.W.-H. and A. MIRZAIAN (1989), A linear reordering algorithm for parallel pivoting of chordal graphs, *SIAM J. Alg. Disc. Meth.* **2**, 100–107.

LIU, J.W.-H., E.G. NG and B.W. PEYTON (1993), On finding supernodes for sparse matrix computations, *SIAM J. Matrix Anal. Appl.* **14** (1), 242–252.

LIU, J.W.-H. and A.H. SHERMAN (1976), Comparative analysis of the Cuthill–McKee and the reverse Cuthill–McKee ordering algorithms for sparse matrices, *SIAM J. Numer. Anal.* **13** (2), 198–213.

LORD, R.E., J.S. KOWALIK and S.P. KUMAR (1983), Solving linear algebraic equations on an MIMD computer, *J. ACM* **30** (1), 103–117.

MARKOWITZ, H.M. (1957), The elimination form of the inverse and its application to linear programming, *Manag. Sci.* **3**, 255–269.

MARRAKCHI, M. and Y. ROBERT (1989), Optimal algorithms for Gaussian elimination on an MIMD computer, *Parallel Comput.* **12** (2), 183–194.

MARTIN, R.S. and J.H. WILKINSON (1965), Symmetric decomposition of positive definite band matrices, *Numer. Math.* **7**, 355–361.

MARTIN, R.S. and J.H. WILKINSON (1967), Solution of symmetric and unsymmetric band equations and the calculation of eigenvectors of band matrices, *Numer. Math.* **9**, 279–301.

MEIER, U. (1985), A parallel partition method for solving banded systems of linear equations, *Parallel Comput.* **2**, 33–43.

MELHEM, R.G. (1988), A modified frontal technique suitable for parallel systems, *SIAM J. Sci. Statist. Comput.* **9** (2), 289–303.

MEURANT, G. (1992), A review on the inverse of symmetric tridiagonal and block tridiagonal matrices, *SIAM J. Matrix Anal. Appl.* **13** (3), 707–728.

MICHIELSE, P.H. and H.A. VAN DER VORST (1988), Data transport in Wang's partition method, *Parallel Comput.* **7**, 87–95.

MITCHISON, G. and R. DURBIN (1986), Optimal numberings of an $N \times N$ array, *SIAM J. Alg. Disc. Meth.* **7** (4), 571–582.

MUNKSGAARD, N. (1979), New factorization codes for sparse, symmetric and positive definite matrices, *BIT* **19**, 43–52.

NASH, S.G., ed. (1990), *A History of Scientific Computing* (ACM Press).

NEAL, L. and G. POOLE (1992), A geometric analysis of Gaussian elimination, II, *Linear Algebra Appl.* **173**, 239–264.

NG, E. (1993), Supernodal symbolic Cholesky factorization on a local memory multiprocessor, *Parallel Comput.* **19**, 153–162.

NG, E. and B.W. PEYTON (1993a), A supernodal Cholesky factorization algorithm for shared memory multiprocessors, *SIAM J. Sci. Statist. Comput.* **14** (4), 761–769.

NG, E. and B.W. PEYTON (1993b), Block sparse Cholesky algorithms on advanced uniprocessor computers, *SIAM J. Sci. Statist. Comput.* **14** (5), 1034–1056.

OETTLI, W. and W. PRAGER (1964), Compatibility of approximate solution of linear equations with given error bounds for coefficients and right hand sides, *Numer. Math.* **6**, 405–409.

OGIELSKI, A.T. and W. AIELLO (1993), Sparse matrix computations on parallel processor arrays, *SIAM J. Sci. Statist. Comput.* **14** (3), 519–530.

O'LEARY, D. and G. STEWART (1986), Assignment and scheduling in parallel matrix factorization, *Linear Algebra Appl.* **77**, 275–299.

OLVER, F.W. and J.H. WILKINSON (1982), A posteriori error bounds for Gaussian elimination, *J. Inst. Math. Appl.* **2**, 377–406.

ORTEGA, J. (1988a), *Introduction to Parallel and Vector Solution of Linear Systems* (Plenum Press).

ORTEGA, J.M. (1988b), The ijk forms of factorization methods I. Vector computers, *Parallel Comput.* **7** (2), 135–148.

ORTEGA, J.M. and C.H. ROMINE (1988), The ijk forms of factorization methods II. Parallel systems, *Parallel Comput.* **7** (2), 149–162.

ORTEGA, J., R.G. VOIGT and C.H. ROMINE (1988), A bibliography on parallel and vector numerical algorithms, ICASE Report, NASA 181764.

ØSTERBY, O. and Z. ZLATEV (1983), *Direct Methods for Sparse Matrices*, Lecture Notes in Comput. Sci. 157 (Springer).

PARTER, S.V. (1961), The use of linear graphs in Gauss elimination, *SIAM Rev.* **3** (2), 119–130.

PEYTON, B. (1986), Some applications of clique trees to the solution of sparse linear systems, Ph.D. thesis, Clemson University.

PISSANETZKY, S. (1984), *Sparse Matrix Technology* (Academic Press).

POOLE, G. and L. NEAL (1991), A geometric analysis of Gaussian elimination, I, *Linear Algebra Appl.* **149**, 249–272.

POOLE, G. and L. NEAL (1992), Gaussian elimination: When is scaling beneficial?, *Linear Algebra Appl.* **162–164**, 309–324.

POTHEN, A. and F.L. ALVARADO (1992), A fast reordering algorithm for parallel sparse triangular solution, *SIAM J. Sci. Statist. Comput.* **13** (2), 645–653.

POTHEN, A., H.D. SIMON and K.P. LIOU (1990), Partitioning sparse matrices with eigenvectors of graphs, *SIAM J. Matrix Anal. Appl.* **11** (3), 430–452.

POTHEN, A. and C. SUN (1993), A mapping algorithm for parallel sparse Cholesky factorization, *SIAM J. Sci. Statist. Comput.* **14** (5), 1253–1257.

RAGHAVAN, P. (1993), Distributed sparse Gaussian elimination and orthogonal factorization, Report UT CS-93-203, University of Tennessee.

REID, J.K., ed. (1971a), *Large Sparse Sets of Linear Equations* (Academic Press).

REID, J.K. (1971b), A note on the stability of Gaussian elimination, *J. Inst. Math. Appl.* **8**, 374–375.

REID, J.K. (1986), Sparse matrices, Report CSS 201, Harwell Lab.

RIGAL, J.L. and J. GACHES (1967), On the compatibility of a given solution with the data of a linear system, *J. ACM* **14**, 543–548.

RODRIGUE, G., ed. (1982), *Parallel Computations* (Academic Press).

RODRIGUE, G., ed. (1989), *Parallel Processing for Scientific Computing* (SIAM).

ROHN, J. (1990), Nonsingularity under data rounding, *Linear Algebra Appl.* **139**, 171–174.

ROMAN, J. (1985), Calcul de complexité relatifs à une méthode de dissection emboîtée, *Numer. Math.* **47**, 175–190.

ROSE, D.J. (1970), Triangulated graphs and the elimination process, *J. Math. Anal. Appl.* **32**, 597–609.

ROSE, D.J. and R.E. TARJAN (1978), Algorithmic aspects of vertex elimination on graphs, *SIAM J. Comput.* **5** (2), 266–283.

ROSE, D.J., R.E. TARJAN and G.S. LUEKER (1976), Algorithmic aspects of vertex elimination on directed graphs, *SIAM J. Appl. Math.* **34** (1), 176–197.

ROSE, D.J. and R.A. WILLOUGHBY, eds. (1972), *Sparse Matrices and Their Applications* (Plenum Press).

ROTHBERG, E. (1993), Exploiting the memory hierarchy in sequential and parallel sparse Cholesky factorization, Ph.D. thesis, Stanford University.

ROTHBERG, E. and A. GUPTA (1994), An efficient block-oriented approach to parallel sparse Cholesky factorization, *SIAM J. Sci. Statist. Comput.* **15** (6), 1413–1439.

SAAD, Y. (1986), Communication complexity of the Gaussian elimination algorithm on multiprocessors, *Linear Algebra Appl.* **77**, 315–340.

SAAD, Y. and M.H. SCHULTZ (1987), Parallel direct methods for solving banded linear systems, *Linear Algebra Appl.* **88**, 623–650.

SAAD, Y. and M.H. SCHULTZ (1989), Data communication in parallel architectures, *Parallel Comput.* **11** (2), 131–150.

SAMEH, A.H. and D.J. KUCK (1978), On stable parallel linear system solvers, *J. ACM* **25** (1), 81–91.

SAMEH, A.H. and R.P. BRENT (1977), Solving triangular systems on a parallel computer, *SIAM J. Numer. Anal.* **14** (6), 1101–1113.

SCHREIBER, R.S. (1982), A new implementation of sparse Gaussian elimination, *ACM Trans. Math. Soft.* **8**, 256–276.

SHAPIRO, A. (1985), Optimal block diagonal l_2-scaling of matrices, *SIAM J. Numer. Anal.* **22** (1), 81–94.

SHERMAN, A.H. (1978), Algorithms for sparse Gaussian elimination with partial pivoting, *ACM Trans. Math. Soft.* **4** (4), 330–338.

SIMON, H.D. (1991), Partitioning of unstructured problems for parallel processing, Report RNR-91-008, NASA Ames.

SKEEL, R.D. (1979), Scaling for numerical stability in Gaussian elimination, *J. ACM* **26** (3), 494–526.

SKEEL, R.D. (1980), Iterative refinement implies numerical stability for Gaussian elimination, *Math. Comp.* **35** (151), 817–832.

STARK, S. and A.N. BERIS (1992), LU decomposition optimized for a parallel computer with a hierarchical memory, *Parallel Comput.* **18** (9), 959–972.

STEWART, G.W. (1973), *Introduction to Matrix Computations* (Academic Press).

STEWART, G.W. (1993), On the perturbation of LU, Cholesky and QR factorizations, *SIAM J. Matrix Anal. Appl.* **14** (4), 1141–1145.

STEWART, G.W. (1990), Communication and matrix computations on large message passing systems, *Parallel Comput.* **16** (1), 27–40.

STEWART, G.W. (1991), Maybe we should call it "Lagrangian elimination", Na-net message of Friday, 21 June 91.

STEWART, G.W. (1995a), The triangular matrices of Gaussian elimination and related decompositions, Report TR-95-91, University of Maryland.

STEWART, G.W. (1995b), On the perturbation of LU and Cholesky factors, Report TR-95-93, University of Maryland.

STEWART, G.W. and J.-G. SUN (1990), *Matrix Perturbation Theory* (Academic Press).

STRANG, G. (1976), *Linear Algebra and Its Applications* (Academic Press).

STRASSEN, V. (1969), Gaussian elimination is not optimal, *Numer. Math.* **13**, 354–356.

STONE, H.S. (1975), Parallel tridiagonal equation solvers, *ACM Trans. Math. Soft.* **1** (4), 289–307.

TEWARSON, R.P. The product form of the inverse of sparse matrices and graph theory, *SIAM Rev.* **9** (1), 91–99.

TEWARSON, R.P. (1970), Computations with sparse matrices, *SIAM Rev.* **12** (4), 527–543.

TEWARSON, R.P. (1973), *Sparse Matrices* (Academic Press).

TINNEY, W.F. and J.W. WALKER (1967), Direct solutions of sparse network equations by optimally ordered triangular factorization, *Proc. IEEE* **55**, 1801–1809.

TINNEY, W.F. and W.S. MEYER (1973), Solution of large sparse systems by ordered triangular factorization, *IEEE Trans. Aut. Control* **AC-18** (4), 333–346.

TISMENETSKY, M. (1986), A direct method for solving linear systems, Tech. Rept. 88.179, IBM Israel.

TRAUB, J.F. (1973), *Complexity of Sequential and Parallel Numerical Algorithms* (Academic Press).

TREFETHEN, L.N. (1985), Three mysteries of Gaussian elimination, *SIGNUM Newsletter* **20** (4), 2–5.

TREFETHEN, L.N. and R.S. SCHREIBER (1990), Average–case stability of Gaussian elimination, *SIAM J. Matrix Anal. Appl.* **11** (3), 335–360.

VARGA, R.S. and D.-Y. CAI (1981), On the LU factorization of M-matrices, *Numer. Math.* **38**, 179–192.

VAN DER SLUIS, A. (1969), Condition numbers and equilibration of matrices, *Numer. Math.* **14**, 14–23.

VAN DER VORST, H.A. (1986), Analysis of a parallel solution method for tridiagonal systems, Report 86-06, Delft University of Technology.

VAN DER VORST, H.A. (1988), Practical aspects of parallel scientific computing, *Fut. Gen. Comp. Sys.* **4**, 285–291.

VON NEUMANN, J. and H.H. GOLDSTINE (1947), Numerical inverting of matrices of high order, *Bull. Amer. Math. Soc.* **53** (11), 1021–1099.

WILKINSON, J.H. (1961), Error analysis of direct methods of matrix inversion, *J. ACM* **10**, 281–330.

WILKINSON, J.H. (1965), *The Algebraic Eigenvalue Problem* (Oxford Univ. Press).

WILKINSON, J.H. (1971), Modern error analysis, *SIAM Rev.* **13**, 548–568.

WRIGHT, S.J. (1991), Parallel algorithms for banded linear systems, *SIAM J. Sci. Statist. Comput.* **12** (4), 824–842.

WRIGHT, S.J. (1993), A collection of problems for which Gaussian elimination with partial pivoting is unstable, *SIAM J. Sci. Statist. Comput.* **14** (1), 231–238.

YANG, W.H. (1977), A method for updating Cholesky factorization of a band matrix, *Comput. Methods Appl. Mech. Engrg.* **12**, 281–288.

YANNAKAKIS, M. (1981), Computing the minimum fill-in is NP-complete, *SIAM J. Alg. Disc. Meth.* **2** (1), 77–79.

ZHANG, G. and H.C. ELMAN (1992), Parallel sparse Cholesky factorization on a shared memory multiprocessor, *Parallel Comput.* **18** (9), 1009–1022.

ZLATEV, Z. (1980) On some pivotal strategies in Gaussian elimination by sparse technique, *SIAM J. Numer. Anal.* **17** (1), 18–30.

ZMIJEWSKI, E. and J.R. GILBERT (1988), A parallel algorithm for sparse symbolic Cholesky factorization on a multiprocessor, *Parallel Comput.* **7**, 199–210.

Subject Index

Techniques of Scientific Computing (Part 3)

The Analysis of Multigrid Methods

James H. Bramble

Xuejun Zhang

Texas A&M University
College Station
TX 77843, USA

HANDBOOK OF NUMERICAL ANALYSIS, VOL. VII
Solution of Equations in $\mathbb{R}^n$ (Part 3)
Techniques of Scientific Computing (Part 3)
edited by P.G. Ciarlet and J.L. Lions

Contents

Preface

This article is an expanded and updated version of the book of BRAMBLE [1993] which, in turn, was based on graduate courses given at Cornell University in 1985, 1988, 1989 and 1992. It is meant to introduce the reader to theoretical considerations involved in the study of multigrid methods keeping the practical applicability of the abstract theory in mind.

The results are presented in a uniform abstract setting. The fundamental ingredients are separated from each other and are presented in the form of conditions. For instance, conditions required on the smoothing operator are kept separate from intrinsic properties (e.g. regularity properties) inherent in the problem to be treated. These conditions are listed in Appendix B as a glossary for convenient reference.

Most problems treated in this article are symmetric and positive definite, except in Section 11 second order nonsymmetric and indefinite problems are considered. References most relevant to the topics treated in each section are briefly discussed in the bibliographical notes at the end of the section. Many papers on related topics which are not directly referred to in this article may be found in the bibliography. No attempt is made to cover the vast literature on the subject of multilevel methods. For subjects that are not treated in this note the reader is referred to the bibliographies in the book of HACKBUSCH [1985], the survey article of BANK, MANDEL and MCCORMICK [1987] and the multigrid bibliography database on the MGnet. Readers interested in the history and the development of multigrid methods may consult the paper of BRAMBLE [1994].

The following is an outline of the topics covered in this article. Chapter I contains two sections. Section 1 is a brief introduction to classical iterative methods for linear systems. This includes the estimate for the rate of convergence of linear iterative methods, the derivation of the preconditioned conjugate gradient (PCG) method, and an estimate for the rate of convergence of the PCG method as an iterative method. In Section 2 we consider a model finite element problem and discuss a two-level method. The model finite element equation serves as a background problem and the two-level method provides a motivation for the multigrid methods to be discussed in later sections.

Chapter II, which consists of Sections 3–8, contains many abstract convergence results for various multigrid algorithms. In Section 3, a V-cycle multigrid algorithm was formulated in a "nested-inherited" setting, i.e., the multiple spaces are assumed to be nested and the bilinear forms are the same on all the spaces. The analysis in this section is based on a two-level recurrence formula. In Section 4, we turn to the analysis of multilevel additive preconditioners. Such preconditioner are of great interest themselves. In addition to that, the technical tools developed in this section also play an essential role in the study of multigrid methods in Section 5. In Section 5 we continue the study of multigrid V-cycle algorithms in the nested-inherited setting. The analysis is however based on a product representation of the error operator of the multigrid algorithm. The results proved in this section are stronger than the results of Section 3 in the cases in which "full elliptic regularity" does not hold. In Section 6 we consider multigrid algorithms in which smoothing is carried out only on subspaces. These results can be used to analyze multigrid algorithms for solutions of elliptic finite element problems on a locally refined mesh. Section 7 is devoted to a more general abstract setting. Nonnested spaces as well as forms which may vary from level to level are covered. General algorithms are formulated which include multiple smoothings which may vary in number from level to level, as well as multiple corrections.

The last section, Section 8, of this chapter concerns the construction of two classes of commonly used smoothing operators (smoothers). We discuss the conditions under which the smoothers satisfy the smoothing conditions required in the abstract multilevel theory. Many examples of smoothers are provided and relevant smoothing conditions are verified.

Applications of the theoretical results are considered in Sections 9–14 which make up Chapter III. Section 9 contains applications to second order elliptic finite element problems on polygonal domains as well as domains with curved boundaries. The case of the multigrid algorithm applied to finite element problems with locally refined meshes is also considered there. Section 10 deals with multigrid algorithms with forms defined via numerical integration. Nonsymmetric and indefinite problems are considered in Section 11, a second order anisotropic problems is discussed in Section 12 and weakly singular boundary integral operators are discussed in Section 13. Finally, applications to the biharmonic Dirichlet problem are discussed in Section 14. These include conforming C^1 finite element approximations, the Morley nonconforming method and the Ciarlet–Raviart mixed finite element method.

The multigrid algorithms in this article are formulated in terms of operators. Section 15 contains a discussion of how the general multigrid algorithm is translated into practical algorithms in specific cases. Recipes are given for their implementation. Estimates on the complexity of the multigrid algorithms are provided and a discussion of the idea of "full multigrid" is presented.

Chapter IV contains two appendices. Appendix A is a brief introduction to the theory of interpolation spaces. Only theorems which are important for our applications are proved there. The last appendix is a "Glossary of Conditions". This is included as a convenient reference for the statements of the earlier theorems and their applications.

We especially want to thank Professors Joseph Pasciak and Vidar Thomée. Professor Pasciak made many suggestions throughout the entire preparation of this article. Professor Thomée's constructive criticism of much of the original manuscript of the book of BRAMBLE [1993] greatly influenced the presentation and is very much appreciated.

CHAPTER I

Introduction

In this chapter we introduce the idea of an iterative method for the solution of systems of linear algebraic equations. Section 1 considers linear iterative methods and discusses their connection with the idea of preconditioning. The preconditioned conjugate gradient method is also treated there. In Section 2 we introduce a model problem and set up a two-level multigrid algorithm for its approximate solution.

1. Introduction to iterative methods

In this section, we provide an introduction to iterative methods for symmetric positive definite problems. We start with the simplest linear iterative method and show that the rate of convergence for such a method can be estimated in terms of the "condition number". Iterative procedures applied to ill conditioned problems converge slowly. In such situations, rapidly converging algorithms are possible provided that effective preconditioners can be developed. In this section, we shall develop algorithms in terms of generic preconditioners. In particular, we present the widely used conjugate gradient method with preconditioning. There are many approaches for developing preconditioners including multigrid, domain decomposition, and incomplete factorization. Much of the remainder of this book is devoted to the development and analysis of multigrid/multilevel algorithms, many of which give rise to effective preconditioners.

1.1. Linear iterative methods

In this subsection, we shall consider an abstract finite-dimensional real inner product space M with inner product $(\bullet, \bullet)$ and corresponding norm $\| \bullet \|$. Let the operator A be linear, symmetric with respect to $(\bullet, \bullet)$ and positive definite (SPD) on M. We shall consider iterative methods for computing the solution $x \in M$ satisfying

$$Ax = f. \tag{1.1}$$

Here $f \in M$ is given.

Through this work the term operator will always mean a linear transformation.

The Picard iteration. We start with the simplest linear iterative method for solving (1.1). Given an initial iterate x^0 (e.g., $x^0 = 0$) set

$$x^{n+1} = x^n + \tau\bigl(f - Ax^n\bigr), \quad \text{for } n = 0, 1, \ldots. \tag{1.2}$$

Let $e^n = x - x^n$. Then it is immediate that

$$e^{n+1} = (I - \tau A)e^n. \tag{1.3}$$

Let $\|I - \tau A\|$ denote the operator norm induced by the norm $\|\bullet\|$, i.e.,

$$\|I - \tau A\| = \sup_{v \in M} \frac{\|(I - \tau A)v\|}{\|v\|}.$$

Since A is symmetric,

$$\|I - \tau A\| = \sup_{v \in M} \frac{|((I - \tau A)v, v)|}{\|v\|^2} = \sup_{\lambda_i \in \sigma(A)} |1 - \tau\lambda_i| = \sup_{\lambda \in [\lambda_1, \lambda_N]} |1 - \tau\lambda|.$$

Here $\sigma(A)$ denotes the spectrum of A and λ_1 and λ_N are, respectively, the smallest and largest eigenvalues of A. It follows from (1.3) that

$$\|e^n\| \leqslant \|I - \tau A\| \, \|e^{n-1}\| \leqslant \cdots \leqslant \|I - \tau A\|^n \, \|e^0\|.$$

The simple choice of $\tau = \lambda_N^{-1}$ results in the bound

$$\|(I - \tau A)\| = 1 - \lambda_1/\lambda_N = 1 - 1/\kappa(A),$$

where $\kappa(A) = \lambda_N/\lambda_1$ is known as the spectral condition number of A. Clearly $\tau = 2/(\lambda_1 + \lambda_N)$ is the best choice and gives rise to the bound

$$\|(I - \tau A)\| = 1 - \frac{2\lambda_1}{\lambda_N + \lambda_1} = \frac{\kappa(A) - 1}{\kappa(A) + 1}.$$

Note that this is not much of an improvement over the simple choice of $\tau = \lambda_N^{-1}$ when $\kappa(A)$ is large.

Preconditioning. From the above discussion, it is clear that even with the best choice of τ, the scheme defined by (1.2) will converge slowly if the condition number of A is large. The idea of preconditioning is to apply a "preconditioner" to (1.1) and then iterate. The rate of convergence can be improved by the use of an appropriately defined preconditioner. More precisely, let B be another symmetric and positive definite operator on M and consider applying the iteration defined by (1.2) to the preconditioned equation

$$BAx = Bf. \tag{1.4}$$

Specifically, we consider the iteration

$$x^{n+1} = x^n + \tau B\big(f - Ax^n\big). \tag{1.5}$$

Let $e^n = x - x^n$ where x^n is defined by (1.5). Then

$$e^n = (I - \tau BA)e^{n-1}.$$

We analyze (1.5) by using the inner product $(\bullet, \bullet)_A$ on M defined by

$$(u, v)_A = (Au, v) \quad \text{for all } u, v \in M$$

along with its corresponding norm $\| \bullet \|_A = (\bullet, \bullet)_A^{1/2}$. Note that the operator BA is symmetric with respect to $(\bullet, \bullet)_A$ and is positive definite. The analysis of the rate of convergence of (1.5) is similar to that of (1.2). Let $\|I - \tau BA\|_A$ denote the operator norm of $I - \tau BA$ induced by the norm $\| \bullet \|_A$, i.e.,

$$\|I - \tau BA\|_A = \sup_{v \in M} \frac{\|(I - \tau BA)v\|_A}{\|v\|_A}.$$

Then

$$\|e^n\|_A \leqslant \|(I - \tau BA)\|_A \|e^{n-1}\|_A \leqslant \cdots \leqslant \|I - \tau BA\|_A^n \|e^0\|_A.$$

Since the operator BA is symmetric with respect to $(\bullet, \bullet)_A$,

$$\|I - \tau BA\|_A = \sup_{v \in M} \frac{|((I - \tau BA)v, v)_A|}{\|v\|_A^2} = \sup_{\hat{\lambda} \in [\hat{\lambda}_1, \hat{\lambda}_N]} |1 - \tau \hat{\lambda}|,$$

where $\hat{\lambda}_1$ and $\hat{\lambda}_N$ are, respectively, the smallest and largest eigenvalue of BA. The rate of convergence of (1.5) for different choices of τ can thus be estimated in terms of the condition number $\kappa(BA)$ as was done previously.

An interesting special case of the above iteration is $\tau = 1$, i.e.,

$$x^{n+1} = x^n - B\big(Ax^n - f\big). \tag{1.6}$$

The errors satisfy

$$\|e^n\|_A \leqslant \rho^n \|e^0\|_A,$$

where $\rho = \rho(I - BA)$ is the spectral radius of $I - BA$ and is given by

$$\rho \equiv \sup_{1 \leqslant i \leqslant N} |1 - \hat{\lambda}_i| = \max\big(|1 - \hat{\lambda}_1|, |1 - \hat{\lambda}_N|\big) = \max(1 - \hat{\lambda}_1, \hat{\lambda}_N - 1).$$

The iteration defined by (1.6) converges for any starting iterate x^0 if and only if,

$$|1-\hat{\lambda}_i| < 1, \quad \text{for } i = 1, \dots, N,$$

or

$$0 < \hat{\lambda}_i < 2, \quad \text{for } i = 1, \dots, N.$$

Since BA is symmetric and positive definite, $\hat{\lambda}_1 > 0$. Consequently, (1.5) converges if and only if $\hat{\lambda}_N < 2$. We shall say that B is a properly scaled preconditioner if $\hat{\lambda}_N$ is less than 2. Note that if $B = A^{-1}$ then $\hat{\lambda}_1 = \hat{\lambda}_N = 1$ and (1.6) converges in one step.

Up to this time, we have considered only a generic preconditioner B. Any SPD operator B provides an example of a preconditioner. To implement the iteration defined in (1.5), one must provide a concrete realization of the preconditioner. The overall cost of an iterative method depends on the total number of iterations and the cost per iteration. Hence a good preconditioner for A should satisfy two basic criteria. First, to reduce significantly the number of iterations, the condition number $\kappa(BA)$ should be much smaller than $\kappa(A)$. Secondly, to keep the cost per iteration low, the computational cost of applying the action of B to functions in M should not be too high. Ideally, we would like to have the cost of applying B to be on the same order as that of applying A (not A^{-1}). The measures of cost may include the number of arithmetic operations and the amount of possible parallelization.

General linear iterative processes. As typically presented, multigrid algorithms are defined as iterative processes. A basic ingredient in the understanding of these algorithms is the interconnection between iterative processes and preconditioning algorithms of the form of (1.6).

A general iterative process for the solution of $Ax = f$ can be defined in terms of a map $\mathcal{I}: M \times M \to M$ as follows. For f and x^0 given in M, the sequence of iterates $x^1, x^2, \dots$ is defined by

$$x^{n+1} = \mathcal{I}(x^n, f). \tag{1.7}$$

The map $\mathcal{I}(\bullet, \bullet)$ is said to be *consistent* (with $Ax = f$) if the solution x is a fixed point of $\mathcal{I}(\bullet, f)$, i.e.,

$$x = \mathcal{I}(x, f). \tag{1.8}$$

The map $\mathcal{I}(\bullet, \bullet)$ is said to be *linear* if it is a linear operator from the product space $M \times M$ into M, i.e.,

$$\mathcal{I}(\alpha x + \beta y, \alpha f + \beta g) = \alpha \mathcal{I}(x, f) + \beta \mathcal{I}(y, g).$$

The following proposition states that the general iterative process defined by (1.7) in terms of a linear consistent map $\mathcal{I}: M \times M \to M$ is the same as the iterative scheme generated by (1.6) with an appropriately defined linear operator B. Furthermore, the errors e^n propagate linearly.

PROPOSITION 1.1. *Let $\mathcal{I}(\bullet, \bullet) : M \times M \to M$. The following are equivalent:*
(i) *The map $\mathcal{I}(\bullet, \bullet)$ is a linear from $M \times M$ into M and is consistent.*
(ii) *The errors $e^n = x - x^n$ of iteration (1.7) are connected by a linear operator E:*

$$e^1 = Ee^0, \quad \textit{for all } f \textit{ and } x^0 \textit{ in } M.$$

(iii) *The map $\mathcal{I}(\bullet, \bullet)$ is given by*

$$\mathcal{I}(x^0, f) = x^0 + B(f - Ax^0),$$

for an appropriately defined linear operator B.
The operators are related by $E\phi = \mathcal{I}(\phi, 0) = (I - BA)\phi$ and $Bf = \mathcal{I}(0, f)$.

PROOF. Suppose $\mathcal{I}(\bullet, \bullet)$ is linear on the product space and is consistent in the sense of (1.8). Then

$$e^1 = x - \mathcal{I}(x^0, f) = \mathcal{I}(x, f) - \mathcal{I}(x^0, f) = \mathcal{I}(e^0, 0) = Ee^0.$$

Hence $E = \mathcal{I}(\bullet, 0)$ which is linear. The shows that (i) implies (ii).
If the errors are connected by a linear operator E, i.e., $(x - x^1) = E(x - x^0)$ for all $x^0 \in M$, then

$$\mathcal{I}(x^0, f) = x^1 = x - E(x - x^0) = x^0 + (I - E)(x - x^0).$$

Set $B = (I - E)A^{-1}$, then

$$\mathcal{I}(x^0, f) = x^0 - B(Ax^0 - f).$$

Hence (ii) implies (iii).
Now let $\mathcal{I}(x^0, f) = x^0 - B(Ax^0 - f)$ for all x^0 and f in M. Clearly, $\mathcal{I}(\bullet, \bullet)$ is linear from $M \times M$ into M. Note that $e^1 = (I - BA)e^0$, the consistency follows immediately. This shows that (iii) implies (i). □

1.2. The conjugate gradient method

We next consider the conjugate gradient (CG) method as a basic iterative scheme for (1.1) and the preconditioned equation (1.4). Although the iterative schemes defined by (1.5) and (1.6) can be used to develop effective algorithms, they share the drawback that one must have a precise bound for the largest eigenvalue of BA. In contrast, the CG method provides a scheme which does not require such an estimate while resulting in a somewhat faster rate of convergence.

The linear equation $Ax = f$ can be rewritten as a minimization problem over M. This minimization with respect to the Krylov subspaces provides a sequence of approximations to the solution x. The conjugate gradient algorithm provides an efficient way to implement this minimization using a simple two term recurrence relation. This minimization property of the CG algorithm also has an important theoretical consequences; it leads to the convergence estimates for the CG algorithm.

A minimization problem on Krylov subspaces. We first consider a minimization problem on Krylov subspaces. The well-known conjugate gradient algorithm is a numerical algorithm for the computation of the minimizers of this problem. Given an operator T on M and a "vector" $r \in M$, the Krylov subspace $K_n(T; r)$ of degree n is defined by

$$K_n(T; r) = \operatorname{span}\{r, Tr, T^2 r, \dots, T^{n-1} r\}.$$

We shall consider the CG algorithm for both (1.1) and (1.4). We shall present the algorithm in a way that can be easily applied to both cases. To this end, let $[\![\cdot, \cdot]\!]$ be an inner product on M. Let $\boldsymbol{A}: M \to M$ be an SPD operator with respect to $[\![\cdot, \cdot]\!]$. Then $|\!|\!| \cdot |\!|\!|_M^2 = [\![\boldsymbol{A}\cdot, \cdot]\!]$ defines a new norm on M. Given $\boldsymbol{f} \in M$ consider the equation

$$\boldsymbol{A}x = \boldsymbol{f}.$$

It is easy to see that this equation is equivalent to the following minimization problem

$$\mathcal{J}(x) = \min_{y \in M} \mathcal{J}(y),$$

where

$$\mathcal{J}(y) = \tfrac{1}{2}[\![\boldsymbol{A}y, y]\!] - [\![\boldsymbol{f}, y]\!] = \tfrac{1}{2}\big[\!\big[\boldsymbol{A}(x-y), (x-y)\big]\!\big] - \tfrac{1}{2}[\![\boldsymbol{A}x, x]\!].$$

Successive approximations to the solution x can be constructed by minimizing $J(y)$ over Krylov subspaces.

Given $x^0 \in M$, let $z^0 = \boldsymbol{f} - \boldsymbol{A}x^0$ and set $K_n = K_n(\boldsymbol{A}; z^0)$. Then the nth conjugate gradient iterate x^n has the form $x^n = x^0 + \theta^n$ with $\theta^n \in K_n$ and is defined to be the minimizer of

$$\mathcal{J}(x^n) = \min_{y \in \{x^0\} + K_n} \mathcal{J}(y).$$

A simple argument shows that $x^n = x^0 + \theta^n$ satisfies

$$\big[\!\big[\boldsymbol{A}x^n, \phi\big]\!\big] = [\![\boldsymbol{f}, \phi]\!], \quad \text{for all } \phi \in K_n. \tag{1.9}$$

Set $e^n = x - x^n$. It follows from (1.9) that

$$\big[\!\big[\boldsymbol{A}e^n, \phi\big]\!\big] = 0, \quad \text{for all } \phi \in K_n.$$

Note also that it follows from (1.9) that $(x^n - x^0) \in K_n$ is the Galerkin approximation of $x - x^0$ in the Krylov subspace K_n; i.e.,

$$[\![\boldsymbol{A}(x^n - x^0), \phi]\!] = [\![\boldsymbol{A}(x - x^0), \phi]\!] = [\![z^0, \phi]\!], \quad \text{for all } \phi \in K_n.$$

Obviously, $x^n = x$ when the error $x - x^0 = \boldsymbol{A}^{-1}z^0$ is in K_n. By the Cayley–Hamilton theorem, $\boldsymbol{A}^{-1}$ is a polynomial in $\boldsymbol{A}$ of degree less than or equal to $N - 1$ where $N = \dim(M)$. Hence $x - x^0 = \boldsymbol{A}^{-1}z^0$ always lies in K_N. This shows that the exact solution x can be obtained in N steps or less and the above scheme mathematically provides a direct solution procedure.

This minimization technique can be applied to (1.1) in a straightforward way by setting $\boldsymbol{A} = A$, $\boldsymbol{f} = f$ and $[\![\bullet,\bullet]\!] = (\bullet,\bullet)$. Note that $[\![\boldsymbol{A}\bullet,\bullet]\!] = (A\bullet,\bullet)$. For the preconditioned equation (1.4), we apply the minimization technique with $\boldsymbol{A} = BA$ and $\boldsymbol{f} = Bf$. The Krylov subspaces in this case are given by $K_n = K_n(BA, z^0)$ with $z^0 = \boldsymbol{f} - \boldsymbol{A}x^0 = Br^0 = B(f - Ax^0)$. We have two choices for the inner product $[\![\bullet,\bullet]\!]$. As already observed, BA is symmetric on M with respect to both the inner product $(A\bullet,\bullet)$ and the inner product defined by $(B^{-1}\bullet,\bullet)$. We can thus set $[\![\bullet,\bullet]\!] = (A\bullet,\bullet)$ or $[\![\bullet,\bullet]\!] = (B^{-1}\bullet,\bullet)$. Note that $[\![\boldsymbol{A}\bullet,\bullet]\!] \equiv (ABA\bullet,\bullet)$ and $[\![\boldsymbol{A}\bullet,\bullet]\!] \equiv (A\bullet,\bullet)$ respectively.

Note that

$$\mathcal{J}(y) = \tfrac{1}{2}[\![\boldsymbol{A}y, y]\!] - [\![\boldsymbol{f}, y]\!] = \tfrac{1}{2}[\![\boldsymbol{A}(x - y), (x - y)]\!] - \tfrac{1}{2}[\![\boldsymbol{A}x, x]\!].$$

Therefore, x^n also minimizes $[\![\boldsymbol{A}(x - y), (x - y)]\!]$ over $y \in \{x^0\} + K_n$. Thus

$$[\![\boldsymbol{A}e^n, e^n]\!] \leqslant [\![\boldsymbol{A}(x - y), (x - y)]\!], \quad \text{for all } y \in \{x^0\} + K_n. \tag{1.10}$$

For the cases of no preconditioning and preconditioning using $[\![\bullet,\bullet]\!] = (B^{-1}\bullet,\bullet)$, the error estimate given by (1.10) reduces to

$$(Ae^n, e^n) \leqslant (A(x - y), (x - y)), \quad \text{for all } y \in \{x^0\} + K_n.$$

The approximation property is determined by the Krylov subspaces. Consequently, the preconditioner B plays a major role in the approximation error.

The conjugate gradient algorithm. We now derive the well-known conjugate gradient algorithm from the minimization property.

We start by choosing x^0 and setting $z^0 = \boldsymbol{f} - \boldsymbol{A}x^0$. Suppose now that $x^n \in \{x^0\} + K_n$ has been computed. Set $z^n = \boldsymbol{f} - \boldsymbol{A}x^n$.

If $z^n = 0$ then $x = x^n$ and the solution was found so the process stops. In fact, because $K_n \subset K_{n+1}$, we also have $x^k = x^n = x$ for $k \geqslant n$.

If $z^n \neq 0$, then $x \neq x^n$. We want to compute $x^{n+1} \in \{x^0\} + K_{n+1}$ from x^n. Note that since $x^n - x^0 \in K_n$,

$$z^n = \boldsymbol{f} - \boldsymbol{A}x^n = \boldsymbol{f} - \boldsymbol{A}x^0 - \boldsymbol{A}(x^n - x^0) = z^0 - \boldsymbol{A}(x^n - x^0) \in K_{n+1}.$$

Furthermore, by (1.9)

$$[\![z^n, \phi]\!] = [\![f - Ax^n, \phi]\!] = 0, \quad \text{for all } \phi \in K_n.$$

Hence $z^n \notin K_n$. Let $K_n^\perp$ be the $[\![A\bullet, \bullet]\!]$-orthogonal complement of K_n in K_{n+1}; i.e.,

$$K_n^\perp = \{y \in K_{n+1} \mid [\![Ay, \phi]\!] = 0, \text{ for all } \phi \in K_n\}.$$

Then since $z^n \in K_{n+1}$ and $z^n \notin K_n$, it follows that $K_n^\perp \neq \{0\}$. Now $K_{n+1} = K_n \oplus K_n^\perp$ and $K_n^\perp$ is one-dimensional. The orthogonality implies that the minimization over $\{x^0\} + K_{n+1}$ can be decoupled into a minimization over $\{x^0\} + K_n$ (whose solution is x^n) and a simple minimization over the one-dimensional space $K_n^\perp$. Now suppose that $p^n \in K_n^\perp$ with $p^n \neq 0$ has been chosen. Then $K_n^\perp = \text{span}\{p^n\}$. We assert that the minimizer in $\{x^0\} + K_{n+1}$ is given by

$$x^{n+1} = x^n + \alpha_n p^n \tag{1.11}$$

with

$$\alpha_n = \frac{[\![z^n, p^n]\!]}{[\![Ap^n, p^n]\!]}. \tag{1.12}$$

Because of uniqueness we only need to check that (1.9) is satisfied. Any element $\chi^{n+1} \in K_{n+1}$ may be written as

$$\chi^{n+1} = \chi^n + \beta p^n \quad \text{with } \chi^n \in K_n.$$

Since $p^n \in K_n^\perp$, $[\![Ap^n, \chi^n]\!] = [\![A\chi^n, p^n]\!] = 0$, we have that

$$\begin{aligned} &[\![A(x^n + \alpha_n p^n) - f, \chi^n + \beta p^n]\!] \\ &\quad = [\![Ax^n - f, \chi^n]\!] - \beta\{[\![f - Ax^n, p^n]\!] - \alpha_n[\![Ap^n, p^n]\!]\} + \alpha_n[\![Ap^n, \chi^n]\!] \\ &\quad = 0, \end{aligned}$$

since each term is zero. Hence (1.11) is proved.

Hence if we can find some $p^n \in K_n^\perp$ with $p^n \neq 0$ then we can compute x^{n+1} from x^n and p^n, using (1.11) and (1.12). Take $p^0 = z^0$. By construction $p^0, \ldots, p^{n-1}$ are orthogonal with respect to $[\![A\bullet, \bullet]\!]$ and hence form a basis for K_n. Since z^n has a component in $K_n^\perp$, we may take p^n to be the $[\![A\bullet, \bullet]\!]$-orthogonal projection of z^n onto $K_n^\perp$; i.e.,

$$p^n = z^n - \sum_{i=0}^{n-1} \frac{[\![Ap^i, z^n]\!]}{[\![Ap^i, p^i]\!]} p^i. \tag{1.13}$$

But if $i \leqslant n-2$ then $\boldsymbol{A}p^i \in K_n$. Hence $[\![\boldsymbol{A}p^i, z^n]\!] = 0$. Thus (1.13) reduces to

$$p^n = z^n - \frac{[\![\boldsymbol{A}p^{n-1}, z^n]\!]}{[\![\boldsymbol{A}p^{n-1}, p^{n-1}]\!]} p^{n-1}.$$

The updated p^n is therefore given by a two term recurrence. We summarize the algorithm as follows.

ALGORITHM 1.0 (version 0). Let x^0 be arbitrary. Set $z^0 = \boldsymbol{f} - \boldsymbol{A}x^0$ and $p^0 = z^0$. For $n = 0, 1, \ldots$ until "convergence", compute

(1) $x^{n+1} = x^n + \alpha_n p^n$, where $\alpha_n = [\![z^n, p^n]\!]/[\![\boldsymbol{A}p^n, p^n]\!]$.
(2) $z^{n+1} = z^n - \alpha_n \boldsymbol{A}p^n$.
(3) $p^{n+1} = z^{n+1} + \beta_n p^n$, where $\beta_n = -[\![\boldsymbol{A}p^n, z^{n+1}]\!]/[\![\boldsymbol{A}p^n, p^n]\!]$.

There are alternative formulas for computing α_n and β_n. Note that $p^{n-1} \in K_n$ and $(z^n, \phi) = 0$ for all $\phi \in K_n$ so that

$$\alpha_n = \frac{[\![z^n, p^n]\!]}{[\![\boldsymbol{A}p^n, p^n]\!]} = \frac{[\![z^n, z^n]\!]}{[\![\boldsymbol{A}p^n, p^n]\!]}.$$

Writing $\boldsymbol{A}p^n = (z^n - z^{n+1})/\alpha_n$ and using $[\![z^n, z^{n+1}]\!] = 0$, we obtain

$$\beta_n = -\frac{1}{\alpha_n} \frac{[\![(z^n - z^{n+1}), z^{n+1}]\!]}{[\![\boldsymbol{A}p^n, p^n]\!]} = \frac{[\![z^{n+1}, z^{n+1}]\!]}{[\![z^n, z^n]\!]}.$$

The implementation of Algorithm 1.0 to (1.1) is easy. Recalling that $\boldsymbol{A} = A$, $\boldsymbol{f} = f$, $[\![\bullet, \bullet]\!] = (\bullet, \bullet)$ and $z^n = r^n = f - Ax^n$, we obtain

ALGORITHM 1.1 (CG). Let x^0 be arbitrary. Set $r^0 = f - Ax^0$ and $p^0 = z^0$. For $n = 0, 1, \ldots$ until "convergence", compute

(1) $x^{n+1} = x^n + \alpha_n p^n$, where $\alpha_n = (r^n, p^n)/(Ap^n, p^n)$ or $\alpha_n = (r^n, r^n)/(Ap^n, p^n)$.
(2) $r^{n+1} = r^n - \alpha_n A p^n$.
(3) $p^{n+1} = r^{n+1} + \beta_n p^n$, where $\beta_n = -(Ap^n, r^{n+1})/(Ap^n, p^n)$ or $\beta_n = (r^{n+1}, r^{n+1})/(r^n, r^n)$.

The preconditioned conjugate gradient (PCG) algorithm refers to the implementation of Algorithm 1.0 applied to the preconditioned equation (1.4). Set $r^n = f - Ax^n$, then $z^n = \boldsymbol{f} - \boldsymbol{A}x^n = Br^n$. If we use inner product $[\![\bullet, \bullet]\!] = (B^{-1}\bullet, \bullet)$, then the formulas for α_n and β_n reduce to

$$\alpha_n = \frac{(r^n, p^n)}{(Ap^n, p^n)}, \qquad \beta_n = -\frac{(Ap^n, r^{n+1})}{(Ap^n, p^n)},$$

and

$$\alpha_n = \frac{(r^n, z^n)}{(Ap^n, p^n)}, \qquad \beta_n = \frac{(r^{n+1}, z^{n+1})}{(r^n, z^n)}.$$

The last set of formulas are preferred in most applications. This way of computing saves one inner product computation and storage space for one vector.

ALGORITHM 1.2 (PCG). Let x^0 be arbitrary. Set $r^0 = f - Ax^0$ and $p^0 = z^0 = Br^0$. For $n = 0, 1, \ldots$ until "convergence", compute

(1) $x^{n+1} = x^n + \alpha_n p^n$, where $\alpha_n = (r^n, p^n)/(Ap^n, p^n)$ or $\alpha_n = (r^n, z^n)/(Ap^n, p^n)$.

(2) $r^{n+1} = r^n - \alpha_n Ap^n$.

(3) $z^{n+1} = Br^{n+1}$, $p^{n+1} = z^{n+1} + \beta_n p^n$, where $\beta_n = -(Ap^n, r^{n+1})/(Ap^n, p^n)$ or $\beta_n = (r^{n+1}, z^{n+1})/(r^n, z^n)$.

If the stopping criterion $(Br^n, r^n)^{1/2} < \varepsilon (Br^0, r^0)^{1/2}$ is used, then it follows that

$$\|x - x^n\|_A \leqslant \sqrt{\kappa}\varepsilon \|x - x^0\|_A,$$

where $\kappa = \kappa(BA)$ is the spectral condition number of BA. In the above algorithm, the evaluations of A and B enter only once per step. Furthermore, the inner product $(B^{-1}\bullet, \bullet)$ never explicitly appears and hence inversion of B is not required. Taking $B = I$ gives the CG algorithm without preconditioning for (1.1).

The matrix form of the PCG algorithm. The above discussion has been in terms of SPD operators A and B on M. We shall show how this abstract algorithm translates into a concrete matrix algorithm assuming that A and B are given in a certain way and that a concrete computational basis is used for representing the elements of the space M.

We shall consider the operator $A : M \to M$ as given by

$$(Au, \phi) = A(u, \phi), \quad \text{for all } \phi \in M.$$

The problem $Au = f$ corresponds to

$$(Au, \phi) = (f, \phi), \quad \text{for all } \phi \in M.$$

If $\{\phi_i\}$ is a computational basis for M, then we seek $u = \sum_{i=1}^N \tilde{u}_i \phi_i$ satisfying

$$\underset{\approx}{A}\tilde{u} = \underset{\sim}{f}, \tag{1.14}$$

where $\tilde{u} = [\tilde{u}_i]$ is the *coefficient vector* of u with respect to the basis $\{\phi_i\}$, $\underset{\sim}{f} = [(f, \phi_i)]$ is the *dual* vector of f and $\underset{\approx}{A} = [A(\phi_i, \phi_i)]$ is the stiffness matrix. Let $\langle \bullet, \bullet \rangle$ be the Euclidean inner product. Then the relation between A and $\underset{\approx}{A}$ is given by

$$(Au, v) = \langle \underset{\approx}{A}\tilde{u}, \tilde{v} \rangle,$$

where $\tilde{u}$ and $\tilde{v}$ are the coefficient vectors of u and v. We shall assume that we are given another SPD matrix $\overset{\approx}{B}$ so that $\overset{\approx}{B}\underset{\approx}{A}\tilde{u} = \widetilde{BAu}$ corresponds to the coefficient vector of

BAu. As shown in Section 15, this is indeed the situation in the construction of $B = B_J$ in the multigrid algorithms.

In terms of $\underset{\approx}{A}$, $\overset{\approx}{B}$ and $\langle \bullet, \bullet \rangle$ the PCG algorithm is as follows.

ALGORITHM 1.3 (Matrix form of PCG). Let $\underset{\sim}{f} = [(f, \phi_i)]$ be a given vector. Let $\tilde{x}^0$ be arbitrary and set $\underset{\sim}{r}^0 = \underset{\sim}{f} - \underset{\approx}{A}\tilde{x}^0$ and $\tilde{p}^0 = \tilde{z}^0 = \overset{\approx}{B}\underset{\sim}{r}^0$. For $n = 0, 1, \ldots$ until "convergence" compute

(1) $\tilde{x}^{n+1} = \tilde{x}^n + \alpha_n \tilde{p}^n$, where $\alpha_n = \langle \underset{\sim}{r}^n, \tilde{z}^n \rangle / \langle \underset{\approx}{A}\tilde{p}^n, \tilde{p}^n \rangle$.

(2) $\underset{\sim}{r}^{n+1} = \underset{\sim}{r}^n - \alpha_n \underset{\approx}{A}\tilde{p}^n$.

(3) $\tilde{z}^{n+1} = \overset{\approx}{B}\underset{\sim}{r}^{n+1}$, $\quad \tilde{p}^{n+1} = \tilde{z}^{n+1} + \beta_n \tilde{p}^n$, where $\beta_n = \langle \underset{\sim}{r}^{n+1}, \tilde{z}^{n+1} \rangle / \langle \underset{\sim}{r}^n, \tilde{z}^n \rangle$.

It is easy to see that Algorithm 1.2 and Algorithm 1.3 are identical provided $\tilde{x}^0$ in Algorithm 1.3 is chosen to be the coefficient vector of x^0 in Algorithm 1.2.

The PCG algorithm is very often presented in terms of matrices by introducing the square root of the matrix $\overset{\approx}{B}$ and then showing that it can be implemented without computing the square root. The matrix algorithm here is mathematically identical to that algorithm.

The crucial step in the PCG algorithm is the computation of the vectors $\tilde{z}^{n+1} = \overset{\approx}{B}\underset{\sim}{r}^{n+1}$. This computation depends on the choice of the preconditioner B. For the special case $B = I$ (no preconditioning), it is straightforward to verify that $\overset{\approx}{B} = \underset{\approx}{G}^{-1}$, where $\underset{\approx}{G} = [(\phi_i, \phi_j)]$ is the mass matrix. Hence $\tilde{z}^{n+1} = \underset{\approx}{G}^{-1}\underset{\sim}{r}^{n+1}$. This means that the CG algorithm applied to (1.1) is in general different from the CG algorithm applied to the corresponding linear system (1.14).

The rate of convergence of the PCG algorithm. We have shown previously that PCG algorithm mathematically provides a direct solution procedure for a linear system of equations and that the exact solution x can be obtained in N steps or less. This, however, is not generally useful for our applications where N is large. Moreover, the PCG algorithm utilizes orthogonality relations which tend to break down in the presence of computer round-off errors and a large number of steps when applied to problems with large condition numbers.

Nevertheless, the PCG algorithm provides an effective iterative scheme when applied to problems with moderate condition numbers.

The efficiency of the PCG algorithm is based on the following well-known theorem.

THEOREM 1.1. *The error $e^n = x - x^n$ with x^n minimizing* (1.10) *satisfies the inequality*

$$[\![\boldsymbol{A}e^n, e^n]\!]^{1/2} \leqslant 2\left(\frac{\sqrt{\kappa} - 1}{\sqrt{\kappa} + 1}\right)^n [\![\boldsymbol{A}e^0, e^0]\!]^{1/2}, \tag{1.15}$$

with $\kappa \equiv \kappa(\boldsymbol{A})$. For both the CG algorithm with the inner product $[\![\bullet, \bullet]\!] = (\bullet, \bullet)$ and the PCG algorithm with the inner product $[\![\bullet, \bullet]\!] = (B^{-1}\bullet, \bullet)$, the estimate reduces to

$$\|e^n\|_A \leqslant 2\left(\frac{\sqrt{\kappa}-1}{\sqrt{\kappa}+1}\right)^n \|e^0\|_A,$$

and $\kappa = \kappa(A)$ for the CG algorithm and $\kappa = \kappa(BA)$ for the PCG algorithm.

PROOF. It follows from (1.10) that

$$[\![\boldsymbol{A}e^n, e^n]\!] \leqslant [\![\boldsymbol{A}(e^0 - \phi), (e^0 - \phi)]\!], \quad \text{for all } \phi \in K_n. \tag{1.16}$$

Now $\phi \in K_n$ means that $\phi = \mathcal{P}_{n-1}(\boldsymbol{A})z^0 = \mathcal{P}_{n-1}(\boldsymbol{A})\boldsymbol{A}e^0$ for some polynomial $\mathcal{P}_{n-1}$ of degree $n-1$ and

$$e^0 - \phi = \big(I - \mathcal{P}_{n-1}(\boldsymbol{A})\boldsymbol{A}\big)e^0 \equiv \mathcal{Q}_n(\boldsymbol{A})e^0$$

for some polynomial $\mathcal{Q}_n$ of degree n with $\mathcal{Q}_n(0) = 1$. Hence (1.16) becomes

$$\begin{aligned}[\![\boldsymbol{A}e^n, e^n]\!] &\leqslant [\![\boldsymbol{A}\mathcal{Q}_n(\boldsymbol{A})e^0, \mathcal{Q}_n(\boldsymbol{A})e^0]\!] \\ &\leqslant |\!|\!|\mathcal{Q}_n(\boldsymbol{A})|\!|\!|_M^2 [\![\boldsymbol{A}e^0, e^0]\!],\end{aligned} \tag{1.17}$$

for any polynomial $\mathcal{Q}_n$ of degree n with $\mathcal{Q}_n(0) = 1$. Here $|\!|\!|\mathcal{Q}_n(\boldsymbol{A})|\!|\!|_M$ denotes the operator norm induced by $[\![\boldsymbol{A}\bullet, \bullet]\!]$.

Since $\boldsymbol{A}$ is SPD with respect to $[\![\boldsymbol{A}\bullet, \bullet]\!]$, its eigenvalues are real and positive. Let $0 < \lambda_1 \leqslant \lambda_2 \leqslant \cdots \leqslant \lambda_N$ be the eigenvalues of $\boldsymbol{A}$. Then

$$|\!|\!|\mathcal{Q}_n(\boldsymbol{A})|\!|\!|_M = \max_{1 \leqslant i \leqslant N} |\mathcal{Q}_n(\lambda_i)| \leqslant \max_{\lambda \in [\lambda_1, \lambda_N]} |\mathcal{Q}_n(\lambda)|. \tag{1.18}$$

For each polynomial $\tilde{\mathcal{Q}}_n$ of degree n with $\tilde{\mathcal{Q}}_n(1) = 1$ set $\mathcal{Q}_n(t) = \tilde{\mathcal{Q}}_n(1 - \frac{2}{\lambda_1+\lambda_N}t)$. Then $\mathcal{Q}_n$ is a polynomial of degree n with $\mathcal{Q}_n(0) = 1$ and

$$\mathcal{Q}_n(\lambda) = \tilde{\mathcal{Q}}_n(\tilde{\lambda}), \quad \text{for } \tilde{\lambda} = 1 - \frac{2\lambda}{\lambda_1 + \lambda_N}.$$

Let $\rho = \frac{\lambda_N - \lambda_1}{\lambda_N + \lambda_1} < 1$. Then $\lambda \in [\lambda_1, \lambda_N]$ if and only if $\tilde{\lambda} \in [-\rho, \rho]$. Hence

$$\max_{\lambda \in [\lambda_1, \lambda_N]} |\mathcal{Q}_n(\lambda)| = \max_{\tilde{\lambda} \in [-\rho, \rho]} |\tilde{\mathcal{Q}}_n(\tilde{\lambda})|. \tag{1.19}$$

From (1.17), (1.18) and (1.19) it follows that

$$[\![\boldsymbol{A}e^n, e^n]\!]^{1/2} \leqslant \max_{\lambda \in [-\rho, \rho]} |\tilde{\mathcal{Q}}_n(\lambda)| [\![\boldsymbol{A}e^0, e^0]\!]^{1/2} \tag{1.20}$$

for any $\tilde{Q}_n$ with $\tilde{Q}_n(1) = 1$. We want to choose Q_n to minimize the right-hand side of (1.20). This problem is well studied in the theory of approximation. It is known that the best choice, with only the knowledge of ρ, is given in terms of the Chebychev polynomials T_n; i.e.,

$$\tilde{Q}_n(\lambda) = T_n(\lambda/\rho)/T_n(1/\rho).$$

Note that any choice and in particular this choice gives an upper bound without using the optimality property of T_n. Now T_n is defined as follows:

$$T_n(t) = \begin{cases} \cos(n\cos^{-1} t), & \text{if } |t| \leqslant 1, \\ \cosh(n\cosh^{-1} t), & \text{if } |t| > 1. \end{cases}$$

Hence

$$\max_{\lambda\in[-\rho,\rho]} \left|\tilde{Q}_n(\lambda)\right| \leqslant \frac{1}{T_n(1/\rho)}.$$

Since $\rho < 1$, setting $\sigma = \cosh^{-1}(1/\rho)$, we have

$$T_n(1/\rho) = \tfrac{1}{2}\left(e^{n\sigma} + e^{-n\sigma}\right) \geqslant \tfrac{1}{2}e^{n\sigma}. \tag{1.21}$$

But $\sigma = \cosh^{-1}(1/\rho) = \ln(1/\rho + \sqrt{1/\rho^2 - 1})$ so that

$$e^{n\sigma} = \left(1/\rho + \sqrt{1/\rho^2 - 1}\right)^n = \left[\frac{1}{\rho}\left(1 + \sqrt{1-\rho^2}\right)\right]^n. \tag{1.22}$$

Now $\kappa = \kappa(\boldsymbol{A}) = \lambda_N/\lambda_1$. Hence

$$\rho \equiv \frac{\lambda_N - \lambda_1}{\lambda_N + \lambda_1} = \frac{\kappa - 1}{\kappa + 1}.$$

Combining (1.21) and (1.22), we find that

$$\frac{1}{T_n(1/\rho)} \leqslant 2\,e^{-n\sigma} = 2\left(\frac{\sqrt{\kappa} - 1}{\sqrt{\kappa} + 1}\right)^n.$$

Thus we have the desired estimate

$$[\![\boldsymbol{A}e^n, e^n]\!]^{1/2} \leqslant 2\left(\frac{\sqrt{\kappa} - 1}{\sqrt{\kappa} + 1}\right)^n [\![\boldsymbol{A}e^0, e^0]\!]^{1/2}.$$

This is (1.15). Note that $[\![\boldsymbol{A}\bullet, \bullet]\!] = (A\bullet, \bullet)$ for the CG method with $[\![\bullet, \bullet]\!] = (\bullet, \bullet)$ and the PCG method with $[\![\bullet, \bullet]\!] = (B^{-1}\bullet, \bullet)$. This completes the proof of the theorem. □

Note that, asymptotically, the convergence rate given by (1.15) is better than that of the simple linear iteration (1.6) since

$$\frac{\sqrt{\kappa}-1}{\sqrt{\kappa}+1}=\frac{\kappa-1}{\kappa+1+2\sqrt{\kappa}}\leqslant\frac{\kappa-1}{\kappa+1},$$

i.e., the PCG method always converges faster than the linear method defined by (1.6) even with optimal choice of the iteration parameter. Note also that n steps of the linear method (1.5) (even varying τ on each step) results in an approximation y^n of the form $x^0+\chi$ for some $\phi\in K_n$. Thus, by (1.10),

$$[\![Ae^n,e^n]\!]\leqslant[\![A(x-y^n),(x-y^n)]\!].$$

1.3. Bibliographical notes

The linear iterative method (1.6) is a special case of the classical Picard method of successive approximations which converges to a unique fixed point when the mapping is a contraction. The material in this subsection is quite standard and can be found in many textbooks (cf. GOLUB and VAN LOAN [1989] and VARGA [1962]). The presentation given here is meant to emphasize the aspect of preconditioning.

The conjugate gradient method was first derived independently by Hestenes and Stiefel. This work was jointly published in HESTENES and STIEFEL [1952]. Their work included the use of a preconditioner. The first use of preconditioning, using a linear iterative algorithm for partial differential equations (the alternating direction method), seems to have been given by WACHSPRESS [1963]. For a history and an extensive annotated bibliography see the survey article by GOLUB and O'LEARY [1989].

2. A two-level multigrid algorithm

We consider a two-level multigrid algorithm applied to a simple model problem in this section. The purpose here is to describe and motivate the use of multiple grids. We will give a complete analysis of this algorithm. For this purpose and for the treatment of multilevel methods, we first provide some preliminary definitions. Next, a model problem and its finite element approximation are described. The two-level method is then defined and an analysis of its properties as a reducer/preconditioner is provided.

2.1. Sobolev spaces

The iterative convergence estimates for multigrid algorithms applied to the computation of the discrete approximations to partial differential equations are most naturally analyzed using Sobolev spaces and their associated norms. To be precise, we shall give the definitions here although a more thorough discussion can be found in, e.g., ADAMS [1975], CIARLET [1978] and LIONS and MAGENES [1972].

Let Ω be a Lebesgue measurable set in d-dimensional Euclidean space $\mathbf{R}^d$ and f be a real valued function defined on Ω. We denote by

$$\|f\|_{L_p(\Omega)} = \left(\int_\Omega |f(x)|^p \,\mathrm{d}x\right)^{1/p}$$

the $L_p(\Omega)$ norm of f. Let $\mathbf{N}$ denote the set of nonnegative integers and let $\alpha = (\alpha_1, \ldots, \alpha_N)$, with $\alpha_i \in \mathbf{N}$, be a multi-index. We set

$$|\alpha| = \alpha_1 + \cdots + \alpha_N$$

and

$$D^\alpha = \left(\frac{\partial}{\partial x_1}\right)^{\alpha_1} \left(\frac{\partial}{\partial x_2}\right)^{\alpha_2} \cdots \left(\frac{\partial}{\partial x_N}\right)^{\alpha_N}.$$

The Sobolev spaces $W_p^s(\Omega)$ for $s \in \mathbf{N}$ are defined to be the set of distributions $f \in \mathcal{D}'(\Omega)$ (cf. HÖRMANDER [1963]) for which the norm

$$\|f\|_{W_p^s(\Omega)} = \left(\sum_{|\alpha| \leqslant s} \|D^\alpha f\|_{L_p(\Omega)}^p\right)^{1/p} < \infty.$$

When $p = 2$, the spaces are Hilbert spaces and are of special interest in this book. We shall denote these by $H^s(\Omega) \equiv W_2^s(\Omega)$. The corresponding norm will be denoted by

$$\|\bullet\|_{s,\Omega} = \|\bullet\|_{W_2^s(\Omega)}.$$

For real s with $i < s < i + 1$, the Sobolev space $H^s(\Omega)$ is defined by interpolation (using the real method) between $H^i(\Omega)$ and $H^{i+1}(\Omega)$ (see, e.g., BUTZER and BERENS [1967], LIONS and MAGENES [1972] or Appendix A). The norm and inner product notation will be further simplified when the domain Ω is clear from the context, in which case we use

$$\|\bullet\|_s = \|\bullet\|_{s,\Omega} \quad \text{and} \quad (\bullet, \bullet) = (\bullet, \bullet)_\Omega.$$

We will also use certain Sobolev spaces with negative indices. These will be defined later as needed.

2.2. *A model problem*

In this subsection, we consider the Dirichlet problem on a bounded domain Ω in $\mathbf{R}^2$. This problem and its finite element approximation can be used to illustrate some of the most fundamental properties of the multigrid algorithms. Let

$$\Delta = \frac{\partial^2}{\partial x^2} + \frac{\partial^2}{\partial y^2}.$$

Given f in an appropriately defined Sobolev space, consider the Dirichlet problem

$$\begin{cases} -\Delta u = f, & \text{in } \Omega, \\ u = 0, & \text{on } \partial\Omega. \end{cases} \tag{2.1}$$

For $v, w \in H^1(\Omega)$, let

$$D(v, w) = \int_\Omega \left(\frac{\partial v}{\partial x}\frac{\partial w}{\partial x} + \frac{\partial v}{\partial y}\frac{\partial w}{\partial y} \right) \mathrm{d}x\,\mathrm{d}y.$$

Denote by $C_0^\infty(\Omega)$ the space of infinitely differentiable functions with compact support in Ω. By Green's identity, for $\phi \in C_0^\infty(\Omega)$,

$$(f, \phi) = (-\Delta u, \phi) = D(u, \phi). \tag{2.2}$$

Let $H_0^1(\Omega)$ be the closure of $C_0^\infty(\Omega)$ with respect to $\|\cdot\|_1$. The Poincaré inequality implies that there is a constant $C > 0$ such that

$$\|v\|_0^2 \leqslant C D(v, v), \quad \text{for all } v \in H_0^1(\Omega).$$

Hence, we can take $D(\cdot,\cdot)^{1/2}$ to be the norm on $H_0^1(\Omega)$. This changes the Hilbert space structure.

In the above inequality, C represents a generic positive constant. Such constants will appear often in this book and will be denoted by C and c, with or without subscript. These constants can take on different values at different occurrences, however, they will always be independent of mesh and grid level parameters.

For a bounded linear functional f on $H_0^1(\Omega)$, the weak solution u of (2.1) satisfies

$$D(u, \phi) = (f, \phi), \quad \text{for all } \phi \in H_0^1(\Omega). \tag{2.3}$$

Here (f, ϕ) is the value of the functional f at ϕ. If $f \in L^2(\Omega)$, it coincides with the L^2-inner product. Existence and uniqueness of the function $u \in H_0^1(\Omega)$ satisfying (2.3) will follow from the Poincaré inequality and the Riesz Representation Theorem.

THEOREM 2.1 (Riesz Representation Theorem). *Let H be a Hilbert space with norm $\|\cdot\|$ and inner product $(\cdot,\cdot)_H$. Let f be a bounded linear functional on H, i.e.,*

$$|f(\phi)| \leqslant C(f)\|\phi\|.$$

Then there exists a unique $u_f \in H$ such that

$$(u_f, \phi)_H = f(\phi), \quad \textit{for all } \phi \in H.$$

To apply the above theorem to (2.3), we take $H = H_0^1(\Omega)$ with $(\cdot,\cdot)_H = D(\cdot,\cdot)$ and set

$$f(\phi) = (f, \phi), \quad \text{for all } \phi \in H_0^1(\Omega).$$

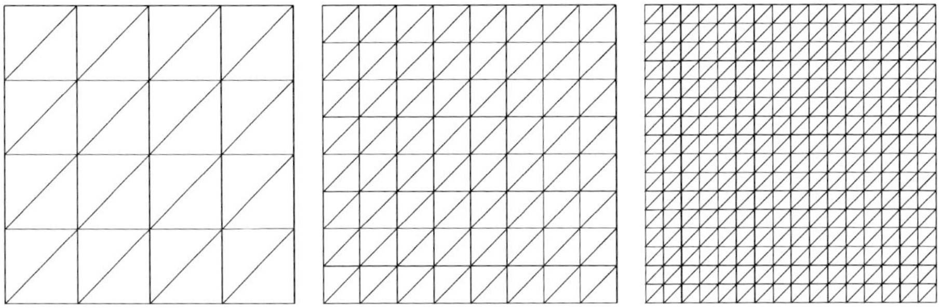

FIG. 2.1. Nested triangulations.

Then, by the definition of $\|\cdot\|_{(H_0^1(\Omega))'}$ and the Poincaré inequality,

$$\left|f(\phi)\right| \leqslant \|f\|_{(H_0^1(\Omega))'}\|\phi\|_1 \leqslant C\|f\|_{(H_0^1(\Omega))'} D(\phi,\phi)^{1/2}.$$

Thus, $f(\cdot)$ is a bounded linear functional on H and Theorem 2.1 implies that there is a unique function $u \in H = H_0^1(\Omega)$ satisfying (2.3).

2.3. *Finite element approximation of the model problem*

In this subsection, we consider the simplest multilevel finite element approximation spaces. We start with the Galerkin method. Let M be a finite-dimensional subspace of $H_0^1(\Omega)$. The Galerkin approximation is the function $u \in M$ satisfying

$$D(u,\phi) = (f,\phi), \qquad \text{for all } \phi \in M. \tag{2.4}$$

As in the continuous case, the existence and uniqueness of solutions to (2.4) follows from the Poincaré inequality and the Riesz Representation Theorem. We shall consider spaces M which result from multilevel finite element constructions.

We first define a nested sequence of triangulations. Let Ω be a domain with polygonal boundary and let $\mathcal{T}_1$ be a given (coarse) triangulation of Ω. Successively finer triangulations $\{\mathcal{T}_k\}$, $k = 2, \ldots, J$, are formed by subdividing the triangles of $\mathcal{T}_{k-1}$. More precisely, for each triangle τ of $\mathcal{T}_k$, $\mathcal{T}_{k+1}$ has four triangles corresponding to those formed by connecting the midpoints of the sides of τ. Note that the angles of the triangles in the finest triangulation are the same as those in the coarsest. Thus, the triangles on all grids are of quasi-uniform shape independent of the mesh parameter k. This construction is illustrated in Fig. 2.1.

Nested approximation spaces are defined in terms of the triangulations. Let M_k, for $k = 1, \ldots, J$, be the space of continuous piecewise linear functions on $\mathcal{T}_k$ which vanish on $\partial\Omega$. Set h_1 to be the length of the side of maximum length in $\mathcal{T}_1$ and $h_k = 2^{-k+1}h_1$. We clearly have that

$$M_1 \subset M_2 \subset \cdots \subset M_J \equiv M \subset H_0^1(\Omega).$$

We also denote $h = h_J$.

The sequence of spaces defined above satisfy the following approximation properties: For $v \in H_0^1(\Omega) \cap H^r(\Omega)$ with $r = 1$ or 2, there exists $\chi \in M_k$ such that

$$\|v-\chi\|_0^2 + h_k^2\|v-\chi\|_1^2 \leqslant Ch_k^{2r}\|v\|_r^2. \tag{2.5}$$

See, e.g., BRAMBLE and XU [1991] for a proof.

2.4. The stiffness matrix and its condition number

The point of our multigrid algorithms is to develop effective iterative techniques for computing approximate solutions satisfying equations exemplified by (2.4). Functions in M are most naturally represented in terms of a basis. Order the interior vertices of $\mathcal{T}_J$, $1 \leqslant i \leqslant \cdots \leqslant N_J$ and let $\phi_i \in M$ be such that

$$\phi_i = \begin{cases} 1 & \text{at } x_i, \\ 0 & \text{at } x_j \neq x_i. \end{cases}$$

Since any continuous, piecewise linear function is determined by its values at the vertices,

$$u = \sum_{i=1}^{N_J} \tilde{u}_i \phi_i, \quad \text{where } \tilde{u}_i = u(x_i).$$

Hence,

$$\sum_{i=1}^{N_J} \tilde{u}_i D(\phi_i, \phi_j) = (f, \phi_j). \tag{2.6}$$

Let $\underset{\approx}{A}_J$ denote the stiffness matrix $[\underset{\approx}{A}_J]_{ij} = D(\phi_i, \phi_j)$. Equation (2.6) is equivalent to

$$\underset{\approx}{A}_J \tilde{u} = \underset{\sim}{f}, \tag{2.7}$$

where $\tilde{u} = (\tilde{u}_1, \ldots, \tilde{u}_N)^T$, $\underset{\sim}{f} = (\underset{\sim}{f}_1, \ldots, \underset{\sim}{f}_N)^T$ and $\underset{\sim}{f}_i = (f, \phi_i)$.

From the earlier discussion, the rate of convergence of simple linear iterative methods or the conjugate gradient method, directly applied to (2.7), can be bounded in terms of the condition number of $\underset{\approx}{A}_J$ which we will now estimate. Let v be in M_J and write

$$v = \sum_{i=1}^{N_J} \tilde{v}_i \phi_i.$$

Let τ be a triangle of $\mathcal{T}_J$. Since τ is of quasi-uniform shape and v is linear on τ, we clearly have that $\|v\|_{0,\tau}^2$ is equivalent to the sum of the squares of the values of v at the vertices of τ times h^2. By summing over the triangles,

$$\|v\|_0^2 \approx h^2 \sum_{i=1}^{N_J} \tilde{v}_i^2, \quad v \in M_J.$$

The notation $\approx$ means equivalence of norms with constants of equivalence independent of h. Hence, by the Poincaré inequality,

$$h^2 \sum_{i=1}^{N_J} \tilde{v}_i^2 \leqslant C\|v\|_0^2 \leqslant CD(v, v) = C \sum_{i,j=1}^{N_J} \left[\underset{\approx}{A}_J\right]_{ij} \tilde{v}_i \tilde{v}_j, \quad v \in M_J.$$

This means that the smallest eigenvalue of $\underset{\approx}{A}_J$ is bounded below by ch^2.

Let τ be a triangle in $\mathcal{T}_J$. Since v is linear on τ and τ is of quasi-uniform shape, we have

$$\int_\tau |\nabla v|^2 \,\mathrm{d}x \approx \left[v(a) - v(b)\right]^2 + \left[v(b) - v(c)\right]^2, \quad v \in M_J.$$

Here a, b, c are the vertices of τ. By summing and using obvious manipulations, it follows that

$$D(v, v) \leqslant C_1 \sum_{i=1}^{N_J} \tilde{v}_i^2, \quad \text{for all } v \in M_J.$$

This means that the largest eigenvalue is bounded and hence the condition number of $\underset{\approx}{A}_J$ is bounded by

$$\kappa\left(\underset{\approx}{A}_J\right) \leqslant Ch^{-2}. \tag{2.8}$$

It is not difficult to show that the estimate (2.8) is sharp, i.e., there is a constant c not depending on J such that

$$\kappa\left(\underset{\approx}{A}_J\right) \geqslant ch^{-2}.$$

We leave this as an exercise for the reader. This means that the problem is rather ill conditioned and, thus, iterative methods applied directly to (2.7) will converge slowly.

2.5. A two-level multigrid method

To motivate the two-level multigrid method, we start by considering the simple linear iterative method

$$\tilde{u}^{n+1} = \tilde{u}^n - \underset{\approx}{\lambda}_J^{-1}\big(\underset{\approx}{A}_J \tilde{u}^n - \underset{\sim}{f}\big). \tag{2.9}$$

Here $\underset{\approx}{\lambda}_J$ denotes the largest eigenvalue of $\underset{\approx}{A}_J$. Since $\underset{\approx}{A}_J$ is symmetric and positive definite, there is a basis $\{\tilde{\psi}_i\}$, for $i = 1, \ldots, N$ of eigenfunctions with corresponding eigenvalues $0 < \eta_1 \leqslant \eta_2 \leqslant \cdots \leqslant \eta_N = \lambda_J$. From (2.9), it immediately follows that the error

$$\tilde{e}^n = \tilde{u} - \tilde{u}^n = \sum_{i=1}^{N} \tilde{e}_i^n \tilde{\psi}_i$$

satisfies

$$\tilde{e}_i^{n+1} = \big(1 - \eta_i/\underset{\approx}{\lambda}_J\big)\tilde{e}_i^n.$$

This means that the rate of reduction of the ith error component is $\rho_i = (1 - \eta_i/\underset{\approx}{\lambda}_J)$. For η_i on the order of $\underset{\approx}{\lambda}_J$, this reduction is bounded away from one and corresponds to a good reduction. However, for η_i near $\eta_1 \approx ch^2$, $\rho_i = 1 - ch^2$ which is near 1. This means that the components of the error corresponding to small eigenvalues will converge to zero rather slowly.

The two-level multigrid algorithm can now be motivated as follows. Since the form $(\underset{\approx}{A}_J \tilde{v}, \tilde{v})$ is equivalent to the sum of the squares of differences between neighboring mesh points, the components corresponding to the larger eigenvalues are highly oscillatory. In contrast, the components corresponding to the smaller eigenvalues are smoother and should be adequately approximated by coarser grid functions. This suggests combining a simple iteration of the form of (2.9) (to reduce the components corresponding to large eigenvalues) with a coarse grid solve (to reduce the components corresponding to smaller eigenvalues).

To describe this procedure, it is most natural to consider the computational problem more abstractly as one of finding functions in finite element subspaces. To this end, we define linear operators on the finite element spaces as follows: Let $A_k : M_k \to M_k$ be defined by

$$(A_k v, \phi) = D(v, \phi), \quad \text{for all } \phi \in M_k. \tag{2.10}$$

Clearly, $A_k v$ is well defined since M_k is finite-dimensional. Moreover, A_k is clearly symmetric and positive definite with respect to the inner product $(\bullet, \bullet)$. Let Q_k denote the $L^2(\Omega)$ projection onto M_k and let P_k denote the orthogonal projector with respect to the $D(\bullet,\bullet)$ inner product. These are defined by

$$(Q_k v, \phi) = (v, \phi), \quad \text{for all } v \in L^2(\Omega),\ \phi \in M_k,$$

and

$$D(P_k v, \phi) = D(v, \phi), \quad \text{for all } v \in H^1(\Omega),\ \phi \in M_k.$$

Now (2.4) can be rewritten as

$$A_J u = f_J \equiv Q_J f.$$

We denote by λ_J the largest eigenvalue of A_J. By the well known inverse properties for M,

$$(A_J v, v) = D(v, v) \leqslant Ch^{-2}(v, v), \quad \text{for all } v \in M.$$

This means that the largest eigenvalue λ_J of A_J is bounded by Ch^{-2}. The Poincaré inequality implies that the smallest eigenvalue of A_J is bounded from below.

In this notation, the two-level multigrid algorithm is given as follows.

ALGORITHM 2.1. Given $u^i \in M$ approximating the solution $u = A_J^{-1} f_J$ of (2.4), define $u^{i+1} \in M$ as follows:

(1) Set $u^{i+1/3} = u^i + \lambda_J^{-1}(f_J - A_J u^i)$.

(2) Define $u^{i+2/3} = u^{i+1/3} + q$, where

$$A_{J-1} q = Q_{J-1}\bigl(f_J - A_J u^{i+1/3}\bigr).$$

(3) Finally, set $u^{i+1} \equiv u^{i+3/3} = u^{i+2/3} + \lambda_J^{-1}(f_J - A_J u^{i+2/3})$.

Steps (1) and (3) above are a simple iteration of the form

$$u^{i+1} = u^i + \lambda_J^{-1}\bigl(f_J - A_J u^i\bigr). \tag{2.11}$$

In terms of the coefficient vectors $\tilde{u}^i$ and the dual vector $\underset{\sim}{f}_J$, it can be written as

$$\tilde{u}^{i+1} = \tilde{u}^i + \lambda_J^{-1} \underset{\approx}{G}_J^{-1}\bigl(\underset{\sim}{f}_J - \underset{\approx}{A}_J \tilde{u}^i\bigr),$$

where $\underset{\approx}{G}_J = [(\phi_i, \phi_j)]$ is the Gram matrix. This matrix is symmetric and positive definite and all its eigenvalues are on the order of Ch_J^2. Note that $\lambda_J \approx h_J^{-2}$ and $\underset{\approx}{\lambda}_J \approx 1$, hence $\lambda_J^{-1} \underset{\approx}{G}_J^{-1}$ is spectrally equivalent to $\underset{\approx}{\lambda}_J^{-1} I$. Consequently iteration (2.11), although slightly different from iteration (2.9), has a smoothing property similar to that of (2.9), i.e., the resulting error after application of (2.11) should be less oscillatory. Steps (1) and (3) above are often referred to as smoothing steps.

The middle step is called a coarse grid correction step. Note that for $\phi \in M_{J-1}$,

$$(Q_{J-1}A_J v, \phi) = (A_J v, \phi) = D(v, \phi) = D(P_{J-1}v, \phi) = (A_{J-1}P_{J-1}v, \phi),$$

i.e., $Q_{J-1}A_J = A_{J-1}P_{J-1}$. Set $e^i = u - u^i$ and $e^{i+j/3} = u - u^{i+j/3}$. We see that

$$q = A_{J-1}^{-1}Q_{J-1}A_J\bigl(u - u^{i+1/3}\bigr) = P_{J-1}e^{i+1/3}.$$

Thus, $e^{i+2/3} = (I - P_{J-1})e^{i+1/3}$. This means that $e^{i+2/3}$ is the $D(\bullet, \bullet)$-orthogonal projection of $e^{i+1/3}$ into the subspace of M_J which is orthogonal to M_{J-1}. We now prove the following theorem.

THEOREM 2.2. *Set $e^i = u - u^i$, where u is the solution of* (2.4) *and u^i is given by Algorithm* 2.1. *Then*

$$|||e^{i+1}||| \leqslant \delta |||e^i|||.$$

Here $\delta < 1$ independently of h and $|||\bullet|||$ denotes the norm defined by

$$|||v||| = D(v, v)^{1/2}.$$

PROOF. From Step (1) and Step (3) of Algorithm 2.1 and the above discussion,

$$\begin{aligned} e^{i+1/3} &= \bigl(I - \lambda_J^{-1}A_J\bigr)e^i, \\ e^{i+2/3} &= (I - P_{J-1})e^{i+1/3}, \end{aligned}$$

and

$$e^{i+1} = e^{i+3/3} = \bigl(I - \lambda_J^{-1}A_J\bigr)e^{i+2/3}.$$

Thus,

$$e^{i+1} = \bigl(I - \lambda_J^{-1}A_J\bigr)(I - P_{J-1})\bigl(I - \lambda_J^{-1}A_J\bigr)e^i \equiv \mathcal{E}e^i.$$

Clearly,

$$|||e^{i+1}||| \leqslant |||\mathcal{E}|||\,|||e^i|||,$$

where $|||\mathcal{E}||| \equiv \sup_{v \in M_J} |||\mathcal{E}v|||/|||v|||$ is the operator norm of $\mathcal{E}$. Note that $\mathcal{E}$ is symmetric with respect to $D(\bullet, \bullet)$. Hence

$$|||\mathcal{E}||| = \sup_{v \in M_J} \frac{|D(\mathcal{E}v, v)|}{|||v|||^2}$$

$$
\begin{aligned}
&= \sup_{v \in M_J} \frac{|||(I - P_{J-1})(I - \lambda_J^{-1} A_J)v|||^2}{|||v|||^2} \\
&= \big|\big|\big|(I - P_{J-1})\big(I - \lambda_J^{-1} A_J\big)\big|\big|\big|^2 \\
&= \big|\big|\big|\big(I - \lambda_J^{-1} A_J\big)(I - P_{J-1})\big|\big|\big|^2.
\end{aligned}
$$

We will estimate the last norm above. Let $w \in M$ and set $\hat{w} = (I - P_{J-1})w$. Then $D(\hat{w}, \theta) = 0$ for all $\theta \in M_{J-1}$. Hence, for any $\theta \in M_{J-1}$,

$$
D(\hat{w}, \hat{w}) = D(\hat{w}, \hat{w} - \theta) = (A_J \hat{w}, \hat{w} - \theta) \leqslant (A_J \hat{w}, A_J \hat{w})^{1/2} \|\hat{w} - \theta\|_0.
$$

Using the approximation property (2.5) gives

$$
D(\hat{w}, \hat{w}) \leqslant C h_{J-1} (A_J \hat{w}, A_J \hat{w})^{1/2} D(\hat{w}, \hat{w})^{1/2}.
$$

Cancelling the common factor and using $\lambda_J \leqslant C h_J^{-2} = 4 C h_{J-1}^{-2}$, we obtain,

$$
D(\hat{w}, \hat{w}) \leqslant C h_{J-1}^2 (A_J \hat{w}, A_J \hat{w}) \leqslant \hat{C} \lambda_J^{-1} (A_J \hat{w}, A_J \hat{w}).
$$

Now $I - \lambda_J^{-1} A_J$ is symmetric with respect to $D(\bullet, \bullet)$ and $\sigma(I - \lambda_J^{-1} A_J) \subseteq [0, 1)$. Hence

$$
\begin{aligned}
\big|\big|\big|\big(I - \lambda_J^{-1} A_J\big)\hat{w}\big|\big|\big|^2 &\leqslant D\big(\big(I - \lambda_J^{-1} A_J\big)\hat{w}, \hat{w}\big) \\
&= D(\hat{w}, \hat{w}) - \lambda_J^{-1} D(A_J \hat{w}, \hat{w}) \\
&\leqslant (1 - 1/\hat{C}) D(\hat{w}, \hat{w}) \\
&\leqslant (1 - 1/\hat{C}) D(w, w) = (1 - 1/\hat{C}) |||w|||^2.
\end{aligned}
$$

This shows that $|||\mathcal{E}||| = |||(I - \lambda_J^{-1} A_J)(I - P_{J-1})|||^2 \leqslant (1 - 1/\hat{C})$. Therefore

$$
\big|\big|\big|e^{i+1}\big|\big|\big| \leqslant \big(1 - 1/\hat{C}\big) \big|\big|\big|e^i\big|\big|\big| = \delta \big|\big|\big|e^i\big|\big|\big|,
$$

where $\delta = 1 - 1/\hat{C}$ is independent of h. □

REMARK 2.1. If we omit Step (1) in Algorithm 2.1, then

$$
e^{i+1} = \big(I - \lambda_J^{-1} A_J\big)(I - P_{J-1}) e^i
$$

and hence

$$
\begin{aligned}
\big|\big|\big|e^{i+1}\big|\big|\big| &= \big|\big|\big|\big(I - \lambda_J^{-1} A_J\big)(I - P_{J-1}) e^i\big|\big|\big| \\
&\leqslant \big|\big|\big|\big(I - \lambda_J^{-1} A_J\big)(I - P_{J-1})\big|\big|\big| \, \big|\big|\big|e^i\big|\big|\big| \\
&\leqslant \delta^{1/2} \big|\big|\big|e^i\big|\big|\big|.
\end{aligned}
$$

We obtain the same bound if we omit, instead, Step (3). One obvious advantage of the symmetric algorithm is that it can be used as a preconditioner in the PCG algorithm.

What have we gained by using the two-level scheme? In this example,

$$\dim M_{J-1} \approx \tfrac{1}{4} \dim M_J .$$

Hence A_{J-1}^{-1} is "cheaper" to compute than A_J^{-1}. This is not a significant improvement for very large problems. This algorithm was described only to illustrate the multilevel technique. More efficient algorithms will be developed in later sections by recursive application of the above idea.

REMARK 2.2. The algorithm described above is not too convenient for direct application. First of all, the largest eigenvalue explicitly appears in Steps (1) and (3). It is not difficult to see that the largest eigenvalue in these algorithms can be replaced by any upper bound λ'_J for λ_J provided that $\lambda'_J \leqslant C\lambda_J$. More importantly, Steps (1) and (3) require that Gram matrix systems be solved at each step of the iteration. Even though these matrices are well conditioned, they result in some unnecessary computational effort. Both of the above mentioned problems will be avoided by the introduction of more appropriate smoothing procedures. Smoothing procedures will be studied in more detail in later sections. The more abstract multigrid algorithms of Sections 3, 4 and 5 will use generic smoothing operators in place of $\lambda_J^{-1} I$.

2.6. *Bibliographical notes*

Estimates of the form of (2.5) are well known (cf. CIARLET [1978]).

The argument given in the proof of Theorem 2.2 for the two-level algorithm was first given by MANDEL[1984]. See also the thesis of XU [1989b]. In contrast to the analysis to be given in Section 3, the two-level analysis here does not use the regularity properties of the continuous problem.

CHAPTER II

Abstract Theory

In this chapter we develop the abstract multilevel theory. The first section generalizes the two-level method of the introduction to the multilevel case and analyzes the V-cycle algorithm under somewhat restrictive conditions. Section 4 introduces the additive multilevel preconditioner and in Section 5 a convergence theory for the V-cycle in the case in which the multilevel spaces are nested and the corresponding forms are inherited from the form on the finest space. Section 6 then treats the case in which smoothing is done only on certain subspaces. This abstract theory covers the case of local mesh refinements discussed in the chapter on applications. A more general setting is introduced in Section 7. There the requirements that the spaces be nested and that the forms be inherited is dropped. The W-cycle is defined as well as the variable V-cycle preconditioner. The final section of this chapter considers carefully the construction of smoothing operators which are an essential ingredient in all of the multigrid algorithms.

3. An abstract V-cycle algorithm

We consider an abstract V-cycle algorithm in this section. We shall provide two theorems which can be used to estimate the rate of convergence of this multigrid algorithm. The first analysis follows that of BRAMBLE and PASCIAK [1987] and requires the so-called *full regularity and approximation* assumption. The fundamental ideas of this analysis are contained in the paper of BRAESS and HACKBUSCH [1983]. However, multigrid methods are applied to many problems which do not satisfy this hypothesis. A second theorem gives some bounds for the rate of convergence in the case of less than full regularity. The first theorem provides a uniform convergence estimate while the second gives rise to a rate of convergence which deteriorates with the number of levels.

For generality, the algorithms and theory presented in this section are given abstractly. However, the usefulness of any abstract theory depends on the possibility of verification of the hypotheses in various applications. We will verify these conditions for many applications in Chapter III. We will also categorize the hypotheses into those which involve the smoothing operator, those which depend on approximation and those which depend on, e.g., some underlying elliptic regularity as well as approximation.

3.1. *The multilevel framework*

In this subsection, we will set up the abstract multilevel framework. Let us consider a nested sequence of finite-dimensional vector spaces

$$M_1 \subset M_2 \subset \cdots \subset M_J.$$

In addition, let $A(\bullet,\bullet)$ and $(\bullet,\bullet)$ be symmetric positive definite bilinear forms on M_J. The norm corresponding to $(\bullet,\bullet)$ will be denoted by $\|\bullet\|$. We shall develop multigrid algorithms for the solution of the following problem: Given $f \in M_J$, find $u \in M_J$ satisfying

$$A(u,\phi) = (f,\phi), \quad \text{for all } \phi \in M_J. \tag{3.1}$$

As in the previous section, we will use auxiliary operators to define the multigrid algorithms. For $k = 1,\ldots,J$, let the operator $A_k : M_k \to M_k$ be defined by

$$(A_k v,\phi) = A(v,\phi), \quad \text{for all } \phi \in M_k.$$

The operator A_k is clearly positive definite and symmetric in both the $A(\bullet,\bullet)$ and $(\bullet,\bullet)$ inner products. Also, we define the projectors $P_k : M_J \to M_k$ and $Q_k : M_J \to M_k$ by

$$A(P_k v,\phi) = A(v,\phi), \quad \text{for all } \phi \in M_k,$$

and

$$(Q_k v,\phi) = (v,\phi), \quad \text{for all } \phi \in M_k.$$

Note that (3.1) can be rewritten in the above notation as

$$A_J u = f. \tag{3.2}$$

We will use generic smoothing operators in our abstract multigrid algorithm. Assume that we are given linear operators, $R_k : M_k \to M_k$ for $k = 2,\ldots,J$. Denote by R_k^t the adjoint of R_k with respect to $(\bullet,\bullet)$. We shall state further conditions concerning these operators as we progress in the development. Section 8 deals with the construction and analysis of effective smoothing operators.

3.2. *The abstract V-cycle algorithm, I*

We describe a V-cycle multigrid algorithm for computing the solution u of (3.2) by means of an iterative process. This involves recursively applying a generalization of Algorithm 2.1. Given an initial iterate $u^0 \in M_J$, we define a sequence approximating u by

$$u^{m+1} = \mathrm{Mg}_J\big(u^m, f\big). \tag{3.3}$$

Here $\text{Mg}_J(\bullet,\bullet)$ is the map of $M_J \times M_J$ into M_J defined by the following algorithm.

ALGORITHM 3.1. Set $\text{Mg}_1(v, g) = A_1^{-1} g$. Let k be greater than one and let v and g be in M_k. Assuming that $\text{Mg}_{k-1}(\bullet,\bullet)$ has been defined, we define $\text{Mg}_k(v, g)$ as follows:
(1) Set $v' = v + R_k^t(g - A_k v)$.
(2) Set $v'' = v' + q$ where $q = \text{Mg}_{k-1}(0, Q_{k-1}(g - A_k v'))$.
(3) Define $\text{Mg}_k(v, g) \equiv v''' = v'' + R_k(g - A_k v'')$.

The above algorithm is a recursive generalization of Algorithm 2.1. In fact, if we take $J = 2$ and $R_J = \lambda_J^{-1} I$ then Algorithm 3.1 coincides with Algorithm 2.1. The above algorithm replaces the solve on the $(J-1)$st subspace (in Algorithm 2.1) with one recursive application of the multilevel procedure.

A straightforward induction argument shows that $\text{Mg}_k(\bullet,\bullet)$ is a linear map of $M_k \times M_k$ into M_k. Moreover, it is obviously consistent. By Proposition 1.1, the linear operator $\mathcal{E}_J v \equiv \text{Mg}_J(v, 0)$ is the error reduction operator for (3.3). That is

$$\mathcal{E}_J\big(u - u^m\big) \equiv u - u^{m+1} = \text{Mg}_J(u, f) - \text{Mg}_J\big(u^m, f\big) = \text{Mg}_J\big(u - u^m, 0\big).$$

Steps (1) and (3) above are referred to as smoothing steps. Step (2) is called a coarse grid correction step. Step (3) is included so that the resulting linear multigrid operator

$$B_J g = \text{Mg}_J(0, g)$$

is symmetric with respect to the inner product $(\bullet,\bullet)$, and hence can be used as a preconditioner for A_J. It is straightforward to show that with $B_k g = \text{Mg}_k(0, g)$,

$$\text{Mg}_k(v, g) = v + B_k(g - A_k v).$$

To use B_k as a preconditioner, it is often convenient to define $B_k g \equiv \text{Mg}_k(0, g)$ directly and the following algorithm gives a recursive definition of B_k.

ALGORITHM 3.2. Let $B_1 = A_1^{-1}$. Assuming $B_{k-1} : M_{k-1} \to M_{k-1}$ has been defined, we define $B_k : M_k \to M_k$ as follows. Let $g \in M_k$.
(1) Set $v' = R_k^t g$.
(2) Set $v'' = v' + q$ where $q = B_{k-1} Q_{k-1}(g - A_k v')$.
(3) Set $B_k g \equiv v''' = v'' + R_k(g - A_k v'')$.

3.3. The two-level error recurrence

We next derive a two-level error recurrence for the multigrid process defined by Algorithm 3.1, i.e., we compute an expression for the error operator $\mathcal{E}_k v = \text{Mg}_k(v, 0)$. Let $u = A_k^{-1} g$. By consistency, $\mathcal{E}_k(u - v) \equiv u - \text{Mg}_k(v, g) = \text{Mg}_k(u - v, 0)$.

We want to express $u - \text{Mg}_k(v, g)$ in terms of $u - v$. We start by considering the effect of the smoothing steps. Steps (1) and (3) are just the simple iterative method (1.6) with $B = R_k^t$ and $B = R_k$ respectively. Hence v''' of Algorithm 3.1 satisfies

$$u - \text{Mg}_k(v, g) = u - v''' = (I - R_k A_k)\big(u - v''\big) \equiv K_k\big(u - v''\big) \tag{3.4}$$

and v' satisfies

$$u - v' = \big(I - R_k^t A_k\big)(u - v) \equiv K_k^*(u - v). \tag{3.5}$$

Note that $K_k^* = (I - R_k^t A_k)$ is the adjoint of K_k with respect to the $A(\bullet, \bullet)$ inner product.

We next consider the effect of the coarse grid correction step. Note that for $\phi \in M_i$ with $i < k$,

$$(Q_i A_k v, \phi) = (A_k v, \phi) = A(v, \phi) = A(P_i v, \phi) = (A_i P_i v, \phi).$$

This means that $Q_i A_k = A_i P_i$. Thus, q of Step (2) is given by

$$\begin{aligned} q &= \mathrm{Mg}_{k-1}\big(0, Q_{k-1}\big(g - A_k v'\big)\big) \\ &= \mathrm{Mg}_{k-1}\big(0, Q_{k-1} A_k \big(u - v'\big)\big) \\ &= \mathrm{Mg}_{k-1}\big(0, A_{k-1} P_{k-1}\big(u - v'\big)\big) \\ &= P_{k-1}\big(u - v'\big) - \mathcal{E}_{k-1} P_{k-1}\big(u - v'\big). \end{aligned}$$

We used consistency for the last equality. Consequently,

$$u - v'' = \big[(I - P_{k-1}) + \mathcal{E}_{k-1} P_{k-1}\big]\big(u - v'\big). \tag{3.6}$$

Combining Eqs. (3.4), (3.5) and (3.6) gives the two-level error recurrence

$$\mathcal{E}_k = K_k\big[(I - P_{k-1}) + \mathcal{E}_{k-1} P_{k-1}\big]K_k^*. \tag{3.7}$$

A simple argument using mathematical induction shows that $\mathcal{E}_k$ is symmetric, positive semidefinite with respect to the $A(\bullet, \bullet)$ inner product, i.e.,

$$0 \leqslant A(\mathcal{E}_k v, v), \quad \text{for all } v \in M_k.$$

In particular,

$$0 \leqslant A(\mathcal{E}_J v, v), \quad \text{for all } v \in M_J. \tag{3.8}$$

3.4. *The Braess–Hackbusch theorem*

We prove a result given by Braess and Hackbusch in this subsection. This result was the first which showed that multigrid algorithms could be expected to converge with very weak conditions on the smoother.

For this result, we require two conditions. The first is on the smoother. Let λ_k be the largest eigenvalue of A_k. Let ω be in $(0, 2)$. For $v, g \in M_k$, Richardson's iteration for the solution of $A_k v = g$ is defined by

$$v^{m+1} = v^m + \omega \lambda_k^{-1}\big(g - A_k v^m\big).$$

The corresponding error reduction operator is

$$K_{k,\omega} = I - \left(\omega\lambda_k^{-1}\right)A_k.$$

Note that $K_k = (I - R_k A_k)$ and

$$A(K_k v, K_k v) = A\left(\left(I - \overline{R}_k A_k\right)v, v\right),$$

where

$$\overline{R}_k = \left(I - K_k^* K_k\right)A_k^{-1} = R_k^t + R_k - R_k^t A_k R_k$$

is the linear smoothing operator corresponding to the symmetrized iterative process with reducer $K_k^* K_k$. Our condition concerning the smoothers R_k, $k = 2, \ldots, J$, is related to Richardson's iteration with an appropriate parameter ω.

CONDITION (SM.1). There exists an $\omega \in (0, 1)$, not depending on J, such that

$$0 \leqslant A(K_k v, K_k v) \leqslant A(K_{k,\omega} v, v), \quad \text{for all } v \in M_k,\ k = 2, 3, \ldots, J,$$

or equivalently

$$\frac{\omega}{\lambda_k}(v, v) \leqslant \left(\overline{R}_k v, v\right) \leqslant \left(A_k^{-1} v, v\right), \quad \text{for all } v \in M_k,\ k = 2, 3, \ldots, J.$$

REMARK 3.1. Note that both $K_k^* K_k$ and $K_{k,\omega}$ are symmetric with respect to $A(\bullet, \bullet)$ and positive semi-definite on M_k. Condition (SM.1) implies that $K_k^* K_k$ is less than or equal to $K_{k,\omega}$. That is, the smoother $\overline{R}_k$ results in a reduction which is at least as good as the result obtained using the Richardson smoother $(\omega/\lambda_k)I$.

The second condition which we will require is the so-called *full regularity and approximation* condition.

CONDITION (A.1). There is a constant C_P not depending on J such that

$$\left\|(I - P_{k-1})v\right\|^2 \leqslant C_P \lambda_k^{-1} A(v, v), \quad \text{for all } v \in M_k,\ k = 2, \ldots, J.$$

In applications, the verification of this condition requires that the underlying elliptic problem satisfy full elliptic regularity (see, e.g., Section 9). In the model problem of Section 2, P_{k-1} is the elliptic projector. The above condition is obtained by application of finite element duality techniques.

We can now state the Braess–Hackbusch theorem.

THEOREM 3.1. *Let* $\mathrm{Mg}_J(\bullet,\bullet)$ *be defined by Algorithm* 3.1. *Assume that Condition* (A.1) *holds and that the smoothers* R_k *satisfy Condition* (SM.1). *Then the error reduction operator* $\mathcal{E}_J v = \mathrm{Mg}_J(v,0)$ *of Algorithm* 3.1 *satisfies*

$$0 \leqslant A(\mathcal{E}_J v, v) \leqslant \frac{C_{\mathrm{P}}}{C_{\mathrm{P}}+\omega} A(v,v), \quad \textit{for all } v \in M_J.$$

This means that the sequence u^m *defined by* (3.3) *satisfies*

$$|||u-u^m||| \leqslant \left(\frac{C_{\mathrm{P}}}{C_{\mathrm{P}}+\omega}\right)^m |||u-u^0|||.$$

Before proving this theorem, we state and prove the following lemma.

LEMMA 3.1. *Assume that Condition* (SM.1) *holds. Then for any* $v \in M_k$,

$$\frac{1}{\lambda_k}\|A_k K_k^* v\|^2 \leqslant \frac{1}{\omega}\big[A(v,v) - A\big(K_k^* v, K_k^* v\big)\big].$$

PROOF. Let v be in M_k. By Condition (SM.1) with v replaced by $A_k K_k^* v$,

$$\begin{aligned}\frac{1}{\lambda_k}\|A_k K_k^* v\|^2 &\leqslant \frac{1}{\omega}\big(\overline{R}_k A_k K_k^* v, A_k K_k^* v\big) = \frac{1}{\omega} A\big(\big(I - K_k^* K_k\big) K_k^* v, K_k^* v\big)\\ &= \frac{1}{\omega} A\big(\big(I-\bar{K}_k\big)\bar{K}_k v, v\big),\end{aligned}$$

where $\bar{K}_k = K_k K_k^*$. The operator $\bar{K}_k$ is symmetric and nonnegative with respect to the $A(\bullet,\bullet)$ inner product. It follows from the symmetry of $\bar{K}_k$ that

$$A\big(\big(I-\bar{K}_k\big)\bar{K}_k v, v\big) \leqslant A\big(\big(I-\bar{K}_k\big)v, v\big) = \big[A(v,v) - A\big(K_k^* v, K_k^* v\big)\big].$$

Combining the above inequalities completes the proof of the lemma. □

PROOF OF THEOREM 3.1. By (3.8), we need only prove the upper inequality. The proof is by induction. We will show that for $i = 1, \ldots, J$,

$$A(\mathcal{E}_i v, v) \leqslant \delta A(v,v), \quad \text{for all } v \in M_i \tag{3.9}$$

holds for $\delta = C_{\mathrm{P}}/(C_{\mathrm{P}}+\omega)$. Since $\mathcal{E}_1 = 0$ is the trivial operator on M_1, (3.9) obviously holds for $i = 1$. Let k be greater than one and assume that (3.9) holds for $i = k-1$. Let v be in M_k. Using the two-level recurrence formula (3.7) gives

$$A(\mathcal{E}_k v, v) = A\big((I - P_{k-1})K_k^* v, K_k^* v\big) + A\big(\mathcal{E}_{k-1} P_{k-1} K_k^* v, P_{k-1} K_k^* v\big).$$

Applying the induction hypothesis we have

$$\begin{aligned} A(\mathcal{E}_k v, v) &\leqslant A\big((I - P_{k-1})K_k^* v, K_k^* v\big) + \delta A\big(P_{k-1} K_k^* v, P_{k-1} K_k^* v\big) \\ &= (1-\delta) A\big((I - P_{k-1})K_k^* v, K_k^* v\big) + \delta A\big(K_k^* v, K_k^* v\big). \end{aligned} \tag{3.10}$$

By the Cauchy–Schwarz inequality and Condition (A.1),

$$\begin{aligned} A\big((I - P_{k-1})K_k^* v, K_k^* v\big) &\leqslant \big\|(I - P_{k-1})K_k^* v\big\| \, \big\|A_k K_k^* v\big\| \\ &\leqslant \big(C_{\mathrm{P}} \lambda_k^{-1} A\big((I - P_{k-1})K_k^* v, K_k^* v\big)\big)^{1/2} \big\|A_k K_k^* v\big\|. \end{aligned}$$

Obvious manipulations give

$$A\big((I - P_{k-1})K_k^* v, K_k^* v\big) \leqslant \frac{C_{\mathrm{P}}}{\lambda_k} \big\|A_k K_k^* v\big\|^2. \tag{3.11}$$

Thus, (3.10), (3.11) and Lemma 3.1 imply that

$$\begin{aligned} A(\mathcal{E}_k v, v) &\leqslant \frac{(1-\delta) C_{\mathrm{P}}}{\omega} \big[A(v, v) - A\big(K_k^* v, K_k^* v\big)\big] + \delta A\big(K_k^* v, K_k^* v\big) \\ &= \delta A(v, v). \end{aligned}$$

This completes the proof of the theorem. □

REMARK 3.2. The original result given by BRAESS and HACKBUSCH [1983] was stated in terms of the particular smoother $R_k = \lambda_k^{-1} I$. They also allowed a fixed number m of applications of this smoother on each level. Theorems with variable numbers of smoothings will appear in Section 7.

3.5. Analysis of multigrid V-cycle with less than full elliptic regularity

In this subsection, we provide an analysis of Algorithm 3.1 with an assumption which is weaker than Condition (A.1). To state this condition, we require scales of discrete norms. The operator A_k is symmetric and positive definite on M_k. Consequently, its fractional powers are well defined. For real s, we consider the scale of discrete norms on M_k defined by

$$|||v|||_{s,k} = \big(A_k^s v, v\big)^{1/2}, \quad \text{for all } v \in M_k.$$

Note that

$$|||v|||_{0,k} = \|v\| \quad \text{and} \quad |||v|||_{1,k} = |||v|||.$$

By expanding in the eigenfunctions of A_k and using the Hölder inequality, it follows easily that

$$|||v|||_{1+\alpha,k} \leqslant |||v|||_{2,k}^{\alpha} |||v|||_{1,k}^{1-\alpha}, \quad 0 \leqslant \alpha \leqslant 1.$$

Our theorem will be based on the following generalization of Condition (A.1).

CONDITION (A.2). There exists a number $\alpha \in (0, 1]$ and a constant C_{P} not depending on J such that

$$\left|\left|\left|(I - P_{k-1})v\right|\right|\right|_{1-\alpha,k}^2 \leqslant C_{\mathrm{P}}^{\alpha} \lambda_k^{-\alpha} A(v, v), \quad \text{for all } v \in M_k,\ k = 2, \ldots, J.$$

In the model problem of Section 2, P_{k-1} is the elliptic projector. It can be shown that for this example and $s \in [0, 1]$, the norm $|||\bullet|||_{s,k}$ is equivalent on M_k (with constants independent of J) to the Sobolev norm $\|\bullet\|_s$. The above condition can be then obtained by application of finite element duality techniques. This will be done explicitly in Section 9.

The following lemma is an easy consequence of Condition (A.2).

LEMMA 3.2. *Assume that Condition* (A.2) *holds. Then, for all* $v \in M_k$,

$$A\big((I - P_{k-1})v, v\big) \leqslant \left(\frac{C_{\mathrm{P}}}{\lambda_k}\right)^{\alpha} \left(|||v|||_{1,k}^{1-\alpha} |||v|||_{2,k}^{\alpha}\right)^2 = C_{\mathrm{P}}^{\alpha} A(v, v)^{1-\alpha} \left(\frac{\|A_k v\|^2}{\lambda_k}\right)^{\alpha}.$$

PROOF. Let $v \in M_k$ be arbitrary. Then Condition (A.2) implies that

$$\begin{aligned} A\big((I - P_{k-1})v, v\big) &\leqslant \left|\left|\left|(I - P_{k-1})v\right|\right|\right|_{1-\alpha,k} |||v|||_{1+\alpha,k} \\ &\leqslant \left(\frac{C_{\mathrm{P}}}{\lambda_k}\right)^{\alpha/2} \left|\left|\left|(I - P_{k-1})v\right|\right|\right| \, |||v|||_{2,k}^{\alpha} |||v|||_{1,k}^{1-\alpha}. \end{aligned}$$

Since $|||(I - P_{k-1})v||| = A((I - P_{k-1})v, v)^{1/2}$, the lemma follows. □

We now give the more general theorem.

THEOREM 3.2. *Let* $\mathrm{Mg}_k(\bullet, \bullet)$ *be defined by Algorithm* 3.1. *Assume that Condition* (A.2) *holds and that the smoothers* R_k *satisfy Condition* (SM.1). *Then there exists a constant* $C_0 > 0$, *independent of* J, *such that the error reduction operator* $\mathcal{E}_k v = \mathrm{Mg}_k(v, 0)$ *satisfies*

$$0 \leqslant A(\mathcal{E}_k v, v) \leqslant \delta_k A(v, v), \quad \textit{for all } v \in M_J$$

with

$$\delta_k = \frac{C_0 k^{(1-\alpha)/\alpha}}{1 + C_0 k^{(1-\alpha)/\alpha}}. \tag{3.12}$$

This means that the sequence u^i *defined by* (3.3) *satisfies*

$$\left|\left|\left|u - u^m\right|\right|\right| \leqslant \delta_J^m \left|\left|\left|u - u^0\right|\right|\right|.$$

PROOF. By (3.8), we need only prove the upper inequality. The proof is by mathematical induction and proceeds initially like that of Theorem 3.1. We will show that for $i = 1, \ldots, J$,

$$A(\mathcal{E}_i w, w) \leqslant \delta_i A(w, w), \quad \text{for all } w \in M_i \tag{3.13}$$

holds for δ_i given by (3.12). Since $\mathcal{E}_1 = 0$ is the trivial operator on M_1, (3.13) obviously holds for $i = 1$. Let k be greater than one and assume that (3.13) holds for $i = k - 1$. Let v be in M_k and $\tilde{v} = K_k^* v$. Following the proof of Theorem 3.1, we see that

$$A(\mathcal{E}_k v, v) \leqslant (1 - \delta_{k-1}) A\big((I - P_{k-1})\tilde{v}, \tilde{v}\big) + \delta_{k-1} A(\tilde{v}, \tilde{v}). \tag{3.14}$$

Combining Lemma 3.2 with the elementary inequality

$$a^\alpha b^{1-\alpha} \leqslant \alpha a + (1 - \alpha) b, \quad \text{for all } a, b > 0$$

gives

$$A\big((I - P_{k-1})\tilde{v}, \tilde{v}\big) \leqslant C_{\mathrm{P}}^\alpha \left[\alpha \gamma_k \frac{\|A_k \tilde{v}\|_k^2}{\lambda_k} + (1 - \alpha)\gamma_k^{-\alpha/(1-\alpha)} A_k(\tilde{v}, \tilde{v})\right],$$

with $\gamma_k > 0$ to be chosen. Applying Lemma 3.1 and combining with (3.14) gives

$$\begin{aligned} A(\mathcal{E}_k v, v) \leqslant\ & (1 - \delta_{k-1}) \frac{C_{\mathrm{P}}^\alpha \alpha \gamma_k}{\omega} \big[A(v, v) - A(\tilde{v}, \tilde{v})\big] \\ & + \big[C_{\mathrm{P}}^\alpha (1 - \alpha)(1 - \delta_{k-1})\gamma_k^{-\alpha/(1-\alpha)} + \delta_{k-1}\big] A_k(\tilde{v}, \tilde{v}). \end{aligned}$$

Note that

$$A(v, v) - A(\tilde{v}, \tilde{v}) \geqslant 0.$$

Thus, to finish the proof, it suffices to choose γ_k and C_0 so that

$$(1 - \delta_{k-1}) \frac{C_{\mathrm{P}}^\alpha \alpha \gamma_k}{\omega} \leqslant \delta_k \tag{3.15}$$

and

$$C_{\mathrm{P}}^\alpha (1 - \alpha)(1 - \delta_{k-1})\gamma_k^{-\alpha/(1-\alpha)} + \delta_{k-1} \leqslant \delta_k. \tag{3.16}$$

We define γ_k so that

$$(1 - \delta_{k-1}) \frac{C_{\mathrm{P}}^\alpha \alpha \gamma_k}{\omega} = \delta_{k-1} \tag{3.17}$$

and hence (3.15) is trivially verified. Inequality (3.16) is equivalent to

$$(1-\delta_{k-1})C_{\mathrm{P}}^{\alpha}(1-\alpha)\gamma_k^{-\alpha/(1-\alpha)} \leqslant \delta_k - \delta_{k-1}. \tag{3.18}$$

Let $\mathcal{D}(k) = 1 + C_0 k^{(1-\alpha)/\alpha}$. Clearly,

$$\delta_k - \delta_{k-1} = \frac{C_0[k^{(1-\alpha)/\alpha} - (k-1)^{(1-\alpha)/\alpha}]}{\mathcal{D}(k)\mathcal{D}(k-1)}. \tag{3.19}$$

With γ_k given by (3.17), the left-hand side of (3.18) can be written as

$$C_{\mathrm{P}}^{\alpha/(1-\alpha)} \frac{(1-\alpha)}{(k-1)\mathcal{D}(k-1)} \left(\frac{\alpha}{\omega C_0}\right)^{\alpha/(1-\alpha)}.$$

A straightforward exercise in calculus, noting that $k \geqslant 2$, gives

$$(k-1)\big[k^{(1-\alpha)/\alpha} - (k-1)^{(1-\alpha)/\alpha}\big] \geqslant \frac{1-\alpha}{\alpha} k^{(1-\alpha)/\alpha} 2^{-s}, \tag{3.20}$$

where

$$s = \begin{cases} 1 & \text{if } \alpha \geqslant 1/2, \\ (1-\alpha)/\alpha & \text{if } \alpha < 1/2. \end{cases}$$

Combining Eqs. (3.19)–(3.20) shows that (3.18) holds if C_0 is such that

$$\big(\alpha C_{\mathrm{P}}^{\alpha}\big)^{1/(1-\alpha)} \omega^{-\alpha/(1-\alpha)} \leqslant C_0^{\alpha/(1-\alpha)} 2^{-s} \frac{C_0 k^{(1-\alpha)/\alpha}}{\mathcal{D}(k)}. \tag{3.21}$$

Let $\tilde{C_0}$ be defined by

$$\big(\alpha C_{\mathrm{P}}^{\alpha}\big)^{1/(1-\alpha)} \omega^{-\alpha/(1-\alpha)} = 2^{-s} \tilde{C_0}^{\alpha/(1-\alpha)}$$

and define C_0 by

$$C_0 = \left(\frac{1+\tilde{C_0}}{\tilde{C_0}}\right)^{(1-\alpha)/\alpha} \tilde{C_0}.$$

Then, since $C_0 \geqslant \tilde{C_0}$,

$$\begin{aligned} \big(\alpha C_{\mathrm{P}}^{\alpha}\big)^{1/(1-\alpha)} \omega^{-\alpha/(1-\alpha)} &= 2^{-s} \tilde{C_0}^{\alpha/(1-\alpha)} = 2^{-s} \frac{\tilde{C_0}}{1+\tilde{C_0}} C_0^{\alpha/(1-\alpha)} \\ &\leqslant 2^{-s} C_0^{\alpha/(1-\alpha)} \frac{C_0}{1+C_0}. \end{aligned}$$

Inequality (3.21) then follows from

$$\frac{C_0}{1+C_0} \leqslant \frac{C_0 k^{(1-\alpha)/\alpha}}{\mathcal{D}(k)}.$$

This completes the proof of the theorem. □

REMARK 3.3. The above theorem provides some bounds for the rate of convergence of the V-cycle algorithm in cases of less than full regularity, i.e., under the assumption of Condition (A.2). These estimates deteriorate as the number of levels increase. In Section 5, we will establish uniform estimates for Algorithm 3.1 under some conditions weaker than Condition (A.1).

3.6. Bibliographical notes

The first papers on multigrid were in the 1960's. FEDORENKO [1964] first introduced the idea in the context of finite differences. Then BAKHVALOV [1966] gave an analysis of the method of Fedorenko. In the early 1970's Brandt studied this method and was a major advocate of it as a powerful tool for solving problems associated with elliptic equations. His work was extremely valuable and he had a lot to do with the popularization of the multigrid techniques (cf. BRANDT [1977]).

BRAESS and HACKBUSCH [1983] were the first to prove that the V-cycle converges uniformly with respect to the number of levels. This proof required "full elliptic regularity". In BRAMBLE and PASCIAK [1987] and DECKER, MANDEL and PARTER [1988] the first estimates were given for the convergence of the V-cycle with less than full regularity, but these estimates showed a deterioration depending on the lack of full regularity. The paper of BRAMBLE and PASCIAK [1987] also introduced the variable V-cycle and proved uniform estimates with some (arbitrary amount of) regularity. The work of Braess and Hackbusch, although requiring full elliptic regularity, led the way to the proofs that the W-cycle gave a uniform reduction under reduced regularity assumptions. The work of BRAMBLE and PASCIAK [1987] and BANK, MANDEL and MCCORMICK [1987] contain proofs of such a result.

Results in this section are based on the two-level error recurrence for the multigrid algorithm. Stronger results which are based on product representation of the error operator for the multigrid algorithm will be discussed in Section 5. More general multigrid algorithms including variable V-cycle and W-cycle algorithms will be discussed in Section 7. We refer to the survey paper of BRAMBLE [1994] for more detailed discussion on the development of the multigrid methods.

4. Multilevel additive preconditioner

In this section, we discuss a multilevel additive preconditioner. Such preconditioners are of interest themselves. In addition, the technical results in this section also play an essential role in Section 5 in the analysis of the standard multigrid V-cycle Algorithm 3.1.

We review here the multilevel framework described in Section 3. We consider a sequence of nested spaces

$$M_1 \subset M_2 \subset \cdots \subset M_J \equiv M.$$

Let $(\bullet,\bullet)$ and $A(\bullet,\bullet)$ be symmetric positive definite bilinear forms on M and the induced norms are denoted by $\|\bullet\|$ and $|||\bullet|||$. Given $f \in M$, we are interested in solving

$$A(u,\phi) = (f,\phi), \quad \text{for all } \phi \in M. \tag{4.1}$$

Let $Q_k : M \to M_k$ and $P_k : M \to M_k$ be the projectors with respect to the inner products $(\bullet,\bullet)$ and $A(\bullet,\bullet)$, i.e.,

$$\begin{aligned} (Q_k v,\phi) &= (v,\phi), \quad \text{for all } \phi \in M_k, \\ A(P_k v,\phi) &= A(v,\phi), \quad \text{for all } \phi \in M_k. \end{aligned}$$

The operators $A_k : M_k \to M_k$ are defined by

$$(A_k v,\phi) = A(v,\phi), \quad \text{for all } v,\phi \in M_k.$$

The operator A_k is symmetric and positive definite in both the $A(\bullet,\bullet)$ and $(\bullet,\bullet)$ inner products. Using the operator A_J, (4.1) can be written as

$$A_J u = f.$$

Let $f_k = Q_k f = Q_k A_J u$, then $u_k = P_k u$ is the solution of

$$A_k u_k = f_k.$$

4.1. *A multilevel additive preconditioner*

To motivate the construction of the multilevel additive preconditioner, we first consider a simple one-level iterative method. Let $R_1 = A_1^{-1}$. For $k \geqslant 2$, let $R_k : M_k \to M_k$ be symmetric, positive definite operators. Let $u^{(0)} \in M_k$. For each k, using R_k, an iterative method for solving $A_k u_k = f_k$ can be defined by

$$u^{(m+1)} = u^{(m)} + \omega R_k\big(f_k - A_k u^{(m)}\big).$$

We define a multilevel solution procedure for solving $A_J u = f$ by combining the above simple one-level iterations as follows.

ALGORITHM 4.1. Let $u^{(m)}$ be an approximation to $u = A_J^{-1} f$. The new approximation $u^{(m+1)}$ is computed as follows:

(1) Project the residual $r = f - Au^{(m)}$ onto subspaces M_k:

$$r_k = Q_k r = Q_k\big(f - Au^{(m)}\big).$$

(2) Correct $u^{(m)}$ in subspace M_k by solving approximately $A_k(P_k e_J) = r_k$ as follows:

$$e_k = R_k r_k.$$

(3) Update $u^{(m)}$: $u^{(m+1)} = u^{(m)} + \omega \sum_k e_k$, where $\omega > 0$ is a damping parameter.

This solution process can be rewritten in more compact form as

$$u^{(m+1)} = u^{(m)} + \omega B_a\big(f - Au^{(m)}\big), \tag{4.2}$$

where

$$B_a = \sum_{k=1}^{J} R_k Q_k. \tag{4.3}$$

The above iteration converges if $0 < \omega < 2[\lambda_{\max}(B_a A)]^{-1}$. The rate of convergence depends on the eigenvalues of $B_a A$ and the choice of the parameter ω. If all the R_k's are symmetric, then B_a is also symmetric. In this case, with the optimal parameter $\omega = 2/(\lambda_{\min}(B_a A) + \lambda_{\max}(B_a A))$, the rate of convergence is estimated by

$$\big|\!\big|\!\big|u - u^{(m)}\big|\!\big|\!\big| \leqslant \left(\frac{\kappa(B_a A) - 1}{\kappa(B_a A) + 1}\right)^m \big|\!\big|\!\big|u - u^{(0)}\big|\!\big|\!\big|.$$

Alternatively, B_a can be used as a preconditioner in the preconditioned conjugate gradient (PCG) method. This method is more efficient than the iteration defined by (4.2). The rate of convergence of the PCG method is determined by the condition number $\kappa(B_a A)$. The interest in B_a is in its use in the PCG algorithm. Hence in the rest of this section, we will concentrate on the estimate for $\kappa(B_a A)$. We will provide a few abstract estimates for the condition number $\kappa(B_a A)$ under various assumptions on the operator A_k, projectors Q_k and P_k, and smoothers R_k.

4.2. *A characterization of the preconditioner*

To estimate $\kappa(B_a A)$, we need to find the constants C_1 and C_2 that satisfy

$$C_1\big(B_a^{-1}v, v\big) \leqslant A(v, v) \leqslant C_2\big(B_a^{-1}v, v\big), \quad \text{for all } v \in M_J. \tag{4.4}$$

Note that (4.4) is equivalent to $C_1 \leqslant \lambda_{\min}(B_a A) \leqslant \lambda_{\max}(B_a A) \leqslant C_2$. Hence if (4.4) holds, then $\kappa(B_a A) \leqslant C_2/C_1$.

In establishing (4.4), we need to compare the two forms $A(v, v)$ and $(B_a^{-1}v, v)$. The following lemma provides a useful characterization for $(B_a^{-1}v, v)$.

LEMMA 4.1. *Let H be a Hilbert space with inner product $(\bullet,\bullet)$. Let $H_k, k=1,\dots,J$, be subspaces of H and $H=\sum_{k=1}^{J} H_k$. Denote by Q_k the orthogonal projectors onto H_k. Let $R_k : H_k \to H_k$ be symmetric and positive definite and $B_a=\sum_{k=1}^{J} R_k Q_k$. Then B_a^{-1} exists and is characterized by*

$$\left(B_a^{-1}v,v\right)=\min_{\{v_k\}}\sum_{k=1}^{J}\left(R_k^{-1}v_k,v_k\right),$$

where the minimum is taken over all $v_k \in H_k$ such that $v=\sum v_k$.

PROOF. Let $v\in H$ and let $v_k\in H_k$ be such that $v=\sum v_k$. Clearly B_a^{-1} exists. By the Cauchy–Schwarz inequality and the definition of B_a, we obtain

$$\begin{aligned}\left(B_a^{-1}v,v\right) &= \sum_{k=1}^{J}\left(B_a^{-1}v,v_k\right)=\sum_{k=1}^{J}\left(Q_kB_a^{-1}v,v_k\right)\\ &\leqslant \left(\sum_{k=1}^{J}\left(R_kQ_kB_a^{-1}v,Q_kB_a^{-1}v\right)\right)^{1/2}\left(\sum_{k=1}^{J}\left(R_k^{-1}v_k,v_k\right)\right)^{1/2}\\ &= \left(B_a^{-1}v,v\right)^{1/2}\left(\sum_{k=1}^{J}\left(R_k^{-1}v_k,v_k\right)\right)^{1/2}.\end{aligned}$$

Thus $(B_a^{-1}v,v)\leqslant \min\sum_k(R_k^{-1}v_k,v_k)$. Equality follows by setting $v_k=R_kQ_kB_a^{-1}v$.

We now briefly discuss how to use Lemma 4.1 to establish (4.4). By Lemma 4.1, the upper estimate $A(v,v)\leqslant C_2(B_a^{-1}v,v)$ holds if and only if

$$A(v,v)\leqslant C_2\sum_{k=1}^{J}\left(R_k^{-1}v_k,v_k\right),\quad \text{for all } v=\sum_{k=1}^{J}v_k \text{ with } v_k\in M_k.$$

For the lower estimate, we have the following equivalent conditions:

(a) $C_1(B_a^{-1}v,v)\leqslant A(v,v)$ for all $v\in M_J$, or $C_1\leqslant \lambda_{\min}(B_aA)$.

(b) For any $v\in M_J$, there exist $v_k\in M_k$ with $v=\sum_{k=1}^{J}v_k$ such that

$$\sum_{k=1}^{J}\left(R_k^{-1}v_k,v_k\right)\leqslant C_1^{-1}A(v,v). \tag{4.5}$$

(c) There exist a sequence of linear operators $\Pi_k : M_J\to M_k$ with $\Pi_J=I$ and $\Pi_0=0$ such that for all $v\in M$,

$$\sum_{k=1}^{J}\left(R_k^{-1}(\Pi_k-\Pi_{k-1})v,(\Pi_k-\Pi_{k-1})v\right)\leqslant C_1^{-1}A(v,v). \tag{4.6}$$

The $(\bullet,\bullet)$-projection Q_k plays a special role in many applications. The lower estimate for $\lambda_{\min}(B_a A)$ is often established by showing that (4.6) holds with $\Pi_k = Q_k$, i.e.,

$$\sum_{k=1}^{J} \bigl(R_k^{-1}(Q_k - Q_{k-1})v, (Q_k - Q_{k-1})v\bigr) \leqslant CA(v,v). \tag{4.7}$$

Clearly (4.7) implies (4.6). It is worth noting that (4.6) also implies (4.7) under some reasonable assumptions on the subspaces M_k. In such cases, (4.7) is equivalent to a uniform lower estimate for $\lambda_{\min}(B_a A)$.

The condition number $\kappa(B_a A)$ depends on the properties of the spaces M_k, the two inner products $(\bullet,\bullet)$ and $A(\bullet,\bullet)$, as well as the smoothers R_k. It is convenient to consider first a multilevel preconditioner B_λ defined in terms of the special smoothers $R_k = \lambda_k^{-1} I$, i.e.

$$B_\lambda = A_1^{-1} Q_1 + \sum_{k=2}^{J} \lambda_k^{-1} Q_k. \tag{4.8}$$

Here λ_k is the largest eigenvalue of A_k. By doing so, we can concentrate on the essential conditions on the two forms $(\bullet,\bullet)$ and $A(\bullet,\bullet)$ first.

By Lemma 4.1,

$$\bigl(B_\lambda^{-1} v, v\bigr) = \min_{\{v_k\}} \left(A(v_1, v_1) + \sum_{k=2}^{J} \lambda_k \|v_k\|^2 \right),$$

where the minimum is taken over all $v_k \in M_k$ such that $v = \sum v_k$. We shall show that under reasonable assumptions, this minimum is essentially achieved for the special decomposition with $v_1 = Q_1 v$ and $v_k = (Q_k - Q_{k-1})v$ for $2 \leqslant k \leqslant J$. That is

$$\bigl(B_\lambda^{-1} v, v\bigr) \approx A(Q_1 v, Q_1 v) + \sum_{k=2}^{J} \lambda_k \bigl\|(Q_k - Q_{k-1})v\bigr\|^2, \tag{4.9}$$

provided that the largest eigenvalue of A_k grows geometrically with k. This geometrical growth property of λ_k can be verified easily for elliptic finite element problems on a sequence of nested meshes. We postulate this property in the following condition. □

CONDITION (A.3). There exist positive constants $\gamma_1 \leqslant \gamma_2 < 1$ such that

$$\gamma_1 \lambda_{k+1} \leqslant \lambda_k \leqslant \gamma_2 \lambda_{k+1}, \quad \text{for } k = 1, 2, \ldots, J-1.$$

When using this condition, it is convenient to introduce the symmetric matrix $\Lambda^{(\alpha)}$, with $0 < \alpha \leqslant 1$, whose lower triangular entries are given by

$$\Lambda_{ij}^{(\alpha)} = (\lambda_j / \lambda_i)^{\alpha/2}, \quad \text{for } j \leqslant i. \tag{4.10}$$

Denote by $\Lambda_{\mathrm{L}}^{(\alpha)}$ and $\Lambda_{\mathrm{U}}^{(\alpha)}$ the lower and upper triangular parts of $\Lambda^{(\alpha)}$. Condition (A.3) then implies that $\Lambda_{ki}^{(\alpha)} = (\lambda_k/\lambda_i)^{\alpha/2} \leqslant \gamma_2^{(i-k)\alpha/2}$. Therefore,

$$\|\Lambda_{\mathrm{L}}^{(\alpha)}\|_{\ell_2} = \|\Lambda_{\mathrm{U}}^{(\alpha)}\|_{\ell_2} \leqslant \left(\frac{\gamma_2^{\alpha/2}}{1-\gamma_2^{\alpha/2}}\right) \quad \text{and} \quad \|\Lambda^{(\alpha)}\|_{\ell_2} \leqslant \left(\frac{1+\gamma_2^{\alpha/2}}{1-\gamma_2^{\alpha/2}}\right).$$

Here $\|\bullet\|_{\ell_2}$ denotes the matrix 2-norm.

It can be shown that (4.9) follows from Condition (A.1). To prove (4.9) under weaker assumptions it is convenient to use the discrete norms defined in Subsection 3.5. Recall that for real s, a scale of discrete norms on M_k is defined by

$$|||v|||_{s,k}^2 = (A_k^s v, v), \quad \text{for } v \in M_k$$

and that $|||v|||_{0,k} = \|v\|$ and $|||v|||_{1,k} = |||v|||$ for $v \in M_k$.

Now the estimate in (4.9) is a direct consequence of the following result.

LEMMA 4.2. *Let $\pi_k : M \to M_k$ be a projection, i.e., $\pi_k v = v$ for any $v \in M_k$. Let $\pi_0 = 0$. Assume that π_k is stable in the sense that there exists a constant $C'_\pi \geqslant 1$ such that, for $0 \leqslant \alpha \leqslant 1$,*

$$\left|\left|\left|(\pi_k - \pi_{k-1})v\right|\right|\right|_{1-\alpha,k}^2 \leqslant C'_\pi |||v|||_{1-\alpha,i}^2, \quad \textit{for all } v \in M_i,\ k \leqslant i.$$

Then with $C = C'_\pi \|I + \Lambda_{\mathrm{U}}^{(\alpha)}\|_{\ell_2}^2$ for the matrix $\Lambda^{(\alpha)}$ defined by (4.10), *we have*

$$|||\pi_1 v|||^2 + \sum_{k=2}^{J} \lambda_k^\alpha \left|\left|\left|(\pi_k - \pi_{k-1})v\right|\right|\right|_{1-\alpha,k}^2 \leqslant C\left(|||v_1|||^2 + \sum_{k=2}^{J} \lambda_k^\alpha |||v_k|||_{1-\alpha,k}^2\right),$$

for any $v = \sum_k v_k$, $v_k \in M_k$. Note that $\|I + \Lambda_{\mathrm{U}}^{(\alpha)}\|_{\ell_2}^2 \leqslant (1/(1-\gamma_2^{\alpha/2}))^2$ if Condition (A.3) *holds.*

PROOF. Let $v \in M_J$ and $v_i \in M_i$ such that $v = \sum v_i$. Since the operators π_k are projections,

$$\pi_1 v_1 = v_1 \quad \text{and} \quad (\pi_k - \pi_{k-1})v_i = 0, \quad \text{for } 1 \leqslant i < k \leqslant J.$$

Using the stability of π_k, we obtain

$$|||\pi_1 v||| \leqslant |||v_1||| + \sum_{i=2}^{J} \lambda_1^{\alpha/2} |||\pi_1 v_i|||_{1-\alpha,1} \leqslant \sqrt{C'_\pi}\left(|||v_1||| + \sum_{i=2}^{J} \lambda_1^{\alpha/2} |||v_i|||_{1-\alpha,i}\right)$$

and

$$\left|\left|\left|(\pi_k - \pi_{k-1})v\right|\right|\right|_{1-\alpha,k} \leqslant \sum_{i=k}^{J} \left|\left|\left|(\pi_k - \pi_{k-1})v_i\right|\right|\right|_{1-\alpha,k} \leqslant \sqrt{C'_\pi} \sum_{i=k}^{J} |||v_i|||_{1-\alpha,i}.$$

Therefore

$$|||\pi_1 v|||^2 + \sum_{k=2}^{J} \lambda_k^{\alpha} \left|\left|\left|(\pi_k - \pi_{k-1})v\right|\right|\right|_{1-\alpha,k}^2$$

$$\leqslant C'_\pi \left[\left(|||v_1||| + \sum_{i=2}^{J} \lambda_1^{\alpha/2} |||v_i|||_{1-\alpha,i} \right)^2 + \sum_{k=2}^{J} \left(\sum_{i=k}^{J} \lambda_k^{\alpha/2} |||v_i|||_{1-\alpha,i} \right)^2 \right].$$

Let x be the J-vector with components $x_1 = |||v_1|||$ and $x_k = \lambda_k^{\alpha/2} |||v_k|||_{1-\alpha,k}$ for $2 \leqslant k \leqslant J$. Then the right-hand side of the above equation is just $C'_\pi \|(I + \Lambda_{\mathrm{U}}^{(\alpha)})x\|_{\ell_2}^2$. The lemma now follows easily. □

We now show that (4.9) holds.

LEMMA 4.3. *If Condition* (A.3) *holds, then*

$$\left(B_\lambda^{-1} v, v\right) \leqslant |||Q_1 v|||^2 + \sum_{k=2}^{J} \lambda_k \left\|(Q_k - Q_{k-1})v\right\|^2 \leqslant \left(\frac{1}{1 - \gamma_2^{1/2}} \right)^2 \left(B_\lambda^{-1} v, v\right).$$

PROOF. Note that $Q_1 v \in M_1$, $(Q_k - Q_{k-1})v \in M_k$ and $v = Q_1 v + \sum_{k=2}^{J} (Q_k - Q_{k-1})v$. Applying Lemma 4.1 with $R_k = \lambda_k I$ we obtain

$$\left(B_\lambda^{-1} v, v\right) \leqslant |||Q_1 v|||^2 + \sum_{k=2}^{J} \lambda_k \left\|(Q_k - Q_{k-1})v\right\|^2, \quad \text{for all } v \in M.$$

On the other hand, applying Lemma 4.2 with $\pi_k = Q_k$ and $\alpha = 1$, we have

$$|||Q_1 v|||^2 + \sum_{k=2}^{J} \lambda_k \left\|(Q_k - Q_{k-1})v\right\|^2 \leqslant \left\|I + \Lambda_{\mathrm{U}}^{(1)}\right\|_{\ell_2}^2 \left(|||v_1|||^2 + \sum_{k=2}^{J} \lambda_k \|v_k\|^2 \right)$$

for any $v = \sum_{k=1}^{J} v_k$, $v_k \in M_k$. Note that $\|I + \Lambda_{\mathrm{U}}^{(1)}\|_{\ell_2}^2 \leqslant (1/(1 - \gamma_2^{1/2}))^2$ if Condition (A.3) holds. The lemma follows from Lemma 4.1. □

4.3. Special smoothers

In this subsection, we consider only the special preconditioner B_λ defined by (4.8). The preconditioner defined in terms of general smoothers R_k will be studied in the next subsection. By Lemma 4.3, the estimate for $(B_\lambda^{-1} v, v)$ is reduced to an estimate for $|||Q_1 v|||^2 + \sum_{k=2}^{J} \lambda_k \|(Q_k - Q_{k-1})v\|^2$ in terms of $A(v, v)$. For this, we also need the following two conditions. The first condition is the regularity-approximation property formulated in Condition (A.2). For convenience we restate it here.

CONDITION (A.2). There exists a number $\alpha \in (0, 1]$ and a constant C_{P}, not depending on J, such that

$$|||(I - P_{k-1})v|||^2_{1-\alpha,k} \leqslant C_{\mathrm{P}}^{\alpha} \lambda_k^{-\alpha} A(v, v), \quad \text{for all } v \in M_k.$$

The second condition is essentially the stability of the $(\cdot, \cdot)$-projection Q_k in the $|||\cdot|||$-norm.

CONDITION (A.4). Let $\alpha \in (0, 1]$ be given. Then there exists a constant C'_{Q} not depending on J such that

$$|||Q_k v|||^2_{1-\alpha,k} \leqslant C'_{\mathrm{Q}} |||v|||^2_{1-\alpha,i}, \quad \text{for all } v \in M_i,\ 1 \leqslant k \leqslant i \leqslant J.$$

The next lemma contains the basic estimates needed in estimating the condition number $\kappa(B_\lambda A)$.

LEMMA 4.4. *Assume that Conditions* (A.2) *and* (A.4) *hold for the same* $\alpha \in (0, 1]$ *and that Condition* (A.3) *holds. Then*

$$\sum_{k=1}^{J} |||(P_k - Q_k)v|||^2 \leqslant C'_b A(v, v), \quad \textit{for all } v \in M_J, \tag{4.11}$$

$$A(v, v) \leqslant C'_c \left(|||Q_1 v|||^2 + \sum_{k=2}^{J} |||(Q_k - Q_{k-1})v|||^2 \right), \quad \textit{for all } v \in M_J, \tag{4.12}$$

and

$$|||Q_1 v|||^2 + \sum_{k=2}^{J} \lambda_k \|(Q_k - Q_{k-1})v\|^2 \leqslant C'_a A(v, v), \quad \textit{for all } v \in M_J. \tag{4.13}$$

Here $C'_b = (C'_{\mathrm{Q}} C_{\mathrm{P}}^{\alpha})(\gamma_2^{\alpha/2}/(1 - \gamma_2^{\alpha/2}))^2$, $C'_c = (\sqrt{C'_b} + \sqrt{C'_b + 1})^2$ *and* $C'_a = 4C'_{\mathrm{Q}} C_{\mathrm{P}}(1/(1 - \gamma_2^{\alpha/2}))^2$.

PROOF. Let v be in M_J. By the definition of λ_k,

$$\begin{aligned} \sum_{k=1}^{J} |||(P_k - Q_k)v|||^2 &\leqslant \sum_{k=1}^{J} \left(\lambda_k^{\alpha/2} |||(P_k - Q_k)v|||_{1-\alpha,k}\right)^2 \\ &= \sum_{k=1}^{J} \left(\lambda_k^{\alpha/2} |||Q_k(P_k - I)v|||_{1-\alpha,k}\right)^2. \end{aligned}$$

Note that $(I - P_k) = \sum_{i=k+1}^{J}(P_i - P_{i-1})$. By the triangle inequality and Condition (A.4),

$$\sum_{k=1}^{J} |||(P_k - Q_k)v|||^2 \leqslant C'_Q \sum_{k=1}^{J}\left(\sum_{i=k+1}^{J} \lambda_k^{\alpha/2} |||(P_i - P_{i-1})v|||_{1-\alpha,i}\right)^2.$$

Note that $(P_i - P_{i-1})$ is a projector and hence Condition (A.2) implies that

$$\begin{aligned}\sum_{k=1}^{J} |||(P_k - Q_k)v|||^2 &\leqslant C'_Q C_P^{\alpha} \sum_{k=1}^{J}\left(\sum_{i=k+1}^{J}\left(\frac{\lambda_k}{\lambda_i}\right)^{\alpha/2} |||(P_i - P_{i-1})v|||\right)^2 \\ &\leqslant C'_Q C_P^{\alpha} \|\Lambda_U^{(\alpha)}\|_{\ell_2}^2 \sum_{k=2}^{J} |||(P_k - P_{k-1})v|||^2.\end{aligned}$$

Condition (A.3) implies that $\|\Lambda_U^{(\alpha)}\|_{\ell_2}^2 \leqslant (\gamma_2^{\alpha/2}/(1-\gamma_2^{\alpha/2}))^2$. Inequality (4.11) now follows from the orthogonality of $(P_i - P_{i-1})$.

Now let $v_k = (Q_k - Q_{k-1})v$. Then $\sum_{i=k+1}^{J} v_i = (I - Q_k)v$. We have

$$\begin{aligned}A(v, v) &= \sum_{k,i=1}^{J} A(v_k, v_i) = \sum_{k=1}^{J} A(v_k, v_k) + 2\sum_{k=1}^{J}\sum_{i=k+1}^{J} A(v_k, v_i) \\ &= \sum_{k=1}^{J} A(v_k, v_k) + 2\sum_{k=1}^{J} A\big(v_k, (I - Q_k)v\big) \\ &= \sum_{k=1}^{J} A(v_k, v_k) + 2\sum_{k=1}^{J} A\big(v_k, (P_k - Q_k)v\big) \\ &\leqslant \sum_{k=1}^{J} |||v_k|||^2 + 2\left(\sum_{k=1}^{J} |||v_k|||^2\right)^{1/2}\left(\sum_{k=1}^{J} |||(P_k - Q_k)v|||^2\right)^{1/2} \\ &\leqslant \sum_{k=1}^{J} |||v_k|||^2 + 2\left(\sum_{k=1}^{J} |||v_k|||^2\right)^{1/2}\big[C'_b A(v, v)\big]^{1/2}.\end{aligned}$$

Elementary algebra shows that

$$A(v, v) \leqslant C'_c \sum_{k=1}^{J} |||v_k|||^2 = C'_c\left(|||Q_1 v|||^2 + \sum_{k=2}^{J} |||(Q_k - Q_{k-1})v|||^2\right)$$

for $C'_c = (\sqrt{C'_b} + \sqrt{C'_b + 1})^2$. This proves (4.12).

By Theorem A.5 of Appendix A, Condition (A.2) implies that

$$\|(I-Q_{k-1})w\|^2 = \inf_{\phi\in M_{k-1}} \|w-\phi\|^2 \leqslant C_{\mathrm{P}}\lambda_k^{-1}|||w|||^2, \quad \text{for all } w \in M_k.$$

By Theorem A.4 of Appendix A, interpolating the above inequality with the trivial result $\|(I-Q_{k-1})w\|^2 \leqslant \|w\|^2$, gives

$$\|(I-Q_{k-1})w\|^2 \leqslant C_{\mathrm{P}}^{(1-\alpha)}\lambda_k^{-(1-\alpha)}|||w|||^2_{1-\alpha,k}.$$

For $v \in M$, setting $w=(Q_k-Q_{k-1})v \in M_k$ in above inequality and using identity $(I-Q_{k-1})w=w$, we obtain

$$\lambda_k\|(Q_k-Q_{k-1})v\|^2 \leqslant C_{\mathrm{P}}^{(1-\alpha)}\lambda_k^{\alpha}|||(Q_k-Q_{k-1})v|||^2_{1-\alpha,k}.$$

By the triangle inequality and Condition (A.4),

$$|||(Q_k-Q_{k-1})\phi|||^2_{1-\alpha,k} \leqslant 2|||Q_k\phi|||^2_{1-\alpha,k} + 2|||Q_{k-1}\phi|||^2_{1-\alpha,k} \leqslant 4C'_{\mathrm{Q}}|||\phi|||^2_{1-\alpha,j},$$

for all $\phi \in M_j$, $k \leqslant j \leqslant J$. In the second inequality, we have used the fact that $|||\psi|||_{1-\alpha,k} \leqslant |||\psi|||_{1-\alpha,k-1}$, for $\psi \in M_{k-1}$. Applying Lemma 4.2 with $\pi_k = Q_k$ gives

$$\begin{aligned}&|||Q_1v|||^2 + \sum_{k=2}^{J}\lambda_k^{\alpha}|||(Q_k-Q_{k-1})v|||^2_{1-\alpha,k}\\ &\quad\leqslant 4C'_{\mathrm{Q}}\|(I+\Lambda_{\mathrm{U}}^{(\alpha)})\|^2_{\ell_2}\left(|||P_1v|||^2 + \sum_{k=2}^{J}\lambda_k^{\alpha}|||(P_k-P_{k-1})v|||^2_{1-\alpha,k}\right).\end{aligned}$$

Now using $P_k - P_{k-1} = (I-P_{k-1})(P_k-P_{k-1})$ and Condition (A.2), we get

$$\lambda_k^{\alpha}|||(P_k-P_{k-1})v|||^2_{1-\alpha,k} \leqslant C_{\mathrm{P}}^{\alpha}|||(P_k-P_{k-1})v|||^2.$$

Using the estimates for $\lambda_k\|(Q_k-Q_{k-1})v\|^2$ and $\lambda_k^{\alpha}|||(P_k-P_{k-1})v|||^2_{1-\alpha,k}$ in the estimate for $\sum_{k=2}^{J}\lambda_k^{\alpha}|||(Q_k-Q_{k-1})v|||^2_{1-\alpha,k}$ we obtain

$$\begin{aligned}&|||Q_1v|||^2 + \sum_{k=2}^{J}\lambda_k\|(Q_k-Q_{k-1})v\|^2\\ &\quad\leqslant 4C'_{\mathrm{Q}}C_{\mathrm{P}}\|(I+\Lambda_{\mathrm{U}}^{(\alpha)})\|^2_{\ell_2}\sum_{k=1}^{J}|||(P_k-P_{k-1})v|||^2.\end{aligned}$$

The estimate in (4.13) now follows from the orthogonality of P_k. □

We can now state and prove the following result.

PROPOSITION 4.1. *Assume that Conditions* (A.2) *and* (A.4) *hold with the same* α *and that* (A.3) *holds. Then*

$$\frac{1}{C'_a}\big(B_\lambda^{-1}v, v\big) \leqslant A(v,v) \leqslant \frac{C'_c}{(1-\gamma_2^{1/2})^2}\big(B_\lambda^{-1}v, v\big), \quad \textit{for all } v \in M_J.$$

Here C'_a *and* C'_c *are defined in Lemma* 4.4.

PROOF. By Lemma 4.4, we have

$$|||Q_1 v|||^2 + \sum_{k=2}^{J} \lambda_k \big\|(Q_k - Q_{k-1})v\big\|^2 \leqslant C'_a A(v,v), \quad \text{for all } v \in M_J$$

and

$$A(v,v) \leqslant C'_c\left(|||Q_1 v|||^2 + \sum_{k=2}^{J} \big|\big|\big|(Q_k - Q_{k-1})v\big|\big|\big|^2\right), \quad \text{for all } v \in M_J.$$

By Lemma 4.1, the first inequality implies the lower estimate. By the second inequality we have

$$A(v,v) \leqslant C'_c\left(|||Q_1 v|||^2 + \sum_{k=2}^{J} \lambda_k \big\|(Q_k - Q_{k-1})v\big\|^2\right), \quad \text{for all } v \in M_J.$$

By Lemma 4.3 and Condition (A.3) we have

$$|||Q_1 v|||^2 + \sum_{k=2}^{J} \lambda_k \big\|(Q_k - Q_{k-1})v\big\|^2 \leqslant \left(\frac{1}{1-\gamma_2^{1/2}}\right)^2 \big(B_\lambda^{-1}v, v\big).$$

This proves the upper estimate. □

COROLLARY 4.1. *If Conditions* (A.1) *and* (A.3) *hold, then*

$$C_{\mathrm{P}}^{-1}\big(B_\lambda^{-1}v, v\big) \leqslant A(v,v) \leqslant C_{\mathrm{P}} \frac{(1+\gamma_2^{1/2})^2}{(1-\gamma_2^{1/2})^4}\big(B_\lambda^{-1}v, v\big).$$

PROOF. By definition, Q_k satisfies Condition (A.4) with $\alpha = 1$, while Condition (A.1) implies that Condition (A.2) holds with $\alpha = 1$. Therefore the upper bound of Corollary 4.1 is a special case of Proposition 4.1. By Condition (A.1), we have

$$|||P_1 v|||^2 + \sum_{k=2}^{J} \lambda_k \big\|(P_k - P_{k-1})v\big\|^2 \leqslant C_{\mathrm{P}} \sum_{k=1}^{J} \big|\big|\big|(P_k - P_{k-1})v\big|\big|\big|^2 = C_{\mathrm{P}} |||v|||^2.$$

It then follows from Lemma 4.1 that $(B_\lambda^{-1}v, v) \leqslant C_P A(v, v)$ for any $v \in M_J$. Note that a direct consequence of the lower estimate in Proposition 4.1 only gives $(B_\lambda^{-1}v, v) \leqslant 4C_P(1/(1-\gamma_2^{1/2}))^2 A(v, v)$ which is slightly weaker than the lower estimate in Corollary 4.1. □

We now consider an alternative way of establishing uniform bounds for the eigenvalues of $B_\lambda A$. We first consider the lower estimate. It follows from Lemma 4.4 that Conditions (A.2), (A.4) (with the same $\alpha \in (0, 1]$) and (A.3) imply that

$$|||Q_1 v|||^2 + \sum_{k=2}^{J} \lambda_k \left\|(Q_k - Q_{k-1})v\right\|^2 \leqslant C_a' A(v, v), \quad \text{for all } v \in M_J, \tag{4.14}$$

which in turn implies a uniform lower bound for $\lambda(B_\lambda A)$. Clearly, such an estimate holds for a form $A(\bullet, \bullet)$ if it holds for an equivalent form $\hat{A}(\bullet, \bullet)$, and consequently it can be established in many applications when Condition (A.2) is not known to hold. This motivates the following assumption.

CONDITION (A.5). There exist linear operators $\Pi_k : M \to M_k$ (not necessarily projectors) with $\Pi_J = I$ and a constant $C_a > 0$ not depending on J, such that

$$|||\Pi_1 v|||^2 + \sum_{k=2}^{J} \lambda_k \left\|(\Pi_k - \Pi_{k-1})v\right\|^2 \leqslant C_a A(v, v), \quad \text{for all } v \in M_J,$$

which is equivalent to

$$A(v, v) \leqslant C_a \left(A(P_1 v, v) + \sum_{k=2}^{J} \lambda_k^{-1} \|A_k P_k v\|^2 \right).$$

Condition (A.5) is a special case of (4.6) with $R_1 = A_1^{-1}$ and $R_k = \lambda_k^{-1} I$. Consequently, it is equivalent to a uniform lower estimate for $\lambda_{\min}(B_\lambda A)$. In many cases, Condition (A.5) is verified by establishing (4.14). Clearly if (4.14) holds then Condition (A.5) holds for $\Pi_k = Q_k$ with $C_a = C_a'$. On the other hand it follows from Lemma 4.3 that if Condition (A.5) holds and if, in addition, Condition (A.3) also holds, then (4.14) holds with $C_a' = (1/(1-\gamma_2^{1/2}))^2 C_a$.

We now consider the upper estimate. The multilevel philosophy suggests that coarser grid functions contain only small components of the highly oscillatory eigenfunctions. The following inverse assumption on M_k is one possible analytic representation of this heuristic idea.

CONDITION (A.6). There is a positive constant C_{cs} and a number $\varepsilon \in (0, 1)$ not depending on J such that for all $1 \leqslant k \leqslant i \leqslant J$

$$A(v_k, v_i) \leqslant C_{cs} \varepsilon^{i-k} |||v_k||| \left(\lambda_i^{1/2} \|v_i\|\right), \quad \text{for all } v_k \in M_k,\ v_i \in M_i,$$

or equivalently for $1 \leqslant k \leqslant i \leqslant J$

$$\|A_i v\|^2 \leqslant \left(C_{\text{cs}}\varepsilon^{i-k}\right)^2 \lambda_i A(v, v), \quad \text{for all } v \in M_k.$$

In many applications we have $\|(Q_i - Q_{i-1})v\|^2 \leqslant C_{\text{Q}}\lambda_i^{-1} A(v, v)$. If we set $v_i = (Q_i - Q_{i-1})v$ in the first inequality in Condition (A.6), we obtain

$$A(v_k, v_i) \leqslant C_{\text{cs}}\varepsilon^{i-k} |||v_k||| \left(\lambda_i^{1/2} \|v_i\|\right) \leqslant C_{\text{Q}}^{1/2} C_{\text{cs}}\varepsilon^{i-k} |||v_k||| \, |||v_i|||.$$

For this reason, (A.6) is often referred to as a *strengthened Cauchy–Schwarz inequality*.

PROPOSITION 4.2. *Assume that Conditions* (A.5) *and* (A.6) *hold. Then*

$$\frac{1}{C_a}\left(B_\lambda^{-1} v, v\right) \leqslant (Av, v) \leqslant C_{\text{cs}} \frac{1+\varepsilon}{1-\varepsilon}\left(B_\lambda^{-1} v, v\right), \quad \textit{for all } v \in M_J.$$

Thus $\kappa(B_\lambda A) \leqslant \frac{C_{\text{cs}}}{C_a}\frac{1+\varepsilon}{1-\varepsilon}$ *so that* $B_\lambda A$ *is uniformly well conditioned.*

PROOF. Let $v_k = (\Pi_k - \Pi_{k-1})v$. Then by Lemma 4.1 and Condition (A.5),

$$\left(B_\lambda^{-1} v, v\right) \leqslant A(v_1, v_1) + \sum_{k=2}^{J} \lambda_k \|v_k\|^2 \leqslant C_a A(v, v).$$

This proves the first inequality. Now let $v = \sum v_k$ with $v_k \in M_k$. Note that $R_1 = A_1^{-1}$ and $R_k = \lambda_k^{-1} I$ for $k \geqslant 2$. By Condition (A.6)

$$A(v_k, v_i) \leqslant C_{\text{cs}}\varepsilon^{|i-k|}\left(R_k^{-1} v_k, v_k\right)^{1/2}\left(R_i^{-1} v_i, v_i\right)^{1/2}.$$

Therefore,

$$A(v, v) = \sum_{k,i=1}^{J} A(v_k, v_i) \leqslant C_{\text{cs}} \frac{1+\varepsilon}{1-\varepsilon} \sum_{k=1}^{J} \left(R_k^{-1} v_k, v_k\right).$$

Since $\{v_k\}$ is an arbitrary decomposition of v, we obtain

$$(Av, v) \leqslant C_{\text{cs}} \frac{1+\varepsilon}{1-\varepsilon}\left(B_\lambda^{-1} v, v\right).$$

This proves the second inequality. □

REMARK 4.1. Condition (A.5) is an approximation property and Condition (A.6) is an inverse property of the subspaces M_k. In approximation theory, similar properties are known as a Jackson type inequality and a Bernstein type inequality, respectively.

4.4. General smoothers

One possible drawback with the preconditioner B_λ is that the implementation of $B_\lambda f$ often requires inverting the mass matrices. This can be avoided by using some other smoother R_k. In this subsection, we consider the additive preconditioner with general smoothers.

The first result only uses an approximation property of subspaces M_k and a very weak assumption on the smoothers R_k. The estimate however deteriorates like J^2.

The stable approximation property is postulated in the following conditions.

CONDITION (A.7). There exist linear operators $\Pi_k : M_J \to M_k$ with $\Pi_J = I$ and positive constants C_π and C'_π not depending on J such that

$$\left\|(\Pi_k - \Pi_{k-1})v\right\|^2 \leqslant C_\pi \lambda_k^{-1} A(v, v), \quad \text{for all } v \in M_J,\ 2 \leqslant k \leqslant J,$$

and

$$A(\Pi_k v, \Pi_k v) \leqslant C'_\pi A(v, v), \quad \text{for all } v \in M_J,\ 1 \leqslant k \leqslant J - 1.$$

REMARK 4.2. In many applications it can be shown that Condition (A.7) holds for $\Pi_k = Q_k$. Clearly if Condition (A.1) holds then Condition (A.7) holds for $\Pi_k = P_k$ with $C_\pi = C_{\mathrm{P}}$ and $C'_\pi = 1$. It is straightforward to show that Condition (A.7) is equivalent to the following stable approximation property of M_{k-1}: There exists a constant $C > 0$, such that

$$\inf_{\phi \in M_{k-1}} \left(\lambda_k \|v - \phi\|^2 + A(\phi, \phi)\right) \leqslant C A(v, v), \quad \text{for all } v \in M_J,\ 2 \leqslant k \leqslant J.$$

For the model problem in Section 2, this is the same as the simultaneous approximation property stated in (2.5). Hence Condition (A.7) holds in many applications when Condition (A.1) is not known to hold.

The condition $R_1 = A_1^{-1}$ used in defining B_λ can be slightly relaxed by assuming that R_1 is symmetric and is uniformly equivalent to A_1^{-1}. For $k \geqslant 2$, it is natural to assume that the iterative method induced by the smoother R_k is at least as good as a weighted Richardson's method. We formulate these more general statements in the following condition.

CONDITION (SA.1). There exist positive constants $\omega_1, \omega_2, \omega_3$ and ω_4, not depending on $1 \leqslant k \leqslant J$, such that the symmetric smoothers, R_k, satisfy

$$\omega_3\left(A_1^{-1} v, v\right) \leqslant (R_1 v, v) \leqslant \omega_4\left(A_1^{-1} v, v\right), \quad \text{for all } v \in M_1$$

and

$$\frac{\omega_1}{\lambda_k}(v, v) \leqslant (R_k v, v) \leqslant \omega_2\left(A_k^{-1} v, v\right), \quad \text{for all } v \in M_k,\ 2 \leqslant k \leqslant J.$$

THEOREM 4.1. *Let* $B_a = \sum_{k=1}^{J} R_k Q_k$ *with* R_k *symmetric and positive definite and satisfying Condition* (SA.1). *Assume that Condition* (A.7) *holds. Then*

$$C_1\big(B_a^{-1}v, v\big) \leqslant (Av, v) \leqslant C_2\big(B_a^{-1}v, v\big), \quad \textit{for all } v \in M_J,$$

where $C_1 = [C'_\pi/\omega_3 + C_\pi(J-1)/\omega_1]^{-1}$ *and* $C_2 = [\omega_4 + (J-1)\omega_2]$.

PROOF. By Condition (A.7), there exist linear operators $\Pi_k : M_J \to M_k$ with $\Pi_J = I$ such that

$$\big\|(\Pi_k - \Pi_{k-1})v\big\|^2 \leqslant C_\pi \lambda_k^{-1} A(v, v), \quad \text{for all } v \in M_J.$$

Therefore,

$$\sum_{k=2}^{J} \lambda_k \big\|(\Pi_k - \Pi_{k-1})v\big\|^2 \leqslant C_\pi(J-1)A(v, v).$$

Set $\Pi_0 = 0$. Then $(\Pi_k - \Pi_{k-1})v \in M_k$ and $v = \sum_{k=1}^{J}(\Pi_k - \Pi_{k-1})v$. By Lemma 4.1 and Condition (SA.1),

$$\begin{aligned}
\big(B_a^{-1}v, v\big) &\leqslant \sum_{k=1}^{J}\big(R_k^{-1}(\Pi_k - \Pi_{k-1})v, (\Pi_k - \Pi_{k-1})v\big)\\
&\leqslant \frac{1}{\omega_3}A(\Pi_1 v, \Pi_1 v) + \frac{1}{\omega_1}\sum_{k=2}^{J}\lambda_k\big\|(\Pi_k - \Pi_{k-1})v\big\|^2\\
&\leqslant \big(C'_\pi/\omega_3 + \omega_1^{-1}C_\pi(J-1)\big)A(v, v).
\end{aligned}$$

This proves the lower estimate. Now suppose $v = \sum v_k$, with $v_k \in M_k$. Then by the Cauchy–Schwarz inequality and Condition (SA.1),

$$\begin{aligned}
A(v, v) &\leqslant \big(\omega_4 + (J-1)\omega_2\big)\left(\omega_4^{-1}A(v_1, v_1) + \omega_2^{-1}\sum_{k=2}^{J}A(v_k, v_k)\right)\\
&\leqslant \big(\omega_4 + (J-1)\omega_2\big)\left(\sum_{k=1}^{J}\big(R_k^{-1}v_k, v_k\big)\right).
\end{aligned}$$

The upper estimate now follows from Lemma 4.1. □

REMARK 4.3. In Theorem 4.1, Condition (SA.1) can be replaced by

$$\omega_1\lambda_k^{-2}(A_k v, v) \leqslant (R_k v, v) \leqslant \omega_2\big(A_k^{-1}v, v\big), \quad \text{for all } v \in M_k,\ k \geqslant 2.$$

Consequently, $R_k = \lambda_k^{-2}A_k$ can be used as a smoother.

The upper estimate in Theorem 4.1 cannot be improved upon under the current assumption (SA.1) on the smoother. Set $R_k = A_k^{-1}$, then Condition (SA.1) is satisfied. However $A(B_a Av, v) = JA(v, v)$ for $v \in M_1$. Thus, to get a better upper estimate on the eigenvalues of $B_a A$, the smoother must not be too good. This is in contrast to multiplicative algorithms such as the multigrid V-cycle. With this motivation, we assume that the smoother $R_k, k \geqslant 2$ performs as well as but not much better than the smoother induced by Richardson's method. As a consequence, the smoother R_k will be effective in reducing the "high frequency" error, while leaving the low frequency error practically unchanged. This is stated more precisely in the following condition.

CONDITION (SA.2). There exist two positive constants ω_1 and ω_2 not depending on $1 \leqslant k \leqslant J$ such that the symmetric smoothers, R_k, satisfy

$$\omega_1 \big(A_1^{-1} v, v\big) \leqslant (R_1 v, v) \leqslant \omega_2 \big(A_1^{-1} v, v\big), \quad \text{for all } v \in M_1$$

and

$$\frac{\omega_1}{\lambda_k}(v, v) \leqslant (R_k v, v) \leqslant \frac{\omega_2}{\lambda_k}(v, v), \quad \text{for all } v \in M_k,\ 2 \leqslant k \leqslant J.$$

To get optimal result, we also use some regularity and approximation property. The following results are direct consequences of Propositions 4.1 and 4.2.

THEOREM 4.2. *Let $B_a = \sum_{k=1}^{J} R_k Q_k$ with R_k symmetric. Assume that the smoothers R_k satisfy Condition* (SA.2). *Assume also that Conditions* (A.2) *and* (A.4) *hold for the same α and that Condition* (A.3) *holds. Then there exist constants C_1 and C_2 not depending on J such that*

$$C_1 \big(B_a^{-1} v, v\big) \leqslant A(v, v) \leqslant C_2 \big(B_a^{-1} v, v\big), \quad \textit{for all } v \in M_J.$$

THEOREM 4.3. *Let $B_a = \sum_{k=1}^{J} R_k Q_k$. Assume that the smoothers R_k are symmetric and satisfy Condition* (SA.2). *Assume in addition that Conditions* (A.5) *and* (A.6) *hold. Then there exist constants C_1 and C_2 not depending on J such that*

$$C_1 \big(B_a^{-1} v, v\big) \leqslant A(v, v) \leqslant C_2 \big(B_a^{-1} v, v\big), \quad \textit{for all } v \in M_J.$$

The smoothing Condition (SA.2) in Theorem 4.2 can be replaced by a slightly weaker condition.

CONDITION (SA.3). There exist positive constants ω_1 and ω_2 and $\beta \in (0, 2]$ such that the symmetric smoothers satisfy

$$\omega_1 \big(A_1^{-1} v, v\big) \leqslant (R_1 v, v) \leqslant \omega_2 \big(A_1^{-1} v, v\big), \quad \text{for all } v \in M_1$$

and

$$\frac{\omega_1}{\lambda_k^2}(A_k v, v) \leqslant (R_k v, v) \leqslant \frac{\omega_2}{\lambda_k^{\beta}} \big(A_k^{-1+\beta} v, v\big), \quad \text{for all } v \in M_k,\ 2 \leqslant k \leqslant J.$$

Condition (SA.3) with $\beta = 1$ is just Condition (SA.2). In general, (SA.3) is weaker than (SA.2). Smoothers derived from the Richardson's method for the normal equation has the form $R_k = \lambda_k^{-2} A_k$. This smoother satisfies Condition (SA.3) but not (SA.2).

THEOREM 4.4. *Let $B_a = \sum_{k=1}^{J} R_k Q_k$ with R_k symmetric and positive definite. Assume that Conditions* (A.2) *and* (A.4) *hold for the same α and that Condition* (A.3) *holds. Assume in addition that the smoothers R_k satisfy Condition* (SA.3) *with $\beta \in [\alpha, 2]$ where α is the constant in Conditions* (A.2) *and* (A.4). *Then there exist positive constants C_1 and C_2, independent of J, such that*

$$C_1\bigl(B_a^{-1}v, v\bigr) \leqslant A(v, v) \leqslant C_2\bigl(B_a^{-1}v, v\bigr), \quad \textit{for all } v \in M_J.$$

PROOF. By Lemma 4.4,

$$|||Q_1 v|||^2 + \sum_{k=2}^{J} \lambda_k \bigl\|(Q_k - Q_{k-1})v\bigr\|^2 \leqslant C_a' A(v, v).$$

By Theorem A.5 of Appendix A, Condition (A.2) implies that

$$\bigl\|(Q_k - Q_{k-1})v\bigr\|^2 \leqslant C_P \lambda_k^{-1} A\bigl((Q_k - Q_{k-1})v, (Q_k - Q_{k-1})v\bigr).$$

By duality, this implies that

$$\bigl|\bigl|\bigl|(Q_k - Q_{k-1})v\bigr|\bigr|\bigr|_{-1,k}^2 \leqslant C_P \lambda_k^{-1} \bigl\|(Q_k - Q_{k-1})v\bigr\|^2.$$

Note that $(R_1^{-1}v, v) \leqslant \omega_1^{-1}(A_1 v, v) = \omega_1^{-1}|||v|||^2$ and $(R_k^{-1}v, v) \leqslant \omega_1^{-1}\lambda_k^2 |||v|||_{-1,k}^2$, and hence

$$\begin{aligned}
&\sum_{k=1}^{J} \bigl(R_k^{-1}(Q_k - Q_{k-1})v, (Q_k - Q_{k-1})v\bigr) \\
&\quad \leqslant \omega_1^{-1}\left(|||Q_1 v|||^2 + \sum_{k=2}^{J} \lambda_k^2 \bigl|\bigl|\bigl|(Q_k - Q_{k-1})v\bigr|\bigr|\bigr|_{-1,k}^2\right) \\
&\quad \leqslant \omega_1^{-1} C_P \left(|||Q_1 v|||^2 + \sum_{k=2}^{J} \lambda_k \bigl\|(Q_k - Q_{k-1})v\bigr\|^2\right) \\
&\quad \leqslant \omega_1^{-1} C_P C_a' A(v, v).
\end{aligned}$$

By Lemma 4.1, $(B_a^{-1}v, v) \leqslant \omega_1^{-1} C_P C_a' A(v, v)$.

By (4.12) in Lemma 4.4,

$$A(v, v) \leqslant \Bigl(\sqrt{C_b'} + \sqrt{C_b' + 1}\Bigr)^2 \left(\sum_{k=1}^{J} \bigl|\bigl|\bigl|(Q_k - Q_{k-1})v\bigr|\bigr|\bigr|^2\right)$$

$$\leqslant \left(\sqrt{C_b'}+\sqrt{C_b'+1}\right)^2\left(|||Q_1v|||^2+\sum_{k=2}^{J}\lambda_k^{\alpha}\,\big|\big|\big|(Q_k-Q_{k-1})v\big|\big|\big|_{1-\alpha,k}^2\right).$$

Condition (A.4) implies that $|||(Q_k-Q_{k-1})v|||_{1-\alpha,k}^2\leqslant 4C_{\mathrm{Q}}'|||v|||_{1-\alpha,i}^2$ for $v\in M_i$ with $i\geqslant k$. We now apply Lemma 4.2 with $\pi_k=Q_k$ and $C_\pi'=4C_{\mathrm{Q}}'$ to obtain

$$\begin{aligned}
&|||Q_1v|||^2+\sum_{k=2}^{J}\lambda_k^{\alpha}\,\big|\big|\big|(Q_k-Q_{k-1})v\big|\big|\big|_{1-\alpha,k}^2\\
&\quad\leqslant 4C_{\mathrm{Q}}'\big\|\Lambda_{\mathrm{U}}^{(\alpha)}\big\|_{\ell_2}^2\left(|||v_1|||^2+\sum_{k=2}^{J}\lambda_k^{\alpha}|||v_k|||_{1-\alpha,k}^2\right)\\
&\quad\leqslant 4C_{\mathrm{Q}}'\big\|\Lambda_{\mathrm{U}}^{(\alpha)}\big\|_{\ell_2}^2\left(|||v_1|||^2+\sum_{k=2}^{J}\lambda_k^{\beta}|||v_k|||_{1-\beta,k}^2\right)\\
&\quad\leqslant 4C_{\mathrm{Q}}'\big\|\Lambda_{\mathrm{U}}^{(\alpha)}\big\|_{\ell_2}^2\omega_2\left(\sum_{k=1}^{J}(R_k^{-1}v_k,v_k)\right)
\end{aligned}$$

for any $v=\sum v_k$. The upper estimate follows from Lemma 4.1. □

4.5. *Uniform perturbation of forms*

We now make a few brief remarks on the validity of various conditions under a uniform perturbation of the form $A(\bullet,\bullet)$.

Two forms $A(\bullet,\bullet)$ and $\mathcal{A}(\bullet,\bullet)$ are said to be uniformly equivalent if there exist positive constants C_1 and C_2 not depending on J such that

$$C_1A(v,v)\leqslant\mathcal{A}(v,v)\leqslant C_2A(v,v),\qquad\text{for all } v\in M_J.$$

Condition (A.7) is insensitive to a uniform perturbation of the form; if Condition (A.7) holds for a form $A(\bullet,\bullet)$, then it holds also for any form $\mathcal{A}(\bullet,\bullet)$ that is uniformly equivalent to $A(\bullet,\bullet)$, with C_π replaced by C_π/C_1 and C_π' by $C_\pi'C_2/C_1$. Similarly, Conditions (A.3), (A.4) and (A.5) are also insensitive to a uniform perturbation of the form.

On the other hand, Conditions (A.1), (A.2) and (A.6) are sensitive to perturbation of the form. It is possible for Condition (A.1) (or (A.2) or (A.6)) to hold for a form $A(\bullet,\bullet)$ but not for a uniformly equivalent form $\mathcal{A}(\bullet,\bullet)$; cf. BRAMBLE [1995].

We have shown that Condition (A.1) (or (A.2) together with (A.3) and (A.4)) implies (A.5), which in turn implies the lower estimate of the eigenvalues of $B_\lambda A$. It is possible that (A.5) holds for $A(\bullet,\bullet)$ while (A.1) (or (A.2)) does not. In applications to elliptic problems, it is often possible to show that (A.1) or (A.2) hold for an equivalent form with smooth coefficients.

REMARK 4.4. It is easy to see that if the conclusion of any of Theorem 4.2, Theorem 4.3, Theorem 4.4, or Theorem 4.1 holds for $A(\bullet, \bullet)$, then it holds also for an equivalent form $\mathcal{A}(\bullet, \bullet)$ and $\kappa(B_a\mathcal{A}) \leqslant \kappa(B_a A)\kappa(A^{-1}\mathcal{A})$. Therefore the multilevel additive preconditioners are robust with respect to such perturbations.

4.6. Bibliographical notes

The multilevel additive preconditioner, B_a, was introduced by BRAMBLE, PASCIAK and XU in [1990]. It is often referred to as the BPX preconditioner in the literature. It proved to be a valuable tool in the study of the convergence property of multigrid methods. The paper by KING [1990], in which he introduced a related two-level preconditioner, had a strong influence on this work. In BRAMBLE, PASCIAK and XU [1990] the technique for analyzing multilevel methods with very general mesh refinements was first introduced.

There are two other similar multilevel preconditioners. The first is the hierarchical basis (HB) preconditioner of YSERENTANT [1986b]. The HB and BPX preconditioners are based on similar (but different) multilevel splittings of the finite element spaces. The cost per iteration for the BPX method is only slightly more than that for the hierarchical basis method. The condition number of the linear system for the HB preconditioner grows quadratically with the number of refinement levels for two-dimensional problems and grows exponentially for three-dimensional problems. On the other hand the condition number of the linear system for the BPX preconditioner is uniformly bounded for elliptic finite element problems in all space dimensions, provided that the corresponding partial differential equation is uniformly elliptic. A comparison of these two preconditioners may be found in YSERENTANT [1990]. Another similar multilevel preconditioner is the multilevel additive Schwarz (MAS) preconditioner proposed in DRYJA and WIDLUND [1991b]. It is known that the MAS preconditioner is equivalent to the BPX preconditioner.

Theorem 4.1 was established in BRAMBLE, PASCIAK and XU [1990]. It is also shown in BRAMBLE, PASCIAK and XU [1990] that the smallest eigenvalue of $B_a A_J$ is bounded from below by $CJ^{1-1/\alpha}$ if Condition (A.2) holds. OSWALD [1992] gave uniform estimates for the condition number of $B_a A_J$ in the case of piecewise linear functions on triangles using approximation theoretic techniques in Besov spaces. Independently ZHANG [1992] essentially proved, in this case, that Condition (A.6) holds. This gave rise to a uniform upper bound for the additive preconditioner.

The results of ZHANG [1992] and OSWALD [1992] motivated the papers by BRAMBLE and PASCIAK [1993, 1994] in which uniform estimates were given for the multigrid V-cycle with only one smoothing step; cf. Theorems 5.2 and 5.1. These results give an extension of the results of BRAESS and HACKBUSCH [1983] to the case of less than full regularity with one smoothing step.

5. Product analysis of V-cycle multigrid algorithm

We now return to the multigrid method defined by Algorithm 3.1. In this section we continue in the framework of the previous two sections and establish six additional theorems which provide bounds for the convergence rate of Algorithm 3.1. The proofs of these results depend on expressing the error propagator for the multigrid algorithm in terms of a product of operators defined on the finest space. The first four theorems give uniform convergence estimates and can be used in many cases without full elliptic regularity. The last two theorems give a result which only depends on ellipticity of the form but gives rise to an iterative convergence estimate which deteriorates rather weakly as J becomes large.

5.1. *The multiplicative error representation*

The theory of Section 3 was based on a two-level error representation. To get strong results in the case of less than full elliptic regularity, we shall need to express the error in a different way. The fine grid error operator $\mathcal{E}_J$ is expressed as a product of factors associated with the smoothings on individual levels.

By (3.7), for $k > 1$ and any $v \in M_J$,

$$(I - P_k)v + \mathcal{E}_k P_k v = (I - P_k)v + K_k\big[(I - P_{k-1}) + \mathcal{E}_{k-1} P_{k-1}\big]K_k^* P_k v.$$

Let $T_k = R_k A_k P_k$ and recall that $R_1 = A_1^{-1}$ so that $T_1 = P_1$. The adjoint of T_k with respect to the inner product $A(\bullet, \bullet)$ is given by $T_k^* = R_k^t A_k P_k$. Note that $I - T_k$ and $I - T_k^*$ respectively extend the operators K_k and K_k^* to operators defined on all of M_J. Clearly $(I - P_k)v = 0$ for all $v \in M_k$. In addition, P_k commutes with T_k. Thus,

$$\begin{aligned}(I - P_k)v + \mathcal{E}_k P_k v &= (I - T_k)\big[(I - P_{k-1} P_k) + \mathcal{E}_{k-1} P_{k-1} P_k\big]\big(I - T_k^*\big)v \\ &= (I - T_k)\big[(I - P_{k-1}) + \mathcal{E}_{k-1} P_{k-1}\big]\big(I - T_k^*\big)v.\end{aligned}$$

The second equality follows from the identity $P_{k-1} P_k = P_{k-1}$. Repeatedly applying the above identity gives

$$\mathcal{E}_J v = (I - T_J)(I - T_{J-1}) \cdots (I - T_1)\big(I - T_1^*\big)\big(I - T_2^*\big) \cdots \big(I - T_J^*\big). \tag{5.1}$$

Equation (5.1) displays the Jth level error operator as a product of factors. These factors show the precise effect of the individual grid smoothings on the fine grid error propagator.

Our goal is to show that

$$0 \leqslant A(\mathcal{E}_J v, v) \leqslant (1 - 1/C_M) A(v, v). \tag{5.2}$$

This means that the sequence u^m defined by (3.3) satisfies

$$|||u - u^m||| \leqslant (1 - 1/C_M)^m\, |||u - u^0|||.$$

Let $E_0 = I$ and set $E_k = (I - T_k)E_{k-1}$, for $k = 1, 2, \ldots, J$. By (5.1), $\mathcal{E}_J = E_J E_J^*$ where E_J^* is the adjoint of E_J with respect to the $A(\bullet, \bullet)$ inner product. We can rewrite (5.2) as

$$A\big(E_J^* v, E_J^* v\big) \leqslant (1 - 1/C_M) A(v, v), \quad \text{for all } v \in M_J,$$

which is equivalent to

$$A(E_J v, E_J v) \leqslant (1 - 1/C_M) A(v, v), \quad \text{for all } v \in M_J.$$

Consequently, (5.2) is equivalent to

$$A(v, v) \leqslant C_M \big[A(v, v) - A(E_J v, E_J v) \big], \quad \text{for all } v \in M_J. \tag{5.3}$$

We will establish a few uniform convergence estimates for the multigrid algorithm by proving (5.3) under various conditions.

5.2. Some technical lemmas

We will prove in this subsection an abstract estimate which is essential to all of our estimates. For this, we will need to use the following identities which we state as lemmas. Recall that $\overline{R}_k = R_k + R_k^t - R_k^t A_k R_k$.

LEMMA 5.1. *The following identity holds.*

$$A(v, v) - A(E_J v, E_J v) = \sum_{k=1}^{J} \big(\overline{R}_k A_k P_k E_{k-1} v, A_k P_k E_{k-1} v \big). \tag{5.4}$$

PROOF. For any $w \in M_J$, a simple calculation shows that

$$A(w, w) - A\big((I - T_k) w, (I - T_k) w\big) = A\big((2I - T_k) w, T_k w\big).$$

Let v be in M_J. Taking $w = E_{k-1} v$ in the above identity and summing gives

$$\begin{aligned} A(v, v) - A(E_J v, E_J v) &= \sum_{k=1}^{J} A\big((2I - T_k) E_{k-1} v, T_k E_{k-1} v\big) \\ &= \sum_{k=1}^{J} \big(\overline{R}_k A_k P_k E_{k-1} v, A_k P_k E_{k-1} v \big). \end{aligned}$$

This is (5.4). □

LEMMA 5.2. *Let $\pi_k : M \to M_k$ be a sequence of linear operators with $\pi_J = I$. Set $\pi_0 = 0$ and $E_0 = I$. Then*

$$\begin{aligned} A(v, v) &= \sum_{k=1}^{J} A\big(E_{k-1} v, (\pi_k - \pi_{k-1}) v\big) + \sum_{k=1}^{J-1} A\big(T_k E_{k-1} v, (P_k - \pi_k) v\big) \\ &= \left(A(P_1 v, v) + \sum_{k=2}^{J} A\big(E_{k-1} v, (\pi_k - \pi_{k-1}) v\big) \right) \end{aligned}$$

$$+\left(\sum_{k=2}^{J-1} A\big(T_k E_{k-1} v, (P_k - \pi_k) v\big)\right). \tag{5.5}$$

PROOF. Expanding v in terms of π_k gives

$$\begin{aligned} A(v, v) &= \sum_{k=1}^{J} A\big(v, (\pi_k - \pi_{k-1}) v\big) \\ &= \sum_{k=1}^{J} A\big(E_{k-1} v, (\pi_k - \pi_{k-1}) v\big) + \sum_{k=1}^{J} A\big((I - E_{k-1}) v, (\pi_k - \pi_{k-1}) v\big). \end{aligned}$$

Let $F_k = I - E_k = \sum_{i=1}^{k} T_i E_{i-1}$. Then since $F_0 = 0$ and $\pi_0 = 0$,

$$\begin{aligned} &\sum_{k=1}^{J} A\big(F_{k-1} v, (\pi_k - \pi_{k-1}) v\big) \\ &\quad = \sum_{k=1}^{J} A(F_{k-1} v, \pi_k v) - \sum_{k=1}^{J-1} A(F_k v, \pi_k v) \\ &\quad = \sum_{k=1}^{J-1} A\big((F_{k-1} - F_k) v, \pi_k v\big) + A(F_{J-1} v, v). \end{aligned}$$

Now $F_{k-1} - F_k = -T_k E_{k-1}$. Rearranging the terms gives

$$\begin{aligned} &\sum_{k=1}^{J} A\big(F_{k-1} v, (\pi_k - \pi_{k-1}) v\big) \\ &\quad = -\sum_{k=1}^{J-1} A(T_k E_{k-1} v, \pi_k v) + \sum_{k=1}^{J-1} A(T_k E_{k-1} v, v) \\ &\quad = \sum_{k=1}^{J-1} A\big(T_k E_{k-1} v, (P_k - \pi_k) v\big). \end{aligned}$$

This gives the first equality in (5.5). Note that $R_1 = A_1^{-1}$ and $T_1 = P_1$ and thus the second equality in (5.5) follows from the first. □

Set $\|w\|_{\overline{R}_k^{-1}} = (\overline{R}_k^{-1} w, w)^{1/2}$ and $\|w\|_{\overline{R}_k} = (\overline{R}_k w, w)^{1/2}$ for all $w \in M_k$. Using Eqs. (5.4) and (5.5) we can prove the following basic estimate.

LEMMA 5.3. *Let $\pi_k : M \to M_k$ be a sequence of linear operators with $\pi_J = I$. Let $\pi_0 = 0$. Then we have the following basic estimate*

$$A(v,v) \leqslant \left[A(v,v) - A(E_J v, E_J v)\right]^{1/2}\left[\sigma_1(v) + \sigma_2(v)\right], \tag{5.6}$$

where

$$\sigma_1(v) = \left(A(P_1 v, v) + \sum_{k=2}^{J} \left\|(\pi_k - \pi_{k-1})v\right\|^2_{\overline{R}_k^{-1}}\right)^{1/2}$$

and

$$\sigma_2(v) = \left(\sum_{k=2}^{J-1} \left\|R_k^t A_k (P_k - \pi_k)v\right\|^2_{\overline{R}_k^{-1}}\right)^{1/2}.$$

PROOF. We will bound the right-hand side of (5.5). By the Cauchy–Schwarz inequality, the first sum of (5.5) can be bounded as

$$\begin{aligned}
&A(P_1 v, v) + \sum_{k=2}^{J} A\big(E_{k-1} v, (\pi_k - \pi_{k-1})v\big) \\
&\quad = A(P_1 v, v) + \sum_{k=2}^{J} \big(A_k P_k E_{k-1} v, (\pi_k - \pi_{k-1})v\big) \\
&\quad \leqslant \left(\sum_{k=1}^{J} \|A_k P_k E_{k-1} v\|^2_{\overline{R}_k}\right)^{1/2} \left(A(P_1 v, v) + \sum_{k=2}^{J} \left\|(\pi_k - \pi_{k-1})v\right\|^2_{\overline{R}_k^{-1}}\right)^{1/2}.
\end{aligned}$$

Similarly the second sum of (5.5) can be bounded as

$$\begin{aligned}
&\sum_{k=2}^{J-1} A\big(T_k E_{k-1} v, (P_k - \pi_k)v\big) \\
&\quad = \sum_{k=2}^{J-1} \big(R_k A_k P_k E_{k-1} v, A_k (P_k - \pi_k)v\big) \\
&\quad \leqslant \left(\sum_{k=2}^{J-1} \|A_k P_k E_{k-1} v\|^2_{\overline{R}_k}\right)^{1/2} \left(\sum_{k=2}^{J-1} \left\|R_k^t A_k (P_k - \pi_k)v\right\|^2_{\overline{R}_k^{-1}}\right)^{1/2}.
\end{aligned}$$

Combining these two estimates and using (5.5) we obtain

$$A(v,v) \leqslant \left(\sum_{k=1}^{J} \big(\overline{R}_k A_k P_k E_{k-1} v, A_k P_k E_{k-1} v\big)\right)^{1/2} \left[\sigma_1(v) + \sigma_2(v)\right].$$

Using (5.4) in the above inequality proves the lemma. □

5.3. *Uniform estimates*

In this subsection, we will show that the uniform convergence estimate of Theorem 3.1 often extends to the case of less than full regularity and approximation. The results are obtained by bounding the second factor in Lemma 5.3 from above by $A(v, v)$.

We shall use the smoother Condition (SM.1), which is restated here for convenience. Recall that $\overline{R}_k = R_k + R_k^t - R_k^t A_k R_k$, $K_k = (I - R_k A_k)$ and $K_{k,\omega} = I - (\omega\lambda_k^{-1})A_k$.

CONDITION (SM.1). There exists a constant $\omega \in (0, 1)$ not depending on J such that

$$A(K_k v, K_k v) \leqslant A(K_{k,\omega} v, v), \qquad \text{for all } v \in M_k,\ k = 2, 3, \ldots, J,$$

or equivalently

$$\frac{\omega}{\lambda_k}(v, v) \leqslant \big(\overline{R}_k v, v\big), \qquad \text{for all } v \in M_k,\ k = 2, 3, \ldots, J.$$

Note that

$$\big(\overline{R}_k v, v\big) \leqslant \big(A_k^{-1} v, v\big), \qquad \text{for } v \in M_k.$$

In addition to the smoother Condition (SM.1), we will need to require that the smoother R_k is properly scaled. More precisely the condition is the following:

CONDITION (SM.2). There exists a constant $\theta \in (0, 2)$ not depending on J such that

$$A(R_k v, R_k v) \leqslant \theta(R_k v, v), \qquad \text{for all } v \in M_k,\ k = 2, 3, \ldots, J.$$

This inequality is the same as any one of the following three inequalities:

$$\big(\overline{R}_k v, v\big) \geqslant (2 - \theta)(R_k v, v) \geqslant \left(\frac{2 - \theta}{\theta}\right)\big((R_k^t A_k R_k)v, v\big), \qquad \text{for all } v \in M_k.$$

A simple change of variable shows that Condition (SM.2) is equivalent to the following condition for $T_k = R_k A_k P_k$.

$$A(T_k v, T_k v) \leqslant \theta A(T_k v, v), \qquad \text{for all } v \in M_J,\ k = 2, 3, \ldots, J.$$

Note that if the smoother R_k satisfies Condition (SM.1), then it also satisfies the inequality in Condition (SM.2) with $\theta < 2$, but possibly depending on k. Therefore smoothers that satisfy (SM.1) will satisfy Condition (SM.2) after a proper scaling.

Our first theorem uses Condition (A.2), a regularity and approximation property of P_k, Condition (A.3), a geometrical growth condition for λ_k, and Condition (A.4), a stability condition on Q_k.

THEOREM 5.1. *Assume that Conditions* (A.2) *and* (A.4) *hold with the same* α *and that Condition* (A.3) *holds. Assume in addition that the smoothers,* R_k, *satisfy Conditions* (SM.1) *and* (SM.2). *Then,*

$$0 \leqslant A(\mathcal{E}_J v, v) \leqslant (1 - 1/C_M)A(v, v), \quad \textit{for all } v \in M_J.$$

Here

$$C_M = \left[\left(\frac{\theta C_b'}{2-\theta}\right)^{1/2} + \left(1 + \frac{C_a'}{\omega}\right)^{1/2}\right]^2$$

with

$$C_a' = 4C_Q' C_P \left(\frac{1}{1 - \gamma_2^{\alpha/2}}\right)^2 \quad \textit{and} \quad C_b' = C_Q' C_P^\alpha \left(\frac{\gamma_2^{\alpha/2}}{1 - \gamma_2^{\alpha/2}}\right)^2.$$

PROOF. The theorem is proved by bounding the second factor in the right-hand side of (5.6) from above by $A(v, v)$. By Lemma 4.4,

$$\sum_{k=1}^{J} |||(P_k - Q_k)v|||^2 \leqslant C_b' A(v, v), \quad \text{for all } v \in M_J$$

and

$$|||Q_1 v|||^2 + \sum_{k=2}^{J} \lambda_k \|(Q_k - Q_{k-1})v\|^2 \leqslant C_a' A(v, v), \quad \text{for all } v \in M_J.$$

By Condition (SM.1), $(\overline{R}_k^{-1} w, w) \leqslant \omega^{-1}\lambda_k \|w\|^2$, for all $w \in M_k$, $k \geqslant 2$. Hence,

$$A(P_1 v, v) + \sum_{k=2}^{J} \|(Q_k - Q_{k-1})v\|_{\overline{R}_k^{-1}}^2 \leqslant \left(1 + \frac{C_a'}{\omega}\right) A(v, v).$$

Condition (SM.2) implies that $(R_k \overline{R}_k^{-1} R_k^t w, w) \leqslant \frac{\theta}{2-\theta}(A_k^{-1} w, w)$ for $w \in M_k$. Hence,

$$\begin{aligned}\sum_{k=2}^{J} \|R_k^t A_k (P_k - Q_k)v\|_{\overline{R}_k^{-1}}^2 &\leqslant \frac{\theta}{2-\theta} \sum_{k=2}^{J} |||(P_k - Q_k)v|||^2 \\ &\leqslant \frac{\theta C_b'}{2-\theta} A(v, v).\end{aligned}$$

Combining these two estimates and applying Lemma 5.3, with $\pi_k = Q_k$, shows that

$$A(v, v) \leqslant C_M\big[A(v, v) - A(E_J v, E_J v)\big].$$

The theorem follows. □

The next theorem shows that if Conditions (A.5) and (A.6) hold, then the multigrid Algorithm 3.1 has a uniform error reduction.

THEOREM 5.2. *Assume that Conditions* (A.5) *and* (A.6) *hold. Assume in addition that the smoothers, R_k, satisfy Conditions* (SM.1) *and* (SM.2). *Then,*

$$A(\mathcal{E}_J v, v) \leqslant (1 - 1/C_M)A(v, v), \quad \textit{for all } v \in M_J,$$

with

$$C_M = \left[\left(1 + \frac{C_a}{\omega}\right)^{1/2} + \frac{C_{\mathrm{cs}}\varepsilon}{1-\varepsilon}\left(\frac{C_a\theta}{2-\theta}\right)^{1/2}\right]^2.$$

PROOF. We again establish this theorem by bounding the second factor in the right-hand side of (5.6) from above by $A(v, v)$. By Condition (A.5), there exists a sequence of linear operators $\Pi_k : M \to M_k$ with $\Pi_J = I$ such that

$$A(\Pi_1 v, \Pi_1 v) + \sum_{k=2}^{J} \lambda_k \left\|(\Pi_k - \Pi_{k-1})v\right\|^2 \leqslant C_a A(v, v).$$

By Condition (SM.1), $(\overline{R}_k^{-1} w, w) \leqslant (\lambda_k/\omega)\|w\|^2$ for $w \in M_k$ and $k \geqslant 2$. Hence,

$$A(P_1 v, v) + \sum_{k=2}^{J} \left\|(\Pi_k - \Pi_{k-1})v\right\|^2_{\overline{R}_k^{-1}} \leqslant \left(1 + \frac{C_a}{\omega}\right)A(v, v). \tag{5.7}$$

Recall that $\Pi_J = I$. By Condition (A.6),

$$\begin{aligned}
\sum_{k=1}^{J} |||(P_k - \Pi_k)v|||^2 &= \sum_{k=1}^{J} A\big((P_k - \Pi_k)v, (I - \Pi_k)v\big) \\
&= \sum_{k=1}^{J} \sum_{i=k+1}^{J} A\big((P_k - \Pi_k)v, (\Pi_i - \Pi_{i-1})v\big) \\
&\leqslant \sum_{k=1}^{J} \sum_{i=k+1}^{J} C_{\mathrm{cs}}\varepsilon^{i-k} |||(P_k - \Pi_k)v||| \left(\lambda_i^{1/2}\left\|(\Pi_i - \Pi_{i-1})v\right\|\right) \\
&\leqslant \left(\frac{C_{\mathrm{cs}}\varepsilon}{1-\varepsilon}\right)\left(\sum_{k=1}^{J} |||(P_k - \Pi_k)v|||^2\right)^{1/2} \left(\sum_{i=2}^{J} \lambda_i \left\|(\Pi_i - \Pi_{i-1})v\right\|^2\right)^{1/2}.
\end{aligned}$$

Cancelling the factor and using Condition (A.5), we obtain

$$\sum_{k=1}^{J} |||(P_k - \Pi_k)v|||^2 \leqslant C_a \left(\frac{C_{\mathrm{cs}}\varepsilon}{1-\varepsilon}\right)^2 A(v, v).$$

Condition (SM.2) implies that $(\overline{R}_k^{-1} R_k^t w, R_k^t w) \leqslant \frac{\theta}{2-\theta}(A_k^{-1} w, w)$ for $k \geqslant 2$, Hence

$$\begin{aligned}\sum_{k=2}^{J} \|R_k^t A_k (P_k - \Pi_k)v\|^2_{\overline{R}_k^{-1}} &\leqslant \frac{\theta}{2-\theta}\sum_{k=2}^{J} |||(P_k - \Pi_k)v|||^2 \\ &\leqslant \frac{C_a\theta}{2-\theta}\left(\frac{C_{\mathrm{cs}}\varepsilon}{1-\varepsilon}\right)^2 A(v, v). \qquad (5.8)\end{aligned}$$

The theorem now follows from Eqs. (5.7) and (5.8) and an application of Lemma 5.3 with $\pi_k = \Pi_k$. □

In Theorems 5.1 and 5.2, Condition (SM.1) can be replaced by the following weaker smoothing condition.

CONDITION (SM.3). There exists a constant $\omega > 0$ not depending on J such that

$$\frac{\omega}{\lambda_k^2}(A_k v, v) \leqslant (\overline{R}_k v, v), \quad \text{for all } v \in M_k,\ k = 2, 3, \ldots, J.$$

THEOREM 5.3. *Assume that Conditions* (A.2) *and* (A.4) *hold with the same* α *and that Condition* (A.3) *holds. Assume in addition that the smoothers,* R_k*, satisfy Conditions* (SM.3) *and* (SM.2). *Then*

$$0 \leqslant A(\mathcal{E}_J v, v) \leqslant (1 - 1/C_M) A(v, v), \quad \textit{for all } v \in M_J.$$

Here

$$C_M = \left[\left(\frac{\theta C_b'}{2-\theta}\right)^{1/2} + \left(1 + \frac{C_{\mathrm{P}} C_a'}{\omega}\right)^{1/2}\right]^2$$

and C_a' *and* C_b' *are defined in Theorem* 5.1.

PROOF. In the proof of Theorem 4.4, we have shown that Condition (A.2) implies that

$$|||(Q_k - Q_{k-1})v|||^2_{-1,k} \leqslant C_{\mathrm{P}} \lambda_k^{-1} \|(Q_k - Q_{k-1})v\|^2.$$

Now by Condition (SM.3), we have

$$\|(Q_k - Q_{k-1})v\|^2_{\overline{R}_k^{-1}} \leqslant \frac{\lambda_k^2}{\omega} |||(Q_k - Q_{k-1})v|||^2_{-1,k} \leqslant \frac{C_{\mathrm{P}}}{\omega}\lambda_k \|(Q_k - Q_{k-1})v\|^2.$$

By Lemma 4.4,

$$|||Q_1 v|||^2 + \sum_{k=2}^{J} \lambda_k \left\|(Q_k - Q_{k-1})v\right\|^2 \leqslant C'_a A(v,v), \quad \text{for all } v \in M_J.$$

Combining these estimates, we obtain

$$A(P_1 v, v) + \sum_{k=2}^{J} \left\|(Q_k - Q_{k-1})v\right\|^2_{\overline{R}_k^{-1}} \leqslant \left(1 + \frac{C_P C'_a}{\omega}\right) A(v,v).$$

The rest of the proof is identical to that of Theorem 5.1. □

THEOREM 5.4. *Assume that the smoothers, R_k, satisfy Conditions* (SM.3) *and* (SM.2). *In addition, assume that*

(i) *either Conditions* (A.3), (A.5) *and* (A.6) *hold or*

(ii) *Condition* (A.5) *holds with $\Pi_k = Q_k$ and Condition* (A.6) *holds.*

Then

$$0 \leqslant A(\mathcal{E}_J v, v) \leqslant (1 - 1/C_M) A(v,v), \quad \textit{for all } v \in M_J$$

for some constant $C_M > 0$ independent of J.

PROOF. By Lemma 4.2, if Conditions (A.3) and (A.5) hold, then Condition (A.5) holds for $\pi_k = Q_k$. The rest of the proof is identical to that of Theorem 5.3. □

Note that if R_k satisfies Condition (SM.3) it also satisfies Condition (SM.1) therefore Theorem 5.1 is a corollary of Theorem 5.3.

5.4. Non-uniform estimates

All the theorems in the previous subsection use some elliptic regularity property. In contrast, the next result uses only the stable approximation property of M_k. The estimate however deteriorates linearly as J increases.

We have the following theorem.

THEOREM 5.5. *Assume that Condition* (A.7) *holds. Assume in addition that the smoothers, R_k, satisfy Conditions* (SM.1) *and* (SM.2). *Then,*

$$0 \leqslant A(\mathcal{E}_J v, v) \leqslant (1 - 1/C_J) A(v,v), \quad \textit{for all } v \in M_J,$$

with

$$C_J = \left[\left(1 + (J-1)\frac{C_\pi}{\omega}\right)^{1/2} + \left(2(J-1)\frac{(1 + C'_\pi)\theta}{2-\theta}\right)^{1/2}\right]^2.$$

PROOF. We need only prove the upper inequality. By Condition (A.7), there exists a sequence of linear operators $\Pi_k : M_J \to M_k$ with $\Pi_J = I$ such that

$$\|(\Pi_k - \Pi_{k-1})v\|^2 \leqslant C_\pi \lambda_k^{-1} A(v, v), \quad \text{for all } v \in M_J,\ 2 \leqslant k \leqslant J.$$

By Condition (SM.1),

$$\|(\Pi_k - \Pi_{k-1})v\|^2_{\overline{R}_k^{-1}} \leqslant \frac{1}{\omega}\lambda_k \|(\Pi_k - \Pi_{k-1})v\|^2 \leqslant \frac{C_\pi}{\omega} A(v, v), \quad \text{for } 2 \leqslant k \leqslant J,$$

and thus

$$A(P_1 v, v) + \sum_{k=2}^{J} \|(\Pi_k - \Pi_{k-1})v\|^2_{\overline{R}_k^{-1}} \leqslant \left(1 + (J-1)\frac{C_\pi}{\omega}\right) A(v, v). \tag{5.9}$$

By the triangle inequality and Condition (A.7) we have

$$|||(P_k - \Pi_k)v|||^2 \leqslant 2\big(|||P_k v|||^2 + |||\Pi_k v|||^2\big) \leqslant 2\big(1 + C'_\pi\big)|||v|||^2.$$

It follows from Condition (SM.2) that

$$\|R_k^t A_k (P_k - \Pi_k)v\|^2_{\overline{R}_k^{-1}} \leqslant \frac{\theta}{2-\theta} |||(P_k - \Pi_k)v|||^2 \leqslant 2\frac{(1 + C'_\pi)\theta}{2-\theta} A(v, v).$$

Summing from 2 to J,

$$\sum_{k=2}^{J} \|R_k^t A_k (P_k - \Pi_k)v\|^2_{\overline{R}_k^{-1}} \leqslant 2(J-1)\frac{(1 + C'_\pi)\theta}{2-\theta} A(v, v). \tag{5.10}$$

The theorem follows from (5.9), (5.10) and Lemma 5.3 with $\pi_k = \Pi_k$. □

In most applications we can verify Condition (A.7) by showing that it holds for $\Pi_k = Q_k$, i.e., $A(Q_k v, Q_k v) \leqslant C'_Q A(v, v)$. It then follows by interpolation that Condition (A.4) holds. Condition (A.2) however depends on the form $A(\bullet, \bullet)$ in a more sensitive way. In contrast to Theorem 5.1, Condition (A.2) is not assumed in Theorem 5.5 and thus only a weaker estimate is derived. On the other hand, if Condition (A.7) holds for a particular form $A(\bullet, \bullet)$, then it holds for an equivalent form $\mathcal{A}(\bullet, \bullet)$. Thus, the analogous Theorem 5.5 holds for the multigrid iteration for the solution of the finite element equations involving $\mathcal{A}(\bullet, \bullet)$. This is not true for Theorem 5.1 in general.

We now show that the convergence estimate in Theorem 5.5 can be improved under a slightly stronger condition on the smoother. We will need two lemmas.

LEMMA 5.4. *Let $A = [A_{ij}]$ be a symmetric positive definite n-by-n block matrix. Denote by $\mathcal{L}(A), \mathcal{U}(A)$ and $\mathcal{D}(A)$ the lower triangular, upper triangular, and diagonal blocks of A, respectively. Then*

$$\left\|\mathcal{L}(A)\right\|_{\ell_2} = \left\|\mathcal{U}(A)\right\|_{\ell_2} \leqslant \tfrac{1}{2}\lceil \log_2 n\rceil \|A\|_{\ell_2},$$

where $\lceil x\rceil = \min\{k \in \mathbf{N}\colon k \geqslant x\}$ denotes the smallest integer upper bound of x.

PROOF. Let $n = 2$, and suppose $A = \left(\begin{smallmatrix} A_{11} & A_{12} \\ A_{21} & A_{22}\end{smallmatrix}\right)$ is a symmetric positive definite 2-by-2 block matrix. It is easy to see that

$$\|A_{11}\|_{\ell_2}, \|A_{22}\|_{\ell_2} \leqslant \|A\|_{\ell_2} \quad \text{and} \quad \|A_{12}\|_{\ell_2} = \|A_{21}\|_{\ell_2} \leqslant \tfrac{1}{2}\|A\|_{\ell_2}.$$

Assume the result holds for any symmetric positive definitive n-by-n block matrix with $n \leqslant 2^k,\ k \geqslant 1$. We now show that the result holds for $2^k < n \leqslant 2^{k+1}$. By grouping blocks of A we can write A as

$$A = \begin{pmatrix} A_{11} & A_{12} \\ A_{21} & A_{22}\end{pmatrix}$$

so that A_{11} contains the first 2^k-by-2^k blocks, A_{12} contains 2^k-by-$(n-2^k)$ blocks, etc. Then the upper triangular block of A is

$$\mathcal{U}(A) = \begin{pmatrix} \mathcal{U}(A_{11}) & A_{12} \\ 0 & \mathcal{U}(A_{22})\end{pmatrix}.$$

It is clear that $\|A_{11}\|_{\ell_2} \leqslant \|A\|_{\ell_2}$ and $\|A_{22}\|_{\ell_2} \leqslant \|A\|_{\ell_2}$ and $\|A_{12}\|_{\ell_2} \leqslant \frac{1}{2}\|A\|_{\ell_2}$. By the induction assumption,

$$\left\|\mathcal{U}(A_{ii})\right\|_{\ell_2} \leqslant \tfrac{1}{2}k\|A_{ii}\|_{\ell_2} \leqslant \tfrac{1}{2}k\|A\|_{\ell_2}, \quad i = 1, 2.$$

As a consequence

$$\left\|\mathcal{U}(A)\right\|_{\ell_2} \leqslant \left\|\begin{pmatrix} \frac{k}{2}\|A\|_{\ell_2} & \frac{1}{2}\|A\|_{\ell_2} \\ 0 & \frac{k}{2}\|A\|_{\ell_2}\end{pmatrix}\right\|_{\ell_2} \leqslant \frac{k+1}{2}\|A\|_{\ell_2}.$$

Note that $k + 1 \leqslant \lceil \log_2 n\rceil$ and hence the result follows. □

LEMMA 5.5. *Let $A(\bullet, \bullet)$ be a symmetric positive definite bilinear form on $M_J \times M_J$ and let $D_k : M_k \to M_k$ be symmetric and positive definite. Assume that*

$$\sum_{k,j=1}^{J} A(u_k, v_j) \leqslant K\left(\sum_{k=1}^{J}(D_k u_k, u_k)\right)^{1/2}\left(\sum_{k=1}^{J}(D_k v_k, v_k)\right)^{1/2}$$

holds for any $u_k, v_k \in M_k$. *Then*

$$\sum_{k=1}^{J}\sum_{j>k}^{J} A(u_k, v_j) \leqslant \tfrac{1}{2}\lceil \log_2 J\rceil K \left(\sum_{k=1}^{J}(D_k u_k, u_k)\right)^{1/2}\left(\sum_{k=1}^{J}(D_k v_k, v_k)\right)^{1/2}$$

holds for any $u_k, v_k \in M_k$.

PROOF. Denote by $\{\phi_s^k\}$ a basis for M_k and represent $u_k, v_k \in M_k$ in terms of the basis functions as $u_k = \sum_s x_s^k \phi_s^k$ and $v_k = \sum_s y_s^k \phi_s^k$. Form the matrices $D_k = [(D_k\phi_s^k, \phi_t^k)]$ and $A_{ij} = [A(\phi_s^i, \phi_t^j)]$. Clearly D_k is nonsingular. Set $x = (x^1, x^2, \ldots, x^J)$ and $y = (y^1, y^2, \ldots, y^J)$ and define the J-by-J block matrices $A = [A_{ij}]$ and $D = \operatorname{diag}(D_1, \ldots, D_J)$. Note that x and y are vectors in $\mathbf{R}^N$ with $N = \sum_{k=1}^{J}\dim(M_k)$. The assumption of the lemma is

$$\langle x, Ay\rangle \leqslant K\langle Dx, x\rangle^{1/2}\langle Dy, y\rangle^{1/2}, \quad \text{for all } x, y \in \mathbf{R}^N,$$

where $\langle\bullet,\bullet\rangle$ is the Euclidean inner product on $\mathbf{R}^N$. In term of the norm,

$$\left\|D^{-1/2}AD^{-1/2}\right\|_{\ell_2} \leqslant K.$$

By Lemma 5.4, $\|\mathcal{U}(D^{-1/2}AD^{-1/2})\|_{\ell_2} \leqslant \frac{1}{2}\lceil\log_2 J\rceil K$. Therefore

$$\langle x, \mathcal{U}(A)y\rangle \leqslant \tfrac{1}{2}\lceil\log_2 J\rceil K\langle Dx, x\rangle\langle Dy, y\rangle, \quad \text{for all } x, y \in \mathbf{R}^N,$$

which is the same as the assertion of the lemma. □

Let $\mathcal{R}_k : M_k \to M_k$ be symmetric and positive definite. Define a multilevel additive preconditioner by

$$B_a = \sum_{k=1}^{J}\mathcal{R}_k Q_k.$$

Denote by $K_1 = \lambda_{\min}(B_a A)$ and $K_2 = \lambda_{\max}(B_a A)$. Thus

$$K_1\left(B_a^{-1}v, v\right) \leqslant A(v, v) \leqslant K_2\left(B_a^{-1}v, v\right).$$

THEOREM 5.6. *Let* $\overline{\overline{R}}_k = R_k\overline{R}_k^{-1}R_k^t$. *Assume that there exist a sequence of symmetric, positive definite operators* $\mathcal{R}_k : M_k \to M_k$ *such that*

$$\alpha_1\left(\overline{\overline{R}}_k v, v\right) \leqslant (\mathcal{R}_k v, v) \leqslant \alpha_2\left(\overline{R}_k v, v\right), \quad \textit{for all } v \in M_k. \tag{5.11}$$

Then

$$A(E_J v, E_J v) \leqslant (1 - 1/C_J)A(v, v), \quad \textit{for all } v \in M_J$$

with

$$C_J = \left[\left(1 + \frac{\alpha_2}{K_1}\right)^{1/2} + \frac{\lceil \log_2 J \rceil K_2}{2(\alpha_1 K_1)^{1/2}} \right]^2.$$

Here $K_1 = \lambda_{\min}(B_a A)$, $K_2 = \lambda_{\max}(B_a A)$ *and* $B_a = \sum_{k=1}^J \mathcal{R}_k Q_k$.

PROOF. By the characterization of B_a, there exist J linear operators $\Pi_k : M \to M_k$ with $\Pi_J = I$ and $\Pi_0 = 0$ such that

$$\sum_{k=1}^J \big(\mathcal{R}_k^{-1}(\Pi_k - \Pi_{k-1})v, (\Pi_k - \Pi_{k-1})v\big) = \big(B_a^{-1}v, v\big) \leqslant \frac{1}{K_1} A(v, v).$$

The smoothing assumption implies that $(\overline{R}_k^{-1} w, w) \leqslant \alpha_2 (\mathcal{R}_k^{-1} w, w)$. Thus

$$A(P_1 v, v) + \sum_{k=2}^J \big(\overline{R}_k^{-1}(\Pi_k - \Pi_{k-1})v, (\Pi_k - \Pi_{k-1})v\big) \leqslant \left(1 + \frac{\alpha_2}{K_1}\right) A(v, v).$$

The theorem will follow from Lemma 5.3 if we can show

$$\sum_{k=2}^J \big\| R_k^t A_k (P_k - \Pi_k) v \big\|_{\overline{R}_k^{-1}}^2 \leqslant \frac{\lceil \log_2 J \rceil^2}{4\alpha_1} \frac{K_2^2}{K_1} A(v, v).$$

By the characterization for B_a (cf. Lemma 4.1),

$$A(v, v) \leqslant K_2 \big(B_a^{-1} v, v\big) \leqslant K_2 \sum_{k=1}^J \big(\mathcal{R}_k^{-1} v_k, v_k\big), \quad \text{for all } v = \sum_{k=1}^J v_k,\ v_k \in M_k.$$

This implies that

$$\sum_{k=1}^J \sum_{i=1}^J A(u_k, v_i) \leqslant K_2 \left(\sum_{k=1}^J \big(\mathcal{R}_k^{-1} u_k, u_k\big) \right)^{1/2} \left(\sum_{k=1}^J \big(\mathcal{R}_k^{-1} v_k, v_k\big) \right)^{1/2},$$

for any $u_k, v_k \in M_k$. It then follows from Lemma 5.5 (with $D_k = \mathcal{R}_k^{-1}$) that

$$\sum_{k=1}^J \sum_{i=k+1}^J A(u_k, v_i) \leqslant \frac{\lceil \log_2 J \rceil}{2} K_2 \left(\sum_{k=1}^J \|u_k\|_{\mathcal{R}_k^{-1}}^2 \right)^{1/2} \left(\sum_{k=1}^J \|v_k\|_{\mathcal{R}_k^{-1}}^2 \right)^{1/2} \tag{5.12}$$

for all $u_k, v_k \in M_k$. Now

$$\sum_{k=1}^J \big\| A_k (P_k - \Pi_k) v \big\|_{\mathcal{R}_k}^2 = \sum_{k=1}^J A\big(\mathcal{R}_k A_k (P_k - \Pi_k) v, (I - \Pi_k) v\big)$$

$$= \sum_{k=1}^{J} \sum_{i=k+1}^{J} A\big(\mathcal{R}_k A_k (P_k - \Pi_k) v, (\Pi_i - \Pi_{i-1}) v\big).$$

Setting $u_k = \mathcal{R}_k A_k (P_k - \Pi_k) v$ and $v_k = (\Pi_k - \Pi_{k-1}) v$ in (5.12) gives

$$\begin{aligned}
&\sum_{k=1}^{J} \left\| A_k (P_k - \Pi_k) v \right\|_{\mathcal{R}_k}^2 \\
&\quad = \sum_{k=1}^{J} \sum_{i=k+1}^{J} A\big(\mathcal{R}_k A_k (P_k - \Pi_k) v, (\Pi_i - \Pi_{i-1}) v\big) \\
&\quad \leqslant \frac{\lceil \log_2 J \rceil}{2} K_2 \left(\sum_{k=1}^{J} \left\| \mathcal{R}_k A_k (P_k - \Pi_k) v \right\|_{\mathcal{R}_k^{-1}}^2 \right)^{1/2} \\
&\qquad \times \left(\sum_{k=1}^{J} \left\| (\Pi_k - \Pi_{k-1}) v \right\|_{\mathcal{R}_k^{-1}}^2 \right)^{1/2} \\
&\quad = \frac{\lceil \log_2 J \rceil}{2} K_2 \left(\sum_{k=1}^{J} \left\| A_k (P_k - \Pi_k) v \right\|_{\mathcal{R}_k}^2 \right)^{1/2} \big(B_a^{-1} v, v\big)^{1/2}.
\end{aligned}$$

Cancelling the common factor, we get

$$\sum_{k=1}^{J} \left\| A_k (P_k - \Pi_k) v \right\|_{\mathcal{R}_k}^2 \leqslant \frac{\lceil \log_2 J \rceil^2}{4} K_2^2 \big(B_a^{-1} v, v\big) \leqslant \frac{\lceil \log_2 J \rceil^2}{4} \frac{K_2^2}{K_1} A(u, u).$$

The smoothing condition says that

$$\big(\overline{R}_k^{-1} R_k^t w, R_k^t w\big) \leqslant \alpha_1^{-1} (\mathcal{R}_k w, w), \quad \text{for all } w \in M_k.$$

Setting $w = A_k (P_k - \Pi_k) v$, we obtain

$$\begin{aligned}
\sum_{k=1}^{J} \left\| R_k^t A_k (P_k - \Pi_k) v \right\|_{\overline{R}_k^{-1}}^2 &\leqslant \frac{1}{\alpha_1} \sum_{k=1}^{J} \left\| A_k (P_k - \Pi_k) v \right\|_{\mathcal{R}_k}^2 \\
&\leqslant \frac{\lceil \log_2 J \rceil^2}{4\alpha_1} \frac{K_2^2}{K_1} A(u, u).
\end{aligned}$$

The theorem now follows from Lemma 5.3 with $\pi_k = \Pi_k$. $\square$

If $R_k = R_k^t$ satisfies Condition (SM.2), then $(2 - \theta)(R_k v, v) \leqslant (\overline{R}_k v, v)$ and thus

$$(2 - \theta)\big(\overline{\overline{R}}_k v, v\big) \leqslant (R_k v, v) \leqslant \frac{1}{2 - \theta} \big(\overline{R}_k v, v\big).$$

Setting $\mathcal{R}_k = R_k$ in Theorem 5.6 and notice that (5.11) in Theorem 5.6 is satisfied with $\alpha_1 = (2-\theta)$ and $\alpha_2 = 1/(2-\theta)$, we obtain the following result.

COROLLARY 5.1. *Assume that* $R_k = R_k^t$ *and satisfies Condition* (SM.2). *Let* $B_a = \sum_{k=1}^J R_k Q_k$ *and denote* $K_1 = \lambda_{\min}(B_a A)$ *and* $K_2 = \lambda_{\max}(B_a A)$. *Then*

$$A(E_J v, E_J v) \leqslant (1 - 1/C_J) A(v, v)$$

with

$$C_J = \left[\left(1 + \frac{1}{(2-\theta)K_1}\right)^{1/2} + \left(\frac{\lceil \log_2 J \rceil K_2}{2[(2-\theta)K_1]^{1/2}}\right)\right]^2.$$

Corollary 5.1 says that if $R_k = R_k^t$ satisfies Condition (SM.2), then the convergence rate of the V-cycle multigrid method (multiplicative algorithm) can be estimated using only the extreme eigenvalues of the corresponding additive operator $B_a A$. In particular, if the additive preconditioner is optimal and properly scaled (e.g., $R_k \approx C\lambda_k^{-1} I$) then the convergence rate of the multiplicative algorithm deteriorates at most like $\lceil \log_2 J \rceil^2$. This result was proved in GRIEBEL and OSWALD [1995].

To get a result for nonsymmetric smoothers, we set $\mathcal{R}_1 = A_1^{-1}$ and $\mathcal{R}_k = \lambda_k^{-1} I$ for $k \geqslant 2$. Then $B_a = B_\lambda \equiv \sum_{k=1}^J \mathcal{R}_k Q_k$. We note that in many applications B_λ is a well scaled optimal preconditioner for A; cf. Section 9. To apply Theorem 5.6, we assume that the smoothers, R_k, also satisfy the following condition.

CONDITION (SM.4). There exist positive constants ω_4 and C such that

$$(R_k v, R_k v) \leqslant \omega_4 \lambda_k^{-1} (R_k v, v), \qquad \text{for all } v \in M_k,\ k = 2, \ldots, J.$$

COROLLARY 5.2. *Let* $R_1 = A_1^{-1}$. *Assume that the smoothers,* R_k, *satisfy Conditions* (SM.1), (SM.2) *and* (SM.4). *Then*

$$A(E_J v, E_J v) \leqslant (1 - 1/C_J) A(v, v)$$

with

$$C_J = \left[\left(1 + \frac{1}{\omega K_1}\right)^{1/2} + \frac{\omega_4^{1/2} \lceil \log_2 J \rceil K_2}{2((2-\theta)K_1)^{1/2}}\right]^2.$$

Here $K_1 = \lambda_{\min}(B_\lambda A)$, $K_2 = \lambda_{\max}(B_\lambda A)$ *and* $B_\lambda = A_1^{-1} Q_1 + \sum_{k=2}^J \lambda_k^{-1} Q_k$.

PROOF. We verify that (5.11) in Theorem 5.6 holds for $\mathcal{R}_1 = A_1^{-1}$ and $\mathcal{R}_k = \lambda_k^{-1} I$. Condition (SM.1) implies that

$$(\mathcal{R}_k v, v) \equiv \frac{1}{\lambda_k} \|v\|^2 \leqslant \frac{1}{\omega} \big(\overline{R}_k v, v\big), \qquad \text{for all } v \in M_k,\ k = 2, \ldots, J.$$

On the other hand, Conditions (SM.2) and (SM.4) imply that

$$(R_k v, R_k v) \leqslant \omega_4 \lambda_k^{-1} (R_k v, v) \leqslant \frac{\omega_4}{2-\theta} \lambda_k^{-1} \big(\overline{R}_k v, v \big).$$

A simple change of variable shows that

$$(v, v) \leqslant \frac{\omega_4}{2-\theta} \lambda_k^{-1} \big(\overline{R}_k R_k^{-1} v, R_k^{-1} v \big) = \frac{\omega_4}{2-\theta} \lambda_k^{-1} \big(\overline{\overline{R}}_k^{-1} v, v \big).$$

Hence

$$\big(\overline{\overline{R}}_k v, v \big) \leqslant \frac{\omega_4}{2-\theta} \lambda_k^{-1} (v, v) = \frac{\omega_4}{2-\theta} (\mathcal{R}_k v, v).$$

Therefore,

$$\frac{2-\theta}{\omega_4} \big(\overline{\overline{R}}_k v, v \big) \leqslant (\mathcal{R}_k v, v) \leqslant \frac{1}{\omega} \big(\overline{R}_k v, v \big), \quad \text{for all } v \in M_k,\ k = 2, \ldots, J.$$

Hence (5.11) in Theorem 5.6 is satisfied with $\alpha_1 = (2-\theta)/\omega_4$ and $\alpha_2 = 1/\omega$. The corollary follows from Theorem 5.6. □

In many applications we can establish uniform estimates for $K_1 = \lambda_{\min}(B_\lambda A)$ and $K_2 = \lambda_{\max}(B_\lambda A)$. In such cases, the reduction rate for the multigrid methods is almost optimal, provided that smoothers satisfying Conditions (SM.1), (SM.2) and (SM.4) are used. Examples of such smoothers include point/line Jacobi and Gauss–Seidel smoothers; cf. Section 8.

5.5. *Bibliographical notes*

The results in this section is based on the product representation of the error operator for the multigrid algorithm. The first successful treatment of the "product type" algorithms was given in BRAMBLE, PASCIAK, WANG and XU [1991b]. This led to BRAMBLE, PASCIAK, WANG and XU [1991a] in which the first multigrid results were given and with no specific regularity properties required. Theorem 5.5 is contained there. The mesh refinement techniques of BRAMBLE, PASCIAK and XU [1990] in the additive case were shown to hold in the standard multigrid algorithms.

The results of OSWALD [1992] and ZHANG [1992] motivated the papers by BRAMBLE and PASCIAK [1993, 1994] in which uniform estimates were given for the multigrid V-cycle with only one smoothing step. These results as in Theorems 5.1 and 5.2 give an extension of the results of BRAESS and HACKBUSCH [1983] to the case of less than full regularity while requiring only one smoothing step.

Corollary 5.1 was proved in GRIEBEL and OSWALD [1995]. This result gives an estimate for the rate of convergence of the multigrid algorithm in term of the extreme eigenvalues of the corresponding multilevel additive preconditioner with the same smoothers. It requires the smoothers to be symmetric. Corollary 5.2 provides a similar estimate allowing nonsymmetric smoothers to be used in the multigrid algorithm.

6. Smoothing on subspaces; the case of local refinement

This section provides theorems which justify smoothing only on subspaces. These results are important in that they can be used to analyze multigrid algorithms for the solution of discrete systems which arise from a locally refined mesh. For motivation, we first extend the model problem of Section 2 to the case of locally refined meshes. We next identify conditions on the smoother which correspond to smoothing on subspaces. Finally, we extend the results of Section 5 under these new conditions.

6.1. Approximation of the model problem with a locally refined mesh

We first describe the finite element approximation of the model problem of Section 2 utilizing a locally refined mesh. Such mesh refinements are convenient for accurate modeling of problems with various types of singular behavior. As in Section 2, we consider, for simplicity, the case where Ω is a polygonal domain and the finite element space consists of piecewise linear functions, although we allow a very general form of mesh refinement.

Following BRAMBLE, PASCIAK and XU [1990], we start with a coarse quasi-uniform triangulation $\mathcal{T}_1$. The refinement triangulation is defined in terms of a sequence of (open) mesh domains with

$$\Omega_J \subseteq \Omega_{J-1} \subseteq \cdots \subseteq \Omega_1 = \Omega.$$

The only restrictions on the mesh domains $\{\Omega_k\}$ are that the boundary of Ω_k, for $k > 1$, consists of edges of mesh triangles in the triangulation $\mathcal{T}_{k-1}$ and that there is at least one edge of $\mathcal{T}_{k-1}$ contained in Ω_k. These mesh domains control the region of refinement. If $\tau_{k-1}^i \in \mathcal{T}_{k-1}$ is a triangle contained in Ω_k, then it is broken into four smaller triangles (in the triangulation $\mathcal{T}_k$) by the lines connecting the midpoints of the edges. Alternatively, if $\tau_{k-1}^i \in \mathcal{T}_{k-1}$ is in the complement of Ω_k, then it is not subdivided but is directly included in the triangulation $\mathcal{T}_k$.

REMARK 6.1. A simple example of this construction is the case of a unit square with local refinement near the upper right hand corner. We define $\Omega_k = \Omega$ for $k = 1, \ldots, j$ and $\Omega_k = (1 - 2^{j-k}, 1) \times (1 - 2^{j-k}, 1)$ for $k = j + 1, \ldots, J$. This results in a uniform triangulation to the mesh size h_j followed by a geometric refinement approaching the point (1,1) when $k > j$.

As in Section 2, the space M_k is defined to be the set of piecewise linear functions with respect to the triangulation $\mathcal{T}_k$ which are continuous on Ω and vanish on $\partial\Omega$. The continuity condition implies that the finer grid nodes on a coarse-fine boundary are "slave nodes" in the sense that the values of the function are completely determined by the values of the function at the nearby coarse grid points (see Fig. 6.1).

Note that the subspaces are still nested, i.e.,

$$M_1 \subset M_2 \subset \cdots \subset M_J.$$

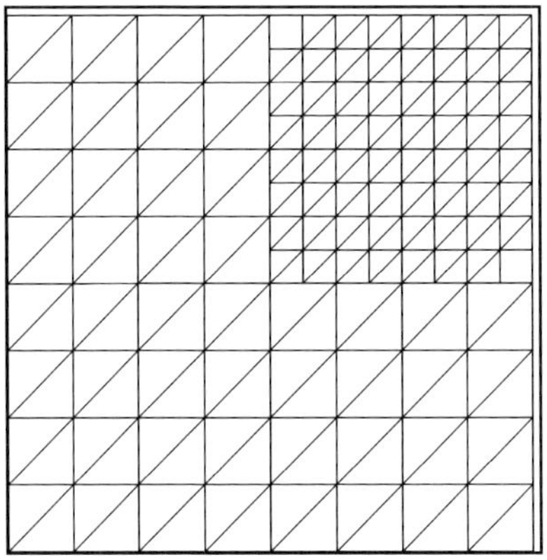

FIG. 6.1. The mesh near a coarse/fine interface.

The cost of multigrid algorithms will be discussed in Section 15. However, one aspect of this cost is the amount of work required to evaluate the action of the smoothing procedure on each level. In the refinement scheme, new nodes are added on level k only in the domain Ω_k. To be efficient, the cost of the smoothing procedure should be proportional to the number of new nodes. This means that the smoother should only have to operate on the subspace $\widehat{M}_k$ generated by the new nodes. In terms of the refinement framework described above,

$$\widehat{M}_k = \{\phi \in M_k \mid \phi = 0 \text{ on } \Omega \setminus \Omega_k\}.$$

REMARK 6.2. Note that λ_k is still defined as being the largest eigenvalue of A_k. Thus λ_k grows inversely proportional to the square of the mesh size of the finest triangles in the composite (coarse/fine) mesh. Obviously, one only has coarser grid approximation in the regions which are not refined. This implies that Conditions (A.1) and (A.2) do not hold with constants independent of the coarse to fine grid size ratios. However, as we shall see in this section, it is possible to generalize the multiplicative analysis of Section 5 to the refinement case.

6.2. *Multigrid algorithm*

As motivated by the previous subsection, smoothing on subspaces involves the introduction of a sequence of subspaces $\{\widehat{M}_k\}$ with $\widehat{M}_k \subseteq M_k$ and such that

$$M_k = M_{k-1} + \widehat{M}_k.$$

Let $\widehat{R}_k : \widehat{M}_k \to \widehat{M}_k$ be a smoother and define $\widehat{R}^t_k : \widehat{M}_k \to \widehat{M}_k$ to be the adjoint of $\widehat{R}_k$ with respect to the inner product $(\cdot,\cdot)$. Let $\widehat{Q}_k : M_J \to \widehat{M}_k$ denote the $(\cdot,\cdot)$ orthogonal projection onto $\widehat{M}_k$. The multigrid algorithm with smoothing on subspaces can be defined as follows.

ALGORITHM 6.1. For $v, g \in M_1$, set $\mathrm{Mg}_1(v, g) = A_1^{-1} g$. Let k be greater than one and v, g be in M_k. Assuming that $\mathrm{Mg}_{k-1}(\bullet, \bullet)$ has been defined, we define $\mathrm{Mg}_k(v, g)$ as follows.

(1) Set $v' = v + \widehat{R}_k^t \widehat{Q}_k(g - A_k v)$.
(2) Set $v'' = v' + q$ where $q = \mathrm{Mg}_{k-1}(0, Q_{k-1}(g - A_k v'))$.
(3) Define $\mathrm{Mg}_k(v, g) \equiv v''' = v'' + \widehat{R}_k \widehat{Q}_k(g - A_k v'')$.

To find the relation between Algorithm 6.1 and Algorithm 3.1, we set $R_k = \widehat{R}_k \widehat{Q}_k$. Then R_k maps M_k into $\widehat{M}_k \subset M_k$. As before, let R_k^t denote the adjoint of the operator R_k on M_k with respect to $(\bullet, \bullet)$, then $R_k^t = \widehat{R}_k^t \widehat{Q}_k$. With this convention, Algorithm 6.1 and Algorithm 3.1 are identical. It is worth noting that Algorithm 6.1 can also be formulated as a subspace correction method with respect to the following space decomposition of M_J:

$$M_J = \widehat{M}_1 + \widehat{M}_2 + \cdots + \widehat{M}_J.$$

Here $\widehat{M}_1 = M_1$. The subspace correction method is defined as follows. Let v be the current approximation to $A_J^{-1} f$, the new approximation u is defined by

(1) Denote $u^{(J+1)} = v$. For $k = J, J-1, \ldots, 2$, set

$$u^{(k)} = u^{(k+1)} + \widehat{R}_k^t \widehat{Q}_k\big(f - A_J u^{(k+1)}\big).$$

(2) Set $u^{(1)} = u^{(2)} + A_1^{-1} Q_1(f - A_J u^{(2)})$.
(3) For $k = 2, \ldots, J$, set

$$u^{(k)} = u^{(k-1)} + \widehat{R}_k \widehat{Q}_k\big(f - A_J u^{(k-1)}\big),$$

and set $u = u^{(J)}$.

6.3. *Conditions on the smoothers*

With $R_k = \widehat{R}_k \widehat{Q}_k$, Algorithm 6.1 and Algorithm 3.1 are identical. We can therefore study Algorithm 3.1 instead. Note that the identities given by Lemma 5.1 and Lemma 5.2 remain valid in this case. However the smoothers $R_k = \widehat{R}_k \widehat{Q}_k$ will not satisfy Condition (SM.1) if $\widehat{M}_k$ is a proper subspace of M_k.

Condition (SM.1) is stated in terms of the symmetrized smoother $\overline{R}_k = R_k + R_k^t - R_k^t A_k R_k$. In this regard, we will need the operator $\overline{\widehat{R}}_k$ defined by

$$\overline{\widehat{R}}_k = \widehat{R}_k + \widehat{R}_k^t - \widehat{R}_k^t \widehat{A}_k \widehat{R}_k.$$

It is natural to assume that the smoothers $\widehat{R}_k$ satisfy Conditions (SM.1) and (SM.2) on the subspaces $\widehat{M}_k$. These conditions are given as follows.

CONDITION $(\widehat{\text{SM.1}})$. Let $\overline{\widehat{R}}_k = \widehat{R}_k + \widehat{R}_k^t - \widehat{R}_k^t \widehat{A}_k \widehat{R}_k$. There exists a positive constant $\omega \in (0, 1]$ not depending on J such that

$$\frac{\omega}{\lambda_k}\|w\|^2 \leqslant \big(\overline{\widehat{R}}_k w, w\big), \quad \text{for all } w \in \widehat{M}_k,\ k = 2, \ldots, J.$$

CONDITION $(\widehat{\text{SM.2}})$. There exists a constant $\theta \in (0, 2)$ not depending on $k = 2, \ldots, J$ such that

$$A\big(\widehat{R}_k v, \widehat{R}_k v\big) \leqslant \theta\big(\widehat{R}_k v, v\big), \quad \text{for all } v \in \widehat{M}_k.$$

This inequality is the same as any one of the following three inequalities:

$$\big(\overline{\widehat{R}}_k v, v\big) \geqslant (2-\theta)\big(\widehat{R}_k v, v\big) \geqslant \left(\frac{2-\theta}{\theta}\right)\big(\big(\widehat{R}_k^t \widehat{A}_k \widehat{R}_k\big)v, v\big), \quad \text{for all } v \in \widehat{M}_k.$$

It is easy to show that

$$\overline{R}_k = \overline{\widehat{R}}_k \widehat{Q}_k.$$

As a consequence if $\widehat{R}_k$ satisfies Condition $(\widehat{\text{SM.2}})$ then R_k satisfies Condition (SM.2), and if $\widehat{R}_k$ satisfies Condition $(\widehat{\text{SM.1}})$ then R_k satisfies Condition (SM.1) when restricted to $\widehat{M}_k$, i.e.,

$$\frac{\omega}{\lambda_k}\|w\|^2 \leqslant \big(\overline{R}_k w, w\big), \quad \text{for all } w \in \widehat{M}_k. \tag{6.1}$$

Clearly, (6.1) is weaker than Condition (SM.1) if $\widehat{M}_k$ is a proper subspace of M_k.

6.4. *V-cycle convergence theory*

Let $\widehat{A}_k$ and $\widehat{P}_k$ be defined analogously to A_k and P_k but with respect to the spaces $\widehat{M}_k$. Specifically, for $k = 1, \ldots, J$, let the operator $\widehat{A}_k : \widehat{M}_k \to \widehat{M}_k$ be defined by

$$\big(\widehat{A}_k w, \phi\big) = A(w, \phi), \quad \text{for all } \phi \in \widehat{M}_k,$$

and the projectors $\widehat{P}_k : M_J \to \widehat{M}_k$ be defined by

$$A\big(\widehat{P}_k w, \phi\big) = A(w, \phi), \quad \text{for all } \phi \in \widehat{M}_k.$$

The error propagation operator of Algorithm 6.1 (or Algorithm 3.1 with $R_k = \widehat{R}_k \widehat{Q}_k$) is given by

$$\mathcal{E}_J v = \mathrm{Mg}_J(v, 0),$$

where $\mathrm{Mg}_J(\bullet, \bullet)$ is defined by Algorithm 6.1 (or Algorithm 3.1). In this subsection, we show, under various conditions that

$$0 \leqslant A(\mathcal{E}_J v, v) \leqslant (1 - 1/C_J) A(v, v), \quad \text{for all } v \in M_J. \tag{6.2}$$

This means that the sequence u^i defined by (3.3) satisfies

$$|||u - u^m||| \leqslant (1 - 1/C_J)^m |||u - u^0|||.$$

Recall that

$$\mathcal{E}_J = E_J E_J^*,$$

where $E_k = (I - T_k) E_{k-1}$, $E_0 = I$ and $T_k = R_k A_k P_k = \widehat{R}_k \widehat{A}_k \widehat{P}_k$. The estimate given by (6.2) is equivalent to

$$A(E_J v, E_J v) \leqslant (1 - 1/C_J) A(v, v), \quad \text{for all } v \in M_J$$

or

$$A(v, v) \leqslant C_J \big[A(v, v) - A(E_J v, E_J v) \big], \quad \text{for all } v \in M_J. \tag{6.3}$$

We first introduce conditions on the subspaces $\{\widehat{M}_k\}$ which lead to convergence estimates for V-cycle algorithms using smoothers which only satisfy Conditions ($\widehat{\text{SM.1}}$) and ($\widehat{\text{SM.2}}$). Although it is possible to generalize the conditions for Theorem 5.1 so that they cover the refinement case, their generalization is somewhat tedious and may only fit the refinement example discussed above.

We start with the generalization of Theorem 5.2. The generalization of Conditions (A.5) and (A.6) to the case of smoothing on subspaces are natural and given below.

CONDITION ($\widehat{\text{A.5}}$). There exist linear operators $\Pi_k : M \to M_k$ (not necessarily projectors) with $\Pi_J = I$ and $(\Pi_k - \Pi_{k-1}) v \in \widehat{M}_k$ for all $v \in M_J$, and a constant $C_a > 0$ not depending on J, such that

$$A(\Pi_1 v, \Pi_1 v) + \sum_{k=2}^{J} \lambda_k \big\| (\Pi_k - \Pi_{k-1}) v \big\|^2 \leqslant C_a A(v, v), \quad \text{for all } v \in M_J.$$

Comparing with Condition (A.5), we see that the only extra requirement in Condition ($\widehat{\text{A.5}}$) is that $\operatorname{range}(\Pi_k - \Pi_{k-1}) \subset \widehat{M}_k$. It is straightforward to show that the following conditions are equivalent.

(a) Condition ($\widehat{\text{A.5}}$) holds.

(b) There is a constant $C_a > 0$ such that

$$A(v, v) \leqslant C_a \left(A\big(\widehat{P}_1 v, v\big) + \sum_{k=2}^{J} \lambda_k^{-1} \big\| \widehat{A}_k \widehat{P}_k v \big\|^2 \right), \quad \text{for all } v \in M_J.$$

(c) There is a constant $C_a > 0$ such that for any $v \in M$, there is a decomposition of v, $v = \sum_{k=1}^{J} v_k$ with $v_k \in \widehat{M}_k$ such that

$$A(v_1, v_1) + \sum_{k=2}^{J} \lambda_k \|v_k\|^2 \leqslant C_a A(v, v).$$

CONDITION $(\widehat{\text{A.6}})$. There is a positive constant C_{cs} and a number $\varepsilon \in (0, 1)$ not depending on J such that

$$A(u, v) \leqslant C_{\text{cs}} \varepsilon^{i-k} |||u||| \left(\lambda_i^{1/2} \|v\|\right), \qquad \text{for all } u \in \widehat{M}_k,\ v \in \widehat{M}_i,\ k \leqslant i \leqslant J,$$

or equivalently

$$\left\| \widehat{A}_i \widehat{P}_i v \right\|^2 \leqslant \lambda_i \left(C_{\text{cs}} \varepsilon^{i-k}\right)^2 A(v, v), \qquad \text{for all } v \in \widehat{M}_k,\ k \leqslant i \leqslant J.$$

The proof of the following identity is similar to that of Lemma 5.2.

LEMMA 6.1. *Let $\pi_k : M \to M_k$ be any sequence of linear operators with $\pi_J = I$. Set $\pi_0 = 0$ and $E_0 = I$. Then*

$$\begin{aligned} A(v, v) &= A(P_1 v, v) + \sum_{k=2}^{J} A\big(E_{k-1} v, (\pi_k - \pi_{k-1}) v\big) \\ &\quad + \sum_{i=2}^{J-1} \sum_{k=i+1}^{J} A\big(T_i E_{i-1} v, (\pi_k - \pi_{k-1}) v\big). \end{aligned}$$

We now prove the following theorem.

THEOREM 6.1. *Assume that Conditions $(\widehat{\text{A.5}})$ and $(\widehat{\text{A.6}})$ hold and that the smoothers $\widehat{R}_k$ satisfy Conditions $(\widehat{\text{SM.1}})$ and $(\widehat{\text{SM.2}})$. Then,*

$$0 \leqslant A(\mathcal{E}_J v, v) \leqslant (1 - 1/C_M) A(v, v), \qquad \textit{for all } v \in M_J,$$

with

$$C_M = \left[\left(1 + \frac{C_a}{\omega}\right)^{1/2} + \frac{C_{\text{cs}} \varepsilon}{(1 - \varepsilon)} \left(\frac{C_a \theta}{2 - \theta}\right)^{1/2} \right]^2.$$

PROOF. Using Lemma 5.2 and the fact that $\widehat{A}_k \widehat{P}_k = \widehat{Q}_k A_k P_k$ we have

$$A(v, v) - A(E_J v, E_J v) = \sum_{k=1}^{J} \|A_k P_k E_{k-1} v\|_{\overline{R}_k}^2 = \sum_{k=1}^{J} \left\| \widehat{A}_k \widehat{P}_k E_{k-1} v \right\|_{\overline{\widehat{R}}_k}^2.$$

Thus the proof of the theorem reduces to showing that

$$A(v,v) \leqslant C_M \sum_{k=1}^{J} \|\widehat{A}_k \widehat{P}_k E_{k-1} v\|^2_{\bar{\bar{R}}_k}.$$

Now let $\{\Pi_k\}$ be a sequence of linear operators satisfying Condition $(\widehat{\text{A.5}})$. Applying Lemma 6.1 with $\pi_k = \Pi_k$ and setting $v_k = (\Pi_k - \Pi_{k-1})v$, we obtain

$$A(v,v) = A(P_1 v, v) + \sum_{k=1}^{J} A(E_{k-1} v, v_k) + \sum_{i=2}^{J-1} \sum_{k=i+1}^{J} A(T_i E_{i-1} v, v_k). \tag{6.4}$$

Note that $v_k = (\Pi_k - \Pi_{k-1})v \in \widehat{M}_k$. By the Cauchy–Schwarz inequality and using Condition $(\widehat{\text{SM.1}})$ we obtain

$$\begin{aligned} A(E_{k-1} v, v_k) &= \left(\widehat{A}_k \widehat{P}_k E_{k-1} v, v_k\right) \leqslant \|\widehat{A}_k \widehat{P}_k E_{k-1} v\|_{\bar{\bar{R}}_k} \|v_k\|_{\bar{\bar{R}}_k^{-1}} \\ &\leqslant \|\widehat{A}_k \widehat{P}_k E_{k-1} v\|_{\bar{\bar{R}}_k} \left(\frac{\lambda_k}{\omega}\right)^{1/2} \|v_k\|. \end{aligned}$$

Recalling that $\widehat{R}_1 = \bar{\bar{R}}_1 = A_1^{-1}$, summing over $2 \leqslant k \leqslant J$ and using the Cauchy–Schwarz inequality and Condition $(\widehat{\text{A.5}})$, we obtain

$$\begin{aligned} &A(P_1 v, v) + \sum_{k=2}^{J} A(E_{k-1} v, v_k) \\ &\quad \leqslant \left(\sum_{k=1}^{J} \|\widehat{A}_k \widehat{P}_k E_{k-1} v\|^2_{\bar{\bar{R}}_k}\right)^{1/2} \left(A(P_1 v, v) + \sum_{k=2}^{J} \left(\frac{\lambda_k}{\omega}\right) \|v_k\|^2\right)^{1/2} \\ &\quad \leqslant \left[\sum_{k=1}^{J} \|\widehat{A}_k \widehat{P}_k E_{k-1} v\|^2_{\bar{\bar{R}}_k}\right]^{1/2} \left[\left(1 + \frac{C_a}{\omega}\right) A(v,v)\right]^{1/2}. \end{aligned} \tag{6.5}$$

Note that $v_k = (\Pi_k - \Pi_{k-1})v \in \widehat{M}_k$ and $T_i E_{i-1} v \in \widehat{M}_i$. By Condition $(\widehat{\text{A.6}})$

$$A(T_i E_{i-1} v, v_k) \leqslant C_{\text{cs}} \varepsilon^{k-i} |||T_i E_{i-1} v||| \left(\lambda_k^{1/2} \|v_k\|\right), \quad \text{for } i \leqslant k. \tag{6.6}$$

By Condition $(\widehat{\text{SM.2}})$

$$|||T_i E_{i-1} v|||^2 = \left|\left|\left|\widehat{R}_i \widehat{A}_i \widehat{P}_i E_{i-1} v\right|\right|\right|^2 \leqslant \frac{\theta}{2-\theta} \|\widehat{A}_i \widehat{P}_i E_{i-1} v\|^2_{\bar{\bar{R}}_i}. \tag{6.7}$$

Summing (6.6) over $i+1 \leqslant k \leqslant J$ and $2 \leqslant i \leqslant J-1$, and using (6.7), we obtain

$$\begin{aligned}
&\sum_{i=2}^{J-1} \sum_{k=i+1}^{J} A(T_i E_{i-1} v, v_k) \\
&\quad \leqslant \left(\frac{\theta}{2-\theta}\right)^{1/2} \sum_{k=2}^{J} \sum_{i=1}^{k-1} C_{\mathrm{cs}} \varepsilon^{k-i} \|\widehat{A}_i \widehat{P}_i E_{i-1} v\|_{\bar{\widehat{R}}_i} \left(\lambda_k^{1/2} \|v_k\|\right) \\
&\quad \leqslant \frac{C_{\mathrm{cs}} \varepsilon}{1-\varepsilon} \left(\frac{\theta}{2-\theta}\right)^{1/2} \left(\sum_{i=2}^{J-1} \|\widehat{A}_i \widehat{P}_i E_{i-1} v\|_{\bar{\widehat{R}}_i}^2\right)^{1/2} \left(\sum_{k=2}^{J} \lambda_k \|v_k\|^2\right)^{1/2} \\
&\quad \leqslant \frac{C_{\mathrm{cs}} \varepsilon}{1-\varepsilon} \left(\frac{\theta}{2-\theta}\right)^{1/2} \left(\sum_{i=2}^{J-1} \|\widehat{A}_i \widehat{P}_i E_{i-1} v\|_{\bar{\widehat{R}}_i}^2\right)^{1/2} \left(C_a A(v, v)\right)^{1/2}. \qquad (6.8)
\end{aligned}$$

Combining Eqs. (6.4), (6.5) and (6.8) it follows that

$$A(v, v) \leqslant C_M \left(\sum_{k=1}^{J} \|\widehat{A}_k \widehat{P}_k E_{k-1} v\|_{\bar{\widehat{R}}_k}^2\right)^{1/2} = C_M \left[A(v, v) - A(E_J v, E_J v)\right].$$

This proves the theorem. $\square$

We next consider the generalization of Theorem 5.5. To do this we start with the generalization of Condition (A.7) given below.

CONDITION $(\widehat{\mathrm{A.7}})$. There exists a sequence of linear operators $\Pi_k : M_J \to M_k$ with $\Pi_J = I$ and $(\Pi_k - \Pi_{k-1})v \in \widehat{M}_k$ for all $v \in M_J$ such that

$$\left\|(\Pi_k - \Pi_{k-1})v\right\|^2 \leqslant C_\pi \lambda_k^{-1} A(v, v), \quad \text{for all } v \in M_J,\ 2 \leqslant k \leqslant J,$$

and

$$A(\Pi_k v, \Pi_k v) \leqslant C'_\pi A(v, v), \quad \text{for all } v \in M_J,\ 1 \leqslant k \leqslant J.$$

The generalization of Theorem 5.5 follows.

THEOREM 6.2. *Assume that Condition* $(\widehat{\mathrm{A.7}})$ *holds and that the smoothers,* $\widehat{R}_k$, *satisfy Conditions* $(\widehat{\mathrm{SM.1}})$ *and* $(\widehat{\mathrm{SM.2}})$. *Then,*

$$0 \leqslant A(\mathcal{E}_J v, v) \leqslant \left(1 - \frac{1}{C_J}\right) A(v, v), \quad \textit{for all } v \in M_J,$$

with

$$C_J = \left[\left(1 + (J-1)\frac{C_\pi}{\omega}\right)^{1/2} + \left(2(J-1)\frac{(1+C'_\pi)\theta}{2-\theta}\right)^{1/2}\right]^2.$$

PROOF. As in the proof of Theorem 6.1, the proof of the theorem reduces to showing that

$$A(v, v) \leqslant C_J \sum_{k=1}^{J} \|\widehat{A}_k \widehat{P}_k E_{k-1} v\|_{\widehat{\bar{R}}_k}^2.$$

Let Π_k, $k = 1, \ldots, J$, be a sequence of linear operators satisfying Condition $(\widehat{\text{A.7}})$. Using Lemma 5.2 with $\pi_k = \Pi_k$, we obtain

$$\begin{aligned} A(v, v) &= A(P_1 v, v) + \sum_{k=2}^{J} A\big(E_{k-1} v, (\Pi_k - \Pi_{k-1}) v\big) \\ &\quad + \sum_{k=2}^{J-1} A\big(T_k E_{k-1} v, (P_k - \Pi_k) v\big). \end{aligned} \tag{6.9}$$

By the Cauchy–Schwarz inequality

$$A\big(E_{k-1} v, (\Pi_k - \Pi_{k-1}) v\big) \leqslant \big\|\widehat{A}_k \widehat{P}_k E_{k-1} v\big\|_{\widehat{\bar{R}}_k} \big\|(\Pi_k - \Pi_{k-1}) v\big\|_{\widehat{\bar{R}}_k^{-1}}. \tag{6.10}$$

It follows from Conditions $(\widehat{\text{SM.1}})$ and $(\widehat{\text{A.7}})$ that

$$\big\|(\Pi_k - \Pi_{k-1}) v\big\|_{\widehat{\bar{R}}_k^{-1}}^2 \leqslant \omega^{-1} \lambda_k \big\|(\Pi_k - \Pi_{k-1}) v\big\|^2 \leqslant \frac{C_\pi}{\omega} A(v, v).$$

Now summing (6.10) and using the Cauchy–Schwarz inequality, we obtain

$$\begin{aligned} &A(P_1 v, v) + \sum_{k=2}^{J} A\big(E_{k-1} v, (\Pi_k - \Pi_{k-1}) v\big) \\ &\quad \leqslant \left[\sum_{k=2}^{J} \|\widehat{A}_k \widehat{P}_k E_{k-1} v\|_{\widehat{\bar{R}}_k}^2\right]^{1/2} \left[\left(1 + (J-1)\frac{C_\pi}{\omega}\right) A(v, v)\right]^{1/2}. \end{aligned} \tag{6.11}$$

For the second sum in (6.9), we have, by the Cauchy–Schwarz inequality

$$A\big(T_k E_{k-1} v, (P_k - \Pi_k) v\big) \leqslant |||T_k E_{k-1} v||| \, \big|\big|\big|(P_k - \Pi_k) v\big|\big|\big|. \tag{6.12}$$

Condition $(\widehat{\text{SM.2}})$ implies that

$$|||T_k E_{k-1} v|||^2 \leqslant \frac{\theta}{2 - \theta} \big\|\widehat{A}_k \widehat{P}_k E_{k-1} v\big\|_{\widehat{\bar{R}}}^2.$$

By the triangle inequality and Condition $(\widehat{\text{A.7}})$, we have

$$\big|\big|\big|(P_k - \Pi_k) v\big|\big|\big|^2 \leqslant 2\big(|||P_k v|||^2 + |||\Pi_k v|||^2\big) \leqslant 2\big(1 + C'_\pi\big) A(v, v).$$

Summing (6.12) we obtain

$$
\begin{aligned}
&\sum_{k=2}^{J-1} A\big(T_k E_{k-1} v, (P_k - \Pi_k) v\big) \\
&\quad \leqslant \left(\sum_{k=2}^{J-1} |||T_k E_{k-1} v|||^2\right)^{1/2} \left(\sum_{k=2}^{J-1} \big|\big|\big|(P_k - \Pi_k) v\big|\big|\big|^2\right)^{1/2} \\
&\quad \leqslant \left(\frac{\theta}{2-\theta} \sum_{k=2}^{J-1} \|\widehat{A}_k \widehat{P}_k E_{k-1} v\|_{\widehat{\bar{R}}}^2\right)^{1/2} \big(2(J-1)\big(1 + C'_\pi\big) A(v, v)\big)^{1/2}. \qquad (6.13)
\end{aligned}
$$

Combining (6.11) and (6.13) gives

$$
A(v, v) \leqslant C_J \sum_{k=1}^{J} \|\widehat{A}_k \widehat{P}_k E_{k-1} v\|_{\widehat{\bar{R}}_k}^2 = C_J\big[A(v, v) - A(E_J v, E_J) v\big].
$$

The theorem follows. □

6.5. Bibliographical notes

The technique for analyzing multilevel methods with very general mesh refinements was introduced in BRAMBLE, PASCIAK and XU [1990]. The multilevel additive preconditioner was discussed in that paper. Multigrid algorithms with locally refined meshes was considered in BRAMBLE and PASCIAK [1993]. The presentation here is based on BRAMBLE and PASCIAK [1993].

7. Multigrid algorithms with nonnested spaces and varying forms

In this section we depart from the framework of the previous sections where we considered only nested sequences of finite-dimensional spaces. More generally, assume that we are given a sequence of finite-dimensional spaces $M_1, M_2, \ldots, M_J = M$. Each space M_k is equipped with an inner product $(\bullet, \bullet)_k$ and denote by $\| \bullet \|_k$ the induced norm. In addition, we assume that we are given symmetric and positive definite forms $A_k(\bullet, \bullet)$ defined on $M_k \times M_k$ and set $A(\bullet, \bullet) = A_J(\bullet, \bullet)$. Each form $A_k(\bullet, \bullet)$ induces an operator $A_k : M_k \to M_k$ defined by

$$
(A_k w, \varphi)_k = A_k(w, \varphi), \qquad \text{for all } \varphi \in M_k.
$$

Denote by λ_k the largest eigenvalue of A_k.

These spaces are connected through $J-1$ linear operators $I_k : M_{k-1} \to M_k$, for $k = 2, \ldots, J$. These operators are often referred to as prolongation operators. The operators $Q_{k-1} : M_k \to M_{k-1}$ and $P_{k-1} : M_k \to M_{k-1}$ are defined by

$$
(Q_{k-1} w, \varphi)_{k-1} = (w, I_k \varphi)_k, \qquad \text{for all } \varphi \in M_{k-1},
$$

and

$$A_{k-1}(P_{k-1}w, \varphi) = A_k(w, I_k\varphi), \quad \text{for all } \varphi \in M_{k-1}.$$

Finally, we assume that smoothers $R_k : M_k \to M_k$ are given and set

$$R_k^{(\ell)} = \begin{cases} R_k & \text{if } \ell \text{ is odd,} \\ R_k^t & \text{if } \ell \text{ is even.} \end{cases}$$

In the case discussed in Sections 3, 4 and 5, $M_k \subset M_{k+1}$, $(\bullet,\bullet)_k = (\bullet,\bullet)$, $A_k(\bullet,\bullet) = A(\bullet,\bullet)$, $I_k = I$, and Q_k and P_k are projectors with respect to $(\bullet,\bullet)$ and $A(\bullet,\bullet)$ respectively. Here Q_k and P_k are not necessarily projectors. Note that the relationship $A_{k-1}P_{k-1} = Q_{k-1}A_k$ still holds.

7.1. *General multigrid algorithms*

Given $f \in M_J$, we are interested in solving

$$A_J u = f.$$

With u^0 given, we will consider the iterative algorithm

$$u^i = \mathrm{Mg}_J\left(u^{i-1}, f\right) \equiv u^{i-1} + B_J\left(f - A_J u^{i-1}\right),$$

where $B_J : M \to M$ is defined recursively by the following general multigrid procedure.

ALGORITHM 7.1. Let p be a positive integer and let m_k be a positive integer depending on k. Set $B_1 = A_1^{-1}$. Assuming that $B_{k-1} : M_{k-1} \to M_{k-1}$ has been defined, we define $B_k : M_k \to M_k$ as follows. Let $g \in M_k$.

(1) Pre-smoothing: Set $x^0 = 0$ and define x^ℓ, $\ell = 1, \dots, m_k$, by

$$x^\ell = x^{\ell-1} + R_k^{(\ell+m_k)}\left(g - A_k x^{\ell-1}\right).$$

(2) Correction: $y^{m_k} = x^{m_k} + I_k q^p$, where $q^0 = 0$ and q^i for $i = 1, \dots, p$ is defined by

$$q^i = q^{i-1} + B_{k-1}\left[Q_{k-1}\left(g - A_k x^{m_k}\right) - A_{k-1} q^{i-1}\right].$$

(3) Post-smoothing: Define y^ℓ for $\ell = m_k + 1, \dots, 2m_k$ by

$$y^\ell = y^{\ell-1} + R_k^{(\ell+m_k)}\left(g - A_k y^{\ell-1}\right).$$

(4) $B_k g = y^{2m_k}$.

The cases $p = 1$ and $p = 2$ correspond to the V-cycle and the W-cycle multigrid algorithms respectively. The case $p = 1$ with the number of smoothings, m_k, varying is known as the variable V-cycle multigrid algorithm. In addition to the V-cycle and the W-cycle multigrid algorithms with the number of smoothings the same on each level, we will consider a variable V-cycle multigrid algorithm in which we will assume that the number of smoothings increases geometrically as k decreases. More precisely, in such a case we assume that there exist two constants β_0 and β_1 with $1 < \beta_0 \leqslant \beta_1$ such that

$$\beta_0 m_k \leqslant m_{k-1} \leqslant \beta_1 m_k.$$

Note that the case $\beta_0 = \beta_1 = 2$ corresponds to doubling the number of smoothings as we proceed from M_k to M_{k-1}, i.e., $m_k = 2^{J-k} m_J$.

Our aim is to study the error reduction operator $\mathcal{E}_k = I - B_k A_k$ and provide conditions under which, we can estimate δ_k between zero and one such that

$$\left|A_k(\mathcal{E}_k u, u)\right| \leqslant \delta_k A_k(u, u), \qquad \text{for all } u \in M_k.$$

We also show that the operator B_k corresponding to the variable V-cycle multigrid algorithm provides a good preconditioner for A_k even in the cases where $\mathcal{E}_k$ is not a reducer.

We first derive a recurrence relation. Let $K_k = I - R_k A_k$, $K_k^* = I - R_k^t A_k$ and set

$$K_k^{(m)} = \begin{cases} (K_k^* K_k)^{m/2} & \text{if } m \text{ is even,} \\ (K_k^* K_k)^{(m-1)/2} K_k^* & \text{if } m \text{ is odd.} \end{cases}$$

By the definition of B_k the following two-level recurrence relation is easily derived:

$$\mathcal{E}_k = \left(K_k^{(m_k)}\right)^* \left[I - I_k P_{k-1} + I_k \mathcal{E}_{k-1}^p P_{k-1}\right] K_k^{(m_k)}. \tag{7.1}$$

Our main assumption relating the spaces M_k and M_{k-1} is the following regularity-approximation property.

CONDITION (A.2″). For some α with $0 < \alpha \leqslant 1$ there exists C_P independent of k such that

$$\left|A_k\big((I - I_k P_{k-1})v, v\big)\right| \leqslant C_\mathrm{P}^{2\alpha} \left(\frac{\|A_k v\|_k^2}{\lambda_k}\right)^\alpha \left[A_k(v, v)\right]^{1-\alpha}.$$

The so-called *variational assumption* refers to the case in which all of the forms $A_k(\bullet, \bullet)$ are defined in terms of the form $A_J(\bullet, \bullet)$ by

$$A_{k-1}(v, v) = A_k(I_k v, I_k v), \qquad \text{for all } v \in M_{k-1}. \tag{7.2}$$

We call this case the *inherited* case. Given the operators I_k, then all of forms come from $A(\bullet, \bullet)$. If the spaces are nested, then I_k can be chosen to be the natural injection operator and the forms can be defined by

$$A_k(v, v) = A(v, v), \quad \text{for all } v \in M_k.$$

We call this case the *nested-inherited* case. It follows from Lemma 3.2 that in the nested-inherited case, Condition (A.2) implies Condition (A.2$''$).

In general, (7.2) does not hold in the nonnested case. We will consider, at first, a weaker condition than (7.2).

CONDITION (I.1). For $k = 2, \dots, J$, the operators I_k satisfy

$$A_k(I_k v, I_k v) \leqslant A_{k-1}(v, v), \quad \text{for all } v \in M_{k-1}.$$

It follows from the definition of P_{k-1} that Condition (I.1) holds if and only if

$$A_{k-1}(P_{k-1} v, P_{k-1} v) \leqslant A_k(v, v), \quad \text{for all } v \in M_k$$

and if and only if

$$A_k\big((I - I_k P_{k-1})v, v\big) \geqslant 0, \quad \text{for all } v \in M_k.$$

LEMMA 7.1. *If Condition* (I.1) *holds then the error propagator $\mathcal{E}_k$ is symmetric and positive semidefinite with respect to $A_k(\bullet, \bullet)$, i.e.,*

$$A_k(\mathcal{E}_k v, v) = A_k\big((I - B_k A_k)v, v\big) \geqslant 0, \quad \textit{for all } v \in M_k.$$

PROOF. Since $A_k(I_k P_{k-1} w, v) = A_{k-1}(P_{k-1} w, P_{k-1} v)$, the operator $I_k P_{k-1}$ is symmetric with respect to $A_k(\bullet, \bullet)$. Hence by induction, since $\mathcal{E}_1 = I - B_1 A_1 = 0$, it follows, using (7.1), that $I - B_k A_k$ is symmetric and positive semidefinite. □

The assumption on the smoother is the same as in Section 3; we restate it here.

CONDITION (SM.1). There exists $\omega > 0$ not depending on J such that

$$\left(\frac{\omega}{\lambda_k}\right) \|v\|_k^2 \leqslant \big(\overline{R}_k v, v\big)_k, \quad \text{for all } v \in M_k,\ k = 2, \dots, J.$$

This is a condition local to M_k and hence does not depend on whether or not the spaces M_k are nested.

The following lemma is a generalization of Lemma 3.1.

LEMMA 7.2. *If Condition* (SM.1) *holds, then*

$$\frac{\|A_k K_k^{(m)} v\|_k^2}{\lambda_k} \leqslant \frac{\omega^{-1}}{m} \big[A_k(v, v) - A_k\big(K_k^{(m)} v, K_k^{(m)} v\big)\big].$$

PROOF. Let $\tilde{v} = K_k^{(m)} v$. By Condition (SM.1),

$$\frac{\|A_k \tilde{v}\|_k^2}{\lambda_k} \leqslant \omega^{-1} A_k\big(\overline{R}_k A_k \tilde{v}, \tilde{v}\big) = \omega^{-1} A_k\big((I - K_k^* K_k)\tilde{v}, \tilde{v}\big).$$

Suppose m is even. Then $\tilde{v} = (K_k^* K_k)^{m/2} v$. Set $\overline{K}_k = K_k^* K_k$. Then

$$\begin{aligned}\frac{\|A_k \tilde{v}\|_k^2}{\lambda_k} &\leqslant \omega^{-1} A_k\big((I - \overline{K}_k)\overline{K}_k^m v, v\big) \leqslant \frac{\omega^{-1}}{m} A_k\big((I - \overline{K}_k^m) v, v\big) \\ &= \frac{\omega^{-1}}{m}\big[A_k(v, v) - A_k(\tilde{v}, \tilde{v})\big].\end{aligned}$$

For m odd, we set $\overline{K}_k = K_k K_k^*$ and the result follows. □

7.2. Multigrid V-cycle as a reducer

We will provide an estimate for the convergence rate of the V-cycle and the variable V-cycle multigrid algorithm under Conditions (A.2″), (I.1) and (SM.1).

It follows from Lemma 7.1 that if Condition (I.1) holds, then $A_k(\mathcal{E}_k u, u) \geqslant 0$. Hence our aim in such a case is to estimate δ_k between zero and one, where δ_k is such that

$$0 \leqslant A_k\big((I - B_k A_k)u, u\big) \leqslant \delta_k A_k(u, u), \quad \text{for all } u \in M_k.$$

THEOREM 7.1 (V-cycle). *Assume that Conditions* (A.2″) *and* (I.1) *hold and that the smoothers satisfy Condition* (SM.1). *Let* $p = 1$. *Then*

$$0 \leqslant A_k(\mathcal{E}_k v, v) \leqslant \delta_k A_k(v, v), \quad \textit{for all } v \in M_k \tag{7.3}$$

with $0 \leqslant \delta_k < 1$ *given in each case as follows*:

(a) *If* $p = 1$ *and* $m_k = m$ *for all* k, *then* (7.3) *holds with*

$$\delta_k = \frac{M k^{(1-\alpha)/\alpha}}{M k^{(1-\alpha)/\alpha} + m^\alpha} \quad \textit{with } M = \mathrm{e}^{1-\alpha}\big(\alpha C_{\mathrm{P}}^2/\omega + 1\big) - 1.$$

(b) *If* $p = 1$ *and* m_k *increases as* k *decreases, then* (7.3) *holds with*

$$\delta_k = 1 - \left(\frac{1}{1 + \frac{\alpha C_{\mathrm{P}}^2}{\omega m_k^\alpha}}\right) \prod_{i=2}^{k} \left(1 - \frac{1-\alpha}{m_i^\alpha}\right).$$

In particular if there exist two constants β_0 *and* β_1 *with* $1 < \beta_0 \leqslant \beta_1$ *such that* $\beta_0 m_k \leqslant m_{k-1} \leqslant \beta_1 m_k$, *then* $\delta_k \leqslant M/(M + m_k^\alpha)$ *for* M *sufficiently large.*

PROOF. By Lemma 7.1, the lower estimate in (7.3) follows from Condition (I.1). For the upper estimate, we claim that Conditions (A.2″) and (SM.1) imply that

$$A_k(\mathcal{E}_k v, v) \leqslant \delta_k(\tau) A_k(v, v), \quad \text{for all } v \in M_k \tag{7.4}$$

for $\tau > 0$ if $0 < \alpha < 1$ and for $\tau \geqslant \max_i(1/m_i^\alpha)$ if $\alpha = 1$. Here $\delta_k(\tau)$ is defined by

$$1 - \delta_k(\tau) = \frac{1}{1 + (\alpha C_{\mathrm{P}}^2/\omega)\tau} \prod_{i=2}^{k} \left(1 - \frac{1-\alpha}{(\tau m_i)^{\alpha/(1-\alpha)}}\right). \tag{7.5}$$

From this the theorem will follow by choosing an appropriate τ.

We shall prove (7.4) by induction. Since $\mathcal{E}_1 = I - B_1 A_1 = 0$, (7.4) holds for $k = 1$. Assume that

$$A_{k-1}(\mathcal{E}_{k-1} v, v) \leqslant \delta_{k-1}(\tau) A_{k-1}(v, v), \quad \text{for all } v \in M_{k-1}.$$

Let $\tilde{v} = K_k^{(m_k)} v$. Using the two-level recurrence relation (7.1), we have

$$\begin{aligned}
A_k(\mathcal{E}_k v, v) &\equiv A_k\big((I - B_k A_k)v, v\big)\\
&= A_k\big((I - I_k P_{k-1})\tilde{v}, \tilde{v}\big) + A_{k-1}\big(\mathcal{E}_{k-1} P_{k-1}\tilde{v}, P_{k-1}\tilde{v}\big)\\
&\leqslant A_k\big((I - I_k P_{k-1})\tilde{v}, \tilde{v}\big) + \delta_{k-1}(\tau) A_{k-1}\big(P_{k-1}\tilde{v}, P_{k-1}\tilde{v}\big)\\
&= A_k\big((I - I_k P_{k-1})\tilde{v}, \tilde{v}\big) + \delta_{k-1}(\tau) A_k\big(I_k P_{k-1}\tilde{v}, \tilde{v}\big)\\
&= \big[1 - \delta_{k-1}(\tau)\big] A_k\big((I - I_k P_{k-1})\tilde{v}, \tilde{v}\big) + \delta_{k-1}(\tau) A_k(\tilde{v}, \tilde{v}).
\end{aligned}$$

By Conditions (A.2″) and (SM.1) and using Lemma 7.2,

$$\begin{aligned}
A_k\big((I - I_k P_{k-1})\tilde{v}, \tilde{v}\big) &\leqslant C_{\mathrm{P}}^{2\alpha} \left(\frac{\|A_k \tilde{v}\|_k^2}{\lambda_k}\right)^{\alpha} \big[A_k(\tilde{v}, \tilde{v})\big]^{1-\alpha}\\
&\leqslant \frac{C_{\mathrm{P}}^{2\alpha}}{(\omega m_k)^\alpha} \big[A_k(v, v) - A_k(\tilde{v}, \tilde{v})\big]^{\alpha} \big[A_k(\tilde{v}, \tilde{v})\big]^{1-\alpha}.
\end{aligned}$$

We now put things together to get

$$\begin{aligned}
&A_k(\mathcal{E}_k v, v)\\
&\quad\leqslant \big[1 - \delta_{k-1}(\tau)\big] \frac{C_{\mathrm{P}}^{2\alpha}}{(\omega m_k)^\alpha} \big(|||v|||^2 - |||\tilde{v}|||^2\big)^{\alpha} \big(|||\tilde{v}|||^2\big)^{1-\alpha} + \delta_{k-1}(\tau) |||\tilde{v}|||^2.
\end{aligned}$$

Let $x = |||\tilde{v}|||^2 / |||v|||^2$. Then

$$A_k(\mathcal{E}_k v, v) \leqslant f\big(\delta_{k-1}(\tau); x, m_k\big) A_k(v, v),$$

where f is defined by

$$f(\delta; x, m) = (1-\delta)\frac{C_{\mathrm{P}}^{2\alpha}}{(\omega m)^{\alpha}}(1-x)^{\alpha}x^{1-\alpha} + \delta x, \quad 0 \leqslant x < 1. \tag{7.6}$$

By the Hölder inequality, we have

$$\frac{C_{\mathrm{P}}^{2\alpha}}{(\omega m)^{\alpha}}(1-x)^{\alpha}x^{1-\alpha} \leqslant \big(\alpha C_{\mathrm{P}}^{2}/\omega\big)\tau(1-x) + (1-\alpha)(\tau m)^{-\alpha/(1-\alpha)}x,$$

for all $\tau > 0$ if $0 < \alpha < 1$ and for all $\tau \geqslant 1/m^{\alpha}$ if $\alpha = 1$. Hence

$$f(\delta; x, m) \leqslant \ell(\delta; x) \equiv \delta x + (1-\delta)\big[\big(\alpha C_{\mathrm{P}}^{2}/\omega\big)\tau(1-x) + (1-\alpha)(\tau m)^{-\alpha/(1-\alpha)}x\big].$$

Note that $\ell(\delta; x)$ is linear in x and therefore

$$f(\delta; x, m) \leqslant \ell(\delta; x) \leqslant \max\big(\ell(\delta; 0), \ell(\delta; 1)\big).$$

If $\frac{(\alpha C_{\mathrm{P}}^{2}/\omega)\tau}{1+(\alpha C_{\mathrm{P}}^{2}/\omega)\tau} \leqslant \delta < 1$, then

$$\ell(\delta; 0) = (1-\delta)\big(\alpha C_{\mathrm{P}}^{2}/\omega\big)\tau \leqslant \delta \leqslant \delta + (1-\delta)\big[(1-\alpha)(\tau m)^{-\alpha/(1-\alpha)}\big] = \ell(\delta; 1),$$

and hence

$$f(\delta; x, m) \leqslant \ell(\delta; 1) = \delta + (1-\delta)\big[(1-\alpha)(\tau m)^{-\alpha/(1-\alpha)}\big].$$

In particular, since $\frac{(\alpha C_{\mathrm{P}}^{2}/\omega)\tau}{1+(\alpha C_{\mathrm{P}}^{2}/\omega)\tau} = \delta_1(\tau) \leqslant \delta_{k-1}(\tau) < 1$,

$$f\big(\delta_{k-1}(\tau); x, m_k\big) \leqslant \delta_{k-1}(\tau) + \big[1-\delta_{k-1}(\tau)\big]\big[(1-\alpha)(\tau m_k)^{-\alpha/(1-\alpha)}\big] \equiv \delta_k(\tau).$$

As a consequence,

$$A_k(\mathcal{E}_k v, v) \leqslant f\big(\delta_{k-1}(\tau); x, m_k\big)A_k(v, v) \leqslant \delta_k(\tau)A_k(v, v)$$

holds for any $\tau > 0$ if $0 < \alpha < 1$ and for any $\tau \geqslant \max_i(1/m_i)$ if $\alpha = 1$. This proves (7.4).

If $m_k \equiv m$, then setting $\tau = k^{(1-\alpha)/\alpha}/m^{\alpha}$ in (7.5) shows that

$$\delta_k(\tau) \leqslant \frac{Mk^{(1-\alpha)/\alpha}}{Mk^{(1-\alpha)/\alpha} + m^{\alpha}}$$

with $M = \mathrm{e}^{1-\alpha}(\alpha C_{\mathrm{P}}^{2}/\omega + 1) - 1 \leqslant C_{\mathrm{P}}^{2}/\omega + (\mathrm{e}^{1-\alpha} - 1)$. This proves part (a) of the theorem.

Part (b) of the theorem follows by setting $\tau = m_k^{-\alpha}$ in (7.5) and noting that the product term is uniformly bounded from below if m_k increases geometrically as k decreases. □

REMARK 7.1. Notice that, for $p = 1$ and $m_k = m$, $\delta_k \to 1$ as $k \to \infty$. This is in contrast to the results in the nested-inherited case under similar hypotheses. The deterioration, however is only like a power of k and hence may not be too serious.

REMARK 7.2. For the variable V-cycle methods, if m_k increases geometrically as k decreases, then clearly the product term is strictly larger than 0 and thus δ_k is strictly less than 1, independently of k. So, e.g., if $m_J = 1$ we have

$$0 \leqslant A_J(\mathcal{E}_J v, v) \leqslant \frac{M}{M+1} A_J(v, v), \qquad \text{for all } v \in M_J.$$

If $m_k = 2^{J-k} m_J$, then the cost per iteration of the variable V-cycle multigrid algorithm is comparable to that of the W-cycle multigrid algorithm.

7.3. Multigrid W-cycle as a reducer

We now provide an estimate for the rate of convergence of the W-cycle multigrid algorithm.

THEOREM 7.2 (W-cycle). *Assume that Conditions* (A.2″) *and* (I.1) *hold and that the smoothers satisfy Condition* (SM.1). *Let $p = 2$ and $m_k = m$ for all k. Then*

$$0 \leqslant A_k(\mathcal{E}_k v, v) \leqslant \delta A_k(v, v), \quad \textit{for all } v \in M_k$$

and

$$\delta = \frac{M}{M + m^\alpha} \quad \textit{with } M \textit{ sufficiently large but independent of } k.$$

PROOF. By Lemma 7.1, the lower estimate follows from Condition (I.1). We obtain the upper estimate inductively. Since $\mathcal{E}_1 = 0$, the estimate holds for $k = 1$. Assume that

$$A_{k-1}(\mathcal{E}_{k-1} v, v) \leqslant \delta A_{k-1}(v, v), \qquad \text{for all } v \in M_{k-1}.$$

Then since $A_{k-1}(\mathcal{E}_{k-1} v, v) \geqslant 0$, we have

$$A_{k-1}\big(\mathcal{E}_{k-1}^2 v, v\big) \leqslant \delta^2 A_{k-1}(v, v), \qquad \text{for all } v \in M_{k-1}.$$

Let $\tilde{v} = K_k^{(m)} v$. Using the two-level recurrence (7.1), we obtain for $p = 2$,

$$A_k(\mathcal{E}_k v, v) \equiv A_k\big((I - B_k A_k) v, v\big)$$

$$
\begin{aligned}
&= A_k\big((I - I_k P_{k-1})\tilde{v}, \tilde{v}\big) + A_{k-1}\big(\mathcal{E}_{k-1}^2 P_{k-1}\tilde{v}, P_{k-1}\tilde{v}\big) \\
&\leqslant A_k\big((I - I_k P_{k-1})\tilde{v}, \tilde{v}\big) + \delta^2 A_{k-1}\big(P_{k-1}\tilde{v}, P_{k-1}\tilde{v}\big) \\
&= A_k\big((I - I_k P_{k-1})\tilde{v}, \tilde{v}\big) + \delta^2 A_k\big(I_k P_{k-1}\tilde{v}, \tilde{v}\big) \\
&= \big(1 - \delta^2\big) A_k\big((I - I_k P_{k-1})\tilde{v}, \tilde{v}\big) + \delta^2 A_k(\tilde{v}, \tilde{v}).
\end{aligned}
$$

By Conditions (A.2$''$) and (SM.1), and using Lemma 7.2,

$$
\begin{aligned}
A_k\big((I - I_k P_{k-1})\tilde{v}, \tilde{v}\big) &\leqslant C_{\mathrm{P}}^{2\alpha}\left(\frac{\|A_k\tilde{v}\|_k^2}{\lambda_k}\right)^{\alpha}\big[A_k(\tilde{v}, \tilde{v})\big]^{1-\alpha} \\
&\leqslant \frac{C_{\mathrm{P}}^{2\alpha}}{(\omega m)^{\alpha}}\big[A_k(v, v) - A_k(\tilde{v}, \tilde{v})\big]^{\alpha}\big[A_k(\tilde{v}, \tilde{v})\big]^{1-\alpha}.
\end{aligned}
$$

We now put things together to get

$$
A_k(\mathcal{E}_k v, v) \leqslant (1 - \delta^2)\frac{C_{\mathrm{P}}^{2\alpha}}{(\omega m)^{\alpha}}\big(|||v|||^2 - |||\tilde{v}|||^2\big)^{\alpha}\big(|||\tilde{v}|||^2\big)^{1-\alpha} + \delta^2|||\tilde{v}|||^2.
$$

Let $x = |||\tilde{v}|||^2/|||v|||^2$. Then

$$
A_k(\mathcal{E}_k v, v) \leqslant f\big(\delta^2; x, m\big) A_k(v, v),
$$

with $f(\delta; x, m)$ defined by (7.6). The theorem will follow if we can show

$$
f\big(\delta^2; x, m\big) \leqslant \delta, \tag{7.7}
$$

for $\delta = M/(M + m^{\alpha})$ with M sufficient large.

We now prove (7.7). We have shown in the proof of Theorem 7.1 that

$$
f\big(\delta^2; x, m\big) \leqslant \max\big(\ell\big(\delta^2; 0\big), \ell\big(\delta^2; 1\big)\big), \quad \text{for all } \tau > 0.
$$

Here $\ell(\delta; x) \equiv \delta x + (1-\delta)[(\alpha C_{\mathrm{P}}^2/\omega)\tau(1-x) + (1-\alpha)(\tau m)^{-\alpha/(1-\alpha)}x]$. It thus suffices to show that there exists a number δ with $0 < \delta < 1$ such that, with an appropriately chosen τ (with the possible restrictions $\tau \geqslant 1/m$ for $\alpha = 1$),

$$
\ell\big(\delta^2; 0\big) = \big(1 - \delta^2\big)\big(\alpha C_{\mathrm{P}}^2/\omega\big)\tau \leqslant \delta
$$

and

$$
\ell\big(\delta^2; 1\big) = \delta^2 + \big(1 - \delta^2\big)(1 - \alpha)(\tau m)^{-\alpha/(1-\alpha)} \leqslant \delta.
$$

These two inequalities can be written as

$$
\left(\frac{1+\delta}{\delta}\right)^{1-\alpha}\left(\frac{1-\alpha}{m^{\alpha/(1-\alpha)}}\right)^{1-\alpha} \leqslant \tau^{\alpha} \leqslant \left(\frac{\delta}{1-\delta^2}\right)^{\alpha}\left(\frac{1}{\alpha C_{\mathrm{P}}^2/\omega}\right)^{\alpha}.
$$

This is equivalent to choosing a $\delta \in (0, 1)$ so that

$$\frac{1+\delta}{\delta}(1-\delta)^{\alpha} \leqslant \frac{m^{\alpha}}{(\alpha C_{\mathrm{P}}^{2}/\omega)^{\alpha}(1-\alpha)^{1-\alpha}} \equiv \frac{m^{\alpha}}{C_{\alpha}}$$

holds uniformly in $m \geqslant 1$. Clearly, we can take $\delta = M/(M+m^{\alpha})$ with M large enough such that above inequality holds. For example, take $M \geqslant \frac{1}{2}(4C_{\alpha})^{1/\alpha} = \frac{1}{2}4^{1/\alpha}(\alpha C_{\mathrm{P}}^{2}/\omega)(1-\alpha)^{(1-\alpha)/\alpha}$. We have thus proved (7.7) and hence the theorem. □

In many applications, Condition (I.1) is not valid. We note that Condition (I.1) is used in Lemma 7.1 to prove

$$A_k(\mathcal{E}_k v, v) \equiv A_k\big((I - B_k A_k)v, v\big) \geqslant 0.$$

Without (I.1), we have to prove

$$\big|A_k(\mathcal{E}_k v, v)\big| \leqslant \delta A_k(v, v) \quad \text{for some } \delta < 1.$$

It is sufficient to assume either the number of smoothings, m, to be sufficiently large (but independent of k), or the following stability condition on I_k.

CONDITION (I.2). For $k = 2, \ldots, J$, the operators I_k satisfy

$$A_k(I_k v, I_k v) \leqslant 2A_{k-1}(v, v), \quad \text{for all } v \in M_{k-1}.$$

THEOREM 7.3 (W-cycle). *Suppose that Condition* (A.2″) *holds and that the smoothers, R_k, satisfy Condition* (SM.1). *Let $p = 2$ and $m_k = m$ for all k. Assume that either*

(a) *Condition* (I.2) *holds or*

(b) *the number of smoothings, m, is sufficiently large.*

Then

$$\big|A_k(\mathcal{E}_k v, v)\big| \leqslant \delta A_k(v, v), \quad \textit{for all } v \in M_k, \tag{7.8}$$

with $\delta = M/(M + m^{\alpha})$ with M sufficiently large.

PROOF. We proceed by induction. Let δ be defined as in Theorem 7.2. Since $\mathcal{E}_k = I - B_1 A_1 = 0$, (7.8) holds for $k = 1$. Assume that

$$\big|A_{k-1}(\mathcal{E}_{k-1} v, v)\big| \leqslant \delta A_{k-1}(v, v), \quad \text{for all } v \in M_{k-1}.$$

Then

$$0 \leqslant A_{k-1}\big(\mathcal{E}_{k-1}^{2} v, v\big) \leqslant \delta^{2} A_{k-1}(v, v), \quad \text{for all } v \in M_{k-1}.$$

We can prove in a way similar to the proof of the previous theorem that

$$A_k(\mathcal{E}_k v, v) \leqslant \delta A_k(v, v),$$

with $\delta = M/(M + m^\alpha)$ given by the previous theorem. We now show that

$$-A_k(\mathcal{E}_k v, v) \leqslant \delta A_k(v, v),$$

with the same δ. We first consider the case in which m is sufficiently large. Note that $\mathcal{E}_k^2$ is always symmetric, positive semidefinite. The two-level recurrence relation (7.1), with $p = 2$, implies that

$$-A_k(\mathcal{E}_k v, v) \leqslant -A_k\big((I - I_k P_{k-1})\tilde{v}, \tilde{v}\big) \leqslant \frac{C_{\mathrm{P}}^{2\alpha}}{(\omega m)^\alpha} A_k(v, v).$$

Without loss of generality, we assume that $M > C_{\mathrm{P}}^{2\alpha}/\omega^\alpha$. Then we choose m so large that $C_{\mathrm{P}}^{2\alpha}/(\omega m)^\alpha \leqslant \delta = M/(M + m^\alpha)$ and then (7.8) follows in this case.

Now if Condition (I.2) holds, then

$$-A_k\big((I - I_k P_{k-1})\tilde{v}, \tilde{v}\big) \leqslant A_k(\tilde{v}, \tilde{v}).$$

This implies that for any $\delta \in (0, 1)$,

$$-A_k(\mathcal{E}_k v, v) \leqslant \big(1 - \delta^2\big)\big|A_k\big((I - I_k P_{k-1})\tilde{v}, \tilde{v}\big)\big| + \delta^2 A_k(\tilde{v}, \tilde{v}).$$

This is the same bound as for $A_k(\mathcal{E}_k v, v)$. Thus the proof proceeds as before to show that

$$\big|A_k(\mathcal{E}_k v, v)\big| \leqslant \delta A_k(v, v),$$

for $\delta = M/(M + m^\alpha)$ with M sufficiently large. This proves the theorem. □

7.4. Multigrid V-cycle as a preconditioner

We now consider using the multigrid operator B_k as a preconditioner for A_k. Any symmetric positive definite operator can be used as a preconditioner for A_k. We will discuss the conditions under which multigrid operator B_k is a good preconditioner.

Theorem 7.1 states that if Conditions (A.2″) and (I.1) hold and the smoothers satisfy Condition (SM.1), then

$$0 \leqslant A_k(\mathcal{E}_k v, v) \equiv A_k\big((I - B_k A_k)v, v\big) \leqslant \delta_k A_k(v, v)$$

which is equivalent to

$$(1 - \delta_k)A_k(v, v) \leqslant A_k(B_k A_k v, v) \leqslant A_k(v, v). \tag{7.9}$$

Therefore, in such a case, B_k is an optimal preconditioner for A_k.

In many situations, Condition (I.1) is not satisfied. In such a case $A_k(\mathcal{E}_k v, v)$ could become negative, and thus $\mathcal{E}_k$ may not be a reducer. For the V-cycle and the variable V-cycle multigrid algorithms, we have proved without using Condition (I.1) that $A_k(\mathcal{E}_k v, v) \leqslant \delta_k A_k(v, v)$, which is equivalent to the lower estimate in (7.9). Condition (I.1) is only used to get the upper estimate in (7.9). We do not need the upper estimate in (7.9) in order to use B_k as a preconditioner. For a good preconditioner, we want to find numbers $\underline{\eta}_k$ and $\overline{\eta}_k$ such that

$$\underline{\eta}_k A_k(v, v) \leqslant A_k(B_k A_k v, v) \leqslant \overline{\eta}_k A_k(v, v), \quad \text{for all } v \in M_k. \tag{7.10}$$

We will provide estimates for $\underline{\eta}_k$ and $\overline{\eta}_k$ for the variable V-cycle multigrid operator under the Conditions (SM.1) and (A.2″). Note again that (SM.1) and (A.2″) imply the lower estimate in (7.9), and hence the lower estimate of (7.10) follows with $\underline{\eta}_k \geqslant 1 - \delta_k$.

THEOREM 7.4 (V-cycle preconditioner). *Let $p = 1$. Assume that Condition* (A.2″) *holds and that the smoothers, R_k, satisfy Condition* (SM.1). *Then* (7.10) *holds with*

$$\underline{\eta}_k = \frac{\omega m_k^\alpha}{\alpha C_{\mathrm{P}}^2 + \omega m_k^\alpha} \prod_{i=2}^{k} \left(1 - \frac{1-\alpha}{m_i^\alpha}\right) \quad \textit{and} \quad \overline{\eta}_k = \prod_{i=2}^{k} (1 + \bar{\delta}_i),$$

where $\bar{\delta}_i = \frac{C_{\mathrm{P}}^{2\alpha}}{\omega^\alpha m_i^\alpha} \alpha^\alpha (1-\alpha)^{1-\alpha}$. *In particular if the number of smoothings, m_k, satisfies*

$$\beta_0 m_k \leqslant m_{k-1} \leqslant \beta_1 m_k$$

for some $1 < \beta_0 \leqslant \beta_1$ *independent of k, then*

$$\underline{\eta}_k \geqslant \frac{m_k^\alpha}{M + m_k^\alpha} \quad \textit{and} \quad \overline{\eta}_k \leqslant 1 + \frac{M}{m_k^\alpha} = \frac{M + m_k^\alpha}{m_k^\alpha} \quad \textit{for some } M.$$

PROOF. We have shown in the proof of Theorem 7.1 that Conditions (A.2″) and (SM.1) imply that $A_k((I - B_k A_k)v, v) = A_k(\mathcal{E}_k v, v) \leqslant \delta_k A_k(v, v)$. Consequently, $(1-\delta_k) A_k(v, v) \leqslant A_k(B_k A_k v, v)$ and the lower estimate in (7.10) holds with

$$\underline{\eta}_k = 1 - \delta_k = \left(\frac{1}{\alpha C_{\mathrm{P}}^2/(\omega m_k^\alpha) + 1}\right) \prod_{i=2}^{k} \left(1 - \frac{1-\alpha}{m_i^\alpha}\right).$$

We prove the upper estimate in (7.10) by induction. For $k = 1$ there is nothing to prove. Assume (7.10) is true for $k-1$. Then

$$-A_{k-1}\big((I - B_{k-1} A_{k-1})w, w\big) \leqslant \big(\overline{\eta}_{k-1} - 1\big) A_{k-1}(w, w), \quad \text{for all } w \in M_{k-1}.$$

Set $\tilde{v} = K_k^{(m_k)} v$. By the induction hypothesis

$$\begin{aligned}
&-A_k\big((I - B_k A_k)v, v\big)\\
&\quad = -A_k\big((I - I_k P_{k-1})\tilde{v}, \tilde{v}\big) - A_{k-1}\big((I - B_{k-1}A_{k-1})P_{k-1}\tilde{v}, P_{k-1}\tilde{v}\big)\\
&\quad \leqslant -A_k\big((I - I_k P_{k-1})\tilde{v}, \tilde{v}\big) + \big(\overline{\eta}_{k-1} - 1\big)A_{k-1}\big(P_{k-1}\tilde{v}, P_{k-1}\tilde{v}\big)\\
&\quad = -\overline{\eta}_{k-1}A_k\big((I - I_k P_{k-1})\tilde{v}, \tilde{v}\big) + \big(\overline{\eta}_{k-1} - 1\big)A_k(\tilde{v}, \tilde{v})\\
&\quad \leqslant -\overline{\eta}_{k-1}A_k\big((I - I_k P_{k-1})\tilde{v}, \tilde{v}\big) + \big(\overline{\eta}_{k-1} - 1\big)A_k(v, v).
\end{aligned} \tag{7.11}$$

It remains to estimate $-A_k((I - I_k P_{k-1})\tilde{v}, \tilde{v})$. By Condition (A.2″)

$$-A_k\big((I - I_k P_{k-1})\tilde{v}, \tilde{v}\big) \leqslant C_{\mathrm{P}}^{2\alpha}\left[\frac{\|A_k\tilde{v}\|_k^2}{\lambda_k}\right]^{\alpha}\big[A_k(\tilde{v}, \tilde{v})\big]^{1-\alpha}.$$

By Lemma 7.2,

$$\frac{\|A_k\tilde{v}\|_k^2}{\lambda_k} \leqslant \frac{1}{\omega m_k}\big[A_k(v, v) - A_k(\tilde{v}, \tilde{v})\big].$$

Hence, since $0 \leqslant A_k(\tilde{v}, \tilde{v}) \leqslant A_k(v, v)$, we have that

$$\begin{aligned}
&-A_k\big((I - I_k P_{k-1})\tilde{v}, \tilde{v}\big)\\
&\quad \leqslant \frac{C_{\mathrm{P}}^{2\alpha}}{(\omega m_k)^{\alpha}}\big[A_k(v, v) - A_k(\tilde{v}, \tilde{v})\big]^{\alpha}\big[A_k(\tilde{v}, \tilde{v})\big]^{1-\alpha}\\
&\quad \leqslant \frac{C_{\mathrm{P}}^{2\alpha}}{(\omega m_k)^{\alpha}}\alpha^{\alpha}(1-\alpha)^{1-\alpha}A_k(v, v)\\
&\quad = \bar{\delta}_i A_k(v, v).
\end{aligned} \tag{7.12}$$

Combining Eqs. (7.11) and (7.12), we obtain

$$\begin{aligned}
-A_k\big((I - B_k A_k)v, v\big) &\leqslant \big[\overline{\eta}_{k-1}\bar{\delta}_k + \big(\overline{\eta}_{k-1} - 1\big)\big]A_k(v, v)\\
&= \big(\overline{\eta}_k - 1\big)A_k(v, v).
\end{aligned}$$

Hence $A_k(B_k A_k v, v) \leqslant \overline{\eta}_k A_k(v, v)$ and the theorem is proved. □

We now prove a general result for the V-cycle multigrid without assuming the regularity-approximation assumption (A.2″).

THEOREM 7.5. *If the operators $\overline{R}_k$, $k = 1, 2, \ldots, J$, are symmetric and positive definite, then the operator B_J corresponding to the V-cycle multigrid method ($p = 1$) is symmetric and positive definite. If in addition, Condition* (SM.1) *holds, then*

$$\frac{\omega}{\lambda_k}A_k(v, v) \leqslant A_k\big((B_k A_k)v, v\big).$$

PROOF. It is the same to prove that $B_k A_k$ is symmetric and positive definite with respect to $A_k(\bullet, \bullet)$. The symmetry is easy to see. Let $\tilde{v} = K_k^{(m_k)} v$. Since $p = 1$, we get

$$A_k\big((I - B_k A_k)v, v\big) = A_k(\tilde{v}, \tilde{v}) - A_{k-1}\big((B_{k-1} A_{k-1}) P_{k-1}\tilde{v}, P_{k-1}\tilde{v}\big).$$

The induction hypothesis implies that

$$A_k\big((I - B_k A_k)v, v\big) < A_k(\tilde{v}, \tilde{v}).$$

If $\overline{R}_k$ is symmetric and positive definite, then $A_k(\tilde{v}, \tilde{v}) < A_k(v, v)$ and B_k is symmetric and positive definite. If in addition, (SM.1) holds, then $A_k(\tilde{v}, \tilde{v}) \leqslant (1 - \omega/\lambda_k) A_k(v, v)$ and thus $\omega\lambda_k^{-1} A_k(v, v) \leqslant A_k(B_k A_k v, v)$. □

The theorem shows that the (variable) V-cycle multigrid operator B_k is symmetric and positive definite, and thus can always be used as a preconditioner if the size of the condition number is not a concern. If in addition we know that Condition (I.1) holds, then by Lemma 7.1, we have

$$0 \leqslant A_k\big((I - B_k A_k)v, v\big), \qquad \text{for all } v \in M_k$$

or equivalently

$$A_k(B_k A_k v, v) \leqslant A_k(v, v), \qquad \text{for all } v \in M_k.$$

Therefore, if Conditions (I.1) and (SM.1) hold, then

$$A_k\big(\overline{R}_k A_k v, v\big) \leqslant A_k(B_k A_k v, v) \leqslant A_k(v, v).$$

7.5. Bibliographical notes

A very important contribution to the multigrid theory was made by BANK and DUPONT [1981] where they presented a rigorous treatment of multigrid as an optimal way to solve equations coming from finite element discretizations of elliptic problems. They assumed nested spaces and proved convergence of the W-cycle with sufficiently many smoothing steps, assuming some regularity in the continuous problem. They proved the part (b) of Theorem 7.3 in the nested case with the same form on all levels with m sufficiently large. This paper led to many subsequent works.

The result of BANK and DUPONT [1981], part (b) of Theorem 7.3, is not a strong result by current standards when applied in the nested-inherited case as they did. However it was an important development in the analysis of multigrid algorithms. Nevertheless, there are situations in which this type of result seems to be the only one within reach; cf. VERFÜRTH [1984].

BRAESS and HACKBUSCH [1983] were the first to prove that the V-cycle converges uniformly with respect to the number of levels. This proof required the "full elliptic regularity". They proved Theorem 7.1 for the nested-inherited case in which Condition (I.1) holds with equality. Theorem 7.1 although weaker than Theorems 5.1 or 5.2 in the case $m = 1$, shows improvement as the number of smoothings steps per iteration is increased. The work of BRAESS and HACKBUSCH [1983] also led the way to the proofs that the W-cycle gave a uniform reduction under reduced regularity assumptions. The work of BRAMBLE and PASCIAK [1987] and BANK, MANDEL and MCCORMICK [1987] contain proofs of such a result. The paper BRAMBLE and PASCIAK [1987] introduced the variable V-cycle and proved uniform estimates with some (arbitrary amount of) regularity. All these papers required the nested-inherited setting.

This section is a condensation of the results of BRAMBLE, PASCIAK and XU [1991] in which non-nested spaces and/or non-inherited forms were considered. Perhaps the most novel of the results is that of Theorem 7.4 which shows the robustness of the variable V-cycle in that, with any amount of regularity, it always gives rise to a uniform preconditioner. Condition (I.1), at first, may seem somewhat artificial, since, in the nested–inherited case (I.1) holds with equality. However, Sections 5, 6 and 8 of BRAMBLE, PASCIAK and XU [1991] contain natural examples in which (I.1) is satisfied without equality. Applications of the theory in this section will be discussed in later sections.

8. Construction and analysis of classes of smoothers

In Sections 3–6 and 7 a generic smoother R_k was introduced and was assumed to satisfy certain smoothing conditions. This section is devoted to the construction and analysis of two classes of smoothers which satisfy the relevant conditions. We shall see in Section 15 that these smoothers are constructed in such a way that the multigrid algorithms are easily implemented. These classes include the smoothers derived from the Jacobi, Gauss–Seidel and symmetric Gauss–Seidel methods as well as the line and block forms of those methods.

8.1. Additive and multiplicative smoothers

Many iterative methods can be considered as additive or multiplicative schemes associated with a decomposition of the solution space. Smoothers deduced from these two classes of iterative schemes are called *additive* and *multiplicative* smoothers respectively. Different smoothers result from using distinct space decompositions. For numerical solutions of partial differential equations, the decomposition of the space often naturally arises from a decomposition of the physical domain. We shall construct two classes of smoothing operators that satisfy the smoothing conditions. We shall do this by decomposing the space M_k into subspaces M_k^i, $i = 1, \ldots, \ell = \ell_k$, with

$$M_k = \sum_{i=1}^{\ell} M_k^i.$$

This may or may not be a direct sum, i.e., the decomposition $v = \sum_{i=1}^{\ell} v_i$ with $v_i \in M_k^i$ may or may not be unique.

Our smoothers are defined in terms of the solution operators associated with problems on the subspaces M_k^i. To this end, we define $A_{k,i}: M_k^i \to M_k^i$, P_k^i, $Q_k^i : M_k \to M_k^i$ by

$$(A_{k,i}v, \phi) = (A_k v, \phi), \quad \text{for all } \phi \in M_k^i,$$

$$\big(A_k P_k^i v, \phi\big) = (A_k v, \phi), \quad \text{for all } \phi \in M_k^i,$$

and

$$\big(Q_k^i v, \phi\big) = (v, \phi), \quad \text{for all } \phi \in M_k^i.$$

We can now define additive and multiplicative smoothers.

DEFINITION 8.1. The additive (Jacobi type) smoother corresponding to the space decomposition of M_k is defined by

$$\mathcal{J}_k = \sum_{i=1}^{\ell} A_{k,i}^{-1} Q_k^i = \left(\sum_{i=1}^{\ell} P_k^i\right) A_k^{-1}. \tag{8.1}$$

The multiplicative (Gauss–Seidel type) smoother is defined by

$$\mathcal{G}_k = \big[I - \big(I - P_k^{\ell}\big)\big(I - P_k^{\ell-1}\big) \cdots \big(I - P_k^1\big)\big] A_k^{-1}. \tag{8.2}$$

We remark that given $f \in M_k$, $\mathcal{G}_k f \in M_k$ can be computed as follows:
(1) Set $v_0 = 0$.
(2) For $i = 1, \ldots, \ell$ define

$$v_i = v_{i-1} + A_{k,i}^{-1} Q_k^i (f - A_k v_{i-1}).$$

(3) $\mathcal{G}_k f = v_\ell$.
To see this, let v be the solution of $A_k v = f$. Note that $A_{k,i} P_k^i = Q_k^i A_k$. Hence

$$v - v_i = v - v_{i-1} - \big(A_{k,i}^{-1} Q_k^i A_k\big)(v - v_{i-1}) = \big(I - P_k^i\big)(v - v_{i-1}).$$

Consequently,

$$(I - \mathcal{G}_k A_k) v \equiv v - v_\ell = \big(I - P_k^{\ell}\big)(v - v_{\ell-1}) = \big(I - P_k^{\ell}\big) \cdots \big(I - P_k^1\big) v,$$

and (8.2) follows.

REMARK 8.1. If the decomposition $M_k = \sum M_k^i$ is a direct sum, then $R_k = \gamma \mathcal{J}_k$ (with $\gamma > 0$) is a weighted block Jacobi smoother and $\mathcal{G}_k$ is a block Gauss–Seidel smoother. In particular, if each M_k^i is given as the span of a single basis function of M_k, then $R_k = \gamma \mathcal{J}_k$ is a weighted Jacobi smoother and $\mathcal{G}_k$ is a Gauss–Seidel smoother.

Note that the inverse A_{ki}^{-1} is used in the above construction of smoothers. In most cases, the dimension of M_k^i is small and thus A_{ki}^{-1} is easy to compute. In some applications, one may still want to replace the inverse A_{ki}^{-1} in Eqs. (8.1) and (8.2) by some other operator R_{ki} which might be more convenient to compute. The operator R_{ki} could be thought of as an approximate inverse of A_{ki}. The sense in which it is approximate will be made more precise by the subsequent conditions placed on R_{ki}. This leads to the following definition of the smoothers.

DEFINITION 8.2. Let $R_{ki} : M_k^i \to M_k^i$ be an approximation of A_{ki}^{-1}. The modified additive smoother is defined by

$$\mathcal{J}_k = \sum_{i=1}^{\ell} R_{ki} Q_k^i, \tag{8.3}$$

and the modified multiplicative smoother is defined by

$$\mathcal{G}_k = \left[I - (I - T_k^{\ell})(I - T_k^{\ell-1}) \cdots (I - T_k^1)\right] A_k^{-1}, \tag{8.4}$$

where $T_k^i = R_{ki} Q_k^i A_k = R_{ki} A_{ki} P_k^i$.

Similar to (8.2), given $f \in M_k$, $\mathcal{G}_k f \in M_k$ can be computed as follows:
(1) Set $v_0 = 0$.
(2) For $i = 1, \ldots, \ell$ define

$$v_i = v_{i-1} + R_{k,i} Q_k^i (f - A_k v_{i-1}).$$

(3) $\mathcal{G}_k f = v_\ell$.
In many applications, the approximate inverse R_{ki} is derived from a few steps of an iterative method for solving $A_{ki} x = b$.

REMARK 8.2. In the case $R_{ki} = \tilde{\theta} A_{ki}^{-1}$, $0 < \tilde{\theta} < 2$, the multiplicative smoother reduces to a block successive overrelaxation (SOR) smoother. A symmetric SOR smoother can also be considered as a special case of the multiplicative smoother by allowing each space in the decomposition to be repeated; i.e., $M_k^i = M_k^{\ell - i}$.

We shall establish that the smoothers defined above satisfy the relevant smoothing conditions. To do this we need to impose certain restrictions on the space decomposition. The *banded* interaction properties of the spaces M_k^i can be characterized by using the *interaction matrices* κ and κ^0 which are defined respectively by

$$\kappa_{ij} = \begin{cases} 0 & \text{if } P_k^i P_k^j = 0, \\ 1 & \text{otherwise} \end{cases} \quad \text{and} \quad \kappa_{ij}^0 = \begin{cases} 0 & \text{if } Q_k^i Q_k^j = 0, \\ 1 & \text{otherwise.} \end{cases}$$

Denote by $\|\bullet\|_2$ and $\|\bullet\|_\infty$ the matrix 2- and ∞-norms respectively. Clearly

$$\|\kappa\|_\infty = \max_i \sum_{j=1}^{\ell} \kappa_{ij} \quad \text{and} \quad \left\|\kappa^0\right\|_\infty = \max_i \sum_{j=1}^{\ell} \kappa_{ij}^0.$$

We make the following hypotheses on the subspaces M_k^i. These hypotheses can be easily verified in applications for elliptic finite element problems.

HYPOTHESIS (i). There is a positive number C_1, independent of k, such that

$$\|\kappa\|_2 \leqslant \|\kappa\|_\infty \leqslant C_1.$$

HYPOTHESIS (i′). There is a positive number C_1', independent of k, such that

$$\left\|\kappa^0\right\|_2 \leqslant \left\|\kappa^0\right\|_\infty \leqslant C_1'.$$

HYPOTHESIS (ii). There exists positive constant C_2, independent of k, such that for each $v \in M_k$ there is a decomposition $v = \sum_{i=1}^{\ell} v_i$, with $v_i \in M_k^i$, satisfying

$$\sum_{i=1}^{\ell} \|v_i\|^2 \leqslant C_2 \|v\|^2.$$

We have the following simple lemma.

LEMMA 8.1. *Let κ and κ^0 be the interaction matrices defined above.*
(a) *Hypothesis* (i) *implies that for all w_i and $v_i \in M_k^i$, $i = 1, \ldots, \ell$,*

$$\sum_{i,j=1}^{\ell} \left|(A_k w_i, v_j)\right| \leqslant C_1 \left(\sum_{i=1}^{\ell} (A_k w_i, w_i)\right)^{1/2} \left(\sum_{j=1}^{\ell} (A_k v_j, v_j)\right)^{1/2}.$$

(b) *Hypothesis* (i′) *implies that*

$$\|v\|^2 \leqslant C_1' \sum_{i=1}^{\ell} \|v_i\|^2, \quad \textit{for all } v = \sum_{i=1}^{\ell} v_i, \ v_i \in M_k^i.$$

PROOF. Note that $\kappa_{ij} \geqslant 0$,

$$\begin{aligned} \sum_{i,j=1}^{\ell} \left|(A_k w_i, v_j)\right| &= \sum_{i,j=1}^{\ell} \kappa_{ij} \left|(A_k w_i, v_j)\right| \leqslant \sum_{i,j=1}^{\ell} \kappa_{ij} |||w_i||| \, |||v_j||| \\ &\leqslant \|\kappa\|_2 \left(\sum_{i=1}^{\ell} (A_k w_i, w_i)\right)^{1/2} \left(\sum_{j=1}^{\ell} (A_k v_j, v_j)\right)^{1/2}. \end{aligned}$$

Since κ is symmetric $\|\kappa\|_2 \leqslant \|\kappa\|_\infty \leqslant C_1$. The proof for the second part is similar. □

Hypotheses (i) and (i′) imply respectively the *uniformly banded interaction* properties of matrices κ and κ^0.

8.2. *Smoothers for multigrid algorithms*

In this subsection, we show that the smoothers defined in the previous subsection satisfy the smoothing conditions used in the analysis of Algorithms 3.1 and 7.1. Recall that $K_k = I - R_k A_k$, $K_k^* = I - R_k^t A_k$ and $\overline{R}_k = R_k + R_k^t - R_k^t A_k R_k$. It is easy to see that

$$A_k(K_k v, K_k v) = A_k\big((I - \overline{R}_k A_k)v, v\big).$$

In the case R_k is an SOR smoother, $\overline{R}_k$ corresponds the the symmetric SOR smoother.

In the convergence analysis of multigrid algorithms given in Sections 3–6 and 7, we have imposed some conditions on the smoother R_k which we recall here.

CONDITION (SM.1). There exists $\omega > 0$, not depending on J, such that

$$\left(\frac{\omega}{\lambda_k}\right)\|v\|^2 \leqslant \big(\overline{R}_k v, v\big), \quad \text{for all } v \in M_k,\ k = 2, \ldots, J.$$

CONDITION (SM.2). There exists $\theta < 2$, not depending on J, such that

$$(A_k R_k v, R_k v) \leqslant \theta(R_k v, v), \quad \text{for all } v \in M_k.$$

CONDITION (SM.4). There exists a constant ω_4, not depending on J, such that

$$(R_k v, R_k v) \leqslant \omega_4 \lambda_k^{-1}(R_k v, v), \quad \text{for all } v \in M_k,\ 2 \leqslant k \leqslant J.$$

It is sometimes convenient to use equivalent forms of these conditions. First note that Condition (SM.1) is equivalent to

$$(A_k K_k v, K_k v) \leqslant (A_k K_{k,\omega} v, v), \quad \text{with } K_{k,\omega} = I - \omega\lambda_k^{-1} A_k.$$

By simple manipulation, it is easy to see that the following conditions are equivalent:

$$\begin{aligned}
\big(R_k^t A_k R_k v, v\big) &\leqslant \theta(R_k v, v), \quad \text{for all } v \in M_k,\\
(2-\theta)(R_k v, v) &\leqslant \big(\overline{R}_k v, v\big), \quad \text{for all } v \in M_k,\\
(A_k R_k v, R_k v) &\leqslant \frac{\theta}{2-\theta}\big(\overline{R}_k v, v\big), \quad \text{for all } v \in M_k.
\end{aligned}$$

A simple change of variable shows that Condition (SM.4) is the same as

$$(v, v) \leqslant \omega_4 \lambda_k^{-1}\big(R_k^{-1} v, v\big), \quad \text{for all } v \in M_k.$$

If R_k is symmetric, then Condition (SM.4) is equivalent to

$$(R_k v, v) \leqslant \omega_4 \lambda_k^{-1}(v, v), \quad \text{for all } v \in M_k.$$

We first show that Conditions (SM.1) and (SM.2) hold for the additive smoother defined by (8.1).

THEOREM 8.1. *Let $\mathcal{J}_k$ be the additive smoother defined by* (8.1). *Suppose that Hypotheses* (i) *and* (ii) *are satisfied. Let $R_k = \gamma \mathcal{J}_k$ and choose γ so that $C_1\gamma \in (0, 2)$. Then Condition* (SM.2) *is satisfied with $\theta = C_1\gamma$ and Condition* (SM.1) *is satisfied with $\omega = \frac{\gamma}{C_2}(2 - \gamma C_1) = \frac{\theta(2-\theta)}{C_2 C_1}$.*

PROOF. We first show that Condition (SM.2) is satisfied. By Hypothesis (i) and Lemma 8.1,

$$\begin{aligned}
A_k(\mathcal{J}_k v, \mathcal{J}_k v) &= \sum_{i=1}^{\ell}\sum_{j=1}^{\ell} A_k\big(A_{ki}^{-1} Q_k^i v, A_{kj}^{-1} Q_k^j v\big) \\
&\leqslant C_1 \sum_{i=1}^{\ell} A_k\big(A_{ki}^{-1} Q_k^i v, A_{ki}^{-1} Q_k^i v\big) \\
&= C_1(\mathcal{J}_k v, v).
\end{aligned}$$

Thus for $R_k = \gamma \mathcal{J}_k$,

$$A_k(R_k v, R_k v) \leqslant \gamma C_1 (R_k v, v) = \theta(R_k v, v).$$

This is equivalent to Condition (SM.2) with $\theta = C_1\gamma$.

For any $v \in M_k$, let $v = \sum_{i=1}^{\ell} v_i$ be a decomposition of v that satisfies Hypothesis (ii). Then

$$\begin{aligned}
\|v\|^2 &= \sum_{i=1}^{\ell} \big(v_i, Q_k^i v\big) \leqslant \left(\sum_{i=1}^{\ell} \|v_i\|^2\right)^{1/2} \left(\sum_{i=1}^{\ell} \big\|Q_k^i v\big\|^2\right)^{1/2} \\
&\leqslant \big(C_2\|v\|^2\big)^{1/2} \left(\sum_{i=1}^{\ell} \big\|Q_k^i v\big\|^2\right)^{1/2}.
\end{aligned}$$

Hence

$$\|v\|^2 \leqslant C_2 \sum_{i=1}^{\ell} \big\|Q_k^i v\big\|^2 \leqslant C_2\lambda_k \sum_{i=1}^{\ell} \big(A_{ki}^{-1} Q_k^i v, v\big) = C_2\lambda_k(\mathcal{J}_k v, v). \tag{8.5}$$

Since $\overline{R}_k = R_k + R_k^t - R_k^t A_k R_k$, by Condition (SM.2) and (8.5), we have

$$\begin{aligned}\left(\overline{R}_k v, v\right) &\geqslant (2-\theta)(R_k v, v) = (2-\theta)\gamma(\mathcal{J}_k v, v)\\ &\geqslant \frac{(2-\theta)\gamma}{C_2}\lambda_k^{-1}\|v\|^2 = \frac{\theta(2-\theta)}{C_2 C_1}\lambda_k^{-1}(v, v).\end{aligned}$$

This proves Condition (SM.1). □

We now show that Conditions (SM.1) and (SM.2) hold for the multiplicative smoother defined by (8.2).

THEOREM 8.2. *Let $\mathcal{G}_k$ be the multiplicative smoother defined in* (8.2). *Suppose that Hypotheses* (i) *and* (ii) *hold with C_1 and C_2 as before. Let $R_k = \mathcal{G}_k$. Then Condition* (SM.1) *holds with $\omega = (C_2 C_1^2)^{-1}$ and Condition* (SM.2) *holds with $\theta = 2C_1/(C_1+1)$, i.e.,*

$$\left(\frac{\omega}{\lambda_k}\right)\|u\|^2 \leqslant \left(\overline{\mathcal{G}}_k u, u\right) \quad \textit{with } \omega = \left(C_2 C_1^2\right)^{-1}$$

and

$$(A_k \mathcal{G}_k u, \mathcal{G}_k u) \leqslant \theta(\mathcal{G}_k u, u) \quad \textit{with } \theta = \frac{2C_1}{C_1+1}.$$

PROOF. Set $E_k^0 = I$ and $E_k^i = (I - P_k^i)E_k^{i-1} = (I - P_k^i)\cdots(I - P_k^1)$. Then

$$I - \mathcal{G}_k A_k = \left(I - P_k^\ell\right)\cdots\left(I - P_k^1\right) = E_k^\ell.$$

Clearly

$$E_k^{i-1} - E_k^i = P_k^i E_k^{i-1}$$

and hence

$$I - E_k^i = \sum_{j=1}^{i} P_k^j E_k^{j-1}.$$

Thus

$$A_k\left(E_k^{i-1}v, E_k^{i-1}v\right) - A_k\left(E_k^i v, E_k^i v\right) = A_k\left(P_k^i E_k^{i-1}v, E_k^{i-1}v\right).$$

Summing gives

$$A_k\left(\overline{\mathcal{G}}_k A_k v, v\right) \equiv (A_k v, v) - \left(A_k E_k^\ell v, E_k^\ell v\right) = \sum_{i=1}^{\ell}\left(A_k P_k^i E_k^{i-1}v, E_k^{i-1}v\right). \quad (8.6)$$

Now for $v \in M_k$, set $w = A_k v$ and assume that the decomposition $w = \sum w_i$ satisfies Hypothesis (ii). Then

$$\begin{aligned}
(A_k v, A_k v) &= A_k(v, w) = \sum_{i=1}^{\ell} A_k(v, w_i) \\
&= \sum_{i=1}^{\ell} \left(A_k\big(E_k^{i-1} v, w_i\big) + \sum_{j=1}^{i-1} A_k\big(P_k^j E_k^{j-1} v, w_i\big) \right) \\
&= \sum_{i=1}^{\ell} \sum_{j=1}^{i} A_k\big(P_k^j E_k^{j-1} v, w_i\big) \\
&\leqslant \|\kappa\|_\infty \left(\sum_{i=1}^{\ell} A_k\big(P_k^i E_k^{i-1} v, E_k^{i-1} v\big) \right)^{1/2} \left(\sum_{i=1}^{\ell} A_k(w_i, w_i) \right)^{1/2} \\
&\leqslant \|\kappa\|_\infty \left(\sum_{i=1}^{\ell} A_k\big(P_k^i E_k^{i-1} v, E_k^{i-1} v\big) \right)^{1/2} \big(\lambda_k C_2 \|w\|^2\big)^{1/2}.
\end{aligned}$$

Now note that $\|w\|^2 = (A_k v, A_k v)$. Therefore, by (8.6)

$$(A_k v, A_k v) \leqslant C_2 C_1^2 \lambda_k \left(\sum_{i=1}^{\ell} A_k\big(P_k^i E_k^{i-1} v, E_k^{i-1} v\big) \right) = C_2 C_1^2 \lambda_k A_k\big(\overline{\mathcal{G}}_k A_k v, v\big).$$

This implies that

$$\|w\|^2 \leqslant \big(C_2 C_1^2\big) \lambda_k \big(\overline{\mathcal{G}}_k w, w\big), \quad \text{for all } w \in M_k.$$

Thus $R_k = \mathcal{G}_k$ satisfies Condition (SM.1) with $\omega = (C_2 C_1^2)^{-1}$.

To show that R_k satisfies Condition (SM.2), we note that

$$\mathcal{G}_k A_k = I - E_k^\ell = I - \big(I - P_k^\ell\big) \cdots \big(I - P_k^1\big) = \sum_{i=1}^{\ell} P_k^i E_k^{i-1}. \tag{8.7}$$

Thus by (8.7), Lemma 8.1 and (8.6)

$$\begin{aligned}
A_k(\mathcal{G}_k A_k v, \mathcal{G}_k A_k v) &= \sum_{i,j=1}^{\ell} A_k\big(P_k^i E_k^{i-1} v, P_k^j E_k^{j-1} v\big) \\
&\leqslant \|\kappa\|_2 \sum_{i=1}^{\ell} A_k\big(P_k^i E_k^{i-1} v, E_k^{i-1} v\big)
\end{aligned}$$

$$= \|\kappa\|_2 A_k(\overline{\mathcal{G}}_k A_k v, v)$$
$$\leqslant C_1 A_k(\overline{\mathcal{G}}_k A_k v, v),$$

for all $v \in M_k$. This shows that

$$(A_k \mathcal{G}_k v, \mathcal{G}_k v) \leqslant \|\kappa\|_2 (\overline{\mathcal{G}}_k v, v) \leqslant C_1 (\overline{\mathcal{G}}_k v, v).$$

Hence Condition (SM.2) holds for $R_k = \mathcal{G}_k$ with $\theta = 2C_1/(1 + C_1) \in (0, 2)$. □

We now consider the smoothing operators $\mathcal{J}_k$ and $\mathcal{G}_k$ defined by Eqs. (8.3) and (8.4), respectively. These two operators are defined in terms of the approximation R_{ki} of A_{ki}^{-1}. If no restriction is imposed on R_{ki}, the smoothers $R_k = \gamma \mathcal{J}_k$ and $R_k = \mathcal{G}_k$ will not satisfy Conditions (SM.1) and (SM.2) in general. In the case when R_{ki} is derived from an iterative scheme for solving $A_{ki} x = b$, it is natural to expect that R_{ki} should satisfy the smoother conditions on subspace M_k^i. Therefore, we assume, in addition to the two hypotheses on the space decomposition, that the approximate inverses R_{ki} satisfy the "smoother conditions" on subspace M_k^i.

HYPOTHESIS (iii). There is a $\tilde{\theta} \in (0, 2)$ such that

$$A_k(R_{ki} v, R_{ki} v) \leqslant \tilde{\theta}(R_{ki} v, v), \qquad \text{for all } v \in M_k^i.$$

HYPOTHESIS (iv). There exists $\tilde{\omega} > 0$ such that $\overline{R}_{ki} = R_{ki} + R_{ki}^t - R_{ki}^t A_{ki} R_{ki}$ satisfies

$$\frac{\tilde{\omega}}{\lambda_k}(v, v) \leqslant (\overline{R}_{ki} v, v), \qquad \text{for all } v \in M_k^i.$$

Note that Hypotheses (iii) and (iv) are identical to the smoother Conditions (SM.1) and (SM.2), except, they are imposed on M_k^i. In the next two theorems, we shall show that if Hypotheses (i)–(iv) hold, then the smoothers defined by Eqs. (8.3) and (8.4) satisfy Conditions (SM.1) and (SM.2). A consequence of these results is the following. In the construction of smoothers, the exact solves A_{ki}^{-1} in Eqs. (8.1) and (8.2) can be replaced by a few iterations of some iterative method for solving $A_{ki} x = b$, provided that the iterative method is derived from a space decomposition of M_k^i and this decomposition of M_k^i satisfies Hypotheses (i) and (ii). This idea can be used in constructing hybrid smoothers of neither additive nor multiplicative type.

THEOREM 8.3. *Let $\mathcal{J}_k$ be the additive smoother defined by* (8.3). *Suppose that Hypotheses* (i) *and* (ii) *hold and that R_{ki} satisfies Hypotheses* (iii) *and* (iv). *Let $R_k = \gamma \mathcal{J}_k$ and choose γ so that $\theta \equiv C_1 \gamma \tilde{\theta} < 2$. Then Condition* (SM.2) *holds with $\theta = C_1 \gamma \tilde{\theta}$ and Condition* (SM.1) *holds with $\omega = \gamma \min((2 - \theta)/(2 - \tilde{\theta}), 1)$, i.e.,*

$$A_k(R_k v, R_k v) \leqslant \theta(R_k v, v), \qquad \text{for all } v \in M_k$$

and

$$\left(\overline{R}_k v, v\right) \geqslant \gamma \min\left(\frac{2-\theta}{2-\tilde{\theta}}, 1\right) \frac{\tilde{\omega}}{C_2} \frac{(v, v)}{\lambda_k}, \quad \textit{for all } v \in M_k.$$

PROOF. By Hypothesis (i) and Lemma 8.1,

$$\begin{aligned} A_k(\mathcal{J}_k v, \mathcal{J}_k v) &= \sum_{i,j=1}^{\ell} A_k\left(R_{ki} Q_k^i v, R_{kj} Q_k^j v\right) \\ &\leqslant C_1 \sum_{i=1}^{\ell} A_k\left(R_{ki} Q_k^i v, R_{ki} Q_k^i v\right). \end{aligned} \tag{8.8}$$

It then follows from Hypothesis (iii) that

$$A_k(\mathcal{J}_k v, \mathcal{J}_k v) \leqslant C_1 \tilde{\theta} \sum_{i=1}^{\ell} \left(R_{ki} Q_k^i v, Q_k^i v\right) = C_1 \tilde{\theta} (\mathcal{J}_k v, v).$$

Thus for $R_k = \gamma \mathcal{J}_k$,

$$A_k(R_k v, R_k v) \leqslant C_1 \tilde{\theta} \gamma (R_k v, v) = \theta (R_k v, v).$$

This proves the first part of the theorem.

It also follows from Hypothesis (iii) and a simple calculation that

$$2(R_{ki} v_i, v_i) - C_1 \gamma A_k(R_{ki} v_i, R_{ki} v_i) \geqslant \min\left(\frac{2 - C_1 \gamma \tilde{\theta}}{2-\tilde{\theta}}, 1\right) \left(\overline{R}_{ki} v_i, v_i\right). \tag{8.9}$$

Now by (8.8), (8.9) and Hypothesis (iv),

$$\begin{aligned} \left(\overline{R}_k v, v\right) &= 2(R_k v, v) - A_k(R_k v, R_k v) = 2\gamma(\mathcal{J}_k v, v) - \gamma^2 A_k(\mathcal{J}_k v, \mathcal{J}_k v) \\ &\geqslant \sum_{i=1}^{\ell} 2\gamma\left(R_{ki} Q_k^i v, Q_k^i v\right) - \sum_{i=1}^{\ell} C_1 \gamma^2 \left(R_{ki}^t A_{ki} R_{ki} Q_k^i v, Q_k^i v\right) \\ &= \gamma \sum_{i=1}^{\ell} \left(2\left(R_{ki} Q_k^i v, Q_k^i v\right) - C_1 \gamma A_k\left(R_{ki} Q_k^i v, R_{ki} Q_k^i v\right)\right) \\ &\geqslant \gamma \min\left(\frac{2 - C_1 \gamma \tilde{\theta}}{2-\tilde{\theta}}, 1\right) \sum_{i=1}^{\ell} \left(\overline{R}_{ki} Q_k^i v, Q_k^i v\right) \\ &\geqslant \gamma \min\left(\frac{2 - C_1 \gamma \tilde{\theta}}{2-\tilde{\theta}}, 1\right) \frac{\tilde{\omega}}{\lambda_k} \sum_{i=1}^{\ell} \left(Q_k^i v, Q_k^i v\right) \end{aligned}$$

$$\geqslant \gamma \min\left(\frac{2 - C_1\gamma\tilde{\theta}}{2 - \tilde{\theta}}, 1\right)\frac{\tilde{\omega}}{\lambda_k} C_2^{-1}(v, v).$$

In the last step, we have used the inequality $\|v\|^2 \leqslant C_2 \sum_{i=1}^{\ell}(Q_k^i v, Q_k^i v)$, which is an easy consequence of Hypothesis (ii). This proves the second part of the theorem. □

In the case $R_{ki} = \tilde{\theta} A_{ki}^{-1}$, the value $\min((2 - C_1\gamma\tilde{\theta})/(2 - \tilde{\theta}), 1)$ in the theorem can be replaced by $(2 - C_1\gamma\tilde{\theta})/(2 - \tilde{\theta})$. In particular, if $R_{ki} = A_{ki}^{-1}$ then $\tilde{\omega} = \tilde{\theta} = 1$ and the estimate reduces to the one given by Theorem 8.1.

THEOREM 8.4. *Assume that Hypotheses* (i) *and* (ii) *hold with constants* C_1 *and* C_2. *If all the operators* R_{ki} *satisfy Hypotheses* (iii) *and* (iv), *then the multiplicative smoother* $\mathcal{G}_k$ *defined by* (8.4) *satisfies Conditions* (SM.1) *and* (SM.2), *i.e.,*

$$\frac{\omega}{\lambda_k}\|v\|^2 \leqslant (\overline{\mathcal{G}}_k v, v) \quad \textit{with } \omega^{-1} = C_2\left[\left(\frac{1}{\tilde{\omega}}\right)^{1/2} + (C_1 - 1)\left(\frac{\tilde{\theta}}{2 - \tilde{\theta}}\right)^{1/2}\right]^2$$

and

$$(A_k\mathcal{G}_k v, \mathcal{G}_k v) \leqslant \theta(\mathcal{G}_k v, v) \quad \textit{where } \theta = \frac{2\tilde{C}_1}{1 + \tilde{C}_1} \textit{ with } \tilde{C}_1 = \frac{C_1\tilde{\theta}}{2 - \tilde{\theta}}.$$

PROOF. Let $T_k^i = R_{ki} A_{ki} P_k^i = R_{ki} Q_k^i A_k$. Set $E_k^i = (I - T_k^i)E_k^{i-1} = (I - T_k^i)\cdots(I - T_k^1)$ and $E_k^0 = I$. Then it is easy to see that

$$I - \mathcal{G}_k A_k = \left(I - T_k^\ell\right)\cdots\left(I - T_k^1\right) = E_k^\ell,$$

and

$$I - \mathcal{G}_k^t A_k = \left(I - T_k^1\right)\cdots\left(I - T_k^\ell\right) = \left(E_k^\ell\right)^*.$$

Notice that $E_k^{i-1} = E_k^i + T_k^i E_k^{i-1}$, we have

$$A_k\left(E_k^{i-1}v, E_k^{i-1}v\right) - A_k\left(E_k^i v, E_k^i v\right) = \left(\overline{R}_{ki} A_{ki} P_k^i E_k^{i-1} v, A_{ki} P_k^i E_k^{i-1} v\right).$$

Summing from 1 to ℓ gives

$$\begin{aligned} A_k\left(\overline{\mathcal{G}}_k A_k v, v\right) &= A_k(v, v) - \left(A_k E_k^\ell v, E_k^\ell v\right) \\ &= \sum_{i=1}^{\ell}\left(\overline{R}_{ki} A_{ki} P_k^i E_k^{i-1} v, A_{ki} P_k^i E_k^{i-1} v\right). \end{aligned} \tag{8.10}$$

Now let $v \in M_k$ and $w \in M_k$ with $w = \sum w_i$ satisfying Hypothesis (ii). Then

$$A_k(v, w) = \sum_{i=1}^{\ell} A_k(v, w_i) = \sum_{i=1}^{\ell} \left(A_k\big(E_k^{i-1} v, w_i\big) + \sum_{j=1}^{i-1} A_k\big(T_k^j E_k^{j-1} v, w_i\big) \right).$$

By Hypotheses (iv) and (ii), the first term can be estimated as

$$\begin{aligned}
&\sum_{i=1}^{\ell} A_k\big(E_k^{i-1} v, w_i\big) \\
&\quad = \sum_{i=1}^{\ell} \big(A_{ki} P_k^i E_k^{i-1} v, w_i\big) \\
&\quad \leqslant \left(\sum_{i=1}^{\ell} \big(\overline{R}_{ki} A_{ki} P_k^i E_k^{i-1} v, A_{ki} P_k^i E_k^{i-1} v\big) \right)^{1/2} \left(\sum_{i=1}^{\ell} \big(\overline{R}_{ki}^{-1} w_i, w_i\big) \right)^{1/2} \\
&\quad \leqslant \big(A_k\big(\overline{\mathcal{G}}_k A_k v, v\big)\big)^{1/2} \left(\tilde{\omega}^{-1} \lambda_k \sum_{i=1}^{\ell} (w_i, w_i) \right)^{1/2} \\
&\quad \leqslant \big(A_k\big(\overline{\mathcal{G}}_k A_k v, v\big)\big)^{1/2} \big(\tilde{\omega}^{-1} \lambda_k C_2 \|w\|^2\big)^{1/2}.
\end{aligned}$$

The second term can be estimated by using Hypotheses (i) and (iii) as

$$\begin{aligned}
&\sum_{i=1}^{\ell} \sum_{j=1}^{i-1} A_k\big(T_k^j E_k^{j-1} v, w_i\big) \\
&\quad \leqslant (C_1 - 1) \left(\sum_{i=1}^{\ell} A_k\big(T_k^i E_k^{i-1} v, T_k^i E_k^{i-1} v\big) \right)^{1/2} \left(\sum_{i=1}^{\ell} A_k(w_i, w_i) \right)^{1/2} \\
&\quad \leqslant (C_1 - 1) \left(\frac{\tilde{\theta}}{2 - \tilde{\theta}} \sum_{i=1}^{\ell} \big(\overline{R}_{ki} A_{ki} P_k^i E_k^{i-1} v, A_{ki} P_k^i E_k^{i-1} v\big) \right)^{1/2} \\
&\qquad \times \left(\lambda_k \sum_{i=1}^{\ell} (w_i, w_i) \right)^{1/2} \\
&\quad \leqslant (C_1 - 1) \left(\frac{\tilde{\theta}}{2 - \tilde{\theta}} A_k\big(\overline{\mathcal{G}}_k A_k v, v\big) \right)^{1/2} \big(C_2 \lambda_k \|w\|^2\big)^{1/2}.
\end{aligned}$$

Combining the two estimates we see that

$$(A_k v, w) \leqslant \left[\left(\frac{1}{\tilde{\omega}}\right)^{1/2} + (C_1 - 1)\left(\frac{\tilde{\theta}}{2-\tilde{\theta}}\right)^{1/2}\right]\left[A_k\left(\overline{\mathcal{G}}_k A_k v, v\right)\right]^{1/2}\left[\lambda_k C_2 \|w\|^2\right]^{1/2}.$$

Since $w \in M_k$ is arbitrary, we obtain

$$\|A_k v\|^2 \leqslant \left[\left(\frac{1}{\tilde{\omega}}\right)^{1/2} + (C_1 - 1)\left(\frac{\tilde{\theta}}{2-\tilde{\theta}}\right)^{1/2}\right]^2 C_2 \lambda_k \left[A_k\left(\overline{\mathcal{G}}_k A_k v, v\right)\right],$$

which is equivalent to

$$\frac{\omega}{\lambda_k}\|v\|^2 \leqslant \left(\overline{\mathcal{G}}_k v, v\right) \quad \text{with } \frac{1}{\omega} = C_2\left[\left(\frac{1}{\tilde{\omega}}\right)^{1/2} + (C_1 - 1)\left(\frac{\tilde{\theta}}{2-\tilde{\theta}}\right)^{1/2}\right]^2.$$

Hence Condition (SM.1) holds for $\mathcal{G}_k$. To show that $\mathcal{G}_k$ satisfies Condition (SM.2), we note that

$$\mathcal{G}_k A_k = I - E_k^\ell = I - \left(I - T_k^\ell\right)\cdots\left(I - T_k^1\right) = \sum_{i=1}^{\ell} T_k^i E_k^{i-1}.$$

Thus by Hypotheses (ii) and (iii),

$$\begin{aligned} A_k(\mathcal{G}_k A_k v, \mathcal{G}_k A_k v) &= \sum_{i,j=1}^{\ell} A_k\left(T_k^i E_k^{i-1} v, T_k^j E_k^{j-1} v\right) \\ &\leqslant \|\kappa\|_2 \sum_{i=1}^{\ell} A_k\left(T_k^i E_k^{i-1} v, T_k^i E_k^{i-1} v\right) \\ &\leqslant \|\kappa\|_2 \frac{\tilde{\theta}}{2-\tilde{\theta}} A_k\left(\overline{\mathcal{G}}_k A_k v, v\right). \end{aligned}$$

This implies that

$$(A_k \mathcal{G}_k v, \mathcal{G}_k v) \leqslant \frac{\|\kappa\|_2 \tilde{\theta}}{2-\tilde{\theta}}\left(\overline{\mathcal{G}}_k v, v\right) \leqslant \frac{C_1 \tilde{\theta}}{2-\tilde{\theta}}\left(\overline{\mathcal{G}}_k v, v\right).$$

Hence Condition (SM.2) holds with $\theta = 2\tilde{C}_1/(1+\tilde{C}_1)$ where $\tilde{C}_1 = C_1\tilde{\theta}/(2-\tilde{\theta})$. Clearly $\theta \in (0, 2)$. □

We now consider Condition (SM.4). In general the smoothers defined in Subsection 8.1 do not satisfy Condition (SM.4), as it can be seen easily by setting $\ell = 1$, $M_k^1 = M_k$ in Eqs. (8.1) and (8.2). We will show that for elliptic finite element problems the smoothers derived from a point or a line relaxation method satisfy Condition

(SM.4). We consider special space decompositions of M_k that correspond to point/line Jacobi and Gauss–Seidel methods. We assume that each M_k^i is either a one-dimensional subspace spanned by a nodal basis function ϕ_k^i or the subspace spanned by the nodal basis functions along a line. For this special decomposition, the following inequality holds:

$$c\lambda_k \|v\|^2 \leqslant A(v, v) \leqslant \lambda_k \|v\|^2, \quad \text{for all } v \in M_k^i.$$

With this in mind, we make the following hypothesis on the operators R_{ki}:

HYPOTHESIS (iv$'$). $C_3\lambda_k^{-1}\|v_i\|^2 \leqslant (R_{ki}v_i, v_i) \leqslant C_4\lambda_k^{-1}\|v_i\|^2$, for all $v_i \in M_k^i$.

Clearly, the point/line Jacobi and Gauss–Seidel smoothers defined by Eqs. (8.1) and (8.2) satisfy this hypothesis. We now show that if the space decomposition of M_k satisfies hypothesis (i$'$) and all the operators R_{ki} satisfy (iii) and (iv$'$), then the smoothers defined by Eqs. (8.3) and (8.4) satisfy Condition (SM.4).

THEOREM 8.5. *Assume that the space decomposition of M_k satisfies Hypothesis* (i$'$). *If all the operators R_{ki} are symmetric and satisfy Hypothesis* (iv$'$), *then the additive smoother $R_k = \mathcal{J}_k$ defined by* (8.3) *satisfies Condition* (SM.4) *with $\omega_4 = C_1'C_4$.*

We will not give a proof for this result. Since $R_k = \mathcal{J}_k$ is symmetric, Condition (SM.4) is equivalent to $(\mathcal{J}_k v, v) \leqslant \omega_4\lambda_k^{-1}\|v\|^2$. This inequality will be proved in the next subsection; cf. Theorem 8.8.

THEOREM 8.6. *Assume that the space decomposition of M_k satisfies Hypothesis* (i$'$). *If all the operators R_{ki} are symmetric and satisfy Hypotheses* (iii) *and* (iv$'$), *then the multiplicative smoother $\mathcal{G}_k$ defined by* (8.4) *satisfies Condition* (SM.4).

PROOF. Set $T_k^i = R_{ki}A_{ki}P_k^i$ and

$$E_k^i = (I - T_k^i)(I - T_k^{i-1})\cdots(I - T_k^1), \quad E_k^0 = I.$$

Note that $I - \mathcal{G}_k A_k = E_k^\ell$ and $E_k^{i-1} = E_k^i + T_k^i E_k^{i-1}$. Hence

$$\mathcal{G}_k A_k = I - E_k^\ell = \sum_{i=1}^{\ell} T_k^i E_k^{i-1} = \sum_{i=1}^{\ell} R_{ki}A_{ki}P_k^i E_k^{i-1}.$$

By Hypothesis (i$'$) and Lemma 8.1,

$$(\mathcal{G}_k A_k v, \mathcal{G}_k A_k v) \leqslant C_1' \sum_{i=1}^{\ell} \left(R_{ki}A_{ki}P_k^i E_k^{i-1}v, R_{ki}A_{ki}P_k^i E_k^{i-1}v\right). \tag{8.11}$$

By Hypotheses (iv′) and (iii), and since R_{ki} is symmetric,

$$(R_{ki}v_i, R_{ki}v_i) \leqslant C_4\lambda_k^{-1}(R_{ki}v_i, v_i) \leqslant \frac{C_4\lambda_k^{-1}}{2-\tilde{\theta}}\big(\overline{R}_{ki}v_i, v_i\big), \quad \text{for all } v_i \in M_k^i. \tag{8.12}$$

Applying (8.12) with $v_i = A_{ki}P_k^i E_k^{i-1}v$ to (8.11), we obtain

$$\begin{aligned}(\mathcal{G}_k A_k v, \mathcal{G}_k A_k v) &\leqslant \frac{C_1' C_4}{2-\tilde{\theta}}\lambda_k^{-1}\big(\overline{R}_{ki}A_{ki}P_{ki}E_k^{i-1}v, A_{ki}P_{ki}E_k^{i-1}v\big) \\ &= \frac{C_1' C_4}{2-\tilde{\theta}}\lambda_k^{-1}\big(\overline{\mathcal{G}}_k A_k v, A_k v\big) \leqslant 2\frac{C_1' C_4}{2-\tilde{\theta}}\lambda_k^{-1}\big(\mathcal{G}_k A_k v, A_k v\big).\end{aligned}$$

This completes the proof of the theorem. □

8.3. Smoothers for additive preconditioners

In the constructions of multilevel additive preconditioners of Section 4, we are mainly interested in symmetric smoothers, R_k. We consider the case in which the smoother R_k is the additive smoother $\mathcal{J}_k$ or the symmetric multiplicative smoother $\overline{\mathcal{G}}_k \equiv \mathcal{G}_k + \mathcal{G}_k^t - \mathcal{G}_k^t A_k \mathcal{G}_k$. We assume that all the R_{ki}'s are symmetric and positive definite. We show that the smoothers constructed Subsection 8.1 satisfy the smoothing conditions required by the multilevel additive preconditioner.

Recall the following smoothing conditions which were used in Section 4 in the study of additive preconditioners.

CONDITION (SA.1). There exist positive constants ω_1 and ω_2, not depending on $1 \leqslant k \leqslant J$, such that

$$\frac{\omega_1}{\lambda_k}(v, v) \leqslant (R_k v, v) \leqslant \omega_2\big(A_k^{-1}v, v\big), \quad \text{for all } v \in M_k,\ k \geqslant 2.$$

CONDITION (SA.2). There exist positive constants ω_1 and ω_2, not depending on $1 \leqslant k \leqslant J$, such that

$$\frac{\omega_1}{\lambda_k}(v, v) \leqslant (R_k v, v) \leqslant \frac{\omega_2}{\lambda_k}(v, v), \quad \text{for all } v \in M_k,\ k \geqslant 2.$$

It is easy to see that Condition (SA.2) implies (SA.1). If R_k is symmetric, then Condition (SA.2) implies Condition (SM.4) with $\omega_4 = \omega_2$ and Conditions (SA.2) and (SM.2) imply Condition (SM.1).

We show first that the additive smoother $\mathcal{J}_k$ and the symmetric multiplicative smoother $\overline{\mathcal{G}}_k$ satisfy Condition (SA.1).

THEOREM 8.7. *Let $\mathcal{J}_k$ and $\mathcal{G}_k$ be defined by Eqs.* (8.3) *and* (8.4), *respectively. Assume that the space decomposition of M_k satisfies Hypotheses* (i) *and* (ii) *and that all the operators R_{ki} are symmetric and satisfy Hypotheses* (iii) *and* (iv). *Then $R_k = \mathcal{J}_k$ and $R_k = \overline{\mathcal{G}}_k \equiv \mathcal{G}_k + \mathcal{G}_k^t - \mathcal{G}_k^t A_k \mathcal{G}_k$ satisfy Condition* (SA.1).

PROOF. We first consider $\mathcal{J}_k$. We have shown in the proof of Theorem 8.3 that Hypotheses (i) and (iii) imply that

$$A_k(\mathcal{J}_k v, \mathcal{J}_k v) \leqslant C_1\tilde{\theta}(\mathcal{J}_k v, v), \quad \text{for all } v \in M_k.$$

Hence $(\mathcal{J}_k v, v) \leqslant C_1\tilde{\theta}(A_k^{-1}v, v)$. Hypothesis (iv) implies that

$$\tilde{\omega}\lambda_k^{-1}\|v_i\|^2 \leqslant \big(\overline{R}_{ki}v_i, v_i\big) \leqslant 2(R_{ki}v_i, v_i), \quad \text{for all } v_i \in M_k^i.$$

Now by Hypothesis (ii),

$$\begin{aligned}\|v\|^2 &\leqslant C_2\sum_{i=1}^{\ell}\big\|Q_k^i v\big\|^2 \leqslant 2C_2\tilde{\omega}^{-1}\lambda_k\sum_{i=1}^{\ell}\big(R_{ki}Q_k^i v, Q_k^i v\big)\\ &= 2C_2\tilde{\omega}^{-1}\lambda_k(\mathcal{J}_k v, v).\end{aligned}$$

Therefore

$$\tfrac{1}{2}C_2^{-1}\tilde{\omega}\lambda_k^{-1}\|v\|^2 \leqslant (\mathcal{J}_k v, v) \leqslant C_1\tilde{\theta}\big(A_k^{-1}v, v\big).$$

This proves that $\mathcal{J}_k$ satisfies Condition (SA.1) with $\omega_1 = \frac{1}{2}\tilde{\omega}/C_2$ and $\omega_2 = C_1\tilde{\theta}$.

By the definition of $\overline{\mathcal{G}}_k$,

$$0 \leqslant A_k\big((I - \mathcal{G}_k A_k)v, (I - \mathcal{G}_k A_k)v\big) = A_k\big((I - \overline{\mathcal{G}}_k A_k)v, v\big).$$

Thus $(\overline{\mathcal{G}}_k v, v) \leqslant (A_k^{-1}v, v)$. On the other hand, by Theorem 8.4,

$$\frac{\omega}{\lambda_k}\|v\|^2 \leqslant \big(\overline{\mathcal{G}}_k v, v\big) \quad \text{for } \omega^{-1} = C_2\left[\left(\frac{1}{\tilde{\omega}}\right)^{1/2} + (C_1 - 1)\left(\frac{\tilde{\theta}}{2-\tilde{\theta}}\right)^{1/2}\right]^2.$$

Thus $\overline{\mathcal{G}}_k$ satisfies Condition (SA.1) with $\omega_1 = \omega$ and $\omega_2 = 1$. □

In general, the smoothers defined in Subsection 8.1 do not satisfy Condition (SA.2), as it can be seen easily by setting $\ell = 1$, $M_k^1 = M_k$ and $R_{k,1} = A_k^{-1}$. We will show that Condition (SA.2) holds for the smoothers derived from a point or a line relaxation method.

THEOREM 8.8. *Let $\mathcal{J}_k$ and $\mathcal{G}_k$ be defined by* (8.3) *and* (8.4). *Assume that all the R_{ki}'s are symmetric. Then*

(a) *If Hypotheses* (i′), (ii) *and* (iv′) *hold, then $R_k = \mathcal{J}_k$ satisfies Condition* (SA.2).

(b) *If Hypotheses* (i)–(iv), (i′) *and* (iv′) *hold, then $R_k = \overline{\mathcal{G}}_k$ satisfies Condition* (SA.2).

PROOF. Given $v \in M_k$, choose $v_i \in M_k^i$ such that Hypothesis (ii) holds. Then by Hypothesis (iv′),

$$
\begin{aligned}
(v, v) &= \sum_{i=1}^{\ell} \left(Q_k^i v, v_i\right) \leqslant \left(\sum_{i=1}^{\ell} \left(R_{ki} Q_k^i v, Q_k^i v\right)\right)^{1/2} \left(\sum_{i=1}^{\ell} \left(R_{ki}^{-1} v_i, v_i\right)\right)^{1/2} \\
&= (\mathcal{J}_k v, v)^{1/2} \left(\sum_{i=1}^{\ell} \left(R_{ki}^{-1} v_i, v_i\right)\right)^{1/2} \\
&\leqslant (\mathcal{J}_k v, v)^{1/2} \left[(C_2/C_3)\lambda_k \|v\|^2\right]^{1/2}.
\end{aligned}
$$

Hence $(v, v) \leqslant (C_2/C_3)\lambda_k(\mathcal{J}_k v, v)$. On the other hand, it follows from Hypotheses (i′) and (iv′) for any $v = \sum v_i$ with $v_i \in M_k^i$,

$$
\begin{aligned}
(v, v) &= \sum_{i=1}^{\ell} \sum_{j=1}^{\ell} (v_i, v_j) \leqslant C_1' \sum_{i=1}^{\ell} \|v_i\|^2 \\
&\leqslant C_1' C_4 \lambda_k^{-1} \sum_{i=1}^{\ell} \left(R_{ki}^{-1} v_i, v_i\right) = C_1' C_4 \lambda_k^{-1} \left(\mathcal{J}_k^{-1} v, v\right).
\end{aligned}
$$

This proves that $R_k = \mathcal{J}_k$ satisfies Condition (SA.2) with $\omega_1 = C_3/C_2$ and $\omega_2 = C_1' C_4$.

To prove part (b) we note that by Theorem 8.4, Hypotheses (i)–(iv) imply that

$$
\frac{\omega}{\lambda_k} \|v\|^2 \leqslant \left(\overline{\mathcal{G}}_k v, v\right) \quad \text{with } \frac{1}{\omega} = C_2 \left[\left(\frac{1}{\tilde{\omega}}\right)^{1/2} + (C_1 - 1)\left(\frac{\tilde{\theta}}{2 - \tilde{\theta}}\right)^{1/2}\right]^2.
$$

We shall show now that

$$
\left(\overline{\mathcal{G}}_k v, v\right) \leqslant \left(\frac{4}{2 - \tilde{\theta}}\right)(\mathcal{J}_k v, v) \leqslant \left(\frac{4}{2 - \tilde{\theta}}\right) C_1' C_4 \lambda_k^{-1} (v, v).
$$

Recall that

$$
\mathcal{G}_k A_k v = \left(I - E_k^{\ell}\right) v = \sum_{i=1}^{\ell} T_k^i E_k^{i-1} v, \quad T_k^i = R_{ki} Q_k^i A_k = R_{ki} A_{ki} P_k^i.
$$

Note that R_{ki} is symmetric and positive definite. By the Cauchy–Schwarz inequality and Hypothesis (iii), we have,

$$
\begin{aligned}
A\left(\overline{\mathcal{G}}_k A_k v, v\right) &\leqslant 2A(\mathcal{G}_k A_k v, v) = 2 \sum_{i=1}^{\ell} A\left(T_k^i E_k^{i-1} v, v\right) \\
&= 2 \sum_{i=1}^{\ell} \left(R_{ki} A_{ki} P_k^i E_k^{i-1} v, A_{ki} P_k^i v\right)
\end{aligned}
$$

$$
\begin{aligned}
&\leqslant 2\Bigg(\sum_{i=1}^{\ell} A\big(T_k^i E_k^{i-1} v, E_k^{i-1} v\big)\Bigg)^{1/2} \Bigg(\sum_{i=1}^{\ell} A\big(T_k^i v, v\big)\Bigg)^{1/2} \\
&= 2\Bigg(\sum_{i=1}^{\ell} \big(R_{ki} A_{ki} P_k^i E_k^{i-1} v, A_{ki} P_k^i E_k^{i-1} v\big)\Bigg)^{1/2} \\
&\quad \times \Bigg(\sum_{i=1}^{\ell} \big(R_{ki} A_{ki} P_k^i v, A_{ki} P_k^i v\big)\Bigg)^{1/2} \\
&\leqslant 2\Bigg(\frac{1}{2-\tilde{\theta}} \sum_{i=1}^{\ell} \big(\overline{R}_{ki} A_{ki} P_k^i E_k^{i-1} v, A_{ki} P_k^i E_k^{i-1} v\big)\Bigg)^{1/2} \\
&\quad \times \big(\mathcal{J}_k A_k v, A_k v\big)^{1/2} \\
&= 2\Bigg(\frac{1}{2-\tilde{\theta}} \big(\overline{\mathcal{G}}_k A_k v, A_k v\big)\Bigg)^{1/2} \big(\mathcal{J}_k A_k v, A_k v\big)^{1/2}.
\end{aligned}
$$

In the last step, we have used identity (8.10). Cancelling the common factor gives

$$
A_k\big(\overline{\mathcal{G}}_k A_k v, v\big) \leqslant \left(\frac{4}{2-\tilde{\theta}}\right)(\mathcal{J}_k A_k v, A_k v)
$$

and hence

$$
\big(\overline{\mathcal{G}}_k v, v\big) \leqslant \left(\frac{4}{2-\tilde{\theta}}\right)(\mathcal{J}_k v, v) \leqslant \left(\frac{4}{2-\tilde{\theta}}\right) C_1' C_4 \lambda_k^{-1} (v, v).
$$

This proves that the smoother $R_k = \overline{\mathcal{G}}_k$ satisfies Condition (SA.2). □

8.4. *Some examples*

In this subsection, we consider some examples of smoothers for second-order elliptic problems (see Sections 2 and 9). We assume that the form $A_k(\bullet, \bullet)$ is uniformly bounded and coercive, i.e., $c\|v\|_1^2 \leqslant A_k(v, v) \leqslant C\|v\|_1^2$, for all $v \in M_k$. We shall denote by $\{\phi_k^i\}$ the standard nodal basis functions of M_k. We denote by $\underset{\approx}{A}_k = [A(\phi_k^i, \phi_k^j)]$ and $\underset{\approx}{G}_k = [(\phi_k^i, \phi_k^j)]$ the corresponding stiffness matrix and mass matrix. For each $u \in M_k$ we denote by $\tilde{u}$ the *coefficient vector* of u with respect to the nodal basis and for each $f \in M_k$, we use $\underset{\sim}{f}$ to denote the *dual vector* $[(f, \phi_k^i)]$ of f with respect to the nodal basis. This notation has already been used in Section 1. We have shown in Section 2 that there exist constants α_1 and α_2 independent of k such that

$$
\alpha_1 \|v\|^2 \leqslant \lambda_k^{-1} \sum_i \big|\tilde{v}_i\big|^2 \leqslant \alpha_2 \|v\|^2, \quad \text{for all } v \in M_k. \tag{8.13}
$$

This implies in particular that the eigenvalues of $\underset{\approx}{G}_k$ are on the order of λ_k^{-1}.

We first consider various types of Richardson smoothers.

EXAMPLE 8.1 (*Richardson smoother* I). The smoother corresponding to the Richardson method applied to the finite element equation $A_k u = f$ is

$$R_k = \bar{\lambda}_k^{-1} I, \quad \text{with } \bar{\lambda}_k \text{ satisfying } \lambda_k \leqslant \bar{\lambda}_k \leqslant C\lambda_k.$$

Here I denotes the identity operator on M_k. It is easy to see that Condition (SM.1) holds with $w = \lambda_k/\bar{\lambda}_k$ and Condition (SM.2) holds with $\theta = 1$. Note that $(R_k v, R_k v) = \bar{\lambda}_k^{-1}(R_k v, v) = (\lambda_k/\bar{\lambda}_k)[\lambda_k^{-1}(R_k v, v)]$, Condition (SM.4) holds with $\omega_4 = \lambda_k/\bar{\lambda}_k$. This smoother clearly satisfies Condition (SA.2) and hence it also satisfies Condition (SA.1). Note that the implementation of this smoother in Algorithm 3.1 requires the inversion of the mass matrix $[(\phi_k^i, \phi_k^j)]$.

EXAMPLE 8.2 (*Richardson smoother* II). The finite element equation $A_k u = f$ is equivalent to the matrix equation $\underset{\approx}{A}_k \tilde{u} = \underset{\sim}{f}$. The smoother corresponding to the Richardson method applied to the matrix equation is $\overset{\approx}{R}_k = \rho^{-1} I$, where I is the identity matrix and ρ satisfies $\underset{\approx}{\lambda}_k \leqslant \rho \leqslant C\underset{\approx}{\lambda}_k$. Here $\underset{\approx}{\lambda}_k$ denotes the largest eigenvalue of $\underset{\approx}{A}_k$. This is equivalent to defining R_k by

$$R_k f = \frac{1}{\rho} \sum_{i=1}^{\ell} \underset{\sim}{f}_i \phi_k^i = \frac{1}{\rho} \sum_{i=1}^{\ell} (f, \phi_k^i) \phi_k^i.$$

Note that

$$(R_k f, f) = \rho^{-1} \langle \underset{\sim}{f}, \underset{\sim}{f} \rangle,$$

$$(R_k f, R_k f) = \rho^{-2} \langle \underset{\approx}{G}_k \underset{\sim}{f}, \underset{\sim}{f} \rangle,$$

and

$$A(R_k f, R_k f) = \rho^{-2} \langle \underset{\approx}{A}_k \underset{\sim}{f}, \underset{\sim}{f} \rangle.$$

It is easy to see that Condition (SM.2) holds with $\theta = 1$. By (8.13),

$$\alpha_2^{-1} \lambda_k^{-1} \langle \underset{\sim}{f}, \underset{\sim}{f} \rangle \leqslant \langle \underset{\approx}{G}_k \underset{\sim}{f}, \underset{\sim}{f} \rangle \leqslant \alpha_1^{-1} \lambda_k^{-1} \langle \underset{\sim}{f}, \underset{\sim}{f} \rangle.$$

This, together with Condition (SM.2), implies that Condition (SM.1) holds with $\omega = 1/(\alpha_2 \rho)$ and (SM.4) holds with $\omega_4 = 1/(\alpha_1 \rho)$. It is clear that (SA.2) also holds.

EXAMPLE 8.3 (*Richardson smoother* III). The smoother corresponding to the Richardson method applied to the normal finite element equation, $A_k^t A_k u = A_k^t f$, is

$$R_k = \bar{\lambda}_k^{-2} A_k, \quad \text{with } \bar{\lambda}_k \text{ satisfying } \lambda_k \leqslant \bar{\lambda}_k \leqslant C\lambda_k.$$

Clearly, Condition (SM.2) holds with $\theta = 1$ and Conditions (SM.3) and (SM.4) hold with $\omega = \omega_4 = (\lambda_k/\bar{\lambda}_k)^2$.

The remaining smoothers correspond to Jacobi and Gauss–Seidel type smoothers including point, line and block methods. The construction is based on the space decomposition of M_k discussed earlier.

EXAMPLE 8.4 (*Domain decomposition smoothers*). Let $\Omega = \bigcup_{i=1}^{\ell} \Omega_k^i$ be a decomposition of Ω, where each subdomain is a union of supports of some nodal basis functions ϕ_k^j. We consider the space decomposition of M_k where each M_k^i is the span of the nodal basis functions associated with the subdomain Ω_k^i. The corresponding smoothing operators $\mathcal{J}_k$ and $\mathcal{G}_k$ are often called (one-level) additive and multiplicative (Schwarz) smoothers and they arise naturally in overlapping domain decomposition methods. When the space decomposition is a direct sum, the operators $\mathcal{J}_k$ and $\mathcal{G}_k$ are the usual block Jacobi and block Gauss–Seidel smoothers.

In general ℓ depends on k and the space decomposition is not a direct sum. We assume that the subdomains $\{\Omega_k^i\}_{i=1}^{\ell}$ have the *limited intersection* property in the following sense. There exists a positive integer n_0, independent of k and ℓ, such that each subdomain Ω_k^i intersects at most n_0 other subdomains. It follows that for every i, the number of indices j for which $(v^i, v^j) \neq 0$ with $v^i \in M_k^i$ and $v^j \in M_k^j$ is bounded above by n_0 and the number of indices j for which $A(v^i, v^j) \neq 0$ with $v^i \in M_k^i$ and $v^j \in M_k^j$ is also bounded above by n_0. These statements can be written as

$$\|\kappa^0\|_\infty \leqslant n_0 \quad \text{and} \quad \|\kappa\|_\infty \leqslant n_0.$$

Hence Hypotheses (i) and (i′) hold with $C_1 = C_1' = n_0$.

We now verify Hypothesis (ii). Denote by Λ_k the set of nodes corresponding to the basis $\{\phi_k^i\}$. Partition Λ_k into ℓ disjoint subsets Λ_k^i such that nodes in Λ_k^i are in Ω_k^i. The way in which this can be done is in general not unique. Now each $v \in M_k$ can be written as

$$v = \sum_{i=1}^{\ell} v_i, \quad \text{where } v_i = \sum_{j \in \Lambda_k^i} \tilde{v}_j \phi_k^j.$$

By (8.13),

$$\|v_i\|^2 \leqslant \alpha_1^{-1} \lambda_k^{-1} \sum_{j \in \Lambda_k^i} |\tilde{v}_j|^2,$$

and thus

$$\sum_{i=1}^{\ell} \|v_i\|^2 \leqslant \alpha_1^{-1}\lambda_k^{-1} \sum_{j\in\Lambda_k} |\tilde{v}_j|^2 \leqslant \frac{\alpha_2}{\alpha_1}\|v\|^2 \equiv C_2\|v\|^2.$$

This is show that Hypothesis (ii) holds.

It follows from Theorems 8.1 and 8.2 that Conditions (SM.1) and (SM.2) hold for the smoothers $\mathcal{J}_k$ and $\mathcal{G}_k$ defined respectively by Eqs. (8.1) and (8.2). Theorem 8.7 shows that $R_k = \mathcal{J}_k$ and $R_k = \overline{\mathcal{G}}_k \equiv \mathcal{G}_k + \mathcal{G}_k^t - \mathcal{G}_k^t A_k \mathcal{G}_k$ satisfy Condition (SA.1).

EXAMPLE 8.5 (*Point/line Jacobi and Gauss–Seidel smoothers*). In this example, we consider the point/line Jacobi and Gauss–Seidel smoothers. These smoothers can be defined respectively by Eqs. (8.1) and (8.2), corresponding to a certain special space decomposition of M_k.

We assume that each M_k^i is either a one-dimensional subspace spanned by a single nodal basis function ϕ_k^i or the subspace spanned by the nodal basis functions along a line. The number of such spaces $\ell = \ell_k$ will often depend on k. Because of the support properties of $\{\phi_k^i\}$, the subspaces $\{M_k^i\}$ satisfy the limited interaction property stated in Example 8.4. Consequently, Hypotheses (i) and (i′) hold. Since this is a special case of smoothers discussed in Example 8.4, Conditions (SM.1) and (SM.2) hold.

For this special case, the following inequality also holds:

$$c\lambda_k\|v\|^2 \leqslant (A_k v, v) \leqslant C\lambda_k\|v\|^2, \quad \text{for all } v \in M_k^i. \tag{8.14}$$

By (8.14), $R_{ki} = A_{ki}^{-1}$ satisfies Hypothesis (iv′). It follows from Theorems 8.5 and 8.6 that the additive smoother $\mathcal{J}_k$ defined by (8.1) and the multiplicative smoother $\mathcal{G}_k$ defined by (8.2) satisfy Conditions (SM.4) and (SA.2). Theorem 8.8 says that Condition (SA.2) holds for the additive smoother $R_k = \mathcal{J}_k$ and the symmetric multiplicative smoother $R_k = \overline{\mathcal{G}}_k \equiv \mathcal{G}_k + \mathcal{G}_k^t - \mathcal{G}_k^t A_k \mathcal{G}_k$.

EXAMPLE 8.6 (*Smoothers using approximate inverses*). We now turn to the additive smoother $\mathcal{J}_k$ and the multiplicative smoother $\mathcal{G}_k$ defined by Eqs. (8.3) and (8.4), respectively. We assume that the underlying space decomposition, $M_k = \sum M_k^i$, satisfies Hypotheses (i), (i′) and (ii) (e.g., a line decomposition). We consider some simple ways to define the approximate inverses R_{ki} of A_{ki}. First we set $R_{ki} = \beta A_{ki}^{-1}$. Then the operators $\mathcal{J}_k$ and $\mathcal{G}_k$ are respectively the block Jacobi and block Gauss–Seidel smoothers corresponding to the given decomposition of M_k. To consider more general cases, we further decompose each M_k^i into a sum of subspaces. We assume that the space decomposition of M_k^i satisfies Hypotheses (i), (i′) and (ii) with constants independent of i. We can then define R_{ki} to be the additive or the multiplicative operators on M_k^i corresponding to the given space decomposition of M_k^i. We note that each decomposition of M_k and each set of decompositions of M_k^i gives rise a new, finer space decomposition of M_k.

We first consider the additive smoother $\mathcal{J}_k$ defined by (8.3).

(i) For each i, let $R_{ki} = \beta A_{ki}^{-1}$ with $\beta > 0$. Then $\mathcal{J}_k$ is the weighted block Jacobi smoother corresponding to the original space decomposition of M_k.

(ii) For each i, let $M_k^i = \operatorname{span}\{\phi_k^i\}$ and $R_{ki} = \rho^{-1}(\phi_k^i, \phi_k^i) = (A(\phi_k^i, \phi_k^i)/\rho)\, A_{ki}^{-1}$. Note that $Q_k^i f = (f, \phi_k^i)/(\phi_k^i, \phi_k^i)\phi_k^i$. Hence the corresponding smoother $\mathcal{J}_k$ is the Richardson smoother discussed in Example 8.1.

(iii) For each i, let R_{ki} be the weighted Jacobi operator on M_k^i corresponding to the space decomposition of M_k^i. Then $\mathcal{J}_k$ is a weighted Jacobi smoother corresponding to the finer space decomposition of M_k.

(iv) For each i, let R_{ki} be the Gauss–Seidel (SOR) operator corresponding to the given decomposition of M_k^i. This is a hybrid smoother.

We now consider the multiplicative smoother $\mathcal{G}_k$.

(i) For each i, let $R_{ki} = \beta A_{ki}^{-1}$ with $\beta \in (0, 2)$. This is the SOR smoother corresponding to the original space decomposition of M_k.

(ii) For each i, let R_{ki} be the Gauss–Seidel operator corresponding to the given decomposition of M_k^i. The resulting operator $\mathcal{G}_k$ is the Gauss–Seidel smoother corresponding to the finer space decomposition of M_k.

(iii) For each i, let R_{ki} be the Jacobi operator corresponding to the given space decomposition of M_k^i. The resulting operator $\mathcal{G}_k$ is a hybrid smoother.

Theorems 8.4–8.8 can be used to verify the relevant smoothing conditions.

8.5. *Bibliographical notes*

Various smoothers have been introduced in the literature. The ones most often introduced correspond to the point-Jacobi or Richardson type iteration; cf. BANK and DUPONT [1981]. Other authors have considered the Gauss–Seidel or symmetric Gauss–Seidel iteration as a smoother; cf. BANK and DOUGLAS [1985] and BANK, MANDEL and MCCORMICK [1987]. Theorems 8.1 and 8.2 taken from BRAMBLE and PASCIAK [1992] and includes point and block Jacobi and Gauss–Seidel as well as smoothers corresponding to overlapping domain decomposition preconditioners (DRYJA and WIDLUND [1987]). Operators defined by Eqs. (8.3) and (8.4) and Theorems 8.3 and 8.4 can be used to construct and analyze some smoothers of neither additive nor multiplicative type.

CHAPTER III

Applications

In this chapter we will consider various applications of the abstract theory of Chapter II. In the first section we will apply the results to standard finite element problems for second-order selfadjoint elliptic boundary value problems, including the case of locally refined meshes. In the second section we treat the case in which numerical integration is used in the approximate problem. Section 11 deals with some cases of nonsymmetric or indefinite problems. Section 13 is devoted to operators of order minus one with the main example being that of a weakly singular integral operator. Finally, we show in Section 14 how the theory applies to many different approximations to the biharmonic Dirichlet problem.

9. Second-order selfadjoint elliptic problems

In this section we shall show how the multilevel methods and the analysis apply to finite element approximations to second-order selfadjoint elliptic boundary value problems. We shall only consider homogeneous Dirichlet boundary conditions. The treatment for nonhomogeneous Dirichlet boundary conditions is standard. Our analysis and results also apply to Neumann and other types of boundary conditions with a straightforward modification. We shall consider polygonal domains as well as domains with curved boundaries. We discuss the case in which the coefficients of the differential equations are smooth as well as the case in which the coefficients are only bounded and measurable. The cases of uniform mesh refinement and local mesh refinement are discussed. For simplicity of exposition, we shall restrict our discussion to polygonal domains in the case of mesh refinements.

Let Ω be a bounded domain in $\mathbf{R}^d$. We shall consider the Dirichlet problem

$$\begin{aligned} \mathcal{L}u &= f \quad \text{in } \Omega, \\ u &= 0 \quad \text{on } \partial\Omega, \end{aligned} \tag{9.1}$$

where

$$\mathcal{L}v = -\sum_{i,j=1}^{d} \frac{\partial}{\partial x_i}\left(a_{ij}\frac{\partial v}{\partial x_j}\right) + av.$$

The Dirichlet problem in weak form is the following: Find $u \in H_0^1(\Omega)$ such that

$$A(u, \phi) = (f, \phi), \quad \text{for all } \phi \in H_0^1(\Omega), \tag{9.2}$$

where the form $A(\bullet, \bullet) : H_0^1(\Omega) \times H_0^1(\Omega) \to \mathbf{R}$ is defined by

$$A(v, w) = \sum_{i,j=1}^{d} \int_\Omega a_{ij} \frac{\partial v}{\partial x_i} \frac{\partial w}{\partial x_j} \, \mathrm{d}x + \int_\Omega a v w \, \mathrm{d}x. \tag{9.3}$$

Throughout this section we shall assume, unless otherwise stated, that the matrix $[a_{ij}(x)]$ is symmetric and uniformly positive definite and $a(x)$ is nonnegative. This assumption on the coefficients implies that the bilinear form $A(\bullet, \bullet)$ is uniformly equivalent to the form corresponding to $-\Delta$.

Denote by $H^{-1}(\Omega)$ the dual of $H_0^1(\Omega)$. The norm $\| \bullet \|_{1+\alpha}$ is the norm on the space defined by interpolation between the space $H_0^1(\Omega)$ and $H^2(\Omega)$. The norm $\| \bullet \|_{-1+\alpha}$ is defined by duality:

$$\|f\|_{-1+\alpha} = \sup_{\phi \in H_0^1(\Omega)} \frac{(f, \phi)}{\|\phi\|_{1-\alpha}}.$$

Here (f, ϕ) is the value of the functional f at ϕ. If $f \in L^2(\Omega)$ then (f, ϕ) is the $L^2(\Omega)$ inner product.

Once we define a finite element space $M \subset H_0^1(\Omega)$, the Galerkin approximation is the solution $u_h \in M$ of the variational equation

$$A(u_h, \phi) = (f, \phi), \quad \text{for all } \phi \in M.$$

We shall consider the multilevel iterative methods of Sections 3 and 4 for solving this finite element problem. We shall apply the multilevel theories of Sections 4 and 5 to four examples: (a) polygonal domains with uniform mesh refinement, (b) equations with rough coefficients, (c) domains with curved boundaries, (d) polygonal domains with local mesh refinement. We shall establish convergence results by applying the theories of Sections 4 and 5.

9.1. Polygonal domains

We first consider the Dirichlet problem (9.1) on a polygonal domain Ω. We assume that a unit-size coarse finite element partition of Ω is given; cf. CIARLET [1978]. Denote this coarse partition by $\mathcal{T}_1 = \{\tau\}$. Associated with the initial mesh partition, we are given a rule for refinement. The triangulation $\mathcal{T}_k$ of Ω is then defined by refining τ in $\mathcal{T}_{k-1}$ by the refinement rule. In the case of two dimensions, we consider the example in which $\mathcal{T}_1$ consists of triangles and $\mathcal{T}_k$ is obtained from $\mathcal{T}_{k-1}$ by breaking each triangle in $\mathcal{T}_{k-1}$ into four triangles by connecting the midpoints of the edges. We assume that the mesh size of the kth triangulation is on the order of δ^k for some fixed $\delta < 1$.

The finite element space M_k is defined to be a space of piecewise polynomial functions with respect to the mesh $\mathcal{T}_k$ which are continuous on Ω and vanish on $\partial\Omega$. For two-dimensional cases, a simple example is that in which M_k consists of continuous functions which are piecewise linear with respect to the triangulation. The finite element spaces are nested, i.e.,

$$M_1 \subset M_2 \subset \cdots \subset M_J = M.$$

The Galerkin approximation to u is the unique function $u_J \in M_J$ satisfying

$$A(u_J, \phi) = (f, \phi), \quad \text{for all } \phi \in M_J. \tag{9.4}$$

We consider approximating the solution of (9.4) by the multigrid V-cycle Algorithm 3.1 and the PCG algorithm with the multilevel additive preconditioner defined by (4.3). For the multigrid algorithm, we assume, unless stated otherwise, that the smoothers satisfy Conditions (SM.1) and (SM.2) (or (SM.3) and (SM.2)). For the multilevel additive preconditioner, we assume that the smoothers satisfy Condition (SA.2). We refer to Section 8 for the construction of examples of such smoothers.

We first establish uniform convergence results by applying Theorems 4.2 and 5.1. Since the smoothers are assumed to satisfy the relevant smoothing conditions, it remains to verify Conditions (A.2), (A.3) and (A.4).

Recall that operators $A_k : M_k \to M_k$ are defined by

$$(A_k v, \phi) = A(v, \phi), \quad \text{for all } v, \phi \in M_k.$$

The maximum eigenvalue λ_k of A_k satisfies $\lambda_k \approx h_k^{-2}$ and hence Condition (A.3) on λ_k is satisfied.

It is well known (cf. BRAMBLE and XU [1991]) that the L^2 projection has the following stable approximation property:

$$\left\|(I - Q_k)v\right\|^2 + \lambda_k^{-1}\|Q_k v\|_1^2 \leqslant C\lambda_k^{-1}\|v\|_1^2, \quad \text{for all } v \in H_0^1(\Omega).$$

In particular,

$$\|Q_k v\|_1 \leqslant C\|v\|_1, \quad \text{for all } v \in H_0^1(\Omega). \tag{9.5}$$

Now recall that the scale of discrete norms on M_k is defined by

$$|||v|||_{s,k} = \left\|A_k^{s/2} v\right\|, \quad \text{for all } v \in M_k.$$

Note that $|||\bullet|||_{0,k} \equiv \|\bullet\|$ and that $|||\bullet|||_{1,k} \equiv A(\bullet,\bullet)^{1/2}$ is uniformly equivalent to $\|\bullet\|_1$. Hence, restricted to $v \in M_i \subset H_0^1(\Omega)$, (9.5) reduces to $|||Q_k v|||_{1,k} \leqslant C|||v|||_{1,i}$ for $v \in M_i$ and $i \geqslant k$. By Theorem A.4, interpolating $|||Q_k v|||_{1,k} \leqslant C|||v|||_{1,i}$ and the trivial inequality $\|Q_k v\| \leqslant \|v\|$, we obtain

$$|||Q_k v|||_{1-\alpha,k} \leqslant C|||v|||_{1-\alpha,i}, \quad \text{for all } v \in M_i, \text{ with } i \geqslant k.$$

This is Condition (A.4). In the verification of Condition (A.4), we have used only the equivalence of the forms $A(\bullet,\bullet)$ and $(\bullet,\bullet)_{H^1}$.

To establish Condition (A.2), we need a certain regularity property of (9.1). We assume that the coefficients $a_{ij}(x)$ and $a(x)$ are sufficiently smooth so that the following a priori estimate is valid: There exists some $\alpha \in (0,1]$ such that for $f \in H^{-1+\alpha}(\Omega)$, the solution u of (9.1) is in $H^{1+\alpha}(\Omega) \cap H_0^1(\Omega)$ and

$$\|u\|_{1+\alpha} \leqslant C\|f\|_{-1+\alpha}. \tag{9.6}$$

It is well known that for $v \in H^{1+\alpha}(\Omega) \cap H_0^1(\Omega)$,

$$\big\|(I - P_{k-1})v\big\|_1 \leqslant Ch_{k-1}^{\alpha}\|v\|_{1+\alpha} \leqslant C\lambda_k^{-\alpha/2}\|v\|_{1+\alpha}. \tag{9.7}$$

The following result shows that Condition (A.2) follows from the regularity assumption in (9.6) and the approximation property in (9.7).

LEMMA 9.1. *The regularity assumption in* (9.6) *and the approximation property in* (9.7) *imply Condition* (A.2); *i.e.,*

$$\big|\big|\big|(I - P_{k-1})v\big|\big|\big|_{1-\alpha,k} \leqslant C\lambda_k^{-\alpha/2}|||v|||, \quad \textit{for all } v \in M_k.$$

PROOF. The proof is based on an application of the duality argument of Aubin and Nitsche; cf. AZIZ and BABUŠKA [1972] or CIARLET [1978]. Set $g = A_k^{1-\alpha}(I - P_{k-1})v \in M_k$. Then

$$\big|\big|\big|(I - P_{k-1})v\big|\big|\big|_{1-\alpha,k}^2 = |||g|||_{-1+\alpha,k}^2 = \big((I - P_{k-1})v, g\big). \tag{9.8}$$

Let $w \in H^{1+\alpha}(\Omega) \cap H_0^1(\Omega)$ satisfy

$$A(w,\phi) = (g,\phi), \quad \text{for all } \phi \in H_0^1(\Omega).$$

Setting $\phi = (I - P_{k-1})v$ and using Eqs. (9.7) and (9.6) we obtain

$$\begin{aligned}
\big((I - P_{k-1})v, g\big) &= A\big(w, (I - P_{k-1})v\big) = A\big(v, (I - P_{k-1})w\big)\\
&\leqslant C\|v\|_1\big\|(I - P_{k-1})w\big\|_1\\
&\leqslant C\lambda_k^{-\alpha/2}\|v\|_1\|w\|_{1+\alpha}\\
&\leqslant C\lambda_k^{-\alpha/2}\|v\|_1\|g\|_{-1+\alpha}.
\end{aligned} \tag{9.9}$$

Now by Theorem A.4 of Appendix A, interpolating $|||Q_k v|||_{1,k} \leqslant C\|Q_k v\|_1 \leqslant C\|v\|_1$ and the trivial inequality $\|Q_k v\| \leqslant \|v\|$ gives

$$|||Q_k v|||_{1-\alpha,k} \leqslant C\|v\|_{1-\alpha}, \quad \text{for all } v \in H_0^1(\Omega).$$

It then follows that for $g \in M_k$ and $\phi \in H_0^1(\Omega)$,

$$(g, \phi) = (g, Q_k\phi) \leqslant |||g|||_{-1+\alpha,k}|||Q_k\phi|||_{1-\alpha,k} \leqslant C|||g|||_{-1+\alpha,k}\|\phi\|_{1-\alpha}.$$

Hence for $g = A_k^{1-\alpha}(I - P_{k-1})v \in M_k$,

$$\|g\|_{-1+\alpha} = \sup_{\phi \in H_0^1(\Omega)} \frac{(g, \phi)}{\|\phi\|_{1-\alpha}} \leqslant C|||g|||_{-1+\alpha,k} = C\Big|\Big|\Big|(I - P_{k-1})v\Big|\Big|\Big|_{1-\alpha,k}. \quad (9.10)$$

Combining Eqs. (9.8), (9.9) and (9.10) proves the lemma. □

Thus we have proved that Conditions (A.2) and (A.4) are satisfied and hence, by Theorem 5.1 (or Theorem 5.3), the multigrid V-cycle has a uniform reduction rate and, from Theorem 4.2, the multilevel additive preconditioner B_a defined by (4.3) is uniformly equivalent to A_J^{-1}, provided that appropriate smoothers are used.

We now establish our second uniform convergence result by verifying Conditions (A.5) and (A.6). In this case, we shall make an explicit assumption on the smoothness of the coefficients. Denote by $W_p^\gamma(\Omega)$ the Sobolev space of order γ defined in terms of the norm $L^p(\Omega)$; cf. GRISVARD [1985] and LIONS and MAGENES [1972]. We assume that the domain $\bar{\Omega} = \bigcup \bar{\Omega}_\ell$, where each Ω_ℓ has a polygonal boundary. For each Ω_ℓ, we shall assume that the coefficients $a_{ij}(x)$ are in $W_p^\gamma(\Omega_\ell)$ with $\gamma \in (0, 1/2)$ and $p\gamma > d$ and $a(x) \in L^\infty(\Omega)$. This condition implies that the coefficients $a_{ij}(x)$ are continuous on each subdomain Ω_ℓ but may have jump discontinuities across the subdomain boundaries. We also assume that the restriction of the coarse triangulation $\mathcal{T}_1$ to each Ω_ℓ is a triangulation of Ω_ℓ.

Since $A(\bullet, \bullet)$ is uniformly equivalent to $(\bullet, \bullet)_{H^1}$, Condition (A.3) holds. By Theorem A.4, Condition (A.4) follows from (9.5) by interpolation as in the previous case. It is well known (cf. GRISVARD [1985] and DAUGE [1988]) that for a polygonal domain Ω, there exists an $\alpha \in (0, 1]$ such that

$$\|u\|_{1+\alpha} \leqslant C\|-\Delta u\|_{-1+\alpha}.$$

This is the regularity assumption (9.6) for the case $\mathcal{L} = -\Delta$. Consequently, by Lemma 9.1, Condition (A.2) holds for $\mathcal{L} = -\Delta$. It follows from Lemma 4.4 that Condition (A.5) holds for $\mathcal{L} = -\Delta$ and hence it holds for a general $\mathcal{L}$ which is uniformly equivalent to $-\Delta$.

The verification of Conditions (A.3) and (A.5) only requires the uniform equivalence of the norms $|||\bullet|||$ and $\|\bullet\|_1$. It is the verification of Condition (A.6) that requires the smoothness assumption of the coefficients of $\mathcal{L}$. For this we need the following technical result.

LEMMA 9.2. *Under the above assumption on the coefficients, there exists a constant C such that for all $\eta > 0$, $v \in H^1(\Omega)$ and $w \in H^{1+\gamma}(\Omega)$*

$$\big|A(v, w)\big| \leqslant C\big(\eta^{-1}\|v\|^2 + \eta^{\gamma/(1-\gamma)}\|v\|_1^2\big)^{1/2}\|w\|_{1+\gamma}. \quad (9.11)$$

PROOF. Let v be in $H^1(\Omega)$ and w be in $H^{1+\gamma}(\Omega)$. There is no problem bounding the lowest order term of (9.3). We need only consider the derivative terms in (9.3). Fix ℓ, and let E_ℓ denote the extension operator defined on Ω_ℓ given by Theorem 1.4.3.1 of GRISVARD [1985]. For a function v defined on Ω_ℓ, let $\tilde{v}$ denote the extension of v by zero to $\mathbf{R}^d$. Since $\gamma < 1/2$, Corollary 1.4.4.5 of GRISVARD [1985] gives that the norm $\|\tilde{v}\|_{W_2^\gamma(\mathbf{R}^d)}$ is equivalent to the norm $\|v\|_{H^\gamma(\Omega_\ell)}$ for all $v \in H^\gamma(\Omega_\ell)$. Thus,

$$\int_{\Omega_\ell} a_{ij} \frac{\partial v}{\partial x_i} \frac{\partial w}{\partial x_i}\, dx = \int_{\mathbf{R}^d} \frac{\partial E_\ell v}{\partial x_i} \widetilde{a_{ij} \frac{\partial w}{\partial x_i}}\, dx = \left(\mathcal{F}\left(\frac{\partial (E_\ell v)}{\partial x_i}\right), \mathcal{F}\left(\widetilde{a_{ij} \frac{\partial w}{\partial x_j}}\right)\right),$$

where $\mathcal{F}$ denotes the Fourier transform. By the Cauchy–Schwarz inequality

$$\begin{aligned}\int_{\Omega_\ell} a_{ij} \frac{\partial v}{\partial x_i} \frac{\partial w}{\partial x_j}\, dx &\leqslant C\left(\int_{\mathbf{R}^d} \frac{|\zeta|^2}{(1+|\zeta|^2)^\gamma} \left|\mathcal{F}(E_\ell v)(\zeta)\right|^2 d\zeta\right)^{1/2} \left\| a_{ij} \frac{\partial w}{\partial x_j} \right\|_{H^\gamma(\Omega_\ell)} \\ &\leqslant C\left(\eta^{-1}\|v\|^2 + \eta^{\gamma/(1-\gamma)}\|v\|_1^2\right)^{1/2} \left\| \frac{\partial w}{\partial x_j} \right\|_{H^\gamma(\Omega_\ell)},\end{aligned}$$

where $\eta > 0$ is arbitrary. For the last inequality, we used Theorem 1.4.4.2 of GRISVARD [1985] which states that multiplication by $a_{ij} \in W_p^\gamma(\Omega_\ell)$ for $p > d/\gamma$ is a bounded operator on $H^\gamma(\Omega_\ell)$. The lemma immediately follows by summing over ℓ. □

Now setting $\eta = (\|v\|^2/\|v\|_1^2)^{1-\gamma}$ in (9.11), we obtain

$$\left|A(v, w)\right| \leqslant C\sqrt{2}\, \|v\|^\gamma\, \|v\|_1^{(1-\gamma)}\, \|w\|_{1+\gamma} \tag{9.12}$$

for all $v \in H^1(\Omega)$ and $w \in H^{1+\gamma}(\Omega)$. It it known that $M_k \subset H^{1+\gamma}(\Omega)$ and the following inverse property holds (see the Appendix of BRAMBLE, PASCIAK and XU [1991] for a proof):

$$\|v_k\|_{1+\gamma} \leqslant C h_k^{-\gamma} \|v_k\|_1, \quad \text{for all } v_k \in M_k.$$

Let $1 \leqslant k \leqslant i \leqslant J$. Setting $w = v_k \in M_k$ and $v = v_i \in M_i$ in (9.12) and using the inverse inequalities, we obtain

$$A(v_i, v_k) \leqslant C\|v_i\|^\gamma \|v_i\|_1^{1-\gamma} \|v_k\|_{1+\gamma} \leqslant C(h_i/h_k)^\gamma \|v_k\|_{1,k} \left(\lambda_i^{1/2} \|v_i\|\right).$$

Since the mesh size h_i is on the order of δ^i, Condition (A.6) holds with $\varepsilon = \delta^\gamma$.

We have thus proved that Conditions (A.5) and (A.6) hold. By Theorem 5.2 or Theorem 5.4, the multigrid V-cycle Algorithm 3.1 has a uniform reduction rate and by Theorem 4.3 the additive preconditioner B_a defined by (4.3) is uniformly equivalent to A_J, provided that appropriate smoothers are used.

REMARK 9.1. A more constructive proof of Condition (A.6) is possible in the case of smooth coefficients and nodal finite element approximation spaces. In this case, one can show, by integration by parts, that

$$\left|A(v_i, v_k)\right| \leqslant C(h_i/h_k)^{1/2}\|v_k\|_{1,k}\big(\lambda_i^{1/2}\|v_i\|\big).$$

Thus Condition (A.6) holds with $\varepsilon = \delta^{1/2}$. As an illustration, we sketch a proof for the model problem of Section 2. Let $v_k \in M_k$ and $v_i \in M_i$ with $k < i$. Let $\tau \in \mathcal{T}_k$ be a triangle of $\mathcal{T}_k$. Integrating by parts gives

$$\begin{aligned}\left|\int_\tau \nabla v_k \cdot \nabla v_i\right| &= \left|\int_{\partial\tau} \partial_\nu v_k v_i \,\mathrm{d}s - \int_\tau \Delta v_k v_i \,\mathrm{d}x\right| \\ &\leqslant \|v_k\|_{H^1(\partial\tau)}\|v_i\|_{L^2(\partial\tau)} + \|v_k\|_{H^2(\tau)}\|v_i\|_{L^2(\tau)}.\end{aligned}$$

Now using the inverse properties

$$\begin{aligned}&\|v_i\|^2_{L^2(\partial\tau)} \leqslant Ch_i^{-1}\|v_i\|^2_{L^2(\tau)},\\ &\|v_k\|^2_{H^1(\partial\tau)} \leqslant Ch_k^1\|v_k\|^2_{W^1_\infty(\tau)} \leqslant Ch_k^{-1}\|v_k\|^2_{H^1(\tau)},\\ &\|v_k\|^2_{H^2(\tau)} \leqslant Ch_k^{-2}\|v_k\|^2_{H^1(\tau)},\end{aligned}$$

we obtain

$$\left|\int_\tau \nabla v_k \cdot \nabla v_i\right| \leqslant Ch_k^{-1/2}h_i^{-1/2}\|v_k\|_{H^1(\tau)}\|v_i\|_{L^2(\tau)}.$$

Summing over $\tau \subset \Omega$ we obtain

$$\left|A(v_k, v_i)\right| \leqslant C(h_i/h_k)^{1/2}\|v_k\|_1\big(h_i^{-1}\|v_i\|\big) \leqslant C(h_i/h_k)^{1/2}\|v_k\|_1\big(\lambda_i^{1/2}\|v_i\|\big).$$

In the case of uniform refinement, $(h_i/h_k)^{1/2} = \varepsilon^{i-k}$ with $\varepsilon = 1/\sqrt{2}$.

9.2. *Equations with rough coefficients*

In establishing the previous results, we have used either explicitly or implicitly the smoothness of the coefficients of the differential operator. In the next example, we consider (9.1) on a polygonal domain with the coefficients $a_{ij}(x)$ being merely uniformly bounded and measurable. The meshes and the finite element spaces M_k are defined as in Subsection 9.1. In such a case $A(\bullet,\bullet)^{1/2}$ defines a norm equivalent to the norm $\|\bullet\|_1$ hence Condition (A.4) holds. However it is not known whether or not Condition (A.2) still holds. Hence we cannot apply Theorem 4.2 or Theorem 5.1.

We first consider the multilevel additive preconditioner B_a defined by (4.3). Since Conditions (A.2) and (A.4) hold for $-\Delta$ and since $|||\bullet|||$ is uniformly equivalent to $\|\bullet\|_1$, the multilevel additive preconditioner B_a is uniformly equivalent to A_J^{-1}, provided that the smoother R_k satisfies the smoothing condition (SA.2).

We now consider the V-cycle multigrid method defined by Algorithm 3.1. The stable approximation property in (9.5) implies that Condition (A.7) holds. It follows from Theorem 5.5 that error propagator of Algorithm 3.1 has a contraction number which deteriorates no worse than linearly in the number of levels, provided that the smoother, R_k, satisfies Conditions (SM.1) and (SM.2).

To get a better estimate on the contraction number of the multigrid V-cycle algorithm, we need to impose on the smoother, R_k, a condition more restrictive than Conditions (SM.1) and (SM.2). We assume that a point (or line) smoother such as a weighted line Jacobi or line Gauss–Seidel smoother is used. We have shown in Section 8 that these smoothers satisfy Conditions (SM.1), (SM.2) and (SM.4). It follows from Corollary 5.2 that the multigrid V-cycle Algorithm 3.1 has an error reduction that deteriorates at a rate no worse than $(1 - C/\log^2 J)$.

We remark that Theorem 5.5 can also be applied to certain interface problems in which there may be large jumps in the coefficients. This situation was treated in BRAMBLE, PASCIAK, WANG and XU [1991a]. In the two-dimensional case, it can be shown that

$$\left\|(I - Q_{k-1})v\right\|^2 + \lambda_k^{-1}\|Q_k v\|_1^2 \leqslant C(J-k)\lambda_k^{-1}\|v\|_1^2, \quad \text{for all } v \in M = M_J.$$

Hence Condition (A.7) holds for $\Pi_k = Q_k$ with constants $C_\pi \leqslant CJ$ and $C'_\pi \leqslant CJ$ depending on the number of levels. An application of Theorem 5.5 then shows that the norm of the error propagation operator of the multigrid V-cycle algorithm is bounded from above by $1 - C/J^2$.

9.3. *Domains with curved boundaries*

We now consider the Dirichlet problem for domains $\Omega \subset \mathbf{R}^2$ with curved boundaries. For simplicity we let $\partial\Omega$ be piecewise smooth and such that each smooth part of the boundary, Γ_i, is convex in the sense that the line joining any two parts on Γ_i lies in $\overline{\Omega}$. We make this convexity assumption for simplicity. It is only technically more involved to treat the more general case and hence we do not do so here.

We assume that $a_{ij}(x)$ and $a(x)$ are smooth enough so that the a priori estimate in (9.6) is valid, i.e., there exists an $\alpha \in (0, 1]$ such that

$$\|u\|_{1+\alpha} \leqslant C\|f\|_{-1+\alpha}.$$

We first set up the finite element problem. Let $\mathcal{T}_1$ be a coarse triangulation in Ω. We take certain points on the boundary (including the end-points of each Γ_i) and in the interior and connect them so that the resulting triangles are in Ω and no triangle has more than two points on $\partial\Omega$.

Now define $\mathcal{T}_k$ from $\mathcal{T}_{k-1}$ by first connecting the midpoints of the sides of each triangle with two vertices in Ω. If a triangle in $\mathcal{T}_{k-1}$ has two vertices on the boundary we assume that they are connected by a smooth arc on the boundary and we take a

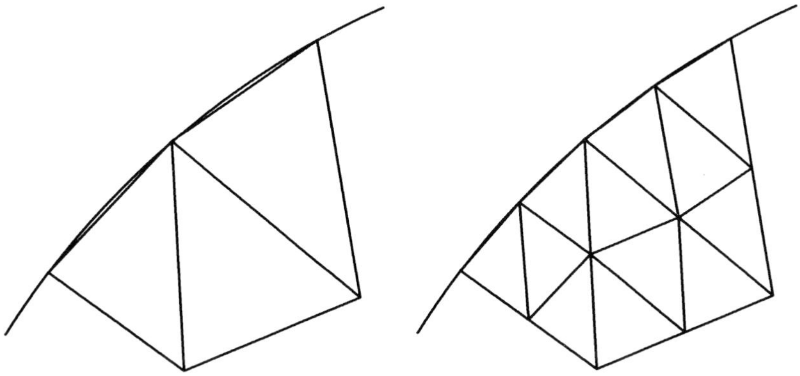

$\{\tau^i_{k-1}\}$ $\{\tau^i_k\}$

FIG. 9.1. The mesh refinement near the boundary.

new boundary point to be the midpoint of this arc. We then form four new triangles by connecting the midpoints of this curved triangle; cf. Fig. 9.1.

We define $\overline{M}_k$ as the piecewise linear subspaces of $H^1_0(\Omega)$ in a natural way. The finite element problem is to find $u_J \in \overline{M}_J$ such that

$$A(u_J, \phi) = (f, \phi), \quad \text{for all } \phi \in \overline{M}_J.$$

Note that in this construction not all triangles in $\mathcal{T}_{k-1}$ are the union of triangles of $\mathcal{T}_k$. As a consequence, the spaces $\overline{M}_k$ are not nested. To remain in the framework of nested spaces, we want to define a family of nested spaces for use in our multigrid algorithms.

Set $M \equiv M_J = \overline{M}_J$. We proceed to construct $\{M_k\}_{k<J}$ in such a way that we have subspaces $M_1 \subset M_2 \subset \cdots \subset M_J = M \subset H^1_0(\Omega)$.

Let Ω°_k, for $k < J$, be the interior of the union of the (closed) triangles of $\mathcal{T}_k$ which do not have vertices on $\partial\Omega$. The spaces M_k, for $k < J$, are defined by

$$M_k = \left\{\phi \in \overline{M}_k \mid \operatorname{supp}(\phi) \subset \overline{\Omega^\circ_k}\right\}. \tag{9.13}$$

Thus $M_k \subset M_{k+1}$. This is clear since each triangle in which some function in M_k is nonzero is the union of four triangles in which some functions in M_{k+1} are nonzero. We have thus defined spaces $M_1 \subset M_2 \subset \cdots \subset M_J$.

We shall use Theorem 5.1 to analyze the convergence properties of Algorithm 3.1. We choose a smoother satisfying (SM.1) and (SM.2). It remains to show that Conditions (A.2), (A.4) and (A.3) hold. We note that the condition on λ_k is satisfied since $\lambda_k \approx h_k^{-2}$. As in our first example, h_k is the length of the longest side of all the triangles in $\mathcal{T}_k$ and, by construction, $h_{k+1} = \frac{1}{2} h_k$.

We have proved in the previous applications that Conditions (A.2) and (A.4) will follow from the Aubin–Nitsche duality argument and interpolation, provided we prove that (9.7) and (9.5) hold; i.e.,

$$\left\|(I-P_{k-1})v\right\|_1 \leqslant Ch_k^{\alpha}\|v\|_{1+\alpha}, \quad \text{for all } v \in H^{1+\alpha}(\Omega)\cap H_0^1(\Omega)$$

and

$$A(Q_k v, Q_k v) \leqslant CA(v,v), \quad \text{for all } v \in H_0^1(\Omega).$$

For this purpose we prove first the following result.

LEMMA 9.3. *Let Ω^{η} denote the strip $\{x \in \Omega \mid \operatorname{dist}(x, \partial\Omega) < \eta\}$ and $0 < s < 1/2$. Then for all $v \in H^{1+s}(\Omega)$,*

$$\|v\|_{1,\Omega^{\eta}} \leqslant C\eta^{s}\|v\|_{1+s}.$$

In addition, for $v \in H_0^1(\Omega)$,

$$\|v\|_{\Omega^{\eta}} \leqslant C\eta\|v\|_1. \tag{9.14}$$

PROOF. Let ω^{η} be a square of side length η. Then

$$\|w\|_{\omega^{\eta}}^2 \leqslant \eta^2/2\int_{\omega^{\eta}}|\nabla\, w|^2\,dx$$

holds for all w vanishing on one edge of ω^{η}. The estimate in (9.14) follows by covering the strip Ω^{η} with subregions Ω_i^{η} each of which can be smoothly mapped onto w^{η}.

By Theorem A.4 of Appendix A, interpolating (9.14) and the trivial inequality

$$\|v\|_{\Omega^{\eta}} \leqslant \|v\|$$

yields

$$\|v\|_{\Omega^{\eta}} \leqslant C\eta^{s}\|v\|_{s}, \quad \text{for all } v \in H_0^{s}(\Omega),\ s \neq 1/2. \tag{9.15}$$

By Corollary 1.4.4.5 of GRISVARD [1985], $H_0^s(\Omega) = H^s(\Omega)$ for $0 < s < 1/2$, provided $\partial\Omega$ is Lipschitz. Thus applying (9.15) to the first derivative of v completes the proof of the lemma. □

We now are in a position to prove (9.7) and (9.5). We first prove (9.7). Fix $w \in H^{1+\alpha}(\Omega)\cap H_0^1(\Omega)$. Let $\chi \in M_{k-1}$ be the interpolant of w at the nodes of Ω_{k-1}° and $\overline{\chi} \in \overline{M}_{k-1}$ interpolate w at the level $k-1$ nodes of Ω. Then

$$\begin{aligned}
&A\big((I-P_{k-1})w, (I-P_{k-1})w\big)\\
&\quad\leqslant A(w-\chi, w-\chi) \leqslant C\big(\|w-\chi\|_{1,\Omega_{k-1}^{\circ}} + \|w\|_{1,\Omega\setminus\Omega_{k-1}^{\circ}}\big)\\
&\quad\leqslant C\big(\|w-\overline{\chi}\|_{1,\Omega_{k-1}^{\circ}}^2 + \|\overline{\chi}-\chi\|_{1,\Omega_{k-1}^{\circ}}^2 + \|w\|_{1,\Omega\setminus\Omega_{k-1}^{\circ}}^2\big).
\end{aligned}$$

It is well known that in two dimensions

$$\|w - \overline{\chi}\|_{1,\Omega}^2 \leqslant C h_{k-1}^{2\alpha} \|w\|_{1+\alpha}^2, \quad 0 < \alpha \leqslant 1.$$

Note that $\overline{\chi} - \chi \in \overline{M}_{k-1}$ and vanishes on all nodes of level $k-1$ except on $\partial\Omega_{k-1}^{\circ}$. Consequently,

$$\|\overline{\chi} - \chi\|_{1,\Omega_{k-1}^{\circ}}^2 \leqslant C \sum_{x_i \in \partial\Omega_{k-1}^{\circ}} \overline{\chi}(x_i)^2.$$

The sum is taken over the level $k-1$ nodes on $\partial\Omega_{k-1}^{\circ}$. It follows from the above two inequalities that

$$\begin{aligned} \|\overline{\chi} - \chi\|_{1,\Omega_{k-1}^{\circ}}^2 &\leqslant C\|\overline{\chi}\|_{1,\Omega\setminus\Omega_{k-1}^{\circ}}^2 \\ &\leqslant C\big(\|w - \overline{\chi}\|_{1,\Omega\setminus\Omega_{k-1}^{\circ}}^2 + \|w\|_{1,\Omega\setminus\Omega_{k-1}^{\circ}}^2\big) \\ &\leqslant C\big(h_{k-1}^{2\alpha}\|w\|_{1+\alpha}^2 + \|w\|_{1,\Omega\setminus\Omega_{k-1}^{\circ}}^2\big). \end{aligned}$$

Applying Lemma 9.3 with $\eta = h_{k-1}$ gives

$$\|w\|_{1,\Omega\setminus\Omega_{k-1}^{\circ}}^2 \leqslant C h_{k-1}^{2\alpha}\|w\|_{1+\alpha}^2.$$

This proves (9.7).

We now prove (9.5). Let $\overline{Q}_k$ denote the $L^2(\Omega)$ orthogonal projector onto the space $\overline{M}_k$. First by using the interpolant in $\overline{M}_k$ we have that

$$\big\|(I - \overline{Q}_k)v\big\| \leqslant C h_k^{1+\alpha}\|v\|_{1+\alpha}.$$

By Theorem A.4, interpolating the above inequality and $\|(I - \overline{Q}_k)v\| \leqslant C\|v\|$, we conclude that

$$\big\|(I - \overline{Q}_k)v\big\|^2 \leqslant C\lambda_k^{-1}\|v\|_1^2, \quad v \in H_0^1(\Omega). \tag{9.16}$$

Let $\theta_k \in M_k$ be equal to $\overline{Q}_k v$ at the interior nodes of Ω_k°. Then

$$\begin{aligned} \big\|(I - Q_k)v\big\|^2 &\leqslant \|v - \theta_k\|^2 = \|v - \theta_k\|_{\Omega_k^{\circ}}^2 + \|v\|_{\Omega\setminus\Omega_k^{\circ}}^2 \\ &\leqslant 2\Big(\big\|(I - \overline{Q}_k)v\big\|_{\Omega_k^{\circ}}^2 + \big\|\overline{Q}_k v - \theta_k\big\|_{\Omega_k^{\circ}}^2 + \|v\|_{\Omega\setminus\Omega_k^{\circ}}^2\Big). \end{aligned}$$

Since $\overline{Q}_k v - \theta_k \in \overline{M}_k$ and vanishes except on $\partial\Omega_k^{\circ}$, we have

$$\big\|\overline{Q}_k v - \theta_k\big\|_{\Omega_k^{\circ}}^2 \leqslant C h_k^2 \sum_{x_i \in \partial\Omega_k^{\circ}} \big|\overline{Q}_k v(x_i)\big|^2$$

$$\begin{aligned}&\leqslant C\|\overline{Q}_k v\|^2_{\Omega\setminus\Omega_k^\circ}\\&\leqslant C\big(\|\overline{Q}_k v - v\|^2_{\Omega\setminus\Omega_k^\circ} + \|v\|^2_{\Omega\setminus\Omega_k^\circ}\big).\end{aligned}$$

This shows that

$$\begin{aligned}\|(I-Q_k)v\|^2 &\leqslant C\big(\|(I-\overline{Q}_k)v\|^2_{\Omega_k^\circ} + \|(I-\overline{Q}_k)v\|^2_{\Omega\setminus\Omega_k^\circ} + \|v\|^2_{\Omega\setminus\Omega_k^\circ}\big)\\&\leqslant C\big(\|(I-\overline{Q}_k)v\|^2 + \|v\|^2_{\Omega\setminus\Omega_k^\circ}\big).\end{aligned}$$

Hence, by Lemma 9.3 and (9.16)

$$\|(I-Q_k)v\|^2 \leqslant C\lambda_k^{-1}\|v\|_1^2.$$

We finally show that

$$A(Q_k v, Q_k v) \leqslant CA(v,v), \quad \text{for all } v \in H_0^1(\Omega).$$

Fix $\tau \in \mathcal{T}_k$. Let θ_τ be the average value of v on τ. By the inverse property and the Poincaré inequality,

$$\begin{aligned}|Q_k v|^2_{1,\tau} &= |Q_k v - \theta_\tau|^2_{1,\tau} \leqslant C\lambda_k\|Q_k v - \theta_\tau\|^2_\tau\\&\leqslant C\lambda_k\big(\|Q_k v - v\|^2_\tau + \|v-\theta_\tau\|^2_\tau\big)\\&\leqslant C\lambda_k\|Q_k v - v\|^2_\tau + C\|v\|^2_{1,\tau}.\end{aligned}$$

Summing over $\tau \in \mathcal{T}_k$ gives

$$|Q_k v|_1^2 \leqslant C\big(\lambda_k\|Q_k v - v\|^2 + \|v\|_1^2\big) \leqslant C\|v\|_1^2.$$

This proves that $A(Q_k v, Q_k v) \leqslant C(\|Q_k v\|^2 + |Q_k v|_1^2) \leqslant CA(v,v)$.

Now we have proved (9.7) and (9.5). Condition (A.4) follows from (9.5) and the trivial inequality $\|Q_k v\| \leqslant \|v\|$ by interpolation. Condition (A.2) follows from Eqs. (9.6) and (9.7) by using Lemma 9.1.

Having established that Conditions (A.2) and (A.4) are valid, we conclude that the multigrid V-cycle Algorithm 3.1 has a uniform reduction rate and the multilevel additive preconditioner is an optimal preconditioner.

9.4. *Mesh refinement*

We now consider finite element approximations that utilize a locally refined mesh. Such mesh refinements are convenient for accurate modeling of problems with various types of singular behavior. As in BRAMBLE and PASCIAK [1993], we consider for simplicity the case where Ω is a polygonal domain and the finite element space consists of piecewise linear functions, although we allow a very general form of refinement.

Following BRAMBLE, PASCIAK and XU [1990], we start with a coarse quasi-uniform triangulation $\mathcal{T}_1$. The refinement triangulation is defined in terms of a sequence of (open) mesh domains with

$$\Omega_J \subseteq \Omega_{J-1} \subseteq \cdots \subseteq \Omega_1 = \Omega.$$

The only restrictions on the mesh domains $\{\Omega_k\}$ are that the boundary of Ω_k, for $k > 1$, consists of edges of mesh triangles in the triangulation $\mathcal{T}_{k-1}$ and that there is at least one edge of $\mathcal{T}_{k-1}$ contained in Ω_k. These mesh domains control the region of refinement. If τ^i_{k-1} is a triangle contained in Ω_k, then it is broken into four smaller triangles (in the triangulation $\mathcal{T}_k$) by the lines connecting the midpoints of the edges. Alternatively, if τ^i_{k-1} is in the complement of Ω_k, then it is not subdivided but is directly included into the triangulation $\mathcal{T}_k$. A simple example of this construction is the case of a unit square with local refinement near the corner. In this case we take $\Omega_k = \Omega$ for $k = 1, \ldots, j$ and $\Omega_k = (1 - 2^{j-k}, 1) \times (1 - 2^{j-k}, 1)$ for $k = j + 1, \ldots, J$.

The space M_k is defined to be the set of piecewise linear functions with respect to the triangulation $\mathcal{T}_k$ which are continuous on Ω and vanish on $\partial\Omega$. The continuity condition implies that the finer grid nodes on a coarse-fine boundary are "slave nodes" in the sense that the values of the function are determined by the values of the function at the nearby coarse grid points. In this case, if $\Omega_{k-1} \neq \Omega_k$, then the subspace $\widehat{M}_k$ on which we smooth is a proper nonzero subspace of M_k. In fact, we define $\widehat{M}_k$ to be the functions in M_k which are zero outside of Ω_k. Thus we smooth on a given level just in the region where new nodes are being added in the refinement scheme. It is easy to see that the mesh corresponding to the nontrivial part of the space $\widehat{M}_k$ is quasi-uniform. Let h_k be the diameter of the smallest triangle in the kth triangulation. By construction, $h_k \equiv 2^{1-k} h_1$.

We assume that smoothers satisfying Conditions $(\widehat{\text{SM.1}})$ and $(\widehat{\text{SM.2}})$ are used. Recall that Conditions $(\widehat{\text{SM.1}})$ and $(\widehat{\text{SM.2}})$ are the same as Conditions (SM.1) and (SM.2), except they are imposed on subspaces $\widehat{M}_k$ instead of M_k. Hence the construction and analysis of smoothers in Section 8 also apply. To apply Theorem 6.1, it remains to verify Conditions $(\widehat{\text{A.5}})$ and $(\widehat{\text{A.6}})$. There is no difficulty in proving Condition $(\widehat{\text{A.6}})$. In fact for $u \in \widehat{M}_k$, $v \in \widehat{M}_i$ with $k \leqslant i$,

$$A(u, v) = A_{\Omega_i}(u, v).$$

Lemma 9.2 remains valid for $A_{\Omega_i}(\bullet, \bullet)$. In addition, when restricted to the domain Ω_i, both u and v are finite element functions on quasi-uniform meshes. Consequently Condition $(\widehat{\text{A.6}})$ holds.

We now construct operators Π_k satisfying Condition $(\widehat{\text{A.5}})$. Let $\overline{M}_k$ denote the finite element space on meshes resulting from a uniform refinement from $\mathcal{T}_1$ over the entire domain, that is, $\overline{M}_k$ is the quasi-uniform space resulting from above construction with $\Omega_1 = \Omega_2 = \cdots = \Omega_J = \Omega$. The finite element spaces $\overline{M}_k$ satisfy the usual approximation and inverse properties. Further $M_k \subset \overline{M}_k$. Let $\overline{Q}_k$ denote the L_2 orthogonal projection onto $\overline{M}_k$. Then

$$A(\overline{Q}_1 v, \overline{Q}_1 v) + \sum_{k=2}^{J} \lambda_k \|(\overline{Q}_k - \overline{Q}_{k-1})v\|^2 \leqslant CA(v,v), \quad \text{for all } v \in M_J \subset \overline{M}_J.$$

Define $\Pi_k : M_J \to M_k$ by

$$\Pi_k v = \begin{cases} \overline{Q}_k v & \text{at nodes of } M_k \text{ in the interior of } \Omega_{k+1}, \\ v & \text{at remaining nodes of } M_k. \end{cases}$$

By construction, $\Pi_J = I$ and $(\Pi_k - \Pi_{k-1})v \in \widehat{M}_k$. It can be shown (see BRAMBLE, PASCIAK, WANG and XU [1991a]) that operators Π_k satisfy

$$\|(I - \Pi_{k-1})v\|^2 \leqslant C\|(I - \overline{Q}_{k-1})v\|^2, \quad \text{for all } v \in M_J,$$

and

$$A(\Pi_k v, \Pi_k v) \leqslant CA(v,v), \quad \text{for all } v \in M_J.$$

By the triangle inequality,

$$\|(\Pi_k - \Pi_{k-1})v\|^2 \leqslant 2\big[\|(I - \Pi_k)v\|^2 + \|(I - \Pi_{k-1})v\|^2\big] \leqslant C\|(I - \overline{Q}_{k-1})v\|^2.$$

Summing over k, we obtain

$$\begin{aligned} \sum_{k=2}^{J} h_k^{-2}\|(\Pi_k - \Pi_{k-1})v\|^2 &\leqslant C\sum_{k=2}^{J} h_k^{-2}\|(I - \overline{Q}_{k-1})v\|^2 \\ &\leqslant C\sum_{k=2}^{J}\left(\sum_{j=k}^{J}\frac{h_j}{h_k}\big(h_j^{-1}\|(\overline{Q}_j - \overline{Q}_{j-1})v\|\big)\right)^2 \\ &\leqslant C\sum_{k=2}^{J} h_k^{-2}\|(\overline{Q}_k - \overline{Q}_{k-1})v\|^2 \leqslant CA(v,v). \end{aligned}$$

Hence Condition $(\widehat{\text{A.5}})$ holds. Thus Theorem 6.1 may be applied to show that the V-cycle algorithm yields a uniform reduction.

9.5. *Bibliographical notes*

The main applications of the multigrid methods are to second-order elliptic problems (cf. BANK, MANDEL and MCCORMICK [1987] and HACKBUSCH [1985]). The particular treatment given in this section is taken from BRAMBLE and PASCIAK [1993] where for the first time a nested multigrid algorithm was formulated for the Dirichlet problem for domains with curved boundaries. In that paper the uniform convergence of the V-cycle with one smoothing step was proved using Theorem 5.2. Here we show that Theorem 5.1 is also applicable. The case of equations with rough coefficients is discussed in BRAMBLE, PASCIAK, WANG and XU [1991a] where it is shown that Theorem 5.5 applies.

10. Perturbation of forms

In applications it is often inconvenient to compute the entries of the matrices coming from the bilinear form $A(\bullet,\bullet)$ defined by (9.3). It may be convenient to replace the form by a new form $\mathcal{A}(\bullet,\bullet)$ which is equivalent, and in a sense close, to $A(\bullet,\bullet)$. For example, for computational purposes, it may be necessary to replace the exact integration over a given triangle by an approximate integration using a quadrature formula. To this end we consider the new approximate problem: Find $u \in M$ such that

$$\mathcal{A}(u,\phi) = f(\phi), \qquad \text{for all } \phi \in M.$$

We shall assume, throughout this section, that the spaces are nested; i.e.,

$$M_1 \subset M_2 \subset \cdots \subset M_J = M.$$

We will consider two settings for formulating multigrid algorithms. In the first setting, we set $\mathcal{A}_k(\bullet,\bullet) = \mathcal{A}(\bullet,\bullet)$ for all k and define a multigrid algorithm by replacing A by $\mathcal{A}$ in Algorithm 3.1. The main step in the analysis of this algorithm is the verification of Conditions (A.2), (A.3) and (A.4) or Conditions (A.5) and (A.6). Alternatively, we may use possibly different forms $\mathcal{A}_k(\bullet,\bullet)$ on each of the spaces M_k, with $\mathcal{A}_J(\bullet,\bullet) = \mathcal{A}(\bullet,\bullet)$. For example, consider the second-order problems of Section 9. Using a different numerical integration rule on each level will lead to the second setting. The corresponding multigrid algorithm is defined by Algorithm 7.1. The main step in the analysis is the verification of Condition (A.2$''$). Since the spaces M_k are nested, we always take I_k to be the natural injection operator.

We will first prove two perturbation lemmas and then give applications.

10.1. Perturbation lemmas

We will make two assumptions relating the forms $\mathcal{A}(\bullet,\bullet)$ and $A(\bullet,\bullet)$. First we assume that $\mathcal{A}(\bullet,\bullet)$ is equivalent to the form $A(\bullet,\bullet)$. That is, there are positive constants C_1 and C_2 such that

$$C_1 A(v,v) \leqslant \mathcal{A}(v,v) \leqslant C_2 A(v,v), \qquad \text{for all } v \in M. \tag{10.1}$$

We also assume that the forms are close in the following sense. There exists $\beta > 0$ such that

$$\left|A(v,w) - \mathcal{A}(v,w)\right| \leqslant C\lambda_J^{-\beta/2} |||v|||\, |||w|||, \qquad \text{for all } v, w \in M. \tag{10.2}$$

The operators A_k, Q_k and P_k are defined as before. Let $\mathcal{A}_k : M_J \to M_k$ be defined by

$$(\mathcal{A}_k v, \phi) = \mathcal{A}(v, \phi), \qquad \text{for all } v, \phi \in M_k.$$

The orthogonal projection $\mathcal{P}_k : M \to M_k$ is defined by

$$\mathcal{A}(\mathcal{P}_k v, w) = \mathcal{A}(v, w), \qquad \text{for all } v \in M,\ w \in M_k.$$

Under the two assumptions in Eqs. (10.1) and (10.2), we can show that if any one of the Conditions (A.2), (A.3), (A.4), (A.5) or (A.6) holds for the form $A(\bullet,\bullet)$ then it also holds for the form $\mathcal{A}(\bullet,\bullet)$.

The cases for Conditions (A.3), (A.4) and (A.5) are simple. We have already remarked in Section 4 that these conditions are not sensitive to a uniform perturbation of the form. Let $\hat{\lambda}_k$ be the largest eigenvalue of $\mathcal{A}_k$. The inequalities (10.1) imply that $\hat{\lambda}_k$ and λ_k are comparable in size. Hence Condition (A.3) holds for $\mathcal{A}(\bullet,\bullet)$ if and only if it holds for $A(\bullet,\bullet)$. It follows also from (10.1) that Condition (A.5) holds for $\mathcal{A}(\bullet,\bullet)$ if and only if it holds for $A(\bullet,\bullet)$. By interpolation, (10.1) implies that, for $0 \leqslant \alpha \leqslant 1$,

$$C_1^{1-\alpha}\big(A_k^{1-\alpha} v, v\big) \leqslant \big(\mathcal{A}_k^{1-\alpha} v, v\big) \leqslant C_2^{1-\alpha}\big(A_k^{1-\alpha} v, v\big), \qquad \text{for all } v \in M_k.$$

It then follows that (A.4) holds for $\mathcal{A}(\bullet,\bullet)$ if and only if it holds for $A(\bullet,\bullet)$.

We now show that under the assumption that Eqs. (10.1) and (10.2) hold, Conditions (A.2) and (A.6) hold for $\mathcal{A}(\bullet,\bullet)$ if they hold for $A(\bullet,\bullet)$. To distinguish the scales of discrete norms derived from the two forms $A(\bullet,\bullet)$ and $\mathcal{A}(\bullet,\bullet)$, we shall use the following notation:

$$|||v|||^2_{A_k^{1-\alpha}} = \big(A_k^{1-\alpha} v, v\big), \qquad |||v|||^2_{\mathcal{A}_k^{1-\alpha}} = \big(\mathcal{A}_k^{1-\alpha} v, v\big), \qquad \text{for all } v \in M_k.$$

LEMMA 10.1. *Assume that* (10.1) *holds and that* (10.2) *holds. If Condition* (A.2) *holds for the form* $A(\bullet,\bullet)$ *with* $\alpha = \beta$, *then it also holds for the form* $\mathcal{A}(\bullet,\bullet)$.

PROOF. If Condition (A.2) holds for $A(\bullet,\bullet)$ then for all $v \in M_k$,

$$\begin{aligned}
&\big|\big|\big|(I - \mathcal{P}_{k-1})v\big|\big|\big|_{\mathcal{A}_k^{1-\alpha}} \\
&\quad \leqslant C\big|\big|\big|(I - \mathcal{P}_{k-1})v\big|\big|\big|_{A_k^{1-\alpha}} \\
&\quad \leqslant C\Big(\big|\big|\big|(I - P_{k-1})v\big|\big|\big|_{A_k^{1-\alpha}} + \big|\big|\big|(P_{k-1} - \mathcal{P}_{k-1})v\big|\big|\big|_{A_k^{1-\alpha}}\Big) \\
&\quad \leqslant C\Big(\lambda_k^{-\alpha/2}|||v|||_{A_k} + \big|\big|\big|(P_{k-1} - \mathcal{P}_{k-1})v\big|\big|\big|_{A_k^{1-\alpha}}\Big).
\end{aligned} \tag{10.3}$$

Now, set $w = A_k^{-\alpha}(P_{k-1} - \mathcal{P}_{k-1})v \in M_{k-1}$. By (10.2),

$$\big|\big|\big|(P_{k-1} - \mathcal{P}_{k-1})v\big|\big|\big|^2_{A_k^{1-\alpha}} = A\big((P_{k-1} - \mathcal{P}_{k-1})v, w\big)$$

$$= A(v,w) - \mathcal{A}(v,w) + \mathcal{A}(\mathcal{P}_{k-1}v,w) - A(\mathcal{P}_{k-1}v,w)$$
$$\leqslant C\lambda_J^{-\alpha/2} |||v|||_{A_k} |||w|||_{A_k}$$
$$= C\lambda_J^{-\alpha/2} |||v|||_{A_k} \left|\left|\left|(P_{k-1} - \mathcal{P}_{k-1})v\right|\right|\right|_{A_k^{1-\alpha}}.$$

This implies that

$$\left|\left|\left|(P_{k-1} - \mathcal{P}_{k-1})v\right|\right|\right|_{A_k^{1-\alpha}} \leqslant C\lambda_J^{-\alpha/2} |||v|||_{A_k}. \tag{10.4}$$

It follows from (10.3) and (10.4) that Condition (A.2) holds for $\mathcal{A}(\bullet,\bullet)$. □

A similar perturbation result holds for Condition (A.6). In addition to Eqs. (10.1) and (10.2), we also need to assume that

$$C(v,v) \leqslant A(v,v), \quad \text{for all } v \in M. \tag{10.5}$$

LEMMA 10.2. *Assume that Eqs.* (10.1), (10.2) *and* (10.5) *hold and that Condition* (A.3) *holds for* λ_k. *If Condition* (A.6) *holds for* $A(\bullet,\bullet)$ *with some* $0<\varepsilon<1$, *then it holds for* $\mathcal{A}(\bullet,\bullet)$ *with some* $0<\hat{\varepsilon}<1$.

PROOF. Clearly, for all $v, w \in M_J$,

$$\mathcal{A}(v,w) \leqslant A(v,w) + \left|A(v,w) - \mathcal{A}(v,w)\right|. \tag{10.6}$$

Inequality (10.2) implies that

$$\left|A(v,w) - \mathcal{A}(v,w)\right| \leqslant C\lambda_J^{-\beta/2} |||v|||_A |||w|||_A, \quad \text{for all } v, w \in M_J. \tag{10.7}$$

Now if Condition (A.6) holds for $A(\bullet,\bullet)$, then there exists $\varepsilon \in (0,1)$ such that for $1 \leqslant k \leqslant i \leqslant J$

$$A(v,w) \leqslant C_{cs}\varepsilon^{(i-k)} |||v|||_A \left(\lambda_i^{1/2} \|w\|\right), \quad \text{for all } v \in M_k, w \in M_i. \tag{10.8}$$

Combining Eqs. (10.6)–(10.8),

$$\mathcal{A}(v,w) \leqslant \left(C_{cs}\varepsilon^{(i-k)} + C\lambda_J^{-\beta/2}\right) |||v|||_A \left(\lambda_i^{1/2}\|w\|\right), \quad \text{for all } v \in M_k, w \in M_i.$$

Inequality (10.5) implies that $\lambda_i > C$ for all i. By Condition (A.3),

$$\lambda_J^{-\beta/2} \leqslant C(\lambda_k/\lambda_i)^{\beta/2} \leqslant C\left(\gamma_2^{\beta/2}\right)^{i-k}.$$

Therefore, with $\hat{\varepsilon} = \max(\varepsilon, \gamma_2^{\beta/2})$,

$$\mathcal{A}(v,w) \leqslant C\hat{\varepsilon}^{i-k} |||v|||_A \left(\lambda_i \|w\|\right), \quad \text{for all } v \in M_k,\ w \in M_i.$$

Note that $\mathcal{A}(\bullet,\bullet)$ is equivalent to $A(\bullet,\bullet)$ and that λ_i is comparable in size to $\hat{\lambda}_i$. Hence Condition (A.6) holds for $\mathcal{A}(\bullet,\bullet)$. □

We now consider a setting for perturbation of the forms on all levels. In some applications, it may be more convenient to replace the form $A(\bullet,\bullet)$ by a new form $\mathcal{A}_k(\bullet,\bullet)$ on each M_k. This is the case when a coarse grid quadrature formula is used on each level. We assume as in the previous case that the form $\mathcal{A}_k(\bullet,\bullet)$ is uniformly equivalent to $A(\bullet,\bullet)$ for each k. That is, there are positive constants C_1 and C_2, not depending on k, such that

$$C_1 A(v,v) \leqslant \mathcal{A}_k(v,v) \leqslant C_2 A(v,v), \qquad \text{for all } v \in M_k. \tag{10.9}$$

We assume that the forms $\mathcal{A}_k(\bullet,\bullet)$ are small perturbations of $A(\bullet,\bullet)$. Thus, instead of (10.2) we assume that

$$\left|\mathcal{A}_k(v,w) - A(v,w)\right| \leqslant C\lambda_k^{-\beta/2} |||v|||_A |||w|||_A, \qquad \text{for all } v, w \in M_k. \tag{10.10}$$

Define $\mathcal{P}_{k-1}: M_k \to M_{k-1}$ by

$$\mathcal{A}_{k-1}(\mathcal{P}_{k-1}v, \phi) = \mathcal{A}_k(v,\phi), \qquad \text{for all } v \in M_k,\ \phi \in M_{k-1}.$$

We now show that if Eqs. (10.9) and (10.10) hold and if Condition (A.2) holds for the form $A(\bullet,\bullet)$, it also holds for the form $\mathcal{A}(\bullet,\bullet)$. Consequently Condition (A.2″) holds for the form $\mathcal{A}(\bullet,\bullet)$.

LEMMA 10.3. *Assume that* (10.9) *holds and that* (10.10) *holds with* $\beta = \alpha$. *If Condition* (A.2) *holds for* $A(\bullet,\bullet)$, *then it holds for* $\mathcal{A}(\bullet,\bullet)$. *Consequently Condition* (A.2″) *holds for* $\mathcal{A}(\bullet,\bullet)$.

PROOF. The proof is similar to the proof of Lemma 10.1. Note that for $v \in M_k$,

$$\begin{aligned} \left|\left|\left|(I - \mathcal{P}_{k-1})v\right|\right|\right|_{\mathcal{A}_k^{1-\alpha}} &\leqslant C\left|\left|\left|(I - \mathcal{P}_{k-1})v\right|\right|\right|_{A_k^{1-\alpha}} \\ &\leqslant C\left(\left|\left|\left|(I - P_{k-1})v\right|\right|\right|_{A_k^{1-\alpha}} + \left|\left|\left|(P_{k-1} - \mathcal{P}_{k-1})v\right|\right|\right|_{A_k^{1-\alpha}}\right). \end{aligned}$$

If Condition (A.2) holds for $A(\bullet,\bullet)$ then

$$\left|\left|\left|(I - P_{k-1})v\right|\right|\right|_{A_k^{1-\alpha}} \leqslant C\lambda_k^{-\alpha/2} |||v|||_A, \qquad \text{for all } v \in M_k.$$

Set $w = A_k^{-\alpha}(P_{k-1} - \mathcal{P}_{k-1})v$. By the definition of $\mathcal{P}_{k-1}$ and (10.10),

$$\begin{aligned} &\left|\left|\left|(P_{k-1} - \mathcal{P}_{k-1})v\right|\right|\right|^2_{A_k^{1-\alpha}} = A\big((P_{k-1} - \mathcal{P}_{k-1})v, w\big) \\ &\quad = \left[A(v,w) - \mathcal{A}_k(v,w) + \mathcal{A}_{k-1}(\mathcal{P}_{k-1}v, w) - A(\mathcal{P}_{k-1}v, w)\right] \\ &\quad \leqslant C\lambda_k^{-\alpha/2} |||v|||_A |||w|||_A = C\lambda_k^{-\alpha/2} |||v|||_A \left|\left|\left|(P_{k-1} - \mathcal{P}_{k-1})v\right|\right|\right|_{A_k^{1-\alpha}}. \end{aligned}$$

This implies that

$$\left|\left|\left|(P_{k-1} - \mathcal{P}_{k-1})v\right|\right|\right|_{A_k^{1-\alpha}} \leqslant C\lambda_k^{-\alpha/2} |||v|||_A.$$

Combining the above estimates, we obtain

$$\left|\left|\left|(I - \mathcal{P}_{k-1})v\right|\right|\right|_{\mathcal{A}_k^{1-\alpha}} \leqslant C\lambda_k^{-\alpha/2} |||v||| \leqslant C\hat{\lambda}_k^{-\alpha/2} |||v|||_{\mathcal{A}_k}.$$

Hence Condition (A.2) holds for form $\mathcal{A}(\bullet, \bullet)$. Since the spaces M_k are nested and $I_k = I$, Condition (A.2$''$) follows from Condition (A.2). □

10.2. Finite elements with quadrature

We consider the Dirichlet problem (9.1) on a polygonal domain Ω in $\mathbf{R}^2$:

$$\mathcal{L}u \equiv -\sum_{i,j=1}^{2} \frac{\partial}{\partial x_i}\left(a_{ij}\frac{\partial u}{\partial x_j}\right) + au = f \quad \text{in } \Omega,$$
$$u = 0 \quad \text{on } \partial\Omega.$$

We assume that the coefficients $a(x)$ and $a_{ij}(x)$ are smooth. The form $A(\bullet, \bullet)$ is defined by (9.3) which we recall here:

$$A(v, w) = \sum_{i,j=1}^{2} \int_\Omega a_{ij} \frac{\partial v}{\partial x_i}\frac{\partial w}{\partial x_j}\,\mathrm{d}x + \int_\Omega avw\,\mathrm{d}x.$$

Let $\{\mathcal{T}_k\}$ be a family of nested quasi-uniform triangulations of Ω. We define on $\{\mathcal{T}_k\}$ the nested set of finite element spaces $M_k,\ k = 1, \ldots, J$, consisting of polynomials on each triangle of degree less than or equal to K, and continuous on Ω. On each fine grid triangle $\tau \in \mathcal{T}_J$ we introduce a quadrature rule

$$\int_\tau \phi(x)\,\mathrm{d}x \cong \sum_{\ell=1}^{L} w_\ell \phi(b_\ell) \equiv \mathcal{Q}_\tau[\phi].$$

We assume that the quadrature formula $\mathcal{Q}_\tau$ is exact for all polynomials of degree less than or equal to $2K - 2$, i.e.,

$$\mathcal{Q}_\tau[\phi] = \int_\tau \phi(x)\,\mathrm{d}x, \quad \text{for all } \phi \in \mathbf{P}_{2K-2}. \tag{10.11}$$

Here $\mathbf{P}_{2K-2}$ denotes the space of polynomials of degree less than or equal to $2K-2$. The usual way to define a quadrature scheme is as follows. Let $\hat{\tau}$ be a closed reference triangle. On $\hat{\tau}$ we introduce the reference quadrature

$$\int_{\hat{\tau}} \hat{\phi}(\hat{x})\,\mathrm{d}\hat{x} \cong \mathcal{Q}_{\hat{\tau}}[\hat{\phi}] \equiv \sum_{\ell=1}^{L} \hat{w}_\ell \hat{\phi}(\hat{b}_\ell),$$

where $\hat{w}_\ell$ are positive weights and $\hat{b}_\ell \in \hat{\tau}$ are the quadrature points. Now the quadrature rule $\mathcal{Q}_\tau$ on each fine grid triangle $\tau \in \mathcal{T}_J$ is defined in terms of $\mathcal{Q}_{\hat{\tau}}$. Let F_τ be an affine mapping from $\hat{\tau}$ onto τ. We set points $b_\ell = F_\tau(\hat{b}_\ell)$ and weights $w_\ell = (|\tau|/|\hat{\tau}|)\hat{w}_\ell$.

The approximate form $\mathcal{A}(\bullet, \bullet)$ is now defined by

$$\mathcal{A}(v, w) = \sum_{\tau \in \mathcal{T}_J} \mathcal{Q}_\tau \left[\left\{ \sum_{i,j=1}^{2} \left(a_{ij} \frac{\partial v}{\partial x_i} \frac{\partial w}{\partial x_j} \right) \right\} + avw \right], \quad \text{for all } v, w \in M_J. \tag{10.12}$$

We set $\mathcal{A}_k(\bullet, \bullet) = \mathcal{A}(\bullet, \bullet)$ for $k = 1, 2, \ldots, J$.

It can be shown that the quadrature rule defined by (10.11) satisfies (10.1) (cf. CIARLET [1978]). Furthermore, the exactness condition also implies that (10.2) holds. We have shown in Section 9 that Conditions (A.2), (A.3), (A.4), (A.5) and (A.6) hold for the form $A(\bullet, \bullet)$ provided that the coefficients $a(x)$ and $a_{ij}(x)$ are sufficiently smooth. By Lemma 10.1 and Lemma 10.2, these conditions also hold for the form $\mathcal{A}(\bullet, \bullet)$. Thus we may apply either Theorem 5.1 or Theorem 5.2 to conclude that the multigrid V-cycle iteration defined by Algorithm 3.1 has a uniform error reduction provided that we use an appropriate smoother.

In the implementation of the above multigrid algorithm, we also need to setup the stiffness matrices $\underset{\approx}{\mathcal{A}}_k$. The matrix $\underset{\approx}{\mathcal{A}}_J = [\mathcal{A}(\phi_J^i, \phi_J^j)]$ is computed by using formula (10.12). Coarse stiffness matrices $\underset{\approx}{\mathcal{A}}_k$ are then computed from $\underset{\approx}{\mathcal{A}}_J$ recursively by expressing coarse basis functions as a linear combination of the finer ones. This can be done recursively and the cost of setting up all the coarse stiffness matrices $\underset{\approx}{\mathcal{A}}_k$ is thus proportional to the dimension of M_J.

An alternative is to use a similar quadrature rule $\mathcal{Q}_\tau$ on each level k. The corresponding multigrid algorithm has varying forms but the spaces are nested. We shall use the natural inclusion operator as the grid transfer operator in the multigrid algorithm. If for each $\tau \in \mathcal{T}_k$ the corresponding the quadrature rule is exact for all polynomials of degree less than or equal to $2K-2$, then Eqs. (10.9) and (10.10) hold. Consequently, Condition (A.2) holds for $\mathcal{A}(\bullet, \bullet)$. Since the spaces M_k are nested and $I_k = I$, Condition (A.2$''$) follows from Condition (A.2). The multigrid theory of Section 7 can be applied to this case. In particular the variable V-cycle multigrid preconditioner is optimal.

10.3. Finite difference example

We consider, for simplicity, a domain $\Omega \subset \mathbf{R}^2$ made up of the union of squares of side $h = h_J$. We define the mesh domain

$$\overline{\Omega}_h = \{x \in \overline{\Omega} \mid x = (\ell h, mh), \text{ for } \ell, m \text{ integers}\}.$$

Set $\Omega_h = \overline{\Omega}_h \cap \Omega$ and $\partial\Omega_h = \overline{\Omega}_h \cap \partial\Omega$. Let S_h denote the set of grid functions on $\overline{\Omega}_h$ which vanish on $\partial\Omega_h$. For $V \in S_h$ set $V_{i,j} = V(ih, jh)$.

We shall consider the special case of (9.1) with

$$\mathcal{L}u \equiv -\left(\frac{\partial}{\partial x_1} a \frac{\partial u}{\partial x_1} + \frac{\partial}{\partial x_2} b \frac{\partial u}{\partial x_2}\right) + cu = f, \tag{10.13}$$

where a and b are positive functions and $c \geqslant 0$. The following is a standard five-point finite difference approximation of (10.13). For $F \in S_h$, find $U \in S_h$ satisfying

$$\mathcal{L}_h U = F,$$

where $\mathcal{L}_h : S_h \to S_h$ is given by

$$\begin{aligned}(\mathcal{L}_h V)_{i,j} &= a_{i+1/2,j}(V_{i,j} - V_{i+1,j}) + a_{i-1/2,j}(V_{i,j} - V_{i-1,j}) \\ &\quad + b_{i,j+1/2}(V_{i,j} - V_{i,j+1}) + b_{i,j-1/2}(V_{i,j} - V_{i,j-1}) \\ &\quad + h^2 c_{i,j} V_{i,j}.\end{aligned} \tag{10.14}$$

Here $\phi_{s,t} = \phi(sh, th)$ for any function ϕ.

Now we can relate the above scheme to a certain finite element scheme by considering it as coming from certain quadratures. We do this as follows:

Consider the triangulation $\mathcal{T}_J$ which results from dividing each square on the finite difference grid into two triangles with positively sloping diagonals. Let M_J be the space of continuous piecewise linear functions on $\mathcal{T}_J$. Assume further that $\mathcal{T}_J$ results from successive subdivisions of an original coarse triangulation $\mathcal{T}_1$.

Let τ be a typical triangle. Denote $\tilde{v}_\ell$, $\ell = 1, 2, 3$, its vertices and m_ℓ the midpoint of the edge parallel to the x_ℓ axis, $\ell = 1, 2$. We use the following three quadrature formulas, respectively, for the three terms coming from (10.13):

$$\int_\tau \phi \, \mathrm{d}x \cong Q_\tau^{(1)}[\phi] \equiv \frac{h^2}{2}\phi(m_1),$$

$$\int_\tau \phi \, \mathrm{d}x \cong Q_\tau^{(2)}[\phi] \equiv \frac{h^2}{2}\phi(m_2)$$

and

$$\int_\tau \phi \, \mathrm{d}x \cong Q_\tau^{(3)}[\phi] \equiv \frac{h^2}{6}\left[\phi(\tilde{v}_1) + \phi(\tilde{v}_2) + \phi(\tilde{v}_3)\right].$$

Let $\mathcal{A}(\bullet,\bullet)$ be the resulting bilinear form on the space M_J. It can be verified that the stiffness matrix associated with $\mathcal{A}(\bullet,\bullet)$ is the same as the matrix of the finite difference scheme (10.14). It is also easy to verify that (10.1) holds for the bilinear form $A(\bullet,\bullet)$ corresponding to (10.13). Furthermore, the above quadrature schemes are exact on constants. Consequently, since $K = 1$ $(2K - 2 = 0)$ it follows easily that (10.2) holds, if a, b and c are Hölder continuous functions with exponent β. Thus with the prolongation coming from the finite element imbedding and using, e.g., Gauss–Seidel as smoother, Theorem 5.1 (or Theorem 5.2) implies that the multigrid iteration defined by Algorithm 3.1 has a uniform error reduction.

We now turn to the computation of the stiffness matrices. It is easy to check that the stiffness matrix corresponding to $\mathcal{A}$ associated with any level k comes from a five point difference operator. The coefficients may be computed recursively from the coefficients on the finest level. The cost of assembling the J stiffness matrices is proportional to the number of unknowns.

The computation of the stiffness matrices can be avoided by using the same finite difference scheme (10.14) on all levels (with appropriate mesh sizes). As in the previous subsection, the corresponding multigrid algorithm has varying forms but the spaces are still nested. Again the theory of Section 7 can be applied.

10.4. Bibliographical notes

A discussion of the use of quadrature in finite element methods may be found in CIARLET [1978].

Convergence results for multigrid method for finite elements with quadrature were given in GOLDSTEIN [1991]. It was assumed there that coarse grid quadrature formulae were used on the coarser levels and that full elliptic regularity (i.e., Condition (A.2) with $\alpha = 1$) was satisfied. This means the Ω must be convex and the coefficients must be smooth. The present method and analysis are based on BRAMBLE, GOLDSTEIN and PASCIAK [1994].

The treatment of finite difference applications may be found in HACKBUSCH [1985]. It is not clear that the theorems there are applicable to our example. The formulation and analysis given in Subsection 10.3 comes from BRAMBLE, GOLDSTEIN and PASCIAK [1994].

11. Second-order nonsymmetric and indefinite problems

In this section, we study a V-cycle multigrid method for second-order elliptic boundary value problems including problems which may be nonsymmetric and/or indefinite. A convergence analysis of the multigrid V-cycle method is presented under rather weak assumptions on the underlying elliptic problems. We show that the uniform convergence results of Section 9 for symmetric positive definite problems carry over to the nonsymmetric and/or indefinite case for a variety of smoothers. More precisely, the multigrid iteration is shown to converge at a uniform rate provided that the coarsest grid in the multilevel iteration is sufficiently fine (but not depending on the number of multigrid levels). The condition that the coarse grid must be fine enough seems to be unavoidable since, in many cases, it is needed even for the finite element approximation to make sense.

Our analysis is based on a perturbation argument. We prove the convergence estimates by comparing the error propagator of the multigrid method for the nonsymmetric and/or indefinite problem with the error propagator of the multigrid method for a symmetric positive definite problem.

In our multigrid method various types of smoothers may be used. One type of smoother which we consider is defined in terms of an associated symmetric problem and includes point and line, Jacobi and Gauss–Seidel iterations. We also study smoothers based entirely on the original operator. One is based on the normal form, i.e., the product of the operator and its transpose. Others studied include point and line, Jacobi and Gauss–Seidel smoothers.

11.1. The differential equation and finite element problems

In this subsection, we set up the model nonsymmetric problem and its finite element approximations on a nested sequence of meshes. We consider, for simplicity, the Dirichlet problem in two dimensions approximated by piecewise linear finite elements. The multigrid convergence results hold for many extensions and generalizations. This is discussed at the end of this section.

We consider as our model problem the following second-order elliptic equation with homogeneous boundary conditions.

$$
\begin{aligned}
&-\sum_{i,j=1}^{2}\frac{\partial}{\partial x_j}\left(a_{ij}\frac{\partial u}{\partial x_i}\right)+\sum_{i=1}^{2}b_i\frac{\partial u}{\partial x_i}+au=f \quad \text{in } \Omega,\\
&u=0 \quad \text{on } \partial\Omega,
\end{aligned}
\tag{11.1}
$$

where Ω is a polygonal domain (possibly nonconvex) in R^2 and $\{a_{ij}(x)\}$ is bounded symmetric, and uniformly positive definite for $x\in\Omega$. We assume that a_{ij} is in the Sobolev space $W_p^\gamma(\Omega)$ for $p>2/\gamma$. Here $W_p^\gamma(\Omega)$ denotes the Sobolev space of order γ defined in terms of the norm $L^p(\Omega)$ (see GRISVARD [1985] for details). Further, we assume that b_i is continuously differentiable on $\bar{\Omega}$ and that $|a|$ is bounded. Finally, we assume that the solution of (11.1) exists.

Let $H^1(\Omega)\equiv W_2^1(\Omega)$ denote the Sobolev space of order one on Ω and let $H_0^1(\Omega)$ denote those functions in $H^1(\Omega)$ whose traces vanish on $\partial\Omega$. For $v, w\in H_0^1(\Omega)$, define

$$
\mathcal{A}(v,w)=\sum_{i,j=1}^{2}\int_\Omega a_{ij}\frac{\partial v}{\partial x_i}\frac{\partial w}{\partial x_j}\,\mathrm{d}x+\sum_{i=1}^{2}\int_\Omega b_i\frac{\partial v}{\partial x_i}w\,\mathrm{d}x+\int_\Omega avw\,\mathrm{d}x. \tag{11.2}
$$

The solution u of (11.1) satisfies

$$
\mathcal{A}(u,v)=(f,v), \quad \text{for all } v\in H_0^1(\Omega),
$$

where $(\bullet,\bullet)$ denotes the inner product in $L^2(\Omega)$.

We define a sequence of nested triangulations of Ω in the usual way. We assume that a coarse triangulation $\{\tau_1^i\}$ of Ω is given. Successively finer triangulations $\{\tau_m^i\}$ for $m>1$ are defined by subdividing each triangle (in a coarser triangulation) into four by connecting the midpoints of the edges. The mesh size of $\{\tau_1^i\}$ will be denoted to be d_1 and can be taken to be the diameter of the largest triangle. By similarity, the mesh size of $\{\tau_m^i\}$ is $2^{1-m}d_1$.

For theoretical and practical purposes, the coarsest grid in the multilevel algorithms must be sufficiently fine. In practice, however, the coarse grid is still considerably coarser than the fine grid. Let J_0 and J be greater than or equal to one. We set $\mathcal{T}_k=\{\tau_{k+J_0}^i\}$. Then the mesh size of $\mathcal{T}_k$ is $h_k=2^{1-J_0-k}d_1=2^{1-k}h_1$. Denote by M_k, for $k=1,\dots,J$, the space of functions which are piecewise linear with respect to the triangulation $\mathcal{T}_k$, continuous on Ω and vanish on $\partial\Omega$. Since the triangulations are nested, it follows that

$$M_1\subset M_2\subset\cdots\subset M_J.$$

We now consider the associated finite element problems: Given a linear functional f defined on M_k, find $u\in M_k$ such that

$$\mathcal{A}(u,\phi)=f(\phi),\qquad \text{for all } \phi\in M_k.$$

We assume that for all k and for every $v\in M_k$,

$$\mathcal{A}(v,\phi)=0,\qquad \text{for all } \phi\in M_k,\qquad \text{implies } v=0. \tag{11.3}$$

This assumption immediately implies the existence and uniqueness of solutions to finite element problem on M_k. In particular, the projection operator $\mathcal{P}_k:H^1(\Omega)\to M_k$ satisfying

$$\mathcal{A}(\mathcal{P}_k u,\phi)=\mathcal{A}(u,\phi),\qquad \text{for all } \phi\in M_k,$$

is well defined.

Clearly, if the form $\mathcal{A}(\bullet,\bullet)$ defined by (11.2) has a positive definite symmetric part then (11.3) holds. More generally, if solutions of (11.1) satisfy regularity estimates of the form

$$\|u\|_{1+\alpha}\leqslant C\|f\|_{-1+\alpha}, \tag{11.4}$$

then, it is well known (cf. SCHATZ [1974]) that there exists a constant h_0 such that for $h_k\leqslant h_0$, (11.3) holds and furthermore

$$\|\mathcal{P}_k u\|_1\leqslant C\|u\|_1 \quad\text{and}\quad \big\|(I-\mathcal{P}_k)u\big\|\leqslant Ch_k^\alpha\big\|(I-\mathcal{P}_k)u\big\|_1.$$

Even if regularity estimates of the form of (11.4) are not known to hold, (11.3) is known from the following result of SCHATZ and WANG [1996].

PROPOSITION 11.1. *There exists an h_0 such that* (11.3) *holds for $h_k \leqslant h_0$. Moreover, given $\varepsilon > 0$, there exists an $h_0(\varepsilon) > 0$ such that for all $h_k \in (0, h_0]$,*

$$\|\mathcal{P}_k v\|_1 \leqslant C\|v\|_1 \quad \textit{and} \quad \big\|(I - \mathcal{P}_k)v\big\| \leqslant C\varepsilon\big\|(I - \mathcal{P}_k)v\big\|_1 \leqslant C\varepsilon\|v\|_1. \tag{11.5}$$

REMARK 11.1. The above ε will appear in our subsequent analysis. We note that ε can be taken arbitrarily small. However, J_0 will be taken large enough so that Eqs. (11.3) and (11.5) hold. Thus, the coarsest grid size (i.e., J_0), for any estimate in which ε appears, will depend on ε.

We seek the solution of

$$\mathcal{A}(u, \phi) = (f, \phi), \quad \text{for all } \phi \in M_J. \tag{11.6}$$

For each k, we define $\mathcal{A}_k : M_k \to M_k$ and $Q_k : L^2(\Omega) \to M_k$ by

$$(\mathcal{A}_k u, \phi) = \mathcal{A}(u, \phi), \quad \text{for all } \phi \in M_k,$$

and

$$(Q_k u, \phi) = (u, \phi), \quad \text{for all } \phi \in M_k.$$

The operator Q_k is the $L^2(\Omega)$ orthogonal projector onto M_k. Using this notation, (11.6) can be rewritten as

$$\mathcal{A}_J u = f_J \equiv Q_J f.$$

For the analysis, we introduce a symmetric positive definite form $A(\bullet, \bullet)$ which has same second order part as $\mathcal{A}(\bullet, \bullet)$. We define $A(\bullet, \bullet)$ by

$$A(v, w) = \sum_{i,j=1}^{2} \int_{\Omega} a_{ij} \frac{\partial v}{\partial x_i} \frac{\partial w}{\partial x_j}\, \mathrm{d}x + \int_{\Omega} vw \,\mathrm{d}x.$$

The difference of $\mathcal{A}(v, w)$ and $A(v, w)$ is denoted by

$$D(v, w) = \mathcal{A}(v, w) - A(v, w).$$

The form $D(\bullet, \bullet)$ satisfies the inequalities

$$\big|D(v, w)\big| \leqslant C\|v\|_1\|w\| \quad \text{and} \quad \big|D(v, w)\big| \leqslant C\|v\|\|w\|_1. \tag{11.7}$$

Here $\|\bullet\|_1$ and $\|\bullet\|$ denote the norms in $H^1(\Omega)$ and $L^2(\Omega)$ respectively. The second inequality above follows from integration by parts.

By the assumptions on the coefficients appearing in the definition of $A(\bullet,\bullet)$, it follows that the norm $A(v,v)^{1/2}$, for $v \in H^1(\Omega)$, is equivalent to the norm on $H^1(\Omega)$. Thus, we take

$$\|v\|_1 = A(v,v)^{1/2}.$$

In our analysis, we shall use the $A(\bullet,\bullet)$ orthogonal projector $P_k : H_0^1(\Omega) \to M_k$ defined by

$$A(P_k u, \phi) = A(u, \phi), \quad \text{for all } \phi \in M_k.$$

We refer to P_k as the elliptic projection. For each k, define $A_k : M_k \to M_k$ by

$$(A_k u, \phi) = A(u, \phi), \quad \text{for all } \phi \in M_k.$$

11.2. Multigrid algorithm and perturbation analysis

We first describe the simplest multigrid algorithm for iteratively computing the solution u of (11.6). We then establish an abstract convergence estimate based on perturbation from the convergence estimates for multigrid algorithms applied to the symmetric problem associated with the form $A(\bullet,\bullet)$.

The multigrid algorithm. Given an initial iterate $u^0 \in M_J$, we define a sequence approximating u by

$$u^{m+1} = \mathrm{Mg}_J\big(u^m, f_J\big). \tag{11.8}$$

Here $\mathrm{Mg}_J(\bullet,\bullet)$ is a map of $M_J \times M_J$ into M_J and is defined recursively as follows.

ALGORITHM 11.1. Set $\mathrm{Mg}_1(v,g) = \mathcal{A}_1^{-1} g$. Let $k > 1$ and v, g be in M_k. Assuming that $\mathrm{Mg}_{k-1}(\bullet,\bullet)$ has been defined, we define $\mathrm{Mg}_k(v,g)$ as follows.

(1) $v' = v + \mathcal{R}_k(g - \mathcal{A}_k v)$.

(2) $\mathrm{Mg}_k(v,g) \equiv v'' = v' + q$, where q is defined by

$$q = \mathrm{Mg}_{k-1}\big(0, Q_{k-1}\big(g - \mathcal{A}_k v'\big)\big).$$

Here $\mathcal{R}_k : M_k \to M_k$ is a linear smoothing operator. Note that in this multigrid algorithm, we smooth only as we proceed to coarser grids. The above algorithm is a special case of more general multigrid algorithms in that we only use pre-smoothing. Alternatively, we could define an algorithm with just post-smoothing or both pre- and post-smoothing. The analysis of these algorithms is similar to that above and will not be presented. Often algorithms with more than one smoothing are considered; cf. HACKBUSCH [1985], and BANK, MANDEL and MCCORMICK [1987]. This is not advised in the above algorithm since the smoothing iteration is generally unstable.

We first derive a product representation for the error propagation operator. We proceed as in Section 5. A straightforward induction argument shows that $\mathrm{Mg}_J(\bullet,\bullet)$ is a linear map from $M_J \times M_J$ into M_J. Moreover, the scheme is consistent in the sense that $v = \mathrm{Mg}_J(v, \mathcal{A}_J v)$ for all $v \in M_J$. It easily follows that the linear operator $\mathcal{E} = \mathrm{Mg}_J(\bullet, 0)$ is the error reduction operator for (11.8), i.e.

$$u - u^{m+1} = \mathrm{Mg}_J\big(u - u^m, 0\big).$$

Now let $\mathcal{T}_k = \mathcal{R}_k \mathcal{A}_k \mathcal{P}_k$ for $k > 1$ and set $\mathcal{T}_1 = \mathcal{P}_1$. Using the facts that $Q_{k-1}\mathcal{A}_k = \mathcal{A}_{k-1}\mathcal{P}_{k-1}$ and $\mathcal{P}_{k-1}\mathcal{P}_k = \mathcal{P}_{k-1}$ and the definition of Algorithm 11.1, a straightforward manipulation gives that for $k > 1$ and any $v \in M_J$,

$$v - \mathrm{Mg}_k(0, \mathcal{A}_k\mathcal{P}_k v) = (I - \mathcal{T}_k)v - \mathrm{Mg}_{k-1}\big(0, \mathcal{A}_{k-1}\mathcal{P}_{k-1}(I - \mathcal{T}_k)v\big).$$

Let $\mathcal{E}_k v = v - \mathrm{Mg}_k(0, \mathcal{A}_k \mathcal{P}_k v)$. In terms of $\mathcal{E}_k$, the above identity is the same as

$$\mathcal{E}_k = \mathcal{E}_{k-1}(I - \mathcal{T}_k).$$

Moreover, by consistency, $\mathcal{E} = \mathcal{E}_J$ and hence

$$\mathcal{E} = (I - \mathcal{T}_1)(I - \mathcal{T}_2)\cdots(I - \mathcal{T}_J). \tag{11.9}$$

This product representation of the error operator is similar to the one given in Section 5 for multigrid algorithms for symmetric problems.

A perturbation analysis. We now provide an abstract analysis of the multigrid iteration (11.8). We analyze the convergence of iteration (11.8) by comparing the error operators of Algorithm 11.1 and the corresponding algorithm applied to the symmetric problem

$$A_J(u, \phi) = (f, \phi), \quad \text{for all } \phi \in M_J. \tag{11.10}$$

The analysis is based on perturbation from the uniform convergence estimates for the multigrid algorithm applied to (11.10) and it uses the product representation of the error operator given in (11.9).

We start with the multigrid algorithm for the symmetric problem corresponding to the form $A(\bullet,\bullet)$. Specifically, we replace $\mathcal{A}_k$ by A_k and $\mathcal{R}_k$ by R_k in Algorithm 11.1. Set $T_1 = P_1$. From the earlier discussion, the error propagation operator associated with this iteration applied to the symmetric problem (11.10) is given by $E = E_J$ where

$$E_k = (I - T_1)(I - T_2)\cdots(I - T_k), \quad k = 1, 2, \ldots, J,$$
$$\text{with } T_1 = P_1 \text{ and } T_k = R_k A_k P_k. \tag{11.11}$$

To analyze the multigrid algorithm for the nonsymmetric problem, we introduce the perturbation operators

$$\mathcal{P}_1 - P_1 \quad \text{and} \quad \mathcal{T}_k - T_k = \mathcal{R}_k\mathcal{A}_k\mathcal{P}_k - R_k A_k P_k, \quad k \geqslant 2.$$

Set $|||v||| = A(v, v)^{1/2}$. For an arbitrary linear operator $O : M_J \to M_J$, let $|||O|||$ denote its operator norm, i.e.,

$$|||O||| = \sup_{v \in M_J} \frac{|||Ov|||}{|||v|||} = \sup_{u,v \in M_J} \frac{A(Ou, v)}{|||u|||\,|||v|||}.$$

We have the following perturbation lemma.

LEMMA 11.1. *Let $\mathcal{E}_k$ and E_k be defined by Eqs.* (11.9) *and* (11.11). *Assume that*

$$|||I - T_k||| \leqslant 1 \quad \text{and} \quad |||\mathcal{T}_k - T_k||| \leqslant Ch_k, \quad \text{for } k = 2, \dots, J.$$

Then given $\varepsilon > 0$, there exist $h_0(\varepsilon) > 0$ and $C > 0$ such that for $h_1 \leqslant h_0(\varepsilon)$,

$$|||\mathcal{E}||| \leqslant |||E||| + C(h_1 + \varepsilon).$$

PROOF. We first derive a bound for $|||\mathcal{T}_1 - T_1|||$. By definition,

$$\begin{aligned} A(\mathcal{P}_1 w, v) &= A(\mathcal{P}_1 w, P_1 v) \\ &= \mathcal{A}(w, P_1 v) - D(\mathcal{P}_1 w, P_1 v) \\ &= A(P_1 w, v) + D\big((I - \mathcal{P}_1)w, P_1 v\big). \end{aligned}$$

Consequently

$$A\big((\mathcal{T}_1 - T_1)w, v\big) = A\big((\mathcal{P}_1 - P_1)w, v\big) = D\big((I - \mathcal{P}_1)w, P_1 v\big). \tag{11.12}$$

Now applying Eqs. (11.7) and (11.5) to (11.12) gives

$$\begin{aligned} \big|A\big((\mathcal{T}_1 - T_1)w, v\big)\big| &= \big|D\big((I - \mathcal{P}_1)w, P_1 v\big)\big| \\ &\leqslant C\big\|(I - \mathcal{P}_1)w\big\|\,\|v\|_1 \leqslant C\varepsilon\|w\|_1\|v\|_1. \end{aligned}$$

This proves that

$$|||\mathcal{T}_1 - T_1||| \leqslant C\varepsilon.$$

Since $|||I - P_1||| \leqslant 1$, the triangle inequality implies that

$$|||I - \mathcal{P}_1||| \leqslant |||I - P_1||| + |||\mathcal{P}_1 - P_1||| \leqslant 1 + C\varepsilon.$$

By the assumption, we have

$$|||\mathcal{T}_k - T_k||| \leqslant Ch_k \quad \text{and} \quad |||I - T_k||| \leqslant 1.$$

The triangle inequality implies that

$$|||I - \mathcal{T}_k||| \leqslant |||I - T_k||| + |||\mathcal{T}_k - T_k||| \leqslant 1 + Ch_k.$$

Hence, it follows that

$$|||\mathcal{E}_k||| \leqslant \prod_{i=1}^{k} |||I - \mathcal{T}_i||| \leqslant (1 + C\varepsilon) \prod_{i=2}^{k} (1 + Ch_i) \leqslant C.$$

It is immediate from the definitions that

$$\mathcal{E}_k - E_k = (\mathcal{E}_{k-1} - E_{k-1})(I - T_k) - \mathcal{E}_{k-1}(\mathcal{T}_k - T_k).$$

We have, for $k > 1$,

$$\begin{aligned} |||\mathcal{E}_k - E_k||| &\leqslant |||\mathcal{E}_{k-1} - E_{k-1}|||\,|||I - T_k||| + |||\mathcal{E}_{k-1}|||\,|||\mathcal{T}_k - T_k||| \\ &\leqslant |||\mathcal{E}_{k-1} - E_{k-1}||| + Ch_k. \end{aligned}$$

Repetitively applying this inequality and using

$$|||\mathcal{E}_1 - E_1||| = |||\mathcal{T}_1 - T_1||| \leqslant C\varepsilon$$

gives that

$$|||\mathcal{E}_J - E_J||| \leqslant C\varepsilon + C\sum_{k=2}^{J} h_k \leqslant C(h_1 + \varepsilon).$$

The lemma follows from the triangle inequality. □

To state the convergence estimate of the multigrid method for symmetric positive definite problems, we first recall the smoother conditions used in the multigrid analysis in Section 5. For $k > 1$, set

$$\overline{R}_k = R_k + R_k^t - R_k^t A_k R_k.$$

CONDITION (SM.1). There is a constant $\omega \in (0, 1)$, not depending on J, such that

$$\frac{\omega}{\lambda_k}\|v\|^2 \leqslant (\overline{R}_k v, v), \quad \text{for all } v \in M_k,\ k = 2, 3, \ldots, J,$$

where λ_k is the largest eigenvalue of A_k.

CONDITION (SM.2). There is a constant $\theta \in (0, 2)$, not depending on J, such that

$$A(R_k v, R_k v) \leqslant \theta(R_k v, v), \quad \text{for all } v \in M_k,\ k = 2, 3, \ldots, J.$$

CONDITION (SM.3). There exists a constant $\omega > 0$, not depending on J, such that

$$\frac{\omega}{\lambda_k^2}(A_k v, v) \leqslant \bigl(\overline{R}_k v, v\bigr), \quad \text{for all } v \in M_k,\ k = 2, 3, \ldots, J.$$

CONDITION (SM.4). There exists a constant C, not depending on J, such that

$$(R_k v, R_k v) \leqslant \omega_4 \lambda_k^{-1} (R_k v, v), \quad \text{for all } v \in M_k,\ k \geqslant 2.$$

When R_k is symmetric with respect to $(\bullet, \bullet)$, Condition (SM.2) implies that $|||T_k||| \leqslant \theta$. Even in the case of nonsymmetric R_k, Condition (SM.2) implies the stability of $(I - T_k)$. In fact, for any $w \in M_J$, Condition (SM.2) implies that

$$\begin{aligned} A\bigl((I - T_k)w, (I - T_k)w\bigr) &= A(w, w) - 2A(T_k w, w) + A(T_k w, T_k w) \\ &\leqslant A(w, w) - (2 - \theta)A(T_k w, w) \leqslant A(w, w). \end{aligned} \tag{11.13}$$

REMARK 11.2. It is easy to verify that if Conditions (SM.1), (SM.2) and (SM.4) (or (SM.3), (SM.2) and (SM.4)) hold for a smoother R_k then they also hold for its adjoint R_k^t with respect to the inner product $(\bullet, \bullet)$. In the case of Conditions (SM.2) and (SM.4), the corresponding inequalities hold with the same constants as those appearing in the original inequalities.

We now recall a convergence result for the multigrid algorithm applied to the symmetric positive definite problem $Au = f$. This result is an application of Theorem 5.2 or Theorem 5.4 to the case of second-order selfadjoint elliptic problems; cf. Section 9.

PROPOSITION 11.2. *Assume that R_k satisfies either Conditions* (SM.1) *and* (SM.2) *or Conditions* (SM.3) *and* (SM.2). *Under the assumptions on the domain Ω and the coefficients of* (11.1) *given in Subsection* 11.1, *there exists a positive constant $\delta < 1$, not depending on J, such that*

$$A(E_J v, E_J v) \leqslant \delta^2 A(v, v), \quad \textit{for all } v \in M_J.$$

11.3. Smoothers based on the symmetric problem

In this subsection, we consider the multigrid Algorithm 11.1 using a smoother associated with the symmetric form $A(\bullet, \bullet)$. We shall denote by R_k any smoother derived from the symmetric form $A_k(\bullet, \bullet)$. We shall discuss five examples of smoothing procedures which satisfy either Conditions (SM.1) and (SM.2) or Conditions (SM.3) and (SM.2). This implies that the error operator of the corresponding multigrid algorithm for the symmetric problem is a uniform contraction; i.e. $|||E_k||| \leqslant \delta < 1$. To show that the perturbation operator $\mathcal{T}_k - T_k$ satisfies the conditions in Lemma 11.1, we assume that the smoothers also satisfy Condition (SM.4).

The following theorem provides an estimate for the multigrid algorithm when the smoothers $\mathcal{R}_k = R_k$ are used in Algorithm 11.1.

THEOREM 11.1. *Assume that the smoothers R_k satisfy either Conditions* (SM.1), (SM.2) *and* (SM.4) *or Conditions* (SM.3), (SM.2) *and* (SM.4). *Let $\mathcal{E}$ be the error operator for Algorithm* 11.1 *with smoothers $\mathcal{R}_k = R_k$ and E be the error operator defined by* (11.11). *Then there exists a constant $C > 0$ such that given $\varepsilon > 0$, there exists $h_0(\varepsilon) > 0$ such that for $h_1 \leqslant h_0(\varepsilon)$,*

$$|||\mathcal{E}||| \leqslant |||E||| + C(h_1 + \varepsilon).$$

Note that Proposition 11.2 *states that $|||E|||$ is bounded by a constant less than one which is independent of J.*

PROOF. To apply Lemma 11.1, we use the perturbation operator $\mathcal{T}_k - T_k$ and derive a bound for $|||\mathcal{T}_k - T_k|||$ and show that $|||I - T_k||| \leqslant 1$. By (11.13),

$$|||I - T_k||| \leqslant 1.$$

Since $\mathcal{R}_k = R_k$, we have

$$\mathcal{T}_k - T_k = R_k \mathcal{A}_k \mathcal{P}_k - R_k A_k P_k.$$

By the definitions of P_k and $\mathcal{P}_k$, we have

$$\begin{aligned} A(\mathcal{T}_k w, v) &= (\mathcal{T}_k w, A_k P_k v) = \big(\mathcal{A}_k \mathcal{P}_k w, R_k^t A_k P_k v\big) \\ &= \big(\mathcal{A}_k \mathcal{P}_k w, T_k^* v\big) = \mathcal{A}\big(\mathcal{P}_k w, T_k^* v\big) = \mathcal{A}\big(w, T_k^* v\big) \\ &= A(w, T_k^* v) + D\big(w, T_k^* v\big) \\ &= A(T_k w, v) + D\big(w, T_k^* v\big). \end{aligned}$$

Hence for $k > 1$,

$$A\big((\mathcal{T}_k - T_k)w, v\big) = D\big(w, T_k^* v\big), \quad \text{for all } w, v \in M_J. \tag{11.14}$$

By (11.14) and using (11.7),

$$\big|A\big((\mathcal{T}_k - T_k)w, v\big)\big| = \big|D\big(w, T_k^* v\big)\big| \leqslant C|||w||| \, \big\|T_k^* v\big\|.$$

By Condition (SM.4), Remark 11.2, and (11.13), we have

$$\big\|T_k^* v\big\| \leqslant C h_k A\big(T_k^* v, v\big)^{1/2} = C h_k A(T_k v, v)^{1/2} \leqslant 2C h_k |||v|||.$$

These two estimates imply that

$$|||\mathcal{T}_k - T_k||| \leqslant C h_k, \quad k = 2, 3, \ldots, J.$$

Therefore, the assumptions of Lemma 11.1 hold and consequently,

$$|||\mathcal{E}_J - E_J||| \leqslant C\varepsilon + C\sum_{k=2}^{J} h_k \leqslant C(h_1 + \varepsilon).$$

The theorem follows from the triangle inequality and Proposition 11.2. □

REMARK 11.3. Note that ε can be made arbitrarily small by taking h_1 small enough. Consequently, Theorem 11.1 shows that the multigrid iteration converges with a rate which is independent of J provided that the coarse grid is fine enough. The coarse grid mesh size can also be taken to be independent of J.

We now consider examples of smoothers which satisfy either Conditions (SM.1), (SM.2) and (SM.4) or (SM.3)), (SM.2) and (SM.4). These examples have already been discussed in Section 8. For the benefit of the reader we repeat these examples briefly.

We first consider Richardson smoothers. As discussed in Section 8, there are a few variants of Richardson smoothers. We denote by $\{\phi_k^i\}$ the standard nodal basis functions of M_k. Denote by $\underset{\approx}{A}_k = [A(\phi_k^i, \phi_k^j)]$ and $\underset{\approx}{G}_k = [(\phi_k^i, \phi_k^j)]$ the corresponding stiffness matrix and mass matrix. For each $u \in M_k$ we denote by $\tilde{u}$ the coefficient vector of u with respect to the nodal basis and for each $f \in M_k$, we use $\underset{\sim}{f}$ to denote the *dual vector* $[(f, \phi_k^i)]$ of f with respect to the nodal basis. This notation was introduced in Sections 1 and 8.

EXAMPLE 11.1 (*Richardson smoother* I). The smoother corresponding to the Richardson method applied to the symmetric positive definite finite element equation $A_k u = f$ is given by

$$R_k = \bar{\lambda}_k^{-1} I, \tag{11.15}$$

Here I denotes the identity operator on M_k and $\bar{\lambda}_k$ satisfies $\lambda_k \leqslant \bar{\lambda}_k \leqslant C\lambda_k$. We have shown in Section 8 that Condition (SM.2) holds with $\theta = 1$ and Conditions (SM.1) and (SM.4) hold with $\omega = \omega_4 = \lambda_k/\bar{\lambda}_k$.

EXAMPLE 11.2 (*Richardson smoother* II). The implementation of the Richardson smoother defined in (11.15) requires the inversion of the mass matrix $[(\phi_k^i, \phi_k^j)]$. The smoother corresponding to the Richardson method applied to the matrix equation, $\underset{\approx}{A}_k \tilde{u} = \underset{\sim}{f}$, is defined by

$$R_k f = \frac{1}{\rho}\sum_{i=1}^{\ell} \underset{\sim}{f}_i \phi_k^i \equiv \frac{1}{\rho}\sum_{i=1}^{\ell} (f, \phi_k^i)\phi_k^i, \tag{11.16}$$

where ρ satisfies $\underset{\approx}{\lambda}_k \leqslant \rho \leqslant C\underset{\approx}{\lambda}_k$ and $\underset{\approx}{\lambda}_k$ denotes the largest eigenvalue of $\underset{\approx}{A}_k$. We have shown in Section 8 that Conditions (SM.1), (SM.2) and (SM.4) hold.

EXAMPLE 11.3 (*Richardson smoother* III). The smoother corresponding to the Richardson method applied to the normal equation of the symmetric finite element problem, $A_k^t A_k u = f$, is given by

$$R_k = \bar{\lambda}_k^{-2} A_k, \tag{11.17}$$

where $\bar{\lambda}_k$ satisfies $\lambda_k \leqslant \bar{\lambda}_k \leqslant C\lambda_k$. We have shown in Section 8 that Condition (SM.2) holds with $\theta = 1$ and Conditions (SM.3) and (SM.4) hold with with $\omega = \omega_4(\lambda_k/\bar{\lambda}_k)^2$. It is possible to define the Richardson smoother for the normal matrix equation to avoid the inversion of the mass matrix.

The remaining smoothers correspond to point (or line) Jacobi and Gauss–Seidel smoothers. These smoothers can be defined in terms of subspace decompositions of M_k. We consider the space decomposition

$$M_k = \sum_{i=1}^{\ell_k} M_k^i,$$

where M_k^i is either the one-dimensional subspace spanned by the nodal basis function ϕ_k^i or the subspace spanned by the nodal basis functions along a line. The number of such spaces $\ell = \ell_k$ will often depend on k. These spaces satisfy the following inequality.

$$c\lambda_k \|v\|^2 \leqslant \|v\|_1^2 \leqslant C\lambda_k \|v\|^2, \quad \text{for all } v \in M_k^i. \tag{11.18}$$

Let $Q_k^i : M_k \to M_k^i$ be the L^2 orthogonal projection onto M_k^i. We also need the operator $A_{k,i} : M_k^i \to M_k^i$ defined by

$$(A_{k,i} v, \phi) = A(v, \phi), \quad \text{for all } \phi \in M_k^i.$$

EXAMPLE 11.4 (*Jacobi*). The smoother corresponding to the weighted point or line Jacobi method applied to the symmetric positive definite finite element problem, $A_k u = f$, is given by

$$R_k = \gamma \sum_{i=1}^{\ell} A_{k,i}^{-1} Q_k^i.$$

The constant γ is a scaling factor which is chosen to ensure that Condition (SM.2) is satisfied; cf. Section 8. It is shown in Theorem 8.8 that this smoother satisfies Condition (SA.2), i.e.,

$$c\lambda_k^{-1}(v, v) \leqslant (R_k v, v) \leqslant C\lambda_k^{-1}(v, v).$$

This, together with Condition (SM.2) implies Conditions (SM.1) and (SM.4).

EXAMPLE 11.5 (*Gauss–Seidel*). The smoother corresponding to the point or line Gauss–Seidel method applied to the symmetric positive definite finite element problem, $A_k u = f$, is given by

$$R_k = \left[I - \left(I - P_k^\ell\right)\left(I - P_k^{\ell-1}\right)\cdots\left(I - P_k^1\right)\right]A_k^{-1}.$$

Note that $E_k^\ell \equiv (I - R_k A_k) = (I - P_k^\ell)(I - P_k^{\ell-1})\cdots(I - P_k^1)$. In Section 8, we have shown that Conditions (SM.1), (SM.2) and (SM.4) hold for this choice of R_k.

REMARK 11.4 (*SOR*). The same analysis could be used for successive overrelaxation type iteration. In that case,

$$R_k = \left[I - \left(I - \beta P_k^\ell\right)\left(I - \beta P_k^{\ell-1}\right)\cdots\left(I - \beta P_k^1\right)\right]A_k^{-1},$$

where $\beta \in (0, 2)$ is the relaxation parameter.

11.4. Smoothers based on the original problem

In this section, we consider smoothing operators $\mathcal{R}_k$ which are defined directly in terms of the nonsymmetric and/or indefinite operator $\mathcal{A}_k$. We will study smoothers case by case.

EXAMPLE 11.6 (*Richardson for the normal equation*). Our first example of a smoother based on $\mathcal{A}_k$ is defined by

$$\mathcal{R}_k = \bar{\lambda}_k^{-2}\mathcal{A}_k^t, \quad \text{with } \bar{\lambda}_k \text{ satisfies } \lambda_k \leqslant \bar{\lambda}_k \leqslant C\lambda_k.$$

Here, $\mathcal{A}_k^t$ is the adjoint of $\mathcal{A}_k$ with respect to the inner product $(\bullet, \bullet)$ and $\bar{\lambda}_k$ is as in Example 11.1.

A possible motivation for such a choice is that, on M_k, the iteration

$$u^{m+1} = u^m + \bar{\lambda}_k^{-2}\mathcal{A}_k^t\left(f - \mathcal{A}_k u^m\right)$$

is stable in the norm $\|\bullet\|$ provided that $\bar{\lambda}_k^2$ is greater than or equal to half the largest eigenvalue of $\mathcal{A}_k^t\mathcal{A}_k$.

For this example, we set $R_k = \bar{\lambda}_k^{-2}A_k$. We analyze this example by comparing the error operator $\mathcal{E}_k$ defined by (11.9) and the error operator E_k for the symmetric problem defined by (11.11). In Example 11.3 we have shown that the smoother $R_k = \bar{\lambda}_k^{-2}A_k$ satisfies Condition (SM.3) with $\omega = (\lambda_k/\bar{\lambda}_k)^2$ and satisfies Condition (SM.2) with $\theta = 1$. Hence Proposition 11.2 implies that $|||E_J||| \leqslant \delta < 1$, with δ independent of J. We can now prove the convergence estimate.

THEOREM 11.2. *Let* $\mathcal{R}_k = \bar{\lambda}_k^{-2} \mathcal{A}_k^t$ *be defined as in Example* 11.6. *Given* $\varepsilon > 0$, *there exists* $h_0(\varepsilon) > 0$ *such that for* $h_1 \leqslant h_0(\varepsilon)$, *the error propagator* $\mathcal{E}$ *of Algorithm* 11.1 *satisfies*

$$|||\mathcal{E}||| \leqslant |||E||| + C(h_1 + \varepsilon).$$

Here E *is defined by* (11.11) *with* $R_k = \bar{\lambda}_k^{-2} A_k$. *Note that Proposition* 11.2 *implies that* $|||E|||$ *is bounded by a constant less than one and independent of* J.

PROOF. Let $R_k = \bar{\lambda}_k^{-2} A_k$ and $T_k = R_k A_k P_k$. Note that Condition (SM.2) implies that $|||I - T_k||| \leqslant 1$ and hence $|||T_k||| \leqslant 2$. For $k > 1$, we consider the perturbation operators

$$\mathcal{T}_k - T_k = \bar{\lambda}_k^{-2}\big(\mathcal{A}_k^t \mathcal{A}_k \mathcal{P}_k - A_k^2 P_k\big).$$

Clearly,

$$\mathcal{T}_k - T_k = \bar{\lambda}_k^{-2}\big[\mathcal{A}_k^t(\mathcal{A}_k \mathcal{P}_k - A_k P_k) + \big(\mathcal{A}_k^t - A_k\big) A_k P_k\big]. \tag{11.19}$$

As in (11.14),

$$\bar{\lambda}_k^{-1} A\big((\mathcal{A}_k \mathcal{P}_k - A_k P_k) w, v\big) = \bar{\lambda}_k^{-1} D(w, A_k P_k v).$$

Using (11.7) it follows that

$$\big|\big|\big|\bar{\lambda}_k^{-1}(\mathcal{A}_k \mathcal{P}_k - A_k P_k)\big|\big|\big| \leqslant C h_k.$$

A similar argument shows that

$$\big|\big|\big|\bar{\lambda}_k^{-1}(\mathcal{A}_k^t - A_k) P_k\big|\big|\big| \leqslant C h_k.$$

It is not difficult to see that

$$\big|\big|\big|\mathcal{A}_k^t\big|\big|\big| \leqslant C \bar{\lambda}_k.$$

Combining the above estimates with (11.19) gives

$$\begin{aligned} |||\mathcal{T}_k - T_k||| \leqslant\ & \big|\big|\big|\bar{\lambda}_k^{-1} \mathcal{A}_k^t\big|\big|\big|\, \big|\big|\big|\bar{\lambda}_k^{-1}(\mathcal{A}_k \mathcal{P}_k - A_k P_k)\big|\big|\big| \\ & + \big|\big|\big|\bar{\lambda}_k^{-1}\big(\mathcal{A}_k^t - A_k\big) P_k\big|\big|\big|\, \big|\big|\big|\bar{\lambda}_k^{-1} A_k P_k\big|\big|\big| \leqslant C h_k. \end{aligned}$$

This proves that the conditions in Lemma 11.1 hold. The theorem follows from Lemma 11.1 and Proposition 11.2. □

The next two examples are point or line Jacobi and Gauss–Seidel methods applied directly to the nonsymmetric indefinite equations. They are closely related to Example 11.4 and Example 11.5. We define the point or line subspaces $\{M_k^i\}$ for $i = 1, \ldots, \ell$. Note that the form $\mathcal{A}(\bullet, \bullet)$ satisfies a Gårding inequality

$$c_1 A(v, v) - c\|v\|^2 \leqslant \mathcal{A}(v, v), \quad \text{for all } v \in H_0^1(\Omega).$$

Consequently, by (11.18),

$$\left(c_1 - Ch_k^2\right)A(v, v) \leqslant \mathcal{A}(v, v), \quad \text{for all } v \in M_k^i.$$

We assume that h_2 is small enough that

$$Ch_2^2 \leqslant c_1/2.$$

This means that $\mathcal{A}(\bullet, \bullet)$ restricted to M_k^i has a positive definite symmetric part. Consequently, the operator $\mathcal{A}_{k,i} : M_k^i \to M_k^i$ defined by

$$(\mathcal{A}_{k,i} v, \phi) = \mathcal{A}(v, \phi), \quad \text{for all } v, \phi \in M_k^i,$$

is invertible and the projector $\mathcal{P}_k^i : M_k \to M_k^i$ satisfying

$$\mathcal{A}\left(\mathcal{P}_k^i v, \phi\right) = \mathcal{A}(v, \phi), \quad \text{for all } \phi \in M_k^i,$$

is well defined and satisfies

$$\left\|\mathcal{P}_k^i u\right\|_1 \leqslant C\|u\|_{1,\Omega_k^i}, \quad \text{for all } u \in M_k. \tag{11.20}$$

The second norm is taken only over the subdomain Ω_k^i which is the set of points of Ω where the functions in M_k^i are nonzero. We assume that there is a constant n_0 independent of k such that each Ω_k^i intersects with at most $n_0 - 1$ other Ω_k^i.

EXAMPLE 11.7 (*Jacobi*). The smoother corresponding to the point or line Jacobi method applied directly to the nonsymmetric/indefinite equations is given by

$$\mathcal{R}_k = \gamma \sum_{i=1}^{\ell} \mathcal{A}_{k,i}^{-1} Q_k^i.$$

We choose γ as in Example 11.4 so that the symmetric smoother defined in Example 11.4 satisfies Condition (SM.2).

We analyze Algorithm 11.1 with this smoother by using perturbation from the multigrid algorithm for A which uses the smoother $R_k = \gamma \sum_{i=1}^{\ell} A_{k,i}^{-1} Q_k^i$ as defined in Example 11.4. Proposition 11.2 provides a uniform estimate for $|||E|||$.

THEOREM 11.3. *Let $\mathcal{R}_k$ be defined by Example* 11.7. *Given $\varepsilon > 0$, there exists an $h_0(\varepsilon) > 0$ such that for $h_1 \leqslant h_0(\varepsilon)$,*

$$|||\mathcal{E}||| \leqslant |||E||| + C(h_1 + \varepsilon).$$

Here E is the error propagator defined in (11.11) *with R_k defined in Example* 11.4. *Note that Proposition* 11.2 *implies that $|||E|||$ is bounded by a constant less than one which is independent of J.*

PROOF. For this case, the perturbation operator is given by

$$\mathcal{T}_k - T_k = \gamma \sum_{i=1}^{\ell} (\mathcal{P}_k^i - P_k^i).$$

As in (11.12),

$$A\big((\mathcal{P}_k^i - P_k^i)w, v\big) = D\big((I - \mathcal{P}_k^i)w, P_k^i v\big).$$

Applying Eqs. (11.7), (11.18) and (11.20) gives

$$A\big((\mathcal{P}_k^i - P_k^i)w, v\big) \leqslant Ch_k \|w\|_{1,\Omega_k^i} \|v\|_{1,\Omega_k^i}. \tag{11.21}$$

Summing over i yields

$$\begin{aligned} A\big((\mathcal{T}_k - T_k)w, v\big) &\leqslant Ch_k \sum_{i=1}^{\ell} \|w\|_{1,\Omega_k^i} \|v\|_{1,\Omega_k^i} \\ &\leqslant Ch_k \left(\sum_{i=1}^{\ell} \|w\|_{1,\Omega_k^i}^2\right)^{1/2} \left(\sum_{i=1}^{\ell} \|v\|_{1,\Omega_k^i}^2\right)^{1/2}. \end{aligned}$$

Using the limited overlap properties of the domains Ω_k^i, we obtain

$$|||\mathcal{T}_k - T_k||| \leqslant Ch_k.$$

The theorem follows from Lemma 11.1. □

EXAMPLE 11.8 (*Gauss–Seidel*). The smoother corresponding to the Gauss–Seidel method applied directly to the nonsymmetric/indefinite equations is defined by

$$\mathcal{R}_k = \big[I - (I - \mathcal{P}_k^l)(I - \mathcal{P}_k^{l-1}) \cdots (I - \mathcal{P}_k^1)\big]\mathcal{A}_k^{-1}.$$

It is easy to see that

$$(I - \mathcal{T}_k) = (I - \mathcal{P}_k^\ell)(I - \mathcal{P}_k^{\ell-1}) \cdots (I - \mathcal{P}_k^1).$$

Note that given $f \in M_k$, $\mathcal{R}_k f$ can be computed by

(1) Set $v_0 = 0 \in M_k$.

(2) Define v_i, for $i = 1, \ldots, l$, by

$$v_i = v_{i-1} + \mathcal{A}_{k,i}^{-1} Q_k^i (f - \mathcal{A}_k v_{i-1}).$$

(3) Set $\mathcal{R}_k f = v_l$.

For this example, we take $\mathcal{R}_k$ to be the Gauss–Seidel smoother for the symmetric form $A(\bullet,\bullet)$ defined in Example 11.5.

We now consider the resulting method as a perturbation of the multigrid algorithm for A which uses the smoother R_k defined by Example 11.5. Proposition 11.2 provides a uniform estimate for $|||E|||$.

THEOREM 11.4. *Let $\mathcal{R}_k$ be the Gauss–Seidel smoother defined by Example* 11.8. *Given $\varepsilon > 0$, there exists an $h_0(\varepsilon) > 0$ such that for $h_1 \leqslant h_0(\varepsilon)$,*

$$|||\mathcal{E}||| \leqslant |||E||| + C(h_1 + \varepsilon).$$

Here E is the error propagator defined by (11.11) *with R_k defined by Example* 11.5.

PROOF. The perturbation operator for this example is

$$\mathcal{T}_k - T_k = (I - T_k) - (I - \mathcal{T}_k) = E_k^\ell - \mathcal{E}_k^\ell,$$

where

$$E_k^0 = I, \qquad E_k^i = (I - P_k^i)(I - P_k^{i-1}) \cdots (I - P_k^1), \quad i = 1, \ldots, \ell,$$

and

$$\mathcal{E}_k^0 = I, \qquad \mathcal{E}_k^i = (I - \mathcal{P}_k^i)(I - \mathcal{P}_k^{i-1}) \cdots (I - \mathcal{P}_k^1), \quad i = 1, \ldots, \ell.$$

It is immediate from the definitions that

$$E_k^i - \mathcal{E}_k^i = (I - P_k^i)(E_k^{i-1} - \mathcal{E}_k^{i-1}) - (P_k^i - \mathcal{P}_k^i)\mathcal{E}_k^{i-1}.$$

Since the last two terms are orthogonal with respect to $A(\bullet,\bullet)$ we have that

$$|||(E_k^i - \mathcal{E}_k^i)v|||^2 = |||(I - P_k^i)(E_k^{i-1} - \mathcal{E}_k^{i-1})v|||^2 + |||(P_k^i - \mathcal{P}_k^i)\mathcal{E}_k^{i-1}v|||^2.$$

Because of (11.21) and the fact that the operator norm of $(I - P_k^i)$ is bounded by one, it follows that

$$|||(E_k^i - \mathcal{E}_k^i)v|||^2 \leqslant |||(E_k^{i-1} - \mathcal{E}_k^{i-1})v|||^2 + Ch_k^2 \|\mathcal{E}_k^{i-1}v\|_{1,\Omega_k^i}^2.$$

Summing over i, since $E_k^0 = \mathcal{E}_k^0 = I$, we obtain

$$|||(E_k^\ell - \mathcal{E}_k^\ell)v|||^2 \leqslant Ch_k^2 \sum_{i=1}^{\ell} \|\mathcal{E}_k^{i-1}v\|_{1,\Omega_k^i}^2. \tag{11.22}$$

We shall show that

$$\sum_{i=1}^{\ell} \|\mathcal{E}_k^{i-1} v\|_{1,\Omega_k^i}^2 \leqslant C|||v|||^2. \tag{11.23}$$

By the arithmetic-geometric mean inequality, the definition $\mathcal{E}_k^i$ and the limited interaction property it follows that

$$\begin{aligned}
\sum_{i=1}^{\ell} \|\mathcal{E}_k^{i-1} v\|_{1,\Omega_k^i}^2 &\leqslant 2\sum_{i=1}^{\ell} \|v\|_{1,\Omega_k^i}^2 + 2\sum_{i=1}^{\ell} \|v - \mathcal{E}_k^{i-1} v\|_{1,\Omega_k^i}^2 \\
&= 2\sum_{i=1}^{\ell} \|v\|_{1,\Omega_k^i}^2 + 2\sum_{i=1}^{\ell} \left\| \sum_{m=1}^{i-1} \mathcal{P}_k^m \mathcal{E}_k^{m-1} v \right\|_{1,\Omega_k^i}^2 \\
&\leqslant \left(2n_0|||v|||^2 + 2n_0 \sum_{i=1}^{\ell}\sum_{m=1}^{\ell} \|\mathcal{P}_k^m \mathcal{E}_k^{m-1} v\|_{1,\Omega_k^i}^2 \right) \\
&\leqslant C\left(|||v|||^2 + \sum_{m=1}^{\ell} |||\mathcal{P}_k^m \mathcal{E}_k^{m-1} v|||^2 \right).
\end{aligned} \tag{11.24}$$

In order to estimate the last term on the right of (11.24) we write

$$\begin{aligned}
|||\mathcal{P}_k^m \mathcal{E}_k^{m-1} v|||^2 &= A(\mathcal{P}_k^m \mathcal{E}_k^{m-1} v, \mathcal{P}_k^m \mathcal{E}_k^{m-1} v) \\
&= A((\mathcal{E}_k^{m-1} - \mathcal{E}_k^m)v, (\mathcal{E}_k^{m-1} - \mathcal{E}_k^m)v) \\
&= A((\mathcal{E}_k^{m-1} - \mathcal{E}_k^m)v, (\mathcal{E}_k^{m-1} + \mathcal{E}_k^m)v) - 2A(\mathcal{P}_k^m \mathcal{E}_k^{m-1} v, \mathcal{E}_k^m v) \\
&= A(\mathcal{E}_k^{m-1} v, \mathcal{E}_k^{m-1} v) - A(\mathcal{E}_k^m v, \mathcal{E}_k^m v) \\
&\quad - 2A(\mathcal{P}_k^m \mathcal{E}_k^{m-1} v, (I - \mathcal{P}_k^m)\mathcal{E}_k^{m-1} v).
\end{aligned} \tag{11.25}$$

Now by (11.21)

$$\begin{aligned}
A(\mathcal{P}_k^m \mathcal{E}_k^{m-1} v, (I - \mathcal{P}_k^m)\mathcal{E}_k^{m-1} v) &= A(\mathcal{P}_k^m \mathcal{E}_k^{m-1} v, (P_k^m - \mathcal{P}_k^m)\mathcal{E}_k^{m-1} v) \\
&\leqslant Ch_k |||\mathcal{P}_k^m \mathcal{E}_k^{m-1} v|||\, \|\mathcal{E}_k^{m-1} v\|_{1,\Omega_k^m}.
\end{aligned} \tag{11.26}$$

Hence combining Eqs. (11.25) and (11.26), we have

$$|||\mathcal{P}_k^m \mathcal{E}_k^{m-1} v|||^2 \leqslant C[A(\mathcal{E}_k^{m-1} v, \mathcal{E}_k^{m-1} v) - A(\mathcal{E}_k^m v, \mathcal{E}_k^m v)] + Ch_k^2 \|\mathcal{E}_k^{m-1} v\|_{1,\Omega_k^m}^2.$$

Summing over m we conclude that

$$\sum_{m=1}^{\ell} |||\mathcal{P}_k^m \mathcal{E}_k^{m-1} v|||^2 \leqslant C|||v|||^2 + Ch_k^2 \sum_{m=1}^{\ell} \|\mathcal{E}_k^{m-1} v\|_{1,\Omega_k^m}^2.$$

This together with (11.24) yields (11.23) when h_k is small enough. Finally, we obtain from Eqs. (11.23) and (11.22) that for $k > 1$,

$$|||\mathcal{T}_k - T_k||| \leqslant Ch_k.$$

The theorem now follows from Lemma 11.1 □

REMARK 11.5. The same analysis could be used for successive overrelaxation type iteration. In that case,

$$\mathcal{R}_k = \big[I - (I - \beta\mathcal{P}_k^{\ell})(I - \beta\mathcal{P}_k^{\ell-1})\cdots(I - \beta\mathcal{P}_k^{1})\big]\mathcal{A}_k^{-1},$$

where $\beta \in (0, 2)$ is the relaxation parameter. We take

$$R_k = \big[I - (I - \beta P_k^{\ell})(I - \beta P_k^{\ell-1})\cdots(I - \beta P_k^{1})\big]A_k^{-1}.$$

as a smoother for the symmetric problem.

11.5. Extensions and generalizations

Many extensions and generalizations of the techniques given above are possible. These techniques lead to uniform estimates for multigrid iteration methods for solving nonsymmetric and/or indefinite problems for the following applications.

(i) Approximations using higher-order nodal finite element spaces.

(ii) Three-dimensional problems.

(iii) Problems with discontinuous coefficients as discussed in BRAMBLE and PASCIAK [1993].

(iv) More general boundary conditions.

(v) Problems with local mesh refinement as described in Sections 6 and 9.

(vi) Finite element approximation of problems on domains with nonpolygonal boundaries as discussed in Section 9.

In addition, the perturbation analysis given above can be combined with results for additive multilevel algorithms, e.g., Theorem 4.2. This leads to new estimates for additive multilevel preconditioning iterations applied to indefinite and nonsymmetric problems. Provided that the coarse grid is sufficiently fine, the operator

$$\mathcal{T} = \sum_{k=1}^{J} \mathcal{T}_k \equiv \sum_{k=1}^{J} \mathcal{R}_k \mathcal{A}_k \mathcal{P}_k = \left(\sum_{k=1}^{J} \mathcal{R}_k \mathcal{Q}_k\right)\mathcal{A}_J$$

has a uniformly (independent of J) positive definite symmetric part with respect to the inner product $A(\bullet, \bullet)$ and has a uniformly bounded operator norm. These results extend to all of the applications discussed in (i)–(vi).

11.6. *Bibliographical notes*

This section is based on BRAMBLE, KWAK and PASCIAK [1994]. Multigrid algorithms for nonsymmetric and indefinite problems were also discussed in BANK [1981], MANDEL [1986], BRAMBLE, PASCIAK and XU [1988] and WANG [1991].

The paper of Bank derives uniform convergence estimates for the W-cycle multigrid iteration with both a standard Jacobi smoother and a smoother which uses the operator times its adjoint. In each case, sufficiently many smoothings are required and a sufficiently fine coarse grid (depending on the number of smoothings) is needed. Some regularity for the elliptic partial differential equation was also required.

Mandel studied the V-cycle iteration and showed that it was effective with only one smoothing and a sufficiently fine coarse grid. His result requires that the underlying partial differential equation satisfy the "full elliptic regularity" hypothesis and generalizes the results of BRAESS and HACKBUSCH [1983] for the symmetric positive definite problem.

Bramble, Pasciak and Xu studied the symmetric smoother introduced by Bank and showed that the W-cycle and variable V-cycle worked without making the undesirable requirement of "sufficiently many smoothings". Somewhat more than minimal regularity was needed.

Wang showed that the "error reduction factor" for the standard V-cycle with one smoothing was bounded by $1 - C/J + C_1 h_1$ where J is the number of levels, h_1 is the size of the coarsest grid and C and C_1 are constants. This estimate deteriorates with the number of levels and will be less than one only if the coarse grid is subsequently finer as the number of levels increase. Minimal elliptic regularity was assumed.

There are also some other techniques for solving nonsymmetric indefinite problems. One approach is to precondition the nonsymmetric problem with a symmetric operator and then solve certain normal equations by the conjugate gradient method; cf. ELMAN [1982], BRAMBLE, LEYK and PASCIAK [1993] and BRAMBLE and PASCIAK [1984]. One possible advantage of such a method is that some nonsymmetric problems which are not "compact perturbations" of symmetric ones may be treated. Of course, the usual normal equations may be formed and then preconditioned; cf. BRAMBLE and PASCIAK [1984] and MANTEUFFEL and PARTER [1990]. This approach seems to be rather restrictive in that good preconditioners may be difficult to construct. Other recent approaches have included Schwarz type methods (CAI [1991] and CAI and WIDLUND [1992]) and two-level methods in which a "coarse space" is introduced to reduce the problem to one with a positive definite symmetric part (cf. BRAMBLE, LEYK and PASCIAK [1993], XU [1992b] and XU and CAI [1992]).

12. Anisotropic problems

In this section, we shall establish a convergence theory for a multigrid algorithm applied to a model anisotropic equation defined on the unit square. We shall consider the standard finite element approximations to this problem. For such an anisotropic problem, the standard finite element solution has a "poor" approximation property and hence the coarse grid solve in the multigrid algorithm is not effective in reducing the smooth components of the errors (in contrast to the cases in Section 9). When a Jacobi or Gauss–Seidel smoother is used, the multigrid algorithm does not provide a uniform reduction in the error. The remedy is to use a smoother, such as the line Jacobi smoother, that is effective in reducing components of the error in a larger part of the spectrum.

Since Conditions (A.1) and (A.2) do not hold for this case, the multigrid theory of Sections 3 and 5 cannot be applied directly. A modification of the theory of Section 3 will be presented. In verifying the hypotheses of the theory, we need a weighted L^2 norm estimate of the error of the finite element approximation and a smoothing property of the weighted line Jacobi smoother. The error estimate is established in Subsection 12.3 by using a duality argument and a regularity property of the anisotropic equation proved in Subsection 12.2. The smoothing property of the line Jacobi smoother is formulated and proved in Subsection 12.4.

12.1. A model equation

Let $\Omega = (0,1)^2$ be the unit square. Consider the equation

$$\begin{cases} Au = -[(au_x)_x + (bu_y)_y] = f & \text{in } \Omega, \\ u = 0 & \text{on } \partial\Omega, \end{cases} \tag{12.1}$$

where $a(x,y)$ and $b(x,y)$ are positive functions.

We are interested in the case in which $a(x,y)$ is of unit size and $b(x,y)$ is possibly small. More precisely, we assume that $a(x,y)$ is uniformly bounded from above and below and $b(x,y)$ is bounded uniformly from above with

$$0 < a_{\min} \leqslant a(x,y) \leqslant a_{\max}, \tag{12.2}$$

and

$$0 < b(x,y) \leqslant b_{\max}. \tag{12.3}$$

We do not assume, however, that $b(x,y)$ has a uniform positive lower bound.

To carry out our analysis for the multigrid algorithm however, we will also make the following technical assumptions on the coefficients. We assume that certain first derivatives of $a(x,y)$ and $b(x,y)$ are uniformly bounded, i.e.,

$$\frac{|\nabla a|}{a} \leqslant \frac{|\nabla a|}{a_{\min}} \leqslant C_a \quad \text{and} \quad \frac{|b_y|}{b} \leqslant C_b. \tag{12.4}$$

Since $a_{\min} \leqslant a \leqslant a_{\max}$, the first inequality in (12.4) is the same as

$$|\nabla a| \leqslant C.$$

On the other hand, since $b(x, y)$ is not assumed to have a uniform positive lower bound, the second inequality in (12.4) states that $b(x, y)$ does not change very much in the y direction relative to its magnitude; i.e., we can write

$$b(x, y) = \varepsilon(x)\tilde{b}(x, y) \quad \text{with } \tilde{c}_1 \leqslant \tilde{b}(x, y) \leqslant \tilde{c}_2 \text{ and } \left|\tilde{b}_y(x, y)\right| \leqslant \tilde{c}_3. \tag{12.5}$$

Clearly, (12.5) implies the second inequality in (12.4). Conversely, if we set $\varepsilon(x) = \max_y b(x, y)$ and $\tilde{b}(x, y) = b(x, y)/\varepsilon(x)$, then the second inequality in (12.4) implies that (12.5) holds with $\mathrm{e}^{-C_b} \leqslant \tilde{b}(x, y) \leqslant 1$ and $|\tilde{b}_y(x, y)| \leqslant C_b$. In our subsequent analysis, the estimate for the rate of convergence of the V-cycle multigrid algorithm will depend on the constants in (12.2), (12.3) and (12.4), but not on a positive lower bound for $b(x, y)$. We will often use (12.5) instead of the second inequality in (12.4). Without loss of generality, we can assume $\varepsilon(x) \leqslant 1$ in (12.5)

The weak form of (12.1) is the following: Find $u \in H_0^1(\Omega)$ such that

$$A(u, \phi) = (f, \phi), \quad \text{for all } \phi \in H_0^1(\Omega),$$

where

$$A(u, v) = \int_\Omega \left[a(x, y)u_x v_x + b(x, y)u_y v_y\right] \mathrm{d}x\, \mathrm{d}y.$$

Here $(\bullet, \bullet)$ is the L^2 inner product. We will also use the notation $|||\bullet|||^2 = A(\bullet, \bullet)$.

12.2. Regularity of the problem

In this subsection, we derive a regularity result for the model anisotropic problem (12.1). This result will be used to derive error estimates of the Galerkin finite element approximation in the energy norm and a weighted L^2 norm.

We first note that if $a(x, y) = 1$ and $b(x, y) = \varepsilon$ is a constant, then, by integration by parts, we have for $u \in H_0^1(\Omega) \cap H^2(\Omega)$

$$\begin{aligned}\int_\Omega \left(u_{xx}^2 + 2u_{xy}^2 + \varepsilon u_{yy}^2\right) \mathrm{d}x\, \mathrm{d}y &= \int_\Omega \left(u_{xx}^2 + 2u_{xx}u_{yy} + \varepsilon u_{yy}^2\right) \mathrm{d}x\, \mathrm{d}y \\ &\leqslant \int_\Omega \frac{1}{\varepsilon}|Au|^2 \,\mathrm{d}x\, \mathrm{d}y.\end{aligned}$$

The following lemma is a generalization of this fact to some variable coefficient cases.

LEMMA 12.1 (Regularity). *If the coefficients $a(x, y)$ and $b(x, y)$ satisfy Eqs.* (12.2)–(12.4), *then the following* a priori *estimate holds*:

$$\int_\Omega \left(u_{xx}^2 + 2u_{xy}^2 + bu_{yy}^2\right) \mathrm{d}x\, \mathrm{d}y \leqslant C \int_\Omega \frac{1}{b}|Au|^2 \,\mathrm{d}x\, \mathrm{d}y.$$

PROOF. Integrating by parts gives

$$
\begin{aligned}
&2\int_\Omega \frac{1}{b}(au_x)_x(bu_y)_y\,\mathrm{d}x\,\mathrm{d}y\\
&\quad = 2\int_\Omega \left((au_x)_x u_{yy} + (au_x)_x \frac{b_y}{b}u_y\right)\mathrm{d}x\,\mathrm{d}y\\
&\quad = 2\int_\Omega \left((au_x)_y u_{xy} + (au_x)_x \frac{b_y}{b}u_y\right)\mathrm{d}x\,\mathrm{d}y\\
&\quad = 2\int_\Omega \left(au_{xy}^2 + (a_yu_x)u_{xy} + (au_x)_x \frac{b_y}{b}u_y\right)\mathrm{d}x\,\mathrm{d}y\\
&\quad \geqslant \int_\Omega \left\{2au_{xy}^2 - \left(\alpha a u_{xy}^2 + \frac{a_y^2}{\alpha a}u_x^2\right) - \left[\frac{\beta}{b}(au_x)_x^2 + \frac{b}{\beta}\left(\frac{b_y}{b}u_y\right)^2\right]\right\}\mathrm{d}x\,\mathrm{d}y\\
&\quad = \int_\Omega \left(-\frac{\beta}{b}(au_x)_x^2 + (2-\alpha)au_{xy}^2 - \frac{a_y^2}{\alpha a}u_x^2 - \frac{b_y^2}{\beta b}u_y^2\right)\mathrm{d}x\,\mathrm{d}y,
\end{aligned}
$$

where α and β are arbitrary positive constants. As a consequence, we have

$$
\begin{aligned}
\int_\Omega \frac{1}{b}|Au|^2\,\mathrm{d}x\,\mathrm{d}y &= \int_\Omega \left(\frac{1}{b}(au_x)_x^2 + \frac{2}{b}(au_x)_x(bu_y)_y + \frac{1}{b}(bu_y)_y^2\right)\mathrm{d}x\,\mathrm{d}y\\
&\geqslant \int_\Omega \left(\frac{1-\beta}{b}(au_x)_x^2 + (2-\alpha)au_{xy}^2\right.\\
&\qquad \left. + \frac{1}{b}(bu_y)_y^2 - \frac{a_y^2}{\alpha a}u_x^2 - \frac{b_y^2}{\beta b}u_y^2\right)\mathrm{d}x\,\mathrm{d}y.
\end{aligned}
$$

Let $\gamma = \min(1, 1/b_{\max}) \leqslant 1$. Then

$$
\begin{aligned}
\int_\Omega \frac{1}{b}(au_x)_x^2\,\mathrm{d}x\,\mathrm{d}y &\geqslant \gamma\int_\Omega (au_x)_x^2\,\mathrm{d}x\,\mathrm{d}y = \gamma\int_\Omega (au_{xx} + a_xu_x)^2\,\mathrm{d}x\,\mathrm{d}y\\
&\geqslant \gamma\int_\Omega \left(\tfrac{1}{2}a^2u_{xx}^2 - a_x^2u_x^2\right)\mathrm{d}x\,\mathrm{d}y
\end{aligned}
$$

and

$$
\int_\Omega \frac{1}{b}(bu_y)_y^2\,\mathrm{d}x\,\mathrm{d}y = \int_\Omega \frac{1}{b}(bu_{yy} + b_yu_y)^2\,\mathrm{d}x\,\mathrm{d}y \geqslant \int_\Omega \left(\tfrac{1}{2}bu_{yy}^2 - \frac{b_y^2}{b}u_y^2\right)\mathrm{d}x\,\mathrm{d}y.
$$

Combining the above estimates, we obtain

$$
\begin{aligned}
&\int_\Omega \frac{1}{b}|Au|^2\,\mathrm{d}x\,\mathrm{d}y\\
&\quad \geqslant (1-\beta)\gamma\int_\Omega \left(\tfrac{1}{2}a^2u_{xx}^2 - a_x^2u_x^2\right)\mathrm{d}x\,\mathrm{d}y + (2-\alpha)\int_\Omega au_{xy}^2\,\mathrm{d}x\,\mathrm{d}y
\end{aligned}
$$

$$+\int_\Omega\left(\frac{b}{2}u_{yy}^2-\frac{b_y^2}{b}u_y^2\right)\mathrm{d}x\,\mathrm{d}y-\int_\Omega\left(\frac{a_y^2}{\alpha a}u_x^2+\frac{b_y^2}{\beta b}u_y^2\right)\mathrm{d}x\,\mathrm{d}y$$

$$=\int_\Omega\left(\tfrac{1}{2}(1-\beta)\gamma a^2u_{xx}^2+(2-\alpha)au_{xy}^2+\left(\frac{b}{2}u_{yy}^2\right)\right)\mathrm{d}x\,\mathrm{d}y$$

$$-\int_\Omega\left((1-\beta)\gamma a_x^2+\frac{a_y^2}{\alpha a}\right)u_x^2\,\mathrm{d}x\,\mathrm{d}y-\int_\Omega\left(\frac{b_y^2}{b}+\frac{b_y^2}{\beta b}\right)u_y^2\,\mathrm{d}x\,\mathrm{d}y.$$

Therefore

$$\int_\Omega\left(u_{xx}^2+2u_{xy}^2+bu_{yy}^2\right)\mathrm{d}x\,\mathrm{d}y$$

$$\leqslant C\int_\Omega\left[\tfrac{1}{2}(1-\beta)\gamma a^2u_{xx}^2+(2-\alpha)au_{xy}^2+\left(\frac{b}{2}u_{yy}^2\right)\right]\mathrm{d}x\,\mathrm{d}y$$

$$\leqslant C\int_\Omega\frac{1}{b}(Au)^2\,\mathrm{d}x\,\mathrm{d}y+C\int_\Omega\left[\left((1-\beta)\gamma a_x^2+\frac{a_y^2}{\alpha a}\right)u_x^2\right.$$

$$\left.+\left(\frac{b_y^2}{b}+\frac{b_y^2}{\beta b}\right)u_y^2\right]\mathrm{d}x\,\mathrm{d}y.$$

The second integral is clearly bounded by $C\int_\Omega|Au|^2$ and hence bounded by $C\int_\Omega b^{-1}|Au|^2$. This proves the lemma. □

12.3. Finite element approximation

Let the domain Ω be partitioned into n by n squares with vertices (ih, jh), $h=1/n$. We consider the linear or the bilinear finite element space M_h associated with this partition. The Galerkin finite element projection $P_h: H_0^1(\Omega)\to M_h$ is defined by

$$A(P_hv,\phi)=A(v,\phi),\qquad \text{for all } \phi\in M_h.$$

We need the following results in proving an approximation property of the finite element solutions.

LEMMA 12.2. *Let $\mathcal{D}=(0,h_1)\times(0,h_2)$ and E be a side of $\mathcal{D}$. If $v\in H^1(\mathcal{D})$ and $\int_E v\,\mathrm{d}s=0$, then*

$$\|v\|_{L^2(\mathcal{D})}^2\leqslant\left(h_1^2\|v_x\|_{L^2(\mathcal{D})}^2+h_2^2\|v_y\|_{L^2(\mathcal{D})}^2\right).$$

Suppose that $b(x,y)=\varepsilon(x)\tilde{b}(x,y)$ with $\tilde{c}_1\leqslant\tilde{b}(x,y)\leqslant\tilde{c}_2$, and $\varepsilon(x)\leqslant 1$. Denote by E a vertical edge of $\mathcal{D}$. If $v\in H^1(\mathcal{D})$ and $\int_E v\,\mathrm{d}s=0$, then

$$\int_{\mathcal{D}}bv^2\,\mathrm{d}x\,\mathrm{d}y\leqslant\tilde{c}_2\left(h_1^2\int_{\mathcal{D}}v_x^2\,\mathrm{d}x\,\mathrm{d}y+\tilde{c}_1^{-1}h_2^2\int_{\mathcal{D}}bv_y^2\,\mathrm{d}x\,\mathrm{d}y\right).$$

PROOF. Without loss of generality, we assume $\int_0^{h_2} u(0,y)\,\mathrm{d}y = 0$. Then

$$u(x,y) = u(0,y_0) + \int_0^x u_x(s,y_0)\,\mathrm{d}s + \int_{y_0}^y u_y(x,t)\,\mathrm{d}t.$$

Integrating y_0 from 0 to h_2 and using $\int_0^{h_2} u(0,y_0)\,\mathrm{d}y_0 = 0$, we obtain

$$u(x,y) = \frac{1}{h_2}\int_0^{h_2} \mathrm{d}y_0 \int_0^x u_x(s,y_0)\,\mathrm{d}s + \frac{1}{h_2}\int_0^{h_2} \mathrm{d}y_0 \int_{y_0}^y u_y(x,t)\,\mathrm{d}t.$$

Squaring and then using the Cauchy–Schwarz inequality, it follows that

$$\begin{aligned}
\left|u(x,y)\right|^2 &\leqslant \frac{2}{h_2^2}\left(\int_0^{h_2} \mathrm{d}y_0 \int_0^x \left|u_x(s,y_0)\right|\mathrm{d}s\right)^2 \\
&\quad + \frac{2}{h_2^2}\left(\int_0^{h_2} \mathrm{d}y_0 \int_0^y \left|u_y(x,t)\right|\mathrm{d}t\right)^2 \\
&\leqslant \frac{2}{h_2^2} h_2 x \int_0^{h_2} \mathrm{d}y_0 \int_0^{h_1} u_x^2(s,y_0)\,\mathrm{d}s + \frac{2}{h_2^2} h_2^2 y \int_0^y u_y^2(x,t)\,\mathrm{d}t \\
&\leqslant \frac{2x}{h_2}\int_{\mathcal{D}} u_x^2\,\mathrm{d}x\,\mathrm{d}y + 2y\int_0^{h_2} u_y^2(x,t)\,\mathrm{d}t.
\end{aligned} \tag{12.6}$$

Integrating over $\mathcal{D}$, we obtain

$$\int_{\mathcal{D}} \left|u(x,y)\right|^2 \mathrm{d}x\,\mathrm{d}y \leqslant h_1^2 \int_{\mathcal{D}} u_x^2\,\mathrm{d}x\,\mathrm{d}y + h_2^2 \int_{\mathcal{D}} u_y^2\,\mathrm{d}x\,\mathrm{d}y.$$

This is the first part of the lemma.

We now prove the second part. Using $\varepsilon(x) \leqslant 1$, we obtain from (12.6)

$$\varepsilon(x)\left|u(x,y)\right|^2 \leqslant \frac{2x}{h_2}\int_{\mathcal{D}} u_x^2\,\mathrm{d}x\,\mathrm{d}y + 2y\int_0^{h_2} \varepsilon(x) u_y^2(x,t)\,\mathrm{d}t.$$

Integrating the above inequality over $\mathcal{D}$ gives

$$\int_{\mathcal{D}} \varepsilon(x) v^2\,\mathrm{d}x\,\mathrm{d}y \leqslant h_1^2 \int_{\mathcal{D}} v_x^2\,\mathrm{d}x\,\mathrm{d}y + h_2^2 \int_{\mathcal{D}} \varepsilon(x) v_y^2\,\mathrm{d}x\,\mathrm{d}y.$$

Since $\tilde{c}_1\varepsilon(x) \leqslant b(x,y) \leqslant \tilde{c}_2\varepsilon(x)$, the second part of the lemma follows from the last inequality. □

Using this lemma, we can prove the following error estimate for the nodal value interpolant.

LEMMA 12.3. *Let $\pi_h : C(\bar{\Omega}) \to M_h$ be the nodal value interpolation operator. Then*

$$|||(I - \pi_h)v|||^2 \leqslant Ch^2 \int_\Omega \left(v_{xx}^2 + v_{xy}^2 + bv_{yy}^2\right) \mathrm{d}x\, \mathrm{d}y, \quad \textit{for all } v \in H^2(\Omega).$$

PROOF. Let $\tau = [x_{i-1}, x_i] \times [y_{j-1}, y_j]$ be arbitrary. Let E_x and E_y be edges in the x and y directions, respectively. For the bilinear element, we have

$$\int_{E_x} (v - \pi_h v)_x \, \mathrm{d}x = \int_{E_y} (v - \pi_h v)_y \, \mathrm{d}y = 0.$$

Applying the first part of Lemma 12.2 to $(v - \pi_h v)_x$ and the second part of Lemma 12.2 to $(v - \pi_h v)_y$ on each element τ we have

$$\int_\tau a\left|(v - \pi_h v)_x\right|^2 \mathrm{d}x\, \mathrm{d}y \leqslant Ca_{\max} h^2 \int_\tau \left(\left|(v - \pi_h v)_{xx}\right|^2 + \left|(v - \pi_h v)_{xy}\right|^2\right) \mathrm{d}x\, \mathrm{d}y,$$

and

$$\int_\tau b\left|(v - \pi_h v)_y\right|^2 \mathrm{d}x\, \mathrm{d}y \leqslant Ch^2 \int_\tau \left(\left|(v - \pi_h v)_{xy}\right|^2 + b\left|(v - \pi_h v)_{yy}\right|^2\right) \mathrm{d}x\, \mathrm{d}y.$$

In the bilinear case, $(\pi_h v)_{xx} = (\pi_h v)_{yy} = 0$ in τ and

$$\begin{aligned}
\left\|(\pi_h v)_{xy}\right\|_{L^2(\tau)}^2 &= h^{-2}\big|\left[v(x_i, y_j) - v(x_{i-1}, y_j)\right] \\
&\quad - \left[v(x_i, y_{j-1}) - v(x_{i-1}, y_{j-1})\right]\big|^2 \\
&\leqslant \|v_{xy}\|_{L^2(\tau)}^2.
\end{aligned}$$

Therefore,

$$\begin{aligned}
&\int_\tau \left(a\left|(v - \pi_h v)_x\right|^2 + b\left|(v - \pi_h v)_y\right|^2\right) \mathrm{d}x\, \mathrm{d}y \\
&\quad \leqslant Ch^2 \int_\tau \left(\left|(v - \pi_h v)_{xx}\right|^2 + 2\left|(v - \pi_h v)_{xy}\right|^2 + b\left|(v - \pi_h v)_{yy}\right|^2\right) \mathrm{d}x\, \mathrm{d}y \\
&\quad \leqslant Ch^2 \int_\tau \left(|v_{xx}|^2 + 2|v_{xy}|^2 + b|v_{yy}|^2\right) \mathrm{d}x\, \mathrm{d}y.
\end{aligned}$$

The result for bilinear elements follows by summing over τ.

In the case of linear elements, we write $\tau = \tau^+ \cup \tau^-$ where τ^+ and τ^- are the two triangles on which $\pi_h v$ is linear. Let ℓ^+ be the linear function on τ which is equal to $\pi_h v$ on τ^+, with ℓ^- similarly defined. Applying the first part of Lemma 12.2 to $(v - \ell^\pm)_x$ and the second part of Lemma 12.2 to $(v - \ell^\pm)_y$ on τ, we obtain

$$\int_\tau a\left|(v - \ell^\pm)_x\right|^2 \mathrm{d}x\, \mathrm{d}y \leqslant Ca_{\max} h^2 \int_\tau \left(\left|(v - \ell^\pm)_{xx}\right|^2 + \left|(v - \ell^\pm)_{xy}\right|^2\right) \mathrm{d}x\, \mathrm{d}y$$

and

$$\int_\tau b\left|\left(v-\ell^\pm\right)_y\right|^2 \mathrm{d}x\,\mathrm{d}y \leqslant Ch^2 \int_\tau \left(\left|\left(v-\ell^\pm\right)_{xy}\right|^2 + b\left|\left(v-\ell^\pm\right)_{yy}\right|^2\right) \mathrm{d}x\,\mathrm{d}y.$$

Since $\ell^\pm$ is linear, its second derivatives all vanish in τ, and hence we obtain

$$\begin{aligned}
&\int_\tau \left(a\left|\left(v-\ell^\pm\right)_x\right|^2 + b\left|\left(v-\ell^\pm\right)_y\right|^2\right) \mathrm{d}x\,\mathrm{d}y \\
&\quad\leqslant Ch^2 \int_\tau \left(|v_{xx}|^2 + 2|v_{xy}|^2 + b|v_{yy}|^2\right) \mathrm{d}x\,\mathrm{d}y.
\end{aligned}$$

Consequently

$$\begin{aligned}
&\int_\tau \left(a\left|(v-\pi_h v)_x\right|^2 + b\left|(v-\pi_h v)_y\right|^2\right) \mathrm{d}x\,\mathrm{d}y \\
&\quad= \int_{\tau^+} \left(a\left|\left(v-\ell^+\right)_x\right|^2 + b\left|\left(v-\ell^+\right)_y\right|^2\right) \mathrm{d}x\,\mathrm{d}y \\
&\qquad+ \int_{\tau^-} \left(a\left|\left(v-\ell^-\right)_x\right|^2 + b\left|\left(v-\ell^-\right)_y\right|^2\right) \mathrm{d}x\,\mathrm{d}y \\
&\quad\leqslant Ch^2 \int_\tau \left(|v_{xx}|^2 + 2|v_{xy}|^2 + b|v_{yy}|^2\right) \mathrm{d}x\,\mathrm{d}y.
\end{aligned}$$

The result now follows by summing over τ. □

The Galerkin projection has the following approximation properties.

LEMMA 12.4. *Let $v \in H_0^1(\Omega) \cap H^2(\Omega)$. Then*

$$|||(I-P_h)v|||^2 \leqslant C_1 h^2 \left(\frac{1}{b} Av, Av\right)$$

and

$$\left(b(I-P_h)v, (I-P_h)v\right) \leqslant C_1 h^2 |||(I-P_h)v|||^2.$$

PROOF. The first inequality is a consequence of Lemma 12.3 and Lemma 12.1; i.e.,

$$\begin{aligned}
|||v-P_h v|||^2 &\leqslant |||v-\pi_h v|||^2 \\
&\leqslant Ch^2 \int_\Omega \left(v_{xx}^2 + 2v_{xy}^2 + bv_{yy}^2\right) \mathrm{d}x\,\mathrm{d}y \\
&\leqslant C_1 h^2 \left(\frac{1}{b} Av, Av\right).
\end{aligned}$$

For the second we use a duality argument. Let $Aw = b(I - P_h)v$. Then

$$\begin{aligned}
&\bigl(b(I - P_h)v, (I - P_h)v\bigr) \\
&\quad = A(w, vP_h v) = A\bigl((I - P_h)w, (I - P_h)v\bigr) \\
&\quad \leqslant |||(I - P_h)w|||\,|||(I - P_h)v||| \\
&\quad \leqslant \sqrt{C_1}h\left(\frac{1}{b}Aw, Aw\right)^{1/2} |||(I - P_h)v||| \\
&\quad = \sqrt{C_1}h\bigl(b(I - P_h)v, (I - P_h)v\bigr)^{1/2} |||(I - P_h)v|||.
\end{aligned}$$

Cancelling the common factor and then squaring, the second inequality follows. □

12.4. The line Jacobi smoother

To define the line Jacobi smoother, we introduce a horizontal stripwise decomposition of Ω:

$$\Omega = \bigcup \Omega_j, \quad \Omega_j = \bigl[(x, y) \in \Omega\colon (j - 1)h < y < (j + 1)h\bigr].$$

We partition the finite element space M_h accordingly as

$$M_h = \sum_{j=1}^{n-1} M_{h,j}, \quad \text{where } M_{h,j} = [v \in M_h\colon v = 0 \text{ in } \Omega \setminus \Omega_j].$$

A line Jacobi smoother $\mathcal{J}_h$ is defined as

$$\mathcal{J}_h = \sum_{j=1}^{n-1} A_{h,j}^{-1} Q_{h,j}, \tag{12.7}$$

where $Q_{h,j}$ is the L^2 projection onto $M_{h,j}$ and $A_{h,j}$ is the "restriction" of A on $M_{h,j}$ defined by

$$(A_{h,j}u, v) = A(u, v), \quad \text{for all } u, v \in M_{h,j}.$$

A smoothing property of the line Jacobi operator is summarized in the following lemma.

LEMMA 12.5. *Let $\mathcal{J}_h$ be the line Jacobi smoother defined by* (12.7). *Then for the piecewise linear and bilinear functions,*

$$\tfrac{1}{2}(Av, v) \leqslant \bigl(\mathcal{J}_h^{-1}v, v\bigr) \leqslant C_2\left[(Av, v) + \frac{1}{h^2}(bv, v)\right], \quad \textit{for all } v \in M_h.$$

PROOF. In either case, we write $v \in M_h$ as $v = \sum_j v_j$, with $v_j \in M_{h,j}$. For linear and bilinear elements, this decomposition is unique. Let $S_j = [(x, y) \in \Omega\colon (j-1)h \leqslant y \leqslant jh]$. Then by Lemma 4.1 $\mathcal{J}_h$ is characterized by

$$(\mathcal{J}_h^{-1} v, v) \equiv \sum_{j=1}^{n-1} A(v_j, v_j) = \sum_{j=1}^{n} \bigl[A_{S_j}(v_{j-1}, v_{j-1}) + A_{S_j}(v_j, v_j)\bigr].$$

On the strip, S_j, $v = v_{j-1} + v_j$ ($v_0 = v_n = 0$). Thus

$$A_{S_j}(v, v) \leqslant 2\bigl[A_{S_j}(v_{j-1}, v_{j-1}) + A_{S_j}(v_j, v_j)\bigr].$$

The first inequality follows by summing the above inequality from 1 to n.

We now prove the upper estimate for $(\mathcal{J}_h^{-1} v, v)$. In the bilinear case, note that

$$v_j(x, y_j) = v(x, y_j) \quad \text{and} \quad v_j(x, y_{j\pm1}) = 0.$$

A simple calculation shows that,

$$\begin{aligned}
&\int_{S_j} \bigl(|D_x v_{j-1}|^2 + |D_x v_j|^2\bigr)\,\mathrm{d}x\,\mathrm{d}y \\
&\quad = \tfrac{1}{3} \sum_{i=1}^{n-1} \Bigl[\bigl|v(x_i, y_{j-1}) - v(x_{i-1}, y_{j-1})\bigr|^2 + \bigl|v(x_i, y_j) - v(x_{i-1}, y_j)\bigr|^2\Bigr] \\
&\quad \leqslant 2 \int_{S_j} |v_x|^2\,\mathrm{d}x\,\mathrm{d}y.
\end{aligned}$$

Consequently,

$$\int_{S_j} a(x, y)\bigl(|D_x v_{j-1}|^2 + |D_x v_j|^2\bigr)\,\mathrm{d}x\,\mathrm{d}y \leqslant 2\frac{a_{\max}}{a_{\min}} \int_{S_j} a(x, y)|v_x|^2\,\mathrm{d}x\,\mathrm{d}y. \tag{12.8}$$

On the other hand, for $(x, y) \in S_j$,

$$\bigl|D_y v_{j-1}(x, y)\bigr|^2 + \bigl|D_y v_j(x, y)\bigr|^2 = \frac{1}{h^2}\bigl(\bigl|v(x, y_{j-1})\bigr|^2 + \bigl|v(x, y_j)\bigr|^2\bigr).$$

Since $v_{yy}(x, \theta) \equiv 0$ for θ between y_{j-1} and y_j,

$$\begin{aligned}
&\bigl|v(x, y_{j-1})\bigr|^2 + \bigl|v(x, y_j)\bigr|^2 \\
&\quad = \bigl|v(x, y) + v_y(x, y)(y_{j-1} - y)\bigr|^2 + \bigl|v(x, y) + v_y(x, y)(y_j - y)\bigr|^2 \\
&\quad \leqslant 4\bigl|v(x, y)\bigr|^2 + 2h^2\bigl|v_y(x, y)\bigr|^2,
\end{aligned}$$

for all $(x, y) \in S_j$. Hence

$$\int_{S_j} b\big(|D_y v_{j-1}|^2 + |D_y v_j|^2\big)\,\mathrm{d}x\,\mathrm{d}y \leqslant \frac{4}{h^2}\int_{S_j} b|v|^2\,\mathrm{d}x\,\mathrm{d}y + 2\int_{S_j} b|v_y|^2\,\mathrm{d}x\,\mathrm{d}y. \tag{12.9}$$

Combining Eqs. (12.8) and (12.9)

$$A_{S_j}(v_{j-1}, v_{j-1}) + A_{S_j}(v_j, v_j) \leqslant \frac{4}{h^2}\int_{S_j} bv^2\,\mathrm{d}x\,\mathrm{d}y + 2\frac{a_{\max}}{a_{\min}}\int_{S_j}\big(a|v_x|^2 + b|v_y|^2\big)\,\mathrm{d}x\,\mathrm{d}y.$$

Summing from 1 to n,

$$\begin{aligned}(\mathcal{J}_h^{-1}v, v) &\equiv \sum_{j=1}^{n}\big[A_{S_j}(v_{j-1}, v_{j-1}) + A_{S_j}(v_j, v_j)\big]\\ &\leqslant \left[2\frac{a_{\max}}{a_{\min}}(Av, v) + \frac{4}{h^2}(bv, v)\right].\end{aligned}$$

This proves the second inequality for the bilinear case.

The proof for the case of linear element is similar. We write $\tau = \tau^+ \cup \tau^-$. Then

$$D_x v_{j-1} = 0,\ D_x v_j = D_x v \text{ on } \tau^+ \quad\text{and}\quad D_x v_j = 0,\ D_x v_{j-1} = D_x v \text{ on } \tau^-.$$

Therefore

$$\int_\tau a\big(|D_x v_{j-1}|^2 + |D_x v_j|^2\big)\,\mathrm{d}x\,\mathrm{d}y = \int_\tau a|D_x v|^2\,\mathrm{d}x\,\mathrm{d}y.$$

On the other hand,

$$|D_y v_{j-1}|^2 + |D_y v_j|^2 = \frac{1}{h^2}\Big(\big|v(x_{i-1}, y_{j-1})\big|^2 + \big|v(x_{i-1}, y_j)\big|^2\Big) \quad \text{in } \tau^+$$

and

$$|D_y v_{j-1}|^2 + |D_y v_j|^2 = \frac{1}{h^2}\Big(\big|v(x_i, y_{j-1})\big|^2 + \big|v(x_i, y_j)\big|^2\Big) \quad \text{in } \tau^-.$$

Since all the second derivatives of v vanish on $\tau^\pm$, we have

$$\big|v(x_{i-1}, y_{j-1})\big|^2 + \big|v(x_{i-1}, y_j)\big|^2$$

$$
\begin{aligned}
&= \left|v(x, y) + \nabla v(x, y) \bullet (x_{i-1} - x, y_{j-1} - y)\right|^2 \\
&\quad + \left|v(x, y) + \nabla v(x, y) \bullet (x_{i-1} - x, y_j - y)\right|^2 \\
&\leqslant 4\left|v(x, y)\right|^2 + 6h^2\left|\nabla v(x, y)\right|^2, \quad \text{for all } (x, y) \in \tau^+.
\end{aligned}
$$

A similar estimate holds for $|v(x_i, y_{j-1})|^2 + |v(x_i, y_j)|^2$. Hence,

$$
\begin{aligned}
&\int_\tau b\left(|D_y v_{j-1}|^2 + |D_y v_j|^2\right) \\
&\quad = \frac{1}{h^2}\left[\int_{\tau^+} b\left(\left|v(x_{i-1}, y_{j-1})\right|^2 + \left|v(x_{i-1}, y_j)\right|^2\right) \mathrm{d}x\, \mathrm{d}y \right. \\
&\qquad \left. + \int_{\tau^-} b\left(\left|v(x_i, y_{j-1})\right|^2 + \left|v(x_i, y_j)\right|^2\right) \mathrm{d}x\, \mathrm{d}y\right] \\
&\quad \leqslant \frac{4}{h^2}\int_\tau bv^2\, \mathrm{d}x\, \mathrm{d}y + 6\int b|\nabla v|^2\, \mathrm{d}x\, \mathrm{d}y.
\end{aligned}
$$

Combining the estimates for $\int_\tau a(|D_x v_{j-1}|^2 + |D_x v_j|^2)$ and $\int_\tau b(|D_y v_{j-1}|^2 + |D_y v_j|^2)$ we obtain

$$
\begin{aligned}
&A_{S_j}(v_{j-1}, v_{j-1}) + A_{S_j}(v_j, v_j) \\
&\quad \leqslant \frac{4}{h^2}\int_{S_j} bv^2\, \mathrm{d}x\, \mathrm{d}y + \int_{S_j} \left[(1 + 6b)v_x^2 + 6bv_y^2\right] \mathrm{d}x\, \mathrm{d}y.
\end{aligned}
$$

The rest of the proof is identical to that for the bilinear elements. □

REMARK 12.1. The line Gauss–Seidel smoother $\mathcal{G}_h$ is defined by

$$
\mathcal{G}_h = \left[I - (I - P_{h,n-1}) \cdots (I - P_{h,1})\right]A_h^{-1}, \quad \text{where } P_{h,j} = A_{h,j}^{-1} Q_{h,j} A_h.
$$

It can be shown that

$$
A(v, v) \leqslant \left(\overline{\mathcal{G}}_h^{-1} v, v\right) \leqslant C\left[A(v, v) + \frac{1}{h^2}(bv, v)\right],
$$

where $\overline{\mathcal{G}}_h \equiv \mathcal{G}_h + \mathcal{G}_h^t - \mathcal{G}_h^t A_h \mathcal{G}_h$ is the symmetric line Gauss–Seidel smoother.

REMARK 12.2. In the proof of Lemma 12.5, we did not make use of (12.4). With minor modification, we can prove that the conclusion of Lemma 12.5 holds for general polynomial elements, provided that (12.4) holds.

12.5. Multigrid algorithm and convergence theory

We now consider the multigrid V-cycle Algorithm 3.1. For simplicity, we only consider linear and bilinear elements. We introduce an initial triangulation $\mathcal{T}_1$ of Ω by partitioning Ω into four smaller equal squares. For linear elements, each square is further decomposed into two triangles by linking the lower-left and upper-right vertices. Let $\{\mathcal{T}_k\}$ be a family of triangulations of Ω, where $\mathcal{T}_k$ is obtained from $\mathcal{T}_{k-1}$ by a halving strategy. Let $\{M_k\}$ be the corresponding family of linear or bilinear finite element spaces defined with respect to $\{\mathcal{T}_k\}$. The multiple finite element spaces are nested, i.e.,

$$M_1 \subset M_2 \subset \cdots \subset M_J.$$

The finite element problem on M_k is the following: Find $u_k \in M_k$ such that

$$A(u_k, \phi) = (f, \phi), \quad \text{for all } \phi \in M_k.$$

The L^2 projection $Q_k : L^2 \to M_k$ and the Galerkin projection $P_k : H_0^1 \to M_k$ are defined by

$$(Q_k w, \phi) = (w, \phi), \quad \text{for all } \phi \in M_k,$$

and

$$A(P_k w, \phi) = A(w, \phi), \quad \text{for all } \phi \in M_k.$$

Let $A_k : M_k \to M_k$ be defined by

$$(A_k w, \phi) = A(w, \phi), \quad \text{for all } \phi \in M_k.$$

Then the finite element equations can be rewritten in the form

$$A_k u_k = f_k \equiv Q_k f.$$

To define the multigrid algorithm, we also need smoothers $R_k : M_k \to M_k$. We denote by R_k^t the adjoint of R_k with respect to the inner product $(\bullet, \bullet)$.

Given an initial iterate $u^0 \in M_k$, a linear multigrid algorithm produces a sequence of approximations to $u_k = A_k^{-1} f_k$ as

$$u^{i+1} = \mathrm{Mg}_k(u^i, f_k) \equiv u^i + B_k(f_k - A_k u^i).$$

The multigrid process $\mathrm{Mg}_k(\bullet, \bullet)$ (or equivalently B_k) is defined in Algorithm 3.1 which we recall here.

ALGORITHM 12.1. With $u^0, g \in M_1$ set $\mathrm{Mg}_1(u^0, g) = A_1^{-1} g$ (or equivalently $B_1 = A_1^{-1}$). For $k > 1$ and $u^0, g \in M_k$, $u^1 = \mathrm{Mg}_k(u^0, g)$ is defined as follows

(1) Pre-smoothing: $u^{1/3} = u^0 + R_k^t(g - A_k u^0)$.

(2) Coarse grid correction:

$$\begin{aligned} u^{2/3} &= u^{1/3} + \mathrm{Mg}_{k-1}\big(0, Q_{k-1}(g - A_k u^{1/3})\big) \\ &= u^{1/3} + B_{k-1} Q_{k-1}\big(g - A_k u^{1/3}\big). \end{aligned}$$

(3) Post-smoothing: $u^1 = u^{2/3} + R_k(g - A_k u^{2/3})$.

Let $u = A_k^{-1} g$. Define $e^0 = u - u^0$, $e^{1/3} = u - u^{1/3}$, $e^{2/3} = u - u^{2/3}$ and $e^1 = u - u^1$. Then by the definition of the multigrid algorithm,

$$\begin{aligned} e^{1/3} &= (I - R_k^t A_k)e^0, \\ e^{2/3} &= (I - B_{k-1}Q_{k-1}A_k)e^{1/3} = (I - B_{k-1}A_{k-1}P_{k-1})e^{1/3}, \\ e^1 &= (I - R_k A_k)e^{2/3} = (I - R_k A_k)(I - B_{k-1}A_{k-1}P_{k-1})(I - R_k^t A_k)e^0. \end{aligned}$$

We thus obtain the following recurrence relation:

$$\begin{aligned} &A\big((I - B_k A_k)v, v\big) \\ &\quad = A\big((I - B_{k-1}A_{k-1}P_{k-1})K_k^* v, K_k^* v\big), \quad \text{for all } v \in M_k. \end{aligned} \tag{12.10}$$

Here $K_k = (I - R_k A_k)$ and $K_k^* = (I - R_k^t A_k)$ are the error propagation operators corresponding to the smoothers R_k and R_k^t.

Since Condition (A.1) and Condition (A.2) do not hold, we cannot apply the theory of Sections 3 and 5. We shall provide a modification of the theory of Section 3 here. For simplicity of presentation, we only consider symmetric smoothers. We shall assume that the spectrum $\sigma(K_k) \subset [0, 1)$.

LEMMA 12.6. *Assume that there is a constant C_M independent of k such that*

$$\begin{aligned} &\big(R_k^{-1}(I - P_{k-1})v, (I - P_{k-1})v\big) \\ &\quad \leqslant C_M A\big((I - P_{k-1})v, v\big), \quad \textit{for all } v \in M_k. \end{aligned} \tag{12.11}$$

Then the multigrid algorithm defined in Algorithm 12.1 *satisfies*

$$0 \leqslant A\big((I - B_k A_k)v, v\big) \leqslant \delta A(v, v), \quad \textit{for all } v \in M_k,$$

with $\delta = C_M/(2 + C_M)$.

PROOF. We only need to prove the upper estimate. We will prove by induction that the estimate holds with $\delta = C_M/(2 + C_M)$. Clearly the assertion holds for $k = 1$. Suppose now that the assertion holds for $k - 1$, i.e.

$$A\big((I - B_{k-1}A_{k-1})v, v\big) \leqslant \delta A(v, v). \tag{12.12}$$

Using the recurrence relation in (12.10), we have, for $v \in M_k$,

$$\begin{aligned} A\big((I - B_k A_k)v, v\big) &= A\big((I - P_{k-1})K_k^* v, K_k^* v\big) \\ &\quad + A\big((I - B_{k-1}A_{k-1})P_{k-1}K_k^* v, P_{k-1}K_k^* v\big). \end{aligned}$$

It is straightforward to show that the induction hypothesis (12.12) implies that

$$A\big((I - B_k A_k)v, v\big) \leqslant (1-\delta)A\big((I - P_{k-1})K_k^* v, K_k^* v\big) + \delta A\big(K_k^* v, K_k^* v\big). \quad (12.13)$$

We now estimate the first term on the right-hand side of (12.13). By the Cauchy–Schwarz inequality and the hypothesis (12.11), we have, for $w \in M_k$,

$$\begin{aligned} A\big((I - P_{k-1})w, w\big) &\leqslant \big(R_k^{-1}(I - P_{k-1})w, (I - P_{k-1})w\big)^{1/2}(R_k A_k w, A_k w)^{1/2} \\ &\leqslant C_M^{1/2} A\big((I - P_{k-1})w, w\big)^{1/2}(R_k A_k w, A_k w)^{1/2}. \end{aligned}$$

Cancelling the common factor, we get

$$A\big((I - P_{k-1})w, w\big) \leqslant C_M(R_k A_k w, A_k w), \qquad \text{for all } w \in M_k.$$

Since R_k is symmetric, $K_k^* = K_k$. Applying the above inequality with $w = K_k^* v$ and recalling that $R_k A_k = I - K_k$ and that $\sigma(K_k) \in [0, 1)$, we obtain

$$\begin{aligned} A\big((I - P_{k-1})K_k^* v, K_k^* v\big) &\leqslant C_M\big(R_k A_k K_k^* v, A_k K_k^* v\big) \\ &= C_M\big((I - K_k)K_k^* v, A_k K_k^* v\big) \\ &\leqslant C_M \tfrac{1}{2}\big[A(v, v) - A\big(K_k^* v, K_k^* v\big)\big]. \end{aligned} \quad (12.14)$$

Combining Eqs. (12.13) and (12.14), we obtain

$$\begin{aligned} A\big((I - B_k A_k)v, v\big) &\leqslant (1-\delta)A\big((I - P_{k-1})K_k^* v, K_k^* v\big) + \delta A\big(K_k^* v, K_k^* v\big) \\ &\leqslant \tfrac{1}{2}C_M(1-\delta)\big[A(v, v) - A\big(K_k^* v, K_k^* v\big)\big] + \delta A\big(K_k^* v, K_k^* v\big) \\ &\leqslant \delta A(v, v), \end{aligned}$$

with $\delta = C_M/(2 + C_M)$.

Lemma 12.6 is “soft”. To apply the lemma, we need to establish (12.11) with C_M independent of J. This will be done by combining the approximation property $(b(I - P_{k-1})v, v) \leqslant Ch_k^2 A((I - P_{k-1})v, v)$ and the smoothing property of the line Jacobi smoother $(R_k^{-1}v, v) \leqslant C[A(v, v) + h_k^{-2}(bv, v)]$. Here h_k is the mesh parameter. □

THEOREM 12.1. *Let $R_k = \frac{1}{2}\mathcal{J}_{h_k}$ be the weighted line Jacobi smoother on M_k. Then there is a positive number δ with $\delta < 1$ independent of k such that the multigrid algorithm defined in Algorithm* 12.1 *satisfies*

$$0 \leqslant A\big((I - B_k A_k)v, v\big) \leqslant \delta A(v, v), \quad \textit{for all } v \in M_k.$$

PROOF. In view of Lemma 12.6, we only need to prove that (12.11) holds for $R_k = \frac{1}{2}\mathcal{J}_{h_k}$. It follows from Lemma 12.4 that

$$\frac{1}{h^2}\big(b(x, y)(I - P_{k-1})v, (I - P_{k-1})v\big) \leqslant C_1 A\big((I - P_{k-1})v, (I - P_{k-1})v\big).$$

By Lemma 12.5, the weighted line Jacobi smoother, R_k, satisfies

$$\left(R_k^{-1}\phi,\phi\right) \leqslant C_2\left[(A_k\phi,\phi)+\frac{1}{h_k^2}(b(x,y)\phi,\phi)\right], \quad \text{for all } \phi \in M_k.$$

Applying the smoothing property of R_k to $\phi = (I - P_{k-1})v$ and using the approximation property of P_{k-1}, we obtain, with $C_M = C_2(1 + C_1)$,

$$\left(R_k^{-1}(I-P_{k-1})v,(I-P_{k-1})v\right) \leqslant C_M\, A\big((I-P_{k-1})v,(I-P_{k-1})v\big).$$

We have thus proved (12.11) and hence the theorem. □

It is desirable to avoid the condition in (12.4) in our theory. However, it is not clear that this is possible even in the case when $b(x, y)$ is uniformly bounded from above and below.

REMARK 12.3. It is not necessary to solve the coarse grid problem exactly. If the approximate coarse grid solve satisfies $0 \leqslant A((I - B_1A_1)\phi, \phi) \leqslant \delta_0 < 1$, for all $\phi \in M_1$, then Lemma 12.6 holds with $\delta = \max(\delta_0, C_M/(2 + C_M))$.

REMARK 12.4. The result of Lemma 12.6 remains valid for nonsymmetric smoothers if we replace the assumption in (12.11) by

$$\left(\overline{R}_k^{-1}(I-P_{k-1})v,(I-P_{k-1})v\right) \leqslant \tfrac{1}{2}C_M A\big((I-P_{k-1})v,v\big).$$

Here $\overline{R}_k \equiv R_k + R_k^t - R_k^t A_k R_k$. This type of estimate holds for the line Gauss–Seidel smoother (see Remark 12.2). Consequently, Theorem 12.1 remains valid for the line Gauss–Seidel smoother.

REMARK 12.5. Note that if $b(x, y) \approx h_k^2$, then the error operator for the line Jacobi or the line Gauss–Seidel method is already a uniform contraction on M_J.

REMARK 12.6. Our analysis remains valid for other polynomial finite elements as well. The approximation property is a consequence of the fact that the linear elements are a subspace of these higher order elements and the smoothing property can be established in a way similar to that of the linear element.

12.6. Bibliographical notes

Some earlier work on multilevel methods for anisotropic problems can be found in HACKBUSCH [1985] and the references therein.

The multigrid algorithm considered in this section is well known and dates back at least to the early eighties. It is also known that the method performs quite well; cf. HACKBUSCH [1985]. The analysis given here is based on BRAMBLE and ZHANG

[2000]. The result can be considered as a generalization of the work of BRAESS and HACKBUSCH [1983] to the case of anisotropic equations.

Some recent work on multilevel methods for anisotropic problems can be found, for example, in WITTUM [1989a, 1989b], STEVENSON [1993b, 1993a, 1994a, 1994b, 1994c], GRIEBEL and OSWALD [1994], and HEMKER [1984].

13. Pseudodifferential operators of order minus one

Weakly singular integral operators arise, e.g., in the approximation of the solution of elliptic boundary value problems by the so-called boundary element method. In this section we shall consider pseudodifferential operators of order minus one which give rise to coercive bilinear forms on $H^{-1/2}(\Omega)$ (to be defined). The aforementioned integral operators will be our main example. We shall develop multigrid algorithms for solving the associated discrete equations.

13.1. Pseudodifferential operators

For purposes of exposition we shall let Ω be a plane polygonal domain which we may think of as being embedded in $\mathbf{R}^3$.

For $s \geqslant 0$, an integer, we denote by $H^s(\Omega)$ the usual Sobolev space with norm given by

$$\|v\|_s^2 = \sum_{|\alpha| \leqslant s} \int_\Omega |D^\alpha v|^2 \, \mathrm{d}x.$$

For $s > 0$, not an integer, $H^s(\Omega)$ is defined by the real method of interpolation between spaces defined for successive integers (cf. Appendix A). The space $H^{-1}(\Omega)$ is defined as the completion of $L^2(\Omega)$ with respect to the norm defined by

$$\|v\|_{-1} = \sup_{\phi \in H^1(\Omega)} \frac{(v, \phi)}{\|\phi\|_1}.$$

For $0 < s < 1$, $H^{-s}(\Omega)$ is defined by interpolation between $L^2(\Omega)$ and $H^{-1}(\Omega)$. These spaces are Hilbert spaces with inner products denoted by $(\bullet, \bullet)_{-s}$. Note that $(\bullet, \bullet)_0 = (\bullet, \bullet)$, the $L^2(\Omega)$ inner product.

Now let $\mathcal{V}(\bullet, \bullet)$ be a positive definite bilinear form on $H^{-1/2}(\Omega) \times H^{-1/2}(\Omega)$ and assume that for positive constants C_0 and C_1

$$C_0 \|v\|_{-1/2}^2 \leqslant \mathcal{V}(v, v) \leqslant C_1 \|v\|_{-1/2}^2, \quad \text{for all } v \in H^{-1/2}(\Omega).$$

Let F be a bounded linear functional on $H^{-1/2}(\Omega)$ and consider the following problem: Find $u \in H^{-1/2}(\Omega)$ such that

$$\mathcal{V}(u, \phi) = F(\phi), \quad \text{for all } \phi \in H^{-1/2}(\Omega). \tag{13.1}$$

By the Riesz Representation Theorem this problem has a unique solution. Our main example of such a form is given by

$$\mathcal{V}(w, v) = \int_\Omega \int_\Omega \frac{w(s_1)v(s_2)}{|s_1 - s_2|}\,\mathrm{d}s_1\,\mathrm{d}s_2. \tag{13.2}$$

We will examine this example in detail later in this section, but for now we do not specialize to this form.

13.2. The multigrid algorithm

As before we consider the spaces M_k with

$$M_1 \subset M_2 \subset \cdots \subset M_J = M \subset H^{-1/2}(\Omega).$$

The approximate problem in M is: Find $u \in M$ such that

$$\mathcal{V}(u, \phi) = F(\phi), \quad \text{for all } \phi \in M. \tag{13.3}$$

In order to motivate and simplify our discussion assume that $M \subset H^1(\Omega)$. This is convenient but not necessary. For the purpose of exposition and analysis, we shall also assume that M_1 is the usual coarse space of continuous piecewise linear functions on Ω. We define M_k as in Section 2. No boundary conditions are imposed.

A naive multigrid approach. To motivate our multigrid algorithm, we first present a naive multigrid approach. Let $\hat{\mathcal{V}}_J\colon M \to M$ be defined by

$$\big(\hat{\mathcal{V}}_J v, \phi\big) = \mathcal{V}(v, \phi), \quad \text{for all } v, \phi \in M. \tag{13.4}$$

The discrete operator $\hat{\mathcal{V}}_J$ is symmetric and positive definite on M and hence has positive eigenvalues $\hat{\mu}_i$ with corresponding orthonormal eigenvectors $\hat{\phi}_i$. Now the extreme eigenvalues, $\hat{\mu}_{\min}$ and $\hat{\mu}_{\max}$, are the extremes of the Rayleigh quotient, i.e.,

$$\hat{\mu}_{\min} \leqslant \frac{\mathcal{V}(v, v)}{(v, v)} \leqslant \hat{\mu}_{\max}, \quad 0 \neq v \in M.$$

This behaves like

$$\frac{\mathcal{V}(v, v)}{(v, v)} \approx \frac{\|v\|^2_{-1/2}}{(v, v)} \approx \frac{(A_J^{-1/2} v, v)_{0,J}}{\|v\|^2}, \tag{13.5}$$

where A_J is analogous to a second-order elliptic operator defined on M by

$$(A_J w, \phi)_{0,J} = D(w, \phi) + (w, \phi) = \int_\Omega (\nabla w \boldsymbol{\cdot} \nabla \phi + w\phi)\,\mathrm{d}x.$$

Here $(\bullet, \bullet)_{0,J}$ is an appropriately defined inner product on M, comparable to the $L^2(\Omega)$ inner product. Following the argument in Section 2, noting that the eigenvalues of $A_J^{1/2}$ are the square roots of those of A_J, we have

$$C_0 h_J \leqslant \frac{(A_J^{-1/2} v, v)_{0,J}}{\|v\|^2} \leqslant C_1. \tag{13.6}$$

Because of Eqs. (13.5) and (13.6)

$$\hat{\mu}_{\min} \geqslant C_0 h_J \quad \text{and} \quad \hat{\mu}_{\max} \leqslant C_1.$$

In order to see what the difficulty is in the present setting we consider for a moment the special case (13.2). The operator $\mathcal{V}: H^{-1/2}(\Omega) \to H^{1/2}(\Omega)$ defined by

$$(\mathcal{V}u)(s) = \int_\Omega \frac{u(s_1)\,\mathrm{d}s_1}{|s_1 - s|}$$

gives rise to the form (13.2) and is compact as an operator from $H^{-1/2}(\Omega)$ to $H^{-1/2}(\Omega)$ because of the compact embedding of $H^{1/2}(\Omega)$ in $H^{-1/2}(\Omega)$ (cf. NEDELEC and PLANCHARD [1973]). The discrete operator $\hat{\mathcal{V}}_J$ defined by (13.4) is symmetric and positive definite and has eigenvalues $\hat{\mu}_i$ which tend to zero with h_J. The highly oscillatory eigenvectors correspond to small eigenvalues so that a simple linear iteration such as

$$u^{n+1} = u^n + \tau \hat{\mathcal{V}}_J (u - u^n)$$

is not convergent for any choice of τ which damps the highly oscillatory components of the error, since such a τ would necessarily be large and hence would amplify the smooth components. Thus the naive multigrid approach fails.

The multigrid algorithm. The remedy for this problem in the general case lies in properly defining the discrete operators $\mathcal{V}_k$. Define $\mathcal{V}_k : M_k \to M_k$ by

$$(\mathcal{V}_k w, \phi)_{-1} = \mathcal{V}(w, \phi), \quad \text{for all } w, \phi \in M_k.$$

Using the operator $\mathcal{V}_J$, (13.3) can be written as

$$\mathcal{V}_J u = f_J,$$

where $f_J \in M$ is defined by

$$(f_J, \phi)_{-1} = F(\phi), \quad \text{for all } \phi \in M.$$

Note that f_J cannot be easily computed, but its action $(f_J, \phi)_{-1}$ can be computed from $[F(\phi)]$.

The discrete operator $\mathcal{V}_k$ is symmetric and positive definite with respect to $(\bullet,\bullet)_{-1}$. Its eigenvalues are positive and the corresponding eigenfunctions can be chosen to be orthonormal in $(\bullet,\bullet)_{-1}$. The extremal eigenvalues are the extremes of the Rayleigh quotient

$$\mu_{\min}^{(k)} \leqslant \frac{\mathcal{V}(\phi,\phi)}{\|\phi\|_{-1}^2} \leqslant \mu_{\max}^{(k)}, \quad 0 \neq \phi \in M_k.$$

The denominator is weaker than the numerator and hence $\mu_{\min}^{(k)} \approx 1$ and $\mu_{\max}^{(k)} \approx h_k^{-1}$.

Now consider the linear iteration

$$u^{n+1} = u^n + \tau \mathcal{V}_J\big(u - u^n\big) = u^n + \tau\big(f_J - \mathcal{V}_J u^n\big).$$

Let $\{\mu_i\}$ be the eigenvalues of $\mathcal{V}_J$ and $\{\psi_i\}$ the corresponding orthonormal eigenfunctions. If we represent the initial error as

$$u - u^0 = \sum_{i=1}^{N_J} a_i \psi_i,$$

then

$$u - u^n = \sum_{i=1}^{N_J} a_i (1 - \tau\mu_i)^n \psi_i.$$

Hence for $c \leqslant \tau\, \mu_{\max}^{(J)} \leqslant 1$ the high frequencies of the error are reduced leaving the low frequencies relatively unchanged. This is the situation in which we expect a multilevel approach to be effective.

To define a multigrid algorithm, we need the orthogonal projectors with respect to the base inner product $(\bullet,\bullet)_{-1}$. We define $\mathcal{Q}_k : H^{-1}(\Omega) \to M_k$ by

$$(\mathcal{Q}_k w, \phi)_{-1} = (w, \phi)_{-1}, \quad \text{for all } \phi \in M_k.$$

Assuming that the smoothers $\mathcal{R}_k : M_k \to M_k$ have been defined, we can solve (13.3) by the multigrid Algorithm 3.1 which can be restated in this case as the following:

ALGORITHM 13.1. Set $\mathcal{B}_1 = \mathcal{V}_1^{-1}$. For $k > 1$, $\mathcal{B}_k$ is defined in terms of $\mathcal{B}_{k-1}$ as follows. Let $g \in M_k$.

(1) $x^1 = \mathcal{R}_k g$,

(2) $x^2 = x^1 + q$, where $q = \mathcal{B}_{k-1}\mathcal{Q}_{k-1}(g - \mathcal{V}_k x^1)$,

(3) $\mathcal{B}_k g = x^2 + \mathcal{R}_k(g - \mathcal{V}_k x^2)$.

13.3. Construction and analysis of smoothers

In order to apply the theorems in Section 5, we need to show that the smoothers $\mathcal{R}_k$ satisfy Conditions (SM.1) and (SM.2). The smoothers defined in Section 8 do not seem to be adaptable to the theory in this section. Thus we need to construct smoothers $\mathcal{R}_k$ satisfying Conditions (SM.1) and (SM.2). To this end, denote by λ_k the largest eigenvalue of $\mathcal{V}_k$, i.e.,

$$\lambda_k = \max_{\phi \in M_k} \frac{\mathcal{V}(\phi, \phi)}{(\phi, \phi)_{-1}}.$$

Conditions (SM.1) states that there exists $\omega \in (0, 1)$ such that

$$\frac{\omega}{\lambda_k} \|v\|_{-1}^2 \leqslant (\mathcal{R}_k v, v)_{-1}, \quad \text{for all } v \in M_k,$$

and Condition (SM.2) states that there exists $\theta \in (0, 2)$ such that

$$\mathcal{V}(\mathcal{R}_k v, \mathcal{R}_k v) \leqslant \theta (\mathcal{R}_k v, v)_{-1}, \quad \text{for all } v \in M_k.$$

A natural choice for the smoother would be $\mathcal{R}_k = \lambda_k^{-1} I$. This smoother clearly satisfies Conditions (SM.1) and (SM.2). However the action of $\mathcal{R}_k$ requires the inversion of the Gram matrix $[(\phi_k^i, \phi_k^j)_{-1}]$, where $\{\phi_k^i\}$ are the nodal basis functions of M_k.

To avoid the inversion of the Gram matrix, we will define the smoother in terms of a discrete minus one inner product $(\bullet, \bullet)_{-1,k}$ on M_k. We assume that $(\bullet, \bullet)_{-1,k}$ is uniformly equivalent to the $(\bullet, \bullet)_{-1}$ on M_k, i.e., there exist positive constants C_1 and C_2, independent of J, such that

$$C_0 \|v\|_{-1}^2 \leqslant (v, v)_{-1,k} \leqslant C_1 \|v\|_{-1}^2, \quad \text{for all } v \in M_k. \tag{13.7}$$

This inner product will be defined a little later. Care must be taken in its choice in order that there are no difficulties in the implementation of the multigrid algorithm. In terms of $(\bullet, \bullet)_{-1,k}$ we define $\mathcal{R}_k : M_k \to M_k$ as

$$(\mathcal{R}_k w, \phi)_{-1,k} = \frac{1}{\bar{\lambda}_k} (w, \phi)_{-1} \tag{13.8}$$

with

$$\hat{\lambda}_k \leqslant \bar{\lambda}_k \leqslant C \hat{\lambda}_k \quad \text{and} \quad \hat{\lambda}_k = \sup_{\phi \in M_k} \frac{\mathcal{V}(\phi, \phi)}{(\phi, \phi)_{-1,k}}.$$

We now verify the smoothing conditions using only the definition of $\mathcal{R}_k$ and the property (13.7). We first verify Condition (SM.2). By the definition of $\mathcal{R}_k$,

$$\begin{aligned}
\left[\mathcal{V}(\mathcal{R}_k \mathcal{V}_k \phi, \phi)\right]^2 &\leqslant \mathcal{V}(\mathcal{R}_k \mathcal{V}_k \phi, \mathcal{R}_k \mathcal{V}_k \phi) \mathcal{V}(\phi, \phi) \\
&\leqslant \bar{\lambda}_k (\mathcal{R}_k \mathcal{V}_k \phi, \mathcal{R}_k \mathcal{V}_k \phi)_{-1,k} \mathcal{V}(\phi, \phi) \\
&= (\mathcal{V}_k \phi, \mathcal{R}_k \mathcal{V}_k \phi)_{-1} \mathcal{V}(\phi, \phi) \\
&= \mathcal{V}(\mathcal{R}_k \mathcal{V}_k \phi, \phi) \mathcal{V}(\phi, \phi).
\end{aligned}$$

Hence

$$\mathcal{V}(\mathcal{R}_k \mathcal{V}_k \phi, \phi) \leqslant \mathcal{V}(\phi, \phi).$$

Thus Condition (SM.2) is satisfied with $\theta = 1$.

We next verify Condition (SM.1). It is easy to check that $\mathcal{R}_k$ is symmetric and positive definite in both $(\bullet, \bullet)_{-1,k}$ and $(\bullet, \bullet)_{-1}$. By the definition of $\bar{\lambda}_k$ and (13.7)

$$\bar{\lambda}_k \leqslant C\hat{\lambda}_k = C \sup_{\theta \in M_k} \frac{\mathcal{V}(\theta, \theta)}{(\theta, \theta)_{-1,k}} \leqslant C\lambda_k$$

and

$$\begin{aligned} \|w\|_{-1}^2 &= \bar{\lambda}_k(\mathcal{R}_k w, w)_{-1,k} = \bar{\lambda}_k\big(\mathcal{R}_k^{1/2} w, \mathcal{R}_k^{1/2} w\big)_{-1,k} \\ &\leqslant C\lambda_k\big(\mathcal{R}_k^{1/2} w, \mathcal{R}_k^{1/2} w\big)_{-1} \\ &= C\lambda_k(\mathcal{R}_k w, w)_{-1}. \end{aligned}$$

This proves the lower inequality of Condition (SA.2). It is easy to see that if $\mathcal{R}_k$ is symmetric, then the lower inequality in Condition (SA.2) implies Condition (SM.1).

The discrete minus one inner product. We now define the discrete minus one inner product. To simplify our presentation we assume as before that $M_k \subset H^1(\Omega)$. We first define a discrete L^2 inner product $(\bullet, \bullet)_{0,k}$ on M_k by

$$(v, v)_{0,k} = \tfrac{1}{3} \sum_{\tau_i \in \mathcal{T}_k} |\tau_i|\big[v(x_{i,1})^2 + v(x_{i,2})^2 + v(x_{i,3})^2\big].$$

Here $|\tau_i|$ denotes the area of τ_i and $x_{i,\ell}$, $\ell = 1, 2, 3$, denote the vertices of τ_i. Now let

$$A(v, w) = \int_\Omega (\nabla v \bullet \nabla w + vw)\,\mathrm{d}x \quad \text{and} \quad \|v\|_1 = A(v, v)^{1/2}.$$

We define $A_k : M_k \to M_k$ by

$$(A_k w, \phi)_{0,k} = A(w, \phi), \quad \text{for all } w, \phi \in M_k.$$

Define on M_k

$$(w, v)_{-1,k} \equiv \big(A_k^{-1} w, v\big)_{0,k}. \tag{13.9}$$

We will see that, with this definition, the multigrid algorithm involves only the action of A_k (not A_k^{-1}).

We prove now that the discrete minus one inner product, $(\bullet, \bullet)_{-1,k}$, is uniformly equivalent to $(\bullet, \bullet)_{-1}$.

LEMMA 13.1. *There exist positive constants C_1 and C_2, independent of J, such that*

$$C_1\|v\|_{-1}^2 \leqslant (v,v)_{-1,k} \leqslant C_2\|v\|_{-1}^2, \quad \textit{for all } v\in M_k.$$

PROOF. Note that $(\bullet,\bullet)_{-1,k}$ is defined in terms of $(\bullet,\bullet)_{0,k}$ and it is known that

$$\left|(v,w)_{0,k}-(v,w)\right| \leqslant Ch_k\|v\|\|w\|_1, \quad \text{for all } v\in M_k \text{ and } w\in M_k. \tag{13.10}$$

Also it follows that there are positive constants c and C, independent of k, such that

$$c(v,v)_{0,k} \leqslant (v,v) \leqslant C(v,v)_{0,k}, \quad \text{for all } v\in M_k.$$

Now $M_k \subset H^1(\Omega)$ so that, for $v\in M_k$,

$$(v,v)_{-1,k} = \left(A_k^{-1}v,v\right)_{0,k} = \sup_{\phi\in M_k}\frac{(v,A_k^{-1/2}\phi)_{0,k}^2}{(\phi,\phi)_{0,k}} = \sup_{\phi\in M_k}\frac{(v,\phi)_{0,k}^2}{A(\phi,\phi)}. \tag{13.11}$$

Using (13.10), we have

$$\left|(v,\phi)_{0,k}\right| \leqslant \left|(v,\phi)_{0,k}-(v,\phi)\right| + \left|(v,\phi)\right| \leqslant Ch_k\|v\|\|\phi\|_1 + \|v\|_{-1}\|\phi\|_1. \tag{13.12}$$

Also $A(\phi,\phi) \leqslant Ch_k^{-2}\|\phi\|^2$ for $\phi\in M_k$. Hence for $v\in M_k$

$$\|v\|^2 = \sup_{\phi\in M_k}\frac{(v,\phi)^2}{\|\phi\|^2} \leqslant Ch_k^{-2}\sup_{\phi\in M_k}\frac{(v,\phi)^2}{A(\phi,\phi)} \leqslant Ch_k^{-2}\|v\|_{-1}^2. \tag{13.13}$$

Similarly

$$\|v\|^2 \leqslant Ch_k^{-2}\sup_{\phi\in M_k}\frac{(v,\phi)_{0,k}^2}{A(\phi,\phi)}. \tag{13.14}$$

Hence, using Eqs. (13.12) and (13.13), we see that

$$\left|(v,\phi)_{0,k}\right| \leqslant C\|v\|_{-1}\|\phi\|_1$$

and therefore, from (13.11)

$$(v,v)_{-1,k} = \sup_{\phi\in M_k}\frac{(v,\phi)_{0,k}^2}{A(\phi,\phi)} \leqslant C\|v\|_{-1}^2,$$

which is the upper inequality of Lemma 13.1.

We next prove that $\|v\|_{-1}^2 \leqslant C(v,v)_{-1,k}$. To this end let $Q_k: L^2(\Omega)\to M_k$ be the $L^2(\Omega)$ orthogonal projection onto M_k. It is known (cf. BRAMBLE and XU [1991]) that

$$\|Q_k\phi\|_1 \leqslant C\|\phi\|_1, \quad \text{for all } \phi\in H^1(\Omega). \tag{13.15}$$

Thus, for $v \in M_k$,

$$\begin{aligned}
\|v\|_{-1} &= \sup_{\phi \in H^1(\Omega)} \frac{(v,\phi)}{\|\phi\|_1} = \sup_{\phi \in H^1(\Omega)} \frac{(v, Q_k\phi)}{\|\phi\|_1} \\
&= \sup_{\phi \in H^1(\Omega)} \frac{(v, Q_k\phi)}{\|Q_k\phi\|_1} \frac{\|Q_k\phi\|_1}{\|\phi\|_1} \\
&\leqslant C \sup_{\phi \in M_k} \frac{(v,\phi)}{\|\phi\|_1}. \qquad (13.16)
\end{aligned}$$

Again we use (13.10) to obtain, for v and ϕ in M_k,

$$\left|(v,\phi)\right| \leqslant Ch_k \|v\| \, \|\phi\|_1 + |(v,\phi)_{0,k}|,$$

or

$$\frac{|(v,\phi)|}{\|\phi\|_1} \leqslant Ch_k\|v\| + \frac{|(v,\phi)_{0,k}|}{\|\phi\|_1}. \qquad (13.17)$$

The result now follows from Eqs. (13.16), (13.17), (13.14) and (13.11). □

13.4. Multigrid convergence analysis

To simplify our discussion, we assume as before that $M_k \subset H^1(\Omega)$. For the convergence analysis we will need the following estimate for the largest eigenvalue λ_k of $\mathcal{V}_k$: There exist positive constants c and C such that

$$ch_k^{-1} \leqslant \lambda_k \leqslant Ch_k^{-1}. \qquad (13.18)$$

We first show the upper estimate. By Theorem A.1 of Appendix A and (13.13),

$$c\mathcal{V}(\phi,\phi) \leqslant \|\phi\|_{-1/2}^2 \leqslant \|\phi\| \|\phi\|_{-1} \leqslant Ch_k^{-1}\|\phi\|_{-1}^2.$$

Hence $\lambda_k \leqslant Ch_k^{-1}$. On the other hand,

$$(v,v)_{0,k} \leqslant C\|v\|^2, \quad \text{for all } v \in M_k,$$

and

$$\left(A_k^{-1}v, v\right)_{0,k} \leqslant C\|v\|_{-1}^2, \quad \text{for all } v \in M_k.$$

By interpolation (Theorem A.4) it follows that

$$\left(A_k^{-1/2}v, v\right)_{0,k} \leqslant C\|v\|_{-1/2}^2 \leqslant C\mathcal{V}(v,v), \quad \text{for all } v \in M_k.$$

This and Lemma 13.1 imply that

$$\lambda_k = \sup_{\phi\in M_k} \frac{\mathcal{V}(\phi,\phi)}{(\phi,\phi)_{-1}} \geqslant C \sup_{\phi\in M_k} \frac{(A_k^{-1/2}\phi,\phi)_{0,k}}{(\phi,\phi)_{-1,k}} = C \sup_{\phi\in M_k} \frac{(A_k^{-1/2}\phi,\phi)_{0,k}}{(A_k^{-1}\phi,\phi)_{0,k}}$$

$$= C \sup_{\psi\in M_k} \frac{(A_k^{1/2}\psi,\psi)}{(\psi,\psi)_{0,k}} \geqslant C h_k^{-1}.$$

We now study the rate of convergence of Algorithm 13.1. We first consider the general case and establish the convergence estimate assuming only that $\mathcal{V}(\bullet,\bullet)$ is uniformly equivalent to $\|\bullet\|_{-1/2}^2$. We then discuss the boundary integral operator with the bilinear form defined by (13.2).

General pseudodifferential operators. We first consider the general case and assume that $\mathcal{V}(\bullet,\bullet)$ is uniformly equivalent to $(\bullet,\bullet)_{-1/2}$. Since we cannot show that Condition (A.6) or (A.2) is satisfied, we apply the weaker result, Theorem 5.5.

LEMMA 13.2. *Let Q_k be the $L^2(\Omega)$ orthogonal projection. Then*

$$\|(Q_k - Q_{k-1})v\|_{-1}^2 \leqslant C\lambda_k^{-1}\mathcal{V}(v,v) \tag{13.19}$$

and

$$\mathcal{V}(Q_k v, Q_k v) \leqslant C\mathcal{V}(v,v). \tag{13.20}$$

PROOF. We first prove (13.19). We use the well known approximation property

$$\|(I-Q_k)\phi\| \leqslant C h_k \|\phi\|_1, \quad \text{for all } \phi\in H^1(\Omega).$$

Now by the definition of $\|\bullet\|_{-1}$, we have

$$\|(I-Q_k)v\|_{-1} = \sup_{\phi\in H^1(\Omega)} \frac{((I-Q_k)v,\phi)}{\|\phi\|_1} \leqslant \sup_{\phi\in H^1(\Omega)} \frac{\|v\|\,\|(I-Q_k)\phi\|}{\|\phi\|_1} \leqslant C h_k \|v\|.$$

Since $M_k \subset H^1(\Omega)$, Q_k is well-defined as a map from $H^{-1}(\Omega)$ to $L^2(\Omega)$. Hence by (13.15), with $u\in H^{-1}(\Omega)$,

$$\|Q_k u\|_{-1} = \sup_{\phi\in H^1(\Omega)} \frac{(Q_k u,\phi)}{\|\phi\|_1} = \sup_{\phi\in H^1(\Omega)} \frac{(u,Q_k\phi)}{\|\phi\|_1} \leqslant C\|u\|_{-1}.$$

Therefore, by interpolation (Theorem A.4),

$$\|(I-Q_k)u\|_{-1}^2 \leqslant C h_k \|u\|_{-1/2}^2 \leqslant C h_k \mathcal{V}(u,u).$$

Since $\lambda_k \leqslant C h_k^{-1}$, (13.19) follows. In a similar way we can prove (13.20). □

Now Eqs. (13.19) and (13.20) imply that Condition (A.7) holds with $\Pi_k = Q_k$. Hence Theorem 5.5 may be applied to conclude that

$$\mathcal{V}\big((I - \mathcal{B}_J \mathcal{V}_J)\phi, \phi\big) \leqslant \left(1 - \frac{1}{CJ}\right)\mathcal{V}(\phi, \phi),$$

where $\mathcal{B}_J$ is defined by Algorithm 13.1 with $\mathcal{R}_k$ given by (13.8).

Single layer potential. We now consider the particular case of the form defined by (13.2). In this case we know more about the form $\mathcal{V}(\bullet, \bullet)$ and hence we are able to prove a stronger result. Recall that the bilinear form is given by

$$\mathcal{V}(u, v) = \int_\Omega \int_\Omega \frac{u(s_1)v(s_2)}{|s_1 - s_2|}\, \mathrm{d}s_1\, \mathrm{d}s_2,$$

where Ω is a polygonal domain in $\mathbf{R}^2$.

We first show that $\mathcal{V}(v, v) \approx \|v\|_{-1/2}$. Let $\mathcal{S}$ be a smooth closed surface in $\mathbf{R}^3$ (a two-dimensional manifold) with $\Omega \subset \mathcal{S}$. Set

$$\mathcal{V}_{(\mathcal{S})}(u, v) = \int_\mathcal{S} \int_\mathcal{S} \frac{u(s_1)v(s_2)}{|s_1 - s_2|}\, \mathrm{d}s_1\, \mathrm{d}s_2.$$

It is shown in NEDELEC and PLANCHARD [1973] that

$$C_0\|v\|^2_{-1/2,\mathcal{S}} \leqslant \mathcal{V}_{(\mathcal{S})}(v, v) \leqslant C_1\|v\|^2_{-1/2,\mathcal{S}}. \tag{13.21}$$

We need the following lemma.

LEMMA 13.3. *Let $\sigma \in L^2(\Omega)$ and $\tilde{\sigma}$ denote the extension by zero of σ to $L^2(\mathcal{S})$. There exist positive constants C_2 and C_3 such that for $s \in [-1, 0]$*

$$C_2\|\sigma\|_s^2 \leqslant \|\tilde{\sigma}\|^2_{s,\mathcal{S}} \leqslant C_3\|\sigma\|_s^2.$$

PROOF. It is obvious for $s = 0$. Furthermore

$$\|\tilde{\sigma}\|_{-1,\mathcal{S}} = \sup_{\phi \in H^1(\mathcal{S})} \frac{(\tilde{\sigma}, \phi)_\mathcal{S}}{\|\phi\|_{1,\mathcal{S}}} \leqslant \sup_{\phi \in H^1(\mathcal{S})} \frac{(\sigma, \phi)}{\|\phi\|_1} \leqslant \|\sigma\|_{-1}.$$

The upper inequality follows by interpolation. To prove the lower inequality let $E : H^s(\Omega) \to H^s(\mathcal{S})$ for $s \in [0, 1]$ be an extension operator which is bounded, uniformly in s (the existence of such an extension is well known, cf. GRISVARD [1985]). Then

$$\|\sigma\|_{-s} = \sup_{\phi \in H^s(\Omega)} \frac{(\sigma, \phi)}{\|\phi\|_s} \leqslant C \sup_{\phi \in H^s(\Omega)} \frac{(\tilde{\sigma}, E\phi)_\mathcal{S}}{\|E\phi\|_{s,\mathcal{S}}} \leqslant C\|\tilde{\sigma}\|_{-s,\mathcal{S}}.$$

This completes the proof of the lemma. □

By Lemma 13.3 and (13.21), we have

$$\mathcal{V}(\sigma,\sigma)=\mathcal{V}_{(\mathcal{S})}(\tilde{\sigma},\tilde{\sigma})\approx\|\tilde{\sigma}\|_{-1/2,\mathcal{S}}^{2}\approx\|\sigma\|_{-1/2}^{2}.$$

Hence the smoother $\mathcal{R}_k$ defined by (13.8) satisfies Conditions (SM.1) and (SM.2).

We now show that Conditions (A.5) and (A.6) hold and hence Theorem 5.2 applies. Define the projection $\mathcal{P}_k : M \to M_k$ by

$$\mathcal{V}(\mathcal{P}_k v,\phi)=\mathcal{V}(v,\phi),\qquad \text{for all } \phi\in M_k.$$

LEMMA 13.4. *Conditions* (A.5) *and* (A.6) *hold, i.e., there exist positive constants* C_0 *and* C_1 *such that*

$$\mathcal{V}(w,w)\leqslant C_0\left(\mathcal{V}(\mathcal{P}_1 w,w)+\sum_{k=2}^{J}\lambda_k^{-1}\|\mathcal{V}_k\mathcal{P}_k w\|_{-1}^{2}\right), \tag{13.22}$$

and for some $0<\varepsilon<1$,

$$\lambda_k^{-1}\|\mathcal{V}_k w\|_{-1}^{2}\leqslant\left(C_1\varepsilon^{k-\ell}\right)^2\mathcal{V}(w,w),\quad \textit{for all } w\in M_\ell \quad \textit{with } \ell\leqslant k. \tag{13.23}$$

PROOF. We first show that Condition (A.6) ((13.23)) holds. The operator

$$(\mathcal{V}_{(\mathcal{S})}u)(s_2)=\int_{\mathcal{S}}\frac{u(s_1)}{|s_1-s_2|}\,\mathrm{d}s_1$$

is an isomorphism of $H^s(\mathcal{S})$ onto $H^{s+1}(\mathcal{S})$ for all real s. That is, for each s,

$$c\|w\|_{s,\mathcal{S}}^{2}\leqslant\|\mathcal{V}_{(\mathcal{S})}w\|_{s+1,\mathcal{S}}^{2}\leqslant C\|w\|_{s,\mathcal{S}}^{2},\qquad \text{for all } w\in H^s(\mathcal{S}). \tag{13.24}$$

Hence

$$\begin{aligned}\|\mathcal{V}_k w\|_{-1}^{2} &= (\mathcal{V}_k w,\mathcal{V}_k w)_{-1}=\mathcal{V}(w,\mathcal{V}_k w)=\mathcal{V}_{(\mathcal{S})}\big(\tilde{w},\widetilde{\mathcal{V}_k w}\big)\\ &= 2\big(\mathcal{V}_{(\mathcal{S})}\tilde{w},\widetilde{\mathcal{V}_k w}\big)_{\mathcal{S}}\\ &\leqslant \|\mathcal{V}_{(\mathcal{S})}\tilde{w}\|_{1,\mathcal{S}}\big\|\widetilde{\mathcal{V}_k w}\big\|_{-1,\mathcal{S}}\\ &\leqslant C\|w\|\|\mathcal{V}_k w\|_{-1}.\end{aligned} \tag{13.25}$$

In the last inequality we used (13.24) and Lemma 13.3. Hence

$$\|\mathcal{V}_k w\|_{-1}^{2}\leqslant C\|w\|^2.$$

By Theorem A.4 of Appendix A, interpolating (13.13) and the trivial result $\|w\| \leqslant \|w\|$ gives

$$\|w\|^2 \leqslant Ch_\ell^{-1}\|w\|_{-1/2}^2, \quad \text{for all } w \in M_\ell.$$

Hence

$$\lambda_k^{-1}\|\mathcal{V}_k w\|_{-1}^2 \leqslant C(\lambda_k h_\ell)^{-1}\|w\|_{-1/2}^2 \leqslant C(\lambda_k h_\ell)^{-1}\mathcal{V}(w,w).$$

The lower inequality in (13.18) states that $\lambda_k \geqslant ch_k^{-1}$, thus

$$\lambda_k^{-1}\|\mathcal{V}_k w\|_{-1}^2 \leqslant C(h_k/h_\ell)\mathcal{V}(w,w) = C\left(\tfrac{1}{2}\right)^{k-\ell}\mathcal{V}(w,w).$$

Thus Condition (A.6) is satisfied with $\varepsilon = 1/\sqrt{2}$.

We next show that Condition (A.5) ((13.22)) holds. Since Condition (A.5) is not sensitive to a uniform perturbation of the form, it is sufficient to show this for the equivalent form

$$\check{\mathcal{V}}(u,v) = (u,v)_{-1/2}.$$

For this purpose define $\check{\mathcal{V}}_k$ and $\check{\mathcal{P}}_k$ by

$$(\check{\mathcal{V}}_k v, \theta)_{-1} = (v,\theta)_{-1/2}, \quad \text{for all } v, \theta \in M_k$$

and

$$\left(\check{\mathcal{P}}_k w, \theta\right)_{-1/2} = (w,\theta)_{-1/2}, \quad \theta \in M_k.$$

It is proved in BRAMBLE, LEYK and PASCIAK [1994] that "full regularity" holds for this operator, i.e.

$$\left((I - \check{\mathcal{P}}_{k-1})v, v\right)_{-1/2} \leqslant C\lambda_k^{-1}\left\|\check{\mathcal{V}}_k v\right\|_{-1}^2, \quad \text{for all } v \in M_k. \tag{13.26}$$

Taking $v = \check{\mathcal{P}}_k w$ in (13.26) (with $\check{\mathcal{P}}_0 = 0$) and summing from 1 to J yields (13.22) for the equivalent operator $\check{\mathcal{V}}$. Consequently Condition (A.5) holds for $\check{\mathcal{V}}(\bullet,\bullet)$ and hence it also holds for $\mathcal{V}(\bullet,\bullet)$. □

Hence all of the conditions of Theorem 5.2 are satisfied and we conclude that the V-cycle multigrid method with one smoothing yields a uniform reduction rate. That is

$$\mathcal{V}\left((I - \mathcal{B}_J\mathcal{V}_J)\phi, \phi\right) \leqslant \left(1 - \frac{1}{C}\right)\mathcal{V}(\phi,\phi).$$

REMARK 13.1. We considered the case in which $M \subset H^1(\Omega)$ only for convenience. There is no fundamental reason why M need be required to consist of continuous functions. In fact, if M is made up of piecewise constant functions then the approximation problem in (13.3) is well defined. An example is given in BRAMBLE, LEYK and PASCIAK [1994] in which $M_1 \subset M_2 \subset \cdots \subset M_J \equiv M$ consist of piecewise constants. The operator A_k (and hence the discrete minus inner product $(\bullet, \bullet)_{-1,k}$) is a little more complicated to define and the proof of Lemma 13.1 a bit more technical. There is also no difficulty in the implementation in this case.

REMARK 13.2. The choice of the operator $\mathcal{R}_k$ is crucial. Since it and the operator $\mathcal{V}_k$ both involve the computationally inconvenient form $(\bullet, \bullet)_{-1}$, it turns out that this form never enters explicitly into the computation. Only the action of A_k is required. This involves only the stiffness matrix which is sparse. The implementation is discussed in detail in Section 15.

13.5. Bibliographical notes

Most of the work on multigrid methods for pseudodifferential operators has been concerned either with Fredholm operators of the second kind or hypersingular operators of the first kind. The second kind of operators are of the form of the identity plus a compact operator. At least in principal, such operators are already well conditioned so that it is not clear that much is to be gained by using a multigrid approach.

The hypersingular operators of the first kind are of positive order so that they fit into a framework similar to that of second-order elliptic differential boundary problems. Work on this subject is contained in VAN PETERSDORFF and STEPHAN [1990] . In contrast, weakly singular operators of potential theory give rise to pseudodifferential operators of negative order and require special consideration.

There has been a lot of work on boundary element methods. The work of NEDELEC and PLANCHARD [1973] contains many results concerning Fredholm first kind equations arising in potential theory. The multigrid method, including the formulation of the method and its analysis, is taken from BRAMBLE, LEYK and PASCIAK [1994].

14. Fourth-order problems

In Sections 9–13 the examples considered all involve nested spaces and inherited forms. In this section we describe a practical setting in which the multigrid theory of Section 7, for nonnested spaces and noninherited forms, may be profitably employed.

Consider, on the plane polygonal domain Ω, the biharmonic Dirichlet problem

$$\begin{cases} \Delta^2 u = f & \text{in } \Omega, \\ u = \frac{\partial u}{\partial n} = 0 & \text{on } \partial\Omega, \end{cases} \tag{14.1}$$

where $\partial/\partial n$ is the outward normal derivative operator on the boundary $\partial\Omega$ of Ω. For convenience we have assumed that the boundary conditions are homogeneous, but our arguments may be extended to include general Dirichlet boundary conditions. The problem (14.1) arises in models for clamped elastic plates and incompressible Stokes flow.

Throughout this section we shall denote by $H^{-s}(\Omega)$, $s>0$, the dual space of $H_0^s(\Omega)$. An important weak formulation of (14.1) is the following. Given $f \in H^{-2}(\Omega)$, find $u \in H_0^2(\Omega)$ such that

$$A(u,v)=(f,v), \quad \text{for all } v \in H_0^2(\Omega), \tag{14.2}$$

where

$$A(u,v)=\int_\Omega \Delta u \Delta v \,\mathrm{d}x=\sum_{i,j}\int_\Omega \frac{\partial^2 u}{\partial x_i \partial x_j}\frac{\partial^2 v}{\partial x_i \partial x_j}\,\mathrm{d}x$$

and

$$(f,v)=\int_\Omega f v \,\mathrm{d}x.$$

It can be shown that $A(\bullet,\bullet)^{1/2}$ induces a norm for $H_0^2(\Omega)$. Hence, it follows from the Riesz Representation Theorem (Theorem 2.1) that (14.2) has a unique solution for each $f \in H^{-2}(\Omega)$.

It is known that the solution u of (14.2) satisfies the following a priori estimate:

$$\|u\|_{2+\beta} \leqslant C\|f\|_{-2+\beta}, \quad \text{for some } \beta>0. \tag{14.3}$$

Note that the estimate also holds for $\beta' \in (0,\beta]$. If Ω is a convex polygon, then the solution u of (14.2) is known to belong to $H^3(\Omega)\cap H_0^2(\Omega)$ and to satisfy (14.3) with $\beta=1$, i.e.,

$$\|u\|_3 \leqslant C\|f\|_{-1}.$$

Full regularity, $u \in H^4(\Omega)$, which can be proved if Ω has a smooth boundary, is not generally obtained on polygonal domains. The regularity of solutions for (14.1) on polygonal domains, including re-entrant corners and cracks, is studied in DAUGE [1988].

By selecting a finite-dimensional (piecewise polynomial) test space M_h one may obtain a finite element method for (14.2) in the usual way. The condition $M_h \subset H_0^2$ leads to the requirement that C^1 finite elements be used. Since such elements are not very convenient to use in practice, other finite element approaches have been developed. Relaxing the condition $M_h \subset H_0^2(\Omega)$ leads to nonconforming finite elements. The Morley element, first described by MORLEY [1968], is the simplest among the nonconforming finite elements. Alternatively, based on different weak formulations of (14.1), certain mixed methods, such as the mixed method of CIARLET and RAVIART [1974], are developed.

In this section we shall consider the behavior of Algorithm 7.1 when it is applied to some conforming finite element equations, Morley nonconforming finite element equations and Ciarlet and Raviart mixed finite element equations. When the Morley nonconforming discretizations as well as many conforming finite element discretizations are constructed with respect to a sequence of nested triangulations, the bilinear forms are inherited but the spaces are nonnested. Alternatively, when certain mixed methods are employed, the spaces are nested, but the bilinear forms are noninherited.

Recall from Section 7 that if Condition (A.2″) is satisfied and if smoothers satisfying Condition (SM.1) are used, then the variable V-cycle multigrid preconditioner, defined by Algorithm 7.1, leads to methods with uniformly bounded condition numbers even if Condition (I.2) fails. The W-cycle multigrid method is a uniform contraction without Condition (I.2), provided that the number of smoothings, m, is sufficiently large. It is interesting to note that calculations performed by HANISCH [1991] showed that it is often necessary to take $m > 1$ when using the W-cycle multigrid methods to solve the Morley or Ciarlet–Raviart equations. The Condition (I.2) fails in each case and the W-cycle multigrid iterations for these methods with small m are observed to diverge. On the other hand, the variable V-cycle multigrid method with $m = 1$ consistently provide excellent preconditioners for these problems.

We shall verify Condition (A.2″) so that Theorem 7.3 and Theorem 7.4 are applicable. The verifications of Condition (A.2″) for conforming elements and the nonconforming Morley element are based on the regularity estimate in (14.3). For the Ciarlet–Raviart method, it is convenient (but not necessary) to assume that Ω is a convex polygon so that (14.3) holds with $\beta = 1$.

14.1. *Conforming finite elements*

We first consider conforming C^1 finite elements approximations to (14.2). Examples of the C^1 elements include the Argyris and Bell triangles, the Hsieh–Clough–Tocher element, the reduced Hsieh–Clough–Tocher element, the singular Zienkiewicz element, the reduced singular Zienkiewicz element, the Birkhoff–Mansfield triangle, the reduced Birkhoff–Mansfield triangle, the Powell–Sabin element, Bogner–Fox–Schmit bicubic element, the Fraeijs de Veubeke–Sander quadrilaterals, the reduced Fraeijs de Veubeke–Sander quadrilaterals, and many others. The definitions and approximation properties can be found in POWELL and SABIN [1977] for the Powell–Sabin element and in CIARLET [1978] for the other elements listed above.

We construct a nested sequence of triangulations $\{\mathcal{T}_k\}$ in the usual way. Let $\mathcal{T}_1$ be an initial triangulation of Ω. We construct $\mathcal{T}_k$ from $\mathcal{T}_{k-1}$ by the halving strategy. Denote by h_k the mesh parameter of $\mathcal{T}_k$. Consequently, $h_{k-1} = 2h_k$. Let $\{M_k\}$ be a family of finite element spaces defined by some conforming C^1 elements with respect to $\mathcal{T}_k$. The finite element solutions $u_k \in M_k$ satisfy

$$A(u_k, \phi) = (f, \phi), \qquad \text{for all } \phi \in M_k.$$

Usually a degree of freedom of $v \in M_k$ is one of the following:

$$v(x_i), \quad D^1 v(x_i) \bullet \xi, \quad D^2 v(x_i)(\xi, \eta).$$

Here ξ and η are some unit vectors representing the directions of the derivatives. The node x_i is usually a vertex of $\mathcal{T}_k$ or the midpoint of an edge. A degree of freedom can also be the average of v or its derivatives over an edge. We denote by ϕ_i^α the nodal basis functions of M_k at node x_i, where α indicates both the order and directions of the derivative of the corresponding degree of freedom. Let Λ_i be the index set for the degrees of freedom at x_i. Using the nodal basis we can represent $v \in M_k$ as

$$v = \sum_i \sum_{\alpha \in \Lambda_i} \partial_\alpha v(x_i) \phi_i^\alpha.$$

With a slight abuse of notation, we will use $|\alpha|$ to denote the order of the corresponding degrees of freedom. Let $\deg(M_k)$ be the maximum order of the derivatives in the degrees of freedom of M_k. For fourth order problems, $\deg(M_k) = 1$ or 2.

Denote by $\|\bullet\|_{s,p,D}$ and $\|\bullet\|_{s,D}$ the standard norm on Sobolev spaces $W^{s,p}(D)$ and $H^s(D) = W^{s,2}(D)$, and by $|\bullet|_{s,p,D}$ and $|\bullet|_{s,D}$ the semi-norms. It is easy to verify that finite elements satisfy the following standard assumptions:

(a) The following local inverse properties hold for $v \in M_k$:

$$|v|_{s,q,\tau} \leqslant C h_k^{2(1/q-1/p)} |v|_{s,p,\tau}, \quad 1 \leqslant p, q \leqslant \infty,\ 0 \leqslant s \leqslant 2,\ \tau \in \mathcal{T}_k. \tag{14.4}$$

(b) The nodal basis functions satisfy

$$|\phi_i^\alpha|_{s,q,\tau} \leqslant C h_k^{2/q+|\alpha|-s}, \quad 1 \leqslant q \leqslant \infty,\ 0 \leqslant s \leqslant 2,\ \tau \in \mathcal{T}_k. \tag{14.5}$$

We now define the operators for the multigrid algorithm. For $1 \leqslant k \leqslant J$, set

$$A_k(v, w) = A(v, w), \quad \text{for all } v \text{ and } w \in M_k.$$

Let $A_k : M_k \to M_k$ be defined as

$$(A_k v, \phi) = A(v, \phi), \quad \text{for all } \phi \in M_k.$$

The L_2 projections $Q_k : L_2 \to M_k$ are defined by

$$(Q_k v, \phi) = (v, \phi), \quad \text{for all } \phi \in M_k.$$

It can be shown (cf., e.g., ZHANG [1996]) that

$$\big|(I - Q_k)v\big|_s \leqslant C h_k^{2-s} |v|_2, \quad \text{for all } v \in H_0^2(\Omega),\ 0 \leqslant s \leqslant 2.$$

Intergrid transfer operators. A sequence of finite element spaces $\{M_k\}$ defined by the Bogner–Fox–Schmit or the Powell–Sabin elements on a nested sequence of triangulations $\{\mathcal{T}_k\}$ are nested, i.e., $M_k \subset M_{k+1}$. For these two elements, a theory which shows uniform convergence for the multigrid V-cycle can be established in a way similar to that of Subsection 9.1; cf. ZHANG [1994] for more details.

For all the other conforming finite elements listed at the beginning of Subsection 14.1, however, the corresponding sequence of finite element spaces are nonnested. We shall consider the behavior of Algorithm 7.1. To define the multigrid algorithm, we need intergrid transfer operators $I_k : M_{k-1} \to M_k$. Clearly $I_k v \in M_k$ can be written as

$$I_k v = \sum_i \sum_{\alpha \in \Lambda_i} v_i^\alpha \phi_i^\alpha = \sum_i \sum_{\alpha \in \Lambda_i} \partial_\alpha (I_k v)(x_i) \phi_i^\alpha .$$

The natural choice of I_k would be the nodal value interpolation operator. This corresponds to setting $\partial_\alpha (I_k v)(x_i) = \partial_\alpha v(x_i)$. However the nodal value interpolation operator is not well defined on M_{k-1}, e.g., for the Argyris, the Bell, and the Birkhoff–Mansfield elements. Many minor modifications are possible. Two possible choices for the intergrid transfer operator are defined by setting

$$\partial_\alpha (I_k v)(x_i) = \begin{cases} \partial_\alpha v(x_i) & \text{if } |\alpha| \leqslant 1,\ \alpha \in \Lambda_i, \\ \text{the average value of } \partial_\alpha v \text{ at } x_i & \text{if } |\alpha| = 2,\ \alpha \in \Lambda_i, \end{cases} \tag{14.6}$$

and

$$\partial_\alpha (I_k v)(x_i) = \begin{cases} \partial_\alpha v(x_i) & \text{if } |\alpha| \leqslant 1,\ \alpha \in \Lambda_k, \\ \sum_{j=1}^3 \lambda_j(x_i) \partial_\alpha v(a_j) & \text{if } |\alpha| = 2,\ \alpha \in \Lambda_k, \end{cases} \tag{14.7}$$

where $\triangle(a_1 a_2 a_3) \in \mathcal{T}_{k-1}$ is a triangle with vertices a_1, a_2 and a_3 containing x_i and $\lambda_j(x)$ is the jth barycentric coordinate of x. Alternatively we can also set

$$\partial_\alpha (I_k v)(x_i) = \begin{cases} \partial_\alpha v(x_i) & \text{if } |\alpha| \leqslant 1,\ \alpha \in \Lambda_i, \\ 0 & \text{if } |\alpha| = 2,\ \alpha \in \Lambda_i. \end{cases} \tag{14.8}$$

If $\deg(M_k) = 1$, e.g., if M_k is defined by the (reduced) Hsieh–Clough–Tocher or the (reduced) singular Zienkiewicz element, then the intergrid operators defined by Eqs. (14.6), (14.7) and (14.8) reduce to the standard nodal value interpolation operator. For the Argyris, Bell, Bogner–Fox–Schmit or the Birkhoff–Mansfield elements, $\deg(M_k) = 2$, and the intergrid transfer operators defined by Eqs. (14.6), (14.8) and (14.7) are different.

Note that for I_k defined by Eqs. (14.6), (14.7) and (14.8), we need to evaluate $\partial_\alpha v(m_i)$ at midpoints m_i of the edges of $\mathcal{T}_{k-1}$, which means that we have to evaluate $\partial_\alpha \Phi(m_i)$ for the basis functions Φ of M_{k-1}. To avoid that, we can use the simplified "$\mathbf{P}_1$ preserving interpolation operator" determined by

$$\partial_\alpha (I_k v)(x_i) = \sum_{j=1}^3 \lambda_j(x_i) \partial_\alpha v(a_j), \quad \alpha \in \Lambda_i, \tag{14.9}$$

where $x_i \in \triangle(a_1 a_2 a_3) \in \mathcal{T}_{k-1}$.

Note that Eqs. (14.8) and (14.9) are well defined on $C^1(\overline{\Omega})$ and Eqs. (14.6) and (14.7) are well defined only on a subset of $C^1(\overline{\Omega})$. If v is also in $H_0^2(\Omega)$, then $I_k v \in M_k \subset H_0^2(\Omega)$, and hence it automatically satisfies the homogeneous boundary conditions.

Note that I_k defined by (14.6) or (14.7) preserves quadratic functions, i.e.,

$$I_k p|_\tau = p, \quad \text{for all } p \in \mathbf{P}_2,\ \tau \in \mathcal{T}_k.$$

In general I_k defined by (14.8) or (14.9) only preserves linear functions, i.e.,

$$I_k p|_\tau = p, \quad \text{for all } p \in \mathbf{P}_1,\ \tau \in \mathcal{T}_k.$$

However, if $\deg(M_k) = 1$, then I_k defined by (14.8) is the standard nodal value interpolation and thus also preserves quadratic functions.

It is known that I_k is not bounded with respect to the $\|\bullet\|_2$ norm, however, restricted to the finite element space M_{k-1} (and thus a local inverse property holds), I_k has the following stable approximation properties.

LEMMA 14.1. *Let* $\tau \in \mathcal{T}_k$ *and* $\hat{\tau} = \bigcup_{\substack{\tau' \in \mathcal{T}_{k-1} \\ \tau' \cap \tau \neq \emptyset}} \tau'$. *There exists* $C > 0$ *independent of* τ *such that* I_k *defined by* (14.6), (14.7), (14.8), *or* (14.9) *satisfies*

$$|I_k v - v|_{s,\tau} \leqslant C h_k^{2-s} |v|_{2,\hat{\tau}}, \quad \text{for all } v \in M_{k-1},\ 0 \leqslant s \leqslant 2.$$

Consequently,

$$|I_k v - v|_{s,\Omega} \leqslant C h_k^{2-s} |v|_{2,\Omega}, \quad \text{for all } v \in M_{k-1},\ 0 \leqslant s \leqslant 2. \tag{14.10}$$

PROOF. Note that $|(\partial_\alpha I_k v)(x_i)| \leqslant |v|_{|\alpha|,\infty,\tau}$ for $|\alpha| \leqslant 1$ and $|(\partial_\alpha I_k v)(x_i)| \leqslant |v|_{|\alpha|,\infty,\hat{\tau}}$ for $|\alpha| = 2$. By the definitions of I_k and (14.5), we have

$$|I_k v|_{s,\tau} \leqslant \sum_i \sum_{|\alpha| \leqslant 2} |v|_{|\alpha|,\infty,\hat{\tau}} |\phi_i^\alpha|_{s,\tau} \leqslant \sum_{r=0}^{2} C h_k^{1+r-s} |v|_{r,\infty,\hat{\tau}}.$$

Since $\hat{\tau}$ is a union of $\tau' \in \mathcal{T}_{k-1}$, the inverse inequality in (14.4) holds for $v \in M_{k-1}$ on $\hat{\tau}$,

$$|v|_{r,\infty,\hat{\tau}} \leqslant C h_{k-1}^{-1} |v|_{r,\hat{\tau}} \leqslant C h_k^{-1} |v|_{r,\hat{\tau}}.$$

Hence

$$|I_k v|_{s,\tau} \leqslant \sum_{r=0}^{2} C h_k^{r-s} |v|_{r,\hat{\tau}}.$$

By the triangle inequality

$$\left|(I - I_k)v\right|_{s,\tau} \leqslant \sum_{r=0}^{2} C h_k^{r-s} |v|_{r,\hat{\tau}}.$$

Now using $I_k p|_\tau = p$ for $p \in \mathbf{P}_1$ and the Poincaré inequality, we obtain

$$\left|(I - I_k)v\right|_{s,\tau} \leqslant \inf_{p\in\mathbf{P}_1} \sum_{r=0}^{2} Ch_k^{r-s}|v+p|_{r,\hat\tau} \leqslant \sum_{r=0}^{2} Ch_k^{r-s}h_{k-1}^{2-r}|v|_{2,\hat\tau} \leqslant Ch_k^{2-s}|v|_{2,\hat\tau}.$$

Squaring and summing the above inequality over $\tau \in \mathcal{T}_k$, we obtain

$$\left|(I - I_k)v\right|_{s,\Omega} \leqslant Ch_k^{2-s}|v|_{2,\Omega}, \quad s = 0, 2.$$

The result for $0 \leqslant s \leqslant 2$ follows from the convexity of the norms. □

REMARK 14.1. In the case in which I_k preserves quadratic functions, it can be shown by applying the Bramble–Hilbert lemma in the fractional-order spaces that

$$\left\|(I - I_k)v\right\|_{s,\Omega} \leqslant Ch_k^{2+r-s}|v|_{2+r,\Omega}, \quad \text{for all } v \in M_{k-1},\ 0 \leqslant s \leqslant 2,\ 0 \leqslant r < 1/2.$$

This holds in particular if I_k is defined by either (14.6) or (14.7).

Multigrid convergence analysis. We now verify Condition (A.2″). For this, it is convenient to use the following discrete norms defined by

$$|||v|||_{s,k}^2 = \left(A_k^{s/2}v, v\right), \quad v \in M_k, \text{ for all } s \in \mathbf{R}.$$

It is trivial to see $|||v|||_{s,k} \leqslant |||v|||_{s+\gamma,k}^{1/2}|||v|||_{s-\gamma,k}^{1/2}$, for all $s, \gamma \in \mathbf{R}$. By expanding in the eigenfunctions of A_k and using the Hölder inequality, it follows easily that

$$|||v|||_{s,k} \leqslant |||v|||_{s_1,k}^{\lambda}|||v|||_{s_2,k}^{1-\lambda}, \quad s = \lambda s_1 + (1-\lambda)s_2,\ 0 \leqslant \lambda \leqslant 1,$$

which is the convexity property of the discrete norms. In particular,

$$|||v|||_{2+r,k} \leqslant |||v|||_{2,k}^{1-r/2}|||v|||_{4,k}^{r/2}, \quad 0 \leqslant r \leqslant 2. \tag{14.11}$$

The norm equivalence

$$|||v|||_{s,k} \approx |v|_s, \quad 0 \leqslant s \leqslant 2,$$

is easy to see as follows. The cases $s = 0$ and $s = 2$ are obvious from the definition of the discrete norms. The result for $0 \leqslant s \leqslant 2$ follows by interpolating the identity operator I and the L^2 projector Q_k (cf. Theorem A.3). For polynomial or piecewise polynomial elements, the result can be extended to the case $2 < s < 5/2$ based on the same reasoning as in BRAMBLE, PASCIAK and XU [1991], where it is shown that $M_k \subset H^{1+s}$ and $|||v|||_{s,k} \asymp |v|_s$ with $0 \leqslant s < 3/2$ for C^0 polynomial elements M_k. We do not know however whether or not this equivalence still holds for singular elements.

Since $A_k(\bullet,\bullet) \equiv A(\bullet,\bullet)$, $P_{k-1}: M_k \to M_{k-1}$ defined in Section 7 satisfies

$$A(P_{k-1}v, \phi) = A(v, I_k\phi), \quad \text{for all } v \in M_k \text{ and } \phi \in M_{k-1}.$$

We assume, without loss of generality, that $0 < \beta \leqslant 1$. We have the following estimate for P_{k-1}.

LEMMA 14.2. *There exists a constant $C > 0$ such that*

$$\left|(I - P_{k-1})v\right|_2 \leqslant Ch_k^\beta |||v|||_{2+\beta,k}, \quad \textit{for all } v \in M_k.$$

PROOF. Let $\bar{P}_{k-1}: H_0^2(\Omega) \to M_{k-1}$ be the Galerkin projection. By the standard finite element error estimate and a duality argument, we have

$$\left|(I - \bar{P}_{k-1})v\right|_{2-\beta} \leqslant Ch_k^\beta |v|_2, \quad \text{for all } v \in H_0^2(\Omega). \tag{14.12}$$

Now for $v \in M_k$,

$$\begin{aligned}
\left|(I - \bar{P}_{k-1})v\right|_2^2 &= A\big(v, (I - \bar{P}_{k-1})v\big) = A\big(v, \bar{P}_k(I - \bar{P}_{k-1})v\big) \\
&\leqslant |||v|||_{2+\beta,k} \big|\big|\big|\bar{P}_k(I - \bar{P}_{k-1})v\big|\big|\big|_{2-\beta,k} \\
&\leqslant C|||v|||_{2+\beta,k} \left|\bar{P}_k(I - \bar{P}_{k-1})v\right|_{2-\beta}.
\end{aligned}$$

Note that $\bar{P}_{k-1}(I - \bar{P}_{k-1}) \equiv 0$. By the triangle inequality and (14.12), we have

$$\begin{aligned}
\left|\bar{P}_k(I - \bar{P}_{k-1})v\right|_{2-\beta} &= \left|(\bar{P}_k - \bar{P}_{k-1})(I - \bar{P}_{k-1})v\right|_{2-\beta} \\
&\leqslant Ch_{k-1}^\beta \left|(I - \bar{P}_{k-1})v\right|_2 \\
&\leqslant Ch_k^\beta \left|(I - \bar{P}_{k-1})v\right|_2.
\end{aligned}$$

Therefore,

$$\left|(I - \bar{P}_{k-1})v\right|_2 \leqslant Ch_k^\beta |||v|||_{2+\beta,k}, \quad \text{for all } v \in M_k. \tag{14.13}$$

By the definition of P_{k-1} and $\bar{P}_{k-1}$, we have for any $\phi \in M_{k-1}$,

$$\begin{aligned}
A\big((P_{k-1} - \bar{P}_{k-1})v, \phi\big) &= A\big(v, (I_k - I)\phi\big) = A\big(v, (I_k - \bar{P}_k)\phi\big) \\
&\leqslant |||v|||_{2+\beta,k} \big|\big|\big|(I_k - \bar{P}_k)\phi\big|\big|\big|_{2-\beta,k} \\
&\leqslant C|||v|||_{2+\beta,k} \left|(I_k - \bar{P}_k)\phi\right|_{2-\beta}.
\end{aligned}$$

Using the triangle inequality, Lemma 14.1 and (14.12) we obtain

$$\left|(I_k - \bar{P}_k)\phi\right|_{2-\beta} \leqslant \left|(I_k - I)\phi\right|_{2-\beta} + \left|(I - \bar{P}_k)\phi\right|_{2-\beta} \leqslant Ch_k^\beta |\phi|_2.$$

Therefore,

$$A\big((P_{k-1}-\bar{P}_{k-1})v,\phi\big)\leqslant Ch_k^{\beta}|||v|||_{2+\beta,k}|\phi|_2.$$

Setting $\phi=(P_{k-1}-\bar{P}_{k-1})v$, we obtain

$$\big|(P_{k-1}-\bar{P}_{k-1})v\big|_2\leqslant Ch_k^{\beta}|||v|||_{2+\beta,k}. \tag{14.14}$$

The lemma follows trivially from Eqs. (14.13) and (14.14). □

THEOREM 14.1. *Assume that* $0<\beta\leqslant 1$. *Then Condition* (A.2″) *holds with* $\alpha=\beta/4$, *i.e.,*

$$A\big((I-I_kP_{k-1})v,v\big)\leqslant C(A_kv,v)^{1-\alpha}\left(\frac{\|A_kv\|_0^2}{\lambda_k}\right)^{\alpha}.$$

PROOF. Let $\alpha=\beta/4$. Then by Lemma 14.2 and (14.11),

$$\begin{aligned}\big|A\big((I-I_kP_{k-1})v,v\big)\big| &= \big|A\big((I-P_{k-1})v,(I+P_{k-1})v\big)\big|\\ &\leqslant \big|(I-P_{k-1})v\big|_2\big|(I+P_{k-1})v\big|_2\\ &\leqslant Ch_k^{\beta}|||v|||_{2+\beta,k}|v|_2\\ &\leqslant Ch_k^{\beta}|||v|||_{2,k}^{2-\beta/2}|||v|||_{4,k}^{\beta/2}\\ &\leqslant C\left(\frac{1}{\lambda_k^{\alpha}}\right)(A_kv,v)^{1-\alpha}\big(\|A_kv\|_0^2\big)^{\alpha}.\end{aligned}$$

This proves the theorem. □

We have thus proved that Condition (A.2″) is satisfied and hence, from Theorem 7.4, the variable V-cycle multigrid preconditioner leads to a problem with a uniformly bounded condition number, and, from Theorem 7.3, the W-cycle multigrid has a uniform reduction in the error provided that the number of smoothings is sufficiently large.

REMARK 14.2. In the cases in which $M_k\subset H^{2+\beta}(\Omega)\cap H_0^2(\Omega)$ and $|||v|||_{2+\beta,k}\approx |v|_{2+\beta}$, the proof of Theorem 14.1 can be simplified slightly. If the spaces $\{M_k\}$ are nested and $I_k=I$, then $P_{k-1}=\bar{P}_{k-1}$ and Theorem 14.1, with $\alpha=\beta/2$, is a direct consequence of (14.13).

REMARK 14.3. Note that if the a priori estimate in (14.3) holds with $\beta>0$ it also holds with $\beta'\in(0,\beta]$. We have assumed (without loss of generality) that $0<\beta\leqslant 1$ in Lemma 14.2 and Theorem 14.1. This assumption is not necessary for the elements that can reproduce cubic polynomials. Without this assumption, however, the estimate in (14.12) is not valid for the reduced Hsieh–Clough–Tocher triangle, the singular Zienkiewicz triangle and the reduced singular Zienkiewicz triangle if $\beta>1$ (since these elements do not reproduce cubic polynomials) and Condition (A.2″) holds for $\alpha = \min(1, \beta)/4$.

14.2. The nonconforming Morley element

Let $\mathcal{T}_h$ denote a triangulation, with mesh diameter h, of a polygon $\Omega \subset \mathbf{R}^2$. The Morley finite element spaces are defined so that $v \in M_h$ if and only if

(a) for each triangle $\tau \in \mathcal{T}_h$, $v|_\tau$ is a quadratic polynomial,

(b) v is continuous at triangle vertices and vanishes at boundary vertices, and

(c) the normal derivative $\partial v/\partial n$ is continuous at the midpoints of each $\tau \in \mathcal{T}_h$ and vanishes at midpoints on $\partial\Omega$.

It is clear that $M_h \not\subset C^0(\Omega)$. Furthermore, it can be seen easily that if $\mathcal{T}_{h/2}$ is defined from $\mathcal{T}_h$ by the usual halving strategy, then $M_h \not\subset M_{h/2}$.

Let $\{\mathcal{T}_k\}$ be a nested sequence of triangulations of Ω and denote by h_k the mesh parameter of $\mathcal{T}_k$. As usual, we assume that $\mathcal{T}_k$ is obtained from $\mathcal{T}_{k-1}$ by the halving strategy. Consequently, $h_k = h_{k-1}/2$ and the Morley spaces $M_k \equiv M_{h_k}$ satisfy familiar inverse properties. We note that Morley spaces $\{M_k\}$ defined on a sequence of nested triangulations are themselves nonnested.

Consider the bilinear forms on $M_k \times M_k$ defined by

$$A_k(u, v) = \sum_{\tau \in \mathcal{T}_k} \sum_{i,j} \int_\tau \frac{\partial^2 u}{\partial x_i \partial x_j} \frac{\partial^2 v}{\partial x_i \partial x_j} \, \mathrm{d}x.$$

The Morley approximation corresponding to (14.2) is the solution of the following problem. Find $u_k \in M_k$ such that for $f \in L^2(\Omega)$

$$A_k(u_k, \phi) = (f, \phi), \quad \text{for all } \phi \in M_k. \tag{14.15}$$

It can be shown that $A_k(\bullet, \bullet)$ is an inner product for the space M_k and consequently there is a unique solution to (14.15).

Denote the induced norm by

$$\|v\|_{2,k} \equiv A_k(v, v)^{1/2}.$$

Note that both the form $A_k(\bullet, \bullet)$ and the norm $\|\bullet\|_{2,k}$ are well defined on the composite space $H_0^2(\Omega) + M_k + M_{k-1} + \cdots + M_1$. Furthermore, for the nested meshes, $A_k(v, w) = A_J(v, w)$ for all v and w in the coarser space M_k and hence the forms $A_k(\bullet, \bullet)$ are "inherited".

As in the case of conforming C^1 elements, we shall assume that (14.3) holds for $0 < \beta \leqslant 1$.

Approximation and error estimates. Let $Q_k : L^2(\Omega) \to M_k$ be the L^2 orthogonal projection onto M_k. Using the nodal value interpolation it can be shown that the Morley spaces, M_k, have optimal approximation properties, i.e.,

$$\|v - Q_k v\|_0 \leqslant C h_k^{2+r} \|v\|_{2+r}, \quad \text{for all } v \in H^{2+r}(\Omega),\ 0 \leqslant r \leqslant 1.$$

By standard arguments it follows that

$$\|Q_k v\|_{2,k} \leqslant C\|v\|_2, \quad \text{for all } v \in H_0^2(\Omega).$$

To see this, let $\pi_k' v$ denote the continuous piecewise $\mathcal{T}_k$-linear interpolation of $v \in H_0^2(\Omega)$. Then using the inverse property of quadratic functions, we obtain

$$\|Q_k v\|_{2,k} \equiv \left\|Q_k v - \pi_k' v\right\|_{2,k} \leqslant C h_k^{-2} \left\|Q_k v - \pi_k' v\right\|_0.$$

The result now follows from the standard estimate $\|v - \pi_k' v\|_0 \leqslant C h_k \|v\|_2$ and the approximation property of $Q_k v$.

Some requirement on f such as $f \in L^2(\Omega)$ is necessary for (f, v) in (14.15) to make sense since the elements of the space M_k are discontinuous. To allow $f \in H^{-1}(\Omega)$, ARNOLD and BREZZI [1985] introduced the following modification to the Morley method: Find $u_k' \in M_k$ such that

$$A_k\left(u_k', v\right) = \left(f, \pi_k' v\right), \quad \text{for all } v \in M_k. \tag{14.16}$$

Here $\pi_k' v$ denotes the continuous piecewise $\mathcal{T}_k$-linear interpolant of $v \in M_k$. Since $\pi_k' v \in H_0^1(\Omega)$, (f, ϕ) in (14.16) makes sense for $f \in H^{-1}(\Omega)$. It is shown in ARNOLD and BREZZI [1985] that the solution u_k' of (14.16) satisfies the error estimate

$$\left\|u - u_k'\right\|_{2,k} \leqslant C h_k \|u\|_3. \tag{14.17}$$

Since $A_k(u_k - u_k', v) = (f, (I - \pi_k')v)$ and $\|v - \pi_k' v\|_0 \leqslant C h_k^2 \|v\|_{2,k}$ for all $v \in M_k$, it follows that the solution u_k of (14.15) satisfies

$$\left\|u_k - u_k'\right\|_{2,k} \leqslant C h_k^2 \|f\|_0,$$

and hence

$$\|u - u_k\|_{2,k} \leqslant C h_k \left(\|u\|_3 + h_k \|f\|_0\right). \tag{14.18}$$

The proof in ARNOLD and BREZZI [1985] is based on a relation between the modified Morley method defined by (14.16) and the mixed method of Hellan, Herrmann and Johnson. Recently SHI [1990] gave a new proof of Eqs. (14.17) and (14.18) based on the standard techniques for nonconforming finite element methods.

To allow $f \in H^{-2}(\Omega)$, we consider the following modification to (14.15):

$$A_k\left(u_k'', v\right) = \left(f, \pi_k'' v\right), \quad \text{for all } v \in M_k \tag{14.19}$$

where $\pi_k'' : M_k \to H_0^2(\Omega)$ is an operator satisfying the following stable approximation property:

$$\left\|v_k - \pi_k'' v_k\right\|_0 + h_k^2 \left\|\pi_k'' v_k\right\|_2 \leqslant C h_k^2 \|v_k\|_{2,k}. \tag{14.20}$$

An operator π_k'' satisfying this property can be constructed by relating M_k to a conforming element (e.g., the Argyris element) on the same grid; cf. BRENNER [1996]. Clearly $(f, \pi_k'' v)$ makes sense for $f \in H^{-2}(\Omega)$. Observe that the proof in SHI [1990] remains valid if we replace π_k' by π_k'', i.e., if $u \in H^3(\Omega)$ then the solution u_k'' of (14.19) satisfies

$$\|u - u_k''\|_{2,k} \leqslant Ch_k \|u\|_3. \tag{14.21}$$

See BRENNER [1995] for details of the proof. Now the stability of π_k'' implies that

$$\|u - u_k''\|_{2,k} \leqslant C\|u\|_2.$$

By Theorem A.4, interpolating these two inequalities gives ($0 < \beta \leqslant 1$)

$$\|u - u_k''\|_{2,k} \leqslant Ch_k^{\beta} \|u\|_{2+\beta}.$$

Since $A_k(u_k - u_k'', v) = (f, (I - \pi_k'')v)$, the approximation property of π_k'' implies that

$$\|u_k - u_k''\|_{2,k} \leqslant Ch_k^2 \|f\|_0.$$

As a consequence, we have the following error estimate for the Morley approximation:

LEMMA 14.3 (error estimate). *Let u_k be the Morley approximation to u in M_k. Then*

$$\|u - u_k\|_{2,k} \leqslant Ch_k^{\beta} \|u\|_{2+\beta} + Ch_k^2 \|f\|_0. \tag{14.22}$$

The operators $A_k : M_k \to M_k$ are defined by

$$(A_k u, v) = A_k(u, v), \qquad \text{for all } v \in M_k.$$

It can be shown that $\kappa(A_k) = \mathrm{O}(h_k^{-4})$ as $h_k \to 0$. Thus the operators A_k are highly ill-conditioned. The bound for the maximum eigenvalue

$$\lambda_k \leqslant Ch_k^{-4}$$

is obtained using an inverse property for piecewise quadratic polynomials. The constant C depends only on the minimum angle of the mesh family.

The intergrid transfer operator. To compute the Morley approximation u_J on the finest mesh, we consider the sequence of Morley discretizations associated with $\{\mathcal{T}_k\}$ and apply the multigrid theory. The resulting bilinear forms, $A_k(\cdot, \cdot)$, are inherited, but the spaces M_k are nonnested. Assume that the smoothers of Section 8 are used. It remains to define an intergrid transfer operator $I_k : M_{k-1} \to M_k$. We follow BRENNER [1989].

DEFINITION *of I_k*. For $v \in M_{k-1}$, $I_k v \in M_k$ is defined so that

(a) if p is a vertex of $\mathcal{T}_{k-1}$, $(I_k v)(p) = v(p)$,

(b) for other vertices p of $\mathcal{T}_k$, v may have a jump at p and $I_k v$ takes the average value of v at p,

(c) if m is a midpoint of an edge of $\mathcal{T}_k$ which is in the interior of $\tau \in \mathcal{T}_{k-1}$, $\partial(I_k v)/\partial n(m) = \partial v/\partial n(m)$, and

(d) for other edge midpoints m associated with $\mathcal{T}_k$, $\partial v/\partial n$ may have a jump and $\partial(I_k v)/\partial n$ takes the average value of $\partial v/\partial n$ at m.

We may now consider Algorithm 7.1. To apply Theorem 7.3 or Theorem 7.4 we need to show that Condition (A.2″) is satisfied for this choice of I_k. We use the following notation:

$$\|v\|_{s,k,D}^2 \equiv \sum_{\substack{\tau \in \mathcal{T}_k \\ \tau \subset D}} |v|_{s,\tau}^2 \quad \text{and} \quad \|v\|_{s,k} \equiv \|v\|_{s,k,\Omega}.$$

Some relevant properties of I_k are summarized in the following lemma:

LEMMA 14.4. *There exists a constant $C > 0$ independent of k such that*

$$\big\|(I - I_k)v\big\|_{s,k} \leqslant C h_k^{2-s} \|v\|_{2,k-1}, \quad s = 0, 2, \tag{14.23}$$

and

$$\|I_k v\|_{s,k} \leqslant C \|v\|_{s,k-1}, \quad s = 0, 2. \tag{14.24}$$

PROOF. Let $\tau \in \mathcal{T}_k$. Then $I_k v \in M_k$ can be written as

$$I_k v = \sum_i (I_k v)(x_i)\phi_i + \sum_i \frac{\partial(I_k v)}{\partial n}(m_i)\psi_i,$$

where x_i and m_i denote the vertices and midpoints of the edges of τ and ϕ_i and ψ_i denote the basis functions of M_k associated with x_i and m_i respectively. By the triangle inequality

$$\|I_k v\|_{s,\tau} \leqslant \sum_i \big|(I_k v)(x_i)\big| \|\phi_i\|_{s,\tau} + \sum_i \left|\frac{\partial(I_k v)}{\partial n}(m_i)\right| \|\psi_i\|_{s,\tau}.$$

It is easy to check (cf. (14.5)) that

$$\|\phi_i\|_{s,\tau} \leqslant C h_k^{1-s} \quad \text{and} \quad \|\psi_i\|_{s,\tau} \leqslant C h_k^{2-s}.$$

Let $\hat{\tau} = \bigcup_{\substack{\tau' \in \mathcal{T}_{k-1} \\ \tau' \cap \tau \neq \emptyset}} \tau'$. By the construction of I_k and the inverse property of M_{k-1},

$$\big|(I_k v)(x_i)\big| \leqslant \|v\|_{L^\infty(\hat{\tau})} \leqslant C h_k^{-1} \|v\|_{0,\hat{\tau}}$$

and

$$\left|\frac{\partial(I_k v)}{\partial n}(m_i)\right| \leqslant \|v\|_{W^{1,\infty}(\hat{\tau})} \leqslant Ch_k^{-1}\|v\|_{1,k-1,\hat{\tau}}.$$

Combining these estimates we have,

$$\|I_k v\|_{s,\tau} \leqslant C\big(h_k^{-s}\|v\|_{0,\hat{\tau}} + h_k^{1-s}\|v\|_{1,k-1,\hat{\tau}}\big) = C\left(\sum_{r=0}^{1} h^{r-s}\|v\|_{r,k-1,\hat{\tau}}\right).$$

By the triangle inequality,

$$\|v - I_k v\|_{s,\tau} \leqslant \sum_{r=0}^{2} Ch_k^{r-s}\|v\|_{r,k-1,\hat{\tau}}.$$

Since $I_k p|_\tau = p$ for $p \in \mathbf{P}_2$, we obtain

$$\|v - I_k v\|_{s,\tau} \leqslant C \inf_{p\in\mathbf{P}_2}\left(\sum_{r=0}^{2} h_k^{r-s}\|v+p\|_{r,k-1,\hat{\tau}}\right).$$

It can be shown (cf. BRAMBLE, PASCIAK and ZHANG [1996, Lemma 2.1]) that if $p \in \mathbf{P}_1 \subset \mathbf{P}_2$ is chosen so that

$$\sum_{\tau'\in\mathcal{T}_k:\tau'\cap\hat{\tau}\neq\emptyset} \int_{\tau'} \partial_\alpha(v+p)\,\mathrm{d}x = 0, \qquad \text{for all } |\alpha| \leqslant 1,$$

then

$$\|v+p\|_{r,k-1,\hat{\tau}} \leqslant Ch_{k-1}^{2-r}\|v\|_{2,k-1,\hat{\tau}}.$$

With this choice of p, we have, for $s = 0, 2$,

$$\|v - I_k v\|_{s,\tau} \leqslant C\left(\sum_{r=0}^{2} h_k^{r-s}\|v+p\|_{r,k-1,\hat{\tau}}\right) \leqslant Ch_k^{2-s}\|v\|_{2,k-1,\hat{\tau}}. \tag{14.25}$$

Inequality (14.23) follows by squaring (14.25) and summing over $\tau \in \mathcal{T}_k$. Inequality (14.24) follows from (14.23) and the inverse property of M_{k-1}. □

Multigrid convergence analysis. We now show, using Lemma 14.4, that Condition (A.2″) holds with $\alpha = \beta/4$ for the Morley elements.

As for the case of conforming elements, we introduce a scale of norms on M_k

$$|||v|||_{s,k}^2 = \big(A_k^{s/2}v, v\big).$$

Note that $|||v|||_{0,k} = \|v\|_0$, $|||v|||_{2,k} = \|v\|_{2,k}$ and $|||v|||_{4,k} = \|A_k v\|_0$. By the convexity of the discrete norms,

$$|||v|||_{2+r,k} \leqslant |||v|||_{2,k}^{1-r/2} |||v|||_{4,h_k}^{r/2}, \quad 0 \leqslant r \leqslant 2.$$

To verify Condition (A.2″), we introduce an auxiliary problem. Let $f_k = A_k u_k$ and denote by u be the solution of the continuous problem (14.2) when $f = f_k$, i.e., $u \in H_0^2(\Omega) \cap H^{2+\beta}(\Omega)$ satisfies

$$A(u, \phi) = (f_k, \phi) \equiv (A_k u_k, \phi), \quad \text{for all } \phi \in H_0^2(\Omega).$$

Then u_k is the finite element approximation to u in M_k. Further denote by u_{k-1} the Morley approximation to u in M_{k-1}.

The following result relates $\| f_k \|_{-s}$ to the discrete norm $||| f_k |||_{-s,k}$.

LEMMA 14.5. *There exists a constant $C > 0$, independent of k, such that*

$$\| f_k \|_{-s} \leqslant C ||| f_k |||_{-s,k}, \quad \textit{for all } f_k \in M_k, \ 0 \leqslant s \leqslant 2.$$

PROOF. Since $f_k \in M_k$, by the Cauchy–Schwarz inequality,

$$(f_k, v) = (f_k, Q_k v) \leqslant ||| f_k |||_{-s,k} ||| Q_k v |||_{s,k}.$$

By Theorem A.4, interpolating

$$||| Q_k v |||_{2,k} \leqslant C \|v\|_2 \quad \text{and} \quad ||| Q_k v |||_{0,k} = \| Q_k v \|_0 \leqslant \|v\|_0,$$

we obtain

$$||| Q_k v |||_{s,k} \leqslant C \|v\|_s, \quad \text{for all } v \in H_0^s(\Omega), \ 0 \leqslant s \leqslant 2.$$

Hence

$$(f_k, v) \leqslant C ||| f_k |||_{-s,k} \|v\|_s, \quad \text{for all } v \in H_0^s(\Omega), \ 0 \leqslant s \leqslant 2.$$

Consequently,

$$\| f_k \|_{-s} = \sup_{v \in H_0^s(\Omega)} \frac{(f_k, v)}{\|v\|_s} \leqslant C ||| f_k |||_{-s,k}, \quad 0 \leqslant s \leqslant 2,$$

and the lemma is proved. □

We now show that Condition (A.2″) holds with $\alpha = \beta/4$. First, since

$$\big| A_k\big((I - I_k P_{k-1}) u_k, u_k\big) \big| \leqslant \big|\big|\big| (I - I_k P_{k-1}) u_k \big|\big|\big|_{2,k} ||| u_k |||_{2,k}, \quad \text{for all } u_k \in M_k$$

and since $h_k^4 \leqslant \lambda_k^{-1}$, it suffices to show that

$$\left|\left|\left|(I - I_k P_{k-1})u_k\right|\right|\right|_{2,k} \leqslant Ch_k^{\beta} |||u_k|||_{4,k}^{\beta/2} |||u_k|||_{2,k}^{2-\beta/2}, \quad \text{for all } u_k \in M_k.$$

LEMMA 14.6. *There exists a constant $C > 0$, independent of k, such that*

$$\left|\left|\left|(I - I_k P_{k-1})u_k\right|\right|\right|_{2,k} \leqslant Ch_k^{\beta} |||u_k|||_{2+\beta,k} + Ch_k^2 |||u_k|||_{4,k}.$$

PROOF. Note that $|||\phi|||_{2,k} = \|\phi\|_{2,k}$ for $\phi \in M_k$. By the triangle inequality,

$$|||u_k - I_k P_{k-1} u_k|||_{2,k} \leqslant \|u_k - I_k u_{k-1}\|_{2,k} + \left\|I_k(u_{k-1} - P_{k-1}u_k)\right\|_{2,k}. \tag{14.26}$$

We first estimate $\|u_k - I_k u_{k-1}\|_{2,k}$. By the triangle inequality,

$$\|u_k - I_k u_{k-1}\|_{2,k} \leqslant \|u_k - u_{k-1}\|_{2,k} + \|u_{k-1} - I_k u_{k-1}\|_{2,k}.$$

Using the error estimate, Lemma 14.3, we have,

$$\|u_k - u_{k-1}\|_{2,k} \leqslant \|u - u_{k-1}\|_{2,k} + \|u - u_k\|_{2,k} \leqslant Ch_k^{\beta} \|u\|_{2+\beta} + Ch_k^2 \|f_k\|_0.$$

By (14.25), $u_{k-1} - I_k u_{k-1}$, on each $\tau \in \mathcal{T}_k$, can be estimated by

$$\|u_{k-1} - I_k u_{k-1}\|_{2,k,\tau} \leqslant C\|u_{k-1}\|_{2,k-1,\hat{\tau}},$$

where $\hat{\tau} = \bigcup_{\substack{\tau' \in \mathcal{T}_{k-1} \\ \tau' \cap \tau \neq \emptyset}} \tau'$. Since $I_k p = p$ for all $p \in \mathbf{P}_2$, replacing u_{k-1} by $u_{k-1} + p$,

$$\begin{aligned} \|u_{k-1} - I_k u_{k-1}\|_{2,k,\tau} &\leqslant C \inf_{p \in \mathbf{P}_2} \|u_{k-1} + p\|_{2,k-1,\hat{\tau}} \\ &\leqslant C\Big(\|u_{k-1} - u\|_{2,k-1,\hat{\tau}} + \inf_{p \in \mathbf{P}_2} |u + p|_{2,\hat{\tau}}\Big) \\ &\leqslant C\|u_{k-1} - u\|_{2,k-1,\hat{\tau}} + Ch_k^{\beta} \|u\|_{2+\beta,\hat{\tau}}. \end{aligned}$$

Summing over $\tau \in \mathcal{T}_k$ and using the error estimate, Lemma 14.3,

$$\begin{aligned} \|u_{k-1} - I_k u_{k-1}\|_{2,k}^2 &\leqslant C\|u_{k-1} - u\|_{2,k-1}^2 + Ch_k^{2\beta} \|u\|_{2+\beta} \\ &\leqslant Ch_k^{2\beta} \|u\|_{2+\beta}^2 + Ch_k^4 \|f_k\|_0^2. \end{aligned}$$

Combining the estimates for $\|u_k - u_{k-1}\|_{2,k}$ and $\|u_{k-1} - I_k u_{k-1}\|_{2,k}$ we obtain,

$$\begin{aligned} \|u_k - I_k u_{k-1}\|_{2,k} &\leqslant \|u_k - u_{k-1}\|_{2,k} + \left\|(I - I_k)u_{k-1}\right\|_{2,k} \\ &\leqslant Ch_k^{\beta} \|u\|_{2+\beta} + Ch_k^2 \|f_k\|_0. \end{aligned} \tag{14.27}$$

We now estimate $\|I_k(u_{k-1} - P_{k-1}u_k)\|_{2,k}$, the second term on the right-hand side of (14.26). Observe that

$$A_{k-1}(u_{k-1}, \phi) = (f_k, \phi), \quad \text{for all } \phi \in M_{k-1}$$

and

$$A_{k-1}(P_{k-1}u_k, \phi) = A_k(u_k, I_k\phi) = (f_k, I_k\phi), \quad \text{for all } v \in M_{k-1}.$$

By (14.23)

$$\begin{aligned} A_{k-1}(u_{k-1} - P_{k-1}u_k, \phi) &= \big(f_k, (I - I_k)\phi\big) \leqslant \|f_k\|_0 \big\|(I - I_k)\phi\big\|_0 \\ &\leqslant Ch_k^2 \|f_k\|_0 \|\phi\|_{2,k-1}, \end{aligned}$$

for all $\phi \in M_{k-1}$. This implies that

$$\big\|(u_{k-1} - P_{k-1}u_k)\big\|_{2,k-1} \leqslant Ch_k^2 \|f_k\|_0.$$

By (14.24)

$$\big\|I_k(u_{k-1} - P_{k-1}u_k)\big\|_{2,k} \leqslant C\|u_{k-1} - P_{k-1}u_k\|_{2,k-1} \leqslant Ch_k^2 \|f_k\|_0. \tag{14.28}$$

Combining Eqs. (14.26), (14.27) and (14.28), it follows that

$$\big\|(I - I_k P_{k-1})u_k\big\|_{2,k} \leqslant Ch_k^{\beta} \|u\|_{2+\beta} + h_k^2 \|f_k\|_0.$$

Clearly $\|f_k\|_0 = |||u_k|||_{4,k}$. By the a priori estimate in (14.3) and Lemma 14.5

$$\|u\|_{2+\beta} \leqslant C\|f_k\|_{-2+\beta} \leqslant C|||f_k|||_{-2+\beta,k} = C|||u_k|||_{2+\beta,k}.$$

The lemma now follows. □

THEOREM 14.2. *Condition* (A.2″) *holds with* $\alpha = \beta/4$.

PROOF. Condition (A.2″) now follows from Lemma 14.5 as follows:

$$\begin{aligned} A_k\big((I - I_k P_{k-1})u_k, u_k\big) &\leqslant \big|\big|\big|(I - I_k P_{k-1})u_k\big|\big|\big|_{2,k} |||u_k|||_{2,k} \\ &\leqslant Ch_k^{\beta} |||u_k|||_{2+\beta,k} |||u_k|||_{2,k} \\ &\leqslant Ch_k^{\beta} |||u_k|||_{2,k}^{2-\beta/2} |||u_k|||_{4,k}^{\beta/2} \\ &\leqslant C|||u_k|||_{2,k}^{2(1-\beta/4)} \left(\frac{|||u_k|||_{4,k}^2}{\lambda_k}\right)^{\beta/4}. \end{aligned}$$

This proves the theorem. □

It now follows from Theorem 7.4 that the variable V-cycle multigrid preconditioner leads to a problem with a uniformly bounded condition number, provided that the number of smoothings, m_k, satisfies $\beta_0 m_k \leqslant m_{k-1} \leqslant \beta_1 m_k$ for some $1 < \beta_0 \leqslant \beta_1$ independent of k. Theorem 7.3 implies that the error operator of the W-cycle multigrid is a uniform contraction, provided that the number of smoothings $m_k = m$ is sufficiently large.

14.3. The mixed method of Ciarlet and Raviart

In order to formulate the mixed method of Ciarlet and Raviart we need an additional weak formulation of (14.1). By introducing an auxiliary variable $\sigma = -\Delta u$, (14.1) can be written as

$$\begin{cases} \sigma + \Delta u = 0, \\ -\Delta\sigma = f. \end{cases}$$

Multiplying by test functions and integrating by parts, we are led to the following mixed-variable weak formulation: Find $(\sigma, u) \in H^1(\Omega) \times H_0^1(\Omega)$ such that

$$\begin{cases} (\sigma, \theta) - D(\theta, u) = 0, & \text{for all } \theta \in H^1(\Omega), \\ -D(\sigma, v) = -(f, v), & \text{for all } v \in H_0^1(\Omega), \end{cases} \tag{14.29}$$

where $D(\bullet, \bullet)$ denotes the Dirichlet form

$$D(\theta, u) = \int_\Omega \nabla\theta \cdot \nabla u \,\mathrm{d}x.$$

It can be shown that when Ω is a convex polygon, (14.29) has a unique solution (σ, u), and that u solves Eqs. (14.1) and (14.2), and that $\sigma = -\Delta u$. (If Ω is such that H^3-regularity is not obtained, e.g., when Ω has a re-entrant corner, it is necessary to change the test spaces in the above formulation as is described in HANISCH [1991].)

A finite element method may be obtained by taking finite-dimensional test spaces $\Sigma_h \subset H^1(\Omega)$ and $M_h \subset H_0^1(\Omega)$ in the above formulation. To obtain uniqueness and good error estimates, the spaces Σ_h and M_h must be "complementary". More precisely, Σ_h and M_h must satisfy the so-called "inf-sup" conditions (cf. AZIZ and BABUŠKA [1972] or BREZZI [1974]). For example, one may choose Σ_h to be the space of continuous piecewise polynomials of degree less than or equal to m, defined with respect to a triangulation $\mathcal{T}_h$ of the convex polygon Ω, with $M_h = \Sigma_h \cap H_0^1(\Omega)$. The resulting finite element methods were first analyzed by Ciarlet and Raviart. If (σ_h, u_h) denotes the mixed finite element solution, and if $m \geqslant 2$, the following error estimates are known

$$\|\sigma - \sigma_h\|_0 \leqslant Ch\|\sigma\|_1$$

and

$$\|u - u_h\|_1 \leqslant Ch^2\|u\|_3.$$

These error estimates are used in a verification of Condition (A.2″) in HANISCH [1991]. Hence we assume that $m \geqslant 2$ which means that continuous piecewise quadratics (at least) are used.

The Schur complement. We next consider the resulting system of linear algebraic equations. Given bases $\{\psi^i\}$ and $\{\phi^i\}$ for Σ_h and M_h, respectively, the Ciarlet–Raviart method reduces to a matrix equation for the coefficient vectors $\tilde{\sigma}_h$ of σ_h and $\tilde{u}_h$ of u_h. If we introduce matrices $G_h = [(\psi^i, \psi^j)]$, $D_h = -[D(\psi^j, \phi^i)]$, and vector $\underset{\sim}{f}_h = [(f, \phi^i)]$, then

$$\mathcal{N}_h \begin{pmatrix} \tilde{\sigma}_h \\ \tilde{u}_h \end{pmatrix} \equiv \begin{pmatrix} G_h & D_h^T \\ D_h & 0 \end{pmatrix} \begin{pmatrix} \tilde{\sigma}_h \\ \tilde{u}_h \end{pmatrix} = \begin{pmatrix} 0 \\ -\underset{\sim}{f}_h \end{pmatrix}. \tag{14.30}$$

The matrix $\mathcal{N}_h$ is symmetric but indefinite. Eliminating $\tilde{\sigma}_h$ in (14.30), we obtain an equation for $\tilde{u}_h$,

$$S_h \tilde{u}_h \equiv D_h G_h^{-1} D_h^T \tilde{u}_h = \underset{\sim}{f}_h. \tag{14.31}$$

This equation is called the Schur complement equation. The vector $\tilde{\sigma}_h$ can be obtained from $\tilde{u}_h$ by back-solving. The Schur complement S_h is symmetric and positive definite but, like $\mathcal{N}_h$, it is ill-conditioned with condition number $\kappa(S_h) = \mathrm{O}(h^{-4})$ as $h \to 0$.

Note that in order to compute the action of the Schur complement S_h it is necessary to "invert" the L^2 Gram matrix G_h. For the usual choices of the basis $\{\psi^i\}$, G_h is well-conditioned and the action of its inverse can be computed with a rapidly converging iterative process. Alternatively, it is possible to replace the L^2 inner product in the mixed method equations with certain approximate L^2 inner products $(\bullet, \bullet)_h$. The solution $\{\bar{u}_h, \bar{\sigma}_h\}$ of this perturbed mixed method exhibits similar approximation properties. Furthermore, a new basis $\{\bar{\psi}^i\}$ can be constructed for which the matrix with entries $(\bar{\psi}^i, \bar{\psi}^j)_h$ is diagonal and $D(\bar{\psi}^j, \phi^i)$ remains sparse. This approach is described in HANISCH [1991].

A multigrid algorithm based on the Schur complement equation. Consider a sequence of Ciarlet–Raviart discretizations, one for each mesh in the usual family of nested triangulations $\{\mathcal{T}_k\}$, $k = 1, \ldots, J$, of Ω. Observe that the resulting spaces $M_k \equiv M_{h_k}$ are nested. Consequently, in defining a multigrid algorithm to solve for $u_J \equiv u_{h_J}$, we may define $I_k : M_{k-1} \to M_k$ to be the natural injection operator.

Let $\langle \bullet, \bullet \rangle_k$ denote the Euclidean inner product on $\mathbf{R}^n \times \mathbf{R}^n$ with $n = N_k$, the dimension of M_k. Corresponding to each space M_k one has the natural bilinear form

$$A_k(u, v) = \langle S_k \tilde{u}, \tilde{v} \rangle_k, \tag{14.32}$$

where $\tilde{u}$, $\tilde{v}$ denote the coefficient vectors of $u, v \in M_k$ with respect to the basis $\{\phi^i\}$. Note, with $D_k \equiv D_{h_k}$ and $G_k \equiv G_{h_k}$, that

$$A_k(v, v) = \langle D_k G_k^{-1} D_k^T \tilde{v}, \tilde{v} \rangle_k$$

$$
\begin{aligned}
&= \langle G_k G_k^{-1} D_k^T \tilde{v}, G_k^{-1} D_k^T \tilde{v} \rangle_k \\
&= \sup_{\tilde{\theta}\in\mathbf{R}^n} \frac{\langle G_k G_k^{-1} D_k^T \tilde{v}, \tilde{\theta}\rangle_k^2}{\langle G_k \tilde{\theta}, \tilde{\theta}\rangle_k} \\
&= \sup_{\theta\in\Sigma_k} \frac{|D(\theta, v)|^2}{\|\theta\|_0^2},
\end{aligned} \tag{14.33}
$$

so that $A_k(\bullet, \bullet)$ is independent of the basis selected for M_k. On the other hand, the forms $A_k(\bullet, \bullet)$ are noninherited as the supremum in (14.33) is k-dependent. In fact, Condition (I.2) seems to fail in this setting.

Multigrid convergence analysis. In HANISCH [1991] Condition (A.2″) is proved for the Ciarlet–Raviart method. (It is quite tedious and will not be included here.) Consequently, Theorem 7.4 and Theorem 7.3 may be applied. Assume that the smoothers of Section 8 are used. It follows from Theorem 7.3 that the W-cycle multigrid algorithm defined by Algorithm 7.1, with $p = 2$, has a uniform reduction in the error provided that the number of smoothings is sufficiently large. Furthermore, it follows from Theorem 7.4 that the variable multigrid preconditioner B_J defined by Algorithm 7.1, with $p = 1$, gives a uniform preconditioner for the Schur complement provided that m_k, the number of smoothings is chosen such that $\beta_0 m_k \leqslant m_{k-1} \leqslant \beta_1 m_k$ for some β_0 and β_1 with $1 < \beta_0 \leqslant \beta_1$.

14.4. Bibliographical notes

For an introduction to finite element methods for the biharmonic Dirichlet problem, see CIARLET [1978]. An extensive catalog of such methods is provided in HRABOK and HRUDEY [1984]. Regularity results for the biharmonic problem are given in DAUGE [1988].

Multigrid W-cycle for some conforming C^1 elements were studied in ZHANG [1989] and in BRAMBLE and ZHANG [1995]. The analysis here is based on BRAMBLE and ZHANG [1995].

The Morley method was introduced in MORLEY [1968]. Error estimates and other analytical details can be found in ARNOLD and BREZZI [1985], LASCAUX and LESAINT [1975], RANNACHER [1979b, 1979a]. A proof of the error estimate (14.22) can be found in BRENNER [1995]. Multigrid analysis for the Morley element can be found in BRENNER [1989]. Alternative multigrid approaches for solving the Morley equations can be found in PEISKER and BRAESS [1987] and HANISCH [1991].

Various analyses of the Ciarlet–Raviart method exist; cf. CIARLET and RAVIART [1974], BREZZI and RAVIART [1978], FALK and OSBORN [1980], BABUŠKA, OSBORN and PITKÄRANTA [1980] and HANISCH [1991]. The multigrid approach considered here is examined in detail in HANISCH [1991]. An alternative multigrid approach which solves (14.30) directly is analyzed by PEISKER [1985] and is based on the work of VERFÜRTH [1984].

15. Implementation, work estimates, and full multigrid

15.1. Implementation issues

In the previous sections the approximate problems have been presented as operator equations on M. In the applications to second and fourth order problems (Sections 9–12 and 14) and to pseudodifferential operators (Section 13), the problems were presented in the following way: Find $u \in M$ such that

$$A(u, \phi) = F(\phi), \quad \text{for all } \phi \in M \tag{15.1}$$

or, find $u \in M$ such that

$$\mathcal{V}(u, \phi) = F(\phi), \quad \text{for all } \phi \in M. \tag{15.2}$$

Most of the subsequent discussion applies to both cases above and we shall use as generic notation that of the first example except when otherwise noted.

Let $f \in M$ be the unique function satisfying

$$(f, \phi) = F(\phi), \quad \text{for all } \phi \in M.$$

In terms of the operator A_J defined by (2.10), (15.1) can be written as

$$A_J u = f. \tag{15.3}$$

Two vector representations of a function. Before proceeding, we briefly examine the way in which finite element equations are described in computer codes. We first discuss the representation of a function in M_k. Let $\{\phi_k^i\}$, $i = 1, \ldots, N_k$, denote a computational basis for the finite element space M_k. Then each function in M_k is associated with two vectors in $\mathbf{R}^{N_k}$. Firstly, a function $v \in M_k$ may be represented by a vector $\tilde{v}$ whose components consist of the coefficients of v with respect to the basis $\{\phi_k^i\}$, i.e., $v = \sum_i \tilde{v}_i \phi_k^i$. We shall call $\tilde{v}$ the *coefficient vector* of v. On the other hand, a function $g \in M_k$ can be considered as a functional on M_k and is characterized by its action on the basis $\{\phi_k^i\}$. Hence g is also associated with the vector $\underset{\sim}{g} = [(g, \phi_k^i)]$, which represents the action of g on the basis. We shall call $\underset{\sim}{g}$ the *dual vector* of $g \in M_k$. The solution u and its approximation v are usually expressed in terms of the coefficient vectors $\tilde{u}$ and $\tilde{v}$, and on the other hand the right-hand side f and the residual $g = f - A_k v$ are often expressed in terms of the dual vectors $\underset{\sim}{f}$ and $\underset{\sim}{g}$.

The the operator A_k is represented by using the *stiffness matrix*

$$[\underset{\approx}{A}_k]_{ij} = A(\phi_k^i, \phi_k^j).$$

Denote by $\langle \cdot, \cdot \rangle$ the Euclidean inner product on $\mathbf{R}^{N_k}$. Then

$$(A_k u, v) = A(u, v) = \langle \underset{\approx}{A}_k \tilde{u}, \tilde{v} \rangle, \quad \text{for all } u, v \in M_k.$$

Furthermore, the dual vector of $A_k v$ is given by $\underset{\approx}{A}_k \tilde{v}$ and the coefficient vector of $A_k^{-1} g$ is given by $\underset{\approx}{A}_k^{-1} \underset{\sim}{g}$.

With the dual vector $\underset{\sim}{f} = [(f, \phi_J^i)] = [F(\phi_J^i)]$ of f computed, (15.1) is reduced to the following matrix equation for the coefficient vector $\tilde{u}$ of u:

$$\underset{\approx}{A}_k \tilde{u} = \underset{\sim}{f}. \tag{15.4}$$

Implementation of the multigrid algorithm. We now discuss the the solution of (15.3) by the multigrid iteration defined by Algorithm 3.1. By the definition of Algorithm 3.1, $\mathrm{Mg}_J(v, g) = v + B_J(g - A_J v)$, and hence the implementation of the multigrid algorithm is the same as that of the action of B_J. We restate Algorithm 3.1 in terms of the computation of $B_J g$ here; see also Algorithm 3.2.

ALGORITHM 15.1. Set $\mathrm{Mg}_1(v, g_1) = A_1^{-1} g_1$. Let $k > 1$. Assume that $\mathrm{Mg}_{k-1}(v, g_{k-1}) \equiv v + B_{k-1}(g_{k-1} - A_{k-1} v)$ has been defined for $v, g_{k-1} \in M_{k-1}$. We define $\mathrm{Mg}_k(v, g_k) \equiv v + B_k(g_k - A_k v)$ for $v, g_k \in M_k$ by

(1) Set $v' = v + R_k^t(g_k - A_k v)$.

(2) Set $v'' = v' + q$, where $q = \mathrm{Mg}_{k-1}(0, Q_{k-1}(g_k - A_k v')) \equiv B_{k-1}[Q_{k-1}(g_k - A_k v')]$.

(3) Define $\mathrm{Mg}_k(v, g_k) \equiv v + B_k(g_k - A_k v) \equiv v''' = v'' + R_k(g_k - A_k v'')$.

The implementation of the multigrid algorithm requires the computation of the coefficient vector of $B_J g$ from the given dual vector $\underset{\sim}{g} = [(g, \phi_J^i)]$. Since $B_1 = A_1^{-1}$, the coefficient vector of $B_1 g_1 = A_1^{-1} g_1$ is just $\underset{\approx}{A}_1^{-1} \underset{\sim}{g}_1$, which can be computed by a direct inversion of the matrix $\underset{\approx}{A}_1$. The multigrid algorithm can now be implemented recursively. Assuming that the computation of the coefficient vector of $B_{k-1} g_{k-1}$ from the dual vector $\underset{\sim}{g}_{k-1}$ has been implemented, we now describe how to compute the coefficient vector of $B_k g_k$ from $\underset{\sim}{g}_k = [(g_k, \phi_k^i)]$, the dual vector of g_k.

The computation of $B_k g_k$ (equivalently, $\mathrm{Mg}_k(v, g_k)$) consists of three steps. Steps 1 and 3 require the computation of the coefficient vectors of

$$v' = R_k^t g_k \quad \text{and} \quad v''' = v'' + R_k(g_k - A_k v'')$$

in terms of the dual vector $\underset{\sim}{g}_k$ and the coefficient vector $\tilde{v}''$. It is desirable to choose a smoother R_k so that for any $g \in M_k$ the coefficient vectors of $R_k g$ and $R_k^t g$ can be computed directly from the dual vector $\underset{\sim}{g} = [(g, \phi_k^i)]$ (without using the coefficient vector $\tilde{g}$). We shall show later that this is the case for the smoothers constructed in Section 8. Note that in Step 1 the dual vector of g_k is $\underset{\sim}{g}_k$ and in Step 3 the dual vector of $g_k - A_k v''$ is given by $\underset{\sim}{g}_k - \underset{\approx}{A}_k \tilde{v}''$.

In Step 2, we need to compute the coefficient vector $\tilde{v}''$ of $v'' = v' + q$ in terms of the coefficient vector $\tilde{v}'$ and dual vector $\underset{\sim}{g}_k$, where q is defined by

$$q = B_{k-1} g_{k-1} \equiv B_{k-1} Q_{k-1}(g_k - A_k v').$$

Two additional ingredients are required in this step. First, to apply B_{k-1}, we need to compute the dual vector of $Q_{k-1}(g_k - A_k v')$ with respect to the basis of M_{k-1}. The dual vector of $(g_k - A_k v')$ with respect to the basis of M_k is just $\underset{\sim}{g}_k - \underset{\approx}{A}_k \tilde{v}'$. Since $M_{k-1} \subset M_k$, the basis functions of M_{k-1} can be expressed in terms of those of M_k, i.e., $\phi_{k-1}^i = \sum_{j=1}^{N_k} \alpha_{ij}^k \phi_k^j$. Hence

$$\begin{aligned}(g_{k-1}, \phi_{k-1}^i) &\equiv (Q_{k-1}(g_k - A_k v'), \phi_{k-1}^i) = (g_k - A_k v', \phi_{k-1}^i) \\ &= \sum_{j=1}^{N_k} \alpha_{ij}^k (g_k - A_k v', \phi_k^j) = \sum_{j=1}^{N_k} \alpha_{ij}^k [\underset{\sim}{g}_k - \underset{\approx}{A}_k \tilde{v}']_j.\end{aligned}$$

Introducing the matrix $\Pi_{k-1} = [\alpha_{ij}^k]$, the dual vector of $g_{k-1} = Q_{k-1}(g_k - A_k v')$ is then given by $\Pi_{k-1}(\underset{\sim}{g}_k - \underset{\approx}{A}_k \tilde{v}')$. This step corresponds to the *restriction* in classical multigrid expositions.

By the induction hypothesis, the coefficient vector of $q = B_{k-1} Q_{k-1}(g_k - A_k v')$ with respect to the basis of M_{k-1} can be computed in terms of $\Pi_{k-1}(\underset{\sim}{g}_k - \underset{\approx}{A}_k \tilde{v}')$, the dual vector of $Q_{k-1}(g_k - A_k v')$.

Finally, we need to compute the coefficient vector of v'' in terms of the coefficient vectors of v' and q with respect to the basis for M_k and M_{k-1} respectively. The coefficient vector of q with respect to the basis for M_k can be computed from the coefficient vector of q with respect to the basis in M_{k-1} by applying the transpose of the matrix Π_{k-1} above. This step corresponds to the *interpolation* in classical multigrid expositions. Note that, in practice, it often happens that for each i, α_{ij}^k is nonzero for only a few values of j and hence Π_{k-1} is sparse.

The matrix form of the algorithm. Instead of working with the multigrid algorithm involving the operators B_J, A_J and functions in M_k, we can define the corresponding preconditioning matrix $\overset{\approx}{B}_J$ of the stiffness matrix $\underset{\approx}{A}_J$ from the multigrid algorithm and work with vectors directly. This is done by introducing the matrices corresponding to the smoothers $\overset{\approx}{R}_k$ and $\overset{\approx}{R}_k^t$. Denote by $\overset{\approx}{R}_k$ and $\overset{\approx}{R}_k^t$ the matrices which map the dual vector of g into the the coefficient vectors of $R_k g$ and $R_k^t g$ respectively. It is straightforward to verify that $\overset{\approx}{R}_k^t$ is the transpose of $\overset{\approx}{R}_k$. We define the matrix $\overset{\approx}{B}_J$ by the the following algorithm.

ALGORITHM 15.2 (*V-cycle in matrix form*). Set $\overset{\approx}{B}_1 = \underset{\approx}{A}_1^{-1}$. Assume that $\overset{\approx}{B}_{k-1}$ has been defined and define $\overset{\approx}{B}_k \underset{\sim}{g}$ for $\underset{\sim}{g} \in R^{N_k}$ as follows:

(1) Pre-smoothing: Set $\tilde{v}' = \overset{\approx}{R}{}^t_k \underset{\sim}{g}$.

(2) Correction: Define $\tilde{v}'' = \tilde{v}' + \Pi^t_{k-1}\tilde{q}$ where

$$\tilde{q} = \overset{\approx}{B}_{k-1}\Pi_{k-1}\big(\underset{\sim}{g} - \underset{\approx}{A}_k\tilde{v}'\big).$$

(3) Post-smoothing: Set $\overset{\approx}{B}_k\underset{\sim}{g} \equiv \tilde{v}''' = \tilde{v}'' + \overset{\approx}{R}_k(\underset{\sim}{g} - \underset{\approx}{A}_k\tilde{v}'')$.

Based on the above discussion, we have the following result.

PROPOSITION 15.1. *Let B_k and $\overset{\approx}{B}_k$ be defined by the above algorithms. Then for any $g_k \in M_k$, the coefficient vector of $B_k g_k$ is $\overset{\approx}{B}_k \underset{\sim}{g}_k$. In particular for any $v \in M_k$ the coefficient vector of $B_k A_k v$ is $\overset{\approx}{B}_k \underset{\approx}{A}_k \tilde{v}$. Furthermore,*

$$\big\langle \underset{\approx}{A}_J\big(I - \overset{\approx}{B}_J\underset{\approx}{A}_J\tilde{v}\big), \tilde{v}\big\rangle = A\big((I - B_J A_J)v, v\big), \quad \textit{for all } v \in M_J.$$

Since $\langle \underset{\approx}{A}_J\tilde{v}, \tilde{v}\rangle = A(v, v)$, the contraction estimates for $I - B_J A_J$ and estimates on the condition number $\kappa(B_J A_J)$ give the same results for their matrix counterparts.

Implementation of the smoothers. We now discuss the implementation of the smoothers. The additive and the multiplicative smoothers introduced in Section 8 are such that the coefficient vector of $R_k g$ is given by a sparse block matrix applied to the dual vector $\underset{\sim}{g} = [(g, \phi^i_k)]$. We shall only discuss the additive and multiplicative smoothers defined by Eqs. (8.1) and (8.2).

For simplicity, let the spaces be given by $M^i_k = \operatorname{span}\{\phi^i_k\}$, i.e., one-dimensional spaces. Then by (8.1), noting that $A_{k,i} : M^i_k \to M^i_k$, we see that

$$R_k g = \sum_{i=1}^{N_k} A_{k,i}^{-1} Q^i_k g = \sum_{i=1}^{N_k} \tilde{\xi}_i \phi^i_k.$$

To determine $\tilde{\xi}_i$, we note that

$$\tilde{\xi}_i A\big(\phi^i_k, \phi^i_k\big) = A\big(A_{k,i}^{-1} Q^i_k g, \phi^i_k\big) = \big(g, \phi^i_k\big).$$

Hence

$$\tilde{\xi}_i = \big[A\big(\phi^i_k, \phi^i_k\big)\big]^{-1}\big(g, \phi^i_k\big) = [\underset{\approx}{A}_k]^{-1}_{ii}[\underset{\sim}{g}]_i.$$

Let $\underset{\approx}{D}_k$ be the diagonal part of the stiffness matrix $\underset{\approx}{A}_k \equiv [A(\phi^i_k, \phi^j_k)]$, then the coefficient vector of $R_k g$ is given by $\widetilde{R_k g} = \tilde{\xi} = \underset{\approx}{D}_k^{-1}\underset{\sim}{g}$. Hence the result of $R_k g$ corresponds to one step of the Jacobi iteration applied to the matrix equation $\underset{\approx}{A}_k\tilde{u} = \underset{\sim}{g}$.

The argument showing that the coefficient vector of $R_k g$ for the multiplicative smoother R_k can be computed from the dual vector $\underset{\sim}{g}$ is similar and can be found in BRAMBLE and PASCIAK [1992]. In particular, in the case in which $M_k^i = \{\phi_k^i\}$, the effect of $R_k g$ for the multiplicative smoother R_k is the same as one-step of the Gauss–Seidel method applied to the matrix equation $\underset{\approx}{A}_k \tilde{u} = \underset{\sim}{g}$.

In a similar way we can have more general subspace decompositions with

$$M_k = \sum_{i=1}^{\ell_k} M_k^i.$$

Thus we could have overlapping spaces, or a direct sum if the spaces are essentially disjoint. This includes the additive and multiplicative "block" methods or, what is the same, the block Jacobi and Gauss–Seidel methods. For easy implementation, we assume that each M_k^i has a basis consisting of some basis functions coming from the basis $\{\phi_k^j\}$. To compute the coefficient vector of $R_k g = \sum_i A_{k,i}^{-1} Q_k^i g$, we note that each term $v \equiv A_{k,i}^{-1} Q_k^i g$ in the sum satisfies

$$A_{k,i} v = Q_k^i g.$$

Note that the stiffness matrix $\underset{\approx}{A}_{k,i}$ that corresponds to $A_{k,i}$ is a principal minor of $\underset{\approx}{A}_k$ and that the dual vector of $Q_k^i g$ is a subvector of $\underset{\sim}{g}$. The coefficient vector of $v = A_{k,i}^{-1} Q_k^i g$ is computed by inverting the stiffness matrix $\underset{\approx}{A}_{k,i}$ and the coefficient vector of $R_k g$ is then obtained from those of $A_{k,i} Q_k^i g$ by superposition.

For a scaled Richardson smoother, $R_k = \tau I$ (corresponding to the operator equation), the coefficient vector of $R_k g = \tau g$ is given by $\tau \tilde{g} = \tau \underset{\approx}{G}_k^{-1} \underset{\sim}{g}$, where $\underset{\approx}{G}_k = [(\phi_k^i, \phi_k^j)]$ is the Gram matrix. Hence the computation of the coefficient vector of $R_k g$ from the dual vector $\underset{\sim}{g}$ requires an application of the inverse of the Gram matrix. In some works, discrete inner products, varying from level to level, are introduced. With respect to such discrete inner products the new Gram matrices are diagonal and hence the Richardson smoother can easily be implemented. The use of the discrete inner products changes the Richardson smoother corresponding the operator equation (15.3) to the Richardson smoother corresponding the matrix equation (15.4). What we have shown here is that, with the smoother properly defined, the same inner product can be used on all levels to define the relevant operators, while still maintaining the ease of implementation.

We finally turn to the implementation of the smoothers for the case of Section 13. Here $\mathcal{V}(\cdot, \cdot)$ and $(\cdot, \cdot)_{-1}$ replace $A(\cdot, \cdot)$ and $(\cdot, \cdot)$ in the earlier discussion. We need only show how to compute the coefficient vector of $R_k g$ with respect to the basis $\{\phi_k^i\}$ from the dual vector of g, which is $[(g, \phi_k^i)_{-1}]$ in this case.

By the definition of R_k,

$$\left(R_k g, \phi_k^i\right)_{-1,k} = \frac{1}{\bar{\lambda}_k}\left(g, \phi_k^i\right)_{-1}. \tag{15.5}$$

Combining Eqs. (13.9) and (15.5), we see that

$$\left(R_k g, \phi_k^i\right)_{0,k} = \left(R_k g, A_k \phi_k^i\right)_{-1,k} = \frac{1}{\bar{\lambda}_k}\left(g, A_k \phi_k^i\right)_{-1}, \quad \text{for } i = 1, \ldots, N_k.$$

Note that

$$\left(\phi_k^i, \phi_k^j\right)_{0,k} = \begin{cases} \tau_i = \frac{1}{3}|\operatorname{supp} \phi_k^i| & \text{if } j = i, \\ 0 & \text{if } j \neq i. \end{cases}$$

Hence,

$$A_k \phi_k^i = \sum_{j=1}^{N_k} \frac{1}{\tau_j} A\left(\phi_k^i, \phi_k^j\right)\phi_k^j.$$

It follows that

$$R_k g = \frac{1}{\bar{\lambda}_k}\sum_{i=1}^{N_k} \frac{1}{\tau_i}\left(g, A_k \phi_k^i\right)_{-1}\phi_k^i = \frac{1}{\bar{\lambda}_k}\sum_{i,j=1}^{N_k} \frac{1}{\tau_i \tau_j} A\left(\phi_k^i, \phi_k^j\right)\left(g, \phi_k^j\right)_{-1}\phi_k^i.$$

Thus the coefficient vector of $R_k g$ can be expressed in terms of the dual vector of g as

$$\widetilde{R_k g} = \frac{1}{\bar{\lambda}_k} \underset{\approx}{G}_k^{-1} \underset{\approx}{A}_k \underset{\approx}{G}_k^{-1} \underset{\sim}{g},$$

where $\underset{\approx}{G}_k = [(\phi_k^i, \phi_k^j)_{0,k}] = \operatorname{diag}(\tau_1, \ldots, \tau_{N_k})$ is the discrete L^2 Gram matrix. Note that the smoothers of Section 8 do not appear to fit into the case treated in Section 13.

15.2. Work estimates

In this subsection we provide some work estimates for multigrid algorithms. As in the previous subsection, we denote by N_k the dimension of M_k and assume that for some $a > 1$

$$N_{k-1} \leqslant \frac{1}{a} N_k. \tag{15.6}$$

We also assume that there is a constant $\beta_1 \geqslant 1$ such that the number of smoothings m_k satisfies

$$m_{k-1} \leqslant \beta_1 m_k. \tag{15.7}$$

Let p be the parameter of the general multigrid algorithm of Section 7. Let W_k be the work to evaluate B_k, i.e., W_k is the number of arithmetic operations required. We assume that the work to evaluate A_k and R_k is proportional to N_k. This is true for the examples in Section 9, but may not hold for the example of Section 13.

By examining the algorithm and ignoring the cost of intergrid transfers we have

$$W_k \leqslant C_0 m_k N_k + p W_{k-1},$$

for some constant C_0. Repeatly applying this inequality we obtain

$$W_k \leqslant C_0\bigl(m_k N_k + p m_{k-1} N_{k-1} + \cdots + p^{k-2} m_2 N_2\bigr) + p^{k-1} W_1. \tag{15.8}$$

We will always assume that

$$\gamma \equiv \frac{p}{a}\beta_1 < 1.$$

Without such a condition the work could increase exponentially. Using Eqs. (15.6) and (15.7) it is easy to see that the first term in (15.8) is bounded by a geometric series with a ratio $\gamma = (pa/\beta_1) < 1$. Consequently

$$W_k \leqslant C_0 m_k N_k \left(1 - \frac{p}{a}\beta_1\right)^{-1} + p^{k-1} W_1. \tag{15.9}$$

We will focus our attention on the examples of Section 9 where the nested triangulations are generated by a halving strategy. Thus a is approximately 4. We will assume that $a = 4$. Now we need $p\beta_1 < 4$.

In the case of the V-cycle with $p = 1$, $m_k = m$ and $\beta_1 = 1$, (15.9) is

$$W_J \leqslant W_1 + \tfrac{4}{3} C_0 m N_J.$$

Now $C_0 m N_J$ is essentially the cost of $2m$ evaluations of A_J and hence, neglecting W_1, the total cost for one V-cycle with m smoothings is less than $3m$ evaluations of A_J.

For the next example we consider the variable V-cycle with $p = 1$, $\beta_1 = 2$, $a = 4$ and $m_J = 1$. Then (15.9) yields

$$W_J \leqslant W_1 + 2C_0 N_J.$$

Hence the cost is approximately that of four evaluations of A_J.

Finally we consider the W-cycle. In this case $p = 2$ and $\beta_1 = 1$. We take $m = 1$. Then (15.9) gives

$$W_J \leqslant 2^{J-1} W_1 + 2C_0 N_J.$$

Note that the second term in this estimate is the same as the second term in the above estimate for the variable V-cycle algorithm with $\beta_1 = 2$. The W-cycle does involve more intergrid transfer operations which we have ignored in this discussion.

15.3. Full multigrid

We shall consider here the example of Section 9. Let u be defined by (9.2). The spaces M_k in this example consist of the piecewise linear functions as defined in Subsection 9.1. On each level k we define the kth level approximate solution by

$$A(u_k, \phi) = (f, \phi), \quad \text{for all } \phi \in M_k.$$

Now the following error estimate is standard:

$$\|u - u_k\|_1 \leqslant C h_k^\alpha \|f\|_{-1+\alpha}, \tag{15.10}$$

where $\alpha \in (0, 1]$ as in (9.6).

The so-called full multigrid algorithm starts with $u_2^{(1)} = u_1 \in M_1 \subset M_2$ as an approximation of u_2. After m steps of some (efficient) algorithm, using u_1 as a starting guess, we produce a better approximation $u_2^{(m)}$ satisfying

$$\left\|u_2 - u_2^{(m)}\right\|_1 \leqslant \delta^m \left\|u_2 - u_2^{(1)}\right\|_1.$$

The two-level multigrid iteration is an example of such a process. Now for $k > 2$ set $u_k^{(1)} = u_{k-1}^{(m)}$ and find an approximate solution $u_k^{(m)}$ to u_k satisfying

$$\left\|u_k - u_k^{(m)}\right\|_1 \leqslant \delta^m \left\|u_k - u_{k-1}^{(m)}\right\|_1. \tag{15.11}$$

A k-level multigrid process could be used here, for example.

Now we may combine the estimates to show that, for some constant C,

$$\left\|u - u_J^{(m)}\right\|_1 \leqslant C h^\alpha \|f\|_{-1+\alpha}. \tag{15.12}$$

To establish (15.12) we proceed as follows. By (15.11)

$$\begin{aligned}
\left\|u - u_J^{(m)}\right\|_1 &\leqslant \|u - u_J\|_1 + \left\|u_J - u_J^{(m)}\right\|_1 \\
&\leqslant \|u - u_J\|_1 + \delta^m \left\|u_J - u_{J-1}^{(m)}\right\|_1 \\
&\leqslant \left(1 + \delta^m\right)\|u - u_J\|_1 + \delta^m \left\|u - u_{J-1}^{(m)}\right\|_1.
\end{aligned}$$

Clearly, continuation of this argument implies that

$$\left\|u - u_J^{(m)}\right\|_1 \leqslant 2 \sum_{k=0}^{J-1} \delta^{mk} \|u - u_{J-k}\|_1.$$

Using (15.10), we have

$$\left\|u - u_J^{(m)}\right\|_1 \leqslant 2C \left(\sum_{k=0}^{J-1} h_{J-k}^\alpha \delta^{mk}\right) \|f\|_{-1+\alpha}.$$

Now $h_{J-k} = 2^k h_J = 2^k h$. Hence

$$\sum_{k=0}^{J-1} h_{J-k}^{\alpha} \delta^{mk} = h^{\alpha} \sum_{k=0}^{J-1} (2^{\alpha} \delta^m)^k.$$

Thus if m is chosen so that $2^{\alpha}\delta^m \leqslant \eta < 1$ then

$$\sum_{k=0}^{J-1} (2^{\alpha}\delta^m)^k \leqslant \frac{1}{1-\eta}.$$

Therefore we have that

$$\left\| u - u_J^{(m)} \right\|_1 \leqslant 2C(1-\eta)^{-1} h^{\alpha} \| f \|_{-1+\alpha},$$

which is (15.12).

The importance of this is that if $\delta < 1$ is independent of h (i.e., independent of the number of levels) then m may be chosen to be a fixed constant independent of h. Now we see that the total cost of arriving at an approximation of u is the sum of the costs on all levels. Hence if $\overline{W}_k$ is the cost on level k of one iterative step (say using the multigrid V-cycle) then the total cost W is

$$W = m \sum_{k=1}^{J} \overline{W}_k.$$

Now suppose $\overline{W}_k \leqslant \overline{C} N_k$. Then

$$W \leqslant \overline{C} m \sum_{k=1}^{J} N_k \leqslant \overline{C} m N_J \sum_{k=0}^{J} 4^{-k} \leqslant \tfrac{4}{3} m \overline{C} N_J.$$

Hence the full multigrid process is said to have "optimal complexity", i.e., the number of operations required to obtain the desired approximation is proportional to the number of unknowns.

15.4. Bibliographical notes

In some works, the multigrid algorithms are defined with a simple scaling of the identity operator as the smoother (cf. BANK and DUPONT [1981]). In such cases, discrete inner products, varying from level to level, are introduced in order to obtain an easily computable implementation of the algorithm. The implementation issues discussed here are discussed in BRAMBLE and PASCIAK [1992] and BRAMBLE, LEYK and PASCIAK [1994] where it was shown that, with the smoother properly defined, the same inner product can be used on all levels to define the relevant operators, while still maintaining the ease of implementation.

The work estimates are essentially the same as those given in BANK and DUPONT [1981]. In that paper they also considered the "full multigrid" process and proved that the W-cycle led to an algorithm of optimal complexity.

CHAPTER IV

Appendix

A. Introduction to interpolation spaces

In this appendix we give a brief introduction to the theory of interpolation spaces using the so-called real method of interpolation of LIONS and MAGENES [1972], LIONS and PEETRE [1964]. We will restrict our attention to certain special cases which are the most relevant to the theory of these notes. We present some basic definitions and some pertinent consequences.

Let B_0 and B_1 be two Banach spaces with B_1 continuously embedded and dense in B_0. An *intermediate space* B is any subspace of B_0 satisfying

$$B_1 \subset B \subset B_0.$$

A.1. Real method of interpolation

We want to define for $0 < s < 1$ a scale of spaces B_s with

$$B_1 \subset B_s \subset B_0,$$

and having nice properties. Define for each $t > 0$ and $u \in B_0$

$$K(t,u) = \inf_{u_0+u_1=u} \left(\|u_0\|_{B_0}^2 + t^2\|u_1\|_{B_1}^2\right)^{1/2}, \tag{A.1}$$

where $u_0 \in B_0$ and $u_1 \in B_1$. Note that we could define $K(t,u)$ by

$$\inf_{u_0+u_1=u} \left(\|u_0\|_{B_0} + t\|u_1\|_{B_1}\right).$$

The definition in (A.1) is necessary in the Hilbert space case in order that the intermediate spaces be Hilbert spaces.

For $0 < s < 1$ and $1 \leqslant p < \infty$ define the quantity

$$|||u|||_{B_{s,p}} = \left(\int_0^\infty t^{-sp} K^p(t,u) \frac{\mathrm{d}t}{t}\right)^{1/p}.$$

The case $p = \infty$ is defined by

$$|||u|||_{B_{s,\infty}} = \sup_{t>0} t^{-s} K(t, u).$$

It follows easily that $|||u|||_{B_{s,p}}$ is a norm on B_1. Now define

$$B_{s,p} = \{u \in B_0;\ |||u|||_{B_{s,p}} < \infty\}.$$

Then $B_{s,p}$ is a Banach space and is intermediate. The following is easily proved.

THEOREM A.1. *For* $u \in B_1$

$$|||u|||_{B_{s,p}} \leqslant C_{s,p} \|u\|_{B_0}^{1-s} \|u\|_{B_1}^{s} \tag{A.2}$$

with $C_{s,p} = (ps(1-s))^{-1/p}$ *if* $p < \infty$ *and* $C_{s,\infty} = 1$.

PROOF. For $u \in B_1$

$$K(t, u) \leqslant \|u\|_{B_0}$$

and

$$K(t, u) \leqslant t \|u\|_{B_1}.$$

Hence, for $p < \infty$ and $\alpha > 0$,

$$\begin{aligned} |||u|||_{B_{s,p}}^p &\leqslant \left[\left(\int_0^\alpha t^{-1+p(1-s)}\,\mathrm{d}t \right) \|u\|_{B_1}^p + \left(\int_\alpha^\infty t^{-1-ps}\,\mathrm{d}t \right) \|u\|_{B_0}^p \right] \\ &= \left[\frac{1}{p(1-s)} \alpha^{p(1-s)} \|u\|_{B_1}^p + \frac{1}{ps} \alpha^{-ps} \|u\|_{B_0}^p \right]. \end{aligned}$$

Now choose $\alpha = \|u\|_{B_0} / \|u\|_{B_1}$. Then

$$|||u|||_{B_{s,p}}^p \leqslant \frac{1}{ps(1-s)} \left(\|u\|_{B_0}^{1-s} \|u\|_{B_1}^{s} \right)^p$$

or

$$|||u|||_{B_{s,p}} \leqslant \|u\|_{B_0}^{1-s} \|u\|_{B_1}^{s} \left(\frac{1}{ps(1-s)} \right)^{1/p}.$$

For $p = \infty$

$$K(t, u) \leqslant \|u\|_{B_0}^{1-s} t^s \|u\|_{B_1}^{s}.$$

This proves Theorem A.1. □

A.2. The Hilbert space case

We shall consider the following situation. Suppose that $B_i = H_i$, $i = 0, 1$, are separable Hilbert spaces with $(\bullet, \bullet)$ the inner product on H_0. It is known that H_1 is the domain of an (unbounded) positive selfadjoint operator $\Lambda : H_1 \to H_0$ connecting the norms as follows:

$$\|u\|_{H_1} = \|\Lambda u\|_{H_0}.$$

For simplicity we shall consider the case where the spectrum of Λ is discrete and the eigenvectors form a complete orthonormal basis for H_0. Then we may expand any element of H_0 as

$$u = \sum_{i=1}^{\infty} (u, \varphi_i)\varphi_i.$$

If $u \in H_1$, then

$$\Lambda u = \sum_{i=1}^{\infty} \lambda_i (u, \varphi_i)\varphi_i$$

and

$$\|u\|^2_{H_1} = \sum_{i=1}^{\infty} \lambda_i^2 (u, \varphi_i)^2.$$

We define the intermediate spaces H_s to consist of those elements of H_0 for which the norm

$$\|u\|_{H_s} = \left(\sum_{i=1}^{\infty} \lambda_i^{2s} (u, \varphi_i)^2 \right)^{1/2} \tag{A.3}$$

is finite. These spaces are clearly intermediate Hilbert spaces. In the finite-dimensional case the sums are finite and as vector spaces they are all the same. In general, the Hilbert space structure is different, however, for different values of s.

Now we want to connect these norms with the norms defined in terms of the real method of interpolation. We define

$$[\![u]\!]_{H_s} = C_s \left(\int_0^{\infty} t^{-2s} K^2(t, u) \frac{\mathrm{d}t}{t} \right)^{1/2}$$

for $0 < s < 1$ and $C_s = (\int_0^{\infty} \frac{t^{1-2s}\,\mathrm{d}t}{t^2+1})^{-1/2} = \sqrt{\frac{2}{\pi}} \sin \pi s$. The following representation result is quite important for our applications.

THEOREM A.2. *For $u \in H_s$, $0 < s < 1$,*

$$[\![u]\!]_{H_s} = \|u\|_{H_s}.$$

PROOF. Now we may write

$$K^2(t,u) = \inf_{u_1 \in H_1} \left(\|u - u_1\|^2_{H_0} + t^2 \|u_1\|^2_{H_1} \right).$$

We solve the minimization problem. Let

$$u = \sum_{i=1}^{\infty} a_i \varphi_i$$

and

$$u_1 = \sum_{i=1}^{\infty} b_i \varphi_i.$$

Then

$$\|u - u_1\|^2_{H_0} + t^2 \|u_1\|^2_{H_1} = \sum_{i=1}^{\infty} \left[(a_i - b_i)^2 + t^2 \lambda_i^2 b_i^2 \right].$$

We choose b_i to minimize $[(a_i - b_i)^2 + t^2 \lambda_i^2 b_i^2]$. The choice is $b_i = a_i (t^2 \lambda_i^2 + 1)^{-1}$. Hence

$$K^2(t,u) = \sum_{i=1}^{\infty} t^2 \lambda_i^2 \left(t^2 \lambda_i^2 + 1\right)^{-1} a_i^2.$$

Now

$$\begin{aligned}
\int_0^{\infty} t^{-2s} K^2(t,u) \frac{\mathrm{d}t}{t} &= \sum_{i=1}^{\infty} \left(\int_0^{\infty} t^{1-2s} \lambda_i^2 \left(t^2 \lambda_i^2 + 1\right)^{-1} \mathrm{d}t \right) a_i^2 \\
&= C_s^{-2} \sum_{i=1}^{\infty} \lambda_i^{2s} a_i^2 = C_s^{-2} \|u\|^2_{H_s}.
\end{aligned}$$

Hence $[\![u]\!]_{H_s} = \|u\|_{H_s}$, $0 < s < 1$. □

The next result is the important property of "interpolation of operators". Suppose that we have spaces $\tilde{B}_0$ and $\tilde{B}_1$ analogous to B_0 and B_1; i.e., $\tilde{B}_1$ is continuously imbedded and dense in $\tilde{B}_0$. Define $\tilde{K}(t,\bullet)$ analogously. Further let L be a linear operator with $L: B_i \to \tilde{B}_i$ and constants C_i such that

$$\|Lu\|_{\tilde{B}_i} \leqslant C_i \|u\|_{B_i} \tag{A.4}$$

for $u \in B_i$, $i = 0, 1$. Here $\|\bullet\|_B$ is the norm on the generic Banach space B.

THEOREM A.3. *Suppose that B_i and $\tilde{B}_i$ are as above and $L: B_i \to \tilde{B}_i$ satisfies* (A.4). *Then*

$$|||Lu|||_{\tilde{B}_{s,p}} \leqslant C_0^{1-s} C_1^s |||u|||_{B_{s,p}}.$$

PROOF. Consider first $p < \infty$ and $0 < s < 1$. Then

$$\begin{aligned} |||Lu|||_{\tilde{B}_{s,p}} &= \left(\int_0^\infty t^{-ps} \tilde{K}^p(t, Lu) \frac{\mathrm{d}t}{t} \right)^{1/p} \\ &\leqslant \left(\int_0^\infty t^{-ps} \inf_{u_0+u_1=u} \left(\|Lu_0\|_{\tilde{B}_0}^2 + t^2 \|Lu_1\|_{\tilde{B}_1}^2 \right)^{p/2} \frac{\mathrm{d}t}{t} \right)^{1/p} \end{aligned}$$

since $Lu = Lu_0 + Lu_1$ is a decomposition of Lu with $Lu_0 \in \tilde{B}_0$ and $Lu_1 \in \tilde{B}_1$. Using (A.4), we have

$$\begin{aligned} |||Lu|||_{\tilde{B}_{s,p}} &\leqslant \left(\int_0^\infty t^{-ps} \inf_{u_0+u_1=u} \left(C_0^2 \|u_0\|_{B_0}^2 + t^2 C_1^2 \|u_1\|_{B_1}^2 \right)^{p/2} \frac{\mathrm{d}t}{t} \right)^{1/p} \\ &= C_0^{1-s} C_1^s \left(\int_0^\infty \left(\frac{C_1 t}{C_0} \right)^{-ps} K^p \left(\frac{C_1 t}{C_0}, u \right) \frac{\mathrm{d}t}{t} \right)^{1/p} \\ &= C_0^{1-s} C_1^s |||u|||_{B_{s,p}}. \end{aligned}$$

The case $p = \infty$ is similar. □

The particular case in which the spaces are Hilbert spaces and the corresponding scale of spaces is defined by the spectral representation is important. The following is a combination of Theorem A.2 and Theorem A.3.

THEOREM A.4. *Let H_s and $\tilde{H}_s$ be Hilbert spaces defined as in* (A.3) *for* $0 \leqslant s \leqslant 1$. *Suppose that L is a linear operator with $L: H_i \to \tilde{H}_i$ and constants C_0 and C_1 such that*

$$\|Lu\|_{\tilde{H}_i} \leqslant C_i \|u\|_{H_i}$$

for $u \in H_i$, $i = 0, 1$. *Then* $L : H_s \to \tilde{H}_s$ *and*

$$\|Lu\|_{\tilde{H}_s} \leqslant C_0^{1-s} C_1^s \|u\|_{H_s}$$

for $0 \leqslant s \leqslant 1$.

PROOF. From the previous definitions

$$[\![u]\!]_{H_s} = C_s |||u|||_{B_{s,2}}, \quad \text{if } B_{s,2} = H_s.$$

Hence Theorem A.4 follows from Theorem A.2 and Theorem A.3. □

A.3. Simultaneous approximation in scales of Banach spaces

Let H_0 and H_1 be two Hilbert spaces as above and the corresponding scale of spaces H_θ is defined by interpolation. Suppose that S is subspace of H_1 which is closed in H_0. We have the following result.

THEOREM A.5. *Suppose that for some* $\varepsilon > 0$ *and* θ *with* $0 < \theta < 1$

$$\inf_{\phi \in S} \|v - \phi\|_{H_\theta} \leqslant \varepsilon^{1-\theta} \|v\|_{H_1}, \quad \text{for all } v \in H_1.$$

Then

$$\inf_{\phi \in S} \|v - \phi\|_{H_0} \leqslant \varepsilon \|v\|_{H_1}, \quad \text{for all } v \in H_1.$$

PROOF. Since H_1 is continuously embedded in H_0, there exists a positive constant K such that $\|v\|_{H_0} \leqslant K\|v\|_{H_1}$. Let $Q_0 : H_0 \to S$ and $Q_\theta : H_\theta \to S$ be the orthogonal projectors with respect to the inner products in H_0 and H_θ, respectively. By definition,

$$\left\|(I - Q_\theta)v\right\|_{H_\theta} = \inf_{\phi \in S} \|v - \phi\|_{H_\theta} \leqslant \varepsilon^{1-\theta} \|v\|_{H_1}, \quad \text{for all } v \in H_1. \tag{A.5}$$

Since $0 < \theta < 1$, the theorem will follow if we prove

$$\left\|(I - Q_0)v\right\|_{H_0} \leqslant K^{\theta^n} \varepsilon^{1-\theta^n} \|v\|_{H_1}, \quad \text{for all } v \in H_1,\ n = 0, 1, \ldots. \tag{A.6}$$

Since Q_0 is the orthogonal projector onto S with respect to the inner product in H_0, we have that $\|(I - Q_0)v\|_{H_0} \leqslant \|v\|_{H_0} \leqslant K\|v\|_{H_1}$ and hence (A.6) holds for $n = 0$. Assume that (A.6) holds for $n = k$. We now show that it also holds for $n = k + 1$. By assumption we have that

$$\left\|(I - Q_0)v\right\|_{H_0} \leqslant K^{\theta^k} \varepsilon^{1-\theta^k} \|v\|_{H_1}, \quad \text{for all } v \in H_1.$$

Noting again the trivial inequality

$$\left\|(I-Q_0)v\right\|_{H_0} \leqslant \|v\|_{H_0}, \quad \text{for all } v \in H_0,$$

and applying Theorem A.4 with $\tilde{H}_0 = \tilde{H}_1 = H_0$ and $L = I - Q_0$, we obtain

$$\left\|(I-Q_0)v\right\|_{H_0} \leqslant K^{\theta^{k+1}} \varepsilon^{(1-\theta^k)\theta} \|v\|_{H_\theta}, \quad \text{for all } v \in H_\theta.$$

Replacing v by $(I - Q_\theta)v$ and noting that $(I - Q_0)(I - Q_\theta) = I - Q_0$, we obtain

$$\begin{aligned}\left\|(I-Q_0)v\right\|_{H_0} &\leqslant K^{\theta^{k+1}} \varepsilon^{(1-\theta^k)\theta} \left\|(I-Q_\theta)v\right\|_{H_\theta} \\ &\leqslant K^{\theta^{k+1}} \varepsilon^{(1-\theta^k)\theta} \varepsilon^{1-\theta} \|v\|_{H_1} \\ &= K^{\theta^{k+1}} \varepsilon^{1-\theta^{k+1}} \|v\|_{H_1}.\end{aligned}$$

In the second inequality above we used (A.5). This shows that (A.6) holds for $n = k + 1$ and hence for all n. Since $0 < \theta < 1$ and n is arbitrary, the theorem follows from (A.6). □

We next present a simultaneous approximation theorem in Banach spaces. This theorem, although not used directly in this book, has many interesting applications.

Let B_0 and B_1 be as in the beginning of this section and suppose that S is a subspace of B_1. Then we have the following result.

THEOREM A.6. *Suppose that for any $p \geqslant 1$ and some θ with $0 < \theta < 1$ and some $\varepsilon > 0$*

$$\inf_{\phi \in S} |||u - \phi|||_{B_{\theta,p}} \leqslant \varepsilon^{1-\theta} \|u\|_{B_1}, \quad \textit{for all } u \in B_1. \tag{A.7}$$

Then there is a constant $C(\theta) > 0$ such that

$$\inf_{\phi \in S} \big(\|u - \phi\|_{B_0} + \varepsilon^\theta |||u - \phi|||_{B_{\theta,p}}\big) \leqslant C(\theta)\varepsilon \|u\|_{B_1}, \quad \textit{for all } u \in B_1.$$

PROOF. For $u \in B_1$, set

$$E(u) = \inf_{\phi \in S} \big(\|u - \phi\|_{B_0} + \varepsilon^\theta |||u - \phi|||_{B_{\theta,p}}\big),$$

and

$$\delta = \sup_{\|u\|_{B_1} = 1} E(u).$$

We shall show that $\delta \leqslant C(\theta)\varepsilon$ which is the desired result. To this end let $v \in B_1$, $v \neq 0$. Then

$$\begin{aligned} E(u) &\leqslant \|u-v\|_{B_0} + \varepsilon^\theta |||u-v|||_{B_{\theta,p}} + E(v) \\ &= \|u-v\|_{B_0} + \varepsilon^\theta |||u-v|||_{B_{\theta,p}} + \frac{E(v)}{\|v\|_{B_1}}\|v\|_{B_1} \\ &\leqslant \|u-v\|_{B_0} + \delta\|v\|_{B_1} + \varepsilon^\theta |||u-v|||_{B_{\theta,p}}. \end{aligned} \tag{A.8}$$

Now choose v so that

$$\|u-v\|_{B_0} + \delta\|v\|_{B_1} \leqslant 2K(\delta, u). \tag{A.9}$$

For such a v we can prove that for all s

$$K(s, u-v) \leqslant \sqrt{10}\, K(s, u), \tag{A.10}$$

from which it follows that

$$|||u-v|||_{B_{\theta,p}} \leqslant \sqrt{10}\, |||u|||_{B_{\theta,p}}. \tag{A.11}$$

To prove (A.10) for $\delta \leqslant s$ we see that, using (A.9),

$$K(s, u-v) \leqslant \|u-v\|_{B_0} \leqslant 2K(\delta, u) \leqslant 2K(s, u),$$

since $K(t, u)$ is an increasing function of t. For $s \leqslant \delta$, and any $w \in B_1$

$$\begin{aligned} K^2(s, u-v) &\leqslant \|u-w\|_{B_0}^2 + s^2\|v-w\|_{B_1}^2 \\ &\leqslant \|u-w\|_{B_0}^2 + 2s^2\|w\|_{B_1}^2 + 2s^2\|v\|_{B_1}^2 \\ &\leqslant 2\big(\|u-w\|_{B_0}^2 + s^2\|w\|_{B_1}^2 + 4s^2\delta^{-2}K^2(\delta, u)\big), \end{aligned}$$

where we used again (A.9). Since $t^{-1}K(t, u)$ is a decreasing function of t, we have

$$K^2(s, u-v) \leqslant 2\big(\|u-w\|_{B_0}^2 + s^2\|w\|_{B_1}^2 + 4K^2(s, u)\big).$$

Taking the infimum over w it follows that for $s \leqslant \delta$

$$K(s, u-v) \leqslant \sqrt{10}\, K(s, u).$$

Hence (A.10) holds. Using Eqs. (A.9) and (A.11) with (A.8), we have

$$\begin{aligned} E(u) &\leqslant 2K(\delta, u) + \sqrt{10}\,\varepsilon^\theta |||u|||_{B_{\theta,p}} \\ &\leqslant 2\delta^\theta |||u|||_{B_{\theta,\infty}} + \sqrt{10}\,\varepsilon^\theta |||u|||_{B_{\theta,p}}. \end{aligned} \tag{A.12}$$

It is an exercise to prove that, for $1 \leqslant p < \infty$,

$$|||u|||_{B_{\theta,\infty}} \leqslant p^{1/p} |||u|||_{B_{\theta,p}}. \tag{A.13}$$

We have from Eqs. (A.12) and (A.13)

$$E(u) \leqslant \left(2p^{1/p}\delta^{\theta} + \sqrt{10}\,\varepsilon^{\theta}\right) |||u|||_{B_{\theta,p}}.$$

But $E(u) = E(u - \phi)$ for any $\phi \in S$ so that, using (A.7),

$$E(u) \leqslant \sqrt{10}\left(\delta^{\theta} + \varepsilon^{\theta}\right) \inf_{\phi \in S} |||u - \phi|||_{B_{\theta,p}} \leqslant \sqrt{10}\left(\delta^{\theta} + \varepsilon^{\theta}\right)\varepsilon^{1-\theta} \|u\|_{B_1}.$$

Hence

$$\delta \leqslant \sqrt{10}\left(\delta^{\theta} + \varepsilon^{\theta}\right)\varepsilon^{1-\theta}.$$

Thus, for appropriate $C(\theta)$,

$$\delta \leqslant C(\theta)\varepsilon.$$

This proves the theorem. □

Theorem A.5 is a special case of Theorem A.6. Notice however that the constant $C(\theta)$ in Theorem A.6, as it is given in the proof, depends on θ and it blows up as $\theta \to 1$. The iteration technique used in the proof of Theorem A.5 shows that the constant $C(\theta)$ can be taken to be one in the Hilbert space case.

A.4. Bibliographical notes

The real method of interpolation of Lions and Peetre was given in LIONS and PEETRE [1964]. It is described in the book of Lions and Magenes where they show the equality of the interpolation norm and the spectrally defined norm on the intermediate spaces. Theorems of the types of Theorem A.1 to Theorem A.4 may be found in many places; cf. BUTZER and BERENS [1967]. The approximation result Theorem A.6 is contained in BRAMBLE and SCOTT [1978]. The iteration technique used in the proof of Theorem A.5 was introduced in BRAMBLE and SCHATZ [1971] in the proof of Lemma 3.2.

B. Glossary of conditions

Conditions on the forms

(A.1) $\|(I - P_{k-1})v\|^2 \leqslant C_{\mathrm{P}}\lambda_k^{-1} A(v, v)$, for all $v \in M_k$.
(A.2) $|||(I - P_{k-1})v|||^2_{1-\alpha,k} \leqslant C_{\mathrm{P}}^{\alpha}\lambda_k^{-\alpha} A(v, v)$, for all $v \in M_k$.

(A.3) $\gamma_1\lambda_{k+1} \leqslant \lambda_k \leqslant \gamma_2\lambda_{k+1}$ for $k = 1, 2, \ldots, J-1$ and $0 < \gamma_1 \leqslant \gamma_2 < 1$.
(A.4) $|||Q_k v|||^2_{1-\alpha,k} \leqslant C'_{\mathrm{Q}}|||v|||^2_{1-\alpha,i}$, for all $v \in M_i, i \geqslant k$.
(A.5) There exist linear operators $\Pi_k : M \to M_k$ with $\Pi_J = I$ such that

$$A(\Pi_1 v, \Pi_1 v) + \sum_{k=2}^{J} \lambda_k \left\|(\Pi_k - \Pi_{k-1})v\right\|^2 \leqslant C_a A(v, v), \quad \forall v \in M_J.$$

(A.6) $A(v_k, v_i) \leqslant C_{\mathrm{cs}}\varepsilon^{i-k}\|v_k\|_A(\lambda_i^{1/2}\|v_i\|)$ for all $v_k \in M_k, v_i \in M_i, 1 \leqslant k \leqslant i \leqslant J$, or equivalently

$$\|A_i v\|^2 \leqslant \left(C_{\mathrm{cs}}\varepsilon^{i-k}\right)^2\lambda_i A(v, v), \quad \text{for all } v \in M_k, 1 \leqslant k \leqslant i \leqslant J.$$

(A.7)

$$\left\|(\Pi_k - \Pi_{k-1})v\right\|^2 \leqslant C_\pi \lambda_k^{-1} A(v, v), \quad \text{for all } v \in M_J, 1 \leqslant k-1 < J,$$

and

$$A(\Pi_k v, \Pi_k v) \leqslant C'_\pi A(v, v), \quad \text{for all } v \in M_J, 1 \leqslant k < J-1.$$

$(\widehat{\mathrm{A.5}})$ There exists a sequence of linear operators $\Pi_k : M \to M_k$ with $\Pi_J = I$ and range$(\Pi_k - \Pi_{k-1}) \subset \widehat{M}_k$, such that

$$A(\Pi_1 v, \Pi_1 v) + \sum_{k=2}^{J} \lambda_k \left\|(\Pi_k - \Pi_{k-1})v\right\|^2 \leqslant C_a A(v, v), \quad \text{for all } v \in M_J.$$

$(\widehat{\mathrm{A.6}})$ There exist $C_{\mathrm{cs}} > 0$ and $\varepsilon \in (0, 1)$ such that

$$A(u, v) \leqslant C_{\mathrm{cs}}\varepsilon^{i-k}|||u|||\left(\lambda_i^{1/2}\|v\|\right), \quad \text{for all } u \in \widehat{M}_k, v \in \widehat{M}_i,\ k \leqslant i \leqslant J.$$

$(\widehat{\mathrm{A.7}})$ There exists a sequence of linear operators $\Pi_k : M_J \to M_k$ with $\Pi_J = I$ and range$(\Pi_k - \Pi_{k-1}) \subset \widehat{M}_k$ such that

$$\left\|(\Pi_k - \Pi_{k-1})v\right\|^2 \leqslant C_\pi \lambda_k^{-1} A(v, v), \quad \text{for all } v \in M_J,\ 2 \leqslant k \leqslant J,$$

and

$$A(\Pi_k v, \Pi_k v) \leqslant C'_\pi A(v, v), \quad \text{for all } v \in M_J,\ 1 \leqslant k \leqslant J.$$

(A.2$''$)

$$\left|A_k\big((I - I_k P_{k-1})v, v\big)\right| \leqslant C_{\mathrm{P}}^{2\alpha}\left(\frac{\|A_k v\|_k^2}{\lambda_k}\right)^{\alpha}\big(A_k(v, v)\big)^{1-\alpha}, \quad 0 < \alpha \leqslant 1.$$

(I.1) $A_k(I_k v, I_k v) \leqslant A_{k-1}(v, v)$, for all $v \in M_{k-1}$.
(I.2) $A_k(I_k v, I_k v) \leqslant 2A_{k-1}(v, v)$, for all $v \in M_{k-1}$.

Conditions on the smoothers

Smoothers for multilevel additive preconditioners
(SA.1)

$$\omega_1 \lambda_k^{-1}(v, v) \leqslant (R_k v, v) \leqslant \omega_2\big(A_k^{-1} v, v\big), \quad \forall v \in M_k,\ 2 \leqslant k \leqslant J.$$

(SA.2)

$$\omega_1 \lambda_k^{-1}(v, v) \leqslant (R_k v, v) \leqslant \omega_2 \lambda_k^{-1}(v, v), \quad \forall v \in M_k, 2 \leqslant k \leqslant J.$$

(SA.3)

$$\omega_1\, \lambda_k^{-2}(A_k v, v) \leqslant (R_k v, v) \leqslant \omega_2\, \lambda_k^{-\beta}\big(A_k^{-1+\beta} v, v\big), \quad \forall v \in M_k,\ 2 \leqslant k \leqslant J.$$

Smoothers in the multigrid algorithms
(SM.1)

$$\omega \lambda_k^{-1}(v, v) \leqslant \big(\overline{R}_k v, v\big), \quad \forall v \in M_k,\ k = 2, 3, \ldots, J,$$

or equivalently $A(K_k v, K_k v) \leqslant A(K_{k,\omega} v, v)$, $\forall v \in M_k$.
(SM.2) There exists a $\theta \in (0, 2)$ not depending on $k = 2, \ldots, J$ such that

$$A(R_k v, R_k v) \leqslant \theta(R_k v, v), \quad \text{for all } v \in M_k.$$

This inequality is the same as any one of the following three inequalities:

$$\big(\overline{R}_k v, v\big) \geqslant (2 - \theta)(R_k v, v) \geqslant \left(\frac{2 - \theta}{\theta}\right)\big((R_k^t A_k R_k) v, v\big), \quad \forall v \in M_k.$$

(SM.3)

$$\frac{\omega}{\lambda_k^2}(A_k v, v) \leqslant \big(\overline{R}_k v, v\big), \quad \forall v \in M_k,\ k = 2, 3, \ldots, J.$$

(SM.4)

$$(R_k v, R_k v) \leqslant \omega_4 \lambda_k^{-1}(R_k v, v) \quad \text{or} \quad \big(R_k^{-1} v, v\big) \geqslant \omega_4^{-1} \lambda_k(v, v), \quad \forall v \in M_k.$$

$(\widehat{\text{SM.1}})$ With $\overline{\widehat{R}}_k = \widehat{R}_k + \widehat{R}_k^t - \widehat{R}_k^t \widehat{A}_k \widehat{R}_k$,

$$\omega \lambda_k^{-1} \|w\|^2 \leqslant \big(\overline{\widehat{R}}_k w, w\big), \quad \text{for all } w \in \widehat{M}_k,\ k = 2, \ldots, J.$$

$(\widehat{\mathrm{SM}.2})$ There exists a $\theta \in (0, 2)$, independent of J, such that

$$A\big(\widehat{R}_k v, \widehat{R}_k v\big) \leqslant \theta\big(\widehat{R}_k v, v\big), \quad \text{for all } v \in \widehat{M}_k,\ 2 \leqslant k \leqslant J.$$

This inequality is the same as any one of the following three inequalities:

$$\big(\overline{\widehat{R}}_k v, v\big) \geqslant (2-\theta)\big(\widehat{R}_k v, v\big) \geqslant \left(\frac{2-\theta}{\theta}\right)\big((\widehat{R}_k^t \widehat{A}_k \widehat{R}_k)v, v\big), \quad \text{for all } v \in \widehat{M}_k.$$

References

ADAMS, R. (1975), *Sobolev Spaces* (Academic Press, New York).

ARNOLD, D. and F. BREZZI (1985), Mixed and nonconforming finite element methods: Implementation, postprocessing and error estimates, *RAIRO Math. Model. Num. Anal.* **19**, 7–32.

ASTRAKHANTSEV, G.P. (1978), Method for fictitious domains for a second-order elliptic equation with natural boundary conditions, *USSR Comput. Math. Math. Phys.* **18**, 114–121.

AUBIN, J. (1990) *Approximation of Elliptic Boundary-Value Problems* (Wiley-Interscience, New York).

AXELSSON, O. (1987), A generalized conjugate gradient, least squares method, *Numer. Math.* **51**, 209–228.

AXELSSON, O. and I. GUSTAFSSON (1983), Preconditioning and two-level multigrid methods of arbitrary degree of approximation, *Math. Comp.* **40**, 219–242.

AXELSSON, O. and P. VASSILEVSKI (1990), Algebraic multilevel preconditioning methods, II, *SIAM J. Numer. Anal.* **27**, 1569–1589.

AZIZ, A.K. and I. BABUŠKA (1972), Part I, survey lectures on the mathematical foundations of the finite element method, in: A. Aziz, ed., *The Mathematical Foundations of the Finite Element Method with Applications to Partial Differential Equations* (Academic Press, New York) 1–362.

BABUŠKA, I. (1958), On the Schwarz algorithm in the theory of differential equations of mathematical physics, *Czechoslovak Math. J.* **8**, 328–342 (in Russian).

BABUŠKA, I., J. OSBORN and J. PITKÄRANTA (1980), Analysis of mixed methods using mesh dependent norms, *Math. Comp.* **35**, 1039–1062.

BAKHVALOV, N. (1966), On the convergence of a relaxation method with natural constraints on the elliptic operator, *USSR Comp. Math. Math. Phys.* **6**, 101–135.

BANK, R.E. (1981), A comparison of two multilevel iterative methods for nonsymmetric and indefinite elliptic finite element equations, *SIAM J. Numer. Anal.* **18**, 724–743.

BANK, R.E. and C. DOUGLAS (1985), Sharp estimates for multigrid rates of convergence with general smoothing and acceleration, *SIAM J. Numer. Anal.* **22**, 617–633.

BANK, R.E. and T. DUPONT (1981), An optimal order process for solving finite element equations, *Math. Comp.* **36**, 35–51.

BANK, R.E., T. DUPONT and H. YSERANTANT (1988), The hierarchical basis multigrid method, *Numer. Math.* **52**, 427–458.

BANK, R.E., J. MANDEL and S. MCCORMICK (1987), Variational multigrid theory, in: S. McCormick, ed., *Multigrid Methods* (SIAM, Philadelphia, PA) 131–178.

BANK, R.E. and D.J. ROSE (1982), Analysis of a multilevel iterative method for nonlinear finite element equations, *Math. Comp.* **39**, 453–465.

BENNET, C. and R. SHARPLEY (1988), *Interpolation of Operators* (Academic Press, Inc., New York).

BJØRSTAD, P. and O. WIDLUND (1984), Solving elliptic problems on regions partitioned into substructures, in: G. Birkhoff and A. Schoenstadt, eds., *Elliptic Problem Solvers* **II** (Academic Press, New York) 245–256.

BJØRSTAD, P. and O. WIDLUND (1986), Iterative methods for the solution of elliptic problems on regions partitioned into substructures, *SIAM J. Numer. Anal.* **23**, 1097–1120.

BRAESS, D. (1984), The convergence rate of a multigrid method with Gauss–Seidel relaxation for the Poisson equation (revised), *Math. Comp.* **42**, 505–519.

BRAESS, D. and W. HACKBUSCH (1983), A new convergence proof for the multigrid method including the V-cycle, *SIAM J. Numer. Anal.* **20**, 967–975.

BRAMBLE, J.H. (1966), A second order finite difference analogue of the first biharmonic boundary value problem, *Numer. Math.* **9**, 236–249.

BRAMBLE, J.H. (1981), The Lagrange multiplier method for Dirichlet's problem, *Math. Comp.* **37**, 1–12.

BRAMBLE, J.H. (1993), *Multigrid Methods*, Pitman Research Notes in Math. 294 (Longman, London).

BRAMBLE, J.H. (1994), On the development of multigrid methods and their analysis, *Proc. Sympos. Appl. Math.* **48**, 5–19.

BRAMBLE, J.H. (1995), Interpolation between Sobolev spaces in Lipschitz domains with an application to multigrid theory, *Math. Comp.* **64**, 1359–1365.

BRAMBLE, J.H., R.E. EWING, R. PARASHKEVOV and J.E. PASCIAK (1991), Domain decomposition methods for problems with uniform local refinement in two dimensions, in: R. Glowinski, Y. Kuznetsov, G. Meurant, J. Périaux and O. Widlund, eds., *Fourth Inter. Symp. on Domain Decomposition Methods for Partial Differential Equations* (SIAM, Philadelphia, PA) 91–100.

BRAMBLE, J.H., R.E. EWING, R. PARASHKEVOV and J.E. PASCIAK (1992), Domain decomposition methods for problems with partial refinement, *SIAM J. Sci. Statist. Comput.* **13**, 397–410.

BRAMBLE, J.H., R.E. EWING, J.E. PASCIAK and A.H. SCHATZ (1988), A preconditioning technique for the efficient solution of problems with local grid refinement, *Comp. Methods Appl. Mech. Engrg.* **67**, 149–159.

BRAMBLE, J.H., C.I. GOLDSTEIN and J.E. PASCIAK (1994), Analysis of V-cycle multigrid algorithms for forms defined by numerical quadrature, *SIAM J. Sci. Statist. Comput.* **15**, 566–576.

BRAMBLE, J.H. and S. HILBERT (1970), Estimation of linear functionals on Sobolev spaces with application to Fourier transforms and spline interpolation, *SIAM J. Numer. Anal.* **7**, 113–124.

BRAMBLE, J.H. and S.R. HILBERT (1971), Bounds for a class linear functionals on with application to Hermite interpolation, *Numer. Math.* **16**, 362–369.

BRAMBLE, J.H., D. KWAK and J.E. PASCIAK (1994), Uniform convergence of multigrid V-cycle iteration for indefinite and nonsymmetric problems, *SIAM J. Numer. Anal.* **31** (6), 1746–1763.

BRAMBLE, J.H., Z. LEYK and J.E. PASCIAK (1993), Iterative schemes for nonsymmetric and indefinite elliptic boundary value problems, *Math. Comp.* **60**, 1–22.

BRAMBLE, J.H., Z. LEYK and J.E. PASCIAK (1994), The analysis of multigrid algorithms for pseudo-differential operators of order minus one, *Math. Comp.* **63**, 461–478.

BRAMBLE, J.H. and J.E. PASCIAK (1983), An efficient numerical procedure for the computation of steady state harmonic currents in flat plates, *IEEE Trans. Mag.* **19**, 2409–2412.

BRAMBLE, J.H. and J.E. PASCIAK (1984), Preconditioned iterative methods for nonselfadjoint or indefinite elliptic boundary value problems, in: H. Kardestuncer, ed., *Unification of Finite Element Methods* (Elsevier Science, New York) 167–184.

BRAMBLE, J.H. and J.E. PASCIAK (1987), New convergence estimates for multigrid algorithms, *Math. Comp.* **49**, 311–329.

BRAMBLE, J.H. and J.E. PASCIAK (1988), A preconditioning technique for indefinite systems resulting from mixed approximations of elliptic problems, *Math. Comp.* **50**, 1–18.

BRAMBLE, J.H. and J.E. PASCIAK (1992), The analysis of smoothers for multigrid algorithms, *Math. Comp.* **58**, 467–488.

BRAMBLE, J.H. and J.E. PASCIAK (1993), New estimates for multigrid algorithms including the V-cycle, *Math. Comp.* **60**, 447–471.

BRAMBLE, J.H. and J.E. PASCIAK (1994), Uniform convergence estimates for multigrid V-cycle algorithms with less than full elliptic regularity, in: A. Quarteroni, Y.A. Kuznetsov, J. Périaux and O.B. Widlund, eds., *Domain Decomposition Methods in Science and Engineering: The Sixth International Conference on Domain Decomposition*, Contemporary Mathematics 157 (Amer. Math. Soc.) 17–26.

BRAMBLE, J.H., J.E. PASCIAK and A.H. SCHATZ (1986a), An iterative method for elliptic problems on regions partitioned into substructures, *Math. Comp.* **46**, 361–369.

BRAMBLE, J.H., J.E. PASCIAK and A.H. SCHATZ (1986b), The construction of preconditioners for elliptic problems by substructuring, I, *Math. Comp.* **47**, 103–134.

BRAMBLE, J.H., J.E. PASCIAK and A.H. SCHATZ (1987), The construction of preconditioners for elliptic problems by substructuring, II, *Math. Comp.* **49**, 1–16.

BRAMBLE, J.H., J.E. PASCIAK and A.H. SCHATZ (1988), The construction of preconditioners for elliptic problems by substructuring, III, *Math. Comp.* **51**, 415–430.

BRAMBLE, J.H., J.E. PASCIAK and A.H. SCHATZ (1989), The construction of preconditioners for elliptic problems by substructuring, IV, *Math. Comp.* **53**, 1–24.

BRAMBLE, J.H., J.E. PASCIAK, J. WANG and J. XU (1991a), Convergence estimates for multigrid algorithms without regularity assumptions, *Math. Comp.* **57**, 23–45.

BRAMBLE, J.H., J.E. PASCIAK, J. WANG and J. XU (1991b), Convergence estimates for product iterative methods with applications to domain decomposition, *Math. Comp.* **57**, 1–21.

BRAMBLE, J.H., J.E. PASCIAK and J. XU (1988), The analysis of multigrid algorithms for nonsymmetric and indefinite elliptic problems, *Math. Comp.* **51**, 389–414.

BRAMBLE, J.H., J.E. PASCIAK and J. XU (1990), Parallel multilevel preconditioners, *Math. Comp.* **55**, 1–22.

BRAMBLE, J.H., J.E. PASCIAK and J. XU (1991), The analysis of multigrid algorithms with nonnested spaces or noninherited quadratic forms, *Math. Comp.* **56** (193), 1–34.

BRAMBLE, J.H., J.E. PASCIAK and J. XU (1992), A multilevel preconditioner for domain decomposition boundary systems, in: *Proceedings of the 10th Inter. Conf. on Comput. Meth. in Appl. Sci. and Engr.* (Nova Sciences, New York).

BRAMBLE, J.H., J.E. PASCIAK and X. ZHANG (1996), Two-level preconditioners for $2m$th order elliptic finite element problems, *East-West J. Numer. Math.* **4** (2), 99–120.

BRAMBLE, J.H. and A.H. SCHATZ (1971), Least squares methods for $2m$th order elliptic boundary-value problems, *Math. Comp.* **25**, 1–32.

BRAMBLE, J.H. and L. SCOTT (1978), Simultaneous approximation in scales of Banach spaces, *Math. Comp.* **32**, 947–954.

BRAMBLE, J.H. and J. XU (1989), A local post-processing technique for improving the accuracy in mixed finite element approximations, *SIAM J. Numer. Anal.* **24**, 1267–1275.

BRAMBLE, J.H. and J. XU (1991), Some estimates for weighted L^2 projections, *Math. Comp.* **56**, 463–476.

BRAMBLE, J.H. and X. ZHANG (1995), Multigrid methods for the biharmonic problem discretized by conforming c^1 finite elements on nonnested meshes, *Numer. Funct. Anal. Optim.* **16**, 835–846.

BRAMBLE, J.H. and X. ZHANG (2000), Uniform convergence of the multigrid V-cycle for an anisotropic problem, *Math. Comp.*

BRAMBLE, J.H. and M. ZLÁMAL (1970), Triangular elements in the finite element method, *Math. Comp.* **24**, 809–820.

BRANDT, A. (1973), Multi-level adaptive technique (MLAT) for fast numerical solution to boundary value problems, in: *Proceedings of the 3rd Inter. Conf. on Numerical Meth. in Fluid Mech.* (Springer-Verlag, Berlin) 82–89.

BRANDT, A. (1977), Multi-level adaptive solutions to boundary value problems, *Math. Comp.* **31**, 333–390.

BRANDT, A. (1982), Guide to multigrid development, in: W. Hackbusch and U. Trottenberg, eds., *Multigrid Methods*, Lecture Notes in Math. 960 (Springer, Berlin).

BRANDT, A. (1986), Algebraic multigrid theory: The symmetric case, *Appl. Math. Comput.* **19**, 23–56.

BRENNER, S.C. (1988), Multigrid methods for nonconforming finite elements, Ph.D. thesis, Department of Mathematics, The University of Michigan.

BRENNER, S.C. (1989a), Multigrid methods for nonconforming finite elements, in: J. Mandel, S.F. McCormick, J.E. Dendy, C. Farhat, G. Lonsdale, S.V. Parter, J.W. Ruge and K. Stüben, eds., *Proceedings of the Fourth Copper Mountain Conference on Multigrid Methods* (SIAM, Philadelphia, PA) 54–65.

BRENNER, S.C. (1989b), An optimal-order multigrid method for P1 nonconforming finite elements, *Math. Comp.* **52**, 1–16.

BRENNER, S.C. (1989), An optimal-order nonconforming multigrid method for the biharmonic equation, *SIAM J. Numer. Anal.* **26**, 1124–1138.

BRENNER, S.C. (1990), A nonconforming multigrid method for the stationary Stokes equation, *Math. Comp.* **55**, 411–437.

BRENNER, S.C. (1995), Convergence of nonconforming multigrid methods without full elliptic regularity, Preprint.

BRENNER, S.C. (1996), Two-level additive Schwarz preconditioners for nonconforming finite element methods, *Math. Comp.* **65**.

BREZZI, F. (1974), On the existence, uniqueness and approximation of saddle-point problems arising from Lagrange multipliers, *R.A.I.R.O.*, 129–151.

BREZZI, F., J. DOUGLAS and L. MARINI (1985), Two families of mixed finite elements for second order elliptic problems, *Numer. Math.* **47**, 217–235.

BREZZI, F. and P.-A. RAVIART (1978), *Mixed Finite Element Methods for 4th Order Elliptic Equations* (Academic Press) 35–56.

BUTZER, P. and H. BERENS (1967), *Semi-Groups of Operators and Approximation* (Springer-Verlag, New York).

BUZBEE, B. and F. DORR (1974), The direct solution of the biharmonic equation on rectangular regions and the Poisson equation on irregular regions, *SIAM J. Numer. Anal.* **11**, 753–763.

BUZBEE, B., F. DORR, J. GEORGE and G.H. GOLUB (1971), The direct solution of the discrete Poisson equation on irregular regions, *SIAM J. Numer. Anal.* **8**, 722–736.

CAI, X.-C. (1991), Additive Schwarz algorithms for parabolic convection–diffusion equations, *Numer. Math.* **60** (1), 41–61.

CAI, X.-C. and O. WIDLUND (1992), Domain decomposition algorithms for indefinite elliptic problems, *SIAM J. Sci. Statist. Comput.* **13**, 243–258.

CHANDRA, R. (1978), Conjugate gradient methods for partial differential equations, Yale Univ., Dept. of Comp. Sci., Rept. 129.

CIARLET, P.G. (1978), *The Finite Element Method for Elliptic Problems* (North-Holland, New York).

CIARLET, P.G. and P.-A. RAVIART (1974), A mixed finite element method for the biharmonic equation, in: *Mathematical Aspects of Finite Elements in Partial Differential Equations* (Academic Press, New York) 125–145.

CONCUS, P., G.H. GOLUB and G. MEURANT (1985), Block preconditioning for the conjugate gradient method, *SIAM J. Sci. Statist. Comput.* **6**, 220–252.

COURANT, R. and D. HILBERT (1962), *Methods of Mathematical Physics* **2** (Interscience, NY).

DAHMEN, W. and A. KUNOTH (1992), Multilevel preconditioning, *Numer. Math.* **63**, 315–344.

DAUGE, M. (1988), *Elliptic Boundary Value Problems on Corner Domains: Smoothness and Asymptotics of Solutions*, Lecture Notes in Math. 1341 (Springer-Verlag, New York).

DECKER, N., J. MANDEL and S. PARTER (1988), On the role of regularity in multigrid methods, in: S. McCormick, ed., *Multigrid Methods: Proceedings of the Third Copper Mountain Conference* (Marcel Dekker, New York).

DIHN, Q., R. GLOWINSKI and J. PÉRIAUX (1984), Solving elliptic problems by domain decomposition methods, in: G. Birkhoff and A. Schoenstadt, eds., *Elliptic Problem Solvers* **II** (Academic Press, New York) 395–426.

DOUGLAS, J. (1956), On the numerical solution of heat conduction problems in two or three space dimensions, *Trans. Amer. Math. Soc.* **82**, 421–439.

DOUGLAS, C. (1984), Multi-grid algorithms with applications to elliptic boundary-value problems, *SIAM J. Numer. Anal.* **21**, 236–254.

DRYJA, M. and O. WIDLUND (1987), An additive variant of the Schwarz alternating method for the case of many subregions, Tech. Rept. 339, Courant Institute of Mathematical Sciences.

DRYJA, M. and O. WIDLUND (1989), Some domain decomposition algorithms for elliptic problems, in: L. Hayes and D. Kincaid, eds., *Iterative Methods for Large Linear Systems* (Academic Press, New York, NY) 273–291.

DRYJA, M. and O. WIDLUND (1991a), Additive Schwarz methods for elliptic finite element problems in three dimensions, Tech. Rept. 570, Courant Institute of Mathematical Sciences.

DRYJA, M. and O.B. WIDLUND (1991b), Multilevel additive methods for elliptic finite element problems, in: W. Hackbusch, ed., *Parallel Algorithms for Partial Differential Equations, Proceedings of the Sixth GAMM-Seminar, Kiel, January 19–21, 1990* (Vieweg & Son, Braunschweig, Germany) 58–69.

DUPONT, T., R. KENDALL and H. RACHFORD (1968), An approximate factorization procedure for solving self-adjoint elliptic difference equations, *SIAM J. Numer. Anal.* **5**, 559–573.

DUPONT, T. and L. SCOTT (1980), Polynomial approximation of functions in Sobolev spaces, *Math. Comp.* **34**, 441–463.

EISENSTAT, S., H. ELMAN, M. SCHULTZ and A. SHERMAN (1984), The (new) Yale sparse matrix package, in: G. Birkhoff and A. Schoenstadt, eds., *Elliptic Problem Solvers* **II** (Academic Press, New York) 45–52.

ELMAN, H. (1982), Iterative methods for large, sparse, nonsymmetric systems of linear equations, Ph.D. thesis, Yale Univ. Dept. of Comp. Sci. Rept. 229.

EWING, R.E., R. LAZAROV, P. LU and P. VASSILEVSKI (1990), Preconditioning indefinite systems arising from mixed finite element discretization of second-order elliptic problems, in: O. Axelsson and L. Kolotilina, eds., *Preconditioned Conjugate Gradient Methods*, Lecture Notes in Math. 1457 (Springer, Berlin) 280–343.

EWING, R.E., R. LAZAROV, T. RUSSELL and P. VASSILEVSKI (1989), Local refinement via domain decomposition techniques for mixed finite element methods with rectangular Raviart–Thomas elements, in: T. Chan, R. Glowinski, J. Periaux and O. Widlund, eds., *Domain Decomposition Methods* (SIAM, Philadelphia, PA) 98–114.

FALK, R. (1976), An analysis of the finite element method using Lagrange multipliers for the stationary Stokes equations, *Math. Comp.* **30**, 241–269.

FALK, R. and J. OSBORN (1980), Error estimates for mixed methods, *R.A.I.R.O. Numer. Anal.* **14**, 249–277.

FEDORENKO, R. (1962), A relaxation method for solving elliptic difference equations, *USSR Comput. Math. Math. Phys.* **1** (4).

FEDORENKO, R. (1964), The speed of convergence of one iterative process, *USSR Comput. Math. Math. Phys.* **4** (3), 227–235 (in Russian).

GEORGE, A. and J. LIU (1981), *Computer Solution of Large Sparse Positive Definite Systems* (Prentice-Hall, Inc., Englewood Cliffs, NJ).

GIRAULT, V. and P.-A. RAVIART (1981), *Finite Element Approximation of the Navier–Stokes Equations*, Lecture Notes in Math. 749 (Springer, New York).

GLOWINSKI, R., W. KINTON and M. WHEELER (1990), Acceleration of domain decomposition algorithms for mixed finite elements by multi-level methods, in: T. Chan, R. Glowinski, J. Periaux and O. Widlund, eds., *Third Inter. Symp. on Domain Decomposition Methods for Partial Differential Equations* (SIAM, Philadelphia, PA) 263–289.

GLOWINSKI, R. and M. WHEELER (1988), Domain decomposition and mixed finite element methods for elliptic problems, in: R. Glowinski, G.H. Golub, G.A. Meurant and J. Periaux, eds., *First Inter. Symp. on Domain Decomposition Methods for Partial Differential Equations* (SIAM, Philadelphia, PA) 144–172.

GLOWINSKI, R. and M. WHEELER (1988), Domain decomposition and mixed methods for elliptic problems, in: *Proceedings, 1'st Inter. Conf. on Domain Decomposition Methods* (SIAM, Philadelphia, PA) 144–172.

GOLDSTEIN, C. (1989), Analysis and application of multigrid preconditioners for singularly perturbed boundary value problems, *SIAM J. Numer. Anal.* **26**, 1090–1123.

GOLDSTEIN, C. (1991), Multigrid analysis of finite element methods with numerical integration, *Math. Comp.* **56**, 409–436.

GOLUB, G.H. and C.F. VAN LOAN (1989), *Matrix Computations* (The Johns Hopkins University Press, Baltimore).

GOLUB, G.H. and D. MEYERS (1983), The use of preconditioning over irregular regions, in: *Proc. 6th. Internl. Conf. Comput. Meth. Sci. and Engng., Versailles, France.*

GOLUB, G.H. and D. O'LEARY (1989), Some history of the the conjugate gradient and Lanczos algorithms: 1948–1976, *SIAM Rev.* **31**, 50–102.

GRIEBEL, M. and P. OSWALD (1994), Tensor-product-type subspace splittings and multilevel methods for anisotropic problems, Tech. Rept. TUM 19434, Technische Universität München.

GRIEBEL, M. and P. OSWALD (1995), On the abstract theory of additive and multiplicative Schwarz algorithms, *Numer. Math.* **70**, 163–180.

GRISVARD, P. (1976), Behavior of the solutions of an elliptic boundary value problem in a polygonal or polyhedral domain, in: B. Hubbard, ed., *Numerical Solution of Partial Differential Equations* **III** (Academic Press, New York) 207–274.

GRISVARD, P. (1985), *Elliptic Problems in Nonsmooth Domains* (Pitman, Boston).

GROPP, W. and D. KEYES (1987), A comparison on domain decomposition techniques for elliptic partial differential equations and the parallel implementation, *SIAM J. Sci. Statist. Comput.* **8**, s166–s203.

HACKBUSCH, W. (1976), Ein iteratives Verfahren zur schnellen Auflösung elliptischer Randwertprobleme, Tech. Rept. 76-12, Mathematisches Institut der Universität zu Köln.

HACKBUSCH, W. (1977), On the convergence of a multigrid iteration applied to finite element equations, Tech. Rept. 77-8, Universität zu Köln.

HACKBUSCH, W. (1978), On the multigrid method applied to difference equations, *Computing* **20**, 291–306.

HACKBUSCH, W. (1980a), Convergence of multi-grid iterations applied to difference equations, *Math. Comp.* **34**, 425–440.

HACKBUSCH, W. (1980b), Survey of convergence proofs for multigrid iterations, in: J. Frehse, D. Pallaschke and U. Trottenberg, eds., *Special Topics of Applied Mathematics* (North-Holland, Amsterdam) 151–164.

HACKBUSCH, W. (1981a), Die schnelle Auflösung der Fredholmschen Integralgleichung zweiter Art, *Beiträge Numer. Math.* **9**, 47–62.

HACKBUSCH, W. (1981b), On the convergence of multi-grid iterations, *Beiträge Numer. Math.* **9**.

HACKBUSCH, W. (1985), *Multi-Grid Methods and Applications*, Springer Ser. Comput. Math. 4 (Springer, New York).

HANISCH, M. (1991), Multigrid preconditioning for mixed finite element methods, Ph.D. thesis, Cornell University.

HANISCH, M.R. (1993), Multigrid preconditioning for the biharmonic Dirichlet problem, *SIAM J. Numer. Anal.* **30**, 184–214.

HEMKER, P. (1984), Multigrid methods for problems with a small parameter in the highest derivative, in: D. Griffiths, ed., *Numerical Analysis. Proceedings Dundee 1983*, Lecture Notes in Math. 1066 (Springer, Berlin) 106–121.

HEMKER, P.W. and H. SCHIPPERS (1981), Multiple grid methods for the solution of Fredholm integral equations of the second kind, *Math. Comp.* **36**, 215–232.

HESTENES, M. (1956), The conjugate gradient method for solving linear systems, *Proc. Symp. Appl. Math.* **6**, 83–102.

HESTENES, M. and E. STIEFEL (1952), Methods of conjugate gradients for solving linear systems, *J. Res. Nat. Bur. Standards* **49**, 409–436.

HÖRMANDER, L. (1963), *Linear Partial Differential Operators* (Springer-Verlag, New York).

HRABOK, M. and T. HRUDEY (1984), A review and catalog of plate bending finite elements, *Comput. & Structures* **19**, 479–495.

JOHNSON, C. and J. PITKÄRANTA (1982), Analysis of some mixed finite element methods related to reduced integration, *Math. Comp.* **38**, 375–400.

KELLOGG, R. (1971), Interpolation between subspaces of a Hilbert space, Tech. Rept. BN-719, Inst. Fluid Dynamics and Appl. Math., Univ. of Maryland.

KING, J. (1990), On the construction of preconditioners by subspace decomposition, *J. Comput. Appl. Math.* **29**, 192–205.

KONDRATÉV, V. (1967), Boundary problems for elliptic equations with conical or angular points, *Trans. Moscow Math. Soc.* **16**, 227–313.

KOČVARA, M. and J. MANDEL (1987), A multigrid method for three-dimensional elasticity and algebraic convergence estimates, *Appl. Math. Comput.* **23**, 121–135.

KREIN, S.G. and Y. PETUNIN (1966), Scales of Banach spaces, *Russian Math. Surveys* **21**, 85–160.

KRONSÖ, L. (1975), A note on the "nested iterations" method, *BIT* **15**, 107–110.

LANCZOS, C. (1952), Solution of systems of linear equations by minimized iterations, *J. Res. Nat. Bur. Standards* **49**, 33–53.

LASCAUX, P. and P. LESAINT (1975), Some nonconforming finite elements for the plate bending problem, *RAIRO Rev. Franc. D'Autom.* **9**, 9–54.

LIONS, J. and E. MAGENES (1972), *Non-Homogeneous Boundary Value Problems and Applications* (Springer-Verlag, New York).

LIONS, J. and J. PEETRE (1964), Sur une classe d'espaces d'interpolation, *Institut des Hautes Etudes Scientifique, Publ. Math.* **19**, 5–68.

LIONS, P. (1987), On the Schwarz alternating method, in: R. Glowinski, G.H. Golub, G.A. Meurant and J. Periaux, eds., *Proceedings of the First International Symposium on Domain Decomposition Methods for Partial Differential Equations* (SIAM, Philadelphia, PA).

MAITRE, J. and F. MUSY (1985), Algebraic formalization of the multigrid method in the symmetric and positive definite case – a convergence estimation for the V-cycle, in: D.J. Paddon and H. Holstien, eds., *Multigrid Methods for Integral and Differential Equations*, The Institute of Mathematics and its Applications Conference Series (Clarendon Press, Oxford) 213–224.

MAITRE, J.F. and F. MUSY (1984), Multigrid methods: Convergence theory in a variational framework, *SIAM J. Numer. Anal.* **21**, 657–671.

MAITRE, J.F. and F. MUSY (1987), Multigrid methods for symmetric variational problems: A general theory and convergence estimates for usual smoothers, *Appl. Math. Comput.* **21**, 21–43.

MANDEL, J. (1984), Étude algébrique d'une méthode multigride pour quelques problèms de frontiére libre, *C. R. Acad. Sci. Paris, Sér. I Math.* **298**, 469–472.

MANDEL, J. (1986), Multigrid convergence for nonsymmetric, indefinite variational problems and one smoothing step, in: *Proc. Copper Mtn. Conf. Multigrid Methods*, Appl. Math. Comput. 19, 201–216.

MANDEL, J. (1988), Algebraic study of multigrid methods for symmetric, definite problems, *Appl. Math. Comput.* **25**, 39–56.

MANDEL, J. and S. MCCORMICK (1989), Iterative solution of elliptic equations with refinement: The two-level case, in: T. Chan, R. Glowinski, J. Periaux and O. Widlund, eds., *Domain Decomposition Methods* (SIAM, Philadelphia, PA) 81–92.

MANDEL, J., S. MCCORMICK and J. RUGE (1988), An algebraic theory for multigrid methods for variational problems, *SIAM J. Numer. Anal.* **25**, 91–110.

MANTEUFFEL, T. and S. PARTER (1990), Preconditioning and boundary conditions, *SIAM J. Numer. Anal.* **27**, 656–694.

MATHEW, T. (1989), Domain decomposition and iterative refinement methods for mixed finite element discretizations of elliptic problems, Ph.D. thesis, New York University.

MCCORMICK, S. (1984), Multigrid methods for variational problems: Further results, *SIAM J. Numer. Anal.* **21**, 255–263.

MCCORMICK, S. (1985), Multigrid methods for variational problems: General theory for the V-cycle, *SIAM J. Numer. Anal.* **22**, 634–643.

MCCORMICK, S.F. and J.W. RUGE (1982), Multigrid methods for variational problems, *SIAM J. Numer. Anal.* **19**, 924–929.

MCCORMICK, S. and J. RUGE (1983), Unigrid for multigrid simulation, *Math. Comp.* **41**, 43–62.

MCCORMICK, S. and J. THOMAS (1986), The fast adaptive composite grid (fac) method for elliptic equations, *Math. Comp.* **46**, 439–456.

MEYERINK, J. and H. VAN DER VORST (1977), Iterative methods for the solution of linear systems of which the coefficient matrix is a symmetric m-matrix, *Math. Comp.* **31**, 148–162.

MILNER, F. (1985), Mixed finite element methods for quasilinear second-order elliptic problems, *Math. Comp.* **44**, 303–320.

MORLEY, L. (1968), The triangular equilibrium element in the solution of plate bending problems, *Aero. Quart.* **19**, 149–169.

NEDELEC, J. (1982), Elements finis mixtes incompressibles pour l'equation de Stokes dans R^3, *Numer. Math.* **39**, 97–112.

NEDELEC, J. and J. PLANCHARD (1973), Une méthod variationnelle d'éléments finis pour la résolution numérique d'un problème extérieur dans R^3, *RAIRO* **7**, 105–129.

NEČAS, J. (1964), Sur la coercivité des formes sesquilinéares elliptiques, *Rev. Roumaine de Meth. Pure Appl.* **9**, 47–69.

NEČAS, J. (1967), *Les Méthodes Directes en Théorie des Équations Elliptiques* (Academia, Prague).

NICOLAIDES, R. (1975), On multiple grid and related techniques for solving discrete elliptic systems, *J. Comput. Phys.* **19**, 418–431.

NICOLAIDES, R. (1977), On the l^2 convergence of an algorithm for solving finite element equations, *Math. Comp.* **31**, 892–906.

NICOLAIDES, R.A. (1978a), On multi-grid convergence in the indefinite case, *Math. Comp.* **32**, 1082–1086.

NICOLAIDES, R.A. (1978b), On the observed rate of convergence of an iterative method applied to a model elliptic difference equation, *Math. Comp.* **32**, 127–133.

NICOLAIDES, R.A. (1979), On some theoretical and practical aspects of multigrid methods, *Math. Comp.* **33**, 933–952.

NITSCHE, J. (1968), Ein Kriterium für die Quasi-optimalität des Ritzchen Verfahrens, *Numer. Math.* **11**, 346–348.

OSWALD, P. (1992), On discrete norm estimates related to multilevel preconditioners in the finite element method, in: P.P.K.G. Ivanov and B. Sendov, eds., *Constructive Theory of Functions*, Proc. Int. Conf. Varna (Bulg. Acad. Sci., Sofia) 203–214.

OSWALD, P. (1995), Multilevel preconditioners for discretizations of the biharmonic equation by rectangular finite elements, *Numer. Linear Algebra Appl.* **2** (6).

PARTER, S.V. (1987), Remarks on multigrid convergence theorems, *Appl. Math. Comput.* **23**, 103–120.

PASCIAK, J.E. (1987), Domain decomposition preconditioners for elliptic problems in two and three dimensions; first approach, in: R. Glowinski, G.H. Golub, G.A. Meurant and J. Périaux, eds., *Proceedings, First International Symposium on Domain Decomposition Methods for Partial Differential Equations* (SIAM, Philadelphia, PA) 62–72.

PASCIAK, J.E. (1988a), Domain decomposition preconditioners for elliptic problems in two and three dimensions, in: M. Schultz, ed., *Numerical Algorithms for Modern Parallel Computer Architectures*, IMA Volumes Math. Appl. 13 (Springer, New York) 163–172.

PASCIAK, J.E. (1988b), Two domain decomposition methods for Stokes equations, in: T. Chan, R. Glowinski, J. Périaux and O. Widlund, eds., *Proceedings, Second International Symposium on Domain Decomposition Methods for Partial Differential Equations* (SIAM, Philadelphia, PA) 419–430.

PEACEMAN, D.W. and H.H. RACHFORD, JR. (1955), The numerical solution of parabolic and elliptic differential equations, *SIAM J. Appl. Math.* **3**, 28–41.

PEISKER, P. (1985), A multilevel algorithm for the biharmonic problem, *Numer. Math.* **46**, 623–634.

PEISKER, P. and D. BRAESS (1987), A conjugate gradient method and a multigrid algorithm for Morley's finite element approximation of the biharmonic equation, *Numer. Math.* **50**, 567–586.

PITKÄRANTA, J. and T. SAARINEN (1985), A multigrid version of a simple finite element method for the Stokes problem, *Math. Comp.* **45**, 1–14.

POWELL, M. and M. SABIN (1977), Piecewise quadratic approximations on triangles, *ACM Trans. Math. Software* **3**, 316–325.

RANNACHER, R. (1979a), Nonconforming finite element methods for eigenvalue problems in linear plate theory, *Numer. Math.* **33**, 23–42.

RANNACHER, R. (1979b), On nonconforming and mixed finite element methods for plate bending problems. The linear case, *RAIRO Anal. Numer.* **13**, 369–387.

RAVIART, P.-A. and J. THOMAS (1977), A mixed finite element method for 2nd order elliptic problems, in: I. Galligani and E. Magenes, eds., *Mathematical Aspects of Finite Element Methods*, Lecture Notes in Math. 606 (Springer, New York) 292–315.

SAAD, Y. and M. SCHULTZ (1986), Gmres: A generalized minimal residual algorithm for solving nonsymmetric linear systems, *SIAM J. Sci. Statist. Comput.* **7**, 856–869.

SCHATZ, A. (1974), An observation concerning Ritz–Galerkin methods with indefinite bilinear forms, *Math. Comp.* **28**.

SCHATZ, A.H. and J. WANG (1996), Some new error estimates for Ritz–Galerkin methods with minimal regularity assumptions, *Math. Comp.* **65** (213), 19–27.

SCHWARZ, H. (1869), Ueber einige Abbildungsaufgaben, *J. Reine Angew. Math.* **70**, 105–120 (*Ges. Math. Abh.* **2**, 65–83 (Springer, 1890)).

SCOTT, L. and M. VOGELIUS (1985), Conforming finite element methods for incompressible and nearly incompressible continua, in: B.E. Engquist, ed., *Lectures in Appl. Math.* **22** (Amer. Math. Soc., Providence), Inst. for Phys. Sci. and Tech., Univ. of Maryland, Tech. Rept. BN-1018.

SCOTT, L. and S. ZHANG (1992), Higher-dimensional nonnested multigrid methods, *Math. Comp.* **58**, 457–466.

SHI, Z. (1990), Error estimates of Morley element, *Chinese J. Numer. Math. Appl.* **12**, 9–15.

SPEKREIJSE, S.P. (1987), Multigrid solution of monotone second-order discretization of hyperbolic conservation laws, *Math. Comp.* **49**, 135–155.

STEVENSON, R. (1993a), New estimates of the contraction number of V-cycle multi-grid with applications to anisotropic equationsm in: W. Hackbusch and G. Wittum, eds., *Incomplete Decompositions, Proceedings of the Eighth GAMM Seminar*, Notes on Numer. Fluid Mech. 41, 159–167.

STEVENSON, R. (1993b), Robustness of multi-grid applied to anisotropic equations on convex domains and domains with re-entrant corners, *Numer. Math.* **66**, 373–398.

STEVENSON, R. (1994a), Modified ILU as a smoother, *Numer. Math.* **68**, 295–309.

STEVENSON, R. (1994b), Robust multi-grid with 7-point ILU smoothing, in: P. Hemker and P.D. Wesseling, eds., *Multigrid Methods IV, Proceedings of the Fourth European Multigrid Conference, Amsterdam* (Birkhäuser) 295–307.

STEVENSON, R. (1994c), Robustness of the additive and multiplicative frequency decomposition multi-level method, Tech. Rept., Department of Mathematics and Computing Science, to appear in *Computing*.

STRANG, G. (1972), Approximation in the finite element method, *Numer. Math.* **19**, 81–93.

SWARZTRAUBER, P. (1977), The methods of cyclic reduction, Fourier analysis and the facr algorithm for the discrete solution of Poisson's equation on a rectangle, *SIAM Rev.* **19**, 490–501.

TEMAM, R. (1977), *Navier–Stokes Equations* (North-Holland, New York).

VARGA, R. (1962), *Matrix Iterative Analysis* (Prentice-Hall, Englewood Cliffs, NJ).

VASSILEVSKI, P. (1987), Iterative methods for solving finite element equations based on, multilevel splitting of the matrix, Bulgar. Acad. Sci., Sofia, Bulgaria.

VERFÜRTH, R. (1984), A multilevel algorithm for mixed problems, *SIAM J. Numer. Anal.* **21**, 264–284.

VERFÜRTH, R. (1991), A posteriori error estimates for the Stokes equations, II. Non-conforming discretizations, *Numer. Math.* **60**, 235–249.

VON PETERSDORFF, T. and E. STEPHAN (1990), On the convergence of the multigrid method for a hypersingular integral equation of the first kind, *Numer. Math.* **57**, 379–391.

WACHSPRESS, E. (1963), Extended application of alternating direction implicit iteration model problem theory, *SIAM J. Appl. Math.* **11**, 994–1016.

WANG, J. (1991), Convergence analysis of multigrid algorithms for non-selfadjoint and indefinite elliptic problems, in: *Proc. 5th Copper Mountain Conference on Multigrid Methods* (Copper Mountain, CO).

WANG, J. (1992), Convergence analysis without regularity assumptions for multigrid algorithms based on SOR smoothing, *SIAM J. Numer. Anal.* **29**, 987–1001.

WESTLAKE, J. (1968), *A Handbook of Numerical Matrix Inversion and Solution of Linear Equations* (John Wiley, New York).

WIDLUND, O. (1988), Optimal iterative refinement methods, Tech. Rept. 391, Courant Institute of Mathematical Sciences.

WITTUM, G. (1989a), Linear iterations as smoothers in multigrid methods: Theory with applications to incomplete decompositions, *IMPACT Comput. Sci. Engrg.* **1**, 180–215.

WITTUM, G. (1989b), On the robustness of ILU smoothing, *SIAM J. Sci. Statist. Comput.* **10** (4), 699–717.

PATTERSON, W.M., R. (1974), *Iterative Methods for the Solution of a Linear Operator Equation in Hilbert Space – A Survey*, Lecture Notes in Math. 394 (Springer, New York).

XU, J. (1989a), Convergence estimates for some multigrid algorithms, in: *Third International Symposium on Domain Decomposition Methods for Partial Differential Equations, held in Houston, Texas (March 20–22)*. Oral presentation.

XU, J. (1989b), Theory of multilevel methods, Ph.D. thesis, Cornell University.

XU, J. (1992a), Iterative methods by space decomposition and subspace correction, *SIAM Rev.* **34**, 581–613.

XU, J. (1992b), A new class of iterative methods for nonsymmetric boundary value problems, *SIAM J. Numer. Anal.* **29**, 303–319.

XU, J. and X.-C. CAI (1992), A preconditioned gmres method for nonsymmetric and indefinite problems, *Math. Comp.* **59**, 311–319.

YOUNG, D.M. (1954), Iterative methods for solving partial difference equations of elliptic type, *Trans. Amer. Math. Soc.* **76**, 92–111.

YSERENTANT, H. (1986a), The convergence of multi-level methods for solving finite-element equations in the presence of singularities, *Math. Comp.* **47**, 399–409.

YSERENTANT, H. (1986b), On the multi-level splitting of finite element spaces, *Numer. Math.* **49**, 379–412.

YSERENTANT, H. (1990), Two preconditioners based on the multi-level splitting of finite element spaces, *Numer. Math.* **58**, 163–184.

YSERENTANT, H. (1993), Old and new convergence proofs for multigrid methods, in: A. Iserles, ed., *Acta Numerica* (Cambridge Univ. Press) 285–326.

ZHANG, S. (1988), Multi-level iterative techniques, Ph.D. thesis, Penn. State Univ.

ZHANG, S. (1989), An optimal order multigrid method for biharmonic C^1 finite element equations, *Numer. Math.* **56**, 613–624.

ZHANG, S. (1990a), Optimal order nonnested multigrid methods for solving finite element equations, I: On quasi-uniform meshes, *Math. Comp.* **55**, 23–36.

ZHANG, S. (1990b), Optimal order nonnested multigrid methods for solving finite element equations, II: On non-quasi-uniform meshes, *Math. Comp.* **55**, 439–450.

ZHANG, X. (1992), Multilevel Schwarz methods, *Numer. Math.* **63**, 521–539.

ZHANG, X. (1994), Multilevel Schwarz methods for the biharmonic Dirichlet problem, *SIAM J. Sci. Comput.* **15** (3), 621–644.

ZHANG, X. (1996), Two-level Schwarz methods for the biharmonic problem discretized by conforming C^1 elements, *SIAM J. Numer. Anal.* **33** (2), 555–570.

Subject Index

Wavelet Methods in Numerical Analysis

Albert Cohen

Laboratoire d'Analyse Numérique
Université Pierre et Marie Curie
16, rue Clisson, 75013 Paris
France

HANDBOOK OF NUMERICAL ANALYSIS, VOL. VII
Solution of Equations in $\mathbb{R}^n$ (Part 3)
Techniques of Scientific Computing (Part 3)
edited by P.G. Ciarlet and J.L. Lions

Contents

Preface

Since the 1960's, multiscale methods have been used in numerous areas of applied mathematics as diverse as signal analysis, statistics, computer aided geometric design, image processing and numerical analysis. The mathematical background underlying these methods was substantially reinforced with the emergence of wavelet bases in the 1980's.

> *The objectives of this article are to survey the theoretical results that are involved in the numerical analysis of wavelet methods, and more generally of multiscale decomposition methods, for numerical simulation problems, and to provide the most relevant examples of such mathematical tools in this particular context.*

Multiscale methods are based on approximations $(f_j)_{j\geqslant 0}$ to the data (or the unknown) f of a given problem, at various resolution levels indexed by j. The corresponding discretization steps h_j are usually chosen to be of order 2^{-j}: the approximation f_j should thus be viewed as a sketchy picture of f that cannot oscillate at a frequency higher than 2^j. As an example, if f is a univariate continuous function, one could choose for f_j the unique function such that $f_j(2^{-j}k) = f(2^{-j}k)$ for all $k \in \mathbb{Z}$ and such that f_j is affine when restricted to $[k2^{-j}, 2^{-j}(k+1)]$.

Formally, one obtains a *multiscale decomposition* by expanding f into the sum of its coarsest approximation and additional details:

$$f = f_0 + \sum_{j\geqslant 0} g_j,$$

where each $g_j = f_{j+1} - f_j$ represents the fluctuations of f between the two successive levels of resolutions j and $j+1$.

In practice, these approximations and decompositions can be defined and implemented in various ways: by contrast with the Fourier transform, a "multiscale transform" stands for a versatility of different mathematical tools. Some of them can furthermore be implemented by means of fast algorithms, and are thus more appealing for numerical applications. In such an applied context, one is often interested in further dividing each fluctuation g_j into local contributions. For specific types of multiresolution approximations, this task can be achieved through a *wavelet basis decomposition*: one uses an appropriate function ψ (in the case where f is a univariate function) that is well localized both in space and frequency, and oscillates in the sense that $\int \psi = 0$, to expand g_j according to

$$g_j = \sum_{k \in \mathbb{Z}} d_{j,k} \psi_{j,k},$$

where the $d_{j,k}$'s are scalar coefficients and each wavelet $\psi_{j,k} := 2^{j/2}\psi(2^j \cdot -k)$ contributes to the fluctuation of f at scale 2^{-j} in a neighborhood of size $2^{-j}|\text{Supp}(\psi)|$ around the point $2^{-j}k$. Here again, there exists a versatile collection of functions ψ that are admissible to generate wavelets bases.

One should thus keep in mind that, for a given problem, the efficiency of a multiscale decomposition method is strongly tied to the choice of a specific tool and associated algorithm.

In the field of numerical analysis, multiscale and wavelets decompositions have been successfully used for three main tasks: the *preconditioning* of large systems arising from the discretization of elliptic partial differential equations, the *adaptive approximation* of functions, allowing to resolve isolated singularities at a low computational cost, and the *sparse representation* of initially full matrices arising from the discretization of integral equations.

It should be noted that in each of these applications, the use of multiscale and wavelet decomposition is very close in essence to other methods that have been priorily introduced: multigrid methods for preconditioning, mesh refinement and moving grids for adaptivity, multipole and pannel clustering algorithms for integral equations.

Do wavelet-based techniques bring specific improvements, when dealing with the above tasks, in comparison to these more classical methods?

Clearly, these techniques enlarge the "library" of available tools that are available to deal with these tasks. As was mentioned previously, an optimal choice among these tools is strongly problem dependent.

More importantly, multiscale decomposition techniques are characterized by a strong theoretical imput from harmonic analysis and approximation theory. In particular, the possibility of characterizing various smoothness classes – e.g., Sobolev, Hölder and Besov spaces – from the numerical properties of multiscale decompositions, turns out to play a key role in the application of wavelet methods to the three previously mentioned tasks. This brings a new and fruitful point of view, which can also be useful in the analysis of more classical, related methods.

Our goal here is not to present all the existing wavelet methods for the numerical simulation of physical processes (the amount of literature is already huge and still growing fast), but rather to give an account of this theoretical imput to the numerical analysis of multiscale methods in the large.

This article is organized as follows:

In Chapter I, we motivate the introduction of a general theory by two very simple examples of multiscale decomposition schemes. These schemes are associated to the most elementary wavelet bases: the Haar system and the Schauder hierarchical basis, that are respectively built from approximations by piecewise constant functions and piecewise affine continuous functions. We show how these schemes can be adapted to the decomposition of multivariate functions and how they can be used for adaptive approximation of functions, sparsification of integral operators, and preconditioning of elliptic equations. We discuss their inherent limitations for these applications. This chapter is descriptive in essence and the results are deliberately not organized according to the usual "lemma-theorem-corollary" sequence. The reader who is prepared to enter directly the general theory can easily skip this chapter.

In Chapter II, we review the general concept of multiresolution approximation, consisting in a nested sequence of approximation spaces,

$$\cdots V_{-1} \subset V_0 \subset V_1 \subset \cdots$$

which are used to define the approximation $f_j \in V_j$. This concept leads in a natural way to the construction of wavelets in terms of *details* complementing the approximation between V_j and V_{j+1}. In practice, there are two main approaches to build such tools. The first one, which is the most intuitive to the numerician, consists in constructing the spaces V_j as a nested sequence of finite element spaces, or as nested discretizations by cell averages or point values equipped with certain inter-scale operators. The second one is based on a single *scaling function* φ generating V_j in the sense that $\varphi(2^{-j}\bullet - k)$, $k \in \mathbb{Z}$, is a Riesz basis of V_j. This last approach yields tools of a different nature than finite elements, that might seem difficult to handle for practical computations: in particular, the function φ is not given explicitly, but as a solution of an equation of the type

$$\varphi(x) = \sum_{n\in\mathbb{Z}} h_n \varphi(2x - n),$$

that expresses the imbedding $V_0 \subset V_1$. A solution of such an equation is also called a *refinable function*. Since they are by far less classical than finite elements in numerical analysis, and since they are a key tool in the construction of wavelet bases, we devote an important part of this chapter to the study of refinable functions. In particular, we show how they can be computed and used for numerical analysis purposes, and we analyze the relations between the properties of the coefficients h_n and properties of φ, such as smoothness, approximation power of the spaces V_j, stability, orthonormality and biorthogonality. We describe the examples of refinable functions and related multiscale decompositions that are relevant in numerical analysis, in particular biorthogonal decompositions into spline functions. We show how these tools can be adapted to multivariate domains with specific boundary conditions. We finally return to the first approach and address the construction of multiscale decompositions associated to the most commonly used discretizations in numerical analysis: point values, cell averages and finite elements.

In Chapter III, we show how smoothness classes can be characterized from the decay properties of $(\|g_j\|_{L^p})_{j\geqslant 0}$ and $(\|f - f_j\|_{L^p})_{j\geqslant 0}$, as j goes to $+\infty$, or equivalently from the numerical properties of the wavelet coefficients $d_{j,k}$. These results turn out to be the key to the theoretical understanding of the multilevel preconditioning of various elliptic operators. The L^p norm which is used here to measure the error is the same as the L^p metric associated to the smoothness class: e.g., in the case of the Sobolev spaces H^s, one will consider the behaviour of $\|f - f_j\|_{L^2}$ or $\|g_j\|_{L^2}$. The quantity $\|f - f_j\|_{L^p}$ is a *linear approximation* error, in the sense that f_j is an approximation of f in the linear space V_j, and is usually obtained through a linear operation applied to f (typically a projection onto V_j). These characterization results are essentially based on the combination of two types of estimates: a *direct* (or Jackson type) inequality and an *inverse* (or Bernstein type) inequality. We show how to establish such inequalities in the case of the multiresolution spaces that were built in Chapter II. We discuss several variants of these results, as well as their adaptation to function spaces defined on bounded domains with prescribed boundary conditions. We finally discuss the applications to multilevel preconditioning.

In Chapter IV, we focus on *nonlinear approximation*: the function f is now approximated by a combination

$$f_N = \sum_{(j,k)\in E_N} c_{j,k}\psi_{j,k},$$

where the set E_N has cardinality N and is allowed to depend on the function f, in order to optimize the approximation. In practice, E_N typically represents the N largest contributions in the wavelet decomposition of f, in the metric where the error is to be measured. As in the case of linear approximation, the decay properties of the error $\|f - f_N\|_{L^p}$ are related to the smoothness properties of f, but the corresponding smoothness classes are now associated to an L^q metric with $q < p$. Similar ideas apply to the wavelet discretization of partial differential and integral operators: sparse approximations can be obtained by simple thresholding operations on the matrix entries. In the practice of numerical simulation, the main difficulty of nonlinear approximation is that one does not know in advance which are the largest coefficients of the solutions. We conclude by discussing the existing practical adaptive strategies that aim to build an optimal set of wavelet coefficients for describing the solution at a low memory and computational cost.

Among the basic references that have influenced the present article, let us mention the two well-known monographs on wavelets by DAUBECHIES [1992] and MEYER [1990], as well as the survey papers by DAHMEN [1997] and DEVORE [1998].

Beside these important references, it goes without saying that many of the results which are presented here have appeared in publications by numerous mathematicians. Most of the time, however, their formulation has been revisited for consistency purpose. We also have chosen to avoid "minimal" or general assumptions in the statement of some of these results, in the cases where these assumptions make the proof more complex without being really useful to numerical applications. In each chapter, several remarks and the last section aim to present an historical account of these contributions, and to give "pointers" toward related theoretical results and algorithmic techniques. The present author is aware of the rapid evolution of the subject and of its multiple connexions with other scientific areas. He has tried to mention the most significant contributions in the context of numerical analysis, together with the constraint of keeping these notes self-contained and of reasonable size.

CHAPTER I

Basic Examples

1. Introduction

Before entering the general theory of multiscale decompositions, we shall study two basic examples and show their potential range of applications in numerical analysis as well as their inherent limitations. Our goal is to identify some important features that will be studied with more details, in the general theory developed in the next chapters. This chapter is descriptive in essence and some of its results are quoted without a detailed proof when they are particular case of theorems that are proved in the next chapters.

In the two examples that we want to study, the approximation f_j of a univariate function f at the scale 2^{-j} will respectively be piecewise constant and piecewise affine on dyadic intervals

$$I_{j,k} := \left[k2^{-j}, (k+1)2^{-j}\right[, \quad k \in \mathbb{Z}. \tag{1.1}$$

In Sections 2 and 3, we introduce these approximations and we show how they are related to decompositions in two elementary wavelet bases: the Haar system and the hierarchical Schauder basis. We describe the decomposition and reconstruction algorithms that can be used to compute the coefficients of a function in these bases, and we show in Section 4 how these schemes can be generalized in a natural way to multivariate functions.

Applications of these elementary tools to numerical analysis are discussed in the next sections: in Section 5, we show the potential of the Haar and Schauder bases for adaptive approximation of functions. The main idea is that a simple thresholding procedure on the coefficients of a function f in these bases amounts in building an adaptive grid for discretizing this function at a low cost. The same idea is applied to sparsify certain integral operators, which usually yield fully populated matrices in their usual finite element discretization. In Section 6, we recall some basic facts on the finite element discretization of a simple model second-order elliptic problem, which is known to yield an ill-conditioned system, and we show how multiscale decompositions can be used to design simple preconditioners and optimal iterative solvers.

We conclude by reviewing the most important features of these schemes that suggest a more general theory. We also point out on their inherent limitations: poor approximation orders and severe restrictions on the class of differential and integral operators that they can handle.

2. The Haar system

Let f be a function in $L^2(\mathbb{R})$. We can define piecewise constant approximations f_j of f at scale 2^{-j} by

$$f_j(x) = 2^j \int_{I_{j,k}} f(t)\,\mathrm{d}t, \quad \text{for all } x \in I_{j,k},\ k \in \mathbb{Z}, \tag{2.1}$$

i.e. f is approximated by its mean value on each interval $I_{j,k}$, $k \in \mathbb{Z}$.

Let us make three simple comments on this particular choice for f_j:

We first remark that the choice of the mean value makes f_j the L^2-orthogonal projection of f onto the space

$$V_j = \{f \in L^2;\ f \text{ is constant on } I_{j,k},\ k \in \mathbb{Z}\}. \tag{2.2}$$

Indeed, an orthogonal basis for V_j is given by the family

$$\varphi_{j,k} := 2^{j/2}\varphi(2^j \bullet - k), \quad k \in \mathbb{Z}, \tag{2.3}$$

where $\varphi := \chi_{[0,1]}$, and clearly f_j can be written

$$f_j = \sum_{k\in\mathbb{Z}} \langle f, \varphi_{j,k}\rangle \varphi_{j,k}, \tag{2.4}$$

with the usual notation $\langle f, g\rangle = \int f(t)\overline{g(t)}\,\mathrm{d}t$. We will thus denote f_j by $P_j f$ where P_j is the orthogonal projector onto V_j. We shall also use the notation $c_{j,k}(f) := \langle f, \varphi_{j,k}\rangle = 2^{j/2}\int_{I_{j,k}} f(t)\,\mathrm{d}t$, for the normalized mean values which are the coordinates of $P_j f$ in the basis $(\varphi_{j,k})_{k\in\mathbb{Z}}$.

We also note that approximation process is *local*: the value of $P_j f$ on $I_{j,k}$ is only influenced by the value of f on the same interval. In particular, we can still use (2.1) to define $P_j f$ when f is only locally integrable, or if f is only defined on a bounded interval such as $[0, 1]$ (in that case, $P_j f$ makes sense only for $j \geqslant 0$). As an example, we display on Fig. 2.1 the function $f(x) = \sin(2\pi x)$ and its approximation $P_4 f$ on $[0, 1]$.

Finally, since $V_j \subset V_{j+1}$, it is clear that $P_{j+1} f$ contains "more information" on f than the coarser approximation: the mean values on the intervals of size $2^{-j-1}k$ entirely determine those on the coarser intervals of double size 2^{-j} by taking their averages. More precisely, we have

$$P_j f|_{I_{j,k}} = (P_{j+1} f|_{I_{j+1,2k}} + P_{j+1} f|_{I_{j+1,2k+1}})/2. \tag{2.5}$$

We can also define the orthogonal projection $Q_j f := P_{j+1} f - P_j f$ onto W_j, the orthogonal complement of V_j into V_{j+1}. From (2.5), it is clear that $Q_j f$ "oscillates" in the sense that

$$Q_j f|_{I_{j+1,2k}} = -Q_j f|_{I_{j+1,2k+1}}. \tag{2.6}$$

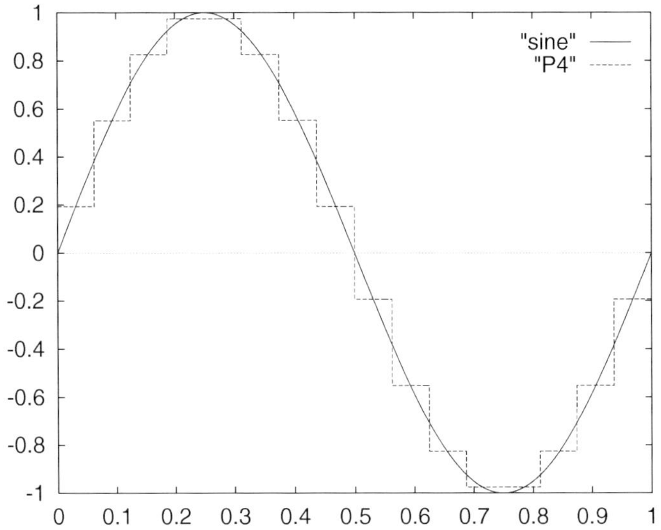

FIG. 2.1. The function $f(x) = \sin(2\pi x)$ and its approximation $P_4 f$.

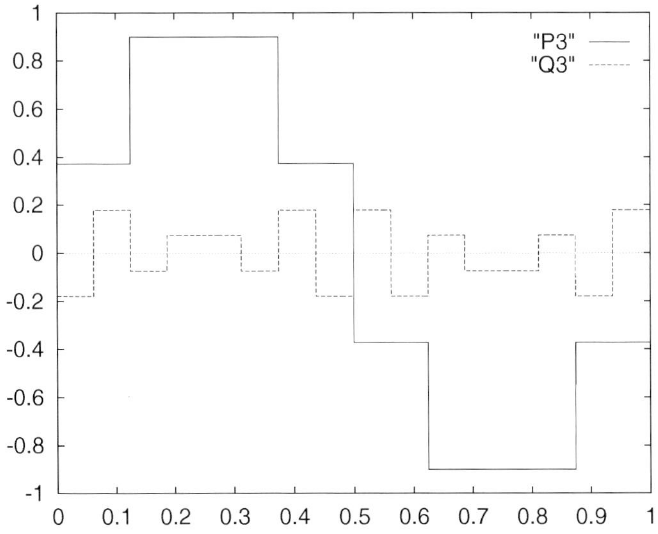

FIG. 2.2. The approximation $P_3 f$ and the fluctuation $Q_3 f$.

As an example, Fig. 2.2 shows the components $P_3 f$ and $Q_3 f$ on $[0, 1]$, for $f(x) = \sin(2\pi x)$: we have thus decomposed $P_4 f$ of Fig. 2.1 into the sum of the coarser approximation $P_3 f$ and the oscillating component $Q_3 f$.

The oscillation property (2.6) allows us to expand $Q_j f$ into

$$Q_j f = \sum_{k\in\mathbb{Z}} d_{j,k}(f)\psi_{j,k}, \tag{2.7}$$

where $\psi_{j,k} := 2^{j/2}\psi(2^j \bullet -k)$ and

$$\psi(x) = \chi_{[0,1/2[} - \chi_{[1/2,1[}. \tag{2.8}$$

Since the $\psi_{j,k}$, $k \in \mathbb{Z}$ are also an orthonormal system, they constitute an orthonormal basis for W_j and we thus have $d_{j,k}(f) = \langle f, \psi_{j,k}\rangle$. We thus have precised the "two-level" decomposition of $P_{j+1} f$ into the coarser approximation $P_j f$ and the additional fluctuations $Q_j f$, according to

$$\sum_k c_{j+1,k}(f)\varphi_{j+1,k} = \sum_k c_{j,k}(f)\varphi_{j,k} + \sum_k d_{j,k}(f)\psi_{j,k}. \tag{2.9}$$

This decomposition can be iterated on an arbitrary number of levels: if $j_0 < j_1$, we can rewrite the orthogonal decomposition

$$P_{j_1} f = P_{j_0} f + \sum_{j_0\leqslant j<j_1} Q_j f, \tag{2.10}$$

into

$$\sum_k c_{j_1,k}(f)\varphi_{j_1,k} = \sum_k c_{j_0,k}(f)\varphi_{j_0,k} + \sum_{j_0\leqslant j<j_1} \sum_k d_{j,k}(f)\psi_{j,k}. \tag{2.11}$$

Both (2.10) and (2.11) express an additive *multiscale decomposition* of $P_{j_1} f$ into a coarser approximation and a succession of fluctuation at intermediate scales. Formula (2.11), however, gives a local description of each contribution and should be viewed an orthonormal change of basis in V_{j_1}: both $\{\varphi_{j_1,k}\}_{k\in\mathbb{Z}}$ and $\{\varphi_{j_0,k}\}_{k\in\mathbb{Z}} \cup \{\psi_{j,k}\}_{j_0\leqslant j<j_1,k\in\mathbb{Z}}$ are orthonormal bases for V_{j_1}, and any function in V_{j_1} has thus a unique decomposition in each of these bases.

Note the different role played by the functions φ and ψ: the former is used to characterize the approximation of a function at different scales, while the latter is needed to represent the "fluctuations" between successive levels of approximations. In particular, we have $\int \psi = 0$, reflecting the oscillatory nature of these fluctuations. In the more general multiresolution context that will be developed in the next chapter, φ will be called *scaling function* and ψ the *mother wavelet*, in the sense all the wavelets $\psi_{j,k}$ are generated from translations and dilations of ψ. We display on Fig. 2.3 the function ψ and two of its translated-dilated versions.

Clearly the union of the approximation spaces V_j is dense in $L^2(\mathbb{R})$, i.e.

$$\lim_{j\to+\infty} \|f - P_j f\|_{L^2} = 0, \tag{2.12}$$

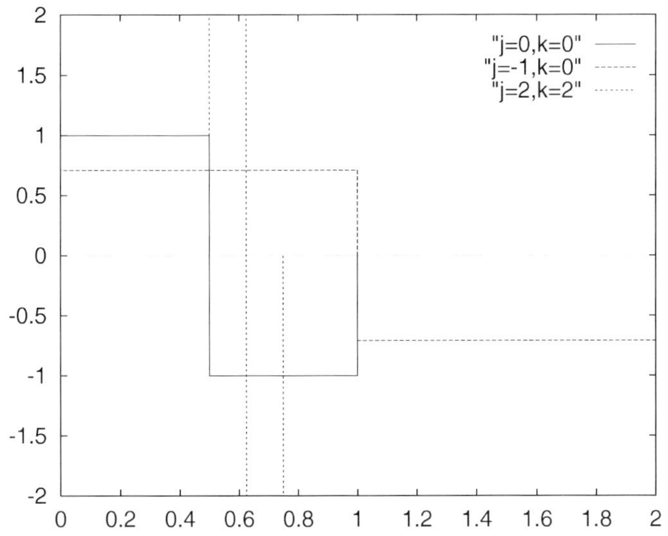

FIG. 2.3. The wavelets $\psi = \psi_{0,0}$, $\psi_{-1,0}$ and $\psi_{2,2}$.

for all $f \in L^2(\mathbb{R})$. Combining (2.12) with (2.11), we obtain that the orthonormal family $\{\varphi_{j_0,k}\}_{k\in\mathbb{Z}} \cup \{\psi_{j,k}\}_{j\geqslant j_0,k\in\mathbb{Z}}$ is a complete system in $L^2(\mathbb{R})$ and thus an orthonormal basis. Any function $f \in L^2(\mathbb{R})$ can thus be decomposed into

$$f = \sum_k c_{j_0,k}(f)\varphi_{j_0,k} + \sum_{j\geqslant j_0}\sum_k d_{j,k}(f)\psi_{j,k}, \tag{2.13}$$

where the series converges in L^2.

REMARK 2.1. Since we have an orthonormal basis, the L^2-convergence in (2.13) is *unconditional*: one can permutate the terms or change their signs without affecting the convergence property. This might not be true for other type of convergence: if f is uniformly continuous, clearly $P_j f$ converges uniformly to f as j goes to $+\infty$, i.e. we can define a summation process by letting j_1 go to $+\infty$ in (2.11). However, a permutation in the terms might lead to a divergent series. The same remark holds for the L^1 convergence of the series of an L^1 function.

REMARK 2.2. We can of course apply the same method to decompose a function that is defined on a bounded interval. In particular, φ and $\psi_{j,k}$, $j \geqslant 0$, $0 \leqslant k < 2^j$, constitute an orthonormal basis for $L^2([0, 1])$, which is known as the *Haar system*, since its introduction by A. Haar in the early twentieth century.

If the function f is defined on the whole of $\mathbb{R}$, we can also let j_0 go to $-\infty$ in (2.13). It turns out that the coarse scale approximation vanishes in the L^2 sense, when passing to this limit: For any f in $L^2(\mathbb{R})$, we have

$$\lim_{j\to-\infty} \|P_j f\|_{L^2} = 0. \tag{2.14}$$

To prove (2.14), we first suppose that $f = \chi_{[a,b]}$, for some $a < b$. Then, when j is negative and large enough so that $2^{-j} \geqslant \max\{|a|, |b|\}$, we have $c_{j,k}(f) = \langle f, \varphi_{j,k}\rangle = 0$ if k is not 0 or -1 (the functions f and $\varphi_{j,k}$ have disjoint supports). We thus have, for this range of j:

$$\|P_j f\|_{L^2}^2 = |c_{j,0}(f)|^2 + |c_{j,-1}(f)|^2 \leqslant 2\big[|a-b|2^{j/2}\big]^2, \tag{2.15}$$

which implies (2.14) for $f = \chi_{[a,b]}$ and thus for any finite linear combination of such characteristic functions.

For an arbitrary $f \in L^2(\mathbb{R})$ and $\varepsilon > 0$, there exists a function g which is of the type

$$g = \sum_{i=0}^{N} c_i \chi_{[a_i,b_i]}, \tag{2.16}$$

such that

$$\|f - g\|_{L^2} \leqslant \varepsilon \tag{2.17}$$

(finite linear combinations of characteristic functions of finite intervals are dense in $L^2(\mathbb{R})$). Since $\|P_j g\|_{L^2}$ tends to zero as j goes to $-\infty$, we have $\|P_j g\|_{L^2} \leqslant \varepsilon$ for j large enough, and thus

$$\|P_j f\|_{L^2} \leqslant \|P_j g\|_{L^2} + \|P_j(f-g)\|_{L^2} \leqslant \|P_j g\|_{L^2} + \|f-g\|_{L^2} \leqslant 2\varepsilon. \tag{2.18}$$

Since ε is arbitrary, this proves (2.14) for a general function $f \in L^2(\mathbb{R})$.

An immediate consequence of (2.14) is the following: *the set* $\{\psi_{j,k}\}_{j,k\in\mathbb{Z}}$ *constitutes an orthonormal basis for* $L^2(\mathbb{R})$. This fact might appear as a paradox: a function f can be expanded in terms of function that all have vanishing integral, even if $\int f \neq 0$. This is possible, simply because the convergence of $P_j f$ to zero as j goes to $-\infty$ holds in L^2 *but not in* L^1, for a general function in $L^1 \cap L^2$. Indeed, according to the definition of P_j by (2.1), we clearly have $\int P_j f = \int f$, so that $P_j f$ cannot go to zero in L^1 as soon as f has not a vanishing integral.

In the context of applications of multiscale decompositions to numerical simulation, one does not make so much use of the full wavelet basis $\{\psi_{j,k}\}_{j,k\in\mathbb{Z}}$ and of the fact that $P_j f$ goes to 0 in L^2 as j tends to $-\infty$. Indeed, most problems are confined to a bounded domain, that limits the coarseness of the analysis scale. We shall thus consider more often multiscale decompositions of the type (2.13) for a fixed minimal j_0. Of more importance to the numerician is the behaviour of $P_j f$ as j grows, in particular approximation results dealing with the measurement of the error $f - P_j f$. This question will be addressed in a more general context in Chapters II and III.

On a computational point of view, if we want to implement the decomposition of a function in the Haar system, we also need to limit the resolution level from above: we shall start our analysis from an approximation $f_{j_1} = \sum_k c_{j_1,j}\varphi_{j,k}$ of a function f in the finest resolution level j_1. We then have to deal with the following problem: how to compute the coefficients $c_{j_0,k}$ and $d_{j,k}$, $j_0 \leqslant j < j_1$, from the coefficients $c_{j_1,k}$ at the finest scale?

Note that this problem makes sense even if f_{j_1} is an approximation of f in V_{j_1} that differs from its projection $P_{j_1} f$: we are then interested in decomposing this particular approximation into multiscale components. This point is crucial, since in numerical analysis applications, the approximate solution of a problem is rarely the L^2-orthogonal projection of the true solution.

To answer this question, we start from the "two level" decomposition (2.9). From (2.5) and the L^2 normalization of the functions $\varphi_{j,k}$ and $\psi_{j,k}$, we derive the simple relations

$$c_{j,k} = (c_{j+1,2k} + c_{j+1,2k+1})/\sqrt{2} \tag{2.19}$$

and

$$d_{j,k} = (c_{j+1,2k} - c_{j+1,2k+1})/\sqrt{2}. \tag{2.20}$$

Another way to derive (2.19) and (2.20) is to inject the identities $\varphi_{j,k} = (\varphi_{j+1,2k} + \varphi_{j+1,2k+1})/\sqrt{2}$ and $\psi_{j,k} = (\varphi_{j+1,2k} - \varphi_{j+1,2k+1})/\sqrt{2}$ in $c_{j,k} = \langle f, \varphi_{j,k}\rangle$ and $d_{j,k} = \langle f, \psi_{j,k}\rangle$. These relations show the local aspect of the change of basis in (2.9): the vectors $(c_{j,k}, d_{j,k})$ and $(c_{j+1,2k}, c_{j+1,2k+1})$ are related by a simple orthogonal transformation.

The inverse transform is given by

$$c_{j+1,2k} = (c_{j,k} + d_{j,k})/\sqrt{2} \tag{2.21}$$

and

$$c_{j+1,2k+1} = (c_{j,k} - d_{j,k})/\sqrt{2}. \tag{2.22}$$

From these four relations, we can derive simple decomposition and reconstruction algorithms:

DECOMPOSITION ALGORITHM.

- Start from finest scale approximation coefficients $c_{j_1,k}$.
- Compute the sequences $c_{j_1-1,k}$ and $d_{j_1-1,k}$ using (2.19) and (2.20).
- Store the details $d_{j_1-1,k}$ and iterate the decomposition on the approximations $c_{j_1-1,k}$.
- Iterate the decomposition from fine to coarse scales.

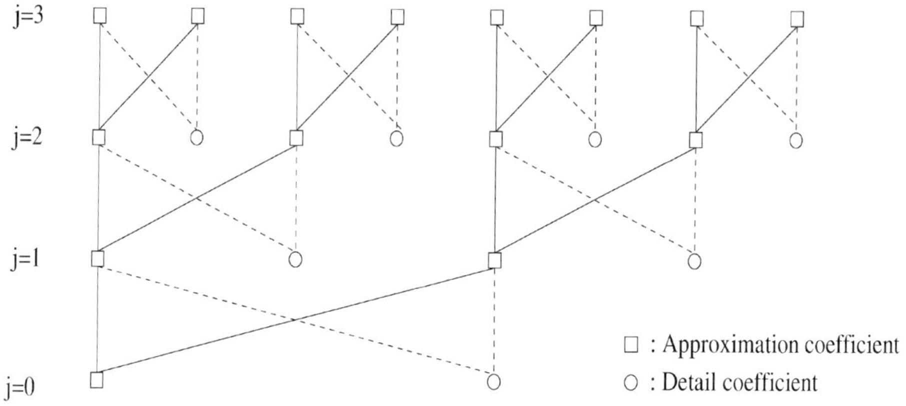

FIG. 2.4. Pyramidal representation of the Haar transform.

- Stop the decomposition when the coarsest level coefficients $c_{j_0,k}$ and $d_{j_0,k}$ are computed.

RECONSTRUCTION ALGORITHM.
- Start from the coarsest level coefficients $c_{j_0,k}$ and $d_{j_0,k}$.
- Compute the sequence $c_{j_0+1,k}$ using (2.21) and (2.22).
- Iterate the reconstruction using the details $d_{j_0+1,k}$.
- Iterate the reconstruction from coarse to fine scales.
- Stop the reconstruction when the finest approximation coefficients $c_{j_1,k}$ are computed.

In practice, the function to be decomposed is defined (and approximated at the finest level) on a bounded domain. In turn, the sequence $c_{j,k}$ has finite length, allowing a computer to perform these algorithms. For example, if the function f is defined on the interval $[0,1]$, these algorithms operate on 2^{j_1} coefficients. We remark that both decomposition and reconstruction algorithms perform 2^{j+1} operations between the level j and $j+1$. This is illustrated by the "pyramidal" representation of these algorithms, in Fig. 2.4, in the case where $j_1 = 3$ and $j_0 = 0$: each "branch" indicates the influence of a coefficient at a given scale j on another one at the level $j+1$ or $j-1$ (the dashed branches correspond to the computations involving the detail coefficients).

The total number of operations for both algorithms is thus $\sum_{j=j_0}^{j_1} 2^{j+1} = \mathcal{O}(N)$ where $N = 2^{j_1}$ is the length of the finest approximation sequence. This complexity is "optimal", in the sense that the number of required operations is of the same order as the number of computed values. A general linear transform would require $\mathcal{O}(N^2)$ operations. From a linear algebra point of view, the Haar decomposition and reconstruction algorithm can thus be viewed as a smart factorisation of a linear transform. A similar situation is encountered when programming the discrete Fourier transform by the Fast Fourier Transform algorithm which has complexity $\mathcal{O}(N \log N)$.

This first example of a multiscale scheme is rarely used in numerical applications, due to the poor approximation properties of piecewise constant functions, as well as their lack of smoothness. However, its particular simplicity allowed us to identify several important features of multiscale decompositions that we shall require in a more general setting: approximations take place in a *nested sequence* of spaces V_j allowing to compute $P_j f$ from $P_{j+1} f$, both approximations and fluctuations are characterized at each scales by *local bases*, and multiscale decompositions and reconstructions in these bases are performed by *fast algorithms* of $\mathcal{O}(N)$ complexity.

3. The Schauder hierarchical basis

We now turn to the approximation of a function f by a continuous, piecewise affine function on the intervals $I_{j,k}$, $k \in \mathbb{Z}$. Such an approximation can be defined in a unique way by its values at the points $2^{-j}k$. Assuming that f is a continuous function, we are allowed to impose

$$f_j(2^{-j}k) = f(2^{-j}k), \quad k \in \mathbb{Z}, \tag{3.1}$$

i.e. choose f_j to be the *linear interpolation* of f at the scale 2^{-j}. If we denote by P_j the interpolation operator that maps f onto f_j, we remark that P_j is a "projector" onto the space

$$V_j = \{ f \in C^0(\mathbb{R});\ f \text{ is affine on } I_{j,k},\ k \in \mathbb{Z} \}, \tag{3.2}$$

in the sense that $P_j f = f$ whenever $f \in V_j$. However, it is by no mean an orthogonal projector, and it is not even bounded in L^p for $p < \infty$ (e.g., if $f(x) = \max\{0, 1 - |x|\}$, the sequence $f_n = f(n\bullet)$, $n > 0$, goes to zero in L^p but is uniformly mapped onto f by P_0).

As in the case of piecewise constant approximation, we start by some simple comments:

A natural "basis" for V_j is given by the functions $\varphi_{j,k}$, $k \in \mathbb{Z}$, where $\varphi(x) = \max\{0, 1 - |x|\}$ and $\varphi_{j,k}(x) = 2^{j/2}\varphi(2^j \bullet - k)$. It allows to expand $P_j f$ in a unique way according to

$$P_j f = \sum_{k \in \mathbb{Z}} c_{j,k}(f)\varphi_{j,k}, \tag{3.3}$$

with $c_{j,k}(f) = 2^{-j/2} f(2^{-j}k)$. This basis is sometimes called a *nodal basis*, since each function $\varphi_{j,k}$ is associated to a "node" of the mesh $\Gamma_j := 2^{-j}\mathbb{Z}$, and the value of a function $f_j \in V_j$ at a node determines its corresponding coordinate. Clearly, this is not an orthonormal basis (due to the overlapping supports of $\varphi_{j,k}$ and $\varphi_{j,k+1}$), and the convergence of (3.3) is meant in a pointwise sense. However, it is easy to check that $\varphi_{j,k}$ is a *Riesz basis* for V_j: if $f_j \in V_j \cap L^2$, then $\|f_j\|_{L^2}$ and $(\sum_k |c_{j,k}(f_j)|^2)^{1/2}$ are equivalent norms and the series in (3.3) converges in L^2. This property will be further analyzed in the more general setting of the next chapter, and we thus postpone any further considerations on nonorthogonal bases in infinite-dimensional spaces.

This approximation process is also local, and (3.1) allows to define $P_j f$ when f is defined on a bounded interval such as [0, 1], if $j \geqslant 0$. Note that an homogeneous Dirichlet condition $f(0) = f(1) = 0$ is then preserved by the interpolation process, which amounts in using the approximation space V_j^D spanned by those $\varphi_{j,k}$ which have their support contained in [0, 1]. As an example, we display on Fig. 3.1 the graphs of the function $f(x) = \sin(2\pi x)$ and its approximation $P_3 f$ on [0, 1].

The approximation $P_j f$ can be derived from $P_{j+1} f$ by interpolation, since we have

$$2^{j/2} c_{j,k}(f) = f\left(2^{-j}k\right) = 2^{(j+1)/2} c_{j+1,2k}(f). \tag{3.4}$$

In other words, we have $P_j P_{j+1} = P_j$, a property which is not guaranteed by general nonorthogonal projectors (in contrast $P_{j+1} P_j = P_j$ is a direct consequence of the nestedness of the spaces V_j).

We can again define the "details" $Q_j f := P_{j+1} f - P_j f$. From (3.4) we see that $Q_j f\left(2^{-j}k\right) = 0$ and

$$Q_j f\left(2^{-j}\left(k + \tfrac{1}{2}\right)\right) = f\left(2^{-j}\left(k + \tfrac{1}{2}\right)\right) - \frac{[f(2^{-j}k) + f(2^{-j}(k+1))]}{2}. \tag{3.5}$$

In other words, $Q_j f$ measures the distance between the real value of f at the fine grid point $2^{-j}(k + 1/2) \in \Gamma_{j+1}$ and the average between its two closest neighbours on the coarser grid Γ_j, or equivalently the second-order finite difference between these three points. As an example, Fig. 3.2 shows the components $P_2 f$ and $Q_2 f$ on [0, 1], for $f(x) = \sin(2\pi x)$: we have decomposed $P_3 f$ of Fig. 3.1 into the sum of the coarser approximation $P_2 f$ and the additional details $Q_2 f$. Note that, in contrast to the case of piecewise constant approximations, we do not have $\int Q_j f = 0$.

From our last remark, we see that we can expand Q_j into local contributions according to

$$Q_j f = \sum_{k \in \mathbb{Z}} d_{j,k} \psi_{j,k}, \tag{3.6}$$

with

$$\psi = \varphi(2x - 1). \tag{3.7}$$

In this case, the "wavelets" $\psi_{j,k}$, $k \in \mathbb{Z}$, are thus simply the nodal basis functions at level $j+1$ that are associated to the grid points in $\Gamma_{j+1} \setminus \Gamma_j$.

For $j_1 > j_0$, we can thus express the multiscale decomposition

$$P_{j_1} f = P_{j_0} f + \sum_{j_0 \leqslant j < j_1} Q_j f, \tag{3.8}$$

in terms of local contributions according to

$$\sum_k c_{j_1,k}(f) \varphi_{j_1,k} = \sum_k c_{j_0,k}(f) \varphi_{j_0,k} + \sum_{j_0 \leqslant j < j_1} \sum_k d_{j,k}(f) \psi_{j,k}. \tag{3.9}$$

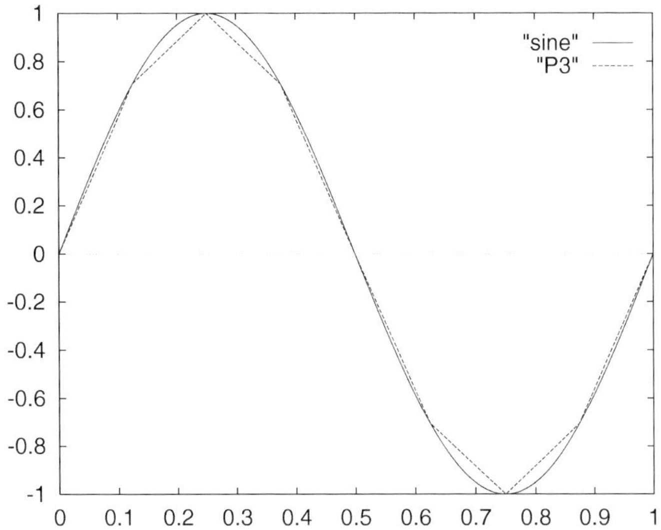

FIG. 3.1. The function $f(x) = \sin(2\pi x)$ and its approximation $P_3 f$.

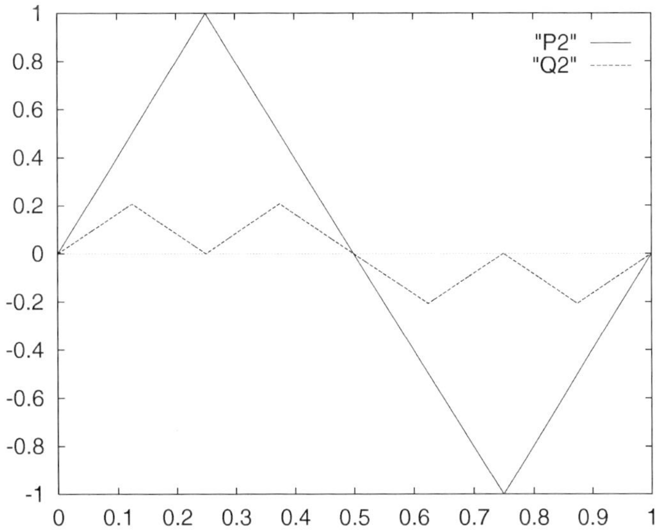

FIG. 3.2. The approximation $P_2 f$ and the fluctuation $Q_2 f$.

As in the case of the Haar system, we have described a change of basis, between the nodal basis $\{\varphi_{j_1,k}\}_{k\in\mathbb{Z}}$ and the *multiscale basis* defined $\{\varphi_{j_0,k}\}_{k\in\mathbb{Z}} \cup \{\psi_{j,k}\}_{j_0\leqslant j<j_1,k\in\mathbb{Z}}$. This nonorthogonal change of basis can easily be adapted to the space V_j^D that incorporates homogeneous Dirichlet boundary conditions, by keeping only the basis functions that are fully supported in $[0, 1]$. As an example, we display the nodal and hierarchical

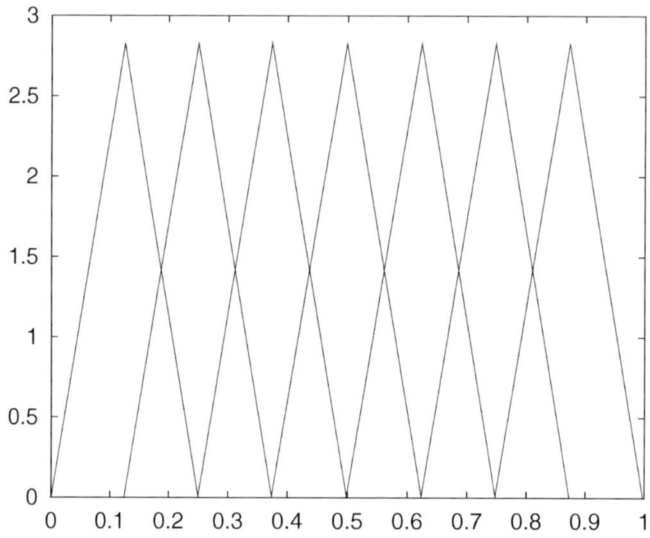

FIG. 3.3. The nodal basis of V_3^D.

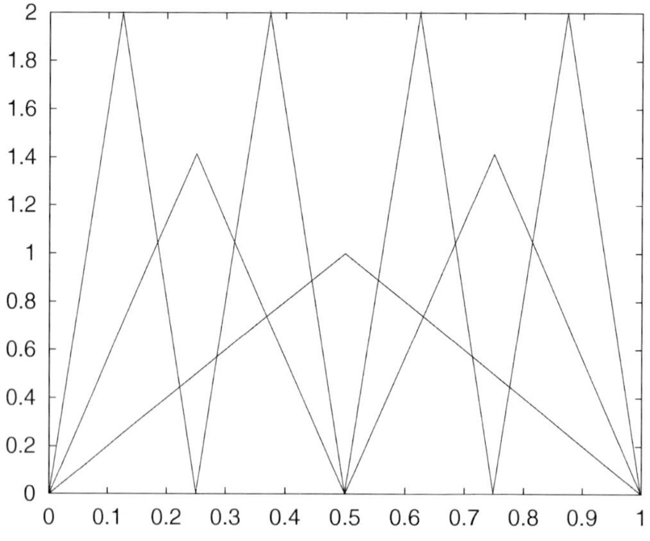

FIG. 3.4. The hierarchical basis of V_3^D.

bases on Figs. 3.3 and 3.4, in the case where $j = 3$.

If we let j_1 go to $+\infty$, the interpolation $P_{j_1} f$ converges uniformly to f, provided that f is uniformly continuous function (in particular on any compact set since we have assumed that f is continuous). In the case of a continuous function f defined on a bounded interval, the summation in k is finite at each resolution level, and this allows us to express f as the uniform limit of the infinite series

$$f = \sum_k c_{j_0,k}\varphi_{j_0,k} + \sum_{j \geqslant j_0} \sum_k d_{j,k}\psi_{j,k}. \tag{3.10}$$

Note that the convergence in (3.10) corresponds to the particular summation process of "piling-up" the details as j_1 go to $+\infty$: a permutation of the terms in the series might ruin it, in contrast to the expansion of an L^2 function f in the orthonormal Haar system which converges to f in L^2 whatever the order of summation.

On a computational point of view, the change of basis in (3.9) can be implemented by the same strategy that was described for the Haar transform, i.e. a "fine to coarse" multiscale decomposition, and a "coarse to fine" reconstruction of the nodal representation.

As in the case of the Haar system, we might operate the multiscale decomposition on any fine scale approximation $f_{j_1} = \sum_k c_{j_1,k}\varphi_{j_1,k}$ of f in V_{j_1} that differs from the interpolation $P_{j_1} f$: e.g., f_{j_1} could be a Galerkin-type approximation to f, in the context of solving a partial differential equation.

These multiscale algorithms are based on the following simple recursions that are derived from (3.5), taking into account the L^2-normalization of $\varphi_{j,k}$ and $\psi_{j,k}$:

$$c_{j,k} = \sqrt{2}c_{j+1,2k} \quad \text{and} \quad d_{j,k} = \sqrt{2}\big[c_{j+1,2k+1} - (c_{j+1,2k} + c_{j+1,2k+2})/2\big], \tag{3.11}$$

for the decomposition, and

$$c_{j+1,2k} = c_{j,k}/\sqrt{2} \quad \text{and} \quad c_{j+1,2k+1} = \big[d_{j,k} + (c_{j,k} + c_{j,k+1})/2\big]/\sqrt{2}, \tag{3.12}$$

for the reconstruction. We display on Fig. 3.5 the "influence" diagram for the decomposition and reconstruction algorithms, between two level $j+1$ and j. In contrast to Fig. 2.4, note the slight difference between the decomposition and reconstruction diagrams, which reflects the nonorthogonality of the discrete transform.

Clearly, these algorithms are of complexity $\mathcal{O}(N)$, similarly to the Haar transform. In the theoretical construction of the Schauder basis, as well as in its practical implementation, we have thus encountered the three main features that we identified on the Haar system in the end of the previous section.

4. Multivariate constructions

If we want to generalize the previous multiscale schemes to multivariate functions, we can imagine many different possibilities. We shall focus on the two most frequently used approaches which are respectively based on tensor product and nested triangulations. For the sake of notational simplicity, we shall describe these generalizations in the case of functions of two variables. The extension to more than two variables does not present any additional difficulty and will be sketched for each approach.

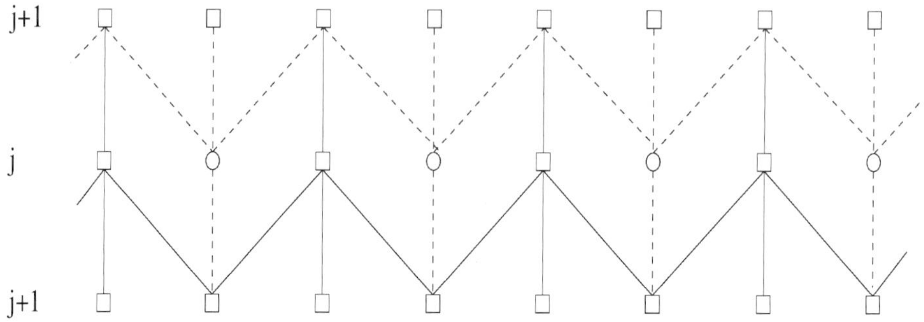

FIG. 3.5. Decomposition $j+1 \mapsto j$ and reconstruction $j \mapsto j+1$ for the Schauder hierarchical basis. □: Approximation coefficient; ○: detail coefficient.

In the case of piecewise constant approximation, a natural generalization of (2.1) for $f \in L^2(\mathbb{R}^2)$ is to define $P_j f$ to be the piecewise constant on the dyadic squares

$$S_{j,k} := I_{j,m} \times I_{j,n}, \quad k = (m,n) \in \mathbb{Z}^2, \tag{4.1}$$

that is equal to the mean value of f on each of these squares.

As in the univariate case, P_j is an orthogonal projection on the space of piecewise constant functions, now generated by the separable orthonormal basis

$$\phi_{j,k}(x,y) = \varphi_{j,m}(x)\varphi_{j,n}(y) = 2^j \phi\big(2^j(x,y) - k\big), \quad k = (m,n) \in \mathbb{Z}^2, \tag{4.2}$$

where $\varphi = \chi_{[0,1]}$ and $\phi = \chi_{[0,1]^2}$. In other words, the approximation space V_j is the *tensor product space* $V_j^x \otimes V_j^y$, where V_j^x and V_j^y represent the univariate space defined in Section 2, in the variable x and y.

If f_{j+1} is a function in V_{j+1}, i.e.

$$f_{j+1} = \sum_{k \in \mathbb{Z}^2} c_{j+1,k} \phi_{j+1,k}, \tag{4.3}$$

we can derive its approximation at the coarser scale

$$P_j f_{j+1} = \sum_{k \in \mathbb{Z}^2} c_{j,k} \phi_{j,k}, \tag{4.4}$$

using $c_{j,(m,n)} = [c_{j+1,(2m,2n)} + c_{j+1,(2m+1,2n)} + c_{j+1,(2m,2n+1)} + c_{j+1,(2m+1,2n+1)}]/2$.

We would like to characterize, as in the univariate case, the detail component $Q_j f = P_{j+1} f - P_j f$ by means of wavelets, that are supported on $S_{j,k}$. However, since $S_{j,k}$ supports one basis function in V_j and four basis functions in V_{j+1}, it is clear that we need three wavelets instead of one to span the complement space. These wavelets can be constructed as follows:

$$\psi^a(x,y) = \psi(x)\varphi(y), \quad \psi^b(x,y) = \varphi(x)\psi(y), \quad \psi^c(x,y) = \psi(x)\psi(y). \tag{4.5}$$

Indeed, one easily checks (using Fubini's theorem), the orthogonality relations

$$\langle \psi^{\alpha}_{j,k}, \phi_{j,k} \rangle = \langle \psi^{\alpha}_{j,k}, \psi^{\beta}_{j,k} \rangle = 0, \quad \alpha, \beta = a, b, c \text{ and } \alpha \neq \beta. \tag{4.6}$$

Thus, $\{\psi^{a}_{j,k}, \psi^{b}_{j,k}, \psi^{c}_{j,k}\}_{k \in \mathbb{Z}^2}$ constitutes an orthonormal basis for the orthogonal complement W_j of V_j into V_{j+1}.

We have in fact used a "distributive" property of tensor product with respect to the orthogonal sum of spaces, in the sense that

$$\begin{aligned} V_{j+1} &= V^x_{j+1} \otimes V^y_{j+1} = (V^x_j \oplus W^x_j) \otimes (V^y_j \oplus W^y_j) \\ &= (V^x_j \otimes V^y_j) \oplus (W^x_j \otimes V^y_j) \oplus (V^x_j \otimes W^y_j) \oplus (W^x_j \otimes W^y_j) \\ &= V_j \oplus W^a_j \oplus W^b_j \oplus W^c_j = V_j \oplus W_j, \end{aligned}$$

each detail subspace W^{α}_j being generated by the shifts and dilates of the corresponding wavelet ψ^{α}, $\alpha = a, b, c$.

This particular representation of the details, allows to implement the decomposition algorithm using a two step recursion to go from level $j+1$ to level j:

- Compute the averages and details along the x-direction using the intermediate values:

$$\begin{aligned} a_{j,(m,n)} &= [c_{j+1,(2m,n)} + c_{j+1,(2m+1,n)}]/\sqrt{2}, \\ b_{j,(m,n)} &= [c_{j+1,(2m,n)} - c_{j+1,(2m+1,n)}]/\sqrt{2}. \end{aligned} \tag{4.7}$$

- Compute averages and details along the y-direction, to derive the wavelets coefficients:

$$\begin{aligned} c_{j,(m,n)} &= [a_{j,(m,2n)} + a_{j,(m,2n+1)}]/\sqrt{2}, \\ d^a_{j,(m,n)} &= [b_{j,(m,2n)} + b_{j,(m,2n+1)}]/\sqrt{2}, \\ d^b_{j,(m,n)} &= [a_{j,(m,2n)} - a_{j,(m,2n+1)}]/\sqrt{2}, \\ d^c_{j,(m,n)} &= [b_{j,(m,2n)} - b_{j,(m,2n+1)}]/\sqrt{2}. \end{aligned} \tag{4.8}$$

Clearly, we can apply the same idea to implement the reconstruction algorithm from level j to $j+1$: recompose successively along the y- and x-directions. We could also have interchanged the order of appearance of the x- and y-directions.

REMARK 4.1. Note that the wavelet coefficients have a directional interpretation: the coefficients $d^a_{j,k}$ (resp. $d^b_{j,k}$) indicate the fluctuations in the x (resp. y) direction, i.e. they are particularly sensitive to a vertical (resp. horizontal) edge on the graph of $f(x, y)$. This fact is directly visualized, if we apply this type of multiscale decomposition on *images*, as it will be explained in the next section.

REMARK 4.2. As in the univariate case, we have obtained that for $j_0 \leqslant j_1$ the functions

$$\{\phi_{j_0,k}\}_{k\in\mathbb{Z}^2} \cup \left\{\psi^{\alpha}_{j,k}\right\}_{j_0\leqslant j<j_1,k\in\mathbb{Z}^2,\alpha=a,b,c}, \tag{4.9}$$

constitute an orthonormal multiscale basis for V_{j_1}.

REMARK 4.3. Note that another basis could be obtained by taking all possible tensor products of two functions in $\{\varphi_{j_0,k}\}_{k\in\mathbb{Z}} \cup \{\psi_{j,k}\}_{j_0\leqslant j<j_1,k\in\mathbb{Z}}$. This leads to a "full tensor product" basis which incorporates functions of the type $\psi_{j,k}(x)\psi_{j',k'}(y)$, which oscillate at different scales in the directions x and y. The corresponding decomposition (resp. reconstruction) algorithm operates in one direction from the finest to coarsest (resp. coarsest to finest) resolution level, and then performs the same operations in the other direction.

REMARK 4.4. As in the univariate case, we can let j_1 go to $+\infty$ (and j_0 to $-\infty$), and derive a multiscale bases for $L^2(\mathbb{R}^2)$.

REMARK 4.5. The generalization of this construction to more than 2 variables is straightforward: the spaces V_j are generated by the orthogonal basis

$$\phi_{j,k}(x) = \varphi_{j,k_1}(x_1)\cdots\varphi_{j,k_d}(x_d), \tag{4.10}$$

and the wavelets are given by

$$\psi^{\varepsilon}(x) = \psi^{\varepsilon_1}(x_1)\cdots\psi^{\varepsilon_d}(x_d), \qquad \varepsilon = (\varepsilon_1,\ldots,\varepsilon_d) \in \{0,1\}^d \setminus (0,\ldots,0), \tag{4.11}$$

where we have set $\psi^0 = \varphi$ and $\psi^1 = \psi$. We can thus characterize the details by $2^d - 1$ wavelets.

We can also apply the above described tensor product technique on piecewise affine univariate approximation, i.e. define the space V_j through a nodal basis $\phi_{j,k}(x,y) = \varphi_{j,m}(x)\varphi_{j,n}(y)$, $k=(m,n)\in\mathbb{Z}^2$, where $\varphi = \max\{1-|x|,0\}$.

We then obtain bivariate approximation spaces which are the well known rectangular Lagrange $\mathbb{Q}_1$ finite elements: on each square $S_{j,k}$, the approximation f_j has the form $a_k + b_k x + c_k y + d_k xy$, and since it is continuous, f_j is completely determined by its values on $2^{-j}\mathbb{Z}^2$: we have $f_j = \sum_{k\in\mathbb{Z}^2} c_{j,k}\phi_{j,k}$, with $c_{j,k} = 2^{-j} f_j(2^{-j}k)$. We can also define in a unique way the interpolation operator P_j that maps a continuous function f onto the space V_j with $P_j f(2^{-j}k) = f(2^{-j}k)$, $k\in\mathbb{Z}^2$.

The same reasoning as in the case of the Haar system shows that formula (4.5) applied to the hat function φ and the Schauder wavelet $\psi = \varphi(2\bullet -1)$ yield a suitable wavelet basis to describe the difference $I_{j+1}f - I_j f$. We obtain a multiscale basis of the form (4.9) and we can also use the same separable technique to implement the decomposition and reconstruction algorithm.

In summary, the tensor product construction inherits the essential features of the corresponding univariate multiresolution approximation and wavelet basis: nested approximation spaces, local bases, fast algorithms.

Another natural multivariate counterpart to the piecewise affine construction consist in preserving the affine nature of the approximation, by using $\mathbb{P}_1$ triangular elements, i.e. approximation by continuous piecewise affine functions on triangles. This approach is well adapted to the approximation of functions defined on domains such as polygons that can easily be triangulated. We shall thus consider functions that are defined on a domain Ω which is either $\mathbb{R}^2$ or a bounded polygonal domain.

In order to build approximation spaces V_j that have the nestedness property $V_j \subset V_{j+1}$, we can start from an initial planar triangulation $\mathcal{T}_0$ of Ω, with two basic properties.

(i) *Conformity*: the edge of a triangle is common with exactly one edge of another triangle, unless it is contained in the boundary of Ω.
(ii) *Regularity*: there exists strictly positive constants c and C such that

$$c \leqslant \inf_{T \in \mathcal{T}_0} r(T) \leqslant \sup_{T \in \mathcal{T}_0} R(T) \leqslant C, \tag{4.12}$$

where $r(T)$ (resp. $R(T)$) is the radius of the largest (resp. smallest) circle contained in (resp. containing) the triangle T.

Note that in the case where the domain is a bounded polygon, the second property simply means that we start with $\mathcal{T}_0$ of finite cardinality. We then build a family of *nested triangulation* by a classical refinement procedure: $\mathcal{T}_{j+1}$ is defined from $\mathcal{T}_j$ by adding a mesh point in the middle of each edge and subdividing each triangle into four subtriangles, as shown on Fig. 4.1 (see also Fig. 21.1 of Chapter II).

This refinement procedures generates conformal triangulations $\mathcal{T}_j$, $j \geqslant 0$, with a property of uniform regularity: the ratio $\rho_j = \sup_{T \in \mathcal{T}_j} R(T)/\inf_{T \in \mathcal{T}_j} r(T)$ is bounded independently of j (it is actually constant by construction). More precisely, we have

$$c2^{-j} = \inf_{T \in \mathcal{T}_j} r(T) \leqslant \sup_{T \in \mathcal{T}_j} R(T) = C2^{-j}. \tag{4.13}$$

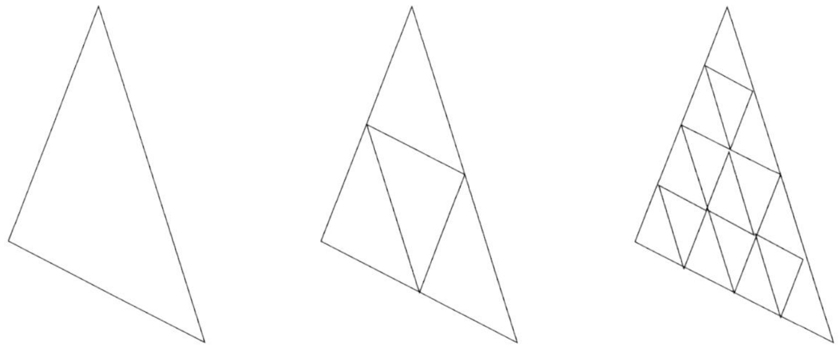

FIG. 4.1. Subdivision of a triangle of $\mathcal{T}_0$.

We also note that the spaces of $\mathbb{P}_1$ finite elements

$$V_j = \{ f \in C^0(\Omega);\ f|_T(x, y) = a_T + b_T x + c_T y,\ T \in \mathcal{T}_j \}, \tag{4.14}$$

are nested, i.e. $V_j \subset V_{j+1}$.

If we denote by Γ_j the set of all vertices of the triangulation $\mathcal{T}_j$, we can again define a simple nodal basis: for $\gamma \in \Gamma_j$, we define $\phi_{j,\gamma}$ to be the unique function in V_j such that

$$\phi_{j,\gamma}(\gamma') = 2^j \delta_{\gamma,\gamma'}, \quad \gamma' \in \Gamma_j. \tag{4.15}$$

As in the univariate case, we have chosen to normalize these functions in such a way that their L^2 norm is controlled independently of j. Any $f_j \in V_j$ can thus be decomposed in a unique way as

$$f_j = \sum_{\gamma \in \Gamma_j} c_{j,\gamma} \phi_{j,\gamma}, \tag{4.16}$$

with

$$c_{j,\gamma} = 2^{-j} f_j(\gamma). \tag{4.17}$$

We can also define in a unique way the interpolation operator P_j that maps a continuous function f onto the space V_j with the constraint $P_j f(\gamma) = f(\gamma)$, $\gamma \in \Gamma_j$.

We remark that, since $[P_{j+1} f - P_j f](\gamma) = 0$, for all $\gamma \in \Gamma_j$, we can use the functions $\phi_{j+1,\gamma}$, $\gamma \in \Gamma_{j+1} \setminus \Gamma_j$, i.e. the fine scale basis functions corresponding to the vertices that have been added in the refinement process from level j to $j+1$, to characterize this difference. Clearly, this representation is also not redundant, since each of these functions is needed to recover the value at the corresponding point of $\Gamma_{j+1} \setminus \Gamma_j$.

Iterating this decomposition on several levels, say between j_1 and $j_0 < j_1$, we thus obtain a multiscale basis for V_{j_1} with the following form:

$$\{\phi_{j_0,\gamma}\}_{\gamma \in \Gamma_{j_0}} \cup \{\phi_{j+1,\gamma}\}_{j_0 \leqslant j < j_1, \gamma \in \Gamma_{j+1} \setminus \Gamma_j}. \tag{4.18}$$

As in all previous constructions, we can use simple relations between two successive level of resolution, in order to derive fine to coarse (decomposition) and coarse to fine (reconstruction) fast algorithms. We denote by $d_{j+1,\gamma}$, $\gamma \in \Gamma_{j+1} \setminus \Gamma_j$, the coefficient of f in the corresponding hierarchical basis function, and by $a(\gamma)$ and $b(\gamma)$, the two points of Γ_j such that γ was built as the mid-point of the edge joining $a(\gamma)$ and $b(\gamma)$. Taking into account the L^2 normalization of the basis functions, we obtain

$$c_{j,\gamma} = 2 c_{j+1,\gamma}, \quad \gamma \in \Gamma_j, \tag{4.19}$$

and

$$d_{j+1,\gamma} = c_{j+1,\gamma} - [c_{j+1,a(\gamma)} + c_{j+1,b(\gamma)}]/2, \quad \gamma \in \Gamma_{j+1} \setminus \Gamma_j, \tag{4.20}$$

for decomposition, and

$$c_{j+1,\gamma} = c_{j,\gamma}/2, \quad \gamma \in \Gamma_j, \tag{4.21}$$

and

$$c_{j+1,\gamma} = d_{j+1,\gamma} + [c_{j,a(\gamma)} + c_{j,b(\gamma)}]/4, \quad \gamma \in \Gamma_{j+1} \setminus \Gamma_j, \tag{4.22}$$

for reconstruction.

REMARK 4.6. As in the case of the univariate Schauder basis, $P_j f$ converges uniformly to f if f is uniformly continuous: any such function can thus be expanded as a uniformly converging series in the hierarchical basis.

REMARK 4.7. A possible approach to generalize this hierarchical basis in a straightforward way to more than two variables is the following: one start from a conformal partition of Ω into simplices, and operate similar refinements, adding a new vertex in the middle of each edges and dividing the simplex into 2^d similar simplices.

REMARK 4.8. As in the univariate case, it is also possible to incorporate homogeneous Dirichlet boundary conditions in V_j, without affecting the construction of the multiscale basis: both nodal and hierarchical basis functions are then associated only to the nodes that are in the interior of Ω.

5. Adaptive approximation of functions and operators

One of the main interest of multiscale decompositions can be stated by the following heuristic: a function f that is smooth, except at some isolated singularities should have a *sparse* representation in a multiscale basis, i.e. only a small number of numerically significant coefficients should carry most of the information on f. Indeed, the wavelet coefficient $d_{j,k}$ measures the local fluctuation of f at scale 2^{-j} near $2^{-j}k$, or equivalently in the support of $\psi_{j,k}$. As j grows, i.e. resolution gets finer, these fluctuations should decay fastly to zero in the smooth regions, and should be larger only when the support of $\psi_{j,k}$ contains a singularity of f, i.e. for a small number of indices k. Such a behaviour also suggests that one could approximate f by keeping only a small number of wavelet coefficients, and reduce significantly the complexity of the description of f without affecting its accuracy.

This *compression property* will be analyzed in details in Chapters III and IV. The goal of this section is simply to illustrate it with some simple examples.

As a first example, we consider the function $f(x) = \sqrt{|\cos(2\pi x)|}$ on $[0, 1]$. A first sampling of f on 8192 points is used to plot its graph on Fig. 5.1, which displays its linear interpolation $P_j f$, as defined in Section 3, at resolution level $j = 13$. From this fine scale approximation, we have applied the decomposition algorithm in the Schauder basis. On Fig. 5.2, we display the points $(2^{-j}k, j)$ corresponding to the wavelet coefficients $|d_{j,k}| \geqslant 5 \times 10^{-3}$. As expected, the contributions from the high scales tend to concentrate near the singularities at $x = 1/4$ and $3/4$.

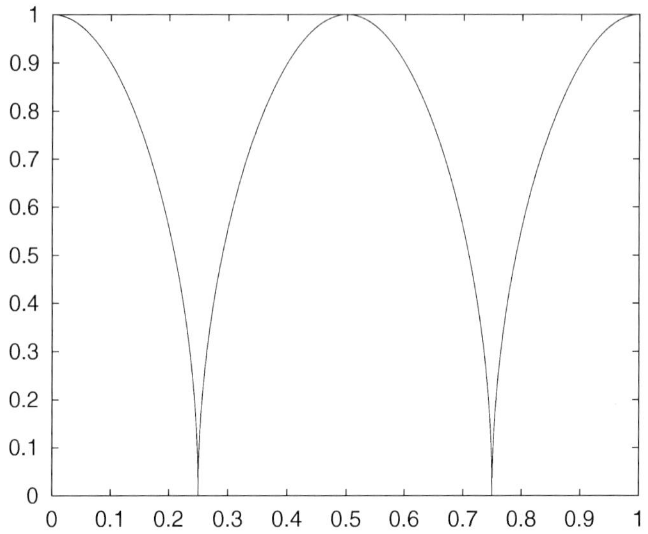

FIG. 5.1. The function $f(x) = \sqrt{|\cos(2\pi x)|}$ at resolution 2^{-13}.

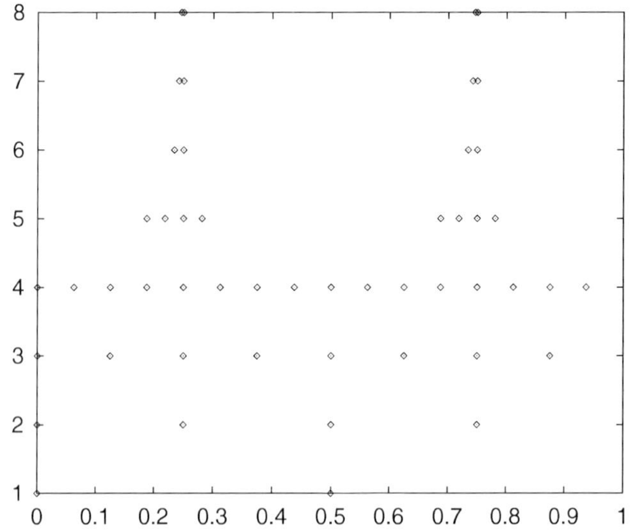

FIG. 5.2. Schauder basis coefficients above threshold 5×10^{-3}.

On Fig. 5.3, we have reconstructed the function f from the coefficients that are marked on Fig. 5.2. The number of preserved coefficients is 50, i.e. a reduction of the initial complexity by a factor above 150. The result of a similar thresholding operation with the Haar system is displayed on Fig. 5.4 (the threshold is kept the same, and the

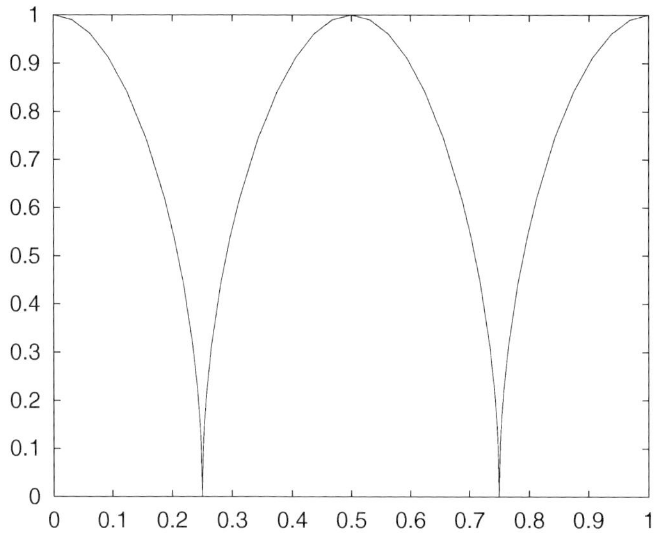

FIG. 5.3. Reconstruction from the 50 Schauder basis coefficients above 5×10^{-3}.

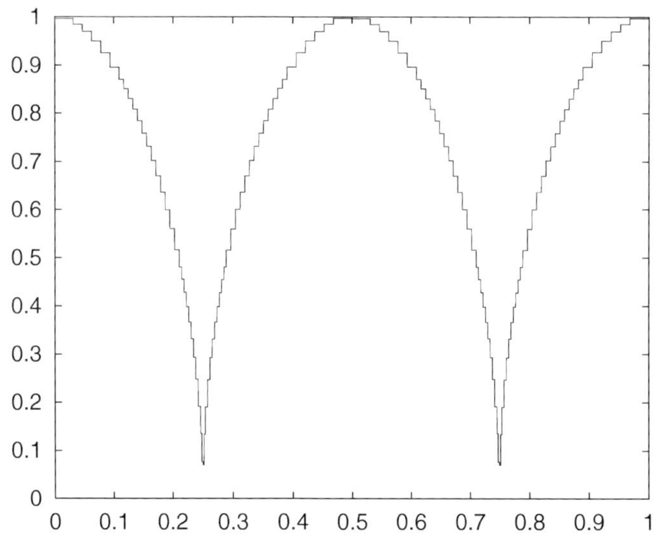

FIG. 5.4. Reconstruction from the 116 Haar basis coefficients above 5×10^{-3}.

number of preserved coefficients is then 116). It is no surprise that the quality of the approximation is visually (and numerically) better with piecewise affine functions.

In both cases, we observe that the approximation is refined near the singularities: an adaptive grid, that takes into account the local fluctuations of the function, is automatically generated by the thresholding of wavelet coefficients.

REMARK 5.1. We can provide a simple analysis of the stronger decay (as $j \to +\infty$) of the wavelet coefficients in the regions where f. In the case of the Haar system, when f is C^1 on the support $I_{j,k}$ of $\psi_{j,k}$, we can use the oscillation property of ψ to derive the estimate

$$\begin{aligned}|d_{j,k}| &= \left|\int_{I_{j,k}} f(x)\psi_{j,k}(x)\,\mathrm{d}x\right| \\ &= \left|\int_{I_{j,k}} \left[f(x) - f\left(2^{-j}k\right)\right]\psi_{j,k}(x)\,\mathrm{d}x\right| \\ &\leqslant \left[\sup_{I_{j,k}}|f'|\right]2^{-j}\int|\psi_{j,k}(x)|\,\mathrm{d}x \leqslant \left[\sup_{I_{j,k}}|f'|\right]2^{-3j/2}.\end{aligned}$$

We have thus combined the smoothness of f with the oscillatory nature of ψ to derive a better estimate than the simple $|d_{j,k}| \leqslant 2^{-j/2}\sup_{I_{j,k}}|f|$. Note that in the case where the function f is only C^α on $I_{j,k}$ with $0 < \alpha < 1$ (i.e. $|f(x) - f(y)| \leqslant C|x-y|^\alpha$), then the same computation will lead to an estimate of $|d_{j,k}|$ in $\mathcal{O}(2^{-(\alpha+1/2)j}$. This explains why more coefficients are preserved at the finest scales near the singularities of our example which have the Hölder exponent $\alpha = 1/2$. The results of Chapter III will actually confirm the intuition that the degree of local smoothness can be *measured* through the decay of the wavelet coefficients. Note that in the case of the Schauder basis, we can obtain a better estimate if $f \in C^2$: using that $d_{j,k}$ is a normalized second-order finite difference, one easily derives $|d_{j,k}| \lesssim 2^{-5j/2}[\sup_{I_{j,k}}|\mathrm{d}^2 f/\mathrm{d}x^2|]$. This better estimate explains why more coefficients have been discarded in the thresholding process with the Schauder basis.

REMARK 5.2. Note that the thresholding procedure is a *nonlinear* operation: the indices (j,k) of the preserved coefficients depend on the function to be approximated. In particular, this means that one need to store both the values of the preserved coefficients and their indices, in order to describe such an adaptive approximation. Another natural way of defining such nonlinear approximations is by prescribing the number of preserved coefficients rather than a threshold, i.e. define f_N to be the approximation of the above f by retaining its N largest wavelet coefficients. The sparsity of the multiscale representation can then be measured through the decay of $\|f_N - f\|_{L^2}$ as N goes to $+\infty$, i.e. the supremum of all s such that

$$\|f_N - f\|_{L^2} \leqslant CN^{-s}. \tag{5.1}$$

In the particular case of the above example, one can easily derive from the estimates on $|d_{j,k}|$ that this error decays like N^{-1} or N^{-2} when using respectively the Haar system or the Schauder basis. General results on nonlinear approximations will be presented in Chapter IV. In particular these results will imply that (5.1) holds with s arbitrarily large for the above example, provided that one uses a sufficiently high order accurate wavelet basis. This is in contrast with *linear approximation* which defines f_N by retaining the N first coefficients of f, i.e. $f_N := P_j f$ when $N = 2^j$. In this case, one essentially cannot improve on the rate N^{-1} in the above example, even when using high order wavelets, due to the presence of the singularities (for the Schauder basis, a "superconvergence" phenomenon still occurs in our example since the singularities are situated at coarse mesh points, which results in the artificially better rate $N^{-2}\log(N)$). This is also in contrast with nonlinear approximation with other basis such as Fourier series: the Fourier coefficients $c_N(f)$ in the above example are not sparse in the sense that they behave like $\mathcal{O}(|N|^{-3/2})$ for all N. In turn, both linear and nonlinear approximation L^2-error behave like N^{-1}.

We shall now illustrate the compression properties of multiscale decompositions in the case of bivariate functions associated to the mathematical representation of *images*. A digital black and white image is a bidimensional array $I(m,n)$ measuring the grey level intensity at each point (or "pixel": picture element) (m,n). As an example, Fig. 5.5 displays a 512×512 image, each pixel being quantized on 8 bits, i.e. 256 possible grey levels (0 for black, 255 for white). The tensor product multiscale decomposition which generalizes the Haar system in 2D is particularly adapted to the representation of such images: we can identify the digital image in Fig. 5.5 to a function in V_9, and proceed to the multiscale decomposition using the separable algorithm that we described in Section 4.

We display the organization of a decomposition on 4 levels on Fig. 5.6: the coefficients of the coarsest approximation (in V_5) appear in the upper left corner, while the rest of the array contains the wavelet coefficients at intermediate resolutions.

In the framework of image processing, one usually normalize the basis functions in L^1 instead of L^2: $\phi_{j,k}(x,y) = 2^{2j}\phi(2^j x - k_x, 2^j y - k_y)$, for $k = (k_x, k_y)$, and similarly for ψ. This normalization allows to visualize the multiscale decomposition of our image

FIG. 5.5. Digitized picture: 512×512 pixels and 256 grey levels.

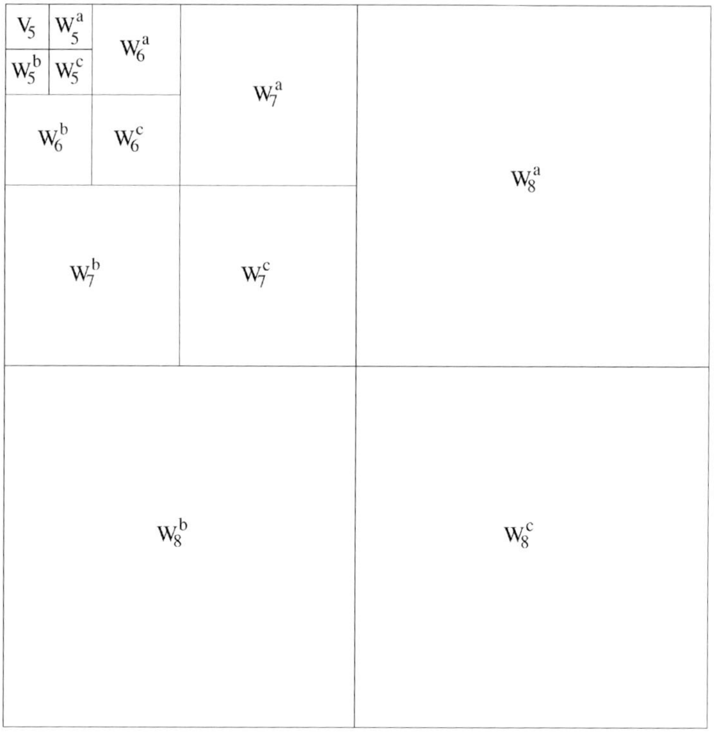

FIG. 5.6. Tensor product multiscale decomposition.

as another image: the approximation coefficients are exactly the averages of the image on squares of pixels, and thus also range between 0 and 255, as well as the absolute values of wavelet coefficients. We display this decomposition image on Fig. 5.7: the coefficients of the coarsest approximation appear as a simplified version of the picture. The rest of the array contains the absolute values of wavelet coefficients: as expected, it is mostly sparse, except near the edges. As it was remarked about tensor product decomposition (Remark 4.1), vertical and horizontal edges are matched by a specific wavelet.

On Fig. 5.8, we have reconstructed the image with the 2000 largest coefficients (after renormalization in L^2), i.e. a parameter reduction above 100.

Clearly, the Haar system is not very well adapted for the task of representing images with a few coefficients: visual artefacts appear, reflecting the square-shaped discontinuities of the generating functions. However, we again observe that the thresholding strategy generates an adaptive approximation of the image, in the sense that the resolution level is increased near the edges.

Finally, we want to show that multiscale decomposition can be also be applied to "compress" operators in integral equations. Such equations arise in numerous contexts, either as a direct modelization of a physical process, or as alternative formulations of a

FIG. 5.7. Tensor product multiscale decomposition.

FIG. 5.8. Reconstruction from 2000 largest coefficients.

partial differential equations. They involve the application or the inversion of an integral operator T defined by a formula of the type

$$Tf(x) = \int K(x, y) f(y)\, \mathrm{d}y, \tag{5.2}$$

where the kernel $K(x,y)$ is a function which is supported over all ranges of x and y. The distributed nature of $K(x,y)$ has the following immediate consequence: the usual discretizations of T – based on finite element methods or straightforward sampling of $K(x,y)$ – result in fully populated matrices that are heavy to store, to apply or to invert.

In order to understand how a multiscale decomposition can "sparsify" the representation of T, let us consider a simple case where x and y range in $I=[0,1]$. We denote by V_J, $J \geqslant 0$, the space of piecewise constant functions defined in Section 2 and adapted to I, and $\varphi_{J,k}$, $k=0,\dots,2^J-1$, its orthonormal basis. We then define a discretization of T on V_J as the matrix

$$T_J = \big(\langle T\varphi_{J,m}, \varphi_{J,n}\rangle\big)_{m,n=0,\dots,2^J-1}. \tag{5.3}$$

This matrix appears naturally in two different situations.

(i) *Approximate the action of* T *on a function*: given f, find an approximation g_J in V_J of $g=Tf$. For this, we can start by approximating f by $f_J \in V_J$ and then define $g_J = P_J T f_J$, where P_J is the orthogonal projection. The computation of the coordinates vector G_J of g_J (in the basis $\varphi_{J,k}$) is then done by applying T_J on the coordinate vector F_J of f_J.

(ii) *Approximate the action of* T^{-1} *on a function*: given f, find an approximation $g_J \in V_J$ of g solution of $Tg=f$. The Galerkin method consists in searching for $g_J \in V_J$ such that $\langle Tg_J, u_J\rangle = \langle f, u_J\rangle$ for all $u_J \in V_J$, i.e. solving $T_J G_J = F_J$, where F_J is the coordinate vector of $f_J = P_J f$, and G_J the coordinate vector of the unknown g_J.

The well-posedness of the system in the second problem, as well as the error estimates that could be obtained in some prescribed norm for both problems, are of course dependent of the specific nature of the operator T and the data f.

Since the matrix elements of T_J are given by

$$T_J(m,n) = \langle T\varphi_{J,m}, \varphi_{J,n}\rangle = \int K(x,y)\varphi_{J,m}(x)\varphi_{J,n}(y)\,\mathrm{d}x\,\mathrm{d}y, \tag{5.4}$$

it is clear that the distributed nature of $K(x,y)$ will result in a full matrix. For example, if we have a uniform bound $|K(x,y)| \leqslant C$, we are ensured that (5.2) defines a bounded operator in $L^2(I)$, and that the operators T_J are bounded independently of J. We also obtain from (5.4) the estimate $|T_J(m,n)| \lesssim 2^{-J}$, but this estimate does not a priori allow us to approximate T_J in operator norm by a sparse matrix.

If we now use the multiscale basis, i.e. $\varphi_{j_0,k}$, $k=0,\dots,2^{j_0}-1$, and $\psi_{j,k}$, $j_0 \leqslant j < J$, $k=0,\dots,2^j-1$, to reformulate both problems, we obtain a new matrix S_J, whose elements are typically given by

$$S_J(j,l,m,n) = \langle T\psi_{j,m}, \psi_{l,n}\rangle = \int K(x,y)\psi_{j,m}(x)\psi_{l,n}(y)\,\mathrm{d}x\,\mathrm{d}y, \tag{5.5}$$

for $j_0 \leqslant j,l \leqslant J-1$, $m=0,\dots,2^j-1$, $n=0,\dots,2^l-1$ (with analogous expressions for those elements involving the basis functions $\varphi_{j_0,k}$). We see from (5.5) that S_J is

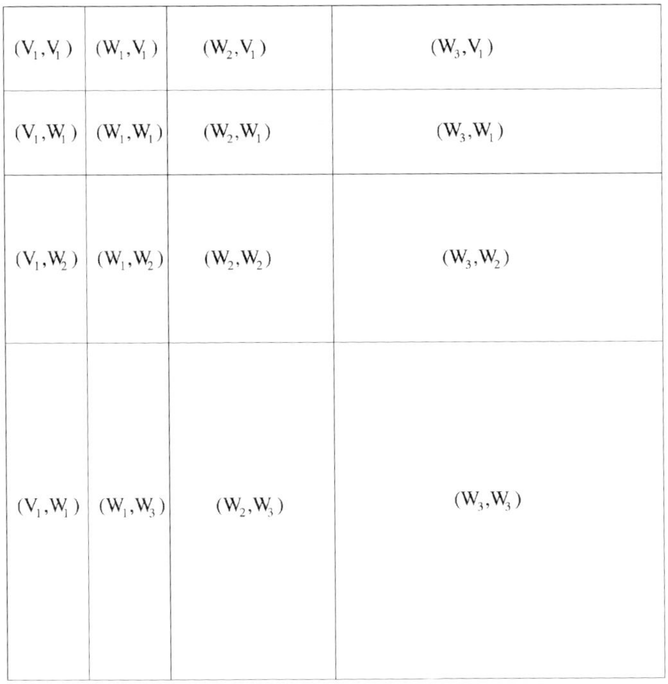

FIG. 5.9. Multiscale discretization of a kernel.

simply obtained by applying on T_J the "full tensor product" wavelet decomposition that was described in Remark 4.3: the discretized kernel is "processed" like a digital picture. The structure of the resulting matrix is displayed on Fig. 5.9, in the case where $J = 4$ and $j_0 = 1$.

We can thus hope to gain some sparsity when the kernel has some smoothness properties. In particular, if $K(x, y)$ is C^1 on the support of $\psi_{j,m}(x)\psi_{l,n}(y)$, we can use the same method as in Remark 5.1, to derive the estimate

$$\left|S_J(j,l,m,n)\right| \leqslant \Big[\sup_{I_{j,m}\times I_{l,n}} |\nabla K|\Big] 2^{-2\max\{j,l\}}, \tag{5.6}$$

which, in contrast to the crude estimate that we had for T_J might allow to discard many coefficients while preserving a good approximation to S_J.

As an example, we consider a single layer logarithmic potential operator that relates the density of electric charge on an infinite cylinder of unit radius $\{(z, e^{i2\pi x}),\ z \in \mathbb{R},\ x \in [0,1]\}$ to the induced potential on the same cylinder, when both functions are independent of the variable z. The associated kernel

$$K(x, y) = \log\left|e^{i2\pi x} - e^{i2\pi y}\right| \tag{5.7}$$

is singular on the diagonal $\{x = y\}$, but integrable.

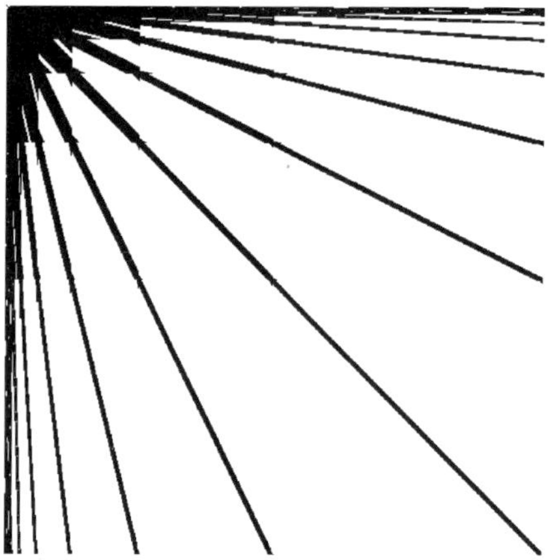

FIG. 5.10. Sparsification of a logarithmic potential kernel.

Starting from the discretization T_9 in V_9, we compute the multiscale matrix S_9 and we define a sparse matrix $\widetilde{S}_9$ by setting to zero all entries of S_9 with modulus less than $10^{-2} \times 2^{-9}$. We display on Fig. 5.10 the location of the preserved coefficients: on each subblock of S_9, that corresponds to a couple of scale (j, l), it is no surprise that we find important coefficients near the diagonal, since it corresponds to the singular part of the kernel.

This approximation of S_9 contains approximately 30000 nonzero entries, i.e. a compression factor near 10. To evaluate its accuracy, we can uses the following simple version of Schur lemma for matrices (see Chapter IV, Lemma 41.1, for the general version)

$$\|A\|^2 \leqslant \left[\sup_j \sum_i |A_{i,j}|\right]\left[\sup_i \sum_j |A_{i,j}|\right], \tag{5.8}$$

which yields the error estimate in operator norm

$$\left\|S_9 - \widetilde{S}_9\right\| \leqslant 10^{-2}. \tag{5.9}$$

REMARK 5.3. The single layer potential is known to be an operator of order -1, which maps $H^{-1/2}$ into $H^{1/2}$. In Chapter IV, we shall see that it is possible to take into account this regularizing property in the compression process, in order to obtain more appropriate error bounds on $\|S_J - \widetilde{S}_J\|_{H^{-1/2} \to H^{1/2}}$.

REMARK 5.4. The singular kernel in (5.7) satisfies $K(x, y) = k(x - y)$ where $k(\bullet)$ is a 1-periodic function, and in turn its discretization T_9 is also a Toepliz cyclic matrix, i.e. $T_9(m, n) = t((m - n)[\text{mod } 2^9])$. A more appropriate change of basis, in that case, would be provided by the discrete Fourier transform, since it would lead to a diagonal matrix. However, it is important to note that the sparsification property of the multiscale decomposition does not rely on a *structural* property of the operator such as translation invariance, but on the *analytical* smoothness properties of the kernel. As a consequence, we can expect more *robustness*, in comparison to Fourier decomposition: a perturbation of the circular shape of the cylinder in the above example should not affect the sparsity of S_9, while it might generate a fully populated matrix when using the Fourier basis. We shall see in Section 41 of Chapter IV that fairly general classes of operators can be sparsified by wavelet techniques.

6. Multilevel preconditioning for elliptic problems

One of the main interest of multiscale methods is that they yield fast solvers for linear elliptic problems. We shall illustrate this point with a simple example involving piecewise affine multiresolution spaces, in order to motivate a more detailed study that will be the object of Chapter III.

Let us first recall some basic results on the numerical approximation of the following model elliptic equation

$$-\Delta u = f \quad \text{in } \Omega, \qquad u = 0 \quad \text{on } \Gamma = \partial\Omega, \tag{6.1}$$

where $\Omega \subset \mathbb{R}^2$ is a convex polygonal domain and $f \in L^2(\Omega)$. With such assumptions, it is known (see, e.g., GRISVARD [1983]), that problem (6.1) has a unique solution $u \in H^2(\Omega) \cap H_0^1(\Omega)$, that satisfies the a priori estimate

$$\|u\|_{H^2} \leqslant C\|f\|_{L^2}, \tag{6.2}$$

where C depends only on Ω. Another classical result is that u is the unique solution in H_0^1 of the variational problem

$$a(u, v) = \langle f, v\rangle, \quad \text{for all } v \in H_0^1(\Omega), \tag{6.3}$$

where $\langle f, v\rangle = \int_\Omega f(x)v(x)\,\mathrm{d}x$ and $a(u, v)$ is the bilinear form

$$a(u, v) = \int_\Omega \nabla u(x)\nabla v(x)\,\mathrm{d}x. \tag{6.4}$$

We recall that existence and unicity of this variational solution follows from the Lax–Milgram lemma, remarking that the bilinear form a is both continuous on $H_0^1 \times H_0^1$ and elliptic, i.e.

$$a(u, u) \geqslant C\|f\|_{H^1}^2, \quad \text{for all } f \in H_0^1(\Omega), \tag{6.5}$$

by Poincaré's inequality. Note that continuity and coercivity imply that the energy norm $\|u\|_a := \sqrt{a(u, u)}$ and the Sobolev norm $\|u\|_{H^1}$ are equivalent norms on H_0^1.

In order to discretize this problem, we now consider the simple $\mathbb{P}_1$ finite element space V_J, $J \geqslant 0$, that was described in the end of Section 4: V_J consists in all H_0^1 piecewise affine functions on a triangulation $\mathcal{T}_J$, built from an initial coarse triangulation $\mathcal{T}_0$ of Ω by J steps of uniform refinements. As it was already remarked, this approximation space is naturally equipped with a nodal basis $\{\phi_{J,\gamma};\, \gamma \in \widetilde{\Gamma}_J\}$, indexed by the set of interior vertices $\widetilde{\Gamma}_J = \Gamma_J \setminus \partial\Omega$. We choose an L^2 normalization for this basis, i.e.

$$\phi_{J,\gamma}(\gamma') = 2^J \delta_{\gamma,\gamma'}, \quad \gamma' \in \Gamma_J. \tag{6.6}$$

Although this basis is not orthonormal, it is not difficult to show the equivalence between the L^2 norm on V_J and the discrete ℓ^2 norm of the coordinates in the nodal basis, i.e.

$$C_1 \sum_{\gamma \in \widetilde{\Gamma}_J} |c_{J,\gamma}|^2 \leqslant \Big\| \sum_{\gamma \in \widetilde{\Gamma}_J} c_{J,\gamma}\phi_{J,\gamma} \Big\|_{L^2}^2 \leqslant C_2 \sum_{\gamma \in \widetilde{\Gamma}_J} |c_{J,\gamma}|^2, \tag{6.7}$$

where the constants $C_1, C_2 > 0$ are independent of the coefficients $c_{J,\gamma}$ and of the resolution level J.

The ellipticity of a on V_J ensures the existence and unicity of the Galerkin approximation to the solution u, i.e. element $u_J \in V_J$ that solves

$$a(u_J, v_J) = \langle f, v_J\rangle, \quad \text{for all } v_J \in V_J, \tag{6.8}$$

or equivalently

$$a(u_J, \phi_{J,\gamma}) = \langle f, \phi_{J,\gamma}\rangle, \quad \gamma \in \widetilde{\Gamma}_j. \tag{6.9}$$

Since we also have $a(u, v_J) = \langle f, v_J\rangle$ for all $v_J \in V_J$, u_J should be viewed as the projection of u onto V_J with respect to the energy norm $\|v\|_a$. Using the equivalence of $\|\cdot\|_a$ and $\|\cdot\|_{H^1}$ and classical finite element approximation results (see, e.g., CIARLET [1978]), we can thus derive the error estimate

$$\|u - u_J\|_{H^1} \lesssim \inf_{v_J \in V_J} \|u - v_J\|_{H^1} \lesssim 2^{-J}\|u\|_{H^2}. \tag{6.10}$$

Such an estimate relates the a priori accuracy $\varepsilon = \varepsilon(J) = \|u - u_J\|_{H^1}$ of the Galerkin approximation to the solution of problem (6.1) with the number of parameters $N_J := \dim(V_J) \sim 2^{2J}$ that are used to describe this approximation: we need (at most) $N(\varepsilon) = \mathcal{O}(\varepsilon^{-2})$ parameters (with a constant proportional to $\|u\|_{H^2}^2$) to reach the accuracy ε.

In the a priori analysis of the finite element method, it is also important to evaluate the *computational cost*, i.e. the number $\mathcal{C}(\varepsilon)$ of basic operations that will be required to reach the accuracy ε. Clearly this cost should be larger than $N(\varepsilon)$, since there is at least $N(\varepsilon)$ unknowns to compute, but we can expect of a good resolution method to make $\mathcal{C}(\varepsilon)$ not substantially larger than $N(\varepsilon)$.

The system resulting from (6.9) has the form

$$A_J U_J = F_J, \tag{6.11}$$

where U_J is the coordinate vector of u_J in the nodal basis, F_J is the vector of coordinates $\langle f, \phi_{j,\gamma}\rangle$, $\gamma \in \widetilde{\Gamma}_J$, and A_J is the positive definite stiffness matrix with entries

$$A_J(\gamma, \gamma') = a(\phi_{j,\gamma}, \phi_{j,\gamma'}), \quad \gamma, \gamma' \in \widetilde{\Gamma}_J. \tag{6.12}$$

For large values of J, direct methods for the computation of U_J becomes unpractical and a more reasonable strategy is to use an iterative algorithm. The simplest algorithm that can be applied to solve (6.12) is the *Richardson iteration*: one starts with an initial guess U_J^0 of U_J (e.g., $U_J^0 = 0$), and define at each step n,

$$U_J^{n+1} = U_J^n + \tau(F_J - A_J U_J^n). \tag{6.13}$$

The parameter $\tau > 0$ aims to stabilize the algorithm: Since U_J is clearly a fixed point of the Richardson iteration, the error $E_J^n = U_J^n - U_J$ satisfies

$$E_J^n = (I - \tau A_J)E_J^n = \cdots = (I - \tau A_J)^n E_J^0, \tag{6.14}$$

and will thus decay to 0 if we tune the parameter τ so that the contraction factor given by the spectral radius $r_J = \rho(I - \tau A_J)$ is strictly less than 1. A natural choice of τ that optimizes this property is $\tau = 2/(\lambda_{J,\max} + \lambda_{J,\min})$ where $\lambda_{J,\max}$ and $\lambda_{J,\min}$ are respectively the largest and smallest eigenvalues of A_J. The contraction factor is thus governed by the condition number $\mathcal{K}(A_J) = \mathcal{K}_J = \lambda_{J,\max}/\lambda_{J,\min}$ since we have $r_J = (\mathcal{K}_J - 1)/(\mathcal{K}_J + 1)$. In our case of interest, the condition number $\mathcal{K}_J$ is well known to depend on J as $\mathcal{O}(2^{2J})$.

Defining u_J^n to be the element in V_J with coordinates U_J^n, i.e. the approximate solution computed after n steps, we can accept an error $\|u_J - u_J^n\|_{H^1}$ of the same order of magnitude as the approximation error $\varepsilon(J) \sim 2^{-J}\|u\|_{H^2}$. This allows us to estimate the number of required iterations, in the case we take 0 as an initial guess: we have then

$$\|u_J - u_J^n\|_{H^1} \leqslant r_J^n \|u_J\|_{H^1} \leqslant C r_J^n \|u\|_{H^1}, \tag{6.15}$$

since the Galerkin projection is stable in H^1. We thus need to choose the number of iteration $n(J)$ so that r_J^n is of the order $\varepsilon(J)$, i.e.

$$\left(1 - C2^{-2J}\right)^{n(J)} \sim 2^{-J}. \tag{6.16}$$

Taking the logarithm of this estimate, we obtain that the number of required iteration is of the order $J2^{2J}$.

The sparsity of A_J allows to perform each iteration in $\mathcal{O}(2^{2J})$ operation. We thus conclude that the accuracy $\varepsilon \sim 2^{-J}$ can be achieved with a computational cost $\mathcal{C}(\varepsilon) \sim J2^{4J} \sim \log(\varepsilon)\varepsilon^{-4}$.

In summary, the bad conditioning of the stiffness matrix results in a computational cost that is much higher than the number of parameters that are used to describe the solution with a prescribed accuracy. In that sense, the method that we have described is far from being optimal.

A natural generalization of the Richardson iteration consists in replacing the scalar factor τ in (6.13) by the action of an invertible operator, i.e.

$$U_J^{n+1} = U_J^n + B_J(F - A_J U_J^n). \tag{6.17}$$

With such a generalization, we can hope to improve the reduction factor which is now given by the spectral radius $r_J = \rho(I - B_J A_J)$.

We call B_J a *preconditioner* for A_J. Note that the simple choice $B_J = A_J^{-1}$ would make the iteration converge in one step, but is meaningless since it amounts in a direct inversion of A_J which is precisely the task that we want to avoid here. A *good* preconditioner B_J should thus have two basic properties.

(i) *Simplicity*: it should be easy to apply. In particular, an iteration of (6.17) should have the same optimal order $\mathcal{O}(N_J)$ of computational cost as the simpler (6.13).
(ii) *Efficiency*: it should keep the reduction factor away from 1, if possible independently of J. This property means that the condition number $\widetilde{\mathcal{K}}_J = \mathcal{K}(B_J A_J)$ is better controlled than $\mathcal{K}(A_J)$ as J grows, if possible bounded independently of J, since up to renormalizing B_J by a proper scalar multiplicative constant we obtain $r_J = (\widetilde{\mathcal{K}}_J - 1)/(\widetilde{\mathcal{K}}_J + 1)$ as the new reduction factor.

In the case where B_J is symmetric, note the second property can also be expressed by $\langle A_J U, U\rangle_d \sim \langle B_J^{-1} U, U\rangle_d$, where $\langle \bullet, \bullet\rangle_d$ stands for the discrete inner product between coordinates vectors, since this is equivalent to $\langle B_J^{1/2} A_J B_J^{1/2} U, U\rangle_d \sim \langle U, U\rangle_d$, i.e. $\mathcal{K}(B_J A_J) = \mathcal{K}(B_J^{1/2} A_J B_J^{1/2}) = \mathcal{O}(1)$. For this reason A_J and B_J^{-1} are said to be *spectrally equivalent*. Note that since $\mathcal{K}(B_J^{-1} A_J^{-1}) = 1/\mathcal{K}(B_J A_J)$, the operators A_J^{-1} and B_J are also spectrally equivalent.

Multiscale decomposition methods provide with such preconditioners: we shall use them to build operators B_J that can be applied in $\mathcal{O}(N_J)$ operations, and such that $\mathcal{K}(B_J A_J)$ is independent of J, i.e. behaves in $\mathcal{O}(1)$. The analysis of such preconditioners makes an important use of ability of characterizing the functions of H_0^1 from the numerical properties of their multiscale decompositions.

Let us first illustrate this point on the one-dimensional case, using the Schauder basis described in Section 3. In that case, we remark that the wavelets $\psi_{j,k}$ defined by (3.7) satisfy $\psi'_{j,k} = 2^{j+1} h_{j,k}$ where $h = \chi_{[0,1/2[} - \chi_{[1/2,1[}$ is the Haar wavelet. In other words, the Schauder hierarchical basis is *orthogonal* with respect to the energy inner product $a(u,v) = \int_0^1 u'v'$ associated to the one-dimensional version of (6.1), i.e. $u'' = f$ with boundary conditions $u(0) = u(1) = 0$. If f is a function in $H_0^1([0,1])$ and if we expand it in the Schauder basis $f = \sum_{j\geqslant 0}\sum_{k=0}^{2^j-1} d_{j,k}\psi_{j,k}$ (which is feasible since such a function is always continuous), we thus have the identity

$$\int |f'|^2 = \int \Big| \sum_{j,k} 2^{j+1} d_{j,k} h_{j,k} \Big|^2 = \sum_{j \geqslant 0} 2^{2j+2} \sum_k |d_{j,k}|^2$$
$$= \sum_{j \geqslant 0} 2^{2j+2} \|P_{j+1} f - P_j f\|^2_{L^2([0,1])},$$

where P_j is the interpolation operator introduced in Section 3 (the term $P_0 f$ is trivial because of the homogeneous boundary condition). A consequence of this fact it that if we use the Schauder basis to discretize the one-dimensional version of (6.1) in the corresponding univariate approximation space V_J, the resulting stiffness matrix $\widetilde{A}_J$ is *diagonal* (still with condition number behaving like $\mathcal{O}(2^{2J})$). In this particular case the inversion of the system becomes trivial, but we can also view this as a way to precondition the stiffness matrix A_J if we want to solve the problem in the nodal basis: we have $\widetilde{A}_J = R_J^* A_J R_J$ where R_J represent the reconstruction operator that maps the coefficients of a function in the Schauder basis to those of the same function in the nodal basis, so that $A_J^{-1} = R_J \widetilde{A}_J^{-1} R_J^*$. The operator $B_J := R_J \widetilde{A}_J^{-1} R_J^*$ qualifies as a good preconditioner according to our previous prescriptions: it clearly result on the best possible condition number $\widetilde{\mathcal{K}} = 1$ since it is the inverse of A_J, and the application of each of his three components has optimal complexity $\mathcal{O}(N_J)$ if we use the fast algorithm described in Section 3 to implement R_J and its transpose.

If we now turn to the multivariate case, a rapid check reveals that the hierarchical basis associated constructed in Section 4 is *not* orthogonal in the sense of the energy inner product. We could still hope more modestly for an equivalence of the type

$$\|f\|^2_{H^1(\Omega)} \sim \|P_0 f\|^2_{L^2(\Omega)} + \sum_{j \geqslant 0} 2^{2j} \|Q_j f\|^2_{L^2(\Omega)}, \tag{6.18}$$

where $Q_j = P_{j+1} - P_j$ and P_j is the corresponding interpolation operator. However, this cannot hold either, since the functions of H_0^1 might be discontinuous and unbounded in more than one dimension, so that $P_j f$ might be meaningless.

The results of Chapter III will nevertheless establish this type of equivalences for more general function spaces and multiscale decompositions in arbitrary dimensions. For the particular case of H_0^1, it will be shown that (6.18) holds in higher dimension, provided that for the definition of P_j one replaces interpolation by some appropriate projection chosen among a fairly general class of projectors onto the finite element space V_j. In particular, we shall see that the L^2-orthogonal projector onto V_j satisfies this property. If in addition we can expand at each level the details according to a wavelet basis $Q_j f = \sum_{\lambda \in \Lambda_j} d_\lambda \psi_\lambda$ with the same uniform stability property as (6.7) for the nodal function, we shall thus have

$$\|f\|^2_{H^1(\Omega)} \sim \sum_{j \geqslant -1} 2^{2j} \sum_{\lambda \in \Lambda_j} |d_\lambda|^2, \tag{6.19}$$

as for the Schauder basis in one dimension. Here by convention, we incorporate the nodal basis of V_0 in the multiscale basis for the indices $\lambda \in \Lambda_{-1}$, in order to include the term $\|P_0 f\|^2_{L^2(\Omega)}$ in the above sum.

The equivalences (6.18) and (6.19) should be at least understood intuitively at this point, by thinking of $P_j f$ as the part f that essentially contains frequencies below 2^j. Let us, e.g., define P_j in the one-dimensional case as the orthogonal projection onto band limited functions such that $\hat{f}(\omega) = 0$ if $|\omega| > 2^j$, i.e. the filtering operator $P_j f = f_j$ with

$$\hat{f}_j(\omega) = \hat{f}(\omega)\chi_{[-2^j\pi, 2^j\pi]}, \tag{6.20}$$

where $\hat{f}(\omega) = \int f(t)\mathrm{e}^{-\mathrm{i}\omega t}\,\mathrm{d}t$ is the usual Fourier transform. Using Parseval's identity, we can decompose the H^1 norm of f according to

$$\begin{aligned}
\|f\|^2_{H^1} &\sim \int \left(1+|\omega|^2\right)\left|\hat{f}(\omega)\right|^2 \mathrm{d}\omega \\
&= \int_{-1}^{1} \left(1+|\omega|^2\right)\left|\hat{f}(\omega)\right|^2 \mathrm{d}\omega + \sum_{j\geqslant 0} \int_{2^j<|\omega|<2^{j+1}} \left(1+|\omega|^2\right)\left|\hat{f}(\omega)\right|^2 \mathrm{d}\omega \\
&\sim \int_{-1}^{1} \left|\hat{f}_0(\omega)\right|^2 \mathrm{d}\omega + \sum_{j\geqslant 0} 2^{2j} \int_{2^j<|\omega|<2^{j+1}} \left|\left(\hat{f}_{j+1}-\hat{f}_j\right)(\omega)\right|^2 \mathrm{d}\omega \\
&\sim \|P_0 f\|^2_{L^2(\Omega)} + \sum_{j\geqslant 0} 2^{2j} \|P_{j+1}f - P_j f\|^2_{L^2(\Omega)}
\end{aligned}$$

and we have thus proved an equivalence of the same type as (6.18). The results of Chapter III will make clear that such equivalences can also be obtained with approximation operators onto finite element spaces, which do not correspond exactly to a "cut-off" in frequency, but are better fitted to numerical simulation problems.

REMARK 6.1. Note that a reformulation of (6.19) is the norm equivalence

$$\|f\|^2_{H^1(\Omega)} \sim \|P_0 f\|^2_{L^2} + \sum_{j\geqslant 0} 2^{2j} \|f - P_j f\|^2_{L^2}. \tag{6.21}$$

Indeed both right-hand side in (6.19) and (6.21) are clearly equivalent, using that $\|f - P_j f\|^2_{L^2} = \sum_{l\geqslant j} \|P_{l+1}f - P_l f\|^2_{L^2}$. We see from (6.21) that $f \in H^1_0$ implies a bit more than the classical error estimate $\|f - P_j f\|_{L^2} \lesssim 2^{-j}\|f\|_{H^1}$, and that moreover, a function f has H^1 smoothness *if and only if* a certain decay property of the finite approximation L^2 error is satisfied.

Let us now explain how (6.19) can be used to build a simple preconditioner for A_J. We again denote by $\widetilde{A}_J = R^*_J A_J R_J$ the stiffness matrix in the multiscale basis

$$\Psi_J := \{\psi_\lambda,\ \lambda \in \Lambda_j;\ j = -1, \ldots, J\}. \tag{6.22}$$

If f is a function in V_J and if F is the coordinate vector corresponding to the expansion $f = \sum_{j<J} \sum_{\lambda \in \Lambda_j} d_\lambda \psi_\lambda$ in the multiscale basis, we thus have by (6.19)

$$\langle \widetilde{A}_J F, F \rangle_d = a(f, f) \sim \|f\|_{H^1}^2 \sim \sum_{j<J} 2^{2j} \sum_{\lambda \in \Lambda_j} |d_\lambda|^2 = \langle D_J F, F \rangle_d, \tag{6.23}$$

where D_J is a simple diagonal operator corresponding to a scaling of d_λ by 2^{2j} if $\lambda \in \Lambda_j$. The operators A_J and D_J are thus spectrally equivalent, i.e. $\mathcal{K}(D_J^{-1}\widetilde{A}_J) = \mathcal{K}(R_J D_J^{-1} R_J^* A_J)$ behaves in $\mathcal{O}(1)$.

We thus see that $B_J := R_J D_J^{-1} R_J^*$ is a good candidate to be a preconditioner for A_J, provided that R_J and its transpose can be applied in $\mathcal{O}(N_J)$ operation, as in the case of the hierarchical basis. Such an optimal complexity for R_J can essentially be obtained if the multiscale basis is local.

The main objective of our next chapter is to address the effective construction of such local multiscale bases that will yield norm equivalences of the type (6.18).

7. Conclusion

In all the applications of multiscale schemes that we were presented here, we have used several important features of multiscale decompositions:

(i) *Nested approximations*: a specific multiresolution approximation process onto a *hierarchy of spaces* $V_j \subset V_{j+1} \subset \cdots$ is defined through projectors P_j.

(ii) *Bases*: the approximations and the fluctuations can be further decomposed into *local* bases.

(iii) *Stability*: certain stability properties are satisfied by these bases, either within one resolution level such as in (6.7), or across the different levels, such as (6.19).

(iv) *Local approximation properties*: we can estimate the accuracy of the multiscale approximations of a function in the regions where it is smooth, and improve this accuracy by local *refinements* near its singularities which are operated by simple *thresholding* procedures.

(v) *Smoothness*: the approximation spaces might also have certain smoothness properties, which are usually required for the discretization of differential problems.

(vi) *Algorithms*: multiscale decompositions and reconstructions are performed by fast $\mathcal{O}(N)$ algorithms.

On the other hand these schemes suffer from numerous drawbacks. The Haar system, based on piecewise constant functions, has no C^0-smoothness and poor approximation properties. In turn, it cannot be used to discretize differential problems. The smoothness and approximation properties of piecewise affine functions are also low. The Schauder hierarchical basis suffers from a lack of stability across level, in the sense that the L^2 norm of a function cannot be estimated from the ℓ^2 norm of its coefficients. As it was indicated in Section 6, it is also unadapted to characterize H^1 smoothness in the multivariate setting. On the other hand, we remarked that an orthogonal decomposition based on piecewise affine functions involves nonlocal computations.

These remarks motivate the development of more general multiscale approximation and decomposition schemes. These generalizations should preserve the important features that we identified, while allowing better smoothness, higher-order approximation and better stability properties.

8. Historical notes

In 1873, Dubois-Reymond proved the existence of a continuous periodic function f such that the Fourier series of f diverges at some point x. This striking result shows how an orthonormal basis that only consists of smooth functions might yet be inadequate for the analysis and the approximation of function spaces different from L^2.

In contrast, the Haar system, introduced in the beginning of the early century (in the Ph.D. dissertation of A. Haar, 1909), has the property that the series of a continuous function on $[0, 1]$ converges uniformly. However, the Haar basis does not constitute a proper basis for the space $C^0([0, 1])$, since its element are not contained in this space.

The Schauder hierarchical basis, which is simply obtained by taking the primitives of the Haar basis functions, was introduced and studied by Faber and Schauder between 1910 and 1920. In contrast to the Haar system, it does constitute a proper basis for the space $C^0([0, 1])$, but cannot be used to decompose a general L^2 function.

Generally speaking, a *Schauder basis* in a Banach space X is a family e_n, $n \geqslant 0$, such that any $x \in X$ has a unique expansion $\sum_{n \geqslant 0} x_n e_n$ that converges in X. The basis is said to be *unconditional* if and only if, the convergence of the series is maintained if x_n is replaced by y_n such that $|y_n| \leqslant |x_n|$. In the setting of function spaces, this means that one can characterize X from the size properties of the coefficients in the basis e_n. Several classical separable Banach spaces, in particular $L^1(I)$ and $C^0(I)$, are known to possess no unconditional basis. A review of these different concepts of basis in a Banach space can be found in YOUNG [1980].

As it will be shown in Chapter III, the Schauder basis is well adapted to the characterization of the Hölder space C^α, $0 < \alpha < 1$, for which it constitutes an unconditional basis. This remarkable property played an important role in the analysis of the Brownian motion by P. Levy in the 1930's, who used the representation in the Schauder basis to estimate (in an almost sure sense) the regularity of its trajectories.

The use of the hierarchical basis in the context of numerical analysis context, for the preconditioning of systems arising from elliptic boundary value problems, is more recent: the idea of replacing the nodal basis by the hierarchical basis was proposed in BABUSHKA, GAGO, KELLY and ZIENKIEWICZ [1982] in the univariate case, and studied in the multivariate case by YSERANTANT [1986]. These contributions, although they should be viewed as part of the important development of multigrid methods, emphasize the point of view of a "multilevel decomposition of finite element spaces", which is not central in the main stream development of multigrid techniques.

CHAPTER II

Multiresolution Approximation

9. Introduction

In this chapter, we shall study in detail the construction and the properties of general multiresolution approximation and decomposition schemes. These schemes are based on hierarchies of nested spaces $V_j \subset V_{j+1}$, that reproduce the main features that we have identified in the simple examples of Chapter I, while allowing additional properties of approximation, smoothness and stability.

A first idea to generalize the constructions of the previous chapter is to use finite element spaces V_j, built on triangulations or quadrangulations with mesh size of the order 2^{-j}, and try to impose the nestedness property, which leads to natural restrictions on the type of mesh and finite element to be used. This approach has the advantage of inheriting the structural simplicity of finite element spaces. We recall below four characteristic features of these spaces.

(i) *Local characterization*: On each triangle or rectangle, the approximating functions have the simple explicit expression of a *polynomial* that depends on a fixed number of nodal values.

(ii) *Bases*: Local *nodal* bases for these spaces can be constructed in a simple way, and the stability properties of these bases is easy to analyze, using affine transformations onto a reference triangle or square.

(iii) *Smoothness*: The smoothness of the functions in the finite element spaces is well understood. It is controlled by the compatibility conditions that are imposed at each node.

(iv) *Accuracy*: The approximation properties of these spaces are also well understood. They are controlled by the maximal degree d such that $\Pi_d \subset V_j$ where Π_d is the space of polynomials of total degree less than d.

However, the construction of multiscale bases, that are both related to nodal bases with fast algorithms and well adapted to characterize smoothness classes, is by no mean a straightforward operation: in many instances of finite element spaces, it is still an open question.

An alternate approach is based on ideas that were introduced and developed more recently: the approximation spaces V_j will be defined through a basis, which is generated (in the univariate case) by the translates and dilates $\varphi(2^j \bullet -k)$, $k \in \mathbb{Z}$, of a single *scal-*

ing function φ. In contrast to the finite element case, the function φ is not necessarily piecewise polynomial, but it needs to satisfy an equation of the type

$$\varphi(x) = \sum_{k\in\mathbb{Z}} h_k \varphi(2x-k), \tag{9.1}$$

which expresses the nestedness of the spaces V_j. Solutions of such equations are called *refinable functions*, and the associated hierarchy of approximation spaces is called a *multiresolution analysis*. We shall require in addition the existence of a dual scaling function $\tilde{\varphi}(x) = \sum_{k\in\mathbb{Z}} \tilde{h}_k \varphi(2x-k)$ that will allow to build local projectors of the form

$$P_j f = \sum_{k\in\mathbb{Z}} \langle f, \tilde{\varphi}_{j,k}\rangle \varphi_{j,k}. \tag{9.2}$$

This second strategy yields in a natural way the construction of biorthogonal wavelet bases and fast algorithms. It is easily generalizable to the multivariate setting by means of tensor product strategies (see Section 4). On the other hand, it is implicitly tied to the setting of uniform tensor product discretization, and it requires a good understanding of the construction, properties and practical manipulation of refinable functions. It also requires particular adaptations, if we want to approximate or decompose functions that are defined on bounded domains, with prescribed boundary conditions.

Since finite elements are classical material that have been documented in numerous textbooks (e.g., CIARLET [1978], STRANG and FIX [1973], BRENNER and SCOTT [1994]), we shall devote a substantial part of this chapter to explain the construction of multiscale decompositions from the second approach.

In the last section, we shall return to the first approach: multiscale decompositions will be directly derived from nested sequences of finite element spaces, or also from nested discretizations by point values or cell averages equipped with certain inter-scale operators. While these last tools are probably the most intuitive and relevant to the numerician, their interpretation in terms of the biorthogonal wavelets framework developed throughout this chapter is very beneficial for the rigorous analysis of their performance in many applications.

This chapter is organized as follows: in Section 10, we introduce the general concept of multiresolution analysis. In Section 11, we focus more precisely on Eq. (9.1). We show how any solution of (9.1) is a potential candidate to generate a multiresolution approximation. In Sections 12 and 13, we show respectively how to compute this solution and how it can be used for numerical approximation, even when it is not known explicitly. We show in Section 14 that these concepts lead in a natural way to the construction of wavelet bases, and fast decomposition and reconstruction algorithms.

The next Sections 15, 16 and 17 are concerned with the problem of analyzing some important properties of refinable functions – smoothness and polynomial reproduction, stability, orthonormality, biorthogonality – from the data of the coefficients h_n in (9.1). We then use these results in Section 18 to build important examples that satisfy interpolation or orthonormality properties, and we discuss examples involving spline functions in Section 19.

For the sake of notational simplicity we have chosen to focus on multiresolution approximation techniques for univariate functions that are defined on the whole of $\mathbb{R}$. We thus postpone to Section 20 the main techniques that are used to adapt these constructions to multivariate functions and to bounded domains with specific boundary conditions.

Finally Section 21 deals with the constructions of multiscale decompositions from the point of view of classical discretizations: point values, cell averages and finite elements.

10. Multiresolution analysis

The following definition was firstly introduced in MALLAT [1989].

DEFINITION 10.1. A multiresolution analysis is a sequence of closed subspaces of $L^2(\mathbb{R})$, such that the following properties are satisfied:

(i) The sequence is nested, i.e. for all j,

$$V_j \subset V_{j+1}. \tag{10.1}$$

(ii) The spaces are related to each other by dyadic scaling, i.e.

$$f \in V_j \Leftrightarrow f(2\bullet) \in V_{j+1} \Leftrightarrow f\left(2^{-j}\bullet\right) \in V_0. \tag{10.2}$$

(iii) The union of the spaces is dense, i.e. for all f in $L^2(\mathbb{R})$

$$\lim_{j\to+\infty} \left\| f - P_j^o f \right\|_{L^2} = 0, \tag{10.3}$$

where P_j^o is the orthonormal projection onto V_j.

(iv) The intersection of the spaces is reduced to the null function, i.e.

$$\lim_{j\to-\infty} \left\| P_j^o f \right\|_{L^2} = 0. \tag{10.4}$$

(v) There exists a function $\varphi \in V_0$ such that the family

$$\varphi(\bullet - k), \quad k \in \mathbb{Z}, \tag{10.5}$$

is a Riesz basis of V_0.

By definition, a family $\{e_n\}_{n\in\mathbb{Z}}$ is a *Riesz basis* of a Hilbert space H, if and only if it spans H, i.e. the finite linear combinations of the e_n are dense in H, and if there exist $0 < C_1 \leqslant C_2$ such that for all finite sequence (x_i), we have

$$C_1 \sum_i |x_i|^2 \leqslant \left\| \sum_i x_i e_i \right\|_H^2 \leqslant C_2 \sum_i |x_i|^2. \tag{10.6}$$

This property expresses the "stability" of the expansion in this basis with respect to the coordinates. It also means that the application

$$T:(c_k)_{k\in\mathbb{Z}}\mapsto\sum_{k\in\mathbb{Z}}c_k e_k, \tag{10.7}$$

defines an isomorphism from $\ell^2(\mathbb{Z})$ to H.

The following three facts concerning Riesz bases are easy to check: firstly, the series $\sum_{i\in\mathbb{Z}}x_i e_i$ converges *unconditionally* in L^2 (i.e. its terms can be permutated without affecting the convergence) if and only if $\sum_{i\in\mathbb{Z}}|x_i|^2$ is finite. Secondly, any $x\in H$ can be decomposed in a unique way according to $x=\sum_{i\in\mathbb{Z}}x_i e_i$ with $(x_i)_{i\in\mathbb{Z}}$ in $\ell^2(\mathbb{Z})$ and the equivalence in (10.6) also hold for such infinite linear combinations. Finally, there exists a unique *biorthogonal* Riesz basis $\{\tilde{e}_n\}_{n\in\mathbb{Z}}$ (defined by $\tilde{e}_n=(TT^*)^{-1}e_n$), such that $\langle e_i,\tilde{e}_j\rangle=\delta_{i,j}$, and the coordinates of x are given by $x_i=\langle x,\tilde{e}_i\rangle$. We also have a dual expansion $x=\sum_{i\in\mathbb{Z}}\langle x,e_i\rangle\tilde{e}_i$.

In this sense, Riesz bases are very close to orthonormal bases, which appear as a particular case where $C_1=C_2=1$.

REMARK 10.1. A consequence of the Riesz basis property is

$$C_1\|x\|_H^2\leqslant\sum_{k\in\mathbb{Z}}\left|\langle x,e_k\rangle\right|^2\leqslant C_2\|x\|_H^2, \tag{10.8}$$

since the adjoint mapping

$$T^*:x\mapsto\big(\langle x,e_k\rangle_H\big)_{k\in\mathbb{Z}}, \tag{10.9}$$

is also bounded and invertible from H to $\ell^2(\mathbb{Z})$. However (10.8) is weaker than (10.6) since it only reflects the left invertibility of T^*. Sequences that satisfy (10.8) are called *frames*. The property (10.8) ensures that a vector x can be reconstructed from its inner products $\langle x,e_k\rangle$. However, in contrast to Riesz bases, this reconstruction is not unique, and frames can be redundant: take the example of $(0,1)$, $(1,0)$, $(-1,0)$ and $(0,-1)$ in $\mathbb{R}^2$ that satisfies (10.8) with $C_1=C_2=2$. Frames were introduced in the 1950's in DUFFIN and SCHAEFFER [1952] in the context of nonharmonic Fourier series and irregular sampling. An introduction to frames and their basic properties can also be found in YOUNG [1980] and in DAUBECHIES [1992].

In our case of interest, we shall sometimes call L^2-*stable* a function φ such that $\varphi(\bullet-k)$, $k\in\mathbb{Z}$, forms a Riesz basis of its L^2-span. We then derive from property (10.2) that the family

$$\varphi_{j,k}=2^{j/2}\varphi\big(2^j\bullet-k\big),\quad k\in\mathbb{Z}, \tag{10.10}$$

is a Riesz basis for the space V_j: any $f_j\in V_j$ can be written $f_j=2^{j/2}f_0(2^j\bullet)$ with $f_0=\sum_{k\in\mathbb{Z}}c_k\varphi(\bullet-k)\in V_0$, and thus $f_j=\sum_{k\in\mathbb{Z}}c_k\varphi_{j,k}$. Moreover, since $\|f_j\|_{L^2}=\|f_0\|_{L^2}$, we see that the constants C_1 and C_2 in the equivalence (10.6) are independent of j. The function φ is called a *scaling function*, since we need to use its scaled version to characterize the approximation at a given resolution.

REMARK 10.2. From the definition of a multiresolution analysis, it is clear that $f \in V_j$ implies that $f(\bullet - kh) \in V_j$ for all $k \in \mathbb{Z}$, where we have set $h = 2^{-j}$. A space V that satisfies such a property for some $h > 0$ is called *shift invariant*, with respect to the shift h. A shift-invariant space that is furthermore generated by the translates $\varphi(\bullet - nh)$, $n \in \mathbb{Z}$ (such as our V_j spaces), of a single function is called *principal shift invariant*. The study of shift-invariant spaces and their property can be done in a more general context than multiresolution approximation (see, e.g., STRANG and FIX [1969], DE BOOR, DEVORE and RON [1993]). In most of these works, as in several results in this chapter, Fourier analysis plays an important role, due to the following observation: if $f = \sum_{n\in\mathbb{Z}} c_n \varphi(\bullet - nh) \in V$, we have

$$\hat{f}(\omega) = \left[\sum_{n\in\mathbb{Z}} c_n \mathrm{e}^{-inh\omega}\right]\hat{\varphi}(\omega). \tag{10.11}$$

The Fourier transform of V can thus be viewed as the simple multiplication of $\hat{\varphi}$ with a space of 2π-periodic functions (which is exactly $L^2([0, 2\pi])$ in the case where φ is L^2-stable since the c_n can be any ℓ^2 sequence), and several properties of V can be "read" on the Fourier transform of φ.

Since $V_0 \subset V_1$, the scaling function φ can be expanded in terms of the basis of V_1, i.e.

$$\varphi(x) = \sum_{n\in\mathbb{Z}} h_n \varphi(2x - n), \tag{10.12}$$

where the sequence $(h_n)_{n\in\mathbb{Z}}$ is in $\ell^2(\mathbb{Z})$ by the stability property. Equation (10.12), known as *refinement equation*, or *two-scale difference equation*, will play a crucial role in the sequel of this chapter. A function that satisfies such an equation is called a *refinable function*, and the coefficients h_n in (10.12) are called the *refinement coefficients*. The terms "refinement" and "refinable" will be justified in Section 12 (Remark 12.1), where it is shown that such functions are related to iterative refinement schemes that are typical tools of computer-aided geometric design.

Our first examples of scaling functions have an explicit expression, while more refined examples will be defined implicitly as solutions of a two-scale difference equation.

EXAMPLE 10.1. The "box function" $\varphi = \chi_{[0,1]}$ has been considered in Section 2 of Chapter I. It generates a multiresolution analysis that consists of the spaces V_j defined in (2.2). It satisfies the two-scale difference equation

$$\varphi(x) = \varphi(2x) + \varphi(2x - 1), \tag{10.13}$$

i.e. $h_0 = h_1 = 1$, else $h_k = 0$. A particular feature of this scaling function is that $\varphi_{j,k}$ is an orthonormal basis for V_j. We shall use the following terminology to summarize this property:

DEFINITION 10.2. A scaling function $\varphi \in L^2$ is said to be orthonormal, when it satisfies

$$\langle \varphi(\bullet - k), \varphi(\bullet - l) \rangle = \delta_{k,l}, \quad k, l \in \mathbb{Z}. \tag{10.14}$$

EXAMPLE 10.2. The "hat function" $\varphi(x) = \max\{1 - |x|, 0\}$ was used in Section 3 of Chapter I. It generates a multiresolution analysis that consists of the spaces

$$V_j = \{f \in L^2 \cap C^0;\ f|_{I_{j,k}} \in \Pi_1\}. \tag{10.15}$$

The proof that its translates forms a Riesz basis is a particular case of Corollary 10.2 below that proves this property for more general spline functions. The associated refinement equation is

$$\varphi(x) = \varphi(2x) + [\varphi(2x-1) + \varphi(2x+1)]/2, \tag{10.16}$$

i.e. $h_0 = 1$, $h_1 = h_{-1} = 1/2$, else $h_k = 0$. A particular feature of this scaling function is that $\varphi(k) = \delta_{0,k}$, $k \in \mathbb{Z}$, so that a function $f_j \in V_j$ satisfies

$$f_j = \sum_{k \in \mathbb{Z}} f_j(2^{-j}k)\varphi(2^j \bullet - k) = \sum_{k \in \mathbb{Z}} 2^{-j/2} f_j(2^{-j}k)\varphi_{j,k}. \tag{10.17}$$

We use the following terminology to summarize this property.

DEFINITION 10.3. A scaling function $\varphi \in L^2 \cap C^0$ is said to be interpolatory, when it satisfies

$$\varphi(k) = \delta_{0,k}, \quad k \in \mathbb{Z}. \tag{10.18}$$

EXAMPLE 10.3. The spaces of *band-limited* functions

$$V_j = \{f \in L^2;\ \mathrm{Supp}(\mathcal{F}f) \subset [-2^j\pi, 2^j\pi]\} \tag{10.19}$$

(with $\mathcal{F}f(\omega) = \hat{f}(\omega)$) constitute a multiresolution analysis: the first four properties in Definition 10.1 are immediate to check (using that the projector P_j is defined by $\mathcal{F}P_j f = \chi_{[-2^j\pi, 2^j\pi]}\mathcal{F}f$). The Fourier transform of V_0 has a natural orthonormal basis $e_n(x) = (2\pi)^{-1/2} \mathrm{e}^{\mathrm{i}nx} \chi_{[-\pi,\pi]}$, $n \in \mathbb{Z}$. Applying the inverse Fourier transform and invoking Parseval identity, we obtain an orthonormal basis for V_0 of the form $\varphi(\bullet - k)$, $k \in \mathbb{Z}$, with

$$\varphi(x) = \frac{\sin(\pi x)}{\pi x}. \tag{10.20}$$

We notice furthermore that the function φ is also interpolatory. This allows to derive the associated scaling equation

$$\varphi(x) = \sum_{n\in\mathbb{Z}} \frac{2\sin(\pi n/2)}{\pi n} \varphi(2x - n). \tag{10.21}$$

This last scaling function strongly differs from the two first in the sense that is has infinite support and slow decay. For these reasons it is not of much use in numerical simulation that often requires local properties. These examples also suggest that the supports of φ and of the sequence h_n are the same. This property will be proved in Section 12.

Our first result is a general characterization of the properties of stability, orthonormality and interpolation for the system $\{\varphi(\bullet - k)\}_{k\in\mathbb{Z}}$ based on the Fourier transform.

THEOREM 10.1. *If $\varphi \in L^2$, the series*

$$\sum_{n\in\mathbb{Z}} \left|\hat{\varphi}(\omega + 2n\pi)\right|^2, \tag{10.22}$$

converges in $L^1(I)$ for any compact set I, to an L^1_{loc}, 2π-periodic function. The function φ is L^2-stable if and only if there exists two constants $C_1, C_2 > 0$, such that

$$C_1 \leqslant \sum_{n\in\mathbb{Z}} \left|\hat{\varphi}(\omega + 2n\pi)\right|^2 \leqslant C_2, \tag{10.23}$$

almost everywhere. Moreover, φ is an orthonormal scaling function if and only if

$$\sum_{n\in\mathbb{Z}} \left|\hat{\varphi}(\omega + 2n\pi)\right|^2 = 1, \tag{10.24}$$

almost everywhere. If we have, for some $\varepsilon > 0$,

$$\left|\hat{\varphi}(\omega)\right| \lesssim \left(1 + |\omega|\right)^{-1-\varepsilon}, \tag{10.25}$$

then φ is interpolatory if and only if

$$\sum_{n\in\mathbb{Z}} \hat{\varphi}(\omega + 2n\pi) = 1, \tag{10.26}$$

almost everywhere.

PROOF. Since $\hat{\varphi}$ is in L^2, the series

$$S_\varphi(\omega) = \sum_{n\in\mathbb{Z}} \left|\hat{\varphi}(\omega + 2n\pi)\right|^2, \tag{10.27}$$

converges in $L^1([-k\pi,(k+1)\pi])$ for all $k \in \mathbb{Z}$, and thus in $L^1(I)$ for any bounded interval I. The limit S_φ is clearly 2π-periodic.

If $f = \sum_{|k|<N} c_k \varphi(\bullet - k)$ is a finite linear combination of the integer shifts of φ, then we can express its L^2 norm as follows:

$$\begin{aligned}
\|f\|_{L^2}^2 &= (2\pi)^{-1}\|\hat{f}(\omega)\|^2 \\
&= (2\pi)^{-1}\int_{-\infty}^{+\infty}\left|\left[\sum_{|k|<N} c_k e^{-ik\omega}\right]\hat{\varphi}(\omega)\right|^2 d\omega \\
&= (2\pi)^{-1}\int_{-\pi}^{\pi}\left|\sum_{|k|<N} c_k e^{-ik\omega}\right|^2 S_\varphi(\omega)\, d\omega \\
&= (2\pi)^{-1}\int_{-\pi}^{\pi}|m(\omega)|^2 S_\varphi(\omega)\, d\omega,
\end{aligned}$$

where $m(\omega) = \sum_{|k|<N} c_k e^{-ik\omega}$. On the other hand, we have that

$$\sum_{|k|<N} |c_k|^2 = (2\pi)^{-1}\int_{-\pi}^{\pi}|m(\omega)|^2\, d\omega. \tag{10.28}$$

Now $\{\varphi(\bullet - k)\}_{k\in\mathbb{Z}}$ is a Riesz basis for its span if and only if the two above quantities are equivalent for any trigonometric polynomial $m(\omega)$ of arbitrary degree N. By density, this also means that these quantities are equivalent for any periodic function $m(\omega)$ which is square integrable on $[0, 2\pi]$, which clearly holds if and only if $S_\varphi(\omega)$ is almost everywhere bounded below and above by the constants C_1 and C_2 of (10.6).

We also remark that the Fourier coefficients of $S_\varphi(\omega)$ are given by

$$\begin{aligned}
S_k &= (2\pi)^{-1}\int_{-\pi}^{\pi} S_\varphi(\omega)e^{-ik\omega}\, d\omega = (2\pi)^{-1}\int_{-\infty}^{+\infty}|\hat{\varphi}(\omega)|^2 e^{-ik\omega}\, d\omega \\
&= \langle\varphi(\bullet - k), \varphi\rangle.
\end{aligned}$$

It follows that $S_\varphi(\omega) = 1$ almost everywhere if and only if φ is an orthonormal scaling function.

It is clear that the decay condition (10.25) implies that φ coincides almost everywhere with a continuous and bounded function that we can identify with φ, and that the series

$$R_\varphi(\omega) = \sum_{n\in\mathbb{Z}} \hat{\varphi}(\omega + 2n\pi), \tag{10.29}$$

converges in L^1 on every compact set. The Fourier coefficients of $R_\varphi(\omega)$ are given by

$$R_k = \frac{1}{2\pi}\int_{-\pi}^{\pi} R_\varphi(\omega)e^{-ik\omega}\, d\omega = \frac{1}{2\pi}\int_{-\infty}^{+\infty}\hat{\varphi}(\omega)e^{-ik\omega}\, d\omega = \varphi(-k). \tag{10.30}$$

It follows that $R_\varphi(\omega) = 1$ almost everywhere if and only if φ is interpolatory. □

REMARK 10.3. It should be noted that under the assumption of the convergence of (10.22), the expansion of the Fourier series

$$S_\varphi(\omega) = \sum_{k\in\mathbb{Z}} \langle \varphi(\cdot - k), \varphi \rangle e^{ik\omega}, \tag{10.31}$$

does not converge better than in $L^2([0, 2\pi])$, unless we make some additional assumptions on the localization properties of φ, in order to derive decay properties on the Fourier coefficients in (10.31). Similar observations can be made for

$$R_\varphi(\omega) = \sum_{k\in\mathbb{Z}} \varphi(k) e^{-ik\omega} \tag{10.32}$$

which is not ensured to converge even in $L^1([0, 2\pi])$ under the sole assumption (10.25).

REMARK 10.4. In the proof of Theorem 10.1, we see that if we only have the right part of (10.23) with some $C_2 > 0$, we still obtain the right part of (10.6), i.e. the mapping T defined by (10.7) is bounded. By duality, we also obtain that $\varphi(\cdot - k)$, $k \in \mathbb{Z}$, is a *Bessel sequence* of L^2, i.e. satisfies

$$\sum_{k\in\mathbb{Z}} \left|\langle f, \varphi(\cdot - k)\rangle\right|^2 \lesssim \|f\|_2^2, \tag{10.33}$$

for all $f \in L^2$.

REMARK 10.5. One can easily check that, under the assumption (10.25), $R_\varphi(\omega) \geqslant C_R > 0$ implies $S_\varphi(\omega) \geqslant C_S > 0$. In particular, an interpolatory scaling function is always L^2-stable.

A simple corollary to Theorem 10.1, deals with the effective construction of orthonormal and interpolatory scaling functions. We recall that $S_\varphi(\omega)$ and $R_\varphi(\omega)$ are the 2π-periodic functions defined by (10.27) and (10.29).

COROLLARY 10.1. *Let $\varphi \in L^2$ be a scaling function that satisfies* (10.23). *Then the function φ^o defined by*

$$\hat{\varphi}^o(\omega) = \left[S_\varphi(\omega)\right]^{-1/2} \hat{\varphi}(\omega), \tag{10.34}$$

is in V_0 and orthonormal. The function φ^d defined by

$$\hat{\varphi}^d(\omega) = \left[S_\varphi(\omega)\right]^{-1} \hat{\varphi}(\omega), \tag{10.35}$$

is in V_0 and generates a dual Riesz basis, i.e. $\langle \varphi(\cdot - k), \varphi^d(\cdot - l)\rangle = \delta_{k,l}$.

If $\varphi \in L^2$ is such that (10.25) *holds and $R(\omega) \geqslant C > 0$, then the function φ^i defined by*

$$\hat{\varphi}^i(\omega) = \left[R_\varphi(\omega)\right]^{-1}\hat{\varphi}(\omega), \tag{10.36}$$

is in V_0 and interpolatory.

When φ has compact support, one can define S_φ and R_φ directly through (10.31) *and* (10.32)*, and if φ is also bounded, the resulting functions φ^o, φ^d and φ^i have exponential decay at infinity.*

PROOF. It is clear that the functions $[S_\varphi(\omega)]^{-1/2}$, $[S_\varphi(\omega)]^{-1}$ and $[R_\varphi(\omega)]^{-1}$ are in $L^2([0, 2\pi])$, when R_φ and S_φ are bounded below by some constant $C > 0$. Based on the characterization of $\mathcal{F}V_0$ (see Remark 10.2), we conclude that the functions φ^o, φ^d and φ^i are in V_0.

By construction it is clear that φ^o and φ^i satisfy respectively the criteria of Theorem 10.1 for orthonormality and interpolation.

Next, we remark that the series

$$U_{\varphi,\varphi^d}(\omega) = \sum_{n\in\mathbb{Z}} \overline{\hat{\varphi}}\hat{\varphi}^d(\omega + 2n\pi)$$

converges in L^1 to 1 on every compact set, and that its Fourier coefficients are given by

$$\begin{aligned} U_k &= (2\pi)^{-1}\int_{-\pi}^{\pi} U_{\varphi,\varphi^d}(\omega)\mathrm{e}^{-\mathrm{i}k\omega}\,\mathrm{d}\omega = (2\pi)^{-1}\int_{-\infty}^{+\infty} \overline{\hat{\varphi}(\omega)}\hat{\varphi}^d(\omega)\mathrm{e}^{-\mathrm{i}k\omega}\,\mathrm{d}\omega \\ &= \left\langle \varphi^d(\bullet - k), \varphi\right\rangle. \end{aligned}$$

Since $U_k = \delta_{0,k}$, we conclude that φ^d generates the dual Riesz basis.

Finally, we remark that if φ has compact support, then both R_φ and S_φ (now directly defined by (10.31) and (10.32)) are trigonometric polynomials. We consider the corresponding meromorphic functions $r_\varphi(z)$ and $s_\varphi(z)$ such that $R_\varphi(\omega) = r_\varphi(\mathrm{e}^{\mathrm{i}\omega})$ and $S_\varphi(\omega) = s_\varphi(\mathrm{e}^{\mathrm{i}\omega})$. Since these functions do not vanish on the unit circle, the functions $[s_\varphi(\mathrm{e}^{\mathrm{i}\omega})]^{-1/2}$, $[s_\varphi(\mathrm{e}^{\mathrm{i}\omega})]^{-1}$ and $[s_\varphi(\mathrm{e}^{\mathrm{i}\omega})]^{-1}$ have analytic extensions on some ring $\{a^{-1} \leqslant |z| \leqslant a\}$, $a > 1$. It follows that the Fourier coefficients of these functions have geometric decay (majorized by $C|a|^{-n}$) at infinity.

Thus, φ^o, φ^d and φ^i are combinations of $\varphi(\bullet - k)$ with coefficients that have geometric decay at infinity. If φ is bounded and compactly supported, it follows that φ^o, φ^d and φ^i have exponential decay at infinity. In this case, one can directly check $\langle\varphi(\bullet - k), \varphi^d\rangle = \langle\varphi^o(\bullet - k), \varphi^o\rangle = \varphi^i(k) = \delta_{0,k}$, by introducing the Grammian matrix $G = (\langle\varphi(\bullet - m), \varphi(\bullet - n)\rangle)_{m,n\in\mathbb{Z}}$ and the collocation matrix $H = (\varphi(m - n))_{m,n\in\mathbb{Z}}$ which have a banded structure. From their definition, the functions $\varphi^d(\bullet - k)$, $\varphi^o(\bullet - k)$ and $\varphi^i(\bullet - k)$, $k \in \mathbb{Z}$, are constructed by applying the matrices G^{-1}, $G^{-1/2}$ and H^{-1} (which are then well defined) on the functions $\varphi(\bullet - k)$, $k \in \mathbb{Z}$, which immediately implies the duality, orthonormality and interpolation properties. □

Let us now apply the Fourier transform on the two-scale difference equation (10.12). For this, we define the *symbol* of the refinable function φ, as the Fourier series

$$m(\omega) = \frac{1}{2}\sum_{n\in\mathbb{Z}} h_n \mathrm{e}^{-in\omega}, \tag{10.37}$$

and we obtain the following reformulation of (10.12):

$$\hat{\varphi}(\omega) = m(\omega/2)\hat{\varphi}(\omega/2). \tag{10.38}$$

An immediate consequence of (10.38), that we shall use in the next example, is that if φ_1 and φ_2 are two refinable functions in $L^1 \cap L^2$ with symbols m_1 and m_2, then the function $\varphi = \varphi_1 * \varphi_2 \in L^1 \cap C^0$ is also refinable, with symbol $m = m_1 m_2$. The refinement coefficients are thus given by $h_n = \frac{1}{2}(h^1 * h^2)_n$ where h_n^1 and h_n^2 are the refinement coefficients of φ_1 and φ_2.

EXAMPLE 10.4. We shall apply the previous results to a slightly more elaborate example. We define the B-spline B_N of degree N by $B_0 = \chi_{[0,1]}$ and

$$B_N = B_0 * B_{N-1} = (*)^{N+1}\chi_{[0,1]}. \tag{10.39}$$

we note that B_1 is the hat function of Example 10.2, up to an integer shift that does not modify the corresponding multiresolution analysis. The function B_N is supported in $[0, N+1]$, and generates the multiresolution spaces

$$V_j = \left\{ f \in L^2 \cap C^{N-1};\ f|_{I_{j,k}} \in \Pi_N \right\}. \tag{10.40}$$

Detailed treatments of spline functions and their remarkable properties can be found in DE BOOR [1978] and SCHUMAKER [1981]. According to the previous remark, $\varphi = B_N$ satisfies (10.12) with

$$h_n = 2^{-N}\binom{N+1}{n}, \quad \text{if } 0 \leqslant n \leqslant N+1, \qquad h_n = 0 \quad \text{else}. \tag{10.41}$$

Figure 10.1 visualizes of the refinement equation in the particular case of the quadratic spline $B_2 = \frac{1}{4}[B_2(2\bullet) + B_2(2\bullet - 3)] + \frac{3}{4}[B_2(2\bullet - 1) + B_2(2\bullet - 2)]$.

COROLLARY 10.2. *The function $\varphi = B_N$ is L^2-stable. The associated orthonormal and dual scaling functions φ^o, φ^d are well defined in V_0 and have exponential decay at infinity. If N is odd, the associated interpolatory function φ^i is well defined in V_0 and has exponential decay.*

PROOF. From (10.39), we obtain

$$\hat{\varphi}(\omega) = \left[\hat{B}_0(\omega)\right]^{N+1} = \left[\frac{\mathrm{e}^{\mathrm{i}\omega} - 1}{\mathrm{i}\omega}\right]^{N+1}, \tag{10.42}$$

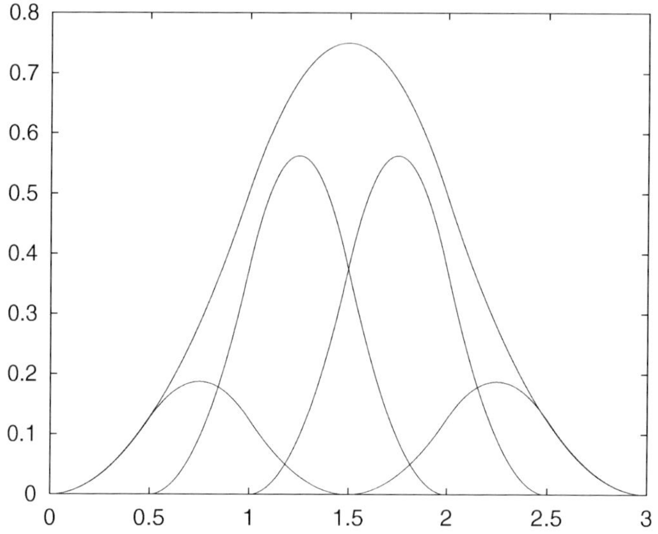

FIG. 10.1. Decomposition of the quadratic B-spline as a combination of its scaled versions.

so that

$$|\hat{\varphi}(\omega)|^2 = \left[\frac{\sin(\omega/2)}{\omega}\right]^{2(N+1)}, \tag{10.43}$$

does not vanish on $[-\pi, \pi]$. Thus $S_\varphi(\omega)$ does not vanish and the existence and properties of φ^o and φ^d follows from the first part of Corollary 10.1.

For odd $N = 2M - 1$, we have

$$\hat{\varphi}(\omega) = \mathrm{e}^{-\mathrm{i}M\omega}\left[\frac{\sin(\omega/2)}{\omega}\right]^{2M} \tag{10.44}$$

which shows that $R_\varphi(\omega) = \mathrm{e}^{-\mathrm{i}M\omega} S_{B_{M-1}}(\omega)$. It follows that $R_\varphi(\omega)$ does not vanish, and the existence and properties of φ^i follows from the second part of Corollary 10.1. □

We have already defined P_j^o to be the L^2-orthogonal projection on V_j. We can use the functions φ^d and φ^o to express P_j^o in two different ways:

$$P_j^o f = \sum_{k\in\mathbb{Z}} \langle f, \varphi_{j,k}^o\rangle \varphi_{j,k}^o, \tag{10.45}$$

or

$$P_j^o f = \sum_{k\in\mathbb{Z}} \langle f, \varphi_{j,k}^d\rangle \varphi_{j,k}. \tag{10.46}$$

In the case of B-splines of degree $N \geqslant 1$, however, both φ^o and φ^d are globally supported on $\mathbb{R}$. As a result, the projector P_j^o has a nonlocal effect and is difficult to manipulate numerically.

In order to circumvent this type of difficulty, we shall allow the use of nonorthogonal projectors, based on the following definition:

DEFINITION 10.4. A refinable function $\tilde{\varphi} = \sum_n \tilde{h}_n \tilde{\varphi}(2\bullet - n)$ in L^2 is dual to φ if it satisfies

$$\langle \varphi(\bullet - k), \tilde{\varphi}(\bullet - l) \rangle = \delta_{k,l}, \quad k, l \in \mathbb{Z}. \tag{10.47}$$

From a pair of dual scaling function, we can define an oblique projector by

$$P_j f = \sum_{k \in \mathbb{Z}} \langle f, \tilde{\varphi}_{j,k} \rangle \varphi_{j,k}. \tag{10.48}$$

Formally, the duality relation (10.47) shows that $f \in V_j$ implies $P_j f = f$. We postpone the study of the convergence of (10.48) and of the L^2-boundedness of P_j to the next section.

In contrast to φ^d, the function $\tilde{\varphi}$ is not forced to be in V_0. In particular, *it is not uniquely determined* for a given φ. We can hope to use this new flexibility to build a compactly supported dual $\tilde{\varphi}$, ensuring that the new projector P_j has a local action. This hope turns out to be fulfilled: a result proved in LEMARIÉ [1997] states that if φ is a compactly supported L^2-stable refinable function, then there always exists a dual scaling function $\tilde{\varphi}$ in L^2 which is also compactly supported.

Note that the dual $P_j^* f = \sum_{k \in \mathbb{Z}} \langle f, \varphi_{j,k} \rangle \tilde{\varphi}_{j,k}$ of P_j is also an oblic projector whose range is the "dual space" $\widetilde{V}_j$ spanned by the Riesz basis $\tilde{\varphi}_{j,k}$, $k \in \mathbb{Z}$. Because we have assumed $\tilde{\varphi}$ to be refinable, the spaces $\widetilde{V}_j$ are also nested. As a consequence, if $j < l$, we have $P_l^* P_j^* = P_j^*$ and thus $P_j P_l = P_j$. This last property is crucial for numerical computations, since it means that the approximations can be computed recursively from fine to coarse scales.

REMARK 10.6. Orthonormal scaling functions can be viewed as a particular case of dual scaling functions for which $\varphi = \tilde{\varphi}$. Interpolatory scaling functions can also be placed in this general setting as a "degenerate case": if φ is interpolatory, we define $\tilde{\varphi}$ to be the Dirac distribution δ centered at the origin, so that (10.47) holds in the sense of the duality $(\mathcal{D}, \mathcal{D}')$. The Dirac distribution satisfies (in a weak sense) $\delta(x) = 2\delta(2x)$ and should thus be considered as refinable. However the resulting "projector" P_j is not L^2-bounded.

11. Refinable functions

The main message of this section is that *any solution φ of a refinement equation* (10.12) *is a potential candidate to generate a multiresolution analysis*, provided that it satisfies some basic properties that we shall study here. By "generate", we mean that we shall directly define V_j as the L^2-span of $\varphi_{j,k}$, $k \in \mathbb{Z}$.

The price to pay in this approach is that, in contrast to finite element bases which are well defined piecewise polynomial functions, the scaling function φ will not in general be given by an explicit formula, but rather as a solution of (10.12), *where only the coefficients h_n will be explicitly given*. In this situation, we need to address the following questions: what are the basic properties of φ which ensure that the spaces V_j satisfy the multiresolution analysis axioms of Definition 10.1? How can we compute φ numerically at some point x? How can we compute accurately inner products such as $\langle f, \varphi_{j,k}\rangle$ or $\langle \varphi'_{j,m}, \varphi'_{j,n}\rangle$, that are needed in Galerkin approximation methods? How can we predict the properties of φ (such as compact support, L^2-stability or smoothness) from the coefficients h_n?

In this section, we shall answer the first question. The answer to the second and third questions are the object of the two next sections, while the last question is the object of Sections 15–17. We shall then use the results of these sections to design the coefficients h_n in (10.12), in such a way that specific properties of φ are fulfilled.

In the rest of this section, as well as in Sections 12–14, φ and $\tilde{\varphi}$ will denote a pair of dual refinable functions that are compactly supported and real valued. Although most of the results that we shall prove are generalizable to noncompactly supported refinable functions, we prefer to keep this assumption since it will be essential for the practical implementation of related algorithms, and because it removes many technicalities in several proofs. Similarly the real-valued assumption is made here to simplify the notations (in particular to avoid complex conjugate quantities appearing in inner products) and because almost all relevant examples are real valued.

A first result is that the existence of $\tilde{\varphi}$ ensures the L^2-stability of φ and the L^2-boundedness of the oblic projector P_j defined by (10.48).

THEOREM 11.1. *Under the above assumptions, the functions φ and $\tilde{\varphi}$ are L^2-stable. Moreover, the oblic projector P_j defined by* (10.48) *is L^2-bounded independently of j.*

PROOF. Let $f = \sum_n c_n \varphi(\bullet - n)$ be a finite linear combination of the functions $\varphi(\bullet - n)$, $n \in \mathbb{Z}$. On the one hand, we have the inequality

$$\|f\|_{L^2}^2 = \sum_{m,n} c_m \overline{c_n} \langle \varphi_{0,m}, \varphi_{0,n}\rangle \leqslant C \sum_n |c_n|^2, \tag{11.1}$$

with $C = [2S - 1]\|\varphi\|_{L^2}^2$, $S = |\mathrm{Supp}(\varphi)|$. On the other hand we have

$$|c_n|^2 = \left|\langle f, \tilde{\varphi}(\bullet - n)\rangle\right|^2 \leqslant \|\tilde{\varphi}\|_{L^2}^2 \int_{\mathrm{Supp}(\tilde{\varphi}_{0,n})} |f|^2, \tag{11.2}$$

by Cauchy–Schwarz inequality, and thus

$$\sum_n |c_n|^2 \leqslant \tilde{C} \|f\|_{L^2}^2, \tag{11.3}$$

with $\tilde{C} = [2\tilde{S} - 1]\|\tilde{\varphi}\|_{L^2}^2$, $\tilde{S} = |\mathrm{Supp}(\tilde{\varphi})|$. A symmetric argument shows that $\tilde{\varphi}$ is also L^2-stable.

The same arguments show that $P_0 f$ is well defined by (10.48) since $(\langle f, \tilde{\varphi}(\bullet - k)\rangle)_{k\in\mathbb{Z}}$ is in ℓ^2 and φ is L^2 stable. Moreover, we have

$$\|P_0 f\|_{L^2}^2 \leqslant C \sum_{n\in\mathbb{Z}} \left|\langle f, \varphi_{0,n}\rangle\right|^2 \leqslant C\tilde{C}\|f\|_{L^2}^2, \tag{11.4}$$

and thus $\|P_0\| \leqslant \sqrt{C\tilde{C}}$. Now we remark that

$$P_j = S_j P_0 S_{-j}, \tag{11.5}$$

where S_j is the scaling operator $f \mapsto 2^{j/2} f(2^j \bullet)$ which preserves the L^2 norm, so that we have $\|P_j\| = \|P_0\| \leqslant \sqrt{C\tilde{C}}$. □

REMARK 11.1. The above results stay valid in weaker assumptions than compact support for φ and $\tilde{\varphi}$. In particular, we can assume that both φ and $\tilde{\varphi}$ are in L^2 and satisfy the right part of (10.23) with constants C_2 and $\tilde{C}_2$. Under these assumptions, according to Remark 10.4, we can use the one-sided L^2-stability and Bessel sequence properties of φ and $\tilde{\varphi}$ to reach the same conclusion. The main interest of the technique that we have used for compactly supported functions is that is it not based on Fourier transform, and will be rather easily adapted to bounded domains and to L^p norms with $p \neq 2$ (Theorem 26.1 of Chapter III).

The properties that remain to be checked in Definition 10.1 are thus (10.3) and (10.4), i.e. the "limit" properties of the V_j spaces as j goes to infinity. For this, we need to make the distinction between P_j and the orthogonal projector P_j^o which is defined by

$$P_j^o f = \sum_{k\in\mathbb{Z}} \langle f, \varphi_{j,k}^o\rangle \varphi_{j,k}^o, \tag{11.6}$$

where φ^o is defined through (10.34).

THEOREM 11.1. *If* $f \in L^2$, *then* $\lim_{j\to-\infty} \|P_j^o f\|_{L^2} = \lim_{j\to-\infty} \|P_j f\|_{L^2} = 0$.

PROOF. Since both P_j and P_j^o are uniformly L^2 bounded in j, by the same reasoning which was used to prove (2.14) in Section 2 of Chapter I, it suffices to prove the result for $f = \chi_{[a,b]}$.

We first show that $\|P_j^o f\|_{L^2}^2 = \sum_{k\in\mathbb{Z}} |\int_a^b \varphi_{j,k}^o(t)\,\mathrm{d}t|^2$ goes to zero as $j \to -\infty$. By Cauchy–Schwartz inequality and a change of variable, for $-j$ large enough so that $2^j(b-a) \leqslant 1$, we obtain

$$\left\|P_j^o f\right\|_{L^2}^2 \leqslant \int_{-\infty}^{+\infty} \chi_{A_j}(t)\left|\varphi^o(t)\right|^2 \mathrm{d}t, \tag{11.7}$$

where $A_j = \bigcup_{k\in\mathbb{Z}}[2^j a + k, 2^j b + k]$. Since $\chi_{A_j}(t)$ tends to zero almost everywhere, we obtain the result by Lebesgue's dominated convergence theorem.

Since $\|P_j f\|^2 \leqslant C \sum_{k\in\mathbb{Z}} |\int_a^b \tilde{\varphi}_{j,k}(t)\,\mathrm{d}t|^2$, with C independent of j, the same reasoning proves the result for P_j. □

We now turn to the behaviour of $P_j f - f$ and $P_j^o f - f$ as $j \to +\infty$. The following observation (originally due to Lebesgue) shows that we only have to treat one of these.

LEMMA 11.1. *If P is a bounded projector from a Banach space X to a closed subspace Y, then for all $f \in X$,*

$$\inf_{g\in Y} \|f-g\|_X \leqslant \|f - Pf\|_X \leqslant \big(1 + \|P\|\big) \inf_{g\in Y} \|f-g\|_X \tag{11.8}$$

with $\|P\| = \sup_{\|f\|_X=1} \|Pf\|_X$.

PROOF. The left inequality is immediate. For the right inequality, we remark that for $g \in Y$,

$$\|f - Pf\|_X \leqslant \|f-g\|_X + \|Pf - Pg\|_X \leqslant \big(1 + \|P\|\big)\|f-g\|_X, \tag{11.9}$$

and we conclude by taking the infimum over all $g \in Y$. □

Since the projectors P_j are uniformly L^2-bounded (by Theorem 11.1), we shall now focus our attention to the behaviour of $P_j f - f$ as $j \to +\infty$.

THEOREM 11.3. *The approximation operator $P_j f = \sum_k \langle f, \tilde{\varphi}_{j,k}\rangle \varphi_{j,k}$ satisfies*

$$\lim_{j\to+\infty} \|P_j f - f\|_{L^2} = 0, \tag{11.10}$$

if and only if

$$\left[\int \tilde{\varphi}\right] \sum_{k\in\mathbb{Z}} \varphi(x-k) = 1, \tag{11.11}$$

almost everywhere on $\mathbb{R}$.

PROOF. Using once again the uniform L^2-boundedness of the projectors P_j and a density argument, we see that the property (11.10) is satisfied for all $f \in L^2$ if and only if it is satisfied for all $f = \chi_{[a,b]}$ where $[a,b]$ is an arbitrary bounded interval. We thus consider $f = \chi_{[a,b]}$ for some $a < b$.

At first, let us remark that the 1-periodic function defined by $A(x) := [\int \tilde{\varphi}] \sum_{k\in\mathbb{Z}} \varphi(x-k)$ is well defined in L^2_{loc}, since φ has compact support.

If s and $\tilde{s}$ are such that $\mathrm{Supp}(\varphi) \subset [-s,s]$ and $\mathrm{Supp}(\tilde{\varphi}) \subset [-\tilde{s},\tilde{s}]$, we remark that we have exactly $P_j f(x) = 0$ for $x < a - 2^{-j}(s+\tilde{s})$ or $x > b + 2^{-j}(s+\tilde{s})$, i.e. the approximation is exact away from the interval at a minimal distance of order 2^{-j}. Moreover, if j is large enough so that $2^{-j}(s+\tilde{s}) \leqslant (b-a)/2$, then for $x \in [a+2^{-j}(s+\tilde{s}), b-2^{-j}(s+\tilde{s})]$, we have exactly

$$P_j f(x) = \sum_k \left[\int \tilde{\varphi}_{j,k}\right] \varphi_{j,k} = \left[\int \tilde{\varphi}\right] \sum_k \varphi\big(2^j x - k\big) = A\big(2^j x\big). \tag{11.12}$$

Assuming that $A(x) = 1$ almost everywhere, it follows that for j large enough, the approximation $P_j f$ is thus equal to f except on the intervals $I_j^a = [a - 2^{-j}(s+\tilde{s}), a + 2^{-j}(s+\tilde{s})]$ and $I_j^b = [b - 2^{-j}(s+\tilde{s}), b + 2^{-j}(s+\tilde{s})]$.

The convergence of $P_j f$ to f in L^2 follows then from the analysis of the contribution of these intervals: clearly $\int_{I_j^a} |f|^2 \leqslant |I_j^a| \to 0$ as $j \to +\infty$, and

$$\begin{aligned}\int_{I_j^a} |P_j f|^2 &\leqslant \int_{\mathbb{R}} \Bigg| \sum_{|k-2^j a| \leqslant 2s+\tilde{s}} \langle f, \tilde{\varphi}_{j,k}\rangle \varphi_{j,k}\Bigg|^2 \\ &\lesssim \sum_{|k-2^j a| \leqslant 2s+\tilde{s}} \big|\langle f, \tilde{\varphi}_{j,k}\rangle\big|^2 \\ &\leqslant (4s+2\tilde{s})2^{-j} \to 0 \quad \text{as } j \to +\infty.\end{aligned}$$

Similar properties hold for I_j^b and we thus conclude that (11.10) is satisfied.

Assuming now that $A(x)$ differs from 1 on a set of nonzero measure, we choose $j_0, k_0 \in \mathbb{Z}$ such that $I_{j_0,k_0} = [2^{-j_0}k_0, 2^{-j_0}(k_0+1)]$ is contained in the interior of $[a,b]$. For $j \geqslant j_0$ large enough, we have $P_j f(x) = A(2^j x)$ on I_{j_0,k_0}, so that

$$\begin{aligned}\|f - P_j f\|_{L^2}^2 &\geqslant \int_{2^{-j_0}k_0}^{2^{-j_0}(k_0+1)} \big|1 - A(2^j x)\big|^2 \,\mathrm{d}x \\ &= 2^{-j} \int_{2^{j-j_0}k_0}^{2^{j-j_0}(k_0+1)} \big|1 - A(x)\big|^2 \,\mathrm{d}x \\ &= 2^{-j_0} \int_0^1 \big|1 - A(x)\big|^2 \,\mathrm{d}x > 0.\end{aligned}$$

We have thus proved the necessity of (11.11). □

REMARK 11.2. In the above proof, we have not made use of the duality property (10.47): $\tilde{\varphi}$ could be any L^2-function with compact support. The operator P_j would still be local and L^2-bounded, but not a projector. Our analysis shows that a minimal requirement for such a local approximation operator is that it "reproduces" the constant functions in the sense that

$$P_j 1 := \sum_k \langle \tilde{\varphi}_{j,k}, 1\rangle \varphi_{j,k} = 1. \tag{11.13}$$

This property implies in particular that the constant functions are generated by the integer translates of φ. Such general approximation operators (when $\tilde{\varphi}$ is not necessarily dual to φ) are known as *quasi-interpolant*. A simple example can be obtained with the choice $\varphi = \tilde{\varphi} = B_1 = \max\{1 - |x|, 0\}$, which clearly satisfies (11.13). Quasi-interpolants have been introduced in DE BOOR and FIX [1973] in order to study the properties of spline approximation. Similar approximation operators have also been proposed for multivariate finite element spaces, see, e.g., CLEMENT [1975].

REMARK 11.3. We also did not exploit the refinability of φ and $\tilde{\varphi}$ in the proof of Theorem 11.3. It will be shown in Section 16 that when φ is both refinable and L^2-stable, the property (11.11) is then automatically fulfilled. For the time being, we shall simply assume that our refinable dual functions φ and $\tilde{\varphi}$ satisfy this property, so that all the multiresolution properties of Definition 10.1 are satisfied.

REMARK 11.4. It is easy to see that if the union of the V_j spaces is dense in L^2, the same property holds for the dual multiresolution approximation spaces $\widetilde{V}_j$ generated by $\tilde{\varphi}$: indeed, if this was not the case, there would exists a nontrivial function f orthogonal to all $\tilde{\varphi}_{j,k}$, $j,k \in \mathbb{Z}$, so that $P_j f = 0$ for all j, which is in contradiction with (11.10). This shows in particular that the dual identity

$$\left[\int \varphi\right] \sum_{k\in\mathbb{Z}} \tilde{\varphi}(x-k) = 1, \tag{11.14}$$

holds almost everywhere.

COROLLARY 11.1. *Both functions φ and $\tilde{\varphi}$ have nonzero integral, and satisfy*

$$\left[\int \varphi\right]\left[\int \tilde{\varphi}\right] = 1. \tag{11.15}$$

Up to a renormalization, we can assume that $\int \varphi = \int \tilde{\varphi} = 1$. The coefficients h_n and $\tilde{h}_n$ satisfy the sum rules

$$\sum h_n = \sum \tilde{h}_n = 2, \qquad \sum (-1)^n h_n = \sum (-1)^n \tilde{h}_n = 0. \tag{11.16}$$

PROOF. Integrating (11.11) over [0, 1], we obtain

$$1 = \left[\int \tilde{\varphi}\right] \int_0^1 \sum_{k\in\mathbb{Z}} \varphi(x-k)\, \mathrm{d}x = \left[\int \varphi\right]\left[\int \tilde{\varphi}\right],$$

i.e. (11.14) which shows that both integrals cannot vanish, and that we can renormalize the functions φ and $\tilde{\varphi}$ such that $\int \varphi = \int \tilde{\varphi} = 1$, without affecting P_j, the duality (10.47) and the refinable nature of these functions.

Integrating both side of the refinement equation (10.12) over $\mathbb{R}$, we obtain

$$1 = \int \varphi = \sum_n h_n \int \varphi(2\bullet - n) = \frac{1}{2}\Big[\sum_n h_n\Big]\int \varphi = \Big[\sum_n h_n\Big]/2,$$

and similarly for the coefficients $\tilde{h}_n$, i.e. the first part of the sum rules (11.16).

For the second part, we inject the refinement equation in (11.11) which gives

$$\begin{aligned}
1 &= \sum_{k\in\mathbb{Z}} \varphi(\bullet - k) \\
&= \sum_{k\in\mathbb{Z}}\Big[\sum_n h_n\Big]\varphi(2\bullet - n - 2k) \\
&= \sum_{k\in\mathbb{Z}}\Big[\sum_n h_{2n}\Big]\varphi(2\bullet - 2k) + \sum_{k\in\mathbb{Z}}\Big[\sum_n h_{2n+1}\Big]\varphi(2\bullet - 2k - 1).
\end{aligned}$$

Taking the inner product with $\tilde{\varphi}(2\bullet)$ and $\tilde{\varphi}(2\bullet - 1)$, we obtain $\sum_n h_{2n} = \sum_n h_{2n+1} = 1$, i.e. the second part of the sum rules (11.16) for the coefficients h_n. The same holds for the coefficients $\tilde{h}_n$ by (11.14). □

REMARK 11.5. From the compact support of φ and $\tilde{\varphi}$, it is also clear that the real valued sequence $h_n = 2\langle\varphi, \tilde{\varphi}(2\bullet - n)\rangle$ has finite support. Moreover, we necessarily have $h_n = 0$ when $n \notin [a, b] = \overline{\mathrm{Supp}(\varphi)}$. Indeed, if we had, e.g., $h_n \neq 0$ for some $n < a$, we reach a contradiction by taking the smallest n with this property and remarking that $\mathrm{Supp}(\varphi(2\bullet - n))$ alone contributes in the refinement equation for $x \in [(n+a)/2, (n+1+a)/2[$, making φ nonzero on this interval, which contradicts its support property.

12. Subdivision schemes

Having identified these diverse properties of "admissible dual refinable functions", now turn the question of how to compute the numerical values of the function φ. The solution to this problem can be obtained by viewing the function φ as a fixed point of the operator

$$f \mapsto Rf = \sum_n h_n f(2\bullet - n), \tag{12.1}$$

and try to obtain it through an iterative method, i.e. as $\varphi = \lim_{j\to+\infty} R^j f_0$ for some well chosen f_0. We note that the property $\sum_n h_n = 2$ implies the invariance by R of the integral of f when it exists, i.e.

$$\int Rf(x)\,\mathrm{d}x = \int f(x)\,\mathrm{d}x, \tag{12.2}$$

so that we need at least $\int f_0 = 1$.

With the simple choice $f_0 = \max\{1 - |x|, 0\}$, we obtain after j iterations a function $f_j = R^j f_0$ which is piecewise affine on each interval $I_{j,k}$, $k \in \mathbb{Z}$, and continuous. Our hope is that f_j constitutes a proper piecewise affine approximation of φ at resolution 2^{-j}. The values $s_{j,k} = f_j(2^{-j}k)$, $k \in \mathbb{Z}$, that completely determine f_j, can be obtained by two different reccursions both applied to the initial data $s_{0,k} = \delta_{0,k}$, $k \in \mathbb{Z}$: on the one hand, if we directly apply $f_{j+1} = Rf_j$, we get $f_j(2^{-j-1}k) = \sum_n h_n f_j(2^{-j}k - n)$, which yields

$$s_{j+1,k} = \sum_n h_n s_{j,k-2^j n}. \tag{12.3}$$

On the other hand, remarking that we have for a more general f $R^j f = \sum_k s_{j,k} f(2^j x - k)$, and applying it to $f = Rf_0$, we obtain

$$f_{j+1} = \sum_k s_{j+1,k} f_0(2^{j+1}x - k) = \sum_k s_{j,k} \left[\sum_n h_n f_0(2^j x - 2k - n) \right],$$

which yields

$$s_{j+1,k} = \sum_n h_{k-2n} s_{j,n}. \tag{12.4}$$

The recursion formulae (12.3) and (12.4) are of different nature. In particular, (12.4) reveals a *local* property of the iteration, in the sense that, if $h_n = 0$ for $|n| > N$, the value of f_j at $2^{-j}k$ is only influenced by the values of f_{j-1} at dyadic points contained in $[2^{-j}k - 2^{-j+1}N, 2^{-j}k + 2^{-j+1}N]$. This property is reflected on Fig. 12.1 that represents the influence of the values from level to level. In contrast, the recursion (12.3) is *global*. Although both recursions produce the same result f_j, the local procedure is often preferred for the following reason: if we are interested in the value of φ at a specific point, or in a specific region, we can use (12.4) to localize the computation of f_j in this region.

The global procedure based on (12.3) is known as the *cascade algorithm*, while the local refinement (12.4) is an instance of a *subdivision scheme*.

Generally speaking, subdivision schemes are a design tool for the fast generation of smooth curves and surfaces from a set of control points by means of iterative refinements. In the most often considered binary univariate case, one starts from a sequence $s_0(k)$ and obtains at step j a sequence $s_j(2^{-j}k)$, generated from the previous one by linear rules:

$$s_j(2^{-j}k) = 2 \sum_{n \in k+2\mathbb{Z}} h_{j,k}(n) s_{j-1}(2^{-j}(k-n)). \tag{12.5}$$

The *masks* $h_{j,k} = \{h_{j,k}(n)\}_{n \in \mathbb{Z}}$ are in general finite sequences, a property that is clearly useful for the practical implementation of (12.5).

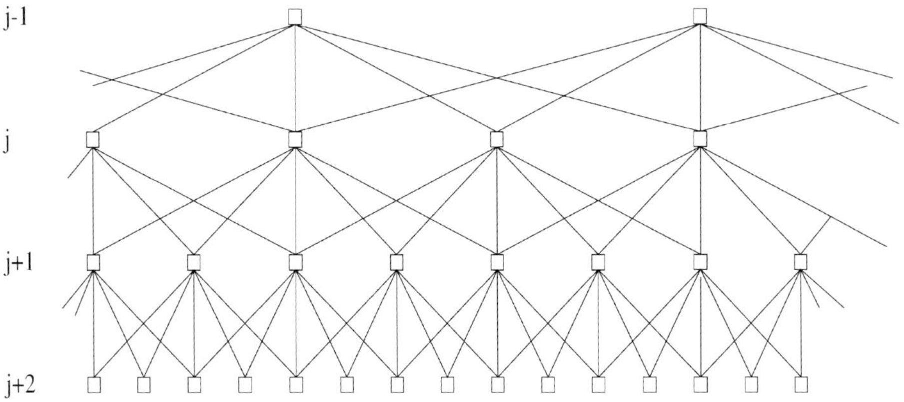

FIG. 12.1. Local subdivision algorithm.

A natural problem is then to study the convergence of such an algorithm to a limit function. In particular, the scheme is said to be strongly convergent if and only if there exists a continuous function $f(x)$ such that

$$\lim_{j\to+\infty}\Big(\sup_k\big|s_j(2^{-j}k)-f(2^{-j}k)\big|\Big)=0. \tag{12.6}$$

A subdivision scheme is said to be stationary and uniform when the masks $h_{j,k}(n)=h_n$ are independent of the parameters j and k. In that case, (12.5) rewrites as (12.4) with the notation $s_{j,k}=s_j(2^{-j}k)$. It is then sufficient to check the convergence of the subdivision on the initial data $s_{0,k}=\delta_{0,k}$: if φ is the limit function with such data, the limit function with more general data is $f=\sum_k s_0(k)\varphi(\bullet-k)$. General treatment of subdivision schemes, with numerous examples, can be found in CAVARETTA, DAHMEN and MICCHELLI [1991] and DYN [1992].

In our case of interest, we precisely start with the fundamental data $s_{0,k}=\delta_{0,k}$. We shall now use the L^2-stability of φ to analyze the uniform convergence of the subdivision scheme.

THEOREM 12.1. *If the scaling function φ is continuous, then the associated subdivision algorithm converges uniformly, i.e. f_j converges uniformly to φ.*

PROOF. We simply remark that, φ being a fixed point of R, we have

$$\varphi(x)=\sum_k s_{j,k}\varphi\big(2^jx-k\big), \tag{12.7}$$

so that

$$s_{j,k}=2^{j/2}\langle\varphi,\tilde\varphi_{j,k}\rangle=2^j\int\varphi(x)\tilde\varphi\big(2^jx-k\big)\,\mathrm{d}x, \tag{12.8}$$

where $\tilde{\varphi}$ is the compactly supported dual scaling function. Since $\tilde{\varphi} \in L^1$ and satisfies $\int \tilde{\varphi} = 1$, and since φ is uniformly continuous (by continuity and compact support), this implies the strong convergence property (12.6). □

REMARK 12.1. This way of obtaining φ through iterative refinements justifies the terms "refinable function". Conversely, if φ is defined as the limit of a converging subdivision scheme with coefficients h_n and initial data $\delta_{0,k}$, one easily checks that it satisfies the refinement equation (10.12) (because it is also the limit of the subdivision of the sequence $s_1(n/2) = h_n$). Thus, we can define a continuous refinable function φ either as a solution of (10.12), or as the fundamental limit function of a subdivision scheme.

The following example, which is of common use in computer graphics, is a good illustration of the relation between subdivision schemes and refinable functions: given a polygonal graph with vertices $p_{0,k} = (x_{0,k}, y_{0,k})$, $k \in \mathbb{Z}$, the *corner cutting algorithm*, introduced by CHAITKIN [1974], consists in a recursive construction of refined polygonal lines $p_{j,k} = (x_{j,k}, y_{j,k})$, $k \in \mathbb{Z}$, $j > 0$, by placing the new points according to

$$p_{j,2k} = \tfrac{3}{4} p_{j,k} + \tfrac{1}{4} p_{j,k+1} \quad \text{and} \quad p_{j,2k+1} = \tfrac{1}{4} p_{j,k} + \tfrac{3}{4} p_{j,k+1}. \tag{12.9}$$

Figure 12.2 reveals the *smoothing* effect of this procedure as j goes to $+\infty$.

FIG. 12.2. Corner cutting algorithm.

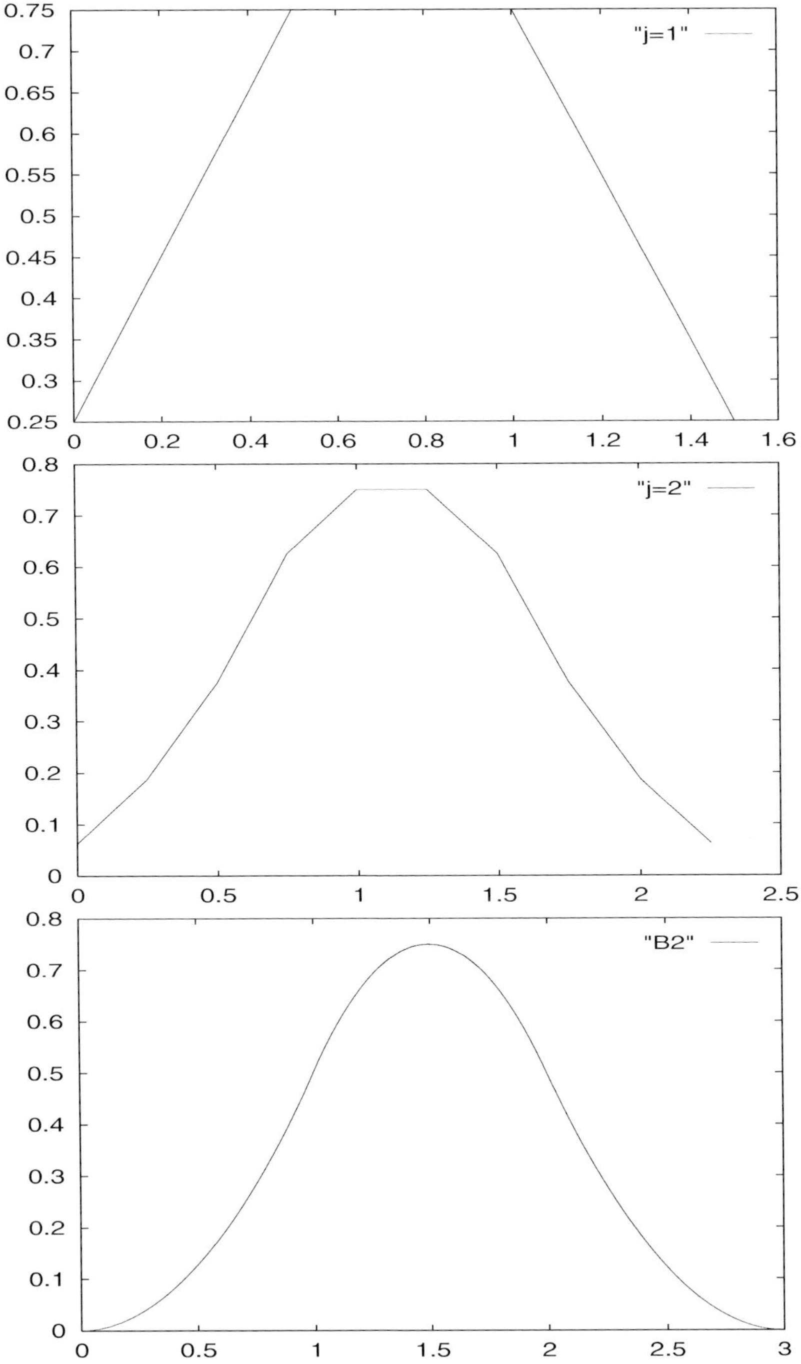

FIG. 12.3. Convergence to the quadratic B-spline.

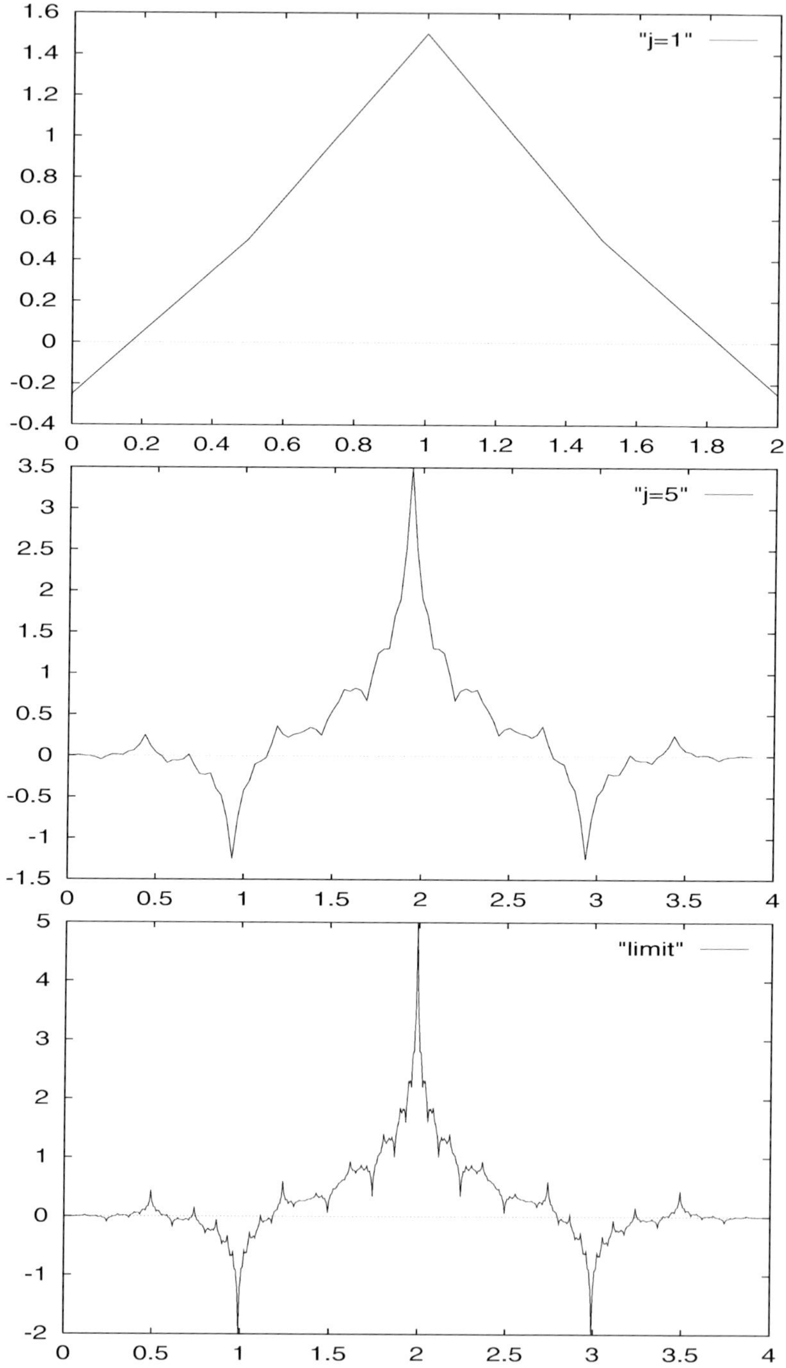

FIG. 12.4. L^p-convergence to a discontinuous limit.

From (12.9), we see that the sequences $x_{j,k}$ and $y_{j,k}$, are obtained by the iteration of a subdivision scheme with coefficients $h_{-1} = h_0 = 3/4$ and $h_{-2} = h_1 = 1/4$. We notice that the associated scaling function is the quadratic B-spline $\varphi = B_2(x+2)$ defined from (10.39). From the stability of this scaling function, we conclude that the corner cutting algorithm converges to the limit curve $p(t) = (\sum_k x_{0,k} B_2(t-k+2), \sum_k y_{0,k} B_2(t-k+2))$. We also display on Fig. 12.3 the subdivision process with fundamental initial data that converges to B_2.

REMARK 12.2. If φ has some additional Hölder regularity C^r, one easily obtains the estimate $\sup_k |s_j(2^{-j}k) - f(2^{-j}k)| \lesssim 2^{-rj}$ for the rate of convergence. Clearly, the uniform convergence of the subdivision is equivalent to the uniform convergence of f_j to f with the choice $f_0 = \max\{1-|x|, 0\}$. It is of course possible to study other types of convergence (for examples in L^p and Sobolev spaces), and to use other choice of initial data f_0. As an example, we display on Fig. 12.4 a subdivision scheme with coefficients $h_0 = h_4 = -1/4$, $h_1 = h_3 = 1/2$, $h_2 = 3/2$, that can be proved to converge in all L^p spaces for $p < +\infty$ to a discontinuous limit which has logarithmic singularities at dyadic points. Results on diverse type of convergence can be found in several references, e.g., JIA and HAN [1998] for L^p convergence or COHEN and RYAN [1995, Chapter 3] for convergence in smoother norms.

REMARK 12.3. The cascade algorithm, introduced in DAUBECHIES [1988], although nonlocal, has the following advantage: if the initial data are chosen to be $s_{0,k} = \varphi(k)$, then the iteration of (12.3) produces the exact values of φ at the dyadic points. We show in the end of the next section (see in particular Remark 13.2) how to compute directly the initial values $\varphi(k)$.

REMARK 12.4. If $h_n = 0$ for $n < a$ and $n > b$, then the support of f_j is contained by construction in $[a_j, b_j]$ with $a_j = (1/2 + 1/4 + \cdots + 2^{-j})a + 2^{-j}$ and $b_j = (1/2 + 1/4 + \cdots + 2^{-j})b - 2^{-j}$. Thus φ is supported in $[a, b]$. Combining this with Remark 11.5, we see that the support of φ is exactly the interval $[a, b]$ with $a = \min\{n;\ h_n \neq 0\}$ and $b = \max\{n;\ h_n \neq 0\}$.

13. Computing with refinable functions

In the practice of using the space V_j to discretize partial differential or integral equations, we might face the problem of computing several quantities. Among others we find:

(i) The coefficients $\langle f, \varphi_{j,k}\rangle$ (or $\langle f, \tilde{\varphi}_{j,k}\rangle$), $k \in \mathbb{Z}$ for some function f.
(ii) The inner products $M_{k,l} = \langle \varphi_{j,k}, \varphi_{j,l}\rangle$, that constitute the mass matrix.
(iii) The elements $R_{k,l} = \langle (\partial/\partial x)^n \varphi_{j,k}, \varphi_{j,l}\rangle$ arising from the discretization of a differential problem.
(iv) The quantities $T_{k,l} = \int K(x, y)\varphi_{j,k}(x)\varphi_{j,l}(y)\, dx\, dy$ arising from the discretization of an integral operator with kernel $K(x, y)$.

In all these different situations, we may think of using the subdivision approximation to φ that we have discussed, in order to evaluate these quantities numerically: one could first compute $\varphi_{j,k}$ on a fine grid and then design quadrature rules on this grid to evaluate the above inner products. However, this approach is unpractical when φ has poor smoothness, in particular when it is not continuous: the subdivision converges too slowly to φ and too many points are then required to obtain a good approximation.

A smarter approach consist in injecting the refinement equation in the computation of these quantities. In this section, we shall explain how this strategy works in the context of the above examples.

We first consider the computation of $c_{j,k} = \int f(t)\varphi_{j,k}(t)\,\mathrm{d}t$, assuming that f has C^r regularity for some $r > 0$. Since $\int \varphi = 1$ and $\varphi_{j,k}$ is localized near the point $2^{-j}k$, we see that a first crude estimate of $c_{j,k}$ is given by

$$c_{j,k} \simeq 2^{-j/2} f\left(2^{-j}k\right). \tag{13.1}$$

This estimate amounts in replacing f by a constant function on the support $S_{j,k}$ of $\varphi_{j,k}$. In order to improve on (13.1), we can think of replacing f on $S_{j,k}$ by a polynomial P_l of degree $l < r$. We can choose for example P_l to be the Lagrange interpolation of f at points $2^{-j}(k+n)$, $n = 0, \ldots, l$. The coefficients of this polynomial depend linearly on the values $f(2^{-j}(k+n))$, $n = 0, \ldots, l$, and the resulting estimate for $c_{j,k}$ is of thus of the type

$$c_{j,k} \simeq \tilde{c}_{j,k} := 2^{-j/2}\left[\sum_{n=0}^{l} w_n f\left(2^{-j}(k+n)\right)\right]. \tag{13.2}$$

The weights w_n are simply given by

$$w_n = \int L_n(x)\varphi(x)\,\mathrm{d}x, \tag{13.3}$$

where L_n are the Lagrange polynomials of degree l such that $L_n(m) = \delta_{m,n}$, $m, n = 0, \ldots, l$. From the classical results on polynomial interpolation, we can then estimate the quadrature error as follows

$$\left|c_{j,k} - \tilde{c}_{j,k}\right| \leqslant \|\varphi_{j,k}\|_{L^1} \sup_{x \in S_{j,k}} \left|f(x) - P_l(x)\right| \lesssim 2^{-j(r+1/2)} \|f\|_{C^r(S_{j,k})},$$

where C is independent of f and j. The computation of the weights w_n amounts in determining the moments

$$M_k = \int x^k \varphi(x)\,\mathrm{d}x. \tag{13.4}$$

At this point, we inject the refinement equation, which leads to

$$M_k = \sum_n h_n \int x^k \varphi(2x-n)\,\mathrm{d}x = 2^{-k-1}\sum_n h_n \int (x+n)^k \varphi(x)\,\mathrm{d}x$$

$$= 2^{-k-1} \sum_{l \leqslant k} \left[\sum_n h_n \binom{k}{l} n^{k-l} \right] M_l,$$

and thus

$$M_k = \left(1 - 2^{-k}\right)^{-1} 2^{-k-1} \sum_{l<k} \left[\sum_n h_n \binom{k}{l} n^{k-l} \right] M_l. \tag{13.5}$$

Since we know the first value $M_0 = 1$, we can use (13.5) to compute the exact values of the moments by induction *without the use of an approximation of φ by a subdivision scheme*. This simple idea was used in SWELDENS and PIESSENS [1994] for the design of economical and accurate quadrature formulae.

Remark that the computation of $T_{k,l} = \int K(x, y)\varphi_{j,k}(x)\varphi_{j,l}(y)\,\mathrm{d}x\,\mathrm{d}y$ can be treated in the same way when $K(x, y)$ is smooth, using the refinable structure of $\phi(x, y) = \varphi(x)\varphi(y) = \sum_{n,m\in\mathbb{Z}} h_n h_m \varphi(2x - n)\varphi(2y - m)$ to compute the multivariate moments.

REMARK 13.1. If one want to improve the accuracy of the quadrature (13.2), several methods can be used: (i) increase the degree l of the polynomial interpolation, (ii) use other type of polynomial interpolation (e.g., at Gauss points) or (iii) apply the subdivision algorithm (12.3) to express the function $\varphi_{j,k}$ according to

$$\varphi_{j,k} = 2^{j/2-J/2} \sum_n s_{J-j,n} \varphi_{J,2^{J-j}k+n}, \tag{13.6}$$

for some $J > j$. Injecting (13.6) in the inner product defining $c_{j,k}$ and using the quadrature rule (13.2) for the functions $\varphi_{J,2^{J-j}k+n}$, we can reduce the error estimate to

$$\begin{aligned}
\varepsilon_J &= 2^{j/2-J/2} \sum_n |s_{J-j,n}| \left| c_{J,2^{J-j}k+n} - \tilde{c}_{J,2^{J-j}k+n} \right| \\
&\lesssim 2^{j/2-J/2} 2^{-J(r+1/2)} \left[\sum_n |s_{J-j,n}| \right] \|f\|_{C^r(S_{j,k})} \\
&\lesssim 2^{j/2} 2^{-Jr} \left[\int |\varphi(x)| \left(\sum_n |\tilde{\varphi}_{J-j,n}(x)| \right) \mathrm{d}x \right] \|f\|_{C^r(S_{j,k})} \\
&\lesssim 2^{-J(r+1/2)} \|f\|_{C^r(S_{j,k})}.
\end{aligned}$$

When f or K have poor regularity at some isolated regions, one can also perform these refinements up to a level J that is allowed vary locally (see COHEN and EZZINE [1996]). A similar idea is proposed in DAHMEN, KUNOTH and SCHNEIDER [1996] for the discretization of singular integral operators.

We now turn to the computation of the mass matrix $M_{k,l} = \langle \varphi_{j,k}, \varphi_{j,l} \rangle$. Clearly, we only need the values $I_k = \langle \varphi, \varphi(\bullet - k) \rangle$, $k \in \mathbb{Z}$, which are finite in number due to the compact support property. Here again, we inject the refinement relation (10.12), so that we obtain

$$I_k = \sum_{m,n} h_n h_m \langle \varphi(2\bullet - m), \varphi(2\bullet - 2k - n)\rangle = \tfrac{1}{2}\sum_{m,n} h_n h_m I_{2k+n-m}$$
$$= \sum_n \Big[\tfrac{1}{2}\sum_m h_{n+m-2k} h_m\Big] I_n.$$

This shows that the I_k's are the coordinates of a eigenvector associated to the eigenvalue $\lambda = 1$ of the matrix $H = (H_{k,n})$ with $H_{k,n} = \frac{1}{2}\sum_m h_{n+m-2k} h_m$. Since I_k is finite-dimensional, we only need to solve this eigenvalue problem in a finite-dimensional space.

We can proceed similarly for the computation of the inner products $R_{k,l} = \langle (\partial/\partial x)^n \varphi_{j,k}, \varphi_{j,l}\rangle$. Again, we only need the $J_k = \langle (\partial/\partial x)^n \varphi, \varphi(\bullet - k)\rangle$. Differentiating the refinement equation, we obtain

$$J_k = 2^n \sum_n \Big[\tfrac{1}{2}\sum_m h_{n+m-2k} h_m\Big] J_n,$$

i.e. the J_k's are the coordinates of a eigenvector associated to the eigenvalue $\lambda = 2^{-n}$ of the same matrix H as above.

In these two examples, two important questions need to be addressed: (i) is there a unique eigenvector (up to a normalization) of H associated to the eigenvalues $1, \ldots, 2^{-n}$? (ii) how should one normalize the eigenvector to obtain the quantities that we search?

These examples turn out to be part of a general theory developed in DAHMEN and MICCHELLI [1993], that deals with the computations of quantities of the type

$$I_{k_1,n_1,\ldots,k_d,n_d} = \int (\partial/\partial x)^{n_1} \varphi_1(x - k_1) \cdots (\partial/\partial x)^{n_d} \varphi_d(x - k_d)\,\mathrm{d}x, \tag{13.7}$$

where $\varphi_1, \ldots, \varphi_d$ are compactly supported refinable functions.

We shall not develop this theory here, but simply summarize its main features.

First, the computation of these quantities always amounts in an eigenvalue problem, by injecting the refinement equation as in the above examples. The existence of an eigenvector is ensured if these quantities are well defined, i.e. when $(\partial/\partial x)^{n_1} \varphi_1(x - k_1) \cdots (\partial/\partial x)^{n_d} \varphi_d(x - k_d)$ is integrable. The unicity of the eigenvector is the hard part of the theory: it depends crucially on certain stability properties of the scaling functions, and (in particular for multivariate refinable functions) on a smart choice of a restricted space in which the eigenvector should be searched. Finally, the normalization of the eigenvector is made possible by sum rules of the type (11.11), that are satisfied by the refinable functions.

In particular, in the case of the mass matrix, it can be shown that the L^2-stability of φ implies that the eigenvalue $\lambda = 1$ is simple (see, e.g., COHEN and DAUBECHIES

[1992]). The normalization of the corresponding eigenvector is made by (11.11) which implies $\sum I_k = 1$.

In the case of a stiffness matrix $J_k = \langle \varphi'', \varphi(\bullet - k)\rangle$, (11.11) is not sufficient since it leads to an homogeneous relation $\sum_k J_k = \int \varphi''(x)\,\mathrm{d}x = 0$. One has to use an additional sum rule of the form

$$\sum_k k^2 \varphi(x-k) = x^2 + ax + b, \tag{13.8}$$

which, as we shall see in Section 16, holds for sufficiently smooth φ. This leads to the normalization rule

$$\sum_k k^2 J_k = \int \big[x^2 + ax + b\big]\varphi''(x)\,\mathrm{d}x = 2\int \varphi(x)\,\mathrm{d}x = 2, \tag{13.9}$$

using integration by part.

REMARK 13.2. The computation of the moments M_k can be viewed as variation on the above theory, since $x^k = 2^{-k}(x/2)^k$ has a refinable structure but is not compactly supported. A particular case of (13.7) is also the computation of the values $\varphi(k)$, $k \in \mathbb{Z}$, since it corresponds to an inner product with Dirac δ functions, which, as we observed in Remark 10.6, also have a refinable structure. Using the refinement equation we obtain the eigenvalue problem $\varphi(k) = \sum_n h_n \varphi(2k-n)$. The normalization of the right solution is ensured by the condition $\sum_k \varphi(k) = 1$ which is imposed by (11.11).

14. Wavelets and multiscale algorithms

Let φ and $\tilde{\varphi}$ be a pair of compactly supported dual scaling functions that satisfy Eq. (10.12) with refinement coefficients h_n and $\tilde{h}_n$, and P_j the oblic projector defined in (10.48).

In this basic setting, we can already perform two simple algorithms that are related to specific tasks:

(i) *Projection*: Given $f_j = \sum_k c_{j,k}\varphi_{j,k} \in V_j$, we want to compute the coefficients of the approximation at a coarser scale $P_{j-1} f_j = \sum_k c_{j-1,k}\varphi_{j-1,k}$. To do so, we inject the refinement equation satisfied by $\tilde{\varphi}$ in the definition of $c_{j-1,k}$ as follows:

$$\begin{aligned} c_{j-1,k} &= 2^{(j-1)/2}\big\langle f_j, \tilde{\varphi}\big(2^{j-1}\bullet - k\big)\big\rangle = 2^{(j-1)/2}\Big\langle f_j, \sum_{n\in\mathbb{Z}} \tilde{h}_n \tilde{\varphi}\big(2^j \bullet - 2k - n\big)\Big\rangle \\ &= \frac{1}{\sqrt{2}} \sum_{n\in\mathbb{Z}} \tilde{h}_n c_{j,2k+n}. \end{aligned}$$

We thus obtain a "fine to coarse" projection algorithm expressed by the formula

$$c_{j-1,k} = \frac{1}{\sqrt{2}} \sum_{n\in\mathbb{Z}} \tilde{h}_{n-2k} c_{j,n}. \tag{14.1}$$

(ii) *Prediction*: Given the projection $P_{j-1}f_j = \sum_k c_{j-1,k}\varphi_{j-1,k}$, we want to express it in terms of the basis functions at finer scale, i.e. obtain an approximation of the real coefficients $c_{j,k}$ that is predicted from the data at resolution $j-1$. To do so, we now exploit the refinement equation of φ, which gives us

$$\begin{aligned} P_{j-1}f_j &= 2^{(j-1)/2}\sum_{k\in\mathbb{Z}} c_{j-1,k}\varphi(2^{j-1} - k) \\ &= 2^{(j-1)/2}\sum_{k\in\mathbb{Z}} c_{j-1,k}\sum_{n\in\mathbb{Z}} h_n\varphi(2^j \bullet -2k-n) \\ &= \frac{1}{\sqrt{2}}\sum_{k\in\mathbb{Z}}\Big[\sum_{n\in\mathbb{Z}} h_{k-2n}c_{j-1,n}\Big]\varphi_{j,k}. \end{aligned}$$

We thus obtain a "coarse to fine" prediction algorithm expressed by the subdivision formula that we already encountered in Section 12:

$$\hat{c}_{j,k} = \frac{1}{\sqrt{2}}\sum_{n\in\mathbb{Z}} h_{k-2n}c_{j-1,n} \tag{14.2}$$

(similar to (12.3) up to a normalization) where $\hat{c}_{j,k}$ is the predicted coefficient associated to the function $\varphi_{j,k}$.

If we now consider the multiscale decomposition of a function $f_{j_1} \in V_{j_1}$ into

$$f_{j_1} = f_{j_0} + \sum_{j=j_0}^{j_1-1} g_j, \tag{14.3}$$

with $f_j = P_j f_{j_1}$ and $g_j = f_{j+1} - f_j$, we see that the details g_j can be expanded into

$$g_j = \sum_{k\in\mathbb{Z}} (c_{j+1,k} - \hat{c}_{j+1,k})\varphi_{j+1,k}. \tag{14.4}$$

A first representation of $f_{j_1} = \sum_k c_{j_1,k}\varphi_{j_1,k}$ is thus provided by the coarse approximation coefficients $c_{j_0,k}$ and the "correction coefficients" $c_{j+1,k} - \hat{c}_{j+1,k}$, $j = j_0, \ldots, j_1 - 1$. The above algorithms allow to compute this multiscale representation from fine to coarse scales and to reconstruct the original coefficients $c_{j_1,k}$ by a coarse to fine procedure.

However, this representation cannot be viewed as a change of basis: it introduces redundancy, since the use of the basis $\varphi_{j+1,k}$, $k \in \mathbb{Z}$, to represent the fluctuation does not exploit the fact that these details are in the kernel of P_j. In order to remove this redundancy, we need to introduce a wavelet basis which can represent these fluctuations in a natural way, as in the case of the Haar basis or the hierarchical Schauder basis that we explored in Chapter I. We shall see that such wavelets can be obtained in a systematic way, once we are given a pair of dual scaling functions. Thus, the main theoretical difficulties in the construction of wavelet bases mostly reside in the selection of an appropriate pair of dual scaling functions.

Recalling the refinement equations satisfied by φ and $\tilde{\varphi}$

$$\varphi(x) = \sum_n h_n \varphi(2x - n) \quad \text{and} \quad \tilde{\varphi}(x) = \sum_n \tilde{h}_n \tilde{\varphi}(2x - n), \tag{14.5}$$

we define new coefficients

$$g_n = (-1)^n \tilde{h}_{1-n} \quad \text{and} \quad \tilde{g}_n = (-1)^n h_{1-n}. \tag{14.6}$$

We then define a pair of dual wavelets by

$$\psi(x) = \sum_n g_n \varphi(2x - n) \quad \text{and} \quad \tilde{\psi}(x) = \sum_n \tilde{g}_n \tilde{\varphi}(2x - n). \tag{14.7}$$

Clearly $\psi \in V_1$ and $\tilde{\psi} \in \widetilde{V}_1$. Before showing how these functions can be used to characterize the fluctuations, we shall first prove some preliminary results on the coefficients h_n, $\tilde{h}_n$, g_n and $\tilde{g}_n$.

LEMMA 14.1. *The coefficients satisfy the identities*

$$\sum_n h_n \tilde{h}_{n+2k} = \sum_n g_n \tilde{g}_{n+2k} = 2\delta_{0,k} \tag{14.8}$$

and

$$\sum_n h_n \tilde{g}_{n+2k} = \sum_n g_n \tilde{h}_{n+2k} = 0. \tag{14.9}$$

PROOF. We inject the refinement equations (14.5) in the duality relation, we find

$$\begin{aligned}
\delta_{0,k} &= \langle \tilde{\varphi}, \varphi(\cdot - k) \rangle \\
&= \sum_{m,n} \tilde{h}_m h_n \langle \tilde{\varphi}(2 \cdot - m), \varphi(2 \cdot - 2k - n) \rangle \\
&= \tfrac{1}{2} \sum_{m,n} \tilde{h}_m h_n \langle \tilde{\varphi}(\cdot), \varphi(\cdot - 2k - n + m) \rangle \\
&= \tfrac{1}{2} \sum_n h_n \tilde{h}_{n+2k}.
\end{aligned}$$

The second identity is immediate, since we have

$$\sum_n g_n \tilde{g}_{n+2k} = \sum_n h_{1-n} \tilde{h}_{1-n-2k} = \sum_m h_m \tilde{h}_{m-2k} = 2\delta_{0,k}.$$

Finally, the last identities are the consequences of the oscillation factor $(-1)^n$ in the definition of g_n and $\tilde{g}_n$. More precisely, we have

$$\sum_n h_n \tilde{g}_{n+2k} = \sum_n (-1)^n h_n h_{1-n-2k}$$
$$= \sum_n h_{2n} h_{1-2n-2k} - \sum_n h_{2n+1} h_{-2n-2k}$$
$$= \sum_n h_{2n} h_{1-2n-2k} - \sum_n h_{-2m-2k+1} h_{2m} = 0,$$

and similarly for $\sum_n g_n \tilde{h}_{n+2k}$. □

REMARK 14.1. The above lemma can be expressed in a matrix formulation: defining the bi-infinite banded matrix M and $\widetilde{M}$ by

$$\begin{aligned} M_{2i,j} &= h_{j-2i}, & M_{2i+1,j} &= g_{j-2i}, \\ \widetilde{M}_{i,2j} &= \tilde{h}_{i-2j}, & \widetilde{M}_{i,2j+1} &= \tilde{g}_{i-2j}, \end{aligned} \tag{14.10}$$

we see that the sum rules (14.8) and (14.9) exactly mean that $M\widetilde{M} = \widetilde{M}M = 2I$, i.e. $\widetilde{M}$ is a twice the inverse of M. Note that the matrices M and $\widetilde{M}$ are have a structure close to Toeplitz, with entries alternating rows by rows for $\widetilde{M}$ and columns by columns for M. In contrast, it is well known that a purely bi-infinite Toeplitz matrix, i.e. of convolution type, cannot be banded with banded inverse, except when it is a multiple of I.

We are now ready to prove the main results of this section. We are interested in characterizing the details $Q_j f = P_{j+1} f - P_j f$. Remark that the operator Q_j satisfies

$$Q_j^2 = P_{j+1}^2 - P_{j+1} P_j - P_j P_{j+1} + P_j^2 = Q_j$$

(using the property $P_j P_{j+1} = P_j$ that we observed in the end of Section 10). Thus Q_j is also a projector on a space W_j which complements V_j in V_{j+1}. The space W_j can also be defined as the kernel of P_j in V_{j+1}.

THEOREM 14.1. *The functions ψ and $\tilde{\psi}$ satisfy*

$$\langle \psi, \tilde{\psi}(\cdot - k) \rangle = \delta_{0,k} \tag{14.11}$$

and

$$\langle \psi, \tilde{\varphi}(\cdot - k) \rangle = \langle \tilde{\psi}, \varphi(\cdot - k) \rangle = 0. \tag{14.12}$$

The projector Q_j can be expanded into

$$Q_j f = \sum_{k\in\mathbb{Z}} \langle f, \tilde{\psi}_{j,k}\rangle \psi_{j,k} \tag{14.13}$$

and the functions $\psi_{j,k}$, $k \in \mathbb{Z}$, constitute a Riesz basis of the complement space W_j.

PROOF. We simply use the definition of ψ and $\tilde{\psi}$ to expand the first inner product according to

$$\begin{aligned}\langle \psi, \tilde{\psi}(\bullet - k)\rangle &= \sum_{m,n} g_m \tilde{g}_n \langle \varphi(2\bullet - m), \tilde{\varphi}(2\bullet - 2k - n)\rangle \\ &= \tfrac{1}{2}\sum_{m,n} g_m \tilde{g}_n \langle \varphi, \tilde{\varphi}(\bullet - 2k - n + m)\rangle \\ &= \tfrac{1}{2}\sum_{n} \tilde{g}_{n+2k} g_n = \delta_{0,k},\end{aligned}$$

by (14.8). In a similar way, we use the refinement equation to obtain

$$\langle \psi, \tilde{\varphi}(\bullet - k)\rangle = \sum_n g_{n+2k}\tilde{h}_n = 0,$$

and

$$\langle \tilde{\psi}, \varphi(\bullet - k)\rangle = \sum_n \tilde{g}_{n+2k} h_n = 0,$$

by (14.9). It is sufficient to prove (14.13) for $j = 0$ since the general case follows by scaling.

We already see from (14.11) that ψ is L^2-stable, using Theorem 11.1, and that $R_0 f = \sum_{k\in\mathbb{Z}} \langle f, \tilde{\psi}_{0,k}\rangle \psi_{0,k}$, is a bounded projector onto the space $S_0 \subset V_1$ spanned by these functions. From (14.12) we also see S_0 is contained in W_0 (since it is the kernel of P_0 in V_1). It remains to prove that S_0 coincides with W_0.

Let $f \in L^2$ and $P_1 f = \sum_{k\in\mathbb{Z}} c_{1,k}\varphi_{1,k} \in V_1$. We can compute the coefficients $d_{0,k}$ of $\psi(\bullet - k)$ in $R_0 f$ from $P_1 f$ by

$$d_{0,k} = \langle f, \tilde{\psi}_{0,k}\rangle = \frac{1}{\sqrt{2}}\sum_n \tilde{g}_n \langle f, \tilde{\varphi}_{1,2k+n}\rangle = \frac{1}{\sqrt{2}}\sum_n \tilde{g}_{n-2k} c_{1,n}. \tag{14.14}$$

From (14.14) and (14.1), we see that the bi-infinite vector X of coordinates $x_{2k} = c_{0,k}$ and $x_{2k+1} = d_{0,k+1}$ is obtained from Y of coordinates $y_k = c_{1,k}$ by the application of $\widetilde{M}$, i.e.

$$X = \frac{1}{\sqrt{2}}\widetilde{M} Y. \tag{14.15}$$

We can then develop $R_0 f + P_0 f$ in the basis $\varphi_{1,k}$, according to

$$\begin{aligned} R_0 f + P_0 f &= \sum_k c_{0,k}\varphi_{0,k} + \sum_k d_{0,k}\psi_{0,k} \\ &= \frac{1}{\sqrt{2}}\sum_k c_{0,k}\Big[\sum_n h_n \varphi_{1,2k+n}\Big] + \sum_k d_{0,k}\Big[\sum_n g_n \varphi_{1,2k+n}\Big] \\ &= \frac{1}{\sqrt{2}}\sum_k \Big[\sum_n h_{k-2n} c_{0,n} + \sum_n g_{k-2n} d_{0,n}\Big]\varphi_{1,k}. \end{aligned}$$

We see that the associated vector Z of coordinates $z_k = \sum_n h_{k-2n}c_{0,n} + \sum_n g_{k-2n}d_{0,n}$, satisfies

$$Z = \frac{1}{\sqrt{2}} M X = \tfrac{1}{2} M \widetilde{M} Y = Y, \tag{14.16}$$

i.e. $P_0 + R_0 = P_1$, and thus $R_0 = Q_0$ and $S_0 = W_0$. $\square$

REMARK 14.2. We clearly have similar results concerning the dual scaling function: $\tilde{\psi}_{j,k}$, $k \in \mathbb{Z}$, is a Riesz basis of the complement space $\widetilde{W}_j$ defined as the kernel of P_j^* in $\widetilde{V}_{j+1}$, and the projection onto this space can be computed by the inner products against the functions $\psi_{j,k}$, $k \in \mathbb{Z}$.

The dual wavelets provide with a multiscale decomposition of V_{j_1} into the (nonorthogonal) sum $V_{j_0} \oplus W_{j_0} \oplus \cdots \oplus W_{j_1-1}$: any function $f_{j_1} \in V_{j_1}$ can be expanded into

$$f_{j_1} = \sum_{k\in\mathbb{Z}} c_{j_1,k}\varphi_{j_1,k} = \sum_{k\in\mathbb{Z}} c_{j_0,k}\varphi_{j_0,k} + \sum_{j=j_0}^{j_1-1}\sum_{k\in\mathbb{Z}} d_{j,k}\psi_{j,k}, \tag{14.17}$$

with $d_{j,k} = \langle f_{j_1}, \tilde{\psi}_{j,k}\rangle$ and $c_{j,k} = \langle f_{j_1}, \tilde{\varphi}_{j,k}\rangle$.

From (14.12), we see that $\widetilde{W}_j$ is orthogonal to V_j and thus to all V_l, $l \leqslant j$, and W_l, $l < j$. By symmetry, we obtain that W_j is orthogonal to $\widetilde{V}_l$ if $l \leqslant j$, and that the spaces W_j and $\widetilde{W}_l$ are orthogonal for all $l \neq j$. In particular, we have the biorthogonality relations

$$\langle \varphi_{j_0,k}, \tilde{\psi}_{j,m}\rangle = \langle \tilde{\varphi}_{j_0,k}, \psi_{j,m}\rangle = 0,\ j \geqslant j_0 \quad \text{and} \quad \langle \psi_{j,k}, \tilde{\psi}_{l,n}\rangle = \delta_{j,l}\delta_{k,n}. \tag{14.18}$$

These relations can be used to show that $\{\varphi_{j_0,k}\}_{k\in\mathbb{Z}} \cup \{\psi_{j,k}\}_{j_0\leqslant j<j_1,k\in\mathbb{Z}}$ is a Riesz basis of V_{j_1}: if a function f_{j_1} has the form (14.17), we have on the one hand

$$\|f_{j_1}\|_{L^2}^2 \leqslant (j_1 - j_0)\left[\Big\|\sum_{k\in\mathbb{Z}} c_{j_0,k}\varphi_{j_0,k}\Big\|_{L^2}^2 + \sum_{j=j_0}^{j_1-1}\Big\|\sum_{k\in\mathbb{Z}} d_{j,k}\psi_{j,k}\Big\|_{L^2}^2\right]$$

$$\lesssim (j_1 - j_0)\left[\sum_{k\in\mathbb{Z}} |c_{j_0,k}|^2 + \sum_{j=j_0}^{j_1-1}\sum_{k\in\mathbb{Z}} |d_{j,k}|^2\right],$$

from the stability at each level. On the other hand, we use the biorthogonality relations to derive

$$\sum_{k\in\mathbb{Z}} |c_{j_0,k}|^2 + \sum_{j=j_0}^{j_1-1}\sum_{k\in\mathbb{Z}} |d_{j,k}|^2 = \sum_{k\in\mathbb{Z}} |\langle f, \tilde{\varphi}_{j_0,k}\rangle|^2 + \sum_{j=j_0}^{j_1-1}\sum_{k\in\mathbb{Z}} |\langle f, \tilde{\psi}_{j,k}\rangle|^2$$

$$\lesssim \left[\|P_{j_0} f\|_{L^2}^2 + \sum_{j=j_0}^{j_1-1} \|Q_j f\|_{L^2}^2\right] \lesssim (j_1 - j_0)\|f_{j_1}\|_{L^2}^2.$$

If we use Theorem 11.3, we derive that any function $f \in L^2$ has a converging expansion of the form

$$f = \lim_{j_1\to+\infty}\left[\sum_{k\in\mathbb{Z}} c_{j_0,k}\varphi_{j_0,k} + \sum_{j=j_0}^{j_1-1}\sum_{k\in\mathbb{Z}} d_{j,k}\psi_{j,k}\right], \tag{14.19}$$

with $d_{j,k} = \langle f_{j_1}, \tilde{\psi}_{j,k}\rangle$ and $c_{j,k} = \langle f_{j_1}, \tilde{\varphi}_{j,k}\rangle$.

However, we cannot conclude from the stability of the multiscale decomposition in V_{j_1} that the families $\{\varphi_{j_0,k}\}_{k\in\mathbb{Z}} \cup \{\psi_{j,k}\}_{j\geqslant j_0,k\in\mathbb{Z}}$ and $\{\tilde{\varphi}_{j_0,k}\}_{k\in\mathbb{Z}} \cup \{\tilde{\psi}_{j,k}\}_{j\geqslant j_0,k\in\mathbb{Z}}$ are biorthogonal Riesz bases of $L^2(\mathbb{R})$: this would mean that we can control the stability constants independently of j_1, which is not the case here since our (crude) estimate involves $(j_1 - j_0)$ in the constant. Note that a particular instance where this global stability always holds is the case of *orthonormal wavelet bases*, i.e. when $\tilde{\varphi} = \varphi$ and $\tilde{\psi} = \psi$. The global stability property for biorthogonal decompositions of L^2 will be proved under mild assumptions on the functions φ and $\tilde{\varphi}$ in Section 31 of Chapter III.

We already described a two-level decomposition and reconstruction procedure (between level 0 and 1) in the proof of Theorem 14.1. The multilevel procedure operates in the same way as for the particular cases that we studied in Chapter I: on the one hand, one *decomposes* from fine to coarse scale, using the formulae

$$c_{j,k} = \frac{1}{\sqrt{2}}\sum_n \tilde{h}_{n-2k}c_{j+1,n} \quad \text{and} \quad d_{j,k} = \frac{1}{\sqrt{2}}\sum_n \tilde{g}_{n-2k}c_{j+1,n}, \tag{14.20}$$

i.e. applying the matrix $\frac{1}{\sqrt{2}}\widetilde{M}$ to the fine scale approximation coefficients to obtain the intertwined sequences of coarser approximation and detail coefficients. On the other hand, one *reconstructs* from coarse to fine scale, using the formula

$$c_{j+1,k} = \frac{1}{\sqrt{2}}\left[\sum_n h_{k-2n}c_{j,n} + \sum_n g_{k-2n}d_{j,n}\right], \tag{14.21}$$

i.e. applying the inverse matrix $\frac{1}{\sqrt{2}}M$.

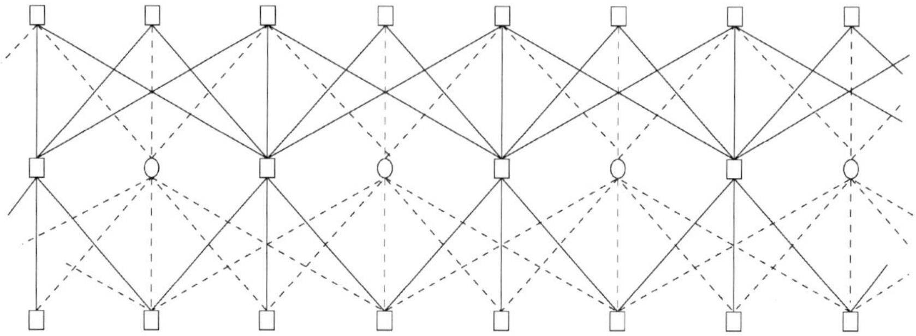

FIG. 14.1. Decomposition (above) and reconstruction (below) algorithms. □: Approximation coefficient; ○: detail coefficient.

As an example, we illustrate the pyramidal structure of these algorithms on Fig. 14.1, in the case where the support of h_n is $[-1, 1]$ and the support of $\tilde{h}_n$ is $[-3, 3]$. The full lines correspond to the coefficients $\tilde{h}_n$ (in the decomposition) and h_n (in the reconstruction), while the dashed line correspond to the coefficients $\tilde{g}_n$ (in the decomposition) and g_n (in the reconstruction). Note that these algorithms are again of complexity $\mathcal{O}(N)$, provided that we know how to deal in a proper way with functions that are defined on a bounded interval such as $[0, 1]$. A simple solution is to extend these functions out of $[0, 1]$ by periodization: this amounts in using the 1-periodizations $(f \mapsto f_{\mathrm{per}} = \sum_k f(\bullet + k))$ of the scaling functions and wavelets at resolution levels $j \geqslant 0$. The terms $Q_j f$ and $P_j f$ are then characterized by 2^j samples, that can be computed by a straightforward adaptation of (14.20) and (14.21). However, this solution has the disadvantage to create an "artificial jump" of f_{per} at 0 and 1 even when f is regular near these points. Better strategies will be discussed in Section 20.

REMARK 14.3. We remark that (14.20) amounts in a *convolution* of the fine scale approximation coefficients by the discrete sequence $\tilde{h}_{-n}$ and $\tilde{g}_{-n}$ (i.e. an autocorrelation by $\tilde{h}_n$ and $\tilde{g}_n$) followed by a *decimation* of one sample out of two in the result. Similarly, (14.21) amounts in convolutions by the discrete sequence h_n and g_n preceded by the insertion of zero values between each samples.

Such an operation is well-known in signal processing as a *subband coding scheme* based on a *two channel filter bank*. Generally speaking, subband coding aims to split a discrete signal $s(n)$, $n \in \mathbb{Z}$, into M components $s_1(n), \ldots, s_M(n)$, each of them representing a certain range of the frequency content of $s(n)$: the s_i are obtained by filtering the initial signals with M filters having complementary frequency characteristics, and by downsampling the resulting signals by a factor of M in order to preserve the total amount of information. Ideally, if the filtering process amounts in a sharp frequency cut-off, i.e. s_i is obtained by preserving the frequencies $|\omega| \in [\frac{i-1}{M}\pi, \frac{i}{M}\pi]$, in the Fourier series $S(\omega) = \sum s_n \mathrm{e}^{\mathrm{i}n\omega}$, then Shannon' sampling theorem states that s_i is perfectly characterized by retaining 1 sample out of M. Since in practice such ideal filters are not implementable (due to the slow decay of their impulse response), one approximates them by finite impulse response filters, i.e. convolutions by finite sequences

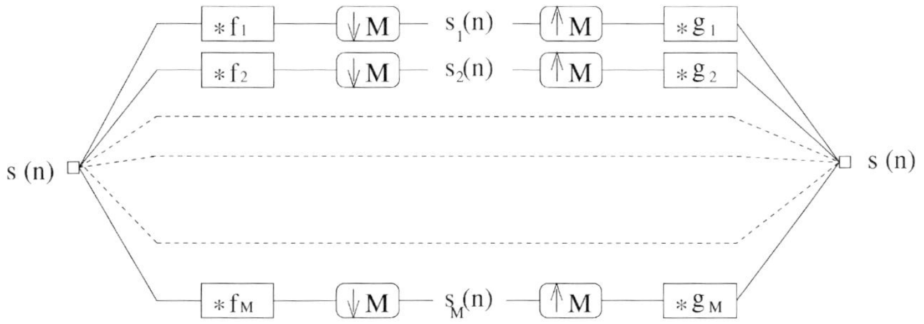

FIG. 14.2. Subband coding scheme.

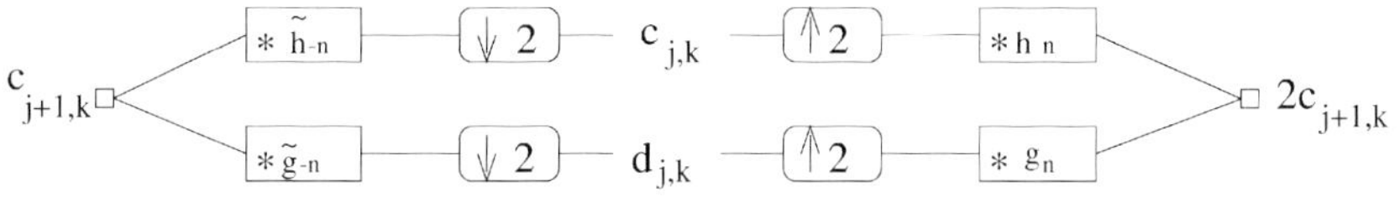

FIG. 14.3. Block diagram for the wavelet two-scale algorithm (without normalization of filters by $1/\sqrt{2}$).

$(f_{1,n})_{n\in\mathbb{Z}}, \ldots, (f_{M,n})_{n\in\mathbb{Z}}$. In order to reconstruct the original signal, the decimated signals are up-sampled by inserting $M-1$ zero values between each samples, and filtered by finite sequences $(g_{1,n})_{n\in\mathbb{Z}}, \ldots, (g_{M,n})_{n\in\mathbb{Z}}$. This ensemble of operations is represented on the so-called "block diagram" of Fig. 14.2: the downsided and upsided arrows respectively stand for downsampling and upsampling operations. In order to ensure perfect reconstruction of the signal, particular properties need to be satisfied by the decomposition filters $f_{i,n}$, and reconstruction filters $g_{i,n}$. In the case of a two-channel filter bank, i.e. $M=2$, these properties are exactly those which are stated in Lemma 14.1, i.e. (14.8) and (14.9).

A general treatment of filter banks and their relation to wavelet bases can be found in the books by VAIDYANATHAN [1992], KOVACEVIC and VETTERLI [1995] and NGUYEN and STRANG [1996].

In our case of interest which is displayed on Fig. 14.3, we notice that the filters h_n and $\tilde{h}_n$ have low-pass frequency characteristics: according to the sum-rules (11.16), their Fourier series $m(\omega)$ and $\tilde{m}(\omega)$ (defined by (10.37)) satisfy

$$m(0)=\tilde{m}(0)=1 \quad \text{and} \quad m(\pi)=\tilde{m}(\pi)=0. \tag{14.22}$$

In contrast, the filters g_n and $\tilde{g}_n$ have "mirror" high pass characteristics since by their definition, they satisfy

$$\frac{1}{2}\sum_n g_n \mathrm{e}^{-\mathrm{i}n\omega} = \mathrm{e}^{-\mathrm{i}\omega}\overline{\tilde{m}(\omega+\pi)} \quad \text{and} \quad \frac{1}{2}\sum_n \tilde{g}_n \mathrm{e}^{-\mathrm{i}n\omega} = \mathrm{e}^{-\mathrm{i}\omega}\overline{m(\omega+\pi)}. \tag{14.23}$$

Note that the oscillation of the sequences g_n and $\tilde{g}_n$, expressed by $\sum g_n = \sum \tilde{g}_n = 0$ implies that ψ and $\tilde{\psi}$ have vanishing integral.

REMARK 14.4. As we already noted in Remark 10.6, interpolatory scaling function can be viewed as a "degenerate" case of dual pair with $\tilde{\varphi} = \delta$ the Dirac distribution centered at the origin. In that case the same derivation of wavelets and multiscale algorithm can be made with $\tilde{h}_n = 2\delta_{0,n}$. As in the particular case of the Schauder hierarchical basis studied in Section 3 of Chapter I, we obtain a wavelet ψ which is simply given by

$$\psi(x) = \varphi(2x - 1), \tag{14.24}$$

and a dual wavelet $\tilde{\psi}$ which is a combination of Dirac distributions. The application of $\tilde{\psi}_{j,k}$ corresponds to the computation of the wavelet coefficient $d_{j,k}$ as the difference between the exact value of f at $2^{-j}(k + 1/2)$ and its predicted value which is obtained by interpolation of the neighbouring values on the coarser grid $2^{-j}\mathbb{Z}$, using the coefficients h_n.

As a conclusion to this section, it is important to note that the decomposition and reconstruction algorithms that we described only involve the sequences h_n, g_n, $\tilde{h}_n$ and $\tilde{g}_n$ as discrete filters and are based on their particular algebraic properties described in Lemma 14.1. These algorithms are purely discrete transformations, and they do not directly involve the underlying functions φ, $\tilde{\varphi}$, ψ and $\tilde{\psi}$, if these exist. In particular, one can apply these transformations on the discretization of a function which was not obtained by approximation in the corresponding space V_j.

15. Smoothness analysis

The goal of this section, as well as the next two sections, is to show how to "guess" from the coefficients h_n, some important properties on the function φ that solves the refinement equation (10.12).

Our starting point is thus a finite set of coefficients h_n and $\tilde{h}_n$ that satisfy the sum rules (11.16), and the corresponding refinement equations with unknowns φ and $\tilde{\varphi}$. A first remark is that we can use the Fourier transform in order to obtain a more explicit expression of φ: assuming that φ is an L^1 solution of (10.12) with the normalization $\int \varphi = 1$, we can iterate Eq. (10.38) as follows

$$\begin{aligned}\hat{\varphi}(\omega) &= m(\omega/2)\hat{\varphi}(\omega/2) = m(\omega/2)m(\omega/4)\hat{\varphi}(\omega/4) \\ &= \cdots = \hat{\varphi}(2^{-j}\omega)\prod_{k=1}^{j} m(2^{-k}\omega).\end{aligned}$$

Letting j tend to $+\infty$, and using the normalization $\hat{\varphi}(0) = 1$, we obtain an explicit expression

$$\hat{\varphi}(\omega) = \prod_{k>0} m(2^{-k}\omega). \tag{15.1}$$

The convergence of the infinite product in (15.1) is uniform on every compact set, since $m(\omega)$ is a trigonometric polynomial (and thus a C^∞ function), so that we have $1 - m(2^{-k}\omega) = \mathcal{O}(2^{-k}\omega)$. It also not difficult to check that uniform convergence on every compact set also holds for all derivatives.

Conversely, we remark that if $m(\omega)$ is a trigonometric polynomial such that $m(0) = 1$, the infinite product (15.1) defines a tempered distribution: $\hat{\varphi}(\omega)$ defined by (15.1) is clearly bounded on any compact set, and for any $j > 0$ and $2^j \leqslant |\omega| < 2^{j+1}$, we have the estimate

$$|\hat{\varphi}(\omega)| = |\hat{\varphi}(2^{-j}\omega)| \prod_{k=1}^{j} |m(2^{-k}\omega)| \leqslant \sup_{1<|\omega|<2} |\hat{\varphi}| [\sup|m|]^j \leqslant C(1 + |\omega|)^r, \quad (15.2)$$

with $C = \sup_{|\omega|<2} |\hat{\varphi}|$ and $r = \log(\sup|m|)/\log 2$. By (10.38), φ is a solution of the refinement equation in the sense of tempered distributions. We can also write

$$\varphi = (*)_{k=1}^{+\infty} \Big[2 \sum_n h_n \delta_{2^{-k}n} \Big], \quad (15.3)$$

which shows that φ is compactly supported in $[N_1, N_2]$ if $h_n = 0$ when $n \notin [N_1, N_2]$, a fact which was already established in Sections 11 and 12, in the setting of dual scaling functions.

Note that the estimate (15.2) does not a priori yield any decay estimate in the Fourier domain, since $\sup_\omega |m(\omega)| \geqslant m(0) = 1$, i.e. $r \geqslant 0$. Such a decay estimate is nevertheless crucial in order to ensure that φ has some regularity. We shall now see that a more refined analysis allows to estimate the smoothness of φ from the properties of the coefficients h_n, or equivalently of the function $m(\omega)$.

In order to measure the smoothness of φ through the properties of its Fourier transform, we use the following definition

DEFINITION 15.1. For $1 \leqslant q \leqslant \infty$, the L^q-Fourier smoothness exponent of φ is the quantity

$$s_q(\varphi) = \sup\{s;\ (1 + |\bullet|)^s \hat{\varphi} \in L^q\}. \quad (15.4)$$

Note that for $q = 2$ this is exactly the classical Sobolev exponent. Another classical indicator of smoothness is the Hölder exponent $\mu(\varphi)$ defined as follows: if φ is in C^n but not in C^{n+1}, $n \in \mathbb{N}$, then $\mu(\varphi) = n + \nu(\varphi)$ where $\nu(\varphi)$ is the supremum of $\alpha \in]0, 1[$ such that $\varphi^{(n)} \in C^\alpha$, i.e.

$$\nu(\varphi) = \inf_x \left(\lim \inf_{|t|\to 0} \frac{\log |\varphi^{(n)}(x + t) - \varphi^{(n)}(x)|}{\log |t|} \right). \quad (15.5)$$

Now if $(1+|\bullet|)^s\hat{\varphi} \in L^1$, for some $s \in]n, n+1[$, we have

$$\begin{aligned}|\varphi^{(n)}(x)-\varphi^{(n)}(x+t)| &\leqslant \int |\omega|^n \big|\big(1-\mathrm{e}^{\mathrm{i}t\omega}\big)\hat{\varphi}(\omega)\big|\,\mathrm{d}\omega \\ &\leqslant 2\int_{|t\omega|\geqslant 1} |\omega|^n \big|\hat{\varphi}(\omega)\big|\,\mathrm{d}\omega + \int_{|t\omega|\leqslant 1} |\omega|^{n+1}\big|\hat{\varphi}(\omega)\big|\,\mathrm{d}\omega \\ &\lesssim |t|^{s-n},\end{aligned}$$

which shows that $\mu(\varphi) \geqslant s_1(\varphi)$. It is also not difficult to show that the exact equality $\mu(\varphi) = s_1(\varphi)$ holds when $\hat{\varphi} \geqslant 0$.

By Hölder's inequality, one can derive

$$s_q(\varphi) \geqslant s_p(\varphi) - 1/q + 1/p \quad \text{if } q \leqslant p, \tag{15.6}$$

which shows that $\mu(\varphi) \geqslant s_1(\varphi) \geqslant s_\infty - 1$. When φ has compact support, we also have an estimate from above:

$$s_\infty(\varphi) - 1 \leqslant \mu(\varphi) \leqslant s_\infty(\varphi). \tag{15.7}$$

To check the upper bound, simply remark that if $\varphi \in C^\mu$ with $\mu \in]0, 1[$, then

$$\big|\hat{\varphi}(\omega)(1-\mathrm{e}^{\mathrm{i}\omega t})\big| \leqslant \int \big|\varphi(x)-\varphi(x+t)\big|\,\mathrm{d}x \lesssim |t|^\mu. \tag{15.8}$$

By taking $\omega = \pi/t$, we see that $|\hat{\varphi}(\omega)| \lesssim |\omega|^{-\mu}$. For $\mu > 1$, we obtain the same result using $|\mathcal{F}\varphi^{(n)}(\omega)| = |\omega|^n|\hat{\varphi}(\omega)|$, where $\mathcal{F}$ is the Fourier transform operator. Thus we can also estimate the Hölder (and the Sobolev) exponent of φ from the rate of decay of its Fourier transform at infinity.

In order to analyze the decay at infinity of the product (15.1), we remark that this product decays like $|\omega|^{-1}$ in the particular case where $m(\omega) = (1+\mathrm{e}^{-\mathrm{i}\omega})/2$ since it corresponds to $\varphi = \chi_{[0,1]}$. More precisely, we have

$$\left|\prod_{k>0}\left(\frac{1+\mathrm{e}^{-\mathrm{i}2^{-k}\omega}}{2}\right)\right| = \big|\hat{\varphi}(\omega)\big| = \left|\frac{1-\mathrm{e}^{\mathrm{i}\omega}}{\mathrm{i}\omega}\right| \lesssim \big(1+|\omega|\big)^{-1}, \tag{15.9}$$

which can also be directly derived from the classical formula $\prod_{k\geqslant 0}\cos(2^{-k}\omega) = \sin(\omega)/\omega$.

We can thus hope that $\hat{\varphi}$ has some decay at infinity if $m(\omega)$ has the factorized form

$$m(\omega) = \left(\frac{1+\mathrm{e}^{-\mathrm{i}\omega}}{2}\right)^L p(\omega), \tag{15.10}$$

where $p(\omega)$ is also a trigonometric polynomial. Note that the sum rules (11.16) imply that such a factorization should hold at least for $L = 1$. This factorization also means that

$\varphi = B_{L-1} * D(\omega)$, where B_{L-1} is the B-spline of degree $L-1$ and D is a distribution defined by

$$\hat{D}(\omega) = \prod_{k>0} p(2^{-k}\omega). \tag{15.11}$$

Applying the estimate (15.2) on $\hat{D}$, we thus obtain

$$|\hat{\varphi}(\omega)| = |\hat{B}_{L-1}(\omega)\hat{D}(\omega)| \leqslant C(1+|\omega|)^{-L+b_1}, \tag{15.12}$$

where $b_1 = \log(\sup|p|)/\log 2$, i.e. a first estimate

$$s_\infty(\varphi) \leqslant L - b_1. \tag{15.13}$$

It is possible to improve on (15.13): if $2^j \leqslant |\omega| \leqslant 2^{j+1}$, we can generalize the estimate (15.2) by grouping the factors in blocks of size $n > 0$ which leads to

$$|\hat{D}(\omega)| \leqslant \sup_{|\omega|\leqslant 1} |\hat{P}|\big[\sup|p|\big]^n \big[\sup|p(\omega)\cdots p(2^{n-1}\omega)|\big]^{j/n} \lesssim (1+|\omega|)^{b_n}, \tag{15.14}$$

with

$$b_n = \frac{\log[\sup_\omega |\prod_{k=0}^{n-1} p(2^k\omega)|]}{n\log 2}. \tag{15.15}$$

We thus obtain

$$s_\infty(\varphi) \geqslant L - b_n, \tag{15.16}$$

which is clearly an improvement on (15.13) since we have $b_n \leqslant b_1$, with a strict inequality in general.

REMARK 15.1. As n goes to $+\infty$, it can be shown (see COHEN and RYAN [1995, Chapter 3]) that the sequence b_n converges to $b_\infty = \inf_{n>0} b_n$, and that moreover, when φ is L^2-stable one has the exact equality

$$s_\infty(\varphi) = L - b_\infty. \tag{15.17}$$

Consequently, (15.16) tends to become an exact estimate as n grows. However, evaluating the maximum of $|\prod_{k=0}^{n-1} p(2^k\omega)|$ usually becomes a difficult task for large values of n. In order to approach b_∞ faster than with the sequence b_n, it often pays to introduce the decreasing sequence

$$d_n = \frac{1}{\log 2}\log\left[\sup_\omega \min_{j=0,\dots,n-1} \left|\prod_{k=0}^{j} p(2^k\omega)\right|^{1/j}\right]. \tag{15.18}$$

Clearly $d_n \leqslant b_n$. Moreover, we also have $b_m \leqslant d_n + \frac{n}{m} b_1$: for a fixed ω, we can group the factors in the product $\prod_{k=0}^{m-1} p(2^k\omega)$ into blocks of various size $1 \leqslant s \leqslant n$, such that each of them, except maybe the last one, is bounded by $[\sup_\omega \min_{j=0,\dots,n-1} |\prod_{k=0}^{j} p(2^k\omega)|^{1/j}]^s$. This decomposition yields the estimate

$$\left|\prod_{k=0}^{m-1} p(2^k\omega)\right| \leqslant \left[\sup_\omega \min_{j=0,\dots,n-1} \left|\prod_{k=0}^{j} p(2^k\omega)\right|^{1/j}\right]^m \left[\sup|p(\omega)|\right]^n.$$

Letting m go to $+\infty$, we obtain $b_\infty \leqslant d_n \leqslant b_n$ so that the sequence d_n also converges to b_∞. In many practical examples the convergence turns out to be much faster than for b_n (in particular in the example of Section 18, it can be shown that $d_n = b_\infty$ for $n \geqslant 2$).

From the evaluation of $s_\infty(\varphi)$, we can derive an estimate for $s_q(\varphi)$ using (15.6), i.e $s_q(\varphi) \geqslant s_i(\varphi) - 1/q$. In general this estimate is not sharp for $q < \infty$. We shall now describe another strategy that allows to obtain directly a sharp estimate for $s_q(\varphi)$, in particular for the case of the Sobolev exponent, i.e. for $q = 2$.

To any 2π-periodic real valued continuous function $u(\omega)$, we associate a *transfer operator* T_u acting in the space of continuous 2π-periodic (complex valued) functions as follows:

$$T_u f(\omega) = u(\omega/2) f(\omega/2) + u(\omega/2 + \pi) f(\omega/2 + \pi). \tag{15.19}$$

We first notice two important properties of T_u: firstly, if $u(\omega) = \sum_{|n| \leqslant N} u_n \mathrm{e}^{in\omega}$, then T_u preserves the corresponding finite-dimensional space spanned by $\mathrm{e}^{in\omega}$, $|n| \leqslant N$. In this basis, T_u is represented by a matrix $(t_{i,j})_{i,j=-N,\dots,N}$, where $t_{i,j} = 2u_{j-2i}$. Secondly, the dual of T_u (viewed as an operator acting in $L^2([-\pi,\pi])$) is given by

$$T_u^* f(\omega) = 2u(\omega) f(2\omega), \tag{15.20}$$

since we have for $f, g \in L^2([-\pi,\pi])$,

$$\begin{aligned}
\langle T_u^* f, \overline{g}\rangle &= \int_{-\pi}^{\pi} T_u g(\omega) f(\omega)\, \mathrm{d}\omega \\
&= 2\int_{-\pi/2}^{\pi/2} \left[u(\omega)g(\omega) + u(\omega+\pi)g(\omega+\pi)\right] f(2\omega) \\
&= 2\int_{-\pi}^{\pi} g(\omega)u(\omega) f(2\omega)\, \mathrm{d}\omega.
\end{aligned}$$

Consequently, if f and g are 2π-periodic continuous functions, we have for all $n \geqslant 0$,

$$\int_{-\pi}^{\pi} g(\omega)(T_u)^n f(\omega)\, \mathrm{d}\omega = \int_{-2^n\pi}^{2^n\pi} g(\omega)\left[\prod_{k=1}^{n} u(2^{-k}\omega)\right] f(2^{-n}\omega)\, \mathrm{d}\omega. \tag{15.21}$$

Recalling the factorized form $m(\omega) = ((1+e^{-i\omega})/2)^L p(\omega)$, we denote by T_q the transfer operator associated to the function $u(\omega) = |p(\omega)|^q$. When q is an even integer, $u = |p|^q$ is a trigonometric polynomial and we can thus study the action of T_q in a finite-dimensional space E_N, according to the first remark.

THEOREM 15.1. *If q is an even integer and if ρ_q is the spectral radius of T_q restricted to E_N, we have the estimate*

$$s_q(\varphi) \geqslant L - \frac{\log(\rho_q)}{q \log 2}. \tag{15.22}$$

PROOF. Applying (15.21) to $f = g = 1$, we obtain that for all $\varepsilon > 0$, there exist a constant $C = C(\varepsilon)$ such that

$$\int_{-2^n\pi}^{2^n\pi} \prod_{k=1}^{n} \left|p\left(2^{-k}\omega\right)\right|^q \mathrm{d}\omega = \left\langle (T_q)^n 1, 1\right\rangle \lesssim (\rho_q + \varepsilon)^n. \tag{15.23}$$

On the interval $[-2^n\pi, 2^n\pi]$, we have

$$\left|\hat{D}(\omega)\right| \leqslant C \prod_{k=1}^{n} \left|p\left(2^{-k}\omega\right)\right|, \tag{15.24}$$

with $C = \sup_{\omega\in[-\pi,\pi]} |\hat{D}(\omega)|$. It follows that

$$\int_{-2^n\pi}^{2^n\pi} \left|\hat{D}(\omega)\right|^q \lesssim (\rho_q + \varepsilon)^n. \tag{15.25}$$

Defining $\Delta_n = \{2^{n-1}\pi \leqslant |\omega| \leqslant 2^n\pi\}$ and using that $\hat{\varphi}(\omega) = \hat{B}_{L-1}(\omega)\hat{D}(\omega)$, we derive from (15.25) the estimate

$$\int_{\Delta_n} \left(1+|\omega|\right)^{sq} \left|\hat{\varphi}(\omega)\right|^q \leqslant C(\rho_q + \varepsilon)^n 2^{-Lqn} 2^{snq}, \tag{15.26}$$

which shows that $(1+|\bullet|)^s \hat{\varphi}$ is in L^q for all $s < L - \frac{\log(\rho_q)}{q \log 2}$, i.e. (15.22). □

REMARK 15.2. Similarly to (15.17), one can also prove the equality in (15.22) in the case where φ is L^2-stable (see VILLEMOES [1994] or EIROLA [1992]). This operator based technique allows thus to obtain a sharp estimate of the Sobolev exponent $s_2(\varphi)$ by checking the spectral radius of a finite matrix representing T_2 in the basis $e^{in\omega}$. Note that, according to the first remark on transition operators, the matrix coefficients of T_2 are given by $t(i, j) = 2r_{j-2i}$, where $(r_k)_{k\in\mathbb{Z}}$ is the (finite) sequence of Fourier coefficients of $|p(\omega)|^2$. We thus have

$$t_{i,j} = 2 \sum_n p_n p_{n+j-2i}, \tag{15.27}$$

where $(p_n)_{n\in\mathbb{Z}}$ is the (finite) sequence of Fourier coefficients of $p(\omega)$. We already have encountered this particular form in Section 13 when we introduced the matrix $H = (H_{k,n})$, $H_{k,n} = 1/2\sum_m h_m h_{m+n-2k}$ for the computation of the mass and stiffness matrix elements: up to a normalization, this matrix exactly represent the transition operator associated to $|m(\omega)|^2$.

REMARK 15.3. When q is not an even integer, if we want to obtain a result similar to Theorem 15.1, we need to address the question of the function space where we look for the spectral radius of T_q. In COHEN and DAUBECHIES [1996], it is shown that the choice of an appropriate space of periodic analytic functions allows to recover a sharp estimate (i.e. an equality in (15.22)) when φ is L^2 stable. The computation of the spectral radius in this infinite-dimensional space is also shown to be feasible through Fredholm determinant techniques.

REMARK 15.4. When $\hat{\varphi}$, $m(\omega)$ and $p(\omega)$ are positive functions, we can use the above technique with $q = 1$ to obtain a sharp estimate of the Hölder exponent $\mu(\varphi) = s_1(\varphi) = L - \log(\rho_1)/\log 2$. In the case where p and m are not positive, a sharp estimate of μ is still possible, using other operator-based techniques: see in particular RIOUL [1992], JIA [1999]. Finally, a technique introduced in DAUBECHIES and LAGARIAS [1991, 1992], also based on matrix analysis, allows to estimate the local pointwise Hölder regularity of scaling functions.

16. Polynomial exactness

In the previous section, we have used the factorized form (15.10) of $m(\omega)$ in order to derive smoothness estimate. The goal of this section is to give an interpretation of this factorization in terms of *polynomial reproduction properties* satisfied by the functions $\varphi_{j,k}$. These properties will be used later on to prove approximation results in the spaces V_j.

We shall also prove in this section that the factorized form (15.10) is necessary to have some smoothness properties for φ, in particular if we want $\varphi \in H^L$. Note that, according to the estimates (15.17) and (15.22), the presence of the factor $p(\omega)$ results in a loss of smoothness, i.e. (15.10) is generally not sufficient to ensure regularity of order L. This will be illustrated in the examples of Section 18 where $p(\omega)$ will be chosen in order to obtain interpolating or orthonormal scaling functions for arbitrary values of the parameter L.

We start by a simple remark: if $m(\omega)$ has the factorized form (15.10), then the function φ defined by (15.1) satisfies

$$\left(\frac{\partial}{\partial\omega}\right)^q \hat{\varphi}(2n\pi) = 0, \quad n \in \mathbb{Z} - \{0\},\ |q| \leqslant L - 1. \tag{16.1}$$

Indeed, if $n \in \mathbb{Z} - \{0\}$, there exist some $k > 0$, such that $2^{-k}(2n) \in 2\mathbb{Z} + 1$, so that $m(2^{-k}\omega)$ vanishes at $2n\pi$ with order L. It follows from (15.1) that $\hat{\varphi}$ vanishes at $2n\pi$ with the same order of flatness.

If φ is L^2 stable, a converse result also hold. Indeed, from the lower inequality in (10.23), there exists $n_0 \in \mathbb{Z}$ such that $\hat{\varphi}((2n_0+1)\pi) \neq 0$. Applying (16.1) to $n = 2n_0 + 1$, we obtain

$$\left|m(\pi + \omega/2)\hat{\varphi}\big((2n_0+1)\pi + \omega\big)\right| = \left|\hat{\varphi}\big((4n_0+2)\pi + \omega\big)\right| = \mathcal{O}\big(|\omega|^L\big), \tag{16.2}$$

so that necessarily, $|m(\pi + \omega)| = \mathcal{O}(|\omega|^L)$, which is equivalent to (15.10).

The cancellations of $\hat{\varphi}$ that are expressed in (16.1) are called the *Strang–Fix conditions* of order $L-1$. These conditions were introduced for more general (not necessarily refinable) functions in STRANG and FIX [1969]. As shown by the following result, the Strang–Fix conditions measure the degree of polynomial exactness for the spaces generated by the functions $\varphi(\bullet - k)$, $k \in \mathbb{Z}$.

THEOREM 16.1. *Let φ be an L^1 function with compact support with $\int \varphi = 1$. The following properties are equivalent*:

(i) *φ satisfies* (16.1), *i.e. the Strang–Fix conditions of order $L-1$.*

(ii) *For all $q = 0, \dots, L-1$, we can expand the polynomial x^q according to*

$$x^q = \sum_{k\in\mathbb{Z}} \big[k^q + p_{q-1}(k)\big]\varphi(x-k), \tag{16.3}$$

where p_{q-1} is a polynomial of degree $q-1$, or equivalently, for all $q = 0, \dots, L-1$,

$$\sum_{k\in\mathbb{Z}} k^q \varphi(x-k) = x^q + r_{q-1}(x), \tag{16.4}$$

where r_{q-1} is a polynomial of degree $q-1$.

PROOF. Remark that since φ is in L^1 with compact support, its Fourier transform is a C^∞ function with bounded derivatives. Applying the Fourier transform on (16.4), we obtain that

$$\hat{\varphi}(\omega)\sum_{k\in\mathbb{Z}} k^q \mathrm{e}^{-ik\omega} = (i)^q \hat{\varphi}(\omega)\Bigg[\sum_{n\in\mathbb{Z}} \delta_{2\pi n}^{(q)}\Bigg],$$

is a singular distribution supported at the origin only. It follows that for all $q = 0, \dots, L$, and $n \in \mathbb{Z} - \{0\}$, the distributions $\hat{\varphi}\delta_{2\pi n}^{(q)}$ are identically zero, which is equivalent to the Strang–Fix conditions of order $L-1$.

Conversely, if (16.1) is satisfied, we remark that the term $\hat{\varphi}\delta_0^{(q)}$ is a combination of derivatives $\delta_0^{(k)}$, $k = 0, \dots, q$, where the highest term is exactly $\delta_0^{(q)}$, since we have assumed that $\int \varphi = 1$. Thus (16.4) is satisfied. □

REMARK 16.1. The properties (16.3) and (16.4) express that the polynomials of degree less or equal to $L-1$ are "contained" in the space V_0 spanned by the functions $\varphi(\bullet - k)$,

$k \in \mathbb{Z}$. From the invariance of Π_{L-1} by a change of scale, we see that these polynomials are also contained in all the spaces V_j, $j \in \mathbb{Z}$. If $\tilde{\varphi}$ is a dual function to φ that satisfies (10.47), and which is sufficiently well localized so that the oblic projector P_j is well defined by (10.48) (e.g., if φ and $\tilde{\varphi}$ are both L^2 with compact support), we then obtain an explicit expression for the coefficients in (16.3) according to

$$k^q + P_{q-1}(k) = \int x^q \tilde{\varphi}(\bullet - k)\,\mathrm{d}x = \int (x+k)^q \tilde{\varphi}(x)\,\mathrm{d}x, \tag{16.5}$$

i.e. in terms of the moments of $\tilde{\varphi}$. We also have that $P_j p = p$ for all $p \in \Pi_{L-1}$.

REMARK 16.2. Theorem 11.3 has already revealed that the polynomial exactness property for constant polynomials was necessary and sufficient to ensure the density of the space V_j. As it will be shown in the next chapter, polynomial exactness at order $L-1$ turns out to be crucial in order to derive *approximation error estimates* of the type

$$\|f - P_j f\|_{L^p} \lesssim 2^{-jL} \big\| f^{(L)} \big\|_{L^p}, \tag{16.6}$$

for functions in the Sobolev space $W^{s,p}(\mathbb{R})$. The Strang–Fix conditions are thus the exact measurement of the *approximation power* of principal shift invariant spaces (see Remark 10.2).

In the case where φ is refinable, the Strang–Fix conditions are also unavoidable to have some smoothness as shown by the following result.

THEOREM 16.2. *If φ is an L^2-stable compactly supported refinable function in H^L for $L \geqslant 0$, then it satisfies the Strang–Fix conditions of order L.*

PROOF. We shall first prove that the Strang–Fix conditions of order L hold when $\varphi \in C^L$. For this, we proceed by induction on L.

Assuming that φ is a continuous compactly supported function, we remark that there exists $n \in \mathbb{Z}$ such that $\varphi(n) \neq 0$: if this was not the case, a simple iteration of the refinement equation (10.12) shows that φ vanishes at all dyadic numbers $2^{-j}k$, $j,k \in \mathbb{Z}$, and is thus identically zero. Up to a redefinition of φ by an integer shift (which does not modify the polynomial reproduction properties), we can thus assume that $\varphi(0) \neq 0$.

For $j \geqslant 0$, we expand $\varphi(2^{-j}\bullet)$ in terms of its scaled version, i.e.

$$\varphi(2^{-j}x) = \sum_{k\in\mathbb{Z}} s_{j,k}\varphi(x-k). \tag{16.7}$$

As it was remarked in the proof of Theorem 12.1, the coefficients $s_{j,k}$ can be expressed as inner product

$$s_{j,k} = \big\langle \varphi(2^{-j}\,\bullet), \tilde{\varphi}(\bullet - k)\big\rangle, \tag{16.8}$$

where $\tilde{\varphi}$ is a compactly supported dual function. Since $\varphi(2^{-j}x)$ tends to $\varphi(0)$ uniformly on any compact set as $j \to +\infty$, and since $\int \tilde{\varphi} = 1$, the expansion (16.7) yields

$$1 = \sum_{k \in \mathbb{Z}} \varphi(x-k), \tag{16.9}$$

and thus, the Strang–Fix conditions at order 0.

Assuming then that the result is proved up to the order $L-1$, we assume that $\varphi \in C^L$. Invoking again the refinement equation, we can assume that the derivative $\varphi^{(L)}$ of order L does not vanish at the origin. Remarking that the function

$$g_L(x) := 2^{jl}\left[\varphi^{(L)}(0)\right]^{-1}\left(\varphi(2^{-j}x) - \sum_{l=0}^{L-1}(2^{-j}x)^l \varphi^{(l)}(0)/l!\right)$$

can be expanded in terms of the $\varphi(\bullet - k)$ (by the induction hypothesis), and converges uniformly to x^L on any compact set as $j \to +\infty$, we obtain that

$$x^L = \sum_{k \in \mathbb{Z}} c_{L,k}\varphi(x-k), \tag{16.10}$$

with $c_{L,k} = \int x^L \tilde{\varphi}(x-k)\,\mathrm{d}x = k^L + p_{l-1}(k)$, $p_{l-1} \in \Pi_{l-1}$.

We have thus proved that a refinable function $\varphi \in C^L$ satisfies the Strang–Fix conditions of order L.

Assume now that $\varphi \in H^L$. It follows that the autocorrelation $\phi(x) = \int \varphi(t)\varphi(t+x)\,\mathrm{d}t$ is in C^{2L}. Since $\hat{\phi} = |\hat{\varphi}|^2$, the symbol of ϕ is given by $M(\omega) = |m(\omega)|^2$, where $m(\omega)$ is the symbol of φ. Now ϕ is L^2-stable (if (10.23) holds for $\hat{\varphi}$, it also hold for $\hat{\phi} = |\hat{\varphi}|^2$), and satisfies the Strang–Fix conditions at order $2L$: according to the first remarks of this section, we thus have $M(\omega+\pi) = \mathcal{O}(|\omega|^{2L+1})$, which immediately implies

$$M(\omega+\pi) = \mathcal{O}\big(|\omega|^{2L+2}\big), \tag{16.11}$$

since $M(\omega) \geqslant 0$. Consequently, we have

$$m(\omega+\pi) = \mathcal{O}\big(|\omega|^{L+1}\big), \tag{16.12}$$

i.e. φ satisfies the Strang–Fix conditions at order L. □

REMARK 16.3. This result shows that the property (11.11) (reproduction of constant functions) is satisfied by any L^2-stable scaling function.

REMARK 16.4. The flatness of $m(\omega)$ at π expressed by (16.12) is equivalent to a *discrete polynomial exactness property*, even when $m(\omega)$ is not related to an L^2-stable refinable function: consider the subdivision scheme associated to the coefficients h_n as it was introduced in Section 12, i.e.

$$s_{j+1,n} = \sum_k h_{n-2k} s_{j,k}. \tag{16.13}$$

Assuming that $s_{j,k} = k^q$ with $q \leqslant L$. Then the refined sequence is given at even points by

$$s_{j+1,2n} = \sum_m h_{2m}(n-m)^q = 2^{-q} \sum_m h_{2m}(2n-2m)^q, \tag{16.14}$$

and at odd points by

$$s_{j+1,2n+1} = \sum_m h_{2m+1}(n-m)^q = 2^{-q} \sum_m h_{2m+1}((2n+1)-(2m+1))^q. \tag{16.15}$$

Both sequences are polynomial in n and we see that the whole sequence $s_{j+1,n}$ is polynomial if and only if

$$\sum_m (2m)^l h_{2m} = \sum_m (2m+1)^l h_{2m+1}, \quad l = 0, \dots, q, \tag{16.16}$$

i.e. $m(\omega+\pi) = \mathcal{O}(|\omega|^{q+1})$. By linearity, we thus find that (16.12) is equivalent to the property that the subdivision scheme maps polynomial sequences of degree less or equal to L into sequences of the same type.

REMARK 16.5. The Strang–Fix conditions of order L can also be expressed in terms of the oscillation of the dual wavelet $\tilde{\psi}$: since we have

$$\hat{\tilde{\psi}}(\omega) = \mathrm{e}^{-\mathrm{i}\omega} m(\omega/2+\pi)\tilde{\varphi}(\hat{\omega}/2), \tag{16.17}$$

the condition (16.12) is equivalent to

$$\big|\hat{\tilde{\psi}}(\omega)\big| = \mathcal{O}\big(|\omega|^{L+1}\big), \tag{16.18}$$

or to the vanishing moment conditions

$$\int x^q \tilde{\psi}(x)\,\mathrm{d}x = 0, \quad q = 0, \dots, L. \tag{16.19}$$

17. Duality, orthogonality and interpolation

The properties of orthonormality (10.14) or interpolation (10.18) of a refinable function φ, or the duality property (10.47) of a pair of refinable functions, induce specific constraints on the corresponding scaling coefficients h_n and $\tilde{h}_n$. More precisely, in the case where φ and $\tilde{\varphi}$ are a pair of dual scaling functions, we have established in Lemma 14.1 the necessity of the identity

$$\sum_n h_n \tilde{h}_{n+2k} = 2\delta_{0,k}. \tag{17.1}$$

For an orthonormal scaling function, i.e. when $\varphi = \tilde{\varphi}$, (17.1) takes the form

$$\sum_n h_n h_{n+2k} = 2\delta_{0,k}. \tag{17.2}$$

If φ is interpolatory, since $h_n = \varphi(n/2)$, we obtain the constraint

$$h_{2k} = \delta_{0,k}. \tag{17.3}$$

Note that (17.3) can also be obtained from (17.1) with $\tilde{h}_n = 2\delta_{0,n}$ corresponding to $\tilde{\varphi} = \delta$ (see Remark 10.6).

In this section we shall prove converse results: if we start from sequences h_n and $\tilde{h}_n$ that satisfy the above conditions, then, under mild conditions, the associated refinable functions φ and $\tilde{\varphi}$ will satisfy the duality, orthonormality or interpolation conditions.

We recall the symbols and their factorized form

$$m(\omega) = \frac{1}{2}\sum_n h_n \mathrm{e}^{-in\omega} = \left(\frac{1+\mathrm{e}^{-\mathrm{i}\omega}}{2}\right)^L p(\omega), \tag{17.4}$$

and

$$\tilde{m}(\omega) = \frac{1}{2}\sum_n \tilde{h}_n \mathrm{e}^{-in\omega} = \left(\frac{1+\mathrm{e}^{-\mathrm{i}\omega}}{2}\right)^{\tilde{L}} \tilde{p}(\omega), \tag{17.5}$$

with $L, \tilde{L} > 0$ and $p(0) = \tilde{p}(0) = 1$, and we assume that φ and $\tilde{\varphi}$ are defined by the infinite product formula (15.1) applied to m and $\tilde{m}$.

We note that the conditions (17.1), (17.2) and (17.3) are respectively equivalent to the following identities

$$\tilde{m}(\omega)\overline{m(\omega)} + \tilde{m}(\omega+\pi)\overline{m(\omega+\pi)} = 1, \tag{17.6}$$

$$|m(\omega)|^2 + |m(\omega+\pi)|^2 = 1, \tag{17.7}$$

and

$$m(\omega) + m(\omega+\pi) = 1. \tag{17.8}$$

We also introduce the band-limited approximants φ_n and $\tilde{\varphi}_n$ for φ and $\tilde{\varphi}$ defined by

$$\hat{\varphi}_n(\omega) = \left[\prod_{k=1}^{n} m(2^{-k}\omega)\right]\chi_{[-\pi,\pi]}(2^{-n}\omega), \tag{17.9}$$

and

$$\hat{\tilde{\varphi}}_n(\omega) = \left[\prod_{k=1}^{n} \tilde{m}(2^{-k}\omega)\right]\chi_{[-\pi,\pi]}(2^{-n}\omega). \tag{17.10}$$

Our first result show that we can recover duality, orthonormality and interpolation properties from the corresponding identities (17.1)–(17.3), if these approximant converge in some specific way to φ and $\tilde{\varphi}$.

THEOREM 17.1. (i) *Duality*: *if m and $\tilde{m}$ satisfy* (17.6), *then the functions φ_n and $\tilde{\varphi}_n$ satisfy the duality property* (10.47) *for all $n \geqslant 0$. The refinable functions φ and $\tilde{\varphi}$ form a dual pair if they are the L^2-limits of the sequences φ_n and $\tilde{\varphi}_n$.*

(ii) *Orthonormality*: *if m satisfies* (17.7), *then the function φ_n has orthonormal integer translates for all $n \geqslant 0$. The refinable function φ is also orthonormal if it is the L^2-limit of the sequence φ_n.*

(iii) *Interpolation*: *if m satisfies* (17.8), *then the function φ_n is interpolatory for all $n \geqslant 0$. The refinable function φ is also interpolatory if it is the uniform limit of the sequence φ_n.*

PROOF. (i) If m and $\tilde{m}$ satisfy (17.6), we obtain for all $n \geqslant 0$,

$$\begin{aligned}
\langle \tilde{\varphi}_n, \varphi_n(\bullet - l)\rangle &= \frac{1}{2\pi}\int \overline{\hat{\varphi}_n(\omega)}\hat{\tilde{\varphi}}_n(\omega) e^{il\omega}\, d\omega \\
&= \frac{1}{2\pi}\int_{-2^n\pi}^{2^n\pi}\left[\prod_{k=1}^{n} \overline{m}\tilde{m}(2^{-k}\omega)\right] e^{il\omega}\, d\omega \\
&= \frac{2^n}{2\pi}\int_{-\pi}^{\pi}\left[\prod_{k=0}^{n-1} \overline{m}\tilde{m}(2^{k}\omega)\right] e^{i2^n l\omega}\, d\omega \\
&= \frac{2^n}{2\pi}\int_{-\pi/2}^{\pi/2}\left(\overline{m}\tilde{m}(\omega) + \overline{m}\tilde{m}(\omega+\pi)\right)\left[\prod_{k=1}^{n-1} \overline{m}\tilde{m}(2^{k}\omega)\right] e^{i2^n l\omega}\, d\omega \\
&= \frac{1}{2\pi}\int_{-2^{n-1}\pi}^{2^{n-1}\pi}\left[\prod_{k=1}^{n-1} \overline{m}\tilde{m}(2^{-k}\omega)\right] e^{il\omega}\, d\omega \\
&= \cdots = \frac{1}{2\pi}\int_{-\pi}^{\pi} e^{il\omega}\, d\omega = \delta_{0,l}.
\end{aligned}$$

We thus obtain that φ_n and $\tilde{\varphi}_n$ satisfy the duality relation (10.47) for all $n \geqslant 0$. By the continuity of the L^2 inner product, this property also holds for φ and $\tilde{\varphi}$ if they are the L^2-limits of their band-limited approximants φ_n and $\tilde{\varphi}_n$.

(ii) Similarly, if m satisfies (17.7), we obtain for all $n \geqslant 0$,

$$\begin{aligned}
\langle \varphi_n, \varphi_n(\bullet - l)\rangle &= \frac{1}{2\pi}\int |\hat{\varphi}_n(\omega)|^2 \mathrm{e}^{\mathrm{i}l\omega}\,\mathrm{d}\omega \\
&= \frac{1}{2\pi}\int_{-2^n\pi}^{2^n\pi}\left[\prod_{k=1}^{n}|m(2^{-k}\omega)|^2\right]\mathrm{e}^{\mathrm{i}l\omega}\,\mathrm{d}\omega \\
&= \frac{2^n}{2\pi}\int_{-\pi}^{\pi}\left[\prod_{k=0}^{n-1}|m(2^{k}\omega)|^2\right]\mathrm{e}^{\mathrm{i}2^n l\omega}\,\mathrm{d}\omega \\
&= \frac{2^n}{2\pi}\int_{-\pi/2}^{\pi/2}\left(|m(\omega)|^2 + |m(\omega+\pi)|^2\right)\left[\prod_{k=1}^{n-1}|m(2^{k}\omega)|^2\right]\mathrm{e}^{\mathrm{i}2^n l\omega}\,\mathrm{d}\omega \\
&= \frac{1}{2\pi}\int_{-2^{n-1}\pi}^{2^{n-1}\pi}\left[\prod_{k=1}^{n-1}|m(2^{-k}\omega)|^2\right]\mathrm{e}^{\mathrm{i}l\omega}\,\mathrm{d}\omega \\
&= \cdots = \frac{1}{2\pi}\int_{-\pi}^{\pi}\mathrm{e}^{\mathrm{i}l\omega}\,\mathrm{d}\omega = \delta_{0,l}.
\end{aligned}$$

We thus obtain that φ_n has orthonormal integer translates for all for all $n \geqslant 0$. By the continuity of the L^2 inner product with respect to the L^2 norm, this property is also satisfied by φ if it is L^2-limit of φ_n.

(iii) Finally, if m satisfies (17.8), we obtain for all $n \geqslant 0$,

$$\begin{aligned}
\varphi_n(l) &= \frac{1}{2\pi}\int \hat{\varphi}_n(\omega)\mathrm{e}^{\mathrm{i}l\omega}\,\mathrm{d}\omega \\
&= \tfrac{1}{2\pi}\int_{-2^n\pi}^{2^n\pi}\left[\prod_{k=1}^{n}m(2^{-k}\omega)\right]\mathrm{e}^{\mathrm{i}l\omega}\,\mathrm{d}\omega \\
&= \frac{2^n}{2\pi}\int_{-\pi}^{\pi}\left[\prod_{k=0}^{n-1}m(2^{k}\omega)\right]\mathrm{e}^{\mathrm{i}2^n l\omega}\,\mathrm{d}\omega \\
&= \frac{2^n}{2\pi}\int_{-\pi/2}^{\pi/2}\left(m(\omega) + m(\omega+\pi)\right)\left[\prod_{k=1}^{n-1}m(2^{k}\omega)\right]\mathrm{e}^{\mathrm{i}2^n l\omega}\,\mathrm{d}\omega \\
&= \frac{1}{2\pi}\int_{-2^{n-1}\pi}^{2^{n-1}\pi}\left[\prod_{k=1}^{n-1}m(2^{-k}\omega)\right]\mathrm{e}^{\mathrm{i}l\omega}\,\mathrm{d}\omega \\
&= \cdots = \frac{1}{2\pi}\int_{-\pi}^{\pi}\mathrm{e}^{\mathrm{i}l\omega}\,\mathrm{d}\omega = \delta_{0,l}.
\end{aligned}$$

We thus obtain that φ_n is interpolatory for all $n \geqslant 0$. By the continuity of the pointwise values with respect to the uniform convergence, this property is also satisfied by φ if it is the uniform limit of φ_n. □

REMARK 17.1. One easily check that $\varphi_n(2^{-n}k) = s_{n,k}$, where $s_{n,k}$ is the result of n iterations of the subdivision algorithm described in Section 12. We can thus view φ_n as the band-limited interpolant of the subdivision values after n-iterations. Recall that in Section 12, we proved the uniform convergence of the linear interpolation of the subdivision, under the assumption that the limit function φ was continuous and L^2 stable. For the L^2 or uniform convergence of φ_n to φ, the situation is very similar: convergence is ensured by the L^2-stability and an assumption on the regularity of the limit. However, we are here interested in placing the assumptions on the function $m(\omega)$ (i.e. on the coefficients h_n) rather than on the function φ.

The following theorem replaces the L^2-stability assumption by a criterion on the cancellations of $m(\omega)$.

THEOREM 17.2. *Assume that $m(\omega)$ does not vanish on $[-\pi/2, \pi/2]$. Then,*
(i) *if $\varphi \in L^2$, it is the L^2-limit of φ_n,*
(ii) *if $\hat{\varphi} \in L^1$, it is the L^1-limit of $\hat{\varphi}_n$ and φ_n converges uniformly to φ.*

PROOF. If $m(\omega)$ does not vanish on $[-\pi/2, \pi/2]$, the infinite product formula (15.1) shows that $\hat{\varphi}(\omega)$ does not vanish on $[-\pi, \pi]$.

We then remark that

$$\hat{\varphi}_n(\omega) = \begin{cases} \hat{\varphi}(\omega)/\hat{\varphi}(2^{-n}\omega) & \text{if } \omega \in [-2^n\pi, 2^n\pi], \\ 0 & \text{elsewhere}, \end{cases} \tag{17.11}$$

so that

$$\left|\hat{\varphi}_n(\omega)\right| \leqslant C\left|\hat{\varphi}(\omega)\right|, \tag{17.12}$$

with $C = \max_{\omega \in [-\pi,\pi]} |\hat{\varphi}(\omega)|^{-1}$. Thus, if $\hat{\varphi}$ is in L^q, $1 \leqslant q < \infty$, the L^q-convergence of $\hat{\varphi}_n$ to $\hat{\varphi}$ is ensured by Lebesgue's dominated convergence theorem (using also that $\hat{\varphi}_n$ converges pointwise to $\hat{\varphi}$).

In the L^1 case, the uniform convergence of φ_n to φ follows, applying the inverse Fourier transform. □

REMARK 17.2. The condition that $m(\omega) \neq 0$ on $[-\pi/2, \pi/2]$ – or equivalently that $\hat{\varphi}(\omega) \neq 0$ on $[-\pi, \pi]$ – is related to the L^2 stability of φ since it clearly implies the lower inequality in (10.23), i.e.

$$\sum_{n \in \mathbb{Z}} \left|\hat{\varphi}(\omega + 2n\pi)\right|^2 \geqslant C_1 > 0. \tag{17.13}$$

However, (17.13) could be achieved without the assumption that $\hat{\varphi}$ does not vanish on $[-\pi, \pi]$. It can be shown (see, e.g., COHEN and RYAN [1995] or COHEN, DAUBECHIES and FEAUVEAU [1992]) that one can use a condition on $m(\omega)$ that is exactly equivalent to (17.13), and reach the same conclusions as in Theorem 17.2: one assumes that there exists a compact set K such that

$$0 \in \operatorname{int}(K), \quad |K| = 2\pi \quad \text{and} \quad \bigcup_{n\in\mathbb{Z}} (K + 2n\pi) = \mathbb{R}, \tag{17.14}$$

for which we have

$$m\big(2^{-k}\omega\big) \neq 0, \quad k \geqslant 0,\ \omega \in K. \tag{17.15}$$

A compact set satisfying the assumption (17.14) is called *congruous to* $[-\pi, \pi]$ *modulo* 2π. Note that when $K = [-\pi, \pi]$, we simply obtain the property that $m(\omega) \neq 0$ on $[-\pi/2, \pi/2]$. In the orthonormal case, it is shown in COHEN and RYAN [1995, Chapter 2] that a smarter choice of K allows to assume only that $m(\omega)$ does not vanish on $[-\pi/3, \pi/3]$. However, such a generalization is not much used in practice: in most relevant examples $m(\omega)$ does not vanish on $[-\pi/2, \pi/2]$.

It remains to discuss the properties that one can impose on $m(\omega)$ and $\tilde{m}(\omega)$ so that $\hat{\varphi}$ and $\tilde{\varphi}$ are ensured to be in L^2 or L^1, in order to obtain the construction of dual, orthonormal or interpolatory scaling functions, directly from some specific assumptions on the refinement coefficients.

In the particular case of orthonormal scaling functions, it is interesting to note that the property (17.7) is enough to ensure that $\varphi \in L^2$: indeed, we remarked in the proof of Theorem 17.1 that (17.7) implies that φ_n is orthonormal, and in particular $\|\varphi_n\|_{L^2} = 1$. Since $\hat{\varphi}_n$ converges pointwise to $\hat{\varphi}$, we obtain by Fatou's lemma that $\varphi \in L^2$ with $\varphi \leqslant 1$. Clearly the same reasoning shows that if $m(\omega)$ satisfies (17.8), and $m(\omega) \geqslant 0$, the limit function $\hat{\varphi}$ is in L^1.

For more general cases, we already discussed in Section 15 the localization properties of $\hat{\varphi}$. In particular, we see that $\hat{\varphi} \in L^q$ whenever $s_q(\varphi) > 0$, so that we can use the estimate (15.17) or (15.22): $\hat{\varphi}$ is in L^q if

$$L - b_\infty > 1/q \quad (\text{i.e. } L - b_n > 1/q \text{ for some } n \geqslant 0), \tag{17.16}$$

or if

$$L - \frac{\log(\rho_q)}{q \log 2} > 0, \tag{17.17}$$

where b_∞, b_n and ρ_q are defined as in Section 15. Additionally, one can easily check that these conditions directly imply the L^q-convergence of $\hat{\varphi}_n$ to $\hat{\varphi}$, so that the assumption that $m(\omega)$ does not vanish on $[-\pi/2, \pi/2]$ can be removed. Introducing the analogous quantities $\tilde{b}_\infty$ and $\tilde{\rho}_q$ related to $\tilde{m}(\omega)$, we can thus summarize the construction of dual, orthonormal and interpolatory scaling function by the following result.

THEOREM 17.3. (i) *Duality*: *if m and $\tilde{m}$ satisfy* (17.6), *and if*

$$\max\left\{L - b_\infty - \tfrac{1}{2}, L - \frac{\log(\rho_2)}{2\log 2}\right\}, \max\left\{\tilde{L} - \tilde{b}_\infty - \tfrac{1}{2}, \tilde{L} - \frac{\log(\tilde{\rho}_2)}{2\log 2}\right\} > 0, \quad (17.18)$$

then φ and $\tilde{\varphi}$ form a pair of dual refinable functions in L^2.

(ii) *Orthonormality*: *if m satisfies* (17.7) *and if*

$$\max\left\{L - b_\infty - \tfrac{1}{2}, L - \frac{\log(\rho_2)}{2\log 2}\right\} > 0, \quad (17.19)$$

then φ is an orthonormal refinable function.

(iii) *Interpolation*: *if m satisfies* (17.8) *and if*

$$\max\left\{L - b_\infty - 1, L - \frac{\log(\rho_1)}{\log 2}\right\} > 0, \quad (17.20)$$

then φ is an interpolatory refinable function.

REMARK 17.3. The question of finding minimal (i.e. necessary and sufficient) conditions on the functions m and $\tilde{m}$, that ensure duality, orthonormality or interpolation, has been addressed in numerous contribution. In the orthonormal case, it is shown in COHEN and RYAN [1995, Chapter 2] that condition (17.15) (together with (17.7)) is minimal. Other sharp criteria (based on the transfer operator introduced in Section 15 or on the arithmetic properties of the zeros of $m(\omega)$) are also discussed there. In the case of dual scaling functions, a minimal criterion is given in COHEN and DAUBECHIES [1992], based on transition operator: it essentially states that the conditions on $L - \log(\rho_2)/(2\log 2)$ and $\tilde{L} - \log(\tilde{\rho}_2)/(2\log 2)$ (together with (17.6)) are necessary and sufficient.

18. Interpolatory and orthonormal scaling functions

We are now ready to describe the construction of an important class of interpolatory and orthonormal scaling functions. This construction will also be used in the next section to obtain scaling functions that are dual to the B-splines.

In order to build regular interpolatory scaling function, we need to look for trigonometric polynomials $m(\omega)$ that can be put in the factorized form (17.4) and that satisfy the condition (17.8). In the case where L is even, we can look for some $m(\omega)$ with real and symmetric Fourier coefficients, i.e. of the form

$$m(\omega) = \frac{1}{2}\left(h_0 + \sum_{n\geqslant 0} h_n\left[e^{in\omega} + e^{-in\omega}\right]\right). \quad (18.1)$$

Since $m(\omega)$ vanishes with order $L = 2N$ at $\omega = \pi$, it should be of the form

$$m(\omega) = \left[\cos^2(\omega/2)\right]^N P\left(\sin^2(\omega/2)\right), \quad (18.2)$$

where P is an algebraic polynomial. Combining this particular form with the constraint (17.8), we obtain the equation

$$(1-x)^N P(x) + x^N P(1-x) = 1, \tag{18.3}$$

that should be satisfied for all $x \in [0,1]$ (and thus for all $x \in \mathbb{C}$ since P is a polynomial). The existence of a minimal degree polynomial solution to (18.2) is ensured by Bezout theorem. We denote by P_N this minimum degree solution. Note that for $N=1$, we have $P_N = 1$ and the corresponding interpolatory refinable function is simply the hat function $\varphi = \max\{1-|x|, 0\}$ that generates linear splines.

For general N, we can relate the minimal degree solution to a simple discrete interpolation scheme: we consider the subdivision scheme associated to the coefficients h_n, i.e.

$$s_{j+1,k} = \sum_n h_{k-2n} s_{j,n}. \tag{18.4}$$

From (17.3), we have $h_0 = 1$ and $h_{2n} = 0$ if $n \neq 0$, so that (18.4) rewrites as an interpolatory scheme

$$s_{j+1,2k} = s_{j,k} \quad \text{and} \quad s_{j+1,2k+1} = \sum_{n \neq 0} h_n s_{j,2k+1+n}. \tag{18.5}$$

As we noted in Remark 16.4, the cancellation of $m(\omega)$ at $\omega = \pi$ with order $2N$ is equivalent to a discrete polynomial reproduction property (up to order $2N-1$) for the subdivision scheme associated to the h_n. Due to the discrete interpolatory property (18.5), we obtain that if $s_{j,k} = q(2^{-j}k)$ for some $q \in \Pi_{2N-1}$, then we should also have $s_{j+1,k} = q(2^{-j-1}k)$. Clearly, the minimally supported sequence h_n allowing this property corresponds to the choice

$$s_{j+1,2k+1} = q_{j,k}\left(2^{-j-1}k\right), \tag{18.6}$$

where $q_{j,k}$ is the unique Lagrangian polynomial interpolant in Π_{2N-1} of the points $(2^{-j}(k+n), s_{j,k+n})$, $n = -N+1, \ldots, N$.

This *iterative Lagrangian interpolation scheme*, firstly introduced in the works of DESLAURIERS and DUBUC [1987], can be viewed as a natural generalization of linear data interpolation viewed as an iterative scheme: when $N=1$, the new points on the fine grid are simply computed as the averages of the two closest neighbours on the coarser grid. Our general scheme simply consists in using the $2N$ closest neighbours and a higher-order polynomial rule.

This interpretation of the minimal degree solution allows to find an explicit expression of the coefficients h_n: $h_{2n} = \delta_{0,n}$ and for $n = 0, \ldots, N-1$,

$$h_{-2n-1} = h_{2n+1} = \frac{\prod_{k=-N,k\neq n}^{N-1}(2k+1)}{\prod_{k=-N,k\neq n}^{N-1}(2k-2n)}. \tag{18.7}$$

Since $m(\omega)$ has degree $2N-1$, it follows that the corresponding minimal degree solution P_N to (18.3) has degree $N-1$.

The exact expression of P_N can also be found directly by the following method: in the expansion of $1=[x+(1-x)]^{2N-1}$, we can factorize x^N in half of the terms, so that we obtain (18.3) with

$$P_N(x)=\sum_{n=0}^{N-1}\binom{2N-1}{N+n}x^n(1-x)^{N-1-n}. \tag{18.8}$$

Since P_N is of degree $N-1$, we conclude that the trigonometric polynomial $m(\omega)$ defined by (18.2) with $P=P_N$ given by (18.8) coincides with $m(\omega)$ defined from the Lagrangian interpolation scheme, i.e. by the coefficients (18.7).

From now on, we shall use the notation

$$m_N(\omega)=\left[\cos^2(\omega/2)\right]^N P_N\left(\sin^2(\omega/2)\right), \tag{18.9}$$

for this minimal solution, and we denote by ϕ_N the associated refinable function defined by the infinite product (15.1). Since $m_N(\omega)$ is positive, as it was remarked in the previous section, we are ensured that $\hat{\phi}_N(\omega)=\prod_{k>0}m_N(2^{-k}\omega)$ is in L^1, so that ϕ_N is at least continuous. We also know that $\phi_1=\max\{1-|x|,0\}$ is $C^{1-\varepsilon}$ for all $\varepsilon>0$.

We display on Fig. 18.1 the iteration of the 4-point iterative interpolation scheme based on cubic interpolation, i.e. corresponding to $N=2$: the new points are given by

$$s_{j+1,2k+1}=\tfrac{9}{16}(s_{j,k}+s_{j,k+1})-\tfrac{1}{16}(s_{j,k-1}+s_{j,k+2}). \tag{18.10}$$

We also display the corresponding interpolatory scaling function ϕ_2 on Fig. 18.2.

Both figures reveal that we have gained some smoothness in comparison to the linear interpolation process corresponding to $N=1$. Using the transition operator technique of Section 15 (see in particular Remark 15.4), one can actually prove that $\phi_2\in C^{2-\varepsilon}$ for all $\varepsilon>0$ and that this estimate is sharp for the Hölder exponent, i.e. $\mu(\phi_2)=2$.

We shall now prove that we can obtain arbitrarily high regularity by increasing the value of N. For this, we shall make use of the following more synthetic formula for P_N, which was derived in DAUBECHIES [1988] using combinatorial arguments:

$$P_N(x)=\sum_{n=0}^{N-1}\binom{N-1+n}{n}x^n. \tag{18.11}$$

Using this form, we first prove two preliminary results.

LEMMA 18.1. *For all $N\geqslant 1$, $x\in[0,1]$, we have*

$$P_N(x)\leqslant[G(x)]^{N-1},\quad \text{with } G(x)=\max\{2,4x\}. \tag{18.12}$$

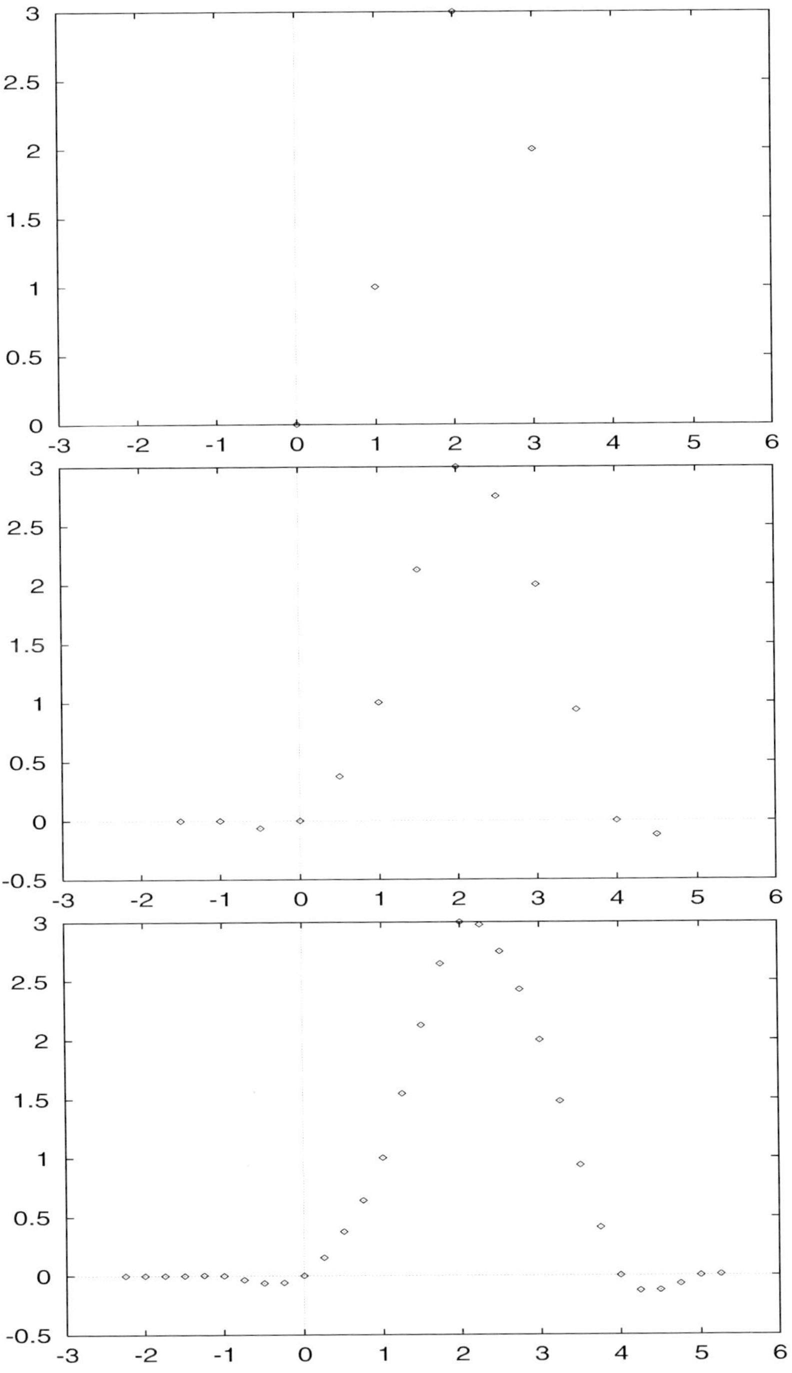

FIG. 18.1. Iteration of the 4-point interpolation scheme ($j = 0, 1, 2$).

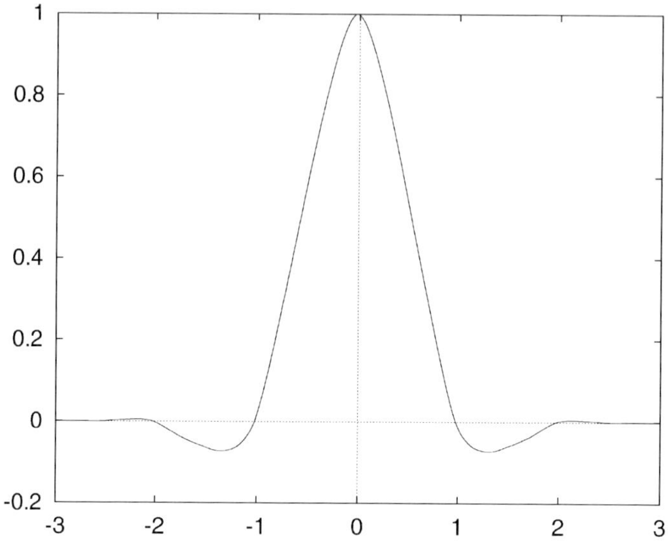

FIG. 18.2. The limit function ϕ_2.

PROOF. We first remark that $P(x)$ is increasing on $[0, 1]$, and that from (18.3), we have $P(1/2) = 2^{N-1}$. The estimate (18.12) follows on $[0, 1/2]$. If $x \in [1/2, 1]$, we write

$$P_N(x) = \sum_{n=0}^{N-1} \binom{N-1+n}{n} 2^{-n}(2x)^n \leqslant (2x)^{N-1} P_N(1/2) = (4x)^{N-1}$$

and thus (18.12) holds on the whole of $[0, 1]$. □

LEMMA 18.2. *The function* $G(x)$ *satisfies*

$$\begin{aligned} &G(x) \leqslant G(3/4) = 3 \quad \text{on } [0, 3/4] \quad \text{and} \\ &G(x)G\big(4x(1-x)\big) \leqslant \big[G(3/4)\big]^2 = 9 \quad \text{on } [3/4, 1]. \end{aligned} \tag{18.13}$$

PROOF. Since G is increasing on $[0, 1]$, we have $G(x) \leqslant G(3/4)$ on $[0, 3/4]$. Since $3/4$ is left unchanged by $x \mapsto 4x(1-x)$, we have $F(3/4) = [G(3/4)]^2$ where $F(x) = G(x)G(4x(1-x))$.

If $y \in [0, 3/4]$ is such that $4y(1-y) = 1/2$, $F(x) = 64x^2(1-x)$ on $[3/4, y]$ and decreases on this interval, while $F(x) = 8x$ on $[y, 1]$ where it increases. Since $F(1) = 8 \leqslant [G(3/4)]^2$, we obtain the second part of (18.13). □

THEOREM 18.1. *The functions* ϕ_N *satisfy*

$$s_\infty(\phi_N) \geqslant 2N\left(1 - \frac{\log 3}{2\log 2}\right) \simeq 0.4\,N. \tag{18.14}$$

Consequently, their regularity can be made arbitrarily high by increasing the value of N.

PROOF. From the results of Section 15, we know that $s_\infty(\phi_N) = 2N - d_2(\phi_N)$ where $d_2(\phi_N)$ is defined through (15.18), i.e.

$$d_2(\phi_N) = \frac{1}{\log 2} \log\Bigl(\max_\omega \min\bigl\{m_N(\omega), \sqrt{m_N(\omega)m_N(2\omega)}\bigr\}\Bigr). \tag{18.15}$$

By a change of variable and using (18.12), we obtain

$$\begin{aligned} d_2(\phi_N) &= [\log 2]^{-1} \log\Bigl[\max_{x\in[0,1]} \min\bigl\{P_N(x), \sqrt{P_N(x)P_N(4x(1-x))}\bigr\}\Bigr] \\ &\leqslant (N-1)[\log 2]^{-1} \log\Bigl[\max_{x\in[0,1]} \min\bigl\{G(x), \sqrt{G(x)G(4x(1-x))}\bigr\}\Bigr]. \end{aligned}$$

From (18.13), we know that $\max_{[0,1]} \min\{G(x), \sqrt{G(x)G(4x(1-x))}\} = 3$, and thus

$$d_2(\phi_N) \leqslant (N-1)\frac{\log 3}{\log 2}, \tag{18.16}$$

which yields (18.14). □

REMARK 18.1. For small values of N, one can obtain better estimates for the Hölder or Sobolev regularity, using the operator based technique described in Section 15 and applying Theorem 15.1. Note that this technique becomes however unpractical for large values of N, due to the important size of the matrix representing the transition operator. As N goes to ∞, it can also be shown (see VOLLKMER [1992], COHEN and RYAN [1995, Chapter 3]) that the estimate (18.14) is sharp *in an asymptotical sense*:

$$\lim_{N\to\infty} \frac{s_\infty(\phi_N)}{2N} = \lim_{N\to\infty} \frac{s_p(\phi_N)}{2N} = \lim_{N\to\infty} \frac{\mu(\phi_N)}{2N} = 1 - \frac{\log 3}{2\log 2} \simeq 0.2. \tag{18.17}$$

This behaviour should be compared with the case of B-splines: the functions ϕ_N are supported in $S_N = [-2N+1, 2N-1]$, i.e. their regularity grows asymptotically like $|S_N|/10$, while the regularity of the splines B_N is asymptotically equivalent to their support. This loss should be viewed as *the price to pay* to the interpolation constraint (17.8). Note that P_N is not the unique solution to (18.3). An open problem is the construction of a family of polynomial solutions of (18.3) that would optimize the ratio between the smoothness of the refinable function and its support, either in an asymptotical sense or for a prescribed support size.

We now turn to the construction of orthonormal scaling functions, following the ideas introduced in DAUBECHIES [1988]. It should be noted that most the techniques that we have used in the previous discussion on the regularity of interpolatory scaling functions, were also introduced in this paper, since, as we shall see, both constructions are closely related.

We are now looking for trigonometric polynomials $m^o(\omega)$ having the factorized form

$$m^o(\omega) = \left(\frac{1+\mathrm{e}^{-\mathrm{i}\omega}}{2}\right)^N p(\omega), \tag{18.18}$$

and such that the orthonormality constraint (17.7) is satisfied. Note that such a polynomial should necessarily satisfy

$$\left|m^o(\omega)\right|^2 = \left[\cos^2(\omega/2)\right]^N P\left(\sin^2(\omega/2)\right), \tag{18.19}$$

where $P(x)$ is a polynomial solution of (18.3) which is positive on $[0, 1]$, or equivalently $|p(\omega)|^2 = P(\sin^2(\omega/2))$.

A natural idea is thus to look for a minimally supported solution

$$m_N^o(\omega) = \left(\frac{1+\mathrm{e}^{-\mathrm{i}\omega}}{2}\right)^N p_N(\omega), \tag{18.20}$$

where $p_N(\omega) = \sum_{n=0}^{N-1} c_n \mathrm{e}^{-in\omega}$ is such that

$$\left|p_N(\omega)\right|^2 = P_N\left(\sin^2(\omega/2)\right), \tag{18.21}$$

or equivalently,

$$\left|m_N^o(\omega)\right|^2 = m_N(\omega). \tag{18.22}$$

The existence of p_N is ensured by the following result, originally due to F. Riesz.

LEMMA 18.3. *Let $A(\omega) = a_0 + \sum_{n=1}^N a_n(\mathrm{e}^{\mathrm{i}n\omega} + \mathrm{e}^{-\mathrm{i}n\omega})$ be a symmetric trigonometric polynomial such that the a_n are real numbers and such that $A(\omega) > 0$. There exists a trigonometric polynomial $B(\omega) = \sum_{n=0}^N b_n \mathrm{e}^{-\mathrm{i}n\omega}$ with real coefficients such that $|B(\omega)|^2 = A(\omega)$.*

PROOF. We introduce an algebraic polynomial

$$P_A(z) = z^N\left[a_0 + \sum_{n=1}^N a_n\left(z^n + z^{-n}\right)\right], \tag{18.23}$$

of degree $2N$, which satisfies $A(\omega) = \mathrm{e}^{-\mathrm{i}N\omega} P(\mathrm{e}^{\mathrm{i}n\omega})$. Since P_A has real coefficients and since $P_A(z^{-1}) = z^{-n} P_A(z)$, we see that

$$P_A(z) = 0 \Leftrightarrow P_A(\bar{z}) = 0 \Leftrightarrow P_A\left(z^{-1}\right) = 0 \Leftrightarrow P_A\left(\bar{z}^{-1}\right) = 0. \tag{18.24}$$

Since P_A does not vanish on the unit circle (because $m(\omega) > 0$), it can thus be factorized in the following way

$$P_A(z) = C\prod_{k=1}^{N_r}(z - r_k)\left(z - r_k^{-1}\right)\prod_{k=1}^{N_z}(z - z_k)\left(z - z_k^{-1}\right)(z - \overline{z}_k)\left(z - \overline{z}_k^{-1}\right), \quad (18.25)$$

where the r_k's are real and the z_k's are complex with nonzero imaginary part.

Since $N = N_r + 2N_z$, we can write

$$A(\omega) = \mathrm{e}^{-\mathrm{i}(N_r+2N_z)\omega} P\left(\mathrm{e}^{\mathrm{i}n\omega}\right) = C A_r(\omega) A_z(\omega),$$

with

$$\begin{aligned} A_r(\omega) &= \prod_{k=1}^{N_r}\left(\mathrm{e}^{\mathrm{i}\omega} - r_k\right)\left(1 - \frac{1}{r_k}\mathrm{e}^{-\mathrm{i}\omega}\right) \\ &= (-1)^{N_r}\left[\prod_{k=1}^{N_r} r_k\right]^{-1}\left|\prod_{k=1}^{N_r}\left(\mathrm{e}^{-\mathrm{i}\omega} - r_k\right)\right|^2, \end{aligned}$$

and

$$\begin{aligned} A_z(\omega) &= \prod_{k=1}^{N_z}\left(\mathrm{e}^{\mathrm{i}\omega} - z_k\right)\left(1 - z_k^{-1}\mathrm{e}^{-\mathrm{i}\omega}\right)\left(\mathrm{e}^{\mathrm{i}\omega} - \overline{z}_k\right)\left(1 - \overline{z}_k^{-1}\mathrm{e}^{-\mathrm{i}\omega}\right) \\ &= \left[\prod_{k=1}^{N_z}|z_k|^2\right]^{-1}\left|\prod_{k=1}^{N_z}\left(\mathrm{e}^{-\mathrm{i}\omega} - z_k\right)\left(\mathrm{e}^{-\mathrm{i}\omega} - \overline{z}_k\right)\right|^2. \end{aligned}$$

It follows that $A(\omega) = |P_B(\mathrm{e}^{-\mathrm{i}\omega})|^2$ with

$$P_B(z) = D\left[\prod_{i=1}^{N_r}(z - r_i)\prod_{i=1}^{N_z}\left(z^2 - (z_i + \overline{z}_i)z + |z_i|^2\right)\right], \quad (18.26)$$

and $D = (C(-1)^{N_r}[\prod_{i=1}^{N_r} r_i]^{-1}[\prod_{i=1}^{N_z}|z_i|^2]^{-1})^{1/2}$ (by the positivity of $A(\omega)$, we are ensured that the quantity in the square root is also positive). □

Note that this lemma is constructive: the coefficients of $m_N^o(\omega)$ can be derived from those of $m_N(\omega)$ through the study of the zeros of the associated algebraic polynomial defined in (18.23). We thus $m_N^o(\omega)$ satisfying (18.22), and the corresponding orthonormal scaling functions φ_N. Note that we have

$$\left|\hat{\varphi}_N(\omega)\right|^2 = \hat{\phi}_N(\omega), \quad (18.27)$$

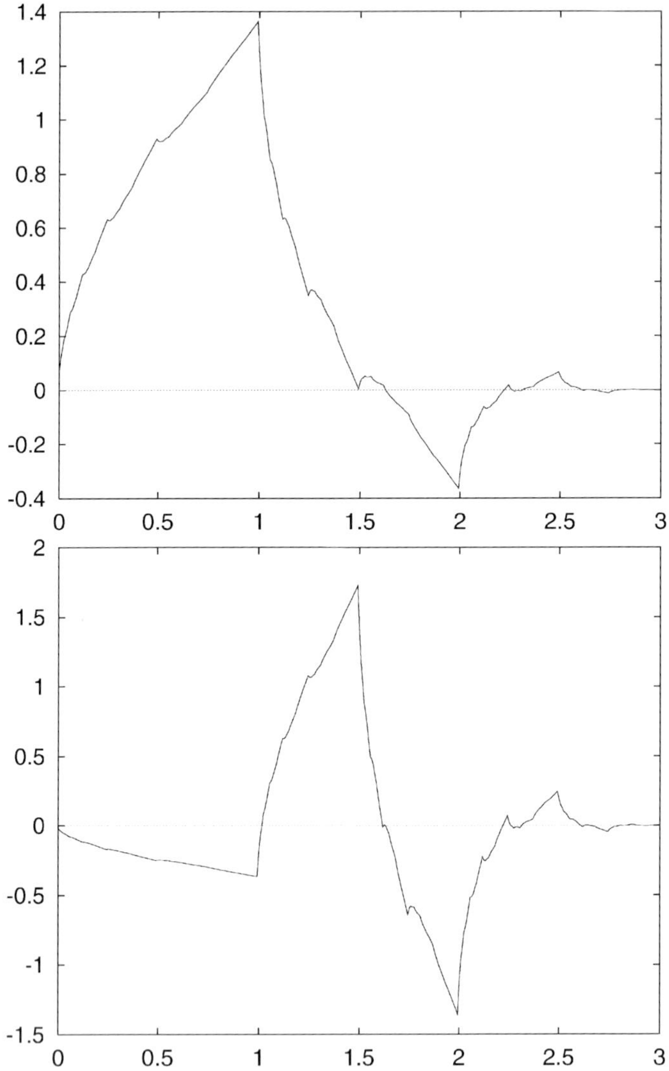

FIG. 18.3. Orthonormal scaling function and wavelet ($N = 2$).

or equivalently

$$\phi_N(x) = \int \varphi_N(x+t)\varphi_N(t)\,\mathrm{d}t. \tag{18.28}$$

As a consequence, Theorem 18.1 also provides an estimate for the regularity of φ_N

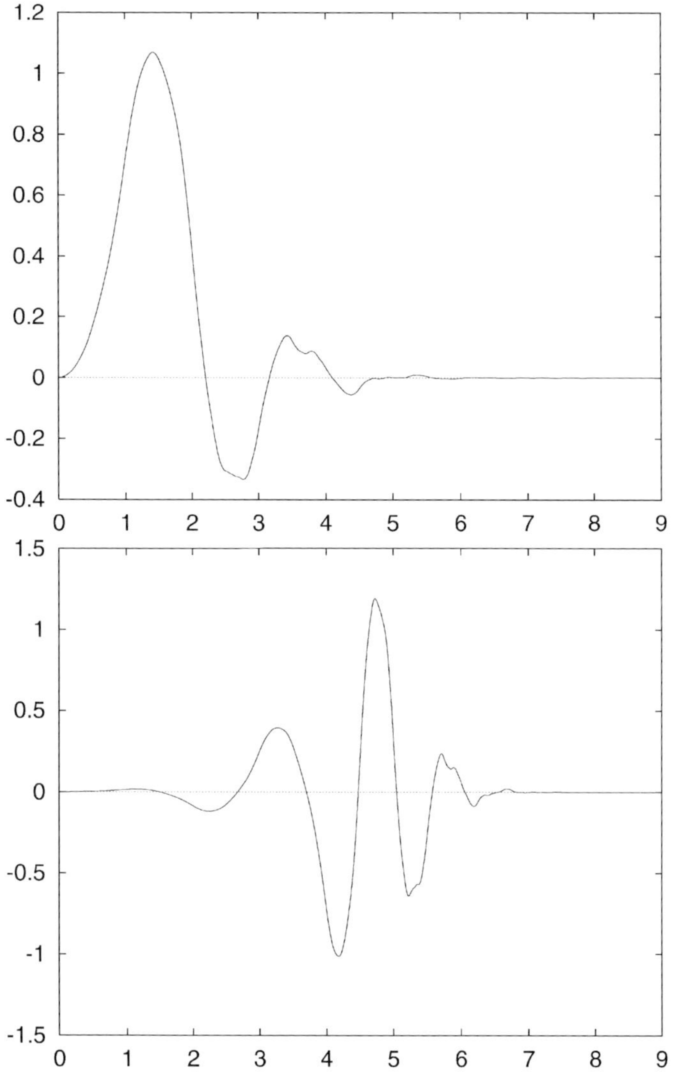

FIG. 18.4. Orthonormal scaling function and wavelet for ($N = 5$).

given by

$$s_\infty(\varphi_N) \geqslant N\left(1 - \frac{\log 3}{2\log 2}\right) \simeq 0.2\,N. \tag{18.29}$$

As an example, we display on Figs. 18.3 and 18.4 the graphs of the functions φ_2 (of Hölder regularity $\mu \simeq 0.55$) and φ_5 (of Hölder regularity $\mu \geqslant 2$) together with their associated wavelets ψ_2 and ψ_5.

19. Wavelets and splines

Spline functions are of common use in the approximation of functions. They are particularly attractive because of their numerical simplicity. We shall now discuss different examples of multiscale decompositions into such functions. Here, we shall denote by φ_N the B-spline of degree $N-1$, i.e.

$$\varphi_N(x) = B_{N-1} = (*)^N \chi_{[0,1]} \tag{19.1}$$

and we denote by

$$m_N(\omega) = \left(\frac{1+\mathrm{e}^{-\mathrm{i}\omega}}{2}\right)^N \tag{19.2}$$

its associated symbol. We finally denote by V_j^N the spline approximation space generated by the functions $\{(\varphi_N)_{j,k}\}_{k\in\mathbb{Z}}$.

Example 19.1. Orthonormal and interpolatory spline wavelets

We recall from Corollary 10.2 that we can build an orthonormal scaling function φ_N^o, applying formula (10.34) to φ_N. For $N>1$, the function φ_N^o is not compactly supported but has exponential decay at infinity and satisfies a refinement equation

$$\varphi_N^o(x) = \sum_{n\in\mathbb{Z}} h_n \varphi_N^o(2x-n), \tag{19.3}$$

where the coefficients $h_n = 2\langle\varphi_N^o(x), \varphi_N^o(2x-n)\rangle$ also have geometric decay at infinity. One way to estimate the coefficients h_n numerically is by computing the Fourier series $H_N(\omega) = \frac{1}{2}\sum_{n\in\mathbb{Z}} h_n \mathrm{e}^{-in\omega}$: combining (10.34) together with $\hat{\varphi}_N(\omega) = m_N(\omega/2)\hat{\varphi}_n(\omega/2)$ and $\hat{\varphi}_N^o(\omega) = H_N(\omega/2)\hat{\varphi}_n^o(\omega/2)$ we obtain

$$H_N(\omega) = \left(\frac{S_{\varphi_N}(\omega)}{S_{\varphi_N}(2\omega)}\right)^{1/2} m_N(\omega). \tag{19.4}$$

We also have

$$\begin{aligned} S_{\varphi_N}(\omega) &= \sum_{n\in\mathbb{Z}} \left|\hat{\varphi}_N(\omega+2n\pi)\right|^2 = \left[2\sin^2(\omega/2)\right]^N \sum_{n\in\mathbb{Z}} \frac{1}{|\omega+2n\pi|^{2N}} \\ &= \left[2\sin^2(\omega/2)\right]^N R_N(\omega), \end{aligned}$$

where R_N can be computed by induction since $R_1(\omega) = [2\sin^2(\omega/2)]^{-1}$ (due to the orthonormality of φ_1) and

$$R_{N+1}(x) = \frac{R_N''(x)}{2N(2N+1)}. \tag{19.5}$$

Orthonormal wavelets ψ_N^o can then be derived according to the formulae of Section 14, which give here

$$\psi_N^o(x) = \sum_{n\in\mathbb{Z}} (-1)^n h_{1-n} \varphi_N^o(2x - n). \tag{19.6}$$

This construction was introduced by LEMARIE [1988]. Note that for practical computations, the sequence h_n need to be truncated and the resulting error has to be taken into account.

In similar way, for even values of N (i.e. for spline of odd order $N-1$), the interpolatory function φ_N^i, defined by applying (10.36) to φ_N, also satisfies a refinement equation involving an infinite number of coefficients (when $N > 2$) that can be estimated through the computation of their Fourier series. The corresponding wavelet is then simply given by $\psi_N^i(x) = \varphi_N^i(2x-1)$, as in the case of the Schauder hierarchical basis (and of all wavelets generated from interpolatory scaling functions).

Example 19.2. Biorthogonal spline wavelets

In order to obtain local multiscale decompositions into spline functions, one needs to give up on orthonormality and work in a biorthogonal setting. The construction of compactly supported refinable functions that are dual to the functions φ_N was addressed in COHEN, DAUBECHIES and FEAUVEAU [1992]. According to the results of Section 17, this construction amounts in finding trigonometric polynomial solutions $\tilde{m}$ to the equation

$$\tilde{m}(\omega)\overline{m_N(\omega)} + \tilde{m}(\omega+\pi)\overline{m_N(\omega+\pi)} = 1, \tag{19.7}$$

and to study the regularity of the corresponding $\tilde{\varphi}$. A family of solution can be immediately derived using the polynomial P_N defined by (18.11): for $L \geqslant 0$ such that $N + L = 2M \in 2\mathbb{Z}$, we have

$$\cos^{2M}(\omega/2) P_M\big(\sin^2(\omega/2)\big) = \mathrm{e}^{\mathrm{i}\frac{N-L}{2}\omega}\left(\frac{1+\mathrm{e}^{-\mathrm{i}\omega}}{2}\right)^L P_M\big(\sin^2(\omega/2)\big)\overline{m_N(\omega)},$$

which implies, since P_N is a solution of (18.3), that the trigonometric polynomials

$$\tilde{m}_{N,L} = \mathrm{e}^{\mathrm{i}\frac{N-L}{2}\omega}\left(\frac{1+\mathrm{e}^{-\mathrm{i}\omega}}{2}\right)^L P_{\frac{N+L}{2}}\big(\sin^2(\omega/2)\big), \tag{19.8}$$

are solutions of (19.7). We denote by $\tilde{\varphi}_{N,L}$ the corresponding scaling functions. The following result gives a criterion for the choice of L in order to build a regular dual function.

THEOREM 19.1. *The functions $\tilde{\varphi}_{N,L}$ satisfy*

$$s_\infty(\tilde{\varphi}_{N,M}) \geqslant \left(L - (N+L-2)\frac{\log 3}{2\log 2}\right) \simeq 0.2\,L - 0.8\,N + 1.6. \tag{19.9}$$

Consequently, their regularity can be made arbitrarily high by increasing the value of L for a fixed value of N. If

$$L > \left(1 - \frac{\log 3}{2\log 2}\right)^{-1}\left(\tfrac{1}{2} + (N-2)\frac{\log 3}{2\log 2}\right), \tag{19.10}$$

the function $\tilde{\varphi}_{N,L}$ is in L^2 and forms a dual pair together with φ_N.

PROOF. By a similar reasoning as in the proof of Theorem 18.1, we obtain the estimate

$$d_2(\tilde{\varphi}_{N,L}) \leqslant \left(\frac{N+L}{2} - 1\right)\frac{\log 3}{\log 2}, \tag{19.11}$$

which yields (19.14) by $s_\infty(\tilde{\varphi}_{N,L}) \geqslant L - d_2(\tilde{\varphi}_{N,L})$. In particular, (19.10) implies that $d_2(\tilde{\varphi}_{N,M}) < L - 1/2$, so that we obtain a dual pair by Theorem 17.3. □

REMARK 19.1. The estimate (19.10) of the minimal value of L is crude and can be improved by operator based techniques (see COHEN and DAUBECHIES [1992]) for small values of N. In particular, the minimal values of L for $N = 2, 3, 4$ are respectively $L = 2, 3, 6$.

From φ_N and $\tilde{\varphi}_{N,L}$ we can then obtain the wavelets $\psi_{N,L}$ and $\tilde{\psi}_{N,L}$ applying formula (14.7). As an example, we display on Fig. 19.1 the graphs of $\tilde{\varphi}_{2,L}$ dual to the (centered) hat function $\varphi_2 = B_1(x+1) = \max\{1 - |x|, 0\}$, and of the wavelets $\psi_{2,2}$ and $\tilde{\psi}_{2,2}$.

REMARK 19.2. From the definition of m_N and $\tilde{m}_{N,L}$, we see that both scaling coefficients of φ_N and $\tilde{\varphi}_{N,L}$ are dyadic numbers, i.e. numbers of the type $2^{-j}k$, $j, k \in \mathbb{Z}$. This property can be of particular interest in the practical implementation of the decomposition and reconstruction algorithms, since such dyadic numbers have by definition a finite binary expansion without round-off error.

REMARK 19.3. From the expression of $\tilde{m}_{N,L}$, we can obtain the recursion formula

$$\tilde{\varphi}_{N,L} = \tilde{\varphi}_{N,L-1} * \chi_{[0,1]}, \tag{19.12}$$

similar to the formula for the B-spline. More generally, if φ and ϕ are refinable functions with symbols $m(\omega)$ $M(\omega) = ((1 + \mathrm{e}^{-\mathrm{i}\omega})/2)m(\omega)$, we have $\phi = \varphi * \chi_{[0,1]}$ which can also be expressed by a differential relation

$$\phi'(x) = \varphi(x) - \varphi(x-1). \tag{19.13}$$

One can then easily check the following facts which were firstly remarked in LEMARIE [1992]): if $\tilde{\phi}$ is a dual function to ϕ with symbol $\widetilde{M}(\omega)$, then $\tilde{\varphi}$ defined through the symbol $\tilde{m}(\omega) = ((1 + \mathrm{e}^{\mathrm{i}\omega})/2)\widetilde{M}(\omega)$ satisfies $\tilde{\varphi}'(x) = \tilde{\phi}(x+1) - \tilde{\phi}(x)$, and it constitutes a

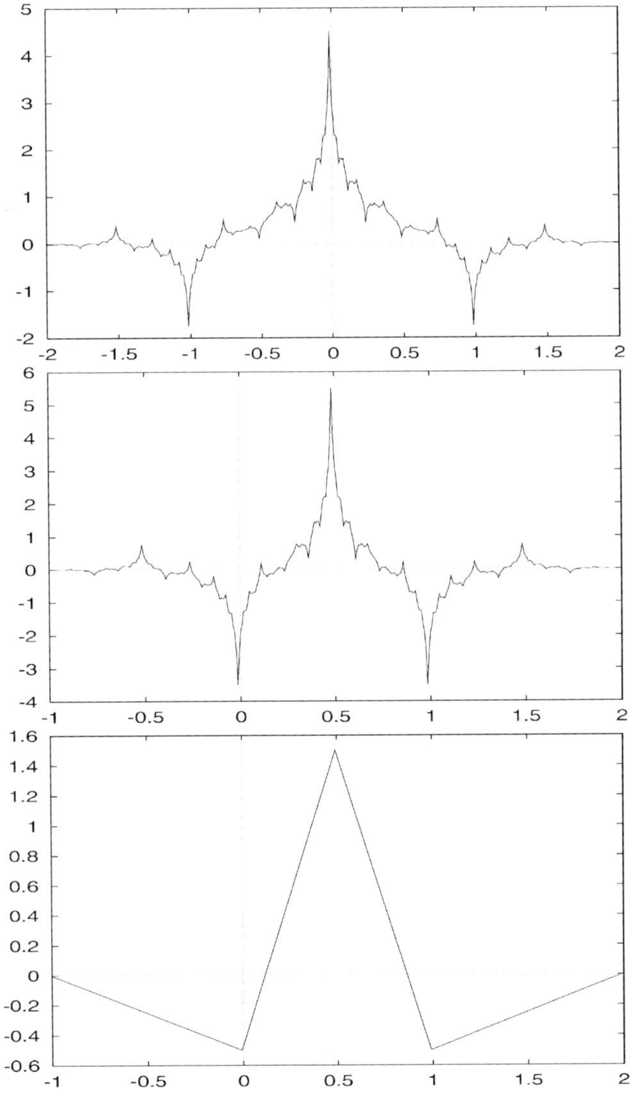

FIG. 19.1. The functions $\tilde{\varphi}_{2,2}$, $\tilde{\psi}_{2,2}$ and $\psi_{2,2}$.

dual scaling function to φ. The associated wavelets are also linked by even simpler relations: if ψ, $\tilde{\psi}$, η and $\tilde{\eta}$ denote the wavelets constructed from the functions φ, $\tilde{\varphi}$, ϕ and $\tilde{\phi}$, we have $\psi = \eta'$ and $\tilde{\eta} = -\psi'$. Finally, the associated projectors satisfy *commutation formulae* of the following type:

$$\left[\sum_k \langle f, \tilde{\phi}_{j,k}\rangle \phi_{j,k}\right]' = \sum_k \langle f', \tilde{\varphi}_{j,k}\rangle \varphi_{j,k}, \tag{19.14}$$

and

$$\Big[\sum_k \langle f, \tilde{\eta}_{j,k}\rangle \eta_{j,k}\Big]' = \sum_k \langle f', \tilde{\psi}_{j,k}\rangle \psi_{j,k}. \tag{19.15}$$

These commutation formulae play a crucial role in certain multivariate constructions where an additional differential constraint is imposed, such as divergence-free wavelets in LEMARIE [1992] and URBAN [1995] or multiscale approximation spaces that satisfy the so-called LBB condition in DAHMEN, KUNOTH and URBAN [1996] for the discretization of velocity and pressure in the mixed formulation of the Stokes problem.

Applied to our biorthogonal splines they mean that the derivative of the oblic projection of f onto V_j^N, relative to the dual index L, can be identified to the oblic projection of f' onto V_j^{N-1}, relative to the index $L+1$.

REMARK 19.4. Figure 19.1 reveals that the function $\psi_{2,2}$ can be obtained from the hierarchical basis generator $\psi = \varphi_2(2x-1)$ by a *coarse grid correction*, i.e. combination of scaling functions at scale $j = 0$. Precisely, we have

$$\psi_{2,2}(x) = 2\psi(x) - \big[\varphi_2(x) - \varphi(x-1)\big]/2. \tag{19.16}$$

More generally, if one is given a set of compactly supported biorthogonal scaling functions and wavelets $(\varphi, \tilde{\varphi}, \psi, \tilde{\psi})$, a new biorthogonal system $(\varphi^n, \tilde{\varphi}^n, \psi^n, \tilde{\psi}^n)$ can be defined (at least in a distributional sense) using such a correction technique. The new wavelet is given by

$$\psi^n(x) = \psi(x) - \sum_{n=N_0}^{N_1} c_n \varphi(x-n), \tag{19.17}$$

while the scaling function is left unchanged, i.e. $\varphi^n = \varphi$. On the dual side, this correspond to a modification of the dual scaling function which is now solution of

$$\tilde{\varphi}^n(x) = \sum_n \Big[\tilde{h}_n + \sum_k c_k \tilde{g}_{2k+n}\Big] \tilde{\varphi}^n(2x-n), \tag{19.18}$$

where $\tilde{h}_n$ and $\tilde{g}_n$ are the scaling coefficients associated to $\tilde{\varphi}$ and $\tilde{\psi}$. The new dual wavelet is given by the same equation $\tilde{\psi}^n(x) = \sum_n \tilde{g}_n \tilde{\varphi}^n(2x-n)$.

This procedure, known as *lifting scheme* was introduced in SWELDENS [1996]. It ensures that the new dual scaling functions and wavelets have also compact support (since $\tilde{\varphi}^n$ is solution of (19.18)). In practice, one often designs the coefficients c_n so that the new wavelet basis has better properties than the old one, e.g., more regularity on the dual side. This is exactly what happens in the case of linear splines that we have discussed: the new dual function $\tilde{\varphi}_{2,2}$ is square integrable (it also has some Sobolev smoothness), in contrast to the Dirac function associated to the decomposition in the hierarchical basis. The main interest of the lifting scheme is its flexibility. In particular, one can easily adapt the above correction process to the hierarchical basis restricted to a bounded interval with some prescribed boundary conditions. Consider, e.g., the spaces V_j^D introduced in Section 3 which are generated by the functions $(\varphi_2)_{j,k}$, $k = 1, \dots, 2^j - 1$, and incorporate the homogeneous Dirichlet conditions $f(0) = f(1) = 0$. For $j \geqslant 2$, we can adapt the correction procedure near the boundary to produce a basis for a complement of V_j^D into V_{j+1}^D in terms of the $(\psi_{2,2})_{j,k}$, $k = 1, \dots, 2^j - 2$, and two boundary adapted functions $2^{j/2}\psi^b(2^j\bullet)$ and $2^{j/2}\psi^b(2^j(1-\bullet))$, where $\psi^b := \psi - \frac{1}{2}\varphi(\bullet - 1)$, which both have one vanishing moment. It is also possible to check that the dual functions corresponding to such a correction process have some positive Sobolev smoothness.

A more general procedure of *stable completion* was introduced in CARNICER, DAHMEN and PEÑA [1996] which also authorizes to correct the function ψ with fine grid functions $\varphi(2\bullet - n)$, resulting however in dual functions which are not compactly supported. In Section 21, we shall describe these procedures in a very general framework of discrete multiresolution decompositions which allows to consider unstructured discretizations, and we shall also use them to build wavelet bases in finite element spaces.

In the next two sections, we discuss examples of wavelets which are built directly as combinations of the fine grid functions $\varphi(2\bullet - n)$ and are instances of stable completion procedures. In particular, these wavelet bases have infinitely compactly supported dual, and the analysis of the properties of these bases are usually more difficult (in particular for Example 19.4, despite the apparent simplicity of the construction). Note however that an infinite support for the dual function is not a real limitation for several practical applications where only the primal basis is needed. This is the case, in particular, when solving a PDE by a Galerkin method, since one uses the primal basis both for expressing the solution and testing the equation.

Example 19.3. Semi-orthogonal spline wavelets

A possible approach to the construction of spline wavelet bases consists in looking directly for a compactly supported function $\psi_N \in V_1^N$ orthogonal to V_0^N and such that $\psi_N(\bullet - k)$, $k \in \mathbb{Z}$, forms a Riesz basis of the orthogonal complement W_0^N of V_0^N into V_1^N.

In CHUI and WANG [1992], it was proved that

$$\psi_N(x) = 2^{1-N} \sum_{k=0}^{2N-2} (-1)^k \varphi_{2N}(k+1)\varphi_{2N}^{(N)}(2x-k), \tag{19.19}$$

is the solution of minimal support, among all generators of W_0^N.

Once we are given ψ_N, the system

$$\{\varphi_N(\bullet - k)\}_{k\in\mathbb{Z}} \cup \{(\psi_N)_{j,k}\}_{j\geqslant 0, k\in\mathbb{Z}}, \tag{19.20}$$

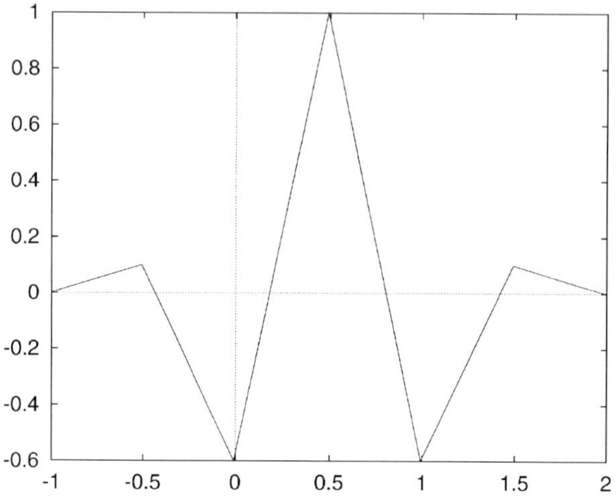

FIG. 19.2. Semi-orthogonal spline wavelet.

is easily proved to be a Riesz basis for the whole of L^2, since the stability across different level is ensured by the orthogonality of the W_j spaces. The biorthogonal system is then given by

$$\{\varphi_N^d(\bullet - k)\}_{k\in\mathbb{Z}} \cup \{(\psi_N^d)_{j,k}\}_{j\geqslant 0, k\in\mathbb{Z}}, \tag{19.21}$$

where φ_N^d and ψ_N^d are constructed by applying formula (10.35) to φ_N and ψ_N. We have thus gained compact support of the basis function and orthogonality across levels, up to the price of having a noncompactly supported dual system.

As an example, we display on Fig. 19.2 the function ψ_2 which is is given by $\psi_2 := \varphi(2\bullet -1) - \frac{3}{5}(\varphi(2\bullet) + \varphi(2\bullet -2)) + \frac{1}{10}(\varphi(2\bullet +1) + \varphi(2\bullet -3))$. Similarly to the previous example, we can adapt this construction near the boundary in the case of multiresolution spaces on a bounded interval with prescribed boundary conditions. Consider again the spaces V_j^D: it is easily seen that for $j \geqslant 2$ a Riesz basis for the orthogonal complement of V_j^D into V_{j+1}^D is given by the $(\psi_2)_{j,k}$, $k = 1, \dots, 2^j - 2$, and two boundary adapted functions $2^{j/2}\psi^b(2^j\bullet)$ and $2^{j/2}\psi^b(2^j(1-\bullet))$, where $\psi^b := \frac{9}{10}\varphi(2\bullet -1) - \frac{3}{5}\varphi(2\bullet -2) + \frac{1}{10}\varphi(2\bullet -3)$.

Example 19.4. Fine grid correction spline wavelets

If we drop the requirement of orthonormality, we can address the more general problem of constructing a compactly supported nonorthogonal wavelet basis by choosing

$$\psi_N(x) = \sum_{n=N_1}^{N_2} a_n \varphi_N(2s - n), \tag{19.22}$$

where a_n is an "oscillating" finite sequence (i.e. $\sum g_n = 0$).

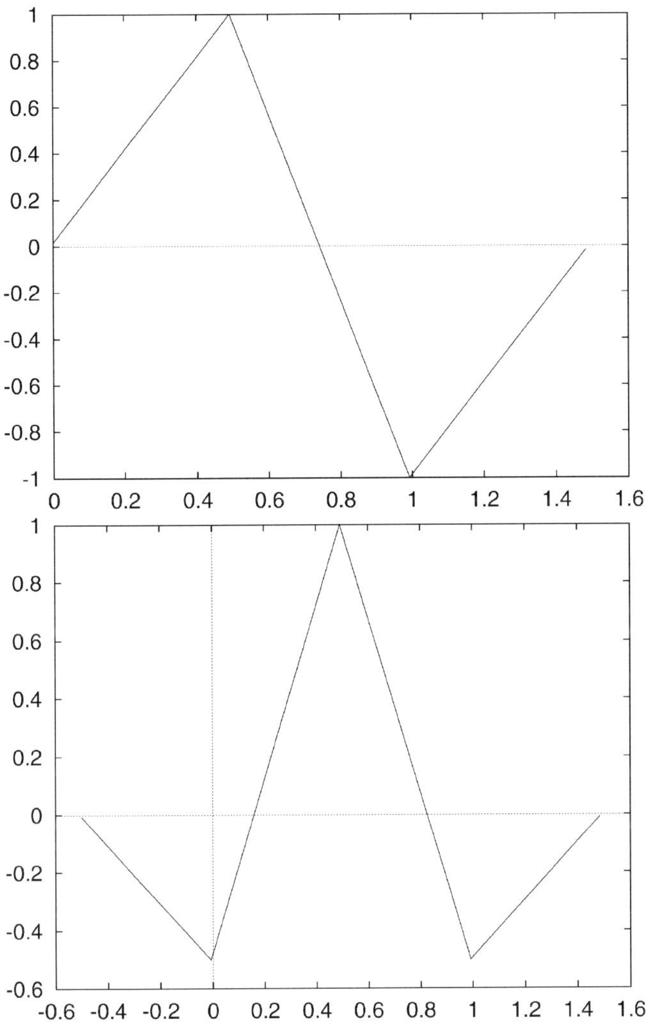

FIG. 19.3. Nonorthogonal spline wavelets.

So far, the general problem of finding a minimal sequence g_n such that the system (19.18) (with ψ_N now defined by (19.19)) is a Riesz basis of $L^2(\mathbb{R})$ is unsolved. In practice, we have in mind simple sequences such as $(a_0, a_1) = (1, -1)$ or $(a_0, a_1, a_2) = (-1, 2, -1)$, etc. This problem was studied in the case of linear splines, i.e. $N = 2$, in COHEN and DAUBECHIES [1996] and OSWALD and LORENTZ [1996] where it is proved that both choices give rise to a multiscale Riesz basis. As in the previous construction, the biorthogonal system is not compactly supported. We display these nonorthogonal wavelets on Fig. 19.3.

More generally, in order to study the stability of the multiscale basis generated by ψ_N, we note that (19.22) rewrites

$$\hat{\psi}_N(\omega) = a(\omega/2)\hat{\varphi}_N(\omega/2), \tag{19.23}$$

where $a(\omega) = \frac{1}{2}\sum_n a_n e^{-in\omega}$. In the case where the construction of $\tilde{\psi}_N$ is related to a dual refinable function (e.g., the biorthogonal spline of Example 19.2), we know that $a(\omega)$ is related to the dual symbol through

$$a(\omega) = e^{-i\omega}\overline{\tilde{m}(\omega+\pi)}. \tag{19.24}$$

The biorthogonality constraint (19.7) leads to the equation

$$e^{i\omega}\big[m_N(\omega)a(\omega+\pi) - m_N(\omega+\pi)a(\omega)\big] = 1, \tag{19.25}$$

that should be solved by $a(\omega)$. This equation is a strong restriction on $a(\omega)$. However, we remark that if $p(\omega)$ is a 2π periodic continuous function that does vanish, then the function $\eta_N(x)$ defined by

$$\hat{\eta}_N(\omega) = p(\omega)\hat{\psi}_N(\omega) = p(\omega)a(\omega/2)\hat{\varphi}_N(\omega/2), \tag{19.26}$$

has the property that $(\psi_N)_{j,k}$, $k \in \mathbb{Z}$, is a Riesz basis of a space W_j if and only if $(\eta_N)_{j,k}$, $k \in \mathbb{Z}$, is a Riesz basis for the same space. We can thus relax Eq. (19.25) by only requiring that it is satisfied by

$$b(\omega) = p(2\omega)a(\omega) \tag{19.27}$$

for some nonvanishing continuous function $p(\omega)$. Note that by combining (19.27) and (19.25), we obtain an explicit expression for $p(\omega)$ as

$$p(2\omega) = e^{-i\omega}\big[m_N(\omega)a(\omega+\pi) - m_N(\omega+\pi)a(\omega)\big]^{-1}, \tag{19.28}$$

and for the dual symbol $\tilde{m}(\omega)$ as

$$\tilde{m}(\omega) = e^{-i\omega}\overline{b(\omega+\pi)} = \frac{\overline{a(\omega+\pi)}}{\overline{m}_N(\omega)\overline{a}(\omega+\pi) - \overline{m}_N(\omega+\pi)\overline{a}(\omega)}. \tag{19.29}$$

The stability of the wavelet basis generated by ψ_N will thus hold if $p(\omega)$ defined by (19.28) is bounded, and if the dual symbol $\tilde{m}(\omega)$ defined by (19.29) satisfies together with $m(\omega)$ the conditions of Theorem 17.3 for generating a pair of dual refinable functions with $\tilde{\varphi} \in L^2$. Since $\tilde{m}$ is not in general a trigonometric polynomial, the dual system to $(\psi_N)_{j,k}$ is not compactly supported.

Example 19.5. Multiwavelets

Multiwavelets are obtained from a natural generalization of multiresolution analysis, where one allows V_0 to be generated by more than one scaling function, i.e. there exists $\varphi_1, \ldots, \varphi_r \in V_0$ such that

$$\varphi_q(\bullet - k), \quad k \in \mathbb{Z},\ q = 1, \ldots, r, \tag{19.30}$$

constitutes a Riesz basis of V_0. The sequence V_j is then called *multiresolution analysis of multiplicity* r. Many of the theoretical results that we have described in Sections 11–17 can be extended to this setting: JIA and LEI [1993] describes the appropriate Strang–Fix conditions, PLONKA [1995] gives the generalization of the factorization of the symbol $m(\omega)$ (which is now a matrix valued trigonometric polynomial), COHEN, DAUBECHIES and PLONKA [1997], SHEN [1998] and JIA, RIEMENSCHNEIDER and ZHOU [1999] generalize the smoothness analysis of for the scaling function vector $(\varphi_1, \ldots, \varphi_r)$.

Several interesting examples of multiwavelets involve spline functions.

First, if V_j consists of the functions which are piecewise polynomial of degree N on the intervals $I_{j,k}$, $k \in \mathbb{Z}$, *with no continuity requirement*, a simple orthonormal basis has the form (19.27) with

$$\varphi_q = \sqrt{2} L_q(2x - 1) \chi_{[0,1]}, \tag{19.31}$$

where L_q is the L^2-normalized Legendre polynomial of degree q. In this case, the derivation of orthonormal wavelets (which are also supported on the unit interval) is very simple. These particular wavelets have been introduced in ALPERT [1993] for the purpose of discretizing integral equation that do not require smoothness of the basis functions. One of their interest is their flexibility: since orthogonality is ensured by disjointness of the supports, one can easily adapt them to unstructured discretizations and to triangulations in higher dimensions. Moreover the derivation of orthogonal wavelets reduces to local problems on each support.

More generally, splines with multiple knots (i.e. piecewise polynomials of degree N on which we do not impose up to C^{N-1} smoothness) can be put in this setting. A very simple example is the case of piecewise cubic C^1 functions, for which a natural basis is given by the nodal functions of Hermite interpolation, i.e.

$$\varphi_1(k) = \varphi_2'(k) = \delta_{0,k} \quad \text{and} \quad \varphi_2(k) = \varphi_1'(k) = 0. \tag{19.32}$$

However, the construction of stable dual refinable functions for this type of bases is still an open problem. Recently, the particular case of cubic Hermite spline was treated in DAHMEN, HAN, JIA and KUNOTH [2000] and a general approach to solve this problem was proposed in DYN [2000].

Finally, it was shown in DONOVAN, GERONIMO and HARDIN [1999] that it is possible to construct a set of spline scaling functions $\varphi_1, \ldots, \varphi_q$ with compact support such that (19.30) is an orthonormal basis for a spline space V_0^i which is *intertwined* between

V_n and V_{n+1} for some $n > 0$ (where V_j is a standard spline multiresolution analysis). Orthonormality can thus be obtained together with compact support, up to a slight modification of the multiresolution approximation spaces.

20. Multivariate bounded domains and boundary conditions

In this section we shall discuss multiresolution decompositions in the multivariate setting, as well as some commonly used strategies to adapt them to a bounded domain $\Omega \in \mathbb{R}^d$. Another frequently used approach, based on multilevel splitting of finite element spaces, will be discussed in Section 21.

The construction of multiresolution analysis and wavelets that we have described for functions defined on $\mathbb{R}$ can easily be extended to multivariate functions on the whole of $\mathbb{R}^d$. The simplest method is by far the *tensor product* strategy that we already discussed in the setting of the basic examples of the previous chapter (Section 4).

In contrast, the adaptation to a bounded domain and the prescription of boundary condition is a delicate task. Several strategies have been proposed to deal with this problem, but so far, none of them has received a full approval in the sense that their efficiency seems to be highly dependent of the type of application.

In a general setting, let us first recall the main principles of the tensor product construction: from a univariate scaling function φ, we define its multivariate version

$$\phi(x) = \varphi(x_1) \cdots \varphi(x_d), \quad \text{for } x = (x_1, \dots, x_d) \in \mathbb{R}^d. \tag{20.1}$$

The multivariate space V_j is then generated by the functions

$$\phi_{j,k}(x) = 2^{dj/2}\phi(2^j - k), \quad k \in \mathbb{Z}^d. \tag{20.2}$$

Given a dual function $\tilde{\varphi}$, we define $\tilde{\phi}$ dual to ϕ in a similar way, which yields the L^2-bounded projectors defined by

$$P_j f = \sum_{k \in \mathbb{Z}^d} \langle f, \tilde{\phi}_{j,k} \rangle \phi_{j,k}. \tag{20.3}$$

In order to characterize the details, we introduce $2^d - 1$ wavelet functions

$$\psi^{\varepsilon}(x) = \psi^{\varepsilon_1}(x_1) \cdots \psi^{\varepsilon_d}(x_d), \quad \varepsilon = (\varepsilon_1, \dots, e_d) \in \{0,1\}^d - \{0\}, \tag{20.4}$$

where we have set $\psi^0(x) = \varphi(x)$ and $\psi^1(x) = \psi(x)$. We define similarly the dual wavelets $\tilde{\psi}^{\varepsilon}$ so that we have

$$Q_j f = P_{j+1} f - P_j f = \sum_{\varepsilon \in \{0,1\}^d - \{0\}} \sum_{k \in \mathbb{Z}^d} \langle f, \tilde{\psi}^{\varepsilon}_{j,k} \rangle \psi^{\varepsilon}_{j,k}, \tag{20.5}$$

where $\psi^{\varepsilon}_{j,k}$ and $\tilde{\psi}^{\varepsilon}_{j,k}$ are defined by a similar scaling as in (20.2). On a computational point of view, the multiscale decomposition and reconstruction algorithms are performed in a separable manner, applying at each stage $j \to j+1$ in the reconstruction (resp. $j+1 \mapsto j$ in the decomposition) the filters h_n and g_n after upsampling (resp. $\tilde{h}_n$ and $\tilde{g}_n$ before downsampling) successively in each direction.

REMARK 20.1. As we already remarked in Section 4, an alternate tensor product multiscale decomposition makes use of the *full tensor product system* defined by

$$\psi_{j,k}(x_1,\dots,x_d) = \psi_{j_1,k_1}(x_1)\cdots\psi_{j_d,k_d}(x_d), \tag{20.6}$$

where $j = (j_1,\dots,j_d)$ and $k = (k_1,\dots,k_d)$. If $j_{\min} < j_{\max}$, using the convention $\psi_{j,k} = \varphi_{j_{\min},k}$ for $j = (j_{\min}-1,\dots,j_{\min}-1)$, we obtain a multiscale decomposition of a function $f \in V_{j_{\max}}$ into

$$f = \sum_{j_{\min}-1\leqslant j_1,\dots,j_d<j_{\max}} \sum_{k\in\mathbb{Z}^d} d_{j,k}\psi_{j,k}, \tag{20.7}$$

where $d_{j,k} = \langle f, \tilde{\psi}_{j,k}\rangle$, with $\tilde{\psi}_{j,k}$ defined in an analogous manner. The associated decomposition and reconstruction algorithm consist in applying the univariate algorithms between the extreme values of the scale, successively in each directions.

An advantage of the tensor product construction is that it inherits all the properties of the associated univariate multiscale decomposition: smoothness and local support of the basis functions, fast $\mathcal{O}(N)$ transformation algorithms, polynomial reproduction properties (if the univariate φ satisfies the Strang–Fix conditions at the order L, then the multivariate polynomials $x^k = x_1^{q_1}\cdots x_d^{q_d}$, $q_1,\dots,q_d \leqslant L$, are reproduced by the projectors P_j).

However, this construction excludes some very classical approximation spaces such as $\mathbb{P}_1$ finite elements defined on a regular grid: if we consider the space V_0 consisting of all bivariate continuous L^2 functions that are affine on the triangles $[(m,n),(m+1,n),(m,n+1)]$ and $[(m,n),(m-1,n),(m,n-1)]$ and its scaled versions $V_j = \{f;\ f(2^{-j}\bullet) \in V_0\}$, we find that the spaces V_j form an embedded sequence of approximation spaces with Riesz bases $\phi_{j,k}$, $k \in \mathbb{Z}^2$, where the generator ϕ is the hat function

$$\phi \in V_0, \quad \phi(m,n) = \delta_{0,m}\delta_{0,n}, \tag{20.8}$$

which is not of tensor product type. This very simple example motivates the study of multiresolution spaces generated by nontensor product functions.

On an algorithmic point of view, this means that we are using nonseparable filters in a multivariate generalization of subband decompositions that were described in Section 14, but we are keeping the same type of downsampling and upsampling operations (one sample out of two in each direction) which can be represented by the dilation matrix $2I$ that maps $2^{-j}\mathbb{Z}^d$ to $2^{-j+1}\mathbb{Z}^d$. Another generalization, which was essentially considered in multivariate signal processing, consists in using for these operations a dilation matrix D that differs from $2I$. In order to define proper downsampling and upsampling operators, this matrix should have integer entries (in order to map $\mathbb{Z}^d$) into itself and be *expanding*, i.e. all eigenvalues λ_i of D satisfy $|\lambda_i| > 1$. It is thus natural to introduce the following generalization of Definition 10.1.

DEFINITION 20.1. A multiresolution analysis of $L^2(\mathbb{R}^d)$ is a ascending dense sequence of closed approximation subspaces

$$\{0\} \cdots \subset V_{-1} \subset V_0 \subset V_1 \subset \cdots L^2(\mathbb{R}^d),$$

that satisfies the scaling relation

$$f \in V_j \Leftrightarrow f(D\bullet) \in V_{j+1}, \tag{20.9}$$

for a fixed expanding matrix D with integer entries, and such that V_0 is generated by a Riesz basis of the type $\{\phi(\bullet - k)\}_{k\in\mathbb{Z}^d}$, for some $\phi \in V_0$.

In this more general setting, the refinability property is expressed by a relation of the type

$$\phi(x) = \sum_{n\in\mathbb{Z}^d} h_n \phi(Dx - n), \tag{20.10}$$

and one can also require the existence of a dual scaling function $\tilde{\phi}$ with similar refinability properties. A projector P_j of the the type (20.3) is then obtained, using the functions

$$\phi_{j,k} = \left|\det(D)\right|^{j/2} \phi\left(D^j \bullet - k\right), \tag{20.11}$$

and $\tilde{\phi}_{j,k}$ defined in an analogous manner.

In the attempt to mimick step by step the construction of such refinable functions and related multiscale decompositions that we have described in the previous sections for univariate functions, one is however faced with several difficulties that we want to point out here. Firstly, the derivation of wavelets from the scaling functions is no more straightforward as in Section 14, except for some particular choice of dilation matrices or refinable functions. Secondly, the relations between the cancellations of the symbol $m(\omega_1, \ldots, \omega_d)$ and the polynomial reproduction properties of the shifts of ϕ, expressed by the Strang–Fix conditions in Section 16, are more intricate. Thirdly, the smoothness analysis cannot be done in the same way as we presented in Section 15 since there is no simple multivariate analog to the factorized form $m(\omega) = ((1 + \mathrm{e}^{-\mathrm{i}\omega})/2)^L p(\omega)$. Finally, for the design of orthonormal scaling function, there is no analogous result to the Riesz Factorisation Lemma 18.3.

Concerning the second and the third difficulties, the reader will find in COHEN, GRÖCHENIG and VILLEMOES [1999] and JIA [1999], other techniques, involving transition operators, to measure the regularity without the help of a factorization of $m(\omega)$. Nevertheless, these difficulties explain the small number of existing examples of nonseparable multiscale decompositions that are of practical use. In the context of numerical simulation, the most relevant nonseparable constructions are related to finite element spaces. They are discussed in the next section. Generally speaking, the tensor product construction remains by far the simplest way to generalize multiscale decomposition to functions defined on $\mathbb{R}^d$.

In most numerical applications, one is faced with the task of decomposing a function which is defined on a *bounded domain* $\Omega \subset \mathbb{R}^d$. We thus need to adapt the ideas of multiscale approximation and decomposition to this setting. A possibility is clearly to extend a function f defined on Ω to the whole of $\mathbb{R}^d$, and apply the techniques that we have described for $L^2(\mathbb{R}^d)$. However, the effectiveness of such a method is strongly dependent on the type of extension which is used: a simple extension by zero, or by periodization in the case of tensor product domains, has the effect of generating an artificial discontinuity on the boundary of Ω even when f is smooth. As a consequence, many wavelets will be needed to resolve this singularity and this might ruin the approximation process. A "smoother" extension process (in the sense that it preserves Hölder continuity of order $\mu < 1$) based on symmetrization can be performed in the case of tensor product domains (see COHEN, DAUBECHIES and VIAL [1993]): such a process is frequently used in image processing where the generation of a singularity in the normal derivative is acceptable. Smoother extensions, although they are known to exist for general Lipschitz domains (STEIN [1970]) are somehow more difficult to perform numerically, making unrealistic this approach for the effective numerical computations of multiscale decompositions.

A more appropriate solution is to build nested approximation spaces V_j^Ω of $L^2(\Omega)$ that reproduce the main features that we have encountered throughout this chapter, in particular the existence of a local basis with a simple structure, satisfying stability, smoothness and polynomial reproduction properties, as well as of a dual basis yielding local projectors P_j and fast decomposition algorithms. An additional difficulty that might arise is the need to incorporate homogeneous boundary conditions in the construction of V_j^Ω.

A very natural strategy for the construction of such spaces is to use a basis that consists of those multivariate functions $\phi_{j,k}$ of (20.2) or (20.8) that are supported in the interior of Ω and of a set of adapted basis functions that are localized against the boundary. The main interest of this approach is the following: in an interior region Ω_j which is distant of $\partial\Omega$ of $C2^{-j}$ for some constant $C > 0$ independent of j, the analysis of a function at resolution 2^{-j} operates exactly as if the function was defined on the whole of $\mathbb{R}^d$. In particular, the quadrature rules and decomposition-reconstruction formulae are the same.

The main problem that we need to address is thus the construction of the basis functions near the boundary. A simple possibility is to generate V_j^Ω by all restrictions to Ω of the basis functions which have a support that is not disjoint with Ω. This choice amounts in defining V_j^Ω as the restriction to Ω of the functions in V_j, i.e.

$$V_j^\Omega = \{ f \in L^2(\Omega) \text{ such that } f = g|_\Omega,\ g \in V_j \}. \tag{20.12}$$

This choice, which was firstly studied in MEYER [1991] in the case of an interval, ensures trivially the nestedness of the spaces V_j.

The restrictions to Ω of the functions $\phi_{j,k}$, $k \in K_j$, where K_j is the set of indices such that $|\Omega \cap \mathrm{Supp}(\phi_{j,k})| \neq 0$, actually form a basis of V_j^{Ω}. This fact is easy to check in the case of a univariate interval I. One first remark that since the restriction does not affect the inner product of the boundary functions with the interior functions, both group generate independent spaces. Finally, the interior ones already form a basis, while the boundary functions are necessarily independent due to their support properties (we also assume here that the scale level j is larger than a minimal j_0 so that the supports of the left side and right side boundary functions do not overlap). For more general domains, this can be shown for tensor product scaling functions, invoking a stronger argument: the functions $\phi_{j,k}$ are *locally linearly independent*, i.e. if for some region $R \in \mathbb{R}^d$ of nonzero Lebesgue measure, we have

$$x \in R \Rightarrow \sum_{k \in \mathbb{Z}^d} c_k \phi_{j,k}(x) = 0, \tag{20.13}$$

then $c_k = 0$ for all k such that $|R \cap \mathrm{Supp}(\phi_{j,k})| \neq 0$. This remarkable property (which was already known for univariate spline functions) was proved in LEMARIE and MALGOUYRES [1991] in the case of an L^2-stable univariate compactly supported refinable function and can easily be extended to tensor product refinable functions (see Theorem 30.1 in the next chapter).

Note that this basis inherits many of the useful features of the basis for $L^2(\mathbb{R}^d)$: the polynomial are reproduced up to the same order, the restricted functions have the same smoothness and they are locally supported. However, the main defect of this construction resides in its lack of stability: the basis functions that overlap Ω on a small portion of Ω, have in general very bad stability constants. This defect will in general result in numerical instabilities in the process of orthonormalizing these functions or of building a proper dual basis. For this reason, the construction of V_j^{Ω} by (20.12) is avoided in most practical computations.

A variant of this construction that yields a more stable basis, was proposed in the case of an interval by COHEN, DAUBECHIES and VIAL [1993]. The basic idea to construct the boundary functions can be described as follows, in the case where $\Omega = [0, 1]$. Suppose that φ is a refinable function with compact support, i.e. $\mathrm{Supp}(\varphi) = [0, L]$ (which we can always assume up to an integer shift in the definition of φ), and such that the Strang–Fix conditions are satisfied up to order N, i.e.

$$x^q = \sum_{k \in \mathbb{Z}} \langle (\cdot)^q, \tilde{\varphi}(\cdot - k) \rangle \varphi(x - k), \quad q = 0, \ldots, N. \tag{20.14}$$

For some fixed $M_1 \geqslant 0$ and $M_2 \geqslant L$, and j large enough so that $2^{-j}(M_1 + M_2 + 2L) \leqslant 1$, we define $V_j^{[0,1]}$ as the space generated by the interior functions

$$\varphi_{j,k}, \quad M_1 \leqslant k \leqslant 2^j - M_2, \tag{20.15}$$

and the left and right edge functions $\varphi^0_{j,q}$ and $\varphi^1_{j,q}$, $q=0,\dots,N$, defined by

$$\begin{aligned}\varphi^0_{j,q}(x) &= 2^{j(1/2-q)} \sum_{k<M_1} \langle (\bullet)^q, \tilde{\varphi}_{j,k}\rangle \varphi_{j,k} \\ &= 2^{j(1/2-q)}\Big[x^q - \sum_{k\geqslant M_1} \langle (\bullet)^q, \tilde{\varphi}_{j,k}\rangle \varphi_{j,k}\Big],\end{aligned} \tag{20.16}$$

and

$$\begin{aligned}\varphi^1_{j,q}(x) &= 2^{j(1/2-q)} \sum_{k>2^j-M_2} \langle (1-\bullet)^q, \tilde{\varphi}_{j,k}\rangle \varphi_{j,k}(x) \\ &= 2^{j(1/2-q)}\Big[(1-x)^q - \sum_{k\leqslant 2^j-M_2} \langle (1-\bullet)^q, \tilde{\varphi}_{j,k}\rangle \varphi_{j,k}(x)\Big].\end{aligned} \tag{20.17}$$

Instead of taking all restriction of the basis functions that are located near the boundary, we are thus selecting the specific linear combinations of these functions that allow to reproduce polynomials up to degree N. These combinations are also independent, due to the particular structure of the coefficients: we recall from the discussion on the Strang–Fix conditions in Section 16 that

$$2^{j(1/2-q)}\langle (\bullet)^q, \tilde{\varphi}_{j,k}\rangle = k^q + P_{q-1}(k), \tag{20.18}$$

where P_{q-1} is a polynomial of degree $q-1$. The linear independence of the left side functions is then an immediate consequence of both linear independence of the polynomials $x^q + P_{q-1}(x)$, $q=0,\dots,N$, and of the restrictions of the functions $\varphi_{j,k}$, $k \leqslant M$. The same holds at the right edge. Note that the scaling factor $2^{j(1/2-q)}$ has been chosen in such way that $\|\varphi^\varepsilon_{j,q}\|_{L^2}$, $\varepsilon = 0, 1$, is independent of j.

REMARK 20.2. There exist variants to the definition of the edges functions by (20.16) and (20.17) which amount in taking other polynomial bases for the coefficients of $\varphi_{j,k}$. The above choice is particularly well fitted to impose *homogeneous boundary conditions* in the construction of $V_j^{[0,1]}$: since the functions $\varphi^0_{j,q}$ (resp. $\varphi^1_{j,q}$) have a smooth extension (up to the order of smoothness of φ) as $2^{j/2}x^q$ on $]-\infty,0]$ (resp. $2^{j/2}(1-x)^q$ on $[1,+\infty[$), we can define an version of $V_j^{[0,1]}$ that satisfies homogeneous Dirichlet conditions by removing $\varphi^0_{j,0}$ and $\varphi^1_{j,0}$ from the set of generators: since these are the only two functions that do not vanish at the boundary, the resulting space V_j^D satisfies the homogeneous Dirichlet boundary condition (we are assuming that φ is at least continuous). More generally, we can impose cancellations of order r at the endpoints by removing the r first edge functions at 0 and 1, assuming that $\varphi \in C^{r-1}$. This idea has been developed in PERRIER and MONASSE [1998] for orthonormal scaling functions, and in MASSON [1996], DAHMEN, KUNOTH and URBAN [1996] for more general biorthogonal scaling functions.

The system defined by (20.15), (20.16) and (20.17) turns out to be better adapted for orthonormalization or construction of a dual basis than the functions obtained by simple restriction, in the sense that these processes do not generate numerical unstabilities (note in particular that the edge functions have a larger support than the interior functions which exclude "small tails" that appear by restriction to the end of the support of φ). On the other hand, the nestedness property of the spaces $V_j^{[0,1]}$ is no more straightforward but still holds as shown by the following result.

THEOREM 20.1. *The spaces* $V_j^{[0,1]}$ *defined by the scaling functions* (20.15), (20.16) *and* (20.17) *are nested, i.e.* $V_j^{[0,1]} \subset V_{j+1}^{[0,1]}$.

PROOF. If $\varphi_{j,k}$ is an interior scaling function, i.e. $M_1 \leqslant k \leqslant 2^j - M_2$, its decomposition as a combination of $\varphi_{j+1,n}$ based the refinable equation, only involves those $\varphi_{j+1,n}$ with support embedded in the support of $\varphi_{j,k}$, and thus n such that $2M_1 \leqslant n \leqslant 2^{j+1} - 2M_2$, i.e. interior functions at the finer level. The interior basis functions of $V_j^{[0,1]}$ are thus contained in $V_{j+1}^{[0,1]}$.

If we now consider the left edge basis function $\varphi_{j,q}^0$ for some $q = 0, \ldots, N$, we remark that the difference

$$\begin{aligned}
&\varphi_{j,q}^0(x) - \frac{1}{\sqrt{2}}\varphi_{j+1,q}^0(x) \\
&\quad = 2^{j(1/2-q)}\Bigg[\sum_{k\geqslant M_1} \langle(\bullet)^q, \tilde{\varphi}_{j,k}\rangle \varphi_{j,k}(x) - \sum_{k\geqslant M_1} \langle(\bullet)^q, \tilde{\varphi}_{j+1,k}\rangle \varphi_{j+1,k}(x)\Bigg],
\end{aligned}$$

is identically zero for $x \geqslant (M_1 + L)2^{-j}$ where we have exactly

$$\sum_{k\geqslant M_1} \langle(\bullet)^q, \tilde{\varphi}_{j,k}\rangle \varphi_{j,k}(x) = \sum_{k\geqslant M_1} \langle(\bullet)^q, \tilde{\varphi}_{j+1,k}\rangle \varphi_{j+1,k}(x) = x^q.$$

Expanding the interior functions $\varphi_{j,k}$, $k \geqslant M_1$, in terms of the interior functions at the finer level, we thus obtain

$$\varphi_{j,q}^0 = \frac{1}{\sqrt{2}}\varphi_{j+1,q}^0 + \sum_{k=M_1}^{2M_1-1} h_{q,k}^0 \varphi_{j+1,k}, \tag{20.19}$$

for some coefficients $h_{q,k}^0$, i.e. $\varphi_{j,q}^0 \in V_{j+1}^{[0,1]}$. A similar reasoning apply to the right edge functions. □

REMARK 20.3. In the case where we start with an orthonormal scaling function φ, the left and right edge scaling functions need to be orthonormalized. The resulting orthonormal bases are presented in COHEN, DAUBECHIES and VIAL [1993], where it is also shown that a natural way to build a wavelet basis is to take some interior $\psi_{j,k}$ and construct "edge wavelets" by an orthonormalization of the functions $\varphi^{\varepsilon}_{j+1,q} - P_j\varphi^{\varepsilon}_{j+1,q}$, $\varepsilon = 0, 1$, $q = 0, \ldots, N$, where P_j is the orthonormal projector onto $V^{[0,1]}_j$.

REMARK 20.4. In the case where we start with a pair of dual scaling functions, there is much more freedom for the construction of the dual system. A basic constraint is still that the parameters (M_1, M_2) and $(\widetilde{M}_1, \widetilde{M}_2)$ (corresponding to the analog construction of $\widetilde{V}_j$) should be linked in such a way that V_j and $\widetilde{V}_j$ have the same dimension. The edge functions $\varphi^{\varepsilon}_{j,q}$, $\varepsilon = 0, 1$, and $\tilde{\varphi}^{\varepsilon}_{j,q}$, $\varepsilon = 0, 1$, should then be "biorthogonalized", i.e. recombined in such a way that they satisfy duality properties. Such a process amounts in a matrix inversion at each edges, but in contrast to the orthonormalization process of the previous remark, the matrix is not guaranteed to be invertible. In the specific case of biorthogonal spline functions (Example 19.2), the well-posedness of this approach could be proved in DAHMEN, KUNOTH and URBAN [1999]. One also has the possibility (by choosing the parameters M_1 and M_2 large enough) to build a dual basis which only consist of interior functions $\tilde{\varphi}_{j,k}$. There is also more flexibility in the derivation of biorthogonal wavelets. In particular, it was shown in COHEN and MASSON [1999] that a basis of $W_j = V_{j+1} \cap \widetilde{V}_j^{\perp}$ can obtained by taking those inner wavelets $\psi_{j,k}$ that can be expressed as combinations of the inner functions $\varphi_{j+1,k}$, $M_1 \leqslant k \leqslant 2^j - M_2$, at the next level, and the projection by $Q_j := P_{j+1} - P_j$ of an appropriate selection of inner functions $\varphi_{j+1,k}$ located near the edges onto W_j. A similar technique is used for generating $\widetilde{W}_j$ and the resulting wavelets bases are then biorthogonalized near the edges by the same process as for the scaling functions.

REMARK 20.5. The edge functions are also "refinable" in the sense that they can have a local expansion of the type (20.19) as a linear combination of the basis functions at the finer level. The coefficients $h^0_{q,k}$ can easily be derived from the standard coefficients h_n of (10.12) and the values of the polynomial coefficients in (10.18). Similarly, one derives particular $h^1_{q,k}$ at the right edges, and coefficients $\tilde{h}^{\varepsilon}_{q,k}$, $g^{\varepsilon}_{q,k}$ and $\tilde{g}^{\varepsilon}_{q,k}$ associated to dual scaling functions, wavelets and dual wavelets at each edge $\varepsilon = 0, 1$, depending on the construction which is chosen. In the fast decomposition and reconstruction algorithm, these particular coefficients correspond to a modification of the filtering process when approaching the boundary.

For a general domain $\Omega \subset \mathbb{R}^d$, the construction of multiscale decomposition is of course a more difficult task, except in the simple case of rectangular domain, i.e.

$$\Omega = I_1 \times \cdots \times I_d, \tag{20.20}$$

where we can simply apply the tensor product strategy to the constructions that we described for the interval [0, 1]. Note that this strategy also allow to prescribe specific homogeneous boundary conditions on a given side of the domain, imposing the corresponding univariate boundary condition in some of the directions.

Such a construction can be further used to treat more general domains Ω, provided that by these domains can be patched into cube-like subdomains Ω_i, $i = 1, \ldots, n$, which are the images of a reference cube $[0,1]^d$ by isoparametric maps T_i. Such maps allow to define successively multiresolution spaces $V_j^{\Omega_i}$ as the image of $V_j^{[0,1]^d}$ (defined by means of tensor product) by T_i, and V_j^{Ω} as

$$V_j^{\Omega} = C^0(\Omega) \cap \left(\bigoplus_{i=1}^{n} V_j^{\Omega_i} \right). \tag{20.21}$$

Since the maps are isoparametric, one can easily build a first basis of V_j^{Ω} by keeping all pullbacks $\varphi_{j,k}(T_i^{-1}\bullet)$ and $\varphi_{j,q}^{\varepsilon}(T_i^{-1}\bullet)$, $q \geqslant 1$, of those basis functions that vanish on the boundary of $[0,1]^d$, and by a simple adaptation on the interface: one simply glues together the symmetric pairs of scaling functions on both side of the interface associated to the edge function $\varphi_{j,0}^{\varepsilon}$, $\varepsilon = 0, 1$, in the normal direction and corresponding to the same function in the tangential direction. A similar process can be performed to construct a basis for $\widetilde{V}_j$ defined the same way as V_j.

The biorthogonalization of these basis is made with respect to a pullback inner product defined by

$$\langle f, g \rangle_P = \sum_{i=1}^{n} \int_{[0,1]^d} f(T_i x) \overline{g(T_i x)} \, \mathrm{d}x. \tag{20.22}$$

The resulting bases $(\varphi_\gamma)_{\gamma \in \Gamma_j}$ of V_j and $(\tilde{\varphi}_\gamma)_{\gamma \in \Gamma_j}$ of $\widetilde{V}_j$ define an oblic projector of the form

$$P_j f = \sum_{\gamma \in \Gamma_j} \langle f, \tilde{\varphi}_\gamma \rangle_p \varphi_\gamma. \tag{20.23}$$

This approach was proposed in DAHMEN and SCHNEIDER [1999b] and further developed in CANUTO, TABACCO and URBAN [1999] and in COHEN and MASSON [1999], where explicit construction of the biorthogonal scaling functions and wavelets are given at the interface between the sub-domains where they need to be adapted. Similarly, the biorthogonality of the wavelet bases is intended in the sense of the pullback inner product $\langle \bullet, \bullet \rangle_P$.

A serious difficulty in such domain decomposition strategies is to allow more smoothness than Lipschitz ($C^{1,1}$) continuity for the basis functions overlapping the interfaces. This is a problem when it comes to the characterization of high order smoothness from the properties of multiscale decompositions, by the techniques that will be developed in the next chapter. A solution to this problem was recently proposed in DAHMEN and SCHNEIDER [1998, 1999a], based on general results of CIESIELSKI and FIGIEL [1983] that allow to describe smoothness spaces on Ω as the direct sum of the analog spaces on the subdomains Ω_i subject to specific boundary conditions.

In summary, the construction of multiresolution approximation and wavelet bases on d-dimensional domains can be derived from the constructions on $\mathbb{R}$ by the following steps:

$$\Omega = \mathbb{R} \to \Omega = [0,1] \to \Omega = [0,1]^d \to \Omega = \bigcup \Omega_i, \quad \Omega_i \sim [0,1]^d,$$

where (i) $\to$ (ii) is done by restriction and adaptation near the boundary, (ii) $\to$ (iii) uses tensor product and (iii) $\to$ (iv) can be performed when domain decomposition is feasible. This general approach is currently the object of active research. It is far from being understood in its full generality, but its success on model geometries (such as L-shaped domains) is promising.

If one wants to avoid domain decomposition techniques and discretization by tensor product grid in each domain, in particular in the setting of triangular finite element discretization, other strategies are also available for the construction of multiscale decomposition on $\Omega \in \mathbb{R}^d$. They will be discussed in the next section.

REMARK 20.6. For a fairly general class of domains, it is also possible to mimic the approach used to deal with the interval, i.e. do a particular adaptation of the scaling functions near the boundary, based on local linear combinations generalizing (20.15) and (20.17). A strategy to build these "boundary scaling functions" is proposed in COHEN, DAHMEN and DEVORE [2000a]. It avoids orthonormalization and biorthogonalization near the boundary, by the choice of a dual system which consists of scaling functions supported in the interior of the domain. This approach is theoretically feasible for any domain with Lipschitz boundary. However, one expects practical difficulties in the implementation of such a general framework, due to the possibly complicated structure of the boundary: this structure enters in the computations of quantities such as mass or stiffness matrices, inner product with boundary scaling functions (since the integrals have to be restricted to the domain). It also complicates the prescription of homogeneous and nonhomogeneous boundary conditions in the discretization of boundary value problems.

21. Point values, cell averages and finite elements

In this last section, we want to adopt a somehow different perspective on the construction of multiscale decompositions, in the sense that these decompositions will be based on specific types of discretizations which are commonly used in numerical analysis. We have in mind three types of discretizations which respectively correspond to the three most frequently used numerical methods.

(i) *Point values* are the unknowns and data of *finite element methods*: given a set of points Γ, a function f is discretized by its values $f(\gamma)$ for $\gamma \in \Gamma$.

(ii) *Cell averages* are the unknown and data of *finite volume methods*: a function f is discretized by its averages $|\Omega_\gamma|^{-1} \int_{\Omega_\gamma} f(x)\,\mathrm{d}x$ on cells Ω_γ, $\gamma \in \Gamma$, which constitute a disjoint partition of the domain of interest.

(iii) *Finite element spaces* are of common use in the context of *Petrov–Galerkin methods*: given a triangulation $\mathcal{T}$, a function f is approximated by another function which is piecewise polynomial on each triangles of $\mathcal{T}$ and characterized by certain local linear functionals, e.g., nodal values.

Note that in the two first examples we are only specifying the discretization process in term of a *functional* applied to the function f (the Dirac distribution δ_γ for point values and the indicator function $|\Omega_\gamma|^{-1}\chi_{\Omega_\gamma}$ for cell averages), while in the last example we also focus on a specific *approximation space* whose elements are characterized by such functionals.

We shall discuss several strategies which allow to build multiscale representations of such discretizations. Such representations constitute a "bridge" between wavelets and more classical numerical methods. They are particularly promising for practical applications since they allow to exploit the advantages of wavelet discretizations which were evoked in the previous chapter (adaptivity and preconditioning) within a specific numerical method which might be well fitted to the resolution of a given problem.

All these strategies can be thought in terms of a very general *discrete framework* which was introduced in HARTEN [1993, 1996]. This framework is based on the concept of interscale operators (which also plays an important role in multigrid methods), as expressed by the following definition.

DEFINITION 21.1. A generalized discrete multiresolution approximation is given by a sequence of index sets Γ_j, $j \geqslant 0$, which are equipped with two types of operators:

(i) Restriction/projection operators P_j^{j+1} acting from $\ell^\infty(\Gamma_{j+1})$ to $\ell^\infty(\Gamma_j)$,

(ii) Prolongation/prediction operators P_{j+1}^j acting from $\ell^\infty(\Gamma_j)$ to $\ell^\infty(\Gamma_{j+1})$.
Such operators should in addition satisfy the property

$$P_j^{j+1} P_{j+1}^j = I, \tag{21.1}$$

i.e. the projection operator is a left inverse to the prediction operator.

Such operators should be thought as a generalization of the filtering processes involving $\tilde{h}_n$ and h_n in (14.1) and (14.2). Note that property (21.1) implies that the image of the projection operator P_j^{j+1} is the entire $\ell^\infty(\Gamma_j)$. If U_{j+1} is a discrete vector in $\ell^\infty(\Gamma_{j+1})$, we can define its "approximation"

$$\hat{U}_{j+1} := P_{j+1}^j P_j^{j+1} U_{j+1}, \tag{21.2}$$

which is obtained from its projection $U_j := P_j^{j+1} U_{j+1}$ at the coarser resolution. Clearly, U_{j+1} can be characterized in a redundant way by the data of U_j and of the prediction error

$$E_{j+1} := U_{j+1} - \hat{U}_{j+1}. \tag{21.3}$$

In order to eliminate the redundancy, we notice that (21.1) implies that the prediction error satisfies $P_j^{j+1} E_{j+1} = 0$, i.e. belongs to the null space N_{j+1} of the projection operator. Let us assume that we have at our disposal a basis for this null space in the sense of a family $(\Psi_{j,\lambda})_{\lambda\in\Lambda_j}$ of vectors in $\ell^\infty(\Gamma_{j+1})$ such that any vector $V\in N_{j+1}$ has the unique expansion

$$V=\sum_{\lambda\in\Lambda_j} v_\lambda \Psi_{j,\lambda} \tag{21.4}$$

for some sequence $(v_\lambda)\in\ell^\infty(\Lambda_j)$ in the sense that for all $\gamma\in\Gamma_{j+1}$ the series $\sum_{\lambda\in\Lambda_j} v_\lambda\Psi_{j,\lambda}(\gamma)$ is absolutely summable and converges to $V(\gamma)$. Note that this such a basis always exists in the case where Γ_{j+1} is a finite set, in which case we clearly have $\#(\Gamma_{j+1})=\#(\Gamma_j)+\#(\Lambda_j)$.

Denoting by D_j the coordinate vector of the error E_{j+1} in the basis $(\Psi_{j,\lambda})_{\lambda\in\Lambda_j}$, we can define two operators associated to the "details" between resolution levels j and $j+1$: (i) A projection operator Q_j^{j+1} acting from $\ell^\infty(\Gamma_{j+1})$ to $\ell^\infty(\Lambda_j)$ which maps U_{j+1} to D_j, and (ii) a prediction operator Q_{j+1}^j acting from $\ell^\infty(\Lambda_j)$ to $\ell^\infty(\Gamma_{j+1})$ which maps D_j to E_{j+1}.

Here again, these operators should be viewed as generalizations of the filtering processes involving the filters $\tilde g_n$ and g_n. From their definition, it is clear that a property similar to (21.1) holds, i.e.

$$Q_j^{j+1}Q_{j+1}^j=I. \tag{21.5}$$

In addition, we also remark that

$$\begin{aligned}U_{j+1}&=\hat U_{j+1}+E_{j+1}=P_{j+1}^jU_j+Q_{j+1}^jD_j\\&=P_{j+1}^jP_j^{j+1}U_{j+1}+Q_{j+1}^jQ_j^{j+1}U_{j+1},\end{aligned}$$

which shows that

$$P_{j+1}^jP_j^{j+1}+Q_{j+1}^jQ_j^{j+1}=I, \tag{21.6}$$

and that U_{j+1} is equivalently characterized by (U_j,D_j). By iteration, we obtain a one to one correspondence between a vector U_J in $\ell^\infty(\Gamma_J)$ and its multiscale decomposition

$$(U_0,D_0,D_1,\dots,D_{J-1})\in\ell^\infty(\Gamma_0\times\Lambda_0\times\cdots\times\Lambda_{J-1}). \tag{21.7}$$

This generalized wavelet representation can be implemented by a (fine to coarse) decomposition algorithm using P_j^{j+1} and Q_j^{j+1}, and a (coarse to fine) reconstruction algorithm using P_{j+1}^j and Q_{j+1}^j, similar to the algorithms discussed in Section 14.

Of course such a tool is much too crude as such, and in practice we always have in mind a more specific situation. In particular, Γ_j and Λ_j correspond to some spatial discretization at resolution 2^{-j} and in this context we are interested that the above operators have some *local properties* in the following sense: there exists a constant C independent of j and γ such that $P_{j+1}^jU(\gamma)$ (resp. $Q_{j+1}^jU(\gamma)$, $P_j^{j+1}U(\gamma)$,

$Q_j^{j+1}U(\gamma))$ depends only of those $\mu \in \Gamma_j$ (resp. $\mu \in \Lambda_j$, $\mu \in \Gamma_{j+1}$, $\mu \in \Gamma_{j+1}$) such that $\mathrm{dist}(\gamma, \mu) \leqslant C2^{-j}$. We also have in mind the typical situation in which the Γ_j are nested meshes, i.e. $\Gamma_j \subset \Gamma_{j+1}$, such as, e.g., $\Gamma_j = 2^{-j}\mathbb{Z}$ corresponding to the algorithms of Section 14, in which case we can simply define Λ_j as $\Gamma_{j+1} \setminus \Gamma_j$. Such prescriptions will typically hold for the examples that we shall present, associated to our three classical discretizations.

Before going to these examples, we shall explain the *lifting scheme* – which was already introduced in Remark 19.4 – in this general context. This procedure consists in building new projection and prediction operators based on the following remark: given a matrix A_j acting from $\ell^\infty(\Lambda_j)$ to $\ell^\infty(\Gamma_j)$, we can define new operators by

$$\tilde{P}_{j+1}^j = P_{j+1}^j, \qquad \tilde{Q}_{j+1}^j = Q_{j+1}^j - P_{j+1}^j A_j, \tag{21.8}$$

and

$$\tilde{P}_j^{j+1} = P_j^{j+1} + A_j Q_j^{j+1}, \qquad \tilde{Q}_j^{j+1} = Q_j^{j+1}. \tag{21.9}$$

It is then easily seen that these new operators also satisfy (21.1), (21.5) and (21.6) and therefore qualify to define a new one to one multiscale transform of the same type as (21.7). We notice that if both A_j and the four initial operators are local, the new operators will also be local. This is clearly the generalization of Remark 19.4 from the algorithmic point of view.

We can give an interpretation of the lifting scheme in matrix terms: if we insert

$$I = \begin{pmatrix} I & -A_j \\ 0 & I \end{pmatrix} \begin{pmatrix} I & A_j \\ 0 & I \end{pmatrix}, \tag{21.10}$$

in between the two factors of the identity

$$\left(P_{j+1}^j, Q_{j+1}^j\right) \begin{pmatrix} P_j^{j+1} \\ Q_j^{j+1} \end{pmatrix} = I. \tag{21.11}$$

which expresses (21.6) as well as (21.1) and (21.5), we obtain a similar identity for the new projectors. In turn, we see that there is a natural other version of the lifting scheme since we also have

$$I = \begin{pmatrix} I & 0 \\ -B_j & I \end{pmatrix} \begin{pmatrix} I & 0 \\ B_j & I \end{pmatrix}, \tag{21.12}$$

for any operator B_j acting from $\ell^\infty(\Gamma_j)$ to $\ell^\infty(\Lambda_j)$. If we use (21.12) instead of (21.10), the new operators are defined by

$$\tilde{P}_{j+1}^j = P_{j+1}^j - Q_{j+1}^j B_j, \qquad \tilde{Q}_{j+1}^j = Q_{j+1}^j, \tag{21.13}$$

and

$$\tilde{P}_j^{j+1} = P_j^{j+1}, \qquad \tilde{Q}_j^{j+1} = Q_j^{j+1} + B_j P_j^{j+1}. \tag{21.14}$$

If the initial operators express the decomposition and reconstruction of a function in some wavelet basis, (21.8) and (21.9) correspond to a correction of the primal wavelets by coarse grid primal scaling functions, as illustrated in Remark 19.4, while (21.13) and (21.14) corresponds to a correction of the dual wavelets by coarse grid dual scaling functions.

We can think of more general modifications of the operators, since (21.10) and (21.12) are not the only possible factorizations of the identity. In particular, the *stable completion process* proposed in CARNICER, DAHMEN and PEÑA [1996], consists in introducing an additional *invertible* operator L_j acting inside $\ell^\infty(\Lambda_j)$, and define the new operators by a more general transformation than (21.8) and (21.9):

$$\tilde{P}_{j+1}^j = P_{j+1}^j, \qquad \tilde{Q}_{j+1}^j = Q_{j+1}^j L_j - P_{j+1}^j A_j, \tag{21.15}$$

and

$$\tilde{P}_j^{j+1} = P_j^{j+1} + A_j L_j^{-1} Q_j^{j+1}, \qquad \tilde{Q}_j^{j+1} = L_j^{-1} Q_j^{j+1}. \tag{21.16}$$

If L_j is local, the resulting reconstruction operators in (21.15) remain local, while the decomposition operators in (21.16) are generally global due to the presence of L_j^{-1}. In turn, the new dual wavelet system is not locally supported. This situation was encountered with the fine grid correction spline wavelets presented in Section 19.

REMARK 21.1. In the particular case corresponding to the subband filtering algorithms of Section 14, the following result was proved in DAUBECHIES and SWELDENS [1998] which reflects the flexibility of the lifting scheme: any arbitrary biorthogonal filter-bank $(h_n, \tilde{h}_n, g_n, \tilde{g}_n)$ can be obtained by applying a finite number of lifting procedures of the two above type to the "lazy wavelet" filters corresponding to $h_n = \tilde{h}_n = \sqrt{2}\delta_{0,n}$ and $g_n = \tilde{g}_n = -\sqrt{2}\delta_{1,n}$.

REMARK 21.2. In the general framework that we have described, it is worth mentioning that one can relax the assumption that the prediction operator is linear: some of the prediction operators proposed in HARTEN [1996] are nonlinearly data dependent for the purpose of a better adapted prediction near the jumps or singularities of the data which usually generate spurious oscillations (i.e. Gibbs phenomenon). These nonlinear predictions are based on *essentially nonoscillatory* (ENO) reconstruction techniques and we shall briefly evoke them in the next two examples. However, the corresponding multiresolution transforms cannot anymore be thought as a change of basis, which results in numerous difficulties when analyzing the stability of these transforms.

We shall now turn to several concrete examples, corresponding to the three types of discretizations. In all these examples, the discretizations will be nested. Although this restriction is not imposed in the above general framework, it is often encountered in practice and it simplifies the construction of multiresolution decompositions in the sense that the restriction operator has a natural definition.

Example 21.1. Point value multiresolution

In this first examples, $(\Gamma_j)_{j\geqslant 0}$ is a sequence of grids which discretizes $\mathbb{R}^d$ or some domain $\Omega \in \mathbb{R}^d$ at different resolutions 2^{-j}: if $\mu(\gamma) \in \Gamma_j$ is the closest neighbour of a point $\gamma \in \Gamma_j$, we have

$$|\gamma - \mu(\gamma)| \sim 2^{-j}, \tag{21.17}$$

with constants independent of γ and j. We assume furthermore that these grids are nested, i.e.

$$\Gamma_j \subset \Gamma_{j+1}. \tag{21.18}$$

In the point value setting, we can interpretate a discrete vector $U_j \in \ell^\infty(\Gamma_j)$ as the values of a continuous function u on the grid Γ_j:

$$U_j = \big(u(\gamma)\big)_{\gamma\in\Gamma_j}. \tag{21.19}$$

This suggest to take for P_j^{j+1} the simple decimation operator defined by

$$P_j^{j+1} U_{j+1}(\gamma) = U_{j+1}(\gamma), \quad \gamma \in \Gamma_j, \tag{21.20}$$

which is the only choice compatible with (21.19) in the sense that we have $U_j = P_j^{j+1} U_{j+1}$.

Let us now turn to the prediction operator. We first notice that by (21.1), the vector $\hat{U}_{j+1} = P_{j+1}^j U_j$ should coincide with U_{j+1} on the coarse grid Γ_j. Therefore, building an prediction operator may be viewed as an interpolation problem: from the values $u(\gamma)$, $\gamma \in \Gamma_j$, we want to reconstruct approximations $\hat{u}(\lambda)$ of the exact values $u(\lambda)$ for λ on the grid Λ_j which is simply defined by

$$\Lambda_j := \Gamma_{j+1} \setminus \Gamma_j. \tag{21.21}$$

Given such an interpolation process, it is natural to define the details by the restriction of the interpolation error E_{j+1} on Λ_j (since E_{j+1} vanishes on Γ_j), i.e.

$$D_j = \big(u(\lambda) - \hat{u}(\lambda)\big)_{\lambda\in\Lambda_j}. \tag{21.22}$$

In the case of univariate structured meshes $\Gamma_j = 2^{-j}\mathbb{Z}$, such multiresolution transforms are typically associated to *interpolatory scaling functions and wavelets*, as already explained in Remark 14.14. In this particular setting, we presented in Section 18 a class of local prediction operators based on *Lagrangian polynomial interpolation*, which are accurate up to some prescribed order N in the sense that the prediction is *exact* if u is a polynomial of degree less than $N-1$.

In a more general setting, let us explain how one follows the same principle in order to build prediction operators which are exact for polynomials of total degree less than $N-1$. A first step is to associate to each $\Lambda \in \Lambda_j$ a *prediction stencil* $S(\lambda)$ consisting of a set of points $\gamma \in \Gamma_j$ in a neighbourhood of λ such that

$$|\lambda - \gamma| \leqslant C2^{-j}, \quad \gamma \in S(\lambda). \tag{21.23}$$

From the values $(u(\gamma))_{\gamma \in S(\lambda)}$, we want to build a polynomial $p_\lambda \in \Pi_{N-1}$ with the property that if the function u is in Π_{N-1} then $p_\lambda = u$. We then simply define

$$\hat{u}(\lambda) = p_\lambda(\lambda). \tag{21.24}$$

The construction of the polynomial p_λ remains to be discussed. Ideally, we would like to define p_λ as the unique polynomial of degree $N-1$ which interpolates the values of u on $S(\lambda)$, i.e.

$$p_\lambda(\gamma) = u(\gamma), \quad \gamma \in S(\lambda). \tag{21.25}$$

Example 21.3 below will discuss a typical instance of such a prediction operator associated to Lagrange finite elements. However, the choice of a stencil for which this problem admits a unique solution is not always possible, in particular for unstructured grids. In such cases, a possibility is to take a sufficiently large stencil such that the application which maps a polynomial $p \in \Pi_{N-1}$ to its values on $S(\lambda)$ is injective, and to define p_λ by solving the least square problem

$$\inf_{p \in \Pi_{N-1}} \sum_{\gamma \in S(\lambda)} \left| u(\gamma) - p(\gamma) \right|^2. \tag{21.26}$$

Such a reconstruction is clearly exact if u is in Π_{N-1}.

REMARK 21.3. Similarly to dyadic Lagrangian interpolation, we can analyze the convergence of the subdivision scheme corresponding to the iterative application of the prediction operator from coarse to fine scales. If such a scheme is uniformly convergent, we can define the associated interpolatory scaling functions $\varphi_{j,\gamma}$, $\gamma \in \Gamma_j$, as the limit of the subdivision scheme applied to the Dirac vector $U_{j,\gamma}(\mu) = \delta_{\gamma,\mu}$, $\mu \in \Gamma_j$. It is easy to check that such functions are refinable in the sense that they are linear combinations of the $\varphi_{j+1,\mu}$, $\mu \in \Gamma_{j+1}$ (using the same arguments as in Section 12), and that they satisfy the interpolatory condition

$$\varphi_{j,\gamma}(\mu) = \delta_{\gamma,\mu}, \quad \mu \in \Gamma_j. \tag{21.27}$$

Such functions can thus be used to generate an interpolatory multiresolution analysis V_j and the corresponding hierarchical wavelet basis is simply given by

$$\{\varphi_{0,\gamma}\}_{\gamma\in\Gamma_0} \cup \{\varphi_{j+1,\lambda}\}_{\lambda\in\Lambda_j}, \tag{21.28}$$

i.e. the wavelets coincide with the fine grid scaling functions on the grid Λ_j. The dual scaling functions are of course the Dirac distributions $\tilde{\varphi}_{j,\gamma} = \delta_\gamma$.

REMARK 21.4. It is possible to apply the lifting scheme on such construction. Note that while (21.13) and (21.14) result in a similar "interpolatory decomposition", the application of (21.8) and (21.9) modify the projection operator which is no more the decimation operator (21.20). As we shall see in the finite element context, one of the interest of such a modification is that it may lead to dual scaling functions which are smoother than the Dirac distribution.

REMARK 21.5. As explained in Remark 21.2, the prediction need not be a linear operator. In the point value context, *essentially nonoscillatory* (ENO) nonlinear prediction schemes of HARTEN [1996] can be explained as follows: we consider several prediction stencils $S_i(\lambda)$, $i = 1, \ldots, N$, resulting in different polynomial $p_{i,\lambda}$ which qualify to predict the value of u at the point λ. We then select by some prescribed numerical criterion the polynomial which is the *least oscillatory* in the neighbourhood of λ. Such a data dependent process aims to minimize the Gibbs effect and improve the accuracy of the prediction in the regions where u is singular. We can then hope for a reduction of the coefficient size near the singularities, and thus better compression properties than with linear reconstruction. However, one should be aware of the difficulties which are specific to such nonlinear transforms: in particular, the convergence and stability of the subdivision process is no more clearly understood in terms of limit functions, as well as the perturbation induced by thresholding the small detail coefficients.

Example 21.2. Cell average multiresolution

In the cell average context, Γ_j represents a partition of $\mathbb{R}^d$ or $\Omega \subset \mathbb{R}^d$ by a set of disjoint *cells* Ω_γ, $\gamma \in \Gamma_j$. Such cells tile the space at various resolution 2^{-j} in the sense that there exists two constants c and C such that

$$c2^{-j} \leqslant \inf_{\gamma\in\Gamma_j} r(\Omega_\gamma) \leqslant \sup_{\gamma\in\Gamma_j} R(\Omega_\gamma) \leqslant C2^{-j}, \tag{21.29}$$

where $r(A)$ (resp. $R(A)$) is the radius of the largest (resp. smallest) circle contained in (resp. containing) a set A. We assume furthermore that these discretizations are nested in the sense that each Ω_γ, $\gamma \in \Gamma_j$, is the union of a finite number of cells Ω_μ, $\mu \in \Gamma_{j+1}$. Note that this number is always bounded by $2^d C/c$ where C and c are the constants in (21.29).

In this context, we interpretate a discrete vector $U_j \in \ell^\infty(\Gamma_j)$ as the average of a locally integrable function over the cell Ω_γ, i.e.

$$U_j = (u_\gamma)_{\gamma\in\Gamma_j} \quad \text{with } u_\gamma := |\Omega_\gamma|^{-1} \int_{\Omega_\gamma} u(x)\,\mathrm{d}x. \tag{21.30}$$

As in the point value setting, this suggests to take for P_j^{j+1} the averaging operator defined by

$$P_j^{j+1} U_{j+1}(\gamma) = |\Omega_\gamma|^{-1} \sum_{\mu\in\Gamma_{j+1},\Omega_\mu\subset\Omega_\gamma} |\Omega_\mu| U_{j+1}(\mu), \tag{21.31}$$

which is compatible with (21.30) in the sense that we have $U_j = P_j^{j+1} U_{j+1}$.

Turning to the prediction operator, we first notice that by (21.1), the vector $\hat{U}_{j+1} = P_{j+1}^j U_j$ should satisfy the compatibility condition

$$\sum_{\mu\in\Gamma_{j+1},\Omega_\mu\subset\Omega_\gamma} |\Omega_\mu| \hat{U}_{j+1}(\mu) = \sum_{\mu\in\Gamma_{j+1},\Omega_\mu\subset\Omega_\gamma} |\Omega_\mu| U_{j+1}(\mu). \tag{21.32}$$

In turn, the prediction error E_{j+1} satisfies the linear constraint

$$\sum_{\mu\in\Gamma_{j+1},\Omega_\mu\subset\Omega_\gamma} |\Omega_\mu| E_{j+1}(\mu) = 0. \tag{21.33}$$

Therefore, we can easily represent E_j in a nonredundant fashion, by defining Λ_j as a subset of Γ_{j+1} obtained as follows: for each $\gamma \in \Gamma_j$, we select a $\mu(\gamma) \in \Gamma_{j+1}$ such that $\Omega_\mu \subset \Omega_\gamma$, and we define

$$\Lambda_j := \Gamma_{j+1} \setminus \{\mu(\gamma);\ \gamma \in \Gamma_j\}. \tag{21.34}$$

We can then define the detail vector D_j as the restriction of E_{j+1} to Λ_j, i.e.

$$D_j = (u_\lambda - \hat{u}_\lambda)_{\lambda\in\Lambda_j}, \tag{21.35}$$

where $\hat{u}_\lambda := \hat{U}_{j+1}(\lambda)$.

In the case of univariate structured meshes $\Gamma_j = 2^{-j}\mathbb{Z}$, such multiresolution transforms are typically associated to biorthogonal scaling functions and wavelets for which the dual scaling function is the box function, i.e. $\tilde{\varphi} = \chi_{[0,1]}$. Some examples of primal functions have been given in Section 19: we can take for φ any of the functions $\tilde{\varphi}_{1,L}$ defined by (19.8). For such examples, the corresponding prediction algorithm has an interpretation which is very similar to point value dyadic Lagrangian interpolation: it consists in reconstructing the unique polynomial $Q_{j,k}$ of degree $L-1$ which admits the same average as u on the intervals $[2^{-j}(k+l), 2^{-j}(k+l+1)]$ for $l = 1-L, \dots, L-1$ and define the predicted averages as those of $Q_{j,k}$ on the half intervals $[2^{-j}k, 2^{-j}(k+1/2)]$ and $[2^{-j}(k+1/2), 2^{-j}k]$.

Similarly to the point value setting, we can generalize this idea by introducing a stencil of cells $S(\gamma)$ for each $\gamma \in \Gamma_j$, using the averages u_μ, $\mu \in S(\gamma)$, to reconstruct a polynomial p_γ, either by interpolation or by least square minimization. We can also derive the same remarks concerning the existence of underlying scaling functions and wavelets, the application of the lifting scheme, and the possibility of using ENO-type reconstructions.

Example 21.3. Hierarchical finite elements

Finite element spaces are of common use for the numerical discretization and simulation of physical processes, in particular when they take place on multidimensional domains that are geometrically less trivial than tensor products of univariate intervals.

For such domains Ω, the construction of a family of finite element spaces V_h, $h > 0$, relies on a decomposition of Ω into subdomains of simple shape – typically triangles or rectangles in the case of a bidimensional domain – and size of order h. The functions that constitute V_h are then piecewise polynomials of a fixed degree on each subdomains, with certain smoothness properties at the interfaces separating these subdomains.

A general question that one can address is the following:

Given a finite element space, can we find natural multiscale decompositions and wavelet bases for this space?

In some sense, we have already given an answer to this question in the case of rectangular finite elements that are obtained by tensor product of univariate spline functions: we can combine the construction of spline wavelet bases described in Section 19 together with the tensor product strategy of Section 20 (provided that Ω can be patched by square-shaped domains), to obtain "rectangular" finite element wavelets.

For many domains however, triangulations – and more generally simplicial decompositions in higher dimensions – are more adapted, in particular for polygons in 2D and polyhedrons in 3D. We thus need to address the construction of multiscale decomposition in this setting.

More general domains Ω are also concerned by the subsequent development provided that they can be patched into simplex-like sub-domains Ω_i, which are the images of a reference simplex by isoparametric maps T_i. Such maps allow to analyze – at least locally – a function f defined on Ω through its pull-back on a reference polygonal or polyhedron geometry, by mean of the multiscale decompositions and wavelet bases that we shall now describe. For the sake of simplicity, we consider the case of a bidimensional polygonal domain.

A natural framework here is a hierarchy of finite element spaces $V_j \subset V_{j+1}$, $V_j = V_{h \simeq 2^{-j}}$, based on *nested triangulations* $\mathcal{T}_j$. The construction of such triangulations, which was already evoked in Section 4, is particularly simple in the case of a polygonal domain: one starts from an initial conformal finite triangulation $\mathcal{T}_0$ and obtains $\mathcal{T}_j$ by a uniform subdivision process, as displayed on Fig. 21.1.

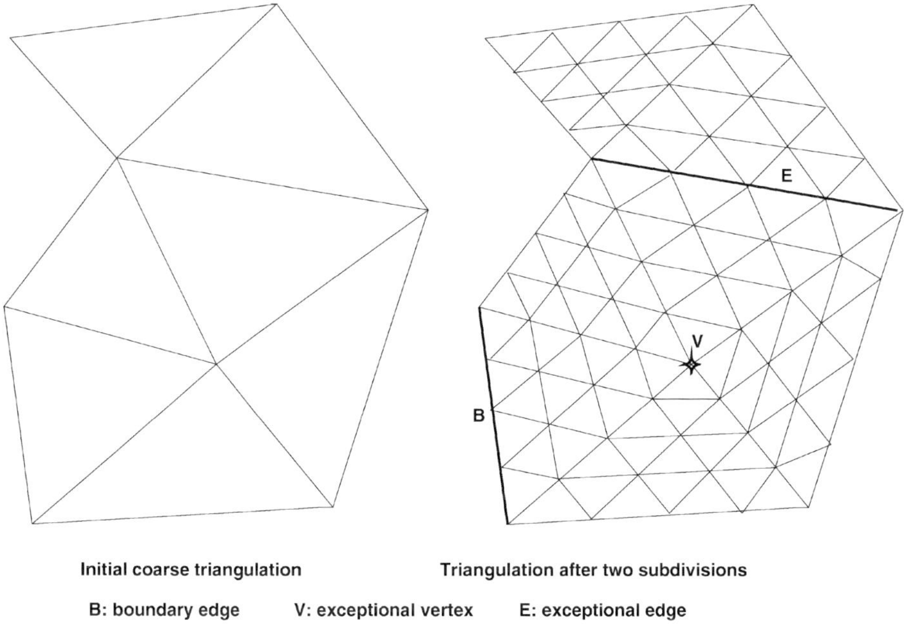

FIG. 21.1. Triangulation of Ω by uniform subdivision: from $\mathcal{T}_0$ to $\mathcal{T}_2$.

Then, a finite element space $V_j = V_{h \simeq 2^{-j}}$ is defined by choosing a polynomial space Π and associating a suitable set of nodes to each triangle $T \in \mathcal{T}_j$, so that each function $f \in V_j$ coincides with an element of Π_k on each triangle, which is uniquely defined by its values (and the values of some of its partial derivatives) at the associated nodes.

Even is the setting of nested triangulations, arbitrary choices of finite elements do not in general provide the property $V_j \subset V_{j+1}$: classical (and useful!) counter-examples are the Argyris C^1 quintic triangle and the Crouzeix–Raviart nonconforming (mid-point) linear triangle. However, an important class that does satisfy this property are the $\mathbb{P}_k$ Lagrange finite elements, i.e. piecewise Π_k functions (i.e. polynomials of total degree k) that are continuous at the interfaces between each triangles.

For such elements, a natural set of nodes associated to each triangle $T = (a, b, c) \in \mathcal{T}_j$ is given by

$$N_T := \left\{ a + \frac{n}{k}(b-a) + \frac{m}{k}(c-a);\ 0 \leqslant n+m \leqslant k \right\}. \tag{21.36}$$

On each triangle $T \in \mathcal{T}_j$, a function of V_j is then uniquely defined by its values at the node points of N_T. We denote by $\Gamma_j = \bigcup_{T \in \mathcal{T}_j} N_T$ the set of nodes globally associated to V_j. In the case of $\mathbb{P}_1$ finite elements this set coincides with the grid of vertices of $\mathcal{T}_j$. More generally, Γ_j is a superset of this grid.

The *nodal* function $\varphi_{j,\gamma}$, $\gamma \in \Gamma_j$, is then defined to be the unique element of V_j such that

$$\varphi_{j,\gamma}(\mu) = \delta_{\gamma,\mu}, \quad \mu \in \Gamma_j. \tag{21.37}$$

A *nodal basis* is then defined as the set $\{\varphi_{j,\gamma};\ \gamma \in \Gamma_j\}$.

The particular choice of equidistributed nodes (21.36) also implies that $\Gamma_j \subset \Gamma_{j+1}$. In this sense, such finite element discretization can be reinterpretated in the setting of point value discretizations of Example 21.1. In particular, while a natural decimation operator is given by (21.20), the choice of a certain finite element also determines a corresponding prediction operator: for $\gamma \in \Gamma_{j+1}$ contained in some triangle $T \in \mathcal{T}_j$ we define its predicted value as

$$\hat{u}(\gamma) = p_T(\gamma), \tag{21.38}$$

where p_T is the unique polynomial of degree N which interpolates u on the set N_T of coarse grid nodes.

The corresponding discrete multiscale transform expresses the decomposition of V_J in the *hierarchical basis*: the set of function $\{\varphi_{j,\gamma};\ \gamma \in \Lambda_j\}$ generates a nonorthogonal complement space W^h_{j-1} of V_{j-1} into V_j, so that the full set

$$\{\varphi_{0,\gamma},\ \gamma \in \Gamma_0\} \cup \{\varphi_{j,\gamma},\ \gamma \in \Lambda_j,\ j = 0, \ldots, J-1\}, \tag{21.39}$$

is a basis for V_J.

As it was remarked in Section 6, the hierarchical basis is intimately associated to the interpolation operator onto V_j (defined by $P_j f \in V_j$ such that $P_j f(\gamma) = f(\gamma)$ for all $\gamma \in \Gamma_j$) in the following sense: in the decomposition of a function f into the hierarchical basis, the layer of details at scale j represented by the $\{\varphi_{j,\gamma};\ \gamma \in \Lambda_j\}$ is exactly given by $P_{j+1}f - P_j f$. This situation reveals a serious defect of the hierarchical basis: the interpolation operator is only well defined for continuous functions. This is also reflected by the fact that the dual scaling functions and wavelets are Dirac distributions. This results in a lack of L^2 stability for the hierarchical basis, as well as the impossibility of preconditioning in $\mathcal{O}(1)$ second order elliptic operators in more than one dimension with such bases.

A natural idea to remedy this drawback is to apply the lifting scheme according to (21.8) and (21.9) or more generally the stable completion procedure, which amounts in *correcting* the hierarchical basis, in order to obtain a more stable multiscale basis: from each hierarchical basis function $\varphi_{j,\gamma}$, $\gamma \in \Lambda_j$, we would like to construct a wavelet $\psi_{j-1,\gamma}$, $\gamma \in \Lambda_j$, by some local perturbation process using neighbouring nodal basis functions at the corresponding scale. Doing so, we expect to obtain a new complement W_j that is "more orthogonal" to V_j than W^h_{j-1}. We already encountered this idea in the different constructions of spline wavelets that were discussed in Section 19. In this task, we are faced with two problems of different nature for choosing an appropriate perturbation:

(i) An *algebra* problem: the choice of the perturbation should preserve the structure of a complement of V_{j-1} into V_j, so that we effectively obtain a multiscale basis.

(ii) An *analysis* problem: the perturbation should be chosen so that the resulting wavelet basis has some improved stability in comparison to the hierarchical basis.

The first problem can be solved by applying properly the procedures of lifting scheme or stable completion which are precisely meant to derive a proper complement space. The second problem will be solved if we can analyze the stability properties of the new approximation operator P_j that has replaced the initial interpolation operator in the sense that the details at scale j in the new decomposition of a function f are represented by $P_{j+1}f - P_j f$. Such an operator will have the general form

$$P_j f = \sum_{\gamma \in \Gamma_j} \langle f, \tilde{\varphi}_{j,\gamma} \rangle \varphi_{j,\gamma}, \tag{21.40}$$

and we thus need to understand the nature (smoothness, support) of the dual functions $\tilde{\varphi}_{j,\gamma}$ which have replaced the Dirac functions δ_γ corresponding to the interpolation operator.

These problems are currently the object of active research. So far, concrete constructions have only been proposed in the case of $\mathbb{P}_1$ finite elements, mostly in 2D. They are particularly well understood in the case of a bi-infinite uniform triangulation (i.e. an hexagonal mesh), in which Fourier methods can be used to solve the two above mentioned problems. This suggests to use these constructions in the interior of each coarse triangle, since at a sufficiently fine scale, the grid coincides with a uniform mesh in the perturbation neighbourhood of a hierarchical basis function. Particular adaptations need to be done at exceptional vertices and edges that belong to the coarse triangulation $\mathcal{T}_0$, as well as near the boundary (see Fig. 21.1).

We shall now describe the three existing constructions in the case of a uniform mesh. These examples should be viewed as the analog for $\mathbb{P}_1$ finite elements of the univariate spline wavelets described in Examples 19.2–19.4.

Example 21.4. Coarse grid correction wavelets

In this example, we apply the lifting scheme according to (21.8) and (21.9) which amounts in defining the new wavelets from the hierarchical basis function by a perturbation formula of the type

$$\psi_{j-1,\gamma} = \varphi_{j,\gamma} + \sum_{\mu \in S_j(\gamma)} c_{\gamma,\mu} \varphi_{j-1,\mu}, \quad \gamma \in \Lambda_{j-1}, \tag{21.41}$$

where the *stencil* $S_j(\gamma)$ is a neighbourhood of γ contained in the coarse grid Γ_{j-1}. Note that this choice always ensures that $\{\psi_{j-1,\gamma};\ \gamma \in \Lambda_j\}$ is a basis for a complement space W_{j-1}^c of V_{j-1}.

The main problem is of course the choice of the coefficients $c_{\gamma,\mu}$ in order to stabilize the basis. A first natural requirement is to impose an oscillation property, i.e.

$$\int \psi_{j,\gamma} = 0, \tag{21.42}$$

since the presence of low frequencies in $\varphi_{j,\gamma}$ can be viewed as the origin of the L^2 instability of the hierarchical basis.

However, a complete understanding of the new system necessarily goes through the analysis of the new dual functions $\tilde{\varphi}_{j,\gamma}$ that appear in (21.40). As we remarked in Section 19 (Remark 19.4), in the case of wavelets defined on the real line, and in the beginning of the present section, the lifting scheme yields a new refinable dual scaling function that has also compact support and satisfies a refinement equation with a finite number of coefficients.

We can then use the techniques of Section 15 (adapted to this multivariate setting) to estimate the smoothness of the new dual function. In this case, one can find simple coarse grid corrections that gives rise to a compactly supported refinable function $\tilde{\varphi}$ (dual to the hat function φ that generates the $\mathbb{P}_1$ elements on a uniform mesh by its translates and dilates) which is square integrable, and has also some positive Sobolev smoothness. The following choice turns out to coincides with a construction proposed by COHEN and SCHLENKER [1993] for which it is proved that $\tilde{\varphi}$ belong to $H^s(\mathbb{R}^2)$ if $s < 0.44$: for $\gamma \in \Lambda_j$, $\varphi_{j,\gamma}$ is perturbated by its two first neighbours on the coarse grid Γ_j with weight $-3/16$ and by its two facing neighbours with weight $1/16$ as shown on Fig. 21.2. Note that this results in three types of wavelets that can be derived from each other by rotation of $2\pi/3$.

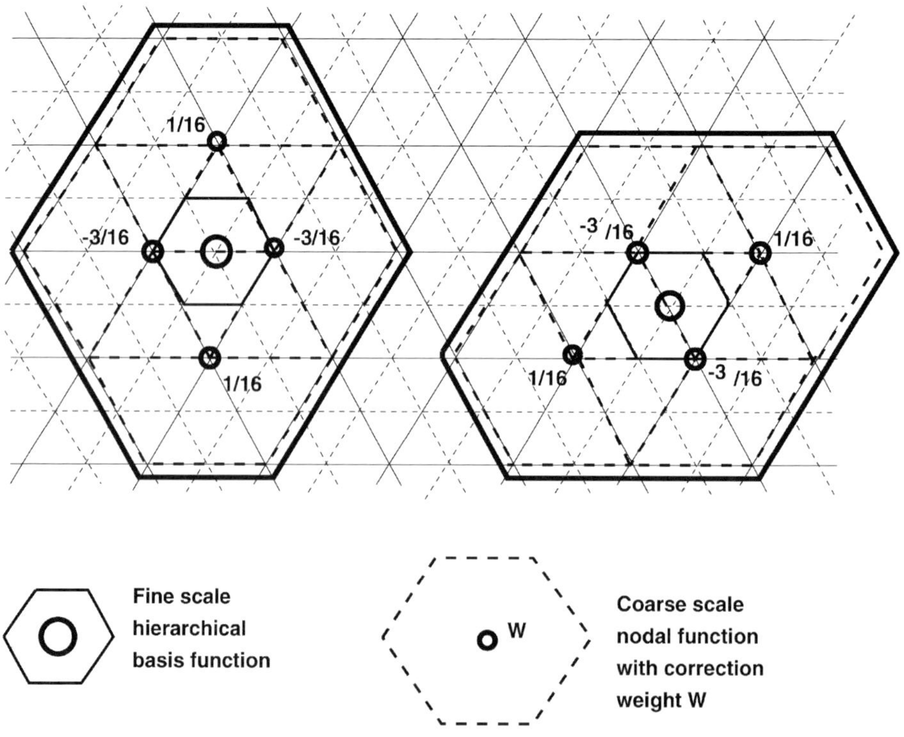

FIG. 21.2. Two coarse grid correction $\mathbb{P}_1$ wavelets.

This particular perturbation gives the maximal Sobolev smoothness for the dual scaling function among all possible weights with such a 4-points stencil. The first numerical results seem to indicate that the use of a a 4-points stencil with modified coefficients near exceptional vertices and edges (in order to preserve (21.42)), and of a modified 3-points stencil near the boundary, allows to preserve the good properties of the resulting multiscale basis. Moreover, the analysis of the corresponding modified dual functions seems feasible.

Example 21.5. Semi-orthogonal wavelets

In this example, the wavelets are obtained from the hierarchical basis function by a combination formula of the type

$$\psi_{j-1,\gamma} = \sum_{\mu \in T_j(\gamma)} d_{\gamma,\mu} \varphi_{j,\mu}, \quad \gamma \in \Lambda_j, \tag{21.43}$$

where the stencil $T_j(\gamma)$ is a neighbourhood of γ contained in the fine grid Γ_j, and where the coefficients are chosen in such a way that $\psi_{j-1,\gamma}$ is orthogonal to V_{j-1}. We also have to check that the functions $\psi_{j-1,\gamma}$, $\gamma \in \Lambda_j$, constitute a Riesz basis for the orthogonal complement W^o_{j-1} of V_{j-1} into V_j.

Note that while the number of coarse grid functions overlapping $\psi_{j-1,\gamma}$ is roughly proportional to $|T_j(\gamma)|/4$, so that taking the stencil $T_j(\gamma)$ large enough we should be able to find coefficients $d_{\gamma,\mu}$ that ensure the orthogonality. The main problem is thus to check the stability of the wavelet basis *within one level of scale*, i.e. that the functions $\psi_{j-1,\gamma}$, $\gamma \in \Lambda_j$, are a Riesz basis for W^o_{j-1}. In the case of a uniform mesh, this can easily be done by the Fourier transform techniques introduced in Section 10 (adapting Theorem 10.1 to this multivariate setting).

We display on Fig. 21.3 a possible choice of stencil that yields such a semi-orthogonal wavelet basis, where as in the previous example, we have three types of wavelets that can be derived from each other by rotation of $2\pi/3$.

This construction seems to be adaptable near exceptional vertices and edges, as well as boundaries. First results going in this direction are presented in STEVENSON [1998]. The main advantage of the semi-orthonormal approach is the relative simplicity of the L^2-stability analysis which can be done separately at each scale level due to the orthogonality between the W^o_j spaces. A generalization of this approach was recently proposed in DAHMEN and STEVENSON [1999]: the corrected wavelets are chosen to be orthogonal to another Lagrange finite element space $\widetilde{V}_j$ which satisfies a minimal angle condition with respect to V_j (for all $f \in V_j$ (resp. $\widetilde{V}_j$), $\|f\|_{L^2} = 1$, there exists $g \in \widetilde{V}_j$ (resp. V_j), $\|g\|_{L^2} = 1$, such that $\langle f, g\rangle > \alpha$, with $\alpha > 0$ independent of j), and such that the $\widetilde{V}_j$ are also nested, i.e. $\widetilde{V}_J \subset \widetilde{V}_{j+1}$. The wavelets can be constructed by an element by element orthogonalization procedure, which allows a proper treatment of exceptional edges and vertices. In both constructions, the dual basis is globally supported in Ω.

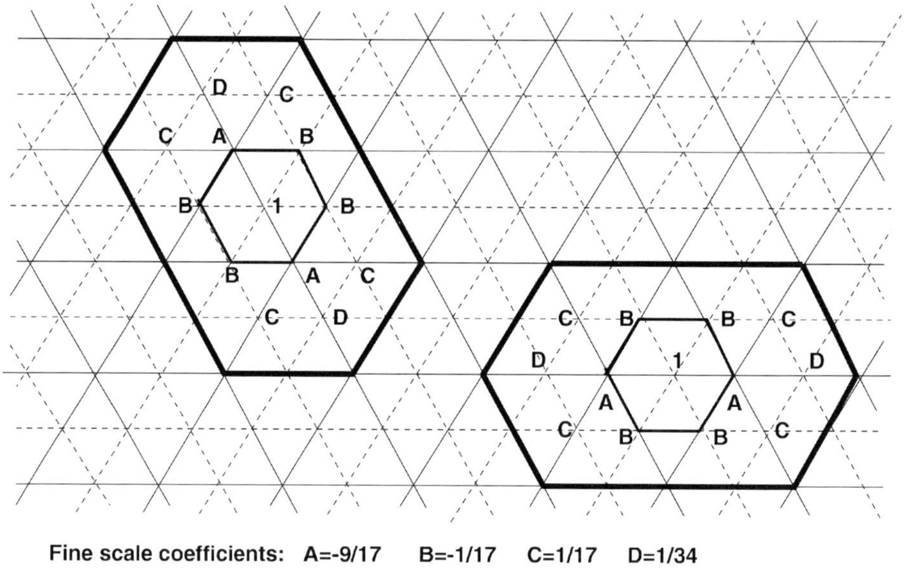

FIG. 21.3. Two semi-orthogonal $\mathbb{P}_1$ wavelets.

Example 21.6. Fine grid correction wavelets

In this example, the wavelets are obtained from the hierarchical basis function by a perturbation formula of the type

$$\psi_{j-1,\gamma} = \varphi_{j,\gamma} + \sum_{\mu \in R_j(\gamma)} c_{\gamma,\mu} \varphi_{j,\mu}, \quad \gamma \in \Lambda_j, \tag{21.44}$$

where the stencil $R_j(\gamma)$ is a neighbourhood of γ contained in the fine grid Γ_j.

In contrast to coarse grid correction, this choice does not always ensures that $\{\psi_{j-1,\gamma};\ \gamma \in \Lambda_j\}$ is a basis for a complement space of V_{j-1} into V_j. In order to solve the algebra problem, we can use a correction with fine scale functions that are centered at points of the coarse grid, i.e. impose that $R_j(\gamma) \in \Gamma_{j-1}$. The following example of such a construction was studied in details in OSWALD and LORENTZ [1996]: the correction consists in removing the two nearest neighbours of $\varphi_{j,\gamma}$, with weight $1/2$. The wavelets are displayed on Fig. 21.4 (after renormalization by a factor of two).

Note that they are orthogonal to the coarse grid nodal functions $\varphi_{j-1,\gamma}$ with respect to the discrete inner product on the fine grid

$$\langle f, g \rangle_{d,j} := \sum_{\gamma \in \Gamma_j} f_\gamma g_\gamma. \tag{21.45}$$

However, one cannot only rely on this "discrete semi-orthogonality" to conclude that this multiscale basis is stable: the dual functions still need to be investigated in order to

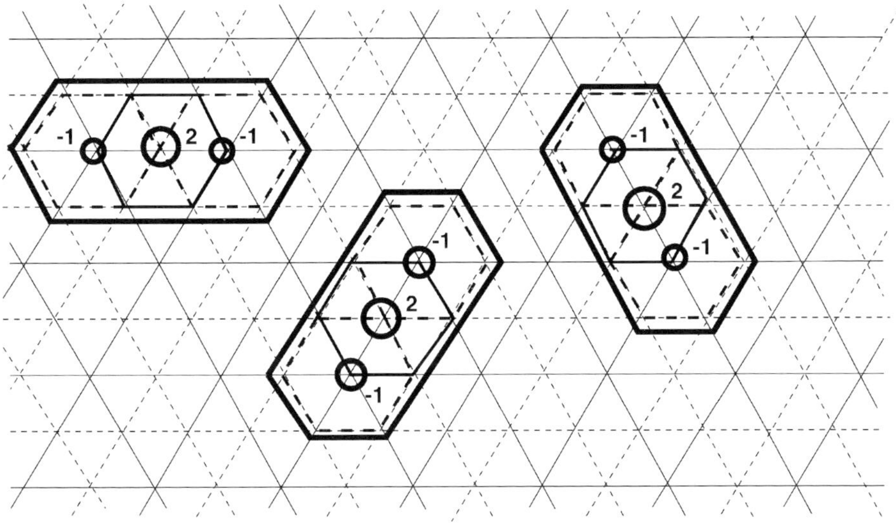

FIG. 21.4. Three fine grid correction $\mathbb{P}_1$ wavelets.

solve the analysis problem. As in Example 19.4, these dual functions are not compactly supported, but Fourier techniques allow to estimate their regularity. For the present example, OSWALD and LORENTZ [1996] have proved that they belong to $H^s(\mathbb{R}^2)$ for $s < 0.99$.

This suggests that such a fine grid correction yields a stabilized basis, although a complete analysis in the realistic situation of a bounded domain is still to be done. Such an analysis seems more difficult than in the two previous cases, due to the combination of two difficulties: the lack of orthogonality between levels and the global support of the dual. It is striking to see how this last example, which has the simplest numerical structure and the smallest support, is by far the most complicated to analyze.

22. Conclusion

Numerous tools are available to the numerician for multiscale approximation and decomposition of functions. These tools can easily be adapted to multivariate functions defined on $\mathbb{R}^d$, and with a little more effort to bounded domains with the possibility of incorporating boundary conditions.

Most of the examples that we have described share the following fundamental features:

(i) *Scaling functions*: approximation takes place in a ladder of embedded spaces V_j, $j \geqslant 0$, equipped with local bases of *scaling functions*. More precisely, these bases have supports of diameter $\mathcal{O}(2^{-j})$ and they satisfy a property of *controlled overlapping*: a point x is contained in at most M supports of basis function at level j with M independent of x and j.

(ii) *Wavelets*: a (not necessarily orthogonal) complement W_j of V_{j+1} onto V_j is generated by a *wavelet bases*, with the same local properties. This results in a multiscale basis consisting of scaling functions at a fixed level and wavelets at all finer levels. The local nature of the scaling functions and wavelets yields a fast reconstruction algorithm to obtain the coefficients of $f \in V_j$ in the scaling function basis from those in the multiscale basis.

(iii) *Dual bases*: the coefficients of the multiscale decomposition of a function in the wavelet basis can be computed by inner products with dual scaling functions and wavelets. The dual scaling functions are not necessarily compactly supported and refinable. However, if they have these two properties, the coefficients of $f \in V_j$ in the multiscale basis can be computed by a fast algorithm from those in the scaling function basis.

(iv) *Approximation and smoothness*: the scaling functions have certain *polynomial exactness properties*, which result in vanishing moments for the dual wavelets. In a shift-invariant setting, these properties are expressed by the Strang–Fix conditions. They also have certain smoothness properties. The dual scaling functions can also be built with such properties. In several constructions, both polynomial exactness and smoothness can be prescribed at an arbitrarily high level, but the price to pay is a support size that grows linearly with these parameters.

These properties play a crucial role in the practical use of multiscale decompositions in numerical analysis, as well as in the proof of most theoretical results underlying these applications, as we shall see in the two remaining chapters.

23. Historical notes

Although the concepts and constructions that we have described throughout this chapter have mostly been developed since the end of the 1980's, they should be viewed as the result of the merging of several independent developments that took place in different area of research:

(i) *Time-frequency and time-scale analysis*: since the 1950's, several ideas had been proposed to "localize" Fourier analysis in the context of signal processing. For example, D. Gabor proposed to compute inner products $Gf(a,b) = \langle f, g_{a,b}\rangle$ of the analyzed function f against localized wave,

$$g_{a,b}(x) = \mathrm{e}^{\mathrm{i}ax} g(x-b), \quad a, b \in \mathbb{R}, \tag{23.1}$$

where g is typically a Gaussian function. In the early 1980's, the geophysicist J. Morlet who was looking for an analysis with arbitrarily high resolution in space (limited in (23.1) by the fixed support of g), proposed to use

$$\psi_{a,b}(x) = a^{-1}\psi\left(\frac{x-b}{a}\right), \quad a > 0,\ b \in \mathbb{R}, \tag{23.2}$$

where ψ is an oscillating function (one assumes that $C = 2\pi \int |\omega|^{-1} |\hat{\psi}(\omega)|^2 \,\mathrm{d}\omega$ is finite so that $\int \psi = 0$) which is well localized both in space and frequency. Moreover, he suggested that a function f can be reconstructed from its continuous wavelet transform $Wf(a,b) = \langle f, \psi_{a,b}\rangle$ by the formula

$$f = C^{-1} \iint Wf(a,b)\psi_{a,b}(x)\frac{\mathrm{d}a\,\mathrm{d}b}{a}. \tag{23.3}$$

This formula was proved to be true by A. Grossmann in 1983. On a more numerical point of view, I. Daubechies proved in 1984 that the sampling

$$\psi_{m,n}(x) = a_0^{m/2}\psi\big(a_0^m x - b_0 n\big), \quad m, n \in \mathbb{Z}, \tag{23.4}$$

generates a *frame* of $L^2(\mathbb{R})$ if $a_0 > 1$ and $b_0 > 0$ are chosen small enough. The question of the existence of an orthonormal basis with the above structure and ψ well localized both in space and frequency (in contrast to the Haar system), was left open until a construction by Y. Meyer in 1985, in which ψ belongs to the Schwartz class $\mathcal{S}(\mathbb{R})$. A detailed and comprehensive survey of these developments will be found in DAUBECHIES [1992].

(ii) *Harmonic analysis*: Since the 1930's, Littlewood and Paley had proposed a systematic way of decomposing a function f into "almost orthogonal blocks". One fixes a positive C^∞ function $r(\omega)$ which has compact support in $[-2,2]$ and equals 1 on $[-1,1]$. Any tempered distribution f can then be decomposed according to

$$f = S_0 f + \sum_{j\geqslant 0} \Delta_j f, \tag{23.5}$$

with $\mathcal{F}S_j f(\omega) = \hat{f}(\omega)r(2^{-j}\omega)$ and $\Delta_j f = S_{j+1}f - S_j f$. The block $\Delta_j f$ thus represents the frequency content of f between 2^j and 2^{j+1}. Such a decomposition has proved to be a powerful tool for the study of linear and multilinear operators, as well as for characterizing various function spaces, in particular smoothness classes, from the size properties of the blocks Δ_j (see FRAZIER, JAWERTH and WEISS [1991] for a review on Littlewood–Paley theory and its applications). Clearly, wavelet decompositions can be viewed as a variant of (23.5) which is more adapted for numerical computations, since the $\Delta_j f$ are replaced by combinations of local functions.

(iii) *Approximation theory*: the study of spline functions began with the work of I.J. Schoenberg in the 1950's (SCHOENBERG [1946]), who is also at the origin of the Strang–Fix conditions described in Section 16. There is an huge amount of existing literature on splines, as well as on finite elements (we already mentioned some general references in this chapter). Let us say that an important step in the understanding of spline approximation was the introduction of local quasi-interpolants operators (which have the same structure as our local projectors P_j, see Remark 11.2) in the 1970's by C. de Boor and G. Fix. The general relations between splines functions and subdivision algorithms were analyzed in the 1980's (DAHMEN and MICCHELLI [1984]).

(iv) *Multiresolution image processing*: many algorithms developed in image processing, as well as in computer-aided geometric design, since the 1960's, are based on a multiresolution processing of the visual information. Beside the subdivision algorithms (that have been discussed in Section 12) for fast curve or surface generation, and the iterated filterbanks (see KOVACEVIC and VETTERLI [1995] for a detailed review) that are principally used for image compression and coding, one should also mention the *Laplacian pyramids* introduced in BURT and ADELSON [1983] which firstly made the connexion between techniques of multiscale refinement with image coding ideas, and can be viewed as a discrete implementation of the quasi-interpolant operators that were evoked in Remark 11.2.

The introduction of the concept of multiresolution approximation by S. Mallat in 1986, and the resulting constructions of compactly supported wavelets by I. Daubechies in 1988, can thus be viewed as a beautiful combination of these ideas.

CHAPTER III

Multiscale Decomposition of Function Spaces

24. Introduction

A basic tool in the numerical analysis of PDE's is approximation theory, which relates the analytical properties of arbitrary functions (in particular their smoothness) with the accuracy of their approximation by simpler functions, such as polynomials, trigonometric series or finite elements.

Perhaps the most well known result for the numerician deals with finite element approximation in L^2-Sobolev spaces: If Ω is a polygonal domain, $\mathcal{T}_h$, $0 < h < h_{\max}$, a family of regular triangulation with mesh size h, and V_h a finite element space built from $\mathcal{T}_h$ that contains polynomials up to degree $n-1$ and is contained in $W^{s,2}$, then for $s \leqslant t \leqslant n$ one has the estimate

$$\inf_{g \in V_h} \|f - g\|_{W^{s,2}} \leqslant C h^{t-s} |f|_{W^{t,2}}, \tag{24.1}$$

where C does not depends of f and h. For the multiresolution spaces $V_j \sim V_{h \approx 2^{-j}}$ that we studied in the previous chapter, (24.1) takes the form

$$\inf_{g \in V_j} \|f - g\|_{W^{s,2}} \lesssim 2^{-(t-s)j} |f|_{W^{t,2}}. \tag{24.2}$$

These results express that a smoothness property implies an approximation rate. If the approximation spaces are themselves constituted of smooth functions, it is reasonable to think of a converse: a certain rate of approximation for a function f might imply that f has some smoothness.

Surprisingly enough, it turns out that a large number of smoothness classes, including L^2-Sobolev spaces, can be *exactly characterized* from the rate of decay of the approximation error in the spaces V_j, or also form the summability and decay properties of the wavelet coefficients. Such an instance was already evoked without proof in Section 6 of Chapter I, in relation with multiscale preconditioning. The goal of this chapter is to prove this type of result in a general context.

As a first step, we introduce in Section 25 the various function spaces that will be considered in this chapter, in particular Hölder, Sobolev and Besov smoothness classes.

We then introduce, in Sections 26 and 27, two important types of inequalities that are essential to build a more general theory: *direct (or Jackson type) inequalities* that are exactly of the type (24.2), and *inverse (or Bernstein type) inequalities* that take into account the smoothness properties of the V_j spaces. In the setting of finite element spaces, such inequalities are classically derived by using the properties of the affine mapping between each element and a reference domain (e.g., the unit square in the case of rectangular elements) and basic results of polynomial approximation on this reference domain (see, e.g., CIARLET [1978]). Since we have in mind approximation spaces V_j which are not necessarily of finite element types we shall proceed in a slightly different way Our proofs are based on local estimates on the dyadic cubes

$$I_{j,k} = 2^{-j}\big(k + [0,1]^d\big), \quad k \in \mathbb{Z}^d, \tag{24.3}$$

and involve the scaling functions generating V_j. They still should be viewed as variants to the proofs of similar results in the finite element context.

In Section 28, we describe a general mechanism, combining direct and inverse estimate with interpolation of function spaces, that yields the characterization of smoothness classes by the rate of approximation.

We apply these ideas in Section 29 to multiscale approximation, and obtain characterization results for L^2-Sobolev, Hölder and more general Besov spaces. In Section 30, we study the possibility of characterizing L^p-Sobolev and Besov spaces by L^p-unstable approximation operators (in particular, we address the case of quasi-normed Besov spaces for $p < 1$ that will be of crucial importance for the nonlinear theory of Chapter IV). We also deal in Section 31 with the characterization of negative smoothness classes and of L^p spaces.

For the sake of simplicity, all these results are at first presented for function spaces defined on the whole of $\mathbb{R}^d$, using multiresolution analysis and wavelets obtained by the tensor product strategy described in Section 20 of Chapter II. While the generalization of our results to other types of multiscale decompositions using other basis functions (nontensor product, multiwavelets, finite element bases) is straightforward, their adaptation to function spaces on bounded domains with specific boundary conditions – which is crucial for practical applications – is a more elaborate task, due to the lack of translation invariance of the multiresolution spaces and the specific adaptations needed near the boundary of the domain. We address these issues in Sections 32 and 33. These two sections are fairly technical and basically reach the same type of norm equivalences that are more easily obtained on $\mathbb{R}^d$.

We end by discussing Section 34 the application of these results to multilevel preconditioning of elliptic operators. The analysis of such techniques only involves the characterization of L^2-Sobolev spaces (possibly with boundary conditions). The characterization of more general Sobolev and Besov spaces will be needed in the analysis of adaptive approximations that will be the object of the next chapter.

Notations. In this chapter, as well as in the next one, we shall frequently adopt notations similar to those used in Section 21, which aim to simplify the representation of a function in a wavelet basis, especially when working on a multivariate domain Ω: for all the constructions in the previous chapter, we can write

$$P_j f := \sum_{\gamma\in\Gamma_j} \langle f, \tilde{\varphi}_\gamma\rangle \varphi_\gamma, \tag{24.4}$$

where $(\varphi_\gamma)_{\gamma\in\Gamma_j}$ is the scaling function basis of V_j and $(\tilde{\varphi}_\gamma)_{\gamma\in\Gamma_j}$ the corresponding dual basis of $\widetilde{V}_j$. We also have

$$f = \sum_{j\geqslant j_0-1} \sum_{\lambda\in\Lambda_j} \langle f, \tilde{\psi}_\lambda\rangle \psi_\lambda, \tag{24.5}$$

where $(\psi_\lambda)_{\lambda\in\Lambda_j}$ denotes the scaling function basis of the coarsest level space V_{j_0} if $j = j_0 - 1$ and the wavelet basis of W_j of $j \geqslant j_0$, and where $\tilde{\psi}_\lambda$ is the corresponding dual function. For sake of simplicity, except in certain specified cases, we shall assume that $j_0 = 0$. We also define $\Lambda = \bigcup_{j\geqslant -1} \Lambda_j$ so that

$$f = \sum_{\lambda\in\Lambda} \langle f, \tilde{\psi}_\lambda\rangle \psi_\lambda. \tag{24.6}$$

If an index λ belongs to Λ_j of Γ_j, we set $|\lambda| := j$. Thus, $\Phi_j := \{\varphi_\gamma;\ |\gamma| = j\}$ and $\Psi_j := \{\psi_\lambda;\ -1 \leqslant |\lambda| < j\}$ are respectively the nodal and multiscale basis of V_j.

We recall some size properties of the functions ψ_λ (the same remarks hold for the functions φ_λ, $\tilde{\psi}_\lambda$ and $\tilde{\varphi}_\lambda$. We assume that the basis functions are normalized in L^2. This means that we have scaling factors when we use another norm to measure these functions. For example, we have

$$\|\psi_\lambda\|_{W^{m,p}} \sim 2^{(m+d/2-d/p)|\lambda|}. \tag{24.7}$$

The supports of ψ_λ are contained in cubes I_λ of sidelength $C2^{-\lambda}$, and do not overlap too much in the following sense: there exists a constant K such that for all $\lambda \in \Lambda_j$,

$$\#\big\{\mu \in \Lambda_j \text{ s.t. } \mathrm{Supp}(\psi_\lambda) \cap \mathrm{Supp}(\psi_\mu) \neq \emptyset\big\} \leqslant K. \tag{24.8}$$

25. Function spaces

There exists many different ways of measuring the smoothness of a function f. The most natural one is certainly the order of differentiability, i.e. the maximal index m such that $f^{(m)} = (\mathrm{d}/\mathrm{d}x)^m f$ (or more generally $\partial^\alpha f$, $|\alpha| := \alpha_1 + \cdots + \alpha_d \leqslant m$, for a multivariate function) is continuous. To this particular measure of smoothness, one associates a class of *function spaces*: for $\Omega \subset \mathbb{R}^d$ an open set, we define $C^m(\Omega)$ to be the space of continuous functions which have bounded and continuous partial derivatives $\partial^\alpha f$, $|\alpha| \leqslant m$. This space is equipped with the norm

$$\|f\|_{C^m(\Omega)} := \sup_{x\in\Omega} \big|f(x)\big| + \sum{}_{|\alpha|=m} \sup_{x\in\Omega} \big|\partial^\alpha f(x)\big|, \tag{25.1}$$

for which it is a Banach space.

In order to measure the smoothness properties of a function in an average sense, it is also natural to introduce the *Sobolev spaces* $W^{m,p}(\Omega)$ consisting of all functions $f \in L^p$ with partial derivatives up to order m in L^p. Here p is a fixed index in $[1, +\infty]$. This space is also a Banach space, when equipped with the norm

$$\|f\|_{W^{m,p}} := \|f\|_{L^p} + |f|_{W^{m,p}}, \quad |f|_{W^{m,p}} := \sum_{|\alpha|=m} \|\partial^\alpha f\|_{L^p}. \tag{25.2}$$

Here, we have used the notation $|\bullet|$ to denote the corresponding semi-norm. Note that the norm (25.1) for C^m spaces coincides with the $W^{m,\infty}$ norm although these spaces are different.

All the above spaces share the common feature that the regularity index is an integer.

In many instances of theoretical and numerical analysis of PDE's, one is interested in describing the regularity of a function in a more precise way through fractional order of smoothness. The question thus arises of *how to fill the gaps between integer smoothness classes*. There are at least two instances where such a generalization is very natural: firstly, in the case of the L^2-Sobolev spaces $H^m := W^{m,2}$ and when $\Omega = \mathbb{R}^d$, we can define an equivalent norm based on the Fourier transform, since by Parseval formula we have

$$\|f\|_{H^m}^2 \sim \int_{\mathbb{R}^d} (1 + |\omega|)^{2m} |\hat{f}(\omega)|^2 \, d\omega. \tag{25.3}$$

For a noninteger $s \geqslant 0$, it is thus natural to define the space H^s as the set of all L^2 functions such that

$$\|f\|_{H^s}^2 := \int_{\mathbb{R}^d} (1 + |\omega|)^{2s} |\hat{f}(\omega)|^2 \, d\omega \tag{25.4}$$

is finite.

Secondly, in the case of C^m spaces, we note that $\sup_{x\in\Omega} |f(x+h) - f(x)| \leqslant [\sup |f'|]|h|$ if $f \in C^1$ for any $h \in \mathbb{R}^d$, whereas for an arbitrary $f \in C^0$, $\sup_{x\in\Omega} |f(x+h) - f(x)|$ might go to zero arbitrarily slow as $|h| \to 0$. This motivates the definition of the Hölder space C^s, $0 < s < 1$, consisting of those $f \in C^0$ such that

$$\sup_{x\in\Omega} |f(x+h) - f(x)| \leqslant C|h|^s. \tag{25.5}$$

If $m < s < m+1$, a natural definition of C^s is given by $f \in C^m$ and $\partial^\alpha f \in C^{s-m}$, $|\alpha| = m$. It is not difficult to prove that this property can also be expressed by

$$\sup_{x\in\Omega} |\Delta_h^n f(x)| \leqslant C|h|^s, \tag{25.6}$$

where $n > s$ and Δ_h^n is the nth order finite difference operator defined recursively by $\Delta_h^1 f(x) = f(x+h) - f(x)$ and $\Delta_h^n f(x) = \Delta_h^1(\Delta_h^{n-1}) f(x)$ (e.g., $\Delta_h^2 f(x) = f(x +$

$2h) - 2f(x+h) + f(x))$. When s is not an integer, the spaces C^s that we have defined are also denoted as $W^{s,\infty}$.

The definition of "s order of smoothness in L^p" for s noninteger and p different of 2 and ∞ is more subject to arbitrary choices. Among others, we find three well-known classes.

Sobolev spaces $W^{s,p}$ are defined (if $m < s < m+1$) by $\|f\|_{W^{s,p}} = \|f\|_{L^p} + |f|_{W^{s,p}}$ with

$$|f|_{W^{s,p}} := \sum_{|\alpha|=m} \int_{\Omega\times\Omega} \frac{|\partial^\alpha f(x) - \partial^\alpha f(y)|^p}{|x-y|^{(s-m)p+d}} \, dx \, dy. \tag{25.7}$$

These spaces coincide with those defined by means of Fourier transform when $p = 2$. They are also a particular case of the Besov spaces defined below when s is not an integer. We refer to ADAMS [1975] for a general introduction.

Potential (or Liouville) spaces $H^{s,p}$ (sometimes also denoted by $L^{p,s}$) are defined by means of the Fourier transform operator $\mathcal{F}$ through

$$\|f\|_{H^{s,p}} = \|f\|_{L^p} + \left\|\mathcal{F}^{-1}\left(1 + |\bullet|^s\right)\mathcal{F}f\right\|_{L^p}. \tag{25.8}$$

These spaces coincides with the Sobolev spaces $W^{m,p}$ when m is an integer and $1 < p < +\infty$ (see TRIEBEL [1983, p. 38]), but their definition by (25.8) requires a priori that $\Omega = \mathbb{R}^d$ in order to apply the Fourier transform. These spaces can of course be defined on arbitrary domains by restriction, but they have no simple inner description for noninteger s. In certain references, they are also referred as fractional Sobolev spaces. However they differ from the $W^{s,p}$ spaces when $p \neq 2$ and s is not an integer.

Besov spaces $B^s_{p,q}$ involve an extra parameter q and can be defined through finite differences. These spaces include most of those that we have listed so far as particular cases for certain ranges of indices. As it will appear in the forthcoming Section 28, these spaces are also produced by general "interpolation techniques" between function spaces of integer smoothness, and they can be exactly characterized by the rate of multiresolution approximation error, as well as from the size properties of the wavelet coefficients. For these reasons, we shall insist a bit more on their definitions and properties.

We define the nth order L^p modulus of smoothness of f by

$$\omega_n(f, t, \Omega)_p = \sup_{|h|\leqslant t} \left\|\Delta_h^n f\right\|_{L^p(\Omega_{h,n})}, \tag{25.9}$$

(h is a vector in $\mathbb{R}^d$ of Euclidean norm less than t) where $\Omega_{h,n} := \{x \in \Omega;\ \text{s.t. } |y-x| \leqslant nh \Rightarrow y \in \Omega\}$. We use the simpler notation $\omega_n(f,t)_p$ if Ω is the whole domain where the function is considered. A basic properties of this quantity are its monotonicity with respect to t: $\omega_n(f,t)_p \leqslant \omega_n(f,s)_p$ if $t < s$. It is also easily checked that if m is a positive integer, we have $\omega_n(f, mt, \Omega)_p \leqslant m^n \omega_n(f, t, \Omega)_p$.

For $p, q \geqslant 1$, $s > 0$, the Besov spaces $B^s_{p,q}(\Omega)$ consists of those functions $f \in L^p(\Omega)$ such that

$$\left(2^{sj}\omega_n\left(f, 2^{-j}\right)_p\right)_{j\geqslant 0} \in \ell^q, \tag{25.10}$$

where n is an integer such that $s < n$. A natural norm for such a space is then given by

$$\|f\|_{B^s_{p,q}} := \|f\|_{L^p} + |f|_{B^s_{p,q}}, \quad |f|_{B^s_{p,q}} := \left\| \left(2^{sj}\omega_n\left(f, 2^{-j}\right)_p\right)_{j\geqslant 0}\right\|_{\ell^q}. \tag{25.11}$$

REMARK 25.1. The Besov semi-norm is often rather defined in its integral form

$$|f|_{B^s_{p,q}} := \left\|t^{-s}\omega_n(f,t)_p\right\|_{L^q([0,1],\,dt/t)}, \tag{25.12}$$

i.e. $(\int_0^1 (t^{-s}\omega_n(f,t)_p)^q \frac{dt}{t})^{1/q}$ if $q < +\infty$. The equivalence between the above expression and its discrete analog of (25.11) is easily derived from the monotonicity of the modulus of smoothness ($\omega_n(f,t)_p \leqslant \omega_n(f,t')_p$ if $t < t'$). Also note that $|f|_{B^s_{p,q}} \sim \|(2^{sj}\omega_n(f,2^{-j})_p)_{j\geqslant J}\|_{\ell^q}$ with constants independent of $J \geqslant 0$ such that $\operatorname{diam}(\Omega) \leqslant 2^{-J}$.

REMARK 25.2. The definition of $B^s_{p,q}$ is independent of n in the sense that two integers $n, n' > s$ yield equivalent norms by (25.11). On the one hand, if $n \geqslant n'$, one clearly has $\omega_n(f,t)_p \lesssim \omega_{n'}(f,t)_p$, and we get a similar inequality for the corresponding expressions in (25.11). To prove the converse part of the norm equivalence, one classically use the Marchaud type (discrete) inequality

$$\omega_{n'}\left(f, 2^{-j}\right)_p \lesssim 2^{-n'j}\left[\|f\|_{L^p} + \left\|\left(2^{n'(j+l)}\omega_n\left(f, 2^{-(j+l)}\right)_p\right)_{l\geqslant 0}\right\|_{\ell^q}\right], \tag{25.13}$$

and conclude by using discrete Hardy's inequalities (see, e.g., BUTZER and BEHRENS [1967, p. 199] or DEVORE and POPOV [1988a, Corollary 4.9], or also DEVORE and LORENTZ [1993, p. 55]).

The space $B^s_{p,q}$ represents "s order of smoothness measured in L^p", with the parameter q allowing a finer tuning on the degree of smoothness – one has $B^s_{p,q_1} \subset B^s_{p,q_2}$ if $q_1 \leqslant q_2$ – but plays a minor role in comparison to s since clearly

$$B^{s_1}_{p,q_1} \subset B^{s_2}_{p,q_2}, \quad \text{if } s_1 > s_2, \tag{25.14}$$

regardless of the values of q_1 and q_2. Roughly speaking, smoothness of order s in L^p is thus expressed here by the fact that, for n large enough, $\omega_n(f,t)_p$ goes to 0 like $\mathcal{O}(t^s)$ as $t \to 0$. Clearly $C^s = B^s_{\infty,\infty}$ when s is not an integer.

REMARK 25.3. More generally, it can be proved (TRIEBEL [1983]) that $W^{s,p} = B^s_{p,p}$ when s is not an integer. One should be aware of the fact that the spaces $W^{m,p}$ are not Besov spaces for $m \in \mathbb{N}$ and $p \neq 2$. In particular, it is well known that $W^{1,p}$ can be characterized by the property $\omega_1(f,t)_p \lesssim t$ if $1 < p < +\infty$, i.e. $W^{1,p} := \operatorname{Lip} 1(L^p)$, but this space differs from $B^1_{p,\infty}$ which involves the second-order differences in its definition.

REMARK 25.4. An alternate definition of Besov space can be obtained through the *Littlewood–Paley decomposition*: one fixes a compactly supported, C^∞ function

$S(\omega) \geqslant 0$ such that $S(0) \neq 0$, and decomposes an arbitrary tempered distribution $f \in \mathcal{S}'(\mathbb{R}^d)$ according to

$$f = S_0 f + \sum_{j \geqslant 0} \Delta_j f, \tag{25.15}$$

where by definition

$$\Delta_j = S_{j+1} - S_j, \qquad \mathcal{F}(S_j f)(\omega) = S(2^{-j}\omega)\hat{f}(\omega). \tag{25.16}$$

The Besov space $B^s_{p,q}$ is then defined as the space of all tempered distributions such that

$$\|S_0 f\|_{L^p} + \left\| \left(2^{sj}\|\Delta_j f\|_{L^p}\right)_{j \geqslant 0} \right\|_{\ell^q}, \tag{25.17}$$

is finite. One easily checks that this property is independent of the choice of the function S. This is actually the definition of Besov spaces which is chosen in some classical references on Besov spaces, e.g., in TRIEBEL [1983], allowing to consider also negative values of s, and all values $p, q > 0$. It can be shown (see, e.g., TRIEBEL [1983]) that (25.17) and (25.11) are equivalent norms when $s > 0$ and $\Omega = \mathbb{R}^d$ (assuming here that $p, q \geqslant 1$). Our results of Sections 28 and 29 will actually provide a proof of this fact. For more general domains, the use of (25.17) obliges to define Besov spaces by restriction, while (25.11) offers an inner description. Nevertheless, these two definitions still coincide for fairly general domains (see Section 30). An immediate consequence of the Littlewood–Paley characterization is that $H^s = W^{s,2} = B^s_{2,2}$ for all $s > 0$.

Sobolev, Besov and potential spaces satisfy two simple imbedding relations: for fixed p (and arbitrary q in the case of Besov spaces, see (25.14)), these spaces get larger as s decreases, and in the case where Ω a bounded domain, for fixed s (and fixed q in the case of Besov spaces), these spaces get larger as p decrease, since $\|f\|_{L^{p_1}} \lesssim \|f\|_{L^{p_2}}$ if $p_1 \leqslant p_2$.

A less trivial type of result is the so-called *Sobolev imbedding* theorem. It states that

$$W^{s_1,p_1} \subset W^{s_2,p_2} \quad \text{if } s_1 - s_2 \geqslant d(1/p_1 - 1/p_2), \tag{25.18}$$

except in the case where $p_2 = +\infty$ and $s_1 - d(1/p_1 - 1/p_2)$ is an integer, for which one needs to assume that $s_1 - s_2 > d(1/p_1 - 1/p_2)$. In the case of Besov spaces, a similar imbedding relation is given by

$$B^{s_1}_{p_1,q} \subset B^{s_2}_{p_2,q} \quad \text{if } s_1 - s_2 \geqslant d(1/p_1 - 1/p_2), \tag{25.19}$$

with no restriction on the indices $s_1, s_2 > 0$, $p_1, p_2 > 0$ and $q > 0$. The proof of these imbedding theorems can be found in ADAMS [1975] for Sobolev spaces and TRIEBEL [1983] for Besov spaces. We shall also give a proof of the less classical Besov imbedding in Section 30.

The Besov spaces can also be defined for p and q less than 1. This extension, which will be of particular importance to us in the study of nonlinear and adaptive approximation of Chapter IV, is the source of additional difficulties. First, the expressions (25.11) or (25.17) do not define norms but only *quasi-norms*: they fail to satisfy the triangle inequality, but still do satisfy

$$\|f+g\| \leqslant C(\|f\|+\|g\|) \quad \text{and} \quad \|f+g\|^{\mu} \leqslant \|f\|^{\mu}+\|g\|^{\mu}, \tag{25.20}$$

for C large enough and $\mu > 0$ small enough. The second property – sometimes called the μ-triangle inequality – clearly holds with $\mu = \min\{p, q\}$ and implies the first property with $C = C(\mu)$. One still has the completeness property, i.e. these spaces are quasi-Banach spaces (in particular $\sum_j f_j$ converges if $\sum_j \|f_j\|^{\mu} < +\infty$). Secondly, for p less than 1, the L^p boundedness of f does not even ensure that it is a distribution, so that the evaluation of inner products $\langle f, g\rangle$ with test functions g is a priori meaningless. One way of turning around this difficulty is to replace L^p-spaces by the Hardy H^p-spaces in the definition of the Besov spaces, as proposed in PEETRE [1974, Chapter 11]. Finally, the equivalence between the expressions (25.11) and (25.17) is only valid under the assumption $s/d > \max\{1/p - 1, 0\}$ (see TRIEBEL [1983, p. 110]). In all the following (in particular in Section 29 and in the next chapter), we shall mainly consider Besov spaces in this range of indices. By the imbedding theorem (25.19), which also holds for $p_1, p_2 < 1$ (see Section 30), this assumption also ensures that the corresponding space is embedded in L^1. In these range of indices, the definition of Besov spaces by (25.11) is thus sufficient to obtain a proper space of distributions.

In the next sections up to Section 30, we shall only consider Besov and Sobolev spaces for $p \geqslant 1$.

We end this brief presentation by recalling without proof two important results dealing with local polynomial approximation. We denote by Π_m the set of polynomials of total degree m, i.e. linear combinations of the basis functions $x_1^{k_1} \cdots x_d^{k_d}$ for $k_1 + \cdots + k_d \leqslant m$.

The first result is a simple version of the Deny–Lions theorem, firstly proved in DENY and LIONS [1954], which is of common use in the numerical analysis of the finite element method (see also CIARLET [1991, Theorem 14.1]). A more elaborate version of this result will be proved in Section 33 in order to treat the case of Sobolev spaces with homogeneous boundary conditions.

THEOREM 25.1. *For $1 \leqslant p \leqslant +\infty$, m a positive integer and $\Omega \subset \mathbb{R}^d$ a connected bounded domain, we have the estimate*

$$\inf_{g \in \Pi_m} \|f-g\|_{L^p(\Omega)} \leqslant C|f|_{W^{m+1,p}(\Omega)}, \tag{25.21}$$

where the constant C only depends on d, m, p and Ω.

The second result is a local variant of a theorem by Whitney, involving the modulus of smoothness. In addition to the previous result, it covers the case $0 < p < 1$, but to our knowledge it is only proved for cubic domains (we shall prove it in Section 32 for more general domains). We refer to BRUDNYI [1970] for a proof of this result in the case $p \geqslant 1$ and to OSWALD and STOROSHENKO [1978] for the generalization to all $p > 0$.

THEOREM 25.2. *For $p > 0$ and $\Omega = [0, 1]^d$, one has*

$$\inf_{g \in \Pi_m} \|f - g\|_{L^p(\Omega)} \leqslant C\omega_{m+1}(f, 1, \Omega)_p, \tag{25.22}$$

where the constant C only depends on d, m and p.

These results have direct consequences on local polynomial approximation error, by a rescaling argument: if I_h is any cube of sidelength $h > 0$, we obtain

$$\inf_{g \in \Pi_m} \|f - g\|_{L^p(I_h)} \lesssim h^{m+1} |f|_{W^{m+1,p}(I_h)}, \tag{25.23}$$

and

$$\inf_{g \in \Pi_m} \|f - g\|_{L^p(I_h)} \lesssim \omega_{m+1}(f, h, I_h)_p \tag{25.24}$$

by applying (25.21) and (25.22) to $f_h := f(T_h \bullet)$, where T_h is an affine transformation mapping Q onto I_h.

26. Direct estimates

In order to prove the direct estimate for the V_j spaces, we shall combine two ingredients: (i) the local polynomial approximation results of Section 25 and (ii) the polynomial reproduction properties of the V_j spaces.

In this section, as well as in Sections 27, 29 and 30, we assume that $(V_j, \widetilde{V}_j)$ is a pair of biorthogonal multiresolution spaces generated by tensor product type, compactly supported, dual scaling functions φ and $\tilde{\varphi}$ as constructed in the previous chapter. We thus have $V_j = \operatorname{Span}\{\varphi_{j,k} = 2^{dj/2}\varphi(2^j \bullet -k);\ k \in \mathbb{Z}^d\}$ and similarly for $\widetilde{V}_j$. We also denote by P_j the corresponding projector defined by

$$P_j f = \sum_{k \in \mathbb{Z}^d} \langle f, \tilde{\varphi}_{j,k} \rangle \varphi_{j,k}. \tag{26.1}$$

As in the previous chapters, we also use the notation $Q_j = P_{j+1} - P_j$ for the projectors onto the detail spaces.

We recall that polynomial exactness up to some order n in the V_j spaces is ensured by designing φ so that the Strang–Fix conditions are fulfilled at order n.

As we already mentioned, we are only concerned here (and up to Section 30) with L^p Sobolev and Besov spaces with $p \geqslant 1$. In the proof of our direct estimates, we shall make use of certain L^p-stability properties of the projector P_j (we already noticed the L^2 stability of this projector in Section 11 of Chapter II). More precisely, we have the following result.

THEOREM 26.1. *Let $1 \leqslant p \leqslant \infty$. Assume that $\varphi \in L^p$ and $\tilde{\varphi} \in L^{p'}$ where $1/p + 1/p' = 1$. Then the projectors P_j are uniformly bounded in L^p. Moreover the basis $\varphi_{j,k}$ is L^p-stable, in the sense that the equivalence*

$$\Big\| \sum_{k \in \mathbb{Z}^d} c_k \varphi_{j,k} \Big\|_{L^p} \sim 2^{dj(1/2-1/p)} \big\| (c_k)_{k \in \mathbb{Z}^d} \big\|_{\ell^p}, \tag{26.2}$$

holds with constants that do not depend on j.

PROOF. We first prove the upper inequality of (26.2), i.e.

$$\Big\| \sum_{k \in \mathbb{Z}^d} c_k \varphi_{j,k} \Big\|_{L^p} \lesssim 2^{dj(1/2-1/p)} \big\| (c_k)_{k \in \mathbb{Z}^d} \big\|_{\ell^p}, \tag{26.3}$$

which is a straightforward consequence of the compact support of φ. On each dyadic cube $I_{j,l}$ defined by (24.3), we have the local estimate

$$\begin{aligned} \Big\| \sum_{k \in \mathbb{Z}^d} c_k \varphi_{j,k} \Big\|_{L^p(I_{j,l})} &\lesssim \sup_{l-k \in \mathrm{Supp}(\varphi)} |c_k| \|\varphi_{j,0}\|_{L^p} \\ &\lesssim 2^{dj(1/2-1/p)} \big\| (c_k)_{l-k \in \mathrm{Supp}(\varphi)} \big\|_{\ell^p}, \end{aligned}$$

where we have used the equivalence of all norms in finite dimensions. Taking the pth power and summing on l (or taking the supremum on l if $p = \infty$) yields the upper inequality.

For the L^p stability of P_j, we use the Hölder estimate to derive

$$\big|\langle f, \tilde{\varphi}_{j,k}\rangle\big| \leqslant \|f\|_{L^p(\mathrm{Supp}(\tilde{\varphi}_{j,k}))} \big\|\tilde{\varphi}_{j,0}\big\|_{L^{p'}} \lesssim 2^{dj(1/2-1/p')} \|f\|_{L^p(\mathrm{Supp}(\tilde{\varphi}_{j,k}))}. \tag{26.4}$$

Taking the pth power and summing on k (or taking the supremum on k if $p = \infty$), we thus obtain

$$\big\| \big(\langle f, \tilde{\varphi}_{j,k}\rangle\big)_{k \in \mathbb{Z}^d} \big\|_{\ell^p} \lesssim 2^{dj(1/2-1/p')} \|f\|_{L^p}, \tag{26.5}$$

which combined with (26.3) gives

$$\|P_j f\|_{L^p} \leqslant C \|f\|_{L^p}, \tag{26.6}$$

where C does not depend on j.

In order to prove the lower inequality in the equivalence (26.2), we simply apply (26.5) to $f = \sum_{k \in \mathbb{Z}^d} c_k \varphi_{j,k}$ and remark that $c_k = \langle f, \tilde{\varphi}_{j,k}\rangle$. □

REMARK 26.1. The L^p stability is *local*, in the sense that

$$\|P_j f\|_{L^p(I_{j,l})} \leqslant C \|f\|_{L^p(\tilde{I}_{j,l})}, \tag{26.7}$$

where

$$\widetilde{I}_{j,l} := \bigcup_{k \in F_l} \operatorname{Supp}(\tilde{\varphi}_{j,k}) \quad \text{with } F_l = \{k \in \mathbb{Z}^d;\ |\operatorname{Supp}(\varphi_{j,k}) \cap I_{j,l}| \neq 0\}, \tag{26.8}$$

(i.e. $\widetilde{I}_{j,l} := I_{j,l} + 2^{-j}[\operatorname{Supp}(\tilde{\varphi}) - \operatorname{Supp}(\varphi)]$) is the larger cube on which the value of f influences the value of $P_j f$ on $I_{j,l}$.

REMARK 26.2. If φ continuous and $\tilde{\varphi}$ is a compactly supported Radon measure, the same reasoning can be used to prove that P_j maps $C(\mathbb{R}^d)$ into itself with an L^∞ bound that does not depend on j. This applies, e.g., to the interpolation operator corresponding to a pair $(\varphi, \tilde{\varphi})$ where φ is interpolatory and $\tilde{\varphi} = \delta_0$.

We now turn to the direct estimate. Note that if P_j is L^p stable uniformly in j, then according to Lemma 11.1, we have

$$\inf_{g \in V_j} \|f - g\|_{L^p} \sim \|f - P_j f\|_{L^p}, \tag{26.9}$$

i.e. the error estimate $\|f - P_j f\|_{L^p}$ is optimal in V_j. We now establish the direct estimate for this particular approximation process.

THEOREM 26.2. *Under the same assumptions as in Theorem* 26.1, *we have*

$$\|f - P_j f\|_{L^p} \lesssim 2^{-nj} |f|_{W^{n,p}}, \tag{26.10}$$

where $n - 1$ is the order of polynomial exactness in V_j.

PROOF. On each dyadic cubes $\widetilde{I}_{j,k}$ as defined by (26.8), we consider a polynomial $p_{j,k} \in \Pi_{n-1}$ such that

$$\|f - p_{j,k}\|_{L^p(\widetilde{I}_{j,k})} \leqslant 2 \inf_{g \in \Pi_{n-1}} \|f - g\|_{L^p(\widetilde{I}_{j,k})} \lesssim 2^{-nj} |f|_{W^{n,p}(\widetilde{I}_{j,k})}, \tag{26.11}$$

where the second inequality follows from Theorem 25.1.

We now estimate the local error by

$$\begin{aligned}
\|f - P_j f\|_{L^p(I_{j,k})} &\leqslant \|f - p_{j,k}\|_{L^p(I_{j,k})} + \|P_j f - p_{j,k}\|_{L^p(I_{j,k})} \\
&= \|f - p_{j,k}\|_{L^p(I_{j,k})} + \big\|P_j(f - p_{j,k})\big\|_{L^p(I_{j,k})} \\
&\lesssim \|f - p_{j,k}\|_{L^p(\widetilde{I}_{j,k})} \\
&\lesssim 2^{-nj} |f|_{W^{n,p}(\widetilde{I}_{j,k})},
\end{aligned}$$

where we have successively used the polynomial reproduction properties of P_j, the local stability estimate (26.7) and the polynomial approximation estimate (26.11).

The global estimate (26.10) follows by taking the pth power and summing on k, or by taking the supremum in k when $p=\infty$. □

REMARK 26.3. In the proof of the direct estimate, as well as for the L^p stability of P_j we make no use of the duality relation $\langle\varphi, \tilde{\varphi}(\bullet - k)\rangle = \delta_{0,k}$. The conclusion would thus be the same for a general operator P_j of the form (26.1) with the sole assumptions that P_j reproduces polynomials up to the order n. This is, e.g., the case of the quasi-interpolant operators introduced by DE BOOR and FIX [1973a] for proving the approximation properties of spline functions. In that particular case, an interesting additional property is that the quasi-interpolant is designed for splines with nonequally spaced knots, while its L^p-norm remains bounded independently of the choice of the knots. In contrast, we have used the existence of dual basis with compact support in order to prove the L^p-stability estimate from below in (26.2). Note however that the refinability of $\tilde{\varphi}$ was not used here, i.e. we could have used any other dual basis with a shift invariant structure. The stability property will be useful for proving the inverse estimate of the next section.

An important variant of the direct estimate (26.10) is the Whitney estimate, which involves the modulus of smoothness.

THEOREM 26.3. *Under the same assumptions as in Theorem* 26.1, *we have*

$$\|f - P_j f\|_{L^p} \lesssim \omega_n\big(f, 2^{-j}\big)_p, \tag{26.12}$$

where $n-1$ *is the order of polynomial exactness in* V_j.

PROOF. We proceed as in the proof of Theorem 26.2, by choosing on each dyadic cubes $\widetilde{I}_{j,k}$ a polynomial $p_{j,k} \in \Pi_{n-1}$ such that

$$\|f - p_{j,k}\|_{L^p(\widetilde{I}_{j,k})} \leqslant 2 \inf_{g\in\Pi_{n-1}} \|f - g\|_{L^p(\widetilde{I}_{j,k})} \lesssim \omega_n\big(f, 2^{-j}, \widetilde{I}_{j,k}\big), \tag{26.13}$$

where the second inequality is easily obtained from (25.25) and the fact that $\omega_n(f, mt, \Omega)_p \leqslant m^n \omega_n(f, t, \Omega)_p$. The local projection error follows by the same arguments as in the proof Theorem 26.2, i.e.

$$\begin{aligned}
\|f - P_j f\|_{L^p(I_{j,k})} &\leqslant \|f - p_{j,k}\|_{L^p(I_{j,k})} + \|P_j f - p_{j,k}\|_{L^p(I_{j,k})} \\
&= \|f - p_{j,k}\|_{L^p(I_{j,k})} + \big\|P_j(f - p_{j,k})\big\|_{L^p(I_{j,k})} \\
&\lesssim \|f - p_{j,k}\|_{L^p(\widetilde{I}_{j,k})} \\
&\lesssim \omega_n\big(f, 2^{-j}, \widetilde{I}_{j,k}\big).
\end{aligned}$$

In order to turn this into a global estimate, we encounter a difficulty since it is not clear that we have

$$\big\|\big(\omega_n\big(f, 2^{-j}, \widetilde{I}_{j,k}\big)_p\big)_{k\in\mathbb{Z}^d}\big\|_{\ell^p} \lesssim \omega_n\big(f, 2^{-j}\big)_p, \tag{26.14}$$

except for the case $p = \infty$. At this point, it is useful to introduce, for $p < +\infty$, the following variant of the modulus of smoothness:

$$\tilde{\omega}_n(f, t, \Omega)_p := \left(t^{-d} \int_{[-t,t]^d} \int_{\Omega_{h,n}} \left| \Delta_h^n f(x) \right|^p \mathrm{d}x \, \mathrm{d}h \right)^{1/p}. \tag{26.15}$$

We clearly have $\tilde{\omega}_n(f, t, \Omega)_p \leqslant 2\omega_n(f, t, \Omega)_p$. A less trivial result is that, assuming that Ω is a cube and $t < K \operatorname{diam}(\Omega)$ with K a fixed constant, one actually has a full equivalence

$$\tilde{\omega}_n(f, t, \Omega)_p \sim \omega_n(f, t, \Omega)_p \tag{26.16}$$

in the sense that

$$\omega_n(f, t, \Omega)_p \leqslant C \tilde{\omega}_n(f, t, \Omega)_p, \tag{26.17}$$

with a constant C that only depends on K, p and n. A proof of this result can be found in DEVORE and LORENTZ [1993, p. 185] in the univariate case, i.e. when Ω is an interval. We adapt this proof below to the case of a multivariate cube. Since the modified modulus clearly satisfies

$$\left\| \left(\tilde{\omega}_n \left(f, 2^{-j}, \tilde{I}_{j,k} \right)_p \right)_{k \in \mathbb{Z}^d} \right\|_{\ell^p} \lesssim \tilde{\omega}_n \left(f, 2^{-j} \right)_p, \tag{26.18}$$

it follows that, if (26.17) is satisfied, (26.14) also holds for all values of p, allowing to conclude the proof of (26.12).

It remains to prove (26.17). By a rescaling argument, it suffices to prove this for $\Omega = [-1, 1]^d$. From the property $\omega_n(f, mt, \Omega)_p \leqslant m^n \omega_n(f, t, \Omega)_p$, it is also sufficient to prove (26.17) for t less than some arbitrary constant. Here, we assume $t \leqslant (2n)^{-1}$. For $|h| \leqslant t$, we want to prove that $\|\Delta_h^n f\|_{L^p(\Omega_{h,n})}$ is less than $C\tilde{\omega}_n(f, t, \Omega)_p$ with a constant C that does not depend on h and t. To do so, we split the cube Ω into 2^d subcubes of half-side length. One of these is $J := [-1, 0]^d$. We shall now estimate $\|\Delta_h^n f\|_{L^p(\Omega_{h,n} \cap J)}$. Our starting point is the identity

$$\Delta_h^n f(x) = \sum_{k=1}^{n} (-1)^k \binom{n}{k} \left[\Delta_{ky}^n f(x + kh) - \Delta_{h+ky}^n f(x) \right], \tag{26.19}$$

which can be proved by simply remarking that

$$\begin{aligned} 0 &= \sum_{k=0}^{n} \sum_{j=0}^{n} (-1)^{k+j} \binom{n}{k} \binom{n}{j} \left[f(x + kh + jky) - f(x + jh + jky) \right] \\ &= \sum_{k=0}^{n} (-1)^k \binom{n}{k} \left[\Delta_{ky}^n f(x + kh) - \Delta_{h+ky}^n f(x) \right], \end{aligned}$$

and remarking that the term corresponding to $k=0$ in the above sum is exactly $-\Delta_h^n f(x)$. From (26.19), we derive

$$\begin{aligned}\|\Delta_h^n f\|_{L^p(\Omega_{h,n}\cap J)} \\ \leqslant \sum_{k=1}^{n}\binom{n}{k}\big[\|\Delta_{ky}^n f(\cdot+kh)\|_{L^p(\Omega_{h,n}\cap J)}+\|\Delta_{h+ky}^n f\|_{L^p(\Omega_{h,n}\cap J)}\big].\end{aligned} \tag{26.20}$$

If we take the pth power of the right-hand side terms and integrate in y over the domain $[0,t/n]^d$, we obtain by changes of variable

$$\begin{aligned}&\int_{[0,t/n]^d}\|\Delta_{ky}^n f(\cdot+kh)\|_{L^p(\Omega_{h,n}\cap J)}^p\,\mathrm{d}y \\ &\quad=\int_{[0,t/n]^d}\|\Delta_{ky}^n f\|_{L^p(kh+\Omega_{h,n}\cap J)}^p\,\mathrm{d}y \\ &\quad\lesssim\int_{[0,t]^d}\int_{\Omega_{y,n}}|\Delta_y^n f(x)|^p\,\mathrm{d}x\,\mathrm{d}y,\end{aligned}$$

and

$$\begin{aligned}\int_{[0,t/n]^d}\|\Delta_{h+ky}^n f\|_{L^p(\Omega_{h,n}\cap J)}^p\,\mathrm{d}y &\lesssim \int_{[h,t+h]^d}\|\Delta_y^n f\|_{L^p(\Omega_{h,n}\cap J)}^p\,\mathrm{d}y \\ &\lesssim \int_{[-t,2t]^d}\int_{\Omega_{y,n}}|\Delta_y^n f(x)|^p\,\mathrm{d}x\,\mathrm{d}y.\end{aligned}$$

It follows that we have the estimate

$$(t/n)^d\|\Delta_h^n f\|_{L^p(\Omega_{h,n}\cap J)}^p \lesssim \int_{[-2t,2t]^d}\int_{\Omega_{y,n}}|\Delta_y^n f(x)|^p\,\mathrm{d}x\,\mathrm{d}y. \tag{26.21}$$

We can proceed in a similar way to obtain estimates of the above type for the 2^d-1 other subcubes of Ω. Summing up these estimates yields

$$\omega_n(f,2t,\Omega)_p \leqslant 2\omega_n(f,t,\Omega)_p \lesssim \tilde{\omega}_n(f,2t,\Omega)_p,$$

and thus (26.17). □

REMARK 26.4. One can easily check that

$$\omega_n(f,t)_p \lesssim t^n|f|_{W^{n,p}}. \tag{26.22}$$

The Whitney estimate of Theorem 26.3 is thus a stronger result than the Jackson estimate of Theorem 26.2.

We end this section by a direct estimate for general Besov spaces, which is a simple corollary of the Whitney estimate.

COROLLARY 26.1. *Under the same assumptions as in Theorem* 26.1, *we have*

$$\|f - P_j f\|_{L^p} \lesssim 2^{-js}|f|_{B^s_{p,q}}, \tag{26.23}$$

for $0 < s < n$, *where* $n - 1$ *is the order of polynomial exactness in* V_j.

PROOF. If $s < n$, it follows from the definition of the Besov semi-norm that

$$\omega_n\big(f, 2^{-j}\big)_p \leqslant 2^{-js}|f|_{B^s_{p,q}}. \tag{26.24}$$

Combining this with the Whitney estimate (26.12), we obtain (26.19). □

REMARK 26.5. From the equivalence (26.16) between ω and $\tilde{\omega}$ and from (26.18), we can also derive

$$\big\|\big(|f|_{B^s_{p,p}(\tilde{I}_{j,k})}\big)_{k\in\mathbb{Z}^d}\big\|_{\ell^p} \lesssim |f|_{B^s_{p,p}(\mathbb{R}^d)}, \tag{26.25}$$

which was not obvious in view of the initial definition of $B^s_{p,p}$ involving only ω. In the case where s is not an integer, (26.16) can also be used to prove the equivalence between the $B^s_{p,p}$ and $W^{s,p}$ semi-norms. Note that it is straightforward from (25.7) that the $W^{s,p}$ semi-norm satisfies a superadditivity property of the type (26.25).

27. Inverse estimates

Inverse (or Bernstein-type) estimate aim to take into account the smoothness properties of the approximation spaces V_j. Our first basic estimate relates the L^p and $W^{n,p}$ norm of a function $f \in V_j$.

THEOREM 27.1. *Under the assumptions of Theorem* 26.1, *and with* $\varphi \in W^{n,p}$ *one has*

$$\|f\|_{W^{n,p}} \leqslant C2^{nj}\|f\|_{L^p} \quad \text{if } f \in V_j, \tag{27.1}$$

with a constant C *that does not depend on* j.

PROOF. We only need to estimate $|f|_{W^{n,p}}$. For this, we use the decomposition into the local basis $f = \sum_{k\in\mathbb{Z}^d} c_k\varphi_{j,k}$. For any multi-index $m = (m_1, \ldots, m_d)$ such that $m_1 + \cdots + m_d = n$, we obtain

$$\begin{aligned}\|\partial^m f\|_{L^p} &= \Big\|\sum_{k\in\mathbb{Z}^d} c_k\partial^m\varphi_{j,k}\Big\|_{L^p} = 2^{nj}\Big\|\sum_{k\in\mathbb{Z}^d} c_k 2^{dj/2}(\partial^m\varphi)(2^j\bullet - k)\Big\|_{L^p}\\ &\lesssim 2^{nj}2^{jd(1/2-1/p)}\big\|(c_k)_{k\in\mathbb{Z}^d}\big\|_{\ell^p},\end{aligned}$$

where we have used the same computation as in the proof of (26.3), replacing φ by $\partial^m\varphi$ (which is also a compactly supported L^p function). We then conclude by using the lower inequality in the equivalence (26.2). □

Another type of inverse estimate involves the modulus of smoothness.

THEOREM 27.2. *Under the assumptions of Theorem* 27.1, *one has*

$$\omega_n(f,t)_p \leqslant C\left[\min\{1,2^jt\}\right]^n \|f\|_{L^p} \quad \text{if } f\in V_j, \tag{27.2}$$

with a constant C that does not depend on j.

PROOF. This estimate is trivial for $t \geqslant 2^{-j}$. From the monotonicity in t of $\omega_n(f,t)_p$, it suffices to prove it for $t=2^{-l}$, $l>j$. For this we proceed as in Theorem 27.1, using the decomposition of f into the scaling function basis, which gives

$$\omega_n(f,t)_p \lesssim \omega_n(\varphi_{j,0},t)_p \left\|(c_k)_{k\in\mathbb{Z}^d}\right\|_{\ell^p}. \tag{27.3}$$

We then remark that, by (26.22), for $l>j$,

$$\omega_n\left(\varphi_{j,0},2^{-l}\right)_p = 2^{jd(1/2-1/p)}\omega_n\left(\varphi,2^{j-l}\right)_p \lesssim |\varphi|_{W^{n,p}}2^{jd(1/2-1/p)}2^{n(j-l)}. \tag{27.4}$$

We conclude the proof, using the lower inequality in the equivalence (26.2). □

Our third inverse estimates deals with general Besov spaces of integer or fractional order.

THEOREM 27.3. *Under the assumptions of Theorem* 26.1, *and with $\varphi\in B^s_{p,q}$ one has*

$$\|f\|_{B^s_{p,q}} \leqslant C2^{sj}\|f\|_{L^p} \quad \text{if } f\in V_j, \tag{27.5}$$

with a constant C that does not depend on j.

PROOF. For $f=\sum_{k\in\mathbb{Z}^d} c_k\varphi_{j,k}\in V_j$, we first apply (27.3) with some $n>s$, and the definition of the Besov semi-norm, to obtain

$$|f|_{B^s_{p,q}} \lesssim \left\|(c_k)_{k\in\mathbb{Z}^d}\right\|_{\ell^p}|\varphi_{j,0}|_{B^s_{p,q}}. \tag{27.6}$$

It remains to evaluate $|\varphi_{j,0}|_{B^s_{p,q}}$. For $l\leqslant j$, we use the crude estimate

$$\omega_n\left(\varphi_{j,0},2^{-l}\right)_p \lesssim \|\varphi_{j,0}\|_{L^p} \lesssim 2^{dj(1/2-1/p)}. \tag{27.7}$$

For $l>j$, we exploit the fact that $\varphi\in B^s_{p,q}$ to obtain the estimate

$$\omega_n\left(\varphi_{j,0},2^{-l}\right)_p = 2^{dj(1/2-1/p)}\omega_n\left(\varphi,2^{j-l}\right)_p \leqslant 2^{s(j-l)}2^{dj(1/2-1/p)}\varepsilon_{l-j}, \tag{27.8}$$

where $(\varepsilon_n)_{n\geqslant 0}$ is an ℓ^q sequence and

$$\|(\varepsilon_n)_{n\geqslant 0}\|_{\ell^q} \leqslant |\varphi|_{B^s_{p,q}}. \tag{27.9}$$

Combining (27.7) and (27.8), we then obtain

$$\begin{aligned}|\varphi_{j,0}|_{B^s_{p,q}} &= \|(2^{ls}\omega_n(\varphi_{j,0}, 2^{-l})_{L^p})_{l\geqslant 0}\|_{\ell^p}\\ &\leqslant \|(2^{ls}\omega_n(\varphi_{j,0}, 2^{-l})_{L^p})_{0\leqslant l\leqslant j}\|_{\ell^p} + \|(2^{ls}\omega_n(\varphi_{j,0}, 2^{-l})_{L^p})_{l>j}\|_{\ell^p}\\ &\lesssim 2^{js}2^{dj(1/2-1/p)}.\end{aligned}$$

We conclude the proof, injecting this estimate in (27.6) and using the lower inequality in the equivalence (26.3). □

We end this section by mentioning more general direct and inverse estimate involving Sobolev and Besov norms on both side of the inequalities. They are corollaries of the direct and inverse estimates that we have obtained so far.

COROLLARY 27.1. *Let $1 \leqslant p, q_1, q_2 \leqslant \infty$ and $0 < s < t$, and assume that $\varphi \in L^p$ and $\tilde{\varphi} \in L^{p'}$. If $\varphi \in B^s_{p,q_1}$ and $t < n$ where $n-1$ is the degree of polynomial reproduction in V_j, one has the direct estimate*

$$\|f - P_j f\|_{B^s_{p,q_1}} \lesssim 2^{-j(t-s)}|f|_{B^t_{p,q_2}}. \tag{27.10}$$

When s and/or t are integers, these estimates also hold with the classical Sobolev spaces $W^{s,p}$ and/or $W^{t,p}$ and t up to n.

PROOF. We use the multiscale decomposition of f to estimate the approximation error by

$$\|f - P_j f\|_{B^s_{p,q_1}} \leqslant \sum_{l\geqslant j} \|P_{l+1}f - P_l f\|_{B^s_{p,q_1}}. \tag{27.11}$$

Combining the inverse estimate of Theorem 27.3 and the direct estimate of Corollary 26.1, we obtain

$$\begin{aligned}\|P_{l+1}f - P_l f\|_{B^s_{p,q_1}} &\lesssim 2^{sl}\|P_{l+1}f - P_l f\|_{L^p}\\ &\lesssim 2^{-l(t-s)}|P_{l+1}f - P_l f|_{B^t_{p,q_2}},\end{aligned} \tag{27.12}$$

where we have used $\|P_{l+1}f - P_l f\| \leqslant \|f - P_l f\| + \|f - P_{l+1}f\|$. This together with (27.11), yields the direct estimate (27.10). The case of classical Sobolev spaces is treated in the same way, using the direct and inverse estimate of Theorem 26.2 and 27.1. □

COROLLARY 27.2. *Let $1 \leqslant p, q_1, q_2 \leqslant \infty$ and $0 < s < t$, and assume that $\varphi \in L^p$ and $\tilde{\varphi} \in L^{p'}$. If $\varphi \in B^t_{p,q_2}$ and $s < n$ where $n-1$ is the degree of polynomial reproduction in V_j, one has the inverse estimate*

$$\|f\|_{B^t_{p,q_2}} \lesssim 2^{j(t-s)} \|f\|_{B^s_{p,q_1}} \quad \text{if } f \in V_j. \tag{27.13}$$

When s and/or t are integers, these estimates also hold with the classical Sobolev spaces $W^{s,p}$ and/or $W^{t,p}$ and s up to n.

PROOF. We use the multiscale decomposition of f to estimate its B^t_{p,q_2} norm as follows

$$\|f\|_{B^t_{p,q_2}} \leqslant \|P_0 f\|_{B^t_{p,q_2}} + \sum_{l=0}^{j-1} \|P_{l+1} f - P_l f\|_{B^t_{p,q_2}}. \tag{27.14}$$

We clearly have $\|P_0 f\|_{B^t_{p,q_2}} \lesssim \|f\|_{L^p} \lesssim \|f\|_{B^s_{p,q_1}}$. For the remaining terms, we combine the inverse estimate of Theorem 27.3 and the direct estimate of Corollary 26.1 and obtain

$$\|P_{l+1} f - P_l f\|_{B^t_{p,q_2}} \lesssim 2^{tl} \|P_{l+1} f - P_l f\|_{L^p} \lesssim 2^{l(t-s)} |f|_{B^s_{p,q_1}}. \tag{27.15}$$

This together with (27.14), yields the inverse estimate (27.13). The case of classical Sobolev spaces is treated in the same way, using the direct and inverse estimate of Theorems 26.2 and 27.1. □

28. Interpolation theory and approximation spaces

The goal of this section is to describe a general mechanism that allows to connect approximation and smoothness properties.

Given a sequence of approximation spaces V_j, we would like to relate the property "$\text{dist}_{L^p}(f, V_j) \leqslant \mathcal{O}(2^{-sj})$" to some classical notion of smoothness satisfied by f. We shall actually be more precise in the description of the approximation rate for the function f, as shown by the following definition.

DEFINITION 28.1. If X is a Banach space and $(V_j)_{j\geqslant 0}$ a nested sequence of subspaces of X such that $\bigcup_{j\geqslant 0} V_j$ is dense in X. For $s > 0$ and $1 \leqslant q \leqslant \infty$, we define the approximation space $\mathcal{A}^s_q(X)$ related to the sequence V_j by

$$\mathcal{A}^s_q(X) := \left\{ f \in X \text{ such that } \left(2^{sj} \,\text{dist}_X(f, V_j)\right)_{j\geqslant 0} \in \ell^q \right\}. \tag{28.1}$$

In the case where X is a Lebesgue space, we use the notation $\mathcal{A}^s_{p,q} := \mathcal{A}^s_q(L^p)$.

Roughly speaking, the spaces $\mathcal{A}^s_q(X)$ describe those functions such that $\text{dist}_X(f, V_j)$ behaves like $\mathcal{O}(2^{-sj})$, with a refined information provided by the extra parameter q.

One easily check that it constitutes a proper subspace of X, and that it is a Banach space when equipped with the norm

$$\|f\|_{\mathcal{A}_q^s(X)} := \|f\|_X + |f|_{\mathcal{A}_q^s(X)},$$
$$|f|_{\mathcal{A}_q^s(X)} := \left\|\left(2^{sj}\operatorname{dist}_X(f, V_j)\right)_{j\geqslant 0}\right\|_{\ell^q}. \tag{28.2}$$

Note also that if we set $V_0 = \{0\}$, then one can remove the term $\|f\|_X$ in the definition of this norm since it is equal to $\operatorname{dist}_X(f, V_0)$. In the next section, we shall prove that under specific assumptions on the multiresolution approximation spaces, the identity

$$\mathcal{A}_{p,q}^s = B_{p,q}^s \tag{28.3}$$

holds, together with the norm equivalences

$$\|f\|_{B_{p,q}^s} \sim \|f\|_{\mathcal{A}_{p,q}^s} \quad \text{and} \quad |f|_{B_{p,q}^s} \sim |f|_{\mathcal{A}_{p,q}^s}. \tag{28.4}$$

As we shall see in Section 29, these equivalences can be proved directly, combining the definition of the Besov spaces together with the inverse and direct estimates involving the modulus of smoothness (Theorems 26.3 and 27.2).

In the present section we shall describe a more general mechanism that allows to identify the approximation spaces $\mathcal{A}_{p,q}^s(X)$ with spaces obtained by *interpolation theory*. Although this mechanism can be avoided when proving (28.3) and (28.4), its usefulness will appear in other instances of relating approximation and smoothness properties, in particular in the nonlinear context of Chapter IV.

Interpolation theory is a useful tool of functional analysis for the study of function spaces and operators. Perhaps, the most well known result of this theory is the oldest one: if T is an operator that is bounded in $L^1(\Omega)$ and in $L^\infty(\Omega)$, then the *Riesz–Thorin theorem* states that T is also bounded in $L^p(\Omega)$ for all $p \in]1, +\infty[$.

More generally the goal of interpolation theory is to build from an initial pair of Banach spaces (X, Y), some "intermediate spaces" Z with similar properties with respect to operators. Of course, such a process will be of particular interest if for some specific X and Y these intermediate spaces coincide with some known function spaces. The two most well known strategies for interpolation of function spaces – both introduced in the 1950's – are the real method of J.-L. Lions and J. Peetre, and the complex method of J.-L. Lions and A.P. Calderon. The first one turns out to be directly related to the concept of approximation spaces. Therefore, we shall focus on this approach and we give below some of its main features. A detailed treatment can be found in the books of BERGH and LÖFSTRÖM [1976], and of BENNETT and SHARPLEY [1988].

Let X and Y be a pair of Banach function spaces. To such a pair, we associate the so-called *K-functional* defined for $f \in X + Y$ and $t \geqslant 0$, by

$$K(f, t) = K(f, t, X, Y) := \inf_{a\in X,\ b\in Y,\ a+b=f} \left[\|a\|_X + t\|b\|_Y\right]. \tag{28.5}$$

This functional has some elementary properties. First, it is continuous, nondecreasing and concave with respect to t. The first two properties are direct consequences of the definition, while the last property can be checked by remarking that

$$\big[K(f,t)+K(f,s)\big]/2 \leqslant \|a\|_X + \frac{t+s}{2}\|b\|_Y$$

if $a+b=f$, and then by minimizing on a and b in this inequality. Secondly, if $X\cap Y$ is dense in Y, then $K(f,0):=0$. Similarly, if $X\cap Y$ is dense in X, then $\lim_{t\to+\infty} K(f,t)/t=0$.

For $\theta\in]0,1[$ and $1\leqslant q\leqslant+\infty$, we define a family of intermediate spaces $X\cap Y\subset [X,Y]_{\theta,q}\subset X+Y$ as follows: $[X,Y]_{\theta,q}$ consists of those functions such that

$$\|f\|_{[X,Y]_{\theta,q}} := \big\|t^{-\theta}K(f,t)\big\|_{L^q(]0,+\infty[,\,\mathrm{d}t/t)}, \tag{28.6}$$

is finite. One easily checks that the above defined intermediate spaces inherit the Banach spaces structure of X and Y.

A first remark is that this process has the expected property with respect to linear operators: assuming that T is continuous from X_1 to X_2 and from Y_1 to Y_2, we obtain that for all $t\geqslant 0$,

$$\begin{aligned} K(Tf,t,X_2,Y_2) &= \inf_{a\in X_2,\ b\in Y_2,\ a+b=Tf}\big[\|a\|_{X_2}+t\|b\|_{Y_2}\big] \\ &\leqslant \inf_{a\in X_1,\ b\in Y_1,\ a+b=f}\big[\|Ta\|_{X_2}+t\|Tb\|_{Y_2}\big] \\ &\leqslant CK(f,t,X_1,Y_1), \end{aligned}$$

with $C=\max\{\|T\|_{X_1\to X_2},\|T\|_{Y_1\to Y_2}\}$. It follows that T is bounded from $[X_1,Y_1]_{\theta,q}$ to $[X_2,Y_2]_{\theta,q}$ for all θ and q.

One of the main result of interpolation theory is the *reiteration theorem* that we state below without proof (see, e.g., BERGH and LÖFSTRÖM [1976]). This theorem confirms the visual intuition of the space $[X,Y]_{\theta,q}$ as barycenter between the "end-points" X and Y, with weight θ in Y and $1-\theta$ in X. It also shows that the parameter q plays a minor role in the definition of intermediate spaces.

THEOREM 28.1. *If $Z_1=[X,Y]_{\theta_1,q_1}$ and $Z_2=[X,Y]_{\theta_2,q_2}$ for $0<\theta_1<\theta_2<1$ and $q_1,q_2\in[1,\infty]$, then for $0<\theta<1$ and $1\leqslant q\leqslant\infty$, one has the identities*

$$[X,Z_1]_{\theta,q}=[X,Y]_{\theta_1\theta,q}, \tag{28.7}$$

$$[Z_1,Y]_{\theta,q}=[X,Y]_{\theta+\theta_1(1-\theta),q}, \tag{28.8}$$

and

$$[Z_1,Z_2]_{\theta,q}=[X,Y]_{\theta_1(1-\theta)+\theta_2\theta,q}. \tag{28.9}$$

As we already noted, the main interest of such a theory is the ability to identify classical function spaces as the results of an interpolation process between two other classical spaces. To do so, one needs to understand more deeply the nature of the K-functional related to the pair of spaces which is considered An important instance is interpolation between Lebesgue spaces, for which the K-functional can be well understood in terms of the decreasing rearrangement of f, and for which one has $[L^p, L^q]_{\theta,r} = L^r$, $1/r = \theta/q + (1-\theta)/p$, if $1 \leqslant p \leqslant q \leqslant \infty$.

In the context of this paper, we shall be interested in interpolation between spaces of representing various degrees of smoothness. In particular, we shall always work in the situation where $Y \subset X$ with a continuous imbedding and Y is dense in X. A typical example is $X = L^p$ and $Y = W^{m,p}$.

In this specific situation, we write

$$K(f,t) = \inf_{g \in Y} \|f-g\|_X + t\|g\|_Y, \tag{28.10}$$

and make a few additional remarks.

Firstly, the K functional is bounded at infinity since $K(f,t) \leqslant \|f\|_X$. Therefore, the finiteness of (28.6) is equivalent to

$$\left\| t^{-\theta} K(f,t) \right\|_{L^q(]0,A[,\,dt/t)} < +\infty \tag{28.11}$$

for some fixed $A > 0$ and we can use this modified expression as an equivalent norm for $[X,Y]_{\theta,q}$.

Secondly, due to the monotonicity of $K(f,t)$ in t, we also have an equivalent discrete norm given by

$$\|f\|_{[X,Y]_{\theta,q}} := \left\| \left(\rho^{j\theta} K\left(f, \rho^{-j}\right)\right)_{j\geqslant 0} \right\|_{\ell^q}, \tag{28.12}$$

for any fixed $\rho > 1$.

Finally, if $\|f\|_Y = \|f\|_X + |f|_Y$ where $|\bullet|_Y$ is a semi-norm, one can build the same intermediate spaces by using the modified K-functional

$$\widetilde{K}(f,t) = \inf_{g \in Y} \|f-g\|_X + t|g|_Y, \tag{28.13}$$

and defining the intermediate semi-norm by

$$|f|_{[X,Y]_{\theta,q}} := \left\| \left(\rho^{j\theta} \widetilde{K}\left(f, \rho^{-j}\right)\right)_{j\geqslant 0} \right\|_{\ell^q}. \tag{28.14}$$

An equivalent norm for $[X,Y]_{\theta,q}$ is then given by $\|f\|_X + |f|_{[X,Y]_{\theta,q}}$.

It should be noted that all the above concepts (including the reiteration theorem) carry over to the case where $q \in]0,1[$ and X is a quasi-Banach spaces with $\|\bullet\|_X$ satisfying the μ-triangle inequality of (25.20). The resulting intermediate spaces are of course quasi-Banach spaces.

We our now ready to state the main result of this section, connecting approximation spaces and interpolation spaces through direct and inverse estimates.

THEOREM 28.2. *Assume that* V_j *is a sequence of approximation spaces*

$$V_j \subset V_{j+1} \subset \cdots \subset Y \subset X, \tag{28.15}$$

such that for some $m > 0$, *one has a Jackson-type estimate*

$$\operatorname{dist}_X(f, V_j) = \inf_{g \in V_j} \|f - g\|_X \lesssim 2^{-mj} \|f\|_Y, \tag{28.16}$$

and a Bernstein-type estimate

$$\|f\|_Y \lesssim 2^{mj} \|f\|_X \quad \text{if } f \in V_j. \tag{28.17}$$

Then, for $s \in]0, m[$, *one has the norm equivalence*

$$\left\| \left(2^{js} K\left(f, 2^{-mj}\right)\right)_{j \geqslant 0} \right\|_{\ell^q} \sim \|f\|_X + \left\| \left(2^{js} \operatorname{dist}_X(f, V_j)\right)_{j \geqslant 0} \right\|_{\ell^q}, \tag{28.18}$$

and thus $[X, Y]_{\theta,q} = \mathcal{A}_q^s(X)$ *for* $s = \theta m$.

PROOF. We need to compare the K-functional $K(f, 2^{-mj})$ and the error of best approximation $\operatorname{dist}_X(f, V_j)$. In one direction, this comparison is simple: for all $f \in X$, $g \in Y$ and $g_j \in V_j$, we have

$$\operatorname{dist}_X(f, V_j) \leqslant \|f - g_j\|_X \leqslant \|f - g\|_X + \|g - g_j\|_X. \tag{28.19}$$

Minimizing $\|g - g_j\|_X$ over $g_j \in V_j$ and using (28.16), we obtain

$$\operatorname{dist}_X(f, V_j) \lesssim \|f - g\|_X + 2^{-mj} \|g\|_Y. \tag{28.20}$$

Finally, we minimize over $g \in Y$ to obtain

$$\operatorname{dist}_X(f, V_j) \lesssim K\left(f, 2^{-mj}\right). \tag{28.21}$$

Since $\|f\|_X \lesssim K(f, 1)$ (by the continuous imbedding of Y into X and the triangle inequality), we have thus proved that $\|f\|_{\mathcal{A}_q^s(X)} \lesssim \|f\|_{[X,Y]_{\theta,q}}$.

In the other direction, we let $f_j \in V_j$ be such that

$$\|f - f_j\|_X \leqslant 2 \operatorname{dist}_X(f, V_j) \tag{28.22}$$

and we write

$$\begin{aligned} K\left(f, 2^{-mj}\right) &\leqslant \|f - f_j\|_X + 2^{-mj} \|f_j\|_Y \\ &\leqslant \|f - f_j\|_X + 2^{-mj} \left[\|f_0\|_Y + \|f_1 - f_0\|_Y + \cdots + \|f_j - f_{j-1}\|_Y\right] \end{aligned}$$

$$\lesssim \|f - f_j\|_X + 2^{-mj}\left[\|f_0\|_X + \sum_{l=0}^{j-1} 2^{ml}\|f_{l+1} - f_l\|_X\right]$$

$$\lesssim 2^{-mj}\|f_0\|_X + 2^{-mj}\left[\sum_{l=0}^{j} 2^{ml}\,\mathrm{dist}_X(f, V_l)\right],$$

where we have used the inverse inequality (28.17) (together with the fact that $f_{l+1} - f_l \in V_{l+1}$) and the crude $\|f_0\|_X \leqslant \|f\|_X + 2\mathrm{dist}_X(f, V_0) \leqslant 3\|f\|_X$. We thus do not have the exact converse inequality to (28.21).

In order to conclude the proof, we first remark that the term $2^{-mj}\|f\|_X$ satisfies

$$\left\|\left(2^{sj}2^{-mj}\|f\|_X\right)_{j\geqslant 0}\right\|_{\ell^q} \lesssim \|f\|_X, \tag{28.23}$$

and we concentrate on the second term. We use here a discrete Hardy inequality: if $(a_j)_{j\geqslant 0}$ is a positive sequence and $b_j := 2^{-mj}\sum_{l=0}^{j} 2^{ml}a_l$ with $0 < s < m$, one has

$$\left\|\left(2^{sj}b_j\right)_{j\geqslant 0}\right\|_{\ell^q} \lesssim \left\|\left(2^{sj}a_j\right)_{j\geqslant 0}\right\|_{\ell^q}, \tag{28.24}$$

for all $q \in [1, \infty]$. Applying this inequality to $a_j = \mathrm{dist}_X(f, V_j)$ allows to estimate the weighted ℓ^q norm of the second term and to conclude that $\|f\|_{[X,Y]_{\theta,q}} \lesssim \|f\|_{\mathcal{A}^s_q(X)}$.

We end by the proof of the Hardy inequality. In the case where $q = \infty$, assuming that $a_j \leqslant C_a 2^{-sj}$, we clearly have

$$b_j \leqslant C_a 2^{mj}\sum_{l=0}^{j} 2^{(m-s)l} \lesssim C_a 2^{-sj}. \tag{28.25}$$

For $q < \infty$, we define q' such that $1/q + 1/q' = 1$ and $\varepsilon = (m - s)/2 > 0$. Using Hölder's inequality, we obtain

$$\begin{aligned}
\sum_{j\geqslant 0}\left(2^{sj}b_j\right)^q &= \sum_{j\geqslant 0} 2^{(s-m)qj}\left(\sum_{l=0}^{j} 2^{ml}a_l\right)^q \\
&\leqslant \sum_{j\geqslant 0} 2^{(s-m)qj}\left[\sum_{l=0}^{j}\left(2^{(m-\varepsilon)l}a_l\right)^q\right]\left[\sum_{l=0}^{j}\left(2^{\varepsilon l}\right)^{q'}\right]^{q/q'} \\
&\lesssim \sum_{j\geqslant 0} 2^{-\varepsilon qj}\left[\sum_{l=0}^{j}\left(2^{(m-\varepsilon)l}a_l\right)^q\right] = \sum_{l\geqslant 0}\left(2^{(m-\varepsilon)l}a_l\right)^q\sum_{j\geqslant l} 2^{-\varepsilon qj} \\
&\lesssim \sum_{l\geqslant 0}\left(2^{ml}a_l\right)^q,
\end{aligned}$$

which proves the Hardy inequality. □

In practice, we are interested in similar norm equivalences involving specific approximation operators, rather than the error of best approximation. This is treated in the following variant of the previous theorem.

THEOREM 28.3. *Under the assumptions of Theorem* 28.2, *suppose that we have*

$$\|P_j f - f\| \lesssim 2^{-mj} \|f\|_Y, \tag{28.26}$$

for a family of linear operator $P_j : X \mapsto V_j$ *which is uniformly bounded in* X. *Then, for* $s = \theta m \in]0, m[$, *the* $\mathcal{A}^s_q(X)$ *and* $[X, Y]_{\theta,q}$ *norms are equivalent to*

$$\|P_0 f\|_X + \big\|\big(2^{sj} \|f - P_j f\|_X\big)_{j \geqslant 0}\big\|_{\ell^q}, \tag{28.27}$$

and to

$$\|P_0 f\|_X + \big\|\big(2^{sj} \|Q_j f\|_X\big)_{j \geqslant 0}\big\|_{\ell^q}, \tag{28.28}$$

where $Q_j = P_{j+1} - P_j$.

PROOF. We first consider (28.27). In one direction, since

$$\operatorname{dist}_X(f, V_j) \leqslant \|f - P_j f\|_X, \tag{28.29}$$

and

$$\|f\|_X \leqslant \|P_0 f\|_X + \|f - P_0 f\|_X, \tag{28.30}$$

we clearly have that the $\mathcal{A}^s_q(X)$ is controlled by (28.27). In the other direction, we operate as in the second part of the proof of the previous theorem: replacing f_j by $P_j f$ proves that the $[X, Y]_{\theta,q}$ norm is controlled by (28.27).

We then turn to (28.28). In one direction, we have

$$\|Q_j f\|_X \leqslant \|f - P_{j+1} f\|_X + \|f - P_j f\|_X, \tag{28.31}$$

which shows that (28.28) is controlled by (28.27). In the other direction, we write

$$\|f - P_j f\|_X \leqslant \sum_{l \geqslant j} \|Q_l f\|_X, \tag{28.32}$$

and we conclude by using a discrete Hardy inequality: if $(a_j)_{j \geqslant 0}$ is a positive sequence and $b_j := \sum_{l \geqslant j} a_l$, then one has

$$\big\|\big(2^{sj} b_j\big)_{j \geqslant 0}\big\|_{\ell^q} \lesssim \big\|\big(2^{sj} a_j\big)_{j \geqslant 0}\big\|_{\ell^q}, \tag{28.33}$$

for all $s>0$ and $q\in[1,\infty]$. Applying this with $a_j=\|Q_jf\|_X$ allows to conclude that (28.27) is controlled by (28.28).

We end by proving the discrete Hardy inequality. For $q=\infty$, it is clear that $a_j\leqslant C_a2^{-sj}$ implies $b_j\lesssim C_a2^{-sj}$. For $q<+\infty$, we define q' such that $1/q+1/q'=1$ and $s'=s/2$. Using Hölder's inequality, we obtain

$$\begin{aligned}\sum_{j\geqslant 0}(2^{sj}b_j)^q &\leqslant \sum_{j\geqslant 0}2^{sqj}\Big[\sum_{l\geqslant j}(2^{s'l}a_l)^q\Big]\Big[\sum_{l\geqslant j}2^{-s'lq'}\Big]^{q/q'}\\ &\lesssim \sum_{j\geqslant 0}2^{s'qj}\Big[\sum_{l\geqslant j}(2^{s'l}a_l)^q\Big]=\sum_{l\geqslant 0}(2^{s'l}a_l)^q\Big[\sum_{j=0}^{l}2^{s'qj}\Big]\\ &\lesssim \sum_{l\geqslant 0}(2^{sl}a_l)^q,\end{aligned}$$

which proves the Hardy inequality. □

REMARK 28.1. The spaces $\mathcal{A}^s_q(X)$ are always interpolation spaces in the following sense: from their definition, we clearly have the direct estimate

$$\mathrm{dist}_X(f,V_j)\lesssim 2^{-sj}\|f\|_{\mathcal{A}^s_\infty(X)}. \tag{28.34}$$

If $f\in V_j$, we also have $\mathrm{dist}_X(f,V_l)=0$ for $l\geqslant j$ and $\mathrm{dist}_X(f,V_l)\lesssim\|f\|_X$ for all l, so that we also have the inverse estimate

$$\|f\|_{\mathcal{A}^s_\infty(X)}\leqslant 2^{sj}\|f\|_X,\quad \text{if } f\in V_j. \tag{28.35}$$

By Theorem 28.2, we thus have for $\theta\in]0,1[$,

$$\mathcal{A}^{\theta s}_q(X)=\big[X,\mathcal{A}^s_\infty(X)\big]_{\theta,q}, \tag{28.36}$$

and thus, by reiteration (Theorem 28.1), we have for all $q_1,q_2\in[1,\infty]$,

$$\mathcal{A}^{\theta s}_{q_2}(X)=\big[X,\mathcal{A}^s_{q_1}(X)\big]_{\theta,q_2}, \tag{28.36}$$

and for $0<t<s$,

$$\big[\mathcal{A}^t_{q_1}(X),\mathcal{A}^s_{q_2}(X)\big]_{\theta,q}=\mathcal{A}^r_q(X)\quad\text{with } r=(1-\theta)t+\theta s. \tag{28.37}$$

REMARK 28.2. Theorem 28.2 and 28.3, as well as the previous remark, easily carry over to the case where $q\in]0,1[$ and X is a quasi-Banach spaces with $\|\bullet\|_X$ satisfying the μ-triangle inequality of (25.20). The proof are exactly similar, to the exception of the evaluation of $K(f, 2^{-mj})$ in terms of the approximation error for the proof of the first theorem: we are forced to use the μ-triangle inequality

$$\|f_j\|_Y \leqslant \left[\|f_0\|_Y^\mu + \sum_{l=0}^{j-1} \|f_{l+1} - f_l\|_Y^\mu\right]^{1/\mu} \tag{28.38}$$

in place of the standard triangle inequality. Similarly, we need to use

$$\|f - P_j f\|_X \leqslant \left[\sum_{l \geqslant j} \|Q_l f\|_Y^\mu\right]^{1/\mu}, \tag{28.39}$$

for the proof of the second theorem. In both cases, a straightforward modification of the Hardy inequalities allows to reach the same conclusions.

REMARK 28.3. The identity (28.37) can also be derived in a direct way, using the basic interpolation theory of the sequence spaces

$$\ell_s^p(\mathbb{N}) := \left\{(a_n)_{n \geqslant 0};\ \left\|(a_n)_{n \geqslant 0}\right\|_{\ell_s^p} := \left\|\left(2^{sj} a_n\right)_{n \geqslant 0}\right\|_{\ell^p} < \infty\right\}, \tag{28.40}$$

for $0 < p \leqslant \infty$ and $s \in \mathbb{R}$. For these weighted spaces, the identities

$$\left[\ell_s^p, \ell_t^p\right]_{\theta,p} = \ell_r^p, \quad r = (1-\theta)s + \theta t, \tag{28.41}$$

can be proved in a straightforward manner, using Hardy's inequalities.

29. Characterization of smoothness classes

We now return to the our specific setting of multiresolution spaces and smoothness classes. We shall first give a direct proof of the equivalence between $\mathcal{A}_{p,q}^s$ and $B_{p,q}^s$, that does not require the general mechanism of the previous section.

THEOREM 29.1. *Under the assumptions of Theorem* 26.3 (*Whitney estimate*) *and of Theorem* 27.3 (*inverse estimate*), *we have the norm equivalences*

$$\|f\|_{B_{p,q}^t} \sim \|P_0 f\|_{L^p} + \left\|\left(2^{tj}\|Q_j f\|_{L^p}\right)_{j \geqslant 0}\right\|_{\ell^q} \tag{29.1}$$

and

$$\|f\|_{\mathcal{A}_{p,q}^t} \sim \|f\|_{B_{p,q}^t}, \tag{29.2}$$

for all $t < \min\{n, s\}$, where $n - 1$ is the order of polynomial reproduction of the V_j spaces and s is such that $\varphi \in B_{p,q_0}^s$ for some q_0.

PROOF. Here, we shall directly compare the modulus of smoothness which is involved in the definition of the Besov spaces $B^t_{p,q}$ and the quantities $\|Q_j f\|_{L^p}$.

In one direction, from Theorem 27.3, we have

$$\mathrm{dist}_{L^p}(f, V_j) \leqslant \|f - P_j f\|_{L^p} \lesssim \omega_n\big(f, 2^{-j}\big)_p, \tag{29.3}$$

and thus $\|Q_j f\|_{L^p} \lesssim \omega_n(f, 2^{-j})_p$. It follows that the $\mathcal{A}^s_q(L^p)$ norm and the right-hand side of (29.1) are both controlled by the $B^s_{p,q}$ norm.

In order to prove the converse result, we remark that the inverse estimate of Theorem 27.3 implies the simpler inverse estimate

$$\omega_n(f, t)_p \lesssim \big[\min\{1, t2^j\}\big]^s \|f\|_{L^p} \quad \text{if } f \in V_j. \tag{29.4}$$

Indeed this property holds for the values $t = 2^{-l}$, $l \geqslant j$, by (27.5), and the other values of t are treated by the monotonicity of $\omega_n(f, t)_p$.

For $f \in L^p$, we let $f_j \in V_j$ be such that

$$\|f - f_j\|_{L^p} \leqslant 2\,\mathrm{dist}_{L^p}(f, V_j). \tag{29.5}$$

We then have

$$\begin{aligned}
\omega_n\big(f, 2^{-j}\big) &\leqslant \omega_n\big(f_0, 2^{-j}\big)_p + \sum_{l=0}^{j} \omega_n\big(f_{l+1} - f_l, 2^{-j}\big)_p + \omega_n\big(f - f_j, 2^{-j}\big)_p \\
&\lesssim 2^{-sj}\|f_0\|_{L^p} + 2^{-sj}\left[\sum_{l=0}^{j-1} 2^{sl}\|f_{l+1} - f_l\|_{L^p}\right] + \|f - f_j\|_{L^p} \\
&\lesssim 2^{-sj}\|f_0\|_{L^p} + 2^{-sj}\left[\sum_{l=0}^{j} 2^{sl}\|f - f_l\|_{L^p}\right] \\
&\lesssim 2^{-sj}\|f\|_{L^p} + 2^{-sj}\left[\sum_{l=0}^{j} 2^{sl}\,\mathrm{dist}_{L^p}(f, V_j)\right],
\end{aligned}$$

where we have used the inverse estimate (29.5). At that point, we find the same estimate as for $K(f, 2^{-mj})$ (with s in place of m) in the second part of the proof of Theorem 28.2, and we can thus conclude that the $B^s_{p,q}$ norm is controlled by the $\mathcal{A}^s_{p,q}$ norm, using the Hardy inequality (28.24).

We can do the same reasoning with $P_j f$ instead of f_j and replace $\mathrm{dist}_{L^p}(f, V_j)$ by $\|f - P_j f\|_{L^p}$ for the characterization of $B^s_{p,q}$. Finally, we can use the Hardy inequality (28.33), as in the proof of Theorem 28.3, to replace $\|f - P_j f\|_{L^p}$ by $\|Q_j f\|_{L^p}$ and conclude the proof. □

Another track to obtain the norm equivalence (29.1) and (29.2) is to use the general mechanism of the previous section. The key observation is that Besov spaces are obtained by the real interpolation applied Sobolev spaces. A more precise result proved in JOHNEN and SCHERRER [1976] is that the quantities $\widetilde{K}(f, 2^{-nj}, L^p, W^{n,p})$ and $\omega_n(f, 2^{-j})_p$ are equivalent (where $\widetilde{K}$ is the modified K-functional of (28.13)). In the above reference, this result is proved to hold on general Lipschitz domains $\Omega \in \mathbb{R}^d$. We give below a simple proof for $\Omega = \mathbb{R}^d$.

THEOREM 29.2. *One has the equivalence*

$$\omega_n(f, t)_p \sim \widetilde{K}\big(f, t^n, L^p, W^{n,p}\big), \tag{29.6}$$

for $t \leqslant 1$. *Consequently* $B^s_{p,q} = [L^p, W^{n,p}]_{\theta,q}$, *with* $s = \theta n$, $\theta \in]0, 1[$.

PROOF. In one direction, we have for all $g \in W^{n,p}$,

$$\omega_n(f, t)_p \leqslant \omega_n(f - g, t)_p + \omega_n(g, t)_p \lesssim \|f - g\|_{L^p} + t^n |g|_{W^{n,p}},$$

using (26.22). Minimizing over $g \in W^{n,p}$ yields

$$\omega_n(f, t)_p \lesssim K\big(f, t^n, L^p, W^{n,p}\big). \tag{29.7}$$

In the other direction, for a given f we shall construct a particular function $g = g(f, t)$ such that

$$\|f - g\|_{L^p} + t^n |g|_{W^{n,p}} \lesssim \omega_n(f, t)_p. \tag{29.8}$$

To do so, we use a pair of dual scaling functions $(\varphi_{n+1}, \tilde{\varphi}_{n+1,m})$ with compact support, as constructed in Section 19 of Chapter II: $\varphi_{n+1} = B_n$ is the spline of degree n and m is supposed to be large enough so that $\tilde{\varphi}_{n+1,m}$ is in L^∞. We define their multidimensional analog $\phi(x) = \varphi_{n+1}(x_1) \cdots \varphi_{n+1}(x_d)$ and $\tilde{\phi}(x) = \tilde{\varphi}_{n+1,m}(x_1) \cdots \tilde{\varphi}_{n+1,m}(x_d)$. For $t \in]2^{-j-1}, 2^{-j}]$, we simply take for g the corresponding projection

$$g := g(f, t) = \sum_{k \in \mathbb{Z}^d} \langle f, \tilde{\phi}_{j,k} \rangle \phi_{j,k}. \tag{29.9}$$

From the Whitney estimate of Theorem 26.3 and the property $\omega_n(f, mt, \Omega)_p \leqslant m^n \omega_n(f, t, \Omega)_p$, we are ensured that

$$\|f - g\|_{L^p} \lesssim \omega_n\big(f, 2^{-j}\big)_p \lesssim \omega_n(f, t)_p. \tag{29.10}$$

It remains to estimate $|g|_{W^{n,p}}$. To do so, we use the property

$$\left(\frac{\mathrm{d}}{\mathrm{d}x}\right)^m B_n = \Delta^m_{-1} B_{n-m}, \quad m \leqslant n, \tag{29.11}$$

which comes directly from the definition of splines. This is also a particular instance of the property (19.13) (Remark 19.3 of Section 19 in Chapter II) iterated m-times. If $\alpha = (\alpha_1, \ldots, \alpha_d)$ such that $|\alpha| = \alpha_1 + \cdots + \alpha_d = n$, we thus have

$$\begin{aligned} \partial^\alpha g &= 2^{nj} \sum_{k \in \mathbb{Z}^d} \langle f, \tilde{\phi}_{j,k} \rangle \widetilde{\Delta}^\alpha_{-2^{-j}} \phi^\alpha_{j,k} \\ &= 2^{nj} \sum_{k \in \mathbb{Z}^d} \langle \widetilde{\Delta}^\alpha_{2^{-j}} f, \tilde{\phi}_{j,k} \rangle \phi^\alpha_{j,k}, \end{aligned}$$

where $\phi^\alpha(x) := \varphi_{n+1-\alpha_1}(x_1) \cdots \varphi_{n+1-\alpha_d}(x_d) \in L^\infty$ and where $\widetilde{\Delta}^\alpha_t$ denotes here the finite difference operator obtained by composing the univariate finite differences operators $\Delta^{\alpha_i}_t$ (in the direction x_i). Since both $\tilde{\phi}$ and ϕ^α are L^∞ and compactly supported, we obtain

$$\|\partial^\alpha g\|_{L^p} \lesssim 2^{nj} \|\widetilde{\Delta}^\alpha_{2^{-j}} f\| \lesssim 2^{nj} \omega_n(f, 2^{-j})_p, \tag{29.12}$$

(by the L^p stability result of Theorem 26.1) and thus $t^n |g|_{W^{n,p}} \lesssim \omega_n(f,t)_p$ which concludes the proof. □

If we combine the above result together with Theorem 28.2 (with $X = L^p$ and $Y = W^{m,p}$), using the direct estimate (26.11) and inverse estimate (27.1), we reach almost the same result as in Theorem 29.1: the norm equivalence (29.1) holds for $s < \min\{n, m\}$ where $n - 1$ is the degree of polynomial reproduction in the V_j spaces and m is such that $\varphi \in W^{m,p}$.

Moreover the reiteration theorem, we also obtain that $B^s_{p,q_1} = [L^p, B^t_{p,q_2}]_{\theta,q_1}$ if $s = \theta t$, $\theta \in]0,1[$, and $B^r_{p,q} = [B^s_{p,q_1}, B^t_{p,q_2}]_{\theta,q}$ if $r = (1-\theta)s + \theta t$. We can thus reduce the above limitation on s to $s < \min\{n, t\}$, where t is such that $\varphi \in B^s_{p,q}$ for some q, using the inverse estimate of Theorem 27.3. Thus, we actually reach exactly the same result as in Theorem 29.1, through interpolation theory. Going further (especially in Chapter IV), we shall encounter instances where the interpolation approach has no simple substitute to reach a specific result.

REMARK 29.1. One should note the robustness of the norm equivalences (29.1) and (29.2): many V_j spaces and approximation operators are allowed, provided that they only satisfy the proper direct and inverse theorem. This can be used in particular to prove the equivalence between the definitions of Besov spaces through modulus of smoothness and Littlewood–Paley decompositions, by viewing the $S_j f$ as approximations of f in spaces V_j of band-limited functions: one simply remarks that these approximation spaces satisfy the direct and inverse estimates that are needed to characterize Besov spaces.

REMARK 29.2. From the proof of Theorem 29.1, we see that the "$\lesssim$" part in the equivalence (29.1) is obtained from the assumption $t < s$ with $\varphi \in B^s_{p,q_0}$, while the converse part is ensured by $t < n$ where $n - 1$ is the degree of polynomial reproduction. In practice, the converse part is obtained for a larger range of indices t, due to the properties of refinable functions which imply that $t \leqslant n$. We have seen such an instance in Theorem 16.2 of Chapter II, in the case where $p = 2$. A similar property is known to holds for $p = 1$ and thus all $p \geqslant 1$ (see, e.g., JIA [1999]). The limitation of $s < \min\{n, t\}$ in Theorem 29.1 can thus be simplified into $s < t$.

REMARK 29.3. The norm equivalence (29.1) shows that P_j is a bounded operator in all the intermediate spaces $B^t_{p,q}$, $0 < t < s$, since we have

$$\begin{aligned}\|P_j f\|_{B^t_{p,q}} &\lesssim \|P_0 f\|_{L^p} + \left\|\left(2^{tj}\|Q_l f\|_{L^p}\right)_{0\leqslant l\leqslant j-1}\right\|_{\ell^q}\\ &\leqslant \|P_0 f\|_{L^p} + \left\|\left(2^{tj}\|Q_l f\|_{L^p}\right)_{l\geqslant 0}\right\|_{\ell^q}\\ &\lesssim \|f\|_{B^t_{p,q}}.\end{aligned}$$

We end this section by two simple corollaries of our results. The first gives a norm equivalence in terms of the error measured in a Besov norm.

COROLLARY 29.1. *Under the same assumptions as in Theorem 29.1, we have the norm equivalences*

$$\|f\|_{B^t_{p,q}} \sim \|P_0 f\|_{B^r_{p,q_1}} + \left\|\left(2^{tj}\|Q_j f\|_{B^r_{p,q_1}}\right)_{j\geqslant 0}\right\|_{\ell^q} \tag{29.13}$$

and

$$\|f\|_{\mathcal{A}^{t-r}_q(B^r_{p,q_1})} \sim \|f\|_{B^t_{p,q}}, \tag{29.14}$$

for all $0 < r < t < \min\{n, s\}$, where $n - 1$ is the order of polynomial reproduction of the V_j spaces and s is such that $\varphi \in B^s_{p,q_0}$ for some q_0.

PROOF. We simply apply Theorem 28.2 for (29.13) and Theorem 28.3 for (29.12) with $X = B^r_{p,q_1}$ and $Y = B^s_{p,q_0}$, using the generalized direct and inverse inequalities of Corollaries 27.1 and 27.2. For the application of Theorem 28.3, we also use that P_j is uniformly stable in B^r_{p,q_1} as shown in Remark 29.3 above. We then identify $B^t_{p,q}$ with the corresponding interpolation space between X and Y. □

The second corollary gives an equivalent norm in terms of the wavelet coefficients. Recall that in the tensor product construction, one needs $2^d - 1$ wavelets ψ^ε, $\varepsilon \in E := \{0,1\}^d - (0,\dots,0)$, to characterize the detail spaces, although this is hidden in the notation (24.5) that we shall use here.

COROLLARY 29.2. *If $f = \sum_{\lambda\in\Lambda} c_\lambda \psi_\lambda$, we have the norm equivalence*

$$\|f\|_{B^t_{p,q}} \sim \left\|\left(2^{tj}2^{d(1/2-1/p)j}\left\|(c_\lambda)_{\lambda\in\Lambda_j}\right\|_{\ell^p}\right)_{j\geqslant -1}\right\|_{\ell^q}, \tag{29.15}$$

under the assumptions of Theorem 29.1.

PROOF. It suffices to remark that for $j \geqslant 0$, we have the equivalence

$$\|Q_j f\|_{L^p} \sim 2^{d(1/2-1/p)j} \|(c_\lambda)_{\lambda\in\Lambda_j}\|_{\ell^p}, \tag{29.16}$$

by the same arguments as for the proof of (26.2). □

This last result also shows that wavelet bases are unconditional bases for all the Besov spaces in the above range: if $f \in B^t_{p,q}$ the convergence of the series holds in the corresponding norm and is not affected by a rearrangement or a change of sign of the coefficients, since it only depends on the finiteness of the right-hand side of (29.15).

REMARK 29.4. In the case $p = q$, note that (29.15) rewrites in a simpler way

$$\|f\|_{B^t_{p,p}} \sim \left\|\left(\|c_\lambda \psi_\lambda\|_{B^t_{p,p}}\right)_{\lambda\in\Lambda}\right\|_{\ell^p}. \tag{29.17}$$

This reveals that the wavelet basis is unconditional in $B^t_{p,p}$. One also says that the multiscale decomposition is *stable* in $B^t_{p,p}$.

REMARK 29.5. The size of a wavelet coefficient $c_\lambda = \langle f, \tilde{\psi}_\lambda\rangle$ can be directly related to the local smoothness of f on the support $\widetilde{S}_\lambda$ of $\tilde{\psi}_\lambda$: if the V_j spaces reproduce polynomials of order up to $n-1$, the wavelets $\tilde{\psi}_\lambda$ are orthogonal to Π_{n-1} for $|\lambda| \geqslant 0$, so that we have

$$\begin{aligned}
\|\langle f, \tilde{\psi}_\lambda\rangle\| &\leqslant \|\tilde{\psi}_\lambda\|_{L^{p'}} \inf_{g\in\Pi_{n-1}} \|f-g\|_{L^p(\widetilde{S}_\lambda)} \\
&\lesssim 2^{|\lambda|d(1/2-1/p')} 2^{-n|\lambda|/p} |f|_{W^{n,p}(\widetilde{S}_\lambda)} \\
&= 2^{|\lambda|[(d-n)/p-d/2]} |f|_{W^{n,p}(\widetilde{S}_\lambda)},
\end{aligned}$$

where p' is such that $1/p + 1/p' = 1$. This suggests that the size of wavelet coefficients also allows to characterize the local smoothness of f at some point x by the size of the wavelet coefficients c_λ such that the support of ψ_λ or of $\tilde{\psi}_\lambda$ contains x. Such a program is not exactly possible in these terms and requires a more elaborate notion of local smoothness, related to the concept of *microlocal spaces* introduced by J.-M. Bony (see JAFFARD [1991]).

30. The case of L^p-unstable approximation and $0 < p < 1$

So far, we have characterized L^p-Besov spaces by mean of approximation operators P_j that are uniformly bounded in L^p. In particular, interpolation operators – corresponding to the case where φ is interpolatory and $\tilde{\varphi} = \delta_0$ – do not fit in this framework, since they are only bounded in L^∞. This limitation is not clearly justified: we already noticed in Section 6 of Chapter I that the univariate Schauder hierarchical basis allows to characterize the Sobolev space $H^1 = B^1_{2,2}$ with a norm equivalence of the type (29.15).

Another limitation is that we have only considered values of p in the range $[1, \infty]$, whereas Besov spaces can be defined for $0 < p < 1$. This second restriction is actually related to the first one in the following sense: for $p < 1$, the projectors $P_j f$ are not bounded in the L^p quasi-norm. Moreover, the inner products $\langle f, \tilde{\varphi}_{j,k}\rangle$ are not well defined if we do not make additional assumptions on f.

The treatment of the case $p < 1$ (which turns out to be crucial for the nonlinear approximation results of Chapter IV) necessitates a particular attention. A first result is that, although we do not have the L^p boundedness of P_j, we still have some L^p stability for the scaling function basis. Here, we continue to assume that $(\varphi, \tilde{\varphi})$ are a pair of compactly supported biorthogonal scaling functions, with $\varphi \in L^r$ and $\tilde{\varphi} \in L^{r'}$ for some $r \geqslant 1$, $1/r' + 1/r = 1$.

THEOREM 30.1. *Assuming that $\varphi \in L^p$, for $p > 0$, one has the L^p stability property*

$$\Big\| \sum_{k\in\mathbb{Z}^d} c_k \varphi_{j,k} \Big\|_{L^p} \sim 2^{dj(1/2-1/p)} \big\| (c_k)_{k\in\mathbb{Z}^d} \big\|_{\ell^p}, \tag{30.1}$$

with constants that do not depend on j.

PROOF. The upper inequality

$$\Big\| \sum_{k\in\mathbb{Z}^d} c_k \varphi_{j,k} \Big\|_{L^p} \lesssim 2^{dj(1/2-1/p)} \big\| (c_k)_{k\in\mathbb{Z}^d} \big\|_{\ell^p}, \tag{30.2}$$

can be proved by the same technique as in Theorem 26.1: on each dyadic cube $I_{j,l}$ defined by (24.3), we still have the local estimate

$$\begin{aligned} \Big\| \sum_{k\in\mathbb{Z}^d} c_k \varphi_{j,k} \Big\|_{L^p(I_{j,l})} &\lesssim \sup_{l-k\in \mathrm{Supp}(\varphi)} |c_k| \|\varphi_{j,0}\|_{L^p} \\ &\lesssim 2^{dj(1/2-1/p)} \big\| (c_k)_{l-k\in\mathrm{Supp}(\varphi)} \big\|_{\ell^p}, \end{aligned}$$

using that all quasi-norms are equivalent in finite dimension.

For the proof of the converse inequality, we cannot use the same argument as in Theorem 26.1, since the estimate (26.5) is no longer valid for general f.

Here we shall use the *local linear independence* of the scaling functions $\varphi_{j,k}$, $k \in \mathbb{Z}^d$, i.e. the fact that if $\sum_{k\in\mathbb{Z}^d} c_k\varphi_{j,k}$ vanishes on some domain Ω, then $c_k = 0$ whenever $|\mathrm{Supp}(\varphi_{j,k}) \cap \Omega| \neq 0$. This property was proved to hold for univariate scaling functions in LEMARIÉ and MALGOUYRES [1991]. Here, we need a simpler result in the sense that we are interested in cubic domains Ω of sidelength 2^{-j}, but we need this result in several space dimension, so we prove it below.

For the time being, we assume that there exists a cube $J=[0,m]^d$, where $m>0$ is an integer, such that the above local linear independence holds on J at scale $j=0$. By a change of scale, defining $J_{j,k}=2^{-j}(k+J)$, we thus obtain that

$$\Bigl\|\sum_{k\in\mathbb{Z}^d} c_k\varphi_{j,k}\Bigr\|_{L^p(J_{j,l})} \sim 2^{dj(1/2-1/p)}\bigl\|(c_k)_{l-k\in\mathrm{Supp}(\varphi)-J}\bigr\|_{\ell^p}, \tag{30.3}$$

using that all quasi-norms are equivalent in finite dimension. Taking the pth power and summing up on $l\in\mathbb{Z}$ gives the global stability (30.1).

We end by proving the local linear independence property. We first address the univariate case, i.e. $d=1$. We assume that $\mathrm{Supp}(\varphi)=[a,b]$ and $\mathrm{Supp}(\tilde{\varphi})=[c,d]$ for some integers $a<b$ and $c<d$ (note that the duality relation implies that these intervals overlap). We set $J=[0,m]$ with $m=\max\{d-c, c+d-2+b-a, b+c-1, d-a-1\}$. With such a choice, we have three properties: (i) since $m\geqslant d-c$, the functions $\tilde{\varphi}(\bullet-k)$ are supported in J for $k=-c,\dots,m-d$, (ii) since $m\geqslant c+d-2+b-a$, the functions $\varphi(x-k)$, $k<c$ have no common support with the functions $\varphi(\bullet-k)$, $k>m-d$ and (iii) $\varphi(x-c+1)$ and $\varphi(\bullet-m+d-1)$ are respectively supported in $]-\infty,m]$ and $[0,+\infty[$. Assume now that $\sum_{k\in\mathbb{Z}} c_k\varphi(x-k)$ is identically zero on $[0,m]$. Integrating over $\tilde{\varphi}(\bullet-k)$, we obtain by (i) that $c_k=0$ for $k=-c,\dots,m-d$. By (ii), it follows that

$$\sum_{k<c} c_k\varphi(x-k)=\sum_{k>m-d} c_k\varphi(x-k)=0, \tag{30.4}$$

on $[0,m]$. Finally, by (iii), the basis functions overlapping $[0,m]$ in the two above sums are linearly independent in $[0,m]$ by their support properties. It follows that $c_k=0$ whenever $|\mathrm{Supp}(\varphi(\bullet-k))\cap[0,m]|\neq 0$.

For the multivariate case with a tensor product scaling function $\phi(x)=\varphi(x_1)\cdots\varphi(x_d)$, we use induction on the dimension d as follows: assuming that the property holds for the dimension $d-1$, let

$$\sum_{k\in\mathbb{Z}^d} c_k\phi(x-k)=0, \tag{30.5}$$

on $[0,m]^d$. Integrating against a test function $g(x_1,\dots,x_{d-1})$ supported on $[0,m]^{d-1}$, we thus obtain

$$\sum_{k_d\in\mathbb{Z}}\Biggl[\int\sum_{l\in\mathbb{Z}^{d-1}} c_{(l,k_d)}\varphi(x_1-l_1)\cdots\varphi(x_{d-1}-l_{d-1})g(x_1,\dots,x_{d-1})\Biggr]\varphi(x_d-k_d)=0 \tag{30.6}$$

for all $x_d\in[0,m]$. From the linear independence in one dimension, it follows that

$$\int\sum_{l\in\mathbb{Z}^{d-1}} c_{(l,k_d)}\varphi(x_1-l_1)\cdots\varphi(x_{d-1}-l_{d-1})g(x_1,\dots,x_{d-1})=0, \tag{30.7}$$

for all k_d such that $|\mathrm{Supp}(\varphi(\bullet - k_d)) \cap [0,m]| \neq 0$. Since g is an arbitrary test function supported in $[0,m]^{d-1}$, we obtain that

$$\sum_{l \in \mathbb{Z}^{d-1}} c_{(l,k_d)} \varphi(x_1 - l_1) \cdots \varphi(x_{d-1} - l_{d-1}) = 0, \tag{30.8}$$

on $[0,m]^{d-1}$ for all k_d such that $|\mathrm{Supp}(\varphi(\bullet - k_d)) \cap [0,m]| \neq 0$. Using the induction hypothesis, we thus obtain that $c_k = 0$ for all k such that $|\phi(\bullet - k)) \cap [0,m]^d| \neq 0$. □

REMARK 30.1. The *local linear independence* proved in LEMARIÉ and MALGOUYRES [1991] is a stronger result in the sense that it actually holds for any nontrivial interval $[a,b]$. We could thus have simply used $J = [0,1]^d$ and the cubes $I_{j,k}$ in place of $J_{j,k}$ in (30.3). Here, we have chosen an interval $[0,m]$ large enough so that local linear independence is almost straightforward. In particular, we can apply the same reasoning for proving the local linear independence of tensor product wavelets in this weaker sense.

An immediate consequence of Theorem 30.1 is that we can extend the inverse inequalities of Theorem 27.2 and 27.3 to the case $p < 1$.

THEOREM 30.2. *Under the assumptions of Theorem* 30.1, *and if* $\varphi \in B^n_{p,q}$ *for some* $q > 0$, *one has*

$$\omega_n(f,t)_p \lesssim \left[\min\{1, t2^j\}\right]^n \|f\|_{L^p} \quad \text{if } f \in V_j. \tag{30.9}$$

If $\varphi \in B^s_{p,q}$ *one has*

$$\|f\|_{B^s_{p,q}} \lesssim 2^{sj} \|f\|_{L^p} \quad \text{if } f \in V_j. \tag{30.10}$$

PROOF. If $f = \sum_{k \in \mathbb{Z}^d} c_k \varphi_{j,k} \in V_j$, we easily obtain

$$\omega_n(f,t)_p \lesssim \omega_n(\varphi_{j,0}, t)_p \left\|(c_k)_{k \in \mathbb{Z}^d}\right\|_{\ell^p} \tag{30.11}$$

with constant 1 if $p \leqslant 1$ (by simply applying the p-triangle inequality). It follows that

$$|f|_{B^s_{p,q}} \lesssim \left\|(c_k)_{k \in \mathbb{Z}^d}\right\|_{\ell^p} |\varphi_{j,0}|_{B^s_{p,q}} \tag{30.12}$$

We then observe that we have for $p < 1$ the same estimates for $|\varphi_{j,0}|_{B^s_{p,q}}$ and $\omega_n(\varphi_{j,0}, t)_p$ as in the proof of Theorems 27.2 and 27.3. Thus, (30.9) and (30.10) are respectively obtained from (30.11) and (30.12) combined together with the L^p-stability of Theorem 30.1. □

If we now want to extend the Whitney estimate of Theorem 26.3 to the case $p < 1$, we are facing the problem that L^p functions are not necessarily distributions and that the operator P_j is not a priori well defined in these spaces (unless we put restrictions on

s). One way to circumvent this problem is to consider the error of best approximation $\mathrm{dist}_{L^p}(f, V_j)$ rather than $\|f - P_j f\|_{L^p}$. In this case, we obtain the following result.

THEOREM 30.3. *Under the same assumptions as in Theorem* 30.1, *we have*

$$\mathrm{dist}_{L^p}(f, V_j) \lesssim \omega_n\big(f, 2^{-j}\big)_p, \tag{30.13}$$

where $n-1$ *is the order of polynomial exactness in* V_j.

PROOF. We consider the cubes $I_{j,k} := 2^j(k + [0,1]^d)$. We denote by E_l (resp. F_k) the set of indices $k \in \mathbb{Z}^d$ (resp. $l \in \mathbb{Z}^d$) such that $|\mathrm{Supp}(\varphi_{j,l}) \cap I_{j,k}| \neq 0$. These sets have the same cardinality N independent of j and k. We then define the larger cubes

$$\widetilde{J}_{j,k} = \bigcup_{l \in F_k} \mathrm{Supp}(\varphi_{j,l}). \tag{30.14}$$

For each of these cubes, we choose a polynomial $p_{j,k} \in \Pi_{n-1}$ such that

$$\|f - p_{j,k}\|_{L^p(\widetilde{J}_{j,k})} \leqslant 2 \inf_{g \in \Pi_{n-1}} \|f - g\|_{L^p(\widetilde{J}_{j,k})} \lesssim \omega_n\big(f, 2^{-j}, \widetilde{J}_{j,k}\big)_p. \tag{30.15}$$

Each of these polynomials admits a (unique) representation

$$p_{j,k} = \sum_{l \in \mathbb{Z}^d} p_{j,k,l}\varphi_{j,l}, \tag{30.16}$$

in the scaling function basis. We build a global approximation $f_j = \sum_{k \in \mathbb{Z}^d} f_{j,l}\varphi_{j,l}$ of f in V_j, with the choice

$$f_{j,l} := N^{-1}\Big(\sum_{k \in E_l} p_{j,k,l}\Big). \tag{30.17}$$

In some sense, f_j is a "blending" of the local polynomial approximations of f.

We now estimate the local approximation error $\|f - f_j\|_{L^p(I_{j,k})}$ by

$$\begin{aligned}
\|f - f_j\|_{L^p(I_{j,k})} &\lesssim \|f - p_{j,k}\|_{L^p(I_{j,k})} + \|f_j - p_{j,k}\|_{L^p(I_{j,k})} \\
&\lesssim \omega_n\big(f, 2^{-j}, I_{j,k}\big)_p + \Big[\sum_{l \in F_k} |p_{j,k,l} - f_{j,l}|^p\Big]^{1/p} \\
&\lesssim \omega_n\big(f, 2^{-j}, I_{j,k}\big)_p + \Big[\sum_{l \in F_k}\sum_{m \in E_l} |p_{j,k,l} - p_{j,m,l}|^p\Big]^{1/p},
\end{aligned}$$

where the first term is obtained by (30.15) and the second one by expanding f_j and $p_{j,k}$. In order to estimate the second term, we use the local linear independence of the polynomials on any domain of nonzero measure which gives

$$\left[\sum_{l\in F_k}\sum_{m\in E_l}|p_{j,k,l}-p_{j,m,l}|^p\right]^{1/p}$$
$$\lesssim\left[\sum_{l\in F_k}\sum_{m\in E_l}\|p_{j,k}-p_{j,m}\|^p_{L^p(\mathrm{Supp}(\varphi_{j,l}))}\right]^{1/p}$$
$$\lesssim\left[\sum_{l\in F_k}\sum_{m\in E_l}\|f-p_{j,m}\|^p_{L^p(\mathrm{Supp}(\varphi_{j,l}))}\right]^{1/p}$$
$$\lesssim\left[\sum_{l\in F_k}\sum_{m\in E_l}\|f-p_{j,m}\|^p_{L^p(\widetilde{J}_{j,m})}\right]^{1/p}$$
$$\lesssim\left[\sum_{l\in F_k}\sum_{m\in E_l}\omega_n\big(f,2^{-j},\widetilde{J}_{j,m}\big)^p_p\right]^{1/p}.$$

We conclude as in Theorem 26.3, by taking the pth power and summing the local estimates on k, using the modified modulus of smoothness $\tilde{\omega}_n(f,2^{-j},\widetilde{J}_{j,m})_p$ of (26.15) (the proof of $\tilde{\omega}\sim\omega$ adapts in a straightforward way to the case $p<1$, with a constant $C\geqslant 1$ appearing in (26.20)). □

Theorems 30.2 and 30.3 contain all the ingredients that are needed to extend the identity $\mathcal{A}^s_{p,q}=B^s_{p,q}$ to all possible values of p and q.

THEOREM 30.4. *Under the assumptions of Theorems* 30.2 *and* 30.3, *we have the norm equivalence*

$$\|f\|_{\mathcal{A}^t_{p,q}}\sim\|f\|_{B^t_{p,q}}, \tag{30.18}$$

for all $t<\min\{n,s\}$, where $n-1$ is the order of polynomial reproduction of the V_j spaces and s is such that $\varphi\in B^s_{p,q_0}$ for some $q_0>0$.

PROOF. We proceed exactly as in the proof of (29.2) in Theorem 29.1, using the p-triangle inequality in the estimate

$$\omega_n\big(f,2^{-j}\big)^p_p\leqslant\omega_n\big(f_0,2^{-j}\big)^p_p+\sum_{l=0}^{j}\omega_n\big(f_{l+1}-f_l,2^{-j}\big)^p_p+\omega_n\big(f-f_j,2^{-j}\big)^p_p.$$

We then conclude by a straightforward adaptation of the Hardy inequality. □

REMARK 30.2. We also have a direct estimate of the same type as in Corollary 26.1. We can thus combine Theorem 30.4 (adapted for quasi-normed spaces) and Theorem 30.2 to obtain that

$$B^s_{p,q_1} = [L^p, B^t_{p,q_2}]_{\theta,q_1}, \tag{30.19}$$

if $s = \theta t$, $\theta \in]0, 1[$. By Remarks 28.1 and 28.2, we also derive

$$B^r_{p,q} = [B^s_{p,q_1}, B^t_{p,q_2}]_{\theta,q}, \tag{30.20}$$

if $r = (1-\theta)s + \theta t$. Here, we have thus used approximation techniques to prove that the Besov spaces $B^s_{p,q}$ defined by moduli of smoothness satisfy the above interpolation identities for all values $0 < p, q \leqslant \infty$ and $s > 0$.

If we now want to use the specific projector P_j – or the wavelet coefficients – to characterize the $B^s_{p,q}$ norm for $p < 1$, we are obliged to impose conditions on s such that P_j will at least be well defined on the corresponding space. As we already mentioned, this problem is similar to the situation where we want to characterize L^p-Besov spaces for $p < \infty$ with an interpolation operator that is only bounded in L^∞. We shall now see that such characterizations are feasible if s is large enough so that $B^s_{p,q}$ is embedded in some L^r, $r \geqslant 1$, such that P_j is bounded on L^r.

For this, we need to understand in more details the imbedding properties of Besov spaces. These imbedding rely on the following basic result, which was firstly proved in DEVORE, RIEMENSCHNEIDER and SHARPLEY [1979], for indices r and p in $[1, \infty]$, and that we prove here also for the range $]0, 1[$.

THEOREM 30.5. *Let Ω be either a cube or $\mathbb{R}^d$. If $0 < p < r \leqslant \infty$, $r \geqslant 1$, and $f \in L^p$, one has*

$$\omega_n(f, 2^{-j}, \Omega)_r \lesssim \sum_{l \geqslant j} 2^{dl(1/p-1/r)} \omega_n(f, 2^{-l}, \Omega)_p, \tag{30.21}$$

with $f \in L^r(\Omega)$ whenever the right-hand side of (30.21) *is finite. For $r < 1$, we have*

$$\omega_n(f, 2^{-j}, \Omega)_r \lesssim \left(\sum_{l \geqslant j} [2^{dl(1/p-1/r)} \omega_n(f, 2^{-l}, \Omega)_p]^r\right)^{1/r}, \tag{30.22}$$

in place of (30.21).

PROOF. We start with the domain $\Omega = [0, 1]^d$ and consider a multiresolution approximation V_j, $j \geqslant 0$, adapted to this setting (by tensor product of the construction on the interval, as explained Section 20 of Chapter II). We assume that it is generated from a scaling function $\varphi \in C^n$ and that polynomials are reproduced up to degree n. Here, we anticipate on the results of Section 32, which show that such an adaptation on the cube (as well as on more general domains) allows to preserve the results of Theorems 30.1–30.4. In particular, we still have the direct estimate

$$\|f - f_j\|_{L^p(\Omega)} \lesssim \omega_n\big(f, 2^{-j}, \Omega\big)_p, \tag{30.23}$$

where $f_j \in V_j$ is such that $\|f - f_j\|_{L^p(\Omega)} \leqslant 2\mathrm{dist}_{L^p}(f, V_j)$, as well as the inverse estimate

$$\omega_n(f, t, \Omega)_t \lesssim \big[\min\{1, t2^j\}\big]^n \|f\|_{L^p(\Omega)}, \quad \text{if } f \in V_j. \tag{30.24}$$

Note that the stability property of Theorem 30.1 shows that any $f_j \in V_j$ also satisfies the inverse estimate

$$\|f_j\|_{L^r(\Omega)} \lesssim 2^{dj(1/p-1/r)}\|f_j\|_{L^p(\Omega)}, \tag{30.25}$$

for $r \geqslant p$ since $\|\bullet\|_{\ell^r} \leqslant \|\bullet\|_{\ell^p}$. We finally impose f_0 to be a polynomial in Π_{n-1} such that

$$\|f - f_0\|_{L^p(\Omega)} \leqslant 2 \inf_{g\in\Pi_{n-1}} \|f - g\|_{L^p(\Omega)} \lesssim \omega_n(f, 1, \Omega)_p. \tag{30.26}$$

Assuming that $r \geqslant 1$, we have for $j \geqslant 0$,

$$\begin{aligned}
\|f\|_{L^r(\Omega)} &\leqslant \|f_j\|_{L^r(\Omega)} + \sum_{l\geqslant j} \|f_{l+1} - f_l\|_{L^r(\Omega)} \\
&\lesssim 2^{dj(1/p-1/r)}\|f_j\|_{L^p(\Omega)} + \sum_{l\geqslant j} 2^{dl(1/p-1/r)}\|f_{l+1} - f_l\|_{L^p(\Omega)} \\
&\lesssim 2^{dj(1/p-1/r)}\|f\|_{L^p(\Omega)} + \sum_{l\geqslant j} 2^{dl(1/p-1/r)}\|f - f_j\|_{L^p(\Omega)} \\
&\lesssim 2^{dj(1/p-1/r)}\|f\|_{L^p(\Omega)} + \sum_{l\geqslant j} 2^{dl(1/p-1/r)}\omega_n\big(f, 2^{-l}, \Omega\big)_p,
\end{aligned}$$

which shows that $f \in L^r(\Omega)$ whenever the right-hand side of (30.21) is finite. In the case $r < 1$, a straightforward modification, using the r-triangle inequality, gives the same result with the right-hand side of (30.22).

We clearly have $\omega_n(f_0, 2^{-j}, \Omega)_r = 0$, since the finite difference operator of order n annihilates P_{n-1}. If $r \geqslant 1$, we thus have for $j \geqslant 0$,

$$\begin{aligned}
\omega_n\big(f, 2^{-j}, \Omega\big)_r &= \omega_n\big(f - f_0, 2^{-j}, \Omega\big)_r \\
&\leqslant \sum_{l\geqslant 0} \omega_n\big(f_{l+1} - f_l, 2^{-j}, \Omega\big)_r \\
&\lesssim \sum_{l=0}^{j-1} 2^{n(l-j)}\|f_{l+1} - f_l\|_{L^r(\Omega)} + \sum_{l\geqslant j} \|f_{l+1} - f_l\|_{L^r(\Omega)},
\end{aligned}$$

where we have used the inverse estimate. We now use the local version of (30.23) and the direct estimate to treat the two sums above. For the low scales, we obtain

$$\begin{aligned}
&\sum_{l=0}^{j-1} 2^{n(l-j)} \| f_{l+1} - f_l \|_{L^r(\Omega)} \\
&\quad \lesssim \sum_{l=0}^{j-1} 2^{n(l-j)} 2^{dl(1/p-1/r)} \| f_{l+1} - f_l \|_{L^p(\Omega)} \\
&\quad \lesssim \sum_{l=0}^{j-1} 2^{n(l-j)} 2^{dl(1/p-1/r)} \| f - f_l \|_{L^p(\Omega)} \\
&\quad \lesssim \sum_{l=0}^{j-1} 2^{n(l-j)} 2^{dl(1/p-1/r)} \omega_n\big(f, 2^{-l}, \Omega\big)_p \\
&\quad \lesssim \left[\sum_{l=0}^{j-1} 2^{(n-1)(l-j)} 2^{dl(1/p-1/r)} \right] \omega_n\big(f, 2^{-j}, \Omega\big)_p \\
&\quad \lesssim 2^{dj(1/p-1/r)} \omega_n\big(f, 2^{-j}, \Omega\big)_p,
\end{aligned}$$

where we have used that n is larger than 1.

For the high scales, we have

$$\begin{aligned}
\sum_{l \geqslant j} \| f_{l+1} - f_l \|_{L^r(\Omega)} &\lesssim \sum_{l \geqslant j} 2^{dl(1/p-1/r)} \| f_{l+1} - f_l \|_{L^p(\Omega)} \\
&\lesssim \sum_{l \geqslant j} 2^{dl(1/p-1/r)} \| f - f_l \|_{L^p(\Omega)} \\
&\lesssim \sum_{l \geqslant j} 2^{dl(1/p-1/r)} \omega_n\big(f, 2^{-l}, \Omega\big)_p.
\end{aligned}$$

We have thus proved

$$\omega_n\big(f, 2^{-j}, \Omega\big)_r \lesssim \sum_{l \geqslant j} 2^{dl(1/p-1/r)} \omega_n\big(f, 2^{-l}, \Omega\big)_p, \tag{30.27}$$

i.e. (30.21) on the unit cube and for $j \geqslant 0$. The same statement for $j \leqslant 0$ is equivalent to the case $j = 0$ due to the size of Ω.

By a change of scale, it follows that (30.27) holds on any cube Ω with a uniform constant and thus also for $\Omega = \mathbb{R}^d$.

In the case $r < 1$, a straightforward modification, using the r-triangle inequality, gives (30.22). □

Note that the first part of the proof shows that $B^s_{p,q}$, $q = \min\{1, p\}$ is continuously embedded in L^r for $1/p = 1/r + s/d$. In fact we also have a continuous imbedding of $B^s_{p,p}$ in L^r, as it will be proved in Section 39 of the next chapter, for all $p > 0$.

An imbedding theorem between Besov spaces also appear as an immediate consequence of the above result, combined with the Hardy inequality (28.33).

COROLLARY 30.1. *For $0 < p < r \leqslant \infty$, and $s, t > 0$ such that $t - s = d(1/p - 1/r)$ we have the imbedding inequality*

$$\|f\|_{B^s_{r,q}} \lesssim \|f\|_{B^t_{p,q}}, \tag{30.28}$$

for all $q > 0$.

Using Theorem 30.4, we shall now establish a direct estimate for a possibly L^p-unstable projector.

THEOREM 30.6. *Assume that $\varphi \in L^r$ and $\tilde{\varphi} \in L^{r'}$ for some $r \in [1, \infty]$, $1/r + 1/r' = 1$, or that $\varphi \in C^0$ and $\tilde{\varphi}$ is a Radon measure in which case we set $r = \infty$. Then, for $0 < p < r$, one has the direct estimate*

$$\|f - P_j f\|_{L^p} \lesssim 2^{-sj} |f|_{B^s_{p,p}}, \tag{30.29}$$

for all $s > 0$ such that $d(1/p - 1/r) < s < n$ if $p > 1$ or $d(1/p - 1/r) \leqslant s < n$ if $p \leqslant 1$, where $n - 1$ is the order of polynomial reproduction in V_j.

PROOF. We shall use the cubes $I_{j,k}$ and $\widetilde{I}_{j,k}$ (defined by (26.8)) that were already used in the proof of Theorems 26.2 and 26.3. We then have

$$\begin{aligned}
\|f - P_j f\|_{L^p(I_{j,k})} &\leqslant 2^{dj(1/r-1/p)} \|f - P_j f\|_{L^r(I_{j,k})} \\
&\lesssim 2^{dj(1/r-1/p)} \omega_n\big(f, 2^{-j}, \widetilde{I}_{j,k}\big)_r \\
&\lesssim 2^{dj(1/r-1/p)} \left[\sum_{l \geqslant j} 2^{dl(1/p-1/r)} \omega_n\big(f, 2^{-l}, \widetilde{I}_{j,k}\big)_p\right],
\end{aligned}$$

where we have used Hölder's inequality, the local L^r estimate (ensured by the L^r-boundedness of P_j) and (30.21) since $r \geqslant 1$. We now take the pth power and sum on k. If $p \leqslant 1$, this gives

$$\begin{aligned}
\|f - P_j f\|^p_{L^p} &\lesssim \sum_{k \in \mathbb{Z}^d} \left(2^{dj(1/r-1/p)} \left[\sum_{l \geqslant j} 2^{dl(1/p-1/r)} \omega_n\big(f, 2^{-l}, \widetilde{I}_{j,k}\big)_p\right]\right)^p \\
&\lesssim \sum_{k \in \mathbb{Z}^d} 2^{dj(p/r-1)} \sum_{l \geqslant j} 2^{dl(1-p/r)} \omega_n\big(f, 2^{-l}, \widetilde{I}_{j,k}\big)^p_p
\end{aligned}$$

$$
\begin{aligned}
&\lesssim \sum_{l\geqslant j} 2^{d(l-j)(1-p/r)}\omega_n\big(f,2^{-l}\big)_p^p\\
&\lesssim 2^{-spj}\sum_{l\geqslant j} 2^{d(l-j)(1-p/r-sp)}2^{spl}\omega_n\big(f,2^{-l}\big)_p^p\\
&\leqslant 2^{-spj}\sum_{l\geqslant j} 2^{spl}\omega_n\big(f,2^{-l}\big)_p^p\\
&\leqslant |f|_{B^s_{p,p}}^p,
\end{aligned}
$$

where we have used that $1-p/r-sp\leqslant 0$. If $p>1$, we define p' such that $1/p+1/p'=1$ and we obtain for any $\varepsilon>0$,

$$
\begin{aligned}
&\|f-P_jf\|_{L^p}^p\\
&\quad\lesssim \sum_{k\in\mathbb{Z}^d}\bigg(2^{dj(1/r-1/p)}\bigg[\sum_{l\geqslant j}2^{dl(1/p-1/r)}\omega_n\big(f,2^{-l},\widetilde{I}_{j,k}\big)_p\bigg]\bigg)^p\\
&\quad\lesssim \sum_{k\in\mathbb{Z}^d}2^{dj(p/r-1)}\bigg[\sum_{l\geqslant j}2^{dl(1-p/r)+lp\varepsilon}\omega_n\big(f,2^{-l},\widetilde{I}_{j,k}\big)_p^p\bigg]\bigg[\sum_{l\geqslant j}2^{-lp'\varepsilon}\bigg]^{p/p'}\\
&\quad\lesssim \sum_{k\in\mathbb{Z}^d}2^{dj(p/r-1)-pj\varepsilon}\sum_{l\geqslant j}2^{dl(1-p/r)+lp\varepsilon}\omega_n\big(f,2^{-l},\widetilde{I}_{j,k}\big)_p^p\\
&\quad\lesssim \sum_{l\geqslant j}2^{d(l-j)(1-p/r+p\varepsilon/d)}\omega_n\big(f,2^{-l}\big)_p^p\\
&\quad\lesssim 2^{-spj}\sum_{l\geqslant j}2^{d(l-j)(1-p/r-sp+p\varepsilon/d)}2^{spl}\omega_n\big(f,2^{-l}\big)_p^p\\
&\quad\leqslant 2^{-spj}\sum_{l\geqslant j}2^{spl}\omega_n\big(f,2^{-l}\big)_p^p\\
&\quad\leqslant |f|_{B^s_{p,p}}^p,
\end{aligned}
$$

by choosing ε small enough so that $1-p/r-sp+p\varepsilon/d\leqslant 0$, which is possible by assumption. □

We are now ready to prove the main result of this section.

THEOREM 30.7. *Assume that $\varphi\in L^r$ and $\tilde\varphi\in L^{r'}$ for some $r\in[1,\infty]$, $1/r+1/r'=1$, or that $\varphi\in C^0$ and $\tilde\varphi$ is a Radon measure in which case we set $r=\infty$. Then, for $0<p\leqslant r$, one has the norm equivalence*

$$
\|f\|_{B^t_{p,q}}\sim\|P_0f\|_{L^p}+\big\|\big(2^{tj}\|Q_jf\|_{L^p}\big)_{j\geqslant 0}\big\|_{\ell^q} \tag{30.30}
$$

for all $t > 0$ such that $d(1/p - 1/r) < t < \min\{s, n\}$, where $n - 1$ is the order of polynomial reproduction in V_j and s is such that $\varphi \in B^s_{p,q_0}$ for some q_0. If $f = \sum_{\lambda\in\Lambda} c_\lambda \psi_\lambda$ is the decomposition of f into the corresponding wavelet basis, we also have the norm equivalence

$$\|f\|_{B^t_{p,q}} \sim \left\| \left(2^{tj} 2^{d(1/2-1/p)j} \left\|(c_\lambda)_{\lambda\in\Lambda_j}\right\|_{\ell^p}\right)_{j\geqslant -1}\right\|_{\ell^q}, \tag{30.31}$$

under the same assumptions.

PROOF. If we denote by $A^t_{p,q}$ the space of L^r functions such that

$$\|f\|_{A^t_{p,q}} := \|P_0 f\|_{L^p} + \left\|\left(2^{tj}\|Q_j f\|_{L^p}\right)_{j\geqslant 0}\right\|_{\ell^q}, \tag{30.32}$$

is bounded, we clearly get from the direct estimate that

$$B^t_{p,p} \subset A^t_{p,\infty}, \tag{30.33}$$

for all $t > 0$ such that $d(1/p - 1/r) < t < n$. By the inverse estimate of Theorem 30.2, we also have

$$\begin{aligned}\|f\|_{B^t_{p,p}} &\leqslant \|P_0 f\|_{B^t_{p,p}} + \sum_{j\geqslant 0} \|Q_j f\|_{B^t_{p,p}} \leqslant \|P_0 f\|_{L^p} + \sum_{j\geqslant 0} 2^{tj}\|Q_j f\|_{L^p} \\ &\leqslant \|f\|_{A^t_{p,1}},\end{aligned}$$

if $p \geqslant 1$ and

$$\|f\|_{B^t_{p,p}} \leqslant \|f\|_{A^t_{p,p}}, \tag{30.34}$$

in the case $p < 1$, for all $t < s$. We thus have

$$A^t_{p,q_p} \subset B^t_{p,p} \subset A^t_{p,\infty}, \tag{30.35}$$

with $q_p := \min\{1, p\}$ and $t > 0$ such that $d(1/p - 1/r) < t < \min\{n, s\}$.

We now remark that the $A^t_{p,q}$ are trivially interpolation spaces, in the same sense that they satisfy identities similar to (28.37), due to the interpolation properties (28.41) of the weighted ℓ^q spaces. We also have a similar structure for Besov spaces, by (30.20). If we consider (30.35) for t_1 and t_2 such that $d(1/p - 1/r) < t_1 < t < t_2 < \min\{n, s\}$, and interpolate with parameters q and θ such that $t = (1-\theta)t_1 + \theta t_2$, we thus obtain, by the reiteration theorem

$$A^t_{p,q} \subset B^t_{p,q} \subset A^t_{p,q}, \tag{30.36}$$

i.e. the announced norm equivalence (30.30).

The wavelet characterization (30.31) is an immediate consequence: we use that

$$\|Q_j f\|_{L^p} \sim 2^{d(1/2-1/p)j} \big\|(c_\lambda)_{\lambda \in \Lambda_j}\big\|_{\ell^p}, \tag{30.37}$$

for $j \geqslant 0$ which is obtained by a similar reasoning as for the scaling functions in Theorem 30.1 (a more general argument, given in Section 32, shows that the L^p stability of the wavelet basis at level j stems from the L^p stability of the scaling function basis at level $j+1$). □

As an example of application, consider the Schauder hierarchical basis obtained from the univariate version of Chapter I by the tensor product method. In this case, polynomials are reproduced up to order 1, i.e. we have $n = 2$. One also easily check that $\varphi \in B^s_{p,p}$ for $s < 1 + 1/p$, with this upper bound independent of the dimension. Finally, we know that the corresponding projector P_j is only bounded in L^∞. Thus according to Theorem 30.6, we can characterize $B^t_{p,q}$ by the Schauder basis for the range

$$d/p < t < \min\{2, 1 + 1/p\}. \tag{30.38}$$

If we are interested in the L^2-Sobolev spaces $H^s = B^s_{2,2}$, this range is $]1/2, 3/2[$ if $d = 1$, $]1, 3/2[$ if $d = 2$ and empty for higher dimensions. If we are interested in Hölder spaces $C^s = B^s_{\infty,\infty}$, this range is $]0, 1[$ in all dimensions.

REMARK 30.3. One can easily extend the result of Theorem 30.6 to the case of operators that are unbounded in all L^p spaces, but bounded in C^m such as *Hermite interpolants*. The restriction by below is of course more severe $(d/p + m < t)$.

REMARK 30.4. We can also use the result of Theorem 30.6 to prove the equivalence the definitions of Besov spaces through modulus of smoothness and Littlewood–Paley decompositions, in the same spirit as explained in Remark 29.1. Here, the restriction $s > \min\{0, d(1/p - 1)\}$ is imposed by the fact that $f \mapsto S_j f$ is unbounded in L^p for $p < 1$. Generally speaking, we cannot hope for a characterization of Besov spaces modelled on L^p for $p < 1$ by the size properties of wavelets coefficients for ranges of smoothness below $d(1/p - 1)$.

REMARK 30.5. Another approach to the characterization of Besov spaces for $p < 1$ is proposed in KYRIASIS [1996]. It is based on the definition of these spaces using the Hardy spaces H^p instead of L^p in the definition of the modulus of smoothness. The interest of such results is that the Hardy spaces are distribution spaces and that the corresponding Besov spaces can be characterized by inner products with sufficiently smooth wavelets even when $s \leqslant d(1/p - 1)$. For larger values of s, the L^p-Besov spaces and H^p-Besov spaces coincide, and the results in the above reference are of similar nature as those of this section, although they are obtained by very different techniques. In particular, the approach of Kyriasis considers the *homogeneous Besov spaces*, for which the scales j in the definition of the Besov norm range from $-\infty$ to $+\infty$. The use of nonhomogeneous spaces, as done in the present approach, seems more appropriate for a straightforward subsequent adaptation of our results to bounded domains.

31. Characterization of negative smoothness and L^p-spaces

So far, we have only considered Hölder, Sobolev and Besov spaces associated with a smoothness index $s > 0$. For $s < 0$, these spaces are usually defined by duality for $p, q \geqslant 1$:

$$B^{-s}_{p',q'} := \left(B^s_{p,q}\right)^*, \tag{31.1}$$

with $1/p + 1/p' = 1$ and $1/q + 1/q' = 1$. The characterization of such dual spaces by means of the multiscale decompositions or wavelet coefficients simply relies on the ability of characterizing the corresponding primal space by the dual multiscale decomposition, as shown by the following result.

THEOREM 31.1. *Assuming that the dual projectors are such that*

$$\|f\|_{B^s_{p,q}} \sim \left\|P_0^* f\right\|_{L^p} + \left\|\left(2^{sj}\left\|Q_j^* f\right\|_{L^p}\right)_{j\geqslant 0}\right\|_{\ell^q}, \tag{31.2}$$

for some $s > 0$ and $p, q \geqslant 1$, we then have

$$\|f\|_{B^{-s}_{p',q'}} \sim \|P_0 f\|_{L^{p'}} + \left\|\left(2^{-sj}\|Q_j f\|_{L^{p'}}\right)_{j\geqslant 0}\right\|_{\ell^{q'}}, \tag{31.3}$$

with $1/p + 1/p' = 1$ and $1/q + 1/q' = 1$. We also have the norm equivalence

$$\|f\|_{B^{-s}_{p',q'}} \sim \left\|\left(2^{-sj}2^{d(1/2-1/p')j}\left\|(c_\lambda)_{\lambda\in\Lambda_j}\right\|_{\ell^{p'}}\right)_{j\geqslant -1}\right\|_{\ell^{q'}}, \tag{31.4}$$

if $f = \sum_{\lambda\in\Lambda} c_\lambda \psi_\lambda$.

PROOF. By definition, we have

$$\|f\|_{B^{-s}_{p',q'}} := \sup_{g\in\mathcal{D},\, \|g\|_{B^s_{p,q}}=1} \langle f, g\rangle, \tag{31.5}$$

where $\langle\bullet,\bullet\rangle$ is meant in the sense of the duality $(\mathcal{D}', \mathcal{D})$. If f is also a test function, we have

$$\begin{aligned}
\langle f, g\rangle &= \left\langle P_0 f + \sum_{j\geqslant 0} Q_j f,\, P_0^* g + \sum_{j\geqslant 0} Q_j^* g\right\rangle \\
&= \left\langle P_0 f, P_0^* g\right\rangle + \sum_{j\geqslant 0}\left\langle Q_j f, Q_j^* g\right\rangle \\
&\leqslant \|P_0 f\|_{L^{p'}}\left\|P_0^* g\right\|_{L^p} + \sum_{j\geqslant 0}\|Q_j f\|_{L^{p'}}\left\|Q_j^* g\right\|_{L^p} \\
&\lesssim \|g\|_{B^s_{p,q}}\left(\|P_0 f\|_{L^{p'}} + \left\|\left(2^{-sj}\|Q_j f\|_{L^{p'}}\right)_{j\geqslant 0}\right\|_{\ell^{q'}}\right),
\end{aligned}$$

where we have used Hölder's inequality on functions and on sequences, and the fact that $V_0 \perp \widetilde{W}_j$ and $\widetilde{V}_0 \perp W_j$ if $j \geqslant 0$ and that $W_j \perp \widetilde{W}_l = \{0\}$ if $l \neq j$.

We have thus proved that

$$\|f\|_{B^{-s}_{p',q'}} \lesssim \|P_0 f\|_{L^{p'}} + \left\| \left(2^{-sj} \|Q_j f\|_{L^{p'}}\right)_{j \geqslant 0} \right\|_{\ell^{q'}}. \tag{31.6}$$

By density of the test functions, this result holds for all f in $B^{-s}_{p',q'}$.

In the other direction, we shall use the duality between ℓ^r and $\ell^{r'}$ if $1/r + 1/r' = 1$ (which holds when $r = 1$ or $r = \infty$). If f is a test function, there exists a sequence $(a_j)_{j \geqslant 0}$ with ℓ^q norm equal to 1 such that

$$\begin{aligned} &\|P_0 f\|_{L^{p'}} + \left\| \left(2^{-sj} \|Q_j f\|_{L^{p'}}\right)_{j \geqslant 0} \right\|_{\ell^{q'}} \\ &\quad = \|P_0 f\|_{L^{p'}} + \sum_{j \geqslant 0} a_j 2^{-sj} \|Q_j f\|_{L^{p'}}. \end{aligned} \tag{31.7}$$

From the norm equivalence

$$\left\| \sum_{\lambda \in \Lambda_j} c_\lambda \psi_\lambda \right\|_{L^p} \sim \left\| \sum_{\lambda \in \Lambda_j} c_\lambda \tilde{\psi}_\lambda \right\|_{L^p} \sim 2^{dj(1/2 - 1/p)} \left\| (c_\lambda)_{\lambda \in \Lambda_j} \right\|_{\ell^p} \tag{31.8}$$

and the identity

$$\left\langle \sum_{\lambda \in \Lambda_j} c_\lambda \psi_\lambda, \sum_{\lambda \in \Lambda_j} d_\lambda \tilde{\psi}_\lambda \right\rangle = \sum_{\lambda \in \Lambda_j} c_\lambda \overline{d_\lambda}, \tag{31.9}$$

we derive that there exists functions $h_j \in \widetilde{W}_j$ such that $\|h_j\|_{L^p} = 1$ and

$$\|Q_j f\|_{L^{p'}} \lesssim \langle Q_j f, h_j \rangle. \tag{31.10}$$

Similarly, there exists $g_0 \in \widetilde{V}_0$ such that $\|g_0\|_{L^p} = 1$ and

$$\|P_0 f\|_{L^{p'}} \lesssim \langle P_0 f, g_0 \rangle. \tag{31.11}$$

If we now define

$$g := g_0 + \sum_{j \geqslant 0} a_j 2^{-sj} h_j, \tag{31.12}$$

it follows from (31.10) and (31.11) that

$$\|P_0 f\|_{L^{p'}} + \left\| \left(2^{-sj} \|Q_j f\|_{L^{p'}}\right)_{j \geqslant 0} \right\|_{\ell^{q'}} \leqslant 4 \langle f, g \rangle. \tag{31.13}$$

Moreover, we have

$$\|g\|_{B^s_{p,q}} \lesssim \|P_0^* g\|_{L^p} + \|(2^{sj}\|Q_j^* g\|_{L^p})_{j\geqslant 0}\|_{\ell^q}$$
$$= \|g_0\|_{L^p} + \|(a_j\|h_j\|_{L^p})_{j\geqslant 0}\|_{\ell^q} = 2.$$

Combining this and (31.13) with the definition of the $B^{-s}_{p',q'}$ norm, we have thus proved

$$\|P_0 f\|_{L^{p'}} + \|(2^{-sj}\|Q_j f\|_{L^{p'}})_{j\geqslant 0}\|_{\ell^{q'}} \lesssim \|f\|_{B^{-s}_{p',q'}}. \tag{31.14}$$

By density the result follows for all $f \in B^{-s}_{p',q'}$. We have thus proved (31.3). The norm equivalence (31.4) is an immediate consequence, using the L^p stability of the wavelet basis at each level, as we already did in the two previous sections. □

REMARK 31.1. The above proof shows that the projectors P_j and Q_j have uniformly bounded extensions in $B^{-s}_{p',q'}$.

REMARK 31.2. A close result to the above is that the Besov spaces of negative smoothness are well defined by (25.17) with the Littlewood–Paley decomposition.

In the case of Sobolev spaces, we have thus proved the norm equivalence

$$\|f\|^2_{H^s} \sim \|P_0 f\|^2_{L^2} + \sum_{j\geqslant 0} 2^{2js}\|Q_j f\|^2_{L^2} \sim \sum_{\lambda\in\Lambda} 2^{2|\lambda|s}|c_\lambda|^2 \tag{31.15}$$

(with $c_\lambda := \langle f, \tilde{\psi}_\lambda\rangle$ the coordinates of f in the multiscale basis), for a regularity index s that is either strictly positive or strictly negative, leaving aside the case $s = 0$ which corresponds to L^2. Note that the above equivalence for $s = 0$ would exactly means that $\{\psi_\lambda\}_{\lambda\in\Lambda}$ constitutes a Riesz basis for $L^2(\mathbb{R})$. By duality, we would also obtain that $\{\tilde{\psi}_\lambda\}_{\lambda\in\Lambda}$ is also a Riesz basis, and thus

$$\|f\|^2_{L^2} \sim \sum_{\lambda\in\Lambda} |c_\lambda|^2 \sim \sum_{\lambda\in\Lambda} |\tilde{c}_\lambda|^2 \tag{31.16}$$

(with $\tilde{c}_\lambda := \langle f, \tilde{\psi}_\lambda\rangle$).

A possible elementary technique to prove (31.16) is to use interpolation theory: L^2 and ℓ^2 can be respectively identified to $[H^{-s}, H^s]_{1/2,2}$ and $[\ell^2_{-s}, \ell^2_s]_{1/2,2}$. Thus (31.16) should hold if both primal and dual multiscale decomposition allow to characterize H^s for some $s > 0$. We shall give here another argument (proposed in DAHMEN [1996]) that will readily adapt to the characterization of $L^2(\Omega)$ if Ω is a bounded domain.

THEOREM 31.2. *Assume that the projectors P_j are uniformly bounded in L^2 and that there exists a normed space Y such that Y is dense in $L^2(\mathbb{R})$ and such that*

$$\|f - P_j f\|_{L^2} \lesssim 2^{-\varepsilon j}\|f\|_Y \quad \text{and} \quad \|f - P_j^* f\|_{L^2} \lesssim 2^{-\varepsilon j}\|f\|_Y \tag{31.17}$$

with $\varepsilon > 0$ and

$$\|f\|_Y \lesssim 2^{\varepsilon j}\|f\|_X \quad \text{if } f \in V_j \text{ or } f \in \widetilde{V}_j. \tag{31.18}$$

Then we have the norm equivalence

$$\|f\|_{L^2}^2 \sim \|P_0 f\|_{L^2}^2 + \sum_{j\geqslant 0}\|Q_j f\|^2 \sim \big\|P_0^* f\big\|_{L^2}^2 + \sum_{j\geqslant 0}\big\|Q_j^* f\big\|^2. \tag{31.19}$$

PROOF. Introducing the dual Y^* of Y, we first remark that we can establish from (31.17) a similar direct estimate between Y^* and L^2: for $f \in L^2$, we have

$$\|f - P_j f\|_{Y^*} = \sup_{\|g\|_Y = 1}\big|\langle f, g - P_j^* g\rangle\big| \lesssim 2^{-\varepsilon j}\|f\|_{L^2}. \tag{31.20}$$

By the symmetry in the assumptions, a similar estimate also holds for P_j^*.

Note that the direct estimate (31.17) implies that the union of the V_j spaces are dense in L^2 and thus for any $f \in L^2$, we have

$$\begin{aligned}
&\lim_{J\to+\infty}\left\|P_0 f + \sum_{j=0}^{J} Q_j f - f\right\|_{L^2} \\
&\quad = \lim_{J\to+\infty}\left\|P_0^* f + \sum_{j=0}^{J} Q_j^* f - f\right\|_{L^2} = 0.
\end{aligned} \tag{31.21}$$

For $f \in V_{J+1}$, we have

$$\begin{aligned}
\|f\|_{L^2}^2 &= \left\|P_0 f + \sum_{j=0}^{J} Q_j f\right\|_{L^2}^2 \\
&\lesssim \|P_0 f\|_{L^2}^2 + \sum_{j\geqslant 0}\|Q_j f\|_{L^2}^2 + \sum_{0\leqslant j_1 < j_2 \leqslant J}\langle Q_{j_1} f, Q_{j_2} f\rangle \\
&\lesssim \|P_0 f\|_{L^2}^2 + \sum_{j\geqslant 0}\|Q_j f\|_{L^2}^2 + \sum_{0\leqslant j_1 < j_2 \leqslant J}\|Q_{j_1} f\|_Y \|Q_{j_2} f\|_{Y^*}.
\end{aligned}$$

The inverse estimate (31.18) gives us $\|Q_{j_1} f\|_Y \lesssim 2^{\varepsilon j_1}\|Q_{j_1} f\|_{L^2}$, while the direct estimate (31.20) yields

$$\|Q_{j_2} f\|_{Y^*} = \|Q_{j_2} f - P_{j_2} f Q_{j_2} f\|_{Y^*} \lesssim 2^{-\varepsilon j_2}\|Q_{j_2} f\|_{L^2}. \tag{31.22}$$

It follows that we have

$$\begin{aligned}\|f\|_{L^2}^2 &\lesssim \|P_0 f\|_{L^2}^2 + \sum_{j\geqslant 0}\|Q_j f\|_{L^2}^2 + \sum_{0\leqslant j_1<j_2} 2^{-\varepsilon(j_2-j_1)}\|Q_{j_1} f\|_{L^2}\|Q_{j_2} f\|_{L^2}\\ &\leqslant \|P_0 f\|_{L^2}^2 + \left[1+2\sum_{l>0}2^{-\varepsilon l}\right]\sum_{j\geqslant 0}\|Q_j f\|_{L^2}^2\\ &\lesssim \|P_0 f\|_{L^2}^2 + \sum_{j\geqslant 0}\|Q_j f\|_{L^2}^2.\end{aligned}$$

By density of the spaces V_j, we have thus proved

$$\|f\|_{L^2}^2 \lesssim \|P_0 f\|_{L^2}^2 + \sum_{j\geqslant 0}\|Q_j f\|_{L^2}^2, \tag{31.23}$$

and by symmetry

$$\|f\|_{L^2}^2 \lesssim \left\|P_0^* f\right\|_{L^2}^2 + \sum_{j\geqslant 0}\left\|Q_j^* f\right\|_{L^2}^2. \tag{31.24}$$

In order to complete the proof, we now remark that

$$\begin{aligned}\left(\|P_0 f\|_{L^2}^2 + \sum_{j\geqslant 0}\|Q_j f\|_{L^2}^2\right)^2 &= \left|\left\langle f, P_0^* P_0 f + \sum_{j\geqslant 0} Q_j^* Q_j f\right\rangle\right|^2\\ &\leqslant \|f\|_{L^2}^2 \left\|P_0^* P_0 f + \sum_{j\geqslant 0} Q_j^* Q_j f\right\|_{L^2}^2\\ &\lesssim \|f\|_{L^2}^2\left(\left\|P_0^* P_0 f\right\|_{L^2}^2 + \sum_{j\geqslant 0}\left\|Q_j^* Q_j f\right\|_{L^2}^2\right)\\ &\lesssim \|f\|_{L^2}^2\left(\|P_0 f\|_{L^2}^2 + \sum_{j\geqslant 0}\|Q_j f\|_{L^2}^2\right),\end{aligned}$$

where we have used (31.24) and the uniform boundedness of the projectors P_j^*. Dividing on both side, we thus obtain

$$\|P_0 f\|_{L^2}^2 + \sum_{j\geqslant 0}\|Q_j f\|_{L^2}^2 \lesssim \|f\|_{L^2}^2, \tag{31.25}$$

and by symmetry

$$\left\|P_0^* f\right\|_{L^2}^2 + \sum_{j\geqslant 0}\left\|Q_j^* f\right\|_{L^2}^2 \lesssim \|f\|_{L^2}^2. \tag{31.26}$$

We have thus proved the full equivalence (31.19). □

The above result shows in particular that L^2 is characterized by (31.19), if both primal and dual multiresolution approximation satisfy direct and inverse estimates between L^2 and H^ε, for some $\varepsilon > 0$. In practice, this amounts in assuming that φ and $\tilde{\varphi}$ have both H^ε smoothness for some $\varepsilon > 0$, since this in turn implies that both V_j and $\widetilde{V}_j$ reproduce constant functions (from the results in Section 16 of Chapter II). Form the L^2-stability of the wavelet bases in each subspace W_j and $\widetilde{W}_j$, we also have (31.16).

REMARK 31.3. Other methods exist for proving (31.19) and (31.16). The approach proposed in COHEN, DAUBECHIES and FEAUVEAU [1992] and in COHEN and DAUBECHIES [1992] gives a direct proof of (31.16), making an important use of the Fourier transform and of the translation invariant structure of the bases $\{\psi_{j,k}\}_{k\in\mathbb{Z}^d}$ at each level j. Such techniques are thus difficult to mimic on a bounded domain Ω. A more general argument for the direct proof of (31.16) is provided by the *vaguelette lemma* (see MEYER [1990, T. 2, Chapter VII]). A vaguelette family is by definition a set of functions $\{g_{j,k}\}_{j\in\mathbb{Z},k\in\mathbb{Z}^d}$ such that $\int g_{j,k} = 0$ and

$$|g_{j,k}| + 2^{-j}\sum_{m=1}^{d} |\partial_{x_m} g_{j,k}| \leqslant 2^{dj/2} g(2^j \cdot - k), \tag{31.27}$$

where g is a positive function such that $g(x) \leqslant C(1+|x|)^{-(d+\varepsilon)/2}$ for some $\varepsilon > 0$. Then the vaguelette lemma states that for all $f \in L^2$, one has

$$\sum_{j\in\mathbb{Z}}\sum_{k\in\mathbb{Z}^d} \left|\langle f, g_{j,k}\rangle\right|^2 \lesssim \|f\|_{L^2}^2. \tag{31.28}$$

Such a result immediately yields one side of the equivalence (31.16) for both primal and dual bases, and the other side is then easily proved by duality arguments. One disadvantage of the vaguelette approach is that it requires a bit more than the minimal smoothness assumptions $\varphi, \tilde{\varphi} \in H^\varepsilon$ (the differentiability of $g_{j,k}$ can still be lowered down to a Hölder regularity assumption).

From the definition of Besov spaces by (25.17) using the Littlewood–Paley decomposition, we see that L^2 identifies with $B^0_{2,2}$. The norm equivalences (31.16) and (31.19) are thus the natural extension to the case $s = 0$ of the characterization of Besov spaces that we have obtained so far. More generally, for $1 \leqslant p \leqslant \infty$, we can use the elementary interpolation properties (28.41) of the weighted sequence spaces ℓ^p_s, to obtain the norm equivalence

$$\|f\|_{B^0_{p,q}} \sim \left\| \left(2^{d(1/2-1/p)j} \left\|(c_\lambda)_{\lambda\in\Lambda_j}\right\|_{\ell^p}\right)_{j\geqslant -1} \right\|_{\ell^q}, \tag{31.29}$$

provided that the wavelet basis allows to characterize B^ε_{p,q_1} and $B^{-\varepsilon}_{p,q_2}$ for some $\varepsilon > 0$ and $q_1, q_2 > 0$. However, we cannot identify L^p with $B^0_{p,q}$ for any $q > 0$ if $p \neq 2$. This reflects the well known fact that the L^p spaces (and more generally the $H^{s,p}$ spaces) do not belong to the scale of Besov spaces when $p \neq 2$.

Concerning such L^p spaces, we can formulate two basic questions: does the wavelet expansion of an arbitrary function $f \in L^p$ converge unconditionally in L^p? Is there a simple characterization of L^p by the size of the wavelets coefficients?

The answer to questions is clearly negative for L^∞ since this space is nonseparable. It is also negative for L^1 which is known to possess no unconditional basis.

For the case $1 < p < \infty$, a positive answer to the first question is provided by the general theory developed since the 1950's by Calderon and Zygmund in order to study the continuity properties of operators. By definition, a Calderon–Zygmund operator is an L^2-bounded operator T with the integral form

$$Tf(x) = \int_{\mathbb{R}^d} K(x, y)\,\mathrm{d}x\,\mathrm{d}y, \tag{31.30}$$

where the the kernel function K satisfies decay estimates of the type

$$\bigl|K(x, y)\bigr| \lesssim |x - y|^{-d}, \tag{31.31}$$

and

$$\sum_{m=1}^{d}\bigl[\bigl|\partial_{x_m} K(x, y)\bigr| + \bigl|\partial_{y_m} K(x, y)\bigr|\bigr] \lesssim |x - y|^{-d-1}, \tag{31.32}$$

away from the diagonal, i.e. for $x \neq y$.

One of the most famous result in the study of such operators states that T is also bounded in L^p for $1 < p < \infty$, with a norm that only depends (for fixed p) of its L^2 norm, p and the constants in the estimates (31.31) and (31.32). A proof of this result can be found in MEYER [1990, Chapter VII] and DAUBECHIES [1992, Chapter IX] (see also STEIN [1970] for a general introduction on singular integral operators).

Returning to the L^p-unconditionality of wavelet bases, we can formulate this property in terms of operators: it holds if and only if the projectors

$$T_\Gamma f = \sum_{\lambda \in \Gamma} \langle f, \tilde{\psi}_\lambda \rangle \psi_\lambda, \tag{31.33}$$

are uniformly bounded in L^p independently of the finite or infinite subset $\Gamma \subset \Lambda$. More generally, to any sequence $(a_\lambda)_{\lambda \in \Lambda}$ such that $|a_\lambda| \leqslant 1$, we associate an operator T_a defined by

$$T_a\Bigl(\sum_{\lambda \in \Lambda} c_\lambda \psi_\lambda\Bigr) = \sum_{\lambda \in \Lambda} a_\lambda c_\lambda \psi_\lambda. \tag{31.34}$$

Clearly T_Γ corresponds to the choice $a_\lambda = 1$ is $\lambda \in \Gamma$, 0 otherwise. Assuming that φ and $\tilde{\varphi}$ have C^1 smoothness, we then remark that the associated kernels

$$K_a(x,y) = \sum_{\lambda\in\Lambda} a_\lambda \psi_\lambda(x)\tilde{\psi}_\lambda(y), \tag{31.35}$$

satisfy the estimates (31.31) and (31.32) with constant that are independent of the eigenvalue sequence a_λ. Indeed, we can estimate

$$\left|K_a(x,y)\right| \leqslant \sum_{\lambda\in\Lambda}\left|\psi_\lambda(x)\tilde{\psi}_\lambda(y)\right| = \sum_{j\geqslant -1}\sum_{\lambda\in\Lambda_j}\left|\psi_\lambda(x)\tilde{\psi}_\lambda(y)\right|,$$

by remarking that due to the support size of the basis function, the nontrivial contributions in the last sum are limited to $j \leqslant J$ if $|x-y| \geqslant C2^{-J}$ for some fixed constant C. Due to the controlled overlapping property (24.8) at each level, we also have

$$\sum_{\lambda\in\Lambda_j}\left|\psi_\lambda(x)\tilde{\psi}_\lambda(y)\right| \lesssim 2^{d|\lambda|}, \tag{31.36}$$

and the estimate (31.31) follows. A similar technique is used to proved (31.32), remarking that

$$\sum_{m=1}^{d}\sum_{\lambda\in\Lambda_j}\left[\left|\partial_{x_m}\psi_\lambda(x)\tilde{\psi}_\lambda(y)\right| + \left|\psi_\lambda(x)\partial_{y_m}\tilde{\psi}_\lambda(y)\right|\right] \lesssim 2^{(d+1)|\lambda|}. \tag{31.37}$$

Assuming that $(\psi_\lambda)_{\lambda\in\Lambda}$ and $(\tilde{\psi}_\lambda)_{\lambda\in\Lambda}$ are biorthogonal Riesz basis for L^2, we also clearly have that the T_a are uniformly bounded in L^2. We can thus derive from the Calderon–Zygmund theory that these operators are uniformly L^p bounded for $1 < p < \infty$, i.e. the wavelet bases are also L^p-unconditional.

It should be noted that the assumption (31.32) can be weakened in such a way that the same conclusions hold with φ and $\tilde{\varphi}$ only in C^ε for some $\varepsilon > 0$ (see, e.g., MEYER [1990, Chapter VII]). We summarize these results below.

THEOREM 31.3. *If the assumptions of Theorem* 31.2 *are satisfied and if the generators of the primal and dual wavelet basis are both in* C^ε *for some* $\varepsilon > 0$, *then these bases are also unconditional for* L^p, $1 < p < \infty$.

REMARK 31.4. In our case of interest, the L^2 boundedness of the operators T_a is a straightforward consequence of the L^2-stability of the multiscale bases. More generally, if $K(x,y)$ only satisfies (31.31) and (31.32), the possible nonintegrability in each variable means that the L^2-boundedness of T should stems from certain cancellation properties in the Kernel K. Such requirements were made more precise by the celebrated $T(1)$ theorem of DAVID and JOURNÉ [1984], which states that the image of the constant function by T and T^* should belong to the space *BMO*.

The characterization of L^p norms from the size properties of the wavelet coefficients is also possible for $1 < p < \infty$, by mean of a *square function*, defined for $f = \sum_{\lambda\in\Lambda} c_\lambda \psi_\lambda$ by

$$Sf(x) = \left[\sum_{\lambda\in\Lambda} |c_\lambda|^2 |\psi_\lambda(x)|^2\right]^{1/2}. \tag{31.38}$$

Clearly, we have $\|f\|_{L^2} \sim \|Sf\|_{L^2}$. A remarkable result is that we also have

$$\|f\|_{L^p} \sim \|Sf\|_{L^p}, \tag{31.39}$$

for $1 < p < \infty$. The proof of such a result relies on the classical Khinchine inequality (see, e.g., STEIN [1970, p. 277]): if $(\varepsilon_\lambda)_{\lambda\in\Lambda}$ is a sequence of independent random variable, with law $(\delta_{-1} + \delta_1)/2$ (i.e. 1 or -1 with equal probability), then one has the equivalence between the expectations

$$\left(E\left(\left|\sum_{\lambda\in\Lambda} \varepsilon_\lambda x_\lambda\right|^p\right)\right)^{1/p} \sim \left(E\left(\left|\sum_{\lambda\in\Lambda} \varepsilon_\lambda x_\lambda\right|^q\right)\right)^{1/q}, \tag{31.40}$$

for any fixed $p, q \in]0, \infty[$. For $q = 2$, the orthogonality of the ε_λ yields

$$\left(E\left(\left|\sum_{\lambda\in\Lambda} \varepsilon_\lambda x_\lambda\right|^p\right)\right)^{1/p} \sim \left(\sum_{\lambda\in\Lambda} |x_\lambda|^2\right)^{1/2}. \tag{31.41}$$

Applying (31.41) to $x_\lambda = c_\lambda \psi_\lambda(x)$ and integrating in x after taking the pth power gives us

$$\|Sf\|_{L^p} \lesssim E\big(\|T_\varepsilon f\|_{L^p}\big), \tag{31.42}$$

where T_ε is defined by (31.34). Since $\|T_\varepsilon f\|_{L^p} \lesssim \|f\|_{L^p}$, we immediately obtain one side of the equivalence (31.38). The other side follows by remarking that since $T_e^2 = I$, we also have $\|f\|_{L^p} \lesssim \|T_\varepsilon f\|_{L^p}$.

REMARK 31.5. Similar characterization are available for general potential spaces $H^{t,p}$ and in particular for $W^{m,p} = H^{m,p}$ is m is an integer, under additional smoothness assumptions on φ. They rely on the modified square function

$$S_t f(x) = \left[\sum_{\lambda\in\Lambda} 4^{t|\lambda|} |c_\lambda|^2 |\psi_\lambda(x)|^2\right]^{1/2} \tag{31.43}$$

and have the form

$$\|f\|_{H^{t,p}} \lesssim \|S_t f\|_{L^p}. \tag{31.44}$$

The proof of such a result can be found in MEYER [1990, p. 165]. The $H^{s,p}$ spaces are an instance of the more general family of *Triebel–Lizorkin spaces* $F^s_{p,q}$ which are are defined through similar expressions involving the Littlewood–Paley decomposition: $f \in F^s_{p,q}$ if and only of $S^q_t f(x) = (|S_0(x)|^q + \sum_{j\geqslant 0} |2^{sj}\Delta_j f(x)|^q)^{1/q}$ belong to L^p. One has in particular $H^{s,p} = F^s_{p,2}$ (see TRIEBEL [1983]).

32. Bounded domains

The goal of this section is to explain how the results obtained so far for function spaces on $\mathbb{R}^r$ in the previous sections can be adapted to the framework of a bounded domains: we are interested in characterizing functions spaces related to a bounded domain $\Omega \subset \mathbb{R}^d$ in terms of their multiscale decomposition.

There are two different approaches to the definition of function spaces $X(\Omega)$ on bounded domains: one can either view them as the *restrictions* of the corresponding function spaces $X(\mathbb{R}^d)$ defined on the whole of $\mathbb{R}^d$, or search for an appropriate *inner description* of $X(\Omega)$. In Section 25 we have given inner descriptions for Hölder, Sobolev and Besov spaces on domains, and we shall continue here with this approach.

REMARK 32.1. For fairly general domains, it is well known that the two approaches yield the same spaces for a wide range of smoothness classes. A classical instance is the case of the Sobolev spaces $W^{m,p}$, for $1 \leqslant p \leqslant \infty$ and $m \in \mathbb{N}$. Clearly, the restriction to Ω of $f \in W^{m,p}(\mathbb{R}^d)$ belongs to $W^{m,p}(\Omega)$ defined as in Section 25. Conversely, if Ω is a Lipschitz domain, it is known from STEIN [1970, p. 181], that any function $f \in W^{m,p}(\Omega)$ can be extended into $\mathcal{E}f \in W^{m,p}(\mathbb{R}^d)$ by a bounded linear extension operator $\mathcal{E}$. As a consequence, we find an analog result for Besov spaces: since $B^s_{p,q}(\Omega)$ can be identified to $[L^p, W^{m,p}]_{\theta,q}$, $\theta = s/m$ for any Lipschitz domain Ω as for $\mathbb{R}^d$ (by the result of JOHNEN and SCHERRER [1976] that we proved in Theorem 29.2 in the special case $\Omega = \mathbb{R}^d$), the extension operator E is also bounded from $B^s_{p,q}(\Omega)$ to $B^s_{p,q}(\mathbb{R}^d)$ by interpolation. Extension results for Besov spaces with $p < 1$ on general Lipschitz domains have also been obtained by DEVORE and SHARPLEY [1993]. As it will be remarked later, the subsequent results will provide with ad hoc concrete extension operator for such spaces.

Of interest to us is also the case of function spaces $X_b(\Omega)$ with prescribed boundary conditions, which are usually defined by the closure in $X(\Omega)$ of a set of smooth functions which vanish up to some order at $\Gamma := \partial\Omega$. Their introduction is natural for the study of boundary value problems, as well as for the proper definition of negative smoothness by duality. The adaptation of our results to such spaces will be the object of the next section.

In Sections 20 and 21 of the previous chapter, we have described the two main approaches to multiscale decompositions and wavelet bases on bounded domains: a domain decomposition strategy, allowing to exploit the simpler construction on the interval, and a strategy based on the multiscale decomposition of finite element spaces.

Both approaches have similar features to the tensor product construction on $\mathbb{R}^d$: local bases, polynomial reproduction up to some prescribed degree, prescribed smoothness properties for the basis functions. This will often allow us to adapt in a straightforward way most of the results that we have stated for $\Omega = \mathbb{R}^d$, to the price of heavier notations. For this reason, we shall only elaborate below when the adaptation is less trivial.

At this stage, we fix some general assumptions on our domain: Ω should have a simple geometry, expressed by a conformal partition

$$\Omega = \bigcup_{i=1}^{n} S_i, \tag{32.1}$$

into simplicial subdomains: each S_i is the image of the unit simplex

$$S := \{0 \leqslant x_1 + \cdots + x_d \leqslant 1\}, \tag{32.2}$$

by an affine transformation. By "conformal" we mean that a face of an S_i is either part of the boundary Γ or coincides with a face of another S_j. We also assume that Ω is connected in the sense that for all $j, l \in \{1, \ldots, n\}$, there exists a sequence $i_0, \ldots, i_m$ such that $i_0 = j$ and $i_m = l$ and such that S_{i_k} and $S_{i_{k+1}}$ have a common face. Clearly, polygons and polyhedrons fall in this category. More general curved domains or manifolds are also concerned here, provided that they can be smoothly parameterized by such simple reference domains, as explained in Sections 20 and 21 of the previous chapter.

For the characterization of Besov spaces $B^s_{p,q}(\Omega)$ with $s > 0$, we directly place ourselves in the general setting of Section 30 where we allow p to be less than 1. Our goal is to show here that all the statements in this section stay valid in the present setting of bounded domains.

Recall that the basic ingredient for proving the results of Section 30 was the covering of $\mathbb{R}^d$ by the dyadic cubes $I_{j,k}$, $\widetilde{I}_{j,k}$, $J_{j,k}$ and $\widetilde{J}_{j,k}$ for $k \in \mathbb{Z}^d$ (the cubes $I_{j,k}$, $\widetilde{I}_{j,k}$ were also used for the direct estimates of Section 26). Also recall (Remark 30.1) that we could actually have taken $J_{j,k} = I_{j,k}$.

The adaptation of these results to Ω thus relies on the ability of covering Ω in an analog manner. More precisely, at each scale j, we need a covering of the type

$$\Omega = \bigcup_{\mu \in \Theta_j} I_\mu, \tag{32.3}$$

with $\operatorname{diam}(I_\mu) \lesssim 2^{-|\mu|}$ (here we also use the convention $|\mu| = j$ if $\mu \in \Theta_j$). The domains I_μ will play the same role as the $I_{j,k}$ and $J_{j,k}$ cubes in the $\mathbb{R}^d$ case. To this covering, we associate $\widetilde{I}_\mu$ and $\widetilde{J}_\mu$ defined if $|\mu| = j$ by

$$\widetilde{I}_\mu = \bigcup_{\lambda \in F_\mu} \operatorname{Supp}(\tilde{\varphi}_\lambda), \tag{32.4}$$

and

$$\widetilde{J}_\mu = \bigcup_{\lambda \in F_\mu} \operatorname{Supp}(\varphi_\lambda), \tag{32.5}$$

with

$$F_\mu = \{\lambda \in \Gamma_j;\ |\mathrm{Supp}(\varphi_\lambda) \cap I_\mu| \neq 0\}. \tag{32.6}$$

These larger subdomains also have diameter of order $2^{-|\mu|}$. They are intended to play the same role as the cubes $\widetilde{I}_{j,k}$ and $\widetilde{J}_{j,k}$ which were used in the proof of Theorems 30.3 and 30.6. In the following, we shall analyze the properties that need to be fulfilled by these covering in order to adapt the proofs of Theorems 30.1–30.7 to Ω, and we shall explain how to design them.

In order to turn local estimates into global estimates, these domains should not overlap too much in the sense that there exists a constant C such that for all $x \in \Omega$,

$$\#\{\mu \in \Theta_j;\ x \in I_\mu\} \leqslant C. \tag{32.7}$$

Clearly, (32.7) implies the same property for the families $\widetilde{I}_\mu$ and $\widetilde{J}_\mu$, $\mu \in \Theta_j$, due to the local supports of the basis functions.

In order to prove Theorems 30.1 and 30.2 the stability estimate

$$\Big\| \sum_{\lambda \in \Gamma_j} c_\lambda \varphi_\lambda \Big\|_{L^p(I_\mu)} \sim \big\| (c_\lambda)_{\lambda \in F_\mu} \big\|_{\ell^p}, \quad \mu \in \Theta_j, \tag{32.8}$$

should hold with a constant that does not depend on μ.

The main headache is to select a covering $(I_\mu)_{\mu \in \Theta_j}$ that satisfies this last prescription, i.e. uniform stability (32.8). This is feasible in a way that depends on the type of multiscale decomposition which is used for Ω.

Domain decomposition approach. Remark first that for $\Omega = [0,1]^d$, we obtain an covering that satisfies the third prescription by simply taking those $I_{j,k} = 2^{-j}(k + [0,1]^d)$, for all $k \in \{0, \ldots, 2^j - 1\}^d$, i.e. such that $I_{j,k} \subset [0,1]^d$. Indeed, we already have proved by (30.3) the uniform stability (30.5) for the $I_{j,k}$ in the interior of $[0,1]^d$ which only meet the supports of unmodified interior scaling functions. For $I_{j,k}$ near the boundary meeting the support of a modified scaling function, we still have local linear independence of these basis functions on $I_{j,k}$. To check this, we can reduce to the univariate case, i.e. $\Omega = [0,1]$, by the same reasoning as in the proof of Theorem 30.1 (induction on the dimension d of the cube). Then, local linear independence near the edges is easily checked, recalling that the edge functions $\varphi^e_{j,q}$ are obtained as linear combinations of the locally independent scaling functions $\varphi_{j,k}$ with Vandermonde type coefficients. The uniformity of the constant in (32.8) is then ensured by change of scale. For a more general Ω, recall (see Section 20 of Chapter II) that Ω is patched into subdomains Ω_i, $i = 0, \ldots, n$, which are the image of the reference cube $[0,1]^d$ by smooth isoparametric maps T_i. The basis functions of $V_j^{\Omega_i}$ are the images of the basis functions of $V_j^{[0,1]^d}$ by the map T_i and V_j^{Ω} is defined by the direct sum of $V_j^{\Omega_i}$ intersected with $C^0(\Omega)$. A natural choice for the covering of Ω is then provided by

$$\{I_\mu\}_{\mu\in\Theta_j} = \bigcup_{i=1}^{n} \{T_i I_{j,k};\ k \in \{0,\dots,2^j - m\}^d\}. \tag{32.9}$$

With such a choice, the uniform stability (32.8) appears as a simple consequence of the same property on the unit cube.

Finite element approach. In this setting, as explained in Section 21 of Chapter II, one starts form a coarse decomposition $\mathcal{T}_0$ of the type (32.1) and obtains a hierarchy of decompositions $\mathcal{T}_j$, $j \geqslant 0$, by iterative uniform subdivision of the simplices S_j using the mid-point rule. The V_j spaces are then simplicial finite element spaces associated to the decomposition $\mathcal{T}_j$. In this setting, local linear independence of the nodal basis functions is well known to hold on each simplices of $\mathcal{T}_j$, and the stability property (32.8) holds uniformly by rescaling. We can thus simply take for $\{I_\mu\}_{\mu\in\Theta_j}$ the set of simplices $\mathcal{T}_j$ (thus a covering by disjoints sets).

With such a covering, we can thus adapt the proof of Theorem 30.1 and obtain a global stability estimate

$$\Big\| \sum_{\lambda\in\Gamma_j} c_\lambda \varphi_\lambda \Big\|_{L^p} \sim 2^{jd(1/2-1/p)} \big\| (c_\lambda)_{\lambda\in\Gamma_j} \big\|_{\ell^p}. \tag{32.10}$$

Theorem 30.2 is then immediately adapted with an unchanged statement. However, it should be remarked that in the case of parameterized curved domains, the smoothness of the map enters in as a limitation of the range of s in the inverse estimate. For example, in the domain decomposition approach, if we only impose C^0 continuity on the interface, we find the limitation $s < 1 + 1/p$: even if the initial generator φ has high smoothness, the functions φ_λ sitting on the interfaces between the subdomains are not in $B^s_{p,q}$ for $s > 1 + 1/p$.

REMARK 32.2. In the case where $p \geqslant 1$, one does not need to work so hard on the design of the I_μ: using a dual scaling function basis $\tilde{\varphi}_\lambda$, $\lambda \in \Lambda_j$ in L^∞, one can directly prove the stability property, as well as the direct estimate, in the same way as it is done in Section 26.

In order to prove Theorem 30.3, we need additional properties for the domains $\widetilde{J}_\mu$: on the one hand, a local Whitney estimate

$$\inf_{g\in\Pi_n} \|f - g\|_{L^p(\widetilde{J}_\mu)} \lesssim \omega_{n+1}\big(f, 2^{-|\mu|}, \widetilde{J}_\mu\big)_p \tag{32.11}$$

should hold with a constant that does not depend on μ. On the other hand, in order to turn the local error estimates into global estimates, we need to use the modified modulus of smoothness $\tilde{\omega}(f, 2^{-|\mu|}, \widetilde{J}_\mu)_p$ defined by (26.15), and thus we should have

$$\tilde{\omega}\big(f, 2^{-|\mu|}, \widetilde{J}_\mu\big)_p \lesssim \omega\big(f, 2^{-|\mu|}, \widetilde{J}_\mu\big)_p, \tag{32.12}$$

with a constant that does not depend on μ.

One of the problems here is that the domains $\widetilde{J}_\mu$ are not cubes, for which these estimates are known to hold according to (25.24) and (26.17). To our knowledge, there is no proof of such results for domains of general shape. Note however that in both domain decomposition and finite element approaches, the domains $\widetilde{J}_\mu$ are the images of a reference domain within a finite set $\{\widetilde{J}_1, \ldots, \widetilde{J}_N\}$ by affine transformations T_μ which are the product of an orthogonal transform with an isotropic scaling by $2^{-|\mu|}$. Such transformation leave invariant the constants in (32.11) and (32.12), which only depend on the shape of the domain, so that we are reduced to prove (25.24) and (26.17) for the reference domains $\{\widetilde{J}_1, \ldots, \widetilde{J}_N\}$.

Once again, these reference domains are no cubes, but have a simple geometry as Ω, i.e. expressed by a partition of the type (32.1). Our work can thus be reduced to prove (25.24) and (25.15) for a domain Ω of the type (32.1).

A simple way of proof is the following: we first remark that since (25.24) and (25.15) hold for cubes, by a change of variable it follows that these estimates also hold on any domain of the form $T[0,1]^d$ where T is an invertible affine transformation. The constants are of course affected by the properties of T. We next remark that for a domain Ω of the type (32.1), we can easily find invertible affine transformations $T_1, \ldots, T_m$ such that

$$\Omega = \bigcup_{j=1}^{m} R_j, \quad R_j := T_j[0,1]^d, \tag{32.13}$$

and such that $|R_j \cap R_{j+1}| \neq 0$. An example of such an overlapping decomposition is shown on Fig. 32.1 in the case of a polygon.

In this case, the overlapping is crucial in order to turn the estimates (25.24) and (25.15) which hold on each R_j into similar estimates for the whole of Ω.

THEOREM 32.1. *If Ω is of the type* (32.13), *we have*

$$\inf_{g \in \Pi_{n-1}} \|f - g\|_{L^p(\Omega)} \lesssim \omega_n(f, 1, \Omega)_p, \tag{32.14}$$

and

$$\omega_n(f, 1, \Omega)_p \lesssim \tilde{\omega}_n(f, 1, \Omega)_p. \tag{32.15}$$

PROOF. On each R_i, there exists a polynomial $p_i \in \Pi_{n-1}$ such that

$$\|f - p_i\|_{L^p(R_i)} \lesssim \omega_n(f, 1, R_i)_p. \tag{32.16}$$

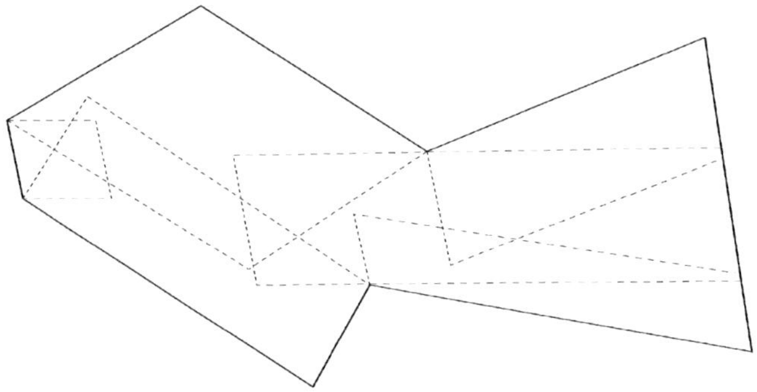

FIG. 32.1. Overlapping partition of Ω by the simpler domains R_i.

Considering first $i = 1, 2$, we then have

$$\begin{aligned}\|p_2 - p_1\|_{L^p(R_1 \cap R_2)} &\lesssim \omega_n(f, 1, R_1)_p + \omega_n(f, 1, R_2)_p \\ &\lesssim \omega_n(f, 1, R_1 \cup R_2)_p,\end{aligned} \tag{32.17}$$

and thus

$$\|p_2 - p_1\|_{L^p(R_1 \cup R_2)} \lesssim \omega_n(f, 1, R_1 \cup R_2)_p, \tag{32.18}$$

from the equivalence of all quasi-norms in finite dimensions. It follows that

$$\|f - p_2\|_{L^p(R_1 \cup R_2)} \lesssim \omega_n(f, 1, R_1 \cup R_2)_p. \tag{32.19}$$

By iteration of this reasoning, we end up with

$$\|f - p_m\|_{L^p(\Omega)} \lesssim \omega_n(f, 1, \Omega)_p, \tag{32.20}$$

i.e. the desired estimate (32.14). In order to prove (32.15), we remark that if $|A \cap B| \neq 0$, we have

$$\omega_n(f, 1, A \cup B)_p \lesssim \omega_n(f, 1, A)_p + \omega_n(f, 1, B)_p, \tag{32.21}$$

so that

$$\omega_n(f, 1, \Omega)_p \lesssim \sum_{i=1}^{m} \omega_n(f, 1, R_i)_p \lesssim \sum_{i=1}^{m} \tilde{\omega}_n(f, 1, R_i)_p. \tag{32.22}$$

Since we clearly have

$$\tilde{\omega}_n(f, 1, A)_p + \tilde{\omega}_n(f, 1, B)_p \lesssim \tilde{\omega}_n(f, 1, A \cup B)_p, \tag{32.23}$$

this concludes the proof. $\square$

From this result, we conclude that the statements of Theorems 30.3 and 30.4 also hold for multiscale decompositions on Ω. Note that this also implies the interpolation results stated in Remark 30.1, and that the argument that we have used in the proof of Theorem 30.5 (based on multiresolution approximation on the cube $[0,1]^d$) is now justified.

The next step is to observe that the estimates (30.21) and (30.22) of Theorem 30.5 also hold for the domains Ω of the type (32.1) in place of a simple cube, with a constant that only depends on the shape of Ω. This allows to mimic the proof of Theorem 30.6, with the I_μ in place of the $I_{j,k}$ and $\widetilde{I}_\mu$ in place of $\widetilde{I}_{j,k}$. Here again, we use that the $\widetilde{I}_\mu$ are rescaled versions of simple reference domains $\widetilde{I}_1, \ldots, \widetilde{I}_N$ in order to have a local estimate of $\|f - P_j f\|_{L^r(\widetilde{I}_\mu)}$ and to apply (30.21) on $\widetilde{I}_\mu$.

The proof of the first statement (30.30) in Theorem 30.7 then a straightforward adaptation. For the second statement, i.e. the wavelet characterization by (30.31), we need an analog of the uniform local stability property (32.10) with wavelets in place of scaling functions.

THEOREM 32.2. *If* (32.10) *holds, then we also have*

$$\Big\| \sum_{\lambda \in \Lambda_j} c_\lambda \psi_\lambda \Big\|_{L^p} \sim 2^{jd(1/2-1/p)} \big\| (c_\lambda)_{\lambda \in \Lambda_j} \big\|_{\ell^p}, \tag{32.24}$$

i.e. the L^p-stability of the wavelet basis $\{\psi_\lambda\}_{\lambda \in \Lambda_j}$ independently of the level j.

PROOF. We remark that we can express a combination of wavelet coefficients at scale j as a combination of scaling functions at scale $j+1$, i.e.

$$\sum_{\lambda \in \Lambda_j} c_\lambda \psi_\lambda = \sum_{\gamma \in \Gamma_{j+1}} d_\gamma \varphi_\gamma. \tag{32.25}$$

We thus have, according to (32.10),

$$\Big\| \sum_{\lambda \in \Lambda_j} c_\lambda \psi_\lambda \Big\|_{L^p} \sim 2^{jd(1/2-1/p)} \big\| (d_\gamma)_{\gamma \in \Gamma_{j+1}} \big\|_{\ell^p}. \tag{32.26}$$

The c_λ and d_γ are linked by local discrete two scale relation. From the fine to coarse decomposition

$$c_\lambda = \sum_{\gamma \in \Gamma_{j+1}} \langle \varphi_\gamma, \tilde{\psi}_\lambda \rangle d_\gamma, \tag{32.27}$$

it follows that $\|(c_\lambda)_{\lambda \in \Lambda_j}\|_{\ell^p} \lesssim \|(d_\gamma)_{\gamma \in \Gamma_{j+1}}\|_{\ell^p}$, while from the coarse to fine reconstruction

$$d_\gamma = \sum_{\lambda \in \Lambda_j} \langle \psi_\lambda, \tilde{\varphi}_\gamma \rangle c_\lambda, \tag{32.28}$$

it follows that $\|(d_\gamma)_{\gamma\in\Gamma_{j+1}}\|_{\ell^p} \lesssim \|(c_\lambda)_{\lambda\in\Lambda_j}\|_{\ell^p}$. We thus obtain here the stability of the wavelet basis as a consequence of the stability of the scaling function basis. □

REMARK 32.3. In both domain decomposition and finite element approaches, the scaling functions that are not supported in the interior of Ω have natural extensions to $\mathbb{R}^d$. The simplest example is the case of $I = [0, 1]$, for which the extension is obtained by removing from the sums (20.16) and (20.17) the scaling functions such that $|\mathrm{Supp}(\varphi_{j,k}) \cap \Omega| = 0$ and viewing the resulting combination as a function defined on $\mathbb{R}$. This allows to build operators $\mathcal{E}_j$ that extend the functions $f \in V_j$ up to a distance of order 2^{-j} from the boundary, by simply keeping the same coefficients with the extended basis. Such an extension at level j satisfies two useful properties: firstly, if the scaling function basis is L^p stable in the sense of Theorem 30.1, then $\|\mathcal{E}_j f\|_{L^p} \lesssim \|f\|_{L^p}$. Secondly, under the assumptions of Theorem 30.2, we also have the inverse estimate (30.9) and (30.10) with $\mathcal{E}_j f$ on the left and f on the right.

From this, on can easily check that

$$\mathcal{E} f = \mathcal{E}_0 P_0 f + \sum_{j\geqslant 0} \mathcal{E}_{j+1} Q_j f \tag{32.29}$$

defines a bounded extension operator from $B^s_{p,q}(\Omega)$ to $B^s_{p,q}(\mathbb{R}^d)$, provided that the first space is characterized by the wavelet decomposition according to (30.30). In the case where $0 < s \leqslant d/p - d$ (which is not covered by Theorem 30.7) it suffices to replace $P_0 f$ by f_0 and $Q_j f$ by $f_{j+1} - f_j$ in (32.29), where f_j is a near best approximation to f in V_j (we then use the characterization of $B^s_{p,q}$ by Theorem 30.4).

33. Boundary conditions and negative smoothness

Before addressing the adaptation of the results in Section 31 (negative smoothness and L^p-spaces) to bounded domains, we need to consider the case of Sobolev and Besov spaces with prescribed homogeneous boundary conditions on $\Gamma := \partial\Omega$. Here, we maintain the same geometrical assumptions on Ω as in the previous section. For the sake of simplicity, we shall only consider $B^s_{p,q}$ and $W^{s,p}$ with $p \geqslant 1$. The case $p < 1$ can be treated by a similar approach, up to some technical modifications, as explained in Remark 33.6.

We denote by $C^\infty_m(\Omega)$ the space of smooth functions defined on Ω which vanish at order m on Γ. This means that $f \in C^\infty_m(\Omega)$ if and only if $f \in C^\infty(\Omega)$ and $|f(x)| \leqslant C[\mathrm{dist}(x, \Gamma)]^{m+1}$. We then define the spaces $W^{s,p}_m(\Omega)$ (resp. $B^s_{p,q,m}(\Omega)$) as the closure of $C^\infty_m(\Omega)$ in $W^{s,p}$ (resp. $B^s_{p,q}$). We recall below some basic facts which are are well-known for Sobolev spaces (see, e.g., GRISVARD [1983, Chapter I]), and are also easy to prove for Besov spaces.

By truncation and regularization near the boundary, one can prove that $C^\infty_m(\Omega)$ is dense in $B^s_{p,q}$ and $W^{s,p}$ for all $m \geqslant 0$ if $s < 1/p$, i.e.

$$W^{s,p}_m(\Omega) = W^{s,p} \quad \text{and} \quad B^s_{p,q,m}(\Omega) = B^s_{p,q}, \quad s < 1/p. \tag{33.1}$$

For values of s above this range, it should also be noticed that for $m_0, m_1 > s - 1/p$, we have

$$W^{s,p}_{m_0}(\Omega) = W^{s,p}_{m_1}(\Omega) \quad \text{and} \quad B^s_{p,q,m_0}(\Omega) = B^s_{p,q,m_1}(\Omega). \tag{33.2}$$

For such values of the flatness parameter m, these spaces can also be obtained as the closure of the space of test functions $\mathcal{D}(\Omega)$. These spaces are classically denoted by $W^{s,p}_0$ or H^s_0. Thus, if $n + 1/p < s \leqslant n + 1/p + 1$, we should only make the distinction between the spaces $W^{s,p}_m(\Omega)$ for $m = 0, \ldots, n$ and similarly for Besov spaces.

There is an alternate definition to such spaces based on traces. If $s > n + 1/p$ and $f \in W^{s,p}$, the traces $\gamma_m f$ of the normal derivatives of f up to order $m = n$ are $W^{s-1/p-m,p}$ functions on the smooth regions of Γ (i.e. on the faces of the S_i in (32.1) that are contained in Γ). Then, for $m \leqslant n$, we have

$$f \in W^{s,p}_m(\Omega) \Leftrightarrow f \in W^{s,p} \quad \text{and} \quad \gamma_k f = 0, \quad k = 0, \ldots, m. \tag{33.3}$$

The same holds with $B^s_{p,q}$ in place of $W^{s,p}$. From this, it follows that

$$W^{s,p}_m(\Omega) = W^{s,p} \cap W^{t,p}_m(\Omega) \quad \text{and} \quad B^s_{p,q,m}(\Omega) = B^s_{p,q} \cap B^t_{p,r,m}(\Omega), \tag{33.4}$$

for t such that $m + 1/p < t < s$. For example, we have $W^{2,2}_0(\Omega) = H^2 \cap H^1_0$, the space of H^2 functions with homogeneous Dirichlet boundary conditions.

REMARK 33.1. More generally, if B is a closed subset of Γ, we can define in a similar way the spaces $B^s_{p,q,m}(\Omega, B)$ and $W^{s,p}_m(\Omega, B)$ as the closure of smooth functions with flatness m near B. We obtain similar properties as the above when B has a nonzero $(d-1)$-dimensional measure. One can also define spaces with mixed boundary conditions involving a linear combination of several normal derivatives, or with different boundary conditions on different regions of Γ. For the sake of simplicity, our subsequent development only deals with the characterization of the spaces $B^s_{p,q,m}(\Omega)$ with the help of multiresolution spaces that are adapted to such boundary conditions. It can be extended to more general boundary conditions provided that one can build the corresponding multiresolution spaces.

In order to characterize the $W^{s,p}_m(\Omega)$ and $B^s_{p,q,m}(\Omega)$ in the same way as we did for function spaces without boundary conditions, we need that the functions in V_j vanish with the proper order near the boundary, while preserving some notion of polynomial exactness. This polynomial exactness cannot be global anymore, since for general domains, only the trivial polynomial vanishes on Γ. A natural intuition is that the characterization is feasible if all polynomials are locally reproduced up to some degree $n-1$ such that $n > s$ by the scaling functions that sit in the interior of the domain, while next to the boundary are only locally reproduced those polynomials of degree $n-1$ which vanish up to order m near Γ. We shall provide here with a precise result which confirms this intuition.

The construction of multiscale approximations with such modified polynomial reproduction properties near the boundary is a classical task in the finite element setting (e.g., $m=0$ with the $\mathbb{P}_n$ Lagrange finite element, one simply impose zero values at the nodes situated on Γ). In the domain decomposition setting, it amounts in imposing such modifications at certain faces of the cube $[0,1]^d$ (which are related to the boundary Γ by the isoparametric maps). By tensor product, this reduces to a similar construction at the edge 0 and/or 1 of the interval $[0,1]$, which is simply obtained by removing the functions $\varphi_{j,q}^{\varepsilon}$, $q=0,\dots,m$, $\varepsilon=0,1$, from the generators of $V_j^{[0,1]}$, i.e. retaining only those which vanish at the edge with order m flatness.

In both approaches, we can build (in a similar way as in the beginning of this section) a covering $(I_\mu)_{\mu\in\Theta_j}$ which satisfies the same basic prescriptions (in particular (32.7) and (32.8)). We can then summarize as follows the modified polynomial reproduction properties of the V_j spaces.

Interior behaviour. There exists a constant C and an integer j_0 such that for all $j\geqslant j_0$, if $\operatorname{dist}(I_\mu,\Gamma)\geqslant C2^{-|\mu|}$, then the restrictions of V_j to $\widetilde{I}_\mu$ and to $\widetilde{J}_\mu$ contain the polynomials of degree n (with $\widetilde{I}_\mu$ and $\widetilde{J}_\mu$ defined by (32.4) and (32.5)).

Boundary behaviour. We denote by $\widetilde{\Theta}_j$ the subset of Θ_j consisting of those μ such that $\operatorname{dist}(I_\mu,\Gamma)\leqslant C2^{-|\mu|}$. From the modifications near the boundary, there exists a family K_μ, $\mu\in\widetilde{\Theta}_j$ such that (i) each K_μ is obtained from a reference domain of the type (32.1) among a finite collection $K_1,\dots,K_N$ by the product of an orthogonal affine transformation with a scaling of $2^{-|\mu|}$ (similarly to the $\widetilde{I}_\mu$ and $\widetilde{J}_\mu$), (ii) $\widetilde{I}_\mu$ and $\widetilde{J}_\mu$ are contained in K_μ and $\partial K_\mu\cap\Gamma$ has nonzero $(d-1)$-dimensional measure, and (iii) the restriction of V_j to K_μ contains those polynomials of degree n that vanish with flatness m at $\partial K_\mu\cap\Gamma$.

We say that such multiresolution approximation spaces reproduces polynomials of degree n with flatness m at the boundary. Note that for $m\geqslant n$ no polynomial is reproduced near the boundary, i.e. we can always assume that $m\leqslant n$. Our goal is now to show that such spaces can be used to characterize the spaces $B_{p,q,m}^s(\Omega)$ by norm equivalences of the type (29.1) and (29.15). Our first step is to establish direct and inverse estimates between L^p and $W_m^{s,p}(\Omega)$.

Inverse estimates operate exactly with the same arguments as when no boundary condition were imposed on the V_j spaces. We thus obtain a straightforward adaptation of the results in Section 27 and of Theorems 30.1 and 30.2.

In order to prove a direct estimate, we cannot proceed exactly as in the previous section. In particular, the local Whitney estimate of Theorem 25.2 cannot be exploited directly, and we shall rather make use of the following generalization of the Deny–Lions theorem.

THEOREM 33.1. *Let Ω be a connected bounded domain and let B be a closed subset of $\Gamma=\partial\Omega$ of nonzero $(d-1)$-dimensional measure. We denote by $\Pi_{n,m}$ the set of polynomial $g\in\Pi_n$ such that $|g(x)|\lesssim[\operatorname{dist}(x,B)]^{m+1}$, we then have*

$$\inf_{g\in\Pi_{n,m}}\|f-g\|_{L^p}\lesssim|f|_{W^{s,p}}, \tag{33.5}$$

for all $f \in W_m^{s,p}(\Omega, B)$ *with* $m + 1/p \leqslant s \leqslant n + 1$.

PROOF. We shall apply a general abstract result of functional analysis: if X, Y and Z are Banach space and if

$$\|f\|_X \sim \|Af\|_Y + \|Kf\|_Z, \tag{33.6}$$

where $A : X \to Y$ and $K : X \to Z$ are respectively bounded and compact operators, then $\mathrm{Ker}(A)$ is finite-dimensional and $\mathrm{Ran}(A)$ is closed. It follows that A is an isomorphism between the quotient space $X/\mathrm{Ker}(A)$ and $\mathrm{Ran}(A)$, and that

$$\inf_{g \in \mathrm{Ker}(A)} \|f - g\|_X \lesssim \|Af\|_Y. \tag{33.7}$$

Before proving this result, let us see how it applies to our specific context. Here, we take $X = W_m^{s,p}(\Omega, B)$, $Z = L^p$ and K the (compact) injection of $W_m^{s,p}(\Omega, B)$ in L^p. Note that for $m = s = 0$ and $p = \infty$ we have $X = Z$ and no compactness, but (33.5) is trivial in this case.

In the case where $s = l \in \mathbb{N} - \{0\}$, we simply take $A : f \mapsto (\partial^\alpha f)_{|\alpha| = l}$ and $Y = (L^p)^{n(l)}$ with $n(l) = \#\{\alpha \in \mathbb{N}^d;\ |\alpha| = l\}$.

In the case where $l - 1 < s < l$ is fractional, we take $A : f \mapsto (g_\alpha)_{|\alpha| = l}$ with $g_\alpha(x, y) = (\partial^\alpha f(x) - \partial^\alpha f(y))/|x - y|^{s - l + 1 + d/p}$ and $Y = (L^p(\Omega \times \Omega))^{n(l)}$.

In both cases, (33.5) holds by definition of the Sobolev norm, and the kernel $\mathrm{Ker}(A)$ consists of the polynomials of degree q that belong to X. Since $l \geqslant m + 1/p$, these are exactly the polynomials $\Pi_{l,m}$. We thus obtain

$$\inf_{g \in \Pi_{l,m}} \|f - g\|_{L^p} \lesssim |f|_{W^{s,p}(B,m)}, \tag{33.8}$$

and thus (33.5) since $\Pi_{l,m} \subset \Pi_{n,m}$.

We end by proving the abstract result. If $(f_n)_{n \geqslant 0}$ is a sequence in $\mathrm{Ker}(A)$ such that $\|f_n\|_X = 1$, then since $\|f_n - f_m\|_X \sim \|Bf_n - Bf_m\|_Z$ and B is compact, there exists a subsequence of $(f_n)_{n \geqslant 0}$ that converges in X. Thus, the unit ball of $\mathrm{Ker}(A)$ is compact, i.e. $\mathrm{Ker}(A)$ is finite-dimensional (note that in our specific situation this statement does not need to be proved since $\mathrm{Ker}(A)$ consists of polynomials of degree less than q).

Now let $(g_n)_{n \geqslant 0}$ be a sequence in $\mathrm{Ran}(A)$ which converges to some $g \in Y$. Since $\mathrm{Ker}(A)$ is finite-dimensional, we can associate $f_n \in X$ such that $g_n = Af_n$ and

$$\|f_n\|_X = \min_{g \in \mathrm{Ker}(A)} \|f_n - g\|_X = \|f_n\|_{X/\mathrm{Ker}(A)}. \tag{33.9}$$

We want to prove that $g \in \mathrm{Ran}(A)$. If $g = 0$ this is trivial. If not, we can assume (up to a subsequence) that g_n does not vanish and define $h_n = f_n/\|f_n\|_X$. In particular, we have $\|h_n\|_X = \|h_n\|_{X/\mathrm{Ker}(A)} = 1$ so that (up to a subsequence) we can assume that Kh_n converges. We can also assume (up to a subsequence) that $\|f_n\|_X$ converges to a finite or infinite limit. If this limit is infinite, then $Ah_n = g_n/\|f_n\|_X$ goes to zero in Y, and thus from (33.6), h_n converges to some limit h in X which should both satisfy $\|Ah\|_Y = 0$

and $\|h\|_{X/\mathrm{Ker}(A)} = 1$, i.e. a contradiction. If $\|f_n\|_X$ goes to some finite limit, then (up a subsequence) Kf_n converges, and thus from (33.6) f_n converges to some f in X. Thus $g = Af \in \mathrm{Ran}(A)$, which concludes the proof. □

We are now ready to prove the direct estimate for Sobolev spaces with boundary conditions.

THEOREM 33.2. *Assume that the V_j spaces reproduce polynomials of degree $n-1$ with flatness $m \leqslant n-1$ at the boundary of Ω. Then, for $s \leqslant n$ we have*

$$\mathrm{dist}_{L^p}(f, V_j) \lesssim 2^{-sj}\|f\|_{W^{s,p}}, \tag{33.10}$$

for all $f \in W_m^{s,p}(\Omega)$.

PROOF. Let us first assume that $s \geqslant m + 1/p$. In the interior of Ω, for $\mu \in \Theta_j \setminus \widetilde{\Theta}_j$, we use the classical Deny–Lions theorem to build polynomials $p_\mu \in \Pi_{n-1}$ such that

$$\|f - p_\mu\|_{L^p(\widetilde{J}_\mu)} \lesssim 2^{-sj}|f|_{W^{s,p}(\widetilde{J}_\mu)}. \tag{33.11}$$

Near the boundary, for $\mu \in \widetilde{\Theta}_j$, we use Theorem 33.1 to build polynomials $p_\mu \in \Pi_{n-1}$ which vanish with flatness m on $\partial K_\mu \cap \Gamma$ and such that

$$\|f - p_\mu\|_{L^p(K_\mu)} \lesssim 2^{-sj}|f|_{W^{s,p}(K_\mu)}. \tag{33.12}$$

We then proceed as in the proof of Theorem 30.3 to build the approximant $f_j \in V_j$, using the expression of the polynomials p_μ in the local basis on $\widetilde{J}_\mu$ and K_μ. By a similar reasoning, we end up with

$$\|f - f_j\|_{L^p(I_\mu)} \lesssim 2^{-sj}\left[\sum_{\nu\in A_\mu}|f|^p_{W^{s,p}(K_\nu)} + \sum_{\nu\in B_\mu}|f|^p_{W^{s,p}(\widetilde{J}_\nu)}\right]^{1/p} \tag{33.13}$$

(replaced by a supremum in ν if $p = \infty$). The sets A_μ and B_μ are of the type

$$\begin{aligned} A_\mu &= \{\nu;\ \mathrm{dist}(I_\mu, K_\nu) \leqslant C2^{-j}\} \quad \text{and} \\ B_\mu &= \{\nu;\ \mathrm{dist}(I_\mu, \widetilde{J}_\nu) \leqslant C2^{-j}\}, \end{aligned} \tag{33.14}$$

with C a fixed constant independent of μ and j.

From the properties of the Sobolev norms with respect to the union of sets with controlled overlapping (observed in Remark 26.5, and which can be generalized to covering with noncubic domains by (32.15) of Theorem 32.1), we obtain the direct estimate (33.10) with the $W^{s,p}$ semi-norm in place of the full norm (i.e. a stronger result).

Assume now that we have $s < m + 1/p$. We claim that the direct estimate (33.10) holds if V_j reproduces polynomials of some degree q with $s \leqslant q$ in the interior, without any requirement on the boundary, and thus under our assumptions. To see this, assume that $l-1+1/p \leqslant s < l+1/p$ for an integer l such that $0 \leqslant l \leqslant m$ (if $l=0$, we only consider the range $0 < s \leqslant 1/p$). We thus cannot directly apply Theorem 33.1, but we remark that $W_m^{s,p}(\Omega) = W_m^{s,l}(\Omega)$.

In the case where $l-1+1/p \leqslant s \leqslant l$, according to the discussion for $s \geqslant m+1/p$, we thus have the direct estimate (33.10) (again with the semi-norm $W^{s,p}$) if V_j reproduces polynomials of degree q with flatness q at the boundary, i.e. no requirement on the boundary. This clearly holds with our assumptions.

In the case where $l < s < l+1/p$, if we again build the approximation f_j in a similar way as in the proof of Theorem 30.3, we still obtain an interior estimate of the type

$$\|f - f_j\|_{L^p(\Omega_j)} \lesssim 2^{-sj}|f|_{W^{s,p}}, \tag{33.15}$$

where Ω_j consists of the union of the I_μ for all μ such that $|A_\mu| = 0$.

We are left with evaluating $\|f\|^p_{L^p(M_j)}$ where $M_j := \Omega \setminus \Omega_j$ is a margin of width 2^{-j} next to the boundary, since we do not assume any polynomial reproduction in this area of Ω. Since for $s < m+1/p$, $W_m^{s,p}(\Omega)$ is also obtained as the closure of test functions, we shall assume that f is in $\mathcal{D}(\Omega)$. By local maps (using the simple geometry of Ω), we can reduce to the following simpler situation: f is a C^∞ function on $[0,1]^d$ which vanishes near the face $x_1 = 0$ and

$$M_j = \left[0, 2^{-j}\right] \times [0,1]^{d-1}. \tag{33.16}$$

We consider r such that $s - l = 1/p - 1/r$. We thus have $p < r < \infty$, and by the Hölder inequality in the x_1 variable,

$$\|f\|^p_{L^p(M_j)} \leqslant 2^{-j(1-p/r)} \int_{[0,1]^{d-1}} \left[\int_0^{2^{-j}} \left|f(x_1,\ldots,x_d)\right|^r \mathrm{d}x_1\right]^{p/r} \mathrm{d}x_2 \cdots \mathrm{d}x_d. \tag{33.17}$$

By integration from $x_1 = 0$, we thus obtain

$$\|f\|^p_{L^p(M_j)} \lesssim 2^{-spj} \int_{[0,1]^{d-1}} \left[\int_0^{2^{-j}} \left|(\partial/\partial x_1)^l f(x_1,\ldots,x_d)\right|^r \mathrm{d}x_1\right]^{p/r} \mathrm{d}x_2 \cdots \mathrm{d}x_d. \tag{33.18}$$

We now apply the Sobolev imbedding of $W^{s,p}$ in $W^{r,l}$ (or L^r if $l=0$) in one variable which gives

$$\begin{aligned}\|f\|^p_{L^p(M_j)} &\lesssim 2^{-spj} \int_{[0,1]^{d-1}} \left\|f(\cdot,x_2,\ldots,x_d)\right\|^p_{W^{s,p}([0,1])} \mathrm{d}x_2 \cdots \mathrm{d}x_d \\ &\leqslant 2^{-sj}\|f\|^p_{W^{s,p}([0,1]^d)},\end{aligned}$$

and we thus finally obtain (33.10) by combining this with the interior estimate (33.15). □

REMARK 33.2. As it can be seen in the proof, the full $W^{s,p}$ norm instead of the semi-norm is needed in (33.10) only for certain values of s, namely $l < s < l + 1/p$ for $0 \leqslant l \leqslant m$. The following simple example (corresponding to $l = 0$) shows that the full norm is indeed necessary: if $0 < s < 1/p$, the constant $f = 1$ belongs to $W_0^{s,p}(\Omega)$ with null semi-norm. Since the V_j spaces reproduce constants only in the interior of Ω, there is no hope for a direct estimate with the semi-norm.

REMARK 33.3. In the case where the generators of V_j are such that P_j is uniformly bounded in L^p we can use $P_j f$ in place of f_j in the direct estimate.

In order to proceed further, we need a better understanding on the interpolation properties of Besov and Sobolev spaces with boundary conditions. We first derive a partial statement.

THEOREM 33.3. *If $s_0 < s_1$ are both in one of the intervals $]m + 1/p, +\infty[$, $]0, 1/p[$ and $]l - 1 + 1/p, l + 1/p[$, $l = 1, \dots, m$, and if $s = (1 - \theta)s_0 + \theta s_1$, $\theta \in]0, 1[$, then*

$$B^s_{p,q,m}(\Omega) = \left[W_m^{s_0,p}(\Omega), W_m^{s_1,p}(\Omega)\right]_{\theta,q}. \tag{33.19}$$

PROOF. The proof of this result is derived from the basic properties that we gave on Sobolev and Besov spaces with boundary conditions (in particular (33.4)), which show that, under our assumptions, we have

$$\begin{aligned} B^s_{p,q,m}(\Omega) &= W_m^{s_0,p}(\Omega) \cap B^s_{p,q}, \quad \text{and} \\ W_m^{s_1,p}(\Omega) &= W_m^{s_0,p}(\Omega) \cap W^{s_1,p}. \end{aligned} \tag{33.20}$$

One concludes by the following observation which is an easy consequence of the definition of real interpolation spaces: if A, $C \subset B$ are Banach spaces and if $A \cap B$ and $A \cap C$ are respectively closed in B and C, then $A \cap [B, C]_{\theta,q}$ is also closed in $[B, C]_{\theta,q}$ and we have

$$[A \cap B, A \cap C]_{\theta,q} = A \cap [B, C]_{\theta,q}. \tag{33.21}$$

Applying (33.21) to $A = W_m^{s_0,p}(\Omega)$, $B = W^{s_0,p}$ and $C = W^{s_1,p}$ yields (33.20). □

Our next result, which should be viewed as the adaptation of Theorem 30.4 to bounded domains with boundary conditions, will also provide additional information on the interpolation of the spaces $B^s_{p,q,m}(\Omega)$ and $W_m^{s,p}(\Omega)$.

THEOREM 33.4. *Assume that V_j reproduces polynomials of degree $n - 1$ with flatness m near the boundary and that the scaling functions which generate V_j belong to*

$B^s_{p,q_0,m}(\Omega)$ for some $q_0 > 0$. Then, for all $t \in]0, \min\{n, s\}]$ such that $t - 1/p$ is not an integer among $0, \ldots, m$, we have $\mathcal{A}^t_{p,q} = B^t_{p,q,m}(\Omega)$ with equivalent norms, i.e.

$$\|f\|_{B^t_{p,q}} \sim \big\| \big(2^{tj} \operatorname{dist}_{L^p}(f, V_j)\big)_{j \geqslant 0} \big\|_{\ell^q}, \tag{33.22}$$

for all $f \in B^t_{p,q,m}(\Omega)$.

PROOF. By assumption, t sits in one of the intervals $]0, 1/p[$, $]m + 1/p, +\infty[$ or $]l - 1 + 1/p, l + 1/p[$, $l = 1, \ldots, m$. Let s_0 and s_1 be in the same interval as t and such that $s_0 < t < s_1 < \min\{n, s\}$.

From the inverse estimate generalizing Theorem 30.2 and the direct estimate established in Theorem 33.2, we can use the same simple arguments as in Corollaries 27.1 and 27.2 to derive direct and inverse estimates between $B^{s_0}_{p,p,m}(\Omega)$ and $B^{s_1}_{p,p,m}(\Omega)$, namely

$$\inf_{f_j \in V_j} \|f - f_j\|_{B^{s_0}_{p,p}} \lesssim 2^{-j(s_1 - s_0)} \|f\|_{B^{s_1}_{p,p}}, \tag{33.23}$$

for all $f \in B^{s_1}_{p,p,m}(\Omega)$ and

$$\|f\|_{B^{s_1}_{p,p}} \lesssim 2^{j(s_1 - s_0)} \|f\|_{B^{s_0}_{p,p}}, \quad \text{if } f \in V_j. \tag{33.24}$$

We can thus combine Theorem 28.2 with Theorem 33.3 to derive that $B^t_{p,q,m}(\Omega) = \mathcal{A}^{t-s_0}_q(B^{s_0}_{p,p})$, i.e. (using Hardy's inequality),

$$\|f\|_{B^t_{p,q}} \lesssim \big\| \big(2^{(t-s_0)j} \|f_j - f_{j-1}\|_{B^{s_0}_{p,p}}\big)_{j \geqslant 0} \big\|_{\ell^q}, \tag{33.25}$$

for any $f \in B^t_{p,q,m}(\Omega)$ where f_j is a near best approximation of f in the $B^{s_0}_{p,p}$ norm and $f_{-1} := 0$.

Since by the inverse estimate

$$\|f_j - f_{j-1}\|_{B^{s_0}_{p,p}} \lesssim 2^{s_0 j} \|f_j - f_{j-1}\|_{L^p}, \tag{33.26}$$

we derive the "$\lesssim$" part of the equivalence (33.22) from (33.25). The converse part is obtained by a direct argument using the inverse estimate as in the proof of Theorem 29.1. □

REMARK 33.4. We can combine the above result together with the known interpolation properties of approximation spaces in order to obtain more information on the interpolation of Sobolev and Besov spaces with boundary conditions: for $t - 1/p$ not an integer among $0, \ldots, m$, we have

$$B^t_{p,q,m}(\Omega) = \big[L^p, W^{s,p}_m(\Omega)\big]_{\theta,q}, \quad t = \theta s, \tag{33.27}$$

as well as

$$B^t_{p,q,m}(\Omega) = \big[B^{s_0}_{p,q_0,m}(\Omega), B^{s_1}_{p,q_1,m}(\Omega)\big]_{\theta,q}, \quad t = (1-\theta)s_0 + \theta s_1. \tag{33.28}$$

Among others, we see that this implies $[L^p, W_0^{s,p}]_{\theta,p} = W_0^{t,p}$, $t = \theta s$ for $t \notin \mathbb{N}$ (except if $p = 2$) and $t - 1/p$ not an integer.

REMARK 33.5. If $t - 1/p$ is an integer among $0, \dots, m$, if and $t = \theta s$, one still has the identity of the space $\mathcal{A}^t_{p,q}$ and of $[L^p, W_m^{s,p}(\Omega)]_{\theta,q}$ but this last space differs from $B^s_{p,q,m}(\Omega)$. For example, if $p = q = 2$, $m \geqslant 0$ and $t = 1/2$, we obtain the space $H^{1/2}_{0,0}$ which is strictly smaller than $H_0^{1/2}$.

The adaptation of Theorems 30.6 and 30.7 (still in the case where $p \geqslant 1$) are easy tasks, with the help of the previous results. For Theorem 30.6, we have the following modifications: (30.28) holds with the full $B^s_{p,p}$ norm, for all $f \in B^s_{p,p,m}(\Omega)$, adding the assumption that V_j reproduces polynomials of degree $n-1$ and flatness m at the boundary. Here we proceed in a slightly different way in evaluating the projection error: for $\mu \in \Theta_j \setminus \widetilde{\Theta}_j$, i.e. in the interior of the domain, we define t such that $d/p - d/r = s - t$, and we write

$$\begin{aligned}
\|f - P_j f\|_{L^p(I_\mu)} &\lesssim 2^{dj(1/r-1/p)} \|f - P_j f\|_{L^r(I_\mu)} \\
&\lesssim 2^{-sj} \inf_{p \in \Pi_{n-1}} |f - p|_{W^{t,r}(\widetilde{I}_\mu)} \\
&\lesssim 2^{-sj} |f|_{W^{s,p}(\widetilde{I}_\mu)}.
\end{aligned}$$

Here, we have first used that P_j reproduces polynomials of degree $n-1$, and then applied the Sobolev imbedding together with Deny–Lions theorem. This yields an interior estimate of the same type as (33.15) for $P_j f$. The estimate on the margin M_j (with possibly the full $W^{s,p}$ norm) is derived by similar modifications as in the proof of Theorem 33.2.

The adaptation of Theorem 30.7 is then straightforward and we simply give its statement below.

THEOREM 33.4. *Assume that the primal scaling functions are in L^r and the dual scaling functions in $L^{r'}$ for some $r \in [1, \infty]$, $1/r + 1/r' = 1$, or that the primal scaling functions are continuous and the dual scaling functions are Radon measures in which case we set $r = \infty$. Also assume that V_j reproduces polynomials of degree $n-1$ with flatness m at the boundary and that $\varphi \in B^s_{p,q_0}$ for some q_0. Then, for $1 \leqslant p \leqslant r$, one has the norm equivalence*

$$\|f\|_{B^t_{p,q}} \sim \|P_0 f\|_{L^p} + \big\|\big(2^{tj} \|Q_j f\|_{L^p}\big)_{j \geqslant 0}\big\|_{\ell^q} \tag{33.29}$$

for all $f \in B^t_{p,q,m}(\Omega)$ *with* $t > 0$ *such that* $d(1/p - 1/r) < t < \min\{s, n\}$ *and* $t - 1/p$ *not an integer among* $0, \dots, m$. *If* $f = \sum_{\lambda \in \Lambda} c_\lambda \psi_\lambda$ *is the decomposition of* f *into the corresponding wavelet basis, we also have the norm equivalence*

$$\|f\|_{B^t_{p,q}} \sim \left\| \left(2^{tj} 2^{d(1/2-1/p)j} \|(c_\lambda)_{\lambda \in \Lambda_j}\|_{\ell^p}\right)_{j \geqslant -1} \right\|_{\ell^q}, \tag{33.30}$$

under the same assumptions.

REMARK 33.6. The case $p < 1$, which was left aside here for the sake of simplicity, can be treated by a similar approach. The spaces $B^s_{p,q,m}(\Omega)$ are defined in a similar manner, and have the same basic properties. In order to reach the same conclusions (in particular Theorem 33.4 above with $p < 1$), one simply need to prove Theorem 33.1 with the seminorm $|f|_{B^s_{p,p}}$ in place of $|f|_{W^{s,p}}$ on the right side of (33.5) and $p < 1$. This is done by similar functional analytic arguments, conveniently adapted to the setting of quasi-Banach spaces: (i) from the equivalence $\omega \sim \tilde{\omega}$ given by Theorem 32.1, we can again write $\|f\|_{B^s_{p,p}} \sim \|f\|_{L^p} + \|Af\|_{L^p}$ where A is an operator involving the finite differences of f and (ii) by the approximation results of the previous section, we still have compacticity of the unit ball of $B^s_{p,p}$ in L^p on a bounded domain if $s > 0$ (in the sense of the metric topology induced by the distance $d(f, g) = \|f - g\|^p_{L^p}$). Here again, the characterization of $B^s_{p,q,m}(\Omega)$ by multiscale decomposition or wavelet coefficients is only feasible if $s - 1/p$ is not an integer among $0, \dots, m$.

We end this section by explaining the adaptation of the results of Section 31, which is much simpler. For $s < 0$, $p, q \geqslant 1$, the space $B^s_{p,q}$ is properly defined as a space of distribution on Ω by by

$$B^s_{p,q}(\Omega) := \left(B^{-s}_{p',q',m}(\Omega)\right)^*, \tag{33.31}$$

for $1/p' + 1/p = 1/q' + 1/q = 1$ and any m such that $m > -s - 1/p$ (recall that, for such values of m, $B^{-s}_{p',q',m}(\Omega)$ is also the closure of $\mathcal{D}(\Omega)$ is $B^{-s}_{p',q'}(\Omega)$).

With such a definition, the adaptation of Theorem 31.1 is straightforward: $B^s_{p,q}$ is characterized by approximation in the primal spaces V_j if $B^{-s}_{p',q',m}(\Omega)$ is characterized by approximation in the dual spaces $\widetilde{V}_j$, and we have the same norm equivalence (31.3) and (31.4).

Theorem 31.2 is independent of the domain Ω. In particular, $L^2(\Omega)$ is characterized by (31.16) if $H^\varepsilon(\Omega)$ is characterized by both approximations in V_j and $\widetilde{V}_j$. Note that since $H^\varepsilon(\Omega) = H^\varepsilon_0(\Omega) = W^{\varepsilon,2}_m(\Omega)$ for $\varepsilon < 1/2$, the characterization of L^2 is feasible with V_j and $\widetilde{V}_j$ constituted of functions that vanish near the boundary.

Finally, the characterization of L^p spaces easily extends to the type of domains that we are considering here, since the Calderon–Zygmund theory (which allows us to establish the uniform L^p-boundedness of the operators $T_E f := \sum_{\lambda \in E} \langle f, \tilde{\psi}_\lambda \rangle \psi_\lambda$ independently of the subset $E \subset \Lambda$) also operates in such domains.

34. Multilevel preconditioning

One of the interest of multiscale discretizations is the possibility of precondition large systems arising from elliptic operator equations. A general setting for such equations is the following: H is a Hilbert space embedded in $L^2(\Omega)$ and $a(\bullet,\bullet)$ is a bilinear form on $H \times H$ such that

$$a(u,u) \sim \|u\|_H^2. \tag{34.1}$$

Given $f \in H^*$, we search for $u \in H$ such that

$$a(u,v) = \langle f, v\rangle, \qquad \text{for all } v \in H. \tag{34.2}$$

From the Lax–Milgram lemma, this problem has a unique solution u. If A is defined by $\langle Au, v\rangle = a(u,v)$ for all $v \in H$, (34.1) implies that A is an isomorphism from H to H^*, so that u is also the unique solution in H of

$$Au = f. \tag{34.3}$$

If V_h is a subspace of H, the Galerkin approximation of u in V_h is classically defined by $u_h \in V_h$ such that

$$a(u_h, v_h) = \langle f, v_h\rangle, \qquad \text{for all } v_h \in V_h. \tag{34.4}$$

If V_h is finite-dimensional, the approximated problem (34.4) amounts in solving a linear system. By its definition, u_h is also the projection of u onto V_h in the sense of the energy norm $\|u\|_a := \sqrt{a(u,u)}$, so that we have the error estimate

$$\|u - u_h\|_H \lesssim \|u - u_h\|_a = \inf_{v_h \in V_h} \|u - v_h\|_a \lesssim \inf_{v_h \in V_h} \|u - v_h\|_H. \tag{34.5}$$

Of interest to us is the practical situation where H is an L^2-Sobolev space. Classical instances are given by the *Poisson* equation $-\Delta u = f$ with Dirichlet boundary condition $u = 0$ on $\Gamma := \partial\Omega$, for which $H = H_0^1$, the *Helmholtz* equation $u - \Delta u = f$ with Neumann boundary condition $\partial u/\partial n = 0$ on Γ, for which $H = H^1$, and the *bi-Laplacian* equation $\Delta^2 u = f$ with condition $u = \partial u/\partial n = 0$ on Γ, for which $H = H_0^2$.

For such equations, it is well known that the matrices resulting from Galerkin discretizations in the finite element spaces are ill-conditioned in the sense that their condition number grows like h^{-2s} where h is the mesh size and s is the order of the corresponding Sobolev space H (i.e. $2s$ is the order of the operator). A particular instance was evoked in Section 6 for the Poisson equation.

REMARK 34.1. Elliptic equations involving integral operators of negative order also enter the above class of problems with the modification that H^* is now embedded in $L^2(\Omega)$. Instances are the single (resp. double) layer potential operators for which $H =$

$H^{-1/2}$ (resp. L^2). This unified point of view was proposed in DAHMEN, PRÖSSDORF and SCHNEIDER [1993, 1994], together with matrix compression techniques that will be evoked in Section 41 of the next chapter.

The use of multilevel methods for preconditioning such matrices is strongly linked with the ability of characterizing the L^2-Sobolev space H by means of multiscale decompositions. In this particular setting, the norm equivalences that were established in the previous sections have the form

$$\|f\|_{H^s}^2 \sim \|P_0 f\|_{L^2}^2 + \sum_{j\geqslant 0} 2^{2sj}\|Q_j f\|_{L^2}^2, \tag{34.6}$$

or in terms of wavelet coefficients

$$\|f\|_{H^s}^2 \sim \sum_{\lambda\in\Lambda} 2^{2s|\lambda|}|c_\lambda|^2. \tag{34.7}$$

In Theorem 30.6, the range of s for which such equivalences hold is described in terms of three basic properties of the multiscale decomposition: smoothness, polynomial reproduction and L^p stability of the associated projector. We have seen that these equivalences also hold for the spaces $H^s(\Omega)$ and $H_0^s(\Omega)$ with appropriate adaptations of the multiscale decomposition near the boundary, as well as for negative values of s. Such norm equivalences will allow us to analyze the performance of several multilevel preconditioning schemes.

Example 34.1. Wavelet diagonal preconditioning

A particular instance of wavelet diagonal preconditioning was already discussed in Section 6 of Chapter I. We also refer to this section for the basic concept of preconditioner for an iterative scheme. Let us consider the Galerkin discretization (34.4) on a multiresolution approximation space $V_J \subset H$ corresponding to a mesh size 2^{-J} and we denote by u_J the corresponding solution. We assume that $\bigcup_{j\geqslant 0} V_j$ is dense in H (which ensure in particular that $\|u_J - u\|_H$ goes to zero as $J \to +\infty$). For the practical computation of u_J, we can use either the nodal basis $\Phi_J = \{\varphi_\gamma;\ |\gamma| = J\}$ or the multiscale basis $\Psi_J = \{\psi_\lambda;\ |\lambda| < J\}$.

With the nodal basis, we obtain a system

$$A_J U_J = F_J, \tag{34.8}$$

where U_J is the coordinate vector of u_J in the basis Φ_J, $F_J = (\langle f, \varphi_\gamma\rangle)_{|\gamma|=J}$, and $A_J = (\langle A\varphi_\gamma, \varphi_\beta\rangle)_{|\gamma|=|\beta|=J}$ is the corresponding stiffness matrix.

Similarly, with the multiscale basis, we obtain a system

$$\widetilde{A}_J \widetilde{U}_J = \widetilde{F}_J, \tag{34.9}$$

where $\widetilde{U}_J$ is the coordinate vector of u_J in the basis Ψ_J, $F_J = (\langle f, \psi_\lambda \rangle)_{|\lambda|<J}$, and $A_J = (\langle A\psi_\lambda, \psi_\lambda \rangle)_{|\lambda|,|\mu|<J}$ is the stiffness matrix.

The link between these two systems is provided by the linear transformation R_J representing the fast wavelet reconstruction algorithm: we clearly have

$$U_J = R_J \widetilde{U}_J, \quad \widetilde{F}_J = R_J^* F_J \quad \text{and} \quad \widetilde{A}_J = R_J^* A_J R_J. \tag{34.10}$$

Recall that R_J is implemented by a coarse to fine $\mathcal{O}(N)$ algorithm. Note that R_J^* has the same fine to coarse structure as the decomposition algorithm, but uses the same low-pass and high-pass filters as the reconstruction R_J: it differs from R_J^{-1}, except if orthonormal wavelets are used.

In the following, we shall denote by $\langle \bullet, \bullet \rangle_d$ and $\| \bullet \|_d$ the discrete inner product and Hilbertian norm.

A first result relates the norm equivalences of the type (34.2) to diagonal preconditioning of A in the multiscale basis.

THEOREM 34.1. *Consider the the diagonal matrix $D_J = (2^{2s|\lambda|}\delta_{\lambda,\mu})_{|\lambda|,|\mu|<J}$, indexed by the multiscale discretization. The two following statements are equivalent*:

(i) *H is characterized by a norm equivalence*

$$\|f\|_H^2 \sim \sum_{\lambda \in \Lambda} 2^{2s|\lambda|} |c_\lambda|^2. \tag{34.11}$$

(ii) *The condition number $\mathcal{K}(D_J^{-1}\widetilde{A}_J) = \mathcal{K}(D_J^{-1/2}\widetilde{A}_J D_J^{-1/2})$ is bounded independently of J.*

PROOF. The property (ii) is equivalent to

$$\langle D_J U, U \rangle_d \sim \langle A_J U, U \rangle_d, \tag{34.12}$$

with constants independent of the vector U and the scale level J. From the definition of A_J, this can also be expressed by

$$\|v_J\|_a^2 \sim \sum_{|\lambda|<J} 2^{2s|\lambda|} |c_\lambda|^2, \tag{34.13}$$

for all $v_J = \sum_{|\lambda|<J} c_\lambda \psi_\lambda$ in V_J. Since $\| \bullet \|_a \sim \| \bullet \|_H$, (34.13) is equivalent to (34.11) for all $f \in V_J$. By density of the V_J spaces, this is equivalent to (34.11) for all $f \in H$. □

From this result, we see that for a suitable choice of τ, the preconditioned iteration

$$\widetilde{U}_J^{n+1} = \widetilde{U}_J^n + \tau D_J^{-1} (\widetilde{F}_J - \widetilde{A}_J \widetilde{U}_J^n), \tag{34.14}$$

decreases the error $E_J^n = U_J^n - U_J$ with a rate $\rho = [1 - \mathcal{K}(D_J^{-1}\widetilde{A}_J)]/[1 + \mathcal{K}(D_J^{-1}\widetilde{A}_J)]$ which is independent of the number of level J.

A particular interest of the nodal basis discretization is the sparse structure of A_J in the case where A is a partial differential operator which has a local action: the cost of one application of A_J is then in $\mathcal{O}(N_J)$ where $N_J = \dim(V_J) \sim 2^{dJ}$. Since $\widetilde{A}_J = R_J^* A_J R_J$ and R_J can also be implemented in $\mathcal{O}(N_J)$ operations, this optimal complexity is preserved in the cost of the above iteration.

Since $D_J^{-1/2}\widetilde{A}_J D_J^{-1/2} = D_J^{-1/2} R_J^* A_J R_J D_J^{-1/2}$, we can also view $B_J := R_J D_J^{-1} R_J^*$ as a preconditioner for A_J. The corresponding preconditioned iteration

$$U_J^{n+1} = U_J^n + \tau B_J\big(F_J - A_J U_J^n\big), \tag{34.15}$$

which directly acts in the nodal discretization, can directly be obtained by application of R_J on (34.14) and produces the same rate of decay for the error.

The norm equivalence (34.11) is also useful to control the progress of these algorithms through their residual $G_J^n := F_J - A_J U_J^n$ and $\widetilde{G}_J^n := \widetilde{F}_J - \widetilde{A}_J \widetilde{U}_J^n$. These quantities satisfy the equations

$$A_J E_J^n = G_J^n \quad \text{and} \quad \widetilde{A}_J \widetilde{E}_J^n = \widetilde{G}_J^n, \tag{34.16}$$

so that we can evaluate the norm of the corresponding error $e_J \in V_J$ as

$$\|e_J^n\|_H^2 \sim \big\langle \widetilde{A}_J \widetilde{E}_J^n, \widetilde{E}_J^n \big\rangle_d \sim \big\langle \widetilde{A}_J^{-1} \widetilde{G}_J^n, \widetilde{G}_J^n \big\rangle_d \sim \big\langle D_J^{-1} \widetilde{G}_J^n, \widetilde{G}_J^n \big\rangle_d$$

for the iteration (30.14) and similarly by $\|e_J^n\|_H^2 \sim \langle B_J G_J^n, G_J^n \rangle_d$ for the iteration (30.15).

REMARK 34.2. Both iterations (34.14) and (34.15) only involve the filters in the primal refinement equations

$$\varphi_\gamma = \sum_{\beta \in \Gamma_{j+1}} h_{\beta,\gamma} \varphi_\beta \quad \text{and} \quad \psi_\lambda = \sum_{\mu \in \Gamma_{j+1}} g_{\beta,\lambda} \varphi_\beta \tag{34.17}$$

(for $\gamma \in \Gamma_j$ and $\lambda \in \Lambda_j$), which are used in the application of R_J and R_J^*. These filters are finitely supported if the primal scaling functions and wavelets have also compact support, in which case R_J and R_J^* apply in $\mathcal{O}(N_J)$ operations. We can thus allow the dual scaling functions and wavelets to have infinite support as in several examples of splines and finite element wavelets presented in Sections 19 and 21.

REMARK 34.3. Further improvement is achieved by using the *preconditioned conjugate gradient algorithm*, which simply amounts in the application of a conjugate gradient algorithm on the preconditioned equation

$$D_J^{-1/2} R_J^* A_J R_J D_J^{-1/2} X_J = Y_J, \tag{34.18}$$

where $Y_J := D_J^{-1/2} R_J^* F_J$ and $U_J = R_J D_J^{-1/2} X_J$. It is well known that the convergence rate for such an algorithm is given by $\tilde{\rho}_J = (\sqrt{\mathcal{K}(B_J A_J)} - 1)/(\sqrt{\mathcal{K}(B_J A_J)} + 1)$, improving on the rate $\rho_J = (\mathcal{K}(B_J A_J) - 1)/(\mathcal{K}(B_J A_J) + 1)$ of the iteration (34.15) (obtained with an optimal choice of τ), although this improvement is less significant when $\mathcal{K}(B_J A_J)$ stays close to 1.

REMARK 34.4. Diagonal wavelet preconditioning was also proposed in the more general setting Petrov–Galerkin discretization in DAHMEN, PRÖSSDORF and SCHNEIDER [1994]: the problem is now to find $u_J \in V_J$ such that

$$a(u_J, v_J) = \langle f, v_J \rangle, \quad v_J \in X_J, \tag{34.19}$$

where X_J has the same dimension as V_J. Such schemes include in particular *collocation* for which the basis functions of X_J are Dirac masses on the grid indexing the nodal basis of V_J. The above reference considers setting of periodic boundary conditions for which the well-posedness of (34.19), the resulting error estimates and the possibility of multiscale preconditioning can be completely analyzed. Such an analysis appears to be more difficult for other types of boundary conditions.

REMARK 34.5. In this example, we are using the multiresolution spaces V_j for two distinct purposes: discretization at the finest level and preconditioning by B_J. In practice, one could consider the idea of decoupling those two tasks, i.e. use a certain space V_J for the Galerkin discretization and precondition with a discrete multiscale transform R'_J which relates to different spaces V'_j. Such a strategy is more difficult to analyze in its full generality and might even fail in some cases. This idea is still particularly attractive in the case where V_J has high approximation order: the filters involved in the application of R_J might then have large supports, so that the computational cost of one iteration will be substantially reduced by using a discrete multiscale transform with shorter filters.

Example 34.2. BPX preconditioning

The wavelet diagonal preconditioning in the iteration (34.15) is by essence a parallel technique: the residual is decomposed into multiscale blocks which are treated independently by the scaling D_J^{-1} and then reassembled.

Similar parallel multilevel preconditioners were introduced in BRAMBLE, PASCIAK and XU [1990], which do not require the existence of a wavelet basis for characterizing the details, but only involve the nodal basis for the multiscale approximation spaces. These construction and analysis of these preconditioners are again related to the norm equivalence

$$\|f\|_H^2 \sim \|P_0 f\|_{L^2}^2 + \sum_{j \geqslant 0} 2^{2js} \|Q_j f\|_{L^2}^2. \tag{34.20}$$

We assume here that $s > 0$ and we recall the variant

$$\|f\|_H^2 \sim \|f\|_{L^2}^2 + \sum_{j\geqslant 0} 2^{2js}\|f - P_j f\|_{L^2}^2. \tag{34.21}$$

If the spaces V_j satisfy have enough smoothness and polynomial reproduction properties, we know from the previous sections that (34.20) and (25.20) are always satisfied with P_j being the L^2-orthogonal projector onto V_j. We also assume that we have an L^2 stable scaling (or nodal) function basis φ_γ, $\gamma \in \Gamma_j$, to characterize the V_j spaces.

For $j \leqslant J$, we also define the discrete spaces $V_{J,j}$ corresponding to the elements of V_j expressed in the nodal basis of V_J, and $P_{J,j}$ the orthogonal projector from $V_{J,J}$ (the full discrete space) onto $V_{j,J}$ with respect to the discrete inner product. We also define $Q_{J,j} := P_{J,j+1} - P_{J,j}$ the projector on the orthogonal complement $W_{J,j}$ of $V_{J,j}$ into $V_{J,j+1}$ with respect to the discrete inner product. If G is the coordinate vector of $g \in V_J$ in the nodal basis, we thus have

$$\begin{aligned}\langle A_J G, G\rangle_d &\sim \|g\|_H^2 \sim \|g\|_{L^2}^2 + \sum_{j=0}^{J-1} 2^{2js}\|g - P_j g\|_{L^2}^2 \\ &\sim \|G\|_d^2 + \sum_{j=0}^{J-1} 2^{2js}\|G - P_{J,j}G\|_d^2 \\ &\sim \|P_{J,0}G\|_d^2 + \sum_{j=0}^{J-1} 2^{2js}\|Q_{J,j}G\|_d^2,\end{aligned}$$

where we have used the ellipticity property and the L^2-stability of the nodal basis of V_J.

It follows that A_J is spectrally equivalent to the self-adjoint discrete operator $Z_J := P_{J,0} + \sum_{j\geqslant 0} 2^{2js} Q_{J,j}G$. As we noticed in Section 6, this also means that A_J^{-1} is spectrally equivalent to Z_J^{-1} which is given by $Z_J^{-1} = P_{J,0} + \sum_{j\geqslant 0} 2^{-2js} Q_{J,j}G$ using the discrete orthogonality of the spaces $V_{J,0}$ and $W_{J,j}$, $j = 0, \ldots, J-1$. We thus have

$$\langle A_J^{-1}G, G\rangle_d \sim \|P_{J,0}G\|_d^2 + \sum_{j=0}^{J-1} 2^{-2js}\|Q_{J,j}G\|_d^2 \sim \sum_{j=0}^{J} 2^{-2js}\|P_{J,j}G\|_d^2,$$

where we have used that $\|Q_{J,j}G\|_d^2 = \|P_{J,j+1}G\|_d^2 - \|P_{J,j}G\|_d^2$.

An $\mathcal{O}(1)$ preconditioner for A_J is thus given by

$$B_J = \sum_{j=0}^{J} 2^{-2js} P_{J,j}. \tag{34.22}$$

The main drawback of this choice is that, except in the case of orthonormal scaling function, the computation of $P_{J,j}$ is not local so that B_J cannot exactly be applied in

$\mathcal{O}(N_J)$ operations: one needs to invert the mass matrix $(\langle\varphi_{J,\gamma},\varphi_{J,\beta}\rangle_d)_{\gamma,\beta\in\Gamma_j}$ where $\varphi_{J,\gamma}$ is the coordinate vector of φ_γ in the nodal basis of V_J.

A possibility to circumvent this drawback is to make the following observation: one has the uniform stability property

$$\Big\|\sum_{\gamma\in\Gamma_j} c_\gamma\varphi_{J,\gamma}\Big\|_d^2 \sim \sum_{\gamma\in\Gamma_j}|c_\gamma|^2 \tag{34.23}$$

from the L^2 stability of the nodal bases. As it was observed in Section 10, this implies the *frame property*

$$\|G\|_d^2 \sim \sum_{\gamma\in\Gamma_j}\big|\langle G,\varphi_{J,\gamma}\rangle_d\big|^2, \tag{34.24}$$

for $G\in V_{J,j}$, with constants also independent of j and J. It follows that

$$\big\langle A_J^{-1}G,G\big\rangle_d \sim \sum_{j=0}^{J}2^{-2js}\sum_{\gamma\in\Gamma_j}\big|\langle G,\varphi_{J,\gamma}\rangle_d\big|^2, \tag{34.25}$$

which shows that an alternate choice for a preconditioner is given by

$$\widetilde{B}_J=\sum_{j=0}^{J}2^{-2js}I_j, \tag{34.26}$$

where $I_j:=\sum_{\gamma\in\Gamma_j}\langle G,\varphi_{J,\gamma}\rangle_d\varphi_{J,\gamma}$ is a discrete quasi-interpolant operator. In comparison to (34.22), this choice is better in the sense that I_j is a local operator. Note that the wavelet preconditioner $R_JD_J^{-1}R_J^*$ of the previous example can also be expressed as

$$R_JD_J^{-1}R_J^*G=A_0G+\sum_{j=0}^{J-1}2^{-2js}\sum_{\lambda\in\Lambda_j}\langle G,\psi_{J,\lambda}\rangle_d\psi_{J,\lambda}, \tag{34.27}$$

with $\psi_{J,\gamma}$ being the coordinates of ψ_λ in the nodal basis of V_J. The so-called BPX preconditioner $\widetilde{B}_J$ is thus implemented in a similar manner as the wavelet diagonal preconditioner, with the distinction that the multiscale transforms only involve the low pass filters and the approximation coefficients.

REMARK 34.6. Note that I_j can also be interpretated as a first order approximation to $P_{J,j}$ in the sense that I_jG is obtained after one iteration in the iterative inversion of the mass matrix $(\langle\varphi_{J,\gamma},\varphi_{J,\beta}\rangle_d)_{\gamma,\beta\in\Gamma_j}$ for the computation of $P_{J,j}G$. Another point of view on the BPX preconditioner is given in OSWALD [1994], in terms of the frame properties of the family $\bigcup_{j\geqslant 0}\{\varphi_\gamma\}_{\gamma\in\Gamma_j}$ in the space H.

REMARK 34.7. In contrast to the wavelet diagonal preconditioning, the constants in the equivalence $\langle A_J^{-1}G, G\rangle_d \sim \langle \widetilde{B}_J^{-1}G, G\rangle_d$ deteriorate as s goes to zero and the equivalence does not hold anymore for the value $s = 0$ (one can still prove an $\mathcal{O}(\log(J)^2)$ condition number). This limitation of the BPX preconditioner to positive order operators makes it inadequate for problems of the type $p(x)u(x) - \nabla[q(x)\nabla u(x)] = f$, if the first coefficient $p(x)$ is numerically dominant, i.e. the L^2 term dominates in the energy norm. A more general remark is that the condition numbers resulting from multiscale preconditioning can be strongly affected by the variations in the coefficients of the operator, even if the space H does not change. We shall discuss in the next example possible solutions circumvent this difficulty.

REMARK 34.8. Both BPX and wavelet preconditioners are *additive*: one iteration treats in parallel the different levels of scale and adds up the result, according to (34.26) or (34.27). In comparison, most algorithms priorily developed in multigrid theory are *multiplicative*, i.e. treat the scales in a sequential way. These techniques also amount in $\mathcal{O}(1)$ preconditioning, although their analysis is less simple than in the additive case. We shall not discuss them here, although it should be mentioned that in several test problems, multiplicative multilevel techniques yield slightly better results in terms of condition numbers than parallel preconditioners. There is now a huge amount of published literature on multigrid methods in various contexts. General treatments can be found in HACKBUSCH [1985] and BRAMBLE [1993]. The reader will also find in OSWALD [1994] an analysis of both multilevel preconditioning and domain decomposition technique from a subspace splitting point of view.

Example 34.3. Adaptive preconditioning

Wavelet preconditioning provides with uniform bounds for the condition number of $D_J^{-1/2}\widetilde{A}_J D_J^{-1/2}$. However, these bounds might be large in practice, in relation with both the multiscale decomposition and the operator.

Indeed, in most constructions of wavelets on bounded domains, the multiscale decomposition cannot be carried out down to the coarsest level of discretization: even in the simple case of the interval, treated in Section 20 of Chapter II, we have seen that the adaptation near the boundary often imposes a minimal level $j_0 \geqslant 0$ allowing to decouple the construction of the modified scaling functions and wavelets at the edges. This affects in practice the ratio between the constants in the norm equivalence (34.11) and thus the resulting condition number of $D_J^{-1}\widetilde{A}_J$. Recall also that there is a certain flexibility in the construction the modified basis functions at the edges (especially in the biorthogonal case) and certain choices might result in better constants than others.

In addition, the ratio between the constants of ellipticity and continuity of the bilinear form in (34.1) also affects the condition number of $D_J^{-1}\widetilde{A}_J$. This ratio deteriorates in particular in the case of nonconstant coefficients operators: if for instance $Au = -\nabla[q(x)\nabla u]$, we find that the quantity $\sup|q(x)|/\inf|q(x)|$ enters in the resulting condition number.

In order to remedy these problems, a good preconditioner should somehow take into account such finer information on the multilevel decomposition and on the operators.

A first basic idea is to eliminate the problem of the coarse scale level by a full inversion of the stiffness matrix. More precisely, in the preconditioning of $\widetilde{A}_J$, we replace the purely diagonal matrix D_J by a block diagonal matrix X_J which consists of a purely diagonal part identical to D_J for the details scales $|\lambda| \geqslant j_0$, and of the coarse scale stiffness matrix A_{j_0} for the basic scale $|\lambda| = j_0 - 1$, for which we have by convention $\{\psi_\lambda\}_{\lambda\in\Lambda_{j_0-1}} = \{\varphi_\gamma\}_{\gamma\in\Gamma_{j_0}}$. The entries of the new matrix X_J are thus given by

$$X_J(\lambda,\mu) = 2^{2s|\lambda|}\delta_{\lambda,\mu} \quad \text{if } |\lambda| \geqslant j_0 \text{ or } |\mu| \geqslant j_0, \tag{34.28}$$

and

$$X_J(\lambda,\mu) = a(\psi_\lambda,\psi_\mu) \quad \text{if } |\lambda| = |\mu| = j_0. \tag{34.29}$$

Note that this matrix is easily inverted since the nondiagonal part is of small size.

Another simple idea is to use replace the simple scaling by $2^{2s|\lambda|}$ in the diagonal part (34.28) by the corresponding diagonal elements of the stiffness matrix $\widetilde{A}_J$. This defines a new block diagonal matrix Y_J which is identical to X_J for the coarse components (34.29) and such that

$$Y_J(\lambda,\mu) = a(\psi_\lambda,\psi_\mu)\delta_{\lambda,\mu} \quad \text{if } |\lambda| \geqslant j_0 \text{ or } |\mu| \geqslant j_0. \tag{34.30}$$

By such a modification, we expect to be more adaptive with respect to the variation of the coefficients of the operator, as well as to the modifications of the scaling functions and wavelets near the edges.

A simple example which reveals the effectiveness of this approach is the singular perturbation problem

$$u - \varepsilon\Delta u = f \quad \text{in } \Omega, \qquad u = 0 \quad \text{on } \partial\Omega. \tag{34.31}$$

The corresponding bilinear form is given by

$$a_\varepsilon(u,u) = \int_\Omega |u(x)|^2\,\mathrm{d}x + \varepsilon\int_\Omega |\nabla u(x)|^2\,\mathrm{d}x. \tag{34.32}$$

On $H = H_0^1$, we still have $a_\varepsilon(u,u) \sim \|u\|^2_{H_0^1}$ but we see that the ratio between the constants in this equivalence is of the order $1/\varepsilon$. As a consequence, we can only expect a condition number $\mathcal{K}(D_J^{-1}A_J)$ of the order $1/\varepsilon$.

We now remark that if $g \in V_J$ has coordinate vector $G = (c_\lambda)_{\lambda\in\Lambda}$ in the wavelet basis, we have

$$\begin{aligned}\langle \widetilde{A}_J G, G\rangle &= a_\varepsilon(g,g)\\ &\sim \sum\nolimits_{|\lambda|<J}\bigl(1+\varepsilon 2^{2s|\lambda|}\bigr)|c_\lambda|^2 \sim \sum\nolimits_{|\lambda|<J} a_\varepsilon(\psi_\lambda,\psi_\lambda)|c_\lambda|^2\\ &\sim \langle Y_J G, G\rangle,\end{aligned}$$

with constants that are independent of both J and ε, provided that the wavelet basis characterizes both L^2 and H_0^1. Thus $R_J Y_J^{-1} R_J^*$ is more adapted to precondition A_J than the simpler B_J of Example 34.1.

In the case of varying coefficients, the effectiveness of such an adaptive method is also observed in practice. However it is not proved in full theoretical generality. Consider an operator of the type $Au := -\nabla[p(x)\nabla u]$ with scalar coefficients $0 < p_{\min} \leqslant p(x) \leqslant p_{\max}$. Ideally, one would like to prove that $\mathcal{K}(Y_J^{-1}\widetilde{A}_J)$ is independent of the ratio $p_{\max}/p_{\min}$, i.e. that the equivalence

$$\int_\Omega p(x)\left|\nabla u(x)\right|^2 \sim \sum\nolimits_{|\lambda|<J} p_\lambda |c_\lambda|^2, \quad p_\lambda := \int_\Omega p(x)\left|\nabla\psi_\lambda\right|^2 \mathrm{d}x, \tag{34.33}$$

holds without dependence on $p_{\max}/p_{\min}$. Such a result is certainly false without any additional assumptions, for instance some bounds on $p(x)$ in a smoother norm than L^∞.

If one wants to improve furthermore on the resulting condition number, a possibility is to drop the diagonal structure of Y_J^{-1} also in the high scales $|\lambda| \geqslant j_0$ and look for a preconditioner which is still "almost diagonal". This approach is proposed in COHEN and MASSON [1999], in which the simple inversion of $a(\psi_\lambda, \psi_\lambda)$ that yields Y_J^{-1} is replaced by a local Petrov–Galerkin problem which amount in the inversion of a small matrix. A close strategy is proposed in LAZAAR, LIANDRAT and TCHAMITCHIAN [1994], using "vaguelettes" and freezing coefficients techniques in the periodic setting. Generally speaking, such preconditioners search for a compromise between the simple wavelet diagonal preconditioner $R_J D_J^{-1} R_J^*$ which has very low cost but might still yields large (although bounded) condition numbers, and the inverse A_J^{-1} of the operator which is too heavy to apply. The best compromise B_J should optimize the overall computational cost to reach an error ε in some prescribed norm, i.e. $\mathcal{C}(\varepsilon) = n(\varepsilon)C_0$ where $n(\varepsilon)$ is the number of needed iterations (which depends on the $\mathcal{K}(B_J A_J)$) and C_0 is the cost of one iteration (which depends on the number of unknown at level J and on the sparsity of B_J). The position of such a compromise clearly depends on the type of problem at hand.

Example 34.4. An optimal complexity multiscale solver

If we want to evaluate the computational cost of the global resolution of (34.4) using the preconditioned iteration (34.15), we need to fix a certain target accuracy ε, that we shall allow in the $\|\bullet\|_H$ norm related to our problem. We then choose a resolution level J such that

$$\|u_J - u\|_H \leqslant \varepsilon, \tag{34.34}$$

where u_J is the Galerkin solution of (34.4) in V_J. Such a scale can be crudely obtained using an a priori information on the smoothness of the solution: e.g., in the case of the

Laplace equation $-\Delta u = f$ on a smooth domain with Dirichlet boundary conditions, we know that $u \in H^{r+2} \cap H_0^1$ if $f \in H^r$, $r \geqslant -1$, so the estimate

$$\|u - u_J\|_{H_0^1} \lesssim \text{dist}_{H_0^1}(u, V_J) \lesssim 2^{-(r+1)J} \|u\|_{H^{r+2}} \tag{34.35}$$

is ensured by the results of this chapter, provided that V_J has the appropriate polynomial reproduction properties. More generally, we assume here that the smoothness of u ensures that

$$\|u_j - u\|_H \leqslant C(u) 2^{-tj}, \quad j \geqslant 0, \tag{34.36}$$

and we denote by $N_J := \dim(V_J) \sim 2^{dJ}$ the dimension of the space V_J which is selected to reach the accuracy $\varepsilon \sim C(u)2^{-tJ}$. Since $C(u)$ is in practice a smoother norm than the H norm, it is reasonable to assume in addition that

$$\|u\|_H \leqslant C(u). \tag{34.37}$$

We are thus allowed to accept a similar error between u_J and the solution u_J^n obtained after n iteration of (34.15). Recall that this error e_J^n can be controlled by the residual $G_J^n = F_J - A_J U_J^n$, according to

$$\|e_J^n\|_H \sim \langle B_J G_J^n, G_J^n \rangle_d. \tag{34.38}$$

The preconditioned iteration decreases the error by a factor $\rho < 1$ which is independent of J. Therefore, if the algorithm is initialized from $U_J^0 = 0$, we can attain the error $\varepsilon \sim C(u)2^{-tJ}$ in a number n_J of iteration such that $\rho^{n_J} \lesssim 2^{-tJ}$ (since $\|e_J^0\|_H = \|u_J\|_H \lesssim C(u)$). Since one iteration has $\mathcal{O}(N_J)$ complexity, the overall computation cost is given by

$$\mathcal{C}(\varepsilon) \sim n_J N_J \lesssim N_J \log(N_J). \tag{34.39}$$

One can actually remove the $\log(N_J)$ factor with a simple elaboration of this multiscale solver, which simply consists in applying the first iterations only on the coarse scale and progressively refining the discretization. More precisely, we proceed as follows:

- Start from a Galerkin discretization at the coarsest scale (here assumed to be $j = 0$ and apply the iteration (34.15) with $J = 0$ and initial condition $u_0^0 = 0$.
- Stop after a number n of iteration such that $\|u_0^n - u_0\|_H \lesssim C(u)$ (using the residual evaluation).
- Discretize the problem at the scale $j = 1$, and apply the iteration (34.15) with $J = 1$ and initial condition $u_1^0 = u_0^n \in V_0 \subset V_1$.
- Stop after a number n of iteration such that $\|u_1^n - u_1\|_H \lesssim C(u)2^{-t}$ (using the residual evaluation).
- Iterate this procedure up from $j = 2$ to $j = J$, requiring at the scale j the error $\|u_j^n - u_j\|_H \lesssim C(u)2^{-tj}$.

Clearly the result of this algorithm satisfies the required error estimate. Let us now compute the computational cost at each scale level: the initial error is estimated by

$$\begin{aligned}\left\|u_j - u_j^0\right\|_H &= \left\|u_j - u_{j-1}^n\right\|_H \\ &\leqslant \left\|u_{j-1} - u_{j-1}^n\right\|_H + \|u_j - u_{j-1}\|_H \\ &\lesssim C(u)2^{-tj} + \|u - u_j\|_H + \|u - u_{j-1}\|_H \\ &\lesssim C(u)2^{-tj},\end{aligned}$$

where we have used the assumption (34.36). Therefore, a fixed number n_0 of iteration is required at each scale level. Each iteration having complexity $\mathcal{O}(N_j)$, we find that the overall computational cost is estimated by

$$\mathcal{C}(\varepsilon) \lesssim n_0 \sum_{j=0}^{J} 2^{dj} \lesssim 2^{dJ} \lesssim N_J. \tag{34.40}$$

We have thus obtained an algorithm of *optimal complexity*: the amount of computation is directly proportional to the number of parameters required to achieve the prescribed accuracy.

REMARK 34.9. This last optimal algorithm should be viewed as a reformulation of the nested iteration in the *full multigrid* algorithm introduced in BANK and DUPONT [1981] which also has optimal complexity.

REMARK 34.10. Throughout this section, we only made use of the characterization of Sobolev spaces of positive order. The characterization of negative order Sobolev spaces can be required for the preconditioning of negative order operators. In several instances, these operators appear as a compact perturbation of the identity, and do not directly require a preconditioner. We shall see in Section 43 of the next chapter that the norm equivalences for negative order Sobolev spaces are needed to develop an a posteriori analysis in the context of adaptive methods for solving elliptic equations. They can also be used to derive stabilization techniques for noncoercive problems, as shown in BERTOLUZZA [1998].

35. Conclusion

The general multiscale approximation tools that we have developed in the previous chapter allow to characterize a large number of smoothness classes, on $\mathbb{R}^d$ or on a bounded domain, with or without prescribed boundary conditions. The basic requirement is that these spaces satisfy proper Jackson and Bernstein inequalities. Such estimates respectively rely on the polynomial reproduction and smoothness properties of the basis functions for the V_j spaces.

The resulting characterizations can be expressed in three different ways: in terms of the size properties (decay and summability) of the approximation error $\text{dist}_X(f, V_j)$ or

$\|f - P_j f\|_X$ in a certain metric X, in terms of the size properties of the detail terms $\|P_{j+1} - P_j f\|_X$ or $\|f_{j+1} - f_j\|_X$, where f_j is a near best approximation in V_j, or in terms of the size properties of the wavelet coefficients $d_\lambda := \langle f, \tilde{\psi}_\lambda \rangle$.

Such result are interesting on a theoretical point of view since they somehow reduce the description of smoothness classes to simple sequence spaces. The multiscale decomposition into wavelet bases can thus be viewed here as a tool of harmonic analysis which is much more flexible than the Fourier series and the Littlewood–Paley decomposition, since it can easily be adapted to general bounded domains.

On an applied point of view, such characterizations (in the case of L^2-Sobolev spaces) can be used to design and analyze multilevel preconditioners for elliptic operator equations. Other types of applications involving more general Besov spaces for $p \neq 2$ will appear in our next chapter.

36. Historical notes

The possibility of characterizing the smoothness of a function by means of approximation properties became clear since the thesis of D. Jackson in 1911. At that time, the zoology of function spaces was limited to a few basic classes, and so were the existing approximation tools (algebraic and trigonometric polynomials).

The theory of function spaces became flourishing in the 1950's and 1960's (a nice historical survey can be found in TRIEBEL [1983]). The $B^s_{p,q}$ spaces (denoted by $\Lambda^s_{p,q}$ in several references) were introduced by BESOV [1959], who immediately pointed out on the characterization of such spaces by approximation procedures based on trigonometric polynomials.

This period also correspond to the development of spline functions and the systematic study of their approximation properties. A characterization of Besov spaces from the rate of approximation by "dyadic splines", i.e. the V_j spaces of spline functions with equally spaced knots at dyadic points $2^{-j}k$, was completed in CIESIELSKI [1973]. The analog result using wavelet bases was proved in MEYER [1990].

Another vision of Besov spaces came through the real interpolation method introduced in LIONS and PEETRE [1964]. One of the main motivation for the introduction of this method was the theoretical study of PDE's. In this setting, interpolation results for spaces with boundary conditions were proved in LIONS and MAGENES [1972] (for smooth domains and L^2-Sobolev spaces).

Approximation spaces were studied in relation with interpolation theory in PEETRE and SPAAR [1972] and BUTZER and SCHERER [1972]. Theorem 28.2 connecting these two concepts is given the same form in Chapter VII of DEVORE and LORENTZ [1993], which also provides with an overview of the identified relations between the analytical properties of functions and their rate of approximation by specific procedures.

The first ideas on multilevel iterative solvers for elliptic PDE's came out in the 1960's, in the work of FEDORENKO [1964] and BAKHVALOV [1966]. Multigrid theory was then developed in the 1970's in the pioneering work of A. Brandt, in the context of finite difference discretizations (see BRANDT [1977]). Important contributions to the theoretical understanding of multigrid methods were made in BANK and DUPONT [1981], which introduces and analyzes the full-multigrid technique, and in BRAESS and HACKBUSCH

[1983] where a proof of the $\mathcal{O}(1)$ preconditioning property for the so-called V-cycle multigrid algorithm was given. There is now an immense literature on multigrid methods in various contexts. We refer to HACKBUSCH [1985] and BRAMBLE [1993] for a general treatment, and to OSWALD [1994] for an analysis of such methods from an approximation theoretical point of view. The use of of multilevel preconditioners based on wavelets was proposed in JAFFARD [1992], while their connexions with the BPX-type preconditioners was analyzed in DAHMEN and KUNOTH [1992].

CHAPTER IV

Nonlinear Approximation and Adaptivity

37. Introduction

The results of the previous chapter reveal that a variety of smoothness classes can be characterized by the rate of decay of the multiscale approximation error. Roughly speaking, we have seen that

$$\operatorname{dist}(f, V_j)_{L^p} \lesssim 2^{-sj}, \tag{37.1}$$

if and only if f has s derivative in L^p. Note that when Ω is a bounded domain of $\mathbb{R}^d$, (37.1) also writes $\operatorname{dist}(f, V_j)_{L^p} = \mathcal{O}(N_j^{-s/d})$, where $N_j := \dim(V_j) \sim 2^{dj}$ represent the number of parameters that describes the approximation of f in V_j.

Such results express that the convergence of uniform approximations to a function f at some prescribed rate, as the mesh size tends to zero, always implies a certain degree of smoothness for this function. In particular, we cannot hope for a high rate of approximation in L^p (more generally in $W^{t,p}$ or $B^t_{p,q}$) if f has poor smoothness in L^p. The presence of singularities in a function f is thus a source of approximation problems, and subsequently of memory size and complexity problems since N_j should be large to reach a prescribed error ε if s is small. This is a concrete difficulty since there are numerous instances of functional equations for which the solution exhibits singularities.

A natural solution to this problem is to search for an *adaptive approximation* of f. In the context of finite element discretizations, this means a certain type of *mesh refinement* near the singularities. In the present context of wavelet discretizations, the term *space refinement* is more appropriate: we look for a space $V_E := \operatorname{Span}\{\psi_\lambda;\ \lambda \in E\}$ (using the general notations proposed in the introduction of the previous chapter) where $E = E(f) \subset \Lambda$ is a finite subset of indices which *depends* on the data (or unknown) f that we want to approximate in V_E. Typically E could be the union of Λ_{j_0} allowing a coarse scale approximation of f in V_{j_0} and of a few indices $|\lambda| > j_0$ that aim to resolve the local singularities of f.

Two main questions need to be addressed at this point: firstly, can we relate the performance of an adaptive approximation strategy to the smoothness properties of f, in the same way as we did in the previous chapter for the approximation of f in the V_j spaces? Secondly, how does one practically build the set $E(f)$ and compute a proper approximation of f in V_E?

The first question can be formulated in a precise way. By "performance", we mean here the behaviour of the error in some norm between f and its approximation in V_E, with respect to the cardinality of E, i.e. the number of parameters that are used to approximate f. It is thus natural to introduce the space

$$S_N = \left\{ \sum_{\lambda \in E} c_\lambda \psi_\lambda;\ |E| \leqslant N \right\} \tag{37.2}$$

of all possible N-term combinations of wavelets, and the error of best N-terms approximation in some norm $\| \bullet \|_X$ defined by

$$\operatorname{dist}_X(f, S_N) = \inf_{E \subset \Lambda, |E| \leqslant N} \inf_{(c_\lambda)_{\lambda \in E}} \left\| f - \sum_{\lambda \in E} c_\lambda \psi_\lambda \right\|_X. \tag{37.3}$$

It should be well understood that S_N is not a linear space: if f and g are in S_N we can only conclude that $f + g \in S_{2N}$. Quite surprisingly, one can relate the decay property $\operatorname{dist}_{L^p}(f, S_N) = \mathcal{O}(N^{-s/d})$ with some notion of smoothness: roughly speaking f should have s derivatives in L^q with q such that $1/q = 1/p + s/d$. Note that this property is a much weaker requirement than s derivatives in L^p, especially when s becomes large (in which case q might be less than 1). It essentially measures the *sparsity* of the wavelet decomposition of f, i.e. the fact that the L^p norm is concentrated on a small number of coefficients. Similar results also hold when the error is measured in a Sobolev or Besov norm.

Concerning the second question, the answer is relatively simple when one has access to the wavelet expansion $f = \sum_{\lambda \in \Lambda} c_\lambda \psi_\lambda$ of the function f to be approximated. In that case, a natural N-term approximation in X is provided by the choice

$$f_N = \sum_{\lambda \in E_N} c_\lambda \psi_\lambda, \tag{37.4}$$

where $E_N = E_N(f, X)$ is the set of indices corresponding to the N largest contributions $\|c_\lambda \psi_\lambda\|_X$. Then, we shall see that for several interesting choice of X, we have

$$\|f - f_N\|_X \lesssim \operatorname{dist}_X(f, S_N), \tag{37.5}$$

i.e. a simple thresholding of the largest contributions in the wavelet decomposition provides a near-optimal N-term approximation.

All these results will be proved in Section 38 for $X = L^2$ or a more general Besov space, and in Section 39 for $X = L^p$ or $W^{m,p}$, $p \neq 2$, They should be viewed as items of a more general nonlinear approximation theory that includes approximation by free knots splines and rational functions. We review in Section 40 some important examples in this general theory and compare them to N-terms wavelet approximation. We also address here the application of nonlinear approximation to data compression.

These results also suggest to reconsider the notion of smoothness for solutions of PDE's that develop singularities: such solutions might still be very smooth in the sense that they have a sparse representation in the multiscale basis allowing high rate of nonlinear approximation.

In order to understand the potential of adaptive discretizations, we should thus turn to an analysis of PDE's in terms of sparsity rather than classical smoothness.

Such an analysis is, at the present time, still at its infancy, and we can only provide here with an incomplete picture. At the start, one needs to understand how the action of the operators involved in partial differential as well as integral equations affects the sparsity of a function. An important observation, that we discuss in Section 41, is that for a large class of such operators the matrices resulting from wavelet discretizations are sparse and have local approximations. In turn, they preserves the sparsity of the function in the multiscale bases. We then review in Section 42 several instances of PDE's for which a theoretical analysis shows that, despite the presence of singularities in the solution, high order nonlinear approximation can be achieved.

In such instances, adaptive methods should thus perform substantially better than linear methods. However, additional difficulties arise from the fact that the object to be compressed is the unknown u of a problem: we are facing both difficulties of finding the indices of the coefficients of u which are numerically significant, and evaluating these coefficients up to a prescribed accuracy. We end this chapter in Section 43 by a brief review of the adaptive strategies that have been so far developed for different types of problems in order to achieve this double task. Such strategies should be viewed as optimal if they succeed in the following ultimate goal: compute the solution within some desired accuracy at a computational cost that stays *proportional* to the number of coefficients needed to represent the function within this accuracy.

We shall use the general notations that were proposed in the introduction of the previous chapter: $\{\psi_\lambda\}_{\lambda\in\Lambda}$ will denote a wavelet basis allowing to characterize function spaces defined on Ω, where Ω is either denote $\mathbb{R}^d$ or a bounded domain of the type addressed in Section 32 of the previous chapter.

38. Nonlinear approximation in Besov spaces

As a starter, we consider N-term approximation in the L^2-norm: we are interested in the behaviour of $\mathrm{dist}_{L^2}(f, S_N)$ as N goes to infinity, where S_N is defined by (37.2). In order to simplify this first example, we assume here that $\{\psi_\lambda\}_{\lambda\in\Lambda}$ is an orthonormal basis for L^2. Thus any $f \in L^2(\Omega)$ can be decomposed into

$$f = \sum_{\lambda\in\Lambda} c_\lambda \psi_\lambda, \quad c_\lambda = \langle f, \psi_\lambda\rangle, \tag{38.1}$$

and we can define the set $E_N = E_N(f) \subset \Lambda$ of the N largest coefficients of f, i.e. such that $|E_N| = N$ and

$$\lambda \in E_N,\ \lambda' \notin E_N \Rightarrow |c_{\lambda'}| \leqslant |c_\lambda|. \tag{38.2}$$

Note that there might be several possible choices for E_N if the modulus of several coefficients of f take the same value. In such a case, we simply take for E_N any of the sets of cardinality N that satisfies (38.2). From the orthonormality of the basis, we clearly have

$$\operatorname{dist}_{L^2}(f, S_N) = \Big\| f - \sum_{\lambda\in E_N} c_\lambda \psi_\lambda \Big\|_{L^2} = \Big(\sum_{\lambda\notin E_N} |c_\lambda|^2 \Big)^{1/2}. \tag{38.3}$$

Let us now consider the spaces $B^s_{q,q}$ where $s > 0$ and q is such that $1/q = 1/2 + s/d$. We assume here that the conditions of Theorem 30.6 (or of its generalization to a bounded domain Ω), are satisfied to that $B^s_{q,q}$ is characterized by (30.31). For such indices, we note that this equivalence can be simplified into

$$\|f\|_{B^s_{q,q}} \sim \big\|(c_\lambda)_{\lambda\in\Lambda}\big\|_{\ell^q}. \tag{38.4}$$

A first immediate consequence of (38.4) is the imbedding of $B^s_{q,q}$ in L^2, since ℓ^q is trivially embedded in ℓ^2. Note that such an imbedding is not compact: the canonical sequence $(s_n)_{n\geqslant 0} = ((\delta_{n,k})_{k\geqslant 0})_{n\geqslant 0}$ is uniformly bounded in ℓ^q but does not contain any subsequence that converges in ℓ^2.

If we now define by $(c_n)_{n\geqslant 1}$ any rearrangement of the coefficients $(c_\lambda)_{\lambda\in\Lambda}$ with decreasing moduli, i.e. such that $|c_{n+1}| \leqslant |c_n|$, we also have

$$n|c_n|^q \leqslant \sum_{k\geqslant 1} |c_k|^q \sim \|f\|^q_{B^s_{q,q}}, \tag{38.5}$$

which yields

$$|c_n| \lesssim \|f\|_{B^s_{q,q}} n^{-1/q}. \tag{38.6}$$

Taking the decreasing rearrangement c_n to be such that $\{c_n;\ n \leqslant N\} = \{c_\lambda;\ \lambda \in E_N\}$, it follows that we have

$$\begin{aligned} \operatorname{dist}_{L^2}(f, S_N) &= \Big\| f - \sum_{\lambda\in E_N} c_\lambda \psi_\lambda \Big\|_{L^2} = \Big(\sum_{n>N} |c_n|^2 \Big)^{1/2} \\ &\lesssim N^{-1/q+1/2} \|f\|^q_{B^s_{q,q}}. \end{aligned}$$

We thus have obtained a Jackson-type estimate

$$\operatorname{dist}_{L^2}(f, S_N) \lesssim N^{-s/d} \|f\|^q_{B^s_{q,q}}, \tag{38.7}$$

with respect to the nonlinear spaces S_N. One the other hand, if $f \in S_N$, we also have by Hölder's inequality

$$\|f\|_{B^s_{q,q}} \lesssim \|(c_\lambda)_{\lambda\in\Lambda}\|_{\ell^q} \leqslant N^{1/q-1/2}\|(c_\lambda)_{\lambda\in\Lambda}\|_{\ell^2} = N^{s/d}\|f\|_{L^2}, \tag{38.8}$$

i.e. a Bernstein-type estimate.

The equivalence (38.4) also shows that for $0 < t < s$ and $1/r = 1/2 + t/d$, we have the interpolation identity

$$B^t_{r,r} = [L^2, B^s_{q,q}]_{\theta,r}, \quad \theta = t/s, \tag{38.9}$$

which is a simple re-expression of $[\ell^2, \ell^q]_{\theta,r} = \ell^r$.

We thus have in hand the three ingredients – direct estimates, inverse estimates and interpolation theory – that were used in the previous chapter to characterize function spaces by the error of linear approximation. At this point, we need to take into account the fact that the spaces S_N are not linear spaces: if these spaces were linear, defining $\Sigma_j := S_{2^{dj}}$ for $j \geqslant 0$, a straightforward application of Theorem 28.2 with $V_j = \Sigma_j$ would give

$$\|f\|_{B^t_{r,r}} \lesssim \|f\|_{L^2} + \|(2^{jt}\mathrm{dist}_{L^2}(f, \Sigma_j))_{j\geqslant 0}\|_{\ell^r}. \tag{38.10}$$

It turns out that such a result also holds in the present nonlinear context, using the following adaptation of Theorem 28.2. Here, as well as in all the following, the approximation spaces $\mathcal{A}^s_q(X)$ and $\mathcal{A}^s_{p,q}$ have the same definition as in Section 28 of Chapter III with the spaces Σ_j in place of V_j.

THEOREM 38.1. *Assume that X and Y are quasi-normed spaces (satisfying the μ-triangle inequality* (25.20)) *and that Σ_j, $j \geqslant 0$, is a sequence of nonlinear approximation spaces*

$$\Sigma_j \subset \Sigma_{j+1} \subset \cdots \subset Y \subset X, \tag{38.11}$$

such that for some $m > 0$, one has a Jackson-type estimate

$$\mathrm{dist}_X(f, \Sigma_j) = \inf_{g\in\Sigma_j} \|f - g\|_X \lesssim 2^{-mj}\|f\|_Y, \tag{38.12}$$

and a Bernstein-type estimate

$$\|f\|_Y \lesssim 2^{mj}\|f\|_X \quad \text{if } f \in \Sigma_j. \tag{38.13}$$

Moreover assume that there exists a fixed integer a such that

$$\Sigma_j + \Sigma_j \subset \Sigma_{j+a}, \quad j \geqslant 0. \tag{38.14}$$

Then, for $s \in]0, m[$, one has the norm equivalence

$$\left\| \left(2^{js} K\left(f, 2^{-mj}\right)\right)_{j \geqslant 0} \right\|_{\ell^q} \sim \|f\|_X + \left\| \left(2^{js} \operatorname{dist}_X(f, \Sigma_j)\right)_{j \geqslant 0} \right\|_{\ell^q}, \tag{38.15}$$

and thus $[X, Y]_{\theta,q} = \mathcal{A}^s_q(X)$ for $s = \theta m$.

PROOF. We mimic the proof of Theorem 28.2 (and its adaptation to quasi-norms as explained in Remark 28.2). In this proof, the linearity of the approximation spaces is only involved in the estimate

$$\|f_j\|_Y \lesssim \left[\|f_0\|_X^\mu + \sum_{l=0}^{j-1} 2^{lm} \|f_{l+1} - f_l\|_X^\mu \right]^{1/\mu},$$

where we use that $f_{l+1} - f_l \in V_{l+1}$ in order to apply the inverse estimate. Here, we can still obtain this result up a change in the constant, since $f_{l+1} - f_l \in V_{l+1+a}$ by the assumption (38.14). □

In the above result, the assumption (38.14) should be viewed as a limitation on the nonlinearity of the spaces Σ_j. In the present context of N-terms approximation, we already noted that $S_n + S_m \subset S_{n+m}$, so that this assumption holds with $a = 1$. We thus concludes that the norm equivalence (38.10) indeed holds.

REMARK 38.1. From the monotonicity of the sequence $\operatorname{dist}_{L^2}(f, S_N)$, we have the equivalence

$$\sum_{j \geqslant 0} \left[2^{jt} \operatorname{dist}_{L^2}(f, \Sigma_j)\right]^r \sim \sum_{N \geqslant 1} N^{-1} \left[N^{t/d} \operatorname{dist}_{L^2}(f, S_N)\right]^r. \tag{38.16}$$

The finiteness of the above quantities for $r < \infty$ is a slightly stronger property than $\operatorname{dist}_{L^2}(f, S_N) \lesssim N^{-t/d}$ which was initially obtained in the direct estimate (38.7). According to the above theorem, this last property characterizes the intermediate space

$$\left[L^2, B^s_{q,q}\right]_{\theta,\infty} = \mathcal{A}^t_{2,\infty}, \tag{38.17}$$

with $t = \theta s$, which cannot be thought as a Besov space. One can easily check that this space is also characterized by the property that $(c_\lambda)_{\lambda \in \Lambda}$ belongs to the weak space ℓ^r_w, i.e.

$$\#\left\{\lambda \in \Lambda;\ |c_\lambda| \geqslant \varepsilon\right\} \lesssim \varepsilon^{-r}. \tag{38.18}$$

More generally, $\mathcal{A}^t_{2,r'}$ identifies with the Lorentz space $\ell^{r,r'} = [\ell^r, \ell^q]_{\theta,r'}$ (one has in particular $\ell^{r,r} = \ell^r$ and $\ell^{r,\infty} = \ell^r_w$). It is not difficult to extend the above results to the case where the error is measured in more general Besov space of the type $B^s_{p,p}$.

Recall that such spaces include Hölder spaces ($p=\infty$), and Sobolev spaces (of noninteger smoothness if $p\neq 2$). Here, we cancel the orthonormality assumption on the basis $\{\psi_\lambda\}_{\lambda\in\Lambda}$ which is irrelevant at this stage of generality. A first result is that a near best N-term approximation in $B^s_{p,p}$ is still provided by a simple thresholding procedure.

LEMMA 38.1. *Assuming that $B^s_{p,p}$ admits a wavelet characterization of the type* (30.31), *and if $f=\sum_{\lambda\in\Lambda}c_\lambda\psi_\lambda$, then*

$$\operatorname{dist}_{B^s_{p,p}}(f,S_N)\lesssim\Big\|f-\sum_{\lambda\in E_N}c_\lambda\psi_\lambda\Big\|_{B^s_{p,p}},\tag{38.19}$$

where $E_N=E_N(f,B^s_{p,p})$ is the set of indices corresponding to the N largest contributions $\|c_\lambda\psi_\lambda\|_{B^s_{p,p}}$ or equivalently the N largest $2^{(s+d/2-d/p)|\lambda|}|c_\lambda|$.

PROOF. The norm equivalence (30.31) shows that $\|\psi_\lambda\|_{B^s_{p,p}}\sim 2^{(s+d/2-d/p)|\lambda|}$. It can thus be reformulated as also writes

$$\|f\|_{B^s_{p,p}}\sim\big\|\big(\|c_\lambda\psi_\lambda\|_{B^s_{p,p}}\big)_{\lambda\in\Lambda}\big\|_{\ell^p}.\tag{38.20}$$

Clearly, the N-term approximation $\sum_{\lambda\in E_N}c_\lambda\psi_\lambda$ minimizes the distance between f and S_N when measured in this equivalent norm for $B^s_{p,p}$. It is thus a near minimizer for the $B^s_{p,p}$ norm in the sense of (38.19). □

We are now ready to prove the general result of nonlinear wavelet approximation in Besov spaces.

THEOREM 38.2. *Assume that the spaces $B^t_{q,q}$, $t-s=d/q-d/p$ admit a wavelet characterization of the type* (30.31) *for $t\in[s,s']$, $s'>s$. Then, for $t\in]s,s'[$, $t-s=d/q-d/p$, we have the norm equivalence*

$$\|f\|_{B^t_{q,q}}\sim\|f\|_{B^s_{p,p}}+\big\|\big(2^{j(t-s)}\operatorname{dist}_{B^s_{p,p}}(f,\Sigma_j)\big)_{j\geqslant 0}\big\|_{\ell^q}.\tag{38.21}$$

PROOF. If $t>s$ and $t-s=d/q-d/p$, the norm equivalence (30.31) can be rewritten as

$$\|f\|_{B^t_{q,q}}\sim\big\|\big(\|c_\lambda\psi_\lambda\|_{B^s_{p,p}}\big)_{\lambda\in\Lambda}\big\|_{\ell^q},\tag{38.22}$$

where $c_\lambda=\langle f,\tilde\psi_\lambda\rangle$ are again the coefficients of f. We can then proceed in a similar way as in the particular case of approximation in L^2. Denoting by $(\varepsilon_n)_{n\geqslant 1}$ the decreasing rearrangement of the sequence $(\|c_\lambda\psi_\lambda\|_{B^s_{p,p}})_{\lambda\in\Lambda}$, we remark that since

$$n\varepsilon_n^q\leqslant\sum_{k\geqslant 1}\varepsilon_k^q\lesssim\|f\|^q_{B^t_{q,q}},\tag{38.23}$$

we have the estimate

$$\varepsilon_n \lesssim n^{-1/q} \|f\|_{B^t_{q,q}}. \tag{38.24}$$

Denoting by E_N a set of indices defined as in Lemma 38.1, we obtain

$$\begin{aligned} \mathrm{dist}_{B^s_{p,p}}(f, S_N) &\leqslant \Big\| f - \sum_{\lambda \in E_N} c_\lambda \psi_\lambda \Big\|_{B^s_{p,p}} \lesssim \Big(\sum_{n \geqslant N} \varepsilon_n^p \Big)^{1/p} \\ &\lesssim N^{1/p - 1/q} \|f\|_{B^t_{q,q}}. \end{aligned}$$

We have thus established the Jackson-type estimate

$$\mathrm{dist}_{B^s_{p,p}}(f, S_N) \leqslant N^{-(t-s)/d} \|f\|_{B^t_{q,q}}. \tag{38.25}$$

If $f \in S_N$, we also have by (38.22) and Hölder's inequality

$$\|f\|_{B^t_{q,q}} \lesssim N^{1/q - 1/p} \big\| \big(\|c_\lambda \psi_\lambda\|_{B^s_{p,p}} \big)_{\lambda \in \Lambda} \big\|_{\ell^p} = N^{(t-s)/d} \|f\|_{B^s_{p,p}}, \tag{38.26}$$

i.e. the corresponding Bernstein-type estimate.

The equivalence (38.22) also shows that the spaces $B^t_{q,q}$, $t - s = d/q - d/p$ are interpolation spaces: for $s < t < s'$, $t - s = d/q - d/p$, and $s' - s = d/p' - d/p$, we have the interpolation identity

$$B^t_{q,q} = \big[B^s_{p,p}, B^{s'}_{p',p'} \big]_{\theta,q}, \quad \theta = (t-s)/(s'-s), \tag{38.27}$$

which is a simple re-expression of $[\ell^p, \ell^{p'}]_{\theta,q} = \ell^q$.

By Theorem 38.1, we thus obtain the norm equivalence (38.21). □

REMARK 38.2. Note that we did not assume here that p or q are larger than 1. This is particularly important for the index q: if we fix the space $B^s_{p,p}$ in which the error is measured, q always become smaller than 1 for large values of t, due to the relation $1/q - 1/p = t/d - s/d$. Thus, the correct description of those functions that can be approximated at a high rate by an adapted combination of N wavelets necessarily involves Besov spaces with $q < 1$.

REMARK 38.3. The analog linear result (Corollary 29.1) tells us that the same error rate $\mathcal{O}(N^{-(t-s)/d})$ in $B^s_{p,p}$ is achieved by a linear method (i.e. with $N = N_j = \dim(V_j)$) for functions in $B^t_{p,p}$, i.e. a smaller space than $B^t_{q,q}$. It should noted that, as t becomes large, the functions in the space $B^t_{p,p}$ become smooth in the classical sense, while $B^t_{q,q}$ might still contain discontinuous functions. As an example, consider a function f defined on a bounded interval which coincides with C^∞ functions except on a finite set of discontinuities. One easily checks that $f \in B^s_{p,p}$ if and only if $s < 1/p$, so that the L^2

approximation rate is limited to $N^{-1/2}$ in the linear case, while in the nonlinear case we have the "spectral estimate"

$$\operatorname{dist}_{L^2}(f, S_N) \leqslant C_s N^{-s}, \tag{38.28}$$

for any $s \leqslant n+1$ where n is the order of polynomial reproduction associated to the wavelet basis.

REMARK 38.4. The space $B^t_{q,q}$ is contained in $B^s_{p,p}$ since these spaces are also defined as ℓ^q and ℓ^p of the sequences $\|c_\lambda \psi_\lambda\|_{B^s_{p,p}}$. Note that this imbedding is not compact, and that neither $B^{t-\varepsilon}_{q,q}$ nor $B^t_{q-\varepsilon,q}$ is contained in $B^s_{p,p}$ for any $\varepsilon > 0$. This critical situation is graphically summarized on Fig. 38.1: any function space corresponding to "s derivative in L^p" is represented by the point $(1/p, s)$, and the space $B^t_{q,q}$ sits on the critical line of slope d initiated from $B^s_{p,p}$.

REMARK 38.5. All the results in this section can be adapted in a straightforward manner to the case of Besov spaces with boundary conditions, using the characterization (33.30) of Theorem 33.4 in place of (30.31). This remark also applies to the results of the next section.

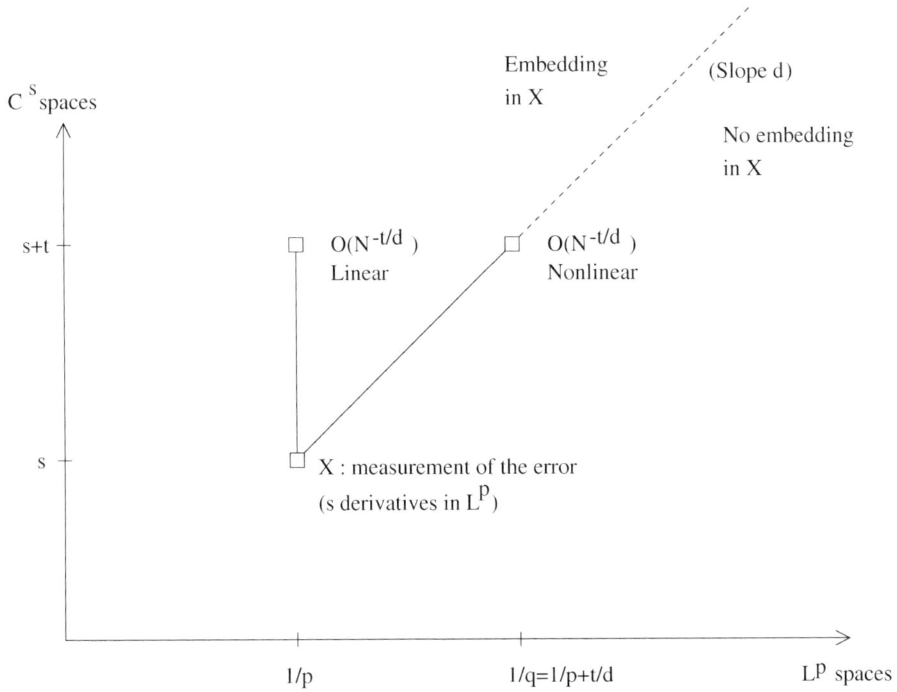

FIG. 38.1. Graphical interpretation of linear and nonlinear approximation.

39. Nonlinear wavelet approximation in L^p

The results of Section 38 are rather easy to prove, due to the simple links that exist between Besov spaces and ℓ^p spaces through wavelet decompositions. For example, Theorem 38.2 can be viewed as the rewriting in terms of Besov spaces of the following simple norm equivalence for sequences: if $(c_n)_{n\geqslant 1}$ is a positive decreasing sequence and $d_n = \|(c_k)_{k\geqslant n}\|_{\ell^p}$, then for $q < p$, one has

$$\|(c_n)_{n\geqslant 1}\|_{\ell^q} \sim \left[\sum_{n\geqslant 1} n^{-1}\left(n^{1/q-1/p} d_n\right)^q\right]^{1/q} = \left\|\left(n^{-1/p} d_n\right)_{n\geqslant 1}\right\|_{\ell^q}. \tag{39.1}$$

The study of best N-terms approximation in L^p norm is more difficult since, as it was pointed out in Section 31 of Chapter III, we cannot identify L^p to $B^0_{p,p}$ for $p \neq 2$. In this section, we shall see that switching from $B^0_{p,p}$ (which is treated by Theorem 38.2) to L^p does not seriously affects the results of N-terms approximation for $1 < p < \infty$.

Let us start with a basic result that allows to estimate the L^p norms of a linear combination of wavelets according to the size of the coefficients.

LEMMA 39.1. *Let $1 \leqslant p < \infty$, and assume that the wavelet basis $\{\psi_\lambda\}_{\lambda\in\Lambda}$ is generated from scaling functions in L^p. If E is a finite subset of Λ of cardinality $|E| < \infty$ and if $(c_\lambda)_{\lambda\in E}$ is a sequence such that $\|c_\lambda\psi_\lambda\|_{L^p} \leqslant 1$, $i = 1, \ldots, N$, then*

$$\left\|\sum_{\lambda\in E} c_\lambda\psi_\lambda\right\|_{L^p} \leqslant C|E|^{1/p}, \tag{39.2}$$

where C is independent of $|E|$.

PROOF. By assumption, we have for $\lambda \in E$

$$\|c_\lambda\psi_\lambda\|_{L^\infty} \lesssim 2^{|\lambda|d/p}\|c_\lambda\psi_\lambda\|_{L^p} \leqslant 2^{|\lambda|d/p}. \tag{39.3}$$

Let us define $E_j = E \cap \Lambda_j$ and $N_j = |E_j|$ the number of wavelets at level j in E. By the controlled overlapping property (24.8), we have

$$\left\|\sum_{\lambda\in E_j} c_\lambda\psi_\lambda\right\|_{L^\infty} \lesssim 2^{jd/p}. \tag{39.4}$$

For $x \in \Omega$, we define

$$j(x) = \max\{j \geqslant -1;\ x \in \mathrm{Supp}(\psi\lambda) \text{ for some } \lambda \in E_j\} \tag{39.5}$$

and

$$\Omega_j := \{x \in \Omega;\ j(x) = j\}. \tag{39.6}$$

Thus $\Omega = \bigcup_{j \geqslant -1} \Omega_j$. We next remark that

$$|\Omega_j| \leqslant \left| \bigcup_{\lambda \in E_j} \mathrm{Supp}(\psi_\lambda) \right| \lesssim N_j 2^{-dj}, \tag{39.7}$$

and that for $x \in \Omega_j$, we have by (39.4)

$$\left| \sum_{\lambda \in E} c_\lambda \psi_\lambda \right| \leqslant \sum_{l \leqslant j} \left| \sum_{\lambda \in E_l} c_\lambda \psi_\lambda \right| \lesssim \sum_{l \leqslant j} 2^{ld/p} \lesssim 2^{jd/p}. \tag{39.8}$$

It follows that

$$\left\| \sum_{\lambda \in E} c_\lambda \psi_\lambda \right\|_{L^p}^p = \sum_{j \geqslant 1} \int_{\Omega_j} \left| \sum_{\lambda \in E} c_\lambda \psi_\lambda \right|^p \lesssim \sum_{j \geqslant 1} |\Omega_j| 2^{jd} \lesssim \sum_{j \geqslant 1} N_j = |E|$$

which concludes the proof. □

Note that, by rescaling, this results gives for a general finite linear combination

$$\left\| \sum_{\lambda \in E} c_\lambda \psi_\lambda \right\|_{L^p} \lesssim |E|^{1/p} \sup_{\lambda \in E} \|c_\lambda \psi_\lambda\|_{L^p}. \tag{39.9}$$

By duality, we shall obtain an analog result with reversed inequalities. We assume here in addition that the dual basis $\{\psi_\lambda\}_{\lambda \in \Lambda}$ is generated from scaling functions in $L^{p'}$, where $1/p + 1/p' = 1$.

LEMMA 39.2. *Let $1 < p \leqslant \infty$. If E is a finite subset of Λ of cardinality $|E| < \infty$ and if $(c_\lambda)_{\lambda \in E}$ is a sequence such that $\|c_\lambda \psi_\lambda\|_{L^p} \geqslant 1$, $i = 1, \ldots, N$, then*

$$\left\| \sum_{\lambda \in E} c_\lambda \psi_\lambda \right\|_{L^p} \geqslant C|E|^{1/p}, \tag{39.10}$$

where C is independent of $|E|$.

PROOF. If $f = \sum_{\lambda \in E} c_\lambda \psi_\lambda$, we define

$$g = \sum_{\lambda \in E} |c_\lambda|^{-2} \overline{c_\lambda} \tilde{\psi}_\lambda, \tag{39.11}$$

so that

$$|E| = \langle f, g \rangle \leqslant \|f\|_{L^p} \|g\|_{L^{p'}}. \tag{39.12}$$

Remarking that

$$\big\|\overline{c_\lambda}|c_\lambda|^{-2}\psi_\lambda\big\|_{L^{p'}} \lesssim |c_\lambda|^{-1}2^{d|\lambda|(1/2-1/p')} \lesssim \|c_\lambda\psi_\lambda\|_{L^p}^{-1} \lesssim 1,$$

we derive from Lemma 39.1 that

$$\|g\|_{L^{p'}} \lesssim |E|^{1/p'}. \tag{39.13}$$

Combining (39.12) and (39.13) yields (39.10). □

We also get, by rescaling, for a general finite linear combination

$$|E|^{1/p} \inf_{\lambda\in E} \|c_\lambda\psi_\lambda\|_{L^p} \lesssim \left\|\sum_{\lambda\in E} c_\lambda\psi_\lambda\right\|_{L^p}. \tag{39.14}$$

A first consequence of (39.9) and (39.14) is the following result originally due to V. Temlyakov: a near best N-term approximation in L^p can be achieved by a simple thresholding procedure.

THEOREM 39.1. *We assume here that the basis $\{\psi_\lambda\}_{\lambda\in\Lambda}$ is generated from scaling functions with some Hölder smoothness C^ε, $\varepsilon > 0$. Let $f = \sum_{\lambda\in\Lambda} c_\lambda\psi_\lambda \in L^p$ with $1 < p < \infty$. Then, we have*

$$\operatorname{dist}_{L^p}(f, S_N) \lesssim \left\|f - \sum_{\lambda\in E_N} c_\lambda\psi_\lambda\right\|_{L^p}, \tag{39.15}$$

where $E_N = E_N(f, L^p)$ is the set of indices corresponding to the N largest contributions $\|c_\lambda\psi_\lambda\|_{L^p}$ or equivalently the N largest $2^{(d/2-d/p)|\lambda|}|c_\lambda|$.

PROOF. We first remark that if V_E denotes the space generated by ψ_λ, $\lambda \in E$, for some $E \subset \Lambda$, then we have

$$\operatorname{dist}_{L^p}(f, V_E) \lesssim \left\|f - \sum_{\lambda\in E} c_\lambda\psi_\lambda\right\|_{L^p}, \tag{39.16}$$

with a constant that does depend on E. This is a straightforward consequence of Lemma 11.1 and the uniform boundedness of the projectors

$$T_E f := \sum_{\lambda\in E} \langle f, \tilde{\psi}_\lambda\rangle\psi_\lambda, \tag{39.17}$$

that we already observed in Section 31 of Chapter III.

It is thus sufficient to prove that if $|E| = N$, one has

$$\|f - T_{E_N} f\|_{L^p} \lesssim \|f - T_E f\|_{L^p}. \tag{39.18}$$

For this, we define $A_N = E \setminus E_N$ and $B_N = E_N \setminus E$. Clearly we have $|A_N| = |B_N|$. By definition of E_N, we have

$$\sup_{\lambda \in A_N} \|c_\lambda \psi_\lambda\|_{L^p} \leqslant \inf_{\lambda \in B_N} \|c_\lambda \psi_\lambda\|_{L^p}. \tag{39.19}$$

Combining this with (39.9) and (39.14), we obtain

$$\Big\| \sum_{\lambda \in A_N} c_\lambda \psi_\lambda \Big\|_{L^p} \lesssim \Big\| \sum_{\lambda \in B_N} c_\lambda \psi_\lambda \Big\|_{L^p}. \tag{39.20}$$

From this, we derive

$$\begin{aligned} \|T_E f - T_{E_N} f\|_{L^p} &\leqslant \|T_{A_N} f\|_{L^p} + \|T_{B_N} f\|_{L^p} \lesssim \|T_{B_N} f\|_{L^p} \\ &= \|T_{B_N} T_{\Lambda \setminus E} f\|_{L^p} \lesssim \|T_{\Lambda \setminus E} f\|_{L^p} \\ &= \|f - T_E f\|_{L^p}, \end{aligned}$$

which implies (39.18). □

We shall now use (39.9) and (39.14) in order to prove Jackson and Bernstein estimate for N-term approximation in L^p.

THEOREM 39.2. *Let $1 < p < \infty$ and $\{\psi_\lambda\}_{\lambda \in \Lambda}$ a wavelet basis constructed from scaling functions in L^p. Assuming that the space $B^s_{q,q}$, $1/q = 1/p + s/d$ admits a wavelet characterization of the type* (30.31), *we have the Jackson estimate*

$$\operatorname{dist}_{L^p}(f, S_N) \lesssim N^{-s/d} \|f\|_{B^s_{q,q}}, \tag{39.21}$$

and, for $f \in S_N$, the Bernstein estimate

$$\|f\|_{B^s_{q,q}} \lesssim N^{s/d} \|f\|_{L^p}. \tag{39.22}$$

PROOF. Let $f \in B^s_{q,q}$. As in the proof of Theorem 38.2, we remark that the norm equivalence (30.31) also writes

$$\|f\|_{B^s_{q,q}} \sim \big\| \big(\|c_\lambda \psi_\lambda\|_{L^p} \big)_{\lambda \in \Lambda} \big\|_{\ell^q}. \tag{39.23}$$

In particular, we have

$$\#\big\{\lambda;\ \|c_\lambda \psi_\lambda\|_{L^p} \geqslant \varepsilon\big\} \lesssim \varepsilon^{-q} \|f\|^q_{B^s_{q,q}}. \tag{39.24}$$

It follows that there exists a constant $C > 0$ (depending on the constant in the equivalence (39.23)) such that if we define

$$A_j = \big\{\lambda;\ C2^{-j/q} \|f\|_{B^s_{q,q}} \leqslant \|c_\lambda \psi_\lambda\|_{L^p} \leqslant C2^{-(j-1)/q} \|f\|_{B^s_{q,q}}\big\}, \tag{39.25}$$

we then have

$$|A_j| \leqslant 2^j. \tag{39.26}$$

From (39.9), we can evaluate the L^p norm of $T_{A_j} f = \sum_{\lambda \in A_j} c_\lambda \psi_\lambda$ by

$$\|T_{A_j}\|_{L^p} \lesssim 2^{-j/q} \|f\|_{B^s_{q,q}} |A_j|^{1/p} \leqslant 2^{j(1/p-1/q)} \|f\|_{B^s_{q,q}}. \tag{39.27}$$

Now define $B_j = \bigcup_{l=0}^{j-1} A_l$. By (39.26), we have $|B_j| \leqslant 2^j$. For $\Sigma_j := S_N$, we thus have

$$\begin{aligned}\operatorname{dist}_{L^p}(f, \Sigma_j) &\leqslant \|f - T_{B_j} f\|_{L^p} \leqslant \sum_{l \geqslant j} \|T_{A_l} f\|_{L^p} \lesssim \sum_{l \geqslant j} 2^{l(1/p-1/q)} \|f\|_{B^s_{q,q}} \\ &\lesssim 2^{j(1/p-1/q)} \|f\|_{B^s_{q,q}} = 2^{-js/d} \|f\|_{B^s_{q,q}}.\end{aligned}$$

By the monotonicity of $\operatorname{dist}_{L^p}(f, S_N)$, this implies the direct estimate (39.21) for all N.

In order to prove the Bernstein estimate, we we distinguish two cases: $p \geqslant 2$ or $p \leqslant 2$. If $p \geqslant 2$, we have

$$\|f\|_{B^0_{p,p}} \sim \big\|\big((\|c_\lambda \psi_\lambda\|\big)_{L^p})_{\lambda \in \Lambda}\big\|_{\ell^p} \lesssim \|f\|_{L^p}. \tag{39.28}$$

One way of checking (39.28) is to use an interpolation argument: the property holds when $p = 2$ for which one actually has the equivalence $\|f\|_{B^0_{2,2}} \sim \|f\|_{L^2}$ and $p = \infty$ since

$$\|c_\lambda \psi_\lambda\|_{L^\infty} \lesssim 2^{|\lambda| d/2} |c_\lambda| = 2^{|\lambda| d/2} \big|\langle f, \tilde{\psi}_\lambda\rangle\big| \lesssim \|f\|_{L^\infty}. \tag{39.29}$$

One can also directly check (39.28) using the characterization of L^p by the square function Sf as shown in (31.39). For this range of p, the Bernstein estimate is thus a straightforward consequence of the same estimate derived for $B^0_{p,p}$ in the previous section.

In the case where $p \leqslant 2$, let $f = \sum_{\lambda \in E} c_\lambda \psi_\lambda \in S_N$, i.e. such that $|E| \leqslant N$. We take here the same definitions of E_j, N_j, $j(x)$ and Ω_j as in the proof of Lemma 39.1. We then estimate $\|f\|_{B^s_{q,q}}$ as follows.

$$\begin{aligned}\|f\|^q_{B^s_{q,q}} &\lesssim \sum_{\lambda \in E} \|c_\lambda \psi_\lambda\|^q_{L^p} = \sum_{\lambda \in E} |c_\lambda|^q \|\psi_\lambda\|^p_{L^p} \|\psi_\lambda\|^{q-p}_{L^p} \\ &\lesssim \int_\Omega \sum_{\lambda \in E} |c_\lambda|^q |\psi_\lambda|^p 2^{d(1/2-1/p)(q-p)|\lambda|} \\ &\lesssim \int_\Omega \sum_{\lambda \in E} |c_\lambda|^q |\psi_\lambda|^q \big[2^{d(1/p-1/2)|\lambda|} |\psi_\lambda|\big]^{p-q} \\ &\lesssim \int_\Omega \big[Sf(x)\big]^q R_E(x)\, \mathrm{d}x,\end{aligned}$$

where we have applied Hölder's inequality on sequences to obtain the last line. Here, $Sf(x)$ is the square function defined by (31.38) and $R_E(x)$ can be estimated by

$$
\begin{aligned}
R_E(x) &= \left(\sum_{\lambda\in E}\left[2^{d(1/p-1/2)|\lambda|}\left|\psi_\lambda(x)\right|\right]^{2(p-q)/(2-q)}\right)^{(2-q)/2}\\
&\lesssim \left(\sum_{\lambda\in E,\,\psi_\lambda(x)\neq 0} 2^{2d|\lambda|(p-q)/(2p-qp)}\right)^{(2-q)/2}\\
&\lesssim 2^{j(x)d(1-q/p)}.
\end{aligned}
$$

Using Hölder's inequality, we thus obtain

$$
\begin{aligned}
\|f\|_{B^s_{q,q}}^q &\lesssim \|Sf\|_{L^p}^{q/p}\left(\int_\Omega 2^{j(x)d}\,\mathrm{d}x\right)^{1-q/p} \lesssim \|f\|_{L^p}\left[\sum_{j\geqslant -1}|\Omega_j|2^{jd}\right]^{1-q/p}\\
&\lesssim \|f\|_{L^p}\left[\sum_{j\geqslant -1}N_j\right]^{1-q/p} = N^{1-q/p}\|f\|_{L^p},
\end{aligned}
$$

i.e. the Bernstein estimate. □

REMARK 39.1. The direct estimate shows in particular that $B^s_{q,q}$ is embedded in L^p, as announced in Section 31 of the previous chapter.

REMARK 39.2. For $p \leqslant 2$, one can also check that $\|f\|_{L^p} \lesssim \|f\|_{B^0_{p,p}}$: the direct estimate is then a straightforward consequence of the same estimate established in Theorem 38.2 for $B^0_{p,p}$.

We are now ready to prove the main result of this section.

THEOREM 39.3. *Let $1 < p < \infty$ and $\{\psi_\lambda\}_{\lambda\in\Lambda}$ a wavelet basis constructed from scaling functions in L^p. Assuming that for $0 < s < t$ the space $B^s_{q,q}$, $1/q = 1/p + s/d$ admits a wavelet characterization of the type* (30.31), *then one also has the norm equivalence*

$$
\|f\|_{B^s_{q,q}} \sim \|f\|_{L^p} + \left\|\left(2^{js}\mathrm{dist}_{L^p}(f,\Sigma_j)\right)_{j\geqslant 0}\right\|_{\ell^q}. \tag{39.30}
$$

PROOF. We want to prove here that $\mathcal{A}^s_{p,q} = B^s_{q,q}$. We first note that the direct estimate in Theorem 39.2 is equivalent to

$$
B^s_{q,q} \subset \mathcal{A}^s_\infty(L^p). \tag{39.31}
$$

On the other hand, if $f \in A^s_{q,q}$ let $f_j \in \Sigma_j$ be such that

$$
\|f - f_j\|_{L^p} \leqslant 2\mathrm{dist}_{L^p}(f,\Sigma_j), \tag{39.32}
$$

and let $r = \min\{1, q\}$. We then have

$$\begin{aligned}\|f\|_{B^s_{q,q}} &\leqslant \Big(\|f_0\|^r_{B^s_{q,q}} + \sum_{j\geqslant 0} \|f_{j+1} - f_j\|^r_{B^s_{q,q}}\Big)^{1/r} \\ &\lesssim \Big(\|f\|^r_{L^p} + \sum_{j\geqslant 0} 2^{js} \|f_{j+1} - f_j\|^r_{L^p}\Big)^{1/r},\end{aligned}$$

where we have used the inverse estimate of Theorem 39.2.

We thus have

$$\mathcal{A}^s_{p,r} \subset B^s_{q,q} \subset \mathcal{A}^s_{p,\infty}. \tag{39.33}$$

Recalling the interpolation properties (28.37) for the $\mathcal{A}^s_{p,r}$ spaces (which is easily extended to the case of nonlinear approximation using Theorem 38.1) and (38.27) for the Besov spaces $B^s_{q,q}$, $1/q = 1/p + s/d$, we conclude by re-iteration (Theorem 28.1) that $B^s_{q,q} = \mathcal{A}^s_{p,q}$ with equivalent norms. □

REMARK 39.3. The above result shows that N-terms approximation in L^p, $1 < p < +\infty$ is also well described by Fig. 38.1. Moreover, all the results of this section can be extended to the case where the error is measured in $W^{m,p}$, $1 < p < \infty$. Here one obtains a norm equivalence of the type

$$\|f\|_{B^s_{q,q}} \sim \|f\|_{W^{m,p}} + \big\|\big(2^{j(s-m)}\mathrm{dist}_{W^{m,p}}(f, \Sigma_j)\big)_{j\geqslant 0}\big\|_{\ell^q}, \tag{39.34}$$

for $s > m$ and $1/q - 1/p = (s-m)/d$. The extension to this case is rather straightforward. One simply notice that, due to the characterization of L^p by the square function Sf and of $W^{m,p}$ by the modified square function $S_m f$, the transformation D_m defined by

$$\langle D_m f, \tilde{\psi}_\lambda\rangle = 2^{m|\lambda|}\langle f, \tilde{\psi}_\lambda\rangle, \tag{39.35}$$

is an isomorphism between $W^{m,p}$ and L^p. From this simple remark, one can easily adapt all the results of this section to the $W^{m,p}$ setting and derive (38.34).

REMARK 39.4. The cases $p = 1$ and $p = \infty$ are not covered by the results of this section. For $p = 1$, a basic difficulty is that the space $B^s_{q,q}$, $1/q = 1 + s/d$, are not simply characterized by the inner products with wavelet coefficients. If one accepts to to measure the error in the Hardy space H^p, then the case $p \leqslant 1$ can treated with the approach of KYRIASIS [1996] which allows to characterize Besov spaces by wavelet decompositions for a wider range of indices. For $p = \infty$, results on N-terms approximation in the class $C^0(\mathbb{R}^d)$ can be found in DEVORE, PETRUSHEV and YU [1992].

40. Other types of nonlinear approximations

Nonlinear approximation is a rich theory that covers a wider range of techniques beside wavelets. In this section, we shall give an account of the main items in this theory.

Generally speaking, we are interested in approximation in spaces S_N (or $\Sigma_j = S_{2^{dj}}$) that are not necessarily linear, but such that $f \in S_N$ can still be parameterized by N (or CN with a fixed C) degrees of freedom, and such that $S_N \subset S_{N+1}$.

We already observed in the abstract Theorem 38.1 that the relation between approximation spaces and interpolation spaces stays valid in this more general context, provided that the nonlinearity of S_N is controlled by an assumption of the type (38.14). Actually, this assumption is also crucial to ensure that the spaces $\mathcal{A}^s_q(X)$ related to S_N are linear, i.e. that if f and g can be approximated at a certain rate then the same holds for $f+g$. We shall now discuss several natural choices for S_N and compare them with N-terms wavelet approximation.

Example 40.1. Free knot splines

We already encountered spline functions with uniformly spaced knots as an example of multiscale approximation in Chapter II. Here we consider spline functions with nonuniformly spaced knots on an interval $[a,b]$. To any subdivision $a = x_0 < x_1 < \cdots < x_N < x_{N+1} = b$ of $[a,b]$ into N intervals, we associate for $0 \leqslant m \leqslant n$ the spaces

$$S\big(m,n,(x_i)\big) := \big\{f \in C^{m-1}\big([a,b]\big)\ f|_{]x_i,x_{i+1}[} \in \Pi_n\big\}, \tag{40.1}$$

of piecewise polynomials functions of degree n with smoothness $m-1$ at the knots x_i. If $a = -\infty$ or $b = +\infty$, we simply modify the definition by imposing that f is identically zero on $]\infty, x_1]$ or $[x_N, +\infty[$. For $m = n$, these are the classical spline functions. For $m = 0$, no smoothness is imposed and we obtain simple piecewise polynomials functions. Functions in the above spaces with $m < n$ can also be obtained as limit of spline functions with $m = n$ by merging knots by groups of at most $n - m + 1$ points. For this reason they are also called splines of degree n and multiplicity $n - m + 1$ (they were already evoked in Example 19.5). We then define the space of free-knot splines of degree n and multiplicity $n - m + 1$ on N intervals as the union

$$S_N(m,n) = \bigcup_{(x_i)\in R_N} S\big(m,n,(x_i)\big), \tag{40.2}$$

where R_N is the set of all possible subdivisions of $[a,b]$ into N-intervals. For fixed parameters m and n, we are thus interested in approximation by functions of the nonlinear space $S_N = S_N(m,n)$. Note that (38.14) does hold for such nonlinear spaces since we clearly have $S_N + S_N \subset S_{2N}$.

The first results on free-knot spline approximation appeared in the 1960's (KAHANE [1961] and BIRMAN and SOLOMJAK [1967]) and a rather complete understanding of this theory (in the case of univariate functions) was reached in the end of the 1980's with works by OSWALD [1990] and PETRUSHEV [1988]. These results were strongly influential in the subsequent study of N-term wavelet approximation as presented in Sections 38 and 39 of this chapter, due to the strong analogies between both tools: in Section 5 of Chapter I, we already observed that the thresholding process of N-term approximation in the Schauder basis amounts in the selection of an adaptive set of N knots. In the univariate case, it is actually known that the approximation spaces are the same for both techniques. More precisely, the following result was proved in PETRUSHEV [1988] for the free knot splines of degree n and multiplicity n (i.e. piecewise polynomials of degree n) on a univariate interval I.

THEOREM 40.1. *For $1 \leqslant p < \infty$, $0 < s < n+1$ and $1/q = 1/p + s$, the approximation space $\mathcal{A}_q^s(L^p)$ for the free knot splines coincides with the Besov space $B_{q,q}^s(I)$.*

As in the case of N-term wavelet approximation, a natural way of proof for such a result consists in establishing first a direct and inverse estimate for the spaces S_N and conclude by interpolation arguments.

It is particularly instructive to follow the arguments of an elementary proof for the direct estimate

$$\inf_{g \in S_N} \|f - g\|_{L^p(I)} \lesssim N^{-s} |f|_{B_{q,q}^s(I)}, \tag{40.3}$$

in the case where $S_N = S_N(0, n)$, i.e. with no smoothness requirement on the spline approximants. From the imbedding of $B_{q,q}^s$ in L^p, we derive that for a given interval J,

$$\inf_{g \in \Pi_n} \|f - g\|_{L^p(J)} \lesssim \inf_{g \in \Pi_n} \|f - g\|_{B_{q,q}^s(J)}. \tag{40.4}$$

Then, using general results on polynomial approximation (for example the local Whitney estimate in Theorem 25.2), we conclude that

$$\inf_{g \in \Pi_n} \|f - g\|_{L^p(J)} \leqslant C |f|_{B_{q,q}^s(J)}. \tag{40.5}$$

In this last estimate, a scaling argument immediately shows that the constant C does not depend on the size of J. From the superadditivity of $|f|_{B_{q,q}^s(J)}^q$ with respect to unions of disjoint intervals (see Remark 26.5), we can find for all N a disjoint partition $I = \bigcup_{k=1}^N I_k$ such that

$$|f|_{B_{q,q}^s(I_k)} \lesssim N^{-1/q} |f|_{B_{q,q}^s(I)}, \tag{40.6}$$

with a constant that does not depend on N. Combining (40.5) and (40.6), we finally obtain

$$\begin{aligned}
\inf_{g \in S_N} \|f - g\|_{L^p(I)} &\leqslant \left[\sum_{k=1}^N \inf_{h \in \Pi_n} \|f - h\|_{L^p(I_k)}^p \right]^{1/p} \\
&\lesssim N^{1/p - 1/q} |f|_{B_{q,q}^s(I)} = N^{-s} |f|_{B_{q,q}^s(I)}.
\end{aligned}$$

This proof shows that a natural subdivision of the interval for the approximation of f by free knot splines corresponds to an well-balanced splitting of the semi-norm $B^s_{q,q}(I)$ between the knots. Note that the critical Sobolev and Besov imbedding, which were already identified in the context of N-term wavelet approximation, now play an explicit role in this proof.

How does one practically implement a free knot spline approximation? The above reasoning suggest to equilibrate the Besov semi-norm $|f|_{B^s_{q,q}}$ over the intervals of the adaptive partition. However, such a strategy is not practically convenient since it relies on our knowledge of the Besov smoothness of the function to be approximated, in contrast to wavelet thresholding. A more realistic approach is to fix some tolerance $\varepsilon > 0$ and design the N intervals I_k in such a way that *the local polynomial approximation error is equilibrated*, i.e. $\inf_{h\in\Pi_n} \|f - h\|_{L^p(I_k)} \sim \varepsilon$ for all k. The resulting free knot spline approximation clearly satisfies $\|f - g\|_{L^p(I)} \leqslant N^{1/p}\varepsilon$, and furthermore if $f \in B^s_{q,q}$, $1/q = 1/p + s$, we can use (40.5) to derive

$$\begin{aligned}\|f - g\|_{L^p(I)} &\leqslant N^{1/p}\varepsilon \leqslant N^{-s}N^{1/q}\varepsilon \\ &\lesssim \left[\sum_{k=1}^{N} \inf_{h\in\Pi_n} \|f - h\|^q_{L^p(I_k)}\right]^{1/q} N^{-s} \\ &\lesssim N^{-s}|f|_{B^s_{q,q}(I)}.\end{aligned}$$

We mention two more practical (however sub-optimal) algorithmic strategies which are of common use for the purpose of equilibrating the local error of polynomial approximation.

(i) *Derefinement strategies*: starting from a spline approximation on a fine uniform discretization, one uses *knot removal* algorithms in order to obtain a (coarser) adaptive discretization with more equilibrated local polynomial approximation error. A survey of such techniques (in particular the famous "Oslo" algorithm) can be found in LYCHE [1993].

(ii) *Adaptive refinement strategies*: starting from a coarse piecewise polynomial approximation, one refines it adaptively by splitting the intervals into half as long as the local error of polynomial approximation remains higher than some prescribed tolerance. The analysis of such algorithms has been done in DEVORE and YU [1990] where it is shown that they reach the rates N^{-s}, as best approximation in S_N, with an slight extra smoothness assumption on the function, e.g., $f \in B^{s+\varepsilon}_{q,q}$. In this sense this procedure almost achieves the approximation rates predicted by Theorem 40.1.

If one disposes of a spline wavelet basis, an alternate approach is to simply apply a thresholding procedure on the decomposition of the function. One can then exploit the results on N-term approximation that were discussed in Sections 38 and 39.

Example 40.2. Rational approximation

Here S_N denotes the set of rational functions of degree N, i.e. functions of the type $r(x) = p(x)/q(x)$ where $p, q \in \Pi_N$. In the case of univariate functions, elements in this set are characterized by $2N$ parameters. As for free knot splines, we have $S_N + S_N \subset S_{2N}$.

Approximation by rational function has been considered since the beginning of century. In particular, the works of Padé provided with a systematic approach to the approximation of an analytic functions f in the complex domain by rational function p/q, $p \in \Pi_N$, $q \in \Pi_M$, which are determined by maximizing the degree of cancellation of $qf - p$ at the origin. We refer to BAKER [1975] and BREZINSKI and VAN ISEGHEM [1994] for detailed surveys on Padé approximants. These methods typically apply to the approximation of analytic functions in the complex domain.

More recently came the idea that rational functions have the ability to produce adaptive approximations for functions that contain singularities in the real domain. One of the first and most striking result in this direction was given in NEWMAN [1964] and deals with the approximation of the function $f(x) = |x|$ in the uniform metric on $I = [-1, 1]$.

THEOREM 40.2. *There exists a constant C and a sequence of polynomials $p_N, q_N \in \Pi_N$ such that*

$$\left\| f - \frac{p_N}{q_N} \right\|_{L^\infty(I)} \leqslant C e^{-\pi\sqrt{N}}, \tag{40.7}$$

and thus $\inf_{g \in S_N} \|f - g\|_{L^\infty(I)} \lesssim e^{-\pi\sqrt{N}}$.

Such a result contrasts deeply with polynomial approximation of f which is limited by the presence of the singularity at the origin: one can easily check that $\inf_{g \in \Pi_N} \|f - g\|_{L^\infty(I)} \sim \mathcal{O}(1/N)$.

This result also reveals the links between rational and free knot spline approximation, since an easy corollary is that a continuous piecewise polynomial function on a bounded interval can be approximated at the rate $e^{-\pi\sqrt{N}}$ in the uniform metric by rational functions in S_N. We can thus expect that a function which is approximated with rate N^{-s} by free knot splines can also be approximated with the same rate by rational functions. Such links were studied in details and exploited in PETRUSHEV [1988] where the following result is proved.

THEOREM 40.3. *For $1 \leqslant p < \infty$, $s > 0$ and $1/q = 1/p + s$, the approximation space $\mathcal{A}_q^s(L^p)$ for rational functions coincides with the Besov space $B_{q,q}^s(I)$.*

An analog result with the metric BMO (bounded mean oscillation) as a close substitute for L^∞ was proved in PELLER [1985]. In other words, the approximation spaces for rational functions coincide with those of free knot splines and wavelets. Here again, the way of proof for such a result consists in establishing direct and inverse estimates for the S_N spaces. However, a simple algorithm that would directly produce an optimal rational approximation (without going through an intermediate free knot spline approximant) is still to be found.

The adaptive potential of rational functions can also be intuitively understood by the following observations: if $\psi(x)$ is a rational function in $L^2(\mathbb{R})$ such that $\int \psi = 0$, e.g., $\psi(x) = x/(x^2+1)$, then the family

$$\psi_{j,k}(x) = 2^{j/2}\psi\big(2^j x - k\big) = 2^{-j/2}\frac{x - 2^{-j}k}{(x - 2^{-j}k)^2 + 2^{-2j}}, \tag{40.8}$$

has the appearance of a wavelet basis, while all its elements are in S_2. This reveals the analogy between nonlinear rational and wavelet approximation since an N-term combination of the above functions is in S_{2N}. Note that as j grows, the poles $2^{-j}k \pm i2^{-j}$ of $\psi_{j,k}$ get closer to the real line, which corresponds to the idea that an adaptive approximation of f by $f_N \in S_N$ requires a particular placement of the poles of f_N in the vicinity of the singularities of f.

REMARK 40.1. It should be well understood that the above analogy between wavelets and rational functions is only heuristic since the family in (40.8) is not known to constitute a basis. A general open problem has been to find a Riesz basis for $L^2(I)$ of a form generalizing (40.8), i.e. consisting of rational functions of degree less than N for some prescribed $N \geqslant 0$, which could provide the optimal approximation orders by simple thresholding procedures. A positive answer to this problem was recently given in PETRUSHEV [1988], which reinforces the links between rational fractions and wavelets. It might also provide with practical algorithms for adaptive rational approximation.

Example 40.3. Highly nonlinear approximation

The two previous examples have demonstrated that, in the univariate setting, free knot splines and rational fractions have approximation properties which are very similar to N-term wavelet approximation. We now want to point out on several instances of nonlinear approximation for which adaptivity is pushed at a higher level. This results in better approximation rates than wavelets for certain types of functions, but also in a more difficult practical implementation as well as a still incomplete theoretical analysis of these methods. We refer to them as "highly nonlinear".

A first instance is provided by free knot splines or adaptive finite element and rational fractions in dimension $d \geqslant 2$. For simplicity, we consider here the bivariate case, i.e. $d = 2$. In order to define approximation families S_N of rational functions depending on $\mathcal{O}(N)$ parameters, we define a polynomial space Π_N^* of dimension N: if M is such that $M(M+1)/2 \leqslant N < (M+1)(M+2)/2$, then Π_N^* is generated by $x^k y^l$ for $k + l < M$ and $x^M y^l$ for $l = 0, \ldots, N - M(M+1)/2 - 1$. We then define S_N as the set of all functions $f(x) = p(x)/q(x)$ with $p, q \in \Pi_N^*$. We also need to choose a specific generalization for the free knot splines. For fixed n and $m < n$, we define the corresponding free knot spline space S_N as the set of all functions f that are piecewise polynomial of degree n and smoothness C^{m-1} on some arbitrary triangulation $\mathcal{T}$ of the polygonal domain Ω under consideration, with prescribed cardinality $|\mathcal{T}| = N$ (with f identically zero outside of $\mathcal{T}$ in the case of an infinite domain Ω, e.g., $\Omega = \mathbb{R}^2$). We can also define such approximants in terms of finite elements of a given type on arbitrary triangulations with a prescribed number N of degrees of freedom.

Such approximation families are "more nonlinear" than the examples that we have considered so far, in the sense that they do not anymore satisfy the property (38.14). In turn, the corresponding approximation spaces $\mathcal{A}_q^s(X)$ are not ensured to be linear (and have been proved not to be linear for simple cases such as $X = L^p$): two functions f and g might be approximated with the rate N^{-s} while this rate is not achieved for $f+g$. There is thus no hope to describe such spaces through classical notion of smoothness.

An important property of rational functions and free knot splines in several dimensions is their ability to capture certain anisotropic singularities in a multivariate function with a small number of parameters. In the case of free knot splines, this is intuitively clear, due to the possibility of using very thin triangles in order to refine the approximation in a single direction. For rational fractions, this is also made possible since the poles are now algebraic curves that can be used to approximate lower-dimensional singularities of certain functions. As a simple example, consider the function $f(x_1, x_2) = |x_1|$ on $[-1, 1]^2$ which has a singularity along the line $x_1 = 0$. For such a function, it is natural to choose the rational approximation $p_N(x_1)/q_N(x_1)$ with p_N and q_N given by Theorem 40.2. With such a choice (which belongs to Σ_{N^2} according to our definition), we inherit of the approximation rate $e^{-\pi\sqrt{N}}$ in the L^∞ norm. In contrast, N-term wavelet approximation in several dimensions fails to produce such high rates, due to the isotropic scaling that defines the basis $(\psi_\lambda)_{\lambda\in\Lambda}$: a singularity of codimension c requires the use of $\mathcal{O}(2^{(d-c)j})$ in order to be captured at resolution 2^{-j}.

REMARK 40.2. One can think of imposing shape constraints in the definition of free knot splines or adaptive finite elements in order to limit their anisotropy, i.e. impose the assumption that the ratio between the diameter of a triangle and the radius of its inscribed circle is bounded by an absolute constant. Such shape constraints allow to recover inverse estimates between L^p and $B_{q,q}^s$ for $1/q = 1/p + s/d$ similar to those of N-term wavelet approximation. With such constraints, it thus seems reasonable that the resulting approximation spaces coincides with those of N-term wavelet approximation: defining $\Sigma_j = S_{2^{dj}}$, we expect $\mathcal{A}_{p,q}^s = B_{q,q}^s$ for $1/q = 1/p + s/d$ and $1 < p < \infty$, assuming $m = 0$ and $n + 1 \geqslant s$ for the parameters of the spline space. This remains so far an open question. Note that most refinement strategies proposed in the finite element context impose such shape constraints. These strategies aim to equilibrate the local error between the different elements, similar to the adaptive splitting strategies that we described for the univariate free knot spline (see, e.g., BABUSHKA and RHEINBOLDT [1978]).

REMARK 40.3. If we impose continuity or higher smoothness across triangles (i.e. $m \geqslant 1$), we should expect that the performance of free knot splines with shape constraints are not as good as those of wavelets: continuity imposes that the triangulation is conformal, which together with shape constraints induces a "grading" property, i.e. a very small triangle cannot be adjacent to a very large one.

A second instance of highly nonlinear approximation is encountered in the following situation: one wants to approximate $f \in L^2(\Omega)$ by an N-term linear combination $\sum_{k \in E} c_k e_k$, where $(e_k)_{k \geqslant 0}$ is a "dictionary" of functions which spans $L^2(\Omega)$ in a redundant fashion. Several choices of such redundant families have been the object of particular interest.

(i) *Time-frequency atoms*: these functions have the general form $g_{m,n}(x) = g(x - ma)\mathrm{e}^{inbx}$, $m, n \in \mathbb{Z}$, where g is a fixed shape function and $a, b > 0$ fixed parameters. A well studied case is the Gabor functions corresponding to the choice of a Gaussian function for g. In such a case (see DAUBECHIES [1992, Chapter IV]), for $a, b > 0$ small enough, these functions constitute a frame of $L^2(\mathbb{R})$ and are in turn a complete family. An interesting variant are the *local cosine bases* of H. Malvar (see, e.g., MEYER [1992]), which in contrast to Gabor functions, allow to generate orthonormal bases.

(ii) *Wavelet packets*: these functions, introduced in COIFMAN, MEYER, QUAKE and WICKERHAUSER [1993], are the natural generalization of wavelets bases, produced by allowing more flexibility in the multiscale decomposition algorithms described in Section 14 of Chapter II. More precisely, one allows to iterate the basic two-band decomposition (the block diagram of Fig. 14.3) on the result of the high-pass filtering of the previous iteration (while wavelet decompositions only iterates this action on the result of the low-pass filtering). Depending on the successive choices of applying this basic decomposition or of leaving the resulting subsignals unchanged, one expands the signal into a specific basis corresponding to a particular subband decomposition, one of them being the wavelet basis (corresponding to the choice of only splitting the low-pass subsignal). Wavelet packets can thus be viewed as a collection of bases, having many common elements.

(iii) *Ridge functions*: these are families of multivariate functions of the type $g(\langle u, x \rangle)$, where g is a univariate function chosen among a finite or infinite set and u a direction vector chosen among a finite or infinite set. Approximation by such functions has been studied in relation with the problem of modelizing neural networks (see BARRON [1994]).

The redundancy of these dictionaries is intended to allow a better level of adaptivity, with the hope of describing complicated functions by very few parameters. However, difficulties arise when it comes to the practical implementation of an optimal N-term approximation: while keeping the N largest coefficients is the natural procedure to apply in the case of a basis, the selection of the best N-term combination within a redundant family is a highly complex problem (it can be shown to have NP-hard complexity in very simple instances, see AVELLANEDA, DAVIS and MALLAT [1997]). For this reason, one practically makes use of sub-optimal algorithms:

(i) *Best bases algorithms*, introduced in COIFMAN and WICKERHAUSER [1992], operate in the context of a dictionary made out of the union of several bases. In a first step, the algorithm compares the action of the different bases on a given function f, using an "entropy criterion", i.e. a measurement of the sparsity of f for each basis. According to the discussion in Section 38, if one is interested in concentrating the L^2 norm on a small number of coefficients, a possibility is to take to ℓ^p norm of the coefficients, for some $p < 2$. The basis which minimizes the criterion is selected and used to compress the function by a simple N-term approximation. In the case of wavelet packets, it is interesting to note that a best basis can be obtained in $\mathcal{O}(N \log N)$ operations, making use of the particular tree-structured algorithm which generates the expansions of a discretized function in these bases.

(ii) *Matching pursuit algorithms*, as studied in AVELLANEDA, DAVIS and MALLAT [1997], operate on any type of dictionary. The simplest (and most often used) of these approximation schemes is the so-called *greedy* algorithm. We assume here that the e_k are normalized in L^2. As a first step, the algorithm compares the values of the coefficients $\langle f, e_k \rangle$ and select the largest one $\langle f, e_{k_0} \rangle$. A first approximation of f is chosen to be $f_0 = \langle f, e_{k_0} \rangle e_{k_0}$. In a second step, the residual $r_0 = f - f_0$ is analyzed in a similar way, yielding an updated approximation of f by $f_1 = f_0 + \langle r_0, e_{k_1} \rangle e_{k_1}$. Iterating this procedure N times yields an N term approximation of f.

Although these algorithms perform very well in several practical situations, most known theoretical results about their nonlinear approximation properties are negative in the following sense: if the function f is such that $\|f - f_N\|_{L^2} \lesssim N^{-s}$ for a sequence of N-term approximations, the algorithm generally fails to produce such a rate of convergence. An interesting open problem is to establish reasonable assumptions on the dictionary or on the class of functions to be analyzed that would restore the optimality of these algorithms.

Example 40.4. Data compression

Among the different techniques of nonlinear approximation that we have discussed, wavelets offer the simplest algorithmic technique: (i) the function is decomposed by a fast algorithm from a fixed finest discretization level (typically imposed by the sampling rate in the case of a digitized signal), (ii) its largest coefficients are retained by imposing either their number or a threshold, (iii) the function can be reconstructed in a compressed form by a fast algorithm up to the finest discretization level.

This particular simplicity explains the interest for wavelets methods when dealing with data compression problems for certain types of signal. In Section 5 of Chapter I, we already made clear that wavelet thresholding behaves particularly well on images, since it corresponds to reduce the number of parameters by focusing on the description of significant structures such as edges. However, another aspect of nonlinear approximation appears in data compression: these parameters (here the unthresholded wavelet coefficients) need to be *quantized*, i.e. encoded by a finite number of bits. We thus cannot expect to encode the exact value of a coefficient c_λ but an approximate value $Q(c_\lambda)$ chosen within a finite set. In turn, the success of a wavelet-based compression is strongly dependent on the chosen quantization strategy.

Without going into details, we shall mention here the two main approaches which have been investigated.

The first idea consists in exploiting certain statistical properties of the wavelet coefficients for a general class of signals in order to derive an optimal quantizer which will process the wavelet coefficients uniformly, regardless of the signal within the class of interest: e.g., a *scalar* quantizer will affect a fixed number of bits $n(\lambda)$, associated to

$2^{n(\lambda)}$ fixed values, to a given coefficient c_λ. A natural generalization is *vector* quantization, which affects a fixed number of bits $n(E)$, associated to $2^{n(E)}$ fixed vectors, to a given block of coefficients $(c_\lambda)_{\lambda \in E}$. In the scalar case, the statistical properties which are exploited are typically the variances of wavelet coefficients: more bits should be affected to coefficients which are larger in average. In the vector case, the intercorrelations between coefficients within a block also influences the bits allocation. Numerous practical algorithms have been designed in order to build optimal vector quantizers in a general setting (see, e.g., GERSHO and GRAY [1992]). This approach was successfully applied to wavelet decompositions of digital pictures in ANTONINI, BARLAUD, DAUBECHIES and MATHIEU [1992]

The second idea is closer in spirit to nonlinear approximation: one chooses to affect to c_λ a number of bits $n(\lambda, |c_\lambda|)$ that will also depend on the size of c_λ *for each signal*. This strategy seems more natural for classes of signal such as images, since the location of the few large coefficients associated to edges vary strongly from one image to the other. Typically, one wants to affect more bits to the large coefficients, and benefit from the fact that there are only few such coefficients. We refer to such a strategy as *adaptive quantization*. Note that it implies the additional encoding of the indices of coefficients within a given order of magnitude. In practice, this encoding is facilitated by imposing some additional *tree structure* on the quantized coefficients, which reflects the properties of real images: if a coefficient d_λ is large, it it then likely that the support of $\tilde{\psi}_\lambda$ contains an edge singularity and that d_μ is also large for the indices μ such that $|\mu| < |\lambda|$ and $\mathrm{Supp}(\tilde{\psi}_\lambda) \subset \mathrm{Supp}(\tilde{\psi}_\mu)$. Therefore d_μ should be quantized with at least as many bits as d_λ (see also Section 43 for a precise definition of a tree-structured sets in the context of adaptive numerical simulation). Such strategies have led to two very competitive image compression algorithms in SHAPIRO [1993] and SAID and PEARLMAN [1996]. On a more mathematical point of view, it has been shown in COHEN, DAHMEN, DAUBECHIES and DEVORE [1999] that for various function classes modelizing the signal, such algorithms lead to optimal asymptotic behaviour of the L^2-error with respect to the number of bits.

As an example we have displayed on Fig. 40.1, the result of an adaptive quantization strategy for the image that we already manipulated in Section 5 of Chapter I, with a compression factor of $1/40$. We have used here the biorthogonal spline wavelet pair $(\psi_{3,5}, \tilde{\psi}_{3,5})$ introduced in Section 19 of Chapter II: reconstruction is ensured by piecewise bilinear functions which allows to avoid the visual artefacts of the Haar system.

REMARK 40.4. Another important application of wavelet-based parameter reduction is statistical estimation and signal denoising: a natural restauration strategy consists in thresholding the coefficients of the noisy signal at a level that will remove most of the noise, but preserve the few significant coefficients in the signal. This is in contrast to classical linear low pass filtering, which tends to blur the edges while removing the noise. Such nonlinear procedures have been the object of important theoretical work and have motivated recent progresses in nonlinear approximation, see, e.g., DONOHO, JOHNSTONE, KERKYACHARIAN and PICARD [1994]. It is fair to say that during the 1990's, wavelets have stimulated a general evolution of signal and image processing

FIG. 40.1. Adaptive compression by a factor 1/40.

toward nonlinear and adaptive methods. The reader interested by a "wavelet tour of signal processing" will certainly enjoy the reading of MALLAT [1998].

41. Adaptive approximation of operators

The goal of this section is both to analyze the sparse structure that results from the multiscale discretization of most operators involved in partial differential and integral equations, and to study the application of such operators on functions that also have a sparse multiscale representation.

We already discussed an instance of "matrix compression" in Section 5 of Chapter I. In order to reach a better theoretical understanding of such a process in more general situations, we need more information on the size of the matrix elements in the wavelet discretizations of general classes of operators. In some sense, we need to analyze the size properties of these elements in the same way as we did in Chapter III for the wavelet coefficients of general classes of functions.

Given an operator A acting on functions defined on a domain $\Omega \subset \mathbb{R}^d$ and two wavelet bases $(\psi_\lambda)_{\lambda\in\Lambda}$ and $(\tilde{\psi}_\lambda)_{\lambda\in\Lambda}$ adapted to such a domain, we are interested in evaluating the entries

$$m_{\lambda,\mu} = \langle A\psi_\lambda, \tilde{\psi}_\mu \rangle. \tag{41.1}$$

When the two bases are biorthogonal, $M = (m_{\lambda,\mu})_{\lambda,\mu\in\Lambda}$ is exactly the matrix of A in the basis $(\psi_\lambda)_{\lambda\in\Lambda}$. Note however that such a coupling between the two bases is not necessary in the general framework of Galerkin and Petrov–Galerkin discretizations of operator equations. We thus leave biorthogonality aside here. We shall first derive some basic estimates for specific classes of operators.

Example 41.1. Partial differential operators

For the sake of simplicity, let us first consider differential operators of the type $Af(x) = \sum_{m\leqslant r} a_m(x)f^{(m)}(x)$, where the coefficients $a_m(x)$ are C^∞ functions. From the local nature of this operator, we can derive the trivial estimate $m_{\lambda,\mu} = 0$ if ψ_λ and $\tilde{\psi}_\mu$ have disjoint support.

If these supports overlap, we can estimate the entries as follows: for $|\lambda| \leqslant |\mu|$, assuming that the wavelets ψ_λ have $C^{r+\alpha}$ smoothness and that the wavelets $\tilde{\psi}_\mu$ are orthogonal to polynomials of degree $n \geqslant \alpha - 1$, we derive the estimate

$$\begin{aligned} |m_{\lambda,\mu}| &= \int A\psi_\lambda(x)\,\tilde{\psi}_\mu(x)\,\mathrm{d}x \leqslant \|\tilde{\psi}_\mu\|_{L^1} \inf_{g\in\Pi_n} \|A\psi_\lambda - g\|_{L^\infty(\mathrm{Supp}(\tilde{\psi}_\mu))} \\ &\lesssim 2^{-|\mu|(1/2+\alpha)}\|A\psi_\lambda\|_{C^\alpha(\mathrm{Supp}(\tilde{\psi}_\mu))} \lesssim 2^{-|\mu|(1/2+\alpha)}2^{|\lambda|(r+\alpha+1/2)} \\ &= 2^{(1/2+\alpha)(|\lambda|-|\mu|)}2^{r|\lambda|}. \end{aligned}$$

In the case where $|\lambda| \geqslant |\mu|$, we can integrate by part and take advantage of the smoothness of the coefficients $a_m(x)$ to obtain a symmetric estimate for $m_{\lambda,\mu}$, under the assumption that the wavelets $\tilde{\psi}_\mu$ have $C^{r+\alpha}$ smoothness and that the wavelets ψ_λ are orthogonal to polynomials of degree $n \geqslant \alpha - 1$. In summary, we have obtained

$$\begin{aligned} |m_{\lambda,\mu}| &\lesssim 2^{-(1/2+\alpha)||\lambda|-|\mu||}2^{r\inf\{|\lambda|,|\mu|\}} \\ &= 2^{-(1/2+r/2+\alpha)||\lambda|-|\mu||}2^{r(|\lambda|+|\mu|)/2}. \end{aligned} \tag{41.2}$$

REMARK 41.1. The case $|\lambda| = -1$ or $|\mu| = -1$ which corresponds to the scaling functions at the coarsest level is not covered by the above considerations since, in contrast to the wavelets, these functions are not orthogonal to polynomials. However, we remark that (41.2) only required the vanishing moment properties for the wavelet of highest level. Thus the only uncovered case remains $|\lambda| = |\mu| = -1$ for which (41.2) becomes trivial.

The case of a more general partial differential operator of order r, i.e. of the type $Af(x) = \sum_{|m|\leqslant r} a_m(x)\partial^m f(x)$ in d dimension can be treated in a similar way: if both wavelet bases have $C^{r+\alpha}$ smoothness and are orthogonal to polynomials of degree $n \geqslant \alpha - 1$, one obtains the estimate

$$|m_{\lambda,\mu}| \lesssim 2^{-(d/2+r/2\alpha)||\lambda|-|\mu||}2^{r(|\lambda|+|\mu|)/2}, \tag{41.3}$$

for entries corresponding to overlapping supports. Note that we can also use Cauchy–Schwarz inequality in the estimation of $|m_{\lambda,\mu}|$ which allows to make the weaker assumption that both wavelet bases have $H^{r+\alpha}$ smoothness, and are orthogonal to polynomials of total degree $n \geqslant \alpha - 1$. We then obtain the estimate

$$|m_{\lambda,\mu}| \lesssim 2^{-(r/2+\alpha)||\lambda|-|\mu||} 2^{r(|\lambda|+|\mu|)/2}. \tag{41.4}$$

One can actually weaken further the assumptions on the smoothness of the wavelet bases: if both wavelet bases have $H^{r/2}$ smoothness, the entries $|m_{\lambda,\mu}|$ still make sense. However, we see from (41.4) that there is a gain in increasing the smoothness of the wavelets, together with their number of vanishing moments, since the factor $2^{-(\alpha+r/2)||\lambda|-|\mu||}$ improves the sparsity of the matrix away from the diagonal blocks corresponding to $|\lambda| = |\mu|$.

Example 41.2. Integral operators

We first consider the very simple example of the integral operator that defines the solution of the two point boundary value problem $-\mathrm{d}^2u/\mathrm{d}x^2 = f$ with $u(0) = u(1) = 0$. This operator has the explicit form

$$\begin{aligned} Af(x) &= \int_0^x (1-x)yf(y)\,\mathrm{d}y + \int_x^1 (1-y)xf(y)\,\mathrm{d}y \\ &= \int_0^1 K(x,y)f(y)\,\mathrm{d}y. \end{aligned} \tag{41.5}$$

In such a case, we remark that the wavelet coefficients

$$m_{\lambda,\mu} = \int_{[0,1]^2} K(x,y)\tilde{\psi}_\mu(x)\psi_\lambda(y)\,\mathrm{d}x\,\mathrm{d}y, \tag{41.6}$$

are zero if ψ_λ and $\tilde{\psi}_\mu$ have disjoint supports, provided that ψ_λ or $\tilde{\psi}_\mu$ is orthogonal to polynomials of degree $n = 1$. In order to estimate the remaining entries, we can proceed in the same way as we did in the previous example: for $|\lambda| \geqslant |\mu|$, assuming that the wavelets $\tilde{\psi}_\mu$ have C^α smoothness and that the wavelets ψ_λ are orthogonal to polynomials of degree $n+2$ with $n \geqslant \alpha - 1$, we derive the estimate

$$\begin{aligned} |m_{\lambda,\mu}| &= \int A\psi_\lambda(x)\,\tilde{\psi}_\mu(x)\,\mathrm{d}x \leqslant \|A\psi_\lambda\|_{L^1} \inf_{g\in\Pi_n} \|\tilde{\psi}_\mu - g\|_{L^\infty(\mathrm{Supp}(A\psi_\lambda))} \\ &\lesssim 2^{-|\lambda|(5/2+\alpha)} \|\tilde{\psi}_\mu\|_{C^\alpha(\mathrm{Supp}(A\psi_\lambda))} \lesssim 2^{-|\lambda|(5/2+\alpha)} 2^{|\mu|(\alpha+1/2)} \\ &= 2^{(1/2+\alpha)(|\mu|-|\lambda|)} 2^{-2|\lambda|}. \end{aligned}$$

By symmetry, we finally obtain the estimate

$$|m_{\lambda,\mu}| \lesssim 2^{-(\alpha+1/2)||\lambda|-|\mu||} 2^{-2\sup\{|\lambda|,|\mu|\}} = 2^{-(\alpha+5/2)||\lambda|-|\mu||} 2^{-(|\lambda|+|\mu|)}. \tag{41.7}$$

This simple example reveals a similarity between the estimates for differential and integral operators. Here, the vanishing moments of ψ_λ are also used to ensure that $A\psi_\lambda$ has a local support of size $\sim 2^{-\lambda}$ (the same as ψ_λ itself). The same comments as in Remark 41.1 apply here to the particular case of scaling functions that do not oscillate.

The case that we have just treated is a very particular instance of a multiscale discretization where most of the matrix entries are zero. One cannot expect this property to hold for more general operators. On the other hand, a fairly large class of integral operator gives rise to "almost sparse" matrices, i.e. containing only few numerically significant entries.

This is the case of the Calderon–Zygmund type operators, i.e. L^2 bounded operators which satisfy estimates of the type (31.31) and (31.32) (see Section 31 of Chapter III). A prototype of Calderon–Zygmund is the univariate Hilbert transform given by the kernel $K(x, y) = (x - y)^{-1}$ if $x \neq y$. In this case, we have the more general estimate

$$\left|\partial_x^m K(x, y)\right| + \left|\partial_y^m K(x, y)\right| \leqslant |x - y|^{-1-m}. \tag{41.8}$$

From (41.8), we can again estimate the entries $m_{\lambda,\mu}$ in the case where the supports of ψ_λ and $\tilde{\psi}_\mu$ are strongly disjoint in the sense that

$$\operatorname{dist}\bigl(\operatorname{Supp}(\psi_\lambda), \operatorname{Supp}(\tilde{\psi}_\mu)\bigr) \geqslant C2^{-\inf\{|\lambda|,|\mu|\}}, \tag{41.9}$$

where C is a fixed constant. We first define a function that will play the role of a spatial distance between indices in Λ by

$$d(\lambda, \mu) := 1 + 2^{\min\{|\lambda|,|\mu|\}}\operatorname{dist}\bigl(\operatorname{Supp}(\psi_\lambda), \operatorname{Supp}(\psi_\mu)\bigr). \tag{41.10}$$

For $|\lambda| \leqslant |\mu|$, assuming that $\tilde{\psi}_\mu$ is orthogonal to all polynomials of degree n, we can exploit (41.8) to derive the estimate

$$\begin{aligned}
|m_{\lambda,\mu}| &= \int K(x, y)\psi_\lambda(y), \tilde{\psi}_\mu(x)\,\mathrm{d}x\,\mathrm{d}y \\
&\leqslant \|\psi_\lambda\|_{L^1}\left\|\tilde{\psi}_\mu\right\|_{L^1} \sup_{y\in\operatorname{Supp}(\psi_\lambda)} \inf_{g\in\Pi_n} \left\|K(\bullet, y) - g\right\|_{L^\infty(\operatorname{Supp}(\tilde{\psi}_\mu))} \\
&\lesssim 2^{-(|\lambda|+|\mu|)/2}2^{-n)|\mu|}\bigl[2^{-|\lambda|} + \operatorname{dist}\bigl(\operatorname{Supp}(\psi_\lambda), \operatorname{Supp}(\tilde{\psi}_\mu)\bigr)\bigr]^{-(n+1)} \\
&\lesssim 2^{-(|\lambda|+|\mu|)/2}2^{-n|\mu|}2^{(n+1)|\lambda|}d(\lambda, \mu)^{-(n+1)} \\
&= 2^{(n+1/2)(|\lambda|-|\mu|)}d(\lambda, \mu)^{-(n+1)}.
\end{aligned}$$

By symmetry, if ψ_λ is also orthogonal to all polynomials of degree n, we finally obtain

$$|m_{\lambda,\mu}| \lesssim 2^{-(n+1/2)||\lambda|-|\mu||}d(\lambda, \mu)^{-(n+1)}. \tag{41.11}$$

In the case where the supports overlap, estimate of the type (41.11) can be preserved if in addition to the $n + 1$ vanishing moments one assumes smoothness properties on the basis functions. We shall illustrate this point in the simple case of the Hilbert transform, with two standard wavelet bases generated from translations and dilations of single functions ψ and $\tilde{\psi}$. In this case, we can use Parseval formula to derive

$$\begin{aligned}
|m_{\lambda,\mu}| &= \left| \int \overline{\hat{\psi}_\lambda(\omega)} \hat{\tilde{\psi}}_\mu(\omega) \mathrm{Sign}(\omega)\, \mathrm{d}\omega \right| \\
&\lesssim 2^{-(|\lambda|+|\mu|)/2} \int \left| \hat{\psi}\left(2^{-|\lambda|}\omega\right) \hat{\tilde{\psi}}\left(2^{-|\mu|}\omega\right) \right| \mathrm{d}\omega \\
&\lesssim 2^{(|\lambda|-|\mu|)/2} \int \left| \hat{\psi}(\omega) \hat{\tilde{\psi}}\left(2^{|\lambda|-|\mu|}\omega\right) \right| \mathrm{d}\omega.
\end{aligned}$$

Assuming that $|\lambda| \leqslant |\mu|$ and that ψ is smooth enough so that $|\hat{\psi}(\omega)| \lesssim (1+|\omega|)^{-n-\varepsilon}$, $\varepsilon > 0$, we can then evaluate the last above integral $I(|\lambda| - |\mu|)$ by splitting it into the domains $|\omega| \leqslant 1$ and $2^j < |\omega| < 2^{j+1}$, $j \geqslant 0$: using the cancellation of $\hat{\tilde{\psi}}$ at the order n near the origin, we obtain

$$\begin{aligned}
I\left(|\lambda| - |\mu|\right) \lesssim & \sum_{j=0}^{|\mu|-|\lambda|} 2(1-n-\varepsilon) j 2^{n(j+|\lambda|-|\mu|)} \\
& + \sum_{j \geqslant |\mu|-|\lambda|} 2^{(1-n-\varepsilon)j} \lesssim 2^{n(|\lambda|-|\mu|)}.
\end{aligned}$$

By a symmetric argument if $|\mu| \leqslant |\lambda|$, we finally obtain that (41.11) also holds for all entries.

Note that the example of the single layer operator acting on 1-periodic functions, which was evoked in Section 5 of Chapter I, is strongly related to the Hilbert transform since a singularity of the type $1/(x-y)$ appears if we differentiate $K(x, y) = \log|\mathrm{e}^{\mathrm{i}2\pi x} - \mathrm{e}^{\mathrm{i}2\pi y}|$. This allows to evaluate the corresponding matrix entries $m_{\lambda,\mu} = \int K(x, y)\psi_\lambda(y), \tilde{\psi}_\mu(x)\, \mathrm{d}x\, \mathrm{d}y$ using integration by part in the variable x (resp. y) if $|\lambda| \leqslant |\mu|$ (resp. $|\lambda| \geqslant |\mu|$), which yields

$$\begin{aligned}
|m_{\lambda,\mu}| &\lesssim 2^{-(n+1/2)||\lambda|-|\mu||} d(\lambda, \mu)^{-(n+1)} 2^{-\sup\{|\lambda|,|\mu|\}} \\
&= 2^{-(n+1)||\lambda|-|\mu||} d(\lambda, \mu)^{-(n+1)} 2^{-(|\lambda|+|\mu|)/2}.
\end{aligned}$$

In the general context of pseudodifferential operator on periodic domains, the reader will find similar estimates in DAHMEN, PRÖSSDORF and SCHNEIDER [1993].

It is interesting to note from these examples that, while finite element discretizations of partial differential and integral operators results in different types of matrices (banded vs. full), multiscale discretizations give rise to the same structure in both case (this is the "finger-like" structure that was illustrated on Fig. 5.10). In order to treat these different examples within a unified framework that describes this structure, we shall now introduce general classes of matrices associated to operators through wavelet bases.

DEFINITION 41.1. Let $s \in \mathbb{R}$ and $\alpha, \beta > 0$. A matrix M belongs to the class $\mathcal{M}^s_{\alpha,\beta}$ if and only if its entries satisfy the estimate

$$|m_{\lambda,\mu}| \leqslant C_M 2^{s(|\lambda|+|\mu|)} 2^{-(d/2+\alpha)||\lambda|-|\mu||} d(\lambda, \mu)^{-(d+\beta)}. \tag{41.12}$$

We denote by $\mathcal{M}_{\alpha,\beta}$ this class when $s = 0$.

Before going further, let us analyze the meaning of the different factors in the right side of (41.12).

The factor $2^{s(|\lambda|+|\mu|)}$ describes the growth or decay (depending on the sign of s) of the entries of M along the diagonal, i.e. the multiplicative effect of the operator on the different range of scales. The parameter s thus indicates the order of the operator: e.g., $s = 2$ if $A = \Delta$, $s = 0$ for the Hilbert transform, $s = -1/2$ for the single layer operator, etc. Note that the diagonal matrix $D_s = (2^{s|\lambda|}\delta_{\lambda,\mu})_{\lambda,\mu}$ allow to renormalize M in the sense that $\widetilde{M} = D_s^{-1} M D_s^{-1}$ satisfies the estimate (41.12) with $s = 0$, i.e. belongs to the class $\mathcal{M}_{\alpha,\beta}$. Note that such a renormalization is exactly the preconditioning process described on finite matrices in Section 34 of Chapter III.

The factor $2^{-(d/2+\alpha)||\lambda|-|\mu||}$ describes the decay of the entries away from the diagonal blocks corresponding to $|\lambda| = |\mu|$. In all examples that we have discussed, the value of α is both influenced by the nature of the operator (e.g., the smoothness of its coefficients in the case of a nonconstant partial differential operator or the regularity of the kernel for an integral operator), as well as by the smoothness and number of vanishing moments of the wavelet basis. As we shall see further, the role of

Finally, the factor $d(\lambda, \mu)^{-(d+\beta)}$ describes the decay of the entries away from the diagonal within each blocks corresponding to fixed values of $|\lambda|$ and $|\mu|$. In the case of a local operator such as a partial differential operator, we note that β can be taken arbitrarily large, up to a modification in the constant C_M in (41.12). In the case of an integral operator, β is influenced like α by the operator and the wavelet basis.

A basic tool for the study of the classes $\mathcal{M}^s_{\alpha,\beta}$ is the Schur lemma that we recall below.

LEMMA 41.1. *Let $M = (m_{\lambda,\mu})_{\lambda,\mu\in\Lambda}$ be a matrix indexed by Λ. Assume that there exists a sequence of positive numbers $(\omega_\lambda)_{\lambda\in\Lambda}$ and a constant C such that*

$$\sum_{\mu\in\Lambda} \omega_\mu |m_{\lambda,\mu}| + \sum_{\mu\in\Lambda} \omega_\mu |m_{\mu,\lambda}| \leqslant C\omega_\lambda, \tag{41.13}$$

for all $\lambda \in \Lambda$. Then M defines a bounded operator in $\ell^2(\Lambda)$ with $||M|| \leqslant C$.

A first application of the Schur lemma is the following simple result.

THEOREM 41.1. *If $\alpha, \beta > 0$, then any $M \in \mathcal{M}_{\alpha,\beta}$ defines a bounded operator in $\ell^2(\Lambda)$. In turn, any matrix $M \in \mathcal{M}_{\alpha,\beta}$ together with a Riesz basis $(\psi_\lambda)_{\lambda\in\Lambda}$ of $L^2(\Omega)$ defines an L^2 bounded operator A represented by M in this basis.*

PROOF. We shall use the Schur lemma with $\omega_\lambda = 2^{-d|\lambda|/2}$. From (41.12), we first obtain

$$\begin{aligned}\omega_\lambda^{-1}\sum_{\mu\in\Lambda}\omega_\mu|m_{\lambda,\mu}| &\lesssim 2^{d|\lambda|/2}\sum_{\mu\in\Lambda}2^{-d|\mu|/2}2^{-(d/2+\alpha)||\lambda|-|\mu||}d(\lambda,\mu)^{-(d+\beta)}\\ &\lesssim 2^{d|\lambda|/2}\sum_{j\geqslant -1}2^{-dj/2}2^{-(d/2+\alpha)||\lambda|-j|}\sum|_{\mu|\in\Lambda_j}d(\lambda,\mu)^{-(d+\beta)}.\end{aligned}$$

Since $\beta > 0$ the last factor $\sum|_{\mu|\in\Lambda_j}d(\lambda,\mu)^{-(d+\beta)}$ is bounded by a uniform constant if $j \leqslant |\lambda|$ and by $2^{d(j-|\lambda|)}$ if $j \geqslant |\lambda|$. Splitting the sum in j according to these two cases, we finally obtain

$$\begin{aligned}\omega_\lambda^{-1}\sum_{\mu\in\Lambda}\omega_\mu|m_{\lambda,\mu}| &\lesssim \sum_{j\geqslant -1}2^{d||\lambda|-j|/2}2^{-(d/2+\alpha)||\lambda|-j|}\\ &\leqslant 2\sum_{l\geqslant 0}2^{-\alpha l} = C < \infty,\end{aligned}$$

which shows that (41.13) holds with such weights. □

An immediate consequence of this result concerns the classes $\mathcal{M}^s_{\alpha,\beta}$ for $s \neq 0$. We already noted that for M in this class, the preconditioned matrix

$$\widetilde{M} = D_s^{-1}MD_s^{-1}, \tag{41.14}$$

with $D_s = (2^{s|\lambda|}\delta_{\lambda,\mu})_{\lambda,\mu}$ belongs to the class $\mathcal{M}_{\alpha,\beta}$. Introducing the weighted spaces

$$\ell^2_t(\Lambda) := \Big\{(c_\lambda)_{\lambda\in\Lambda};\ \big\|(c_\lambda)\big\|^2_{\ell^2_t} := \sum_{\lambda\in\Lambda}2^{2t|\lambda|}|c_\lambda|^2 < \infty\Big\} \tag{41.15}$$

(which are associated to the Sobolev spaces H^s by the results of Chapter III, e.g., Theorem 33.4), we remark that D_s defines an isomorphism from $\ell^2_t(\Lambda)$ to $\ell^2_{t+s}(\Lambda)$. Combining these remarks with Theorem 41.1, we can describe the action of $M = D_s\widetilde{M}D_s$ as follows.

COROLLARY 41.1. *If $\alpha, \beta > 0$, then any $M \in \mathcal{M}^s_{\alpha,\beta}$ defines a bounded operator from $\ell^2_s(\Lambda)$ to $\ell^2_{-s}(\Lambda)$. In turn, any matrix $M \in \mathcal{M}^s_{\alpha,\beta}$ together with a wavelet basis $(\psi_\lambda)_{\lambda\in\Lambda}$ that both characterizes $H^s(\Omega)$ and $H^{-s}(\Omega)$ (possibly with boundary conditions) defines a bounded operator A from H^s to H^{-s}, represented by M in this basis.*

REMARK 41.2. One can address the problem of the multiplication of two matrices in the classes $\mathcal{M}^s_{\alpha,\beta}$. With a slight modification – appending the additional factor $[1 + (|\lambda| - |\mu|)^2]^{-1}$ in the estimate (41.12) – it can be proved by direct matrix product computations (see, e.g., MEYER [1990]) that the resulting classes $\widetilde{\mathcal{M}}_{\alpha,\beta}$ (again for

$s = 0$) are algebras of operators. In contrast, these classes are not stable by inversion: it is false in general that an invertible matrix in $\mathcal{M}^s_{\alpha,\beta}$ or $\widetilde{\mathcal{M}}_{\alpha,\beta}$ has its inverse in the same class (see TCHAMITCHIAN [1996a] for some explicit counter-examples).

Our next step is to show that the estimate (41.12) allows to compress the matrices in the class $\mathcal{M}^s_{\alpha,\beta}$ by discarding certain entries. Again, we first consider the case $s = 0$.

THEOREM 41.2. *Let $M \in \mathcal{M}_{\alpha,\beta}$ and $t < \inf\{\alpha/d, \beta/d\}$. For all $N \geqslant 0$ one can discard the entries of M in such a way that the resulting matrix M_N has N nonzero entries per rows and columns and satisfies*

$$\|M - M_N\| \lesssim N^{-t} \tag{41.16}$$

in the operator norm of $\ell^2(\Lambda)$.

PROOF. We first truncate the matrix M in scale: for a given $J > 0$, we discard $m_{\lambda,\mu}$ if $||\lambda| - |\mu|| \geqslant J$. Denoting by A_J the resulting matrix, we can use the same technique as in the proof of Theorem 41.1 (Schur lemma with weights $2^{d|\lambda|/2}$) to measure the error $\|M - A_J\|$ in operator norm. By a very similar computation, we obtain

$$\|M - A_J\| \lesssim \sum_{l \geqslant J} 2^{-\alpha l} \lesssim 2^{-\alpha J}. \tag{41.17}$$

We next truncate A_J in space, by preserving in each remaining block of A_J the entries $m_{\lambda,\mu}$ such that $d(\lambda,\mu) \geqslant k(||\lambda| - |\mu||)$ where the function k is to be determined. We denote by B_J the resulting matrix. Using again the Schur lemma in the same way as in the proof of Theorem 41.1, we evaluate the error $\|A_J - B_J\|$ by the supremum in λ of

$$\omega_\lambda^{-1} \sum_{j=|\lambda|-J}^{|\lambda|+J} \sum_{\mu \in \Lambda_j} \omega_\mu \left| b^J_{\lambda,\mu} - m_{\lambda,\mu} \right|, \tag{41.18}$$

and we obtain an estimated contribution of $2^{-\alpha J}$ for each term in j by taking $k(l) = 2^{J\alpha/\beta} 2^{l(1-\alpha/\beta)}$. The total error is thus estimated by

$$\|M - B_J\| \lesssim J 2^{-\alpha J} \tag{41.19}$$

while the number of nonzero entries per rows and columns in B_J is estimated by $N(J) \lesssim \sum_{l=0}^{J} k(l)^d$. In the case where $\alpha > \beta$ (resp. $\beta < \alpha$), this sum is dominated by the first term $k(0)^d = 2^{Jd\alpha/\beta}$ (resp. last term $K(J) = 2^{dJ}$). In the case $\alpha = \beta$ we obtain $N(J) \lesssim J 2^{dJ}$. In all cases, it follows from the evaluation of $N(J)$ and of the error (41.19) that

$$\|M - B_J\| \lesssim N(J)^{-t}. \tag{41.20}$$

if t is such that $t < \inf\{\alpha/d, \beta/d\}$. Since J ranges over all positive integers, this is enough to conclude the proof. □

We can derive simple consequences of this result concerning the sparsity of the operators in the classes $\mathcal{M}^s_{\alpha,\beta}$, by the same considerations as for the study of boundedness properties: for $M \in \mathcal{M}^s_{\alpha,\beta}$, we apply the compression process of Theorem 41.2 to the preconditioned matrix $\widetilde{M}$ defined by (41.14). Denoting by $\widetilde{M}_N$ the compressed matrix, we then define $M_N = D_s \widetilde{M}_N D_s$. This new matrix has also N entries per rows and columns and approximates M in the sense expressed by the following corollary.

COROLLARY 41.2. *Let $M \in \mathcal{M}^s_{\alpha,\beta}$ and $t < \inf\{\alpha/d, \beta/d\}$. For all $N \geqslant 0$ one can discard the entries of M in such a way that the resulting matrix M_N has N nonzero entries per rows and columns and satisfies*

$$\|M - M_N\| \lesssim N^{-t} \tag{41.21}$$

in the norm of operators from $\ell^2_s(\Lambda)$ to $\ell^2_{-s}(\Lambda)$.

Our last result concerns the application of sparse matrices of the type that we have introduced in this section on sparse vectors that result of the multiscale discretization of functions. In the context of nonlinear approximation theory, the sparsity of such a vector is precisely described by the rate of decay of the error of N-term approximation: an infinite vector U has a degree of sparsity $t > 0$ in some metric X, if there exists a sequence of vectors $(U_N)_{N\geqslant 0}$ such that U_N has N nonzero coordinates and such that

$$\|U - U_N\|_X \lesssim N^{-t}. \tag{41.22}$$

In the case where $X = \ell^2$, the vectors U_N are simply obtained by keeping the N largest coordinate in U. In this case, we have also remarked that the property (41.22) is equivalent to $U \in \ell^p_w$ with $1/p = 1/2 + t$.

THEOREM 41.3. *The matrices $M \in \mathcal{M}_{\alpha,\beta}$ define bounded operator in $\ell^2 \cap \ell^p_w$ for $1/p = 1/2 + t$ and $t < \min\{\alpha/d, \beta/d\}$. In other words a vector U of sparsity t in ℓ^2 is mapped by M onto a vector $V = MU$ with the same property.*

PROOF. Our proof will directly construct an N-term approximation to $V = MU$ from the N-term approximation of U. For $j \geqslant 0$, we denote by U_j the vector that consists of the 2^j largest coordinates of U. From the assumptions we know that

$$\|U - U_j\|_X \lesssim 2^{-tj}. \tag{41.23}$$

Fixing $r \in]t, \min\{\alpha/d, \beta/d\}[$, we can define according to Theorem 41.2 truncated operators M_j such that M_j has at most $2^{(1-\varepsilon)j}$ nonzero entries per rows and columns with $\varepsilon > 0$ and

$$\|M - M_j\| \lesssim 2^{-rj}, \tag{41.24}$$

according to Theorem 41.2. We define an approximation to $V = MU$ by

$$V_j := A_j U_0 + A_{j-1}(U_1 - U_0) + \cdots + A_0(U_j - U_{j-1}) \\ = A_j U_0 + \sum_{l=1}^{j} A_{j-l}(U_l - U_{l-1}). \tag{41.25}$$

We can thus evaluate the number of nonzero entries of V_j by

$$N(j) \leqslant 2^{(1-\varepsilon)j} + \sum_{l=1}^{j} 2^{(1-\varepsilon)j-l} 2^{l-1} \lesssim 2^j. \tag{41.26}$$

Finally, we can evaluate the error of approximation as follows.

$$\begin{aligned} \|V - V_j\| &= \|M(U - U_j) + \sum_{l=0}^{j-1} (M - M_l)(U_{j-l} - U_{j-l-1}) + (M - M_j)U_0 \\ &\leqslant \|M\|\,\|U - U_j\| + \sum_{l=0}^{j-1} \|M - M_l\|\,\|U_{j-l} - U_{j-l-1}\| \\ &\quad + \|M - M_j\|\,\|U_0\| \\ &\lesssim 2^{-tj} + 2^{-tj} \sum_{l=0}^{j-1} 2^{(t-r)l} + 2^{-rj} \lesssim 2^{-tj} \lesssim N(j)^{-t}. \end{aligned}$$

Since j ranges over all possible integers, we have thus proved that $V \in \ell^p_w$. □

It is interesting to note that the simple choice $V_j = M_j U_j$ in the above proof does not give the optimal decay estimate. The particular construction of the approximation by (41.25) corresponds to the intuitive idea that the operator should be truncated according to the size of the component of U to which it is applied.

We can again derive immediate corollary, obtained by the same considerations as Corollaries 41.1 and 41.2.

COROLLARY 41.3. *Let $M \in \mathcal{M}^s_{\alpha,\beta}$ and U a vector of sparsity t in ℓ^2_s with $s < \inf\{\alpha/d, \beta/d\}$. Then $V = MU$ has sparsity t in the dual space ℓ^2_{-s}.*

REMARK 41.3. In this section, we only have addressed sparsity in the sense where the error is measured in ℓ^2 or ℓ^2_s. In a similar way it is possible to derive similar results with ℓ^p and ℓ^p_s spaces (which corresponds to analyze the error in Besov spaces that differ from H^s). The Hilbertian theory is of particular interest, due to its direct applications to the discretization of elliptic problems (in particular for the a posteriori analysis which will be evoked in Section 43).

REMARK 41.4. Another technique of matrix compression is based on the so-called nonstandard representation, which was introduced in BEYLKIN, COIFMAN and ROKHLIN [1991], in relation with multipole algorithms (GREENGARD and ROKHLIN [1987]) and pannel clustering techniques (HACKBUSCH and NOWAK [1989]). The nonstandard representation is derived from an initial (usually full) matrix M_J representing the discretization of an operator in a nodal basis at resolution 2^{-J}. One views this matrix as the "restriction" of an operator A onto a space V_J, in the sense that its entries $m^J_{\lambda,\mu}$ are identified to the inner products $\langle A\varphi_\mu, \tilde{\varphi}_\lambda\rangle$, $\lambda, \mu \in \Gamma_J$. Thus M_J represents the operator $P_J A P_J$, which can be decomposed according to

$$\begin{aligned} P_J A P_J &= (P_{J-1} + Q_{J-1})A(P_{J-1} + Q_{J-1}) \\ &= P_{J-1}AP_{J-1} + P_{J-1}AQ_{J-1} + Q_{J-1}AP_{J-1} + Q_{J-1}AQ_{J-1} \\ &= \cdots = P_0AP_0 + \sum_{j=0}^{J-1}[P_{j-1}AQ_{j-1} + Q_{j-1}AP_{j-1} + Q_{j-1}AQ_{j-1}]. \end{aligned}$$

The members of this decomposition are represented by the coarse matrix

$$\big(\langle A\varphi_\mu, \tilde{\varphi}_\lambda\rangle\big)_{\lambda,\mu\in\Gamma_0}$$

and by the matrices

$$\big(\langle A\psi_\mu, \tilde{\varphi}_\lambda\rangle\big)_{\lambda\in\Gamma_j,\mu\in\Lambda_j}, \quad \big(\langle A\varphi_\mu, \tilde{\psi}_\lambda\rangle\big)_{\lambda\in\Lambda_j,\mu\in\Gamma_j} \quad \text{and} \quad \big(\langle A\psi_\mu, \tilde{\psi}_\lambda\rangle\big)_{\lambda\in\Lambda_j,\mu\in\Lambda_j}$$

at intermediate scales. Note that these matrices are obtained from M_J by applying a standard (isotropic) bivariate wavelet decomposition: the resulting representation has the same organization as the decomposition of an image (see Figs. 5.6 and 5.7). We can thus expect that, similar to the standard wavelet discretization that was discussed in the present section, many entries are numerically small and can be thresholded in order to compress the action of A, provided that its has some smoothness properties. It should be well understood that such a representation is not a change of basis since its application on a vector requires the redundant data of both its approximation and wavelet coefficients at levels $0, \ldots, J-1$. In turn one cannot hope to combine this type of operator compression with a nonlinear approximation of the vector (as we did in the case of standard multiscale discretization) since the approximation coefficients are in general fully populated. On the other hand, a particular advantage of the nonstandard form is that in many practical cases, e.g., Calderon–Zygmund operators, the structure of the compressed matrices consists in a set of uniformly banded square matrix, simpler and sparser than the finger-like structure encountered with standard wavelet discretizations. We refer to MEYER [1997] for a detailed analysis of operator compression algorithms based on the nonstandard representation.

REMARK 41.5. Wavelet-based multiscale representations of operators can also be applied to matrix reduction problems in the context of homogenization: an operator acting on coarse scale is derived from a fine scale operator by multilevel reduction technique, in order to solve a problem on a coarse grid with a "memory" of fine grid information that may affect the behaviour of the coarse solution, e.g., highly oscillating coefficients. These methods have been introduced in BEYLKIN and BREWSTER [1995] (see also GILBERT [1998] for a comparison with classical homogenization schemes).

42. Besov regularity theory for PDE's

The goal of this section is to present instances of PDE's for which the solution can be proved to have a high smoothness in the sense of nonlinear approximation rate, while its classical smoothness is poor, due to the presence of isolated singularities. In such instances, there is an important potential gain in switching from uniform to adaptive discretization.

We insist however on the substantial gap between the evidence of high order sparsity of the solution in the wavelet basis and the effective implementation of an adaptive scheme which indeed provides the predicted nonlinear approximation rate. The results that we mention here should thus be viewed as "challenges for numericians", since the error rates that they predict are theoretically achievable by smart adaptive schemes.

Example 42.1. Hyperbolic conservation laws

It is well known that the solutions $u(x,t) = (u_1(x,t),\dots,u_m(x,t))$ of initial value problems of the type

$$\begin{aligned} &\frac{\partial u_i}{\partial t} + \operatorname{div}\big[A(u)\big] = 0, \quad x \in \mathbb{R}^n,\ t \geqslant 0, \\ &u_i(x,0) = u_{i,0}(x), \quad i = 1,\dots,m, \end{aligned} \tag{42.1}$$

can develop singularities in finite time even when the functions $u_{i,0}$ and A are smooth, if the flux A is not a simple linear transformation (in which case the solution is simply advected). The simplest prototype of such a situation is furnished by the Burger equation for which $m = n = 1$ and $A(u) = u^2/2$.

When $u(x,t)$ is no more C^1, the solution of (42.1) can only be meant in the distribution sense and is not necessarily unique. In the case of scalar conservation laws, i.e. when $m = 1$, uniqueness can be restaured by appending suitable *entropy conditions* to the equation. We refer to the monograph by LAX [1972] for a synthetic survey of these fundamental issues.

The generation of shocks is clearly a problem for the approximation of solutions by numerical schemes (see LEVEQUE [1992] for an introduction to this aspect of conservation laws): a uniform discretization with mesh size h can at most yield $\|u(\cdot,t) - u_h(\cdot,t)\|_{L^1} \lesssim h$ (with $u_h(\cdot,t)$ the approximate solution at time t), since solutions generally fail to have $W^{1,1}$ smoothness, but still lie in the space BV (functions with bounded variation, i.e. such that $\|u(\cdot,t) - u(\cdot+h,t)\|_{L^1} \leqslant C|h|$) under reasonable assumptions in the case of scalar conservation laws. If we now intend to approximate u on an adaptive grid, smoothness should be reconsidered in $B^s_{p,p}$, $1/p = 1 + s/n$. In the case of scalar univariate conservation laws, i.e. $m = n = 1$, this issue was addressed in DEVORE and LUCIER [1990] with the following striking result.

THEOREM 42.1. *Assume that the flux function A is C^∞ and strictly convex, and that the initial data u_0 is in $B^s_{p,p}$ for some $s > 0$ and $1/p = 1 + s$. Then for all time $t > 0$, the function $u(\bullet, t)$ remains in the space $B^s_{p,p}$.*

The way of proof to the theorem of DeVore and Lucier can be understood at the intuitive level by considering the simple case of the Burger's equation (i.e. $A(u) = u^2/2$) with the smoothness parameter $s \leqslant 2$. For such range of smoothness, $B^s_{p,p}$ can exactly be characterized by the decay property of approximation by free knot piecewise affine functions (without smoothness constraints). One then remarks that the evolution operator E_t associated to the entropy solution of Burger's equation is an L^1 contraction, and that E_t maps piecewise affine function with N pieces into another piecewise affine function with at most $C(t)N$ pieces. The second property is due to the fact that the speed $A'(u) = u$ is itself affine in u, and the constant $C(t)$ takes account of the possible development of shocks and rarefaction waves. Thus if u_0 is approximated with a certain rate by piecewise affine functions $v_N \in S_N$, one obtains the same rate for $u(\bullet, t)$ with the affine functions $E_t v_N \in S_{C(t)N}$, i.e. $u(\bullet, t)$ has the same smoothness as u_0 in the scale Besov spaces associated to nonlinear approximation by piecewise polynomials in L^1.

It is interesting to note here that the proof goes through the study of a specific approximation procedure. From the remarks of Section 40, one can also conclude from Theorem 42.1 that other approximation tools, such as wavelets or rational functions, might yield high order schemes for conservation laws. A multivariate analog of this result is still an open problem (one should expect a less optimistic statement, due to the anisotropic structures of shocks in several dimensions).

Example 42.2. Elliptic operators equations

We already saw in Section 34 of Chapter III that a large class of elliptic equations can be preconditioned through multiscale Galerkin discretization. We again place ourselves in this general setting, recalling (34.1) to (34.7). In particular, we assume that the energy norm $\|\bullet\|_a$ is equivalent to a Sobolev norm $\|\bullet\|_H$, with the multiscale characterizations (34.6) and (34.7).

The use of adaptive multiscale methods raises the following questions: (i) what are the possible sources of singularities in elliptic equations? (ii) Are these singularities highly regular in the scale of nonlinear approximation for some specific metric?

Since the Galerkin method produces a near best approximation of u by a function of the trial space in the norm $\|\bullet\|_H$, it is natural to compare linear and nonlinear approximation spaces associated to the specific Sobolev space H. In the most familiar cases of a second-order elliptic operator A, e.g., $A = \Delta$, the results that we have stated in Chapter III as well as in Section 38 of the present chapter allow us to give a precise description of these spaces.

(i) *Uniform case*: if u has smoothness H^{1+t} then uniform discretizations produce solutions u_h described by $N = N(h)$ parameters such that

$$\|u - u_h\|_{H^1} \lesssim N^{-t/d}. \tag{42.2}$$

(ii) *Adaptive case*: if u has smoothness $B^{1+t}_{p,p}$, $1/p = 1/2 + t/d$, then for all N, there exist an adaptive space $V_E = \text{Span}\{\psi_\lambda;\ \lambda \in E\}$ with $|E| = N$ such that

$$\|u - u_E\|_{H^1} \lesssim N^{-t/d}. \tag{42.3}$$

In the second case, an additional problem is to construct an adaptive index set $E = E(N)$ such that (42.3) holds. We postpone this crucial point to Section 43. For the time being, our main point is to compare the smoothness of the solution u in the two above senses. We list below the various possible sources of singularities for the solutions of linear elliptic equations and the known facts about their linear and nonlinear approximation properties.

(i) *Singularities generated by the interior data*: if the data f has poor smoothness, say f does not belong to H^t for $t \geqslant t_0$, the solution of $Au = f$ will not be in H^t for $t \geqslant t_0 + 2s$ where $2s$ is the order of the operator A. It may occur that the data f has higher smoothness in the sense of nonlinear approximation in some metric, in which case one might also expect a higher smoothness for the solution in this sense. In other words, we raise the question: do sparse data generate sparse solutions? The answer to this question relies on the sparsity of the inverse operator A^{-1} in its multiscale discretization. In the most favorable cases, estimates on the kernel of A^{-1} show that the matrix of this operator in a wavelet basis belongs to the class $\mathcal{M}^{-s}_{\alpha,\beta}$ (defined in Section 41), where α and β can be made arbitrarily large by increasing the smoothness and number of vanishing moments of the basis functions. This is the case (with $s = 1$) when $A = \Delta$ with homogeneous Dirichlet conditions on $\partial\Omega$ and Ω is a C^∞ domain. In such cases, we derive from Corollary 41.3 that if the data f is sparse in H^* in the sense that $\text{dist}_{H^*}(f, S_N) \lesssim N^{-t}$, then the solution u will be sparse in H in the sense that $\text{dist}_H(f, S_N) \lesssim N^{-t}$ (assuming here that the wavelet basis has enough smoothness and oscillations to ensure the estimate (41.12) on the coefficients of A^{-1} with $t < \inf\{\alpha/d, \beta/d\}$). In less favorable cases, other sources of singularities (that we describe below) might perturbate the sparsity of A^{-1}.

(ii) *Singularities generated by the boundary data*: if we append a non homogeneous boundary condition $Bu = g$ on $\partial\Omega$, a singular behaviour of g can perturbate the smoothness of the solution u. Intuitively, one expects to remedy this problem by local refinement of the interior mesh near the location of the singularity on the boundary. This approach can be precisely analyzed in simple situation: consider the example of the Laplace equation $-\Delta u = f$ on the square $\Omega = [0, 1]^2$ with Dirichlet data $u = g$ on $\partial\Omega$. In such a case, it is shown in COHEN and MASSON [1999] that a tensor product multiscale decomposition of $[0, 1]$ can be coupled together with a multiscale decomposition adapted to the boundary, in the sense where each wavelet ψ^b_λ is the trace of an interior wavelet $\psi^i_\mu(\lambda)$. Such a coupling allows to derive a natural lifting of $g = \sum_{\lambda \in \Lambda^b} g_\lambda \psi_\lambda^b$ into $Lg = \sum_{\lambda \in \Lambda^b} g_\lambda \psi_\lambda^i$. From the function space characterization results for both interior and boundary bases, one can derive that L is stable from H^s to $H^{s+1/2}$ (for a range of s that depends on the wavelet bases), and that L preserves sparsity: if $\|g - g_N\|_{H^{1/2}} \lesssim N^{-t}$ for some N term approximation on the boundary, then Lg_N is an N-term approximation of Lg which satisfies $\|Lg - Lg_N\|_{H^1} \lesssim N^{-t}$. Since $u = Lg + v$ where v is solution of $-\Delta v = f + \Delta(Lg)$ with homogeneous Dirichlet boundary condition, we are back to the previous case, i.e. facing a (sparse) singularity in the interior data.

(iii) *Singularities generated by the coefficients in the operator*: the presence of singularities in the coefficients, e.g., a jump in $a(x)$ for an operator $u \mapsto \nabla(a(x)\nabla u)$ also perturbates the smoothness of u. The analysis of the sparsity of u, in particular when the coefficients have themselves a sparse wavelet decomposition is so far not well understood, although some first results are available in DAHLKE [1998]. An additional difficulty is that such singularities in the coefficients also affects the sparsity of the operator A and the possibility of compressing it (by reducing the coefficient α in the estimate (41.12)).

(iv) *Singularities generated by the geometry of the domain*: it is well known that the solution u of an elliptic problem with smooth data f and g and coefficients $a_i(x)$ in A can still be singular if the geometry of the domain is not smooth. Are these singularities still highly regular in the sense of nonlinear approximation? In certain cases for which the nature of the singularity can be precisely elucidated, one can give a clear answer to this question. As an example, consider the Laplace equation $-\Delta u = f$ on a polyhedral domain by measuring the smoothness of these singularities in the scale of Besov spaces related to nonlinear approximation in H^1. Then it is known (see, e.g., GRISVARD [1983]) that for a data $f \in H^m$, the solution can be decomposed into $u = u_r + u_s$, where the regular part u_r belongs to H^{m+2} while u_s is a finite linear combination of functions s_{n,v_i}, $n = 1, \dots, n_{\max} = n_{\max}(m)$, which are singular at the corresponding vertex v_i of $\partial\Omega$ in the sense that they do not belong to H^{m+2}. With a conformal change of coordinate that maps v_i to the origin and such that Ω locally agrees with the cone $C = \{(r, \omega);\ r > 0 \text{ and } 0 < \omega < \theta\}$, the singular functions are given in polar coordinates by $s_n(r, \omega) = r^{n\pi/\theta} \sin(n\pi\omega/\theta)$ near the origin. One easily checks that such functions always belong to $B^{3/2+t}_{p,p}$ for all $t > 0$ and p such that $1/p = 1/2 + t/2$. In turn they also belong to all spaces $B^{1+t}_{p,p}$, $1/p = 1/2 + s/2$, and thus do not perturbate the rate of nonlinear approximation of the solution in H^1 (see DAHLKE [1999] for a more detailed analysis). Less favorable case include polyhedrons which develop anisotropic singularities along the edges (and thus require better adapted anisotropic refinements), as well as general Lipschitz domains which may develop singularities along the whole boundary. Several results dealing with this last case can be found in DAHLKE and DEVORE [1996].

43. Toward wavelet adaptive schemes

In this last section, we shall give a brief account of several existing wavelet-based adaptive schemes, which are currently the object of intensive theoretical and practical study.

Nonlinear approximation provides a natural *benchmark* for an adaptive scheme: if $\|\cdot\|_X$ is the norm where in which we measure the error between the solution of a PDE and its numerical approximation, and if the solution u is such that $\|u - u_N\|_X \lesssim N^{-s}$ for a family of N-term wavelet approximations $u_N \in \Sigma_N$ (e.g., obtained by thresholding the coefficients of u), then an *optimal adaptive scheme* should produce N-terms approximate solutions $\tilde{u}_N \in \Sigma_N$ such that one also has $\|u - \tilde{u}_N\|_X \lesssim N^{-s}$. On a computational point of view, the operation cost in evaluating $\tilde{u}_N$ should remain within order N. Note that this last property was fulfilled for multiscale elliptic solver described in Example 34.4, however in the context of uniform discretizations. Our present goal is more ambitious since the algorithm should also provide the optimal adaptive discretization. It should be noted that the norm X in which one can hope for an optimal error estimate is often dictated by the problem at hand: e.g., in the case of an elliptic problem, this will typically be a Sobolev norm equivalent to the energy norm (e.g., the H^1 norm for a second-order problem).

In most instances, the above ideal goal is far from being reached, either by the wavelet-based schemes that we shall evoke here, or by more classical finite differences or finite element adaptive schemes. In the latter, a crucial tool for adaptivity is *a posteriori analysis*, i.e. the derivation of *local error indicators* from the currently computed solution. These indicators are then used to locally refine (or derefine) the discretization at the next computation step.

Wavelet-based schemes have been developed following the empirical idea that local error indicators are directly given by the size of the currently computed wavelet coefficients: a large coefficient indicates important fluctuations of the solution on the support of the corresponding wavelet, and suggests to refine the approximation by adding wavelets at finer scales in this region. This idea was firstly introduced in MADAY, PERRIER and RAVEL [1991] for the discretization of nonstationary initial value problems. More recently, it was also applied to stationary problems for which the possibility of computing a residual allows to derive more precise a posteriori error indicators from the computed wavelet coefficients.

Most existing wavelet adaptive schemes have in common the following general structure. At some step n of the computation, a set E_n is used to represent the numerical solution

$$u_{E_n} = \sum_{\lambda \in E_n} d_\lambda^n \psi_\lambda. \tag{43.1}$$

In the context of a *nonstationary initial value problem* of the type

$$\partial_t u = F(u), \qquad u(x, 0) = u_0(x), \tag{43.2}$$

the numerical solution at step n is typically an approximation to u at time $n\Delta t$ where Δt is the time step of the resolution scheme. In the context of an *stationary problem* of the type

$$F(u) = 0, \tag{43.3}$$

the numerical solution at step n is typically an approximation to u which should converge to the exact solution as n tends to $+\infty$. In both cases, the derivation of $(E_{n+1}, u_{E_{n+1}})$ from (E_n, u_{E_n}) goes typically in three basic steps:

(i) *Refinement*: a larger set $\tilde{E}_{n+1}$ with $E_n \subset \tilde{E}_{n+1}$ is derived from an *a posteriori* analysis of the computed coefficients d_λ^n, $\lambda \in E_n$.

(ii) *Computation*: an intermediate numerical solution $u_{\tilde{E}_{n+1}} = \sum_{\lambda \in \tilde{E}_{n+1}} d_\lambda^{n+1} \psi_\lambda$ is computed from u_n and the data of the problem.

(iii) *Coarsening*: the smallest coefficients of $\tilde{u}_{n+1}$ are thresholded, resulting in the new approximation $u_{E_{n+1}} = \sum_{\lambda \in E_{n+1}} d_\lambda^{n+1} \psi_\lambda$ supported on the smaller set $E_{n+1} \subset \tilde{E}_{n+1}$.

We shall now describe these strategies in more details, for problems of the type (43.2) and (43.3).

Example 43.1. Nonstationary problems: dynamical adaptivity

The idea of dynamical adaptivity proposed in MADAY, PERRIER and RAVEL [1991] applies to general class of initial value problems of the type (43.2) where F is possibly a nonlinear operator. For the sake of simplicity we shall describe this idea in the context of an explicit one step forward Euler scheme: the numerical solution $u_{E_{n+1}}$ approximating $u(\bullet, (n+1)\Delta t)$ will be computed explicitly from the numerical solution u_{E_n} at the previous time step.

Before discussing the derivation of $(u_{E_{n+1}}, E_{n+1})$ from (u_{E_n}, E_n), a few words are in order concerning the initialization of the scheme: ideally, we can obtain an optimal adaptive expansion u_{E_0} of the initial value data u_0 into a linear combination of wavelets by a thresholding procedure on its global expansion, i.e.

$$u_{E_0} = \sum_{\lambda \in E_0} d_\lambda^0 \psi_\lambda, \quad E_0 := \{\lambda \text{ s.t. } \|d_\lambda^0 \psi_\lambda\|_X \geqslant \eta\}, \tag{43.4}$$

where X is some prescribed norm in which we are interested in controlling the error, η a prescribed threshold and $d_\lambda^0 := \langle u_0, \tilde{\psi}_\lambda \rangle$ are the wavelet coefficients of u_0. In practice, we cannot compute all the values of these coefficients, and one thus needs a more reasonable access to a compressed representation. This is typically done through some *a priori* analysis of the initial value u_0. In particular, if u_0 is provided by an analytic expression, or if we have some information on the local size of its derivatives, estimates on the decay of wavelet coefficients, such as in Remark 29.5, can be used to avoid the computation of most details which are below threshold. With such a strategy, we expect to obtain E_0 and $(u_\lambda^0)_{\lambda \in E_0}$ with a memory and computational cost which is proportional to $\#(E_0)$.

Then, assuming that at time $n\Delta t$ the approximate solution u_{E_n} has the form (43.1) for some set E_n of coefficients, the problem is thus both to select a correct set of indices E_{n+1} as well as to compute the new coefficients d_λ^{n+1} for $\lambda \in E_{n+1}$. As we already explained, this is done by (i) refining E_n into an intermediate set $\tilde{E}_{n+1}$ which is well fitted to describe the solution at time $(n+1)\Delta t$, (ii) computing $u_{\tilde{E}_{n+1}}$ supported by $\tilde{E}_{n+1}$ and

(iii) deriving $(u_{E_{n+1}}, E_{n+1})$ from $u_{\tilde{E}_{n+1}}$ by a thresholding process similar to (43.4). The selection of the intermediate set $\tilde{E}_{n+1}$ should thus take into account the effect of the nonlinear transformation F on the sparse expansion (43.1), integrated between $n\Delta t$ and $(n+1)\Delta t$. A procedure being prescribed for the refinement of E_n into $\tilde{E}_{n+1}$, several strategies are available for computing $u_{\tilde{E}_{n+1}}$ from u_{E_n}, such as Petrov–Galerkin methods in MADAY, PERRIER and RAVEL [1991] or collocation methods in BERTOLUZZA [1995b, 1997]. The relevance and efficiency of such strategies are strongly dependent of the given problem at hand. In the following we want to focus on a systematic approach based on the *discrete framework* described in Section 21.

Recall that such a framework allows to consider wavelet decompositions of point values or cell averages discretizations. In this context, one can hope to combine the above ideas of dynamical adaptivity with existing finite difference or finite volume schemes which are proved or known to perform well: the main idea is to start from such a scheme which operates on discretizations U_J^n of $u(\bullet, n\Delta t)$ at scale 2^{-J}, and accelerate the scheme by applying dynamical adaptivity on a discrete multiscale decomposition of U_J^n at intermediate scales $j = 0, \ldots, J$. Before explaining this strategy in more detail, let us point out that imposing a limitation on the scale in explicit dynamically adaptive scheme is usually necessary for *stability*: in the context of linear or nonlinear convection terms, the time-step Δt is tied to the highest level J of refinement contained in the sets E_n according to the so-called CFL condition which imposes that $\Delta t \lesssim c\Delta x \sim c2^{-J}$ where c is the local speed of convection. For a given time step, this constraint typically imposes a limitation on the scales contained in E_n, i.e. fixes some highest discretization level J.

Our starting point is thus a given one-step explicit scheme which computes discretized versions $U_J^n = (u_\gamma^n)_{\gamma \in \Gamma_j}$ of $u(\bullet, n\Delta t)$ at scale 2^{-J} according to a general formula of the type

$$U_J^{n+1} = U_J^n + B_J^n, \tag{43.5}$$

where the increment $B_J^n = (b_\gamma^n)_{\gamma \in \Gamma_j}$ depends locally on the numerical solution, i.e. can be expressed as

$$b_\gamma^n := \tilde{F}\big(u_\mu^n;\ \mu \in I(\gamma)\big), \tag{43.6}$$

where $I(\gamma)$ represents the computation stencil attached to γ, and $\tilde{F}$ is a discrete approximation to the action of F integrated between $n\Delta t$ and $(n+1)\Delta t$. We denote by F_J the corresponding discrete evolution operator (possibly nonlinear) associated to this scheme, in the sense that we have

$$U_J^n = F_J U_J^{n-1} = \cdots = F_J^n U_J^0. \tag{43.7}$$

Denoting by $\overline{U}_J^n$ the discretization of the exact solution $u(\bullet, n\Delta t)$, we can measure the error in some prescribed metric X by

$$e_n := \big\| U_J^n - \overline{U}_J^n \big\|_X. \tag{43.8}$$

The idea of now to use a dynamically adaptive strategy in order to produce adaptive solutions of the form

$$V_J^n = \sum_{\lambda \in E_n} d_\lambda^n \Psi_{J,\lambda}, \tag{43.9}$$

where the $\Psi_{J,\lambda}$ are the discrete wavelets introduced in (21.4). Such solutions will be computed and described in terms of their wavelet representations $(d_\lambda^n)_{\lambda \in E_n}$. If the exact solution has a sparse wavelet representation, we can hope for significant computation and memory saving in comparison to the reference scheme (43.5), while the additional error

$$a_n := \|U_J^n - V_J^n\|_X \tag{43.10}$$

remains within a prescribed accuracy. It is natural to fix this accuracy of the same order of magnitude as the error estimate available for e_n.

Assume that we are given V_J^n through the data of $(d_\lambda^n)_{\lambda \in E_n}$, and that we have fixed the intermediate set $\tilde{E}_{n+1}$. The question is now to define and compute the coefficients $(d_\lambda^{n+1})_{\lambda \in \tilde{E}_{n+1}}$ of the adaptive solution at the next time step. A natural choice is to define them as the coefficients of $F_J V_J^n$ restricted to $\tilde{E}_{n+1}$. Therefore $(d_\lambda^{n+1})_{\lambda \in \tilde{E}_{n+1}}$ could be obtained from $(d_\lambda^n)_{\lambda \in E_n}$ as follows: (i) Reconstruct V_J^n on the fine grid, (ii) apply the discrete evolution operator F_J, and (iii) restrict the multiscale decomposition of $F_J V_J^n$ to $\tilde{E}_{n+1}$. However, applying such a procedure at each time step would the useless since its computational and memory cost is at least similar to the reference scheme (43.5). We therefore seek for a more direct mapping between $(d_\lambda^n)_{\lambda \in E_n}$ and $(d_\lambda^{n+1})_{\lambda \in \tilde{E}_{n+1}}$.

Such a mapping is particularly easy to derive in the context of *point value* multiscale decompositions based on the decimation operator (21.20). In this case, the wavelet indices λ identify with points of Γ_J and a set $E \subset \Gamma_J$ of such indices thus identifies to an "adaptive mesh". Moreover, if we impose that the set E at least contain the coarsest grid Γ_0, it is easily checked that there exists there is a one to one correspondence between the restricted details $(d_\lambda)_{\lambda \in E}$ and the restricted point values $(v_\gamma)_{\gamma \in E}$ of a vector $V_J = (v_\gamma)_{\gamma \in \Gamma_J}$. This fact is illustrated, e.g., on Fig. 5.3, in the case of the pointvalue decomposition associated to the Schauder hierarchical basis.

This allows to propose the following strategy for a more economical derivation of $(d_\lambda^{n+1})_{\lambda \in \tilde{E}_{n+1}}$: (i) From $(d_\lambda^n)_{\lambda \in E_n}$ the point values v_μ^n of V_J^n restricted to those μ such that $\mu \in I(\gamma)$ for some $\gamma \in \tilde{E}_{n+1}$, where $I(\gamma)$ is the computation stencil associated to γ in the reference scheme, (ii) use these values to compute the point values of $F_J V_J^n$ on the restricted set $\tilde{E}_{n+1}$ according to (43.5) and (43.6), and (iii) from these point values, compute the restricted details $(d_\lambda^{n+1})_{\lambda \in \tilde{E}_{n+1}}$. In contrast to the first procedure, the computational and memory cost is then no more tied to the cardinality of the finest mesh but essentially proportional to $\#(\tilde{E}_{n+1})$.

Once the $(d_\lambda^{n+1})_{\lambda\in\tilde{E}_{n+1}}$ are computed, V_J^{n+1} and E_{n+1} can be derived by a thresholding procedure of the same type as (43.4), namely

$$\begin{aligned} V_J^{n+1} &= \sum_{\lambda\in E_{n+1}} d_\lambda^{n+1}\Psi_{J,\lambda},\\ E_{n+1} &:= \big\{\lambda\in\tilde{E}_{n+1} \text{ s.t. } \big\|d_\lambda^{n+1}\Psi_{J,\lambda}\big\|_X \geqslant \eta\big\}. \end{aligned} \tag{43.11}$$

Denoting by T_E the truncation operator which maps a vector $V_J=\sum_{|\lambda|\leqslant J} d_\lambda\Psi_{J,\lambda}$ to $T_E V_J := \sum_{\lambda\in E} d_\lambda \Psi_{J,\lambda}$, we can thus summarize the evolution of the adaptive solution by

$$V_J^{n+1} = T_{E_{n+1}} T_{\tilde{E}_{n+1}} F_J V_J^n. \tag{43.12}$$

In the case where the scheme is X-stable in the sense that for some $\alpha > 0$,

$$\|F_J U - F_J V\|_X \leqslant (1+\alpha\Delta t)\|U-V\|_X, \tag{43.12}$$

this identity allows a "cumulative analysis" of the additional error a_n. For this purpose we define the truncation errors

$$b_n := \big\|V_J^{n+1} - T_{\tilde{E}_{n+1}} F_J V_J^n\big\|_X \tag{43.13}$$

and

$$c_n := \big\|T_{\tilde{E}_{n+1}} F_J V_J^n - F_J V_J^n\big\|_X. \tag{43.14}$$

Assuming that these truncation errors are majorized by some prescribed $\varepsilon > 0$, we then obtain at time $T=(n+1)\Delta t$ the estimate

$$\begin{aligned} a_{n+1} &= \big\|U_J^{n+1} - V_J^{n+1}\big\|_X \leqslant \big\|F_J U_J^n - F_J V_J^n\big\|_X + b_n + c_n\\ &\leqslant (1+\alpha\Delta t)a_n + b_n + c_n \leqslant (1+\alpha\Delta t)a_n + 2\varepsilon\\ &\leqslant 2\varepsilon\sum_{k=0}^{n}(1+\alpha\Delta t)^k \leqslant 2[\alpha\Delta t]^{-1}(1+\alpha\Delta t)^{T/\Delta t}\varepsilon\\ &\leqslant C(T)(n+1)\varepsilon, \end{aligned}$$

where $C(T)$ behaves like $e^{\alpha T}$. In order to control a_n, it thus suffices to ensure that the truncation errors b_n and c_n are bounded by some prescribed ε. This is a simple task for b_n which correspond to the error produced by the last thresholding step: on fixes η to some value which ensures that

$$\sum_{\lambda\in\tilde{E}_{n+1},\ \|d_\lambda^{n+1}\Psi_{J,\lambda}\|_X\leqslant\eta} \big\|d_\lambda^{n+1}\Psi_{J,\lambda}\big\|_X \leqslant \varepsilon, \tag{43.15}$$

e.g., by taking the crude threshold $\eta=\eta(\varepsilon):=\varepsilon/N_J$ where $N_J=\#(\Gamma_J)$. It is a more delicate task to ensure a similar estimate on c_n since this corresponds to neglecting the effect of F_J outside the intermediate set $\tilde{E}_{n+1}$. In particular, such an estimate will hold if we can ensure that for $\lambda\notin\tilde{E}_{n+1}$, we have $\|d_\lambda^{n+1}\Psi_{J,\lambda}\|_X\leqslant\eta(\varepsilon)$. A proper refinement of E_n into $\tilde{E}_{n+1}$ requires thus a good understanding of the effect of F_J in the coefficient domain.

In most existing approaches, this effect is assumed to be *local*: the intermediate set has the form $\tilde{E}_{n+1}=E_n\cup F_n$ where F_n is a set of indices neighbouring E_n both in scale and in space. Typically, one fixes two "inflation" parameters C_1 and C_2 and defines the new set by

$$\tilde{E}_{n+1}=\{\lambda\in\Lambda\ ||\lambda|-|\mu||\leqslant C_1\text{ and }d(\lambda,\mu)\leqslant C_2\text{ for some }\mu\in E_n\},\tag{43.16}$$

where $d(\lambda,\mu)$ is the spatial distance defined by (41.9). We already observed in Section 41 that a large class of *linear operators* have indeed a local action in the wavelet domain, in the sense that this action can be approximated by the sparse matrix-vector multiplication (41.25) in Theorem 41.3. For such linear operators, one can invoke similar matrix compression arguments to design a local refinement procedure which ensures the required estimate on c_n. Such a procedure will be slightly more elaborate than (43.16) in the sense that the inflation parameter will depend on the *size* of the coefficients d_μ^n: in analogy with the sparse matrix-vector multiplication (41.25), large details will have a stronger influence on the refinement procedure. In this sense, the size of wavelet coefficients serve as local error indicators. Unfortunately, this type of strategy fails for most *nonlinear operators* which appear in classical PDE's. Consider as an example the simple nonlinear map $u\mapsto u^2$ which appears e.g., in Burger's equation. If ψ_λ is a wavelet, then $\psi_\lambda^2=\sum_\mu g_{\lambda\mu}\psi_\mu$ has the same support as ψ_λ and mostly consists of scales $|\mu|\leqslant|\lambda|+1$, since the effect of squaring is a self convolution in the frequency domain. However, we can expect that all scales from $|\mu|=0$ to $|\lambda|+1$ are represented since squaring destroys the oscillation in ψ_λ and thus introduces low frequencies. In order to circumvent this problem, a standard practical strategy is to impose a *tree structure* on the set $\tilde{E}_{n+1}$ which can be summarized as follows.

DEFINITION 43.1. Let $C>0$ be a fixed constant. A set E of wavelet indices is a C-tree if for all $\lambda\in E$, the indices μ such that $|\mu|\leqslant|\lambda|$ and $d(\lambda,\mu)\leqslant C$ are also contained in E.

Note that such tree structures are actually quite natural since they usually appear in the wavelet thresholding of functions with isolated singularities, as illustrated on Fig. 5.2. It should be noted that the tree structure is also necessary to derive fast transformation algorithms between restricted details $(d_\lambda)_{\lambda\in E}$ and restricted point values $(v_\gamma)_{\gamma\in E}$, *with complexity proportional to the size of the compressed data*, i.e. $\mathcal{O}(\#(E))$.

REMARK 43.1. In the case of hyperbolic equations derived from conservation laws (as considered in Section 42), the multiresolution approach developed in HARTEN [1995] can be qualified as "semi-adaptive" in the following sense: while the solution U_J^n is still described and evolved on a uniform grid, the increment B_J^n (corresponding here to the numerical flux balance vector) is replaced by its adaptive approximation $T_{E_n} B_J^n$ where E_n now represents the set of wavelet coefficients of U_J^n above some threshold. Such an approach allows some substantial saving on the computational cost in the case where the evaluation of B_J^n is the heavy part of the reference scheme.

Example 43.2. Stationary problems: a-posteriori analysis

Let us now turn to stationary problems of the type (43.3). For such problems, a classical approach to adaptive strategy is by a posteriori analysis based on the *residual* of the equation corresponding to the approximate solution: assume that $u_E = \sum_{\lambda \in E} d_\lambda \psi_\lambda$ is a Galerkin solution to (43.3) in the sense that

$$\langle F(u_E), \psi_\lambda \rangle = 0, \quad \lambda \in E. \tag{43.17}$$

We can then view each of the quantities

$$c_\lambda := |\langle F(u_E), \psi_\lambda \rangle|, \quad \lambda \notin E, \tag{43.18}$$

as reflecting the contribution to the Galerkin approximation error (expressed by the residual $F(u_E)$) of the wavelet ψ_λ which was not used in the Galerkin approximation. This suggest to use such quantities as *error indicators* and refine the set E by those $\lambda \notin E$ corresponding to the largest c_λ. Such a strategy cannot be operated as such, since the c_λ for $\lambda \notin E$ constitute an infinite sequence which should necessarily be truncated in practice, based on some *a priori* knowledge on their values.

In the framework of linear elliptic PDE's, this approach leads to a more rigorous numerical analysis of the wavelet adaptive strategy, as introduced in BERTOLUZZA [1995a] and further developed in DAHLKE, DAHMEN, HOCHMUTH and SCHNEIDER [1997]. For the sake of simplicity, we shall present these ideas in the case of a second-order elliptic equation $Au = f$ for which the underlying energy norm $\|u\|_a = \langle Au, u\rangle$ is equivalent to the H^1 norm, having in mind that they can be applied to the general setting of elliptic operator equations described in Section 34 of Chapter III.

Assuming that we have selected a first set E of indices and that we have computed the Galerkin approximation u_E of u in $V_E = \mathrm{Span}\{\psi_\lambda;\ \lambda \in E\}$, we consider the residual

$$r_E := f - Au_E, \tag{43.19}$$

which satisfies the equation $Ar_E = u - u_E$. Using the fact that A is an isomorphism from H^1 to H^{-1}, and the characterization of H^{-1} by the dual wavelet decomposition (derived from the characterization of H^1 by the primal decomposition as shown in

Section 31 of Chapter III), we now evaluate the error by

$$\|u-u_E\|_{H^1} \sim \|r_E\|_{H^{-1}} \sim \Big[\sum_{\lambda\in\Lambda} 2^{-2|\lambda|}\big|\langle r_E,\psi_\lambda\rangle\big|^2\Big]^{1/2}$$
$$=\Big[\sum_{\lambda\in\Lambda\setminus E} 2^{-2|\lambda|}\big|\langle f,\psi_\lambda\rangle-\langle Au_E,\psi_\lambda\rangle\big|^2\Big]^{1/2},$$

where we have used the fact that the residual $f-Au_E$ is by construction orthogonal to V_E. In the above sum, each term $a_\lambda=2^{-2|\lambda|}|\langle f,\psi_\lambda\rangle-\langle Au_E,\psi_\lambda\rangle|^2$ represents the contribution of the indices $\lambda\notin E$ to the error, and we would like to refine the set E into $\widetilde{E}=E\cup F$ where F corresponds to the largest a_λ.

Again, such a strategy is practically infeasible since the set $\Lambda\setminus E$ is infinite, so that the evaluation of the ℓ^2 norm of the sequence $(a_\lambda)_{\lambda\notin E}$ and of its largest component is irrealistic. Before presenting a "realistic variant" of this strategy, we want to analyze its effect on the error.

The Galerkin method can be viewed as an orthogonal projection in the energy norm so that, if we consider the new solution $u_{\widetilde{E}}$, we have

$$\|u-u_{\widetilde{E}}\|_a^2=\|u-u_E\|_a^2-\|u_E-u_{\widetilde{E}}\|_a^2. \tag{43.20}$$

Assuming that we define $\widetilde{E}$ by selecting the largest contributions a_λ in such way that

$$\sum_{\lambda\in F}|a_\lambda|^2\geqslant\sigma\Big(\sum_{\lambda\notin E}|a_\lambda|^2\Big), \tag{43.21}$$

for some fixed $\sigma\in\,]0,1[$, then we derive that

$$\begin{aligned}
\|u_E-u_{\widetilde{E}}\|_a^2 &\sim \|Au_E-Au_{\widetilde{E}}\|_{H^{-1}}^2\\
&\sim \sum_{\lambda\in\Lambda}2^{-2|\lambda|}\big|\langle Au_E,\psi_\lambda\rangle-\langle Au_{\widetilde{E}},\psi_\lambda\rangle\big|^2\\
&= \sum_{\lambda\in\Lambda\setminus E}2^{-2|\lambda|}\big|\langle Au_E,\psi_\lambda\rangle-\langle Au_{\widetilde{E}},\psi_\lambda\rangle\big|^2\\
&\geqslant \sum_{\lambda\in\widetilde{E}\setminus E}2^{-2|\lambda|}\big|\langle Au_E,\psi_\lambda\rangle-\langle Au_{\widetilde{E}},\psi_\lambda\rangle\big|^2\\
&= \sum_{\lambda\in\widetilde{E}\setminus E}2^{-2|\lambda|}\big|\langle Au_E,\psi_\lambda\rangle-\langle f,\psi_\lambda\rangle\big|^2\geqslant\sigma\Big(\sum_{\lambda\notin E}|a_\lambda|^2\Big)\\
&\sim \|u-u_E\|_a^2.
\end{aligned}$$

In summary, (43.21) implies the existence of another constant $\tilde{\sigma}$ depending only on σ and on the equivalence constants involved in the above computation, such that

$$\|u_E - u_{\widetilde{E}}\|_a^2 \geqslant \tilde{\sigma}\|u - u_E\|_a^2, \tag{43.22}$$

and thus

$$\|u - u_{\widetilde{E}}\|_a^2 \leqslant \theta\|u - u_E\|_a^2, \tag{43.23}$$

with $\theta = \sqrt{1-\tilde{\sigma}}$. We thus see that the refinement strategy based on extracting a fixed portion in the ℓ^2 norm of the indicators a_λ allows to reduce the error of a fixed amount.

REMARK 43.2. The inequality (43.22) which ensures such a fixed reduction is called the *saturation property*. In the finite element context, this property is usually assumed but not proved to hold for most automatic refinement procedures based on a posteriori error indicators, although it was recently proved to hold for some specific mesh refinement procedures in DOERFLER [1996].

As we already mentioned, one needs a more realistic strategy based on computable error indicators. The main idea is to fix a tolerance $\varepsilon > 0$ and to select first a set E such that

$$\sum_{\lambda \notin E} 2^{-2|\lambda|}\left|\langle f, \psi_\lambda\rangle\right|^2 \leqslant \varepsilon^2, \tag{43.24}$$

i.e. E is selected by a thresholding procedure on the data f. This allows to eliminates of the part containing f in the coefficients a_λ, in the sense that we now have

$$\|u - u_E\|_a^2 \sim \left[\sum_{\lambda \in \Lambda\setminus E} 2^{-2|\lambda|}\left|\langle Au_E, \psi_\lambda\rangle\right|^2\right]^{1/2} + K(\varepsilon), \tag{43.25}$$

with $K(\varepsilon) \lesssim \varepsilon$. The next step is to localize the action of the stiffness matrix R involved in the computation of the coefficients $\langle Au_E, \psi_\lambda\rangle$: using Corollary 41.2, we can obtain local approximations R_N in the sense of the operator norm from $\ell_1^2(\Lambda)$ to $\ell_{-1}^2(\Lambda)$. Since $\|u_E\|_{H^1} \lesssim \|u\|_{H^1}$ independently of E, the components of u_E are uniformly in $\ell_1^2(\Lambda)$ and we can choose $R_N = R_{N(\varepsilon)}$ in such a way that the corresponding operator $A_{N(\varepsilon)}$ satisfies

$$\sum_{\lambda \in \Lambda} 2^{-2|\lambda|}\left|\langle (A - A_{N(\varepsilon)})u_E, \psi_\lambda\rangle\right|^2 \leqslant \varepsilon^2. \tag{43.26}$$

We have now a realistic error indicator, up to the tolerance ε given by

$$\|u - u_E\|_a^2 \sim \left[\sum_{\lambda \in \Lambda\setminus E} 2^{-2|\lambda|}\left|\langle A_{N(\varepsilon)}u_E, \psi_\lambda\rangle\right|^2\right]^{1/2} + K(\varepsilon), \tag{43.27}$$

with $K(\varepsilon) \lesssim \varepsilon$. Since $A_{N(\varepsilon)}$ is local, the above sum can be limited to a finite range of indices which represent a margin $M(E,\varepsilon)$ of coefficients surrounding the initial adaptive set E. The new error estimator has now the form

$$I(E) = \left[\sum_{\lambda \in M(E,\varepsilon)} b_\lambda^2 \right]^{1/2}, \tag{43.28}$$

with $b_\lambda := 2^{-2|\lambda|} |\langle A_{N(\varepsilon)} u_E, \psi_\lambda \rangle|^2$. One then adopt the strategy of selecting the largest contribution b_λ above a fixed portion of this estimator, in the same way as it was described for the a_λ. Due to the presence of $K(\varepsilon)$ in (43.13), error reduction by a constant factor is ensured as long as this error is not of the same order as ε. We can thus start from an initial set E_0 and use this refinement technique to construct an increasing sequence $E_1, \ldots, E_n$ of adaptive multiscale discretizations with associated Galerkin solution u_{E_i}, so that we finally obtain $\|u - u_{E_n}\| \lesssim \varepsilon$.

Of course, the crucial question remains to understand how the cardinality of the sets E_j grows as the error decreases, and in particular if the relation between ε and $|E_n|$ is in accordance with the optimal rate predicted by nonlinear approximation theory. In order to reach this goal, several elaborations are needed, which improve the performance of an adaptive scheme based on such a refinement strategy.

(i) Firstly, one can start with a large tolerance ε_0 (say comparable to $\|u\|_{H^1}$) in place of the desired accuracy ε, apply the refinement strategy and reduce ε_0 by half once the error has attained this order. The procedure is then repeated until the prescribed accuracy $\varepsilon_J = 2^{-J}\varepsilon_0 \sim \varepsilon$ is reached. In turn, the computational cost is reduced since the first iterations do not involve as many term as if one started at once with the final value of ε.

(ii) Secondly, the Galerkin solutions u_{E_n} should not be computed exactly but only up to an accuracy of the same order as the current tolerance ε_j.

(iii) A natural idea is also that the truncation of the action of A should actually vary according to the size of the component of u_E to which it is applied, in a similar way as the sparse matrix-vector multiplication (41.25) in the proof of Theorem 41.3: the smallest components of the currently computed solution should not affect too much the extension of the margin $M(E)$.

(iv) Finally, derefinement procedures – in terms of thresholding – should also be regularly applied in order to ensure an optimal compression of the solution.

By the time we complete these notes, a theoretical result proved in COHEN, DAHMEN and DEVORE [2000a, 2000b] (illustrated by numerical experiments in COHEN and MASSON [1999]), reveals that the combination of these elaborations allows to obtain the rate predicted by nonlinear approximation theory, to the expense of a number of operations that stays proportional to the final number of parameters. In the case of stationary elliptic problems, one can thus achieve the optimal benchmark that we defined for an adaptive strategy: optimal N-term approximation to the expense of $\mathcal{O}(N)$ computational cost.

44. Conclusions

Multiscale decomposition into wavelet bases allow to generate optimal adaptive approximations of functions by simple thresholding procedures. The accuracy of such nonlinear approximations can be predicted from the smoothness of the function in certain scales of quasi-normed Besov spaces.

While wavelet adaptive techniques have been successfully applied to the reduction of parameters in the description of data (e.g., compression, curve and surface design, statistical estimation and denoising), their ability to compute the unknown solution of a physical process in an optimally compressed form is still to be confirmed in many instances.

The potential of these methods for a given problem can be understood through a fine analysis of the smoothness of solutions in the scales of Besov spaces characterized by the rate of nonlinear approximation. Their practical implementation requires in addition the development and the numerical analysis of efficient resolution schemes. These schemes should benefit from the many positive features of wavelets discretizations that have been described throughout this paper: high order and local approximations, fast decomposition and reconstruction algorithms, diagonal preconditioning of elliptic operators, operator compression and adaptive approximation of functions.

45. Historical notes

Since the 1960's, the study of nonlinear methods became a central issue in approximation theory. In his famous monograph on Besov spaces, J. Peetre mentions the study of nonlinear approximation spaces as a motivation to consider the case $p < 1$ (PEETRE [1974]). In the 1980's, a systematic treatment of nonlinear approximation spaces in relation with interpolation theory was achieved by R. DeVore, V. Popov and their collaborators. In particular, the nonlinear version of Theorem 28.2 was given in DEVORE and POPOV [1988a, 1988b, 1988c] with applications in the context of free knot spline approximation. Results specific to wavelets were firstly proved in DEVORE, JAWERTH and POPOV [1992], including Theorem 41.2 with a slightly different proof.

On a more practical point of view, since the pioneering work in the 1970's, e.g., BABUSHKA and RHEINBOLDT [1978], adaptive methods have been intensively developed and analyzed by numericians in the context of finite elements and finite differences. These methods has been documented in numerous contributions (see, e.g., VERFURTH [1994] and ERIKSSON, ESTEP, HANSBO and JOHNSON [1995]). A key ingredient is the derivation of *local a posteriori error indicators* from the currently computed solutions. Finite elements indicators are more "local" than the wavelet indicators of Section 43, in the sense that each of them aims to reflect the error on a specific element and is computed from the current solution in the immediate neighbourhood of this element. However it is not clear that they lead to optimal convergence in the sense of nonlinear approximation (i.e. generation of an optimal mesh).

The first adaptive wavelet method for PDE's was proposed in MADAY, PERRIER and RAVEL [1991], in the context of initial value problems. In the same period, the idea of using wavelets in order to sparsify integral operators was introduced in BEYLKIN,

COIFMAN and ROKHLIN [1991]. Since then, wavelets adaptive discretizations have been extensively implemented and analyzed in various situations. As explained in Section 43, a common point to most approaches is the idea that a basic multiscale error indicator is provided by the size of the computed wavelet coefficients. The derivation of more rigorous error indicators was proposed (for elliptic equations) in BERTOLUZZA [1995a], as well as in DAHLKE, DAHMEN, HOCHMUTH and SCHNEIDER [1997] which also proves the convergence of the refinement strategy based on such estimates.

Notations

We list below symbols that appear frequently throughout the article. If not below, their definition appears at the first place where they are introduced.

Spaces

$C^m(\Omega)$ (integer m): Spaces of m-times continuously differentiable functions
$C^s(\Omega)$ (non-integer $s > 0$): Hölder spaces
$L^p(\Omega)$: Lebesgue spaces
$W^{m,p}(\Omega)$: Sobolev spaces ($H^s := W^{s,2}$)
$B^s_{p,q}(\Omega)$: Besov spaces
Π_n: polynomials of degree n
V_j and $\widetilde{V}_j$: primal and dual multiresolution spaces at scale j
W_j and $\widetilde{W}_j$: primal and dual detail (or wavelet) spaces at scale j

Functions

$f_{j,k} := 2^{dj/2} f(2^j \bullet -k)$: L^2-scaling of f defined on $\mathbb{R}^d$
$\psi_{j,k}$ and $\tilde{\psi}_{j,k}$: primal and dual wavelet of resolution 2^{-j}
$\varphi_{j,k}$ and $\tilde{\varphi}_{j,k}$: primal and dual scaling functions of resolution 2^{-j}
φ_λ and $\tilde{\varphi}_\lambda$ for $\lambda \in \Gamma_j$: compact notations for the primal and dual scaling functions of resolution 2^{-j}
ψ_λ and $\tilde{\psi}_\lambda$ for $\lambda \in \Lambda_j$: compact notations for the primal and dual wavelets of resolution 2^{-j}
$|\lambda|$: resolution level of the index λ ($|\lambda| = j$ if $\lambda \in \Gamma_j$ or $\lambda \in \Lambda_j$)
$\langle f, g\rangle$: duality product ($\int f\overline{g}$ if $f \in L^p$ and $g \in L^{p'}$, $1/p + 1/p' = 1$)

Operators

P_j, $\widetilde{P}_j$, Q_j, $\widetilde{Q}_j$: projectors onto V_j, $\widetilde{V}_j$, W_j, $\widetilde{W}_j$
$\mathcal{F}f(\omega) = \hat{f}(\omega) := \int_{\mathbb{R}^d} f(x)\mathrm{e}^{-\mathrm{i}\langle\omega,x\rangle}\,\mathrm{d}x$: Fourier transform of f

Sets

$\#E$ (or $|E|$): cardinality of a finite set E
$\mathrm{meas}(E)$ (or $|E|$): Lebesgue's measure of a domain $E \subset \mathbb{R}^d$
$\mathrm{diam}(E)$: diameter of a domain $E \subset \mathbb{R}^d$
$\mathrm{dist}(A, B) = \inf_{x\in A, y\in B} |x - y|$: distance between the sets A and B
$\mathrm{Supp}(f)$: support of the function f

Equivalences
If $A(u)$ and $B(u)$ are positive functions of a set u of parameters, we shall often use the notation

$$A(u) \lesssim B(u),$$

to express that there exist a constant $C > 0$ such that $A(u) \leqslant CB(u)$ independently of the parameters. For example $\|P_j f\| \lesssim \|f\|$ means that the operator P_j is bounded independently of j. We also use the notation

$$A(u) \sim B(u),$$

to express that $A(u) \lesssim B(u)$ and $B(u) \lesssim A(u)$.

References

ABGRALL, R. (1995), Multiresolution analysis on unstructured meshes: Application to CFD, in: K.W. Morton and M.J. Baines, eds., *Numerical Methods for Fluid Dynamics* **V** (Oxford Science Publications).

ADAMS, R. (1975), *Sobolev Spaces* (Academic Press, New York).

ALPERT, B. (1993), A class of bases in L^2 for the sparse representation of integral operators, *SIAM J. Math. Anal.* **24**, 246–262.

ANTONINI, M., M. BARLAUD, I. DAUBECHIES and P. MATHIEU (1992), Image coding using wavelet transforms, *IEEE Trans. Image Process.* **1**, 205–220.

AVELLANEDA, M., G. DAVIS and S. MALLAT (1997), Adaptive greedy approximations, *Constr. Approx.* **13**, 57–98.

BABUSHKA, I., J. GAGO, D. KELLY and O. ZIENKIEWICZ (1982), Hierarchical finite element approaches, error estimates and adaptive refinements, in: J.R. Whiteman, ed., *The Mathematics of Finite Elements and Applications* **IV** (Academic Press, London).

BABUSHKA, I. and W.C. RHEINBOLDT (1978), Error estimates for adaptive finite element computations, *SIAM J. Numer. Anal.* **15**, 736–754.

BAKER, G.A. (1975), *Essentials of Padé Approximants* (Academic Press, New York).

BAKHVALOV, N.S. (1966), On the convergence of a relaxation method with natural constraints on the elliptic operator, *USSR Comp. Math. and Math. Phys.* **6**, 101–135.

BANK, R.E. and T. DUPONT (1981), An optimal order process for solving finite element equations, *Math. Comp.* **36**, 35–51.

BANK, R.E., A.H. SHERMAN and A. WEISER (1983), Refinement algorithms and data structures for regular local mesh refinement, in: R. Stepleman et al., eds., *Scientific Computing*, Amsterdam: IMACS (Noth-Holland, Amsterdam) 3–17.

BANK, R.E. and A. WEISER (1985), Some a posteriori error estimates for elliptic partial differential equations, *Math. Comp.* **44**, 283–301.

BARRON, A.R. (1994), Approximation and estimation bounds for artificial neural networks, *Machine Learning* **14**, 115–133.

BENNETT, C. and R. SHARPLEY (1988), *Interpolation of Operators* (Academic Press, New York).

BERGH, J. and J. LÖFSTRÖM (1976), *Interpolation Spaces* (Springer-Verlag, Berlin).

BERTOLUZZA, S. (1995a), A posteriori error estimates for wavelet Galerkin methods, *Appl. Math. Lett.* **8**, 1–6.

BERTOLUZZA, S. (1995b), Adaptive wavelet collocation for the solution of steady state equation, in: *SPIE Proc. Wavelet Appl.* **II 2491**.

BERTOLUZZA, S. (1997), An adaptive collocation method based on interpolating wavelets, in: W. Dahmen, A.J. Kurdila and P. Oswald, eds., *Multiscale Wavelet Methods for PDE's* (Academic Press, New York).

BERTOLUZZA, S. (1998), Stabilization by multiscale decompositions, *Appl. Math. Lett.* **6**, 129–134.

BESOV, O.V. (1959), On a family of functions spaces. Embedding theorems and applications, *Dokl. Akad. Nauk USSR* **126**, 1163–1165.

BEYLKIN, G., R. COIFMAN and V. ROKHLIN (1991), Fast wavelet transforms and numerical algorithms, *Comm. Pure Appl. Math.* **44**, 141–183.

BEYLKIN, G. and J.M. KEISER (1997), An adaptive pseudo-wavelet approach for solving nonlinear partial differential equations, in: W. Dahmen, A.J. Kurdila and P. Oswald, eds., *Multiscale Wavelet Methods for PDE's* (Academic Press, New York).

BEYLKIN, G. and M. BREWSTER (1995), A multiresolution strategy for numerical homogenization, *Appl. Comp. Harm. Anal.* **2**, 327–349.

BIRMAN, M.S. and M. SOLOMJAK (1967), Piecewise polynomial approximation of functions of the class W_p^α, *Math. USSR-Sb.* **2**, 295–317.

DE BOOR, C. (1978), *A Practical Guide to Splines* (Springer-Verlag, Berlin).

DE BOOR, C. (1973), The quasi-interpolant as a tool in elementary polynomial spline theory, in: G.G. Lorentz, ed., *Approximation Theory* (Academic Press, New York).

DE BOOR, C., R. DEVORE and A. RON (1993), On the construction of multivariate pre-wavelets, *Constr. Approx.* **9**, 123–166.

DE BOOR, C. and G. FIX (1973a), Approximation from shift-invariant subspaces of $L_2(\mathbb{R}^d)$, *Trans. Amer. Math. Soc.* **341**, 787–806.

DE BOOR, C. and G. FIX (1973b), Spline approximation by quasi-interpolants, *J. Approx. Theory* **8**, 19–45.

BRAESS, D. and W. HACKBUSCH (1983), A new convergence proof for the multigrid method including the V-cycle, *SIAM J. Numer. Anal.* **20**, 967–975.

BRAMBLE, J.H. and S.R. HILBERT (1970), Estimation of linear functionals on Sobolev spaces with applications to Fourier transform and spline interpolation, *SIAM J. Numer. Anal.* **7**, 112–124.

BRAMBLE, J.H. and S.R. HILBERT (1971), Bounds for a class of linear functionals with applications to Hermite interpolation, *Numer. Math.* **16**, 362–369.

BRAMBLE, J.H., J.E. PASCIAK and J. XU (1990), Parallel multilevel preconditioners, *Math. Comp.* **55**, 1–22.

BRAMBLE, J.H. (1993), *Multigrid Methods* (Longman Scientific and Technical, Harlow, England).

BRANDT, A. (1977), Multilevel adaptive solutions to boundary value problems, *Math. Comp.* **31**, 333–390.

BRENNER, S.C. and R.L. SCOTT (1994), *The Mathematical Theory of Finite Element Methods* (Springer-Verlag, New York).

BREZINSKI, C. and J. VAN ISEGHEM (1994), Padé approximations, in: P.G. Ciarlet and J.L. Lions, eds., *Handbook of Numerical Analysis* **III** (North-Holland, Amsterdam).

BRUDNYI, Y. (1970), Approximation of functions of n-variables by quasi-polynomials, *Math. USSR Izv.* **4**, 568–586.

BURT, P. and E. ADELSON (1983), The Laplacian pyramid as a compact image code, *IEEE Trans. Comm.* **31**, 482–540.

BUTZER, P.L. and H. BEHRENS (1967), *Semi-Groups of Operators and Approximation* (Springer-Verlag, Berlin).

BUTZER, P.L. and K. SCHERER (1972), Jackson and Bernstein type inequalities for families of commutative operators in Banach spaces, *J. Approx. Theory* **5**, 308–342.

CANUTO, C. and I. CRAVERO (1997), Wavelet-based adaptive methods for advection-diffusion problems, *Math. Mod. Meth. Appl. Sci.* **7**, 265–289.

CANUTO, C., A.TABACCO and K. URBAN (1999), The wavelet element method. Part I: construction and analysis, *Appl. Comp. Harm. Anal.* **6**, 1–52.

CARNICER, J.M., W. DAHMEN and J.M. PEÑA (1996), Local decomposition of refinable spaces and wavelets, *Appl. Comp. Harm. Anal.* **3**, 127–153.

CAVARETTA, A., W. DAHMEN and C.A. MICCHELLI (1991), *Stationary Subdivision*, Mem. Amer. Math. Soc. 453.

CHAITKIN, G.M. (1974), An algorithm for high speed curve generation, *Comp. Graph. and Image Process.* **3**, 346–349.

CHUI, C.K. (1988), *Multivariate Splines* (SIAM, Philadelphia).

CHUI, C.K. (1992), *An Introduction to Wavelets* (Academic Press, Boston).

CHUI, C.K. and E. QUAK (1992), Wavelets on a bounded interval, in: D. Braess and L. Schumaker, eds., *Numerical Methods of Approximation Theory* (Birkhäuser, Basel) 57–76.

CHUI, C.K. and Y. WANG (1992), A general framework for compactly supported splines and wavelets, *J. Approx. Theory* **71**, 263–304.

CIARLET, P.G. (1978), *The Finite Element Method for Elliptic Problems* (North-Holland, Amsterdam).

CIARLET, P.G. (1991), Basic error estimate for the finite element method, in: P.G. Ciarlet and J.L. Lions, eds., *Handbook of Numerical Analysis* **II** (Elsevier, Amsterdam).

CIESIELSKI, Z. (1973), Constructive function theory and spline systems, *Studia Math.* **52**, 277–302.

CIESIELSKI, Z. and J. FIGIEL (1983), Spline bases in classical function spaces on compact C^∞ manifolds, *Studia Math.* **76**, 1–58.

CLEMENT, P. (1975), Approximation by finite element functions using local regularizations, *RAIRO Mod. Math. Anal. Num.* **2**, 77–84.

COHEN, A., W. DAHMEN, I. DAUBECHIES and R. DEVORE (1999), Tree-structured approximation and optimal encoding, Preprint LAN/UPMC, to appear in *App. Comp. Harm. Anal.* (2000).

COHEN, A., W. DAHMEN and R. DEVORE (2000a), Multiscale methods on bounded domains, Preprint LAN/UPMC, to appear in *Trans. Amer. Math. Soc.* (2000).

COHEN, A., W. DAHMEN and R. DEVORE (2000b), Adaptive wavelet methods for elliptic equations – Convergence rates, Preprint LAN/UPMC 98049, to appear in *Math. Comp.* (2000).

COHEN, A., I. DAUBECHIES and P. VIAL (1993), Wavelets and fast wavelet transforms on an interval, *Appl. Comp. Harm. Anal.* **1**, 54–81.

COHEN, A. and I. DAUBECHIES (1992), A stability criterion for biorthogonal wavelets and their related subband coding schemes, *Duke Math. J.* **68**, 313–335.

COHEN, A. and I. DAUBECHIES (1996), A new technique to estimate the regularity of refinable functions, *Rev. Mat. Iberoamericana* **12**, 527–591.

COHEN, A., I. DAUBECHIES and J.C. FEAUVEAU (1992), Biorthogonal bases of compactly supported wavelets, *Comm. Pure Appl. Math.* **45**, 485–560.

COHEN, A., I. DAUBECHIES and G. PLONKA (1997), Regularity of refinable function vectors, *J. Fourier Anal. Appl.* **3**, 295–324.

COHEN and EZZINE (1996), Quadratures singulieres et fonctions d'echelle, *C. R. Acad. Sci. Paris Sér. I* **323**, 829–834.

COHEN, A., K. GRÖCHENIG and L. VILLEMOES (1999), On the regularity of multivariate refinable functions, *Constr. Approx.* **15**, 241–255.

COHEN, A. and R. MASSON (1999), Wavelet adaptive methods for elliptic problems – preconditioning and adaptivity, *SIAM J. Sci. Comput.* **21**, 1006–1026.

COHEN, A. and R. MASSON (2000), Wavelet methods for elliptic problems – boundary conditions and domain decomposition, Preprint LAN/UPMC 98007, to appear in *Numer. Math.* (2000).

COHEN, A. and R. RYAN (1995), *Wavelets and Multiscale Signal Processing* (Chapman and Hall, London).

COHEN, A. and J.M. SCHLENKER (1993), Compactly supported wavelets with hexagonal symmetry, *Constr. Approx.* **9**, 209–236.

COIFMAN, R., Y. MEYER, S. QUAKE and M.V. WICKERHAUSER (1993), Signal processing and compression with wavelet packets, in: S. Roques and Y. Meyer, eds., *Progresses in Wavelet Analysis and Applications* (Frontieres, Gif sur Yvette, France).

COIFMAN, R. and M.V. WICKERHAUSER (1992), Entropy-based algorithms for best basis selection, *IEEE Trans. Inform. Theory* **38**, 713–718.

DAHLKE, S. (1998), Besov regularity for elliptic boundary value problems with variable coefficients, *Manuscripta Math.* **95**, 59–77.

DAHLKE, S. (1999), Besov regularity for elliptic boundary value problems on polygonal domains, *Appl. Math. Lett.* **12**, 31–36.

DAHLKE, S., W. DAHMEN, R. HOCHMUTH and R. SCHNEIDER (1997), Stable multiscale bases and local error estimation for elliptic problems, *Appl. Numer. Math.* **23**, 21–48.

DAHLKE, S. and R. DEVORE (1996), Besov regularity for elliptic boundary value problems, *Comm. in PDE's* **22**, 1–16.

DAHMEN, W. (1996), Stability of multiscale transforms, *J. Fourier Anal. Appl.* **2**, 341–361.

DAHMEN, W. (1997), Wavelets and multiscale methods for operator equations, *Acta Numerica*, 55–228.

DAHMEN, W., B. HAN, R.-Q. JIA and A. KUNOTH (2000), Biorthogonal multiwavelets on the interval: Cubic Hermite splines, *Constr. Approx.* **16**, 221–259.

DAHMEN, W. and A. KUNOTH (1992), Multilevel preconditioning, *Numer. Math.* **63**, 315–344.

DAHMEN, W., A. KUNOTH and R. SCHNEIDER (1996), Operator equations, multiscale concepts and complexity, *1995 AMS-SIAM Summer Seminar, Math. of Numer. Anal.*, Amer. Math. Soc., Lectures in Appl. Math. **32**, 225–261.

DAHMEN, W., A. KUNOTH and K. URBAN (1996), A wavelet Galerkin method for the Stokes problem, *Computing* **56**, 259–302.

DAHMEN, W., A. KUNOTH and K. URBAN (1999), Biorthogonal spline-wavelets on the interval. Stability and moment conditions, *Appl. Comp. Harm. Anal.* **6**, 132–196.

DAHMEN, W. and C.A. MICCHELLI (1984), Subdivision algorithms for the generation of box spline surfaces, *Computer Aided Geom. Design* **1**, 115–129.

DAHMEN, W. and C.A. MICCHELLI (1993), Using the refinement equation for evaluating integrals of wavelets, *SIAM J. Numer. Anal.* **30**, 507–537.

DAHMEN, W., S. PRÖSSDORF and R. SCHNEIDER (1993), Wavelet methods for pseudodifferential equations II: Matrix compression and fast solution, *Adv. Comput. Math.* **1**, 259–335.

DAHMEN, W., S. PRÖSSDORF and R. SCHNEIDER (1994), Wavelet approximation methods for pseudodifferential equations I: Stability and convergence, *Math. Z.* **215**, 583–620.

DAHMEN, W. and R. SCHNEIDER (1998), Wavelets with complementary boundary conditions – function spaces on the cube, *Results in Math.* **34**, 255–293.

DAHMEN, W. and R. SCHNEIDER (1999a), Composite wavelet bases for operator equations, *Math. Comp.* **68**, 1533–1567.

DAHMEN, W. and R. SCHNEIDER (1999b), Wavelets on manifold I. Construction and domain decomposition, *SIAM J. Math. Anal.* **31**, 184–230.

DAHMEN, W. and R. STEVENSON (1999), Element by element construction of wavelets satisfying stability and moment conditions, *SIAM J. Numer. Anal.* **37**, 319–352.

DAUBECHIES, I. (1988), Orthonormal bases of compactly supported wavelets, *Comm. Pure Appl. Math.* **41**, 909–996.

DAUBECHIES, I. (1992), *Ten Lectures on Wavelets* (SIAM, Philadelphia).

DAUBECHIES, I. and J. LAGARIAS (1991), Two scale difference equations I. Existence and global regularity of solutions, *SIAM J. Math. Anal.* **22**, 1388–1410.

DAUBECHIES, I. and J. LAGARIAS (1992), Two scale difference equations II. Local regularity, infinite products of matrices and fractals, *SIAM J. Math. Anal.* **23**, 1031–1079.

DAUBECHIES, I. and W. SWELDENS (1998), Factoring wavelet transform into lifting steps, *J. Fourier Anal. Appl.* **4**, 247–269.

DAUGE, M. (1988), *Elliptic Boundary Value Problems on Corner Domains* (Springer-Verlag, Berlin).

DAVID, G. and J.L. JOURNÉ (1984), A boundedness criterion for generalized Calderon–Zygmund operators, *Ann. of Math.* **120**, 371–397.

DENY, J and J.L. LIONS (1954), Les espaces de type de Beppo Levi, *Ann. Inst. Fourier* **5**, 305–370.

DESLAURIERS, G. and S. DUBUC (1987), Interpolation dyadique, in: G. Cherbit, ed., *Fractals, Dimensions Non Entières et Applications* (Masson, Paris).

DEVORE, R. (1998), Non-linear approximation, *Acta Numerica* **7**, 51–150.

DEVORE, R., B. JAWERTH and V. POPOV (1992), Compression of wavelet decompositions, *Amer. J. Math.* **114**, 737–785.

DEVORE, R. and G.G. LORENTZ (1993), *Constructive Approximation* (Springer-Verlag, Berlin).

DEVORE, R. and B. LUCIER (1990), High order regularity for conservation laws, *Indiana Math. J.* **39**, 413–430.

DEVORE, R., P. PETRUSHEV and X.M. YU (1992), Wavelet approximation in the space C, in: *Progress in Approximation Theory* (Springer-Verlag, New York).

DEVORE, R. and V. POPOV (1988a), Free multivariate splines, *Constr. Approx.* **3**, 239–248.

DEVORE, R. and V. POPOV (1988b), Interpolation of Besov spaces, *Trans. Amer. Math. Soc.* **305**, 397–414.

DEVORE, R. and V. POPOV (1988c), Interpolation spaces and nonlinear approximation, in: M. Cwikel, J. Peetre, Y. Sagher and H. Wallin, eds., *Function Spaces and Applications*, Lecture Notes in Math. 1302 (Springer-Verlag, Berlin) 191–205.

DEVORE, R., S. RIEMENSCHNEIDER and R. SHARPLEY (1979), Weak interpolation in Banach spaces, *J. Funct. Anal.* **33**, 58–91.

DEVORE, R.A and R.C. SHARPLEY (1993), Besov spaces on domains in $\mathbb{R}^d$, *Trans. Amer. Math. Soc.* **335**, 843–864.

DEVORE, R. and X.M. YU (1990), Degree of adaptive approximation, *Math. Comp.* **55**, 625–635.

DONOHO, D., I. JOHNSTONE, G. KERKYACHARIAN and D. PICARD (1994), Wavelet shrikage: Asymptotia? (with discussion), *J. Roy. Statist. Soc. Ser. B* **57**, 301–369.

DONOVAN G., J. GERONIMO and D. HARDIN (1999), Orthogonal polynomials and the construction of piecewise polynomial smooth wavelets, *SIAM J. Math. Anal.* **30**, 1029–1056.

DOERFLER, W. (1996), A convergent adaptive algorithm for Poisson's equation, *SIAM J. Numer. Anal.* **33**, 1106–1124.

DUFFIN, R.J. and A.C. SCHAEFFER (1952), A class of nonharmonic Fourier series, *Trans. Amer. Math. Soc.* **72**, 341–366.

DYN, N. (1992), Subdivision schemes in computer-aided geometric design, in: W.A. Light, ed., *Advances in Numerical Analysis* **II**, *Wavelets, Subdivision Algorithms, and Radial Basis Functions* (Clarendon Press, Oxford).

DYN, N. (2000), A construction of biorthogonal functions to B-splines with multiple knots, *Appl. Comp. Harm. Anal.* **8**, 24–31.

DYN, N., J. GREGORY and D. LEVIN (1991), Analysis of uniform binary subdivision schemes for curve design, *Constr. Approx.* **7**, 127–147.

EIROLA, T. (1992), Sobolev characterization of solutions of dilation equations, *SIAM J. Math. Anal.* **23**, 1015–1030.

ERIKSSON, K., D. ESTEP, P. HANSBO and C. JOHNSON (1995), Introduction to adaptive methods for differential equations, in: *Acta Numerica* (Cambridge University Press) 105–158.

FEDORENKO, R.P. (1964), The speed of convergence of one iterative process, *USSR Comp. Math. Math. Phys.* **4**, 1092–1096.

FRAZIER, M, B. JAWERTH and G. WEISS (1991), *Littlewood–Paley Theory and the Study of Function Spaces*, CBMS Conference Lecture Notes 79 (Amer. Math. Soc., Providence, RI).

GERSHO, A. and R.M. GRAY (1992), *Vector Quantization and Signal Compression* (Kluwer Acad. Publ., Boston).

GILBERT, A. (1998), A comparison of multiresolution and classical one-dimensional homogeneization schemes, *Appl. Comp. Harm. Anal.* **5**, 1–35.

GREENGARD, L. and V. ROKHLIN (1987), A fast algorithm for particle simulations, *J. Comp. Phys.* **325**.

GRISVARD, P. (1983), *Elliptic Problems on Non-Smooth Domains* (Pittman, New York).

HACKBUSCH, W. (1985), *Multi-Grid Methods and Applications* (Springer-Verlag, New York).

HACKBUSCH, W. and Z.P. NOWAK (1989), On the fast matrix multiplication in the boundary element method by panel clustering, *Numer. Math.* **54**, 463–491.

HARTEN, A. (1993), Discrete multiresolution and generalized wavelets, *J. Appl. Numer. Math.* **12**, 153–193.

HARTEN, A. (1995), Multiresolution algorithms for the numerical solution of hyperbolic conservation laws, *Comm. Pure Appl. Math.* **48**, 1305–1342.

HARTEN, A. (1996), Multiresolution representation of data II: Generalized framework, *SIAM J. Numer. Anal.* **33**, 1205–1256.

JAFFARD, S. (1991), Pointwise smoothness, two-microlocalization and wavelet coefficients, *Publ. Mat.* **35**, 155–168.

JAFFARD, S. (1992), Wavelet methods for fast resolution of elliptic problems, *SIAM J. Numer. Anal.* **29**, 965–986.

JIA, R.Q (1999), Smoothness of multivariate refinable functions in Sobolev spaces, *Trans. Amer. Math. Soc.* **351**, 4089–4112.

JIA, R.Q. and B. HAN (1998), Multivariate refinement equations and convergence of subdivision schemes *SIAM J. Math. Anal.* **29**, 1177–1199.

JIA, R.Q. and J.J. LEI (1993), Approximation by multi-integer translates of functions having global support, *J. Approx. Theory* **72**, 2–23.

JIA, R.Q., S. RIEMENSCHNEIDER and D.X. XHOU (1999), Smoothness of multiple refinable functions and multiple wavelets, *SIAM J. Matrix Anal. Appl.* **21**, 1–28.

JOHNEN, H. and K. SCHERER (1976), On the equivalence of the K-functional and moduli of continuity and some applications, in: *Constructive Theory of Functions of Several Variables*, Lecture Notes in Math. 571 (Springer-Verlag, Berlin).

KAHANE, J.P. (1961), *Teoria Constructiva de Functiones*, Course Notes (University of Buenos Aires).

KOVACEVIC, J. and M. VETTERLI (1995), *Wavelets and Subband Coding* (Prentice-Hall, Englewood Cliff, NJ).

KUNOTH, A. (1995), Multiscale preconditioning – Appending boundary conditions by Lagrange multipliers, *Adv. Comput. Math.* **4**, 145–170.

KYRIASIS, G (1996), Wavelet coefficients measuring smoothness in $H_p(\mathbb{R}^d)$, *Appl. Com. Harm. Anal.* **2**, 100–119.

LAX, P. (1972), *Hyperbolic System of Conservation Laws and the Mathematical Theory of Shock Waves* (SIAM, Philadelphia).

LAZAAR, S., J. LIANDRAT and P. TCHAMITCHIAN (1994), Algorithme à base d'ondelettes pour la résolution numérique d'équations aux dérivées partielles à coefficients variables, *C. R. Acad. Sci. Paris Sér. I* **319**, 1101–1107.

LEMARIÉ, P.G. (1988), Ondelettes à localisation exponentielle, *J. Math. Pures Appl.* **67**, 227–236.

LEMARIÉ, P.G. (1992), Analyses multirésolutions nonorthogonales, commutation entre projecteurs et dérivation et ondelettes vecteurs à divergence nulle, *Rev. Mat. Iberoamericana* **8**, 221–236.

LEMARIÉ, P.G. (1997), On the existence of compactly supported dual wavelets, *Appl. Comp. Harm. Anal.* **3**, 117–118.

LEMARIÉ, P.G. and G. MALGOUYRES (1991), Support des fonctions de base dans une analyse multiresolution, *C. R. Acad. Sci. Paris Sér. I* **213**, 377–380.

LEVEQUE, R.J. (1992), *Numerical Methods for Conservation Laws* (Birkhäuser, Basel).

LIONS, J.L and E. MAGENES (1972), *Non-Homogeneous Boundary Value Problems and Applications* (Springer-Verlag, New York)

LIONS, J.L. and J. PEETRE (1964), Sur une classe d'espaces d'interpolation, *Publ. Math. Inst. Hautes Etudes Sci.* **19**, 5–68.

LYCHE, T. (1993), Knot removal for spline curves and surfaces, in: E.W. Cheney, C.K. Chui and L. Schumaker, eds., *Approximation Theory* **VII** (Academic Press, Boston) 207–227.

MADAY, Y., V. PERRIER and J.C. RAVEL (1991), Adaptativité dynamique sur bases d'ondelettes pour l'approximation d'équations aux dérivées partielles, *C. R. Acad. Sci. Paris Sér. I* **1**, 405–410.

MALLAT, S. (1989), Multiresolution approximation and wavelet orthonormal bases of $L^2(\mathbb{R})$, *Trans. Amer. Math. Soc.* **315**, 69–88.

MALLAT, S. (1998), *A Wavelet Tour of Signal Processing* (Academic Press, New York).

MASSON, R. (1996), Biorthogonal spline wavelets on the interval for the resolution of boundary problems, *M3As* **6**, 749–791.

MEYER, Y. (1990), *Ondelettes et Opérateurs* (Hermann, Paris); English translation by D.H. Salinger (Cambridge University Press, Cambridge, 1992).

MEYER, Y. (1991), Ondelettes sur l'intervalle, *Rev. Mat. Iberoamericana* **7**, 115–134.

MEYER, Y. (1992), *Ondelettes, Algorithmes et Applications* (Armand Colin, Paris); English translation by R. Ryan (SIAM, Philadelphia, 1994).

MEYER, Y. (1997), Wavelets and fast numerical algorithms, in: P.G. Ciarlet and J.L. Lions, eds., *Handbook of Numerical Analysis* **V** (Elsevier, Amsterdam).

NEWMAN, D.J. (1964), Rational approximation to $|x|$, *Michigan Math. J.* **11**, 11–14.

NGUYEN, V. and G. STRANG (1996), *Wavelets and Filter Banks* (Wellesley–Cambridge Press, Cambridge).

OSWALD, P. (1990), On the degree of nonlinear spline approximation in Besov–Sobolev spaces, *J. Approx. Theory* **61**, 131–157.

OSWALD, P. (1994), *Multilevel Finite Element Approximation: Theory and Applications* (Teubner, Stuttgart).

OSWALD, P. and R. LORENTZ (1996), Multilevel finite element Riesz bases in Sobolev spaces, in: P. Bjorstad, ed., *Proceeding of the Ninth International Conf. on Domain Decomposition* (Bergen).

OSWALD, P. and E.A. STOROZHENKO (1978), Jackson's theorem in the space $L_p(\mathbb{R}^k)$, $0 < p < 1$, *Siberian Math J.* **19**, 630–639.

PEETRE, J. (1974), *New Thoughts on Besov Spaces* (Duke University Math. Series, Duham).

PEETRE, J. and G. SPARR (1972), Interpolation of normed Abelian groups, *Ann. Mat. Pura Appl.* **92**, 217–252.

PELLER, V. (1985), Description of Hankel operators of the class σ_p for $p < 1$, investigation of the order of rational approximation, *Math. USSR Sb.* **50**, 465–494.

PERRIER, V. and P. MONASSE (1998), Orthogonal wavelet bases adapted for partial differential equations with boundary conditions, *SIAM J. Math. Anal.* **29**, 1040–1065.

VON PETERSDORFF, T. and C. SCHWAB (1997), Fully discrete multiscale Galerkin BEM, in: W. Dahmen, A.J. Kurdila and P. Oswald, eds., *Multiscale Wavelet Methods for PDE's* (Academic Press, New York).

PETRUSHEV, P. (1988), Direct and converse theorems for spline and rational approximation and Besov spaces, in: M. Cwikel, J. Peetre, Y. Sagher and H. Wallin, eds., *Function Spaces and Applications*, Lecture Notes in Math. 1302 (Springer-Verlag, Berlin) 363–377.

PETRUSHEV, P. (2000), Bases consisting of rational functions of uniformly bounded degrees or more general functions, Preprint Univ. of South. Carolina, to appear in *J. Funct. Anal.* (2000).

PLONKA, G. (1995), Factorization of refinement masks for function vectors, in: C.K. Chui and L.L. Schumaker, eds., *Wavelets and Multilevel Approximation* (World Scientific, Singapore).

RIOUL, O. (1992), Simple criteria for subdivision scheme, *SIAM J. Math. Anal.* **23**, 1544–1576.

SAID, A. and W.A. PEARLMAN (1996), An image multiresolution representation for lossless and lossy compression, *IEEE Trans. Image Process.* **5**, 1303–1310.

SCHOENBERG, I.J. (1946), Contributions to the problem of approximation of equidistant data by analytic functions, *Quart. Appl. Math.* **4**, 45–99 (part A), 112–141 (part B).

SCHUMAKER, L. (1981), *Spline Functions: Basic Theory* (Wiley, New York).

SHAPIRO, J. (1993), Embedded image coding using zerotrees of wavelet coefficients, *IEEE Signal Process.* **41**, 3445–3462.

SHEN, Z. (1998), Refinable function vectors, *SIAM J. Math. Anal.* **29**, 235–250.

STEIN, E.M. (1970), *Singular Integrals and Differentiability Properties of Functions* (Princeton University Press, Princeton).

STEVENSON, R. (1998), Piecewise linear (pre)-wavelets on nonuniform meshes, in: W. Hackbusch and G. Wittum, eds., *Multigrid Methods* **V**, Lecture Notes in Comput. Sci. Eng. 3 (Springer-Verlag, Heidelberg) 306–319.

STEVENSON, R. (1997), Experiments in 3D with a three point hierarchical basis preconditioner, *Appl. Numer. Math.* **23**, 159–175.

STRANG, G. and G. FIX (1969), Fourier analysis of the finite element method in Ritz–Galerkin theory, *Stud. Appl. Math.* **48**, 265–273.

STRANG, G. and G. FIX (1973), *An Analysis of the Finite Element Method* (Wellesley–Cambridge Press, Cambridge).

SWELDENS, W. and R. PIESSENS (1994), Quadrature formulae and asymptotic error expansions for wavelet approximation of smooth functions, *SIAM J. Numer. Anal.* **31**, 2140–2164.

SWELDENS, W. (1996), The lifting scheme: A custom design construction of biorthogonal wavelets, *Appl. Comp. Harm. Anal.* **3**, 186–200.

SWELDENS, W. (1998), The lifting scheme: A construction of second generation wavelets, *SIAM J. Math. Anal.* **29**, 511–546.

TCHAMITCHIAN, P. (1996a), Inversion explicite de certains operateurs elliptiques, *SIAM J. Math. Anal.* **27**, 1680–1703.

TCHAMITCHIAN, P (1996b), Wavelets functions and operators, in: G. Erlebacher, Y. Hussaini and L. Jameson, eds., *Wavelets, Theory and Applications* (Oxford University Press) 87–178.

TRIEBEL, H. (1983), *Theory of Function Spaces* (Birkhäuser, Basel).

URBAN, K. (1995), On divergence-free wavelets, *Advances in Comp. Math.* **4**, 51–82.

VAIDYANATHAN, P.P. (1992), *Multirate Systems and Filter Banks* (Prentice-Hall, Englewood Cliff, NJ).

VERFURTH, R. (1994), A posteriori error estimation and adaptive mesh refinement techniques, *J. Comput. Appl. Math.* **50**, 67–83.

VILLEMOES, L. (1994), Wavelet analysis of two-scale refinement equations, *SIAM J. Math. Anal.* **25**, 1433–1460.

VOLKMER, H. (1992), On the regularity of wavelets, *IEEE Trans. Inform. Theory* **38**, 872–876.

YOUNG, R.M. (1980), *An Introduction to Non-Harmonic Fourier Series* (Academic Press, New York).

YSERANTANT, H. (1986), On the multi-level splitting of finite element spaces, *Numer. Math.* **49**, 379–412.

YSERANTANT, H. (1993), Old and new convergence proof for multigrid methods, in: *Acta Numerica* (Cambridge University Press) 285–326.

Subject Index

Finite Volume Methods

Robert Eymard

Ecole Nationale des Ponts et Chaussées
Marne-la-Vallée
et Université de Paris XIII
France

Thierry Gallouët

Ecole Normale Supérieure de Lyon
France

Raphaèle Herbin

Université de Provence, Marseille
France
e-mail: Raphaele.Herbin@cmi.univ-mrs.fr

HANDBOOK OF NUMERICAL ANALYSIS, VOL. VII
Solution of Equations in $\mathbb{R}^n$ (Part 3)
Techniques of Scientific Computing (Part 3)
edited by P.G. Ciarlet and J.L. Lions

Contents

CHAPTER I

Introduction

The finite volume method is a discretization method which is well suited for the numerical simulation of various types (elliptic, parabolic or hyperbolic, for instance) of conservation laws; it has been extensively used in several engineering fields, such as fluid mechanics, heat and mass transfer or petroleum engineering. Some of the important features of the finite volume method are similar to those of the finite element method, see ODEN [1991]: it may be used on arbitrary geometries, using structured or unstructured meshes, and it leads to robust schemes. An additional feature is the local conservativity of the numerical fluxes, that is the numerical flux is conserved from one discretization cell to its neighbour. This last feature makes the finite volume method quite attractive when modelling problems for which the flux is of importance, such as in fluid mechanics, semi-conductor device simulation, heat and mass transfer.... The finite volume method is locally conservative because it is based on a " balance" approach: a local balance is written on each discretization cell which is often called "control volume"; by the divergence formula, an integral formulation of the fluxes over the boundary of the control volume is then obtained. The fluxes on the boundary are discretized with respect to the discrete unknowns.

Let us introduce the method more precisely on simple examples, and then give a description of the discretization of general conservation laws.

1. Examples

Two basic examples can be used to introduce the finite volume method. They will be developed in details in the following chapters.

EXAMPLE 1.1 (*Transport equation*). Consider first the linear transport equation

$$\begin{cases} u_t(x,t) + \operatorname{div}(\boldsymbol{v}u)(x,t) = 0, & x \in \mathbb{R}^2, t \in \mathbb{R}_+, \\ u(x,0) = u_0(x), & x \in \mathbb{R}^2, \end{cases} \tag{1.1}$$

where u_t denotes the time derivative of u, $\boldsymbol{v} \in C^1(\mathbb{R}^2, \mathbb{R}^2)$, and $u_0 \in L^\infty(\mathbb{R}^2)$. Let $\mathcal{T}$ be a mesh of $\mathbb{R}^2$ consisting of polygonal bounded convex subsets of $\mathbb{R}^2$ and let $K \in \mathcal{T}$ be a "control volume", that is an element of the mesh $\mathcal{T}$. Integrating the first equation of (1.1) over K yields the following "balance equation" over K:

$$\int_K u_t(x,t)\,\mathrm{d}x + \int_{\partial K} \boldsymbol{v}(x,t)\boldsymbol{n}_K(x)u(x,t)\,\mathrm{d}x = 0, \quad \forall t \in \mathbb{R}_+, \tag{1.2}$$

where $\boldsymbol{n}_K$ denotes the normal vector to ∂K, outward to K. Let $k \in \mathbb{R}_+^*$ be a constant time discretization step and let $t_n = nk$, for $n \in \mathbb{N}$. Writing Eq. (1.2) at time t_n, $n \in \mathbb{N}$ and discretizing the time partial derivative by the Euler explicit scheme suggests to find an approximation $u^{(n)}(x)$ of the solution of (1.1) at time t_n which satisfies the following semi-discretized equation:

$$\begin{aligned}\frac{1}{k}\int_K \big(u^{(n+1)}(x) - u^{(n)}(x)\big)\,\mathrm{d}x \\ + \int_{\partial K} \boldsymbol{v}(x,t_n)\cdot\boldsymbol{n}_K(x)u^{(n)}(x)\,\mathrm{d}\gamma(x) = 0, \quad \forall \boldsymbol{n} \in \mathbb{N},\ \forall K \in \mathcal{T},\end{aligned} \tag{1.3}$$

where $\mathrm{d}\gamma$ denotes the one-dimensional Lebesgue measure on ∂K and $u^{(0)}(x) = u(x,0) = u_0(x)$. We need to define the discrete unknowns for the (finite volume) space discretization. We shall be concerned here principally with the so-called "cell-centered" finite volume method in which each discrete unknown is associated with a control volume. Let $(u_K^{(n)})_{K\in\mathcal{T},n\in\mathbb{N}}$ denote the discrete unknowns. For $K \in \mathcal{T}$, let $\mathcal{E}_K$ be the set of edges which are included in ∂K, and for $\sigma \subset \partial K$, let $\boldsymbol{n}_{K,\sigma}$ denote the unit normal to σ outward to K. The second integral in (1.3) may then be split as:

$$\int_{\partial K} \boldsymbol{v}(x,t_n)\cdot\boldsymbol{n}_K(x)u^{(n)}(x)\,\mathrm{d}\gamma(x) = \sum_{\sigma\in\mathcal{E}_K}\int_\sigma \boldsymbol{v}(x,t_n)\cdot\boldsymbol{n}_{K,\sigma}u^{(n)}(x)\,\mathrm{d}\gamma(x); \tag{1.4}$$

for $\sigma \subset \partial K$, let

$$v_{K,\sigma}^{(n)} = \int_\sigma \boldsymbol{v}(x,t_n)\boldsymbol{n}_{K,\sigma}(x)\,\mathrm{d}\gamma(x).$$

Each term of the sum in the right-hand side of (1.4) is then discretized as

$$F_{K,\sigma}^{(n)} = \begin{cases} v_{K,\sigma}^{(n)}u_K^{(n)} & \text{if } v_{K,\sigma}^{(n)} \geqslant 0, \\ v_{K,\sigma}^{(n)}u_L^{(n)} & \text{if } v_{K,\sigma}^{(n)} < 0, \end{cases} \tag{1.5}$$

where L denotes the neighbouring control volume to K with common edge σ. This "upstream" or "upwind" choice is classical for transport equations; it may be seen, from the mechanical point of view, as the choice of the "upstream information" with respect to the location of σ. This choice is crucial in the mathematical analysis; it ensures the stability properties of the finite volume scheme (see Chapters V and VI). We have therefore derived the following finite volume scheme for the discretization of (1.1):

$$\begin{cases} \dfrac{\mathrm{m}(K)}{k}\big(u_K^{(n+1)} - u_K^{(n)}\big) + \sum_{\sigma\in\mathcal{E}_K} F_{K,\sigma}^{(n)} = 0, \quad \forall K \in \mathcal{T}, \forall n \in \mathbb{N}, \\ u_K^{(0)} = \displaystyle\int_K u_0(x)\,\mathrm{d}x, \end{cases} \tag{1.6}$$

where $\mathrm{m}(K)$ denotes the measure of the control volume K and $F_{K,\sigma}^{(n)}$ is defined in (1.5). This scheme is locally conservative in the sense that if σ is a common edge to the control volumes K and L, then $F_{K,\sigma} = -F_{L,\sigma}$. This property is important in several application fields; it will later be shown to be a key ingredient in the mathematical proof of convergence. Similar schemes for the discretization of linear or nonlinear hyperbolic equations will be studied in Chapters V and VI.

EXAMPLE 1.2 (*Stationary diffusion equation*). Consider the basic diffusion equation

$$\begin{cases} -\Delta u = f & \text{on } \Omega =]0,1[\times]0,1[, \\ u = 0 & \text{on } \partial\Omega. \end{cases} \tag{1.7}$$

Let $\mathcal{T}$ be a rectangular mesh. Let us integrate the first equation of (1.7) over a control volume K of the mesh; with the same notations as in the previous example, this yields:

$$\sum_{\sigma\in\mathcal{E}_K} \int_\sigma -\nabla u(x)\cdot \boldsymbol{n}_{K,\sigma}\,\mathrm{d}\gamma(x) = \int_K f(x)\,\mathrm{d}x. \tag{1.8}$$

For each control volume $K \in \mathcal{T}$, let x_K be the center of K. Let σ be the common edge between the control volumes K and L. The flux $-\int_\sigma \nabla u(x)\cdot \boldsymbol{n}_{K,\sigma}\,\mathrm{d}\gamma(x)$, may be approximated by the following finite difference approximation:

$$F_{K,\sigma} = -\frac{\mathrm{m}(\sigma)}{d_\sigma}(u_L - u_K), \tag{1.9}$$

where $(u_K)_{K\in\mathcal{T}}$ are the discrete unknowns and d_σ is the distance between x_K and x_L. This finite difference approximation of the first-order derivative $\nabla u \cdot \boldsymbol{n}$ on the edges of the mesh (where n denotes the unit normal vector) is consistent: the truncation error on the flux is of order h, where h is the maximum length of the edges of the mesh. It is necessary for this to be true that the points x_K be the intersections of the orthogonal bisectors of the edges of K. Indeed, this is the case here since the control volumes are rectangular. This property is satisfied by other meshes which will be studied hereafter. It is crucial for the discretization of diffusion operators.

In the case where the edge σ is part of the boundary, then d_σ denotes the distance between the center x_K of the control volume K to which σ belongs and the boundary. The flux $-\int_\sigma \nabla u(x)\cdot \boldsymbol{n}_{K,\sigma}\,\mathrm{d}\gamma(x)$, is then approximated by

$$F_{K,\sigma} = \frac{\mathrm{m}(\sigma)}{d_\sigma} u_K. \tag{1.10}$$

Hence the finite volume scheme for the discretization of (1.7) is:

$$\sum_{\sigma\in\mathcal{E}_K} F_{K,\sigma} = \mathrm{m}(K) f_K, \quad \forall K \in \mathcal{T}, \tag{1.11}$$

where $F_{K,\sigma}$ is defined by (1.9) and (1.10), and f_K denotes (an approximation of) the mean value of f on K. We shall see later (see Chapters II, III and IV) that the finite volume scheme is easy to generalize to a triangular mesh, whereas the finite difference method is not. As in the previous example, the finite volume scheme is locally conservative, since for any edge σ separating K from L, one has $F_{K,\sigma} = -F_{L,\sigma}$.

2. The finite volume principles for general conservation laws

The finite volume method is used for the discretization of conservation laws. We gave in the above section two examples of such conservation laws. Let us now present the discretization of general conservation laws by finite volume schemes. As suggested by its name, a conservation law expresses the conservation of a quantity $q(x,t)$. For instance, the conserved quantities may be the energy, the mass, or the number of moles of some chemical species. Let us first assume that the local form of the conservation equation may be written as

$$q_t(x,t) + \operatorname{div} \boldsymbol{F}(x,t) = f(x,t), \tag{2.1}$$

at each point x and each time t where the conservation of q is to be written. In Eq. (2.1), $(\bullet)_t$ denotes the time partial derivative of the entity within the parentheses, div represents the space divergence operator: $\operatorname{div}\boldsymbol{F} = \partial F_1/\partial x_1 + \cdots + \partial F_d/\partial x_d$, where $\boldsymbol{F} = (F_1, \ldots, F_d)^t$ denotes a vector function depending on the space variable x and on the time t, x_i is the ith space coordinate, for $i = 1, \ldots, d$, and d is the space dimension, i.e. $d = 1, 2$ or 3; the quantity $\boldsymbol{F}$ is a flux which expresses a transport mechanism of q; the "source term" f expresses a possible volumetric exchange, due, e.g., to chemical reactions between the conserved quantities.

Thanks to the physicist's work, the problem can be closed by introducing constitutive laws which relate q, $\boldsymbol{F}$, f with some scalar or vector unknown $u(x,t)$, function of the space variable x and of the time t. For example, the components of u can be pressures, concentrations, molar fractions of the various chemical species by unit volume.... The quantity q is often given by means of a known function $\bar{q}$ of $u(x,t)$, of the space variable x and of the time t, i.e. $q(x,t) = \bar{q}(x,t,u(x,t))$. The quantity $\boldsymbol{F}$ may also be given by means of a function of the space variable x, the time variable t and of the unknown $u(x,t)$ and (or) by means of the gradient of u at point (x,t) .... The transport equation of Example 1.1 is a particular case of (2.1) with $q(x,t) = u(x,t)$ and $\boldsymbol{F}(x,t) = vu(x,t)$; so is the stationary diffusion equation of Example 1.2 with $q(x,t) = u(x)$, $\boldsymbol{F}(x,t) = -\nabla u(x)$ and $f(x,t) = f(x)$. The source term f may also be given by means of a function of x, t and $u(x,t)$.

EXAMPLE 2.1 (*The one-dimensional Euler equations*). Let us consider as an example of a system of conservation laws the 1D Euler equations for equilibrium real gases; these equations may be written under the form (2.1), with

$$q = \begin{pmatrix} \rho \\ \rho u \\ E \end{pmatrix} \quad \text{and} \quad \boldsymbol{F} = \begin{pmatrix} \rho u \\ \rho u^2 + p \\ u(E+p) \end{pmatrix},$$

where ρ, u, E and p are functions of the space variable x and the time t, and refer respectively to the density, the velocity, the total energy and the pressure of the particular gas under consideration. The system of equations is closed by introducing the constitutive laws which relate p and E to the specific volume τ, with $\tau = 1/\rho$ and the entropy s, through the constitutive laws:

$$p = \frac{\partial \varepsilon}{\partial \tau}(\tau, s) \quad \text{and} \quad E = \rho\left(\varepsilon(\tau, s) + \frac{u^2}{2}\right),$$

where ε is the internal energy per unit mass, which is a given function of τ and s.

Equation (2.1) may be seen as the expression of the conservation of q in an infinitesimal domain; it is formally equivalent to the equation

$$\int_K q(x, t_2)\,\mathrm{d}x - \int_K q(x, t_1)\,\mathrm{d}x + \int_{t_1}^{t_2}\int_{\partial K} \boldsymbol{F}(x, t) \bullet \boldsymbol{n}_K(x)\,\mathrm{d}\gamma(x)\,\mathrm{d}t$$
$$= \int_{t_1}^{t_2}\int_K f(x, t)\,\mathrm{d}x\,\mathrm{d}t, \tag{2.2}$$

for any subdomain K and for all times t_1 and t_2, where $\boldsymbol{n}_K(x)$ is the unit normal vector to the boundary ∂K, at point x, outward to K. Equation (2.2) expresses the conservation law in subdomain K between times t_1 and t_2. Here and in the sequel, unless otherwise mentioned, $\mathrm{d}x$ is the integration symbol for the d-dimensional Lebesgue measure in $\mathbb{R}^d$ and $\mathrm{d}\gamma$ is the integration symbol for the $(d-1)$-dimensional Hausdorff measure on the considered boundary.

2.1. *Time discretization*

The time discretization of Eq. (2.1) is performed by introducing an increasing sequence $(t_n)_{n\in\mathbb{N}}$ with $t_0 = 0$. For the sake of simplicity, only constant time steps will be considered here, keeping in mind that the generalization to variable time steps is straightforward. Let $k \in \mathbb{R}_+^\star$ denote the time step, and let $t_n = nk$, for $n \in \mathbb{N}$. It can be noted that Eq. (2.1) could be written with the use of a space-time divergence. Hence, Eq. (2.1) could be either discretized using a space-time finite volume discretization or a space finite volume discretization with a time finite difference scheme (the explicit Euler scheme, for instance). In the first case, the conservation law is integrated over a time interval and a space "control volume" as in the formulation (2.1). In the latter case, it is only integrated space wise, and the time derivative is approximated by a finite difference scheme; with the explicit Euler scheme, the term $(q)_t$ is therefore approximated by the differential quotient $(q^{(n+1)} - q^{(n)})/k$, and $q^{(n)}$ is computed with an approximate value of u at time t_n, denoted by $u^{(n)}$. Implicit and higher-order schemes may also be used.

2.2. *Space discretization*

In order to perform a space finite volume discretization of Eq. (2.1), a mesh $\mathcal{T}$ of the domain Ω of $\mathbb{R}^d$, over which the conservation law is to be studied, is introduced. The mesh is such that $\overline{\Omega} = \bigcup_{K\in\mathcal{T}} \overline{K}$, where an element of $\mathcal{T}$, denoted by K, is an open subset of Ω and is called a control volume. Assumptions on the meshes will be needed for the definition of the schemes; they also depend on the type of equation to be discretized.

For the finite volume schemes considered here, the discrete unknowns at time t_n are denoted by $u_K^{(n)}$, $K \in \mathcal{T}$. The value $u_K^{(n)}$ is expected to be some approximation of u on the cell K at time t_n. The basic principle of the classical finite volume method is to integrate Eq. (2.1) over each cell K of the mesh $\mathcal{T}$. One obtains a conservation law under a nonlocal form (related to Eq. (2.2)) written for the volume K. Using the Euler time discretization, this yields

$$\int_K \frac{q^{(n+1)}(x) - q^{(n)}(x)}{k}\,\mathrm{d}x + \int_{\partial K} \boldsymbol{F}(x, t_n) \bullet \boldsymbol{n}_K(x)\,\mathrm{d}\gamma(x) = \int_K f(x, t_n)\,\mathrm{d}x, \quad (2.3)$$

where $\boldsymbol{n}_K(x)$ is the unit normal vector to ∂K at point x, outward to K.

The remaining step in order to define the finite volume scheme is therefore the approximation of the "flux", $\boldsymbol{F}(x, t_n) \bullet \boldsymbol{n}_K(x)$, across the boundary ∂K of each control volume, in terms of $\{u_L^{(n)}, L \in \mathcal{T}\}$ (this flux approximation has to be done in terms of $\{u_L^{n+1}, L \in \mathcal{T}\}$ if one chooses the implicit Euler scheme instead of the explicit Euler scheme for the time discretization). More precisely, omitting the terms on the boundary of Ω, let $K|L = \overline{K} \cap \overline{L}$, with K, $L \in \mathcal{T}$, the exchange term (from K to L), $\int_{K|L} \boldsymbol{F}(x, t_n) \bullet \boldsymbol{n}_K(x)\,\mathrm{d}\gamma(x)$, between the control volumes K and L during the time interval $[t_n, t_{n+1})$ is approximated by some quantity, $F_{K,L}^{(n)}$, which is a function of $\{u_M^{(n)}, M \in \mathcal{T}\}$ (or a function of $\{u_M^{n+1}, M \in \mathcal{T}\}$ for the implicit Euler scheme, or more generally a function of $\{u_M^{(n)}, M \in \mathcal{T}\}$ and $\{u_M^{n+1}, M \in \mathcal{T}\}$ if the time discretization is a one-step method). Note that $F_{K,L}^{(n)} = 0$ if the Hausdorff dimension of $\overline{K} \cap \overline{L}$ is less than $d - 1$ (e.g., $\overline{K} \cap \overline{L}$ is a point in the case $d = 2$ or a line segment in the case $d = 3$).

Let us point out that two important features of the classical finite volume method are

(1) the conservativity, that is $F_{K,L}^{(n)} = -F_{L,K}^{(n)}$, for all K and $L \in \mathcal{T}$ and for all $n \in \mathbb{N}$,

(2) the "consistency" of the approximation of $\boldsymbol{F}(x, t_n) \bullet \boldsymbol{n}_K(x)$, which has to be defined for each relation type between $\boldsymbol{F}$ and the unknowns.

These properties, together with adequate stability properties which are obtained by estimates on the approximate solution, will give some convergence properties of the finite volume scheme.

3. Comparison with other discretization techniques

The finite volume method is quite different from (but sometimes related to) the finite difference method or the finite element method. On these classical methods see, e.g., DAHLQUIST and BJÖRCK [1974], THOMÉE [1991], CIARLET [1978, 1991], ROBERTS and THOMAS [1991].

Roughly speaking, the principle of the finite difference method is, given a number of discretization points which may be defined by a mesh, to assign one discrete unknown per discretization point, and to write one equation per discretization point. At each discretization point, the derivatives of the unknown are replaced by finite differences through the use of Taylor expansions. The finite difference method becomes difficult to use when the coefficients involved in the equation are discontinuous (e.g., in the case of heterogeneous media). With the finite volume method, discontinuities of the coefficients will not be any problem if the mesh is chosen such that the discontinuities of the coefficients occur on the boundaries of the control volumes (see Sections 7 and 11, for elliptic problems). Note that the finite volume scheme is often called "finite difference scheme" or "cell centered difference scheme". Indeed, in the finite volume method, the finite difference approach can be used for the approximation of the fluxes on the boundary of the control volumes. Thus, the finite volume scheme differs from the finite difference scheme in that the finite difference approximation is used for the flux rather than for the operator itself.

The finite element method (see, e.g., CIARLET [1978]) is based on a variational formulation, which is written for both the continuous and the discrete problems, at least in the case of conformal finite element methods which are considered here. The variational formulation is obtained by multiplying the original equation by a "test function". The continuous unknown is then approximated by a linear combination of "shape" functions; these shape functions are the test functions for the discrete variational formulation (this is the so called "Galerkin expansion"); the resulting equation is integrated over the domain. The finite volume method is sometimes called a "discontinuous finite element method" since the original equation is multiplied by the characteristic function of each grid cell which is defined by $1_K(x) = 1$, if $x \in K$, $1_K(x) = 0$, if $x \notin K$, and the discrete unknown may be considered as a linear combination of shape functions. However, the techniques used to prove the convergence of finite element methods do not generally apply for this choice of test functions. In the following chapters, the finite volume method will be compared in more detail with the classical and the mixed finite element methods.

From the industrial point of view, the finite volume method is known as a robust and cheap method for the discretization of conservation laws (by robust, we mean a scheme which behaves well even for particularly difficult equations, such as nonlinear systems of hyperbolic equations and which can easily be extended to more realistic and physical contexts than the classical academic problems). The finite volume method is cheap thanks to short and reliable computational coding for complex problems. It may be more adequate than the finite difference method (which in particular requires a simple geometry). However, in some cases, it is difficult to design schemes which give enough precision. Indeed, the finite element method can be much more precise than the finite volume method when using higher-order polynomials, but it requires an adequate functional framework which is not always available in industrial problems. Other more precise methods are, e.g., particle methods or spectral methods but these methods can be more expensive and less robust than the finite volume method.

4. General guideline

The mathematical theory of finite volume schemes has recently been undertaken. Even though we choose here to refer to the class of scheme which is the object of our study as the "finite volume" method, we must point out that there are several methods with different names (box method, control volume finite element methods, balance method to cite only a few) which may be viewed as finite volume methods. The name "finite difference" has also often been used referring to the finite volume method. We shall mainly quote here the works regarding the mathematical analysis of the finite volume method, keeping in mind that there exist numerous works on applications of the finite volume methods in the applied sciences, some references to which may be found in the books which are cited below.

Finite volume methods for convection–diffusion equations seem to have been first introduced in the early sixties by TICHONOV and SAMARSKII [1962], SAMARSKII [1965, 1971].

The convergence theory of such schemes in several space dimensions has only recently been undertaken. In the case of vertex-centered finite volume schemes, studies were carried out by SAMARSKII, LAZAROV and MAKAROV [1987] in the case of Cartesian meshes, HEINRICH [1986], BANK and ROSE [1986], CAI [1991], CAI, MANDEL and MCCORMICK [1991] and VANSELOW [1996] in the case of unstructured meshes; see also MORTON and SÜLI [1991], SÜLI [1989], MACKENZIE and MORTON [1992], MORTON, STYNES and SÜLI [1997] and SHASHKOV [1987] in the case of quadrilateral meshes. Cell-centered finite volume schemes are addressed in MANTEUFFEL and WHITE [1986], FORSYTH and SAMMON [1988], WEISER and WHEELER [1988] and LAZAROV, MISHEV and VASSILEVSKI [1996] in the case of Cartesian meshes and in VASSILESKI, PETROVA and LAZAROV [1992], HERBIN [1995, 1996], LAZAROV and MISHEV [1996], MISHEV [1998] in the case of triangular or Voronoi meshes; let us also mention COUDIÈRE, VILA and VILLEDIEU [1996] and COUDIÈRE, VILA and VILLEDIEU [1999] where more general meshes are treated, with, however, a somewhat technical geometrical condition. In the pure diffusion case, the cell centered finite volume method has also been analyzed with finite element tools: AGOUZAL, BARANGER, MAITRE and OUDIN [1995], BARANGER, MAITRE and OUDIN [1996], ANGERMANN [1996], ARBOGAST, WHEELER and YOTOV [1997]. Semilinear convection–diffusion are studied in FEISTAUER, FELCMAN and LUKACOVA-MEDVIDOVA [1997] with a combined finite element-finite volume method, EYMARD, GALLOUËT and HERBIN [1999] with a pure finite volume scheme.

Concerning nonlinear hyperbolic conservation laws, the one-dimensional case is now classical; let us mention the following books on numerical methods for hyperbolic problems: GODLEWSKI and RAVIART [1991, 1996], LEVEQUE [1990], KRÖNER [1997], and references therein. In the multidimensional case, let us mention the convergence results which where obtained in CHAMPIER, GALLOUËT and HERBIN [1993], KRÖNER and ROKYTA [1994], COCKBURN, COQUEL and LEFLOCH [1995] and the error estimates of COCKBURN, COQUEL and LEFLOCH [1994] and VILA [1994] in the case of an explicit scheme and EYMARD, GALLOUËT, GHILANI and HERBIN [1998] in the case of explicit and implicit schemes.

The purpose of the following chapters is to lay out a mathematical framework for the convergence and error analysis of the finite volume method for the discretization of elliptic, parabolic or hyperbolic partial differential equations under conservative form, following the philosophy of the works of CHAMPIER, GALLOUËT and HERBIN [1993], HERBIN [1995], EYMARD, GALLOUËT, GHILANI and HERBIN [1998], and EYMARD, GALLOUËT and HERBIN [1999]. In order to do so, we shall describe the implementation of the finite volume method on some simple (linear or nonlinear) academic problems, and develop the tools which are needed for the mathematical analysis. This approach will help to determine the properties of finite volume schemes which lead to "good" schemes for complex applications.

Chapter II introduces the finite volume discretization of an elliptic operator in one space dimension. The resulting numerical scheme is compared to finite difference, finite element and mixed finite element methods in this particular case. An error estimate is given; this estimate is in fact contained in results shown later in the multidimensional case; however, with the one-dimensional case, one can already understand the basic principles of the convergence proof, and understand the difference with the proof of MANTEUFFEL and WHITE [1986] or FORSYTH and SAMMON [1988], which does not seem to generalize to the unstructured meshes. In particular, it is made clear that, although the finite volume scheme is not consistent in the finite difference sense since the truncation error does not tend to 0, the conservativity of the scheme, together with a consistent approximation of the fluxes and some "stability" allow the proof of convergence. The scheme and the error estimate are then generalized to the case of a more general elliptic operator allowing discontinuities in the diffusion coefficients. Finally, a semilinear problem is studied, for which a convergence result is proved. The principle of the proof of this result may be used for nonlinear problems in several space dimensions. It will be used in Chapter III in order to prove convergence results for linear problems when no regularity on the exact solution is known.

In Chapter III, the discretization of elliptic problems in several space dimensions by the finite volume method is presented. Structured meshes are shown to be an easy generalization of the one-dimensional case; unstructured meshes are then considered, for Dirichlet and Neumann conditions on the boundary of the domain. In both cases, admissible meshes are defined, and, following EYMARD, GALLOUËT and HERBIN [1999], convergence results (with no regularity on the data) and error estimates assuming a C^2 or H^2 regular solution to the continuous problems are proved. As in the one-dimensional case, the conservativity of the scheme, together with a consistent approximation of the fluxes and some "stability" are used for the proof of convergence. In addition to the properties already used in the one-dimensional case, the multidimensional estimates require the use of a "discrete Poincaré" inequality which is proved in both Dirichlet and Neumann cases, along with some compactness properties which are also used and are given in the last section. It is then shown how to deal with matrix diffusion coefficients and more general boundary conditions. Singular sources and mesh refinement are also studied.

Chapter IV deals with the discretization of parabolic problems. Using the same concepts as in Chapter III, an error estimate is given in the linear case. A nonlinear degenerate parabolic problem is then studied, for which a convergence result is proved, thanks to a uniqueness result which is proved at the end of the chapter.

Chapter V introduces the finite volume discretization of a hyperbolic operator in one space dimension. Some basics on entropy weak solutions to nonlinear hyperbolic equations are recalled. Then the concept of stability of a scheme is explained on a simple linear advection problem, for which both finite difference and finite volume schemes are considered. Some well known schemes are presented with a finite volume formulation in the nonlinear case. A proof of convergence using a "weak BV inequality" which was found to be crucial in the multidimensional case (Chapter VI) is given in the one-dimensional case for the sake of clarity. For the sake of completeness, the proof of convergence based on "strong BV estimates" and the Lax–Wendroff theorem is also recalled, although it does not seem to extend to the multidimensional case with general meshes.

In Chapter VI, finite volume schemes for the discretization of multidimensional nonlinear hyperbolic conservation equations are studied. Under suitable assumptions, which are satisfied by several well known schemes, it is shown that the considered schemes are L^∞ stable (this is classical) but also satisfy some "weak BV inequality". This "weak BV" inequality is the key estimate to the proof of convergence of the schemes. Following EYMARD, GALLOUËT, GHILANI and HERBIN [1998], both time implicit and explicit discretizations are considered. The existence of the solution to the implicit scheme is proved. The approximate solutions are shown to satisfy some discrete entropy inequalities. Using the weak BV estimate, the approximate solution is also shown to satisfy some continuous entropy inequalities. Introducing the concept of "entropy process solution" to the nonlinear hyperbolic equations (which is similar to the notion of measure valued solutions of DIPERNA [1985]), the approximate solutions are proved to converge towards an entropy process solution as the mesh size tends to 0. The entropy process solution is shown to be unique, and is therefore equal to the entropy weak solution, which concludes the convergence of the approximate solution towards the entropy weak solution. Finally error estimates are proved for both the explicit and implicit schemes.

The last chapter is concerned with systems of equations. In the case of hyperbolic systems which are considered in the first part, little is known concerning the continuous problem, so that the schemes which are introduced are only shown to be efficient by numerical experimentation. These "rough" schemes seem to be efficient for complex cases such as the Euler equations for real gases. The incompressible Navier–Stokes equations are then considered; after recalling the classical staggered grid finite volume formulation (see, e.g., PATANKAR [1980]), a finite volume scheme defined on a triangular mesh for the Stokes equation is studied. In the case of equilateral triangles, the tools of Chapter III allow to show that the approximate velocities converge to the exact velocities. Systems arising from modelling multiphase flow in porous media are then considered.

The convergence of the approximate finite volume solution for a simplified case is then proved with the tools introduced in Chapter VI.

More precise references to recent works on the convergence of finite volume methods will be made in the following chapters. However, we shall not quote here the numerous works on applications of the finite volume methods in the applied sciences.

CHAPTER II

A One-Dimensional Elliptic Problem

The purpose of this chapter is to give some developments of the Example 1.2 of the introduction in the one-dimensional case. The formalism needed to define admissible finite volume meshes is first given and applied to the Dirichlet problem. After some comparisons with other relevant schemes, convergence theorems and error estimates are provided. Then, the case of general linear elliptic equations is handled and, finally, a first approach of a nonlinear problem is studied. It yields an application of compactness theorems, which will be useful in further chapters, in a quite simple framework.

5. A finite volume method for the Dirichlet problem

5.1. Formulation of a finite volume scheme

The principle of the finite volume method will be shown here on the academic Dirichlet problem, namely a second-order differential operator without time dependent terms and with homogeneous Dirichlet boundary conditions. Let f be a given function from $(0, 1)$ to $\mathbb{R}$, consider the following differential equation:

$$\begin{aligned} -u_{xx}(x) &= f(x), \quad x \in (0,1), \\ u(0) &= 0, \\ u(1) &= 0. \end{aligned} \tag{5.1}$$

If $f \in C([0,1],\mathbb{R})$, there exists a unique solution $u \in C^2([0,1],\mathbb{R})$ to problem (5.1). In the sequel, this exact solution will be denoted by u. Note that the equation $-u_{xx} = f$ can be written in the conservative form $\operatorname{div}(\boldsymbol{F}) = f$ with $\boldsymbol{F} = -u_x$.

In order to compute a numerical approximation to the solution of this equation, let us define a mesh, denoted by $\mathcal{T}$, of the interval $(0, 1)$ consisting of N cells (or control volumes), denoted by K_i, $i = 1, \dots, N$, and N points of $(0, 1)$, denoted by x_i, $i = 1, \dots, N$, satisfying the following assumptions:

DEFINITION 5.1 (*Admissible one-dimensional mesh*). An admissible mesh of $(0, 1)$, denoted by $\mathcal{T}$, is given by a family $(K_i)_{i=1,\dots,N}$, $N \in \mathbb{N}^\star$, such that $K_i = (x_{i-1/2}, x_{i+1/2})$, and a family $(x_i)_{i=0,\dots,N+1}$ such that

$$\begin{aligned} x_0 = x_{1/2} = 0 &< x_1 < x_{3/2} < \cdots < x_{i-1/2} < x_i < x_{i+1/2} < \cdots \\ &< x_N < x_{N+1/2} = x_{N+1} = 1. \end{aligned}$$

One sets

$$h_i = \mathrm{m}(K_i) = x_{i+1/2} - x_{i-1/2}, \quad i = 1, \ldots, N, \text{ and therefore } \sum_{i=1}^{N} h_i = 1,$$

$$h_i^- = x_i - x_{i-1/2}, \qquad h_i^+ = x_{i+1/2} - x_i, \quad i = 1, \ldots, N,$$

$$h_{i+1/2} = x_{i+1} - x_i, \quad i = 0, \ldots, N,$$

$$\mathrm{size}(\mathcal{T}) = h = \max\{h_i,\ i = 1, \ldots, N\}.$$

The discrete unknowns are denoted by u_i, $i = 1, \ldots, N$, and are expected to be some approximation of u in the cell K_i (the discrete unknown u_i can be viewed as an approximation of the mean value of u over K_i, or of the value of $u(x_i)$, or of other values of u in the control volume K_i ...). The first equation of (5.1) is integrated over each cell K_i, as in (2.3) and yields

$$-u_x(x_{i+1/2}) + u_x(x_{i-1/2}) = \int_{K_i} f(x)\,\mathrm{d}x, \quad i = 1, \ldots, N.$$

A reasonable choice for the approximation of $-u_x(x_{i+1/2})$ (at least, for $i = 1, \ldots, N - 1$) seems to be the differential quotient

$$F_{i+1/2} = -\frac{u_{i+1} - u_i}{h_{i+1/2}}.$$

This approximation is consistent in the sense that, if $u \in C^2([0, 1], \mathbb{R})$, then

$$F_{i+1/2}^{\star} = -\frac{u(x_{i+1}) - u(x_i)}{h_{i+1/2}} = -u_x(x_{i+1/2}) + \mathrm{o}(h), \tag{5.2}$$

where $|\mathrm{o}(h)| \leqslant Ch$, $C \in \mathbb{R}_+$ only depending on u.

REMARK 5.1. Assume that x_i is the center of K_i. Let $\tilde{u}_i$ denote the mean value over K_i of the exact solution u to problem (5.1). One may then remark that $|\tilde{u}_i - u(x_i)| \leqslant Ch_i^2$, with some C only depending on u; it follows easily that $(\tilde{u}_{i+1} - \tilde{u}_i)/h_{i+1/2} = u_x(x_{i+1/2}) + \mathrm{o}(h)$ also holds, for $i = 1, \ldots, N - 1$ (recall that $h = \max\{h_i,\ i = 1, \ldots, N\}$). Hence the approximation of the flux is also consistent if the discrete unknowns u_i, $i = 1, \ldots, N$, are viewed as approximations of the mean value of u in the control volumes.

The Dirichlet boundary conditions are taken into account by using the values imposed at the boundaries to compute the fluxes on these boundaries. Taking these boundary conditions into consideration and setting $f_i = \frac{1}{h_i} \int_{K_i} f(x)\,\mathrm{d}x$ for $i = 1, \ldots, N$ (in an actual computation, an approximation of f_i by numerical integration can be used), the finite volume scheme for problem (5.1) writes

$$F_{i+1/2} - F_{i-1/2} = h_i f_i, \quad i = 1, \ldots, N, \tag{5.3}$$

$$F_{i+1/2} = -\frac{u_{i+1} - u_i}{h_{i+1/2}}, \quad i = 1, \dots, N-1, \tag{5.4}$$

$$F_{1/2} = -\frac{u_1}{h_{1/2}}, \tag{5.5}$$

$$F_{N+1/2} = \frac{u_N}{h_{N+1/2}}. \tag{5.6}$$

Note that (5.4), (5.5), (5.6) may also be written

$$F_{i+1/2} = -\frac{u_{i+1} - u_i}{h_{i+1/2}}, \quad i = 0, \dots, N, \tag{5.7}$$

setting

$$u_0 = u_{N+1} = 0. \tag{5.8}$$

The numerical scheme (5.3)–(5.6) may be written under the following matrix form:

$$AU = b, \tag{5.9}$$

where $U = (u_1, \dots, u_N)^t$, $b = (b_1, \dots, b_N)^t$, with (5.8) and with A and b defined by

$$(AU)_i = \frac{1}{h_i}\left(-\frac{u_{i+1} - u_i}{h_{i+1/2}} + \frac{u_i - u_{i-1}}{h_{i-1/2}}\right), \quad i = 1, \dots, N, \tag{5.10}$$

$$b_i = \frac{1}{h_i}\int_{K_i} f(x)\,\mathrm{d}x, \quad i = 1, \dots, N. \tag{5.11}$$

REMARK 5.2. There are other finite volume schemes for problem (5.1).

(1) For instance, it is possible, in Definition 5.1, to take $x_1 \geqslant 0$, $x_N \leqslant 1$ and, for the definition of the scheme (that is (5.3)–(5.6)), to write (5.3) only for $i = 2, \dots, N-1$ and to replace (5.5) and (5.6) by $u_1 = u_N = 0$ (note that (5.4) does not change). For this so called "modified finite volume" scheme, it is also possible to obtain an error estimate as for the scheme (5.3)–(5.6) (see Remark 6.2). Note that, with this scheme, the union of all control volumes for which the "conservation law" is written is slightly different from [0, 1] (namely $[x_{3/2}, x_{N-1/2}] \neq [0, 1]$).

(2) Another possibility is to take (primary) unknowns associated to the boundaries of the control volumes. We shall not consider this case here (cf. KELLER [1971], COURBET and CROISILLE [1998]).

5.2. *Comparison with a finite difference scheme*

With the same notations as in Section 5.1, consider that u_i is now an approximation of $u(x_i)$. It is interesting to notice that the expression

$$\frac{1}{h_i}\left(-\frac{u_{i+1}-u_i}{h_{i+1/2}}+\frac{u_i-u_{i-1}}{h_{i-1/2}}\right)$$

is not a consistent approximation of $-u_{xx}(x_i)$ in the finite difference sense, i.e. the error made by replacing the derivative by a difference quotient (the truncation error DAHLQUIST and BJÖRCK [1974]) does not tend to 0 as h tends to 0. Indeed, let $\overline{U}=(u(x_1),\dots,u(x_N))^t$; with the notations of (5.9)–(5.11), the truncation error may be defined as

$$r=A\overline{U}-b,$$

with $r=(r_1,\dots,r_N)^t$. Note that for f regular enough, which is assumed in the sequel, $b_i=f(x_i)+\mathrm{o}(h)$. An estimate of r is obtained by using Taylor's expansion:

$$u(x_{i+1})=u(x_i)+h_{i+1/2}u_x(x_i)+\tfrac{1}{2}h_{i+1/2}^2u_{xx}(x_i)+\tfrac{1}{6}h_{i+1/2}^3u_{xxx}(\xi_i),$$

for some $\xi_i\in(x_i,x_{i+1})$, which yields

$$r_i=-\frac{1}{h_i}\frac{h_{i+1/2}+h_{i-1/2}}{2}u_{xx}(x_i)+u_{xx}(x_i)+\mathrm{o}(h),\quad i=1,\dots,N,$$

which does not, in general tend to 0 as h tends to 0 (except in particular cases) as may be seen on the simple following example:

EXAMPLE 5.1. Let $f\equiv 1$ and consider a mesh of (0, 1), in the sense of Definition 5.1, satisfying $h_i=h$ for even i, $h_i=h/2$ for odd i and $x_i=(x_{i+1/2}t+x_{i-1/2})/2$, for $i=1,\dots,N$. An easy computation shows that the truncation error r is such that

$$\begin{aligned} r_i &= -\tfrac{1}{4}, && \text{for even } i,\\ r_i &= +\tfrac{1}{2}, && \text{for odd } i. \end{aligned}$$

Hence $\sup\{|r_i|,\ i=1,\dots,N\}\nrightarrow 0$ as $h\to 0$.

Therefore, the scheme obtained from (5.3)–(5.6) is not consistent in the finite difference sense, even though it is consistent in the finite volume sense, i.e., the numerical approximation of the fluxes is conservative and the truncation error on the fluxes tends to 0 as h tends to 0.

If, e.g., x_i is the center of K_i, for $i = 1, \dots, N$, it is well known that for problem (5.1), the consistent finite difference scheme would be, omitting boundary conditions,

$$\frac{4}{2h_i + h_{i-1} + h_{i+1}}\left[-\frac{u_{i+1} - u_i}{h_{i+1/2}} + \frac{u_i - u_{i-1}}{h_{i-1/2}}\right] = f(x_i), \quad i = 2, \dots, N-1. \tag{5.12}$$

REMARK 5.3. Assume that x_i is, for $i = 1, \dots, N$, the center of K_i and that the discrete unknown u_i of the finite volume scheme is considered as an approximation of the mean value $\tilde{u}_i$ of u over K_i (note that $\tilde{u}_i = u(x_i) + (h_i^2/24)u_{xx}(x_i) + \mathrm{o}(h^3)$, if $u \in C^3([0,1], \mathbb{R})$) instead of $u(x_i)$, then again, the finite volume scheme, considered once more as a finite difference scheme, is not consistent in the finite difference sense. Indeed, let $\tilde{R} = A\tilde{U} - b$, with $\tilde{U} = (\tilde{u}_1, \dots, \tilde{u}_N)^t$, and $\tilde{R} = (\tilde{R}_1, \dots, \tilde{R}_N)^t$, then, in general, $\tilde{R}_i$ does not go to 0 as h goes to 0. In fact, it will be shown later that the finite volume scheme, when seen as a finite difference scheme, is consistent in the finite difference sense if u_i is considered as an approximation of $u(x_i) - (h_i^2/8)u_{xx}(x_i)$. This is the idea upon which the first proof of convergence by Forsyth and Sammon in 1988 is based, see FORSYTH and SAMMON [1988] and Section 6.2.

In the case of problem (5.1), both the finite volume and finite difference schemes are convergent. The finite difference scheme (5.12) is convergent since it is stable, in the sense that $\|X\|_\infty \leqslant C\|AX\|_\infty$, for all $X \in \mathbb{R}^N$, where C is a constant and $\|X\|_\infty = \sup(|X_1|, \dots, |X_N|)$, $X = (X_1, \dots, X_N)^t$, and consistent in the usual finite difference sense. Since $A(\overline{U} - U) = R$, the stability property implies that $\|\overline{U} - U\|_\infty \leqslant C\|R\|_\infty$ which goes to 0, as h goes to 0, by definition of the consistency in the finite difference sense. The convergence of the finite volume scheme (5.3)–(5.6) needs some more work and is described in Section 6.1.

5.3. *Comparison with a mixed finite element method*

The finite volume method has often be thought of as a kind of mixed finite element method. Nevertheless, we show here that, on the simple Dirichlet problem (5.1), the two methods yield two different schemes. For problem (5.1), the discrete unknowns of the finite volume method are the values u_i, $i = 1, \dots, N$. However, the finite volume method also introduces one discrete unknown at each of the control volume extremities, namely the numerical flux between the corresponding control volumes. Hence, the finite volume method for elliptic problems may appear closely related to the mixed finite element method. Recall that the mixed finite element method consists in introducing in problem (5.1) the auxiliary variable $q = -u_x$, which yields the following system:

$$q + u_x = 0,$$
$$q_x = f;$$

assuming $f \in L^2((0,1))$, a variational formulation of this system is:

$$q \in H^1\big((0,1)\big), \qquad u \in L^2\big((0,1)\big), \tag{5.13}$$

$$\int_0^1 q(x)p(x)\,\mathrm{d}x = \int_0^1 u(x)p_x(x)\,\mathrm{d}x, \quad \forall p \in H^1\big((0,1)\big), \tag{5.14}$$

$$\int_0^1 q_x(x)v(x)\,\mathrm{d}x = \int_0^1 f(x)v(x)\,\mathrm{d}x, \quad \forall v \in L^2\big((0,1)\big). \tag{5.15}$$

Considering an admissible mesh of (0, 1) (see Definition 5.1), the usual discretization of this variational formulation consists in taking the classical piecewise linear finite element functions for the approximation H of $H^1((0,1))$ and the piecewise constant finite element for the approximation L of $L^2((0,1))$. Then, the discrete unknowns are $\{u_i,\ i=1,\dots,N\}$ and $\{q_{i+1/2},\ i=0,\dots,N\}$ (u_i is an approximation of u in K_i and $q_{i+1/2}$ is an approximation of $-u_x(x_{i+1/2})$). The discrete equations are obtained by performing a Galerkin expansion of u and q with respect to the natural basis functions $\psi_l, l=1,\dots,N$ (spanning L), and $\varphi_{j+1/2}$, $j=0,\dots,N$ (spanning H), and by taking $p=\varphi_{i+1/2}$, $i=0,\dots,N$, in (5.14) and $v=\psi_k$, $k=1,\dots,N$, in (5.15). Let $h_0 = h_{N+1}=0$, $u_0=u_{N+1}=0$ and $q_{-1/2}=q_{N+3/2}=0$; then the discrete system obtained by the mixed finite element method has $2N+1$ unknowns. It writes

$$q_{i+1/2}\left(\frac{h_i+h_{i+1}}{3}\right)+q_{i-1/2}\left(\frac{h_i}{6}\right)+q_{i+3/2}\left(\frac{h_{i+1}}{6}\right)=u_i-u_{i+1}, \quad i=0,\dots,N,$$

$$q_{i+1/2}-q_{i-1/2}=\int_{K_i} f(x)\,\mathrm{d}x, \quad i=1,\dots,N.$$

Note that the unknowns $q_{i+1/2}$ cannot be eliminated from the system. The resolution of this system of equations does not give the same values $\{u_i,\ i=1,\dots,N\}$ than those obtained by using the finite volume scheme (5.3)–(5.6). In fact it is easily seen that, in this case, the finite volume scheme can be obtained from the mixed finite element scheme by using the following numerical integration for the left-hand side of (5.14):

$$\int_{K_i} g(x)\,\mathrm{d}x = \frac{g(x_{i+1})+g(x_i)}{2}h_i.$$

This is also true for some two-dimensional elliptic problems and therefore the finite volume error estimates for these problems may be obtained via the mixed finite element theory, see AGOUZAL, BARANGER, MAITRE and OUDIN [1995], BARANGER, MAITRE and OUDIN [1996].

6. Convergence theorems and error estimates for the Dirichlet problem

6.1. *A finite volume error estimate in a simple case*

We shall now prove the following error estimate, which will be generalized to more general elliptic problems, and in higher dimensions of space.

THEOREM 6.1. *Let $f \in C([0,1],\mathbb{R})$ and let $u \in C^2([0,1],\mathbb{R})$ be the (unique) solution of problem (5.1). Let $\mathcal{T} = (K_i)_{i=1,\dots,N}$ be an admissible mesh in the sense of Definition 5.1. Then, there exists a unique vector $U = (u_1,\dots,u_N)^t \in \mathbb{R}^N$ solution to (5.3)–(5.6) and there exists $C \geqslant 0$, only depending on u, such that*

$$\sum_{i=0}^{N} \frac{(e_{i+1}-e_i)^2}{h_{i+1/2}} \leqslant C^2h^2, \tag{6.1}$$

and

$$|e_i| \leqslant Ch, \quad \forall i \in \{1,\dots,N\}, \tag{6.2}$$

with $e_0 = e_{N+1} = 0$ and $e_i = u(x_i) - u_i$, for all $i \in \{1,\dots,N\}$.

This theorem is in fact a consequence of Theorem 7.1, which gives an error estimate for the finite volume discretization of a more general operator. However, we now give the proof of the error estimate in this first simple case.

PROOF. First remark that there exists a unique vector $U = (u_1,\dots,u_N)^t \in \mathbb{R}^N$ solution to (5.3)–(5.6). Indeed, multiplying (5.3) by u_i and summing for $i = 1,\dots,N$ gives

$$\frac{u_1^2}{h_{1/2}} + \sum_{i=1}^{N-1} \frac{(u_{i+1}-u_i)^2}{h_{i+1/2}} + \frac{u_N^2}{h_{N+1/2}} = \sum_{i=1}^{N} u_i h_i f_i.$$

Therefore, if $f_i = 0$ for any $i \in \{1,\dots,N\}$, then the unique solution to (5.3) is obtained by taking $u_i = 0$, for any $i \in \{1,\dots,N\}$. This gives existence and uniqueness of $U = (u_1,\dots,u_N)^t \in \mathbb{R}^N$ solution to (5.3) (with (5.4)–(5.6)).

One now proves (6.1). Let

$$\overline{F}_{i+1/2} = -u_x(x_{i+1/2}), \quad i = 0,\dots,N.$$

Integrating the equation $-u_{xx} = f$ over K_i yields

$$\overline{F}_{i+1/2} - \overline{F}_{i-1/2} = h_i f_i, \quad i = 1,\dots,N.$$

By (5.3), the numerical fluxes $F_{i+1/2}$ satisfy

$$F_{i+1/2} - F_{i-1/2} = h_i f_i, \quad i = 1,\dots,N.$$

Therefore, with $G_{i+1/2} = \overline{F}_{i+1/2} - F_{i+1/2}$,

$$G_{i+1/2} - G_{i-1/2} = 0, \quad i = 1,\dots,N.$$

Using the consistency of the fluxes (5.2), there exists $C > 0$, only depending on u, such that

$$F^{\star}_{i+1/2} = \overline{F}_{i+1/2} + R_{i+1/2} \quad \text{and} \quad |R_{i+1/2}| \leqslant Ch, \tag{6.3}$$

Hence with $e_i = u(x_i) - u_i$, for $i = 1, \ldots, N$, and $e_0 = e_{N+1} = 0$, one has

$$G_{i+1/2} = -\frac{e_{i+1} - e_i}{h_{i+1/2}} - R_{i+1/2}, \quad i = 0, \ldots, N,$$

so that $(e_i)_{i=0,\ldots,N+1}$ satisfies

$$-\frac{e_{i+1} - e_i}{h_{i+1/2}} - R_{i+1/2} + \frac{e_i - e_{i-1}}{h_{i-1/2}} + R_{i-1/2} = 0, \quad \forall i \in \{1, \ldots, N\}. \tag{6.4}$$

Multiplying (6.4) by e_i and summing over $i = 1, \ldots, N$ yields

$$-\sum_{i=1}^{N} \frac{(e_{i+1} - e_i)e_i}{h_{i+1/2}} + \sum_{i=1}^{N} \frac{(e_i - e_{i-1})e_i}{h_{i-1/2}} = -\sum_{i=1}^{N} R_{i-1/2}e_i + \sum_{i=1}^{N} R_{i+1/2}e_i.$$

Noting that $e_0 = 0$, $e_{N+1} = 0$ and reordering by parts, this yields (with (6.3))

$$\sum_{i=0}^{N} \frac{(e_{i+1} - e_i)^2}{h_{i+1/2}} \leqslant Ch \sum_{i=0}^{N} |e_{i+1} - e_i|. \tag{6.5}$$

The Cauchy–Schwarz inequality applied to the right-hand side gives

$$\sum_{i=0}^{N} |e_{i+1} - e_i| \leqslant \left(\sum_{i=0}^{N} \frac{(e_{i+1} - e_i)^2}{h_{i+1/2}} \right)^{1/2} \left(\sum_{i=0}^{N} h_{i+1/2} \right)^{1/2}. \tag{6.6}$$

Since $\sum_{i=0}^{N} h_{i+1/2} = 1$ in (6.6) and from (6.5), one deduces (6.1).

Since, for all $i \in \{1, \ldots, N\}$, $e_i = \sum_{j=1}^{i} (e_j - e_{j-1})$, one can deduce, from (6.6) and (6.1) that (6.2) holds. □

REMARK 6.1. The error estimate given in this section does not use the discrete maximum principle (that is the fact that $f_i \geqslant 0$, for all $i = 1, \ldots, N$, implies $u_i \geqslant 0$, for all $i = 1, \ldots, N$), which is used in the proof of error estimates by the finite difference techniques, but the coerciveness of the elliptic operator, as in the proof of error estimates by the finite element techniques.

REMARK 6.2. (1) The above proof of convergence gives an error estimate of order h. It is sometimes possible to obtain an error estimate of order h^2. Indeed, this is the case, at least if $u \in C^4([0, 1], \mathbb{R})$, if x_i is the center of K_i for all $i = 1, \ldots, N$. One obtains, in this case, $|e_i| \leqslant Ch^2$, for all $i \in \{1, \dots, N\}$, where C only depends on u (see FORSYTH and SAMMON [1988] or BARANGER, MAITRE and OUDIN [1996]).

(2) It is also possible to obtain an error estimate for the modified finite volume scheme described in the first item of Remark 5.2. It is even possible to obtain an error estimate of order h^2 in the case $x_1 = 0$, $x_N = 1$ and assuming that $x_{i+1/2} = (1/2)(x_i + x_{i+1})$, for all $i = 1, \dots, N-1$. In fact, in this case, one obtains $|R_{i+1/2}| \leqslant C_1 h^2$, for all $i = 1, \dots, N-1$. Then, the proof of Theorem 6.1 gives (6.1) with h^4 instead of h^2 which yields $|e_i| \leqslant C_2 h^2$, for all $i \in \{1, \dots, N\}$ (where C_1 and C_2 are only depending on u). Note that this modified finite volume scheme is also consistent in the finite difference sense. Then, the finite difference techniques yield also an error estimate on $|e_i|$, but only of order h.

(3) It could be tempting to try and find error estimates with respect to the mean value of the exact solution on the control volumes rather than with respect to its value at some point of the control volumes. This is not such a good idea: indeed, if x_i is not the center of K_i (this will be the general case in several space dimensions), then one does not have (in general) $|\tilde{e}_i| \leqslant C_3 h^2$ (for some C_3 only depending on u) with $\tilde{e}_i = \tilde{u}_i - u_i$ where $\tilde{u}_i$ denotes the mean value of u over K_i.

REMARK 6.3. (1) If the assumption $f \in C([0,1], \mathbb{R})$ is replaced by the assumption $f \in L^2((0,1))$ in Theorem 6.1, then $u \in H^2((0,1))$ instead of $C^2([0,1], \mathbb{R})$, but the estimates of Theorem 6.1 still hold. Then, the consistency of the fluxes must be obtained with a Taylor expansion with an integral remainder. This is feasible for C^2 functions, and since the remainder only depends on the H^2 norm, a density argument allows to conclude; see also Theorem 9.4 and EYMARD, GALLOUËT and HERBIN [1999].

(2) If the assumption $f \in C([0,1], \mathbb{R})$ is replaced by the assumption $f \in L^1((0,1))$ in Theorem 6.1, then $u \in C^2([0,1], \mathbb{R})$ no longer holds (and even $u \in H^2((0,1))$), but the convergence still holds; indeed there exists $C(u,h)$, only depending on u and h, such that $C(u,h) \to 0$, as $h \to 0$, and $|e_i| \leqslant C(u,h)$, for all $i = 1, \dots, N$. The proof is similar to the one above, except that the estimate (6.3) is replaced by $|R_{i+1/2}| \leqslant C_1(u,h)$, for all $i = 0, \dots, N$, with some $C_1(u,h)$, only depending on u and h, such that $C(u,h) \to 0$, as $h \to 0$.

REMARK 6.4. Estimate (6.1) can be interpreted as a "discrete H_0^1" estimate on the error. A theoretical result which underlies the L^∞ (6.2) is the fact that if Ω is an open bounded subset of $\mathbb{R}$, then $H_0^1(\Omega)$ is imbedded in $L^\infty(\Omega)$. This is no longer true in higher dimension. In two space dimensions, for instance, a discrete version of the imbedding of H_0^1 in L^p allows to obtain (see, e.g., FIARD [1994]) $\|e\|_p \leqslant Ch$, for all finite p, which in turn yields $\|e\|_\infty \leqslant Ch \ln h$ for convenient meshes (see Corollary 9.1).

The important features needed for the above proof seem to be the consistency of the approximation of the fluxes and the conservativity of the scheme; this conservativity is natural the fact that the scheme is obtained by integrating the equation over each cell, and the approximation of the flux on any interface is obtained by taking into account the flux balance (continuity of the flux in the case of no source term on the interface).

The above proof generalizes to other elliptic problems, such as a convection–diffusion equation of the form $-u_{xx}+au_x+bu=f$, and to equations of the form $-(\lambda u_x)_x=f$ where $\lambda \in L^\infty$ may be discontinuous, and is such that there exist α and β in $\mathbb{R}_+^\star$ such that $\alpha \leqslant \lambda \leqslant \beta$. These generalizations are studied in the next section. Other generalizations include similar problems in 2 (or 3) space dimensions, with meshes consisting of rectangles (parallelepipeds), triangles (tetrahedra), or general meshes of Voronoï type, and the corresponding evolutive (parabolic) problems. These generalizations will be addressed in further chapters.

Let us now give a proof of estimate (6.2), under slightly different conditions, which uses finite difference techniques.

6.2. *An error estimate using finite difference techniques*

Convergence can be obtained via a method similar to that of the finite difference proof of convergence (following, e.g., FORSYTH and SAMMON [1988], MANTEUFFEL and WHITE [1986], FAILLE [1992a]). Most of these methods, are, however, limited to the finite volume method for problem (5.1). Using the notations of Section 5.2 (recall that $\overline{U}=(u(x_1),\dots,u(x_N))^t$, and $r=A\overline{U}-b=\mathrm{o}(1)$), the idea is to find $\overline{\overline{U}}$ "close" to $\overline{U}$, such that

$$A\overline{\overline{U}}=b+\overline{\overline{r}}, \quad \text{with } \overline{\overline{r}}=\mathrm{o}(h).$$

This value of $\overline{\overline{U}}$ was found in FORSYTH and SAMMON [1988] and is such that $\overline{\overline{U}}=\overline{U}-V$, where

$$V=(v_1,\dots,v_N)^t \quad \text{and} \quad v_i=\frac{h_i^2 u_{xx}(x_i)}{8}, \qquad i=1,\dots,N.$$

Then, one may decompose the truncation error as

$$r=A(\overline{U}-U)=AV+\overline{\overline{r}} \quad \text{with } \|V\|_\infty=\mathrm{o}\big(h^2\big) \text{ and } \overline{\overline{r}}=\mathrm{o}(h).$$

The existence of such a V is given in Lemma 6.1. In order to prove the convergence of the scheme, a stability property is established in Lemma 6.2.

LEMMA 6.1. *Let $\mathcal{T}=(K_i)_{i=1,\dots,N}$ be an admissible mesh of $(0,1)$, in the sense of Definition 5.1, such that x_i is the center of K_i for all $i=1,\dots,N$. Let $\alpha_\mathcal{T}>0$ be such that $h_i>\alpha_\mathcal{T} h$ for all $i=1,\dots,N$ (recall that $h=\max\{h_1,\dots,h_N\}$). Let $\overline{U}=(u(x_1),\dots,u(x_N))^t\in\mathbb{R}^N$, where u is the solution to (5.1), and assume $u\in C^3([0,1],\mathbb{R})$. Let A be the matrix defining the numerical scheme, given in (5.10). Then there exists a unique $U=(u_1,\dots,u_N)$ solution of (5.3)–(5.6) and there exists $\overline{\overline{r}}$ and $V\in\mathbb{R}^N$ such that*

$$r=A(\overline{U}-U)=AV+\overline{\overline{r}}, \quad \textit{with } \|V\|_\infty\leqslant Ch^2 \textit{ and } \big\|\overline{\overline{r}}\big\|_\infty\leqslant Ch,$$

where C only depends on u and $\alpha_\mathcal{T}$.

PROOF. The existence and uniqueness of U is classical (it is also proved in Theorem 6.1).

For $i = 0, \dots N$, define

$$R_{i+1/2} = -\frac{u(x_{i+1}) - u(x_i)}{h_{i+1/2}} + u_x(x_{i+1/2}).$$

Remark that

$$r_i = \frac{1}{h_i}(R_{i+1/2} - R_{i-1/2}), \quad \text{for } i = 0, \dots, N, \tag{6.7}$$

where r_i is the ith component of $r = A(\overline{U} - U)$.

The computation of $R_{i+1/2}$ yields

$$R_{i+1/2} = -\tfrac{1}{4}(h_{i+1} - h_i)u_{xx}(x_{i+1/2}) + \mathrm{o}(h^2), \quad i = 1, \dots, N-1,$$
$$R_{1/2} = -\tfrac{1}{4}h_1 u_{xx}(0) + \mathrm{o}(h^2), \qquad R_{N+1/2} = \tfrac{1}{4}h_N u_{xx}(1) + \mathrm{o}(h^2).$$

Define $V = (v_1, \dots, v_N)^t$ with $v_i = h_i^2 u_{xx}(x_i)/8$, $i = 1, \dots, N$. Then,

$$-\frac{v_{i+1} - v_i}{h_{i+1/2}} = R_{i+1/2} + \mathrm{o}(h^2), \quad i = 1, \dots, N-1,$$

$$-\frac{2v_1}{h_1} = R_{1/2} + \mathrm{o}(h^2),$$

$$\frac{2v_N}{h_N} = R_{N+1/2} + \mathrm{o}(h^2).$$

Since $h_i \geqslant \alpha_{\mathcal{T}} h$, for $i = 1, \dots, N$, replacing $R_{i+1/2}$ in (6.7) gives that $r_i = (AV)_i + \mathrm{o}(h)$, for $i = 1, \dots, N$, and $\|V\|_\infty = \mathrm{o}(h^2)$. Hence the lemma is proved. □

LEMMA 6.2 (Stability). *Let $\mathcal{T} = (K_i)_{i=1,\dots,N}$ be an admissible mesh of* $[0, 1]$ *in the sense of Definition* 5.1. *Let A be the matrix defining the finite volume scheme given in* (5.10). *Then A is invertible and*

$$\|A\|_\infty^{-1} \leqslant \tfrac{1}{4}. \tag{6.8}$$

PROOF. First we prove a discrete maximum principle; indeed if $b_i \geqslant 0$, for all $i = 1, \dots, N$, and if U is solution of $AU = b$ then we prove that $u_i \geqslant 0$ for all $i = 1, \dots, N$.

Let $a = \min\{u_i, i = 0, \dots, N+1\}$ (recall that $u_0 = u_{N+1} = 0$) and $i_0 = \min\{i \in \{0, \dots, N+1\};\ u_i = a\}$.

If $i_0 \neq 0$ and $i_0 \neq N+1$, then

$$\frac{1}{h_{i_0}}\left(\frac{u_{i_0}-u_{i_0-1}}{h_{i_0-1/2}}-\frac{u_{i_0+1}-u_{i_0}}{h_{i_0+1/2}}\right)=b_{i_0}\geqslant 0,$$

this is impossible since $u_{i_0+1}-u_{i_0}\geqslant 0$ and $u_{i_0}-u_{i_0-1}<0$, by definition of i_0. Therefore, $i_0=0$ or $N+1$. Then, $a=0$ and $u_i\geqslant 0$ for all $i=1,\dots,N$.

Note that, by linearity, this implies that A is invertible.

Next, we shall prove that there exists $M>0$ such that $\|A^{-1}\|_\infty\leqslant M$ (indeed, $M=1/4$ is convenient). Let ϕ be defined on $[0,1]$ by $\phi(x)=\frac{1}{2}x(1-x)$. Then $-\phi_{xx}(x)=1$ for all $x\in[0,1]$. Let $\Phi=(\phi_1,\dots,\phi_N)$ with $\phi_i=\phi(x_i)$; if A represented the usual finite difference approximation of the second-order derivative, then we would have $A\Phi=1$, since the difference quotient approximation of the second-order derivative of a second-order polynomial is exact ($\phi_{xxx}=0$). Here, with the finite volume scheme (5.3)–(5.6), we have $A\Phi-\mathbf{1}=AW$ (where $\mathbf{1}$ denotes the vector of $\mathbb{R}^N$ the components of which are all equal to 1), with $W=(w_1,\dots,w_N)\in\mathbb{R}^N$ such that $W_i=-h_i^2/8$ (see proof of Lemma 6.1). Let $b\in\mathbb{R}^N$ and $AU=b$, since $A(\Phi-W)=\mathbf{1}$, we have

$$A\big(U-\|b\|_\infty(\Phi-W)\big)\leqslant 0,$$

this last inequality being meant componentwise. Therefore, by the above maximum principle, assuming, without loss of generality, that $h\leqslant 1$, one has

$$u_i\leqslant\|b\|_\infty(\phi_i-w_i),\qquad\text{so that}\quad u_i\leqslant\frac{\|b\|_\infty}{4}$$

(note that $\phi(x)\leqslant 1/8$). But we also have

$$A\big(U+\|b\|_\infty(\Phi-W)\big)\geqslant 0,$$

and again by the maximum principle, we obtain

$$u_i\geqslant-\frac{\|b\|_\infty}{4}.$$

Hence $\|U\|_\infty\leqslant\frac{1}{4}\|b\|_\infty$. This shows that $\|A^{-1}\|_\infty\leqslant 1/4$. □

This stability result, together with the existence of V given by Lemma 6.1, yields the convergence of the finite volume scheme, formulated in the next theorem.

THEOREM 6.2. *Let $\mathcal{T}=(K_i)_{i=1,\dots,N}$ be an admissible mesh of* $[0,1]$ *in the sense of Definition* 5.1. *Let $\alpha_{\mathcal{T}}\in\mathbb{R}_+^\star$ be such that $h_i\geqslant\alpha_{\mathcal{T}}h$, for all $i=1,\dots,N$ (recall that $h=\max\{h_1,\dots,h_N\}$). Let $\overline{U}=(u(x_1),\dots,u(x_N))^t\in\mathbb{R}^N$, and assume $u\in C^3([0,1],\mathbb{R})$ (recall that u is the solution to* (5.1)*). Let $U=(u_1,\dots,u_N)$ be the solution given by the numerical scheme* (5.3)–(5.6). *Then there exists $C>0$, only depending on $\alpha_{\mathcal{T}}$ and u, such that $\|U-\overline{U}\|_\infty\leqslant Ch$.*

REMARK 6.5. In the proof of Lemma 6.2, it was shown that $A(\overline{U} - V) = b + o(h)$; therefore, if, once again, the finite volume scheme is considered as a finite difference scheme, it is consistent, in the finite difference sense, when u_i is considered to be an approximation of $u(x_i) - (1/8)h_i^2 u_{xx}(x_i)$.

REMARK 6.6. With the notations of Lemma 6.1, let r be the function defined by

$$r(x) = r_i, \quad \text{if } x \in K_i,\ i = 1, \dots, N,$$

the function r does not necessarily go to 0 (as h goes to 0) in the L^∞ norm (and even in the L^1 norm), but, thanks to the conservativity of the scheme, it goes to 0 in $L^\infty((0,1))$ for the weak-$\star$ topology, i.e.

$$\int_0^1 r(x)\varphi(x)\,\mathrm{d}x \to 0, \quad \text{as } h \to 0,\ \forall \varphi \in L^1\big((0,1)\big).$$

This property will be called "weak consistency" in the sequel and may also be used to prove the convergence of the finite volume scheme (see FAILLE [1992a]).

The proof of convergence described above may be easily generalized to the two-dimensional Laplace equation $-\Delta u = f$ in two and three space dimensions if a rectangular or a parallelepipedic mesh is used, provided that the solution u is of class C^3. However, it does not seem to be easily generalized to other types of meshes.

7. General 1D elliptic equations

7.1. Formulation of the finite volume scheme

This section is devoted to the formulation and to the proof of convergence of a finite volume scheme for a one-dimensional linear convection–diffusion equation, with a discontinuous diffusion coefficient. The scheme can be generalized in the two-dimensional and three-dimensional cases (for a space discretization which uses, e.g., simplices or parallelepipeds or a "Voronoï mesh", see Section 9.2) and to other boundary conditions.

Let $\lambda \in L^\infty((0,1))$ such that there exist $\underline{\lambda}$ and $\overline{\lambda} \in \mathbb{R}_+^\star$ with $\underline{\lambda} \leqslant \lambda \leqslant \overline{\lambda}$ a.e. and let $a, b, c, d \in \mathbb{R}$, with $b \geqslant 0$, and $f \in L^2((0,1))$. The aim, here, is to find an approximation to the solution, u, of the following problem:

$$-(\lambda u_x)_x(x) + a u_x(x) + b u(x) = f(x), \quad x \in [0,1], \tag{7.1}$$

$$u(0) = c, \qquad u(1) = d. \tag{7.2}$$

The discontinuity of the coefficient λ may arise, e.g. for the permeability of a porous medium, the ratio between the permeability of sand and the permeability of clay being of an order of 10^3; heat conduction in a heterogeneous medium can also yield such discontinuities, since the conductivities of the different components of the medium may be quite different. Note that the assumption $b \geqslant 0$ ensures the existence of the solution to the problem.

REMARK 7.1. Problem (7.1)–(7.2) has a unique solution u in the Sobolev space $H^1((0, 1))$. This solution is continuous (on $[0, 1]$) but is not, in general, of class C^2 (even if $\lambda(x) = 1$, for all $x \in [0, 1]$). Note that one has

$$-\lambda u_x(x) = \int_0^x g(t)\,\mathrm{d}t + C,$$

where C is some constant and $g = f - au_x - bu \in L^1((0, 1))$, so that λu_x is a continuous function and $u_x \in L^\infty((0, 1))$.

Let $\mathcal{T} = (K_i)_{i=1,\dots,N}$ be an admissible mesh, in the sense of Definition 5.1, such that the discontinuities of λ coincide with the interfaces of the mesh. It is assumed that $\lambda|_{K_i} \in C^1(\overline{K_i})$, for all $i \in \{1, \dots, N\}$.

The notations being the same as in Section 5, integrating Eq. (7.1) over K_i yields

$$-(\lambda u_x)(x_{i+1/2}) + (\lambda u_x)(x_{i-1/2}) + au(x_{i+1/2}) - au(x_{i-1/2}) + \int_{K_i} bu(x)\,\mathrm{d}x$$
$$= \int_{K_i} f(x)\,\mathrm{d}x, \quad i = 1, \dots, N.$$

Let $(u_i)_{i=1,\dots,N}$ be the discrete unknowns. In the case $a \geqslant 0$, which will be considered in the sequel, the convective term $au(x_{i+1/2}t)$ is approximated by au_i ("upstream") because of stability considerations. Indeed, this choice always yields a stability result whereas the approximation of $au(x_{i+1/2}t)$ by $(a/2)(u_i + u_{i+1})$ (with the approximation of the other terms as it is done below) yields a stable scheme if $ah \leqslant 2\lambda$, for a uniform mesh of size h and a constant diffusion coefficient λ. The case $a \leqslant 0$ is easily handled in the same way by approximating $au(x_{i+1/2})$ by au_{i+1}. The term $\int_{K_i} bu(x)\,\mathrm{d}x$ is approximated by bh_iu_i. Let us now turn to the approximation $H_{i+1/2}$ of $-\lambda u_x(x_{i+1/2})$. Let $\lambda_i = \frac{1}{h_i}\int_{K_i} \lambda(x)\,\mathrm{d}x$; since $\lambda|_{K_i} \in C^1(\overline{K_i})$, there exists $c_\lambda \in \mathbb{R}_+$, only depending on λ, such that $|\lambda_i - \lambda(x)| \leqslant c_\lambda h$, $\forall x \in K_i$. In order that the scheme be conservative, the discretization of the flux at $x_{i+1/2}$ should have the same value on K_i and K_{i+1}. To this purpose, we introduce the auxiliary unknown $u_{i+1/2}$ (approximation of u at $x_{i+1/2}$). Since on K_i and K_{i+1}, λ is continuous, the approximation of $-\lambda u_x$ may be performed on each side of $x_{i+1/2}$ by using the finite difference principle:

$$H_{i+1/2} = -\lambda_i \frac{u_{i+1/2} - u_i}{h_i^+} \quad \text{on } K_i,\ i = 1, \dots, N,$$

$$H_{i+1/2} = -\lambda_{i+1} \frac{u_{i+1} - u_{i+1/2}}{h_{i+1}^-} \quad \text{on } K_{i+1},\ i = 0, \dots, N-1,$$

with $u_{1/2} = c$, and $u_{N+1/2} = d$, for the boundary conditions. (Recall that $h_i^+ = x_{i+1/2} - x_i$ and $h_i^- = x_i - x_{i-1/2}$.) Requiring the two above approximations of $\lambda u_x(x_{i+1/2})$ to be equal (conservativity of the flux) yields the value of $u_{i+1/2}$ (for $i = 1, \dots, N-1$):

$$u_{i+1/2} = \frac{u_{i+1}\frac{\lambda_{i+1}}{h_{i+1}^-} + u_i\frac{\lambda_i}{h_i^+}}{\frac{\lambda_{i+1}}{h_{i+1}^-} + \frac{\lambda_i}{h_i^+}} \tag{7.3}$$

which, in turn, allows to give the expression of the approximation $H_{i+1/2}$ of $\lambda u_x(x_{i+1/2})$:

$$\begin{aligned} H_{i+1/2} &= -\tau_{i+1/2}(u_{i+1} - u_i), \quad i = 1, \dots, N-1, \\ H_{1/2} &= -\frac{\lambda_1}{h_1^-}(u_1 - c), \\ H_{N+1/2} &= -\frac{\lambda_N}{h_N^+}(d - u_N) \end{aligned} \tag{7.4}$$

with

$$\tau_{i+1/2} = \frac{\lambda_i \lambda_{i+1}}{h_i^+ \lambda_{i+1} + h_{i+1}^- \lambda_i}, \quad i = 1, \dots, N-1. \tag{7.5}$$

EXAMPLE 7.1. If $h_i = h$, for all $i \in \{1, \dots, N\}$, and x_i is assumed to be the center of K_i, then $h_i^+ = h_i^- = h/2$, so that

$$H_{i+1/2} = -\frac{2\lambda_i \lambda_{i+1}}{\lambda_i + \lambda_{i+1}} \frac{u_{i+1} - u_i}{h},$$

and therefore the mean harmonic value of λ is involved.

The numerical scheme for the approximation of problem (7.1)–(7.2) is therefore,

$$F_{i+1/2} - F_{i-1/2} + bh_i u_i = h_i f_i, \quad \forall i \in \{1, \dots, N\}, \tag{7.6}$$

with $f_i = \frac{1}{h_i}\int_{x_{i-1/2}}^{x_{i+1/2}} f(x)\,\mathrm{d}x$, for $i = 1, \dots, N$, and where $(F_{i+1/2})_{i\in\{0,\dots,N\}}$ is defined by the following expressions

$$F_{i+1/2} = -\tau_{i+1/2}(u_{i+1} - u_i) + au_i, \quad \forall i \in \{1, \dots, N-1\}, \tag{7.7}$$

$$F_{1/2} = -\frac{\lambda_1}{h_1^-}(u_1 - c) + ac, \qquad F_{N+1/2} = -\frac{\lambda_N}{h_N^+}(d - u_N) + au_N. \tag{7.8}$$

REMARK 7.2. In the case $a \geqslant 0$, the choice of the approximation of $au(x_{i+1/2})$ by au_{i+1} would yield an unstable scheme, except for h small enough (when $a \leqslant 0$, the unstable scheme is au_i).

Taking (7.5), (7.7) and (7.8) into account, the numerical scheme (7.6) yields a system of N equations with N unknowns $u_1, \ldots, u_N$.

7.2. Error estimate

THEOREM 7.1. *Let $a, b \geqslant 0$, $c, d \in \mathbb{R}$, $\lambda \in L^\infty((0,1))$ such that $\underline{\lambda} \leqslant \lambda \leqslant \overline{\lambda}$ a.e. with some $\underline{\lambda}$, $\overline{\lambda} \in \mathbb{R}_+^\star$ and $f \in L^1((0,1))$. Let u be the (unique) solution of* (7.1)–(7.2). *Let $\mathcal{T} = (K_i)_{i=1,\ldots,N}$ be an admissible mesh, in the sense of Definition* 5.1, *such that $\lambda \in C^1(\overline{K_i})$ and $f \in C(\overline{K_i})$, for all $i = 1, \ldots, N$. Let $\gamma = \max\{\|u_{xx}\|_{L^\infty(K_i)},\ i = 1, \ldots, N\}$ and $\delta = \max\{\|\lambda\|_{L^\infty(K_i)},\ i = 1, \ldots, N\}$. Then,*

(1) *there exists a unique vector $U = (u_1, \ldots, u_N)^t \in \mathbb{R}^N$ solution to* (7.5)–(7.8),

(2) *there exists C, only depending on $\underline{\lambda}$, $\overline{\lambda}$, γ and δ, such that*

$$\sum_{i=0}^{N} \tau_{i+1/2}(e_{i+1} - e_i)^2 \leqslant Ch^2, \tag{7.9}$$

where $\tau_{i+1/2}$ is defined in (7.5), *and*

$$|e_i| \leqslant Ch, \quad \forall i \in \{1, \ldots, N\}, \tag{7.10}$$

with $e_0 = e_{N+1} = 0$ and $e_i = u(x_i) - u_i$, for all $i \in \{1, \ldots, N\}$.

PROOF.

Step 1. *Existence and uniqueness of the solution to* (7.5)–(7.8). Multiplying (7.6) by u_i and summing for $i = 1, \ldots, N$ yields that if $c = d = 0$ and $f_i = 0$ for any $i \in \{1, \ldots, N\}$, then the unique solution to (7.5)–(7.8) is obtained by taking $u_i = 0$, for any $i \in \{1, \ldots, N\}$. This yields existence and uniqueness of the solution to (7.5)–(7.8).

Step 2. *Consistency of the fluxes.* Recall that $h = \max\{h_1, \ldots, h_N\}$. Let us first show the consistency of the fluxes.

Let $\overline{H}_{i+1/2} = -(\lambda u_x)(x_{i+1/2})$ and $H^\star_{i+1/2} = -\tau_{i+1/2}(u(x_{i+1}) - u(x_i))$, for $i = 0, \ldots, N$, with $\tau_{1/2} = \lambda_1 / h_1^-$ and $\tau_{N+1/2} = \lambda_N / h_N^+$. Let us first show that there exists $C_1 \in \mathbb{R}_+^\star$, only depending on $\underline{\lambda}$, $\overline{\lambda}$, γ and δ, such that

$$\begin{aligned} &H^\star_{i+1/2} = \overline{H}_{i+1/2} + T_{i+1/2}, \\ &|T_{i+1/2}| \leqslant C_1 h, \quad i = 0, \ldots, N. \end{aligned} \tag{7.11}$$

In order to show this, let us introduce

$$\begin{aligned} &H^{\star,-}_{i+1/2} = -\lambda_i \frac{u(x_{i+1/2}) - u(x_i)}{h_i^+} \quad \text{and} \\ &H^{\star,+}_{i+1/2} = -\lambda_{i+1} \frac{u(x_{i+1}) - u(x_{i+1/2})}{h_{i+1}^-}; \end{aligned} \tag{7.12}$$

since $\lambda \in C^1(\overline{K}_i)$, one has $u \in C^2(\overline{K}_i)$; hence, there exists $C \in \mathbb{R}_+^\star$, only depending on γ and δ, such that

$$H_{i+1/2}^{\star,-} = \overline{H}_{i+1/2} + R_{i+1/2}^{-}, \quad \text{where } |R_{i+1/2}^{-}| \leqslant Ch,\ i = 1, \ldots, N, \tag{7.13}$$

and

$$H_{i+1/2}^{\star,+} = \overline{H}_{i+1/2} + R_{i+1/2}^{+}, \quad \text{where } |R_{i+1/2}^{+}| \leqslant Ch,\ i = 0, \ldots, N-1. \tag{7.14}$$

This yields (7.11) for $i = 0$ and $i = N$.

The following equality:

$$\overline{H}_{i+1/2} = H_{i+1/2}^{\star,-} - R_{i+1/2}^{-} = H_{i+1/2}^{\star,+} - R_{i+1/2}^{+}, \quad i = 1, \ldots, N-1, \tag{7.15}$$

yields that

$$u(x_{i+1/2}) = \frac{\frac{\lambda_{i+1}}{h_{i+1}^{-}} u(x_{i+1}) + \frac{\lambda_i}{h_i^{+}} u(x_i)}{\frac{\lambda_i}{h_i^{+}} + \frac{\lambda_{i+1}}{h_{i+1}^{-}}} + S_{i+1/2}, \quad i = 1, \ldots, N-1, \tag{7.16}$$

where

$$S_{i+1/2} = \frac{R_{i+1/2}^{+} - R_{i+1/2}^{-}}{\frac{\lambda_i}{h_i^{+}} + \frac{\lambda_{i+1}}{h_{i+1}^{-}}}$$

so that

$$|S_{i+1/2}| \leqslant \frac{1}{\underline{\lambda}} \frac{h_i^{+} h_{i+1}^{-}}{h_i^{+} + h_{i+1}^{-}} \left| R_{i+1/2}^{+} - R_{i+1/2}^{-} \right|.$$

Let us replace the expression (7.16) of $u(x_{i+1/2})$ in $H_{i+1/2}^{\star,-}$ defined by (7.12) (note that the computation is similar to that performed in (7.3)–(7.4)); this yields

$$H_{i+1/2}^{\star,-} = -\tau_{i+1/2}\big(u(x_{i+1}) - u(x_i)\big) - \frac{\lambda_i}{h_i^{+}} S_{i+1/2}, \quad i = 1, \ldots, N-1. \tag{7.17}$$

Using (7.15), this implies that $H_{i+1/2}^{\star} = \overline{H}_{i+1/2} + T_{i+1/2}$ where

$$|T_{i+1/2}| \leqslant |R_{i+1/2}^{-}| + \left| R_{i+1/2}^{+} - R_{i+1/2}^{-} \right| \frac{\overline{\lambda}}{2\underline{\lambda}}.$$

Using (7.13) and (7.14), this last inequality yields that there exists C_1, only depending on $\overline{\lambda}$, $\underline{\lambda}$, γ, δ, such that

$$\left| H_{i+1/2}^{\star} - \overline{H}_{i+1/2} \right| = |T_{i+1/2}| \leqslant C_1 h, \quad i = 1, \ldots, N-1.$$

Therefore (7.11) is proved.

Define now the total exact fluxes;

$$\overline{F}_{i+1/2} = -(\lambda u_x)(x_{i+1/2}) + au(x_{i+1/2}), \quad \forall i \in \{0, \dots, N\},$$

and define

$$F^\star_{i+1/2} = -\tau_{i+1/2}\big(u(x_{i+1}) - u(x_i)\big) + au(x_i), \quad \forall i \in \{1, \dots, N-1\},$$

$$F^\star_{1/2} = -\frac{\lambda_1}{h_1^-}\big(u(x_1) - c\big) + ac, \qquad F^\star_{N+1/2} = -\frac{\lambda_N}{h_N^+}\big(d - u(x_N)\big) + au_N.$$

Then, from (7.11) and the regularity of u, there exists C_2, only depending on $\underline{\lambda}, \overline{\lambda}, \gamma$ and δ, such that

$$F^\star_{i+1/2} = \overline{F}_{i+1/2} + R_{i+1/2}, \quad \text{with } |R_{i+1/2}| \leqslant C_2 h,\ i = 0, \dots, N. \tag{7.18}$$

Hence the numerical approximation of the flux is consistent .

Step 3. *Error estimate.* Integrating Eq. (7.1) over each control volume yields that

$$\overline{F}_{i+1/2} - \overline{F}_{i-1/2} + bh_i\big(u(x_i) + S_i\big) = h_i f_i, \quad \forall i \in \{1, \dots, N\}, \tag{7.19}$$

where $S_i \in \mathbb{R}$ is such that there exists C_3 only depending on u such that $|S_i| \leqslant C_3 h$, for $i = 1, \dots, N$. Using (7.18) yields that

$$\begin{aligned} &F^\star_{i+1/2} - F^\star_{i-1/2} + bh_i\big(u(x_i) + S_i\big) \\ &\quad = h_i f_i + R_{i+1/2} - R_{i-1/2}, \quad \forall i \in \{1, \dots, N\}. \end{aligned} \tag{7.20}$$

Let $e_i = u(x_i) - u_i$, for $i = 1, \dots, N$, and $e_0 = e_{N+1} = 0$. Subtracting (7.6) from (7.20) yields

$$\begin{aligned} &-\tau_{i+1/2}(e_{i+1} - e_i) + \tau_{i-1/2}(e_i - e_{i-1}) + a(e_i - e_{i-1}) + bh_i e_i \\ &\quad = -bh_i S_i + R_{i+1/2} - R_{i-1/2}, \quad \forall i \in \{1, \dots, N\}. \end{aligned}$$

Let us multiply this equation by e_i, sum for $i = 1, \dots, N$, reorder the summations. Remark that

$$\sum_{i=1}^{N} e_i(e_i - e_{i-1}) = \tfrac{1}{2}\sum_{i=1}^{N+1}(e_i - e_{i-1})^2$$

and therefore

$$\sum_{i=0}^{N} \tau_{i+1/2}(e_{i+1} - e_i)^2 + \frac{a}{2}\sum_{i=1}^{N+1}(e_i - e_{i-1})^2 + \sum_{i=1}^{N} bh_i e_i^2$$

$$= -\sum_{i=1}^{N} bh_i S_i e_i - \sum_{i=0}^{N} R_{i+1/2}(e_{i+1} - e_i).$$

Since $|S_i| \leqslant C_3 h$ and thanks to (7.18), one has

$$\sum_{i=0}^{N} \tau_{i+1/2}(e_{i+1} - e_i)^2 \leqslant \sum_{i=1}^{N} bC_3 h_i h |e_i| + \sum_{i=1}^{N} C_2 h |e_{i+1} - e_i|.$$

Remark that $|e_i| \leqslant \sum_{j=1}^{N} |e_j - e_{j-1}|$. Denote by $A = (\sum_{i=0}^{N} \tau_{i+1/2}(e_{i+1} - e_i)^2)^{1/2}$ and $B = (\sum_{i=0}^{N} \frac{1}{\tau_{i+1/2}})^{1/2}$. The Cauchy–Schwarz inequality yields

$$A^2 \leqslant \sum_{i=1}^{N} bC_3 h_i h AB + C_2 h AB.$$

Now, since

$$\frac{1}{\tau_{i+1/2}} \leqslant \frac{\overline{\lambda}}{\underline{\lambda}^2}(h_{i+1}^- + h_i^+),$$

$$\sum_{i=0}^{N} (h_{i+1}^- + h_i^+) = 1, \quad \text{with } h_0^+ = h_{N+1}^- = 0, \quad \text{and} \quad \sum_{i=1}^{N} h_i = 1,$$

one obtains that $A \leqslant C_4 h$, with C_4 only depending on $\underline{\lambda}$, $\overline{\lambda}$, γ and δ, which yields estimate (7.9). Applying once again the Cauchy–Schwarz inequality yields estimate (7.10). □

7.3. The case of a point source term

In many physical problems, some discontinuous or point source terms appear. In the case where a source term exists at the interface $x_{i+1/2}$, the fluxes relative to K_i and K_{i+1} will differ because of this source term. The computation of the fluxes is carried out in a similar way, writing that the sum of the approximations of the fluxes must be equal to the source term at the interface. Consider again the one-dimensional conservation problem (7.1), (7.2) (with, for the sake of simplification, $a = b = c = d = 0$, we use below the notations of the previous section), but assume now that at $\underline{x} \in (0, 1)$, a point source of intensity α exists. In this case, the problem may be written in the following way:

$$-\big(\lambda u_x(x)\big)_x = f(x), \quad x \in (0, \underline{x}) \cup (\underline{x}, 1), \tag{7.21}$$

$$u(0) = 0, \tag{7.22}$$

$$u(1) = 0, \tag{7.23}$$

$$(\lambda u_x)^+(\underline{x}) - (\lambda u_x)^-(\underline{x}) = -\alpha, \tag{7.24}$$

where

$$(\lambda u_x)^+(\underline{x}) = \lim_{x\to\underline{x},\ x>\underline{x}} (\lambda u_x)(x) \quad \text{and} \quad (\lambda u_x)^-(\underline{x}) = \lim_{x\to\underline{x},\ x<\underline{x}} (\lambda u_x)(x).$$

Equation (7.24) states that the flux is discontinuous at point $\underline{x}$. Another formulation of the problem is the following:

$$-(\lambda u_x)_x = g \quad \text{in } \mathcal{D}'\big((0,1)\big), \tag{7.25}$$

$$u(0) = 0, \tag{7.26}$$

$$u(1) = 0, \tag{7.27}$$

where $g = f + \alpha\delta_{\underline{x}}$, where $\delta_{\underline{x}}$ denotes the Dirac measure, which is defined by $\langle\delta_{\underline{x}}, \varphi\rangle_{\mathcal{D}',\mathcal{D}} = \varphi(\underline{x})$, for any $\varphi \in \mathcal{D}((0,1)) = C_c^\infty((0,1),\mathbb{R})$, and $\mathcal{D}'((0,1))$ denotes the set of distributions on (0,1), i.e. the set of continuous linear forms on $\mathcal{D}((0,1))$.

Assuming the mesh to be such that $\underline{x} = x_{i+1/2}$ for some $i \in 1,\dots,N-1$, the equation corresponding to the unknown u_i is $F_{i+1/2}^- - F_{i-1/2} = \int_{K_i} f(x)\,\mathrm{d}x$, while the equation corresponding to the unknown u_{i+1} is $F_{i+3/2} - F_{i+1/2}^+ = \int_{K_{i+1}} f(x)\,\mathrm{d}x$. In order to compute the values of the numerical fluxes $F_{i+1/2}^\pm$, one must take the source term into account while writing the conservativity of the flux; hence at $x_{i+1/2}$, the two numerical fluxes at $x = \underline{x}$, namely $F_{i+1/2}^+$ and $F_{i+1/2}^-$, must satisfy, following Eq. (7.24),

$$F_{i+1/2}^+ - F_{i+1/2}^- = \alpha. \tag{7.28}$$

Next, the fluxes $F_{i+1/2}^+$ and $F_{i+1/2}^-$ must be expressed in terms of the discrete variables u_k, $k = 1,\dots,N$; in order to do so, introduce the auxiliary variable $u_{i+1/2}$ (which will be eliminated later), and write

$$\begin{aligned} F_{i+1/2}^+ &= -\lambda_{i+1}\frac{u_{i+1} - u_{i+1/2}}{h_{i+1}^-}, \\ F_{i+1/2}^- &= -\lambda_i\frac{u_{i+1/2} - u_i}{h_i^+}. \end{aligned}$$

Replacing these expressions in (7.28) yields

$$u_{i+1/2} = \frac{h_i^+ h_{i+1}^-}{h_{i+1}^-\lambda_i + h_i^+\lambda_{i+1}}\left[\frac{\lambda_{i+1}}{h_{i+1}^-}u_{i+1} + \frac{\lambda_i}{h_i^+}u_i + \alpha\right]$$

and therefore

$$F_{i+1/2}^+ = \frac{h_i^+\lambda_{i+1}}{h_{i+1}^-\lambda_i + h_i^+\lambda_{i+1}}\alpha - \frac{\lambda_i\lambda_{i+1}}{h_{i+1}^-\lambda_i + h_i^+\lambda_{i+1}}(u_{i+1} - u_i),$$

$$F^-_{i+1/2} = \frac{-h^-_{i+1}\lambda_i}{h^-_{i+1}\lambda_i + h^+_i\lambda_{i+1}}\alpha - \frac{\lambda_i\lambda_{i+1}}{h^-_{i+1}\lambda_i + h^+_i\lambda_{i+1}}(u_{i+1} - u_i).$$

Note that the source term α is distributed on either side of the interface proportionally to the coefficient λ, and that, when $\alpha = 0$, the above expressions lead to

$$F^+_{i+1/2} = F^-_{i+1/2} = -\frac{\lambda_i\lambda_{i+1}}{h^-_{i+1}\lambda_i + h^+_i\lambda_{i+1}}(u_{i+1} - u_i).$$

Note that the error estimate given in Theorem 7.1 still holds in this case (under adequate assumptions).

8. A semilinear elliptic problem

8.1. Problem and scheme

This section is concerned with the proof of convergence for some nonlinear problems. We are interested, as an example, by the following problem:

$$-u_{xx}(x) = f\big(x, u(x)\big), \quad x \in (0, 1), \tag{8.1}$$

$$u(0) = u(1) = 0, \tag{8.2}$$

with a function $f : (0, 1) \times \mathbb{R} \to \mathbb{R}$ such that

$$\begin{aligned} &f(x, s) \text{ is measurable with respect to } x \in (0, 1) \text{ for all } s \in \mathbb{R} \\ &\quad \text{and continuous with respect to } s \in \mathbb{R} \text{ for a.e. } x \in (0, 1), \end{aligned} \tag{8.3}$$

$$f \in L^\infty\big((0, 1) \times \mathbb{R}\big). \tag{8.4}$$

It is possible to prove that there exists at least one weak solution to (8.1), (8.2), that is a function u such that

$$\begin{aligned} &u \in H^1_0\big((0, 1)\big), \quad \int_0^1 u_x(x)v_x(x)\,\mathrm{d}x = \int_0^1 f\big(x, u(x)\big)v(x)\,\mathrm{d}x, \\ &\quad \forall v \in H^1_0\big((0, 1)\big). \end{aligned} \tag{8.5}$$

Note that (8.5) is equivalent to "$u \in H^1_0((0, 1))$ and $-u_{xx} = f(\bullet, u)$ in the distribution sense in $(0, 1)$".

The proof of the existence of such a solution is possible by using, e.g., the Schauder's fixed point theorem (see, e.g., DEIMLING [1980]) or by using the Convergence Theorem 8.1 which is proved in the sequel.

Let $\mathcal{T}$ be an admissible mesh of $[0,1]$ in the sense of Definition 5.1. In order to discretize (8.1), (8.2), let us consider the following (finite volume) scheme

$$F_{i+1/2} - F_{i-1/2} = h_i f_i(u_i), \quad i = 1, \ldots, N, \tag{8.6}$$

$$F_{i+1/2} = -\frac{u_{i+1} - u_i}{h_{i+1/2}}, \quad i = 0, \ldots, N, \tag{8.7}$$

$$u_0 = u_{N+1} = 0, \tag{8.8}$$

with $f_i(u_i) = \frac{1}{h_i}\int_{K_i} f(x, u_i)\,\mathrm{d}x$, $i = 1, \ldots, N$. The discrete unknowns are therefore $u_1, \ldots, u_N$.

In order to give a convergence result for this scheme (Theorem 8.1), one first proves the existence of a solution to (8.6)–(8.8), a stability result, that is, an estimate on the solution of (8.6)–(8.8) (Lemma 8.1) and a compactness lemma (Lemma 8.2).

LEMMA 8.1 (Existence and stability result). *Let* $f:(0,1)\times\mathbb{R}\to\mathbb{R}$ *satisfying* (8.3), (8.4) *and* $\mathcal{T}$ *be an admissible mesh of* $(0,1)$ *in the sense of Definition* 5.1. *Then, there exists* $(u_1, \ldots, u_N)^t \in \mathbb{R}^N$ *solution of* (8.6)–(8.8) *and which satisfies*:

$$\sum_{i=0}^{N} \frac{(u_{i+1} - u_i)^2}{h_{i+1/2}} \leqslant C, \tag{8.9}$$

for some $C \geqslant 0$ *only depending on* f.

PROOF. Define $M = \|f\|_{L^\infty((0,1)\times\mathbb{R})}$. The proof of estimate (8.9) is given in a first step, and the existence of a solution to (8.6)–(8.8) in a second step.

Step 1 (*Estimate*). Let $V = (v_1, \ldots, v_N)^t \in \mathbb{R}^N$, there exists a unique $U = (u_1, \ldots, u_N)^t \in \mathbb{R}^N$ solution of (8.6)–(8.8) with $f_i(v_i)$ instead of $f_i(u_i)$ in the right-hand side (see Theorem 6.1). One sets $U = F(V)$, so that F is a continuous application from $\mathbb{R}^N$ to $\mathbb{R}^N$, and $(u_1, \ldots, u_N)$ is a solution to (8.6)–(8.8) if and only if $U = (u_1, \ldots, u_N)^t$ is a fixed point to F.

Multiplying (8.6) by u_i and summing over i yields

$$\sum_{i=0}^{N} \frac{(u_{i+1} - u_i)^2}{h_{i+1/2}} \leqslant M \sum_{i=1}^{N} h_i |u_i|, \tag{8.10}$$

and from the Cauchy–Schwarz inequality, one has

$$|u_i| \leqslant \left(\sum_{j=0}^{N} \frac{(u_{j+1} - u_j)^2}{h_{j+1/2}}\right)^{1/2}, \quad i = 1, \ldots, N,$$

then (8.10) yields, with $C = M^2$,

$$\sum_{i=0}^{N} \frac{(u_{i+1} - u_i)^2}{h_{i+1/2}} \leqslant C. \tag{8.11}$$

This gives, in particular, estimate (8.9) if $(u_1, \dots, u_N)^t \in \mathbb{R}^N$ is a solution of (8.6)–(8.8) (i.e. $u_i = v_i$ for all i).

Step 2 (*Existence*). The application $F : \mathbb{R}^N \to \mathbb{R}^N$ defined above is continuous and, taking in $\mathbb{R}^N$ the norm

$$\|V\| = \left(\sum_{i=0}^{N} \frac{(v_{i+1} - v_i)^2}{h_{i+1/2}}\right)^{1/2}, \quad \text{for } V = (v_1, \dots, v_N)^t, \text{ with } v_0 = v_{N+1} = 0,$$

one has $F(B_M) \subset B_M$, where B_M is the closed ball of radius M and center 0 in $\mathbb{R}^N$. Then, F has a fixed point in B_M thanks to the Brouwer fixed point theorem (see, e.g., DEIMLING [1980]). This fixed point is a solution to (8.6)–(8.8). □

8.2. Compactness results

LEMMA 8.2 (Compactness). *For an admissible mesh $\mathcal{T}$ of* (0, 1) (*see Definition* 5.1), *let $(u_1, \dots, u_N)^t \in \mathbb{R}^N$ satisfy* (8.9) *for some $C \in \mathbb{R}$ (independent of $\mathcal{T}$) and let $u_{\mathcal{T}} : (0,1) \to \mathbb{R}$ be defined by $u_{\mathcal{T}}(x) = u_i$ if $x \in K_i$, $i = 1, \dots, N$.*

Then, the set $\{u_{\mathcal{T}}, \mathcal{T}$ admissible mesh of $(0,1)\}$ is relatively compact in $L^2((0,1))$. Furthermore, if $u_{\mathcal{T}_n} \to u$ in $L^2((0,1))$ and $\text{size}(\mathcal{T}_n) \to 0$, as $n \to \infty$, then, $u \in H_0^1((0,1))$.

PROOF. A possible proof is to use "classical" compactness results, replacing $u_{\mathcal{T}}$ by a continuous function, say $\overline{u}_{\mathcal{T}}$, piecewise affine, such that $\overline{u}_{\mathcal{T}}(x_i) = u_i$ for $i = 1, \dots, N$, and $\overline{u}_{\mathcal{T}}(0) = \overline{u}_{\mathcal{T}}(1) = 0$. The set $\{\overline{u}_{\mathcal{T}}, \mathcal{T}$ admissible mesh of $(0,1)\}$ is then bounded in $H_0^1((0,1))$, see Remark 9.8.

Another proof is given here, the interest of which is its simple generalization to multi-dimensional cases (such as the case of one unknown per triangle in 2 space dimensions, see Section 9.2 and Section 14) when the construction of such a function, $\overline{u}_{\mathcal{T}}$, "close" to $u_{\mathcal{T}}$ and bounded in $H_0^1((0,1))$ (independently of $\mathcal{T}$), is not so easy.

In order to have $u_{\mathcal{T}}$ defined on $\mathbb{R}$, one sets $u_{\mathcal{T}}(x) = 0$ for $x \notin [0,1]$. The proof may be decomposed into four steps.

Step 1. First remark that the set $\{u_{\mathcal{T}}, \mathcal{T}$ an admissible mesh of $(0,1)\}$ is bounded in $L^2(\mathbb{R})$. Indeed, this an easy consequence of (8.9), since one has, for all $x \in [0,1]$ (since $u_0 = 0$ and by the Cauchy–Schwarz inequality),

$$|u_{\mathcal{T}}(x)| \leqslant \sum_{i=0}^{N} |u_{i+1} - u_i| \leqslant \left(\sum_{i=0}^{N} \frac{(u_{i+1} - u_i)^2}{h_{i+1/2}}\right)^{1/2} \leqslant C.$$

Step 2. Let $0 < \eta < 1$. One proves, in this step, that

$$\left\| u_{\mathcal{T}}(\cdot + \eta) - u_{\mathcal{T}} \right\|_{L^2(\mathbb{R})}^2 \leqslant C\eta(\eta + 2h). \tag{8.12}$$

(Recall that $h = \text{size}(\mathcal{T})$.)

Indeed, for $i \in \{0, \ldots, N\}$ define $\chi_{i+1/2}: \mathbb{R} \to \mathbb{R}$, by $\chi_{i+1/2}(x) = 1$, if $x_{i+1/2} \in [x, x + \eta]$ and $\chi_{i+1/2}(x) = 0$, if $x_{i+1/2} \notin [x, x + \eta]$. Then, one has, for all $x \in \mathbb{R}$,

$$\begin{aligned}
&\left(u_{\mathcal{T}}(x + \eta) - u_{\mathcal{T}}(x)\right)^2 \\
&\quad \leqslant \left(\sum_{i=0}^{N} |u_{i+1} - u_i| \chi_{i+1/2}(x) \right)^2 \\
&\quad \leqslant \left(\sum_{i=0}^{N} \frac{(u_{i+1} - u_i)^2}{h_{i+1/2}} \chi_{i+1/2}(x) \right) \left(\sum_{i=0}^{N} \chi_{i+1/2}(x) h_{i+1/2} \right).
\end{aligned} \tag{8.13}$$

Since $\sum_{i=0}^{N} \chi_{i+1/2}(x) h_{i+1/2} \leqslant \eta + 2h$, for all $x \in \mathbb{R}$, and $\int_{\mathbb{R}} \chi_{i+1/2}(x)\, \mathrm{d}x = \eta$, for all $i \in \{0, \ldots, N\}$, integrating (8.13) over $\mathbb{R}$ yields (8.12).

Step 3. For $0 < \eta < 1$, estimate (8.12) implies that

$$\left\| u_{\mathcal{T}}(\cdot + \eta) - u_{\mathcal{T}} \right\|_{L^2(\mathbb{R})}^2 \leqslant 3C\eta.$$

This gives (with Step 1), by the Kolmogorov compactness theorem (recalled in Section 14, see Theorem 14.1), the relative compactness of the set $\{u_{\mathcal{T}}, \mathcal{T}$ an admissible mesh of $(0, 1)\}$ in $L^2((0, 1))$ and also in $L^2(\mathbb{R})$ (since $u_{\mathcal{T}} = 0$ on $\mathbb{R} \setminus [0, 1]$).

Step 4. In order to conclude the proof of Lemma 8.2, one may use Theorem 14.2, which we prove here in the one-dimensional case for the sake of clarity. Let $(\mathcal{T}_n)_{n \in \mathbb{N}}$ be a sequence of admissible meshes of $(0, 1)$ such that $\text{size}(\mathcal{T}_n) \to 0$ and $u_{\mathcal{T}_n} \to u$, in $L^2((0, 1))$, as $n \to \infty$. Note that $u_{\mathcal{T}_n} \to u$, in $L^2(\mathbb{R})$, with $u = 0$ on $\mathbb{R} \setminus [0, 1]$. For a given $\eta \in (0, 1)$, let $n \to \infty$ in (8.12), with $u_{\mathcal{T}_n}$ instead of $u_{\mathcal{T}}$ (and $\text{size}(\mathcal{T}_n)$ instead of h). One obtains

$$\left\| \frac{u(\cdot + \eta) - u}{\eta} \right\|_{L^2(\mathbb{R})}^2 \leqslant C. \tag{8.14}$$

Since $(u(\cdot + \eta) - u)/\eta$ tends to Du (the distribution derivative of u) in the distribution sense, as $\eta \to 0$, estimate (8.14) yields that $Du \in L^2(\mathbb{R})$. Furthermore, since $u = 0$ on $\mathbb{R} \setminus [0, 1]$, the restriction of u to $(0, 1)$ belongs to $H_0^1((0, 1))$. The proof of Lemma 8.2 is complete. □

8.3. Convergence

The following convergence result follows from Lemmata 8.1 and 8.2.

THEOREM 8.1. *Let* $f:(0,1)\times\mathbb{R}\to\mathbb{R}$ *satisfying* (8.3), (8.4). *For an admissible mesh,* $\mathcal{T}$, *of* $(0,1)$ (*see Definition* 5.1), *let* $(u_1,\dots,u_N)^t\in\mathbb{R}^N$ *be a solution to* (8.6)–(8.8) (*the existence of which is given by Lemma* 8.1), *and let* $u_{\mathcal{T}}:(0,1)\to\mathbb{R}$ *by* $u_{\mathcal{T}}(x)=u_i$, *if* $x\in K_i$, $i=1,\dots,N$.

Then, for any sequence $(\mathcal{T}_n)_{n\in\mathbb{N}}$ *of admissible meshes such that* $\mathrm{size}(\mathcal{T}_n)\to 0$, *as* $n\to\infty$, *there exists a subsequence, still denoted by* $(\mathcal{T}_n)_{n\in\mathbb{N}}$, *such that* $u_{\mathcal{T}_n}\to u$, *in* $L^2((0,1))$, *as* $n\to\infty$, *where* $u\in H_0^1((0,1))$ *is a weak solution to* (8.1), (8.2) (*i.e., a solution to* (8.5)).

PROOF. Let $(\mathcal{T}_n)_{n\in\mathbb{N}}$ be a sequence of admissible meshes of $(0,1)$ such that $\mathrm{size}(\mathcal{T}_n)\to 0$, as $n\to\infty$. By Lemmata 8.1 and 8.2, there exists a subsequence, still denoted by $(\mathcal{T}_n)_{n\in\mathbb{N}}$, such that $u_{\mathcal{T}_n}\to u$, in $L^2((0,1))$, as $n\to\infty$, where $u\in H_0^1((0,1))$. In order to conclude, it only remains to prove that $-u_{xx}=f(\bullet,u)$ in the distribution sense in $(0,1)$.

To prove this, let $\varphi\in C_c^\infty((0,1))$. Let $\mathcal{T}$ be an admissible mesh of $(0,1)$, and $\varphi_i=\varphi(x_i)$, $i=1,\dots,N$, and $\varphi_0=\varphi_{N+1}=0$. If $(u_1,\dots,u_N)$ is a solution to (8.6)–(8.8), multiplying (8.6) by φ_i and summing over $i=1,\dots,N$ yields

$$\int_0^1 u_{\mathcal{T}}(x)\psi_{\mathcal{T}}(x)\,\mathrm{d}x=\int_0^1 f_{\mathcal{T}}(x)\varphi_{\mathcal{T}}(x)\,\mathrm{d}x, \tag{8.15}$$

where

$$\begin{aligned}\psi_{\mathcal{T}}(x)&=\frac{1}{h_i}\left(\frac{\varphi_i-\varphi_{i-1}}{h_{i-1/2}}-\frac{\varphi_{i+1}-\varphi_i}{h_{i+1/2}}\right),\\ f_{\mathcal{T}}(x)&=f(x,u_i)\quad\text{and}\quad\varphi_{\mathcal{T}}(x)=\varphi_i,\quad\text{if } x\in K_i.\end{aligned}$$

Note that, thanks to the regularity of the function φ,

$$\frac{\varphi_{i+1}-\varphi_i}{h_{i+1/2}}=\varphi_x(x_{i+1/2})+R_{i+1/2},\quad |R_{i+1/2}|\leqslant C_1 h,$$

with some C_1 only depending on φ, and therefore

$$\begin{aligned}\int_0^1 u_{\mathcal{T}}(x)\psi_{\mathcal{T}}(x)\,\mathrm{d}x&=\sum_{i=1}^N\int_{K_i}\frac{u_i}{h_i}\big(\varphi_x(x_{i-1/2})-\varphi_x(x_{i+1/2})\big)\,\mathrm{d}x\\&\quad+\sum_{i=1}^N u_i(R_{i-1/2}-R_{i+1/2})\end{aligned}$$

$$= \int_0^1 -u_{\mathcal{T}} \,\mathrm{d}x \theta_{\mathcal{T}}(x)\,\mathrm{d}x + \sum_{i=0}^{N} R_{i+1/2}(u_{i+1} - u_i),$$

with $u_0 = u_{N+1} = 0$, where the piecewise constant function

$$\theta_{\mathcal{T}} = \sum_{i=1,N} \frac{\varphi_x(x_{i+1/2}) - \varphi_x(x_{i-1/2})}{h_i} 1_{K_i}$$

tends to $\varphi_x x$ as h tends to 0.

Let us consider (8.15) with $\mathcal{T}_n$ instead of $\mathcal{T}$; thanks to the Cauchy–Schwarz inequality, a passage to the limit as $n \to \infty$ gives, thanks to (8.9),

$$-\int_0^1 u(x)\varphi_{xx}(x)\,\mathrm{d}x = \int_0^1 f\big(x, u(x)\big)\varphi(x)\,\mathrm{d}x,$$

and therefore $-u_{xx} = f(\bullet, u)$ in the distribution sense in $(0, 1)$. This concludes the proof of Theorem 8.1. Note that the crucial idea of this proof is to use the property of consistency of the fluxes on the regular test function φ. □

REMARK 8.1. It is possible to give some extensions of the results of this section. For instance, Theorem 8.1 is true with an assumption of "sublinearity" on f instead of (8.4). Furthermore, in order to have both existence and uniqueness of the solution to (8.5) and a rate of convergence (of order h) in Theorem 8.1, it is sufficient to assume, instead of (8.3) and (8.4), that $f \in C^1([0, 1] \times \mathbb{R}, \mathbb{R})$ and that there exists $\gamma < 1$, such that $(f(x, s) - f(x, t))(s - t) \leqslant \gamma(s - t)^2$, for all $(x, s) \in [0, 1] \times \mathbb{R}$.

CHAPTER III

Elliptic Problems in Two or Three Dimensions

The topic of this chapter is the discretization of elliptic problems in several space dimensions by the finite volume method. The one-dimensional case which was studied in Chapter II is easily generalized to nonuniform rectangular or parallelepipedic meshes. However, for general shapes of control volumes, the definition of the scheme (and the proof of convergence) requires some assumptions which define an "admissible mesh". Dirichlet and Neumann boundary conditions are both considered. In both cases, a discrete Poincaré inequality is used, and the stability of the scheme is proved by establishing estimates on the approximate solutions. The convergence of the scheme without any assumption on the regularity of the exact solution is proved; this result may be generalized, under adequate assumptions, to nonlinear equations. Then, again in both the Dirichlet and Neumann cases, an error estimate between the finite volume approximate solution and the C^2 or H^2 regular exact solution to the continuous problems are proved. The results are generalized to the case of matrix diffusion coefficients and more general boundary conditions. Section 12 is devoted to finite volume schemes written with unknowns located at the vertices. Some links between the finite element method, the "classical" finite volume method and the "control volume finite element" method introduced by FORSYTH [1989] are given. Section 13 is devoted to the treatment of singular sources and to mesh refinement; under suitable assumption, it can be shown that error estimates still hold for "atypical" refined meshes. Finally, Section 14 is devoted to the proof of compactness results which are used in the proofs of convergence of the schemes.

9. Dirichlet boundary conditions

Let us consider here the following elliptic equation

$$-\Delta u(x) + \mathrm{div}(\boldsymbol{v}u)(x) + bu(x) = f(x), \quad x \in \Omega, \tag{9.1}$$

with Dirichlet boundary condition:

$$u(x) = g(x), \quad x \in \partial\Omega, \tag{9.2}$$

where

ASSUMPTION 9.1.
(1) Ω is an open bounded polygonal subset of $\mathbb{R}^d$, $d = 2$ or 3,
(2) $b \geqslant 0$,
(3) $f \in L^2(\Omega)$,
(4) $\boldsymbol{v} \in C^1(\overline{\Omega}, \mathbb{R}^d)$; $\operatorname{div} \boldsymbol{v} \geqslant 0$,
(5) $g \in C(\partial\Omega, \mathbb{R})$ is such that there exists $\tilde{g} \in H^1(\Omega)$ such that $\overline{\gamma}(\tilde{g}) = g$ a.e. on $\partial\Omega$.

Here, and in the sequel, "polygonal" is used for both $d = 2$ and $d = 3$ (meaning polyhedral in the latter case) and $\overline{\gamma}$ denotes the trace operator from $H^1(\Omega)$ into $L^2(\partial\Omega)$. Note also that "a.e. on $\partial\Omega$" is a.e. for the $(d-1)$-dimensional Lebesgue measure on $\partial\Omega$.

Under Assumption 9.1, by the Lax–Milgram theorem, there exists a unique variational solution $u \in H^1(\Omega)$ of problem (9.1)–(9.2). (For the study of elliptic problems and their discretization by finite element methods, see, e.g., CIARLET [1978] and references therein.) This solution satisfies $u = w + \tilde{g}$, where $\tilde{g} \in H^1(\Omega)$ is such that $\overline{\gamma}(\tilde{g}) = g$, a.e. on $\partial\Omega$, and w is the unique function of $H_0^1(\Omega)$ satisfying

$$\begin{aligned}
&\int_\Omega \big(\nabla w(x) \cdot \nabla \psi(x) + \operatorname{div}(\boldsymbol{v} w)(x)\psi(x) + b w(x)\psi(x)\big)\,\mathrm{d}x \\
&\quad = \int_\Omega \big(-\nabla \tilde{g}(x) \cdot \nabla \psi(x) - \operatorname{div}(\boldsymbol{v}\tilde{g})(x)\psi(x) \\
&\qquad - b\tilde{g}(x)\psi(x) + f(x)\psi(x)\big)\,\mathrm{d}x, \quad \forall \psi \in H_0^1(\Omega).
\end{aligned} \tag{9.3}$$

9.1. *Structured meshes*

If Ω is a rectangle $(d = 2)$ or a parallelepiped $(d = 3)$, it may then be meshed with rectangular or parallelepipedic control volumes. In this case, the one-dimensional scheme may easily be generalized.

9.1.1. *Rectangular meshes for the Laplace operator*

Let us, e.g., consider the case $d = 2$, let $\Omega = (0, 1) \times (0, 1)$, and $f \in C^2(\Omega, \mathbb{R})$ (the three-dimensional case is similar). Consider problem (9.1)–(9.2) and assume here that $b = 0$, $\boldsymbol{v} = 0$ and $g = 0$ (the general case is considered later, on general unstructured meshes). The problem reduces to the pure diffusion equation:

$$\begin{aligned}
&-\Delta u(x, y) = f(x, y), \quad (x, y) \in \Omega, \\
&u(x, y) = 0, \quad (x, y) \in \partial\Omega.
\end{aligned} \tag{9.4}$$

In this section, it is convenient to denote by (x, y) the current point of $\mathbb{R}^2$ (elsewhere, the notation x is used for a point or a vector of $\mathbb{R}^d$).

Let $\mathcal{T} = (K_{i,j})_{i=1,\dots,N_1;\ j=1,\dots,N_2}$ be an admissible mesh of $(0,1)\times(0,1)$, i.e., satisfying the following assumptions (which generalize Definition 5.1).

ASSUMPTION 9.2. Let $N_1 \in \mathbb{N}^\star$, $N_2 \in \mathbb{N}^\star$, $h_1,\dots,h_{N_1} > 0$, $k_1,\dots,k_{N_2} > 0$ such that

$$\sum_{i=1}^{N_1} h_i = 1, \qquad \sum_{i=1}^{N_2} k_i = 1,$$

and let $h_0 = 0$, $h_{N_1+1} = 0$, $k_0 = 0$, $k_{N_2+1} = 0$. For $i = 1,\dots,N_1$, let $x_{1/2} = 0$, $x_{i+1/2} = x_{i-1/2} + h_i$ (so that $x_{N_1+1/2} = 1$), and for $j = 1,\dots,N_2$, $y_{1/2} = 0$, $y_{j+1/2} = y_{j-1/2} + k_j$ (so that $y_{N_2+1/2} = 1$) and

$$K_{i,j} = [x_{i-1/2}, x_{i+1/2}] \times [y_{j-1/2}, y_{j+1/2}].$$

Let $(x_i)_{i=0,N_1+1}$, and $(y_j)_{j=0,N_2+1}$, such that

$$x_{i-1/2} < x_i < x_{i+1/2}, \quad \text{for } i = 1,\dots,N_1,\ x_0 = 0,\ x_{N_1+1} = 1,$$
$$y_{j-1/2} < y_j < y_{j+1/2}, \quad \text{for } j = 1,\dots,N_2,\ y_0 = 0,\ y_{N_2+1} = 1,$$

and let $x_{i,j} = (x_i, y_j)$, for $i = 1,\dots,N_1$, $j = 1,\dots,N_2$; set

$$h_i^- = x_i - x_{i-1/2}, \qquad h_i^+ = x_{i+1/2} - x_i, \quad \text{for } i = 1,\dots,N_1,$$
$$h_{i+1/2} = x_{i+1} - x_i, \quad \text{for } i = 0,\dots,N_1,$$
$$k_j^- = y_j - y_{j-1/2}, \qquad k_j^+ = y_{j+1/2} - y_j, \quad \text{for } j = 1,\dots,N_2,$$
$$k_{j+1/2} = y_{j+1} - y_j, \quad \text{for } j = 0,\dots,N_2.$$

Let $h = \max\{(h_i, i = 1,\dots,N_1), (k_j, j = 1,\dots,N_2)\}$.

As in the 1D case, the finite volume scheme is found by integrating the first equation of (9.4) over each control volume $K_{i,j}$, which yields

$$-\int_{y_{j-1/2}}^{y_{j+1/2}} u_x(x_{i+1/2}, y)\,\mathrm{d}y + \int_{y_{i-1/2}}^{y_{i+1/2}} u_x(x_{i-1/2}, y)\,\mathrm{d}y$$
$$+\int_{x_{i-1/2}}^{x_{i+1/2}} u_y(x, y_{j-1/2})\,\mathrm{d}x - \int_{x_{i-1/2}}^{x_{i+1/2}} u_y(x, y_{j+1/2})\,\mathrm{d}x = \int_{K_{ij}} f(x,y)\,\mathrm{d}x\,\mathrm{d}y.$$

The fluxes are then approximated by differential quotients with respect to the discrete unknowns $(u_{i,j}, i = 1,\dots,N_1, j = 1,\dots,N_2)$ in a similar manner to the 1D case; hence the numerical scheme writes

$$F_{i+1/2,j} - F_{i-1/2,j} + F_{i,j+1/2} - F_{i,j-1/2} = h_{i,j} f_{i,j},$$
$$\forall (i,j) \in \{1,\dots,N_1\} \times \{1,\dots,N_2\}, \tag{9.5}$$

where $h_{i,j} = h_i \times k_j$, $f_{i,j}$ is the mean value of f over $K_{i,j}$, and

$$\begin{aligned} F_{i+1/2,j} &= -\frac{k_j}{h_{i+1/2}}(u_{i+1,j} - u_{i,j}), \quad \text{for } i = 0, \ldots, N_1,\ j = 1, \ldots, N_2, \\ F_{i,j+1/2} &= -\frac{h_i}{k_{j+1/2}}(u_{i,j+1} - u_{i,j}), \quad \text{for } i = 1, \ldots, N_1,\ j = 0, \ldots, N_2, \end{aligned} \tag{9.6}$$

$$u_{0,j} = u_{N_1+1,j} = u_{i,0} = u_{i,N_2+1} = 0, \quad \text{for } i = 1, \ldots, N_1,\ j = 1, \ldots, N_2. \tag{9.7}$$

The numerical scheme (9.5)–(9.7) is therefore clearly conservative and the numerical approximations of the fluxes can easily be shown to be consistent.

PROPOSITION 9.1 (Error estimate). *Let $\Omega = (0,1) \times (0,1)$ and $f \in L^2(\Omega)$. Let u be the unique variational solution to (9.4). Under Assumptions 9.2, let $\zeta > 0$ be such that $h_i \geqslant \zeta h$ for $i = 1, \ldots, N_1$ and $k_j \geqslant \zeta h$ for $j = 1, \ldots, N_2$. Then, there exists a unique solution $(u_{i,j})_{i=1,\ldots,N_1, j=1,\ldots,N_2}$ to (9.5)–(9.7). Moreover, there exists $C > 0$ only depending on u, Ω and ζ such that*

$$\sum_{i,j} \frac{(e_{i+1,j} - e_{i,j})^2}{h_{i+1/2}} k_j + \sum_{i,j} \frac{(e_{i,j+1} - e_{i,j})^2}{k_{j+1/2}} h_i \leqslant Ch^2 \tag{9.8}$$

and

$$\sum_{i,j} (e_{i,j})^2 h_i k_j \leqslant Ch^2, \tag{9.9}$$

where $e_{i,j} = u(x_{i,j}) - u_{i,j}$, for $i = 1, \ldots, N_1$, $j = 1, \ldots, N_2$.

In the above proposition, since $f \in L^2(\Omega)$ and Ω is convex, it is well known that the variational solution u to (9.4) belongs to $H^2(\Omega)$. We do not give here the proof of this proposition since it is in fact included in Theorem 9.4 (see also LAZAROV, MISHEV and VASSILEVSKI [1996] where the case $u \in H^s$, $s \geqslant 3/2$ is also studied).

In the case $u \in C^2(\overline{\Omega})$, the estimates (9.8) and (9.9) can be shown with the same technique as in the 1D case (see, e.g., FIARD [1994]). If $u \in C^2$ then the above estimates are a consequence of Theorem 9.3; in this case, the value C in (9.8) and (9.9) independent of ζ, and therefore the assumption $h_i \geqslant \zeta h$ for $i = 1, \ldots, N_1$ and $k_j \geqslant \zeta h$ for $j = 1, \ldots, N_2$ is no longer needed.

Relation (9.8) can be seen as an estimate of a "discrete H_0^1 norm" of the error, while relation (9.9) gives an estimate of the L^2 norm of the error.

REMARK 9.1. Some slight modifications of the scheme (9.5)–(9.7) are possible, as in the first item of Remark 5.2. It is also possible to obtain, sometimes, an "h^2" estimate on the L^2 (or L^∞) norm of the error (i.e., "h^4" instead of "h^2" in (9.9)), exactly as in the 1D case, see Remark 6.2. In the case equivalent to the second case of Remark 6.2, the point $x_{i,j}$ is not necessarily the center of $K_{i,j}$.

When the mesh is no longer rectangular, the scheme (9.5)–(9.6) is not easy to generalize if keeping to a 5 points scheme. In particular, the consistency of the fluxes or the conservativity can be lost, see FAILLE [1992a], which yields a bad numerical behaviour of the scheme. One way to keep both properties is to introduce a 9-points scheme.

9.1.2. Quadrangular meshes: a nine-point scheme
Let Ω be an open bounded polygonal subset of $\mathbb{R}^2$, and f be a regular function from $\overline{\Omega}$ to $\mathbb{R}$. We still consider problem (9.4), turning back to the usual notation x for the current point of $\mathbb{R}^2$,

$$\begin{aligned} -\Delta u(x) &= f(x), \quad x \in \Omega, \\ u(x) &= 0, \quad x \in \partial\Omega. \end{aligned} \tag{9.10}$$

Let $\mathcal{T}$ be a mesh defined over Ω; then, integrating the first equation of (9.10) over any cell K of the mesh yields

$$-\int_{\partial K} \mathbf{grad}\, u \bullet \boldsymbol{n}_K = \int_K f,$$

where $\boldsymbol{n}_K$ is the normal to the boundary ∂K, outward to K. Let u_K denote the discrete unknown associated to the control volume $K \in \mathcal{T}$. In order to obtain a numerical scheme, if σ is a common edge to $K \in \mathcal{T}$ and $L \in \mathcal{T}$ (denoted by $K|L$) or if σ is an edge of $K \in \mathcal{T}$ belonging to $\partial\Omega$, the expression $\mathbf{grad}\, u \bullet \boldsymbol{n}_K$ must be approximated on σ by using the discrete unknowns. The study of the finite volume scheme in dimension 1 and the above straightforward generalization to the rectangular case showed that the fundamental properties of the method seem to be

(1) Conservativity: in the absence of any source term on $K|L$, the approximation of $\mathbf{grad}\, u \bullet \boldsymbol{n}_K$ on $K|L$ which is used in the equation associated with cell K is equal to the approximation of $-\mathbf{grad}\, u \bullet \boldsymbol{n}_L$ which is used in the equation associated with cell L. This property is naturally obtained when using a finite volume scheme.

(2) Consistency of the fluxes: taking for u_K the value of u in a fixed point of K (e.g., the center of gravity of K), where u is a regular function, the difference between $\mathbf{grad}\, u \bullet \boldsymbol{n}_K$ and the chosen approximation of $\mathbf{grad}\, u \bullet \boldsymbol{n}_K$ is of an order less or equal to that of the mesh size. This need of consistency will be discussed in more detail: see Remarks 9.2 and 9.7

Several computer codes use the following "natural" extension of (9.6) for the approximation of $\mathbf{grad}\, u \bullet \boldsymbol{n}_K$ on $\overline{K} \cap \overline{L}$:

$$\mathbf{grad}\, u \bullet \boldsymbol{n}_K = \frac{u_L - u_K}{d_{K|L}},$$

where $d_{K|L}$ is the distance between the center of the cells K and L. This choice, however simple, is far from optimal, at least in the case of a general (nonrectangular) mesh, because the fluxes thus obtained are not consistent; this yields important errors, especially in the case where the mesh cells are all oriented in the same direction, see

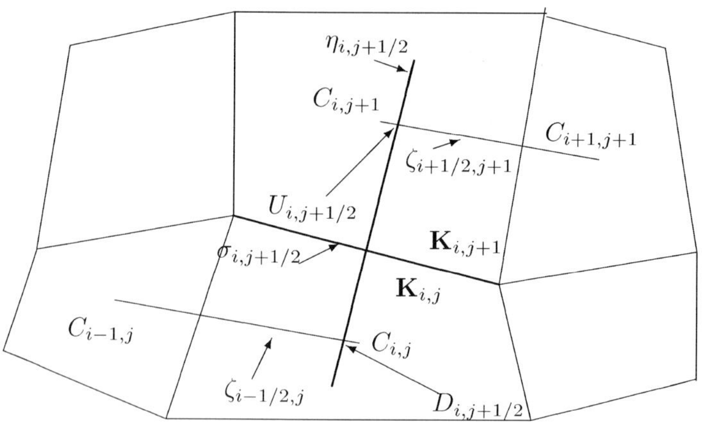

FIG. 9.1. FV9 scheme.

FAILLE [1992a, 1992b]. This problem may be avoided by modifying the approximation of $\mathbf{grad}\, u \cdot \boldsymbol{n}_K$ so as to make it consistent. However, one must be careful, in doing so, to maintain the conservativity of the scheme. To this purpose, a 9-points scheme was developed, which is denoted by FV9.

Let us describe now how the flux $\mathbf{grad}\, u \cdot \boldsymbol{n}_K$ is approximated by the FV9 scheme. Assume here, for the sake of clarity, that the mesh $\mathcal{T}$ is structured; indeed, it consists in a set of quadrangular cells $\{K_{i,j},\ i = 1, \ldots, N;\ j = 1, \ldots, M\}$. As shown in Fig. 9.1, let $C_{i,j}$ denote the center of gravity of the cell $K_{i,j}$, $\sigma_{i,j-1/2}, \sigma_{i+1/2,j}, \sigma_{i,j+1/2}, \sigma_{i-1/2,j}$ the four edges to $K_{i,j}$ and $\eta_{i,j-1/2}, \eta_{i+1/2,j}, \eta_{i,j+1/2}, \eta_{i-1/2,j}$ their respective orthogonal bisectors. Let $\zeta_{i,j-1/2}$, (resp. $\zeta_{i+1/2,j}, \zeta_{i,j+1/2}, \zeta_{i-1/2,j}$) be the lines joining points $C_{i,j}$ and $C_{i,j-1}$ (resp. $C_{i,j}$ and $C_{i+1,j}$, $C_{i,j}$ and $C_{i,j+1}$, $C_{i-1,j}$ and $C_{i,j}$).

Consider, e.g., the edge $\sigma_{i,j+1/2}$ which lies between the cells $K_{i,j}$ and $K_{i,j+1}$ (see Fig. 9.1). In order to find an approximation of $\mathbf{grad}\, u \cdot \boldsymbol{n}_K$, for $K = K_{i,j}$, at the center of this edge, we shall first derive an approximation of u at the two points $U_{i,j+1/2}$ and $D_{i,j+1/2}$ which are located on the orthogonal bisector $\eta_{i,j+1/2}$ of the edge $\sigma_{i,j+1/2}$, on each side of the edge. Let $\phi_{i,j+1/2}$ be the approximation of $-\mathbf{grad}\, u \cdot \boldsymbol{n}_K$ at the center of the edge $\sigma_{i,j+1/2}$. A natural choice for $\phi_{i,j+1/2}$ consists in taking

$$\phi_{i,j+1/2} = -\frac{u^U_{i,j+1/2} - u^D_{i,j+1/2}}{d(U_{i,j+1/2}, D_{i,j+1/2})}, \tag{9.11}$$

where $u^U_{i,j+1/2}$ and $u^D_{i,j+1/2}$ are approximations of u at $U_{i,j+1/2}$ and $D_{i,j+1/2}$, and $d(U_{i,j+1/2}, D_{i,j+1/2})$ is the distance between points $U_{i,j+1/2}$ and $D_{i,j+1/2}$.

The points $U_{i,j+1/2}$ and $D_{i,j+1/2}$ are chosen so that they are located on the lines ζ which join the centers of the neighbouring cells. The points $U_{i,j+1/2}$ and $D_{i,j+1/2}$ are therefore located at the intersection of the orthogonal bisector $\eta_{i,j+1/2}$ with the adequate

ζ lines, which are chosen according to the geometry of the mesh. More precisely,

$$\begin{aligned} U_{i,j+1/2} &= \eta_{i,j+1/2} \cap \zeta_{i-1/2,j+1} && \text{if } \eta_{i,j+1/2} \text{ is to the left of } C_{i,j+1}, \\ &= \eta_{i,j+1/2} \cap \zeta_{i+1/2,j+1} && \text{otherwise}, \\ D_{i,j+1/2} &= \eta_{i,j+1/2} \cap \zeta_{i-1/2,j} && \text{if } \eta_{i,j+1/2} \text{ is to the left of } C_{i,j}, \\ &= \eta_{i,j+1/2} \cap \zeta_{i+1/2,j} && \text{otherwise}. \end{aligned}$$

In order to satisfy the property of consistency of the fluxes, a second-order approximation of u at points $U_{i,j+1/2}$ and $D_{i,j+1/2}$ is required. In the case of the geometry which is described in Fig. 9.1, the following linear approximations of $u^U_{i,j+1/2}$ and $u^D_{i,j+1/2}$ can be used in (9.11);

$$\begin{aligned} u^U_{i,j+1/2} &= \alpha u_{i+1,j+1} + (1-\alpha) u_{i,j+1} \quad \text{where } \alpha = \frac{d(C_{i,j+1}, U_{i,j+1/2})}{d(C_{i,j+1}, C_{i+1,j+1})}, \\ u^D_{i,j+1/2} &= \beta u_{i-1,j} + (1-\beta) u_{i,j} \quad \text{where } \beta = \frac{d(C_{i,j}, D_{i,j+1/2})}{d(C_{i-1,j}, C_{i,j})}. \end{aligned}$$

The approximation of $\mathbf{grad}\, u \bullet \boldsymbol{n}_K$ at the center of a "vertical" edge $\sigma_{i+1/2,j}$ is performed in a similar way, by introducing the points $R_{i+1/2,j}$ intersection of the orthogonal bisector $\eta_{i+1/2,j}$ and, according to the geometry, of the line $\zeta_{i,j-1/2}$ or $\zeta_{i,j+1/2}$, and $L_{i+1/2,j}$ intersection of $\eta_{i+1/2,j}$ and $\zeta_{i+1,j-1/2}$ or $\zeta_{i+1,j+1/2}$.

Note that the outmost grid cells require a particular treatment (see FAILLE [1992a]).

The scheme which is described above is stable under a geometrical condition on the family of meshes which is considered. Since the fluxes are consistent and the scheme is conservative, it also satisfies a property of "weak consistency", i.e., as in the one-dimensional case (see Remark 6.6), the exact solution of (9.10) satisfies the numerical scheme with an error which tends to 0 in $L^\infty(\Omega)$ for the weak-$\star$ topology. Under adequate restrictive assumptions, the convergence of the scheme can be deduced, see FAILLE [1992a].

Numerical tests were performed for the Laplace operator and for operators of the type $-\mathrm{div}(\Lambda\, \mathbf{grad}.)$, where Λ is a variable and discontinuous matrix (see FAILLE [1992a]); the discontinuities of Λ are treated in a similar way as in the 1D case (see Section 7). Comparisons with solutions which were obtained by the bilinear finite element method, and with known analytical solutions, were performed. The results given by the VF9 scheme and by the finite element scheme were very similar.

The two drawbacks of this method are the fact that it is a 9-points scheme, and therefore computationally expensive, and that it yields a nonsymmetric matrix even if the original continuous operator is symmetric. Also, its generalization to three dimensions is somewhat complex.

REMARK 9.2. The proof of convergence of this scheme is hindered by the lack of consistency for the discrete adjoint operator (see Section 9.4). An error estimate is also difficult to obtain because the numerical flux at an interface $K|L$ cannot be written under the form $\tau_{K|L}(u_K - u_L)$ with $\tau_{K|L} > 0$. Note, however, that under some geometrical assumptions on the mesh, see FAILLE [1992a] and COUDIÈRE, VILA and VILLEDIEU [1999], error estimates may be obtained.

9.2. General meshes and schemes

Let us now turn to the discretization of convection–diffusion problems on general structured or nonstructured grids, consisting of any polygonal (recall that we shall call "polygonal" any polygonal domain of $\mathbb{R}^2$ or polyhedral domain or $\mathbb{R}^3$) control volumes (satisfying adequate geometrical conditions which are stated in the sequel) and not necessarily ordered in a Cartesian grid. The advantage of finite volume schemes using nonstructured meshes is clear for convection–diffusion equations. On one hand, the stability and convergence properties of the finite volume scheme (with an upstream choice for the convective flux) ensure a robust scheme for any admissible mesh as defined in Definitions 9.1 and 10.1 below, without any need for refinement in the areas of a large convection flux. On the other hand, the use of a nonstructured mesh allows the computation of a solution for any shape of the physical domain.

We saw in the previous section that a consistent discretization of the normal flux $-\nabla u \cdot \boldsymbol{n}$ over the interface of two control volumes K and L may be performed with a differential quotient involving values of the unknown located on the orthogonal line to the interface between K and L, on either side of this interface. This remark suggests the following definition of admissible finite volume meshes for the discretization of diffusion problems. We shall only consider here, for the sake of simplicity, the case of polygonal domains. The case of domains with a regular boundary does not introduce any supplementary difficulty other than complex notations.

DEFINITION 9.1 (*Admissible meshes*). Let Ω be an open bounded polygonal subset of $\mathbb{R}^d$, $d = 2$, or 3. An admissible finite volume mesh of Ω, denoted by $\mathcal{T}$, is given by a family of "control volumes", which are open polygonal convex subsets of Ω, a family of subsets of $\overline{\Omega}$ contained in hyperplanes of $\mathbb{R}^d$, denoted by $\mathcal{E}$ (these are the edges (two-dimensional) or sides (three-dimensional) of the control volumes), with strictly positive $(d-1)$-dimensional measure, and a family of points of Ω denoted by $\mathcal{P}$ satisfying the following properties (in fact, we shall denote, somewhat incorrectly, by $\mathcal{T}$ the family of control volumes):

(i) The closure of the union of all the control volumes is $\overline{\Omega}$.

(ii) For any $K \in \mathcal{T}$, there exists a subset $\mathcal{E}_K$ of $\mathcal{E}$ such that $\partial K = \overline{K} \setminus K = \bigcup_{\sigma \in \mathcal{E}_K} \overline{\sigma}$. Furthermore, $\mathcal{E} = \bigcup_{K \in \mathcal{T}} \mathcal{E}_K$.

(iii) For any $(K, L) \in \mathcal{T}^2$ with $K \neq L$, either the $(d-1)$-dimensional Lebesgue measure of $\overline{K} \cap \overline{L}$ is 0 or $\overline{K} \cap \overline{L} = \overline{\sigma}$ for some $\sigma \in \mathcal{E}$, which will then be denoted by $K|L$.

(iv) The family $\mathcal{P} = (x_K)_{K \in \mathcal{T}}$ is such that $x_K \in \overline{K}$ (for all $K \in \mathcal{T}$) and, if $\sigma = K|L$, it is assumed that $x_K \neq x_L$, and that the straight line $\mathcal{D}_{K,L}$ going through x_K and x_L is orthogonal to $K|L$.

(v) For any $\sigma \in \mathcal{E}$ such that $\sigma \subset \partial\Omega$, let K be the control volume such that $\sigma \in \mathcal{E}_K$. If $x_K \notin \sigma$, let $\mathcal{D}_{K,\sigma}$ be the straight line going through x_K and orthogonal to σ, then the condition $\mathcal{D}_{K,\sigma} \cap \sigma \neq \emptyset$ is assumed; let $y_\sigma = \mathcal{D}_{K,\sigma} \cap \sigma$.

In the sequel, the following notations are used.

The mesh size is defined by: $\text{size}(\mathcal{T}) = \sup\{\text{diam}(K),\ K \in \mathcal{T}\}$.

For any $K \in \mathcal{T}$ and $\sigma \in \mathcal{E}$, $\text{m}(K)$ is the d-dimensional Lebesgue measure of K (it is the area of K in the two-dimensional case and the volume in the three-dimensional case) and $\text{m}(\sigma)$ the $(d-1)$-dimensional measure of σ.

The set of interior (resp. boundary) edges is denoted by $\mathcal{E}_{\text{int}}$ (resp. $\mathcal{E}_{\text{ext}}$), i.e., $\mathcal{E}_{\text{int}} = \{\sigma \in \mathcal{E};\ \sigma \not\subset \partial\Omega\}$ (resp. $\mathcal{E}_{\text{ext}} = \{\sigma \in \mathcal{E};\ \sigma \subset \partial\Omega\}$).

The set of neighbours of K is denoted by $\mathcal{N}(K)$, i.e., $\mathcal{N}(K) = \{L \in \mathcal{T};\ \exists \sigma \in \mathcal{E}_K,\ \overline{\sigma} = \overline{K} \cap \overline{L}\}$.

If $\sigma = K|L$, we denote by d_σ or $d_{K|L}$ the Euclidean distance between x_K and x_L (which is positive) and by $d_{K,\sigma}$ the distance from x_K to σ.

If $\sigma \in \mathcal{E}_K \cap \mathcal{E}_{\text{ext}}$, let d_σ denote the Euclidean distance between x_K and y_σ (then, $d_\sigma = d_{K,\sigma}$).

For any $\sigma \in \mathcal{E}$; the "transmissibility" through σ is defined by $\tau_\sigma = \text{m}(\sigma)/d_\sigma$ if $d_\sigma \neq 0$.

In some results and proofs given below, there are summations over $\sigma \in \mathcal{E}_0$, with $\mathcal{E}_0 = \{\sigma \in \mathcal{E};\ d_\sigma \neq 0\}$. For simplicity (in these results and proofs), $\mathcal{E} = \mathcal{E}_0$ is assumed.

REMARK 9.3. (i) The definition of y_σ for $\sigma \in \mathcal{E}_{\text{ext}}$ requires that $y_\sigma \in \overline{\sigma}$. However, in many cases, this condition may be relaxed. The condition $x_K \in \overline{K}$ may also be relaxed as described, e.g., in Example 9.1 below.

(ii) The condition $x_K \neq x_L$ if $\sigma = K|L$, is in fact quite easy to satisfy: two neighbouring control volumes K, L which do not satisfy it just have to be collapsed into a new control volume M with $x_M = x_K = x_L$, and the edge $K|L$ removed from the set of edges. The new mesh thus obtained is admissible.

EXAMPLE 9.1 (*Triangular meshes*). Let Ω be an open bounded polygonal subset of $\mathbb{R}^2$. Let $\mathcal{T}$ be a family of open triangular disjoint subsets of Ω such that two triangles having a common edge have also two common vertices. Assume that all angles of the triangles are less than $\pi/2$. This last condition is sufficient for the orthogonal bisectors to intersect inside each triangle, thus naturally defining the points $x_K \in K$. One obtains an admissible mesh. In the case of an elliptic operator, the finite volume scheme defined on such a grid using differential quotients for the approximation of the normal flux yields a 4-point scheme (HERBIN [1995]). This scheme does not lead to a finite difference scheme consistent with the continuous diffusion operator (using a Taylor expansion). The consistency is only verified for the approximation of the fluxes, but this, together with the conservativity of the scheme yields the convergence of the scheme, as it is proved below.

Note that the condition that all angles of the triangles are less than $\pi/2$ (which yields $x_K \in K$) may be relaxed (at least for the triangles the closure of which are in Ω) to the so called "strict Delaunay condition" which is that the closure of the circumscribed circle to each triangle of the mesh does not contain any other triangle of the mesh. For such a mesh, the point x_K (which is the intersection of the orthogonal bisectors of the edges of K) is not always in K, but the scheme (9.17)–(9.19) is convenient since (9.18) yields a consistent approximation of the diffusion fluxes and since the transmissibilities (denoted by $\tau_{K|L}$) are positive.

EXAMPLE 9.2 (*Voronoï meshes*). Let Ω be an open bounded polygonal subset of $\mathbb{R}^d$. An admissible finite volume mesh can be built by using the so called "Voronoï" technique. Let $\mathcal{P}$ be a family of points of $\overline{\Omega}$. For example, this family may be chosen as $\mathcal{P} = \{(k_1 h, \dots, k_d h),\ k_1, \dots, k_d \in \mathbb{Z}\} \cap \Omega$, for a given $h > 0$. The control volumes of the Voronoï mesh are defined with respect to each point x of $\mathcal{P}$ by

$$K_x = \{y \in \Omega,\ |x - y| < |z - y|,\ \forall z \in \mathcal{P},\ z \neq x\}.$$

Recall that $|x - y|$ denotes the Euclidean distance between x and y.

Voronoï meshes are admissible in the sense of Definition 9.1 if the assumption "on the boundary", namely part (v) of Definition 9.1, is satisfied. Indeed, this is true, in particular, if the number of points $x \in \mathcal{P}$ which are located on $\partial\Omega$ is "large enough". Otherwise, the assumption (v) of Definition 9.1 may be replaced by the weaker assumption "$d(y_\sigma, \sigma) \leqslant \mathrm{size}(\mathcal{T})$ for any $\sigma \in \mathcal{E}_{\mathrm{ext}}$" which is much easier to satisfy. Note also that a slight modification of the treatment of the boundary conditions in the finite volume scheme (9.20)–(9.23) allows us to obtain convergence and error estimates results (as in Theorems 9.1 and 9.3) for all Voronoï meshes. This modification is the obvious generalization of the scheme described in the first item of Remark 5.2 for the 1D case. It consists in replacing, for $K \in \mathcal{T}$ such that $\mathcal{E}_K \cap \mathcal{E}_{\mathrm{ext}} \neq \emptyset$, Eq. (9.20), associated to this control volume, by the equation $u_K = g(z_K)$, where z_K is some point on $\partial\Omega \cap \partial K$. In fact, Voronoï meshes often satisfy the following property:

$$\mathcal{E}_K \cap \mathcal{E}_{\mathrm{ext}} \neq \emptyset \Rightarrow x_K \in \partial\Omega$$

and the mesh is therefore admissible in the sense of Definition 9.1 (then, the scheme (9.20)–(9.23) yields $u_K = g(x_K)$ if $K \in \mathcal{T}$ is such that $\mathcal{E}_K \cap \mathcal{E}_{\mathrm{ext}} \neq \emptyset$).

An advantage of the Voronoï method is that it easily leads to meshes on nonpolygonal domains Ω.

Let us now introduce the space of piecewise constant functions associated to an admissible mesh and some "discrete H_0^1" norm for this space. This discrete norm will be used to obtain stability properties which are given by some estimates on the approximate solution of a finite volume scheme.

DEFINITION 9.2. Let Ω be an open bounded polygonal subset of $\mathbb{R}^d$, $d = 2$ or 3, and $\mathcal{T}$ an admissible mesh. Define $X(\mathcal{T})$ as the set of functions from Ω to $\mathbb{R}$ which are constant over each control volume of the mesh.

DEFINITION 9.3 (*discrete H_0^1 norm*). Let Ω be an open bounded polygonal subset of $\mathbb{R}^d$, $d = 2$ or 3, and $\mathcal{T}$ an admissible finite volume mesh in the sense of Definition 9.1.

For $u \in X(\mathcal{T})$, define the discrete H_0^1 norm by

$$\|u\|_{1,\mathcal{T}} = \left(\sum_{\sigma \in \mathcal{E}} \tau_\sigma (D_\sigma u)^2\right)^{1/2}, \tag{9.12}$$

where $\tau_\sigma = \mathrm{m}(\sigma)/d_\sigma$ and

$$D_\sigma u = |u_K - u_L| \quad \text{if } \sigma \in \mathcal{E}_{\text{int}},\ \sigma = K|L,$$
$$D_\sigma u = |u_K| \quad \text{if } \sigma \in \mathcal{E}_{\text{ext}} \cap \mathcal{E}_K,$$

where u_K denotes the value taken by u on the control volume K and the sets $\mathcal{E}$, $\mathcal{E}_{\text{int}}$, $\mathcal{E}_{\text{ext}}$ and $\mathcal{E}_K$ are defined in Definition 9.1.

The discrete H_0^1 norm is used in the following sections to prove the convergence of finite volume schemes and, under some regularity conditions, to give error estimates. It is related to the H_0^1 norm, see the convergence of the norms in Theorem 9.1. One of the tools used below is the following "discrete Poincaré inequality" which may also be found in TEMAM [1977]:

LEMMA 9.1 (Discrete Poincaré inequality). *Let Ω be an open bounded polygonal subset of $\mathbb{R}^d$, $d = 2$ or 3, $\mathcal{T}$ an admissible finite volume mesh in the sense of Definition* 9.1 *and $u \in X(\mathcal{T})$ (see Definition* 9.2*), then*

$$\|u\|_{L^2(\Omega)} \leqslant \operatorname{diam}(\Omega)\|u\|_{1,\mathcal{T}}, \tag{9.13}$$

where $\|\bullet\|_{1,\mathcal{T}}$ is the discrete H_0^1 norm defined in Definition 9.3.

PROOF. For $\sigma \in \mathcal{E}$, define χ_σ from $\mathbb{R}^d \times \mathbb{R}^d$ to $\{0,1\}$ by $\chi_\sigma(x,y) = 1$ if $\sigma \cap [x,y] \neq \emptyset$ and $\chi_\sigma(x,y) = 0$ otherwise.

Let $u \in X(\mathcal{T})$. Let $\boldsymbol{d}$ be a given unit vector. For all $x \in \Omega$, let $\mathcal{D}_x$ be the semi-line defined by its origin, x, and the vector $\boldsymbol{d}$. Let $y(x)$ such that $y(x) \in \mathcal{D}_x \cap \partial\Omega$ and $[x, y(x)] \subset \overline{\Omega}$, where $[x, y(x)] = \{tx + (1-t)y(x),\ t \in [0,1]\}$ (i.e. $y(x)$ is the first point where $\mathcal{D}_x$ meets $\partial\Omega$).

Let $K \in \mathcal{T}$. For a.e. $x \in K$, one has

$$|u_K| \leqslant \sum_{\sigma\in\mathcal{E}} D_\sigma u \chi_\sigma\big(x, y(x)\big),$$

where the notations $D_\sigma u$ and u_K are defined in Definition 9.3. We write the above inequality for a.e. $x \in \Omega$ and not for all $x \in \Omega$ in order to account for the cases where an edge or a vertex of the mesh is included in the semi-line $[x, y(x)]$; in both cases one may not write the above inequality, but there are only a finite number of edges and vertices, and since $\boldsymbol{d}$ is fixed, the above inequality may be written almost everywhere.

Let $c_\sigma = |\boldsymbol{d} \bullet \boldsymbol{n}_\sigma|$ (recall that $\xi \bullet \eta$ denotes the usual scalar product of ξ and η in $\mathbb{R}^d$). By the Cauchy–Schwarz inequality, the above inequality yields:

$$|u_K|^2 \leqslant \sum_{\sigma\in\mathcal{E}} \frac{(D_\sigma u)^2}{d_\sigma c_\sigma} \chi_\sigma\big(x, y(x)\big) \sum_{\sigma\in\mathcal{E}} d_\sigma c_\sigma \chi_\sigma\big(x, y(x)\big), \quad \text{for a.e. } x \in K. \tag{9.14}$$

Let us show that, for a.e. $x \in \Omega$,

$$\sum_{\sigma\in\mathcal{E}} d_\sigma c_\sigma \chi_\sigma\big(x, y(x)\big) \leqslant \operatorname{diam}(\Omega). \tag{9.15}$$

Let $x \in K$, $K \in \mathcal{T}$, such that $\sigma \cap [x, y(x)]$ contains at most one point, for all $\sigma \in \mathcal{E}$, and $[x, y(x)]$ does not contain any vertex of $\mathcal{T}$ (proving (9.15) for such points x leads to (9.15) a.e. on Ω, since $\boldsymbol{d}$ is fixed).

There exists $\sigma \in \mathcal{E}_{\text{ext}}$ such that $y(x) \in \sigma$. Then, using the fact that the control volumes are convex, one has:

$$\sum_{\sigma\in\mathcal{E}} \chi_\sigma\big(x, y(x)\big) d_\sigma c_\sigma = \big|(x_K - y_\sigma)\cdot \boldsymbol{d}\big|.$$

Since x_K and $y_\sigma \in \overline{\Omega}$ (see Definition 9.1), this gives (9.15).

Let us integrate (9.14) over Ω; (9.15) gives

$$\sum_{K\in\mathcal{T}} \int_K |u_K|^2 \,\mathrm{d}x \leqslant \operatorname{diam}(\Omega) \sum_{\sigma\in\mathcal{E}} \frac{(D_\sigma u)^2}{d_\sigma c_\sigma} \int_\Omega \chi_\sigma\big(x, y(x)\big)\,\mathrm{d}x.$$

Since $\int_\Omega \chi_\sigma(x, y(x))\,\mathrm{d}x \leqslant \operatorname{diam}(\Omega)\mathrm{m}(\sigma)c_\sigma$, this last inequality yields

$$\sum_{K\in\mathcal{T}} \int_K |u_K|^2 \,\mathrm{d}x \leqslant \big(\operatorname{diam}(\Omega)\big)^2 \sum_{\sigma\in\mathcal{E}} |D_\sigma u|^2 \frac{\mathrm{m}(\sigma)}{d_\sigma}\,\mathrm{d}x.$$

Hence the result. □

Let $\mathcal{T}$ be an admissible mesh. Let us now define a finite volume scheme to discretize (9.1), (9.2).

Let

$$f_K = \frac{1}{\mathrm{m}(K)} \int_K f(x)\,\mathrm{d}x, \quad \forall K \in \mathcal{T}. \tag{9.16}$$

Let $(u_K)_{K\in\mathcal{T}}$ denote the discrete unknowns. In order to describe the scheme in the most general way, one introduces some auxiliary unknowns (as in the 1D case, see Section 7), namely the fluxes $F_{K,\sigma}$, for all $K \in \mathcal{T}$ and $\sigma \in \mathcal{E}_K$, and some (expected) approximation of u in σ, denoted by u_σ, for all $\sigma \in \mathcal{E}$. For $K \in \mathcal{T}$ and $\sigma \in \mathcal{E}_K$, let $\boldsymbol{n}_{K,\sigma}$ denote the normal unit vector to σ outward to K and $v_{K,\sigma} = \int_\sigma \boldsymbol{v}(x)\cdot\boldsymbol{n}_{K,\sigma}\,\mathrm{d}\gamma(x)$. Note that $\mathrm{d}\gamma$ is the integration symbol for the $(d-1)$-dimensional Lebesgue measure on the considered hyperplane. In order to discretize the convection term $\operatorname{div}(\boldsymbol{v}(x)u(x))$ in a stable way (see Section 7), let us define the upstream choice $u_{\sigma,+}$ of u on an edge σ with respect to $\boldsymbol{v}$ in the following way. If $\sigma = K|L$, then $u_{\sigma,+} = u_K$ if $v_{K,\sigma} \geqslant 0$, and $u_{\sigma,+} = u_L$ otherwise; if $\sigma \subset K \cap \partial\Omega$, then $u_{\sigma,+} = u_K$ if $v_{K,\sigma} \geqslant 0$ and $u_{\sigma,+} = g(y_\sigma)$ otherwise.

Let us first assume that the points x_K are located in the interior of each control volume, and are therefore not located on the edges, hence $d_{K,\sigma} > 0$ for any $\sigma \in \mathcal{E}_K$, where $d_{K,\sigma}$ is the distance from x_K to σ. A finite volume scheme can be defined by the following set of equations:

$$\sum_{\sigma\in\mathcal{E}_K} F_{K,\sigma} + \sum_{\sigma\in\mathcal{E}_K} v_{K,\sigma} u_{\sigma,+} + b\mathrm{m}(K)u_K = \mathrm{m}(K) f_K, \quad \forall K \in \mathcal{T}, \tag{9.17}$$

$$F_{K,\sigma} = -\tau_{K|L}(u_L - u_K), \quad \forall \sigma \in \mathcal{E}_{\mathrm{int}}, \text{ if } \sigma = K|L, \tag{9.18}$$

$$F_{K,\sigma} = -\tau_\sigma\big(g(y_\sigma) - u_K\big), \quad \forall \sigma \in \mathcal{E}_{\mathrm{ext}} \text{ such that } \sigma \in \mathcal{E}_K. \tag{9.19}$$

In the general case, the center of the cell may be located on an edge. This is the case, e.g., when constructing Voronoï meshes with some of the original points located on the boundary $\partial\Omega$. In this case, the following formulation of the finite volume scheme is valid, and is equivalent to the above scheme if no cell center is located on an edge:

$$\sum_{\sigma\in\mathcal{E}_K} F_{K,\sigma} + \sum_{\sigma\in\mathcal{E}_K} v_{K,\sigma} u_{\sigma,+} + b\mathrm{m}(K)u_K = \mathrm{m}(K) f_K, \quad \forall K \in \mathcal{T}, \tag{9.20}$$

$$F_{K,\sigma} = -F_{L,\sigma}, \quad \forall \sigma \in \mathcal{E}_{\mathrm{int}}, \text{ if } \sigma = K|L, \tag{9.21}$$

$$F_{K,\sigma} d_{K,\sigma} = -\mathrm{m}(\sigma)(u_\sigma - u_K), \quad \forall \sigma \in \mathcal{E}_K, \ \forall K \in \mathcal{T}, \tag{9.22}$$

$$u_\sigma = g(y_\sigma), \quad \forall \sigma \in \mathcal{E}_{\mathrm{ext}}. \tag{9.23}$$

Note that (9.20)–(9.23) always lead, after an easy elimination of the auxiliary unknowns, to a linear system of N equations with N unknowns, namely the $(u_K)_{K\in\mathcal{T}}$, with $N = \mathrm{card}(\mathcal{T})$.

REMARK 9.4. (1) Note that one may have, for some $\sigma \in \mathcal{E}_K$, $x_K \in \sigma$, and therefore, thanks to (9.22), $u_\sigma = u_K$.

(2) The choice $u_\sigma = g(y_\sigma)$ in (9.23) needs some discussion. Indeed, this choice is possible since g is assumed to belong to $C(\partial\Omega, \mathbb{R})$ and then is everywhere defined on $\partial\Omega$. In the case where the solution to (9.1), (9.2) belongs to $H^2(\Omega)$ (which yields $g \in C(\partial\Omega, \mathbb{R})$), it is clearly the "good choice" since it yields the consistency of fluxes (even though an error estimate also holds with other choices for u_σ, the choice given below is, e.g., possible). If $g \in H^{1/2}$ (and not continuous), the value $g(y_\sigma)$ is not necessarily defined. Then, another choice for u_σ is possible, e.g.,

$$u_\sigma = \frac{1}{\mathrm{m}(\sigma)} \int_\sigma g(x)\,\mathrm{d}\gamma(x).$$

With this latter choice for u_σ, a convergence result also holds, see Theorem 9.2.

For the sake of simplicity, it is assumed in Definition 9.1 that $x_K \neq x_L$, for all $K, L \in \mathcal{T}$. This condition may be relaxed; it simply allows an easy expression of the numerical flux $F_{K,\sigma} = -\tau_{K|L}(u_L - u_K)$ if $\sigma = K|L$.

9.3. *Existence and estimates*

Let us first prove the existence of the approximate solution and an estimate on this solution. This estimate ensures the stability of the scheme and will be obtained by using the discrete Poincaré inequality (9.13) and will yield convergence thanks to a compactness theorem given in Section 14.

LEMMA 9.2 (Existence and estimate). *Under Assumptions* 9.1, *let $\mathcal{T}$ be an admissible mesh in the sense of Definition* 9.1; *there exists a unique solution $(u_K)_{K\in\mathcal{T}}$ to Eqs.* (9.20)–(9.23).

Furthermore, assuming $g = 0$ and defining $u_{\mathcal{T}} \in X(\mathcal{T})$ (see Definition 9.2) *by $u_{\mathcal{T}}(x) = u_K$ for a.e. $x \in K$, and for any $K \in \mathcal{T}$, the following estimate holds*:

$$\|u_{\mathcal{T}}\|_{1,\mathcal{T}} \leqslant \operatorname{diam}(\Omega)\|f\|_{L^2(\Omega)}, \tag{9.24}$$

where $\|\bullet\|_{1,\mathcal{T}}$ is the discrete H_0^1 norm defined in Definition 9.3.

PROOF. Equations (9.20)–(9.23) lead, after an easy elimination of the auxiliary unknowns, to a linear system of N equations with N unknowns, namely the $(u_K)_{K\in\mathcal{T}}$, with $N = \operatorname{card}(\mathcal{T})$.

Step 1 (*Existence and uniqueness*). Assume that $(u_K)_{K\in\mathcal{T}}$ satisfies this linear system with $g(y_\sigma) = 0$ for any $\sigma \in \mathcal{E}_{\text{ext}}$, and $f_K = 0$ for all $K \in \mathcal{T}$. Let us multiply (9.20) by u_K and sum over K; from (9.21) and (9.22) one deduces

$$b\sum_{K\in\mathcal{T}} \mathrm{m}(K)u_K^2 + \sum_{K\in\mathcal{T}}\sum_{\sigma\in\mathcal{E}_K} F_{K,\sigma}u_K + \sum_{K\in\mathcal{T}}\sum_{\sigma\in\mathcal{E}_K} v_{K,\sigma}u_{\sigma,+}u_K = 0, \tag{9.25}$$

which gives, reordering the summation over the set of edges

$$b\sum_{K\in\mathcal{T}} \mathrm{m}(K)u_K^2 + \sum_{\sigma\in\mathcal{E}} \tau_\sigma(D_\sigma u)^2 + \sum_{\sigma\in\mathcal{E}} v_\sigma(u_{\sigma,+} - u_{\sigma,-})u_{\sigma,+} = 0, \tag{9.26}$$

where

- $|D_\sigma u| = |u_K - u_L|$, if $\sigma = K|L$ and $|D_\sigma u| = |u_K|$, if $\sigma \in \mathcal{E}_K \cap \mathcal{E}_{\text{ext}}$;
- $v_\sigma = |\int_\sigma \boldsymbol{v}(x) \bullet \boldsymbol{n}\, \mathrm{d}\gamma(x)|$, $\boldsymbol{n}$ being a unit normal vector to σ;
- $u_{\sigma,-}$ is the downstream value to σ with respect to $\boldsymbol{v}$, i.e. if $\sigma = K|L$, then $u_{\sigma,-} = u_K$ if $v_{K,\sigma} \leqslant 0$, and
- $u_{\sigma,-} = u_L$ otherwise; if $\sigma \in \mathcal{E}_K \cap \mathcal{E}_{\text{ext}}$, then $u_{\sigma,-} = u_K$ if $v_{K,\sigma} \leqslant 0$ and $u_{\sigma,-} = u_\sigma$ if $v_{K,\sigma} > 0$.

Note that $u_\sigma = 0$ if $\sigma \in \mathcal{E}_{\text{ext}}$.

Now, remark that

$$\sum_{\sigma\in\mathcal{E}} v_\sigma u_{\sigma,+}(u_{\sigma,+}-u_{\sigma,-}) = \frac{1}{2}\sum_{\sigma\in\mathcal{E}} v_\sigma\left((u_{\sigma,+}-u_{\sigma,-})^2 + \left(u_{\sigma,+}^2 - u_{\sigma,-}^2\right)\right) \tag{9.27}$$

and, thanks to the assumption $\operatorname{div}\boldsymbol{v} \geqslant 0$,

$$\begin{aligned}\sum_{\sigma\in\mathcal{E}} v_\sigma\left(u_{\sigma,+}^2 - u_{\sigma,-}^2\right) &= \sum_{K\in\mathcal{T}}\left(\int_{\partial K}\boldsymbol{v}(x)\cdot\boldsymbol{n}_K\,\mathrm{d}\gamma(x)\right)u_K^2\\ &= \int_\Omega\left(\operatorname{div}\boldsymbol{v}(x)\right)u_{\mathcal{T}}^2(x)\,\mathrm{d}x \geqslant 0.\end{aligned} \tag{9.28}$$

Hence,

$$b\|u_{\mathcal{T}}\|_{L^2(\Omega)}^2 + \|u_{\mathcal{T}}\|_{1,\mathcal{T}}^2 = b\sum_{K\in\mathcal{T}}\mathrm{m}(K)u_K^2 + \sum_{\sigma\in\mathcal{E}}\tau_\sigma(D_\sigma u)^2 \leqslant 0. \tag{9.29}$$

One deduces, from (9.29), that $u_K = 0$ for all $K\in\mathcal{T}$.

This proves the existence and the uniqueness of the solution $(u_K)_{K\in\mathcal{T}}$, of the linear system given by (9.20)–(9.23), for any $\{g(y_\sigma),\ \sigma\in\mathcal{E}_{\text{ext}}\}$ and $\{f_K,\ K\in\mathcal{T}\}$.

Step 2 (*Estimate*). Assume $g=0$. Multiply (9.20) by u_K, sum over K; then, thanks to (9.21), (9.22), (9.27) and (9.28) one has

$$b\|u_{\mathcal{T}}\|_{L^2(\Omega)}^2 + \|u_{\mathcal{T}}\|_{1,\mathcal{T}}^2 \leqslant \sum_{K\in\mathcal{T}}\mathrm{m}(K)f_K u_K.$$

By the Cauchy–Schwarz inequality, this inequality yields

$$\|u_{\mathcal{T}}\|_{1,\mathcal{T}}^2 \leqslant \left(\sum_{K\in\mathcal{T}}\mathrm{m}(K)u_K^2\right)^{1/2}\left(\sum_{K\in\mathcal{T}}\mathrm{m}(K)f_K^2\right)^{1/2} \leqslant \|f\|_{L^2(\Omega)}\|u_{\mathcal{T}}\|_{L^2(\Omega)}.$$

Thanks to the discrete Poincaré inequality (9.13), this yields

$$\|u_{\mathcal{T}}\|_{1,\mathcal{T}} \leqslant \|f\|_{L^2(\Omega)}\operatorname{diam}(\Omega),$$

which concludes the proof of the lemma. □

Let us now state a discrete maximum principle which is satisfied by the scheme (9.20)–(9.23); this is an interesting stability property, even though it will not be used in the proofs of the convergence and error estimate.

PROPOSITION 9.2. *Under Assumption* 9.1, *let* $\mathcal{T}$ *be an admissible mesh in the sense of Definition* 9.1, *let* $(f_K)_{K\in\mathcal{T}}$ *be defined by* (9.16). *If* $f_K \geqslant 0$ *for all* $K\in\mathcal{T}$, *and* $g(y_\sigma)\geqslant 0$, *for all* $\sigma\in\mathcal{E}_{\text{ext}}$, *then the solution* $(u_K)_{K\in\mathcal{T}}$ *of* (9.20)–(9.23) *satisfies* $u_K \geqslant 0$ *for all* $K\in\mathcal{T}$.

PROOF. Assume that $f_K \geqslant 0$ for all $K \in \mathcal{T}$ and $g(y_\sigma) \geqslant 0$ for all $\sigma \in \mathcal{E}_{\text{ext}}$. Let $a = \min\{u_K, K \in \mathcal{T}\}$. Let K_0 be a control volume such that $u_{K_0} = a$. Assume first that K_0 is an "interior" control volume, in the sense that $\mathcal{E}_K \subset \mathcal{E}_{\text{int}}$, and that $u_{K_0} \leqslant 0$. Then, from (9.20),

$$\sum_{\sigma \in \mathcal{E}_{K_0}} F_{K_0,\sigma} + \sum_{\sigma \in \mathcal{E}_{K_0}} v_{K_0,\sigma} u_{\sigma,+} \geqslant 0; \tag{9.30}$$

since for any neighbour L of K_0 one has $u_L \geqslant u_{K_0}$, then, noting that $\operatorname{div} \boldsymbol{v} \geqslant 0$, one must have $u_L = u_{K_0}$ for any neighbour L of K. Hence, setting $B = \{K \in \mathcal{T}, u_K = a\}$, there exists $K \in B$ such that $\mathcal{E}_K \not\subset \mathcal{E}_{\text{int}}$, i.e. K is a control volume "neighbouring the boundary".

Assume then that K_0 is a control volume neighbouring the boundary and that $u_{K_0} = a < 0$. Then, for an edge $\sigma \in \mathcal{E}_{\text{ext}} \cap \mathcal{E}_K$, relations (9.22) and (9.23) yield $g(y_\sigma) < 0$, which is in contradiction with the assumption. Hence Proposition 9.2 is proved. $\square$

REMARK 9.5. The maximum principle immediately yields the existence and uniqueness of the solution of the numerical scheme (9.20)–(9.23), which was proved directly in Lemma 9.2.

9.4. Convergence

Let us now show the convergence of approximate solutions obtained by the above finite volume scheme when the size of the mesh tends to 0. One uses Lemma 9.2 together with the Compactness Theorem 14.2 given at the end of this chapter to prove the convergence result. In order to use Theorem 14.2, one needs the following lemma.

LEMMA 9.3. *Let Ω be an open bounded set of $\mathbb{R}^d$, $d = 2$ or 3. Let $\mathcal{T}$ be an admissible mesh in the sense of Definition 9.1 and $u \in X(\mathcal{T})$ (see Definition 9.2). One defines $\tilde{u}$ by $\tilde{u} = u$ a.e. on Ω, and $\tilde{u} = 0$ a.e. on $\mathbb{R}^d \setminus \Omega$. Then there exists $C > 0$, only depending on Ω, such that*

$$\left\|\tilde{u}(\cdot + \eta) - \tilde{u}\right\|^2_{L^2(\mathbb{R}^d)} \leqslant \|u\|^2_{1,\mathcal{T}} |\eta| \big(|\eta| + C \operatorname{size}(\mathcal{T})\big), \quad \forall \eta \in \mathbb{R}^d. \tag{9.31}$$

PROOF. For $\sigma \in \mathcal{E}$, define χ_σ from $\mathbb{R}^d \times \mathbb{R}^d$ to $\{0, 1\}$ by $\chi_\sigma(x, y) = 1$ if $[x, y] \cap \sigma \neq \emptyset$ and $\chi_\sigma(x, y) = 0$ if $[x, y] \cap \sigma = \emptyset$.

Let $\eta \in \mathbb{R}^d$, $\eta \neq 0$. One has

$$\left|\tilde{u}(x + \eta) - \tilde{u}(x)\right| \leqslant \sum_{\sigma \in \mathcal{E}} \chi_\sigma(x, x + \eta) |D_\sigma u|, \quad \text{for a.e. } x \in \Omega$$

(see Definition 9.3 for the definition of $D_\sigma u$).

This gives, using the Cauchy–Schwarz inequality,

$$\left|\tilde{u}(x+\eta)-\tilde{u}(x)\right|^2 \leqslant \sum_{\sigma\in\mathcal{E}} \chi_\sigma(x,x+\eta)\frac{|D_\sigma u|^2}{d_\sigma c_\sigma} \sum_{\sigma\in\mathcal{E}} \chi_\sigma(x,x+\eta) d_\sigma c_\sigma,$$

$$\text{for a.e. } x\in\mathbb{R}^d, \tag{9.32}$$

where $c_\sigma = |\boldsymbol{n}_\sigma \bullet \frac{\eta}{|\eta|}|$, and $\boldsymbol{n}_\sigma$ denotes a unit normal vector to σ.

Let us now prove that there exists $C > 0$, only depending on Ω, such that

$$\sum_{\sigma\in\mathcal{E}} \chi_\sigma(x,x+\eta) d_\sigma c_\sigma \leqslant |\eta| + C\,\text{size}(\mathcal{T}), \tag{9.33}$$

for a.e. $x\in\mathbb{R}^d$.

Let $x\in\mathbb{R}^d$ such that $\sigma\cap[x,x+\eta]$ contains at most one point, for all $\sigma\in\mathcal{E}$, and $[x,x+\eta]$ does not contain any vertex of $\mathcal{T}$ (proving (9.33) for such points x gives (9.33) for a.e. $x\in\mathbb{R}^d$, since η is fixed). Since Ω is not assumed to be convex, it may happen that the line segment $[x,x+\eta]$ is not included in $\overline{\Omega}$. In order to deal with this, let $y,z\in[x,x+\eta]$ such that $y\neq z$ and $[y,z]\subset\overline{\Omega}$; there exist K, $L\in\mathcal{T}$ such that $y\in\overline{K}$ and $z\in\overline{L}$. Hence,

$$\sum_{\sigma\in\mathcal{E}} \chi_\sigma(y,z) d_\sigma c_\sigma = \left|(y_1-z_1)\bullet\frac{\eta}{|\eta|}\right|,$$

where $y_1 = x_K$ or y_σ with $\sigma\in\mathcal{E}_{\text{ext}}\cap\mathcal{E}_K$ and $z_1 = x_L$ or $y_{\tilde{\sigma}}$ with $\tilde{\sigma}\in\mathcal{E}_{\text{ext}}\cap\mathcal{E}_L$, depending on the position of y and z in $\overline{K}$ or $\overline{L}$ respectively.

Since $y_1 = y + y_2$, with $|y_2|\leqslant\text{size}(\mathcal{T})$, and $z_1 = z+z_2$, with $|z_2|\leqslant\text{size}(\mathcal{T})$, one has

$$\left|(y_1-z_1)\bullet\frac{\eta}{|\eta|}\right| \leqslant |y-z|+|y_2|+|z_2| \leqslant |y-z| + 2\,\text{size}(\mathcal{T})$$

and

$$\sum_{\sigma\in\mathcal{E}} \chi_\sigma(y,z) d_\sigma c_\sigma \leqslant |y-z| + 2\,\text{size}(\mathcal{T}). \tag{9.34}$$

Note that this yields (9.33) with $C=2$ if $[x,x+\eta]\subset\overline{\Omega}$.

Since Ω has a finite number of sides, the line segment $[x,x+\eta]$ intersects $\partial\Omega$ a finite number of times; hence there exist $t_1,\dots,t_n$ such that $0\leqslant t_1<t_2<\cdots<t_n\leqslant 1$, $n\leqslant N$, where N only depends on Ω (indeed, it is possible to take $N=2$ if Ω is convex and N equal to the number of sides of Ω for a general Ω) and such that

$$\sum_{\sigma\in\mathcal{E}} \chi_\sigma(x,x+\eta) d_\sigma c_\sigma = \sum_{\substack{i=1\\ \text{odd } i}}^{n-1} \sum_{\sigma\in\mathcal{E}} \chi_\sigma(x_i,x_{i+1}) d_\sigma c_\sigma,$$

with $x_i = x+t_i\eta$, for $i=1,\dots,n$, $x_i\in\partial\Omega$ if $t_i\notin\{0,1\}$ and $[x_i,x_{i+1}]\subset\overline{\Omega}$ if i is odd.

Then, thanks to (9.34) with $y = x_i$ and $z = x_{i+1}$, for $i = 1, \ldots, n-1$, one has (9.33) with $C = 2(N-1)$ (in particular, if Ω is convex, $C = 2$ is convenient for (9.33) and therefore for (9.31) as we shall see below).

In order to conclude the proof of Lemma 9.3, remark that, for all $\sigma \in \mathcal{E}$,

$$\int_{\mathbb{R}^d} \chi_\sigma(x, x+\eta)\,\mathrm{d}x \leqslant \mathrm{m}(\sigma) c_\sigma |\eta|.$$

Therefore, integrating (9.32) over $\mathbb{R}^d$ yields, with (9.33),

$$\|\tilde{u}(\cdot + \eta) - \tilde{u}\|^2_{L^2(\mathbb{R}^d)} \leqslant \Bigl(\sum_{\sigma\in\mathcal{E}} \frac{\mathrm{m}(\sigma)}{d_\sigma} |D_\sigma u|^2\Bigr) |\eta| \bigl(|\eta| + C\,\mathrm{size}(\mathcal{T})\bigr). \qquad \square$$

We are now able to state the convergence theorem. We shall first prove the convergence result in the case of homogeneous Dirichlet boundary conditions, i.e. $g = 0$; the nonhomogeneous case is then considered in the two-dimensional case (see Theorem 9.2), following EYMARD, GALLOUËT and HERBIN [1999].

THEOREM 9.1 (Convergence, homogeneous Dirichlet boundary conditions). *Under Assumption* 9.1 *with* $g = 0$, *let* $\mathcal{T}$ *be an admissible mesh* (*in the sense of Definition* 9.1). *Let* $(u_K)_{K\in\mathcal{T}}$ *be the solution of the system given by Eqs.* (9.20)–(9.23) (*existence and uniqueness of* $(u_K)_{K\in\mathcal{T}}$ *are given in Lemma* 9.2). *Define* $u_\mathcal{T} \in X(\mathcal{T})$ *by* $u_\mathcal{T}(x) = u_K$ *for a.e.* $x \in K$, *and for any* $K \in \mathcal{T}$. *Then* $u_\mathcal{T}$ *converges in* $L^2(\Omega)$ *to the unique variational solution* $u \in H_0^1(\Omega)$ *of problem* (9.1), (9.2) *as* $\mathrm{size}(\mathcal{T}) \to 0$. *Furthermore* $\|u_\mathcal{T}\|_{1,\mathcal{T}}$ *converges to* $\|u\|_{H_0^1(\Omega)}$ *as* $\mathrm{size}(\mathcal{T}) \to 0$.

REMARK 9.6. (1) In Theorem 9.1, the hypothesis $f \in L^2(\Omega)$ is not necessary. It is used essentially to obtain a bound on $\|u_\mathcal{T}\|_{1,\mathcal{T}}$. In order to pass to the limit, the hypothesis "$f \in L^1(\Omega)$" is sufficient. Then, in Theorem 9.1, the hypothesis $f \in L^2(\Omega)$ can be replaced by $f \in L^p(\Omega)$ for some $p > 1$, if $d = 2$, and for $p \geqslant 6/5$, if $d = 3$, provided that the meshes satisfy, for some fixed $\zeta > 0$, $d_{K,\sigma} \geqslant \zeta d_\sigma$, for all $\sigma \in \mathcal{E}_K$ and for all control volumes K. Indeed, one obtains, in this case, a bound on $\|u_\mathcal{T}\|_{1,\mathcal{T}}$ by using a "discrete Sobolev inequality" (proved in Lemma 9.5).

It is also possible to obtain convergence results, towards a "very weak solution" of problem (9.1), (9.2), with only $f \in L^1(\Omega)$, by working with some discrete equivalent of the $W_0^{1,q}$-norm, with $q < d/(d-1)$. This is not detailed here.

(2) In Theorem 9.1, it is also possible to prove convergence results when $f(x)$ is replaced by some nonlinear function $f(x, u(x))$ as in Theorem 8.1. The proof is an easy adaptation of that of Theorem 8.1.

PROOF OF THEOREM 9.1. Let Y be the set of approximate solutions, i.e., the set of $u_\mathcal{T}$ where $\mathcal{T}$ is an admissible mesh in the sense of Definition 9.1. First, we want to prove that $u_\mathcal{T}$ tends to the unique solution (in $H_0^1(\Omega)$) to (9.3) as $\mathrm{size}(\mathcal{T}) \to 0$.

Thanks to Lemma 9.2 and to the discrete Poincaré inequality (9.13), there exists $C_1 \in \mathbb{R}$, only depending on Ω and f, such that $\|u_\mathcal{T}\|_{1,\mathcal{T}} \leqslant C_1$ and $\|u_\mathcal{T}\|_{L^2(\Omega)} \leqslant C_1$

for all $u_{\mathcal{T}} \in Y$. Then, thanks to Lemma 9.3 and to the compactness result given in Theorem 14.2, the set Y is relatively compact in $L^2(\Omega)$ and any possible limit (in $L^2(\Omega)$) of a sequence $(u_{\mathcal{T}_n})_{n\in\mathbb{N}} \subset Y$ (such that $\text{size}(\mathcal{T}_n) \to 0$) belongs to $H_0^1(\Omega)$. Therefore, thanks to the uniqueness of the solution (in $H_0^1(\Omega)$) of (9.3), it is sufficient to prove that if $(u_{\mathcal{T}_n})_{n\in\mathbb{N}} \subset Y$ converges towards some $u \in H_0^1(\Omega)$, in $L^2(\Omega)$, and $\text{size}(\mathcal{T}_n) \to 0$ (as $n \to \infty$), then u is the solution to (9.3). We prove this result below, omitting the index n, i.e., assuming $u_{\mathcal{T}} \to u$ in $L^2(\Omega)$ as $\text{size}(\mathcal{T}) \to 0$.

Let $\psi \in C_c^\infty(\Omega)$ and let $\text{size}(\mathcal{T})$ be small enough so that $\psi(x) = 0$ if $x \in K$ and $K \in \mathcal{T}$ is such that $\partial K \cap \partial\Omega \neq \emptyset$. Multiplying (9.20) by $\psi(x_K)$, and summing the result over $K \in \mathcal{T}$ yields

$$T_1 + T_2 + T_3 = T_4, \tag{9.35}$$

with

$$T_1 = b \sum_{K\in\mathcal{T}} \mathrm{m}(K) u_K \psi(x_K),$$

$$T_2 = -\sum_{K\in\mathcal{T}} \sum_{L\in\mathcal{N}(K)} \tau_{K|L}(u_L - u_K)\psi(x_K),$$

$$T_3 = \sum_{K\in\mathcal{T}} \sum_{\sigma\in\mathcal{E}_K} v_{K,\sigma} u_{\sigma,+} \psi(x_K),$$

$$T_4 = \sum_{K\in\mathcal{T}} \mathrm{m}(K)\psi(x_K) f_K.$$

First remark that, since $u_{\mathcal{T}}$ tends to u in $L^2(\Omega)$,

$$T_1 \to b \int_\Omega u(x)\psi(x)\,\mathrm{d}x \quad \text{as size}(\mathcal{T}) \to 0.$$

Similarly,

$$T_4 \to \int_\Omega f(x)\psi(x)\,\mathrm{d}x \quad \text{as size}(\mathcal{T}) \to 0.$$

Let us now turn to the study of T_2;

$$T_2 = -\sum_{K|L\in\mathcal{E}_{\text{int}}} \tau_{K|L}(u_L - u_K)\big(\psi(x_K) - \psi(x_L)\big).$$

Consider the following auxiliary expression:

$$T_2' = \int_\Omega u_{\mathcal{T}}(x)\Delta\psi(x)\,\mathrm{d}x$$

$$= \sum_{K\in\mathcal{T}} u_K \int_K \Delta\psi(x)\,\mathrm{d}x$$

$$= \sum_{K|L\in\mathcal{E}_{\text{int}}} (u_K - u_L) \int_{K|L} \nabla\psi(x)\cdot\boldsymbol{n}_{K,L}\,\mathrm{d}\gamma(x).$$

Since $u_{\mathcal{T}}$ converges to u in $L^2(\Omega)$, it is clear that T_2' tends to $\int_\Omega u(x)\Delta\psi(x)\,\mathrm{d}x$ as $\mathrm{size}(\mathcal{T})$ tends to 0.

Define

$$R_{K,L} = \frac{1}{\mathrm{m}(K|L)} \int_{K|L} \nabla\psi(x)\cdot\boldsymbol{n}_{K,L}\,\mathrm{d}\gamma(x) - \frac{\psi(x_L)-\psi(x_K)}{d_{K|L}},$$

where $\boldsymbol{n}_{K,L}$ denotes the unit normal vector to $K|L$, outward to K, then

$$\begin{aligned}
\left|T_2 + T_2'\right| &= \left| \sum_{K|L\in\mathcal{E}_{\text{int}}} \mathrm{m}(K|L)(u_K - u_L)R_{K,L} \right| \\
&\leqslant \left[\sum_{K|L\in\mathcal{E}_{\text{int}}} \mathrm{m}(K|L)\frac{(u_K-u_L)^2}{d_{K|L}} \sum_{K|L\in\mathcal{E}_{\text{int}}} \mathrm{m}(K|L) d_{K|L}(R_{K,L})^2 \right]^{1/2}.
\end{aligned}$$

Regularity properties of the function ψ give the existence of $C_2\in\mathbb{R}$, only depending on ψ, such that $|R_{K,L}| \leqslant C_2 \mathrm{size}(\mathcal{T})$. Therefore, since

$$\sum_{K|L\in\mathcal{E}_{\text{int}}} \mathrm{m}(K|L) d_{K|L} \leqslant d\,\mathrm{m}(\Omega),$$

from estimate (9.24), we conclude that $T_2 + T_2' \to 0$ as $\mathrm{size}(\mathcal{T}) \to 0$.

Let us now show that T_3 tends to $-\int_\Omega \boldsymbol{v}(x)u(x)\nabla\psi(x)\,\mathrm{d}x$ as $\mathrm{size}(\mathcal{T}) \to 0$. Let us decompose $T_3 = T_3' + T_3''$ where

$$T_3' = \sum_{K\in\mathcal{T}} \sum_{\sigma\in\mathcal{E}_K} v_{K,\sigma}(u_{\sigma,+} - u_K)\psi(x_K)$$

and

$$T_3'' = \sum_{K\in\mathcal{T}} \sum_{\sigma\in\mathcal{E}_K} v_{K,\sigma} u_K \psi(x_K) = \int_\Omega \mathrm{div}\,\boldsymbol{v}(x) u_{\mathcal{T}}(x)\psi_{\mathcal{T}}(x)\,\mathrm{d}x,$$

where $\psi_{\mathcal{T}}$ is defined by $\psi_{\mathcal{T}}(x) = \psi(x_K)$ if $x\in K$, $K\in\mathcal{T}$. Since $u_{\mathcal{T}} \to u$ and $\psi_{\mathcal{T}} \to \psi$ in $L^2(\Omega)$ as $\mathrm{size}(\mathcal{T}) \to 0$ (indeed, $\psi_{\mathcal{T}} \to \psi$ uniformly on Ω as $\mathrm{size}(\mathcal{T}) \to 0$) and since $\mathrm{div}\,\boldsymbol{v} \in L^\infty(\Omega)$, one has

$$T_3'' \to \int_\Omega \mathrm{div}\,\boldsymbol{v}(x) u(x)\psi(x)\,\mathrm{d}x \quad \text{as } \mathrm{size}(\mathcal{T}) \to 0.$$

Let us now rewrite T_3' as $T_3' = T_3''' + r_3$ with

$$T_3''' = \sum_{K \in \mathcal{T}} \sum_{\sigma \in \mathcal{E}_K} (u_{\sigma,+} - u_K) \int_\sigma \boldsymbol{v}(x) \cdot \boldsymbol{n}_{K,\sigma} \psi(x) \, \mathrm{d}\gamma(x)$$

and

$$r_3 = \sum_{K \in \mathcal{T}} \sum_{\sigma \in \mathcal{E}_K} (u_{\sigma,+} - u_K) \int_\sigma \boldsymbol{v}(x) \cdot \boldsymbol{n}_{K,\sigma} \big(\psi(x_K) - \psi(x)\big) \, \mathrm{d}\gamma(x).$$

Thanks to the regularity of $\boldsymbol{v}$ and ψ, there exists C_3 only depending on $\boldsymbol{v}$ and ψ such that

$$|r_3| \leqslant C_3 \, \mathrm{size}(\mathcal{T}) \sum_{K|L \in \mathcal{E}_{\mathrm{int}}} |u_K - u_L| \mathrm{m}(K|L),$$

which yields, with the Cauchy–Schwarz inequality,

$$|r_3| \leqslant C_3 \, \mathrm{size}(\mathcal{T}) \Big(\sum_{K|L \in \mathcal{E}_{\mathrm{int}}} \tau_{K|L} |u_K - u_L|^2 \Big)^{1/2} \Big(\sum_{K|L \in \mathcal{E}_{\mathrm{int}}} \mathrm{m}(K|L) d_{K|L} \Big)^{1/2},$$

from which one deduces, with estimate (9.24), that $r_3 \to 0$ as $\mathrm{size}(\mathcal{T}) \to 0$.

Next, remark that

$$\begin{aligned} T_3''' &= -\sum_{K \in \mathcal{T}} u_K \sum_{\sigma \in \mathcal{E}_K} \int_\sigma \boldsymbol{v}(x) \cdot \boldsymbol{n}_{K,\sigma} \psi(x) \, \mathrm{d}\gamma(x) \\ &= -\sum_{K \in \mathcal{T}} u_K \int_K \mathrm{div}\big(\boldsymbol{v}(x) \psi(x)\big) \, \mathrm{d}x. \end{aligned}$$

This implies (since $u_{\mathcal{T}} \to u$ in $L^2(\Omega)$) that $T_3''' \to -\int_\Omega \mathrm{div}(\boldsymbol{v}(x)\psi(x)) u(x) \, \mathrm{d}x$, so that T_3' has the same limit and $T_3 \to -\int_\Omega \boldsymbol{v}(x) \cdot \nabla\psi(x) u(x) \, \mathrm{d}x$.

Hence, letting $\mathrm{size}(\mathcal{T}) \to 0$ in (9.35) yields that the function $u \in H_0^1(\Omega)$ satisfies

$$\int_\Omega \big(bu(x)\psi(x) - u(x)\Delta\psi(x) - \boldsymbol{v}(x)u(x)\nabla\psi(x) - f(x)\psi(x)\big) \, \mathrm{d}x = 0, \\ \forall \psi \in C_c^\infty(\Omega),$$

which, in turn, yields (9.3) thanks to the fact that $u \in H_0^1(\Omega)$, and to the density of $C_c^\infty(\Omega)$ in $H_0^1(\Omega)$.

This concludes the proof of $u_{\mathcal{T}} \to u$ in $L^2(\Omega)$ as $\mathrm{size}(\mathcal{T}) \to 0$, where u is the unique solution (in $H_0^1(\Omega)$) to (9.3).

Let us now prove that $\|u_{\mathcal{T}}\|_{1,\mathcal{T}}$ tends to $\|u\|_{H_0^1(\Omega)}$ in the pure diffusion case, i.e. assuming $b = 0$ and $\boldsymbol{v} = 0$. Since

$$\|u_{\mathcal{T}}\|_{1,\mathcal{T}}^2 = \int_\Omega f_{\mathcal{T}}(x)u_{\mathcal{T}}(x)\,\mathrm{d}x \to \int_\Omega f(x)u(x)\,\mathrm{d}x \quad \text{as size}(\mathcal{T}) \to 0,$$

where $f_{\mathcal{T}}$ is defined from Ω to $\mathbb{R}$ by $f_{\mathcal{T}}(x) = f_K$ a.e. on K for all $K \in \mathcal{T}$, it is easily seen that

$$\|u_{\mathcal{T}}\|_{1,\mathcal{T}}^2 \to \int_\Omega f(x)u(x)\,\mathrm{d}x = \|u\|_{H_0^1(\Omega)}^2 \quad \text{as size}(\mathcal{T}) \to 0.$$

This concludes the proof of Theorem 9.1. □

REMARK 9.7 (*Consistency for the adjoint operator*). The proof of Theorem 9.1 uses the property of consistency of the (diffusion) fluxes on the test functions. This property consists in writing the consistency of the fluxes for the adjoint operator to the discretized Dirichlet operator. This consistency is achieved thanks to that of fluxes for the discretized Dirichlet operator and to the fact that this operator is self adjoint. In fact, any discretization of the Dirichlet operator giving "L^2-stability" and consistency of fluxes on its adjoint, yields a convergence result (see also Remark 9.2). On the contrary, the error estimates proved in Sections 9.5 and 9.6 directly use the consistency for the discretized Dirichlet operator itself.

REMARK 9.8 (*Finite volume schemes and H^1 approximate solutions*). In the above proof, we showed that a sequence of approximate solutions (which are piecewise constant functions) converges in $L^2(\Omega)$ to a limit which is in $H_0^1(\Omega)$. An alternative to the use of Theorem 14.2 is the construction of a bounded sequence in $H^1(\mathbb{R}^d)$ from the sequence of approximate solutions. This can be performed by convoluting the approximate solution with a mollifier "of size size($\mathcal{T}$)". Using Rellich's compactness theorem and the weak sequential compactness of the bounded sets of H^1, one obtains that the limit of the sequence of approximate solutions is in H_0^1.

Let us now deal with the case of nonhomogeneous Dirichlet boundary conditions, in which case $g \in H^{1/2}(\partial\Omega)$ is no longer assumed to be 0. The proof uses the following preliminary result:

LEMMA 9.4. *Let Ω be an open bounded polygonal subset of $\mathbb{R}^2$, $\tilde{g} \in H^1(\Omega)$ and $g = \overline{\gamma}(\tilde{g})$ (recall that $\overline{\gamma}$ is the "trace" operator from $H^1(\Omega)$ to $H^{1/2}(\partial\Omega)$). Let $\mathcal{T}$ be an admissible mesh (in the sense of Definition 9.1) such that, for some $\zeta > 0$, the inequality $d_{K,\sigma} \geqslant \zeta\,\mathrm{diam}(K)$ holds for all control volumes $K \in \mathcal{T}$ and for all $\sigma \in \mathcal{E}_K$, and let $M \in \mathbb{N}$ be such that $\mathrm{card}(\mathcal{E}_K) \leqslant M$ for all $K \in \mathcal{T}$. Let us define $\tilde{g}_K$ for all $K \in \mathcal{T}$ by*

$$\tilde{g}_K = \frac{1}{\mathrm{m}(K)} \int_K \tilde{g}(x)\,\mathrm{d}x$$

and $\tilde{g}_\sigma$ for all $\sigma \in \mathcal{E}_{\rm ext}$ by

$$\tilde{g}_\sigma = \frac{1}{\mathrm{m}(\sigma)} \int_\sigma g(x)\,\mathrm{d}\gamma(x).$$

Let us define

$$\mathcal{N}(\tilde{g}, \mathcal{T}) = \Bigg(\sum_{\sigma = K|L \in \mathcal{E}_{\rm int}} \tau_{K|L}\big(\tilde{g}_K - \tilde{g}_L\big)^2 + \sum_{\sigma \in \mathcal{E}_{\rm ext}} \tau_\sigma\big(\tilde{g}_{K(\sigma)} - \tilde{g}_\sigma\big)^2 \Bigg)^{1/2}, \tag{9.36}$$

where $K(\sigma) = K$ if $\sigma \in \mathcal{E}_{\rm ext} \cap \mathcal{E}_K$. Then there exists $C \in \mathbb{R}_+$, only depending on ζ and M, such that

$$\mathcal{N}(\tilde{g}, \mathcal{T}) \leqslant C \|\tilde{g}\|_{H^1(\Omega)}. \tag{9.37}$$

PROOF. Lemma 9.4 is given in the two-dimensional case, an analogous result is possible in the three-dimensional case. Let Ω, $\tilde{g}$, $\mathcal{T}$, ζ, M satisfying the hypotheses of Lemma 9.4. By a classical argument of density, one may assume that $\tilde{g} \in C^1(\overline{\Omega}, \mathbb{R})$.

A first step consists in proving that there exists $C_1 \in \mathbb{R}_+$, only depending on ζ, such that

$$\big(\tilde{g}_K - \tilde{g}_\sigma\big)^2 \leqslant C_1 \frac{\mathrm{diam}(K)}{\mathrm{m}(\sigma)} \int_K \big|\nabla \tilde{g}(x)\big|^2\,\mathrm{d}x, \quad \forall K \in \mathcal{T},\ \forall \sigma \in \mathcal{E}_K, \tag{9.38}$$

where $\tilde{g}_K$ (resp. $\tilde{g}_\sigma$) is the mean value of $\tilde{g}$ on K (resp. σ), for $K \in \mathcal{T}$ (resp. $\sigma \in \mathcal{E}$). Indeed, without loss of generality, one assumes that $\sigma = \{0\} \times J_0$, with J_0 is a closed interval of $\mathbb{R}$ and $K \subset \mathbb{R}_+ \times \mathbb{R}$.

Let $\alpha = \max\{x_1, x = (x_1, x_2)^t \in \overline{K}\}$ and $a = (\alpha, \beta)^t \in \overline{K}$. In the following, a is fixed. For all $x_1 \in (0, \alpha)$, let $J(x_1) = \{x_2 \in \mathbb{R}$, such that $(x_1, x_2)^t \in \overline{K}\}$, so that $J_0 = J(0)$.

For a.e. $x = (x_1, x_2)^t \in K$ and a.e., for the 1-Lebesgue measure, $y = (0, \overline{y})^t \in \sigma$ (with $\overline{y} \in J_0$), one sets $z(x, y) = ta + (1 - t)y$ with $t = x_1/\alpha$. Note that, since $\overline{K}$ is convex, $z(x, y) \in \overline{K}$ and $z(x, y) = (x_1, z_2(x_1, \overline{y}))^t$, with $z_2(x_1, \overline{y}) = \frac{x_1}{\alpha}\beta + (1 - \frac{x_1}{\alpha})\overline{y}$.

One has, using the Cauchy–Schwarz inequality,

$$\big(\tilde{g}_K - \tilde{g}_\sigma\big)^2 \leqslant \frac{2}{\mathrm{m}(K)\mathrm{m}(\sigma)}(A + B), \tag{9.39}$$

where

$$A = \int_K \int_\sigma \big(\tilde{g}(x) - \tilde{g}\big(z(x, y)\big)\big)^2\,\mathrm{d}\gamma(y)\,\mathrm{d}x,$$

and

$$B = \int_K \int_\sigma \big(\tilde{g}\big(z(x, y)\big) - \tilde{g}(y)\big)^2\,\mathrm{d}\gamma(y)\,\mathrm{d}x.$$

Let us now obtain a bound of A. Let $D_i\tilde{g}$, $i = 1$ or 2, denote the partial derivative of $\tilde{g}$ w.r.t. the components of $x = (x_1, x_2)^t \in \mathbb{R}^2$. Then,

$$A = \int_0^\alpha \int_{J(x_1)} \int_{J(0)} \left(\int_{z_2(x_1,\overline{y})}^{x_2} D_2\tilde{g}(x_1, s)\,\mathrm{d}s \right)^2 \mathrm{d}\overline{y}\,\mathrm{d}x_2\,\mathrm{d}x_1.$$

The Cauchy–Schwarz inequality yields

$$A \leqslant \mathrm{diam}(K) \int_0^\alpha \int_{J(x_1)} \int_{J(0)} \int_{J(x_1)} \big(D_2\tilde{g}(x_1, s)\big)^2 \mathrm{d}s\,\mathrm{d}\overline{y}\,\mathrm{d}x_2\,\mathrm{d}x_1$$

and therefore

$$A \leqslant \mathrm{diam}(K)^3 \int_K \big(D_2\tilde{g}(x)\big)^2 \mathrm{d}x. \tag{9.40}$$

One now turns to the study of B, which can be rewritten as

$$B = \int_0^\alpha \int_{J(x_1)} \int_{J(0)} \left(\int_0^{x_1} \left[D_1\tilde{g}\big(s, z_2(s,\overline{y})\big) + \frac{\beta - \overline{y}}{\alpha} D_2\tilde{g}\big(s, z_2(s,\overline{y})\big) \right] \mathrm{d}s \right)^2 \mathrm{d}\overline{y}\,\mathrm{d}x_2\,\mathrm{d}x_1.$$

The Cauchy–Schwarz inequality and the fact that $\alpha \geqslant \zeta\,\mathrm{diam}(K)$ give that

$$B \leqslant 2\,\mathrm{diam}(K)\left(B_1 + \frac{1}{\zeta^2} B_2 \right), \tag{9.41}$$

with

$$B_i = \int_0^\alpha \int_{J(x_1)} \int_{J(0)} \int_0^{x_1} \big(D_i\tilde{g}\big(s, z_2(s,\overline{y})\big)\big)^2 \mathrm{d}s\,\mathrm{d}\overline{y}\,\mathrm{d}x_2\,\mathrm{d}x_1, \quad i = 1, 2.$$

First, using Fubini's theorem, one has

$$B_i = \int_{J(0)} \int_0^\alpha \big(D_i\tilde{g}\big(s, z_2(s,\overline{y})\big)\big)^2 \int_s^\alpha \int_{J(x_1)} \mathrm{d}x_2\,\mathrm{d}x_1\,\mathrm{d}s\,\mathrm{d}\overline{y}.$$

Therefore

$$B_i \leqslant \mathrm{diam}(K) \int_0^\alpha \int_{J(0)} \big(D_i\tilde{g}\big(s, z_2(s,\overline{y})\big)\big)^2 (\alpha - s)\,\mathrm{d}\overline{y}\,\mathrm{d}s.$$

Then, with the change of variables $z_2 = z_2(s, \overline{y})$, one gets

$$B_i \leqslant \mathrm{diam}(K) \int_0^\alpha \int_{J(s)} \big(D_i\tilde{g}(s, z_2)\big)^2 \frac{\alpha - s}{1 - \frac{s}{\alpha}}\,\mathrm{d}z_2\,\mathrm{d}s.$$

Hence

$$B_i \leqslant \operatorname{diam}(K)^2 \int_K \big(D_i \tilde{g}(x)\big)^2 \,\mathrm{d}x. \tag{9.42}$$

Using the fact that $\mathrm{m}(K) \geqslant \pi\zeta^2(\operatorname{diam}(K))^2$, (9.39), (9.40), (9.41) and (9.42), one concludes (9.38).

In order to conclude the proof of (9.37), one remarks that

$$\big(\mathcal{N}(\tilde{g},\mathcal{T})\big)^2 \leqslant 2\sum_{K\in\mathcal{T}}\sum_{\sigma\in\mathcal{E}_K} \tau_\sigma\big(\tilde{g}_K - \tilde{g}_\sigma\big)^2.$$

Because, for all $K \in \mathcal{T}$ and $\sigma \in \mathcal{E}_K$, $d_\sigma \geqslant \zeta \operatorname{diam}(K)$, one gets thanks to (9.38), that

$$\big(\mathcal{N}(\tilde{g},\mathcal{T})\big)^2 \leqslant 2\sum_{K\in\mathcal{T}}\sum_{\sigma\in\mathcal{E}_K} \frac{C_1}{\zeta}\int_K \big|\nabla\tilde{g}(x)\big|^2 \,\mathrm{d}x.$$

The above inequality shows that

$$\big(\mathcal{N}(\tilde{g},\mathcal{T})\big)^2 \leqslant 2M\frac{C_1}{\zeta}\int_\Omega \big|\nabla\tilde{g}(x)\big|^2 \,\mathrm{d}x,$$

which implies (9.37). □

THEOREM 9.2 (Convergence, nonhomogeneous Dirichlet boundary condition). *Assume items* (1), (2), (3) *and* (4) *of Assumption* 9.1 *and* $g \in H^{1/2}(\partial\Omega)$. *Let* $\zeta \in \mathbb{R}_+$ *and* $M \in \mathbb{N}$ *be given values. Let* $\mathcal{T}$ *be an admissible mesh (in the sense of Definition* 9.1) *such that* $d_{K,\sigma} \geqslant \zeta \operatorname{diam}(K)$ *for all control volumes* $K \in \mathcal{T}$ *and for all* $\sigma \in \mathcal{E}_K$, *and* $\operatorname{card}(\mathcal{E}_K) \leqslant M$ *for all* $K \in \mathcal{T}$. *Let* $(u_K)_{K\in\mathcal{T}}$ *be the solution of the system given by Eqs.* (9.20)–(9.22) *and*

$$u_\sigma = \frac{1}{\mathrm{m}(\sigma)}\int_\sigma g(x)\,\mathrm{d}\gamma(x), \quad \forall\sigma \in \mathcal{E}_{\mathrm{ext}} \tag{9.43}$$

(note that the proofs of existence and uniqueness of $(u_K)_{K\in\mathcal{T}}$ *which were given in Lemma* 9.2 *remain valid). Define* $u_\mathcal{T} \in X(\mathcal{T})$ *by* $u_\mathcal{T}(x) = u_K$ *for a.e.* $x \in K$ *and for any* $K \in \mathcal{T}$. *Then,* $u_\mathcal{T}$ *converges, in* $L^2(\Omega)$, *to the unique variational solution* $u \in H^1(\Omega)$ *of problem* (9.1), (9.2) *as* $\operatorname{size}(\mathcal{T}) \to 0$.

PROOF. The proof is only detailed for the case $b = 0$ and $\boldsymbol{v} = 0$ (the extension of the proof to the general case is straightforward using the proof of Theorem 9.1). Let $\tilde{g} \in H^1(\Omega)$ be such that the trace of $\tilde{g}$ on $\partial\Omega$ is equal to g. One defines $\tilde{u}_\mathcal{T} \in X(\mathcal{T})$ by $\tilde{u}_\mathcal{T}(x) = \tilde{u}_K = u_K - \frac{1}{\mathrm{m}(K)}\int_K \tilde{g}(y)\,\mathrm{d}y$ for all $x \in K$ and all $K \in \mathcal{T}$. Then $(\tilde{u}_K)_{K\in\mathcal{T}}$ satisfies

$$\sum_{\sigma\in\mathcal{E}_K} \tilde{F}_{K,\sigma} = \mathrm{m}(K) f_K - \sum_{\sigma\in\mathcal{E}_K} \overline{F}_{K,\sigma}, \quad \forall K \in \mathcal{T}, \tag{9.44}$$

$$\tilde{F}_{K,\sigma} = -\tau_{K|L}\big(\tilde{u}_L - \tilde{u}_K\big), \quad \forall \sigma \in \mathcal{E}_{\text{int}}, \text{ if } \sigma = K|L, \tag{9.45}$$

$$\tilde{F}_{K,\sigma} = \tau_\sigma\big(\tilde{u}_K\big), \quad \forall \sigma \in \mathcal{E}_{\text{ext}} \text{ such that } \sigma \in \mathcal{E}_K. \tag{9.46}$$

$$\overline{F}_{K,\sigma} = -\tau_{K|L}\left(\frac{1}{\mathrm{m}(L)}\int_L \tilde{g}(y)\,\mathrm{d}y - \frac{1}{\mathrm{m}(K)}\int_K \tilde{g}(y)\,\mathrm{d}y\right),$$
$$\forall \sigma \in \mathcal{E}_{\text{int}}, \text{ if } \sigma = K|L, \tag{9.47}$$

$$\overline{F}_{K,\sigma} = -\tau_\sigma\left(\frac{1}{\mathrm{m}(\sigma)}\int_\sigma g(x)\,\mathrm{d}\gamma(x) - \frac{1}{\mathrm{m}(K)}\int_K \tilde{g}(y)\,\mathrm{d}y\right),$$
$$\forall \sigma \in \mathcal{E}_{\text{ext}} \text{ such that } \sigma \in \mathcal{E}_K. \tag{9.48}$$

Multiplying (9.44) by $\tilde{u}_K$, summing over $K \in \mathcal{T}$, gathering by edges in the right-hand side and using the Cauchy–Schwarz inequality yields

$$\|\tilde{u}_{\mathcal{T}}\|_{1,\mathcal{T}}^2 \leqslant \sum_{K\in\mathcal{T}} \mathrm{m}(K) f_K \tilde{u}_K + \mathcal{N}(\tilde{g},\mathcal{T})\|\tilde{u}_{\mathcal{T}}\|_{1,\mathcal{T}},$$

from the definition (9.36) of $\mathcal{N}(\tilde{g},\mathcal{T})$ and Definition 9.3 of $\|\cdot\|_{1,\mathcal{T}}$. Therefore, thanks to Lemma 9.4 and the discrete Poincaré inequality (9.13), there exists $C_1 \in \mathbb{R}$, only depending on Ω, $\|\tilde{g}\|_{H^1(\Omega)}$, ζ, M and f, such that $\|\tilde{u}_{\mathcal{T}}\|_{1,\mathcal{T}} \leqslant C_1$ and $\|\tilde{u}_{\mathcal{T}}\|_{L^2(\Omega)} \leqslant C_1$. Let us now prove that $\tilde{u}_{\mathcal{T}}$ converges in $L^2(\Omega)$, as $\mathrm{size}(\mathcal{T}) \to 0$, towards the unique solution in $H_0^1(\Omega)$ to (9.3). We proceed as in Theorem 9.1. Using Lemma 9.3, the compactness result given in Theorem 14.2 and the uniqueness of the solution (in $H_0^1(\Omega)$) of (9.3), it is sufficient to prove that if $\tilde{u}_{\mathcal{T}}$ converges towards some $\tilde{u} \in H_0^1(\Omega)$, in $L^2(\Omega)$ as $\mathrm{size}(\mathcal{T}) \to 0$, then $\tilde{u}$ is the solution to (9.3). In order to prove this result, let us introduce the function $\tilde{g}_{\mathcal{T}}$ defined by

$$\tilde{g}_{\mathcal{T}}(x) = \frac{1}{\mathrm{m}(K)}\int_K \tilde{g}(y)\,\mathrm{d}y, \quad \forall x \in K,\ \forall K \in \mathcal{T},$$

which converges to $\tilde{g}$ in $L^2(\Omega)$, as $\mathrm{size}(\mathcal{T}) \to 0$. Then the function $u_{\mathcal{T}}$ converges in $L^2(\Omega)$, as $\mathrm{size}(\mathcal{T}) \to 0$ to $u = \tilde{u} + \tilde{g} \in H^1(\Omega)$ and the proof that $\tilde{u}$ is the unique solution of (9.3) is identical to the corresponding part in the proof of Theorem 9.1. This completes the proof of Theorem 9.2. □

REMARK 9.9. A more simple proof of convergence for the finite volume scheme with nonhomogeneous Dirichlet boundary condition can be made if g is the trace of a Lipschitz-continuous function $\tilde{g}$. In that case, ζ and M do not have to be introduced and Lemma 9.4 is not used. The scheme is defined with $u_\sigma = g(y_\sigma)$ instead of the average value of g on σ, and the proof uses $\tilde{g}(x_K)$ instead of the average value of $\tilde{g}$ on K.

9.5. C^2 error estimate

Under adequate regularity assumptions on the solution of problem (9.1)–(9.2), one may prove that the error between the exact solution and the approximate solution given by the finite volume scheme (9.20)–(9.23) is of order $\mathrm{size}(\mathcal{T}) = \sup_{K\in\mathcal{T}} \mathrm{diam}(K)$, in a certain sense which we give in the following theorem:

THEOREM 9.3. *Under Assumption 9.1, let $\mathcal{T}$ be an admissible mesh as defined in Definition 9.1 and $u_{\mathcal{T}} \in X(\mathcal{T})$ (see Definition 9.2) be defined a.e. in Ω by $u_{\mathcal{T}}(x) = u_K$ for a.e. $x \in K$, for all $K \in \mathcal{T}$, where $(u_K)_{K\in\mathcal{T}}$ is the solution to (9.20)–(9.23). Assume that the unique variational solution u of problem (9.1)–(9.2) satisfies $u \in C^2(\overline{\Omega})$. Let, for each $K \in \mathcal{T}$, $e_K = u(x_K) - u_K$, and $e_{\mathcal{T}} \in X(\mathcal{T})$ defined by $e_{\mathcal{T}}(x) = e_K$ for a.e. $x \in K$, for all $K \in \mathcal{T}$.*

Then, there exists $C > 0$ only depending on u, $\boldsymbol{v}$ and Ω such that

$$\|e_{\mathcal{T}}\|_{1,\mathcal{T}} \leqslant C\mathrm{size}(\mathcal{T}), \tag{9.49}$$

where $\|\bullet\|_{1,\mathcal{T}}$ is the discrete H_0^1 norm defined in Definition 9.3,

$$\|e_{\mathcal{T}}\|_{L^2(\Omega)} \leqslant C\mathrm{size}(\mathcal{T}) \tag{9.50}$$

and

$$\begin{aligned}
&\sum_{\substack{\sigma\in\mathcal{E}_{\mathrm{int}}\\ \sigma=K|L}} \mathrm{m}(\sigma) d_\sigma \left(\frac{u_L - u_K}{d_\sigma} - \frac{1}{\mathrm{m}(\sigma)}\int_\sigma \nabla u(x)\bullet \boldsymbol{n}_{K,\sigma}\,\mathrm{d}\gamma(x)\right)^2 \\
&\quad + \sum_{\substack{\sigma\in\mathcal{E}_{\mathrm{ext}}\\ \sigma\in\overline{K}\cap\partial\Omega}} \mathrm{m}(\sigma) d_\sigma \left(\frac{g(y_\sigma) - u_K}{d_\sigma} - \frac{1}{\mathrm{m}(\sigma)}\int_\sigma \nabla u(x)\bullet \boldsymbol{n}_{K,\sigma}\,\mathrm{d}\gamma(x)\right)^2 \\
&\leqslant C\mathrm{size}(\mathcal{T})^2.
\end{aligned} \tag{9.51}$$

REMARK 9.10. (1) Inequality (9.49) (resp. (9.50)) yields an estimate of order 1 for the discrete H_0^1 norm (resp. L^2 norm) of the error on the solution. Note also that, since $u \in C^1(\overline{\Omega})$, one deduces, from (9.50), the existence of C only depending on u and Ω such that $\|u - u_{\mathcal{T}}\|_{L^2(\Omega)} \leqslant C\mathrm{size}(\mathcal{T})$. Inequality (9.51) may be seen as an estimate of order 1 for the L^2 norm of the flux.

(2) In BARANGER, MAITRE and OUDIN [1996], finite element tools are used to obtain error estimates of order $\mathrm{size}(\mathcal{T})^2$ in the case $d = 2$, $\boldsymbol{v} = b = g = 0$ and if the elements of $\mathcal{T}$ are triangles of a finite element mesh satisfying the Delaunay condition (see Section 12). Note that this result is quite different of those of Remarks 6.2 and 9.1, which are obtained by using a higher-order approximation of the flux.

(3) The proof of Theorem 9.3 given below is close to that of error estimates for finite elements schemes in the sense that it uses the coerciveness of the operator (the discrete Poincaré inequality) instead of the discrete maximum principle of Proposition 9.2 (which is used for error estimates with finite difference schemes).

PROOF OF THEOREM 9.3. Let $u_{\mathcal{T}} \in X(\mathcal{T})$ be defined a.e. in Ω by $u_{\mathcal{T}}(x) = u_K$ for a.e. $x \in K$, for all $K \in \mathcal{T}$, where $(u_K)_{K\in\mathcal{T}}$ is the solution to (9.20)–(9.23). Let us write the flux balance for any $K \in \mathcal{T}$;

$$\sum_{\sigma\in\mathcal{E}_K} \left(\overline{F}_{K,\sigma} + \overline{V}_{K,\sigma}\right) + b\int_K u(x)\,\mathrm{d}x = \int_K f(x)\,\mathrm{d}x, \tag{9.52}$$

where $\overline{F}_{K,\sigma} = -\int_\sigma \nabla u(x)\cdot \boldsymbol{n}_{K,\sigma}\,\mathrm{d}\gamma(x)$, and $\overline{V}_{K,\sigma} = \int_\sigma u(x)\boldsymbol{v}(x)\cdot\boldsymbol{n}_{K,\sigma}\,\mathrm{d}\gamma(x)$ are respectively the diffusion and convection fluxes through σ outward to K.

Let $F^\star_{K,\sigma}$ and $V^\star_{K,\sigma}$ be defined by

$$F^\star_{K,\sigma} = -\tau_{K|L}\big(u(x_L) - u(x_K)\big), \quad \forall\sigma = K|L \in \mathcal{E}_K\cap\mathcal{E}_{\mathrm{int}},\ \forall K\in\mathcal{T},$$

$$F^\star_{K,\sigma} d(x_K,\sigma) = -\mathrm{m}(\sigma)\big(u(y_\sigma) - u(x_K)\big), \quad \forall\sigma\in\mathcal{E}_K\cap\mathcal{E}_{\mathrm{ext}}, \quad \forall K\in\mathcal{T},$$

$$V^\star_{K,\sigma} = v_{K,\sigma}u(x_{\sigma,+}), \forall\sigma\in\mathcal{E}_K, \quad \forall K\in\mathcal{T},$$

where $x_{\sigma,+} = x_K$ (resp. x_L) if $\sigma\in\mathcal{E}_{\mathrm{int}}$, $\sigma = K|L$ and $v_{K,\sigma}\geqslant 0$ (resp. $v_{K,\sigma}\leqslant 0$) and $x_{\sigma,+} = x_K$ (resp. y_σ) if $\sigma = \mathcal{E}_K\cap\mathcal{E}_{\mathrm{ext}}$ and $v_{K,\sigma}\geqslant 0$ (resp. $v_{K,\sigma}\leqslant 0$). Then, the consistency error on the diffusion and convection fluxes may be defined as

$$R_{K,\sigma} = \frac{1}{\mathrm{m}(\sigma)}\left(\overline{F}_{K,\sigma} - F^\star_{K,\sigma}\right), \tag{9.53}$$

$$r_{K,\sigma} = \frac{1}{\mathrm{m}(\sigma)}\left(\overline{V}_{K,\sigma} - V^\star_{K,\sigma}\right), \tag{9.54}$$

Thanks to the regularity of u and $\boldsymbol{v}$, there exists $C_1\in\mathbb{R}$, only depending on u and $\boldsymbol{v}$, such that $|R_{K,\sigma}| + |r_{K,\sigma}| \leqslant C_1\mathrm{size}(\mathcal{T})$ for any $K\in\mathcal{T}$ and $\sigma\in\mathcal{E}_K$. For $K\in\mathcal{T}$, let

$$\rho_K = u(x_K) - \big(1/\mathrm{m}(K)\big)\int_K u(x)\,\mathrm{d}x,$$

so that $|\rho_K|\leqslant C_2\mathrm{size}(\mathcal{T})$ with some $C_2\in\mathbb{R}_+$ only depending on u.

Subtract (9.20) to (9.52); thanks to (9.53) and (9.54), one has

$$\begin{aligned}&\sum_{\sigma\in\mathcal{E}_K}(G_{K,\sigma} + W_{K,\sigma}) + b\mathrm{m}(K)e_K\\ &\quad = b\mathrm{m}(K)\rho_K - \sum_{\sigma\in\mathcal{E}_K}\mathrm{m}(\sigma)(R_{K,\sigma} + r_{K,\sigma}),\end{aligned} \tag{9.55}$$

where $G_{K,\sigma} = F^\star_{K,\sigma} - F_{K,\sigma}$ is such that

$$G_{K,\sigma} = -\tau_{K|L}(e_L - e_K), \quad \forall K \in \mathcal{T}, \forall \sigma \in \mathcal{E}_K \cap \mathcal{E}_{\text{int}}, \sigma = K|L,$$

$$G_{K,\sigma} d(x_K, \sigma) = \mathrm{m}(\sigma) e_K, \quad \forall K \in \mathcal{T}, \forall \sigma \in \mathcal{E}_K \cap \mathcal{E}_{\text{ext}},$$

with $e_K = u(x_K) - u_K$, and $W_{K,\sigma} = V^{\star}_{K,\sigma} - V_{K,\sigma} = v_{K,\sigma}(u(x_{\sigma,+}) - u_{\sigma,+})$.

Multiply (9.55) by e_K, sum for $K \in \mathcal{T}$, and note that

$$\sum_{K \in \mathcal{T}} \sum_{\sigma \in \mathcal{E}_K} G_{K,\sigma} e_K = \sum_{\sigma \in \mathcal{E}} |D_\sigma e|^2 \frac{\mathrm{m}(\sigma)}{d_\sigma} = \|e\|^2_{1,\mathcal{T}}.$$

Hence

$$\|e_\mathcal{T}\|^2_{1,\mathcal{T}} + \sum_{K \in \mathcal{T}} \sum_{\sigma \in \mathcal{E}_K} v_{K,\sigma} e_{\sigma,+} e_K + b \|e_\mathcal{T}\|^2_{L^2(\Omega)}$$
$$\leqslant b \sum_{K \in \mathcal{T}} \mathrm{m}(K) \rho_K e_K - \sum_{K \in \mathcal{T}} \sum_{\sigma \in \mathcal{E}_K} \mathrm{m}(\sigma)(R_{K,\sigma} + r_{K,\sigma}) e_K, \tag{9.56}$$

where

- $e_\mathcal{T} \in X(\mathcal{T})$, $e_\mathcal{T}(x) = e_K$ for a.e. $x \in K$ and for all $K \in \mathcal{T}$,
- $|D_\sigma e| = |e_K - e_L|$, if $\sigma \in \mathcal{E}_{\text{int}}$, $\sigma = K|L$, $|D_\sigma e| = |e_K|$, if $\sigma \in \mathcal{E}_K \cap \mathcal{E}_{\text{ext}}$,
- $e_{\sigma,+} = u(x_{\sigma,+}) - u_{\sigma,+}$.

By Young's inequality, the first term of the left-hand side satisfies:

$$\left| \sum_{K \in \mathcal{T}} \mathrm{m}(K) \rho_K e_K \right| \leqslant \tfrac{1}{2} \|e_\mathcal{T}\|^2_{L^2(\Omega)} + \tfrac{1}{2} C_2^2 \big(\text{size}(\mathcal{T})\big)^2 \mathrm{m}(\Omega). \tag{9.57}$$

Thanks to the assumption $\operatorname{div} \boldsymbol{v} \geqslant 0$, one obtains, through a computation similar to (9.27)–(9.28) that

$$\sum_{K \in \mathcal{T}} \sum_{\sigma \in \mathcal{E}_K} v_{K,\sigma} e_{\sigma,+} e_K \geqslant 0.$$

Hence, (9.56) and (9.57) yield that there exists C_3 only depending on u, b and Ω such that

$$\|e_\mathcal{T}\|^2_{1,\mathcal{T}} + \tfrac{1}{2} b \|e_\mathcal{T}\|^2_{L^2(\Omega)}$$
$$\leqslant C_3 \big(\text{size}(\mathcal{T})\big)^2 - \sum_{K \in \mathcal{T}} \sum_{\sigma \in \mathcal{E}_K} \mathrm{m}(\sigma)(R_{K,\sigma} + r_{K,\sigma}) e_K. \tag{9.58}$$

Thanks to the property of conservativity, one has $R_{K,\sigma} = -R_{L,\sigma}$ and $r_{K,\sigma} = -r_{L,\sigma}$ for $\sigma \in \mathcal{E}_{\text{int}}$ such that $\sigma = K|L$. Let $R_\sigma = |R_{K,\sigma}|$ and $r_\sigma = |r_{K,\sigma}|$ if $\sigma \in \mathcal{E}_K$. Reordering the summation over the edges and from the Cauchy–Schwarz inequality, one then obtains

$$\begin{aligned}
&\left|\sum_{K\in\mathcal{T}}\sum_{\sigma\in\mathcal{E}_K}\mathrm{m}(\sigma)(R_{K,\sigma}+r_{K,\sigma})e_K\right| \\
&\quad\leqslant\sum_{\sigma\in\mathcal{E}}\mathrm{m}(\sigma)(D_\sigma e)(R_\sigma+r_\sigma) \\
&\quad\leqslant\left(\sum_{\sigma\in\mathcal{E}}\frac{\mathrm{m}(\sigma)}{d_\sigma}(D_\sigma e)^2\right)^{1/2}\left(\sum_{\sigma\in\mathcal{E}}\mathrm{m}(\sigma)d_\sigma(R_\sigma+r_\sigma)^2\right)^{1/2}.
\end{aligned}\tag{9.59}$$

Now, since $|R_\sigma+r_\sigma|\leqslant C_1\mathrm{size}(\mathcal{T})$ and since $\sum_{\sigma\in\mathcal{E}}\mathrm{m}(\sigma)d_\sigma=d\,\mathrm{m}(\Omega)$, (9.58) and (9.59) yield the existence of $C_4\in\mathbb{R}_+$ only depending on u, $\boldsymbol{v}$ and Ω such that

$$\|e_{\mathcal{T}}\|_{1,\mathcal{T}}^2+\tfrac{1}{2}b\|e_{\mathcal{T}}\|_{L^2(\Omega)}^2\leqslant C_3\big(\mathrm{size}(\mathcal{T})\big)^2+C_4\mathrm{size}(\mathcal{T})\|e\|_{1,\mathcal{T}}.$$

Using again Young's inequality, there exists C_5 only depending on u, $\boldsymbol{v}$, b and Ω such that

$$\|e_{\mathcal{T}}\|_{1,\mathcal{T}}^2+b\|e_{\mathcal{T}}\|_{L^2(\Omega)}^2\leqslant C_5\big(\mathrm{size}(\mathcal{T})\big)^2.\tag{9.60}$$

This inequality yields estimate (9.49) and, in the case $b>0$, estimate (9.50). In the case where $b=0$, one uses the discrete Poincaré inequality (9.13) and the inequality (9.60) to obtain

$$\|e_{\mathcal{T}}\|_{L^2(\Omega)}^2\leqslant\mathrm{diam}(\Omega)^2C_5\big(\mathrm{size}(\mathcal{T})\big)^2,$$

which yields (9.50).

Remark now that (9.49) can be written

$$\begin{aligned}
&\sum_{\substack{\sigma\in\mathcal{E}_{\mathrm{int}}\\ \sigma=K|L}}\mathrm{m}(\sigma)d_\sigma\left(\frac{u_L-u_K}{d_\sigma}-\frac{u(x_L)-u(x_K)}{d_\sigma}\right)^2 \\
&\quad+\sum_{\substack{\sigma\in\mathcal{E}_{\mathrm{ext}}\\ \sigma\in\overline{K}\cap\partial\Omega}}\mathrm{m}(\sigma)d_\sigma\left(\frac{g(y_\sigma)-u_K}{d_\sigma}-\frac{u(y_\sigma)-u(x_K)}{d_\sigma}\right)^2\leqslant\big(C\mathrm{size}(\mathcal{T})\big)^2.
\end{aligned}\tag{9.61}$$

From definition (9.53) and the consistency of the fluxes, one has

$$\sum_{\substack{\sigma\in\mathcal{E}_{\mathrm{int}}\\ \sigma=K|L}}\mathrm{m}(\sigma)d_\sigma\left(\frac{u(x_L)-u(x_K)}{d_\sigma}-\frac{1}{\mathrm{m}(\sigma)}\int_\sigma\nabla u(x)\cdot\boldsymbol{n}_{K,\sigma}\,\mathrm{d}\gamma(x)\right)^2$$

$$+\sum_{\substack{\sigma\in\mathcal{E}_{\mathrm{ext}}\\ \sigma\in\overline{K}\cap\partial\Omega}} \mathrm{m}(\sigma)d_\sigma\left(\frac{u(y_\sigma)-u(x_K)}{d_\sigma}-\frac{1}{\mathrm{m}(\sigma)}\int_\sigma \nabla u(x)\cdot \boldsymbol{n}_{K,\sigma}\,\mathrm{d}\gamma(x)\right)^2$$

$$=\sum_{\sigma\in\mathcal{E}}\mathrm{m}(\sigma)d_\sigma R_\sigma^2 \leqslant d\mathrm{m}(\Omega)C_1^2\big(\mathrm{size}(\mathcal{T})\big)^2. \tag{9.62}$$

Then (9.61) and (9.62) give (9.51). □

9.6. H^2 error estimate

In Theorem 9.3, the hypothesis $u\in C^2(\overline{\Omega})$ has been used. In the following theorem (Theorem 9.4), one obtains estimates (9.49) and (9.50), in the case $b=\boldsymbol{v}=0$ and assuming some additional assumption on the mesh (see Definition 9.4 below), under the weaker assumption $u\in H^2(\Omega)$. This additional assumption on the mesh is not completely necessary (see Remark 9.12). It is also possible to obtain estimates (9.49) and (9.50) in the cases $b\neq 0$ or $\boldsymbol{v}\neq 0$ assuming $u\in H^2(\Omega)$ (see Remark 9.12 and GALLOUET, HERBIN, VIGNAL). Some similar results are also in LAZAROV, MISHEV and VASSILEVSKI [1996] and COUDIÈRE, VILA and VILLEDIEU [1999].

DEFINITION 9.4 (*Restricted admissible meshes*). Let Ω be an open bounded polygonal subset of $\mathbb{R}^d$, $d=2$ or 3. A restricted admissible finite volume mesh of Ω, denoted by $\mathcal{T}$, is an admissible mesh in the sense of Definition 9.1 such that, for some $\zeta>0$, one has $d_{K,\sigma}\geqslant\zeta\,\mathrm{diam}(K)$ for all control volumes K and for all $\sigma\in\mathcal{E}_K$.

THEOREM 9.4 (H^2 regularity). *Under Assumption* 9.1 *with* $b=\boldsymbol{v}=0$, *let* $\mathcal{T}$ *be a restricted admissible mesh in the sense of Definition* 9.4 *and* $u_\mathcal{T}\in X(\mathcal{T})$ *(see Definition* 9.2*) be the approximate solution defined in* Ω *by* $u_\mathcal{T}(x)=u_K$ *for a.e.* $x\in K$, *for all* $K\in\mathcal{T}$, *where* $(u_K)_{K\in\mathcal{T}}$ *is the (unique) solution to* (9.20)–(9.23) *(existence and uniqueness of* $(u_K)_{K\in\mathcal{T}}$ *are given by Lemma* 9.2*). Assume that the unique solution, u, of* (9.3) *(with* $b=\boldsymbol{v}=0$*) belongs to* $H^2(\Omega)$. *For each control volume K, let* $e_K=u(x_K)-u_K$, *and* $e_\mathcal{T}\in X(\mathcal{T})$ *defined by* $e_\mathcal{T}(x)=e_K$ *for a.e.* $x\in K$, *for all* $K\in\mathcal{T}$.

Then, there exists C, only depending on u, ζ *and* Ω, *such that* (9.49), (9.50) *and* (9.51) *hold.*

REMARK 9.11. (1) In Theorem 9.4, the function $e_\mathcal{T}$ is still well defined, and so is the quantity "$\nabla u\cdot\boldsymbol{n}_\sigma$" on σ, for all $\sigma\in\mathcal{E}$. Indeed, since $u\in H^2(\Omega)$ (and $d\leqslant 3$), one has $u\in C(\overline{\Omega})$ (and then $u(x_K)$ is well defined for all control volumes K) and $\nabla u\cdot\boldsymbol{n}_\sigma$ belongs to $L^2(\sigma)$ (for the $(d-1)$-dimensional Lebesgue measure on σ) for all $\sigma\in\mathcal{E}$.

(2) Note that, under Assumption 9.1 with $b=\boldsymbol{v}=g=0$ the (unique) solution of (9.3) is necessarily in $H^2(\Omega)$ provided that Ω is convex.

PROOF OF THEOREM 9.4. Let K be a control volume and $\sigma\in\mathcal{E}_K$. Define $\mathcal{V}_{K,\sigma}=\{tx_K+(1-t)x,\ x\in\sigma,\ t\in[0,1]\}$. For $\sigma\in\mathcal{E}_{\mathrm{int}}$, let $\mathcal{V}_\sigma=\mathcal{V}_{K,\sigma}\cup\mathcal{V}_{L,\sigma}$, if K and L are the control volumes such that $\sigma=K|L$. For $\sigma\in\mathcal{E}_{\mathrm{ext}}\cap\mathcal{E}_K$, let $\mathcal{V}_\sigma=\mathcal{V}_{K,\sigma}$.

The main part of the proof consists in proving the existence of some C, only depending on the space dimension d and ζ (given in Definition 9.4), such that, for all control volumes K and for all $\sigma \in \mathcal{E}_K$,

$$|R_{K,\sigma}|^2 \leqslant C\frac{(\text{size}(\mathcal{T}))^2}{\text{m}(\sigma)d_\sigma}\int_{\mathcal{V}_\sigma}\left|H(u)(z)\right|^2 \text{d}z, \tag{9.63}$$

where H is the Hessian matrix of u and

$$\left|H(u)(z)\right|^2 = \sum_{i,j=1}^{d}\left|D_i D_j u(z)\right|^2,$$

and D_i denotes the (weak) derivative with respect to the component z_i of $z = (z_1,\dots,z_d)^t \in \mathbb{R}^d$. Recall that $R_{K,\sigma}$ is the consistency error on the diffusion flux (see (9.53)), i.e.:

$$R_{K,\sigma} = \frac{u(x_L)-u(x_K)}{d_\sigma} - \frac{1}{\text{m}(\sigma)}\int_\sigma \nabla u(x)\cdot \boldsymbol{n}_{K,\sigma}\,\text{d}\gamma(x),$$
$$\text{if } \sigma \in \mathcal{E}_{\text{int}} \text{ and } \sigma = K|L,$$
$$R_{K,\sigma} = \frac{u(y_\sigma)-u(x_K)}{d_\sigma} - \frac{1}{\text{m}(\sigma)}\int_\sigma \nabla u(x)\cdot \boldsymbol{n}_{K,\sigma}\,\text{d}\gamma(x), \quad \text{if } \sigma \in \mathcal{E}_{\text{ext}} \cap \mathcal{E}_K.$$

Note that $R_{K,\sigma}$ is well defined, thanks to $u \in H^2(\Omega)$, see Remark 9.11.

In Step 1, one proves (9.63), and, in Step 2, we conclude the proof of estimates (9.49) and (9.50).

Step 1. *Proof of* (9.63). Let $\sigma \in \mathcal{E}$. Since $u \in H^2(\Omega)$, the restriction of u to $\mathcal{V}_\sigma$ belongs to $H^2(\mathcal{V}_\sigma)$. The space $C^2(\overline{\mathcal{V}_\sigma})$ is dense in $H^2(\mathcal{V}_\sigma)$ (see, e.g., NEČAS [1967], this can be proved quite easily be a regularization technique). Then, by a density argument, one needs only to prove (9.63) for $u \in C^2(\overline{\mathcal{V}_\sigma})$. Therefore, in the remainder of Step 1, it is assumed $u \in C^2(\overline{\mathcal{V}_\sigma})$.

First, one proves (9.63) if $\sigma \in \mathcal{E}_{\text{int}}$. Let K and L be the 2 control volumes such that $\sigma = K|L$.

It is possible to assume, for simplicity of notations and without loss of generality, that $\sigma = 0 \times \tilde{\sigma}$, with some $\tilde{\sigma} \subset \mathbb{R}^{d-1}$, and $x_K = (-\alpha, 0)^t$, $x_L = (\beta, 0)^t$, with some $\alpha > \zeta\,\text{diam}(K)$, $\beta > \zeta\,\text{diam}(L)$ (ζ is defined in Definition 9.4).

Since $u \in C^2(\overline{\mathcal{V}_\sigma})$ a Taylor expansion gives for a.e. (for the $(d-1)$-dimensional Lebesgue measure on σ) $x = (0, \tilde{x})^t \in \sigma$,

$$\begin{aligned} u(x_L) - u(x) = {} & \nabla u(x)\cdot(x_L - x) \\ & + \int_0^1 H(u)\big(tx + (1-t)x_L\big)(x_L - x)\cdot(x_L - x)t\,\text{d}t, \end{aligned}$$

and

$$u(x_K) - u(x) = \nabla u(x)\cdot(x_K - x)$$

$$+\int_0^1 H(u)\big(tx+(1-t)x_K\big)(x_K-x)\cdot(x_K-x)t\,\mathrm{d}t,$$

where $H(u)(z)$ denotes the Hessian matrix of u at point z.

Subtracting one equation to the other and integrating over σ yields (note that $x_L - x_K = \boldsymbol{n}_{K,\sigma} d_\sigma$) $|R_{K,\sigma}| \leqslant B_{K,\sigma} + B_{L,\sigma}$, with, for some C_1 only depending on d,

$$B_{K,\sigma} = \frac{C_1}{\mathrm{m}(\sigma)d_\sigma}\int_\sigma\int_0^1 \big|H(u)\big(tx+(1-t)x_K\big)\big||x_K-x|^2 t\,\mathrm{d}t\,\mathrm{d}\gamma(x). \tag{9.64}$$

The quantity $B_{L,\sigma}$ is obtained with $B_{K,\sigma}$ by changing K in L.

One uses a change of variables in (9.64). Indeed, one sets $z = tx + (1-t)x_K$. Since $|x_K - x| \leqslant \mathrm{diam}(K)$ and $\mathrm{d}z = t^{d-1}\alpha\,\mathrm{d}t\,\mathrm{d}\gamma(x)$, one obtains, since $z_1 = (t-1)\alpha$, $z = (z_1, \overline{z})^t$,

$$B_{K,\sigma} \leqslant \frac{C_1(\mathrm{diam}(K))^2}{\mathrm{m}(\sigma)d_\sigma}\int_{\mathcal{V}_{K,\sigma}} \big|H(u)(z)\big|\frac{\alpha^{d-2}}{\alpha(z_1+\alpha)^{d-2}}\,\mathrm{d}z.$$

This gives, with the famous Cauchy–Schwarz inequality,

$$B_{K,\sigma} \leqslant \frac{C_1\alpha^{d-3}(\mathrm{diam}(K))^2}{\mathrm{m}(\sigma)d_\sigma}\left(\int_{\mathcal{V}_{K,\sigma}} \big|H(u)(z)\big|^2\,\mathrm{d}z\right)^{1/2}$$
$$\times\left(\int_{\mathcal{V}_{K,\sigma}} \frac{1}{(z_1+\alpha)^{(d-2)2}}\,\mathrm{d}z\right)^{1/2}. \tag{9.65}$$

For $d = 2$, (9.65) gives

$$B_{K,\sigma} \leqslant \frac{C_1(\mathrm{diam}(K))^2}{\alpha\mathrm{m}(\sigma)d_\sigma}\left(\frac{\alpha\mathrm{m}(\sigma)}{2}\right)^{1/2}\left(\int_{\mathcal{V}_{K,\sigma}} \big|H(u)(z)\big|^2\,\mathrm{d}z\right)^{1/2},$$

and therefore

$$B_{K,\sigma} \leqslant \frac{C_1(\mathrm{diam}(K))^2}{2^{1/2}(\mathrm{m}(\sigma)d_\sigma)^{1/2}(d_\sigma\alpha)^{1/2}}\left(\int_{\mathcal{V}_{K,\sigma}} \big|H(u)(z)\big|^2\,\mathrm{d}z\right)^{1/2}.$$

A similar estimate holds on $B_{L,\sigma}$ by changing K in L and α in β. Since $\alpha, \beta \geqslant \zeta\,\mathrm{diam}(K)$ and $d_\sigma = \alpha + \beta \geqslant \zeta\,\mathrm{diam}(K)$, these estimates on $B_{K,\sigma}$ and $B_{L,\sigma}$ yield (9.63) for some C only depending on d and ζ.

For $d = 3$, denoting by $\mathcal{V}_{K,\sigma}^{z_1} = \{x \in \mathcal{V}_{K,\sigma}$ such that $x_1 = z_1\}$, it can be remarked that

$$\int_{\mathcal{V}_{K,\sigma}^{z_1}} \mathrm{d}z_2\,\mathrm{d}z_3 = \mathrm{m}\big(\mathcal{V}_{K,\sigma}^{z_1}\big) = \left(\frac{z_1+\alpha}{\alpha}\right)^2\mathrm{m}(\sigma)$$

and therefore (9.65) gives:

$$B_{K,\sigma} \leqslant \frac{C_1(\mathrm{diam}(K))^2}{\mathrm{m}(\sigma)d_\sigma}\left(\int_{-\alpha}^{0}\frac{\mathrm{m}(\sigma)}{\alpha^2}\,\mathrm{d}z_1\right)^{1/2}\left(\int_{\mathcal{V}_{K,\sigma}}\left|H(u)(z)\right|^2\mathrm{d}z\right)^{1/2},$$

and then

$$B_{K,\sigma} \leqslant \frac{C_1(\mathrm{diam}(K))^2}{(\mathrm{m}(\sigma)d_\sigma)^{1/2}(d_\sigma\alpha)^{1/2}}\left(\int_{\mathcal{V}_{K,\sigma}}\left|H(u)(z)\right|^2\mathrm{d}z\right)^{1/2}.$$

With a similar estimate on $B_{L,\sigma}$, this yields (9.63) for some C only depending on d and ζ.

Now, one proves (9.63) if $\sigma \in \mathcal{E}_{\mathrm{ext}}$. Let K be the control volume such that $\sigma \in \mathcal{E}_K$. One can assume, without loss of generality, that $x_K = 0$ and $\sigma = \{2\alpha\} \times \tilde{\sigma}$ with $\tilde{\sigma} \subset \mathbb{R}^{d-1}$ and some $\alpha \geqslant \frac{1}{2}\zeta\,\mathrm{diam}(K)$. The above proof gives (see Definition 9.1 for the definition of y_σ), with some C_2 only depending on d,

$$\begin{aligned}&\left|\frac{u(y_\sigma)-u(x_K)}{2\alpha} - \frac{1}{\mathrm{m}(\hat{\sigma})}\int_{\hat{\sigma}}\nabla u(x)\cdot\boldsymbol{n}_{K,\sigma}\,\mathrm{d}\gamma(x)\right|^2\\ &\quad\leqslant C_2\frac{(\mathrm{size}(\mathcal{T}))^2}{\mathrm{m}(\sigma)d_\sigma}\int_{\mathcal{V}_{\hat{\sigma}}}\left|H(u)(z)\right|^2\mathrm{d}z,\end{aligned} \tag{9.66}$$

with $\hat{\sigma} = \{(\alpha\,\frac{x}{2}), x \in \tilde{\sigma}\}$, and $\mathcal{V}_{\hat{\sigma}} = \{ty_\sigma + (1-t)x,\ x \in \hat{\sigma}, t \in [0,1]\} \cup \{tx_K + (1-t)x,\ x \in \hat{\sigma}, t \in [0,1]\}$.

Note that $\mathrm{m}(\hat{\sigma}) = \mathrm{m}(\sigma)/2^{d-1}$ and that $\mathcal{V}_{\hat{\sigma}} \subset \mathcal{V}_\sigma$.

One has now to compare $I_\sigma = \frac{1}{\mathrm{m}(\sigma)}\int_\sigma \nabla u(x)\cdot\boldsymbol{n}_{K,\sigma}\,\mathrm{d}\gamma(x)$ with $I_{\hat{\sigma}} = \frac{1}{\mathrm{m}(\hat{\sigma})}\int_{\hat{\sigma}} \nabla u(x)\cdot\boldsymbol{n}_{K,\sigma}\,\mathrm{d}\gamma(x)$.

A Taylor expansion gives

$$I_\sigma - I_{\hat{\sigma}} = \frac{1}{\mathrm{m}(\sigma)}\int_\sigma\int_{1/2}^{1} H(u)\big(x_K + t(x-x_K)\big)(x-x_K)\cdot\boldsymbol{n}_{K,\sigma}\,\mathrm{d}t\,\mathrm{d}\gamma(x).$$

The change of variables in this last integral $z = x_K + t(x - x_K)$, which gives $\mathrm{d}z = 2\alpha t^{d-1}\,\mathrm{d}t\,\mathrm{d}\gamma(x)$, yields, with $E_\sigma = \{tx + (1-t)x_K,\ x \in \sigma, t \in [1/2, 1]\}$ and some C_3 only depending on d (note that $t \geqslant 1/2$),

$$|I_\sigma - I_{\hat{\sigma}}| \leqslant \frac{C_3}{\mathrm{m}(\sigma)\alpha}\int_{E_\sigma}\left|H(u)(z)\right||x - x_K|\,\mathrm{d}z.$$

Then, from the Cauchy–Schwarz inequality and since $|x - x_K| \leqslant \mathrm{diam}(K)$,

$$|I_\sigma - I_{\hat{\sigma}}|^2 \leqslant \frac{C_4(\mathrm{diam}(K))^2}{\mathrm{m}(\sigma)d_\sigma}\int_{E_\sigma}\left|H(u)(z)\right|^2\mathrm{d}z, \tag{9.67}$$

with some C_4 only depending on d and ζ.

Inequalities (9.66) and (9.67) yield (9.63) for some C only depending on d and ζ.

One may therefore choose $C \in \mathbb{R}_+$ such that (9.63) holds for $\sigma \in \mathcal{E}_{\text{int}}$ or $\sigma \in \mathcal{E}_{\text{ext}}$. This concludes Step 1.

Step 2. *Proof of estimates* (9.49), (9.50) *and* (9.51). In order to obtain estimate (9.49) (and therefore (9.50) from the discrete Poincaré inequality (9.13)), one proceeds as in Theorem 9.3. Inequality (9.56) writes here, since $R_{K,\sigma} = -R_{L,\sigma}$, if $\sigma = K|L$,

$$\|e_{\mathcal{T}}\|_{1,\mathcal{T}}^2 \leqslant \sum_{\sigma\in\mathcal{E}} R_\sigma |D_\sigma e| \mathrm{m}(\sigma),$$

with $R_\sigma = |R_{K,\sigma}|$, if $\sigma \in \mathcal{E}_K$. Recall also that $|D_\sigma e| = |e_K - e_L|$ if $\sigma \in \mathcal{E}_{\text{int}}$, $\sigma = K|L$ and $|D_\sigma e| = |e_K|$, if $\sigma \in \mathcal{E}_{\text{ext}} \cap \mathcal{E}_K$. Cauchy and Schwarz strike again:

$$\|e_{\mathcal{T}}\|_{1,\mathcal{T}}^2 \leqslant \left(\sum_{\sigma\in\mathcal{E}} R_\sigma^2 \mathrm{m}(\sigma) d_\sigma\right)^{1/2} \left(\sum_{\sigma\in\mathcal{E}} |D_\sigma e|^2 \frac{\mathrm{m}(\sigma)}{d_\sigma}\right)^{1/2}.$$

The main consequence of (9.63) is that

$$\begin{aligned}\sum_{\sigma\in\mathcal{E}} \mathrm{m}(\sigma) d_\sigma R_\sigma^2 &\leqslant C\big(\mathrm{size}(\mathcal{T})\big)^2 \sum_{\sigma\in\mathcal{E}} \int_{\mathcal{V}_\sigma} \big|H(u)(z)\big|^2 \,\mathrm{d}z \\ &= C\big(\mathrm{size}(\mathcal{T})\big)^2 \int_\Omega \big|H(u)(z)\big|^2 \,\mathrm{d}z. \end{aligned} \tag{9.68}$$

Then, one obtains

$$\|e_{\mathcal{T}}\|_{1,\mathcal{T}} \leqslant \sqrt{C}\,\mathrm{size}(\mathcal{T}) \left(\int_\Omega \big|H(u)(z)\big|^2 \,\mathrm{d}z\right)^{1/2}.$$

This concludes the proof of (9.49) since $u \in H^2(\Omega)$ implies $\int_\Omega |H(u)(z)|^2 \,\mathrm{d}z < \infty$.

Estimate (9.51) follows from (9.68) in a similar manner as in the proof of Theorem 9.3. This concludes the proof of Theorem 9.4. □

REMARK 9.12 (*Generalizations*). (1) By developing the method used to bound the consistency error on the flux on the elements of $\mathcal{E}_{\text{ext}}$, it is possible to replace, in Theorem 9.4, the hypothesis $d_{K,\sigma} \geqslant \zeta\,\mathrm{diam}(K)$ in Definition 9.4 by the weaker hypothesis $d_\sigma \geqslant \zeta\,\mathrm{diam}(\sigma)$ provided that $\mathcal{V}_\sigma$ is convex. Note also that, in this case, the hypothesis $x_K \in K$ is not necessary, it suffices that $x_L - x_K = d_\sigma \boldsymbol{n}_{K,\sigma}$, for all $\sigma \in \mathcal{E}_{\text{int}}$, $\sigma = K|L$ (for $\sigma \in \mathcal{E}_{\text{ext}}$, one always needs $y_\sigma - x_K = d_\sigma \boldsymbol{n}_{K,\sigma}$).

(2) It is also possible to prove Theorem 9.4 if $b \neq 0$ or $\boldsymbol{v} \neq 0$ (or, of course, $b \neq 0$ and $\boldsymbol{v} \neq 0$). Indeed, if the solution, u, to (9.3) is not only in $H^2(\Omega)$ but is also Lipschitz continuous on $\overline{\Omega}$ (this is the case if, e.g., there exists $p > d$ such that $u \in W^{2,p}(\Omega)$), the treatment of the consistency error terms due to the terms involving b and $\boldsymbol{v}$ are exactly as in Theorem 9.3. If u is not Lipschitz continuous on $\overline{\Omega}$, one has to deal with the

consistency error terms due to b and $\boldsymbol{v}$ similarly as in the proof of Theorem 9.4 (see EYMARD, GALLOUËT and HERBIN [1999]).

It is also possible, essentially under Assumption 9.1, to obtain an L^q estimate of the error, for $2 \leqslant q < +\infty$ if $d = 2$, and for $1 \leqslant q \leqslant 6$ if $d = 3$, see COUDIÈRE, GALLOUËT and HERBIN [1998]. The error estimate for the L^q norm is a consequence of the following lemma:

LEMMA 9.5 (Discrete Sobolev inequality). *Let Ω be an open bounded polygonal subset of $\mathbb{R}^d$ and $\mathcal{T}$ be a general finite volume mesh of Ω in the sense of Definition* 10.1, *and let $\zeta > 0$ be such that*

$$\forall K \in \mathcal{T},\ \forall \sigma \in \mathcal{E}_K, \quad d_{K,\sigma} \geqslant \zeta d_\sigma, \text{ and } d_{K,\sigma} \geqslant \zeta \operatorname{diam}(K). \tag{9.69}$$

Let be $u \in X(\mathcal{T})$ (see Definition 9.2*), then, there exists $C > 0$ only depending on Ω and ζ, such that for all $q \in [2, +\infty)$, if $d = 2$, and $q \in [2, 6]$, if $d = 3$,*

$$\|u\|_{L^q(\Omega)} \leqslant Cq \|u\|_{1,\mathcal{T}}, \tag{9.70}$$

where $\|\bullet\|_{1,\mathcal{T}}$ is the discrete H_0^1 norm defined in Definition 9.3.

PROOF. Let us first prove the two-dimensional case. Assume $d = 2$ and let $q \in [2, +\infty)$. Let $\boldsymbol{d}_1 = (1, 0)^t$ and $\boldsymbol{d}_2 = (0, 1)^t$; for $x \in \Omega$, let $\mathcal{D}_x^1$ and $\mathcal{D}_x^2$ be the straight lines going through x and defined by the vectors $\boldsymbol{d}_1$ and $\boldsymbol{d}_2$.

Let $v \in X(\mathcal{T})$. For all control volume K, one denotes by v_K the value of v on K. For any control volume K and a.e. $x \in K$, one has

$$v_K^2 \leqslant \sum_{\sigma \in \mathcal{E}} D_\sigma v\, \chi_\sigma^{(1)}(x) \sum_{\sigma \in \mathcal{E}} D_\sigma v\, \chi_\sigma^{(2)}(x), \tag{9.71}$$

where $\chi_\sigma^{(1)}$ and $\chi_\sigma^{(2)}$ are defined by

$$\chi_\sigma^{(i)}(x) = \begin{cases} 1 & \text{if } \sigma \cap \mathcal{D}_x^i \neq \emptyset, \\ 0 & \text{if } \sigma \cap \mathcal{D}_x^i = \emptyset, \end{cases} \quad \text{for } i = 1, 2.$$

Recall that $D_\sigma v = |v_K - v_L|$, if $\sigma \in \mathcal{E}_{\text{int}}$, $\sigma = K|L$ and $D_\sigma v = |v_K|$, if $\sigma \in \mathcal{E}_{\text{ext}} \cap \mathcal{E}_K$. Integrating (9.71) over K and summing over $K \in \mathcal{T}$ yields

$$\int_\Omega v^2(x)\,\mathrm{d}x \leqslant \int_\Omega \left(\sum_{\sigma \in \mathcal{E}} D_\sigma v\, \chi_\sigma^{(1)}(x) \sum_{\sigma \in \mathcal{E}} D_\sigma v\, \chi_\sigma^{(2)}(x) \right) \mathrm{d}x.$$

Note that $\chi_\sigma^{(1)}$ (resp. $\chi_\sigma^{(2)}$) only depends on the second component x_2 (resp. the first component x_1) of x and that both functions are nonzero on a region the width of which is less than $\mathrm{m}(\sigma)$; hence

$$\int_\Omega v^2(x)\,\mathrm{d}x \leqslant \Bigl(\sum_{\sigma\in\mathcal{E}} \mathrm{m}(\sigma) D_\sigma v\Bigr)^2. \tag{9.72}$$

Applying the inequality (9.72) to $v = |u|^\alpha \operatorname{sign}(u)$, where $u \in X(\mathcal{T})$ and $\alpha > 1$ yields

$$\int_\Omega \bigl|u(x)\bigr|^{2\alpha}\,\mathrm{d}x \leqslant \Bigl(\sum_{\sigma\in\mathcal{E}} \mathrm{m}(\sigma) D_\sigma v\Bigr)^2.$$

Now, since $|v_K - v_L| \leqslant \alpha(|u_K|^{\alpha-1} + |u_L|^{\alpha-1})|u_K - u_L|$, if $\sigma \in \mathcal{E}_{\mathrm{int}}$, $\sigma = K|L$ and $|v_K| \leqslant \alpha(|u_K|^{\alpha-1})|u_K|$, if $\sigma \in \mathcal{E}_{\mathrm{ext}} \cap \mathcal{E}_K$,

$$\Bigl(\int_\Omega \bigl|u(x)\bigr|^{2\alpha}\,\mathrm{d}x\Bigr)^{1/2} \leqslant \alpha \sum_{K\in\mathcal{T}} \sum_{\sigma\in\mathcal{E}_K} \mathrm{m}(\sigma)|u_K|^{\alpha-1} D_\sigma u.$$

Using Hölder's inequality with $p, p' \in \mathbb{R}_+$ such that $1/p + 1/p' = 1$ yields that

$$\Bigl(\int_\Omega \bigl|u(x)\bigr|^{2\alpha}\,\mathrm{d}x\Bigr)^{1/2} \leqslant \alpha\Bigl(\sum_{K\in\mathcal{T}} \sum_{\sigma\in\mathcal{E}_K} |u_K|^{p(\alpha-1)} \mathrm{m}(\sigma) d_{K,\sigma}\Bigr)^{1/p} \times \Bigl(\sum_{K\in\mathcal{T}} \sum_{\sigma\in\mathcal{E}_K} \frac{|D_\sigma u|^{p'}}{d_{K,\sigma}{}^{p'}} \mathrm{m}(\sigma) d_{K,\sigma}\Bigr)^{1/p'}.$$

Since $\sum_{\sigma\in\mathcal{E}_K} \mathrm{m}(\sigma) d_{K,\sigma} = 2\mathrm{m}(K)$, this gives

$$\Bigl(\int_\Omega \bigl|u(x)\bigr|^{2\alpha}\,\mathrm{d}x\Bigr)^{1/2} \leqslant \alpha 2^{1/p}\Bigl(\int_\Omega \bigl|u(x)\bigr|^{p(\alpha-1)}\,\mathrm{d}x\Bigr)^{1/p} \Bigl(\sum_{K\in\mathcal{T}} \sum_{\sigma\in\mathcal{E}_K} \frac{|D_\sigma u|^{p'}}{d_{K,\sigma}{}^{p'}} \mathrm{m}(\sigma) d_{K,\sigma}\Bigr)^{1/p'},$$

which yields, choosing p such that $p(\alpha - 1) = 2\alpha$, i.e. $p = 2\alpha/(\alpha - 1)$ and $p' = 2\alpha/(\alpha + 1)$,

$$\|u\|_{L^q(\Omega)} = \Bigl(\int_\Omega \bigl|u(x)\bigr|^{2\alpha}\,\mathrm{d}x\Bigr)^{1/2\alpha} \leqslant \alpha 2^{1/p}\Bigl(\sum_{K\in\mathcal{T}} \sum_{\sigma\in\mathcal{E}_K} \frac{|D_\sigma u|^{p'}}{d_{K,\sigma}{}^{p'}} \mathrm{m}(\sigma) d_{K,\sigma}\Bigr)^{1/p'}, \tag{9.73}$$

where $q = 2\alpha$. Let $r = 2/p'$ and $r' = 2/(2-p')$, Hölder's inequality yields

$$\sum_{K\in\mathcal{T}}\sum_{\sigma\in\mathcal{E}_K}\frac{|D_\sigma u|^{p'}}{d_{K,\sigma}{}^{p'}}\mathrm{m}(\sigma)d_{K,\sigma}$$
$$\leqslant\left(\sum_{K\in\mathcal{T}}\sum_{\sigma\in\mathcal{E}_K}\frac{|D_\sigma u|^{2}}{d_{K,\sigma}{}^{2}}\mathrm{m}(\sigma)d_{K,\sigma}\right)^{p'/2}\left(\sum_{K\in\mathcal{T}}\sum_{\sigma\in\mathcal{E}_K}\mathrm{m}(\sigma)d_{K,\sigma}\right)^{1/r'},$$

replacing in (9.73) gives

$$\|u\|_{L^q(\Omega)}\leqslant\alpha 2^{1/p}\left(\frac{2}{\zeta}\right)^{1/2}\bigl(2\mathrm{m}(\Omega)\bigr)^{1/(p'r')}\|u\|_{1,\mathcal{T}}$$

and then (9.70) with, e.g., $C = (2/\zeta)^{1/2}((2\mathrm{m}(\Omega))^{1/2}+1)$.

Let us now prove the three-dimensional case. Let $d = 3$. Using the same notations as in the two-dimensional case, let $\boldsymbol{d}_1 = (1,0,0)^t$, $\boldsymbol{d}_2 = (0,1,0)^t$ and $\boldsymbol{d}_3 = (0,0,1)^t$; for $x\in\Omega$, let $\mathcal{D}_x^1$, $\mathcal{D}_x^2$ and $\mathcal{D}_x^3$ be the straight lines going through x and defined by the vectors $\boldsymbol{d}_1$, $\boldsymbol{d}_2$ and $\boldsymbol{d}_3$. Let us again define the functions $\chi_\sigma^{(1)}$, $\chi_\sigma^{(2)}$ and $\chi_\sigma^{(3)}$ by

$$\chi_\sigma^{(i)}(x)=\begin{cases}1 & \text{if } \sigma\cap\mathcal{D}_x^i\neq\emptyset,\\ 0 & \text{if } \sigma\cap\mathcal{D}_x^i=\emptyset,\end{cases}\qquad\text{for } i=1,2,3.$$

Let $v\in X(\mathcal{T})$ and let $A\in\mathbb{R}_+$ such that $\Omega\subset[-A,A]^3$; we also denote by v the function defined on $[-A,A]^3$ which equals v on Ω and 0 on $[-A,A]^3\setminus\Omega$. By the Cauchy–Schwarz inequality, one has:

$$\int_{-A}^{A}\int_{-A}^{A}\bigl|v(x_1,x_2,x_3)\bigr|^{3/2}\,\mathrm{d}x_1\,\mathrm{d}x_2$$
$$\leqslant\left(\int_{-A}^{A}\int_{-A}^{A}\bigl|v(x_1,x_2,x_3)\bigr|\,\mathrm{d}x_1\,\mathrm{d}x_2\right)^{1/2}$$
$$\times\left(\int_{-A}^{A}\int_{-A}^{A}\bigl|v(x_1,x_2,x_3)\bigr|^{2}\,\mathrm{d}x_1\,\mathrm{d}x_2\right)^{1/2}. \tag{9.74}$$

Now remark that

$$\int_{-A}^{A}\int_{-A}^{A}\bigl|v(x_1,x_2,x_3)\bigr|\,\mathrm{d}x_1\,\mathrm{d}x_2\leqslant\sum_{\sigma\in\mathcal{E}}D_\sigma v\int_{-A}^{A}\int_{-A}^{A}\chi_\sigma^{(3)}(x)\,\mathrm{d}x_1\,\mathrm{d}x_2$$
$$\leqslant\sum_{\sigma\in\mathcal{E}}\mathrm{m}(\sigma)D_\sigma v.$$

Moreover, computations which were already performed in the two-dimensional case give that

$$\begin{aligned}
&\int_{-A}^{A}\int_{-A}^{A}\left|v(x_1,x_2,x_3)\right|^2\mathrm{d}x_1\,\mathrm{d}x_2\\
&\quad\leqslant\int_{-A}^{A}\int_{-A}^{A}\sum_{\sigma\in\mathcal{E}}D_\sigma v\chi_\sigma^{(1)}(x)\sum_{\sigma\in\mathcal{E}}D_\sigma v\chi_\sigma^{(2)}(x)\,\mathrm{d}x_1\,\mathrm{d}x_2\\
&\quad\leqslant\left(\sum_{\sigma\in\mathcal{E}}\mathrm{m}(\sigma_{x_3})D_\sigma v\right)^2,
\end{aligned}$$

where σ_{x_3} denotes the intersection of σ with the plane which contains the point $(0,0,x_3)$ and is orthogonal to $\boldsymbol{d}_3$. Therefore, integrating (9.74) in the third direction yields:

$$\int_\Omega\left|v(x)\right|^{3/2}\mathrm{d}x\leqslant\left(\sum_{\sigma\in\mathcal{E}}\mathrm{m}(\sigma)D_\sigma v\right)^{3/2}. \tag{9.75}$$

Now let $v=|u|^4\mathrm{sign}(u)$, since $|v_K-v_L|\leqslant 4(|u_K|^3+|u_L|^3)|u_K-u_L|$, inequality (9.75) yields:

$$\int_\Omega\left|u(x)\right|^6\mathrm{d}x\leqslant\left[4\sum_{K\in\mathcal{T}}\sum_{\sigma\in\mathcal{E}_K}|u_K|^3D_\sigma u\,\mathrm{m}(\sigma)\right]^{3/2}.$$

By Cauchy–Schwarz' inequality and since $\sum_{\sigma\in\mathcal{E}_K}\mathrm{m}(\sigma)d_{K,\sigma}=3\mathrm{m}(K)$, this yields

$$\|u\|_{L^6}\leqslant 4\sqrt{3}\sum_{K\in\mathcal{T}}\sum_{\sigma\in\mathcal{E}_K}(D_\sigma u)^2\frac{\mathrm{m}(\sigma)}{d_{K,\sigma}},$$

and since $d_{K,\sigma}\geqslant\zeta d_\sigma$, this yields (9.70) with, e.g., $C=4\sqrt{3}/\sqrt{\zeta}$. □

REMARK 9.13 (*Discrete Poincaré inequality*). In the above proof, inequality (9.72) leads to another proof of some discrete Poincaré inequality (as in Lemma 9.1) in the two-dimensional case. Indeed, let Ω be an open bounded polygonal subset of $\mathbb{R}^2$. Let $\mathcal{T}$ be an admissible finite volume mesh of Ω in the sense of Definition 9.1 (but more general meshes are possible). Let $v\in X(\mathcal{T})$. Then, (9.72), the Cauchy–Schwarz inequality and the fact that $\sum_{\sigma\in\mathcal{E}}\mathrm{m}(\sigma)d_\sigma=2\,\mathrm{m}(\Omega)$ yield

$$\|v\|_{L^2(\Omega)}^2\leqslant 2\,\mathrm{m}(\Omega)\|v\|_{1,\mathcal{T}}^2.$$

A similar result holds in the three-dimensional case.

COROLLARY 9.1. *Under the same assumptions and with the same notations as in Theorem 9.3, or as in Theorem 9.4, and assuming that the mesh satisfies, for some $\zeta > 0$, $d_{K,\sigma} \geqslant \zeta d_\sigma$, for all $\sigma \in \mathcal{E}_K$ and for all control volume K, there exists $C > 0$ only depending on u, ζ and Ω such that*

$$\|e_{\mathcal{T}}\|_{L^q(\Omega)} \leqslant Cq\,\mathrm{size}(\mathcal{T}); \quad \textit{for any } q \in \begin{cases} [1,6] & \textit{if } d=3, \\ [1,+\infty) & \textit{if } d=2, \end{cases} \tag{9.76}$$

furthermore, there exists $C \in \mathbb{R}_+$ only depending on u, ζ, $\zeta_{\mathcal{T}} = \min\{\mathrm{m}(K)/\mathrm{size}(\mathcal{T})^2, K \in \mathcal{T}\}$, and Ω, such that

$$\|e_{\mathcal{T}}\|_{L^\infty(\Omega)} \leqslant C\mathrm{size}(\mathcal{T})\big(\big|\ln\big(\mathrm{size}(\mathcal{T})\big)\big| + 1\big), \quad \textit{if } d=2. \tag{9.77}$$

$$\|e_{\mathcal{T}}\|_{L^\infty(\Omega)} \leqslant C\mathrm{size}(\mathcal{T})^{2/3}, \quad \textit{if } d=3. \tag{9.78}$$

PROOF. Estimate (9.49) of Theorem 9.3 (or Theorem 9.4) and inequality (9.70) of Lemma 9.5 immediately yield estimate (9.76) in the case $d = 2$. Let us now prove (9.77). Remark that

$$\|e_{\mathcal{T}}\|_{L^\infty(\Omega)} = \max\big\{|e_K|, K \in \mathcal{T}\big\} \leqslant \left(\frac{1}{\zeta_{\mathcal{T}}\mathrm{size}(\mathcal{T})^2}\right)^{1/q} \|e_{\mathcal{T}}\|_{L^q}. \tag{9.79}$$

For $d = 2$, a study of the real function defined, for $q \geqslant 2$, by $q \mapsto \ln q + (1 - 2/q)\ln h$ (with $h = \mathrm{size}(\mathcal{T})$) shows that its minimum is attained for $q = -2\ln h$, if $\ln h \leqslant -1/2$. And therefore (9.76) and (9.79) yield (9.77).

The 3-dimensional case is an immediate consequence of (9.76) with $q = 6$. □

10. Neumann boundary conditions

This section is devoted to the convergence proof of the finite volume scheme when Neumann boundary conditions are imposed. The discretization of a general convection–diffusion equation with Dirichlet, Neumann and Fourier boundary conditions is considered in Section 11 below, and the convection term is largely studied in the previous section. Hence we shall limit here the presentation to the pure diffusion operator. Consider the following elliptic problem:

$$-\Delta u(x) = f(x), \quad x \in \Omega, \tag{10.1}$$

with Neumann boundary conditions:

$$\nabla u(x) \cdot \boldsymbol{n}(x) = g(x), \quad x \in \partial\Omega, \tag{10.2}$$

where $\partial\Omega$ denotes the boundary of Ω and $\boldsymbol{n}$ its unit normal vector outward to Ω.

The following assumptions are made on the data:

ASSUMPTION 10.1. (1) Ω is an open bounded polygonal connected subset of $\mathbb{R}^d$, $d = 2$ or 3,

(2) $g \in L^2(\partial\Omega)$, $f \in L^2(\Omega)$ and $\int_{\partial\Omega} g(x)\,\mathrm{d}\gamma(x) + \int_\Omega f(x)\,\mathrm{d}x = 0$.

Under Assumption 10.1, problem (10.1), (10.2) has a unique (variational) solution, u, belonging to $H^1(\Omega)$ and such that $\int_\Omega u(x)\,\mathrm{d}x = 0$. It is the unique solution of the following problem:

$$u \in H^1(\Omega), \quad \int_\Omega u(x)\,\mathrm{d}x = 0, \tag{10.3}$$

$$\begin{aligned}&\int_\Omega \nabla u(x)\nabla\psi(x)\\ &\quad = \int_\Omega f(x)\psi(x)\,\mathrm{d}x + \int_{\partial\Omega} g(x)\overline{\gamma}(\psi)(x)\,\mathrm{d}\gamma(x), \quad \forall \psi \in H^1(\Omega).\end{aligned} \tag{10.4}$$

Recall that $\overline{\gamma}$ is the "trace" operator from $H^1(\Omega)$ to $L^2(\partial\Omega)$ (or to $H^{1/2}(\partial\Omega)$).

10.1. Meshes and schemes

Admissible meshes

The definition of the scheme in the case of Neumann boundary conditions is easier, since the finite volume scheme naturally introduces the fluxes on the boundaries in its formulation. Hence the class of admissible meshes considered here is somewhat wider than the one considered in Definition 9.1, thanks to the Neumann boundary conditions and the absence of convection term.

DEFINITION 10.1 (*Admissible meshes*). Let Ω be an open bounded polygonal connected subset of $\mathbb{R}^d$, $d = 2$, or 3. An admissible finite volume mesh of Ω for the discretization of problem (10.1), (10.2), denoted by $\mathcal{T}$, is given by a family of "control volumes", which are open disjoint polygonal convex subsets of Ω, a family of subsets of $\overline{\Omega}$ contained in hyperplanes of $\mathbb{R}^d$, denoted by $\mathcal{E}$ (these are the "sides" of the control volumes), with strictly positive $(d-1)$-dimensional Lebesgue measure, and a family of points of Ω denoted by $\mathcal{P}$ satisfying properties (i), (ii), (iii) and (iv) of Definition 9.1.

The same notations as in Definition 9.1 are used in the sequel.

One defines the set $X(\mathcal{T})$ of piecewise constant functions on the control volumes of an admissible mesh as in Definition 9.2.

DEFINITION 10.2 (*Discrete H^1 seminorm*). Let Ω be an open bounded polygonal subset of $\mathbb{R}^d$, $d = 2$ or 3, and $\mathcal{T}$ an admissible finite volume mesh in the sense of Definition 10.1.

For $u \in X(\mathcal{T})$, the discrete H^1 seminorm of u is defined by

$$|u|_{1,\mathcal{T}} = \left(\sum_{\sigma \in \mathcal{E}_{\text{int}}} \tau_\sigma (D_\sigma u)^2 \right)^{1/2},$$

where $\tau_\sigma = \mathrm{m}(\sigma)/d_\sigma$ and $\mathcal{E}_{\text{int}}$ are defined in Definition 9.1, u_K is the value of u in the control volume K and $D_\sigma u = |u_K - u_L|$ if $\sigma \in \mathcal{E}_{\text{int}}$, $\sigma = K|L$.

The finite volume scheme
Let $\mathcal{T}$ be an admissible mesh in the sense of Definition 10.1 . For $K \in \mathcal{T}$, let us define:

$$f_K = \frac{1}{\mathrm{m}(K)} \int_K f(x)\,\mathrm{d}x, \tag{10.5}$$

$$\begin{aligned} g_K &= \frac{1}{\mathrm{m}(\partial K \cap \partial\Omega)} \int_{\partial K \cap \partial\Omega} g(x)\,\mathrm{d}\gamma(x) \quad \text{if } \mathrm{m}(\partial K \cap \partial\Omega) \neq 0, \\ g_K &= 0 \quad \text{if } \mathrm{m}(\partial K \cap \partial\Omega) = 0. \end{aligned} \tag{10.6}$$

Recall that, in formula (10.5), $\mathrm{m}(K)$ denotes the d-dimensional Lebesgue measure of K, and, in (10.6), $\mathrm{m}(\partial K \cap \partial\Omega)$ denotes the $(d-1)$-dimensional Lebesgue measure of $\partial K \cap \partial\Omega$. Note that $g_K = 0$ if the dimension of $\partial K \cap \partial\Omega$ is less than $d-1$. Let $(u_K)_{K \in \mathcal{T}}$ denote the discrete unknowns; the numerical scheme is defined by (9.20)–(9.22), with $b = 0$ and $\boldsymbol{v} = 0$. This yields:

$$-\sum_{L \in \mathcal{N}(K)} \tau_{K|L}(u_L - u_K) = \mathrm{m}(K) f_K + \mathrm{m}(\partial K \cap \partial\Omega) g_K, \quad \forall K \in \mathcal{T} \tag{10.7}$$

(see the notations in Definitions 9.1 and 10.1). Condition (10.3) is discretized by:

$$\sum_{K \in \mathcal{T}} \mathrm{m}(K) u_K = 0. \tag{10.8}$$

Then, the approximate solution, $u_\mathcal{T}$, belongs to $X(\mathcal{T})$ (see Definition 9.2) and is defined by

$$u_\mathcal{T}(x) = u_K, \quad \text{for a.e. } x \in K,\ \forall K \in \mathcal{T}.$$

The following lemma gives existence and uniqueness of the solution of (10.7) and (10.8).

LEMMA 10.1. *Under Assumption* 10.1. *let $\mathcal{T}$ be an admissible mesh* (*see Definition* 10.1) *and* $\{f_K,\ K \in \mathcal{T}\}$, $\{g_K,\ K \in \mathcal{T}\}$ *defined by* (10.5), (10.6). *Then, there exists a unique solution* $(u_K)_{K \in \mathcal{T}}$ *to* (10.7)–(10.8).

PROOF. Let $N = \mathrm{card}(\mathcal{T})$. Equations (10.7) are a system of N equations with N unknowns, namely $(u_K)_{K\in\mathcal{T}}$. Ordering the unknowns (and the equations), this system can be written under a matrix form with a $N \times N$ matrix A. Using the convexity of Ω, the null space of this matrix is the set of "constant" vectors (that is $u_K = u_L$, for all $K, L \in \mathcal{T}$). Indeed, if $f_K = g_K = 0$ for all $K \in \mathcal{T}$ and $\{u_K, K \in \mathcal{T}\}$ is solution of (10.7), multiplying (10.7) (for $K \in \mathcal{T}$) by u_K and summing over $K \in \mathcal{T}$ yields

$$\sum_{\sigma\in\mathcal{E}_{\mathrm{int}}} \tau_\sigma (D_\sigma u)^2 = 0,$$

where $D_\sigma u = |u_K - u_L|$ if $\sigma \in \mathcal{E}_{\mathrm{int}}$, $\sigma = K|L$. This gives, thanks to the positivity of τ_σ and the convexity of Ω, $u_K = u_L$, for all $K, L \in \mathcal{T}$.

For general $(f_K)_{K\in\mathcal{T}}$ and $(g_K)_{K\in\mathcal{T}}$, a necessary condition, in order that (10.7) has a solution, is that

$$\sum_{K\in\mathcal{T}} \big(\mathrm{m}(K) f_K + \mathrm{m}(\partial K \cap \partial\Omega) g_K\big) = 0. \tag{10.9}$$

Since the dimension of the null space of A is one, this condition is also a sufficient condition. Therefore, system (10.7) has a solution if and only if (10.9) holds, and this solution is unique up to an additive constant. Adding condition (10.8) yields uniqueness. Note that (10.9) holds thanks to the second item of Assumption 10.1; this concludes the proof of Lemma 10.1. □

10.2. Discrete Poincaré inequality

The proof of an error estimate, under a regularity assumption on the exact solution, and of a convergence result, in the general case (under Assumption 10.1), requires a "discrete Poincaré" inequality as in the case of the Dirichlet problem.

LEMMA 10.2 (Discrete mean Poincaré inequality). *Let Ω be an open bounded polygonal connected subset of $\mathbb{R}^d$, $d = 2$ or 3. Then, there exists $C \in \mathbb{R}_+$, only depending on Ω, such that for all admissible meshes (in the sense of Definition* 10.1*), $\mathcal{T}$, and for all $u \in X(\mathcal{T})$ (see Definition* 9.2*), the following inequality holds*:

$$\|u\|^2_{L^2(\Omega)} \leqslant C|u|^2_{1,\mathcal{T}} + 2\big(\mathrm{m}(\Omega)\big)^{-1}\left(\int_\Omega u(x)\,\mathrm{d}x\right)^2, \tag{10.10}$$

where $|\bullet|_{1,\mathcal{T}}$ is the discrete H^1 seminorm defined in Definition 10.2.

PROOF. The proof given here is a "direct proof"; another proof, by contradiction, is possible (see Remark 10.2). Let $\mathcal{T}$ be an admissible mesh and $u \in X(\mathcal{T})$. Let $m_\Omega(u)$ be the mean value of u over Ω, i.e.

$$m_\Omega(u) = \frac{1}{\mathrm{m}(\Omega)} \int_\Omega u(x)\,\mathrm{d}x.$$

Since

$$\|u\|^2_{L^2(\Omega)} \leqslant 2\|u - m_\Omega(u)\|^2_{L^2(\Omega)} + 2\big(m_\Omega(u)\big)^2 \mathrm{m}(\Omega),$$

proving Lemma 10.2 amounts to proving the existence of $D \geqslant 0$, only depending on Ω, such that

$$\|u - m_\Omega(u)\|^2_{L^2(\Omega)} \leqslant D|u|^2_{1,\mathcal{T}}. \tag{10.11}$$

The proof of (10.11) may be decomposed into three steps (indeed, if Ω is convex, the first step is sufficient).

Step 1 (*Estimate on a convex part of* Ω). Let ω be an open convex subset of Ω, $\omega \neq \emptyset$ and $m_\omega(u)$ be the mean value of u on ω. In this step, one proves that there exists C_0, depending only on Ω, such that

$$\|u(x) - m_\omega(u)\|^2_{L^2(\omega)} \leqslant \frac{1}{\mathrm{m}(\omega)} C_0 |u|^2_{1,\mathcal{T}}. \tag{10.12}$$

(Taking $\omega = \Omega$, this proves (10.11) and Lemma 10.2 in the case where Ω is convex.)

Noting that

$$\int_\omega \big(u(x) - m_\omega(u)\big)^2 \,\mathrm{d}x \leqslant \frac{1}{\mathrm{m}(\omega)} \int_\omega \left(\int_\omega \big(u(x) - u(y)\big)^2 \,\mathrm{d}y \right) \mathrm{d}x,$$

(10.12) is proved provided that there exists $C_0 \in \mathbb{R}_+$, only depending on Ω, such that

$$\int_\omega \int_\omega \big(u(x) - u(y)\big)^2 \,\mathrm{d}x\,\mathrm{d}y \leqslant C_0 |u|^2_{1,\mathcal{T}}. \tag{10.13}$$

For $\sigma \in \mathcal{E}_{\mathrm{int}}$, let the function χ_σ from $\mathbb{R}^d \times \mathbb{R}^d$ to $\{0, 1\}$ be defined by

$$\begin{aligned} &\chi_\sigma(x, y) = 1, \quad \text{if } x, y \in \overline{\Omega},\ [x, y] \cap \sigma \neq \emptyset, \\ &\chi_\sigma(x, y) = 0, \quad \text{if } x \notin \overline{\Omega} \text{ or } y \notin \overline{\Omega} \text{ or } [x, y] \cap \sigma = \emptyset. \end{aligned}$$

(Recall that $[x, y] = \{tx + (1-t)y,\ t \in [0, 1]\}$.) For a.e. $x, y \in \omega$, one has, with $D_\sigma u = |u_K - u_L|$ if $\sigma \in \mathcal{E}_{\mathrm{int}}$, $\sigma = K|L$,

$$\big(u(x) - u(y)\big)^2 \leqslant \left(\sum_{\sigma \in \mathcal{E}_{\mathrm{int}}} |D_\sigma u| \chi_\sigma(x, y) \right)^2$$

(note that the convexity of ω is used here) which yields, thanks to the Cauchy–Schwarz inequality,

$$\big(u(x) - u(y)\big)^2 \leqslant \sum_{\sigma \in \mathcal{E}_{\mathrm{int}}} \frac{|D_\sigma u|^2}{d_\sigma c_{\sigma,y-x}} \chi_\sigma(x, y) \sum_{\sigma \in \mathcal{E}_{\mathrm{int}}} d_\sigma c_{\sigma,y-x} \chi_\sigma(x, y), \tag{10.14}$$

with

$$c_{\sigma,y-x} = \left| \frac{y-x}{|y-x|} \bullet \boldsymbol{n}_\sigma \right|,$$

recall that $\boldsymbol{n}_\sigma$ is a unit normal vector to σ, and that $x_K - x_L = \pm d_\sigma \boldsymbol{n}_\sigma$ if $\sigma \in \mathcal{E}_{\text{int}}$, $\sigma = K|L$. For a.e. $x, y \in \omega$, one has

$$\sum_{\sigma \in \mathcal{E}_{\text{int}}} d_\sigma c_{\sigma,y-x} \chi_\sigma(x,y) = \left| (x_K - x_L) \bullet \frac{y-x}{|y-x|} \right|,$$

for some convenient control volumes K and L, depending on x, y and σ (the convexity of ω is used again here). Therefore,

$$\sum_{\sigma \in \mathcal{E}_{\text{int}}} d_\sigma c_{\sigma,y-x} \chi_\sigma(x,y) \leqslant \operatorname{diam}(\Omega).$$

Thus, integrating (10.14) with respect to x and y in ω,

$$\int_\omega \int_\omega \big(u(x) - u(y)\big)^2 \,\mathrm{d}x\,\mathrm{d}y \leqslant \operatorname{diam}(\Omega) \int_\omega \int_\omega \sum_{\sigma \in \mathcal{E}_{\text{int}}} \frac{|D_\sigma u|^2}{d_\sigma c_{\sigma,y-x}} \chi_\sigma(x,y)\,\mathrm{d}x\,\mathrm{d}y,$$

which gives, by a change of variables,

$$\int_\omega \int_\omega \big(u(x) - u(y)\big)^2 \,\mathrm{d}x\,\mathrm{d}y \leqslant \operatorname{diam}(\Omega) \int_{\mathbb{R}^d} \left(\sum_{\sigma \in \mathcal{E}_{\text{int}}} \frac{|D_\sigma u|^2}{d_\sigma c_{\sigma,z}} \int_\omega \chi_\sigma(x, x+z)\,\mathrm{d}x \right) \mathrm{d}z. \tag{10.15}$$

Noting that, if $|z| > \operatorname{diam}(\Omega)$, $\chi_\sigma(x, x+z) = 0$, for a.e. $x \in \Omega$, and

$$\int_\Omega \chi_\sigma(x, x+z)\,\mathrm{d}x \leqslant \mathrm{m}(\sigma)|z \bullet \boldsymbol{n}_\sigma| = \mathrm{m}(\sigma)|z| c_{\sigma,z} \quad \text{for a.e. } z \in \mathbb{R}^d,$$

therefore, with (10.15):

$$\int_\omega \int_\omega \big(u(x) - u(y)\big)^2 \,\mathrm{d}x\,\mathrm{d}y \leqslant \big(\operatorname{diam}(\Omega)\big)^2 \mathrm{m}(B_\Omega) \sum_{\sigma \in \mathcal{E}_{\text{int}}} \frac{\mathrm{m}(\sigma)|D_\sigma u|^2}{d_\sigma},$$

where B_Ω denotes the ball of $\mathbb{R}^d$ of center 0 and radius $\operatorname{diam}(\Omega)$.

This inequality proves (10.13) and then (10.12) with $C_0 = (\operatorname{diam}(\Omega))^2 \mathrm{m}(B_\Omega)$ (which only depends on Ω). Taking $\omega = \Omega$, it concludes the proof of Lemma 10.2 in the case where Ω is convex.

Step 2 (*Estimate with respect to the mean value on a part of the boundary*). In this step, one proves the same inequality than (10.12) but with the mean value of u on a (arbitrary) part I of the boundary of ω instead of $m_\omega(u)$ and with a convenient C_1 depending on I, Ω and ω instead of C_0.

More precisely, let ω be a polygonal open convex subset of Ω and let $I \subset \partial\omega$, with $m(I) > 0$ ($m(I)$ is the $(d-1)$-Lebesgue measure of I). Assume that I is included in a hyperplane of $\mathbb{R}^d$. Let $\overline{\gamma}(u)$ be the "trace" of u on the boundary of ω, that is $\overline{\gamma}(u)(x) = u_K$ if $x \in \partial\omega \cap \overline{K}$, for $K \in \mathcal{T}$. (If $x \in \overline{K} \cap \overline{L}$, the choice of $\overline{\gamma}(u)(x)$ between u_K and u_L does not matter.) Let $m_I(u)$ be the mean value of $\overline{\gamma}(u)$ on I. This step is devoted to the proof that there exists C_1, only depending on Ω, ω and I, such that

$$\left\|u(x) - m_I(u)\right\|^2_{L^2(\omega)} \leqslant C_1 |u|^2_{1,\mathcal{T}}. \tag{10.16}$$

For the sake of simplicity, only the case $d = 2$ is considered here. Since I is included in a hyperplane, it may be assumed, without loss of generality, that $I = \{0\} \times J$, with $J \subset \mathbb{R}$ and $\omega \subset \mathbb{R}_+ \times \mathbb{R}$ (one uses here the convexity of ω).

Let $\alpha = \max\{x_1, x = (x_1, x_2)^t \in \overline{\omega}\}$ and $a = (\alpha, \beta)^t \in \overline{\omega}$. In the following, a is fixed. For a.e. $x = (x_1, x_2)^t \in \omega$ and for a.e. (for the 1-Lebesgue measure) $y = (0, \overline{y})^t \in I$ (with $\overline{y} \in J$), one sets $z(x, y) = ta + (1-t)y$ with $t = x_1/\alpha$. Note that, thanks to the convexity of ω, $z(x, y) = (z_1, z_2)^t \in \overline{\omega}$, with $z_1 = x_1$. The following inequality holds:

$$\pm\big(u(x) - \overline{\gamma}(u)(y)\big) \leqslant \big|u(x) - u\big(z(x, y)\big)\big| + \big|u\big(z(x, y) - \overline{\gamma}(u)(y)\big)\big|.$$

In the following, the notation C_i, $i \in \mathbb{N}^\star$, will be used for quantities only depending on Ω, ω and I.

Let us integrate the above inequality over $y \in I$, take the power 2, from the Cauchy–Schwarz inequality, an integration over $x \in \omega$ leads to

$$\begin{aligned}\int_\omega \big(u(x) - m_I(u)\big)^2 \,\mathrm{d}x \leqslant{}& \frac{2}{m(I)} \int_\omega \int_I \big(u(x) - u\big(z(x, y)\big)\big)^2 \,\mathrm{d}\gamma(y)\,\mathrm{d}x \\ &+ \frac{2}{m(I)} \int_\omega \int_I \big(u\big(z(x, y)\big) - u(y)\big)^2 \,\mathrm{d}\gamma(y)\,\mathrm{d}x.\end{aligned}$$

Then,

$$\int_\omega \big(u(x) - m_I(u)\big)^2 \,\mathrm{d}x \leqslant \frac{2}{m(I)}(A + B),$$

with, since ω is convex,

$$A = \int_\omega \int_I \Big(\sum_{\sigma \in \mathcal{E}_{\mathrm{int}}} |D_\sigma u| \chi_\sigma\big(x, z(x, y)\big)\Big)^2 \,\mathrm{d}\gamma(y)\,\mathrm{d}x,$$

and

$$B = \int_\omega \int_I \Big(\sum_{\sigma \in \mathcal{E}_{\text{int}}} |D_\sigma u| \chi_\sigma\big(z(x, y), y\big)\Big)^2 \, \mathrm{d}\gamma(y) \, \mathrm{d}x.$$

Recall that, for $\xi, \eta \in \overline{\Omega}$, $\chi_\sigma(\xi, \eta) = 1$ if $[\xi, \eta] \cap \sigma \neq \emptyset$ and $\chi_\sigma(\xi, \eta) = 0$ if $[\xi, \eta] \cap \sigma = \emptyset$. Let us now look for some bounds of A and B of the form $C|u|_{1,\mathcal{T}}^2$.

The bound for A is easy. Using the Cauchy–Schwarz inequality and the fact that

$$\sum_{\sigma \in \mathcal{E}_{\text{int}}} c_{\sigma, x - z(x,y)} d_\sigma \chi_\sigma\big(x, z(x, y)\big) \leqslant \operatorname{diam}(\Omega)$$

(recall that $c_{\sigma,\eta} = |\frac{\eta}{|\eta|} \cdot \boldsymbol{n}_\sigma|$ (for $\eta \in \mathbb{R}^2 \setminus 0$) gives

$$A \leqslant C_2 \int_\omega \int_I \sum_{\sigma \in \mathcal{E}_{\text{int}}} \frac{|D_\sigma u|^2 \chi_\sigma(x, z(x, y))}{c_{\sigma, x - z(x,y)} d_\sigma} \, \mathrm{d}x \, \mathrm{d}\gamma(y).$$

Since $z_1 = x_1$, one has $c_{\sigma, x - z(x,y)} = c_{\sigma, e}$, with $e = (0, 1)^t$. Let us perform the integration of the right-hand side of the previous inequality, with respect to the first component of x, denoted by x_1, first. The result of the integration with respect to x_1 is bounded by $|u|_{1,\mathcal{T}}^2$. Then, integrating with respect to x_2 and $y \in I$ gives $A \leqslant C_3 |u|_{1,\mathcal{T}}^2$.

In order to obtain a bound B, one remarks, as for A, that

$$B \leqslant C_4 \int_\omega \int_I \sum_{\sigma \in \mathcal{E}_{\text{int}}} \frac{|D_\sigma u|^2 \chi_\sigma(z(x, y), y)}{c_{\sigma, y - z(x,y)} d_\sigma} \, \mathrm{d}x \, \mathrm{d}\gamma(y).$$

In the right-hand side of this inequality, the integration with respect to $y \in I$ is transformed into an integration with respect to $\xi = (\xi_1, \xi_2)^t \in \sigma$, this yields (note that $c_{\sigma, y - z(x,y)} = c_{\sigma, a - y}$)

$$B \leqslant C_4 \sum_{\sigma \in \mathcal{E}_{\text{int}}} \frac{|D_\sigma u|^2}{d_\sigma} \int_\omega \int_\sigma \frac{\psi_\sigma(x, \xi)}{c_{I, a - y(\xi)}} \frac{|a - y(\xi)|}{|a - \xi|} \, \mathrm{d}x \, \mathrm{d}\gamma(\xi),$$

where $y(\xi) = s\xi + (1 - s)a$, with $s\xi_1 + (1 - s)\alpha = 0$, and where ψ_σ is defined by

$$\begin{aligned} \psi_\sigma(x, \xi) &= 1, \quad \text{if } y(\xi) \in I \text{ and } \xi_1 \leqslant x_1 \\ \psi_\sigma(x, \xi) &= 0, \quad \text{if } y(\xi) \notin I \text{ or } \xi_1 > x_1. \end{aligned}$$

Noting that $c_{I, a - y(\xi)} \geqslant C_5 > 0$, one deduces that

$$B \leqslant C_6 \sum_{\sigma \in \mathcal{E}_{\text{int}}} \frac{|D_\sigma u|^2}{d_\sigma} \int_\sigma \Big(\int_\omega \psi_\sigma(x, \xi) \frac{|a - y(\xi)|}{|a - \xi|} \, \mathrm{d}x \Big) \mathrm{d}\gamma(\xi) \leqslant C_7 |u|_{1,\mathcal{T}}^2,$$

with, e.g., $C_7 = C_6(\mathrm{diam}(\omega))^2$. The bounds on A and B yield (10.16).

Step 3. *Proof of* (10.11). Let us now prove that there exists $D \in \mathbb{R}_+$, only depending on Ω such that (10.11) hold. Since Ω is a polygonal set ($d = 2$ or 3), there exists a finite number of disjoint convex polygonal sets, denoted by $\{\Omega_1, \ldots, \Omega_n\}$, such that $\overline{\Omega} = \bigcup_{i=1}^n \overline{\Omega_i}$. Let $I_{i,j} = \overline{\Omega_i} \cap \overline{\Omega_j}$, and B be the set of couples $(i, j) \in \{1, \ldots, n\}^2$ such that $i \neq j$ and the $(d-1)$-dimensional Lebesgue measure of $I_{i,j}$, denoted by $m(I_{i,j})$, is positive.

Let m_i denote the mean value of u on Ω_i, $i \in \{1, \ldots, n\}$, and $m_{i,j}$ denote the mean value of u on $I_{i,j}$, $(i, j) \in B$. (For $\sigma \in \mathcal{E}_{\mathrm{int}}$, in order that u be defined on σ, a.e. for the $(d-1)$-dimensional Lebesgue measure, let $K \in \mathcal{T}$ be a control volume such that $\sigma \in \mathcal{E}_K$, one sets $u = u_K$ on σ.) Note that $m_{i,j} = m_{j,i}$ for all $(i, j) \in B$.

Step 1 gives the existence of C_i, $i \in \{1, \ldots, n\}$, only depending on Ω (since the Ω_i only depend on Ω), such that

$$\|u - m_i\|^2_{L^2(\Omega_i)} \leqslant C_i |u|^2_{1,\mathcal{T}}, \quad \forall i \in \{1, \ldots, n\}. \tag{10.17}$$

Step 2 gives the existence of $C_{i,j}$, $i, j \in B$, only depending on Ω, such that

$$\|u - m_{i,j}\|^2_{L^2(\Omega_i)} \leqslant C_{i,j} |u|^2_{1,\mathcal{T}}, \quad \forall (i, j) \in B.$$

Then, one has $(m_i - m_{i,j})^2 \mathrm{m}(\Omega_i) \leqslant 2(C_i + C_{i,j})|u|^2_{1,\mathcal{T}}$, for all $(i, j) \in B$. Since Ω is connected, the above inequality yields the existence of M, only depending on Ω, such that $|m_i - m_j| \leqslant M|u|_{1,\mathcal{T}}$ for all $(i, j) \in \{1, \ldots, n\}^2$, and therefore $|m_\Omega(u) - m_i| \leqslant M|u|_{1,\mathcal{T}}$ for all $i \in \{1, \ldots, n\}$. Then, (10.17) yields the existence of D, only depending on Ω, such that (10.11) holds. This completes the proof of Lemma 10.2. □

An easy consequence of the proof of Lemma 10.2 is the following lemma. Although this lemma is not used in the sequel, it is interesting in its own sake.

LEMMA 10.3 (Mean boundary Poincaré inequality). *Let Ω be an open bounded polygonal connected subset of $\mathbb{R}^d$, $d = 2$ or 3. Let $I \subset \partial\Omega$ such that the $(d-1)$-Lebesgue measure of I is positive. Then, there exists $C \in \mathbb{R}_+$, only depending on Ω and I, such that for all admissible mesh* (*in the sense of Definition* 10.1) *$\mathcal{T}$ and for all $u \in X(\mathcal{T})$* (*see Definition* 9.2), *the following inequality holds*:

$$\left\|u - m_I(u)\right\|^2_{L^2(\Omega)} \leqslant C|u|^2_{1,\mathcal{T}},$$

where $|\bullet|_{1,\mathcal{T}}$ is the discrete H^1 seminorm defined in Definition 10.2 *and $m_I(u)$ is the mean value of $\overline{\gamma}(u)$ on I with $\overline{\gamma}(u)$ defined a.e. on $\partial\Omega$ by $\overline{\gamma}(u)(x) = u_K$ if $x \in \sigma$, $\sigma \in \mathcal{E}_{\mathrm{ext}} \cap \mathcal{E}_K$, $K \in \mathcal{T}$.*

Finally, let us point out that a continuous version of Lemmata 10.2 and 10.3 holds and that the proof is similar and rather easier. Let us state this continuous version which can be proved by contradiction or with a technique similar to Lemma 9.4. The advantage of the latter is that it gives a more explicit bound.

LEMMA 10.4. *Let Ω be an open bounded polygonal connected subset of $\mathbb{R}^d$, $d=2$ or 3. Let $I \subset \partial\Omega$ such that the $(d-1)$-Lebesgue measure of I is positive.*

Then, there exists $C \in \mathbb{R}_+$, only depending on Ω, and $\tilde{C} \in \mathbb{R}_+$, only depending on Ω and I, such that, for all $u \in H^1(\Omega)$, the following inequalities hold:

$$\|u\|^2_{L^2(\Omega)} \leqslant C|u|^2_{H^1(\Omega)} + 2\big(\mathrm{m}(\Omega)\big)^{-1}\left(\int_\Omega u(x)\,\mathrm{d}x\right)^2$$

and

$$\big\|u - m_I(u)\big\|^2_{L^2(\Omega)} \leqslant \tilde{C}|u|^2_{H^1(\Omega)},$$

where $|\bullet|_{H^1(\Omega)}$ is the H^1 seminorm defined by

$$|v|^2_{H^1(\Omega)} = \|\nabla u\|^2_{(L^2(\Omega))^d} = \int_\Omega \big|\nabla v(x)\big|^2\,\mathrm{d}x$$

for all $v \in H^1(\Omega)$, and $m_I(u)$ is the mean value of $\overline{\gamma}(u)$ on I. Recall that $\overline{\gamma}$ is the trace operator from $H^1(\Omega)$ to $H^{1/2}t(\partial\Omega)$.

10.3. Error estimate

Under Assumption 10.1, let $\mathcal{T}$ be an admissible mesh (see Definition 10.1) and $\{f_K, K \in \mathcal{T}\}$, $\{g_K, K \in \mathcal{T}\}$ defined by (10.5), (10.6). By Lemma 10.1, there exists a unique solution $(u_K)_{K\in\mathcal{T}}$ to (10.7)–(10.8). Under an additional regularity assumption on the exact solution, the following error estimate holds:

THEOREM 10.1. *Under Assumption* 10.1, *let $\mathcal{T}$ be an admissible mesh (see Definition* 10.1) *and $h = \mathrm{size}(\mathcal{T})$. Let $(u_K)_{K\in\mathcal{T}}$ be the unique solution to* (10.7) *and* (10.8) (*thanks to* (10.5) *and* (10.6), *existence and uniqueness of $(u_K)_{K\in\mathcal{T}}$ is given in Lemma* 10.1). *Let $u_\mathcal{T} \in X(\mathcal{T})$ (see Definition* 9.2) *be defined by $u_\mathcal{T}(x) = u_K$ for a.e. $x \in K$, for all $K \in \mathcal{T}$. Assume that the unique solution, u, to problem* (10.3), (10.4) *satisfies $u \in C^2(\overline{\Omega})$.*

Then there exists $C \in \mathbb{R}_+$ which only depends on u and Ω such that

$$\|u_\mathcal{T} - u\|_{L^2(\Omega)} \leqslant Ch, \tag{10.18}$$

$$\sum_{\sigma=K|L\in\mathcal{E}_{\mathrm{int}}} \mathrm{m}(\sigma)d_\sigma\left(\frac{u_L - u_K}{d_\sigma} - \frac{1}{\mathrm{m}(\sigma)}\int_\sigma \nabla u(x)\bullet \boldsymbol{n}_{K,\sigma}\,\mathrm{d}\gamma(x)\right)^2 \leqslant Ch^2. \tag{10.19}$$

Recall that, in the above theorem, $K|L$ denotes the element σ of $\mathcal{E}_{\mathrm{int}}$ such that $\overline{\sigma} = \partial K \cap \partial L$, with $K, L \in \mathcal{T}$.

PROOF OF THEOREM 10.1. Let $C_{\mathcal{T}} \in \mathbb{R}$ be such that

$$\sum_{K\in\mathcal{T}} \overline{u}(x_K)\mathrm{m}(K) = 0,$$

where $\overline{u} = u + C_{\mathcal{T}}$.

Let, for each $K \in \mathcal{T}$, $e_K = \overline{u}(x_K) - u_K$, and $e_{\mathcal{T}} \in X(\mathcal{T})$ defined by $e_{\mathcal{T}}(x) = e_K$ for a.e. $x \in K$, for all $K \in \mathcal{T}$. Let us first prove the existence of C only depending on u and Ω such that

$$|e_{\mathcal{T}}|_{1,\mathcal{T}} \leqslant Ch \quad \text{and} \quad \|e_{\mathcal{T}}\|_{L^2(\Omega)} \leqslant Ch. \tag{10.20}$$

Integrating (10.1) over $K \in \mathcal{T}$, and taking (10.2) into account yields:

$$\sum_{\sigma\in\mathcal{E}_K} \int_\sigma \nabla u(x) \cdot \boldsymbol{n}_{K,\sigma}\,\mathrm{d}\gamma(x) = \int_K f(x)\,\mathrm{d}x + \int_{\partial K\cap\partial\Omega} g(x)\,\mathrm{d}\gamma(x). \tag{10.21}$$

For $\sigma \in \mathcal{E}_{\mathrm{int}}$ such that $\sigma = K|L$, let us define the consistency error on the flux from K through σ by:

$$R_{K,\sigma} = \frac{1}{\mathrm{m}(\sigma)} \int_\sigma \nabla u(x) \cdot \boldsymbol{n}_{K,\sigma}\,\mathrm{d}\gamma(x) - \frac{u(x_L) - u(x_K)}{d_\sigma}. \tag{10.22}$$

Note that the definition of $R_{K,\sigma}$ remains with $\overline{u}$ instead of u in (10.22).

Thanks to the regularity of the solution u, there exists $C_1 \in \mathbb{R}_+$, only depending on u, such that $|R_{K,L}| \leqslant C_1 h$. Using (10.21), (10.22) and (10.7) yields

$$\sum_{K|L\in\mathcal{E}_{\mathrm{int}}} \tau_{K|L}(e_L - e_K)^2 \leqslant d\mathrm{m}(\Omega)(C_1 h)^2,$$

which gives the first part of (10.20).

Thanks to the discrete Poincaré inequality (10.10) applied to the function $e_{\mathcal{T}}$, and since

$$\sum_{K\in\mathcal{T}} \mathrm{m}(K)e_K = 0$$

(which is the reason why $e_{\mathcal{T}}$ was defined with $\overline{u}$ instead of u) one obtains the second part of (10.20), i.e. the existence of C_2 only depending on u and Ω such that

$$\sum_{K\in\mathcal{T}} \mathrm{m}(K)(e_K)^2 \leqslant C_2 h^2.$$

From (10.20), one deduces (10.18) from the fact that $u \in C^1(\overline{\Omega})$. Indeed, let C_2 be the maximum value of $|\nabla u|$ in Ω. One has $|u(x) - u(y)| \leqslant C_2 h$, for all $x, y \in K$, for all $K \in \mathcal{T}$. Then, from $\int_\Omega u(x)\,\mathrm{d}x = 0$, one deduces $C_\mathcal{T} \leqslant C_2 h$. Furthermore, one has

$$\sum_{K\in\mathcal{T}} \int_K \big(u(x_K) - u(x)\big)^2\,\mathrm{d}x \leqslant \sum_{K\in\mathcal{T}} \mathrm{m}(K)(C_2 h)^2 = \mathrm{m}(\Omega)(C_2 h)^2.$$

Then, noting that

$$\begin{aligned}\|u_\mathcal{T} - u\|^2_{L^2(\Omega)} &= \sum_{K\in\mathcal{T}} \int_K \big(u_K - u(x)\big)^2\,\mathrm{d}x \\ &\leqslant 3\sum_{K\in\mathcal{T}} \mathrm{m}(K)(e_K)^2 + 3(C_\mathcal{T})^2\mathrm{m}(\Omega) + 3\sum_{K\in\mathcal{T}} \int_K \big(u(x_K) - u(x)\big)^2\,\mathrm{d}x\end{aligned}$$

yields (10.18).

The proof of estimate (10.19) is exactly the same as in the Dirichlet case. This property will be useful in the study of the convergence of finite volume methods in the case of a system consisting of an elliptic equation and a hyperbolic equation (see Section 35.6). □

As for the Dirichlet problem, the hypothesis $u \in C^2(\overline{\Omega})$ is not necessary to obtain error estimates. Assuming an additional assumption on the mesh (see Definition 10.3), estimates (10.20) and (10.19) hold under the weaker assumption $u \in H^2(\Omega)$ (see Theorem 10.2 below). It is therefore also possible to obtain (10.18) under the additional assumption that u is Lipschitz continuous.

DEFINITION 10.3 (*Neumann restricted admissible meshes*). Let Ω be an open bounded polygonal connected subset of $\mathbb{R}^d$, $d = 2$ or 3. A restricted admissible mesh for the Neumann problem, denoted by $\mathcal{T}$, is an admissible mesh in the sense of Definition 10.1 such that, for some $\zeta > 0$, one has $d_{K,\sigma} \geqslant \zeta\,\mathrm{diam}(K)$ for all control volume K and for all $\sigma \in \mathcal{E}_K \cap \mathcal{E}_{\mathrm{int}}$.

THEOREM 10.2 (H^2 regularity, Neumann problem). *Under Assumption* 10.1, *let* $\mathcal{T}$ *be an admissible mesh in the sense of Definition* 10.3 *and* $h = \mathrm{size}(\mathcal{T})$. *Let* $u_\mathcal{T} \in X(\mathcal{T})$ (*see Definition* 9.2) *be the approximated solution defined in* Ω *by* $u_\mathcal{T}(x) = u_K$ *for a.e.* $x \in K$, *for all* $K \in \mathcal{T}$, *where* $(u_K)_{K\in\mathcal{T}}$ *is the (unique) solution to* (10.7) *and* (10.8) (*thanks to* (10.5) *and* (10.6), *existence and uniqueness of* $(u_K)_{K\in\mathcal{T}}$ *is given in Lemma* 10.1). *Assume that the unique solution, u, of* (10.3), (10.4) *belongs to* $H^2(\Omega)$. *Let* $C_\mathcal{T} \in \mathbb{R}$ *be such that*

$$\sum_{K\in\mathcal{T}} \overline{u}(x_K)\mathrm{m}(K) = 0 \quad \textit{where } \overline{u} = u + C_\mathcal{T}.$$

Let, for each control volume $K \in \mathcal{T}$, $e_K = \overline{u}(x_K) - u_K$, *and* $e_\mathcal{T} \in X(\mathcal{T})$ *defined by* $e_\mathcal{T}(x) = e_K$ *for a.e.* $x \in K$, *for all* $K \in \mathcal{T}$.

Then there exists C, only depending on u, ζ and Ω, such that (10.20) *and* (10.19) *hold.*

Note that, in Theorem 10.2, the function $e_{\mathcal{T}}$ is well defined, and the quantity "$\nabla u \cdot \boldsymbol{n}_\sigma$" is well defined on σ, for all $\sigma \in \mathcal{E}$ (see Remark 9.11).

PROOF. The proof is very similar to that of Theorem 9.4, from which the same notations are used.

There exists some C, depending only on the space dimension (d) and ζ (given in Definition 10.3), such that, for all $\sigma \in \mathcal{E}_{\text{int}}$,

$$|R_\sigma|^2 \leqslant C \frac{h^2}{\mathrm{m}(\sigma) d_\sigma} \int_{\mathcal{V}_\sigma} \left|H(u)(z)\right|^2 \mathrm{d}z, \tag{10.23}$$

and therefore

$$\sum_{\sigma \in \mathcal{E}_{\text{int}}} \mathrm{m}(\sigma) d_\sigma R_\sigma^2 \leqslant C h^2 \int_\Omega \left|H(u)(z)\right|^2 \mathrm{d}z. \tag{10.24}$$

The proof of (10.23) (from which (10.24) is an easy consequence) was already done in the proof of Theorem 9.4 (note that, here, there is no need to consider the case of $\sigma \in \mathcal{E}_{\text{ext}}$). In order to obtain estimate (10.20), one proceeds as in Theorem 9.4. Recall

$$|e_{\mathcal{T}}|_{1,\mathcal{T}}^2 \leqslant \sum_{\sigma \in \mathcal{E}_{\text{int}}} R_\sigma |D_\sigma e| \mathrm{m}(\sigma),$$

where $|D_\sigma e| = |e_K - e_L|$ if $\sigma \in \mathcal{E}_{\text{int}}$ is such that $\sigma = K|L$; hence, from the Cauchy–Schwarz inequality, one obtains that

$$|e_{\mathcal{T}}|_{1,\mathcal{T}}^2 \leqslant \left(\sum_{\sigma \in \mathcal{E}_{\text{int}}} R_\sigma^2 \mathrm{m}(\sigma) d_\sigma \right)^{1/2} \left(\sum_{\sigma \in \mathcal{E}_{\text{int}}} |D_\sigma e|^2 \frac{\mathrm{m}(\sigma)}{d_\sigma} \right)^{1/2}.$$

Then, one obtains, with (10.24),

$$|e_{\mathcal{T}}|_{1,\mathcal{T}} \leqslant \sqrt{C} h \left(\int_\Omega \left|H(u)(z)\right|^2 \mathrm{d}z \right)^{1/2}.$$

This concludes the proof of the first part of (10.20). The second part of (10.20) is a consequence of the discrete Poincaré inequality (10.10). Using (10.24) also easily leads (10.19).

Note also that, if u is Lipschitz continuous, inequality (10.18) follows from the second part of (10.20) and the definition of $\overline{u}$ as in Theorem 10.1.

This concludes the proof of Theorem 10.2. □

Some generalizations of Theorem 10.2 are possible, as for the Dirichlet case, see Remark 9.12.

10.4. Convergence

A convergence result, under Assumption 10.1, may be proved without any regularity assumption on the exact solution.

The proof of convergence uses the following preliminary inequality on the "trace" of an element of $X(\mathcal{T})$ on the boundary:

LEMMA 10.5 (Trace inequality). *Let Ω be an open bounded polygonal connected subset of $\mathbb{R}^d$, $d=2$ or 3 (indeed, the connexity of Ω is not used in this lemma). Let $\mathcal{T}$ be an admissible mesh, in the sense of Definition 10.1, and $u \in X(\mathcal{T})$ (see Definition 9.2). Let u_K be the value of u in the control volume K. Let $\overline{\gamma}(u)$ be defined by $\overline{\gamma}(u) = u_K$ a.e. (for the $(d-1)$-dimensional Lebesgue measure) on σ, if $\sigma \in \mathcal{E}_{\text{ext}}$ and $\sigma \in \mathcal{E}_K$. Then, there exists C, only depending on Ω, such that*

$$\left\|\overline{\gamma}(u)\right\|_{L^2(\partial\Omega)} \leqslant C\left(|u|_{1,\mathcal{T}} + \|u\|_{L^2(\Omega)}\right). \tag{10.25}$$

REMARK 10.1. The result stated in this lemma still holds if Ω is not assumed connected. Indeed, one needs only modify (in an obvious way) the definition of admissible meshes (Definition 10.1) so as to take into account nonconnected subsets.

PROOF OF LEMMA 10.5. By compactness of the boundary of $\partial\Omega$, there exists a finite number of open hyper-rectangles ($d=2$ or 3), $\{R_i,\ i=1,\dots,N\}$, and normalized vectors of $\mathbb{R}^d$, $\{\eta_i, i=1,\dots,N\}$, such that

$$\begin{cases} \partial\Omega \subset \bigcup_{i=1}^N R_i, \\ \eta_i \cdot \boldsymbol{n}(x) \geqslant \alpha > 0 \quad \text{for all } x \in R_i \cap \partial\Omega,\ i \in \{1,\dots,N\}, \\ \{x + t\eta_i, x \in R_i \cap \partial\Omega, t \in \mathbb{R}_+\} \cap R_i \subset \Omega, \end{cases}$$

where α is some positive number and $\boldsymbol{n}(x)$ is the normal vector to $\partial\Omega$ at x, inward to Ω. Let $\{\alpha_i, i=1,\dots,N\}$ be a family of functions such that $\sum_{i=1}^N \alpha_i(x) = 1$, for all $x \in \partial\Omega$, $\alpha_i \in C_c^\infty(\mathbb{R}^d, \mathbb{R}_+)$ and $\alpha_i = 0$ outside of R_i, for all $i=1,\dots,N$. Let $\Gamma_i = R_i \cap \partial\Omega$; let us prove that there exists C_i only depending on α and α_i such that

$$\left\|\alpha_i\overline{\gamma}(u)\right\|_{L^2(\Gamma_i)} \leqslant C_i\left(|u|_{1,\mathcal{T}} + \|u\|_{L^2(\Omega)}\right). \tag{10.26}$$

The existence of C, only depending on Ω, such that (10.25) holds, follows easily (taking $C = \sum_{i=1}^N C_i$, and using $\sum_{i=1}^N \alpha_i(x) = 1$, note that α and α_i depend only on Ω). It remains to prove (10.26).

Let us introduce some notations. For $\sigma \in \mathcal{E}$ and $K \in \mathcal{T}$, define χ_σ and χ_K from $\mathbb{R}^d \times \mathbb{R}^d$ to $\{0, 1\}$ by $\chi_\sigma(x,y) = 1$, if $[x,y] \cap \sigma \neq \emptyset$, $\chi_\sigma(x,y) = 0$, if $[x,y] \cap \sigma = \emptyset$, and $\chi_K(x,y) = 1$, if $[x,y] \cap K \neq \emptyset$, $\chi_K(x,y) = 0$, if $[x,y] \cap K = \emptyset$.

Let $i \in \{1,\dots,N\}$ and let $x \in \Gamma_i$. There exists a unique $t > 0$ such that $x + t\eta_i \in \partial R_i$, let $y(x) = x + t\eta_i$. For $\sigma \in \mathcal{E}$, let $z_\sigma(x) = [x, y(x)] \cap \sigma$ if $[x, y(x)] \cap \sigma \neq \emptyset$ and is reduced to one point. For $K \in \mathcal{T}$, let $\xi_K(x), \eta_K(x)$ be such that $[x, y(x)] \cap K = [\xi_K(x), \eta_K(x)]$ if $[x, y(x)] \cap K \neq \emptyset$.

One has, for a.e. (for the $(d-1)$-dimensional Lebesgue measure) $x \in \Gamma_i$,

$$\begin{aligned}\left|\alpha_i \overline{\gamma}(u)(x)\right| \leqslant & \sum_{\sigma=K|L\in\mathcal{E}_{\mathrm{int}}} \left|\alpha_i\big(z_\sigma(x)\big)(u_K - u_L)\right| \chi_\sigma\big(x, y(x)\big) \\ & + \sum_{K\in\mathcal{T}} \left|\alpha_i\big(\xi_K(x)\big) - \alpha_i\big(\eta_K(x)\big)u_K\right| \chi_K\big(x, y(x)\big),\end{aligned}$$

i.e.,

$$\left|\alpha_i \overline{\gamma}(u)(x)\right|^2 \leqslant A(x) + B(x) \tag{10.27}$$

with

$$\begin{aligned}A(x) &= 2\Bigg(\sum_{\sigma=K|L\in\mathcal{E}_{\mathrm{int}}} \left|\alpha_i\big(z_\sigma(x)\big)(u_K - u_L)\right| \chi_\sigma\big(x, y(x)\big)\Bigg)^2, \\ B(x) &= 2\Bigg(\sum_{K\in\mathcal{T}} \left|\big(\alpha_i\big(\xi_K(x)\big) - \alpha_i\big(\eta_K(x)\big)\big)u_K\right| \chi_K\big(x, y(x)\big)\Bigg)^2.\end{aligned}$$

A bound on $A(x)$ is obtained for a.e. $x \in \Gamma_i$, by remarking that, from the Cauchy–Schwarz inequality:

$$A(x) \leqslant D_1 \sum_{\sigma\in\mathcal{E}_{\mathrm{int}}} \frac{|D_\sigma u|^2}{d_\sigma c_\sigma} \chi_\sigma\big(x, y(x)\big) \sum_{\sigma\in\mathcal{E}_{\mathrm{int}}} d_\sigma c_\sigma \chi_\sigma\big(x, y(x)\big),$$

where D_1 only depends on α_i and $c_\sigma = |\eta_i \cdot \boldsymbol{n}_\sigma|$. (Recall that $D_\sigma u = |u_K - u_L|$.) Since

$$\sum_{\sigma\in\mathcal{E}_{\mathrm{int}}} d_\sigma c_\sigma \chi_\sigma\big(x, y(x)\big) \leqslant \mathrm{diam}(\Omega),$$

this yields:

$$A(x) \leqslant \mathrm{diam}(\Omega) D_1 \sum_{\sigma\in\mathcal{E}_{\mathrm{int}}} \frac{|D_\sigma u|^2}{d_\sigma c_\sigma} \chi_\sigma\big(x, y(x)\big).$$

Then, since

$$\int_{\Gamma_i} \chi_\sigma\big(x, y(x)\big)\, \mathrm{d}\gamma(x) \leqslant \frac{1}{\alpha} c_\sigma \mathrm{m}(\sigma),$$

there exists D_2, only depending on Ω, such that

$$A = \int_{\Gamma_i} A(x)\, \mathrm{d}\gamma(x) \leqslant D_2 |u|_{1,\mathcal{T}}^2.$$

A bound $B(x)$ for a.e. $x \in \Gamma_i$ is obtained with the Cauchy–Schwarz inequality:

$$B(x) \leqslant D_3 \sum_{K \in \mathcal{T}} u_K^2 \chi_K\big(x, y(x)\big)\big|\xi_K(x) - \eta_K(x)\big| \\ \times \sum_{K \in \mathcal{T}} \big|\xi_K(x) - \eta_K(x)\big|\chi_K\big(x, y(x)\big),$$

where D_3 only depends on α_i. Since

$$\sum_{K \in \mathcal{T}} \big|\xi_K(x) - \eta_K(x)\big|\chi_K\big(x, y(x)\big) \leqslant \operatorname{diam}(\Omega) \quad \text{and}$$

$$\int_{\Gamma_i} \chi_K\big(x, y(x)\big)\big|\xi_K(x) - \eta_K(x)\big| \,\mathrm{d}\gamma(x) \leqslant \frac{1}{\alpha}\mathrm{m}(K),$$

there exists D_4, only depending on Ω, such that

$$B = \int_{\Gamma_i} B(x)\,\mathrm{d}\gamma(x) \leqslant D_4 \|u\|^2_{L^2(\Omega)}.$$

Integrating (10.27) over Γ_i, the bounds on A and B lead (10.26) for some convenient C_i and it concludes the proof of Lemma 10.5. □

REMARK 10.2. Using this "trace inequality" (10.25) and the Kolmogorov theorem (see Theorem 14.1, it is possible to prove Lemma 10.2 (Discrete Poincaré inequality) by way of contradiction. Indeed, assume that there exists a sequence $(u_n)_{n\in\mathbb{N}}$ such that, for all $n \in \mathbb{N}$, $\|u_n\|_{L^2(\Omega)} = 1$, $\int_\Omega u_n(x)\,\mathrm{d}x = 0$, $u_n \in X(\mathcal{T}_n)$ (where $\mathcal{T}_n$ is an admissible mesh in the sense of Definition 10.1) and $|u_n|_{1,\mathcal{T}_n} \leqslant 1/n$. Using the trace inequality, one proves that $(u_n)_{n\in\mathbb{N}}$ is relatively compact in $L^2(\Omega)$, as in Theorem 10.3. Then, one can assume that $u_n \to u$, in $L^2(\Omega)$, as $n \to \infty$. The function u satisfies $\|u\|_{L^2(\Omega)} = 1$, since $\|u_n\|_{L^2(\Omega)} = 1$, and $\int_\Omega u(x)\,\mathrm{d}x = 0$, since $\int_\Omega u_n(x)\,\mathrm{d}x = 0$. Using $|u_n|_{1,\mathcal{T}_n} \leqslant 1/n$, a proof similar to that of Theorem 14.3, yields that $D_i u = 0$, for all $i \in \{1, \ldots, n\}$ (even if $\operatorname{size}(\mathcal{T}_n) \nrightarrow 0$, as $n \to \infty$), where $D_i u$ is the derivative in the distribution sense with respect to x_i of u. Since Ω is connected, one deduces that u is constant on Ω, but this is impossible since $\|u\|_{L^2(\Omega)} = 1$ and $\int_\Omega u(x)\,\mathrm{d}x = 0$.

Let us now prove that the scheme (10.7) and (10.8), where $(f_K)_{K\in\mathcal{T}}$ and $(g_K)_{K\in\mathcal{T}}$ are given by (10.5) and (10.6) is stable: the approximate solution given by the scheme is bounded independently of the mesh, as we proceed to show.

LEMMA 10.6 (Estimate for the Neumann problem). *Under Assumption* 10.1, *let* $\mathcal{T}$ *be an admissible mesh* (*in the sense of Definition* 10.1). *Let* $(u_K)_{K\in\mathcal{T}}$ *be the unique solution to* (10.7) *and* (10.8), *where* $(f_K)_{K\in\mathcal{T}}$ *and* $(g_K)_{K\in\mathcal{T}}$ *are given by* (10.5) *and* (10.6); *the existence and uniqueness of* $(u_K)_{K\in\mathcal{T}}$ *is given in Lemma* 10.1. *Let* $u_\mathcal{T} \in X(\mathcal{T})$ (*see*

Definition 9.2) *be defined by* $u_{\mathcal{T}}(x) = u_K$ *for a.e.* $x \in K$, *for all* $K \in \mathcal{T}$. *Then, there exists* $C \in \mathbb{R}_+$, *only depending on* Ω, g *and* f, *such that*

$$|u_{\mathcal{T}}|_{1,\mathcal{T}} \leqslant C, \tag{10.28}$$

where $|\bullet|_{1,\mathcal{T}}$ *is defined in Definition* 10.2.

PROOF. Multiplying (10.7) by u_K and summing over $K \in \mathcal{T}$ yields

$$\sum_{K|L \in \mathcal{E}_{\mathrm{int}}} \tau_{K|L}(u_L - u_K)^2 = \sum_{K \in \mathcal{T}} \mathrm{m}(K) f_K u_K + \sum_{\sigma \in \mathcal{E}_{\mathrm{ext}}} u_{K_\sigma} g_{K_\sigma} \mathrm{m}(\sigma), \tag{10.29}$$

where, for $\sigma \in \mathcal{E}_{\mathrm{ext}}$, $K_\sigma \in \mathcal{T}$ is such that $\sigma \in \mathcal{E}_{K_\sigma}$.

We get (10.28) from (10.29) using (10.25), (10.10) and the Cauchy–Schwarz inequality. □

Using the estimate (10.28) on the approximate solution, a convergence result is given in the following theorem.

THEOREM 10.3 (Convergence in the case of the Neumann problem). *Under Assumption* 10.1, *let* u *be the unique solution to* (10.3), (10.4). *For an admissible mesh (in the sense of Definition* 10.1) $\mathcal{T}$, *let* $(u_K)_{K \in \mathcal{T}}$ *be the unique solution to* (10.7) *and* (10.8) (*where* $(f_K)_{K \in \mathcal{T}}$ *and* $(g_K)_{K \in \mathcal{T}}$ *are given by* (10.5) *and* (10.6), *the existence and uniqueness of* $(u_K)_{K \in \mathcal{T}}$ *is given in Lemma* 10.1) *and define* $u_{\mathcal{T}} \in X(\mathcal{T})$ (*see Definition* 9.2) *by* $u_{\mathcal{T}}(x) = u_K$ *for a.e.* $x \in K$, *for all* $K \in \mathcal{T}$. *Then,*

$$u_{\mathcal{T}} \to u \quad \textit{in } L^2(\Omega) \textit{ as } \mathrm{size}(\mathcal{T}) \to 0,$$

$$|u_{\mathcal{T}}|_{1,\mathcal{T}}^2 \to \int_\Omega |\nabla u(x)|^2 \,\mathrm{d}x \quad \textit{as } \mathrm{size}(\mathcal{T}) \to 0$$

and

$$\overline{\gamma}(u_{\mathcal{T}}) \to \overline{\gamma}(u) \quad \textit{in } L^2(\Omega) \textit{ for the weak topology as } \mathrm{size}(\mathcal{T}) \to 0,$$

where the function $\overline{\gamma}(u)$ *stands for the trace of* u *on* $\partial\Omega$ *in the sense given in Lemma* 10.5 *when* $u \in X(\mathcal{T})$ *and in the sense of the classical trace operator from* $H^1(\Omega)$ *to* $L^2(\partial\Omega)$ (*or* $H^{1/2}(\partial\Omega)$) *when* $u \in H^1(\Omega)$.

PROOF.

Step 1 (*Compactness*). Denote by Y the set of approximate solutions $u_{\mathcal{T}}$ for all admissible meshes $\mathcal{T}$. Thanks to Lemma 10.6 and to the discrete Poincaré inequality (10.10), the set Y is bounded in $L^2(\Omega)$. Let us prove that Y is relatively compact in $L^2(\Omega)$, and that, if $(\mathcal{T}_n)_{n \in \mathbb{N}}$ is a sequence of admissible meshes such that $\mathrm{size}(\mathcal{T}_n)$ tends to 0 and $u_{\mathcal{T}_n}$ tends to u, in $L^2(\Omega)$, as n tends to infinity, then u belongs to $H^1(\Omega)$. Indeed, these results follow from Theorems 14.1 and 14.3, provided that there exists a real positive number C only depending on Ω, f and g such that

$$\left\|\tilde{u}_{\mathcal{T}}(\bullet+\eta)-\tilde{u}_{\mathcal{T}}\right\|_{L^2(\mathbb{R}^d)}^2 \leqslant C|\eta|, \quad \text{for any admissible mesh } \mathcal{T}$$
$$\text{and for any } \eta \in \mathbb{R}^d,\ |\eta| \leqslant 1, \tag{10.30}$$

and that, for any compact subset $\bar{\omega}$ of Ω,

$$\left\|u_{\mathcal{T}}(\bullet+\eta)-u_{\mathcal{T}}\right\|_{L^2(\bar{\omega})}^2 \leqslant C|\eta|\big(|\eta|+2\,\mathrm{size}(\mathcal{T})\big), \quad \text{for any admissible mesh } \mathcal{T}$$
$$\text{and for any } \eta \in \mathbb{R}^d \text{ such that } |\eta| < d\big(\bar{\omega}, \Omega^c\big). \tag{10.31}$$

Recall that $\tilde{u}_{\mathcal{T}}$ is defined by $\tilde{u}_{\mathcal{T}}(x)=u_{\mathcal{T}}(x)$ if $x \in \Omega$ and $\tilde{u}_{\mathcal{T}}(x)=0$ otherwise. In order to prove (10.30) and (10.31), define χ_σ from $\mathbb{R}^d \times \mathbb{R}^d$ to $\{0,1\}$ by $\chi_\sigma(x,y)=1$ if $[x,y]\cap\sigma \neq \emptyset$ and $\chi_\sigma(x,y)=0$ if $[x,y]\cap\sigma=\emptyset$. Let $\eta \in \mathbb{R}^d \setminus \{0\}$. Then:

$$\begin{aligned}\big|\tilde{u}(x+\eta)-\tilde{u}(x)\big| \leqslant{} & \sum_{\sigma\in\mathcal{E}_{\mathrm{int}}} \chi_\sigma(x,x+\eta)|D_\sigma u| \\ & + \sum_{\sigma\in\mathcal{E}_{\mathrm{ext}}} \chi_\sigma(x,x+\eta)|u_\sigma|, \quad \text{for a.e. } x \in \Omega,\end{aligned} \tag{10.32}$$

where, for $\sigma \in \mathcal{E}_{\mathrm{ext}}$, $u_\sigma = u_K$, and K is the control volume such that $\sigma \in \mathcal{E}_K$. Recall also that $D_\sigma u = |u_K - u_L|$, if $\sigma = K|L$. Let us first prove inequality (10.31). Let $\bar{\omega}$ be a compact subset of Ω. If $x \in \bar{\omega}$ and $|\eta| < d(\bar{\omega}, \Omega^c)$, the second term of the right-hand side of (10.32) is 0, and the same proof as in Lemma 9.3 gives, from an integration over $\bar{\omega}$ instead of Ω and from (9.33) with $C=2$ since $[x, x+\eta] \subset \Omega$ for $x \in \bar{\omega}$,

$$\left\|u_{\mathcal{T}}(\bullet+\eta)-u_{\mathcal{T}}\right\|_{L^2(\bar{\omega})}^2 \leqslant |u|_{1,\mathcal{T}}^2|\eta|\big(|\eta|+2\,\mathrm{size}(\mathcal{T})\big). \tag{10.33}$$

In order to prove (10.30), remark that the number of nonzero terms in the second term of the right-hand side of (10.32) is, for a.e. $x \in \Omega$, bounded by some real positive number, which only depends on Ω, which can be taken, e.g., as the number of sides of Ω, denoted by N. Hence, with $C_1=(N+1)^2$ (which only depends on Ω. Indeed, if Ω is convex, $N=2$ is also convenient), one has

$$\begin{aligned}\big|\tilde{u}(x+\eta)-\tilde{u}(x)\big|^2 \leqslant{} & C_1\Big(\sum_{\sigma\in\mathcal{E}_{\mathrm{int}}} \chi_\sigma(x,x+\eta)|D_\sigma u|\Big)^2 \\ & + C_1 \sum_{\sigma\in\mathcal{E}_{\mathrm{ext}}} \chi_\sigma(x,x+\eta)u_\sigma^2, \quad \text{for a.e. } x \in \Omega.\end{aligned} \tag{10.34}$$

Let us integrate this inequality over $\mathbb{R}^d$. As seen in the proof of Lemma 9.3,

$$\int_{\mathbb{R}^d}\Big(\sum_{\sigma\in\mathcal{E}_{\text{int}}}\chi_\sigma(x,x+\eta)|D_\sigma u|\Big)^2\,\mathrm{d}x \leqslant |u|_{1,\mathcal{T}}^2|\eta|\big(|\eta|+2(N-1)\operatorname{size}(\mathcal{T})\big);$$

hence, by Lemma 10.6, there exists a real positive number C_2, only depending on Ω, f and g, such that (if $|\eta|\leqslant 1$)

$$\int_{\mathbb{R}^d}\Big(\sum_{\sigma\in\mathcal{E}_{\text{int}}}\chi_\sigma(x,x+\eta)|D_\sigma u|\Big)^2\,\mathrm{d}x \leqslant C_2|\eta|.$$

Let us now turn to the second term of the right-hand side of (10.34) integrated over $\mathbb{R}^d$;

$$\int_{\mathbb{R}^d}\Big(\sum_{\sigma\in\mathcal{E}_{\text{ext}}}\chi_\sigma(x,x+\eta)u_\sigma^2\Big)\,\mathrm{d}x \leqslant \sum_{\sigma\in\mathcal{E}_{\text{ext}}}\mathrm{m}(\sigma)|\eta|u_\sigma^2 \leqslant \big\|\overline{\gamma}(u_\mathcal{T})\big\|_{L^2(\partial\Omega)}^2|\eta|;$$

therefore, thanks to Lemma 10.5, Lemma 10.6 and to the discrete Poincaré inequality (10.10), there exists a real positive number C_3, only depending on Ω, f and g, such that

$$\int_{\mathbb{R}^d}\Big(\sum_{\sigma\in\mathcal{E}_{\text{ext}}}\chi_\sigma(x,x+\eta)u_\sigma^2\Big)\,\mathrm{d}x \leqslant C_3|\eta|.$$

Hence (10.30) is proved for some real positive number C only depending on Ω, f and g.

Step 2 (*Passage to the limit*). In this step, the convergence of $u_\mathcal{T}$ to the solution of (10.3), (10.4) (in $L^2(\Omega)$ as $\operatorname{size}(\mathcal{T})\to 0$) is first proved.

Since the solution to (10.3), (10.4) is unique, and thanks to the compactness of the set Y described in Step 1, it is sufficient to prove that, if $u_{\mathcal{T}_n}\to u$ in $L^2(\Omega)$ and $\operatorname{size}(\mathcal{T}_n)\to 0$ as $n\to 0$, then u is a solution to (10.3)–(10.4).

Let $(\mathcal{T}_n)_{n\in\mathbb{N}}$ be a sequence of admissible meshes and $(u_{\mathcal{T}_n})_{n\in\mathbb{N}}$ be the corresponding solutions to (10.7)–(10.8) with $\mathcal{T}=\mathcal{T}_n$. Assume $u_{\mathcal{T}_n}\to u$ in $L^2(\Omega)$ and $\operatorname{size}(\mathcal{T}_n)\to 0$ as $n\to 0$. By Step 1, one has $u\in H^1(\Omega)$ and since the mean value of $u_{\mathcal{T}_n}$ is zero, one also has $\int_\Omega u(x)\,\mathrm{d}x=0$. Therefore, u is a solution of (10.3). It remains to show that u satisfies (10.4). Since $(\overline{\gamma}(u_{\mathcal{T}_n}))_{n\in\mathbb{N}}$ is bounded in $L^2(\partial\Omega)$, one may assume (up to a subsequence) that it converges to some v weakly in $L^2(\partial\Omega)$. Let us first prove that

$$\begin{aligned}&-\int_\Omega u(x)\Delta\varphi(x)\,\mathrm{d}x+\int_{\partial\Omega}\nabla\varphi(x)\cdot\boldsymbol{n}(x)v(x)\,\mathrm{d}\gamma(x)\\&\quad=\int_\Omega f(x)\varphi(x)\,\mathrm{d}x+\int_{\partial\Omega}g(x)\varphi(x)\,\mathrm{d}\gamma(x),\quad\forall\varphi\in C^2(\overline{\Omega}),\end{aligned}\tag{10.35}$$

and then that u satisfies (10.4).

Let $\mathcal{T}$ be an admissible mesh, $u_{\mathcal{T}}$ the corresponding approximate solution to the Neumann problem, given by (10.7) and (10.8), where $(f_K)_{K\in\mathcal{T}}$ and $(g_K)_{K\in\mathcal{T}}$ are given by (10.5) and (10.6) and let $\varphi \in C^2(\overline{\Omega})$. Let $\varphi_K = \varphi(x_K)$, define $\varphi_{\mathcal{T}}$ by $\varphi_{\mathcal{T}}(x) = \varphi_K$, for a.e. $x \in K$ and for any control volume K, and $\overline{\gamma}(\varphi_{\mathcal{T}})(x) = \varphi_K$ for a.e. $x \in \sigma$ (for the $(d-1)$-dimensional Lebesgue measure), for any $\sigma \in \mathcal{E}_{\text{ext}}$ and control volume K such that $\sigma \in \mathcal{E}_K$.

Multiplying (10.7) by φ_K, summing over $K \in \mathcal{T}$ and reordering the terms yields

$$\sum_{K\in\mathcal{T}} u_K \sum_{L\in\mathcal{N}(K)} \tau_{K|L}(\varphi_L - \varphi_K)$$
$$= \int_\Omega f(x)\varphi_{\mathcal{T}}(x)\,\mathrm{d}x + \int_{\partial\Omega} \overline{\gamma}(\varphi_{\mathcal{T}})(x) g(x)\,\mathrm{d}\gamma(x). \tag{10.36}$$

Using the consistency of the fluxes and the fact that $\varphi \in C^2(\overline{\Omega})$, there exists C only depending on φ such that

$$\sum_{L\in\mathcal{N}(K)} \tau_{K|L}(\varphi_L - \varphi_K)$$
$$= \int_K \Delta\varphi(x)\,\mathrm{d}x - \int_{\partial\Omega\cap\partial K} \nabla\varphi(x)\cdot \boldsymbol{n}(x)\,\mathrm{d}\gamma(x) + \sum_{L\in\mathcal{N}(K)} R_{K,L}(\varphi),$$

with $R_{K,L} = -R_{L,K}$, for all $L \in \mathcal{N}(K)$ and $K \in \mathcal{T}$, and $|R_{K,L}| \leqslant C_4 \mathrm{m}(K|L)\mathrm{size}(\mathcal{T})$, where C_4 only depends on φ. Hence (10.36) may be rewritten as

$$-\int_\Omega u_{\mathcal{T}}(x)\Delta\varphi(x)\,\mathrm{d}x + \int_{\partial\Omega} \nabla\varphi(x)\cdot\boldsymbol{n}(x)\overline{\gamma}(u_{\mathcal{T}})(x)\,\mathrm{d}\gamma(x) + r(\varphi,\mathcal{T})$$
$$= \int_\Omega f(x)\varphi_{\mathcal{T}}(x)\,\mathrm{d}x + \int_{\partial\Omega} \overline{\gamma}(\varphi_{\mathcal{T}})(x) g(x)\,\mathrm{d}\gamma(x), \tag{10.37}$$

where

$$\begin{aligned}\bigl|r(\varphi,\mathcal{T})\bigr| &= C_4 \sum_{\sigma\in\mathcal{E}_{\text{int}}} |D_\sigma u| \mathrm{m}(\sigma)\mathrm{size}(\mathcal{T}) \\ &\leqslant C_4 \Bigl(\sum_{\sigma\in\mathcal{E}_{\text{int}}} |D_\sigma u|^2 \frac{\mathrm{m}(\sigma)}{d_\sigma}\Bigr)^{1/2} \Bigl(\sum_{\sigma\in\mathcal{E}_{\text{int}}} \mathrm{m}(\sigma) d_\sigma\Bigr)^{1/2} \mathrm{size}(\mathcal{T}) \\ &\leqslant C_5 \mathrm{size}(\mathcal{T}),\end{aligned}$$

where C_5 is a real positive number only depending on f, g, Ω and φ (thanks to Lemma 10.6).

Writing (10.37) with $\mathcal{T} = \mathcal{T}_n$ and passing to the limit as n tends to infinity yields (10.35).

Let us now prove that u satisfies (10.4). Since $u \in H^1(\Omega)$, an integration by parts in (10.35) yields

$$\int_\Omega \nabla u(x) \cdot \nabla \varphi(x)\,\mathrm{d}x + \int_{\partial\Omega} \nabla\varphi(x) \cdot \boldsymbol{n}(x)\big(v(x) - \overline{\gamma}(u)(x)\big)\,\mathrm{d}\gamma(x)$$
$$= \int_\Omega f(x)\varphi(x)\,\mathrm{d}x + \int_{\partial\Omega} g(x)\varphi(x)\,\mathrm{d}\gamma(x), \quad \forall\varphi \in C^2(\overline{\Omega}), \tag{10.38}$$

where $\overline{\gamma}(u)$ denotes the trace of u on $\partial\Omega$ (which belongs to $L^2(\partial\Omega)$). In order to prove that u is solution to (10.4) (this will conclude the proof of Theorem 10.3), it is sufficient, thanks to the density of $C^2(\overline{\Omega})$ in $H^1(\Omega)$, to prove that $v = \overline{\gamma}(u)$ a.e. on $\partial\Omega$ (for the $(d-1)$-dimensional Lebesgue measure on $\partial\Omega$). Let us now prove that $v = \overline{\gamma}(u)$ a.e. on $\partial\Omega$ by first remarking that (10.38) yields

$$\int_\Omega \nabla u(x) \cdot \nabla\varphi(x)\,\mathrm{d}x = \int_\Omega f(x)\varphi(x)\,\mathrm{d}x, \quad \forall\varphi \in C_c^\infty(\Omega),$$

and therefore, by density of $C_c^\infty(\Omega)$ in $H_0^1(\Omega)$,

$$\int_\Omega \nabla u(x) \cdot \nabla\varphi(x)\,\mathrm{d}x = \int_\Omega f(x)\varphi(x)\,\mathrm{d}x, \quad \forall\varphi \in H_0^1(\Omega).$$

With (10.38), this yields

$$-\int_{\partial\Omega} \nabla\varphi(x) \cdot \boldsymbol{n}(x)\big(v(x) - \overline{\gamma}(u)(x)\big)\mathrm{d}\gamma(x) = 0,$$
$$\forall\varphi \in C^2(\overline{\Omega}) \text{ such that } \varphi = 0 \text{ on } \partial\Omega. \tag{10.39}$$

There remains to show that the wide choice of φ in (10.39) allows to conclude $v = \overline{\gamma}(u)$ a.e. on $\partial\Omega$ (for the $(d-1)$-dimensional Lebesgue measure of $\partial\Omega$). Indeed, let I be a part of the boundary $\partial\Omega$, such that I is included in a hyperplane of $\mathbb{R}^d$. Assume that $I = \{0\} \times J$, where J is an open ball of $\mathbb{R}^{d-1}$ centered on the origin. Let $z = (a, \tilde{z}) \in \mathbb{R}^d$ with $a \in \mathbb{R}_+^\star$, $\tilde{z} \in \mathbb{R}^{d-1}$ and $B = \{(t, \frac{a-|t|}{a}y + \frac{|t|}{a}\tilde{z});\, t \in (-a, a),\, y \in J\}$; assume that, for a convenient a, one has

$$B \cap \Omega = \left\{\left(t, \frac{a-|t|}{a}y + \frac{|t|}{a}\tilde{z}\right);\ t \in (0, a),\ y \in J\right\}.$$

Let $\psi \in C_c^\infty(J)$, and for $x = (x_1, y) \in \mathbb{R} \times J$, define $\varphi_1(x) = -x_1\psi(y)$. Then,

$$\varphi_1 \in C^\infty(\mathbb{R}^d) \quad \text{and} \quad \frac{\partial\varphi_1}{\partial n} = \psi \quad \text{on } I.$$

(Recall that $\boldsymbol{n}$ is the normal unit vector to $\partial\Omega$, outward to Ω.) Let $\varphi_2 \in C_c^\infty(B)$ such that $\varphi_2 = 1$ on a neighborhood of $\{0\} \times \{\psi \neq 0\}$, where $\{\psi \neq 0\} = \{x \in J;\ \psi(x) \neq 0\}$, and set $\varphi = \varphi_1\varphi_2$; φ is an admissible test function in (10.39), and therefore

$$\int_J \psi(y)\big(\overline{\gamma}(u)(0, y) - v(0, y)\big)\,\mathrm{d}y = 0,$$

which yields, since ψ is arbitrary in $C_c^\infty(J)$, $v = \overline{\gamma}(u)$ a.e. on I. Since J is arbitrary, this implies that $v = \overline{\gamma}(u)$ a.e. on $\partial\Omega$.

This conclude the proof of $u_{\mathcal{T}} \to u$ in $L^2(\Omega)$ as $\mathrm{size}(\mathcal{T}) \to 0$, where u is the solution to (10.3), (10.4).

Note also that the above proof gives (by way of contradiction) that $\overline{\gamma}(u_{\mathcal{T}}) \to \overline{\gamma}(u)$ weakly in $L^2(\partial\Omega)$, as $\mathrm{size}(\mathcal{T}) \to 0$.

Then, a passage to the limit in (10.29) together with (10.4) yields

$$|u_{\mathcal{T}}|_{1,\mathcal{T}}^2 \to \big\||\nabla u|\big\|_{L^2(\Omega)}^2, \qquad \text{as } \mathrm{size}(\mathcal{T}) \to 0.$$

This concludes the proof of Theorem 10.3. □

Note that, with some discrete Sobolev inequality (similar to (9.70)), the hypothesis "$f \in L^2(\Omega)$, $g \in L^2(\partial\Omega)$" may be relaxed in some way similar to that of item (2) of Remark 9.6.

11. General elliptic operators

11.1. Discontinuous matrix diffusion coefficients

Meshes and schemes

Let Ω be an open bounded polygonal subset of $\mathbb{R}^d$, $d = 2$ or 3. We are interested here in the discretization of an elliptic operator with discontinuous matrix diffusion coefficients, which may appear in real case problems such as electrical or thermal transfer problems or, more generally, diffusion problems in heterogeneous media. In this case, the mesh is adapted to fit the discontinuities of the data. Hence the definition of an admissible mesh given in Definition 9.1 must be adapted. As an illustration, let us consider here the following problem, which was studied in Section 7 in the one-dimensional case:

$$-\mathrm{div}(\Lambda\nabla u)(x) + \mathrm{div}(\boldsymbol{v}u)(x) + bu(x) = f(x), \quad x \in \Omega, \tag{11.1}$$

$$u(x) = g(x), \quad x \in \partial\Omega, \tag{11.2}$$

with the following assumptions on the data (one denotes by $\mathbb{R}^{d\times d}$ the set of $d \times d$ matrices with real coefficients):

ASSUMPTION 11.1. (1) Λ is a bounded measurable function from Ω to $\mathbb{R}^{d\times d}$ such that for any $x \in \Omega$, $\Lambda(x)$ is symmetric, and that there exists $\underline{\lambda}$ and $\overline{\lambda} \in \mathbb{R}_+^\star$ such that $\underline{\lambda}\xi \cdot \xi \leqslant \Lambda(x)\xi \cdot \xi \leqslant \overline{\lambda}\xi \cdot \xi$ for any $x \in \Omega$ and any $\xi \in \mathbb{R}^d$.

(2) $\boldsymbol{v} \in C^1(\overline{\Omega}, \mathbb{R}^d)$, $\operatorname{div} \boldsymbol{v} \geqslant 0$ on Ω, $b \in \mathbb{R}_+$.

(3) f is a bounded piecewise continuous function from Ω to $\mathbb{R}$.

(4) g is such that there exists $\tilde{g} \in H^1(\Omega)$ such that $\overline{\gamma}(\tilde{g}) = g$ (a.e. on $\partial\Omega$) and is a bounded piecewise continuous function from $\partial\Omega$ to $\mathbb{R}$.

(Recall that $\overline{\gamma}$ denotes the trace operator from $H^1(\Omega)$ into $L^2(\partial\Omega)$.) As in Section 9, under Assumption 11.1, there exists a unique variational solution $u \in H^1(\Omega)$ of problem (11.1), (11.2). This solution satisfies $u = w + \tilde{g}$, where $\tilde{g} \in H^1(\Omega)$ is such that $\overline{\gamma}(\tilde{g}) = g$, a.e. on $\partial\Omega$, and w is the unique function of $H_0^1(\Omega)$ satisfying

$$\begin{aligned}
&\int_\Omega \big(\Lambda(x)\nabla w(x) \cdot \nabla\psi(x) + \operatorname{div}(\boldsymbol{v}w)(x)\psi(x) + bw(x)\psi(x)\big)\,\mathrm{d}x \\
&\quad = \int_\Omega \big(-\Lambda(x)\nabla\tilde{g}(x) \cdot \nabla\psi(x) - \operatorname{div}(\boldsymbol{v}\tilde{g})(x)\psi(x) \\
&\qquad - b\tilde{g}(x)\psi(x) + f(x)\psi(x)\big)\,\mathrm{d}x, \quad \forall\psi \in H_0^1(\Omega).
\end{aligned}$$

Let us now define an admissible mesh for the discretization of problem (11.1)–(11.2).

DEFINITION 11.1 (*Admissible mesh for a general diffusion operator*). Let Ω be an open bounded polygonal subset of $\mathbb{R}^d$, $d = 2$ or 3. An admissible finite volume mesh for the discretization of problem (11.1)–(11.2) is an admissible mesh $\mathcal{T}$ of Ω in the sense of Definition 9.1 where items (iv) and (v) are replaced by the two following conditions:

(iv′) The set $\mathcal{T}$ is such that the restriction of g to each edge $\sigma \in \mathcal{E}_{\text{ext}}$ is continuous. For any $K \in \mathcal{T}$, let Λ_K denote the mean value of Λ on K, i.e.

$$\Lambda_K = \frac{1}{\mathrm{m}(K)} \int_K \Lambda(x)\,\mathrm{d}x.$$

There exists a family of points

$$\mathcal{P} = (x_K)_{K\in\mathcal{T}} \quad \text{such that } x_K = \bigcap_{\sigma\in\mathcal{E}_K} \mathcal{D}_{K,\sigma} \in \overline{K},$$

where $\mathcal{D}_{K,\sigma}$ is a straight line perpendicular to σ with respect to the scalar product induced by Λ_K^{-1} such that $\mathcal{D}_{K,\sigma} \cap \sigma = \mathcal{D}_{L,\sigma} \cap \sigma \neq \emptyset$ if $\sigma = K|L$. Furthermore, if $\sigma = K|L$, let $y_\sigma = \mathcal{D}_{K,\sigma} \cap \sigma$ $(= \mathcal{D}_{L,\sigma} \cap \sigma)$ and assume that $x_K \neq x_L$.

(v′) For any $\sigma \in \mathcal{E}_{\text{ext}}$, let K be the control volume such that $\sigma \in \mathcal{E}_K$ and let $\mathcal{D}_{K,\sigma}$ be the straight line going through x_K and orthogonal to σ with respect to the scalar product induced by Λ_K^{-1}; then, there exists $y_\sigma \in \sigma \cap \mathcal{D}_{K,\sigma}$; let $g_\sigma = g(y_\sigma)$.

In the sequel, the following notations are used. $\operatorname{size}(\mathcal{T}) = \sup\{\operatorname{diam}(K),\ K \in \mathcal{T}\}$. For any $K \in \mathcal{T}$ and $\sigma \in \mathcal{E}$, $\mathrm{m}(K)$ is the d-dimensional Lebesgue measure of K (i.e. area if $d = 2$, volume if $d = 3$) and $\mathrm{m}(\sigma)$ the $(d-1)$-dimensional Lebesgue measure of σ. The set of interior (resp. boundary) edges is denoted by $\mathcal{E}_{\text{int}}$ (resp. $\mathcal{E}_{\text{ext}}$), that is $\mathcal{E}_{\text{int}} = \{\sigma \in \mathcal{E};\ \sigma \not\subset \partial\Omega\}$ (resp. $\mathcal{E}_{\text{ext}} = \{\sigma \in \mathcal{E};\ \sigma \subset \partial\Omega\}$). The set of neighbours of K is denoted by

$\mathcal{N}(K)$, that is $\mathcal{N}(K) = \{L \in \mathcal{T};\ \exists\sigma \in \mathcal{E}_K,\ \sigma = K|L\}$. If $\sigma = K|L$, we denote by $d_{K,\sigma}$ the Euclidean distance between x_K and $y_\sigma = \mathcal{D}_{K,\sigma} \cap \sigma$, and $d_{K|L} = d_{K,\sigma} + d_{L,\sigma}$; if $\sigma = \mathcal{E}_{\text{ext}} \cap \mathcal{E}_K$, d_σ denotes the Euclidean distance between x_K and $y_\sigma = \mathcal{D}_{K,\sigma} \cap \sigma$. For any $\sigma \in \mathcal{E}$, the transmissibility through σ is defined by $\tau_\sigma = \mathrm{m}(\sigma)/d_\sigma$ (if $d_\sigma \neq 0$).

We shall now define the discrete unknowns of the numerical scheme, with the same notations as in Section 9.2. As in the case of the Dirichlet problem, the primary unknowns $(u_K)_{K\in\mathcal{T}}$ will be used, which aim to be approximations of the values $u(x_K)$, and some auxiliary unknowns, namely the fluxes $F_{K,\sigma}$, for all $K \in \mathcal{T}$ and $\sigma \in \mathcal{E}_K$, and some (expected) approximation of u in σ, say u_σ, for all $\sigma \in \mathcal{E}$. Again, these auxiliary unknowns are helpful to write the scheme, but they can be eliminated locally so that the discrete equations will only be written with respect to the primary unknowns $(u_K)_{K\in\mathcal{T}}$. For any $\sigma \in \mathcal{E}_{\text{ext}}$, set $u_\sigma = g(y_\sigma)$. The finite volume scheme for the numerical approximation of the solution to problem (11.1)–(11.2) is obtained by integrating Eq. (11.1) over each control volume K, and approximating the fluxes over each edge σ of K. This yields

$$\sum_{\sigma\in\mathcal{E}_K} F_{K,\sigma} + \sum_{\sigma\in\mathcal{E}_K} v_{K,\sigma} u_{\sigma,+} + \mathrm{m}(K) b u_K = f_K, \quad \forall K \in \mathcal{T}, \tag{11.3}$$

where $v_{K,\sigma} = \int_\sigma \boldsymbol{v}(x) \cdot \boldsymbol{n}_{K,\sigma}\, \mathrm{d}\gamma(x)$ (where $\boldsymbol{n}_{K,\sigma}$ denotes the normal unit vector to σ outward to K); if $\sigma = K_{\sigma,+}|K_{\sigma,-}$, $u_{\sigma,+} = u_{K_{\sigma,+}}$, where $K_{\sigma,+}$ is the upstream control volume, i.e. $v_{K,\sigma} \geqslant 0$, with $K = K_{\sigma,+}$; if $\sigma \in \mathcal{E}_{\text{ext}}$, then $u_{\sigma,+} = u_K$ if $v_{K,\sigma} \geqslant 0$ (i.e. K is upstream to σ with respect to v), and $u_{\sigma,+} = u_\sigma$ otherwise.

$F_{K,\sigma}$ is an approximation of $\int_\sigma -\Lambda_K \nabla u(x) \cdot \boldsymbol{n}_{K,\sigma}\, \mathrm{d}\gamma(x)$; the approximation $F_{K,\sigma}$ is written with respect to the discrete unknowns $(u_K)_{K\in\mathcal{T}}$ and $(u_\sigma)_{\sigma\in\mathcal{E}}$. For $K \in \mathcal{T}$ and $\sigma \in \mathcal{E}_K$, let $\lambda_{K,\sigma} = |\Lambda_K \boldsymbol{n}_{K,\sigma}|$ (recall that $|\cdot|$ denote the Euclidean norm).

- If $x_K \notin \sigma$, a natural expression for $F_{K,\sigma}$ is then

$$F_{K,\sigma} = -\mathrm{m}(\sigma)\lambda_{K,\sigma} \frac{u_\sigma - u_K}{d_{K,\sigma}}.$$

Writing the conservativity of the scheme, i.e. $F_{L,\sigma} = -F_{K,\sigma}$ if $\sigma = K|L \subset \Omega$, yields the value of u_σ, if $x_L \notin \sigma$, with respect to $(u_K)_{K\in\mathcal{T}}$;

$$u_\sigma = \frac{1}{\frac{\lambda_{K,\sigma}}{d_{K,\sigma}} + \frac{\lambda_{L,\sigma}}{d_{L,\sigma}}} \left(\frac{\lambda_{K,\sigma}}{d_{K,\sigma}} u_K + \frac{\lambda_{L,\sigma}}{d_{L,\sigma}} u_L \right).$$

Note that this expression is similar to that of (7.3) in the 1D case.

- If $x_K \in \sigma$, one sets $u_\sigma = u_K$.

Hence the value of $F_{K,\sigma}$;

- internal edges:

$$F_{K,\sigma} = -\tau_\sigma (u_L - u_K), \quad \text{if } \sigma \in \mathcal{E}_{\text{int}},\ \sigma = K|L, \tag{11.4}$$

where

$$\tau_\sigma = \mathrm{m}(\sigma)\frac{\lambda_{K,\sigma}\lambda_{L,\sigma}}{\lambda_{K,\sigma}d_{L,\sigma}+\lambda_{L,\sigma}d_{K,\sigma}} \quad \text{if } y_\sigma \neq x_K \text{ and } y_\sigma \neq x_L$$

and

$$\tau_\sigma = \mathrm{m}(\sigma)\frac{\lambda_{K,\sigma}}{d_{K,\sigma}} \quad \text{if } y_\sigma \neq x_K \text{ and } y_\sigma = x_L;$$

- boundary edges:

$$F_{K,\sigma} = -\tau_\sigma(g_\sigma - u_K), \quad \text{if } \sigma \in \mathcal{E}_{\text{ext}} \text{ and } x_K \notin \sigma, \tag{11.5}$$

where

$$\tau_\sigma = \mathrm{m}(\sigma)\frac{\lambda_{K,\sigma}}{d_{K,\sigma}};$$

if $x_K \in \sigma$, then the equation associated to u_K is $u_K = g_\sigma$ (instead of that given by (11.3)) and the numerical flux $F_{K,\sigma}$ is an unknown which may be deduced from (11.3).

REMARK 11.1. Note that if $\Lambda = \mathrm{Id}$, then the scheme (11.3)–(11.5) is the same scheme than the one described in Section 9.2.

Error estimate

THEOREM 11.1. *Let Ω be an open bounded polygonal subset of $\mathbb{R}^d$, $d = 2$ or 3. Under Assumption* 11.1, *let u be the unique variational solution to problem* (11.1)–(11.2). *Let $\mathcal{T}$ be an admissible mesh for the discretization of problem* (11.1)–(11.2), *in the sense of Definition* 11.1. *Let ζ_1 and $\zeta_2 \in \mathbb{R}_+$ such that*

$$\zeta_1\big(\mathrm{size}(\mathcal{T})\big)^2 \leqslant \mathrm{m}(K) \leqslant \zeta_2\big(\mathrm{size}(\mathcal{T})\big)^2,$$
$$\zeta_1\mathrm{size}(\mathcal{T}) \leqslant \mathrm{m}(\sigma) \leqslant \zeta_2\mathrm{size}(\mathcal{T}),$$
$$\zeta_1\mathrm{size}(\mathcal{T}) \leqslant d_\sigma \leqslant \zeta_2\mathrm{size}(\mathcal{T}).$$

Assuming moreover that

- *the restriction of f to K belongs to $C(\overline{K})$, for any $K \in \mathcal{T}$;*
- *the restriction of Λ to K belongs to $C^1(\overline{K}, \mathbb{R}^{d\times d})$, for any $K \in \mathcal{T}$;*
- *the restriction of u (unique variational solution of problem* (11.1)–(11.2)) *to K belongs to $C^2(\overline{K})$, for any $K \in \mathcal{T}$.*

(Recall that $C^m(\overline{K}, \mathbb{R}^N) = \{v_{|K},\ v \in C^m(\mathbb{R}^d, \mathbb{R}^N)\}$ and $C^m(\bullet) = C^m(\bullet, \mathbb{R})$.)

Then, there exists a unique family $(u_K)_{K\in\mathcal{T}}$ satisfying (11.3)–(11.5); *furthermore, denoting by $e_K = u(x_K) - u_K$, there exists $C \in \mathbb{R}_+$ only depending on ζ_1, ζ_2, $\gamma = \sup_{K\in\mathcal{T}}(\|D^2u\|_{L^\infty(K)})$ and $\delta = \sup_{K\in\mathcal{T}}(\|D\Lambda\|_{L^\infty(K)})$ such that*

$$\sum_{\sigma\in\mathcal{E}} \frac{(D_\sigma e)^2}{d_\sigma}\mathrm{m}(\sigma) \leqslant C\big(\mathrm{size}(\mathcal{T})\big)^2 \tag{11.6}$$

and

$$\sum_{K\in\mathcal{T}} e_K^2 \mathrm{m}(K) \leqslant C\big(\mathrm{size}(\mathcal{T})\big)^2. \tag{11.7}$$

Recall that $D_\sigma e = |e_L - e_K|$ for $\sigma \in \mathcal{E}_{\mathrm{int}}$, $\sigma = K|L$ and $D_\sigma e = |e_K|$ for $\sigma \in \mathcal{E}_{\mathrm{ext}} \cap \mathcal{E}_K$.

PROOF. First, one may use Taylor expansions and the same technique as in the 1D case (see Step 2 of the proof of Theorem 7.1) to show that the expressions (11.4) and (11.5) are consistent approximations of the exact diffusion flux $\int_\sigma -\Lambda(x)\nabla u(x)\cdot \boldsymbol{n}_{K,\sigma}\,\mathrm{d}\gamma(x)$, i.e. there exists C_1 only depending on u and Λ such that, for all $\sigma \in \mathcal{E}$, with $F^\star_{K,\sigma} = \tau_\sigma(u(x_L) - u(x_K))$, if $\sigma = K|L$, and $F^\star_{K,\sigma} = \tau_\sigma(u(y_\sigma) - u(x_K))$, if $\sigma \in \mathcal{E}_{\mathrm{ext}} \cap \mathcal{E}_K$,

$$F^\star_{K,\sigma} - \int_\sigma -\Lambda(x)\nabla u(x)\cdot \boldsymbol{n}_{K,\sigma}\,\mathrm{d}\gamma(x) = R_{K,\sigma},$$
$$\text{with } |R_{K,\sigma}| \leqslant C_1 \mathrm{size}(\mathcal{T})\mathrm{m}(\sigma).$$

There also exists C_2 only depending on u and $\boldsymbol{v}$ such that, for all $\sigma \in \mathcal{E}$,

$$v_{K,\sigma}u(x_{K_{\sigma,+}}) - \int_\sigma \boldsymbol{v}\cdot\boldsymbol{n}_{K,\sigma}u = r_{K,\sigma},$$
$$\text{with } |r_{K,\sigma}| \leqslant C_2 \mathrm{size}(\mathcal{T})\mathrm{m}(\sigma).$$

Let us then integrate Eq. (11.1) over each control volume, subtract to (11.3) and use the consistency of the fluxes to obtain the following equation on the error:

$$\begin{aligned} &-\sum_{\sigma\in\mathcal{E}_K} G_{K,\sigma} + \sum_{\sigma\in\mathcal{E}_K} v_{K,\sigma}e_{\sigma,+} + \mathrm{m}(K)be_K \\ &\quad = \sum_{\sigma\in\mathcal{E}_K}(R_{K,\sigma} + r_{K,\sigma}) + S_K, \quad \forall K \in \mathcal{T}, \end{aligned}$$

where $G_{K,\sigma} = \tau_\sigma(e_L - e_K)$, if $\sigma = K|L$, and $G_{K,\sigma} = \tau_\sigma(-e_K)$, if $\sigma \in \mathcal{E}_{\mathrm{ext}} \cap \mathcal{E}_K$, $e_{\sigma,+} = e_{K_{\sigma,+}}$ is the error associated to the upstream control volume to σ and $S_K = b(\mathrm{m}(K)u(x_K) - \int_K u(x)\,\mathrm{d}x)$ is such that $|S_K| \leqslant \mathrm{m}(K)C_3h$, where $C_3 \in \mathbb{R}_+$ only depends on u and b. Then, similarly to the proof of Theorem 9.3, let us multiply by the e_K, sum over $K \in \mathcal{T}$, and use the conservativity of the scheme, which yields that if $\sigma = K|L$

then $R_{K,\sigma} = -R_{L,\sigma}$. A reordering of the summation over $\sigma \in \mathcal{E}$ yields the "discrete H_0^1 estimate" (11.6). Then, following HERBIN [1995], one shows the following discrete Poincaré inequality:

$$\sum_{K \in \mathcal{T}} e_K^2 \mathrm{m}(K) \leqslant C_4 \sum_{\sigma \in \mathcal{E}} \frac{(D_\sigma e)^2}{d_\sigma} \mathrm{m}(\sigma), \tag{11.8}$$

where C_4 only depends on Ω, ζ_1 and ζ_2, which in turn yields the L^2 estimate (11.7). □

REMARK 11.2. In the case where Λ is constant, or more generally, in the case where $\Lambda(x) = \lambda(x)\Lambda$, where Λ is a constant symmetric positive definite matrix, the proof of Lemma 9.1 is easily extended. However, for a general matrix Λ, the generalization of this proof is not so clear; this is the reason of the dependency of the estimates (11.6) and (11.7) on ζ_1 and ζ_2, which arises when proving (11.8) as in HERBIN [1995].

11.2. Other boundary conditions

The finite volume scheme may be used to discretize elliptic problems with Dirichlet or Neumann boundary conditions, as we saw in the previous sections. It is also easily implemented in the case of Fourier (or Robin) and periodic boundary conditions. The case of interface conditions between two geometrical regions is also generally easy to implement; the purpose here is to present the treatment of some of these boundary and interface conditions. One may also refer to ANGOT [1989] and references therein, FIARD and HERBIN [1994] for the treatment of more complex boundary conditions and coupling terms in a system of elliptic equations.

Let Ω be (for the sake of simplicity) the open rectangular subset of $\mathbb{R}^2$ defined by $\Omega = (0,1) \times (0,2)$, let $\Omega_1 = (0,1) \times (0,1)$, $\Omega_2 = (0,1) \times (1,2)$, $\Gamma_1 = [0,1] \times \{0\}$, $\Gamma_2 = \{1\} \times [0,2]$, $\Gamma_3 = [0,1] \times \{2\}$, $\Gamma_4 = \{0\} \times [0,2]$ and $I = [0,1] \times \{1\}$. Let λ_1 and $\lambda_2 > 0$, $f \in C(\overline{\Omega})$, $\alpha > 0$, $\bar{u} \in \mathbb{R}$, $g \in C(\Gamma_4)$, θ and $\Phi \in C(I)$. Consider here the following problem (with some "natural" notations):

$$-\mathrm{div}(\lambda_i \nabla u)(x) = f(x), \quad x \in \Omega_i,\ i = 1, 2, \tag{11.9}$$

$$-\lambda_i \nabla u(x) \cdot \boldsymbol{n}(x) = \alpha\big(u(x) - \bar{u}\big), \quad x \in \Gamma_1 \cup \Gamma_3, \tag{11.10}$$

$$\nabla u(x) \cdot \boldsymbol{n}(x) = 0, \quad x \in \Gamma_2, \tag{11.11}$$

$$u(x) = g(x), \quad x \in \Gamma_4, \tag{11.12}$$

$$\big(\lambda_2 \nabla u(x) \cdot \boldsymbol{n}_I(x)\big)_{|2} = \big(\lambda_1 \nabla u(x) \cdot \boldsymbol{n}_I(x)\big)_{|1} + \theta(x), \quad x \in I, \tag{11.13}$$

$$u_{|2}(x) - u_{|1}(x) = \Phi(x), \quad x \in I, \tag{11.14}$$

where $\boldsymbol{n}$ denotes the unit normal vector to $\partial\Omega$ outward to Ω and $\boldsymbol{n}_I = (0,1)^t$ (it is a unit normal vector to I).

Let $\mathcal{T}$ be an admissible mesh for the discretization of (11.9)–(11.14) in the sense of Definition 11.1. For the sake of simplicity, let us assume here that $d_{K,\sigma} > 0$ for all $K \in \mathcal{T}, \sigma \in \mathcal{E}_K$. Integrating Eq. (11.9) over each control volume K, and approximating the fluxes over each edge σ of K yields the following finite volume scheme:

$$\sum_{\sigma\in\mathcal{E}_K} F_{K,\sigma} = f_K, \quad \forall K \in \mathcal{T}, \tag{11.15}$$

where $F_{K,\sigma}$ is an approximation of $\int_\sigma -\lambda_i \nabla u(x)\boldsymbol{\cdot}\boldsymbol{n}_{K,\sigma}\,\mathrm{d}\gamma(x)$, with i such that $K \subset \Omega_i$.

Let $N_{\mathcal{T}} = \mathrm{card}(\mathcal{T})$, $N_{\mathcal{E}} = \mathrm{card}(\mathcal{E})$, $N_{\mathcal{E}}^0 = \mathrm{card}(\{\sigma \in \mathcal{E}; \sigma \not\subset \partial\Omega \cup I\})$, $N_{\mathcal{E}}^i = \mathrm{card}(\{\sigma \in \mathcal{E}; \sigma \subset \Gamma_i\})$, and $N_{\mathcal{E}}^I = \mathrm{card}(\{\sigma \in \mathcal{E}; \sigma \subset I\})$ (note that $N_{\mathcal{E}} = N_{\mathcal{E}}^0 + \sum_{i=1}^4 N_{\mathcal{E}}^i + N_{\mathcal{E}}^I$). Introduce the $N_{\mathcal{T}}$ (primary) discrete unknowns $(u_K)_{K\in\mathcal{T}}$; note that the number of (auxiliary) unknowns of the type $F_{K,\sigma}$ is $2(N_{\mathcal{E}}^0 + N_{\mathcal{E}}^I) + \sum_{i=1}^4 N_{\mathcal{E}}^i$; let us introduce the discrete unknowns $(u_\sigma)_{\sigma\in\mathcal{E}}$, which aim to be approximations of u on σ. In order to take into account the jump condition (11.14), two unknowns of this type are necessary on the edges $\sigma \subset I$, namely $u_{\sigma,1}$ and $u_{\sigma,2}$. Hence the number of (auxiliary) unknowns of the type u_σ is $N_{\mathcal{E}}^0 + \sum_{i=1}^4 N_{\mathcal{E}}^i + 2N_{\mathcal{E}}^I$. Therefore, the total number of discrete unknowns is

$$N_{\mathrm{tot}} = N_{\mathcal{T}} + 3N_{\mathcal{E}}^0 + 4N_{\mathcal{E}}^I + 2\sum_{i=1}^4 N_{\mathcal{E}}^i.$$

Hence, it is convenient, in order to obtain a well-posed system, to write N_{tot} discrete equations. We already have $N_{\mathcal{T}}$ equations from (11.15). The expression of $F_{K,\sigma}$ with respect to the unknowns u_K and u_σ is

$$F_{K,\sigma} = -\mathrm{m}(\sigma)\lambda_i \frac{u_\sigma - u_K}{d_{K,\sigma}}, \quad \forall K \in \mathcal{T};\ K \subset \Omega_i\ (i = 1,2),\ \forall\sigma \in \mathcal{E}_K; \tag{11.16}$$

which yields $2(N_{\mathcal{E}}^0 + N_{\mathcal{E}}^I) + \sum_{i=1}^4 N_{\mathcal{E}}^i$. (In (11.16), u_σ stands for $u_{\sigma,i}$ if $\sigma \subset I$.)

Let us now take into account the various boundary and interface conditions:

- *Fourier boundary conditions.* Discretizing condition (11.10) yields

$$F_{K,\sigma} = \alpha\mathrm{m}(\sigma)(u_\sigma - \bar{u}), \quad \forall K \in \mathcal{T},\ \forall\sigma \in \mathcal{E}_K;\ \sigma \subset \Gamma_1 \cup \Gamma_3, \tag{11.17}$$

  that is $N_{\mathcal{E}}^1 + N_{\mathcal{E}}^3$ equations.
- *Neumann boundary conditions.* Discretizing condition (11.11) yields

$$F_{K,\sigma} = 0, \quad \forall K \in \mathcal{T},\ \forall\sigma \in \mathcal{E}_K;\ \sigma \subset \Gamma_2, \tag{11.18}$$

  that is $N_{\mathcal{E}}^2$ equations.

- *Dirichlet boundary conditions.* Discretizing condition (11.12) yields

$$u_\sigma = g(y_\sigma), \quad \forall \sigma \in \mathcal{E};\ \sigma \subset \Gamma_4, \tag{11.19}$$

that is $N_{\mathcal{E}}^4$ equations.
- *Conservativity of the flux.* Except at interface I, the flux is continuous, and therefore

$$F_{K,\sigma} = -F_{L,\sigma}, \quad \forall \sigma \in \mathcal{E};\ \sigma \not\subset \left(\bigcup_{i=1}^{4} \Gamma_i \cup I\right) \text{ and } \sigma = K|L, \tag{11.20}$$

that is $N_{\mathcal{E}}^0$ equations.
- *Jump condition on the flux.* At interface I, condition (11.13) is discretized into

$$F_{K,\sigma} + F_{L,\sigma} = \int_\sigma \theta(x)\,\mathrm{d}s,$$
$$\forall \sigma \in \mathcal{E};\ \sigma \subset I \text{ and } \sigma = K|L;\ K \subset \Omega_2, \tag{11.21}$$

that is $N_{\mathcal{E}}^I$ equations.
- *Jump condition on the unknown.* At interface I, condition (11.14) is discretized into

$$u_{\sigma,2} = u_{\sigma,1} + \Phi(y_\sigma), \quad \forall \sigma \in \mathcal{E};\ \sigma \subset I \text{ and } \sigma = K|L. \tag{11.22}$$

that is another $N_{\mathcal{E}}^I$ equations.

Hence the total number of equations from (11.15) to (11.22) is N_{tot}, so that the numerical scheme can be expected to be well posed.

The finite volume scheme for the discretization of equations (11.9)–(11.14) is therefore completely–defined by (11.15)–(11.22). Particular cases of this scheme are the schemes (9.20)–(9.23) (written for Dirichlet boundary conditions) and (10.7)–(10.8) (written for Neumann boundary conditions and no convection term) which were thoroughly studied in the two previous sections.

12. Dual meshes and unknowns located at vertices

One of the principles of the classical finite volume method is to associate the discrete unknowns to the grid cells. However, it is sometimes useful to associate the discrete unknowns with the vertices of the mesh; e.g., the finite volume method may be used for the discretization of a hyperbolic equation coupled with an elliptic equation (see Chapter VII). Suppose that an existing finite element code is implemented for the elliptic equation and yields the discrete values of the unknown at the vertices of the mesh. One might then want to implement a finite volume method for the hyperbolic equation with the values of the unknowns at the vertices of the mesh. Note also that for some physical problems, e.g., the modelling of two phase flow in porous media, the conservativity principle is easier to respect if the discrete unknowns have the same location. For these various reasons, we introduce here some finite volume methods where the discrete unknowns are located at the vertices of an existing mesh.

For the sake of simplicity, the treatment of the boundary conditions will be omitted here. Recall that the construction of a finite volume method is carried out (in particular) along the following principles:

(1) Divide the spatial domain in control volumes.

(2) Associate to each control volume and, for time dependent problems, to each discrete time, one discrete unknown.

(3) Obtain the discrete equations (at each discrete time) by integration of the equation over the control volume and the definition of one exchange term between two (adjacent) control volumes.

Recall, in particular, that the definition of one (and one only) exchange term between two control volumes is important; this is called the property of conservativity of a finite volume method. The aim here is to present finite volume methods for which the discrete unknowns are located at the vertices of the mesh. Hence, to each vertex must correspond a control volume. Note that these control volumes may be somehow "fictive" (see the next section); the important issue is to respect the principles given above in the construction of the finite volume scheme. In the three following sections, we shall deal with the two-dimensional case; the generalization to the three-dimensional case is the purpose of Section 12.4.

12.1. The piecewise linear finite element method viewed as a finite volume method

We consider here the Dirichlet problem. Let Ω be a bounded open polygonal subset of $\mathbb{R}^2$, f and g be some "regular" functions (from Ω or $\partial\Omega$ to $\mathbb{R}$). Consider the following problem:

$$\begin{cases} -\Delta u(x) = f(x), & x \in \Omega, \\ u(x) = g(x), & x \in \partial\Omega. \end{cases} \tag{12.1}$$

Let us show that the "piecewise linear" finite element method for the discretization of (12.1) may be viewed as a kind of finite volume method. Let $\mathcal{M}$ be a finite element mesh of Ω, consisting of triangles (see, e.g., CIARLET [1978] for the conditions on the triangles), and let $\mathcal{V} \subset \overline{\Omega}$ be the set of vertices of $\mathcal{M}$. For $K \in \mathcal{V}$ (note that here K denotes a point of $\overline{\Omega}$), let φ_K be the shape function associated to K in the piecewise linear finite element method for the mesh $\mathcal{M}$. We remark that

$$\sum_{K\in\mathcal{V}} \varphi_K(x) = 1, \quad \forall x \in \Omega,$$

and therefore

$$\sum_{K\in\mathcal{V}} \int_\Omega \varphi_K(x)\,\mathrm{d}x = \mathrm{m}(\Omega) \tag{12.2}$$

that the whole spatial domain is the "disjoint union" of the control volumes), a possibility is to construct a "dual mesh" which will be denoted by $\mathcal{T}$. In order for this mesh to be admissible in the sense of Definition 9.1, a simple way is to use the Voronoï mesh defined with $\mathcal{V}$ (see Example 9.2). In order to write the "classical" finite volume scheme with this mesh (see (9.20)–(9.23)), a slight modification is necessary at the boundary for some particular $\mathcal{M}$ (see Example 9.2); this method is denoted CFV/DM (classical finite volume on dual mesh); it is conservative, the numerical fluxes are consistent, and the transmissibilities are nonnegative. Hence, the convergence results and error estimates which were studied in previous sections hold (see, in particular, Theorems 9.1 and 9.3).

A case of particular interest is found when the primal mesh (that is $\mathcal{M}$) consists in triangles with acute angles. One uses, as dual mesh, the Voronoï mesh defined with $\mathcal{V}$. Then, the dual mesh is admissible in the sense of Definition 9.1 and is constructed with the orthogonal bisectors of the edges of the elements of $\mathcal{M}$, parts of these orthogonal bisectors (and parts of $\partial\Omega$) give the boundaries to the control volumes of the dual mesh. In this case, the CFV/DM scheme is "close" to the piecewise linear finite element scheme on the primal mesh. Let us elaborate on this point.

For $K \in \mathcal{V}$, let K also denote the control volume (of the dual mesh) associated to K (in the sequel, the sense of "K", which denotes vertex or control volume, will not lead any confusion) and let φ_K be the shape function associated to the vertex K (in the piecewise linear finite element associated to $\mathcal{M}$). The term $\tau_{K|L}$ (ratio between the length of the edge $K|L$ and the distance between vertices), which is used in the finite volume scheme, verifies

$$\tau_{K|L} = -\int_\Omega \nabla\varphi_K(x) \cdot \nabla\varphi_L(x)\,\mathrm{d}x.$$

The CFV/DM scheme (finite volume scheme on the dual mesh) writes

$$-\sum_{L\in\mathcal{N}(K)} \tau_{K|L}(u_L - u_K) = \int_K f(x)\,\mathrm{d}x, \quad \text{if } K \in \mathcal{V}\cap\Omega,$$

$$u_K = g(K), \quad \text{if } K \in \mathcal{V}\cap\partial\Omega,$$

where K stands for an element of $\mathcal{V}$ or for the control volume (of the dual mesh) associated to this point.

The finite element scheme (on the primal mesh) writes

$$-\sum_{L\in\mathcal{N}(K)} \tau_{K|L}(u_L - u_K) = \int_\Omega f(x)\varphi_K(x)\,\mathrm{d}x, \quad \text{if } K \in \mathcal{V}\cap\Omega,$$

$$u_K = g(K), \quad \text{if } K \in \mathcal{V}\cap\partial\Omega.$$

Therefore, the only difference between the finite element and finite volume schemes is in the definition of the right-hand sides. Note that these right-hand sides may be quite different. Consider, e.g., a node K which is the vertex of four identical triangles featuring an angle of $\pi/2$ at the vertex K, and denote by a the area of each of these triangles. Then, for $f \equiv 1$, the right-hand side computed for the discrete equation associated to the node K is equal to a in the case of the finite element (piecewise linear finite element) scheme, and equal to $2a$ for the dual mesh finite volume (CFV/DM) scheme. Both schemes may be shown to converge, by using finite volume techniques for the CFV/DM scheme (see previous sections), and finite element techniques for the piecewise linear finite element (see, e.g., CIARLET [1978]).

Let us now weaken the hypothesis that all angles of the triangles of the primal mesh $\mathcal{M}$ are acute to the so called Delaunay condition and the additional assumption that an angle of an element of $\mathcal{M}$ is less or equal $\pi/2$ if its opposite edge lies on $\partial\Omega$ (see, e.g., VANSELOW [1996]). Under this new assumption the schemes (piecewise linear FE and CFV/DM with the Voronoï mesh defined with $\mathcal{V}$) still lead to the same transmissibilities and still differ in the definition of the right-hand sides.

Recall that the Delaunay condition states that no neighboring element (of $\mathcal{M}$) is included in the circumscribed circle of an arbitrary element of $\mathcal{M}$. This is equivalent to saying that the sum of two opposite angles to an edge is less or equal π. The dual mesh is still admissible in the sense of Definition 9.1 and is still constructed with the orthogonal bisectors of the edges of the elements of $\mathcal{M}$, parts of these orthogonal bisectors (and parts of $\partial\Omega$) give the boundaries to the control volumes of the dual mesh. This is not the case when $\mathcal{M}$ does not satisfy the Delaunay condition.

Consider now a primal mesh, $\mathcal{M}$, consisting of triangles, but which does not satisfy the Delaunay condition and let the dual mesh be the Voronoï mesh defined with $\mathcal{V}$. Then, the two schemes, piecewise linear FE and CFV/DM are quite different. If the Delaunay condition does not hold say between the angles $\widehat{KAL}$ and $\widehat{KBL}$ (the triplets (K, A, L) and (K, B, L) defining two elements of $\mathcal{M}$), the sum of these two angles is greater than π and the transmissibility $\tau_{K|L} = -\int_\Omega \nabla\varphi_K(x) \cdot \nabla\varphi_L(x)\,\mathrm{d}x$ between the two control volumes associated respectively to K and L becomes negative with the piecewise linear FE scheme; there is no transmissibility between A and B (since A and B do not belong to a common element of $\mathcal{M}$). Hence the maximum principle is no longer respected for the FE scheme, while it remains valid for the CFV/DM finite volume scheme. This is due to the fact that the CFV/DM scheme allows an exchange term between A and B, with a positive transmissibility (and leads to no exchange term between K and L), while the FE scheme does not. Also note that the common edge to the control volumes (of the dual mesh) associated to A and B is not a part of an orthogonal bisector of an edge of an element of $\mathcal{M}$ (it is a part of the orthogonal bisector of the segment $[A, B]$).

To conclude this section, note that an admissible mesh for the classical finite volume is generally not a dual mesh of a primal triangular mesh consisting of triangles (e.g., the general triangular meshes which are considered in HERBIN [1995] are not dual meshes to triangular meshes).

12.3. *"Finite volume finite element" methods*

The "finite volume finite element" method for elliptic problems also uses a dual mesh $\mathcal{T}$ constructed from a finite element primal mesh, such that each cell of $\mathcal{T}$ is associated with a vertex of the primal mesh $\mathcal{M}$. Let $\mathcal{V}$ again denote the set of vertices of $\mathcal{M}$. As in the classical finite volume method, the conservation law is integrated over each cell of the (dual) mesh. Indeed, this integration is performed only if the cell is associated to a vertex (of the primal mesh) belonging to Ω.

Let us consider problem (12.1). Integrating the conservation law over K_P, where $P \in \mathcal{V} \cap \Omega$ and K_P is the control volume (of the dual mesh) associated to P yields

$$-\int_{\partial K_P} \nabla u(x) \cdot \boldsymbol{n}_P(x)\,\mathrm{d}\gamma(x) = \int_{K_P} f(x)\,\mathrm{d}x.$$

(Recall that $\boldsymbol{n}_P$ is the unit normal vector to ∂K_P outward to K_P.) Now, following the idea of finite element methods, the function u is approximated by a Galerkin expansion $\sum_{M\in\mathcal{V}} u_M \varphi_M$, where the functions φ_M are the shape functions of the piecewise linear finite element method. Hence, the discrete unknowns are $\{u_P,\ P \in \mathcal{V}\}$ and the scheme writes

$$-\sum_{M\in\mathcal{V}} \left(\int_{\partial K_P} \nabla \varphi_M(x) \cdot \boldsymbol{n}_P(x)\,\mathrm{d}\gamma(x)\right) u_M = \int_{K_P} f(x)\,\mathrm{d}x, \quad \forall P \in \mathcal{V}\cap\Omega, \tag{12.4}$$

$$u_P = g(P), \quad \forall P \in \mathcal{V} \cap \partial\Omega.$$

Equations (12.4) may also be written under the conservative form

$$\sum_{Q\in\mathcal{V}} E_{P,Q} = \int_{K_P} f(x)\,\mathrm{d}x, \quad \forall P \in \mathcal{V}\cap\Omega, \tag{12.5}$$

$$u_P = g(P), \quad \forall P \in \mathcal{V}\cap\partial\Omega, \tag{12.6}$$

where

$$E_{P,Q} = -\sum_{M\in\mathcal{V}} \int_{\partial K_P \cap \partial K_Q} \nabla\varphi_M(x) \cdot \boldsymbol{n}_P(x)\,\mathrm{d}\gamma(x). \tag{12.7}$$

Note that $E_{Q,P} = -E_{P,Q}$. Unfortunately, the exchange term $E_{P,Q}$ between P and Q is not, in general, a function of the only unknowns u_P and u_Q (this property was used, in the previous sections, to obtain convergence results of finite volume schemes). Another way to write (12.4) is, thanks to (12.3),

$$-\sum_{Q\in\mathcal{V}} \left(\int_{\partial K_P} \nabla\varphi_Q(x) \cdot \boldsymbol{n}_P(x)\,\mathrm{d}\gamma(x)\right)(u_Q - u_P) = \int_{K_P} f(x)\,\mathrm{d}x, \quad \forall P \in \mathcal{V}\cap\Omega.$$

Hence a new exchange term from P to Q might be

$$\bar{E}_{P,Q} = -\left(\int_{\partial K_P} \nabla\varphi_Q(x) \cdot \boldsymbol{n}_P(x)\,\mathrm{d}\gamma(x)\right)(u_Q - u_P)$$

and the scheme is therefore conservative if $\bar{E}_{P,Q} = -\bar{E}_{Q,P}$. Unfortunately, this is not the case for a general dual mesh.

There are several ways of constructing a dual mesh from a primal mesh. A common way (see, e.g., FEZOUI, LANTERI, LARROUTUROU and OLIVIER [1989]) is to take a primal mesh ($\mathcal{M}$) consisting of triangles and to construct the dual mesh with the medians (of the triangles of $\mathcal{M}$), joining the centers of gravity of the triangles to the midpoints of the edges of the primal mesh. The main interest of this way is that the resulting scheme (called FVFE/M below, Finite Volume Finite Element with Medians) is very close to the piecewise linear finite element scheme associated to $\mathcal{M}$. Indeed the FVFE/M scheme is defined by (12.5)–(12.7) while the piecewise linear finite element scheme writes

$$\sum_{Q\in\mathcal{V}} E_{P,Q} = \int_{\Omega} f(x)\varphi_P(x)\,\mathrm{d}x, \quad \forall P \in \mathcal{V}\cap\Omega,$$

$$u_P = g(P), \quad \forall P \in \mathcal{V}\cap\partial\Omega,$$

where $E_{P,Q}$ is defined by (12.7).

These two schemes only differ by the right-hand sides and, in fact, these right-hand sides are "close" since

$$\mathrm{m}(K_P) = \int_{\Omega} \varphi_P(x)\,\mathrm{d}x, \quad \forall P \in \mathcal{V}.$$

This is due to the fact that $\int_T \varphi_P(x)\,\mathrm{d}x = \mathrm{m}(T)/3$ and $\mathrm{m}(K_P \cap T) = \mathrm{m}(T)/3$, for all $T \in \mathcal{M}$ and all vertex P of T.

Thus, convergence properties of the FVFE/M scheme can be proved by using the finite element techniques. Recall however that the piecewise linear finite element scheme (and the FVFE/M scheme) does not satisfy the (discrete) maximum principle if $\mathcal{M}$ does not satisfy the Delaunay condition.

There are other means to construct a dual mesh starting from a primal triangular mesh. One of them is the Voronoï mesh associated to the vertices of the primal mesh, another possibility is to join the centers of gravity; in the latter case, the control volume associated to a vertex, say S, of the primal mesh is then limited by the lines joining the centers of gravity of the neighboring triangles of which S is a vertex (with some convenient modification for the vertices which are on the boundary of Ω). See also BARTH [1994] for descriptions of dual meshes.

Note that the proof of convergence which we designed for finite volume with admissible meshes does not generalize to any "FVFE" (Finite Volume Finite Element) method for several reasons. In particular, since the exchange term between P and Q (denoted by $E_{P,Q}$) is not, in general, a function of the only unknowns u_P and u_Q (and even if it is the transmissibilities may become negative) and also since, as in the case of the finite element method, the concept of consistency of the fluxes is not clear with the FVFE schemes.

12.4. Generalization to the three-dimensional case

The methods described in the three above sections generalize to the three-dimensional case, in particular when the primal mesh is a tetrahedral mesh. With such a mesh, the Delaunay condition no longer ensures the nonnegativity of the transmissibilities in the case of the piecewise linear finite element method. It is however possible to construct a dual mesh (the "three-dimensional Voronoï" mesh) to a Delaunay triangulation such that the FVFE scheme leads to positive transmissibilities, and therefore such that the maximum principle holds, see CORDES and PUTTI [1998].

Note that the theoretical results (convergence and error estimate) which were shown for the classical finite volume method on an admissible mesh (Sections 9.2 and 10) still hold for CFV/DM in three-dimensional, since the dual mesh is admissible.

13. Mesh refinement and singularities

Some problems involve singular source terms. In the case of petroleum engineering, e.g., one may model (in two space dimensions) the well with a Dirac measure. Other problems may require a better precision of some unknown in certain areas. This section is devoted to the treatment of this kind of problem, either with an adequate treatment of the singularity or by mesh refinement.

13.1. Singular source terms and finite volumes

It is possible to take into account, in the discretization with the finite volume method, the singularities of the solution of an elliptic problem. A common example is the study of wells in petroleum engineering. As a model example we can consider the following problem, which appears, e.g., in the study of a two phase flow in a porous medium. Let B be the ball of $\mathbb{R}^2$ of center 0 and radius r_p (B represents a well of radius r_p). Let $\Omega = (-R, R)^2$ be the whole domain of simulation; r_p is of the order of 10 cm while R can be of the order of 1 km for instance. An approximation to the solution of the following problem is sought:

$$\begin{aligned} &-\operatorname{div}(\nabla u)(x) = 0, \quad x \in \Omega \setminus B, \\ &u(x) = P_p, \quad x \in \partial B, \\ &\text{“BC”} \quad \text{on } \partial\Omega, \end{aligned} \tag{13.1}$$

where "BC" stands for some "smooth" boundary conditions on $\partial\Omega$ (e.g., Dirichlet or Neumann condition). This system is a mathematical model (under convenient assumptions ...) of the two phase flow problem, with u representing the pressure of the fluid and P_p an imposed pressure at the well. In order to discretize (13.1) with the finite volume method, a mesh $\mathcal{T}$ of Ω is introduced. For the sake of simplicity, the elements of $\mathcal{T}$ are assumed to be squares of length h (the method is easily generalized to other meshes). It is assumed that the well, represented by B, is located in the middle of one cell, denoted by K_0, so that the origin 0 is the center of K_0. It is also assumed that the mesh size, h, is large with respect to the radius of the well, r_p (which is the case in real applications, where, e.g., h ranges between 10 and 100 m). Following the principle of the finite volume method, one discrete unknown u_K per cell K ($K \in \mathcal{T}$) is introduced in order to discretize the following system:

$$\begin{aligned}&\int_{\partial K} \nabla u(x) \cdot \boldsymbol{n}_K(x)\,\mathrm{d}\gamma(x) = 0, \quad K \in \mathcal{T},\ K \neq K_0,\\ &\int_{\partial K_0} \nabla u(x) \cdot \boldsymbol{n}_{K_0}(x)\,\mathrm{d}\gamma(x) = \int_{\partial B} \nabla u(x) \cdot \boldsymbol{n}_B(x)\,\mathrm{d}\gamma(x),\end{aligned} \tag{13.2}$$

where $\boldsymbol{n}_P$ denotes the normal to ∂P, outward to P (with $P = K$, K_0 or B).

Hence, we have to discretize $\nabla u \cdot \boldsymbol{n}_K$ on ∂K (and $\nabla u \cdot \boldsymbol{n}_B$ on ∂B) in terms of $\{u_L, L \in \mathcal{T}\}$ (and "BC" and P_p).

The problems arise in the discretization of $\nabla u \cdot \boldsymbol{n}_{K_0}$ and $\nabla u \cdot n_B$. Indeed, if $\sigma = K|L$ is the common edge to K and L (elements of $\mathcal{T}$), with $K \neq K_0$ and $L \neq K_0$, since the solution of (13.1) is "smooth" enough with respect to the mesh size, except "near" the well, $\nabla u \cdot n_K$ can be discretized by $\frac{1}{h}(u_L - u_K)$ on σ.

In order to discretize ∇u near the well, it is assumed that $\nabla u \cdot n_B$ is constant on ∂B. Let $q(x) = -2\pi r_p \nabla u \cdot n_B$ for $x \in \partial B$ (recall that $\boldsymbol{n}_B$ is the normal to ∂B, outward to B). Then $q \in \mathbb{R}$ is a new unknown, which satisfies

$$\int_{\partial B} -\nabla u \cdot n_B \,\mathrm{d}\gamma(x) = q.$$

Denoting by $|\cdot|$ the Euclidean norm in $\mathbb{R}^2$, and u the solution to (13.1), let v be defined by

$$v(x) = \frac{q}{2\pi} \ln\big(|x|\big) + u(x), \quad x \in \Omega \setminus B, \tag{13.3}$$

$$v(x) = \frac{q}{2\pi} \ln(r_p) + P_p, \quad x \in B. \tag{13.4}$$

Thanks to the boundary conditions satisfied by u on ∂B, the function v satisfies $-\mathrm{div}(\nabla v) = 0$ on the whole domain Ω, and therefore v is regular on the whole domain Ω. Note that, if we set

$$u(x) = -\frac{q}{2\pi} \ln\big(|x|\big) + v(x), \quad \text{a.e. } x \in \Omega,$$

then

$$-\mathrm{div}(\nabla u) = q\delta_0 \quad \text{on } \Omega,$$

where δ_0 is the Dirac mass at 0. A discretization of $\nabla u \cdot n_{K_0}$ is now obtained in the following way. Let σ be the common edge to $K_1 \in \mathcal{T}$ and K_0, since v is smooth, it is possible to approximate $\nabla v \cdot \boldsymbol{n}_{K_0}$ on σ by $\frac{1}{h}(v_{K_1} - v_{K_0})$, where v_{K_i} is some approximation of v in K_i (e.g., the value of v at the center of K_i). Then, by (13.4), it is natural to set

$$v_{K_0} = \frac{q}{2\pi} \ln(r_p) + P_p,$$

and by (13.3),

$$v_{K_1} = \frac{q}{2\pi} \ln(h) + u_{K_1}.$$

By (13.3) and from the fact that the integral over σ of $\nabla(\frac{q}{2\pi} \ln(|x|)) \cdot \boldsymbol{n}_{K_0}$ is equal to $q/4$, we find the following approximation for $\int_\sigma \nabla u \cdot n_{K_0}\, d\gamma$:

$$-\frac{q}{4} + \frac{q}{2\pi} \ln\left(\frac{h}{r_p}\right) + u_{K_1} - P_p.$$

The discretization is now complete, there are as many equations as unknowns. The discrete unknowns appearing in the discretized problem are $\{u_K, K \in \mathcal{T}, K \neq K_0\}$ and q. Note that, up to now, the unknown u_{K_0} has not been used. The discrete equations are given by (13.2) where each term of (13.2) is replaced by its approximation in terms of $\{u_K, K \in \mathcal{T}, K \neq K_0\}$ and q. In particular, the discrete equation "associated" to the unknown q is the discretization of the second equation of (13.2), which is

$$\sum_{i=1}^{4} \left(\frac{q}{2\pi} \ln\left(\frac{h}{r_p}\right) + u_{K_i} - P_p \right) = 0, \tag{13.5}$$

where $\{K_i,\ i = 1, 2, 3, 4\}$ are the four neighbouring cells to K_0.

It is possible to replace the unknown q by the unknown u_{K_0} (as it is done in petroleum engineering) by setting

$$u_{K_0} = \frac{q}{4} - \frac{q}{2\pi} \ln\left(\frac{h}{r_p}\right) + P_p, \tag{13.6}$$

the interest of which is that it yields the usual formula for the discretization of $\nabla u \cdot n_{K_0}$ on σ if σ is the common edge to K_1 and K_0, namely $1/h\ (u_{K_1} - u_{K_0})$; the discrete equation associated to the unknown u_{K_0} is then (from (13.5))

$$\sum_{i=1}^{4} (u_{K_i} - u_{K_0}) = -q$$

and (13.6) may be written as:

$$q = i_p(P_p - u_{K_0}), \quad \text{with } i_p = \frac{1}{-\frac{1}{4} + \frac{1}{2\pi} \ln(\frac{h}{r_p})}.$$

This last equation defines i_p, the so called "well-index" in petroleum engineering. With this formula for i_p, the discrete unknowns are now $\{u_K, K \in \mathcal{T}\}$. The discrete equations associated to $\{u_K, K \in \mathcal{T}, K \neq K_0\}$ are given by the first part of (13.2) where each terms of (13.2) is replaced by its approximation in terms of $\{u_K, K \in \mathcal{T}\}$ (using also "BC" on $\partial\Omega$). The discrete equation associated to the unknown u_{K_0} is

$$\sum_{i=1}^{4}(u_{K_i} - u_{K_0}) = -i_p(P_p - u_{K_0}),$$

where $\{K_i, i = 1, 2, 3, 4\}$ are the four neighbouring cells to K_0.

Note that the discrete unknown u_{K_0} is somewhat artificial, it does not really represent the value of u in K_0. In fact, if $x \in K_0$, the "approximate value" of $u(x)$ is $-\frac{q}{2\pi}\ln(\frac{|x|}{r_p}) + P_p$ and $u_{K_0} = \frac{q}{4} - \frac{q}{2\pi}\ln(\frac{h}{r_p}) + P_p$.

13.2. Mesh refinement

Mesh refinement consists in using, in certain areas of the domain, control volumes of smaller size than elsewhere. In the case of triangular grids, a refinement may be performed, e.g., by dividing each triangle in the refined area into four subtriangles, and those at the boundary of the refined area in two triangles. Then, with some additional technique (e.g., change of diagonal), one may obtain an admissible mesh in the sense of Definitions 9.1, 10.1 and 11.1; therefore the error estimates 9.3, 10.1 and 11.1 hold under the same assumptions.

In the case of rectangular grids, the same refining procedure leads to "atypical" nodes and edges, i.e. an edge σ of a given control volume K may be common to two other control volumes, denoted by L and M. This is also true in the triangular case if the triangles of the boundary of the refined area are left untouched.

Let us consider, e.g., the same problem as in Section 9.1, with the same assumptions and notations, namely the discretization of

$$-\Delta u(x, y) = f(x, y), \quad (x, y) \in \Omega = (0, 1) \times (0, 1),$$
$$u(x, y) = 0, \quad (x, y) \in \partial\Omega.$$

It is easily seen that, in this case, if the approximation of the fluxes is performed using differential quotients such as in (9.6), the fluxes on the "atypical" edge σ cannot be consistent, since the lines joining the centers of K and L and the centers of K and M are not orthogonal to σ. However, the error which results from this lack of consistency can be controlled if the number of atypical edges is not too large.

In the case of rectangular grids (with a refining procedure), denoting by $\mathcal{E}_\dashv$ the set of "atypical" edges of a given mesh $\mathcal{T}$, i.e. edges with separate more than two control volumes, and $\mathcal{T}_\dashv$ the set of "atypical" control volumes, i.e. the control volumes containing an atypical edge in their boundaries; let e_K denote the error between $u(x_K)$ and u_K

for each control volume K, and $e_{\mathcal{T}}$ denote the piecewise constant function defined by $e(x) = e_K$ for any $x \in K$, then one has

$$\|e\|_{L^2(\Omega)} \leqslant C\Big(\operatorname{size}(\mathcal{T}) + \sum_{K \in \mathcal{T}_\dashv} \mathrm{m}(K)\Big).$$

The proof is similar to that of Theorem 9.3. It is detailed in BELMOUHOUB [1996].

14. Compactness results in L^2

This section is devoted to some functional analysis results which were used in the previous section. Let Ω be a bounded open set of $\mathbb{R}^d$, $d \geqslant 1$. Two relative compactness results in $L^2(\Omega)$ for sequences "almost" bounded in $H^1(\Omega)$ which were used in the proof of convergence of the schemes are presented here. Indeed, they are variations of the Rellich theorem (relative compactness in $L^2(\Omega)$ of a bounded sequence in $H^1(\Omega)$ or $H_0^1(\Omega)$). The originality of these results is not the fact that the sequences are relatively compact in $L^2(\Omega)$, which is an immediate consequence of the Kolmogorov theorem (see below), but the fact that the eventual limit, in $L^2(\Omega)$, of the sequence (or of a subsequence) is necessarily in $H^1(\Omega)$ (or in $H_0^1(\Omega)$ for Theorem 14.2), a space which does not contain the elements of the sequence.

We shall make use in this section of the Kolmogorov compactness theorem in $L^2(\Omega)$ which we now recall. The essential part of the proof of this theorem may be found in BREZIS [1983].

THEOREM 14.1. *Let ω be an open bounded set of $\mathbb{R}^N$, $N \geqslant 1$, $1 \leqslant q < \infty$ and $A \subset L^q(\omega)$. Then, A is relatively compact in $L^q(\omega)$ if and only if there exists $\{p(u),\ u \in A\} \subset L^q(\mathbb{R}^N)$ such that*

(1) *$p(u) = u$ a.e. on ω, for all $u \in A$,*
(2) *$\{p(u),\ u \in A\}$ is bounded in $L^q(\mathbb{R}^N)$,*
(3) *$\|p(u)(\bullet + \eta) - p(u)\|_{L^q(\mathbb{R}^N)} \to 0$, as $\eta \to 0$, uniformly with respect to $u \in A$.*

Let us now state the compactness results used in this chapter.

THEOREM 14.2. *Let Ω be an open bounded set of $\mathbb{R}^d$ with a Lipschitz continuous boundary, $d \geqslant 1$, and $\{u_n,\ n \in \mathbb{N}\}$ a bounded sequence of $L^2(\Omega)$. For $n \in \mathbb{N}$, one defines $\tilde{u}_n$ by $\tilde{u}_n = u_n$ a.e. on Ω and $\tilde{u}_n = 0$ a.e. on $\mathbb{R}^d \setminus \Omega$. Assume that there exist $C \in \mathbb{R}$ and $\{h_n,\ n \in \mathbb{N}\} \subset \mathbb{R}_+$ such that $h_n \to 0$ as $n \to \infty$ and*

$$\big\|\tilde{u}_n(\bullet + \eta) - \tilde{u}_n\big\|^2_{L^2(\mathbb{R}^d)} \leqslant C|\eta|\big(|\eta| + h_n\big), \quad \forall n \in \mathbb{N},\ \forall \eta \in \mathbb{R}^d. \tag{14.1}$$

Then, $\{u_n,\ n \in \mathbb{N}\}$ is relatively compact in $L^2(\Omega)$. Furthermore, if $u_n \to u$ in $L^2(\Omega)$ as $n \to \infty$, then $u \in H_0^1(\Omega)$.

PROOF. Since $\{h_n,\ n \in \mathbb{N}\}$ is bounded, the fact that $\{u_n,\ n \in \mathbb{N}\}$ is relatively compact in $L^2(\Omega)$ is an immediate consequence of Theorem 14.1, taking $N = d$, $\omega = \Omega$, $q = 2$

and $p(u_n) = \tilde{u}_n$. Then, assuming that $u_n \to u$ in $L^2(\Omega)$ as $n \to \infty$, it is only necessary to prove that $u \in H_0^1(\Omega)$. Let us first remark that $\tilde{u}_n \to \tilde{u}$ in $L^2(\mathbb{R}^d)$, as $n \to \infty$, with $\tilde{u} = u$ a.e. on Ω and $\tilde{u} = 0$ a.e. on $\mathbb{R}^d \setminus \Omega$.

Then, for $\varphi \in C_c^\infty(\mathbb{R}^d)$, one has, for all $\eta \in \mathbb{R}^d$, $\eta \neq 0$ and $n \in \mathbb{N}$, using the Cauchy–Schwarz inequality and thanks to (14.1),

$$\int_{\mathbb{R}^d} \frac{\tilde{u}_n(x+\eta) - \tilde{u}_n(x)}{|\eta|} \varphi(x)\,\mathrm{d}x \leqslant \frac{\sqrt{C|\eta|(|\eta| + h_n)}}{|\eta|} \|\varphi\|_{L^2(\mathbb{R}^d)},$$

which gives, letting $n \to \infty$, since $h_n \to 0$,

$$\int_{\mathbb{R}^d} \frac{\tilde{u}(x+\eta) - \tilde{u}(x)}{|\eta|} \varphi(x)\,\mathrm{d}x \leqslant \sqrt{C} \|\varphi\|_{L^2(\mathbb{R}^d)},$$

and therefore, with a trivial change of variables in the integration,

$$\int_{\mathbb{R}^d} \frac{\varphi(x-\eta) - \varphi(x)}{|\eta|} \tilde{u}(x)\,\mathrm{d}x \leqslant \sqrt{C} \|\varphi\|_{L^2(\mathbb{R}^d)}. \tag{14.2}$$

Let $\{e_i,\ i = 1, \ldots, d\}$ be the canonical basis of $\mathbb{R}^d$. For $i \in \{1, \ldots, d\}$ fixed, taking $\eta = he_i$ in (14.2) and letting $h \to 0$ (with $h > 0$, for instance) leads to

$$-\int_{\mathbb{R}^d} \frac{\partial \varphi(x)}{\partial x_i} \tilde{u}(x)\,\mathrm{d}x \leqslant \sqrt{C} \|\varphi\|_{L^2(\mathbb{R}^d)},$$

for all $\varphi \in C_c^\infty(\mathbb{R}^d)$.

This proves that $D_i \tilde{u}$ (the derivative of $\tilde{u}$ with respect to x_i in the sense of distributions) belongs to $L^2(\mathbb{R}^d)$, and therefore that $\tilde{u} \in H^1(\mathbb{R}^d)$. Since u is the restriction of $\tilde{u}$ on Ω and since $\tilde{u} = 0$ a.e. on $\mathbb{R}^d \setminus \Omega$, therefore $u \in H_0^1(\Omega)$. This completes the proof of Theorem 14.2. □

THEOREM 14.3. *Let Ω be an open bounded set of $\mathbb{R}^d$, $d \geqslant 1$, and $\{u_n,\ n \in \mathbb{N}\}$ a bounded sequence of $L^2(\Omega)$. For $n \in \mathbb{N}$, one defines $\tilde{u}_n$ by $\tilde{u}_n = u_n$ a.e. on Ω and $\tilde{u}_n = 0$ a.e. on $\mathbb{R}^d \setminus \Omega$. Assume that there exist $C \in \mathbb{R}$ and $\{h_n,\ n \in \mathbb{N}\} \subset \mathbb{R}_+$ such that $h_n \to 0$ as $n \to \infty$ and such that*

$$\left\|\tilde{u}_n(\bullet + \eta) - \tilde{u}_n\right\|_{L^2(\mathbb{R}^d)}^2 \leqslant C|\eta|, \quad \forall n \in \mathbb{N},\ \forall \eta \in \mathbb{R}^d, \tag{14.3}$$

and, for all compact $\bar{\omega} \subset \Omega$,

$$\left\|u_n(\bullet + \eta) - u_n\right\|_{L^2(\bar{\omega})}^2 \leqslant C|\eta|\big(|\eta| + h_n\big),$$
$$\forall n \in \mathbb{N},\ \forall \eta \in \mathbb{R}^d,\ |\eta| < d\big(\bar{\omega}, \Omega^c\big). \tag{14.4}$$

(The distance between $\bar{\omega}$ and $\mathbb{R}^d \setminus \Omega$ is denoted by $d(\bar{\omega}, \Omega^c)$.)

Then $\{u_n,\ n \in \mathbb{N}\}$ is relatively compact in $L^2(\Omega)$. Furthermore, if $u_n \to u$ in $L^2(\Omega)$ as $n \to \infty$, then $u \in H^1(\Omega)$.

and

$$\sum_{K\in\mathcal{V}} \nabla\varphi_K(x) = 0, \quad \text{for a.e. } x \in \Omega. \tag{12.3}$$

Using the latter equality, the discrete finite element equation associated to the unknown u_K, if $K \in \Omega$, can therefore be written as

$$\sum_{L\in\mathcal{V}} \int_\Omega (u_L - u_K)\nabla\varphi_L(x)\cdot\nabla\varphi_K(x)\,\mathrm{d}x = \int_\Omega f(x)\varphi_K(x)\,\mathrm{d}x.$$

Then the finite element method may be written as

$$\sum_{L\in\mathcal{V}} -\tau_{K|L}(u_L - u_K) = \int_\Omega f(x)\varphi_K(x)\,\mathrm{d}x, \quad \text{if } K \in \mathcal{V}\cap\Omega,$$

$$u_K = g(K), \quad \text{if } K \in \mathcal{V}\cap\partial\Omega,$$

with

$$\tau_{K|L} = -\int_\Omega \nabla\varphi_L(x)\cdot\nabla\varphi_K(x)\,\mathrm{d}x.$$

Under this form, the finite element method may be viewed as a finite volume method, except that there are no "real" control volumes associated to the vertices of $\mathcal{M}$. Indeed, thanks to (12.2), the control volume associated to K may be viewed as the support of φ_K "weighted" by φ_K. This interpretation of the finite element method as a finite volume method was also used in FORSYTH [1989, 1991] and EYMARD and GALLOUËT [1993] in order to design a numerical scheme for a transport equation for which the velocity field is the gradient of the pressure, which is itself the solution to an elliptic equation (see also HERBIN and LABERGERIE [1997] for numerical tests). This method is often referred to as the "control volume finite element" method.

In this finite volume interpretation of the finite element scheme, the notion of "consistency of the fluxes" does not appear. This notion of consistency, however, seems to be an interesting tool in the study of the "classical" finite volume schemes.

Note that the (discrete) maximum principle is satisfied with this scheme if and only if the transmissibilities $\tau_{K|L}$ are nonnegative (for all $K, L \in \mathcal{V}$ with $K \in \Omega$); this is the case under the classical Delaunay condition; this condition states that the (interior of the) circumscribed circle (or sphere in the three-dimensional case) of any triangle (tetrahedron in the three-dimensional case) of the mesh does not contain any element of $\mathcal{V}$. This is equivalent, in the case of two-dimensional triangular meshes, to the fact that the sum of two opposite angles facing a common edge is less or equal π.

12.2. Classical finite volumes on a dual mesh

Let $\mathcal{M}$ be a mesh of Ω ($\mathcal{M}$ may consist of triangles, but it is not necessary) and $\mathcal{V}$ be the set of vertices of $\mathcal{M}$. In order to associate to each vertex (of $\mathcal{M}$) a control volume (such

PROOF. The proof is very similar to that of Theorem 14.2. Using assumption (14.3), Theorem 14.1 yields that $\{u_n,\ n \in \mathbb{N}\}$ is relatively compact in $L^2(\Omega)$. Assuming now that $u_n \to u$ in $L^2(\Omega)$, as $n \to \infty$, one has to prove that $u \in H^1(\Omega)$.

Let $\varphi \in C_c^\infty(\Omega)$ and $\varepsilon > 0$ such that $\varphi(x) = 0$ if the distance from x to $\mathbb{R}^d \setminus \Omega$ is less than ε. Assumption (14.4) yields

$$\int_\Omega \frac{u_n(x+\eta) - u_n(x)}{|\eta|}\varphi(x)\,\mathrm{d}x \leqslant \frac{\sqrt{C|\eta|(|\eta| + h_n)}}{|\eta|}\|\varphi\|_{L^2(\Omega)},$$

for all $\eta \in \mathbb{R}^d$ such that $0 < |\eta| < \varepsilon$.

From this inequality, it may be proved, as in the proof of Theorem 14.2 (letting $n \to \infty$ and using a change of variables in the integration),

$$\int_\Omega \frac{\varphi(x-\eta) - \varphi(x)}{|\eta|}u(x)\,\mathrm{d}x \leqslant \sqrt{C}\|\varphi\|_{L^2(\Omega)},$$

for all $\eta \in \mathbb{R}^d$ such that $0 < |\eta| < \varepsilon$.

Then, taking $\eta = he_i$ and letting $h \to 0$ (with $h > 0$, for instance) one obtains, for all $i \in \{1, \ldots, d\}$,

$$-\int_\Omega \frac{\partial \varphi(x)}{\partial x_i}u(x)\,\mathrm{d}x \leqslant \sqrt{C}\|\varphi\|_{L^2(\Omega)},$$

for all $\varphi \in C_c^\infty(\Omega)$.

This proves that $D_i u$ (the derivative of u with respect to x_i in the sense of distributions) belongs to $L^2(\Omega)$, and therefore that $u \in H^1(\Omega)$. This completes the proof of Theorem 14.3. □

CHAPTER IV

Parabolic Equations

15. Introduction

The aim of this chapter is the study of finite volume schemes applied to a class of linear or nonlinear parabolic problems. We consider the following transient diffusion–convection equation:

$$u_t(x,t) - \Delta\varphi(u)(x,t) + \operatorname{div}(\boldsymbol{v}u)(x,t) + bu(x,t) = f(x,t),$$
$$x \in \Omega,\ t \in (0,T), \tag{15.1}$$

where Ω is an open polygonal bounded subset of $\mathbb{R}^d$, with $d = 2$ or $d = 3$, $T > 0$, $b \geqslant 0$, $\boldsymbol{v} \in \mathbb{R}^d$ is, for the sake of simplicity, a constant velocity field, f is a function defined on $\Omega \times \mathbb{R}_+$ which represents a volumetric source term. The function φ is a nondecreasing Lipschitz continuous function, which arises in the modelling of general diffusion processes. A simplified version of Stefan's problem may be expressed with the formulation (15.1) where φ is a continuous piecewise linear function, which is constant on an interval. The porous medium equation is also included in Eq. (15.1), with $\varphi(u) = u^m$, $m > 1$. However, the linear case, i.e. $\varphi(u) = u$, is of full interest and the error estimate of Section 17 will be given in such a case. In Section 18, we study the convergence of the explicit and of the implicit Euler scheme for the nonlinear case with $v = 0$ and $b = 0$.

REMARK 15.1. One could also consider a nonlinear convection term of the form $\operatorname{div}(\boldsymbol{v}\psi(u))(x,t)$ where $\psi \in C^1(\mathbb{R},\mathbb{R})$. Such a nonlinear convection term will be largely studied in the framework of nonlinear hyperbolic equations (Chapters V and VI) and we restrain here to a linear convection term for the sake of simplicity.

An initial condition is given by

$$u(x,0) = u_0(x), \quad x \in \Omega. \tag{15.2}$$

Let $\partial\Omega$ denote the boundary of Ω, and let $\partial\Omega_d \subset \partial\Omega$ and $\partial\Omega_n \subset \partial\Omega$ such that $\partial\Omega_d \cup \partial\Omega_n = \partial\Omega$ and $\partial\Omega_d \cap \partial\Omega_n = \emptyset$. A Dirichlet boundary condition is specified on $\partial\Omega_d \subset$

$\partial\Omega$. Let g be a real function defined on $\partial\Omega_d \times \mathbb{R}_+$, the Dirichlet boundary condition states that

$$u(x,t) = g(x,t), \quad x \in \partial\Omega_d,\ t \in (0,T). \tag{15.3}$$

A Neumann boundary condition is given with a function $\tilde{g}$ defined on $\partial\Omega_n \times \mathbb{R}_+$:

$$-\nabla\varphi(u)(x,t) \cdot \boldsymbol{n}(x) = \tilde{g}(x,t), \quad x \in \partial\Omega_n,\ t \in (0,T), \tag{15.4}$$

where $\boldsymbol{n}$ is the unit normal vector to $\partial\Omega$, outward to Ω.

REMARK 15.2. Note that, formally, $\Delta\varphi(u) = \operatorname{div}(\varphi'(u)\nabla u)$. Then, if $\varphi'(u)(x,t) = 0$ for some $(x,t) \in \Omega \times (0,T)$, the diffusion coefficient vanishes, so that Eq. (15.1) is a "degenerate" parabolic equation. In this case of degeneracy, the choice of the boundary conditions is important in order for the problem to be well-posed. In the case where φ' is positive, the problem is always parabolic.

In the next section, a finite volume scheme for the discretization of (15.1)–(15.4) is presented. An error estimate in the linear case (that is $\varphi(u) = u$) is given in Section 17. Finally, a nonlinear (and degenerate) case is studied in Section 18; a convergence result is given for subsequences of sequences of approximate solutions, and, when the weak solution is unique, for the whole set of approximate solutions. A uniqueness result is therefore proved for the case of a smooth boundary.

16. Meshes and schemes

In order to perform a finite volume discretization of system (15.1)–(15.4), admissible meshes are used in a similar way to the elliptic cases. Let $\mathcal{T}$ be an admissible mesh of Ω in the sense of Definition 9.1 with the additional assumption that any $\sigma \in \mathcal{E}_{\mathrm{ext}}$ is included in the closure of $\partial\Omega_d$ or included in the closure of $\partial\Omega_n$. The time discretization may be performed with a variable time step; in order to simplify the notations, we shall choose a constant time step $k \in (0,T)$. Let $N_k \in \mathbb{N}^\star$ such that $N_k = \max\{n \in \mathbb{N}, nk < T\}$, and we shall denote $t_n = nk$, for $n \in \{0,\dots,N_k+1\}$. Note that with a variable time step, error estimates and convergence results similar to that which are given in the next sections hold.

Denote by $\{u_K^n,\ K \in \mathcal{T},\ n \in \{0,\dots,N_k+1\}\}$ the discrete unknowns; the value u_K^n is an expected approximation of $u(x_K, nk)$.

In order to obtain the numerical scheme, let us integrate formally Eq. (15.1) over each control volume K of $\mathcal{T}$, and time interval $(nk,(n+1)k)$, for $n \in \{0,\dots,N_k\}$:

$$\int_K \big(u(x,t_{n+1}) - u(x,t_n)\big)\,\mathrm{d}x - \int_{nk}^{(n+1)k} \int_{\partial K} \nabla\varphi(u)(x,t) \cdot \boldsymbol{n}_K(x)\,\mathrm{d}\gamma(x)\,\mathrm{d}t$$
$$+ \int_{nk}^{(n+1)k} \int_{\partial K} \boldsymbol{v} \cdot \boldsymbol{n}_K(x) u(x,t)\,\mathrm{d}\gamma(x)\,\mathrm{d}t + b \int_{nk}^{(n+1)k} \int_K u(x,t)\,\mathrm{d}x\,\mathrm{d}t$$

$$= \int_{nk}^{(n+1)k} \int_K f(x,t)\,\mathrm{d}x\,\mathrm{d}t, \tag{16.1}$$

where $\boldsymbol{n}_K$ is the unit normal vector to ∂K, outward to K.

Recall that, as usual, the stability condition for an explicit discretization of a parabolic equation requires the time step to be limited by a power two of the space step, which is generally too strong a condition in terms of computational cost. Hence the choice of an implicit formulation in the left-hand side of (16.1) which yields

$$\begin{aligned}
&\frac{1}{k}\int_K \big(u(x,t_{n+1}) - u(x,t_n)\big)\,\mathrm{d}x - \int_{\partial K} \nabla\varphi(u)(x,t_{n+1}) \bullet \boldsymbol{n}_K(x)\,\mathrm{d}\gamma(x) \\
&\quad + \int_{\partial K} \boldsymbol{v} \bullet \boldsymbol{n}_K(x) u(x,t_{n+1})\,\mathrm{d}\gamma(x) + b\int_K u(x,t_{n+1})\,\mathrm{d}x\,\mathrm{d}t \\
&= \frac{1}{k}\int_{nk}^{(n+1)k} \int_K f(x,t)\,\mathrm{d}x\,\mathrm{d}t,
\end{aligned} \tag{16.2}$$

There now remains to replace in Eq. (16.1) each term by its approximation with respect to the discrete unknowns (and the data). Before doing so, let us remark that another way to obtain (16.2) is to integrate (in space) formally Eq. (15.1) over each control volume K of $\mathcal{T}$, at time $t \in (0,T)$. This gives

$$\begin{aligned}
&\int_K u_t(x,t)\,\mathrm{d}x - \int_{\partial K} \nabla\varphi(u)(x,t) \bullet \boldsymbol{n}_K(x)\,\mathrm{d}\gamma(x) \\
&\quad + \int_{\partial K} \boldsymbol{v} \bullet \boldsymbol{n}_K(x) u(x,t)\,\mathrm{d}\gamma(x) + b\int_K u(x,t)\,\mathrm{d}x \\
&= \int_K f(x,t)\,\mathrm{d}x.
\end{aligned} \tag{16.3}$$

An implicit time discretization is then obtained by taking $t = t_{n+1}$ in the left-hand side of (16.3), and replacing $u_t(x,t_{n+1})$ by $(u(x,t_{n+1}) - u(x,t_n))/k$. For the right-hand side of (16.3) a mean value of f between t_n and t_{n+1} may be used. This gives (16.2). It is also possible to take $f(x,t_{n+1})$ in the right-hand side of (16.3). This latter choice is simpler for the proof of some error estimates (see Section 17).

Writing the approximation of the various terms in Eq. (16.2) with respect to the discrete unknowns (namely, $\{u_K^n,\ K \in \mathcal{T},\ n \in \{0,\dots,N_k+1\}\}$) and taking into account the initial and boundary conditions yields the following implicit finite volume scheme for the discretization of (15.1)–(15.4), using the same notations and introducing some auxiliary unknowns as in Chapter III (see Eqs. (9.20)–(9.23)):

$$\begin{aligned}
&\mathrm{m}(K)\frac{u_K^{n+1} - u_K^n}{k} + \sum_{\sigma\in\mathcal{E}_K} F_{K,\sigma}^{n+1} + \sum_{\sigma\in\mathcal{E}_K} v_{K,\sigma} u_{\sigma,+}^{n+1} + \mathrm{m}(K) b u_K^{n+1} = \mathrm{m}(K) f_K^n, \\
&\forall K \in \mathcal{T},\ \forall n \in \{0,\dots,N_k\},
\end{aligned} \tag{16.4}$$

with

$$d_{K,\sigma} F^n_{K,\sigma} = -\mathrm{m}(\sigma)\big(\varphi\big(u^n_\sigma\big) - \varphi\big(u^n_K\big)\big) \quad \text{for } \sigma \in \mathcal{E}_K, \\ \text{for } n \in \{1, \dots, N_k + 1\}, \tag{16.5}$$

$$F^n_{K,\sigma} = -F^n_{L,\sigma} \quad \text{for all } \sigma \in \mathcal{E}_{\mathrm{int}} \text{ such that } \sigma = K|L, \\ \text{for } n \in \{1, \dots, N_k + 1\}, \tag{16.6}$$

$$F^n_{K,\sigma} = \frac{1}{k} \int_{(n-1)k}^{nk} \int_\sigma \tilde{g}(x,t)\,\mathrm{d}\gamma(x)\,\mathrm{d}t \quad \text{for } \sigma \in \mathcal{E}_K \text{ such that } \sigma \subset \partial\Omega_n, \\ \text{for } n \in \{1, \dots, N_k + 1\}, \tag{16.7}$$

and

$$u^n_\sigma = g(y_\sigma, nk) \quad \text{for } \sigma \subset \partial\Omega_d, \ \text{for } n \in \{1, \dots, N_k + 1\}. \tag{16.8}$$

The upstream choice for the convection term is performed as in the elliptic case (recall that $v_{K,\sigma} = \mathrm{m}(\sigma)\boldsymbol{v} \bullet \boldsymbol{n}_{K,\sigma}$),

$$u^n_{\sigma,+} = \begin{cases} u^n_K, & \text{if } \boldsymbol{v} \bullet \boldsymbol{n}_{K,\sigma} \geqslant 0, \\ u^n_L, & \text{if } \boldsymbol{v} \bullet \boldsymbol{n}_{K,\sigma} < 0, \end{cases} \quad \text{for all } \sigma \in \mathcal{E}_{\mathrm{int}} \text{ such that } \sigma = K|L, \tag{16.9}$$

$$u^n_{\sigma,+} = \begin{cases} u^n_K, & \text{if } \boldsymbol{v} \bullet \boldsymbol{n}_{K,\sigma} \geqslant 0, \\ u^n_\sigma & \text{if } \boldsymbol{v} \bullet \boldsymbol{n}_{K,\sigma} < 0, \end{cases} \quad \text{for all } \sigma \in \mathcal{E}_K \text{ such that } \sigma \subset \partial\Omega. \tag{16.10}$$

Note that, in the same way as in the elliptic case, the unknowns u^{n+1}_σ may be eliminated using (16.5)–(16.8). There remains to define the right-hand side, which may be defined by:

$$f^n_K = \frac{1}{k\,\mathrm{m}(K)} \int_{nk}^{(n+1)k} \int_K f(x,t)\,\mathrm{d}x\,\mathrm{d}t, \quad \forall K \in \mathcal{T}, \ \forall n \in \{0, \dots, N_k\}, \tag{16.11}$$

or by:

$$f^n_K = \frac{1}{\mathrm{m}(K)} \int_K f(x, t_{n+1})\,\mathrm{d}x, \quad \forall K \in \mathcal{T}, \ \forall n \in \{0, \dots, N_k\}. \tag{16.12}$$

Initial conditions can be taken into account by different ways, depending on the regularity of the data u_0. For example, it is possible to take

$$u^0_K = \frac{1}{\mathrm{m}(K)} \int_K u_0(x)\,\mathrm{d}x, \quad K \in \mathcal{T}, \tag{16.13}$$

or

$$u_K^0 = u_0(x_K), \quad K \in \mathcal{T}. \tag{16.14}$$

REMARK 16.1. It is not obvious to prove that the implicit finite volume scheme (16.4)–(16.10) (with (16.11) or (16.12) and (16.13) or (16.14)) has a solution. Once the unknowns $F_{K,\sigma}^{n+1}$ are eliminated, a nonlinear system of equations has to be solved. A proof of the existence and uniqueness of a solution to this system is proved in the next section for the linear case, and is sketched in Remark 18.4 for the nonlinear case.

REMARK 16.2 (*Comparison with finite difference and finite element*). Assume here that $\boldsymbol{v} = 0$, $b = 0$, $\varphi(s) = s$ for all $s \in \mathbb{R}$, $\partial\Omega_d = \partial\Omega$ (Dirichlet condition on the whole boundary) and that the mesh consists in rectangular control volumes with constant space step in each direction. The discretization obtained with the finite volume method then gives, as in the case of elliptic problems, the same scheme than the one obtained with the finite difference method (for which the discretization points are the centers of the elements of $\mathcal{T}$) except at the boundary. Finite difference methods have been used in, e.g., ATTEY [1974], KAMENOMOSTSKAJA [1995] and MEYER [1973]. Finite elements methods have been classically used in, e.g., AMIEZ and GREMAUD [1991] and CIAVALDINI [1975]. Consider now the dual mesh to a triangular mesh satisfying the Delaunay condition, such that the dual edges are the orthogonal bisectors of the primal edges. (The cells of this dual mesh are called "co-volumes" in BAUGHMAN and WALKINGTON [1993].) The classical piecewise bilinear finite element method on the dual mesh, for rectangles, or piecewise linear, for triangles (forgetting here the problem of the boundary conditions) using a "mass-lumping" process yields the same scheme as the finite volume scheme described above.

A finite element formulation for (15.1), with the implicit Euler scheme in time, yields

$$\frac{1}{k}\left(\int_\Omega \big(u^{n+1}(x) - u^n(x)\big)X_i(x)\,\mathrm{d}x\right) + \int_\Omega \nabla u^{n+1}(x) \cdot \nabla X_i(x)\,\mathrm{d}x$$
$$= \int_\Omega f(x, t_{n+1})X_i(x)\,\mathrm{d}x,$$

where X_i is the shape function of the finite element basis, associated with node i, for $i = 1, \dots, N$ (where N is the number of nodes where the unknown u is to be determined at time t_{n+1}). Let us approximate u^n by the following Galerkin expansion:

$$u^{n+1} = \sum_{j=1}^{\overline{N}} u_j^{n+1} X_j \quad \text{and} \quad u^n = \sum_{j=1}^{\overline{N}} u_j^n X_j$$

(where $\overline{N}$ is the total number of nodes, and u_j^n is expected to be an approximation of u at time t_n and node j, for all j and n); replacing in the above equation, this yields:

$$\frac{1}{k}\sum_{j=1}^{\overline{N}} \int_\Omega \big(u_j^{n+1} - u_j^n\big)X_j(x)X_i(x)\,\mathrm{d}x + \sum_{j=1}^{\overline{N}} \int_\Omega u_j^{n+1} \nabla X_j(x) \cdot \nabla X_i(x)\,\mathrm{d}x$$

$$= \int_{\Omega} f(x, t_{n+1}) X_i(x)\,\mathrm{d}x. \tag{16.15}$$

Hence, the finite element formulation yields, at each time step, a linear system of the form $CU^{n+1} + AU^{n+1} = B$ (where $U^{n+1} = (u_1, \dots, u_N)^t$, and A and C are $N \times N$ matrices); this scheme, however, is generally used after a mass-lumping, i.e. by assigning to the diagonal term of C the sum of the coefficients of the corresponding line and transforming it into a diagonal matrix; in fact, the mass lumping technique transforms, for the left-hand side of (16.15), the finite element scheme into a finite volume scheme.

17. Error estimate for the linear case

We consider, in this section, the linear case, $\varphi(s) = s$ for all $s \in \mathbb{R}$, and assume $\partial\Omega_d = \partial\Omega$, i.e. that a Dirichlet boundary condition is given on the whole boundary, in which case problem (15.1)–(15.4) becomes

$$u_t(x,t) - \Delta u(x,t) + \operatorname{div}(\boldsymbol{v}u)(x,t) + bu(x,t) = f(x,t), \quad x \in \Omega,\ t \in (0,T),$$

$$u(x,0) = u_0(x), \quad x \in \Omega,$$

$$u(x,t) = g(x,t), \quad x \in \partial\Omega,\ t \in (0,T);$$

the finite volume scheme (16.4)–(16.10) then becomes, assuming, for the sake of simplicity, that $x_K \in K$ for all $K \in \mathcal{T}$,

$$\mathrm{m}(K)\frac{u_K^{n+1} - u_K^n}{k} + \sum_{\sigma \in \mathcal{E}_K} F_{K,\sigma}^{n+1} + \sum_{\sigma \in \mathcal{E}_K} v_{K,\sigma} u_{\sigma,+}^{n+1} + \mathrm{m}(K) b u_K^{n+1} = \mathrm{m}(K) f_K^n,$$
$$\forall K \in \mathcal{T},\ \forall n \in \{0, \dots, N_k\}, \tag{17.1}$$

with

$$F_{K,\sigma}^n = -\tau_{K|L}\big(u_L^n - u_K^n\big) \quad \text{for all } \sigma \in \mathcal{E}_{\text{int}} \text{ such that } \sigma = K|L,$$
$$\text{for } n \in \{1, \dots, N_k + 1\}, \tag{17.2}$$

$$F_{K,\sigma}^n = -\tau_\sigma\big(g(y_\sigma, nk) - u_K^n\big) \quad \text{for all } \sigma \in \mathcal{E}_K \text{ such that } \sigma \subset \partial\Omega,$$
$$\text{for } n \in \{1, \dots, N_k + 1\}, \tag{17.3}$$

and

$$\begin{cases} u_{\sigma,+}^n = u_K^n, & \text{if } \boldsymbol{v} \cdot \boldsymbol{n}_{K,\sigma} \geqslant 0, \\ u_{\sigma,+}^n = u_L^n, & \text{if } \boldsymbol{v} \cdot \boldsymbol{n}_{K,\sigma} < 0, \end{cases} \quad \text{for all } \sigma \in \mathcal{E}_{\text{int}} \text{ such that } \sigma = K|L, \tag{17.4}$$

$$\begin{cases} u_{\sigma,+}^n = u_K^n, & \text{if } \boldsymbol{v} \cdot \boldsymbol{n}_{K,\sigma} \geqslant 0, \\ u_{\sigma,+}^n = g(y_\sigma, nk), & \text{if } \boldsymbol{v} \cdot \boldsymbol{n}_{K,\sigma} < 0, \end{cases} \quad \text{for all } \sigma \in \mathcal{E}_K \text{ such that } \sigma \subset \partial\Omega. \tag{17.5}$$

The source term and initial condition f and u_0, are discretized by (16.12) and (16.14).

A convergence analysis of a one-dimensional vertex-centered scheme was performed in GUO and STYNES [1997] by writing the scheme in a finite element framework. Here we shall use direct finite volume techniques which also handle the multi-dimensional case.

The following theorem gives an L^∞ estimate (on the approximate solution) and an error estimate. Some easy generalizations are possible (e.g., the same theorem holds with $b < 0$, the only difference is that in the L^∞ estimate (17.6) the bound c also depends on b).

THEOREM 17.1. *Let Ω be an open polygonal bounded subset of $\mathbb{R}^d$, $T > 0$, $u \in C^2(\overline{\Omega} \times \mathbb{R}_+, \mathbb{R})$, $b \geqslant 0$ and $\mathbf{v} \in \mathbb{R}^d$. Let $u_0 \in C^2(\overline{\Omega}, \mathbb{R})$ be defined by $u_0 = u(\bullet, 0)$, let $f \in C^0(\overline{\Omega} \times \mathbb{R}_+, \mathbb{R})$ be defined by $f = u_t - \operatorname{div}(\nabla u) + \operatorname{div}(\mathbf{v}u) + bu$ and $g \in C^0(\partial\Omega \times \mathbb{R}_+, \mathbb{R})$ defined by $g = u$ on $\partial\Omega \times \mathbb{R}_+$. Let $\mathcal{T}$ be an admissible mesh in the sense of Definition* 9.1 *and $k \in (0, T)$. Then there exists a unique vector $(u_K)_{K \in \mathcal{T}}$ satisfying* (17.1)–(17.5) (*or* (16.4)–(16.10)) *with* (16.12) *and* (16.14). *There exists c only depending on u_0, T, f and g such that*

$$\sup\{|u_K^n|,\ K \in \mathcal{T},\ n \in \{1, \ldots, N_k + 1\}\} \leqslant c. \tag{17.6}$$

Furthermore, let $e_K^n = u(x_K, t_n) - u_K^n$, for $K \in \mathcal{T}$ and $n \in \{1, \ldots, N_k + 1\}$, and $h = \operatorname{size}(\mathcal{T})$. Then there exists $C \in \mathbb{R}_+$ only depending on b, u, $\mathbf{v}$, Ω and T such that

$$\left(\sum_{K \in \mathcal{T}} (e_K^n)^2 \mathrm{m}(K)\right)^{1/2} \leqslant C(h + k), \quad \forall n \in \{1, \ldots, N_k + 1\}. \tag{17.7}$$

PROOF. For simplicity, let us assume that $x_K \in K$ for all $K \in \mathcal{T}$. Generalization without this condition is straightforward.

(i) *Existence, uniqueness, and L^∞ estimate.* For a given $n \in \{0, \ldots, N_k\}$, set $f_K^n = 0$ and $u_K^n = 0$ in (17.1), and $g(y_\sigma, (n+1)k) = 0$ for all $\sigma \in \mathcal{E}$ such that $\sigma \subset \partial\Omega$. Multiplying (17.1) by u_K^{n+1} and using the same technique as in the proof of Lemma 9.2 yields that $u_K^{n+1} = 0$ for all $K \in \mathcal{T}$. This yields the uniqueness of the solution $\{u_K^{n+1}, K \in \mathcal{T}\}$ to (17.1)–(17.5) for given $\{u_K^n, K \in \mathcal{T}\}$, $\{f_K^n, K \in \mathcal{T}\}$ and $\{g(y_\sigma, (n+1)k), \sigma \in \mathcal{E}, \sigma \subset \partial\Omega_d\}$. The existence follows immediately, since (17.1)–(17.5) is a finite-dimensional linear system with respect to the unknown $\{u_K^{n+1}, K \in \mathcal{T}\}$ (with as many unknowns as equations).

Let us now prove the estimate (17.6).

Set $m_f = \min\{f(x,t),\ x \in \overline{\Omega},\ t \in [0, 2T]\}$ and $m_g = \min\{g(x,t),\ x \in \partial\Omega,\ t \in [0, 2T]\}$.

Let $n \in \{0, \ldots, N_k\}$. Then, we claim that

$$\min\{u_K^{n+1},\ K \in \mathcal{T}\} \geqslant \min\{\min\{u_K^n,\ K \in \mathcal{T}\} + k m_f, 0, m_g\}. \tag{17.8}$$

Indeed, if $\min\{u_K^{n+1},\ K\in\mathcal{T}\}<\min\{0,m_g\}$, let $K_0\in\mathcal{T}$ such that $u_{K_0}^{n+1}=\min\{u_K^{n+1},\ K\in\mathcal{T}\}$. Since $u_{K_0}^{n+1}<0$ and $u_{K_0}^{n+1}<m_g$ writing (17.1) with $K=K_0$ and n leads to

$$u_{K_0}^{n+1}\geqslant u_{K_0}^n+kf_{K_0}^n\geqslant\min\{u_K^n,\ K\in\mathcal{T}\}+km_f,$$

this proves (17.8), which yields, by induction, that:

$$\min\{u_K^n,\ K\in\mathcal{T}\}\geqslant\min\{\min\{u_K^0,\ K\in\mathcal{T}\},0,m_g\}+nk\min\{m_f,0\},$$
$$\forall n\in\{0,\dots,N_k+1\}.$$

Similarly,

$$\max\{u_K^n,\ K\in\mathcal{T}\}\leqslant\max\{\max\{u_K^0,\ K\in\mathcal{T}\},0,M_g\}+nk\max\{M_f,0\},$$
$$\forall n\in\{0,\dots,N_k+1\},$$

with $M_f=\max\{f(x,t),\ x\in\overline{\Omega},\ t\in[0,2T]\}$ and $M_g=\max\{g(x,t),\ x\in\partial\Omega,\ t\in[0,2T]\}$.

This proves (17.6) with $c=\|u_0\|_{L^\infty(\Omega)}+\|g\|_{L^\infty(\partial\Omega\times(0,2T))}+2T\|f\|_{L^\infty(\Omega\times(0,2T))}$.

(ii) *Error estimate.* As in the stationary case (see the proof of Theorem 9.3), one uses the regularity of the data and the solution to write an equation for the error $e_K^n=u(x_K,t_n)-u_K^n$, defined for $K\in\mathcal{T}$ and $n\in\{0,\dots,N_k+1\}$. Note that $e_K^0=0$ for $K\in\mathcal{T}$. Let $n\in\{0,\dots,N_k\}$. Integrating (in space) Eq. (15.1) over each control volume K of $\mathcal{T}$, at time $t=t_{n+1}$, gives, thanks to the choice of f_K^n (see (16.12)),

$$\int_K u_t(x,t_{n+1})\,\mathrm{d}x-\int_{\partial K}\big(\nabla u(x,t)-\boldsymbol{v}u(x,t_{n+1})\big)\boldsymbol{\cdot}\boldsymbol{n}_K(x)\,\mathrm{d}\gamma(x)$$
$$+b\int_K u(x,t_{n+1})\,\mathrm{d}x=\mathrm{m}(K)f_K^n. \tag{17.9}$$

Note that, for all $x\in K$ and all $K\in\mathcal{T}$, a Taylor expansion yields, thanks to the regularity of u:

$$u_t(x,t_{n+1})=(1/k)\big(u(x_K,t_{n+1})-u(x_K,t_n)\big)+s_K^n(x)$$
$$\text{with } \left|s_K^n(x)\right|\leqslant C_1(h+k)$$

with some C_1 only depending on u and T. Therefore, defining $S_K^n=\int_K s_K^n(x)\,\mathrm{d}x$, one has: $|S_K^n|\leqslant C_1\mathrm{m}(K)(h+k)$.

One follows now the lines of the proof of Theorem 9.3, adding the terms due to the time derivative u_t. Subtracting (17.1) to (17.9) yields

$$\mathrm{m}(K)\frac{e_K^{n+1}-e_K^n}{k}+\sum_{\sigma\in\mathcal{E}_K}\big(G_{K,\sigma}^{n+1}+W_{K,\sigma}^{n+1}\big)+b\mathrm{m}(K)e_K^{n+1}$$

$$= b\mathrm{m}(K)\rho_K^n - \sum_{\sigma\in\mathcal{E}_K} \mathrm{m}(\sigma)\big(R_{K,\sigma}^n + r_{K,\sigma}^n\big) - S_K^n, \quad \forall K \in \mathcal{T}, \tag{17.10}$$

where (with the notations of Definition 9.1),

$$\begin{aligned} G_{K,\sigma}^{n+1} &= -\tau_\sigma\big(e_L^{n+1} - e_K^{n+1}\big), \quad \forall K \in \mathcal{T},\ \forall\sigma \in \mathcal{E}_K \cap \mathcal{E}_{\mathrm{int}},\ \sigma = K|L, \\ G_{K,\sigma}^{n+1} &= \tau_\sigma e_K^{n+1}, \quad \forall K \in \mathcal{T},\ \forall\sigma \in \mathcal{E}_K \cap \mathcal{E}_{\mathrm{ext}}, \\ W_{K,\sigma}^{n+1} &= \mathrm{m}(\sigma)\boldsymbol{v}\cdot\boldsymbol{n}_{K,\sigma}\big(u(x_{\sigma,+}, t_{n+1}) - u_{\sigma,+}^{n+1}\big), \end{aligned}$$

where $x_{\sigma,+} = x_K$ (resp. x_L) if $\sigma \in \mathcal{E}_{\mathrm{int}}$, $\sigma = K|L$ and $\boldsymbol{v}\cdot\boldsymbol{n}_{K,\sigma} \geqslant 0$ (resp. $\boldsymbol{v}\cdot\boldsymbol{n}_{K,\sigma} < 0$) and $x_{\sigma,+} = x_K$ (resp. y_σ) if $\sigma = \mathcal{E}_K \cap \mathcal{E}_{\mathrm{ext}}$ and $\boldsymbol{v}\cdot\boldsymbol{n}_{K,\sigma} \geqslant 0$ (resp. $\boldsymbol{v}\cdot\boldsymbol{n}_{K,\sigma} < 0$),

$$\begin{aligned} &\rho_K^n = u(x_K, t_{n+1}) - \frac{1}{\mathrm{m}(K)}\int_K u(x, t_{n+1})\,\mathrm{d}x, \\ &m(\sigma)R_{K,\sigma}^n = \tau_\sigma\big(u(x_K, t_{n+1}) - u(x_L, t_{n+1})\big) + \int_\sigma \nabla u(x, t_{n+1})\cdot\boldsymbol{n}_{K,\sigma}\,\mathrm{d}\gamma(x) \\ &\quad \text{if } \sigma = K|L \in \mathcal{E}_{\mathrm{int}}, \\ &m(\sigma)R_{K,\sigma}^n = \tau_\sigma\big(u(x_K, t_{n+1}) - g(y_\sigma, t_{n+1})\big) + \int_\sigma \nabla u(x, t_{n+1})\cdot\boldsymbol{n}_{K,\sigma}\,\mathrm{d}\gamma(x) \\ &\quad \text{if } \sigma \in \mathcal{E}_K \cap \mathcal{E}_{\mathrm{int}}, \end{aligned}$$

and

$$m(\sigma)r_{K,\sigma}^n = \boldsymbol{v}\cdot\boldsymbol{n}_{K,\sigma}\left(m(\sigma)u(x_{\sigma,+}, t_{n+1}) - \int_\sigma u(x, t_{n+1})\,\mathrm{d}\gamma(x)\right), \quad \text{for any } \sigma \in \mathcal{E}.$$

As in Theorem 9.3, thanks to the regularity of u, there exists C_2, only depending on u, $\boldsymbol{v}$ and T, such that $|R_{K,\sigma}^n| + |r_{K,\sigma}^n| \leqslant C_2 h$ and $|\rho_K^n| \leqslant C_2 h$, for any $K \in \mathcal{T}$ and $\sigma \in \mathcal{E}_K$.

Multiplying (17.10) by e_K^{n+1}, summing for $K \in \mathcal{T}$, and performing the same computations as in the proof of Theorem 9.3 between (9.56) to (9.60) yields, with some C_3 only depending on u, $\boldsymbol{v}$, b, Ω and T,

$$\begin{aligned} &\frac{1}{k}\sum_{K\in\mathcal{T}} \mathrm{m}(K)\big(e_K^{n+1}\big)^2 + \tfrac{1}{2}\big\|e_{\mathcal{T}}^{n+1}\big\|_{1,\mathcal{T}}^2 + \tfrac{1}{2}b\big\|e_{\mathcal{T}}^{n+1}\big\|_{L^2(\Omega)}^2 \\ &\quad \leqslant C_3 h^2 + C_1(h+k)\sum_{K\in\mathcal{T}} \mathrm{m}(K)\big|e_K^{n+1}\big| + \frac{1}{k}\sum_{K\in\mathcal{T}} \mathrm{m}(K)e_K^{n+1}e_K^n, \end{aligned} \tag{17.11}$$

where the second term of the right-hand side is due to the bound on S_K^n and where $e_{\mathcal{T}}^{n+1}$ is a piecewise constant function defined by

$$e_{\mathcal{T}}^{n+1}(x) = e_K^{n+1}, \quad \text{for } x \in K,\ K \in \mathcal{T}.$$

Inequality (17.11) yields

$$\|e_{\mathcal{T}}^{n+1}\|_{L^2(\Omega)}^2 \leqslant 2kC_3h^2 + 2kC_1\mathrm{m}(\Omega)(k+h)\|e_{\mathcal{T}}^{n+1}\|_{L^2(\Omega)} + \|e_{\mathcal{T}}^{n}\|_{L^2(\Omega)}^2,$$

which gives

$$\|e_{\mathcal{T}}^{n+1}\|_{L^2(\Omega)}^2 \leqslant \|e_{\mathcal{T}}^{n}\|_{L^2(\Omega)}^2 + C_4\big(kh^2 + k(k+h)\|e_{\mathcal{T}}^{n+1}\|_{L^2(\Omega)}\big), \tag{17.12}$$

where $C_4 \in \mathbb{R}_+$ only depends on u, $\boldsymbol{v}$, b, Ω and T. Remarking that for $\varepsilon > 0$, the following inequality holds:

$$C_4k(k+h)\|e_{\mathcal{T}}^{n+1}\|_{L^2(\Omega)} \leqslant \varepsilon^2\|e_{\mathcal{T}}^{n+1}\|_{L^2(\Omega)}^2 + (1/\varepsilon^2)C_4^2k^2(k+h)^2,$$

taking $\varepsilon^2 = k/(k+1)$, (17.12) yields

$$\begin{aligned}\|e_{\mathcal{T}}^{n+1}\|_{L^2(\Omega)}^2 \leqslant\ & (1+k)\|e_{\mathcal{T}}^{n}\|_{L^2(\Omega)}^2 + C_4kh^2(1+k)\\ & + (1+k)^2C_4^2k(k+h)^2.\end{aligned} \tag{17.13}$$

Then, if $\|e_{\mathcal{T}}^{n}\|_{L^2(\Omega)}^2 \leqslant c_n(h+k)^2$, with $c_n \in \mathbb{R}_+$, one deduces from (17.13), using $h \leqslant h+k$ and $k < T$, that

$$\|e_{\mathcal{T}}^{n+1}\|_{L^2(\Omega)}^2 \leqslant c_{n+1}(h+k)^2$$
$$\text{with } c_{n+1} = (1+k)c_n + C_5k \text{ and } C_5 = C_4(1+T) + C_4^2(1+T)^2.$$

(Note that C_5 only depends on u, $\boldsymbol{v}$, b, Ω and T.)

Choosing $c_0 = 0$ (since $\|e_{\mathcal{T}}^{0}\|_{L^2(\Omega)} = 0$), the relation between c_n and c_{n+1} yields (by induction) $c_n \leqslant C_5e^{2kn}$. Estimate (17.7) follows with $C^2 = C_5e^{4T}$. □

REMARK 17.1. The error estimate given in Theorem 17.1 may be generalized to the case of discontinuous coefficients. The admissibility of the mesh is then redefined so that the data and the solution are piecewise regular on the control volumes as in Definition 11.1, see also HERBIN [1996].

18. Convergence in the nonlinear case

18.1. Solutions to the continuous problem

We consider problem (15.1)–(15.4) with $\boldsymbol{v} = 0$, $b = 0$, $\partial\Omega_n = \partial\Omega$ and $\tilde{g} = 0$, that is a homogeneous Neumann condition on the whole boundary, in which case the problem becomes

$$u_t(x,t) - \Delta\varphi(u)(x,t) = f(x,t), \quad \text{for } (x,t) \in \Omega \times (0,T), \tag{18.1}$$

with

$$\nabla\varphi(u)(x,t)\cdot\boldsymbol{n}(x)=0,\quad \text{for } (x,t)\in\partial\Omega\times(0,T), \tag{18.2}$$

and the initial condition

$$u(x,0)=u_0(x),\quad \text{for all } x\in\Omega. \tag{18.3}$$

We suppose that the following hypotheses are satisfied:

ASSUMPTION 18.1.
(i) Ω is an open bounded polygonal subset of $\mathbb{R}^d$ and $T>0$.
(ii) The function $\varphi\in C(\mathbb{R},\mathbb{R})$ is a nondecreasing locally Lipschitz continuous function.
(iii) The initial data u_0 satisfies $u_0\in L^\infty(\Omega)$.
(iv) The right-hand side f satisfies $f\in L^\infty(\Omega\times\mathbb{R}_+^\star)$.

Equation (18.1) is a degenerate parabolic equation. Formally, $\Delta\varphi(u)=\operatorname{div}(\varphi'(u)\nabla u)$, so that, if $\varphi'(u)=0$, the diffusion coefficient vanishes. Let us give a definition of a weak solution u to problem (18.1)–(18.3) (the proof of the existence of such a solution is given in KAMENOMOSTSKAJA [1995], LADYŽENSKAJA, SOLONNIKOV and URAL'CEVA [1968], MEIRMANOV [1992], OLEINIK [1960]).

DEFINITION 18.1. Under Assumption 18.1, a measurable function u is a weak solution of (18.1)–(18.3) if

$$\begin{aligned}
&u\in L^\infty(\Omega\times(0,T)),\\
&\int_0^T\int_\Omega\big(u(x,t)\psi_t(x,t)+\varphi\big(u(x,t)\big)\Delta\psi(x,t)+f(x,t)\psi(x,t)\big)\,\mathrm{d}x\,\mathrm{d}t\\
&\quad+\int_\Omega u_0(x)\psi(x,0)\,\mathrm{d}x=0,\quad \text{for all } \psi\in\mathcal{A}_T,
\end{aligned} \tag{18.4}$$

where $\mathcal{A}_T=\{\psi\in C^{2,1}(\overline{\Omega}\times[0,T]),\ \nabla\psi\cdot\boldsymbol{n}=0 \text{ on } \partial\Omega\times[0,T], \text{ and } \psi(\cdot,T)=0\}$, and $C^{2,1}(\overline{\Omega}\times[0,T])$ denotes the set of functions which are restrictions on $\overline{\Omega}\times[0,T]$ of functions from $\mathbb{R}^d\times\mathbb{R}$ into $\mathbb{R}$ which are twice (resp. once) continuously differentiable with respect to the first (resp. second) variable. (Recall that, as usual, $\boldsymbol{n}$ is the unit normal vector to $\partial\Omega$, outward to Ω.)

REMARK 18.1. It is possible to use a solution in a stronger sense, using only one integration by parts for the space term. It then leads to a larger test function space than $\mathcal{A}_T$.

REMARK 18.2. Note that the function u formally satisfies the conservation law

$$\int_\Omega u(x,t)\,\mathrm{d}x=\int_\Omega u_0(x)\,\mathrm{d}x+\int_0^t\int_\Omega f(x,t)\,\mathrm{d}x\,\mathrm{d}t, \tag{18.5}$$

for all $t\in[0,T]$. This property is also satisfied by the finite volume approximation.

18.2. Definition of the finite volume approximate solutions

As in Sections 9.2 and 10.1, an admissible mesh of Ω is defined, with respect to which a functional space is introduced: this space contains the approximate solutions obtained from the finite volume discretization over the admissible mesh.

DEFINITION 18.2. Let Ω be an open bounded polygonal subset of $\mathbb{R}^d$, $\mathcal{T}$ be an admissible mesh in the sense of Definition 10.1, $T > 0$, $k \in (0, T)$ and $N_k = \max\{n \in \mathbb{N};\ nk < T\}$. Let $X(\mathcal{T}, k)$ be the set of functions u from $\Omega \times (0, (N_k + 1)k)$ to $\mathbb{R}$ such that there exists a family of real values $\{u_K^n, K \in \mathcal{T}, n \in \{0, \dots, N_k\}\}$, with $u(x, t) = u_K^n$ for a.e., $x \in K$, $K \in \mathcal{T}$ and for a.e. $t \in [nk, (n + 1)k)$, $n \in \{0, \dots, N_k\}$.

Since we only consider, for the sake of simplicity, a Neumann boundary condition, we can easily eliminate the unknowns $F_{K,\sigma}^n$ located at the edges in Eq. (16.4) using Eqs. (16.5), (16.6), and (16.7). An explicit version of the scheme can then be written in the following way:

$$\mathrm{m}(K)\frac{u_K^{n+1} - u_K^n}{k} - \sum_{L \in \mathcal{N}(K)} \tau_{K|L}\big(\varphi(u_L^n) - \varphi(u_K^n)\big) = \mathrm{m}(K) f_K^n,$$
$$\forall K \in \mathcal{T},\ \forall n \in \{0, \dots, N_k\}, \tag{18.6}$$

$$u_K^0 = \frac{1}{\mathrm{m}(K)} \int_K u_0(x)\, \mathrm{d}x, \quad \forall K \in \mathcal{T}, \tag{18.7}$$

$$f_K^n = \frac{1}{k\mathrm{m}(K)} \int_{nk}^{(n+1)k} \int_K f(x, t)\, \mathrm{d}x\, \mathrm{d}t, \quad \forall K \in \mathcal{T},\ \forall n \in \{0, \dots, N_k\}. \tag{18.8}$$

(Recall that $\tau_{K|L} = \mathrm{m}(K|L)/d_{K|L}$, see Definition 10.1.)

REMARK 18.3. The definition using the mean value in (18.7) is motivated by the lack of regularity assumed on the data u_0.

The scheme (18.6)–(18.8) is then used to build an approximate solution, $u_{\mathcal{T},k} \in X(\mathcal{T}, k)$ by

$$u_{\mathcal{T},k}(x, t) = u_K^n, \qquad \forall x \in K,\ \forall t \in \big[nk, (n + 1)k\big),\ \forall K \in \mathcal{T},$$
$$\forall n \in \{0, \dots, N_k\}. \tag{18.9}$$

REMARK 18.4. The implicit finite volume scheme is defined by

$$\mathrm{m}(K)\frac{u_K^{n+1} - u_K^n}{k} - \sum_{L \in \mathcal{N}(K)} \tau_{K|L}\big(\varphi(u_L^{n+1}) - \varphi(u_K^{n+1})\big) = \mathrm{m}(K) f_K^n,$$
$$\forall K \in \mathcal{T},\ \forall n \in \{0, \dots, N_k\}. \tag{18.10}$$

The proof of the existence of u_K^{n+1}, for any $n \in \{0, \dots, N_k\}$, can be obtained using the following fixed point method:

$$u_K^{n+1,0} = u_K^n, \quad \text{for all } K \in \mathcal{T}, \tag{18.11}$$

and

$$\mathrm{m}(K)\frac{u_K^{n+1,m+1} - u_K^n}{k} - \sum_{L\in\mathcal{N}(K)} \tau_{K|L}\big(\varphi\big(u_L^{n+1,m}\big) - \varphi\big(u_K^{n+1,m+1}\big)\big) = \mathrm{m}(K) f_K^n,$$
$$\forall K \in \mathcal{T},\ \forall m \in \mathbb{N}. \tag{18.12}$$

Equation (18.12) gives a contraction property, which leads first to prove that for all $K \in \mathcal{T}$, the sequence $(\varphi(u_K^{n+1,m}))_{m\in\mathbb{N}}$ converges. Then we deduce that $(u_K^{n+1,m})_{m\in\mathbb{N}}$ also converges.

We shall see further that all results obtained for the explicit scheme are also true, with convenient adaptations, for the implicit scheme. The function $u_{\mathcal{T},k}$ is then defined by $u_{\mathcal{T},k}(x,t) = u_K^{n+1}$, for all $x \in K$, for all $t \in [nk, (n+1)k)$.

The mathematical problem is to study, under Assumption 18.1 and with a mesh in the sense of Definition 10.1, the convergence of $u_{\mathcal{T},k}$ to a weak solution of problem (18.1)–(18.3), when $h = \mathrm{size}(\mathcal{T}) \to 0$ and $k \to 0$. Exactly in the same manner as for the elliptic case, we shall use estimates on the approximate solutions which are discrete versions of the estimates which hold on the solution of the continuous problem and which ensure the stability of the scheme. We present the proofs in the case of the explicit scheme and show in several remarks how they can be extended to the case of the implicit scheme (which is significantly easier to study). The proof of convergence of the scheme uses a weak-$\star$ convergence property, as in CIAVALDINI [1975], which is proved in a general setting in Section 18.5. For the sake of completeness, the proof of uniqueness of the weak solution of problem (18.1)–(18.3) is given for the case of a regular boundary; this allows to prove that the whole sequence of approximate solutions converges to the weak solution of problem (18.1)–(18.3), in which case an admissible mesh for a smooth domain can easily be defined (see Definition 18.4).

18.3. Estimates on the approximate solution

Maximum principle

LEMMA 18.1. *Under Assumption* 18.1, *let* $\mathcal{T}$ *be an admissible mesh in the sense of Definition* 10.1 *and* $k \in (0,T)$. *Let* $U = \|u_0\|_{L^\infty(\Omega)} + T\|f\|_{L^\infty(\Omega\times(0,T))}$, $B = \sup_{-U\leqslant x<y\leqslant U} \frac{\varphi(x)-\varphi(y)}{x-y}$. *Assume that the condition*

$$k \leqslant \frac{\mathrm{m}(K)}{B\sum_{L\in\mathcal{N}(K)} \tau_{K|L}}, \quad \textit{for all } K \in \mathcal{T}, \tag{18.13}$$

is satisfied. Then the function $u_{\mathcal{T},k}$ defined by (18.6)–(18.9) verifies

$$\|u_{\mathcal{T},k}\|_{L^\infty(\Omega\times(0,T))} \leqslant U. \tag{18.14}$$

PROOF. Let $n \in \{0,\ldots,N_k-1\}$ and assume $u_K^n \in [-U,+U]$ for all $K \in \mathcal{T}$.

Let $K \in \mathcal{T}$, Eq. (18.6) can be written as

$$u_K^{n+1} = \left(1 - \frac{k}{\mathrm{m}(K)} \sum_{L\in\mathcal{N}(K)} \tau_{K|L} \frac{\varphi(u_L^n)-\varphi(u_K^n)}{u_L^n - u_K^n}\right) u_K^n$$
$$+ \frac{k}{\mathrm{m}(K)} \sum_{L\in\mathcal{N}(K)} \left(\tau_{K|L} \frac{\varphi(u_L^n)-\varphi(u_K^n)}{u_L^n - u_K^n}\right) u_L^n + k f_K^n,$$

with the convention that $\frac{\varphi(u_L^n)-\varphi(u_K^n)}{u_L^n-u_K^n} = 0$ if $u_L^n - u_K^n = 0$.

Thanks to the condition (18.13) and since φ is nondecreasing, the following inequality can be deduced:

$$\left|u_K^{n+1}\right| \leqslant \sup_{L\in\mathcal{T}} \left|u_L^n\right| + k\|f\|_{L^\infty(\Omega\times(0,T))}.$$

Then, since K is arbitrary in $\mathcal{T}$,

$$\sup_{K\in\mathcal{T}} \left|u_K^{n+1}\right| \leqslant \sup_{L\in\mathcal{T}} \left|u_L^n\right| + k\|f\|_{L^\infty(\Omega\times(0,T))}. \tag{18.15}$$

Using (18.15), an induction on n yields, for $n \in \{0,\ldots,N_k\}$, $\sup_{K\in\mathcal{T}} |u_K^n| \leqslant \|u_0\|_{L^\infty(\Omega)} + nk\|f\|_{L^\infty(\Omega\times(0,T))}$, which leads to inequality (18.14) since $N_k k \leqslant T$. □

REMARK 18.5. Assume that there exist $\alpha, \beta, \gamma \in \mathbb{R}_+^\star$ such that $\mathrm{m}(K) \geqslant \alpha h^d$, $\mathrm{m}(\partial K) \leqslant \beta h^{d-1}$, for all $K \in \mathcal{T}$, and $d_{K|L} \geqslant \gamma h$, for all $K|L \in \mathcal{E}_{\mathrm{int}}$ (recall that $h = \mathrm{size}(\mathcal{T})$). Then, $k \leqslant Ch^2$ with $C = (\alpha\gamma)/(B\beta)$ yields (18.13).

REMARK 18.6. Let $(\mathcal{T}_n, k_n)_{n\in\mathbb{N}}$ be a sequence of admissible meshes and time steps, and $(u_{\mathcal{T}_n,k_n})_{n\in\mathbb{N}}$ the associated sequence of approximate finite volume solutions; then, thanks to (18.14), there exists a function $u \in L^\infty(\Omega\times(0,T))$ and a subsequence of $(u_{\mathcal{T}_n,k_n})_{n\in\mathbb{N}}$ which converges to u for the weak-$\star$ topology of $L^\infty(\Omega\times(0,T))$.

REMARK 18.7. Estimate (18.14) is also true, with

$$U = \|u_0\|_{L^\infty(\Omega)} + 2T\|f\|_{L^\infty(\Omega\times(0,2T))},$$

for the implicit scheme, because the fixed point method guarantees (18.15) (with $\|f\|_{L^\infty(\Omega\times(0,2T))}$ instead of $\|f\|_{L^\infty(\Omega\times(0,T))}$ and until $n = N_k$), without any condition on k.

Space translates of approximate solutions
Let us now define a seminorm, which is the discrete version of the seminorm in the space $L^2(0,T;H^1(\Omega))$.

DEFINITION 18.3 (*Discrete $L^2(0,T;H^1(\Omega))$ seminorm*). Let Ω be an open bounded polygonal subset of $\mathbb{R}^d$, $\mathcal{T}$ an admissible finite volume mesh in the sense of Definition 10.1, $T>0$, $k\in(0,T)$ and $N_k=\max\{n\in\mathbb{N};\, nk<T\}$. For $u\in X(\mathcal{T},k)$, let the following seminorms be defined by:

$$\left|u(\bullet,t)\right|_{1,\mathcal{T}}^2=\sum_{K|L\in\mathcal{E}_{\text{int}}}\tau_{K|L}\left(u_L^n-u_K^n\right)^2,$$
$$\text{for a.e. } t\in(0,T) \text{ and } n=\max\{n\in\mathbb{N};\; nk\leqslant t\},\tag{18.16}$$

and

$$|u|_{1,\mathcal{T},k}^2=\sum_{n=0}^{N_k}k\sum_{K|L\in\mathcal{E}_{\text{int}}}\tau_{K|L}\left(u_L^n-u_K^n\right)^2.\tag{18.17}$$

Let us now state some preliminary lemmata to the use of Kolmogorov's theorem (compactness properties in $L^2(\Omega\times(0,T))$) in the proof of convergence of the approximate solutions.

LEMMA 18.2. *Let Ω be an open bounded polygonal subset of $\mathbb{R}^d$, $\mathcal{T}$ an admissible mesh in the sense of Definition* 10.1, $T>0$, $k\in(0,T)$ *and* $u\in X(\mathcal{T},k)$. *For all $\eta\in\mathbb{R}^d$, let Ω_η be defined by $\Omega_\eta=\{x\in\Omega,\ [x,x+\eta]\subset\Omega\}$. Then*:

$$\left\|u(\bullet+\eta,\bullet)-u(\bullet,\bullet)\right\|_{L^2(\Omega_\eta\times(0,T))}^2\leqslant|u|_{1,\mathcal{T},k}^2|\eta|\left(|\eta|+2\,\text{size}(\mathcal{T})\right),$$
$$\forall\eta\in\mathbb{R}^d.\tag{18.18}$$

PROOF. Reproducing the proof of Lemma 9.3 (see also the proof of (10.31)), we get, for a.e. $t\in(0,T)$:

$$\left\|u(\bullet+\eta,t)-u(\bullet,t)\right\|_{L^2(\Omega_\eta)}^2\leqslant\left|u(\bullet,t)\right|_{1,\mathcal{T}}^2|\eta|\left(|\eta|+2\,\text{size}(\mathcal{T})\right),\quad\forall\eta\in\mathbb{R}^d.\tag{18.19}$$

Integrating (18.19) on $t\in(0,T)$ gives (18.18). □

The set Ω_η defined in Lemma 18.2 verifies $\Omega\setminus\Omega_\eta\subset\bigcup_{\sigma\in\mathcal{E}_{\text{ext}}}\overline{\omega_{\eta,\sigma}}$, with $\omega_{\eta,\sigma}=\{y-t\eta,\ y\in\sigma,\ t\in[0,1]\}$. Then, $\mathrm{m}(\Omega\setminus\Omega_\eta)\leqslant|\eta|\ \mathrm{m}(\partial\Omega)$, since $\mathrm{m}(\bar{\omega}_\eta)\leqslant\eta\mathrm{m}(\sigma)$. Then, an immediate corollary of Lemma 18.2 is the following:

LEMMA 18.3. *Let Ω be an open bounded polygonal subset of $\mathbb{R}^d$, $\mathcal{T}$ an admissible mesh in the sense of Definition* 10.1, $T>0$, $k\in(0,T)$ *and* $u\in X(\mathcal{T},k)$. *Let $\tilde{u}$ be* *defined by* $\tilde{u} = u$ *a.e. on* $\Omega \times (0, T)$, *and* $\tilde{u} = 0$ *a.e. on* $\mathbb{R}^{d+1} \setminus \Omega \times (0, T)$. *Then*:

$$\begin{aligned}\|\tilde{u}(\cdot + \eta, \cdot) - \tilde{u}(\cdot, \cdot)\|^2_{L^2(\mathbb{R}^{d+1})} &\leqslant |\eta|\big(|u|^2_{1,\mathcal{T},k}\big(|\eta| + 2\,\mathrm{size}(\mathcal{T})\big) \\ &\quad + 2\mathrm{m}(\partial\Omega)\|u\|^2_{L^\infty(\Omega\times(0,T))}\big), \quad \forall \eta \in \mathbb{R}^d. \qquad (18.20)\end{aligned}$$

REMARK 18.8. Estimate (18.20) makes use of the $L^\infty(\Omega \times (0, T))$-norm of $u \in X(\mathcal{T}, k)$. A similar estimate may be proved with the $L^2(\Omega \times (0, T))$-norm of u (instead of the $L^\infty(\Omega \times (0, T))$-norm). Indeed, the right-hand side of (18.20) may be replaced by $C\eta(|u|^2_{1,\mathcal{T},k} + \|u\|^2_{L^2(\Omega\times(0,T))})$, where C only depends on Ω. This estimate is proved in Theorem 10.3 where it is used for the convergence of numerical schemes for the Neumann problem (for which no L^∞ estimate on the approximate solutions is available). The key to its proof is the "Trace lemma" 10.5.

Let us now state the following lemma, which gives an estimate of the discrete $L^2(0, T; H^1(\Omega))$ seminorm of the nonlinearity.

LEMMA 18.4. *Under Assumption* 18.1, *let* $\mathcal{T}$ *be an admissible mesh in the sense of Definition* 10.1. *Let* $\xi \in (0, 1)$ *and* $k \in (0, T)$ *such that*

$$k \leqslant (1 - \xi)\frac{\mathrm{m}(K)}{B\sum_{L\in\mathcal{N}(K)}\tau_{K|L}}, \quad \textit{for all } K \in \mathcal{T}. \qquad (18.21)$$

Let $u_{\mathcal{T},k} \in X(\mathcal{T}, k)$ *be given by* (18.6)–(18.9).

Let $U = \|u_0\|_{L^\infty(\Omega)} + T\|f\|_{L^\infty(\Omega\times(0,T))}$ *and* B *be the Lipschitz constant of* φ *on* $[-U, U]$. *Then there exists* $F_1 \geqslant 0$, *which only depends on* Ω, T, φ, u_0, f *and* ξ *such that*

$$\big|\varphi(u_{\mathcal{T},k})\big|^2_{1,\mathcal{T},k} \leqslant F_1. \qquad (18.22)$$

PROOF. Let us first remark that the condition (18.21) is stronger than (18.13). Therefore, the result of Lemma 18.1 holds, i.e. $|u^n_K| \leqslant U$, for all $K \in \mathcal{T}$, $n \in \{0, \dots, N_k\}$. Multiplying Eq. (18.6) by ku^n_K, and summing the result over $n \in \{0, \dots, N_k\}$ and $K \in \mathcal{T}$ yields:

$$\begin{aligned}&\sum_{n=0}^{N_k}\sum_{K\in\mathcal{T}}\mathrm{m}(K)\big(u^{n+1}_K - u^n_K\big)u^n_K - \sum_{n=0}^{N_k}k\sum_{K\in\mathcal{T}}\sum_{L\in\mathcal{N}(K)}\tau_{K|L}\big(\varphi\big(u^n_L\big) - \varphi\big(u^n_K\big)\big)u^n_K \\ &\quad = \sum_{n=0}^{N_k}k\sum_{K\in\mathcal{T}}\mathrm{m}(K)u^n_K f^n_K. \qquad (18.23)\end{aligned}$$

In order to obtain a lower bound on the first term on the left-hand side of (18.23), let us first remark that:

$$\big(u^{n+1}_K - u^n_K\big)u^n_K = \tfrac{1}{2}\big(u^{n+1}_K\big)^2 - \tfrac{1}{2}\big(u^n_K\big)^2 - \tfrac{1}{2}\big(u^{n+1}_K - u^n_K\big)^2. \qquad (18.24)$$

Now, applying (18.6), using Young's inequality, the following inequality is obtained:

$$\begin{aligned}&\left(u_K^{n+1}-u_K^n\right)^2\\&\quad\leqslant k^2(1+\xi)\left[\left(\frac{1}{\mathrm{m}(K)}\sum_{L\in\mathcal{N}(K)}\tau_{K|L}\left(\varphi\left(u_L^n\right)-\varphi\left(u_K^n\right)\right)\right)^2+\frac{\left(f_K^n\right)^2}{\xi}\right].\end{aligned}\tag{18.25}$$

which yields in turn, using the Cauchy–Schwarz inequality:

$$\begin{aligned}\left(u_K^{n+1}-u_K^n\right)^2\leqslant{}&\frac{k^2}{\mathrm{m}(K)^2}(1+\xi)\left[\sum_{L\in\mathcal{N}(K)}\tau_{K|L}\right]\left[\sum_{L\in\mathcal{N}(K)}\tau_{K|L}\left(\varphi\left(u_L^n\right)-\varphi\left(u_K^n\right)\right)^2\right]\\&+\frac{(1+\xi)(k\,f_K^n)^2}{\xi}.\end{aligned}\tag{18.26}$$

Taking condition (18.21) into account gives:

$$\begin{aligned}\left(u_K^{n+1}-u_K^n\right)^2\leqslant{}&\left(1-\xi^2\right)\frac{k}{B\mathrm{m}(K)}\left[\sum_{L\in\mathcal{N}(K)}\tau_{K|L}\left(\varphi\left(u_L^n\right)-\varphi\left(u_K^n\right)\right)^2\right]\\&+\frac{(1+\xi)(k\,f_K^n)^2}{\xi}.\end{aligned}\tag{18.27}$$

Using (18.24) and (18.27) leads to the following lower bound on the first term of the left-hand side of (18.23):

$$\begin{aligned}&\sum_{n=0}^{N_k}\sum_{K\in\mathcal{T}}\mathrm{m}(K)\left(u_K^{n+1}-u_K^n\right)u_K^n\\&\quad\geqslant\tfrac12\sum_{K\in\mathcal{T}}\mathrm{m}(K)\left(\left(u_K^{N_k+1}\right)^2-\left(u_K^0\right)^2\right)\\&\qquad-\frac{1-\xi^2}{2B}\sum_{n=0}^{N_k}k\sum_{K\in\mathcal{T}}\left[\sum_{L\in\mathcal{N}(K)}\tau_{K|L}\left(\varphi\left(u_L^n\right)-\varphi\left(u_K^n\right)\right)^2\right]\\&\qquad-\frac{k(1+\xi)}{2\xi}\sum_{n=0}^{N_k}k\sum_{K\in\mathcal{T}}\mathrm{m}(K)\left(f_K^n\right)^2.\end{aligned}\tag{18.28}$$

Let us now handle the second term on the left-hand side of (18.23). Let $\phi\in C(\mathbb{R},\mathbb{R})$ be defined by $\phi(x)=x\varphi(x)-\int_{x_0}^x\varphi(y)\,\mathrm{d}y$, where $x_0\in\mathbb{R}$ is an arbitrary given real value. Then the following equality holds:

$$\phi\left(u_L^n\right)-\phi\left(u_K^n\right)=u_K^n\left(\varphi\left(u_L^n\right)-\varphi\left(u_K^n\right)\right)-\int_{u_K^n}^{u_L^n}\left(\varphi(x)-\varphi\left(u_L^n\right)\right)\mathrm{d}x.\tag{18.29}$$

The following technical lemma is used here and several times in the sequel:

LEMMA 18.5. *Let $g:\mathbb{R}\to\mathbb{R}$ be a monotone Lipschitz continuous function, with a Lipschitz constant $G>0$. Then*:

$$\left|\int_c^d \big(g(x)-g(c)\big)\,\mathrm{d}x\right| \geqslant \frac{1}{2G}\big(g(d)-g(c)\big)^2, \quad \forall c,d\in\mathbb{R}. \tag{18.30}$$

PROOF. In order to prove Lemma 18.5, we assume, e.g., that g is nondecreasing and $c<d$ (the other cases are similar). Then, one has $g(s)\geqslant h(s)$, for all $s\in[c,d]$, where $h(s)=g(c)$ for $s\in[c,d-l]$ and $h(s)=g(c)+(s-d+l)G$ for $s\in[d-l,d]$, with $lG=g(d)-g(c)$, and therefore:

$$\begin{aligned}\int_c^d \big(g(s)-g(c)\big)\,\mathrm{d}s &\geqslant \int_c^d \big(h(s)-g(c)\big)\,\mathrm{d}s = \frac{l}{2}\big(g(d)-g(c)\big)\\ &= \frac{1}{2G}\big(g(d)-g(c)\big)^2,\end{aligned}$$

this completes the proof of Lemma 18.5. □

Using Lemma 18.5, (18.29) and the equality $\sum_{K\in\mathcal{T}}\sum_{L\in\mathcal{N}(K)}\tau_{K|L}(\phi(u_L^n)-\phi(u_K^n))=0$ yields:

$$\begin{aligned}&-\sum_{n=0}^{N_k} k\sum_{K\in\mathcal{T}}\sum_{L\in\mathcal{N}(K)}\tau_{K|L}\big(\varphi(u_L^n)-\varphi(u_K^n)\big)u_K^n\\ &\quad\geqslant \frac{1}{2B}\sum_{n=0}^{N_k} k\sum_{K\in\mathcal{T}}\sum_{L\in\mathcal{N}(K)}\tau_{K|L}\big(\varphi(u_L^n)-\varphi(u_K^n)\big)^2.\end{aligned} \tag{18.31}$$

Since $k<T$ we deduce from (18.14) that the right-hand side of Eq. (18.23) satisfies

$$\left|\sum_{n=0}^{N_k} k\sum_{K\in\mathcal{T}}\mathrm{m}(K)u_K^n f_K^n\right| \leqslant 2T\mathrm{m}(\Omega)U\|f\|_{L^\infty(\Omega\times(0,2T))}. \tag{18.32}$$

Relations $k<T$, (18.23), (18.28), (18.31) and (18.32) lead to

$$\begin{aligned}&\frac{\xi^2}{2B}\sum_{n=0}^{N_k} k\sum_{K\in\mathcal{T}}\sum_{L\in\mathcal{N}(K)}\tau_{K|L}\big(\varphi(u_L^n)-\varphi(u_K^n)\big)^2\\ &\quad\leqslant 2T\mathrm{m}(\Omega)\|f\|_{L^\infty(\Omega\times(0,2T))}\left(U+\frac{1+\xi}{2\xi}\|f\|_{L^\infty(\Omega\times(0,2T))}T\right)\\ &\qquad+\tfrac{1}{2}\mathrm{m}(\Omega)\|u_0\|^2_{L^\infty(\Omega)}\end{aligned} \tag{18.33}$$

which concludes the proof of the lemma. □

REMARK 18.9. Estimate (18.22) also holds for the implicit scheme, without any condition on k. One multiplies (18.10) by u_K^{n+1}: the last term on the right-hand side of (18.24) appears with the opposite sign, which considerably simplifies the previous proof.

Time translates of approximate solutions
In order to fulfill the hypotheses of Kolmogorov's theorem, the study of time translates must now be performed. The following estimate holds:

LEMMA 18.6. *Under Assumption* 18.1, *let $\mathcal{T}$ be an admissible mesh in the sense of Definition* 10.1 *and $k \in (0, T)$. Let $u_{\mathcal{T},k} \in X(\mathcal{T}, k)$ be given by* (18.6)–(18.9). *Let $U = \|u_{\mathcal{T},k}\|_{L^\infty(\Omega\times(0,T))}$ and B be the Lipschitz constant of φ on $[-U, U]$. Then*:

$$\begin{aligned} &\big\|\varphi(u_{\mathcal{T},k}(\bullet,\bullet+\tau)) - \varphi(u_{\mathcal{T},k}(\bullet,\bullet))\big\|^2_{L^2(\Omega\times(0,T-\tau))} \\ &\quad\leqslant 2B\tau\big(|\varphi(u_{\mathcal{T},k})|^2_{1,\mathcal{T},k} + BT\mathrm{m}(\Omega)U\|f\|_{L^\infty(\Omega\times(0,T))}\big), \quad \forall\tau\in(0,T). \end{aligned} \tag{18.34}$$

PROOF. Let $\tau \in (0, T)$. Since B is the Lipschitz constant of φ on $[-U, U]$, $U = \|u_{\mathcal{T},k}\|_{L^\infty(\Omega\times(0,T))}$ and φ is nondecreasing, the following inequality holds:

$$\begin{aligned} &\int_{\Omega\times(0,T-\tau)} \big(\varphi\big(u_{\mathcal{T},k}(x,t+\tau)\big) - \varphi\big(u_{\mathcal{T},k}(x,t)\big)\big)^2\,\mathrm{d}x\,\mathrm{d}t \\ &\quad\leqslant B\int_0^{T-\tau} A(t)\,\mathrm{d}t, \end{aligned} \tag{18.35}$$

where, for almost every $t \in (0, T - \tau)$,

$$A(t) = \int_\Omega \big(\varphi\big(u_{\mathcal{T},k}(x,t+\tau)\big) - \varphi\big(u_{\mathcal{T},k}(x,t)\big)\big)\big(u_{\mathcal{T},k}(x,t+\tau) - u_{\mathcal{T},k}(x,t)\big)\,\mathrm{d}x.$$

Let $t \in (0, T - \tau)$. Using the definition of $u_{\mathcal{T},k}$ (18.9), this may also be written:

$$A(t) = \sum_{K\in\mathcal{T}} \mathrm{m}(K)\big(\varphi\big(u_K^{n_1(t)}\big) - \varphi\big(u_K^{n_0(t)}\big)\big)\big(u_K^{n_1(t)} - u_K^{n_0(t)}\big), \tag{18.36}$$

with $n_0(t), n_1(t) \in \{0, \ldots, N_k\}$ such that $n_0(t)k \leqslant t < (n_0(t)+1)k$ and $n_1(t)k \leqslant t + \tau < (n_1(t)+1)k$. Equality (18.36) may be written as

$$A(t) = \sum_{K\in\mathcal{T}} \big(\varphi\big(u_K^{n_1(t)}\big) - \varphi\big(u_K^{n_0(t)}\big)\big)\left(\sum_{n=n_0(t)+1}^{n_1(t)} \mathrm{m}(K)\big(u_K^n - u_K^{n-1}\big)\right),$$

which also writes

$$A(t) = \sum_{K\in\mathcal{T}} \big(\varphi\big(u_K^{n_1(t)}\big) - \varphi\big(u_K^{n_0(t)}\big)\big)\left(\sum_{n=1}^{N_k} \chi_n(t,t+\tau)\mathrm{m}(K)\big(u_K^n - u_K^{n-1}\big)\right), \tag{18.37}$$

with $\chi_n(t,t+\tau)=1$ if $nk \in (t,t+\tau]$ and $\chi_n(t,t+\tau)=0$ if $nk \notin (t,t+\tau]$.

In (18.37), the order of summation between n and K is changed and the scheme (18.6) is used. Hence,

$$\begin{aligned} A(t) &= k\sum_{n=1}^{N_k}\chi_n(t,t+\tau)\Bigg[\sum_{K\in\mathcal{T}}\big(\varphi\big(u_K^{n_1(t)}\big)-\varphi\big(u_K^{n_0(t)}\big)\big) \\ &\quad\times\Bigg(\sum_{L\in\mathcal{N}(K)}\tau_{K|L}\big(\varphi\big(u_L^{n-1}\big)-\varphi\big(u_K^{n-1}\big)\big)+\mathrm{m}(K)f_K^{n-1}\Bigg)\Bigg]. \end{aligned}$$

Gathering by edges, this yields:

$$\begin{aligned} A(t) &= k\sum_{n=1}^{N_k}\Bigg[\sum_{K|L\in\mathcal{E}_{\rm int}}\tau_{K|L}\big(\varphi\big(u_K^{n_1(t)}\big)-\varphi\big(u_L^{n_1(t)}\big)-\varphi\big(u_K^{n_0(t)}\big)+\varphi\big(u_L^{n_0(t)}\big)\big) \\ &\quad\times\big(\varphi\big(u_L^{n-1}\big)-\varphi\big(u_K^{n-1}\big)\big)+\sum_{K\in\mathcal{T}}\big(\varphi\big(u_K^{n_1(t)}\big) \\ &\quad-\varphi\big(u_K^{n_0(t)}\big)\big)\mathrm{m}(K)f_K^{n-1}\Bigg]\chi_n(t,t+\tau). \end{aligned}$$

Using the inequality $2ab \leqslant a^2+b^2$, this yields:

$$A(t)\leqslant \tfrac{1}{2}A_0(t)+\tfrac{1}{2}A_1(t)+A_2(t)+A_3(t), \tag{18.38}$$

with

$$\begin{aligned} A_0(t) &= k\sum_{n=1}^{N_k}\chi_n(t,t+\tau)\Bigg(\sum_{K|L\in\mathcal{E}_{\rm int}}\tau_{K|L}\big(\varphi\big(u_L^{n_0(t)}\big)-\varphi\big(u_K^{n_0(t)}\big)\big)^2\Bigg), \\ A_1(t) &= k\sum_{n=1}^{N_k}\chi_n(t,t+\tau)\Bigg(\sum_{K|L\in\mathcal{E}_{\rm int}}\tau_{K|L}\big(\varphi\big(u_L^{n_1(t)}\big)-\varphi\big(u_K^{n_1(t)}\big)\big)^2\Bigg), \\ A_2(t) &= k\sum_{n=1}^{N_k}\chi_n(t,t+\tau)\Bigg(\sum_{K|L\in\mathcal{E}_{\rm int}}\tau_{K|L}\big(\varphi\big(u_L^{n-1}\big)-\varphi\big(u_K^{n-1}\big)\big)^2\Bigg), \end{aligned}$$

and

$$A_3(t)=k\sum_{n=1}^{N_k}\chi_n(t,t+\tau)\Bigg(\sum_{K\in\mathcal{T}}\big(\varphi\big(u_K^{n_1(t)}\big)-\varphi\big(u_K^{n_0(t)}\big)\big)\mathrm{m}(K)f_K^{n-1}\Bigg).$$

Note that, since $t \in (0, T-\tau)$, $n_0(t) \in \{0, \dots, N_k\}$, and, for $m \in \{0, \dots, N_k\}$, $n_0(t) = m$ if and only if $t \in [mk, (m+1)k)$. Therefore,

$$\int_0^{T-\tau} A_0(t)\,\mathrm{d}t \leqslant \sum_{m=0}^{N_k} \int_{mk}^{(m+1)k} k \sum_{n=1}^{N_k} \chi_n(t, t+\tau) \times \left(\sum_{K|L \in \mathcal{E}_{\mathrm{int}}} \tau_{K|L} \left(\varphi\left(u_L^m\right) - \varphi\left(u_K^m\right)\right)^2 \right) \mathrm{d}t,$$

which also writes

$$\int_0^{T-\tau} A_0(t)\,\mathrm{d}t \leqslant \sum_{m=0}^{N_k} k \int_{mk}^{(m+1)k} \left(\sum_{n=1}^{N_k} \chi_n(t, t+\tau) \right) \mathrm{d}t \times \sum_{K|L \in \mathcal{E}_{\mathrm{int}}} \tau_{K|L} \left(\varphi\left(u_L^m\right) - \varphi\left(u_K^m\right)\right)^2. \tag{18.39}$$

The change of variable $t = s + (n-m)k$ yields

$$\begin{aligned} \int_{mk}^{(m+1)k} \chi_n(t, t+\tau)\,\mathrm{d}t &= \int_{2mk-nk}^{2mk-nk+k} \chi_n\big(s + (n-m)k, s + (n-m)k + \tau\big)\,\mathrm{d}s \\ &= \int_{2mk-nk}^{2mk-nk+k} \chi_m(s, s+\tau)\,\mathrm{d}s, \end{aligned}$$

then, for all $m \in \{0, \dots, N_k\}$,

$$\int_{mk}^{(m+1)k} \left(\sum_{n=1}^{N_k} \chi_n(t, t+\tau) \right) \mathrm{d}t \leqslant \int_{\mathbb{R}} \chi_m(s, s+\tau)\,\mathrm{d}s = \tau,$$

since $\chi_m(s, s+\tau) = 1$ if and only if $mk \in (s, s+\tau]$ which is equivalent to $s \in [mk - \tau, mk)$.

Therefore (18.39) yields

$$\int_0^{T-\tau} A_0(t)\,\mathrm{d}t \leqslant \tau \left|\varphi(u_{\mathcal{T},k})\right|_{1,\mathcal{T},k}^2. \tag{18.40}$$

Similarly:

$$\int_0^{T-\tau} A_1(t)\,\mathrm{d}t \leqslant \tau \left|\varphi(u_{\mathcal{T},k})\right|_{1,\mathcal{T},k}^2. \tag{18.41}$$

Let us now study the term $\int_0^{T-\tau} A_2(t)\,\mathrm{d}t$:

$$\int_0^{T-\tau} A_2(t)\,\mathrm{d}t \leqslant \sum_{n=1}^{N_k} k \sum_{K|L\in\mathcal{E}_{\mathrm{int}}} \tau_{K|L}\big(\varphi\big(u_L^{n-1}\big)-\varphi\big(u_K^{n-1}\big)\big)^2 \times \int_0^{T-\tau} \chi_n(t,t+\tau)\,\mathrm{d}t. \tag{18.42}$$

Since $\int_0^{T-\tau} \chi_n(t,t+\tau) \leqslant \tau$ (recall that $\chi_n(t,t+\tau)=1$ if and only if $t\in[nk-\tau,nk)$), the following inequality holds:

$$\int_0^{T-\tau} A_2(t)\,\mathrm{d}t \leqslant \tau\big|\varphi(u_{\mathcal{T},k})\big|_{1,\mathcal{T},k}^2. \tag{18.43}$$

In the same way:

$$\begin{aligned}\int_0^{T-\tau} A_3(t)\,\mathrm{d}t &\leqslant \sum_{n=1}^{N_k} k\bigg(\sum_{K\in\mathcal{T}} \mathrm{m}(K)2BU\|f\|_{L^\infty(\Omega\times(0,T))}\bigg)\int_0^{T-\tau}\chi_n(t,t+\tau)\,\mathrm{d}t\\ &\leqslant \tau T\mathrm{m}(\Omega)2BU\|f\|_{L^\infty(\Omega\times(0,T))}.\end{aligned} \tag{18.44}$$

Using inequalities (18.35), (18.38) and (18.40)–(18.44), (18.34) is proved. □

REMARK 18.10. Estimate (18.34) is again true for the implicit scheme, with $\|f\|_{L^\infty(\Omega\times(0,2T))}$ instead of $\|f\|_{L^\infty(\Omega\times(0,T))}$.

An immediate corollary of Lemma 18.6 is the following.

LEMMA 18.7. *Under Assumption* 18.1, *let $\mathcal{T}$ be an admissible mesh in the sense of Definition* 10.1 *and $k\in(0,T)$. Let $u_{\mathcal{T},k}\in X(\mathcal{T},k)$ be given by* (18.6)–(18.9). *Let $U=\|u_{\mathcal{T},k}\|_{L^\infty(\Omega\times(0,T)}$ and B be the Lipschitz constant of φ on $[-U,U]$. One defines $\tilde u$ by $\tilde u=u_{\mathcal{T},k}$ a.e. on $\Omega\times(0,T)$, and $\tilde u=0$ a.e. on $\mathbb{R}^{d+1}\setminus\Omega\times(0,T)$. Then*:

$$\begin{aligned}&\big\|\varphi\big(\tilde u(\bullet,\bullet+\tau)\big)-\varphi\big(\tilde u(\bullet,\bullet)\big)\big\|_{L^2(\mathbb{R}^{d+1})}^2\\ &\quad\leqslant 2|\tau|B\big(\big|\varphi(u_{\mathcal{T},k})\big|_{1,\mathcal{T},k}^2+BT\mathrm{m}(\Omega)U\|f\|_{L^\infty(\Omega\times(0,T))}+B\mathrm{m}(\Omega)U^2\big),\\ &\quad\forall\tau\in\mathbb{R}.\end{aligned}$$

18.4. Convergence

THEOREM 18.1. *Under Assumption* 18.1, *let $U=\|u_0\|_{L^\infty(\Omega)}+T\|f\|_{L^\infty(\Omega\times(0,T))}$ and*

$$B=\sup_{-U\leqslant x<y\leqslant U}\frac{\varphi(x)-\varphi(y)}{x-y}.$$

Let $\xi \in (0,1)$ be a given real value. For $m \in \mathbb{N}$, let $\mathcal{T}_m$ be an admissible mesh in the sense of Definition 10.1 *and $k_m \in (0,T)$ satisfying the condition* (18.21) *with $\mathcal{T} = \mathcal{T}_m$ and $k = k_m$. Let $u_{\mathcal{T}_m,k_m}$ be given by* (18.6)–(18.9) *with $\mathcal{T} = \mathcal{T}_m$ and $k = k_m$. Assume that* $\mathrm{size}(\mathcal{T}_m) \to 0$ *as $m \to \infty$.*

Then, there exists a subsequence of the sequence of approximate solutions, still denoted by $(u_{\mathcal{T}_m,k_m})_{m\in\mathbb{N}}$, which converges to a weak solution u of problem (18.1)–(18.3), *as $m \to \infty$, in the following sense:*

(i) *$u_{\mathcal{T}_m,k_m}$ converges to u in $L^\infty(\Omega \times (0,T))$, for the weak-$\star$ topology as m tends to $+\infty$,*

(ii) *$\varphi(u_{\mathcal{T}_m,k_m})$ converges to $\varphi(u)$ in $L^1(\Omega \times (0,T))$ as m tends to $+\infty$,*

where $u_{\mathcal{T}_m,k_m}$ and $\varphi(u_{\mathcal{T}_m,k_m})$ also denote the restrictions of these functions to $\Omega \times (0,T)$.

PROOF. Let us set $u_m = u_{\mathcal{T}_m,k_m}$ and assume, without loss of generality, that $\varphi(0) = 0$. First remark that, by (18.21), $k_m \to 0$ as $m \to 0$. Thanks to Lemma 18.1, the sequence $(u_m)_{m\in\mathbb{N}}$ is bounded in $L^\infty(\Omega \times (0,T)$. Then, there exists a subsequence, still denoted by $(u_m)_{m\in\mathbb{N}}$, such that u_m converges, as $m \to \infty$, to u in $L^\infty(\Omega \times (0,T))$, for the weak-$\star$ topology.

For the study of the sequence $(\varphi(u_m))_{m\in\mathbb{N}}$, we shall apply Theorem 14.1 with $N = d+1$, $q = 2$, $\omega = \Omega \times (0,T)$ and $p(v) = \tilde{v}$ with $\tilde{v}$ defined, as usual, by $\tilde{v} = v$ on $\Omega \times (0,T)$ and $\tilde{v} = 0$ on $\mathbb{R}^{d+1} \setminus \Omega \times (0,T)$.

The first and second items of Theorem 14.1 are clearly satisfied; let us prove hereafter that the third is also satisfied. By Lemma 18.4, the sequence $(|\varphi(u_m)|_{1,\mathcal{T}_m,k_m})_{m\in\mathbb{N}}$ is bounded. Let $\eta \in \mathbb{R}^d$ and $\tau \in \mathbb{R}$, since

$$\begin{aligned}
&\big\|\varphi\big(\tilde{u}_m(\bullet+\eta,\bullet+\tau)\big) - \varphi\big(\tilde{u}_m(\bullet,\bullet)\big)\big\|_{L^2(\mathbb{R}^{d+1})} \\
&\quad\leqslant \big\|\varphi(\tilde{u}_m(\bullet+\eta,\bullet)) - \varphi(\tilde{u}_m(\bullet,\bullet))\big\|_{L^2(\mathbb{R}^{d+1})} \\
&\qquad + \big\|\varphi\big(\tilde{u}_m(\bullet,\bullet+\tau)\big) - \varphi\big(\tilde{u}_m(\bullet,\bullet)\big)\big\|_{L^2(\mathbb{R}^{d+1})},
\end{aligned}$$

Lemmata 18.3 and 18.7 give the third item of Theorem 14.1 and this yields the compactness of the sequence $(\varphi(u_m))_{m\in\mathbb{N}}$ in $L^2(\Omega \times (0,T))$.

Therefore, there exists a subsequence, still denoted by $(\varphi(u_m))_{m\in\mathbb{N}}$, and there exists $\chi \in L^2(\Omega \times (0,T))$ such that $\varphi(u_{\mathcal{T}_m,k_m})$ converges, as $m \to \infty$, to χ in $L^2(\Omega \times (0,T))$. Indeed, since $(\varphi(u_m))_{m\in\mathbb{N}}$ is bounded in $L^\infty(\Omega \times (0,T))$, this convergence holds in $L^q(\Omega \times (0,T))$ for all $1 \leqslant q < \infty$. Furthermore, since φ is nondecreasing, Theorem 18.2 (see p. 863) gives that $\chi = \varphi(u)$.

Up to now, the following properties have been shown to be satisfied by a convenient subsequence:

(i) $(u_m)_{m\in\mathbb{N}}$ converges to u, as $m \to \infty$, in $L^\infty(\Omega \times (0,T))$ for the weak-$\star$ topology,

(ii) $(\varphi(u_m))_{m\in\mathbb{N}}$ converges to $\varphi(u)$ in $L^1(\Omega \times (0,T))$ (and even in $L^p(\Omega \times (0,T))$ for all $p \in [0,\infty)$).

There remains to show that u is a weak solution of problem (18.1)–(18.3), which concludes the proof of Theorem 18.1.

Let $m \in \mathbb{N}$. For the sake of simplicity, we shall use the notations $\mathcal{T} = \mathcal{T}_m$, $h = \text{size}(\mathcal{T})$ and $k = k_m$. Let $\psi \in \mathcal{A}_T$. We multiply (18.6) by $k\psi(x_K, nk)$, and sum the result on $n \in \{0, \dots, N_k\}$ and $K \in \mathcal{T}$. We obtain

$$T_{1m} + T_{2m} = T_{3m}, \tag{18.45}$$

with

$$T_{1m} = \sum_{n=0}^{N_k} \sum_{K \in \mathcal{T}} \mathrm{m}(K)\big(u_K^{n+1} - u_K^n\big)\psi(x_K, nk),$$

$$T_{2m} = -\sum_{n=0}^{N_k} k \sum_{K \in \mathcal{T}} \sum_{L \in \mathcal{N}(K)} \tau_{K|L}\big(\varphi(u_L^n) - \varphi(u_K^n)\big)\psi(x_K, nk),$$

and

$$T_{3m} = \sum_{n=0}^{N_k} k \sum_{K \in \mathcal{T}} \psi(x_K, nk)\mathrm{m}(K) f_K^n.$$

We first consider T_{1m}.

$$\begin{aligned} T_{1m} &= \sum_{n=1}^{N_k} \sum_{K \in \mathcal{T}} \mathrm{m}(K) u_K^n \big(\psi(x_K, (n-1)k) - \psi(x_K, nk)\big) \\ &\quad + \sum_{K \in \mathcal{T}} \mathrm{m}(K)\big(u_K^{N_k+1}\psi(x_K, kN_k) - u_K^0 \psi(x_K, 0)\big). \end{aligned}$$

Performing one more step of the induction in Lemma 18.1, it is clear that $|u_K^{N_k+1}| < U + 2T\|f\|_{L^\infty(\Omega\times(0,2T))}$, for all $K \in \mathcal{T}$.

Since $0 < T - N_k k \leqslant k$, there exists $C_{1,\psi}$ which only depends on ψ, T and Ω, such that $|\psi(x_K, N_k k)| \leqslant kC_{1,\psi}$. Hence,

$$\sum_{K \in \mathcal{T}} \mathrm{m}(K) u_K^{N_k+1} \psi(x_K, kN_k) \to 0 \quad \text{as } m \to \infty.$$

Since

$$\bigg\| \sum_{K \in \mathcal{T}} u_K^0 1_K - u_0 \bigg\|_{L^1(\Omega)} \to 0, \quad \text{as } m \to \infty$$

(where $1_K(x) = 1$ if $x \in K$, 0 otherwise), one has

$$\sum_{K \in \mathcal{T}} \mathrm{m}(K) u_K^0 \psi(x_K, 0) \to \int_\Omega u_0(x)\psi(x, 0)\,\mathrm{d}x \quad \text{as } m \to \infty.$$

Since $(u_m)_{m\in\mathbb{N}}$ converges, as $m \to +\infty$, to u in $L^\infty(\Omega \times (0,T))$, for the weak-$\star$ topology, and since $|u_K^{N_k}| < U + T\|f\|_{L^\infty(\Omega\times(0,T))}$, for all $K \in \mathcal{T}$, the following property also holds:

$$\sum_{n=1}^{N_k}\sum_{K\in\mathcal{T}} \mathrm{m}(K)u_K^n\big(\psi(x_K,(n-1)k) - \psi(x_K,nk)\big)$$
$$\to -\int_0^T\int_\Omega u(x,t)\psi_t(x,t)\,\mathrm{d}x\,\mathrm{d}t \quad \text{as } m \to \infty.$$

Therefore,

$$T_{1m} \to -\int_0^T\int_\Omega u(x,t)\psi_t(x,t)\,\mathrm{d}x\,\mathrm{d}t - \int_\Omega u_0(x)\psi(x,0)\,\mathrm{d}x, \quad \text{as } m \to \infty.$$

We now study T_{2m}. This term can be rewritten as

$$T_{2m} = -\sum_{n=0}^{N_k} k \sum_{K|L\in\mathcal{E}_{\mathrm{int}}} m(K|L)\big(\varphi(u_L^n) - \varphi(u_K^n)\big)\frac{\psi(x_K,nk) - \psi(x_L,nk)}{d_{K|L}}.$$

It is useful to introduce the following expression:

$$\begin{aligned}
T'_{2m} &= \sum_{n=0}^{N_k}\int_{nk}^{(n+1)k}\int_\Omega \varphi\big(u_{\mathcal{T},k}(x,t)\big)\Delta\psi(x,nk)\,\mathrm{d}x\,\mathrm{d}t \\
&= \sum_{n=0}^{N_k} k \sum_{K\in\mathcal{T}} \varphi(u_K^n)\int_K \Delta\psi(x,nk)\,\mathrm{d}x \\
&= \sum_{n=0}^{N_k} k \sum_{K|L\in\mathcal{E}_{\mathrm{int}}} \big(\varphi(u_K^n) - \varphi(u_L^n)\big)\int_{K|L}\nabla\psi(x,nk)\cdot \boldsymbol{n}_{K,L}\,\mathrm{d}\gamma(x).
\end{aligned}$$

The sequence $(\varphi(u_m))_{m\in\mathbb{N}}$ converges to $\varphi(u)$ in $L^1(\Omega \times (0,T))$; furthermore, it is bounded in L^∞ so that the integral between T and $(N_k+1)k$ tends to 0. Therefore:

$$T'_{2m} \to \int_0^T\int_\Omega \varphi\big(u(x,t)\big)\Delta\psi(x,t)\,\mathrm{d}x\,\mathrm{d}t, \quad \text{as } m \to \infty.$$

The term $T_{2m} + T'_{2m}$ can be written as

$$T_{2m} + T'_{2m} = \sum_{n=0}^{N_k} k \sum_{K|L\in\mathcal{E}} m(K|L)\big(\varphi(u_K^n) - \varphi(u_L^n)\big)R_{K,L}^n,$$

with

$$R_{K,L}^n = \frac{1}{m(K|L)} \int_{K|L} \nabla\psi(x, nk) \cdot \boldsymbol{n}_{K,L} \, \mathrm{d}\gamma(x) - \frac{\psi(x_L, nk) - \psi(x_K, nk)}{d_{K|L}}.$$

Thanks to the regularity properties of ψ there exists C_ψ, which only depends on ψ, such that $|R_{K,L}^n| \leqslant C_\psi h$. Then, using the estimate (18.22), we conclude that $T_{2m} + T'_{2m} \to 0$ as $m \to \infty$. Therefore,

$$T_{2m} \to -\int_0^T \int_\Omega \varphi\big(u(x,t)\big) \Delta\psi(x,t) \, \mathrm{d}x \, \mathrm{d}t, \quad \text{as } m \to \infty.$$

Let us now study T_{3m}.

Define $f_{\mathcal{T},k} \in X(\mathcal{T}, k)$ by $f_{\mathcal{T},k}(x,t) = f_K^n$ if $(x,t) \in K \times (nk, nk+k)$. Since $f_{\mathcal{T},k} \to f$ in $L^1(\Omega \times (0,T)$ and since $f \in L^\infty(\Omega \times (0, 2T)$,

$$T_{3m} \to \int_\Omega \int_0^T f(x,t)\psi(x,t) \, \mathrm{d}t \, \mathrm{d}x, \quad \text{as } m \to \infty.$$

Passing to the limit in Eq. (18.45) gives that u is a weak solution of problem (18.1)–(18.3). This concludes the proof of Theorem 18.1. □

REMARK 18.11. This convergence proof is quite similar in the case of the implicit scheme, with the additional condition that $(k_m)_{m\in\mathbb{N}}$ converges to zero, since condition (18.21) does not have to be satisfied.

REMARK 18.12. The above convergence result was shown for a subsequence only. A convergence theorem is obtained for the full set of approximate solutions, if a uniqueness result is valid. Such a result can be easily obtained in the case of a smooth boundary and is given in Section 18.6 below. For this case, an extension to the Definition 10.1 of admissible meshes is given hereafter.

DEFINITION 18.4 (*Admissible meshes for regular domains*). Let Ω be an open bounded connected subset of $\mathbb{R}^d$, $d = 2$ or 3 with a C^2 boundary $\partial\Omega$. An admissible finite volume mesh of Ω is given by an open bounded polygonal set Ω' containing Ω, and an admissible mesh $\mathcal{T}'$ of Ω' in the sense of Definition 10.1. The set of control volumes of the mesh of Ω are $\{K' \cap \Omega,\ K' \in \mathcal{T}'$ such that $m_d(K' \cap \Omega) > 0\}$ and the set of edges of the mesh is $\mathcal{E} = \{\sigma \cap \overline{\Omega},\ \sigma \in \mathcal{E}'$ such that $m_{d-1}(\sigma \cap \overline{\Omega}) > 0\}$, where $\mathcal{E}'$ denotes the set of edges of $\mathcal{T}'$ and m_N denotes the N-dimensional Lebesgue measure.

REMARK 18.13. For smooth domains Ω, the set of edges $\mathcal{E}$ of an admissible mesh of Ω does not contain the parts of the boundaries of the control volumes which are included in the boundary $\partial\Omega$ of Ω.

18.5. Weak convergence and nonlinearities

We show here a property which was used in the proof of Theorem 18.1.

THEOREM 18.2. *Let $U > 0$ and $\varphi \in C([-U, U])$ be a nondecreasing function. Let ω be an open bounded subset of $\mathbb{R}^N$, $N \geqslant 1$. Let $(u_n)_{n\in\mathbb{N}} \subset L^\infty(\omega)$ such that*

(i) *$-U \leqslant u_n \leqslant U$ a.e. in ω, for all $n \in \mathbb{N}$;*

(ii) *there exists $u \in L^\infty(\omega)$ such that $(u_n)_{n\in\mathbb{N}}$ converges to u in $L^\infty(\omega)$ for the weak-$\star$ topology;*

(iii) *there exists a function $\chi \in L^1(\omega)$ such that $(\varphi(u_n))_{n\in\mathbb{N}}$ converges to χ in $L^1(\omega)$. Then $\chi(x) = \varphi(u(x))$, for a.e. $x \in \omega$.*

PROOF. First we extend the definition of φ by $\varphi(v) = \varphi(-U) + v + U$ for all $v < -U$ and $\varphi(v) = \varphi(U) + v - U$ for all $v > U$, and denote again by φ this extension of φ which now maps $\mathbb{R}$ into $\mathbb{R}$, is continuous and nondecreasing. Let us define $\alpha_\pm$ from $\mathbb{R}$ to $\mathbb{R}$ by $\alpha_-(t) = \inf\{v \in \mathbb{R}, \varphi(v) = t\}$ and $\alpha_+(t) = \sup\{v \in \mathbb{R}, \varphi(v) = t\}$, for all $t \in \mathbb{R}$.

Note that the functions $\alpha_\pm$ are increasing and that

(i) α_- is left continuous and therefore lower semi-continuous, i.e.

$$t = \lim_{n\to\infty} t_n \Rightarrow \alpha_-(t) \leqslant \liminf_{n\to\infty} \alpha_-(t_n),$$

(ii) α_+ is right continuous and therefore upper semi-continuous, i.e.

$$t = \lim_{n\to\infty} t_n \Rightarrow \alpha_+(t) \geqslant \limsup_{n\to\infty} \alpha_+(t_n).$$

Thus, since we may assume, up to a subsequence, that $\varphi(u_n) \to \chi$ a.e. in ω,

$$\alpha_-(\chi(x)) \leqslant \liminf_{n\to\infty} \alpha_-\big(\varphi\big(u_n(x)\big)\big) \leqslant \limsup_{n\to\infty} \alpha_+\big(\varphi\big(u_n(x)\big)\big) \leqslant \alpha_+\big(\chi(x)\big), \tag{18.46}$$

for a.e. $x \in \omega$.

A direct application of the definition of the functions α_- and α_+ gives

$$\alpha_-\big(\varphi\big(u_n(x)\big)\big) \leqslant u_n(x) \leqslant \alpha_+\big(\varphi\big(u_n(x)\big)\big). \tag{18.47}$$

Let $L^1_+ = \{\psi \in L^1(\omega),\ \psi \geqslant 0 \text{ a.e.}\}$. Let $\psi \in L^1_+$. We multiply (18.47) by $\psi(x)$ and integrate over ω, it yields

$$\begin{aligned}\int_\omega \alpha_-\big(\varphi\big(u_n(x)\big)\big)\psi(x)\,\mathrm{d}x &\leqslant \int_\omega u_n(x)\psi(x)\,\mathrm{d}x \\ &\leqslant \int_\omega \alpha_+\big(\varphi\big(u_n(x)\big)\big)\psi(x)\,\mathrm{d}x.\end{aligned} \tag{18.48}$$

Applying Fatou's lemma to the sequences of L^1 positive functions $\alpha_-(\varphi(u_n))\psi - \alpha_-(\varphi(-U))\psi$ and $\alpha_+(\varphi(U))\psi - \alpha_+(\varphi(u_n))\psi$ yields, with (18.46),

$$\int_\omega \alpha_-\big(\chi(x)\big)\psi(x)\,\mathrm{d}x \leqslant \liminf_{n\to\infty} \int_\omega \alpha_-\big(\varphi\big(u_n(x)\big)\big)\psi(x)\,\mathrm{d}x,$$

and

$$\limsup_{n\to\infty} \int_\omega \alpha_+\big(\varphi\big(u_n(x)\big)\big)\psi(x)\,\mathrm{d}x \leqslant \int_\omega \alpha_+\big(\chi(x)\big)\psi(x)\,\mathrm{d}x.$$

Then, passing to the lim inf and lim sup in (18.48) and using the convergence of $(u_n)_{n\in\mathbb{N}}$ to u in $L^\infty(\omega)$ for the weak-$\star$ topology gives

$$\int_\omega \alpha_-\big(\chi(x)\big)\psi(x)\,\mathrm{d}x \leqslant \int_\omega u(x)\psi(x)\,\mathrm{d}x \leqslant \int_\omega \alpha_+\big(\chi(x)\big)\psi(x)\,\mathrm{d}x.$$

Thus, since ψ is arbitrary in L^1_+, the following inequality holds for a.e. $x \in \omega$:

$$\alpha_-\big(\chi(x)\big) \leqslant u(x) \leqslant \alpha_+\big(\chi(x)\big),$$

which implies in turn that $\chi(x) = \varphi(u(x))$ for a.e. $x \in \omega$. This completes the proof of Theorem 18.2. □

REMARK 18.14. Another proof of Theorem 18.2 is possible by passing to the limit in the inequality

$$0 \leqslant \int_\omega \big(\varphi(u_n)(x) - \varphi\big(v(x)\big)\big)\big(u_n(x) - v(x)\big)\,\mathrm{d}x, \quad \forall v \in L^\infty(\omega),$$

which leads to

$$0 \leqslant \int_\omega \big(\chi(x) - \varphi\big(v(x)\big)\big)\big(u(x) - v(x)\big)\,\mathrm{d}x, \quad \forall v \in L^\infty(\omega).$$

From this inequality, one deduces that $\chi = \varphi(u)$ a.e. on ω.

A third proof is possible by using the concept of nonlinear weak-$\star$ convergence, see Definition 30.1.

18.6. A uniqueness result for nonlinear diffusion equations

The uniqueness of the weak solution to variations of problem (18.1)–(18.3) has been proved by several authors. For precise references we refer to MEIRMANOV [1992]. Also rather similar proofs have been given in BERTSCH, KERSNER and PELETIER [1995] and GUEDDA, HILHORST and PELETIER [1997]. Recall that this uniqueness result allows to obtain a convergence result on the whole set of finite volume approximate solutions to problem (15.1)–(15.4) (see Remark 18.12).

The uniqueness of the weak solution to problem (18.1)–(18.3) immediately results from the following property.

THEOREM 18.3. *Let Ω be an open bounded subset of $\mathbb{R}^d$ with a C^2 boundary, and suppose that items* (ii), (iii) *and* (iv) *of Assumption* 18.1 *are satisfied. Let u_1 and u_2 be two solutions of problem* (18.1)–(18.3) *in the sense of Definition* 18.1, *with initial conditions $u_{0,1}$ and $u_{0,2}$ and source terms v_1 and v_2 respectively, that is, for u_1 (resp. u_2), $u_0 = u_{0,1}$ (resp. $u_0 = u_{0,2}$) in* (18.3) *and $f = v_1$ (resp. v_2) in* (18.1).

Then for all $T > 0$,

$$\int_0^T \int_\Omega \big|u_1(x,t) - u_2(x,t)\big| \,\mathrm{d}x\,\mathrm{d}t$$
$$\leqslant T \int_\Omega \big|u_{0,1}(x) - u_{0,2}(x)\big| \,\mathrm{d}x + \int_0^T \int_\Omega (T-t)\big|v_1(x,t) - v_2(x,t)\big| \,\mathrm{d}x\,\mathrm{d}t.$$

Before proving Theorem 18.3, let us first show the following auxiliary result.

The existence of regular solutions to the adjoint problem

LEMMA 18.8. *Let Ω be an open bounded subset of $\mathbb{R}^d$ with a C^2 boundary, and suppose that φ is a nondecreasing locally Lipschitz-continuous function. Let $T > 0$, $w \in C_c^\infty(\Omega \times (0,T))$ such that $|w| \leqslant 1$, and $g \in C^\infty(\overline{\Omega} \times [0,T])$ such that there exists $r \in \mathbb{R}$ with $0 < r \leqslant g(x,t)$, for all $(x,t) \in \Omega \times (0,T)$.*

Then there exists a unique function $\psi \in C^{2,1}(\overline{\Omega} \times [0,T])$ such that

$$\psi_t(x,t) + g(x,t)\Delta\psi(x,t) = w(x,t), \quad \textit{for all } (x,t) \in \Omega \times (0,T), \tag{18.49}$$

$$\nabla\psi \cdot \boldsymbol{n}(x,t) = 0, \quad \textit{for all } (x,t) \in \partial\Omega \times (0,T), \tag{18.50}$$

$$\psi(x,T) = 0, \quad \textit{for all } x \in \Omega. \tag{18.51}$$

Moreover, the function ψ satisfies

$$\big|\psi(x,t)\big| \leqslant T - t, \quad \textit{for all } (x,t) \in \Omega \times (0,T), \tag{18.52}$$

and

$$\int_0^T \int_\Omega g(x,t)\big(\Delta\psi(x,t)\big)^2 \,\mathrm{d}x\,\mathrm{d}t \leqslant 4T \int_0^T \int_\Omega \big|\nabla w(x,t)\big|^2 \,\mathrm{d}x\,\mathrm{d}t. \tag{18.53}$$

PROOF. It will be useful in the following to point out that the right-hand side of (18.53) does not depend on g. Since the function g is bounded away from zero, Eqs. (18.49)–(18.51) define a boundary value problem for a usual heat equation with an initial condition, in which the time variable is reversed. Since Ω, g and w are sufficiently smooth, this problem has a unique solution $\psi \in \mathcal{A}_T$, see LADYŽENSKAJA, SOLONNIKOV and URAL'CEVA [1968]. Since $|w| \leqslant 1$, the functions $T - t$ and $-(T - t)$ are respectively upper and lower solutions of problem (18.49)–(18.50). Hence we get (18.52) (see LADYŽENSKAJA, SOLONNIKOV and URAL'CEVA [1968]).

In order to show (18.53), multiply (18.49) by $\Delta\psi(x,t)$, integrate by parts on $\Omega \times (0,\tau)$, for $\tau \in (0,T]$. This gives

$$\frac{1}{2}\int_\Omega \left|\nabla\psi(x,0)\right|^2 \mathrm{d}x - \frac{1}{2}\int_\Omega \left|\nabla\psi(x,\tau)\right|^2 \mathrm{d}x + \int_0^\tau \int_\Omega g(x,t)\big(\Delta\psi(x,t)\big)^2 \mathrm{d}x\,\mathrm{d}t$$
$$= -\int_0^\tau \int_\Omega \nabla w(x,t)\cdot\nabla\psi(x,t)\,\mathrm{d}x\,\mathrm{d}t. \tag{18.54}$$

Since $\nabla\psi(\cdot,T) = 0$, letting $\tau = T$ in (18.54) leads to

$$\frac{1}{2}\int_\Omega \left|\nabla\psi(x,0)\right|^2 \mathrm{d}x + \int_0^T \int_\Omega g(x,t)\big(\Delta\psi(x,t)\big)^2 \mathrm{d}x\,\mathrm{d}t$$
$$= -\int_0^T \int_\Omega \nabla w(x,t)\cdot\nabla\psi(x,t)\,\mathrm{d}x\,\mathrm{d}t. \tag{18.55}$$

Integrating (18.54) with respect to $\tau \in (0,T)$ leads to

$$\frac{1}{2}\int_0^T \int_\Omega \left|\nabla\psi(x,\tau)\right|^2 \mathrm{d}x d\tau$$
$$\leqslant \frac{T}{2}\int_\Omega \left|\nabla\psi(x,0)\right|^2 \mathrm{d}x + T\int_0^T \int_\Omega g(x,t)\big(\Delta\psi(x,t)\big)^2 \mathrm{d}x\,\mathrm{d}t$$
$$+ T\int_0^T \int_\Omega \left|\nabla w(x,t)\cdot\nabla\psi(x,t)\right| \mathrm{d}x\,\mathrm{d}t. \tag{18.56}$$

Using (18.55) and (18.56), we get

$$\frac{1}{2}\int_0^T \int_\Omega \left|\nabla\psi(x,\tau)\right|^2 \mathrm{d}x\,\mathrm{d}\tau \leqslant 2T\int_0^T \int_\Omega \left|\nabla w(x,t)\cdot\nabla\psi(x,t)\right| \mathrm{d}x\,\mathrm{d}t. \tag{18.57}$$

Thanks to the Cauchy–Schwarz inequality, the right-hand side of (18.57) may be estimated as follows:

$$\left[\int_0^T \int_\Omega \left|\nabla w(x,t)\cdot\nabla\psi(x,t)\right| \mathrm{d}x\,\mathrm{d}t\right]^2$$
$$\leqslant \int_0^T \int_\Omega \left|\nabla\psi(x,t)\right|^2 \mathrm{d}x\,\mathrm{d}t \int_0^T \int_\Omega \left|\nabla w(x,t)\right|^2 \mathrm{d}x\,\mathrm{d}t.$$

With (18.57), this implies

$$\left[\int_0^T \int_\Omega \left|\nabla w(x,t)\cdot\nabla\psi(x,t)\right| \mathrm{d}x\,\mathrm{d}t\right]^2$$

$$\leqslant 4T \int_0^T \int_\Omega \left|\nabla w(x,t) \cdot \nabla \psi(x,t)\right| \mathrm{d}x\,\mathrm{d}t \int_0^T \int_\Omega \left|\nabla w(x,t)\right|^2 \mathrm{d}x\,\mathrm{d}t.$$

Therefore,

$$\int_0^T \int_\Omega \left|\nabla w(x,t) \cdot \nabla \psi(x,t)\right| \mathrm{d}x\,\mathrm{d}t \leqslant 4T \int_0^T \int_\Omega \left|\nabla w(x,t)\right|^2 \mathrm{d}x\,\mathrm{d}t,$$

which, together with (18.55), yields (18.53). □

Proof of the uniqueness theorem

Let u_1 and u_2 be two solutions of problem (18.4), with initial conditions $u_{0,1}$ and $u_{0,2}$ and source terms v_1 and v_2 respectively. We set $u_d = u_1 - u_2$, $v_d = v_1 - v_2$ and $u_{0,d} = u_{0,1} - u_{0,2}$. Let us also define, for all $(x,t) \in \Omega \times \mathbb{R}_+^\star$, $q(x,t) = \frac{\varphi(u_1(x,t)) - \varphi(u_2(x,t))}{u_1(x,t) - u_2(x,t)}$ if $u_1(x,t) \neq u_2(x,t)$, else $q(x,t) = 0$. For all $T \in \mathbb{R}_+^\star$ and for all $\psi \in \mathcal{A}_T$, we deduce from (18.4) that

$$\int_0^T \int_\Omega \left[u_d(x,t)\big(\psi_t(x,t) + q(x,t)\Delta\psi(x,t)\big) + v_d(x,t)\psi(x,t)\right] \mathrm{d}x\,\mathrm{d}t \\ + \int_\Omega u_{0,d}(x)\psi(x,0)\,\mathrm{d}x = 0. \tag{18.58}$$

Let $w \in C_c^\infty(\Omega \times (0,T))$, such that $|w| \leqslant 1$. Since φ is locally Lipschitz continuous, we can define its Lipschitz constant, say B_M, on $[-M, M]$, where $M = \max\{\|u_1\|_{L^\infty(\Omega\times(0,T))}, \|u_2\|_{L^\infty(\Omega\times(0,T))}\}$ so that $0 \leqslant q \leqslant B_M$ a.e. on $\Omega \times (0,T)$.

Using mollifiers, functions $q_{1,n} \in C_c^\infty(\Omega \times (0,T))$ may be constructed such that $\|q_{1,n} - q\|_{L^2(\Omega\times(0,T))} \leqslant \frac{1}{n}$ and $0 \leqslant q_{1,n} \leqslant B_M$, for $n \in \mathbb{N}^\star$. Let $q_n = q_{1,n} + 1/n$. Then

$$\frac{1}{n} \leqslant q_n(x,t) \leqslant B_M + \frac{1}{n}, \quad \text{for all } (x,t) \in \Omega \times (0,T),$$

and

$$\int_0^T \int_\Omega \frac{(q_n(x,t) - q(x,t))^2}{q_n(x,t)} \mathrm{d}x\,\mathrm{d}t \leqslant 2\left(\int_0^T \int_\Omega \frac{(q_n(x,t) - q_{1,n}(x,t))^2}{q_n(x,t)} \mathrm{d}x\,\mathrm{d}t \right. \\ \left. + \int_0^T \int_\Omega \frac{(q_{1,n}(x,t) - q(x,t))^2}{q_n(x,t)} \mathrm{d}x\,\mathrm{d}t\right),$$

which shows that

$$\int_0^T \int_\Omega \frac{(q_n(x,t) - q(x,t))^2}{q_n(x,t)} \mathrm{d}x\,\mathrm{d}t \leqslant 2n\left(\frac{T\,\mathrm{m}(\Omega)}{n^2} + \frac{1}{n^2}\right).$$

It leads to

$$\left\| \frac{q_n - q}{\sqrt{q_n}} \right\|_{L^2(\Omega\times(0,T))} \to 0 \quad \text{as } n \to \infty. \tag{18.59}$$

Let $\psi_n \in \mathcal{A}_T$ be given by Lemma 18.8, with $g = q_n$. Substituting ψ by ψ_n in (18.58), using (with $g = q_n$ and $\psi = \psi_n$) (18.49) and (18.52) give

$$\begin{aligned}
&\left| \int_0^T \int_\Omega u_d(x,t)\big(w(x,t) + \big(q(x,t) - q_n(x,t)\big)\Delta\psi_n(x,t)\big)\,\mathrm{d}x\,\mathrm{d}t \right| \\
&\quad \leqslant \int_0^T \int_\Omega \big|v_d(x,t)\big|(T-t)\,\mathrm{d}x\,\mathrm{d}t + T\int_\Omega \big|u_{0,d}(x)\big|\,\mathrm{d}x.
\end{aligned} \tag{18.60}$$

The Cauchy–Schwarz inequality yields

$$\begin{aligned}
&\left[\int_0^T \int_\Omega \big|u_d(x,t)\big|\big|\big(q(x,t) - q_n(x,t)\big)\Delta\psi_n(x,t)\big|\,\mathrm{d}x\,\mathrm{d}t \right]^2 \\
&\quad \leqslant 4M^2 \int_0^T \int_\Omega \left(\frac{q(x,t) - q_n(x,t)}{\sqrt{q_n(x,t)}} \right)^2 \mathrm{d}x\,\mathrm{d}t \\
&\quad\quad \times \int_0^T \int_\Omega q_n(x,t)\big(\Delta\psi_n(x,t)\big)^2\,\mathrm{d}x\,\mathrm{d}t.
\end{aligned} \tag{18.61}$$

We deduce from (18.53) and (18.59) that the right-hand side of (18.61) tends to zero as $n \to \infty$. Hence the left-hand side of (18.61) also tends to zero as $n \to \infty$. Therefore letting $n \to \infty$ in (18.60) gives

$$\begin{aligned}
&\left| \int_0^T \int_\Omega u_d(x,t) w(x,t)\,\mathrm{d}x\,\mathrm{d}t \right| \\
&\quad \leqslant \int_0^T \int_\Omega \big|v_d(x,t)\big|(T-t)\,\mathrm{d}x\,\mathrm{d}t + T\int_\Omega \big|u_{0,d}(x)\big|\,\mathrm{d}x.
\end{aligned} \tag{18.62}$$

Inequality (18.62) holds for any function $w \in C_c^\infty(\Omega \times (0,T))$, with $|w| \leqslant 1$. Let us take as functions w the elements of a sequence $(w_m)_{m\in\mathbb{N}}$ such that $w_m \in C_c^\infty(\Omega \times (0,T))$ and $|w_m| \leqslant 1$ for all $m \in \mathbb{N}$, and the sequence $(w_m)_{m\in\mathbb{N}}$ converges to $\operatorname{sign}(u_d(\bullet,\bullet))$ in $L^1(\Omega \times (0,T))$. Letting $m \to \infty$ yields

$$\int_0^T \int_\Omega \big|u_d(x,t)\big|\,\mathrm{d}x\,\mathrm{d}t \leqslant \int_0^T \int_\Omega \big|v_d(x,t)\big|(T-t)\,\mathrm{d}x\,\mathrm{d}t + T\int_\Omega \big|u_{0,d}(x)\big|\,\mathrm{d}x,$$

which concludes the proof of Theorem 18.3.

CHAPTER V

Hyperbolic Equations in the One-Dimensional Case

This chapter is devoted to the numerical schemes for one-dimensional hyperbolic conservation laws. Some basics on the solution to linear or nonlinear hyperbolic equations with initial data and without boundary conditions will first be recalled. We refer to GODLEWSKI and RAVIART [1991, 1996], KRÖNER [1997], LEVEQUE [1990] and SERRE [1996] for extensive studies of theoretical and/or numerical aspects; we shall highlight here the finite volume point of view for several well known schemes, comparing them with finite difference schemes, either for the linear and the nonlinear case. Convergence results for numerical schemes are presented, using a "weak BV inequality" which will be used later in the multidimensional case. We also recall the classical proof of convergence which uses a "strong BV estimate" and the Lax–Wendroff theorem. The error estimates which can also be obtained will be given later in the multidimensional case (Chapter VI).

Throughout this chapter, we shall focus on explicit schemes. However, all the results which are presented here can be extended to implicit schemes (this requires a bit of work). This will be detailed in the multidimensional case (see (24.6) for the scheme).

19. The continuous problem

Consider the nonlinear hyperbolic equation with initial data:

$$\begin{cases} u_t(x,t) + (f(u))_x(x,t) = 0 & x \in \mathbb{R}, t \in \mathbb{R}_+, \\ u(x,0) = u_0(x), & x \in \mathbb{R}, \end{cases} \tag{19.1}$$

where f is a given function from $\mathbb{R}$ to $\mathbb{R}$, of class C^1, $u_0 \in L^\infty(\mathbb{R})$ and where the partial derivatives of u with respect to time and space are denoted by u_t and u_x.

EXAMPLE 19.1 (*Bürgers equation*). A simple flow model was introduced by Bürgers and yields the following equation:

$$u_t(x,t) + u(x,t)u_x(x,t) - \varepsilon u_{xx}(x,t) = 0. \tag{19.2}$$

Bürgers studied the limit case which is obtained when ε tends to 0; the resulting equation is (19.1) with $f(s)=s^2/2$, i.e.

$$u_t(x,t)+\tfrac{1}{2}\left(u^2\right)_x(x,t)=0.$$

DEFINITION 19.1 (*Classical solution*). Let $f\in C^1(\mathbb{R},\mathbb{R})$ and $u_0\in C^1(\mathbb{R},\mathbb{R})$; a classical solution to problem (19.1) is a function $u\in C^1(\mathbb{R}\times\mathbb{R}_+,\mathbb{R})$ such that

$$\begin{cases} u_t(x,t)+f'\big(u(x,t)\big)u_x(x,t)=0, & \forall x\in\mathbb{R}, \forall t\in\mathbb{R}_+,\\ u(x,0)=u_0(x), & \forall x\in\mathbb{R}.\end{cases}$$

Recall that in the linear case, i.e. $f(s)=cs$ for all $s\in\mathbb{R}$, for some $c\in\mathbb{R}$, there exists (for $u_0\in C^1(\mathbb{R},\mathbb{R})$) a unique classical solution. It is $u(x,t)=u_0(x-ct)$, for all $x\in\mathbb{R}$ and for all $t\in\mathbb{R}_+$. In the nonlinear case, the existence of such a solution depends on the initial data u_0; in fact, the following result holds:

PROPOSITION 19.1. *Let $f\in C^1(\mathbb{R},\mathbb{R})$ be a nonlinear function, i.e. such that there exist s_1, $s_2\in\mathbb{R}$ with $f'(s_1)\neq f'(s_2)$; then there exists $u_0\in C_c^\infty(\mathbb{R},\mathbb{R})$ such that problem* (19.1) *has no classical solution.*

Proposition 19.1 is an easy consequence of the following remark.

REMARK 19.1. If u is a classical solution to (19.1), then u is constant along the characteristic lines which are defined by

$$x(t)=f'\big(u_0(x_0)\big)t+x_0,\quad t\in\mathbb{R}_+,$$

where $x_0\in\mathbb{R}$ is the origin of the characteristic. This is the equation of a straight line issued from the point $(x_0,0)$ (in the (x,t) coordinates). Note that if f depends on x and u (rather than only on u), the characteristics are no longer straight lines.

The concept of weak solution is introduced in order to define solutions of (19.1) when classical solutions do not exist.

DEFINITION 19.2 (*Weak solution*). Let $f\in C^1(\mathbb{R},\mathbb{R})$ and $u_0\in L^\infty(\mathbb{R})$; a weak solution to problem (19.1) is a function u such that

$$\begin{cases} u\in L^\infty\big(\mathbb{R}\times\mathbb{R}_+^\star\big),\\ \displaystyle\int_{\mathbb{R}}\int_{\mathbb{R}_+}u(x,t)\varphi_t(x,t)\,\mathrm{d}t\,\mathrm{d}x+\int_{\mathbb{R}}\int_{\mathbb{R}_+}f\big(u(x,t)\big)\varphi_x(x,t)\,\mathrm{d}t\,\mathrm{d}x\\ \quad\displaystyle+\int_{\mathbb{R}}u_0(x)\varphi(x,0)\,\mathrm{d}x=0,\\ \forall\varphi\in C_c^1(\mathbb{R}\times\mathbb{R}_+,\mathbb{R}).\end{cases}\tag{19.3}$$

REMARK 19.2. (1) If $u \in C^1(\mathbb{R} \times \mathbb{R}_+, \mathbb{R}) \cap L^\infty(\mathbb{R} \times \mathbb{R}_+^\star)$ then u is a weak solution if and only if u is a classical solution.

(2) Note that in the above definition, we require the test function φ to belong to $C_c^1(\mathbb{R} \times \mathbb{R}_+, \mathbb{R})$, so that φ may be non zero at time $t = 0$.

One may show that there exists at least one weak solution to (19.1). In the linear case, i.e. $f(s) = cs$, for all $s \in \mathbb{R}$, for some $c \in \mathbb{R}$, this solution is unique (it is $u(x,t) = u_0(x - ct)$ for a.e. $(x,t) \in \mathbb{R} \times \mathbb{R}_+$). However, the uniqueness of this weak solution in the general nonlinear case is no longer true. Hence the concept of entropy weak solution, for which an existence and uniqueness result is known.

DEFINITION 19.3 (*Entropy weak solution*). Let $f \in C^1(\mathbb{R}, \mathbb{R})$ and $u_0 \in L^\infty(\mathbb{R})$; the entropy weak solution to problem (19.1) is a function u such that

$$\begin{cases} u \in L^\infty\big(\mathbb{R} \times \mathbb{R}_+^\star\big), \\ \displaystyle\int_{\mathbb{R}}\int_{\mathbb{R}_+} \eta\big(u(x,t)\big)\varphi_t(x,t)\,\mathrm{d}t\,\mathrm{d}x + \int_{\mathbb{R}}\int_{\mathbb{R}_+} \Phi\big(u(x,t)\big)\varphi_x(x,t)\,\mathrm{d}t\,\mathrm{d}x \\ \quad \displaystyle + \int_{\mathbb{R}} \eta\big(u_0(x)\big)\varphi(x,0)\,\mathrm{d}x \geqslant 0, \\ \forall \varphi \in C_c^1(\mathbb{R} \times \mathbb{R}_+, \mathbb{R}_+), \text{ for all convex function } \eta \in C^1(\mathbb{R}, \mathbb{R}) \text{ and} \\ \Phi \in C^1(\mathbb{R}, \mathbb{R}) \text{ such that } \Phi' = \eta' f'. \end{cases} \tag{19.4}$$

REMARK 19.3. The solutions of (19.4) are necessarily solutions of (19.3). This can be shown by taking in (19.4) $\eta(s) = s$ for all $s \in \mathbb{R}$, $\eta(s) = -s$, for all $s \in \mathbb{R}$, and regularizations of the positive and negative parts of the test functions of the weak formulation.

THEOREM 19.1. *Let $f \in C^1(\mathbb{R}, \mathbb{R})$, $u_0 \in L^\infty(\mathbb{R})$, then there exists a unique entropy weak solution to problem* (19.1).

The proof of this result was first given by VOL'PERT [1967], introducing the space $BV(\mathbb{R})$ which is defined hereafter and assuming $u_0 \in BV(\mathbb{R})$, see also OLEINIK [1963] for the convex case. KRUSHKOV [1970], proved the theorem of existence and uniqueness in the general case $u_0 \in L^\infty(\mathbb{R})$, using a regularization of u_0 in $BV(\mathbb{R})$, under the slightly stronger assumption $f \in C^3(\mathbb{R}, \mathbb{R})$. Krushkov also proved that the solution is in the space $C(\mathbb{R}_+, L^1_{\mathrm{loc}}(\mathbb{R}))$. Krushkov's proof uses particular entropies, namely the functions $|\bullet - \kappa|$ for all $\kappa \in \mathbb{R}$, which are generally referred to as "Krushkov's entropies". The "entropy flux" associated to $|\bullet - \kappa|$ may be taken as $f(\bullet \top \kappa) - f(\bullet \bot \kappa)$, where $a \top b$ denotes the maximum of a and b and $a \bot b$ denotes the minimum of a and b, for all real values a, b (recall that $f(a \top b) - f(a \bot b) = \mathrm{sign}(a - b)(f(a) - f(b))$).

DEFINITION 19.4 ($BV(\mathbb{R})$). A function $v \in L^1_{\mathrm{loc}}(\mathbb{R})$ is of bounded variation, i.e. $v \in BV(\mathbb{R})$, if

$$\begin{aligned} |v|_{BV(\mathbb{R})} &= \sup\left\{\int_{\mathbb{R}} v(x)\varphi_x(x)\,\mathrm{d}x,\ \varphi \in C_c^1(\mathbb{R}, \mathbb{R}),\ \big|\varphi(x)\big| \leqslant 1\ \forall x \in \mathbb{R}\right\} \\ &< +\infty. \end{aligned} \tag{19.5}$$

REMARK 19.4. (1) If $v:\mathbb{R}\to\mathbb{R}$ is piecewise constant, i.e. if there exists an increasing sequence $(x_i)_{i\in\mathbb{Z}}$ with $\mathbb{R}=\bigcup_{i\in\mathbb{Z}}[x_i,x_{i+1}]$ and a sequence $(v_i)_{i\in\mathbb{Z}}$ such that $v|_{(x_i,x_{i+1})}=v_i$, then $|v|_{BV(\mathbb{R})}=\sum_{i\in\mathbb{Z}}|v_{i+1}-v_i|$.

(2) If $v\in C^1(\mathbb{R},\mathbb{R})$ then $|v|_{BV(\mathbb{R})}=\|v_x\|_{L^1(\mathbb{R})}$.

(3) The space $BV(\mathbb{R})$ is included in the space $L^\infty(\mathbb{R})$; furthermore, if $u\in BV(\mathbb{R})\cap L^1(\mathbb{R})$ then $\|u\|_{L^\infty(\mathbb{R})}\leqslant|u|_{BV(\mathbb{R})}$.

(4) Let $u\in BV(\mathbb{R})$ and let $(x_{i+1/2})_{i\in\mathbb{Z}}$ be an increasing sequence of real values such that $\mathbb{R}=\bigcup_{i\in\mathbb{Z}}[x_{i-1/2},x_{i+1/2}]$. For $i\in\mathbb{Z}$, let $K_i=(x_{i-1/2},x_{i+1/2})$ and u_i be the mean value of u over K_i. Then, choosing conveniently φ in the definition of $|u|_{BV(\mathbb{R})}$, it is easy to show that

$$\sum_{i\in\mathbb{Z}}|u_{i+1}-u_i|\leqslant|u|_{BV(\mathbb{R})}. \tag{19.6}$$

Inequality (19.6) is used for the classical proof of "BV estimates" for the approximate solutions given by finite volume schemes (see Lemma 21.5 and Corollary 21.1).

Note that (19.6) is also true when u_i is the mean value of u over a subinterval of K_i instead of the mean value of u over K_i.

Krushkov used a characterization of entropy weak solutions which is given in the following proposition.

PROPOSITION 19.2 (Entropy weak solution using "Krushkov's entropies"). *Let $f\in C^1(\mathbb{R},\mathbb{R})$ and $u_0\in L^\infty(\mathbb{R})$, u is the unique entropy weak solution to problem* (19.1) *if and only if u is such that*

$$\begin{cases} u\in L^\infty(\mathbb{R}\times\mathbb{R}_+^\star),\\ \displaystyle\int_{\mathbb{R}}\int_{\mathbb{R}_+}\big|u(x,t)-\kappa\big|\varphi_t(x,t)\,\mathrm{d}t\,\mathrm{d}x\\ \displaystyle\quad+\int_{\mathbb{R}}\int_{\mathbb{R}_+}\big(f\big(u(x,t)\top\kappa\big)-f\big(u(x,t)\bot\kappa\big)\big)\varphi_x(x,t)\,\mathrm{d}t\,\mathrm{d}x\\ \displaystyle\quad+\int_{\mathbb{R}}\big|u_0(x)-\kappa\big|\varphi(x,0)\,\mathrm{d}x\geqslant 0,\\ \forall\varphi\in C_c^1(\mathbb{R}\times\mathbb{R}_+,\mathbb{R}_+),\ \forall\kappa\in\mathbb{R}. \end{cases} \tag{19.7}$$

The result of existence of an entropy weak solution defined by (19.4) was already proved by passing to the limit on the solutions of an appropriate numerical scheme, see, e.g., OLEINIK [1963], and may also be obtained by passing to the limit on finite volume approximations of the solution (see Theorem 21.1 in the one-dimensional case and Theorem 28.2 in the multidimensional case).

REMARK 19.5. An entropy weak solution is sometimes defined as a function u satisfying:

$$\int_{\mathbb{R}}\int_{\mathbb{R}_+} u(x,t)\varphi_t(x,t)\,\mathrm{d}t\,\mathrm{d}x + \int_{\mathbb{R}}\int_{\mathbb{R}_+} f\big(u(x,t)\big)\varphi_x(x,t)\,\mathrm{d}t\,\mathrm{d}x$$
$$+\int_{\mathbb{R}} u_0(x)\varphi(x,0)\,\mathrm{d}x = 0, \quad \forall \varphi \in C_c^1(\mathbb{R}\times\mathbb{R}_+,\mathbb{R}). \tag{19.8}$$

$$\int_{\mathbb{R}}\int_{\mathbb{R}_+} \eta\big(u(x,t)\big)\varphi_t(x,t)\,\mathrm{d}t\,\mathrm{d}x + \int_{\mathbb{R}}\int_{\mathbb{R}_+} \Phi\big(u(x,t)\big)\varphi_x(x,t)\,\mathrm{d}t\,\mathrm{d}x \geqslant 0,$$
$$\forall \varphi \in C_c^1(\mathbb{R}\times\mathbb{R}_+^\star,\mathbb{R}_+), \text{ for all convex function } \eta \in C^1(\mathbb{R},\mathbb{R})$$
$$\text{and } \Phi \in C^1(\mathbb{R},\mathbb{R}) \text{ such that } \Phi' = \eta' f'. \tag{19.9}$$

The uniqueness of an entropy weak solution thus defined depends on the functional space to which u is chosen to belong. Indeed, the uniqueness result given in Theorem 19.1 is no longer true with u defined by (19.8) such that

$$u,\ f(u) \in L^1_{\mathrm{loc}}(\mathbb{R}\times\mathbb{R}_+),\ u \in L^\infty(\mathbb{R}\times(\varepsilon,\infty)),\ \forall \varepsilon \in \mathbb{R}_+^\star. \tag{19.10}$$

Under assumption (19.10), every term in (19.8) makes sense. Note that (19.10)–(19.8) is weaker than (19.4). An easy counterexample to a uniqueness result of the solution to (19.8)–(19.10) is obtained with $f(s) = s^2$ for all $s \in \mathbb{R}$ and $u_0(x) = 0$ for a.e. $x \in \mathbb{R}$. In this case, a first solution to (19.8)–(19.10) is $u(x,t) = 0$ for a.e. $(x,t) \in \mathbb{R}\times\mathbb{R}_+$ (it is the entropy weak solution). A second solution to (19.8)–(19.10) is defined for a.e. $(x,t) \in \mathbb{R}\times\mathbb{R}_+$ by

$$\begin{cases} u(x,t) = 0, & \text{if } x < -\sqrt{t} \text{ or } x > \sqrt{t},\\ u(x,t) = \dfrac{x}{2t}, & \text{if } -\sqrt{t} < x < \sqrt{t}. \end{cases}$$

This second solution is not an entropy weak solution: it does not satisfy (19.4). Also note that this second solution is not in the space $C(\mathbb{R}_+, L^1_{\mathrm{loc}}(\mathbb{R}))$ nor in the space $L^\infty(\mathbb{R}\times\mathbb{R}_+)$ (it belongs to $L^\infty(\mathbb{R}_+, L^1(\mathbb{R}))$). Indeed, under the assumption $u \in L^\infty(\mathbb{R}\times\mathbb{R}_+)\cap C(\mathbb{R}_+, L^1_{\mathrm{loc}}(\mathbb{R}))$, the solution of (19.8) is unique.

The entropy weak solution to (19.1) satisfies the following L^∞ and BV stability properties:

PROPOSITION 19.3. *Let $f \in C^1(\mathbb{R},\mathbb{R})$ and $u_0 \in L^\infty(\mathbb{R})$. Let u be the entropy weak solution to* (19.1). *Then, $u \in C(\mathbb{R}_+, L^1_{\mathrm{loc}}(\mathbb{R}))$; furthermore, the following estimates hold*:

(1) $\|u(\cdot,t)\|_{L^\infty(\mathbb{R})} \leqslant \|u_0\|_{L^\infty(\mathbb{R})}$, *for all $t \in \mathbb{R}_+$.*
(2) *If $u_0 \in BV(\mathbb{R})$, then $|u(\cdot,t)|_{BV(\mathbb{R})} \leqslant |u_0|_{BV(\mathbb{R})}$, for all $t \in \mathbb{R}_+$.*

20. Numerical schemes in the linear case

We shall first introduce the numerical schemes in the linear case $f(u) = u$ in (19.1). The problem considered in this section is therefore

$$\begin{cases} u_t(x,t) + u_x(x,t) = 0 & x \in \mathbb{R}, t \in \mathbb{R}_+, \\ u(x,0) = u_0(x), & x \in \mathbb{R}. \end{cases} \tag{20.1}$$

Assume that $u_0 \in C^1(\mathbb{R}, \mathbb{R})$; problem (20.1) has a unique classical solution, as defined in Definition 19.1, which is $u(x,t) = u_0(x-t)$ for all $(x,t) \in \mathbb{R} \times \mathbb{R}_+$. If $u_0 \in L^\infty(\mathbb{R})$, then problem (20.1) has a unique weak solution, as defined in Definition 19.2, which is again $u(x,t) = u_0(x-t)$ for a.e. $(x,t) \in \mathbb{R} \times \mathbb{R}_+$. Therefore, if $u_0 \geqslant 0$, the solution u is also nonnegative. Hence, it is advisable for many problems that the solution given by the numerical scheme should preserve the nonnegativity of the solution.

20.1. The centered finite difference scheme

Assume $u_0 \in C(\mathbb{R}, \mathbb{R})$. Let $h \in \mathbb{R}_+^\star$ and $x_i = ih$ for all $i \in \mathbb{Z}$. Let $k \in \mathbb{R}_+^\star$ be the time step. With the explicit Euler scheme for the time discretization (the implicit Euler scheme could also be used), the centered finite difference scheme associated to points x_i and k is

$$\begin{cases} \dfrac{u_i^{n+1} - u_i^n}{k} + \dfrac{u_{i+1}^n - u_{i-1}^n}{2h} = 0, & \forall n \in \mathbb{N},\ \forall i \in \mathbb{Z}, \\ u_i^0 = u_0(x_i), & \forall i \in \mathbb{Z}. \end{cases} \tag{20.2}$$

The discrete unknown u_i^n is expected to be an approximation of $u(x_i, nk)$ where u is the solution to (20.1).

It is well known that this scheme should be avoided. In particular, for the following reasons:

(1) it does not preserve positivity, i.e. $u_i^0 \geqslant 0$ for all $i \in \mathbb{Z}$ does not imply $u_i^1 \geqslant 0$ for all $i \in \mathbb{Z}$; take, e.g., $u_i^0 = 0$ for $i \leqslant 0$ and $u_i^0 = 1$ for $i > 0$, then $u_0^1 = -k/(2h) < 0$;

(2) it is not "L^∞-diminishing", i.e. $\max\{|u_i^0|, i \in \mathbb{Z}\} = 1$ does not imply that $\max\{|u_i^1|, i \in \mathbb{Z}\} \leqslant 1$; e.g., in the previous example, $\max\{|u_i^0|, i \in \mathbb{Z}\} = 1$ and $\max\{|u_i^1|, i \in \mathbb{Z}\} = 1 + k/(2h)$;

(3) it is not "L^2-diminishing", i.e. $\sum_{i\in\mathbb{Z}} (u_i^0)^2 = 1$ does not imply that $\sum_{i\in\mathbb{Z}} (u_i^1)^2 \leqslant 1$; take, e.g., $u_i^0 = 0$ for $i \neq 0$ and $u_i^0 = 1$ for $i = 0$, then $u_0^1 = 1$, $u_1^1 = k/(2h)$, $u_{-1}^1 = -k/(2h)$, so that $\sum_{i\in\mathbb{Z}} (u_i^1)^2 = 1 + k^2/(2h^2) > 1$;

(4) it is unstable in the von Neumann sense: if the initial condition is taken under the form $u_0(x) = \exp(\mathrm{i}px)$, where p is given in $\mathbb{Z}$, then $u(x,t) = \exp(-\mathrm{i}pt)\exp(\mathrm{i}px)$ (i is, here, the usual complex number, u_0 and u take values in $\mathbb{C}$). Hence $\exp(-\mathrm{i}pt)$ can be seen as an amplification factor, and its modulus is 1. The numerical scheme is stable in the von Neumann sense if the amplification factor for the discrete solution is less than or equal to 1. For the scheme (20.2), we have $u_j^1 = u_j^0 - (u_{j+1}^0 - u_{j-1}^0)k/(2h)$

$= \exp(\mathrm{i}pjh)\xi_{p,h,k}$, with $\xi_{p,h,k} = 1 - (\exp(\mathrm{i}ph) - \exp(-\mathrm{i}ph))k/(2h)$. Hence $|\xi_{p,h,k}|^2 = 1 + (k^2/h^2)\sin^2 ph > 1$ if $ph \neq q\pi$ for any q in $\mathbb{Z}$.

In fact, one can also show that there exists $u_0 \in C_c^1(\mathbb{R}, \mathbb{R})$ such that the solution given by the numerical scheme does not tend to the solution of the continuous problem when h and k tend to 0 (whatever the relation between h and k).

REMARK 20.1. The scheme (20.2) is also a finite volume scheme with the (spatial) mesh $\mathcal{T}$ given by $x_{i+1/2} = (i + 1/2)h$ in Definition 20.1 below and with a centered choice for the approximation of $u(x_{i+1/2}, nk)$ (that is $u(x_{i+1/2}, nk)$ approached by $(u_i^n + u_{i+1}^n)/2$, see (20.5) where an upstream choice for $u(x_{i+1/2}, nk)$ is performed). In fact, the choice of u_i^0 is different in (20.5) and in (20.2) but this does not change the unstability of the centered scheme.

20.2. *The upstream finite difference scheme*

Consider now a nonuniform distribution of points x_i, i.e. an increasing sequence of real values $(x_i)_{i\in\mathbb{Z}}$ such that $\lim_{i\to\pm\infty} x_i = \pm\infty$. For all $i \in \mathbb{Z}$, we set $h_{i-1/2} = x_i - x_{i-1}$. The time discretization is performed with the explicit Euler scheme with time step $k > 0$. Still assuming $u_0 \in C(\mathbb{R}, \mathbb{R})$, consider the upwind (or upstream) finite difference scheme defined by

$$\begin{cases} \dfrac{u_i^{n+1} - u_i^n}{k} + \dfrac{u_i^n - u_{i-1}^n}{h_{i-1/2}} = 0, & \forall n \in \mathbb{N}, \forall i \in \mathbb{Z}, \\ u_i^0 = u_0(x_i), & \forall i \in \mathbb{Z}. \end{cases} \tag{20.3}$$

Rewriting the scheme as

$$u_i^{n+1} = \left(1 - \frac{k}{h_{i-1/2}}\right)u_i^n + \frac{k}{h_{i-1/2}}u_{i-1}^n,$$

it appears that if $\inf_{i\in\mathbb{Z}} h_{i-1/2} > 0$ and if k is such that $k \leqslant \inf_{i\in\mathbb{Z}} h_{i-1/2}$ then u_i^{n+1} is a convex combination of u_i^n and u_{i-1}^n; by induction, this proves that the scheme (20.3) is stable, in the sense that if u_0 is such that $U_m \leqslant u_0(x) \leqslant U_M$ for a.e. $x \in \mathbb{R}$, where U_m, $U_M \in \mathbb{R}$, then $U_m \leqslant u_i^n \leqslant U_M$ for any $i \in \mathbb{Z}$ and $n \in \mathbb{N}$.

Moreover, if $u_0 \in C^2(\mathbb{R}, \mathbb{R}) \cap L^\infty(\mathbb{R})$ and u_0' and u_0'' belong to $L^\infty(\mathbb{R})$, it is easily shown that the scheme is consistent in the finite difference sense, i.e. the consistency error defined by

$$R_i^n = \frac{u(x_i, (n+1)k) - u(x_i, nk)}{k} + \frac{u(x_i, nk) - u(x_{i-1}, nk)}{h_{i-1/2}} \tag{20.4}$$

is such that $|R_i^n| \leqslant Ch$, where $h = \sup_{i\in\mathbb{Z}} h_i$ and $C \geqslant 0$ only depends on u_0 (recall that u is the solution to problem (20.1)). Hence the following error estimate holds:

PROPOSITION 20.1 (Error estimate for the upwind finite difference scheme). *Let $u_0 \in C^2(\mathbb{R}, \mathbb{R}) \cap L^\infty(\mathbb{R})$, such that u_0' and $u_0'' \in L^\infty(\mathbb{R})$. Let $(x_i)_{i\in\mathbb{Z}}$ be an increasing sequence of real values such that $\lim_{i\to\pm\infty} x_i = \pm\infty$. Let $h = \sup_{i\in\mathbb{Z}} h_{i-1/2}$, and assume that $h < \infty$ and $\inf_{i\in\mathbb{Z}} h_{i-1/2} > 0$. Let $k > 0$ such that $k \leqslant \inf_{i\in\mathbb{Z}} h_{i-1/2}$. Let u denote the unique solution to* (20.1) *and $\{u_i^n,\ i \in \mathbb{Z},\ n \in \mathbb{N}\}$ be given by* (20.3); *let $e_i^n = u(x_i, nk) - u_i^n$, for any $n \in \mathbb{N}$ and $i \in \mathbb{Z}$, and let $T \in]0, +\infty[$ (note that $u(x_i, nk)$ is well defined since $u \in C^2(\mathbb{R} \times \mathbb{R}_+, \mathbb{R})$).*

Then there exists $C \in \mathbb{R}_+$, only depending on u_0, such that $|e_i^n| \leqslant ChT$, for any $n \in \mathbb{N}$ such that $nk \leqslant T$, and for any $i \in \mathbb{Z}$.

PROOF. Let $i \in \mathbb{Z}$ and $n \in \mathbb{N}$. By definition of the consistency error R_i^n in (20.4), the error e_i^n satisfies

$$\frac{e_i^{n+1} - e_i^n}{k} + \frac{e_i^n - e_{i-1}^n}{h_{i-1/2}} = R_i^n.$$

Hence

$$e_i^{n+1} = e_i^n\left(1 - \frac{k}{h_{i-1/2}}\right) + \frac{k}{h_{i-1/2}} e_{i-1}^n + kR_i^n.$$

Using $|R_i^n| \leqslant Ch$ (for some C only depending on u_0) and the assumption $k \leqslant \inf_{i\in\mathbb{Z}} h_{i-1/2}$, this yields

$$\left|e_i^{n+1}\right| \leqslant \sup_{j\in\mathbb{Z}}\left|e_j^n\right| + Ckh.$$

Since $e_i^0 = 0$ for any $i \in \mathbb{Z}$, an induction yields

$$\sup_{i\in\mathbb{Z}}\left|e_i^n\right| \leqslant Cnkh$$

and the result follows. □

Note that in the above proof, the linearity of the equation and the regularity of u_0 are used. The next questions to arise are what to do in the case of a nonlinear equation and in the case $u_0 \in L^\infty(\mathbb{R})$.

20.3. The upwind finite volume scheme

Let us first give a definition of the admissible meshes for the finite volume schemes.

DEFINITION 20.1 (*One-dimensional admissible mesh*). An admissible mesh $\mathcal{T}$ of $\mathbb{R}$ is given by an increasing sequence of real values $(x_{i+1/2})_{i\in\mathbb{Z}}$, such that $\mathbb{R} = \bigcup_{i\in\mathbb{Z}}[x_{i-1/2}, x_{i+1/2}]$. The mesh $\mathcal{T}$ is the set $\mathcal{T} = \{K_i,\ i \in \mathbb{Z}\}$ of subsets of $\mathbb{R}$ defined by $K_i = (x_{i-1/2}, x_{i+1/2})$ for all $i \in \mathbb{Z}$. The length of K_i is denoted by h_i, so that $h_i = x_{i+1/2} - x_{i-1/2}$ for all $i \in \mathbb{Z}$. It is assumed that $h = \text{size}(\mathcal{T}) = \sup\{h_i, i \in \mathbb{Z}\} < +\infty$ and that, for some $\alpha \in \mathbb{R}_+^\star$, $\alpha h \leqslant \inf\{h_i, i \in \mathbb{Z}\}$.

Consider an admissible mesh in the sense of Definition 20.1. Let $k \in \mathbb{R}_+^\star$ be the time step. Assume $u_0 \in L^\infty(\mathbb{R})$ (this is a natural hypothesis for the finite volume framework). Integrating (20.1) on each control volume of the mesh, approximating the time derivatives by differential quotients and using an upwind choice for $u(x_{i+1/2}, nk)$ yields the following (time explicit) scheme:

$$\begin{cases} h_i \dfrac{u_i^{n+1} - u_i^n}{k} + u_i^n - u_{i-1}^n = 0, & \forall n \in \mathbb{N}, \forall i \in \mathbb{Z}, \\ u_i^0 = \dfrac{1}{h_i} \displaystyle\int_{K_i} u_0(x)\,\mathrm{d}x, & \forall i \in \mathbb{Z}. \end{cases} \tag{20.5}$$

The value u_i^n is expected to be an approximation of u (solution to (20.1)) in K_i at time nk. It is easily shown that this scheme is not consistent in the finite difference sense if u_i^n is considered to be an approximation of $u(x_i, nk)$ with, e.g., $x_i = (x_{i-1/2} + x_{i+1/2})/2$ for all $i \in \mathbb{Z}$. Even if $u_0 \in C_c^\infty(\mathbb{R}, \mathbb{R})$, the quantity R_i^n defined by (20.4) does not satisfy (except in particular cases) $|R_i^n| \leqslant Ch$, with some C only depending on u_0.

It is however possible to interpret this scheme as another expression of the upwind finite difference scheme (20.3) (except for the minor modification of u_i^0, $i \in \mathbb{Z}$). One simply needs to consider u_i^n as an approximation of $u(x_{i+1/2}, nk)$ which leads to a consistency property in the finite difference sense. Indeed, taking $x_j = x_{j+1/2}$ (for $j = i$ and $i - 1$) in definition (20.4) of R_i^n yields $|R_i^n| \leqslant Ch$, where C only depends on u_0. Therefore, a convergence result for this scheme is given by Proposition 20.1. This analogy cannot be extended to the general case of "monotone flux schemes" (see Definition 21.1 below) for a nonlinear equation for which there may be no value of x_i (independent of u) leading to such a consistency property, see Remark 21.1 for a counterexample (the analogy holds however for the scheme (21.7), convenient for a nondecreasing function f, see Remark 21.3).

The approximate finite volume solution $u_{\mathcal{T},k}$ may be defined on $\mathbb{R} \times \mathbb{R}_+$ from the discrete unknowns u_i^n, $i \in \mathbb{Z}$, $n \in \mathbb{N}$ which are computed in (20.5):

$$u_{\mathcal{T},k}(x, t) = u_i^n \quad \text{for } x \in K_i \text{ and } t \in \big[nk, (n+1)k\big). \tag{20.6}$$

The following L^∞ estimate holds:

LEMMA 20.1 (L^∞ estimate in the linear case). *Let $u_0 \in L^\infty(\mathbb{R})$ and U_m, $U_M \in \mathbb{R}$ such that $U_m \leqslant u_0(x) \leqslant U_M$ for a.e. $x \in \mathbb{R}$. Let $\mathcal{T}$ be an admissible mesh in the sense of Definition* 20.1 *and let $k \in \mathbb{R}_+^\star$ satisfying the Courant–Friedrichs–Levy* (*CFL*) *condition*

$$k \leqslant \inf_{i \in \mathbb{Z}} h_i$$

(*note that taking $k \leqslant \alpha h$ implies the above condition*). *Let $u_{\mathcal{T},k}$ be the finite volume approximate solution defined by* (20.5) *and* (20.6).

Then,

$$U_m \leqslant u_{\mathcal{T},k}(x,t) \leqslant U_M \quad \textit{for a.e. } x \in \mathbb{R} \textit{ and a.e. } t \in \mathbb{R}_+.$$

PROOF. The proof that $U_m \leqslant u_i^n \leqslant U_M$, for all $i \in \mathbb{Z}$ and $n \in \mathbb{N}$, as in the case of the upwind finite difference scheme (see (20.3)), consists in remarking that Eq. (20.5) gives, under the CFL condition, an expression of u_i^{n+1} as a linear convex combination of u_i^n and u_{i-1}^n, for all $i \in \mathbb{Z}$ and $n \in \mathbb{N}$. □

The following inequality will be crucial for the proof of convergence.

LEMMA 20.2 (Weak BV estimate, linear case). *Let $\mathcal{T}$ be an admissible mesh in the sense of Definition* 20.1 *and let $k \in \mathbb{R}_+^\star$ satisfying the CFL condition*

$$k \leqslant (1-\xi) \inf_{i \in \mathbb{Z}} h_i, \tag{20.7}$$

for some $\xi \in (0,1)$ (taking $k \leqslant (1-\xi)\alpha h$ implies this condition).

Let $\{u_i^n,\ i \in \mathbb{Z},\ n \in \mathbb{N}\}$ be given by the finite volume scheme (20.5). *Let $R \in \mathbb{R}_+^\star$ and $T \in \mathbb{R}_+^\star$ and assume $h = \operatorname{size}(\mathcal{T}) < R$, $k < T$. Let $i_0 \in \mathbb{Z}$, $i_1 \in \mathbb{Z}$ and $N \in \mathbb{N}$ be such that $-R \in \overline{K}_{i_0}$, $R \in \overline{K}_{i_1}$ and $T \in (Nk, (N+1)k]$ (note that $i_0 < i_1$).*

Then there exists $C \in \mathbb{R}_+^\star$, only depending on R, T, u_0, α and ξ, such that

$$\sum_{i=i_0}^{i_1} \sum_{n=0}^{N} k \left| u_i^n - u_{i-1}^n \right| \leqslant C h^{-1/2}. \tag{20.8}$$

PROOF. Multiplying the first equation of (20.5) by $k u_i^n$ and summing on $i = i_0, \ldots, i_1$ and $n = 0, \ldots, N$ yields $A + B = 0$ with

$$A = \sum_{i=i_0}^{i_1} \sum_{n=0}^{N} h_i \left(u_i^{n+1} - u_i^n\right) u_i^n$$

and

$$B = \sum_{i=i_0}^{i_1} \sum_{n=0}^{N} k \left(u_i^n - u_{i-1}^n\right) u_i^n.$$

Noting that

$$A = -\tfrac{1}{2} \sum_{i=i_0}^{i_1} \sum_{n=0}^{N} h_i \left(u_i^{n+1} - u_i^n\right)^2 + \tfrac{1}{2} \sum_{i=i_0}^{i_1} h_i \left[\left(u_i^{N+1}\right)^2 - \left(u_i^0\right)^2\right]$$

and using the scheme (20.5) gives

$$A = -\tfrac{1}{2}\sum_{i=i_0}^{i_1}\sum_{n=0}^{N}\frac{k^2}{h_i}\left(u_i^n - u_{i-1}^n\right)^2 + \tfrac{1}{2}\sum_{i=i_0}^{i_1} h_i\left[\left(u_i^{N+1}\right)^2 - \left(u_i^0\right)^2\right];$$

therefore, using the CFL condition (20.7),

$$A \geqslant -(1-\xi)\tfrac{1}{2}\sum_{i=i_0}^{i_1}\sum_{n=0}^{N} k\left(u_i^n - u_{i-1}^n\right)^2 - \tfrac{1}{2}\sum_{i=i_0}^{i_1} h_i\left(u_i^0\right)^2.$$

We now study the term B, which may be rewritten as

$$B = \tfrac{1}{2}\sum_{i=i_0}^{i_1}\sum_{n=0}^{N} k\left(u_i^n - u_{i-1}^n\right)^2 + \tfrac{1}{2}\sum_{n=0}^{N} k\left[\left(u_{i_1}^n\right)^2 - \left(u_{i_0-1}^n\right)^2\right].$$

Thanks to the L^∞ estimate of Lemma 20.1, this last equality implies that

$$B \geqslant \tfrac{1}{2}\sum_{i=i_0}^{i_1}\sum_{n=0}^{N} k\left(u_i^n - u_{i-1}^n\right)^2 - T\max\{-U_m, U_M\}^2.$$

Therefore, since $A + B = 0$ and $\sum_{i=i_0}^{i_1} h_i \leqslant 4R$, the following inequality holds:

$$0 \geqslant \xi\sum_{i=i_0}^{i_1}\sum_{n=0}^{N} k\left(u_i^n - u_{i-1}^n\right)^2 - (4R + 2T)\max\{-U_m, U_M\}^2,$$

which, in turn, gives the existence of $C_1 \in \mathbb{R}_+^\star$, only depending on R, T, u_0 and ξ such that

$$\sum_{i=i_0}^{i_1}\sum_{n=0}^{N} k\left(u_i^n - u_{i-1}^n\right)^2 \leqslant C_1. \tag{20.9}$$

Finally, using

$$\sum_{i=i_0}^{i_1} 1 \leqslant \sum_{i=i_0}^{i_1}\frac{h_i}{\alpha h} \leqslant \frac{4R}{\alpha h},$$

the Cauchy–Schwarz inequality leads to

$$\left[\sum_{i=i_0}^{i_1}\sum_{n=0}^{N} k\left|u_i^n - u_{i-1}^n\right|\right]^2 \leqslant C_1 2T\frac{4R}{\alpha h},$$

which concludes the proof of the lemma. □

Contrary to the discrete H_0^1 estimates which were obtained on the approximate finite volume solutions of elliptic equations, see, e.g., (9.24), the weak BV estimate (20.8) is not related to an a priori estimate on the solution to the continuous problem (20.1). It does not give any compactness property in the space $L^1_{\mathrm{loc}}(\mathbb{R})$ (there are some counterexamples); such a compactness property is obtained thanks to a "strong BV estimate" (with, e.g., an L^∞ estimate) as it is recalled below (see Lemma 21.4). In the one-dimensional case which is studied here such a "strong BV estimate" can be obtained if $u_0 \in BV(\mathbb{R})$, see Corollary 21.1; this is no longer true in the multidimensional case with general meshes, for which only the above weak BV estimate is available.

REMARK 20.2. The weak BV estimate is a crucial point for the proof of convergence. Indeed, the property which is used in the proof of convergence (see Proposition 20.2 below) is, with the notations of Lemma 20.2,

$$h\sum_{i=i_0}^{i_1}\sum_{n=0}^{N} k\left|u_i^n - u_{i-1}^n\right| \to 0, \quad \text{as } h \to 0, \tag{20.10}$$

for R, T, u_0, α and ξ fixed.

If a piecewise constant function $u_{\mathcal{T},k}$, such as given by (20.6) (with some u_i^n in $\mathbb{R}$, not necessarily given by (20.5)), is bounded in (e.g.) $L^\infty(\mathbb{R}\times\mathbb{R}_+)$ and converges in $L^1_{\mathrm{loc}}(\mathbb{R}\times\mathbb{R}_+)$ as $h \to 0$ and $k \to 0$ (with a possible relation between k and h) then (20.10) holds. This proves that the hypothesis (20.10) is included in the hypotheses of the classical Lax–Wendroff theorem of convergence (see Theorem 21.2).

We show in the following remark how the "weak" and "strong" BV estimates may "formally" be obtained on the "continuous equation"; this gives a hint of the reason why this estimate may be obtained even if the exact solution does not belong to the space $BV(\mathbb{R}\times\mathbb{R}_+)$. A similar remark also holds in the nonlinear case (i.e. for problem (19.1)).

REMARK 20.3 (*Formal derivations of the strong and weak BV estimates*). When approximating the solution to (20.1) by the finite volume scheme (20.5) (with $h_i = h$ for all i, for the sake of simplicity), the equation to which an approximation of a solution is sought is "close" to the equation

$$u_t + u_x - \varepsilon u_{xx} = 0, \tag{20.11}$$

where $\varepsilon = (h-k)/2$ is positive under the CFL condition (20.7), which ensures that the scheme is diffusive.

We assume that u is regular enough, with null limits for $u(x,t)$ and its derivatives as $x \to \pm\infty$.

(i) "*Strong*" *BV estimate*. Derivating Eq. (20.11) with respect to the variable x, multiplying by $\text{sign}_r(u_x(x,t))$, where sign_r denotes a nondecreasing regularization of the function sign, and integrating over $\mathbb{R}$ yields

$$\left(\int_{\mathbb{R}} \phi_r\big(u_x(x,t)\big)\,\mathrm{d}x\right)_t + \int_{\mathbb{R}} u_{xx}(x,t)\text{sign}_r\big(u_x(x,t)\big)\,\mathrm{d}x$$
$$= -\varepsilon \int_{\mathbb{R}} \text{sign}_r'\big(u_x(x,t)\big)\big(u_{xx}(x,t)\big)^2\,\mathrm{d}x \leqslant 0,$$

where $\phi_r' = \text{sign}_r$ and $\phi_r(0) = 0$. Since

$$\int_{\mathbb{R}} u_{xx}(x,t)\text{sign}_r\big(u_x(x,t)\big)\,\mathrm{d}x = \int_{\mathbb{R}} \big(\phi_r\big(u_x(x,t)\big)\big)_x\,\mathrm{d}x = 0,$$

this yields, passing to the limit on the regularization, that $\|u_x(\bullet,t)\|_{L^1(\mathbb{R})}$ is nonincreasing with respect to t. Copying this formal proof on the numerical scheme yields a strong BV estimate, which is an a priori estimate giving compactness properties in $L^1_{\text{loc}}(\mathbb{R}\times\mathbb{R}_+)$, see Lemma 21.5, Corollary 21.1 and Lemma 21.4.

(ii) "*Weak*" *BV estimate.* Multiplying (20.11) by u and summing over $\mathbb{R}\times(0,T)$ yields

$$\frac{1}{2}\int_{\mathbb{R}} u^2(x,T)\,\mathrm{d}x - \frac{1}{2}\int_{\mathbb{R}} u^2(x,0)\,\mathrm{d}x + \int_0^T\int_{\mathbb{R}} \varepsilon u_x^2(x,t)\,\mathrm{d}x\,\mathrm{d}t = 0,$$

which yields in turn

$$\varepsilon\int_0^T\int_{\mathbb{R}} u_x^2(x,t)\,\mathrm{d}x\,\mathrm{d}t \leqslant \frac{1}{2}\|u_0\|^2_{L^2(\mathbb{R})}.$$

This is the continuous analogous of (20.9). Hence if $h-k=\varepsilon \geqslant \xi h$ (this is condition (20.7), note that this condition is more restrictive than the usual CFL condition required for the L^∞ stability), the discrete equivalent of this formal proof yields (20.9) (and then (20.8)).

In the first case, we derivate the equation and we use some regularity on u_0 (namely $u_0 \in BV(\mathbb{R})$). In the second case, it is sufficient to have $u_0 \in L^\infty(\mathbb{R})$ but we need the diffusion term to be large enough in order to obtain the estimate which, by the way, does not yield any estimate on the solution of (20.11) with $\varepsilon = 0$. This formal derivation may be carried out similarly in the nonlinear case.

Let us now give a convergence result for the scheme (20.5) in $L^\infty(\mathbb{R}\times\mathbb{R}_+^\star)$ for the weak-$\star$ topology. Recall that a sequence $(v_n)_{n\in\mathbb{N}} \subset L^\infty(\mathbb{R}\times\mathbb{R}_+^\star)$ converges to $v \in L^\infty(\mathbb{R}\times\mathbb{R}_+^\star)$ in $L^\infty(\mathbb{R}\times\mathbb{R}_+^\star)$ for the weak-$\star$ topology if

$$\int_{\mathbb{R}_+}\int_{\mathbb{R}} \big(v_n(x,t)-v(x,t)\big)\varphi(x,t)\,\mathrm{d}x\,\mathrm{d}t \to 0 \quad \text{as } n\to\infty,\ \forall\varphi\in L^1\big(\mathbb{R}\times\mathbb{R}_+^\star\big).$$

A stronger convergence result is available, and comes from the nonlinear study given in Section 21.

PROPOSITION 20.2 (Convergence in the linear case). *Let $u_0 \in L^\infty(\mathbb{R})$ and u be the unique weak solution to problem* (20.1) *in the sense of Definition* 19.2, *with $f(s) = s$ for all $s \in \mathbb{R}$. Let $\xi \in (0,1)$ and $\alpha > 0$ be given. Let $\mathcal{T}$ be an admissible mesh in the sense of Definition* 20.1 *and let $k \in \mathbb{R}_+^\star$ satisfying the CFL condition* (20.7) (*taking $k \leqslant (1-\xi)\alpha h$ implies this condition, note that ξ and α do not depend on $\mathcal{T}$*).

Let $u_{\mathcal{T},k}$ be the finite volume approximate solution defined by (20.5) *and* (20.6). *Then $u_{\mathcal{T},k} \to u$ in $L^\infty(\mathbb{R}\times\mathbb{R}_+^\star)$ for the weak-$\star$ topology as $h = \mathrm{size}(\mathcal{T}) \to 0$.*

PROOF. Let $(\mathcal{T}_m, k_m)_{m\in\mathbb{N}}$ be a sequence of meshes and time steps satisfying the hypotheses of Proposition 20.2 and such that $\mathrm{size}(\mathcal{T}_m) \to 0$ as $m \to \infty$.

Lemma 20.1 gives the existence of a subsequence, still denoted by $(\mathcal{T}_m, k_m)_{m\in\mathbb{N}}$, and of a function $u \in L^\infty(\mathbb{R}\times\mathbb{R}_+^\star)$ such that $u_{\mathcal{T}_m,k_m} \to u$ in $L^\infty(\mathbb{R}\times\mathbb{R}_+^\star)$ for the weak-$\star$ topology, as $m \to +\infty$. There remains to show that u is the solution of (19.3) (with $f(s) = s$ for all $s \in \mathbb{R}$). The uniqueness of the weak solution to problem (20.1) will then imply that the full sequence converges to u.

Let $\varphi \in C_c^1(\mathbb{R}\times\mathbb{R}_+, \mathbb{R})$. Let $m \in \mathbb{N}$ and $\mathcal{T} = \mathcal{T}_m$, $k = k_m$ and $h = \mathrm{size}(\mathcal{T})$. Let us multiply the first equation of (20.5) by $(k/h_i)\varphi(x, nk)$, integrate over $x \in K_i$ and sum for all $i \in \mathbb{Z}$ and $n \in \mathbb{N}$. This yields

$$A_m + B_m = 0$$

with

$$A_m = \sum_{i\in\mathbb{Z}}\sum_{n\in\mathbb{N}} \bigl(u_i^{n+1} - u_i^n\bigr) \int_{K_i} \varphi(x, nk)\,\mathrm{d}x$$

and

$$B_m = \sum_{i\in\mathbb{Z}}\sum_{n\in\mathbb{N}} k\bigl(u_i^n - u_{i-1}^n\bigr)\frac{1}{h_i}\int_{K_i} \varphi(x, nk)\,\mathrm{d}x.$$

Let us remark that $A_m = A_{1,m} - A'_{1,m}$ with

$$A_{1,m} = -\int_k^\infty \int_{\mathbb{R}} u_{\mathcal{T},k}(x,t)\varphi_t(x, t-k)\,\mathrm{d}x\,\mathrm{d}t - \int_{\mathbb{R}} u_0(x)\varphi(x,0)\,\mathrm{d}x$$

and

$$A'_{1,m} = \sum_{i\in\mathbb{Z}} u_i^0 \int_{K_i} \varphi(x,0)\,\mathrm{d}x - \int_{\mathbb{R}} u_0(x)\varphi(x,0)\,\mathrm{d}x.$$

Using the fact that $\sum_{i\in\mathbb{Z}} u_i^0 1_{K_i} \to u_0$ in $L^1_{\mathrm{loc}}(\mathbb{R})$ as $m \to \infty$, we get that $A'_{1,m} \to 0$ as $m \to \infty$. (Recall that $1_{K_i}(x) = 1$ if $x \in K_i$ and $1_{K_i}(x) = 0$ if $x \notin K_i$.)

Therefore, since, as $m \to \infty$, $u_{\mathcal{T},k} \to u$ in $L^\infty(\mathbb{R} \times \mathbb{R}_+^\star)$ for the weak-$\star$ topology and $\varphi_t(\bullet, \bullet - k)1_{\mathbb{R}\times(k,\infty)} \to \varphi_t$ in $L^1(\mathbb{R} \times \mathbb{R}_+^\star)$ (note that $k \to 0$ thanks to (20.7)),

$$\lim_{m\to+\infty} A_m = \lim_{m\to+\infty} A_{1,m} = -\int_{\mathbb{R}_+}\int_{\mathbb{R}} u(x,t)\varphi_t(x,t)\,\mathrm{d}x\,\mathrm{d}t - \int_{\mathbb{R}} u_0(x)\varphi(x,0)\,\mathrm{d}x.$$

Let us now turn to the study of B_m. We compare B_m with

$$B_{1,m} = -\sum_{n\in\mathbb{N}} \int_{nk}^{(n+1)k} \int_{\mathbb{R}} u_{\mathcal{T},k}(x,t)\varphi_x(x,nk)\,\mathrm{d}x\,\mathrm{d}t,$$

which tends to $-\int_{\mathbb{R}_+}\int_{\mathbb{R}} u(x,t)\varphi_x(x,t)\,\mathrm{d}x\,\mathrm{d}t$ as $m \to \infty$. The term $B_{1,m}$ can be rewritten as

$$B_{1,m} = \sum_{i\in\mathbb{Z}}\sum_{n\in\mathbb{N}} k\big(u_i^n - u_{i-1}^n\big)\varphi(x_{i-1/2}, nk).$$

Let $R > 0$ and $T > 0$ be such that $\varphi(x,t) = 0$ if $|x| \geqslant R$ or $t \geqslant T$. Then, there exists $C \in \mathbb{R}_+^\star$, only depending on φ, such that, if $h < R$ and $k < T$ (which is true for h small enough, thanks to (20.7)),

$$|B_m - B_{1,m}| \leqslant Ch \sum_{i=i_0}^{i_1}\sum_{n=0}^{N} k\big|u_i^n - u_{i-1}^n\big|, \tag{20.12}$$

where $i_0 \in \mathbb{Z}$, $i_1 \in \mathbb{Z}$ and $N \in \mathbb{N}$ are such that $-R \in \overline{K}_{i_0}$, $R \in \overline{K}_{i_1}$ and $T \in (Nk, (N+1)k]$.

Using (20.12) and Lemma 20.2, we get that $B_m - B_{1,m} \to 0$ and then

$$B_m \to -\int_{\mathbb{R}_+}\int_{\mathbb{R}} u(x,t)\varphi_x(x,t)\,\mathrm{d}x\,\mathrm{d}t \quad \text{as } m \to \infty,$$

which completes the proof that u is the weak solution to problem (20.1) (note that here the useful consequence of Lemma 20.2 is (20.10)). □

REMARK 20.4. In Proposition 20.2, a simpler proof of convergence could be achieved, with $\xi = 0$, using a multiplication of the first equation of (20.5) by $(k/h_i)\varphi(x_{i-1/2}, nk)$. However, this proof does not generalize to the general case of nonlinear hyperbolic problems.

REMARK 20.5. Proving the convergence of the finite difference method (with the scheme (20.3)) with $u_0 \in L^\infty(\mathbb{R})$ can be done using the same technique as the proof of the finite volume method (i.e. considering the finite difference scheme as a finite volume scheme on a convenient mesh).

21. The nonlinear case

In this section, finite volume schemes for the discretization of problem (19.1) are presented and a theorem of convergence is given (Theorem 21.1) which will be generalized to the multidimensional case in the next chapter. We also recall the classical proof of convergence which uses a "strong BV estimate" and the Lax–Wendroff theorem. This proof, however, does not seem to extend to the multidimensional case for general meshes. The following properties are assumed to be satisfied by the data of problem (19.1).

ASSUMPTION 21.1. The flux function f belongs to $C^1(\mathbb{R},\mathbb{R})$, the initial data u_0 belongs to $L^\infty(\mathbb{R})$ and $U_m, U_M \in \mathbb{R}$ are such that $U_m \leqslant u_0 \leqslant U_M$ a.e. on $\mathbb{R}$.

21.1. Meshes and schemes

Let $\mathcal{T}$ be an admissible mesh in the sense of Definition 20.1 and $k \in \mathbb{R}_+^\star$ be the time step. In the general nonlinear case, the finite volume scheme for the discretization of problem (19.1) writes

$$\begin{cases} \dfrac{h_i}{k}\left(u_i^{n+1} - u_i^n\right) + f_{i+1/2}^n - f_{i-1/2}^n = 0, & \forall n \in \mathbb{N}, \forall i \in \mathbb{Z}, \\ u_i^0 = \dfrac{1}{h_i}\displaystyle\int_{x_{i-1/2}}^{x_{i+1/2}} u_0(x)\,\mathrm{d}x, & \forall i \in \mathbb{Z}, \end{cases} \tag{21.1}$$

where u_i^n is expected to be an approximation of u at time $t_n = nk$ in cell K_i. The quantity $f_{i+1/2}^n$ is often called the numerical flux at point $x_{i+1/2}$ and time t_n (it is expected to be an approximation of $f(u)$ at point $x_{i+1/2}$ and time t_n). Note that a common expression of $f_{i+1/2}^n$ is used for both equations i and $i+1$ in (21.1); therefore the scheme (21.1) satisfies the property of conservativity, common to all finite volume schemes. In the case of a so called "scheme with $2p+1$ points" ($p \in \mathbb{N}^\star$), the numerical flux may be written

$$f_{i+1/2}^n = g\left(u_{i-p+1}^n, \ldots, u_{i+p}^n\right), \tag{21.2}$$

where g is the numerical flux function, which determines the scheme. It is assumed to be a locally Lipschitz continuous function.

As in the linear case (20.6), the approximate finite volume solution is defined by

$$u_{\mathcal{T},k}(x,t) = u_i^n \quad \text{for } x \in K_i \text{ and } t \in \left[nk, (n+1)k\right). \tag{21.3}$$

The property of consistency for the finite volume scheme (21.1), (21.2) with $2p+1$ points, is ensured by writing the following condition:

$$g(s, \ldots, s) = f(s), \quad \forall s \in \mathbb{R}. \tag{21.4}$$

This condition is equivalent to writing the consistency of the approximation of the flux (as in the elliptic and parabolic cases, which were described in the previous chapters, see, e.g., Section 5).

REMARK 21.1 (*Finite volumes and finite differences*). We can remark that, as in the elliptic case, the condition (21.4) does not generally give the consistency of the scheme (21.1) when it is considered as a finite difference scheme. For instance, assume $f(s) = s^2$ for all $s \in \mathbb{R}$, $p = 1$ and $g(a, b) = f_1(a) + f_2(b)$ for all $a, b \in \mathbb{R}$ with $f_1(s) = \max\{s, 0\}^2$, $f_2(s) = \min\{s, 0\}^2$ (which is shown below to be a "good choice", see Example 21.1). Assume also $h_{2i} = h$ and $h_{2i+1} = h/2$ for all $i \in \mathbb{Z}$. In this case, there is no choice of points $x_i \in \mathbb{R}$ such that the quantity $(f_{i+1/2}^n - f_{i-1/2}^n)/h_i$ is an approximation of order 1 of $(f(u))_x(x_i, nk)$, for any regular function u, when $u_i^n = u(x_i, nk)$ for all $i \in \mathbb{Z}$. Indeed, up to second-order terms, this property of consistency is achieved if and only if $f_2'(a)|x_{i+1} - x_i| + f_1'(a)|x_{i-1} - x_i| = f'(a)h_i$ for all $i \in \mathbb{Z}$ and for all $a \in \mathbb{R}$. Choosing $a > 0$ and $a < 0$, this condition leads to $|x_{i+1} - x_i| = h_i$ and $|x_{i+1} - x_i| = h_{i+1}$ for all $i \in \mathbb{Z}$, which is impossible.

Examples of convenient choices for the function g will now be given. An interesting class of schemes is the class of 3-points schemes with a monotone flux, which we now define.

DEFINITION 21.1 (*Monotone flux schemes*). Under Assumption 21.1, the finite volume scheme (21.1)–(21.2) is said to be a "monotone flux scheme" if $p = 1$ and if the function g, only depending on f, U_m and U_M, satisfies the following assumptions:

- g is locally Lipschitz continuous from $\mathbb{R}^2$ to $\mathbb{R}$,
- $g(s, s) = f(s)$, for all $s \in [U_m, U_M]$,
- $(a, b) \mapsto g(a, b)$, from $[U_m, U_M]^2$ to $\mathbb{R}$, is nondecreasing with respect to a and nonincreasing with respect to b.

The monotone flux schemes are worthy of consideration for they are consistent in the finite volume sense, they are L^∞-stable under a condition (the so called Courant–Friedrichs–Levy condition) of the type $k \leqslant C_1 h$, where C_1 only depends on g and u_0 (see Section 21.2 below), and they are "consistent with the entropy inequalities" also under a condition of the type $k \leqslant C_2 h$, where C_2 only depends on g and u_0 (but C_2 may be different of C_1, see Section 21.3).

REMARK 21.2. A monotone flux scheme is a monotone scheme, under a Courant–Friedrichs–Levy condition, which means that the scheme can be written under the form

$$u_i^{n+1} = H\left(u_{i-1}^n, u_i^n, u_{i+1}^n\right),$$

with H nondecreasing with respect to its three arguments.

EXAMPLE 21.1 (*Examples of monotone flux schemes*). (See also GODLEWSKI and RAVIART [1996], LEVEQUE [1990] and references therein.) Under Assumption 21.1, here are some numerical flux functions g for which the finite volume scheme (21.1)–(21.2) is a monotone flux scheme (in the sense of Definition 21.1):

- the flux splitting scheme: assume $f = f_1 + f_2$, with $f_1, f_2 \in C^1(\mathbb{R}, \mathbb{R})$, $f_1'(s) \geqslant 0$ and $f_2'(s) \leqslant 0$ for all $s \in [U_m, U_M]$ (such a decomposition for f is always possible, see the modified Lax–Friedrichs scheme below), and take

$$g(a,b) = f_1(a) + f_2(b).$$

Note that if $f' \geqslant 0$, taking $f_1 = f$ and $f_2 = 0$, the flux splitting scheme boils down to the upwind scheme, i.e. $g(a,b) = f(a)$.
- the Godunov scheme: the Godunov scheme, which was introduced in GODUNOV [1976], may be summarized by the following expression.

$$g(a,b) = \begin{cases} \min\{f(\xi), \xi \in [a,b]\} & \text{if } a \leqslant b, \\ \max\{f(\xi), \xi \in [b,a]\} & \text{if } b \leqslant a. \end{cases} \tag{21.5}$$

- the modified Lax–Friedrichs scheme: take

$$g(a,b) = \frac{f(a) + f(b)}{2} + D(a - b), \tag{21.6}$$

with $D \in \mathbb{R}$ such that $2D \geqslant \max\{|f'(s)|, s \in [U_m, U_M]\}$. Note that in this modified version of the Lax–Friedrichs scheme, the coefficient D only depends on f, U_m and U_M, while the original Lax–Friedrichs scheme consists in taking $D = h/(2k)$, in the case $h_i = h$ for all $i \in \mathbb{N}$, and therefore satisfies the three items of Definition 21.1 under the condition $h/k \geqslant \max\{|f'(s)|, s \in [U_m, U_M]\}$. However, an inverse CFL condition appears to be necessary for the convergence of the original Lax–Friedrichs scheme (see Remark 29.1); such a condition is not necessary for the modified version.
Note also that the modified Lax–Friedrichs scheme consists in a particular flux splitting scheme with $f_1(s) = (1/2)f(s) + Ds$ and $f_2(s) = (1/2)f(s) - Ds$ for $s \in [U_m, U_M]$.

REMARK 21.3. In the case of a nondecreasing (resp. nonincreasing) function f, the Godunov monotone flux scheme (21.5) reduces to $g(a,b) = f(a)$ (resp. $f(b)$). Then, in the case of a nondecreasing function f, the scheme (21.1), (21.2) reduces to

$$h_i \frac{u_i^{n+1} - u_i^n}{k} + f(u_i^n) - f(u_{i-1}^n) = 0, \tag{21.7}$$

i.e. the upstream (or upwind) finite volume scheme. The scheme (21.7) is sometimes called "upstream finite difference" scheme. In that particular case (f monotone and 1D) it is possible to find points x_i in order to obtain a consistent scheme in the finite difference sense (if f is nondecreasing, take $x_i = x_{i+1/2}$ as for the scheme (20.5)).

21.2. L^∞-stability for monotone flux schemes

LEMMA 21.1 (L^∞ estimate in the nonlinear case). *Under Assumption 21.1, let $\mathcal{T}$ be an admissible mesh in the sense of Definition 20.1 and let $k \in \mathbb{R}_+^\star$ be the time step.*

Let $u_{\mathcal{T},k}$ be the finite volume approximate solution defined by (21.1)–(21.3) and assume that the scheme is a monotone flux scheme in the sense of Definition 21.1. Let g_1 and g_2 be the Lipschitz constants of g on $[U_m, U_M]^2$ with respect to its two arguments.

Under the Courant–Friedrichs–Levy (CFL) condition

$$k \leqslant \frac{\inf_{i\in\mathbb{Z}} h_i}{g_1 + g_2} \tag{21.8}$$

(note that taking $k \leqslant \alpha h/(g_1 + g_2)$ implies (21.8)), the approximate solution $u_{\mathcal{T},k}$ satisfies

$$U_m \leqslant u_{\mathcal{T},k}(x,t) \leqslant U_M \quad \text{for a.e. } x \in \mathbb{R} \text{ and a.e. } t \in \mathbb{R}_+.$$

PROOF. Let us prove that

$$U_m \leqslant u_i^n \leqslant U_M, \quad \forall i \in \mathbb{Z},\ \forall n \in \mathbb{N}, \tag{21.9}$$

by induction on n, which proves the lemma. Assertion (21.9) holds for $n = 0$ thanks to the definition of u_i^0 in (21.1). Suppose that it holds for $n \in \mathbb{N}$.

For all $i \in \mathbb{Z}$, scheme (21.1), (21.2) (with $p = 1$) gives

$$u_i^{n+1} = \left(1 - b_{i+1/2}^n - a_{i-1/2}^n\right)u_i^n + b_{i+1/2}^n u_{i+1}^n + a_{i-1/2}^n u_{i-1}^n,$$

with

$$b_{i+1/2}^n = \begin{cases} \dfrac{k}{h_i}\dfrac{g(u_i^n, u_{i+1}^n) - f(u_i^n)}{u_i^n - u_{i+1}^n} & \text{if } u_i^n \neq u_{i+1}^n, \\ 0 & \text{if } u_i^n = u_{i+1}^n, \end{cases}$$

and

$$a_{i-1/2}^n = \begin{cases} \dfrac{k}{h_i}\dfrac{g(u_{i-1}^n, u_i^n) - f(u_i^n)}{u_{i-1}^n - u_i^n} & \text{if } u_i^n \neq u_{i-1}^n, \\ 0 & \text{if } u_i^n = u_{i-1}^n. \end{cases}$$

Since $f(u_i^n) = g(u_i^n, u_i^n)$ and thanks to the monotonicity of g, $0 \leqslant b_{i+1/2}^n \leqslant g_2 k/h_i$ and $0 \leqslant a_{i-1/2}^n \leqslant g_1 k/h_i$, for all $i \in \mathbb{Z}$. Therefore, under condition (21.8), the value u_i^{n+1} may be written as a convex linear combination of the values u_i^n and u_{i-1}^n. Assertion (21.9) is thus proved for $n + 1$, which concludes the proof of the lemma. $\square$

21.3. Discrete entropy inequalities

LEMMA 21.2 (Discrete entropy inequalities). *Under Assumption* 21.1, *let $\mathcal{T}$ be an admissible mesh in the sense of Definition* 20.1 *and let $k \in \mathbb{R}_+^\star$ be the time step.*

Let $u_{\mathcal{T},k}$ be the finite volume approximate solution defined by (21.1)–(21.3) *and assume that the scheme is a monotone flux scheme in the sense of Definition* 21.1. *Let g_1 and g_2 be the Lipschitz constants of g on $[U_m, U_M]^2$ with respect to its two arguments. Under the CFL condition* (21.8), *the following inequation holds*:

$$\begin{aligned}
&\frac{h_i}{k}\left(\left|u_i^{n+1}-\kappa\right|-\left|u_i^n-\kappa\right|\right)\\
&\quad + g\left(u_i^n\top\kappa, u_{i+1}^n\top\kappa\right) - g\left(u_i^n\bot\kappa, u_{i+1}^n\bot\kappa\right) - g\left(u_{i-1}^n\top\kappa, u_i^n\top\kappa\right)\\
&\quad + g\left(u_{i-1}^n\bot\kappa, u_i^n\bot\kappa\right) \leqslant 0, \quad \forall n\in\mathbb{N},\ \forall i\in\mathbb{Z},\ \forall\kappa\in\mathbb{R}.
\end{aligned} \tag{21.10}$$

Recall that $a\top b$ (resp. $a\bot b$) denotes the maximum (resp. the minimum) of the two real numbers a and b.

PROOF. Thanks to the monotonicity properties of g and to the condition (21.8) (see Remark 21.2),

$$u_i^{n+1} = H\left(u_{i-1}^n, u_i^n, u_{i+1}^n\right), \quad \forall i\in\mathbb{Z},\ \forall n\in\mathbb{N},$$

where H is a function from $\mathbb{R}^3$ to $\mathbb{R}$ which is nondecreasing with respect to all its arguments and such that $\kappa = H(\kappa,\kappa,\kappa)$ for all $\kappa\in\mathbb{R}$.

Hence, for all $\kappa\in\mathbb{R}$,

$$u_i^{n+1} \leqslant H\left(u_{i-1}^n\top\kappa, u_i^n\top\kappa, u_{i+1}^n\top\kappa\right),$$

and

$$\kappa \leqslant H\left(u_{i-1}^n\top\kappa, u_i^n\top\kappa, u_{i+1}^n\top\kappa\right),$$

which yields

$$u_i^{n+1}\top\kappa \leqslant H\left(u_{i-1}^n\top\kappa, u_i^n\top\kappa, u_{i+1}^n\top\kappa\right).$$

In the same manner, we get

$$u_i^{n+1}\bot\kappa \geqslant H\left(u_{i-1}^n\bot\kappa, u_i^n\bot\kappa, u_{i+1}^n\bot\kappa\right),$$

and therefore, by subtracting the last two equations,

$$\left|u_i^{n+1}-\kappa\right| \leqslant H\left(u_{i-1}^n\top\kappa, u_i^n\top\kappa, u_{i+1}^n\top\kappa\right) - H\left(u_{i-1}^n\bot\kappa, u_i^n\bot\kappa, u_{i+1}^n\bot\kappa\right),$$

i.e. (21.10). □

In the two next sections, we study the convergence of the schemes defined by (21.1), (21.2) with $p = 1$ (see Remarks 21.4 and 21.5 and Section 22 for the schemes with $2p + 1$ points).

We first develop a proof of convergence for the monotone flux schemes; this proof is based on a weak BV estimate similar to (20.8) like the proof of Proposition 20.2 in the linear case. It will be generalized in the multidimensional case studied in Chapter VI. We then briefly describe the BV framework which gave the first convergence results; its generalization to the multidimensional case is not so easy, except in the case of Cartesian meshes.

21.4. Convergence of the upstream scheme in the general case

A proof of convergence similar to the proof of convergence given in the linear case can be developed. For the sake of simplicity, we shall consider only the case of a nondecreasing function f and of the classical upstream scheme (the general case for f and for the monotone flux schemes being handled in Chapter VI). We shall first prove a "weak BV" estimate.

LEMMA 21.3 (Weak BV estimate for the nonlinear case). *Under Assumption* 21.1, *assume that f is nondecreasing. Let $\xi \in (0, 1)$ be a given value. Let $\mathcal{T}$ be an admissible mesh in the sense of Definition* 20.1, *let M be the Lipschitz constant of f in $[U_m, U_M]$ and let $k \in \mathbb{R}_+^\star$ satisfying the CFL condition*

$$k \leqslant (1-\xi)\frac{\inf_{i\in\mathbb{Z}} h_i}{M}. \tag{21.11}$$

(The condition $k \leqslant (1-\xi)\alpha h/M$ implies the above condition.) Let $\{u_i^n,\ i \in \mathbb{Z},\ n \in \mathbb{N}\}$ be given by the finite volume scheme (21.1), (21.2) *with $p = 1$ and $g(a, b) = f(a)$. Let $R \in \mathbb{R}_+^\star$ and $T \in \mathbb{R}_+^\star$ and assume $h < R$ and $k < T$. Let $i_0 \in \mathbb{Z}$, $i_1 \in \mathbb{Z}$ and $N \in \mathbb{N}$ be such that $-R \in \overline{K}_{i_0}$, $R \in \overline{K}_{i_1}$, and $T \in (Nk, (N+1)k]$. Then there exists $C \in \mathbb{R}_+^\star$, only depending on R, T, u_0, α, f and ξ, such that*

$$\sum_{i=i_0}^{i_1}\sum_{n=0}^{N} k\left|f(u_i^n) - f(u_{i-1}^n)\right| \leqslant Ch^{-1/2}. \tag{21.12}$$

PROOF. We multiply the first equation of (21.1) by ku_i^n, and we sum on $i = i_0, \ldots, i_1$ and $n = 0, \ldots, N$. We get $A + B = 0$, with

$$A = \sum_{i=i_0}^{i_1}\sum_{n=0}^{N} h_i\left(u_i^{n+1} - u_i^n\right)u_i^n,$$

and

$$B = \sum_{i=i_0}^{i_1}\sum_{n=0}^{N} k\left(f(u_i^n) - f(u_{i-1}^n)\right)u_i^n.$$

We have

$$A=-\tfrac{1}{2}\sum_{i=i_0}^{i_1}\sum_{n=0}^{N}h_i\bigl(u_i^{n+1}-u_i^n\bigr)^2+\tfrac{1}{2}\sum_{i=i_0}^{i_1}h_i\bigl[\bigl(u_i^{N+1}\bigr)^2-\bigl(u_i^0\bigr)^2\bigr].$$

Using the scheme (21.1), we get

$$A=-\tfrac{1}{2}\sum_{i=i_0}^{i_1}\sum_{n=0}^{N}\frac{k^2}{h_i}\bigl(f\bigl(u_i^n\bigr)-f\bigl(u_{i-1}^n\bigr)\bigr)^2+\tfrac{1}{2}\sum_{i=i_0}^{i_1}h_i\bigl[\bigl(u_i^{N+1}\bigr)^2-\bigl(u_i^0\bigr)^2\bigr],$$

and therefore, using the CFL condition (21.11),

$$A\geqslant-\frac{1}{2M}(1-\xi)\sum_{i=i_0}^{i_1}\sum_{n=0}^{N}k\bigl(f\bigl(u_i^n\bigr)-f\bigl(u_{i-1}^n\bigr)\bigr)^2-\tfrac{1}{2}\sum_{i=i_0}^{i_1}h_i\bigl(u_i^0\bigr)^2. \tag{21.13}$$

We now study the term B.

Denoting by Φ the function $\Phi(a)=\int_{U_m}^{a}sf'(s)\,\mathrm{d}s$, for all $a\in\mathbb{R}$, an integration by parts yields, for all $(a,b)\in\mathbb{R}^2$,

$$\Phi(b)-\Phi(a)=b\bigl(f(b)-f(a)\bigr)-\int_a^b\bigl(f(s)-f(a)\bigr)\,\mathrm{d}x.$$

Using the technical Lemma 18.5 which states $\int_a^b(f(s)-f(a))\,\mathrm{d}x\geqslant\frac{1}{2M}(f(b)-f(a))^2$, we obtain

$$b\bigl(f(b)-f(a)\bigr)\geqslant\frac{1}{2M}\bigl(f(b)-f(a)\bigr)^2+\Phi(b)-\Phi(a).$$

The above inequality with $a=u_{i-1}^n$ and $b=u_i^n$ yields

$$B\geqslant\frac{1}{2M}\sum_{i=i_0}^{i_1}\sum_{n=0}^{N}k\bigl(f\bigl(u_i^n\bigr)-f\bigl(u_{i-1}^n\bigr)\bigr)^2+\sum_{n=0}^{N}k\bigl[\Phi\bigl(u_{i_1}^n\bigr)-\Phi\bigl(u_{i_0-1}^n\bigr)\bigr].$$

Thanks to the L^∞ estimate of Lemma 20.1, there exists $C_1>0$, only depending on u_0 and f such that

$$B\geqslant\frac{1}{2M}\sum_{i=i_0}^{i_1}\sum_{n=0}^{N}k\bigl(f\bigl(u_i^n\bigr)-f\bigl(u_{i-1}^n\bigr)\bigr)^2-TC_1.$$

Therefore, since $A+B=0$ and $\sum_{i=i_0}^{i_1} h_i \leqslant 4R$, the following inequality holds:

$$0 \geqslant \xi \sum_{i=i_0}^{i_1} \sum_{n=0}^{N} k\big(f(u_i^n) - f(u_{i-1}^n)\big)^2 - 4RM \max\{-U_m, U_M\}^2 - 2MTC_1,$$

which gives the existence of $C_2 \in \mathbb{R}_+^\star$, only depending on R, T, u_0, f and ξ such that

$$\sum_{i=i_0}^{i_1} \sum_{n=0}^{N} k\big(f(u_i^n) - f(u_{i-1}^n)\big)^2 \leqslant C_2.$$

The Cauchy–Schwarz inequality yields

$$\left[\sum_{i=i_0}^{i_1} \sum_{n=0}^{N} k\big|f(u_i^n) - f(u_{i-1}^n)\big|\right]^2 \leqslant C_2 2T \frac{4R}{\alpha h},$$

which concludes the proof of the lemma. □

We can now state the convergence theorem.

THEOREM 21.1 (Convergence in the nonlinear case). *Assume Assumption* 21.1 *and f nondecreasing. Let $\xi \in (0,1)$ and $\alpha > 0$ be given. Let M be the Lipschitz constant of f in $[U_m, U_M]$. For an admissible mesh $\mathcal{T}$ in the sense of Definition* 20.1 *and for a time step $k \in \mathbb{R}_+^\star$ satisfying the CFL condition* (21.11) *(taking $k \leqslant (1-\xi)\alpha h/M$ is a sufficient condition, note that ξ and α do not depend of $\mathcal{T}$), let $u_{\mathcal{T},k}$ be the finite volume approximate solution defined by* (21.1)–(21.3) *with $p=1$ and $g(a,b) = f(a)$.*

Then the function $u_{\mathcal{T},k}$ converges to the unique entropy weak solution u of (19.1) *in $L^1_{\mathrm{loc}}(\mathbb{R} \times \mathbb{R}_+)$ as* $\mathrm{size}(\mathcal{T})$ *tends to* 0.

PROOF. Let Y be the set of approximate solutions, that is the set of $u_{\mathcal{T},k}$, defined by (21.1)–(21.3) with $p=1$ and $g(a,b) = f(a)$, for all $(\mathcal{T}, k)$ where $\mathcal{T}$ is an admissible mesh in the sense of Definition 20.1 and $k \in \mathbb{R}_+^\star$ satisfies the CFL condition (21.11). Thanks to Lemma 21.1, the set Y is bounded in $L^\infty(\mathbb{R} \times \mathbb{R}_+)$.

The proof of Theorem 21.1 is performed in three steps. In the first step, a compactness result is given for Y, only using the boundedness of Y in $L^\infty(\mathbb{R} \times \mathbb{R}_+)$. In the second step, it is proved that the eventual limit (in a convenient sense) of a sequence of approximate solutions is a solution (in a convenient sense) of problem (19.1). In the third step a uniqueness result yields the conclusion. For Steps 1 and 3, we refer to Chapter VI for a complete proof.

Step 1 (*Compactness result*). Let us first use a compactness result in $L^\infty(\mathbb{R} \times \mathbb{R}_+)$ which is stated in Proposition 30.1. Since Y is bounded in $L^\infty(\mathbb{R} \times \mathbb{R}_+)$, for any sequence $(u_m)_{m\in\mathbb{N}}$ of Y there exists a subsequence, still denoted by $(u_m)_{m\in\mathbb{N}}$, and there exists $\mu \in L^\infty(\mathbb{R} \times \mathbb{R}_+ \times (0,1))$ such that $(u_m)_{m\in\mathbb{N}}$ converges to μ in the "nonlinear weak-$\star$ sense", i.e.

$$\int_{\mathbb{R}}\int_{\mathbb{R}_+} \theta\big(u_m(x,t)\big)\varphi(x,t)\,\mathrm{d}t\,\mathrm{d}x$$
$$\to \int_{\mathbb{R}}\int_{\mathbb{R}_+}\int_0^1 \theta\big(\mu(x,t,\alpha)\big)\varphi(x,t)\,\mathrm{d}\alpha\,\mathrm{d}t\,\mathrm{d}x, \quad \text{as } m \to \infty,$$

for all $\varphi \in L^1(\mathbb{R} \times \mathbb{R}_+)$ and all $\theta \in C(\mathbb{R}, \mathbb{R})$. In other words, for any $\theta \in C(\mathbb{R}, \mathbb{R})$,

$$\theta(u_m) \to \mu_\theta \quad \text{in } L^\infty(\mathbb{R} \times \mathbb{R}_+) \text{ for the weak-}\star\text{ topology as } m \to \infty, \tag{21.14}$$

where μ_θ is defined by

$$\mu_\theta(x,t) = \int_0^1 \theta\big(\mu(x,t,\alpha)\big)\,\mathrm{d}\alpha, \quad \text{for a.e. } (x,t) \in \mathbb{R} \times \mathbb{R}_+.$$

Step 2 (*Passage to the limit*). Let $(u_m)_{m\in\mathbb{N}}$ be a sequence of Y. Assume that $(u_m)_{m\in\mathbb{N}}$ converges to μ in the nonlinear weak-$\star$ sense and that $u_m = u_{\mathcal{T}_m,k_m}$ (for all $m \in \mathbb{N}$) with $\text{size}(\mathcal{T}_m) \to 0$ as $m \to \infty$ (note that $k_m \to 0$ as $m \to \infty$, thanks to (21.11)).

Let us prove that μ is a "solution" to problem (19.1) in the following sense (we shall say that μ is "an entropy process solution" to problem (19.1)):

$$\begin{cases} \mu \in L^\infty\big(\mathbb{R} \times \mathbb{R}_+ \times (0,1)\big), \\ \displaystyle\int_{\mathbb{R}}\int_{\mathbb{R}_+}\int_0^1 \Big(\big|\mu(x,t,\alpha) - \kappa\big|\varphi_t(x,t) + \big(f(\mu(x,t,\alpha)\top\kappa) \\ \quad - f(\mu(x,t,\alpha)\bot\kappa)\big)\varphi_x(x,t)\Big)\,\mathrm{d}\alpha\,\mathrm{d}t\,\mathrm{d}x \\ \quad + \displaystyle\int_{\mathbb{R}} \big|u_0(x) - \kappa\big|\varphi(x,0)\,\mathrm{d}x \geqslant 0, \quad \forall\varphi \in C_c^1(\mathbb{R} \times \mathbb{R}_+, \mathbb{R}_+),\ \forall\kappa \in \mathbb{R}. \end{cases} \tag{21.15}$$

Let $\kappa \in \mathbb{R}$. Setting

$$v(x,t) = \int_0^1 \big|\mu(x,t,\alpha) - \kappa\big|\,\mathrm{d}\alpha, \quad \text{for a.e. } (x,t) \in \mathbb{R} \times \mathbb{R}_+$$

and

$$w(x,t) = \int_0^1 \Big(f\big(\mu(x,t,\alpha)\top\kappa\big) - f\big(\mu(x,t,\alpha)\bot\kappa\big)\Big)\,\mathrm{d}\alpha,$$
$$\text{for a.e. } (x,t) \in \mathbb{R} \times \mathbb{R}_+,$$

the inequality in (21.15) writes

$$\int_{\mathbb{R}}\int_{\mathbb{R}_+}\big[v(x,t)\varphi_t(x,t)+w(x,t)\varphi_x(x,t)\big]\,\mathrm{d}t\,\mathrm{d}x+\int_{\mathbb{R}}\big|u_0(x)-\kappa\big|\varphi(x,0)\,\mathrm{d}x\geqslant 0,$$

$$\forall\varphi\in C_c^1(\mathbb{R}\times\mathbb{R}_+,\mathbb{R}_+). \tag{21.16}$$

Let us prove that (21.16) holds; for $m\in\mathbb{N}$ we shall denote by $\mathcal{T}=\mathcal{T}_m$ and $k=k_m$. We use the result of Lemma 21.2, which writes in the present particular case $f'\geqslant 0$,

$$h_i\frac{v_i^{n+1}-v_i^n}{k}+w_i^n-w_{i-1}^n\leqslant 0,\quad \forall i\in\mathbb{Z},\ \forall n\in\mathbb{N},$$

where $v_i^n=|u_i^n-\kappa|$ and $w_i^n=f(u_i^n\top\kappa)-f(u_i^n\bot\kappa)=|f(u_i^n)-f(\kappa)|$.

The functions $v_{\mathcal{T}_m,k_m}$ and $w_{\mathcal{T}_m,k_m}$ are defined in the same way as the function $u_{\mathcal{T}_m,k_m}$, i.e. with constant values v_i^n and w_i^n in each control volume K_i during each time step $(nk,(n+1)k)$. Choosing θ equal to the continuous functions $|\bullet-\kappa|$ and $|f(\bullet)-f(\kappa)|$ in (21.14) yields that the sequences $(v_{\mathcal{T}_m,k_m})_{m\in\mathbb{N}}$ and $(w_{\mathcal{T}_m,k_m})_{m\in\mathbb{N}}$ converge to v and w in $L^\infty(\mathbb{R}\times\mathbb{R}_+^\star)$ for the weak-$\star$ topology.

Applying the method which was used in the proof of Proposition 20.2, taking v_i^l instead of u_i^l in the definition of A_m (for $l=n$ and $n+1$) and w_j^n instead of u_j^n in the definition of B_m (for $j=i$ and $i-1$), we conclude that (21.16) holds.

Indeed, a weak BV inequality holds on the values w_i^n (i.e. (20.8) holds with w_j^n instead of u_j^n for $j=i$ and $i-1$), thanks to Lemma 21.3 and the relation

$$\big||f(u_i^n)-\kappa|-|f(u_{i-1}^n)-\kappa|\big|\leqslant\big|f(u_i^n)-f(u_{i-1}^n)\big|,\quad \forall i\in\mathbb{Z},\ \forall n\in\mathbb{N}.$$

(Note that here, as in the linear case, the useful consequence of the weak BV inequality, is (20.10) with w_j^n instead of u_j^n for $j=i$ and $i-1$.)

This concludes Step 2.

Step 3 (*Uniqueness result for* (21.15) *and conclusion*). Theorem 28.1 states that there exists at most one solution to (21.15) and that there exists $u\in L^\infty(\mathbb{R}\times\mathbb{R}_+)$ such that μ solution to (21.15) implies $\mu(x,t,\alpha)=u(x,t)$ for a.e. $(x,t,\alpha)\in\mathbb{R}\times\mathbb{R}_+\times(0,1)$. Then, u is necessarily the entropy weak solution to (19.1).

Furthermore, if $(u_m)_{m\in\mathbb{N}}$ converges to u in the nonlinear weak-$\star$ sense, an easy argument shows that $(u_m)_{m\in\mathbb{N}}$ converges to u in $L^1_{\mathrm{loc}}(\mathbb{R}\times\mathbb{R}_+)$ (and even in $L^p_{\mathrm{loc}}(\mathbb{R}\times\mathbb{R}_+)$ for all $1\leqslant p<\infty$), see Remark (30.2).

Then, the conclusion of Theorem 21.1 follows easily from Step 2 and Step 1 by way of contradiction (in order to prove the convergence of a sequence $u_{\mathcal{T}_m,k_m}\subset Y$ to u, if $\mathrm{size}(\mathcal{T}_m)\to 0$ as $m\to\infty$, without any extraction of a "subsequence"). □

REMARK 21.4. In Theorem 21.1, we only consider the case $f'\geqslant 0$ and the so called "upstream scheme". It is quite easy to generalize the result for any $f\in C^1(\mathbb{R},\mathbb{R})$ and any monotone flux scheme (see the following chapter). It is also possible to consider other schemes (e.g., some 5-points schemes, as in Section 22). For a given scheme, the proof of convergence of the approximate solution towards the entropy weak solution contains 2 steps:

(1) prove an L^∞ estimate on the approximate solutions, which allows to use the compactness result of Step 1 of the proof of Theorem 21.1,

(2) prove a "weak BV" estimate and some "discrete entropy inequality" in order to have the following property:

If $(u_m)_{m\in\mathbb{N}}$ is a sequence of approximate solutions which converges in the nonlinear weak-$\star$ sense, then

$$\lim_{m\to\mathbb{N}} \int_{\mathbb{R}}\int_{\mathbb{R}_+} \Big(\big|u_m(x,t)-\kappa\big|\varphi_t(x,t) + \big(f\big(u_m(x,t)\top\kappa\big)$$
$$- f\big(u_m(x,t)\bot\kappa\big)\big)\varphi_x(x,t)\Big)\,\mathrm{d}t\,\mathrm{d}x + \int_{\mathbb{R}} \big|u_0(x)-\kappa\big|\varphi(x,0)\,\mathrm{d}x \geqslant 0,$$
$$\forall\varphi\in C_c^1(\mathbb{R}\times\mathbb{R}_+,\mathbb{R}_+),\ \forall\kappa\in\mathbb{R}.$$

21.5. Convergence proof using BV

We now give the details of the classical proof of convergence (considering only 3 points schemes), which requires regularizations of u_0 in $BV(\mathbb{R})$. It consists in using Helly's compactness theorem (which may also be used in the linear case to obtain a strong convergence of $u_{\mathcal{T},k}$ to u in $L^1_{\mathrm{loc}}(\mathbb{R}\times\mathbb{R}_+)$). This theorem is a direct consequence of Kolmogorov's theorem (Theorem 14.1). We give below the definition of $BV(\Omega)$ where Ω is an open subset of $\mathbb{R}^p(\Omega)$, $p\geqslant 1$ (already given in Definition 20.1 for $\Omega=\mathbb{R}$) and we give a straightforward consequence of Helly's theorem for the case of interest here.

DEFINITION 21.2 ($BV(\Omega)$). Let $p\in\mathbb{N}^\star$ and let Ω be an open subset of $\mathbb{R}^p$. A function $v\in L^1_{\mathrm{loc}}(\Omega)$ has a bounded variation, i.e., $v\in BV(\Omega)$, if $|v|_{BV(\Omega)}<\infty$ where

$$|v|_{BV(\Omega)} = \sup\Big\{\int_\Omega v(x)\operatorname{div}\varphi(x)\,\mathrm{d}x,\ \varphi\in C_c^1\big(\Omega,\mathbb{R}^p\big),$$
$$\big|\varphi(x)\big|\leqslant 1,\ \forall x\in\Omega\Big\}. \tag{21.17}$$

LEMMA 21.4 (Consequence of Helly's theorem). *Let $\mathcal{A}\subset L^\infty(\mathbb{R}^2)$. Assume that there exists $C\in\mathbb{R}_+$ and, for all $T>0$, there exists $C_T\in\mathbb{R}_+$ such that*

$$\|v\|_{L^\infty(\mathbb{R}^2)}\leqslant C,\quad \forall v\in\mathcal{A},$$

and

$$|v|_{BV(\mathbb{R}\times(-T,T))}\leqslant C_T,\quad \forall v\in\mathcal{A},\ \forall T>0.$$

Then for any sequence $(v_n)_{n\in\mathbb{N}}$ of elements of $\mathcal{A}$, there exists a subsequence, still denoted by $(v_n)_{n\in\mathbb{N}}$, and there exists $v\in L^\infty(\mathbb{R}^2)$, with $\|v\|_{L^\infty(\mathbb{R}^2)}\leqslant C$ and $|v|_{BV(\mathbb{R}\times(-T,T))}\leqslant C_T$ for all $T>0$, such that $v_n\to v$ in $L^1_{\mathrm{loc}}(\mathbb{R}^2)$ as $n\to\infty$, that is $\int_{\bar\omega}|v_n(x)-v(x)|\,\mathrm{d}x\to 0$, as $n\to\infty$ for any compact set $\bar\omega$ of $\mathbb{R}^2$.

In order to use Lemma 21.4, one first shows the following BV stability estimate for the approximate solution:

LEMMA 21.5 (Discrete space BV estimate). *Under Assumption* 21.1, *assume that* $u_0 \in BV(\mathbb{R})$; *let* $\mathcal{T}$ *be an admissible mesh in the sense of Definition* 20.1 *and let* $k \in \mathbb{R}_+^\star$ *be the time step. Let* $\{u_i^n,\ i \in \mathbb{Z},\ n \in \mathbb{N}\}$ *be given by* (21.1), (21.2) *and assume that the scheme is a monotone flux scheme in the sense of Definition* 21.1. *Let* g_1 *and* g_2 *be the Lipschitz constants of* g *on* $[U_m, U_M]^2$ *with respect to its two arguments. Then, under the CFL condition* (21.8), *the following inequality holds*:

$$\sum_{i\in\mathbb{Z}} |u_{i+1}^{n+1} - u_i^{n+1}| \leqslant \sum_{i\in\mathbb{Z}} |u_{i+1}^n - u_i^n|, \quad \forall n \in \mathbb{N}. \tag{21.18}$$

PROOF. First remark that, for $n = 0$, $\sum_{i\in\mathbb{Z}} |u_{i+1}^0 - u_i^0| \leqslant |u_0|_{BV(\mathbb{R})}$ (see Remark 19.4).

For all $i \in \mathbb{Z}$, the scheme (21.1), (21.2) (with $p = 1$) leads to

$$u_i^{n+1} = u_i^n + b_{i+1/2}^n(u_{i+1}^n - u_i^n) + a_{i-1/2}^n(u_{i-1}^n - u_i^n),$$

and

$$u_{i+1}^{n+1} = u_{i+1}^n + b_{i+3/2}^n(u_{i+2}^n - u_{i+1}^n) + a_{i+1/2}^n(u_i^n - u_{i+1}^n),$$

where $a_{i+1/2}$ and $b_{i+1/2}$ are defined (for all $i \in \mathbb{Z}$) in Lemma 21.1. Subtracting one equation to the other leads to

$$\begin{aligned} u_{i+1}^{n+1} - u_i^{n+1} &= (u_{i+1}^n - u_i^n)(1 - b_{i+1/2}^n - a_{i+1/2}^n) + b_{i+3/2}^n(u_{i+2}^n - u_{i+1}^n) \\ &\quad + a_{i-1/2}^n(u_i^n - u_{i-1}^n). \end{aligned}$$

Under the condition (21.8), we get

$$\begin{aligned} |u_{i+1}^{n+1} - u_i^{n+1}| &\leqslant |u_{i+1}^n - u_i^n|(1 - b_{i+1/2}^n - a_{i+1/2}^n) \\ &\quad + b_{i+3/2}^n|u_{i+2}^n - u_{i+1}^n| + a_{i-1/2}^n|u_i^n - u_{i-1}^n|. \end{aligned}$$

Summing the previous equation over $i \in \mathbb{Z}$ gives (21.18). □

COROLLARY 21.1 (Discrete BV estimate). *Under Assumption* 21.1, *let* $u_0 \in BV(\mathbb{R})$; *let* $\mathcal{T}$ *be an admissible mesh in the sense of Definition* 20.1 *and let* $k \in \mathbb{R}_+^\star$ *be the time step. Let* $u_{\mathcal{T},k}$ *be the finite volume approximate solution defined by* (21.1)–(21.3) *and assume that the scheme is a monotone flux scheme in the sense of Definition* 21.1. *Let* g_1 *and* g_2 *be the Lipschitz constants of* g *on* $[U_m, U_M]^2$ *with respect to its two arguments and assume that* k *satisfies the CFL condition* (21.8). *Let* $u_{\mathcal{T},k}(x,t) = u_i^0$ *for a.e.* $(x,t) \in K_i \times \mathbb{R}_-$, *for all* $i \in \mathbb{Z}$ (*hence* $u_{\mathcal{T},k}$ *is defined a.e. on* $\mathbb{R}^2$). *Then, for any* $T > 0$, *there exists* $C \in \mathbb{R}_+^\star$, *only depending on* u_0, g *and* T *such that*:

$$|u_{\mathcal{T},k}|_{BV(\mathbb{R}\times(-T,T))} \leqslant C. \tag{21.19}$$

PROOF. As in Lemma 21.5, remark that $\sum_{i\in\mathbb{Z}} |u^0_{i+1} - u^0_i| \leqslant |u_0|_{BV(\mathbb{R})}$.

Let us first assume that $T \leqslant k$. Then, the BV semi-norm of $u_{\mathcal{T},k}$ satisfies

$$|u_{\mathcal{T},k}|_{BV(\mathbb{R}\times(-T,T))} \leqslant 2T \sum_{i\in\mathbb{Z}} \left|u^0_{i+1} - u^0_i\right|.$$

Hence the estimate (21.19) is true for $C = 2T|u_0|_{BV(\mathbb{R})}$.

Let us now assume that $k < T$. Let $N \in \mathbb{N}^\star$ such that $Nk < T \leqslant (N+1)k$. The definition of $|\bullet|_{BV(\mathbb{R}\times(-T,T))}$ yields

$$\begin{aligned}
&|u_{\mathcal{T},k}|_{BV(\mathbb{R}\times(-T,T))} \\
&\quad \leqslant T \sum_{i\in\mathbb{Z}} \left|u^0_{i+1} - u^0_i\right| + \sum_{n=0}^{N-1} \sum_{i\in\mathbb{Z}} k\left|u^n_{i+1} - u^n_i\right| + (T - Nk) \sum_{i\in\mathbb{Z}} \left|u^N_{i+1} - u^N_i\right| \\
&\qquad + \sum_{n=0}^{N-1} \sum_{i\in\mathbb{Z}} h_i \left|u^{n+1}_i - u^n_i\right|. \qquad (21.20)
\end{aligned}$$

Lemma 21.5 gives $\sum_{i\in\mathbb{Z}} |u^n_{i+1} - u^n_i| \leqslant |u_0|_{BV(\mathbb{R})}$ for all $n \in \mathbb{N}$, and therefore,

$$\sum_{n=0}^{N-1} \sum_{i\in\mathbb{Z}} k\left|u^n_{i+1} - u^n_i\right| + (T - Nk) \sum_{i\in\mathbb{Z}} \left|u^N_{i+1} - u^N_i\right| \leqslant T|u_0|_{BV(\mathbb{R})}. \qquad (21.21)$$

In order to bound the last term of (21.20), using the scheme (21.1) yields, for all $i \in \mathbb{Z}$ and all $n \in \mathbb{N}$,

$$\left|u^{n+1}_i - u^n_i\right| \leqslant \frac{k}{h_i} g_1 \left|u^n_i - u^n_{i-1}\right| + \frac{k}{h_i} g_2 \left|u^n_i - u^n_{i+1}\right|.$$

Therefore,

$$\sum_{i\in\mathbb{Z}} h_i \left|u^{n+1}_i - u^n_i\right| \leqslant k(g_1 + g_2) \sum_{i\in\mathbb{Z}} \left|u^n_i - u^n_{i+1}\right|, \quad \text{for all } n \in \mathbb{N},$$

which yields, since $Nk < T$,

$$\sum_{n=0}^{N-1} \sum_{i\in\mathbb{Z}} h_i \left|u^{n+1}_i - u^n_i\right| \leqslant T(g_1 + g_2)|u_0|_{BV(\mathbb{R})}. \qquad (21.22)$$

Therefore inequality (21.19) follows from (21.20), (21.21) and (21.22) with $C = T(2 + g_1 + g_2)|u_0|_{BV(\mathbb{R})}$. □

Consider a sequence of admissible meshes and time steps verifying the CFL condition, and the associated sequence of approximate solutions (prolonged on $\mathbb{R} \times \mathbb{R}_-$ as in Corollary 21.1). By Lemma 21.1 and Corollary 21.1, the sequence of approximate solutions satisfies the hypotheses of Lemma 21.4. It is therefore possible to extract a subsequence which converges in $L^1_{\mathrm{loc}}(\mathbb{R}\times\mathbb{R}_+)$ to a function $u \in L^\infty(\mathbb{R}\times\mathbb{R}_+^\star)$. It must still be shown that the function u is the unique weak entropy solution of problem (19.1). This may be proven by using the discrete entropy inequalities (21.10) and the strong BV estimate (21.18) or the classical Lax–Wendroff theorem recalled below.

THEOREM 21.2 (Lax–Wendroff). *Under Assumption* 21.1, *let $\alpha > 0$ be given and let $(\mathcal{T}_m)_{m\in\mathbb{N}}$ be a sequence of admissible meshes in the sense of Definition* 20.1 (*note that, for all $m \in \mathbb{N}$, the mesh $\mathcal{T}_m$ satisfies the hypotheses of Definition* 20.1 *where $\mathcal{T} = \mathcal{T}_m$ and α is independent of m*). *Let $(k_m)_{m\in\mathbb{N}}$ be a sequence of* (*positive*) *time steps. Assume that* $\mathrm{size}(\mathcal{T}_m) \to 0$ *and $k_m \to 0$ as $m \to \infty$.*

For $m \in \mathbb{N}$, setting $\mathcal{T} = \mathcal{T}_m$ and $k = k_m$, let $u_m = u_{\mathcal{T},k}$ be the solution of (21.1)–(21.3) *with $p = 1$ and some g from $\mathbb{R}^2$ to $\mathbb{R}$, only depending on f and u_0, locally Lipschitz continuous and such that $g(s,s) = f(s)$ for all $s \in \mathbb{R}$.*

Assume that $(u_m)_{m\in\mathbb{N}}$ is bounded in $L^\infty(\mathbb{R}\times\mathbb{R}_+)$ and that $u_m \to u$ a.e. on $\mathbb{R}\times\mathbb{R}_+$. Then, u is a weak solution to problem (19.1) (*i.e. u satisfies* (19.3)).

Furthermore, assume that for any $\kappa \in \mathbb{R}$ there exists some locally Lipschitz continuous function G_κ from $\mathbb{R}^2$ to $\mathbb{R}$, only depending on f, u_0 and κ, such that $G_\kappa(s,s) = f(s\top\kappa) - f(s\bot\kappa)$ for all $s \in \mathbb{R}$ and such that for all $m \in \mathbb{N}$

$$\frac{1}{k}\left(\left|u_i^{n+1}-\kappa\right| - \left|u_i^n-\kappa\right|\right) + \frac{1}{h_i}\left(G_\kappa\left(u_i^n, u_{i+1}^n\right) - G_\kappa\left(u_{i-1}^n, u_i^n\right)\right) \leqslant 0,$$
$$\forall i \in \mathbb{Z},\ \forall n \in \mathbb{N}, \tag{21.23}$$

where $\{u_i^n,\ i \in \mathbb{Z},\ n \in \mathbb{N}\}$ is the solution to (21.1)–(21.2) *for $\mathcal{T} = \mathcal{T}_m$ and $k = k_m$. Then, u is the entropy weak solution to problem* (19.1) (*i.e., u is the unique solution of* (19.4)).

PROOF. Since $(u_m)_{m\in\mathbb{N}}$ is bounded in $L^\infty(\mathbb{R}\times\mathbb{R}_+)$ and $u_m \to u$ a.e. on $\mathbb{R}\times\mathbb{R}_+$, the sequence $(u_m)_{m\in\mathbb{N}}$ converges to u in $L^1_{\mathrm{loc}}(\mathbb{R}\times\mathbb{R}_+)$. This implies in particular (from Kolmogorov's theorem, see Theorem 14.1) that, for all $R > 0$ and all $T > 0$,

$$\sup_{m\in\mathbb{N}} \int_0^{2T}\int_{-2R}^{2R} \left|u_m(x,t) - u_m(x-\eta,t)\right| \mathrm{d}x\,\mathrm{d}t \to 0 \quad \text{as } \eta \to 0.$$

Then, taking $\eta = \alpha\,\mathrm{size}(\mathcal{T}_m)$ (for $m \in \mathbb{N}$) and letting $m \to \infty$ yields, in particular,

$$\int_0^{2T}\int_{-2R}^{2R} \left|u_m(x,t) - u_m\big(x-\alpha\,\mathrm{size}(\mathcal{T}_m),t\big)\right| \mathrm{d}x\,\mathrm{d}t \to 0 \quad \text{as } m \to \infty. \tag{21.24}$$

For $m \in \mathbb{N}$, let $\{u_i^n,\ i \in \mathbb{Z},\ n \in \mathbb{N}\}$ be the solution to (21.1)–(21.2) for $\mathcal{T} = \mathcal{T}_m$ and $k = k_m$ (note that u_i^n depends on m, even though this dependency is not so clear in the notation). We also set $k_m = k$ and $\mathrm{size}(\mathcal{T}_m) = h$, so that k and h are depending on m (but recall that α is not depending on m).

Let $R > 0$ and $T > 0$. Let $i_0 \in \mathbb{Z}$, $i_1 \in \mathbb{Z}$ and $N \in \mathbb{N}$ be such that $-R \in \overline{K}_{i_0}$, $R \in \overline{K}_{i_1}$ and $T \in (Nk, (N+1)k]$. Then, for $h < R$ and $k < T$ (which is true for m large enough),

$$\alpha h \sum_{i=i_0}^{i_1} \sum_{n=0}^{N} k \left| u_i^n - u_{i-1}^n \right| \leqslant \int_0^{2T} \int_{-2R}^{2R} \left| u_m(x,t) - u_m(x - \alpha h, t) \right| \mathrm{d}x \, \mathrm{d}t.$$

Therefore, inequality (21.24) leads to (20.10), i.e.

$$h \sum_{i=i_0}^{i_1} \sum_{n=0}^{N} k \left| u_i^n - u_{i-1}^n \right| \to 0 \quad \text{as } m \to \infty. \tag{21.25}$$

Using (21.25), the remainder of the proof of Theorem 21.2 is very similar to the proof of Proposition 20.2 and to Step 2 in the proof of Theorem 21.1 (inequality (21.25) replaces the weak BV inequality).

In order to prove that u is solution to (19.3), let us multiply the first equation of (21.1) by $(k/h_i)\varphi(x, nk)$, integrate over $x \in K_i$ and sum for all $i \in \mathbb{Z}$ and $n \in \mathbb{N}$. This yields

$$A_m + B_m = 0$$

with

$$A_m = \sum_{i \in \mathbb{Z}} \sum_{n \in \mathbb{N}} \left(u_i^{n+1} - u_i^n \right) \int_{K_i} \varphi(x, nk) \, \mathrm{d}x$$

and

$$B_m = \sum_{i \in \mathbb{Z}} \sum_{n \in \mathbb{N}} k \left(g\left(u_i^n, u_{i+1}^n\right) - g\left(u_{i-1}^n, u_i^n\right) \right) \frac{1}{h_i} \int_{K_i} \varphi(x, nk) \, \mathrm{d}x.$$

As in the proof of Proposition 20.2, one has

$$\lim_{m \to +\infty} A_m = -\int_{\mathbb{R}_+} \int_{\mathbb{R}} u(x,t) \varphi_t(x,t) \, \mathrm{d}x \, \mathrm{d}t - \int_{\mathbb{R}} u_0(x) \varphi(x, 0) \, \mathrm{d}x.$$

Let us now turn to the study of B_m. We compare B_m with

$$B_{1,m} = -\sum_{n \in \mathbb{N}} \int_{nk}^{(n+1)k} \int_{\mathbb{R}} f\left(u_{\mathcal{T},k}(x,t)\right) \varphi_x(x, nk) \, \mathrm{d}x \, \mathrm{d}t,$$

which tends to $-\int_{\mathbb{R}_+} \int_{\mathbb{R}} f(u(x,t)) \varphi_x(x,t) \, \mathrm{d}x \, \mathrm{d}t$ as $m \to \infty$ since $f(u_{\mathcal{T},k}) \to f(u)$ in $L^1_{\mathrm{loc}}(\mathbb{R} \times \mathbb{R}_+)$ as $m \to \infty$.

The term $B_{1,m}$ can be rewritten as

$$B_{1,m} = \sum_{i\in\mathbb{Z}}\sum_{n\in\mathbb{N}} k\big(f(u_i^n) - f(u_{i-1}^n)\big)\varphi(x_{i-1/2}, nk),$$

which yields, introducing $g(u_{i-1}^n, u_i^n)$,

$$\begin{aligned} B_{1,m} &= \sum_{i\in\mathbb{Z}}\sum_{n\in\mathbb{N}} k\big(f(u_i^n) - g(u_{i-1}^n, u_i^n)\big)\varphi(x_{i-1/2}, nk) \\ &\quad + \sum_{i\in\mathbb{Z}}\sum_{n\in\mathbb{N}} k\big(g(u_{i-1}^n, u_i^n) - f(u_{i-1}^n)\big)\varphi(x_{i-1/2}, nk). \end{aligned}$$

Similarly, introducing $f(u_i^n)$ in B_m,

$$\begin{aligned} B_m &= \sum_{i\in\mathbb{Z}}\sum_{n\in\mathbb{N}} k\big(f(u_i^n) - g(u_{i-1}^n, u_i^n)\big)\frac{1}{h_i}\int_{K_i}\varphi(x, nk)\,\mathrm{d}x \\ &\quad + \sum_{i\in\mathbb{Z}}\sum_{n\in\mathbb{N}} k\big(g(u_i^n, u_{i+1}^n) - f(u_i^n)\big)\frac{1}{h_i}\int_{K_i}\varphi(x, nk)\,\mathrm{d}x. \end{aligned}$$

In order to compare B_m and $B_{1,m}$, let $R > 0$ and $T > 0$ be such that $\varphi(x,t) = 0$ if $|x| \geqslant R$ or $t \geqslant T$. Let $A > 0$ be such that $\|u_m\|_{L^\infty(\mathbb{R}\times\mathbb{R}_+)} \leqslant A$ for all $m \in \mathbb{N}$. Then there exists $C > 0$, only depending on φ and the Lipschitz constants on g on $[-A, A]^2$, such that, if $h < R$ and $k < T$ (which is true for m large enough),

$$|B_m - B_{1,m}| \leqslant Ch \sum_{i=i_0}^{i_1}\sum_{n=0}^{N} k\big|u_i^n - u_{i-1}^n\big|, \tag{21.26}$$

where $i_0 \in \mathbb{Z}$, $i_1 \in \mathbb{Z}$ and $N \in \mathbb{N}$ are such that $-R \in \overline{K}_{i_0}$, $R \in \overline{K}_{i_1}$ and $T \in (Nk, (N+1)k]$.

Using (21.26) and (21.25), we get $|B_m - B_{1,m}| \to 0$ and then

$$B_m \to -\int_{\mathbb{R}}\int_{\mathbb{R}_+} f\big(u(x,t)\big)\varphi_x(x,t)\,\mathrm{d}t\,\mathrm{d}x \quad \text{as } m \to \infty,$$

which completes the proof that u is a solution to problem 19.3.

Under the additional assumption that u_m satisfies (21.23), one proves that u satisfies (19.7) (and therefore that u satisfies (19.4)) and is the entropy weak solution to problem (19.1) by a similar method.

Indeed, let $\kappa \in \mathbb{R}$. One replaces u_i^l by $|u_i^l - \kappa|$ in A_m (for $l = n$ and $n+1$) and one replaces g by G_κ in B_m. Then, passing to the limit in $A_m + B_m \leqslant 0$ (which is a consequence of the inequation (21.23)) leads the desired result.

This concludes the proof of Theorem 21.2 □

REMARK 21.5. Theorem 21.2 still holds with $(2p+1)$-points schemes $(p > 1)$. The generalization of the first part of Theorem 21.2 (the proof that u is a solution to (19.3)) is quite easy. For the second part of Theorem 21.2 (entropy inequalities) the discrete entropy inequalities may be replaced by some weaker ones (in order to handle interesting schemes such as those which are described in the following section).

However, the use of Theorem 21.2 needs a compactness property of sequences of approximate solutions in the space $L^1_{\mathrm{loc}}(\mathbb{R} \times \mathbb{R}_+)$. Such a compactness property is generally achieved with a "strong BV estimate" (similar to (21.18)). Hence an extensive literature on "TVD schemes" (see HARTEN [1983]), "ENO schemes"... (see GODLEWSKI and RAVIART [1991, 1996] and references therein). The generalization of this method in the multidimensional case (studied in the following chapter) does not seem so clear except in the case of Cartesian meshes.

22. Higher order schemes

Consider a monotone flux scheme in the sense of Definition 21.1. By definition, the considered scheme is a 3-points scheme; recall that the numerical flux function is denoted by g. The approximate solution obtained with this scheme converges to the entropy weak solution of problem (19.1) as the mesh size tends to 0 and under a so called CFL condition (it is proved in Theorem 21.1 for a particular case and in the next chapter for the general case). However, 3-points schemes are known to be diffusive, so that the approximate solution is not very precise near the discontinuities. An idea to reduce the diffusion is to go to a 5-points scheme by introducing "slopes" on each discretization cell and limiting the slopes in order for the scheme to remain stable. A classical way to do this is the "MUSCL" (Monotonic Upwind Scheme for Conservation Laws, see VAN LEER [1979]) technique.

We briefly describe, with the notations of Section 21.1, an example of such a scheme, see, e.g., GODLEWSKI and RAVIART [1991] and GODLEWSKI and RAVIART [1996] for further details. Let $n \in \mathbb{N}$.

- Computation of the slopes

$$\tilde{p}_i^n = \frac{u_{i+1}^n - u_{i-1}^n}{h_i + \frac{h_{i-1}}{2} + \frac{h_{i+1}}{2}}, \qquad i \in \mathbb{Z}.$$

- Limitation of the slopes. $p_i^n = \alpha_i^n \tilde{p}_i^n$, $i \in \mathbb{Z}$, where α_i^n is the largest number in $[0, 1]$ such that

$$u_i^n + \frac{h_i}{2}\alpha_i^n \tilde{p}_i^n \in \left[u_i^n \perp u_{i+1}^n, u_i^n \top u_{i+1}^n\right] \quad \text{and}$$
$$u_i^n - \frac{h_i}{2}\alpha_i^n \tilde{p}_i^n \in \left[u_i^n \perp u_{i-1}^n, u_i^n \top u_{i-1}^n\right].$$

In practice, other formulas giving smaller values of α_i^n are sometimes needed for stability reasons.

- Computation of u_i^{n+1} for $i \in \mathbb{Z}$. One replaces $g(u_i^n, u_i^{n+1})$ in (21.2) by:

$$\overline{g}(u_{i-1}^n, u_i^n, u_{i+1}^n, u_{i+2}^n) = g\left(u_i^n + \frac{h_i}{2} p_i^n, u_{i+1}^n - \frac{h_{i+1}}{2} p_i^{n+1}\right).$$

The scheme thus constructed is less diffusive than the original one and it remains stable thanks to the limitation of the slope. Indeed, if the limitation of the slopes is not active (that is $\alpha_i^n = 1$), the space diffusion term disappears from this new scheme, while the time "antidiffusion" term remains. Hence it seems appropriate to use a higher-order scheme for the time discretization. This may be done by using, e.g., an RK2 (Runge–Kutta order 2, or Heun) method for the discretization of the time derivative. The MUSCL scheme may be written as

$$\frac{U^{n+1} - U^n}{k} = \overline{H}(U^n) \quad \text{for } n \in \mathbb{N},$$

where $U^n = (u_i^n)_{i \in \mathbb{Z}}$; hence it may be seen as the explicit Euler discretization of

$$U_t = \overline{H}(U);$$

therefore, the RK2 time discretization yields to the following scheme:

$$\frac{U^{n+1} - U^n}{k} = \tfrac{1}{2}\overline{H}(U^n) + \tfrac{1}{2}\left(\overline{H}\left(U^n + k\overline{H}(U^n)\right)\right) \quad \text{for } n \in \mathbb{N}.$$

Going to a second-order discretization in time allows larger time steps, without loss of stability.

Results of convergence are possible with these new schemes (with eventually some adaptation of the slope limitations to obtain convenient discrete entropy inequalities, see VILA [1986]. It is also possible to obtain error estimates in the spirit of those given in the following chapter, in the multidimensional case, see, e.g., CHAINAIS-HILLAIRET [1996], NOËLLE [1996], KRÖNER, NOELLE and ROKYTA [1995]. However these error estimates are somewhat unsatisfactory since they are of a similar order to that of the original 3-points scheme (although these schemes are numerically more precise that the original 3-points schemes).

The higher-order schemes are nonlinear even if problem (19.1) is linear, because of the limitation of the slopes.

Implicit versions of these higher-order schemes are more or less straightforward. However, the numerical implementation of these implicit versions requires the solution of nonlinear systems. In many cases, the solutions to these nonlinear systems seem impossible to reach for large k; in fact, the existence of the solutions is not so clear, see PFERTZEL [1987]. Since the advantage of implicit schemes is essentially the possibility to use large values of k, the above flaw considerably reduces the opportunity of their use. Therefore, although implicit 3-points schemes are very diffusive, they remain the basic schemes in several industrial environments. See also Section 33.3 for some clues on implicit schemes applied to complex industrial applications.

CHAPTER VI

Multidimensional Nonlinear Hyperbolic Equations

The aim of this chapter is to define and study finite volume schemes for the approximation of the solution to a nonlinear scalar hyperbolic problem in several space dimensions. Explicit and implicit time discretizations are considered. We prove the convergence of the approximate solution towards the entropy weak solution of the problem and give an error estimate between the approximate solution and the entropy weak solution with respect to the discretization mesh size.

23. The continuous problem

We consider here the following nonlinear hyperbolic equation in d space dimensions ($d \geqslant 1$), with initial condition

$$u_t(x,t) + \operatorname{div}\big(\boldsymbol{v} f(u)\big)(x,t) = 0, \quad x \in \mathbb{R}^d,\ t \in \mathbb{R}_+, \tag{23.1}$$

$$u(x,0) = u_0(x), \quad x \in \mathbb{R}^d, \tag{23.2}$$

where u_t denotes the time derivative of u ($t \in \mathbb{R}_+$), and div the divergence operator with respect to the space variable (which belongs to $\mathbb{R}^d$). Recall that $|x|$ denotes the Euclidean norm of x in $\mathbb{R}^d$, and $x \bullet y$ the usual scalar product of x and y in $\mathbb{R}^d$.

The following hypotheses are made on the data:

ASSUMPTION 23.1.
(i) $u_0 \in L^\infty(\mathbb{R}^d)$, $U_m, U_M \in \mathbb{R}$, $U_m \leqslant u_0 \leqslant U_M$ a.e.,
(ii) $\boldsymbol{v} \in C^1(\mathbb{R}^d \times \mathbb{R}_+, \mathbb{R}^d)$,
(iii) $\operatorname{div} \boldsymbol{v}(x,t) = 0$, $\forall (x,t) \in \mathbb{R}^d \times \mathbb{R}_+$,
(iv) $\exists V < \infty$ such that $|\boldsymbol{v}(x,t)| \leqslant V$, $\forall (x,t) \in \mathbb{R}^d \times \mathbb{R}_+$,
(v) $f \in C^1(\mathbb{R}, \mathbb{R})$.

REMARK 23.1. Note that part (iv) of Assumption 23.1 is crucial. It ensures the property of "propagation in finite time" which is needed for the uniqueness of the solution of (23.3) and for the stability (under a "Courant–Friedrichs–Levy" (CFL) condition) of the time explicit numerical scheme. Part (iii) of Assumption 23.1, on the other hand, is only considered for the sake of simplicity; the results of existence and uniqueness of the entropy weak solution and convergence (including error estimates as in Theorems 29.1 and 29.2) of the numerical schemes presented below may be extended to the case div $\boldsymbol{v} \neq 0$. However, part (iii) of Assumption 23.1 is natural in many "applications" and avoids several technical complications. Note, in particular, that, e.g., if div $\boldsymbol{v} \neq 0$, the L^∞-bound on the solution of (23.3) and the L^∞ estimate (in Lemma 25.1 and Proposition 26.1) on the approximate solution depend on $\boldsymbol{v}$ and T. The case $F(x,t,u)$ instead of $\boldsymbol{v}(x,t)f(u)$ is also feasible, but somewhat more technical, see CHAINAIS-HILLAIRET [1996, 1997].

Problem (23.1)–(23.2) has a unique entropy weak solution, which is the solution to the following equation (which is the multidimensional extension of the one-dimensional Definition 19.3).

$$\begin{cases} u \in L^\infty(\mathbb{R}^d \times \mathbb{R}_+^\star), \\ \displaystyle\int_{\mathbb{R}_+}\int_{\mathbb{R}^d} \big[\eta(u(x,t))\varphi_t(x,t) + \Phi(u(x,t))\boldsymbol{v}(x,t)\cdot\nabla\varphi(x,t)\big]\,\mathrm{d}x\,\mathrm{d}t \\ \quad \displaystyle + \int_{\mathbb{R}^d} \eta(u_0(x))\varphi(x,0)\,\mathrm{d}x \geqslant 0, \quad \forall \varphi \in C_c^\infty(\mathbb{R}^d \times \mathbb{R}_+, \mathbb{R}_+), \\ \forall \eta \in C^1(\mathbb{R},\mathbb{R}),\ \text{convex function, and } \Phi \in C^1(\mathbb{R},\mathbb{R}) \text{ such that } \Phi' = f'\eta', \end{cases}$$

where $\nabla\varphi$ denotes the gradient of the function φ with respect to the space variable (which belongs to $\mathbb{R}^d$). Recall that $C_c^m(E,F)$ denotes the set of functions C^m from E to F, with compact support in E.

The characterization of the entropy weak solution by the Krushkov entropies (Proposition 19.2) still holds in the multidimensional case. Let us define again, for all $\kappa \in \mathbb{R}$, the Krushkov entropies $(|\cdot - \kappa|)$ for which the entropy flux is $f(\cdot \top \kappa) - f(\cdot \bot \kappa)$ (for any pair of real values a, b, we denote again by $a \top b$ the maximum of a and b, and by $a \bot b$ the minimum of a and b). The unique entropy weak solution is also the unique solution to the following problem:

$$\begin{cases} u \in L^\infty(\mathbb{R}^d \times \mathbb{R}_+^\star), \\ \displaystyle\int_{\mathbb{R}_+}\int_{\mathbb{R}^d} \big[|u(x,t) - \kappa|\varphi_t(x,t) + \big(f(u(x,t)\top\kappa) \\ \quad -f(u(x,t)\bot\kappa)\big)\boldsymbol{v}(x,t)\cdot\nabla\varphi(x,t)\big]\,\mathrm{d}x\,\mathrm{d}t \\ \quad \displaystyle + \int_{\mathbb{R}^d} |u_0(x) - \kappa|\varphi(x,0)\,\mathrm{d}x \geqslant 0, \quad \forall \kappa \in \mathbb{R},\ \forall \varphi \in C_c^\infty(\mathbb{R}^d \times \mathbb{R}_+, \mathbb{R}_+). \end{cases} \tag{23.3}$$

As in the one-dimensional case (Theorem 19.1), existence and uniqueness results are also known for the entropy weak solution to problem (23.1)–(23.2) under assumptions which differ slightly from Assumption 23.1 (see, e.g., KRUSHKOV [1970], VOL'PERT [1967]). In particular, these results are obtained with a nonlinearity F (in our case $F =$

$\boldsymbol{v}f$) of class C^3. We recall that the methods which were used in KRUSHKOV [1970] require a regularization in $BV(\mathbb{R}^d)$ of the function u_0, in order to take advantage, for any $T > 0$, of compactness properties which are similar to those given in Lemma 21.4 for the case $d = 1$. Recall that the space $BV(\Omega)$ where Ω is an open subset of $\mathbb{R}^p$, $p \geqslant 1$, was defined in Definition 21.2; it will be used later with $\Omega = \mathbb{R}^d$ or $\Omega = \mathbb{R}^d \times (-T, T)$.

The existence of solutions to similar problems to (23.1)–(23.2) was already proved by passing to the limit on solutions of an appropriate numerical scheme, see CONWAY and SMOLLER [1966]. The work of CONWAY and SMOLLER [1966] uses a finite difference scheme on a uniform rectangular grid, in two space dimensions, and requires that the initial condition u_0 belongs to $BV(\mathbb{R}^d)$ (and thus, the solution to problem (23.1)–(23.2) also has a locally bounded variation). These assumptions (on meshes and on u_0) yield, as in Lemma 21.4, a (strong) compactness property in $L^1_{\text{loc}}(\mathbb{R}^d \times \mathbb{R}_+)$ on a family of approximate solutions. In the following, however, we shall only require that $u_0 \in L^\infty(\mathbb{R}^d)$ and we shall be able to deal with more general meshes. We may use, e.g., a triangular mesh in the case of two space dimensions. For each of these reasons, the BV framework may not be used and a (strong) compactness property in L^1_{loc} on a family of approximate solutions is not easy to obtain (although this compactness property does hold and results from this chapter). In order to prove the existence of a solution to (23.1)–(23.2) by passing to the limit on the approximate solutions given by finite volume schemes on general meshes (in the sense used below) in two or three space dimensions, we shall work with some "weak" compactness result in L^∞, namely Proposition 30.1, which yields the "nonlinear weak-$\star$ convergence" (see Definition 30.1) of a family of approximate solutions. When doing so, passing to the limit with the approximate solutions will give the existence of an "entropy process solution" to problem (23.1)–(23.2), see Definition 28.1. A uniqueness result for the entropy process solution to problem (23.1)–(23.2) is then proven. This uniqueness result proves that the entropy process solution is indeed the entropy weak solution, hence the existence and uniqueness of the entropy weak solution. This uniqueness result also allows us to conclude to the convergence of the approximate solution given by the numerical scheme (i.e. (24.4), (24.2)) towards the entropy weak solution to (23.1)–(23.2) (this convergence holds in $L^p_{\text{loc}}(\mathbb{R}^d \times \mathbb{R}_+)$ for any $1 \leqslant p < \infty$).

Note that uniqueness results for "generalized" solutions (namely measure valued solutions) to (23.1)–(23.2) have recently been proved (see DIPERNA [1985], SZEPESSY [1989], GALLOUËT and HERBIN [1994]). The proofs of these results rely on the one hand on the concept of measure valued solutions and on the other hand on the existence of an entropy weak solution. The direct proof of the uniqueness of a measure valued solution (i.e. without assuming any existence result of entropy weak solutions) leads to a difficult problem involving the application of the theorem of continuity in mean. This difficulty is easier to deal within the framework of entropy process solutions (but in fact, measure valued solutions and entropy process solutions are two presentations of the same concept).

Developing the above analysis gives a (strong) convergence result of approximate solutions towards the entropy weak solution. But moreover, we also derive some error estimates depending on the regularity of u_0.

In the case of a Cartesian grid, the convergence and error analysis reduces essentially to a one-dimensional discretization problem for which results were proved some time ago, see, e.g., KUZNETSOV [1976], CRANDALL and MAJDA [1980], SANDERS [1983]. In the case of general meshes, the numerical schemes are not generally "TVD" (Total Variation Diminishing) and therefore the classical framework of the 1D case (see Section 21.5) may not be used. More recent works deal with several convergence results and error estimates for time explicit finite volume schemes, see, e.g., COCKBURN, COQUEL and LEFLOCH [1994], CHAMPIER, GALLOUËT and HERBIN [1993], VILA [1994], KRÖNER and ROKYTA [1994], KRÖNER, NOELLE and ROKYTA [1995], KRÖNER [1997]: following Szepessy's work on the convergence of the streamline diffusion method (see SZEPESSY [1989]), most of these works use DiPerna's uniqueness theorem, see DIPERNA [1985] (or an adaptation of it, see GALLOUËT and HERBIN [1994] and EYMARD, GALLOUËT and HERBIN [1995]), and the error estimates generalize the work by KUZNETSOV [1976]. Here we use the framework of CHAMPIER, GALLOUËT and HERBIN [1993], EYMARD, GALLOUËT, GHILANI and HERBIN [1998]; we prove directly that any monotone flux scheme (defined below) satisfies a "weak BV" estimate (see Lemmata 25.2 and 26.1). This inequality appears to be a key for the proof of convergence and for the error estimate. Some convergence results and error estimates are also possible with some so called "higher-order schemes" which are not monotone flux schemes (briefly presented for the 1D case in Section 22). These results are not presented here, see NOËLLE [1996] and CHAINAIS-HILLAIRET [1996] for some of them.

Note that the nonlinearity considered here is of the form $\mathbf{v}(x,t)f(u)$. This kind of flux is often encountered in porous medium modelling, where the hyperbolic equation may then be coupled with an elliptic or parabolic equation (see, e.g., EYMARD and GALLOUËT [1993], VIGNAL [1996a, 1996b], HERBIN and LABERGERIE [1997]). It adds an extra difficulty to the case $F(u)$ because of the dependency on x and t. Note again (see Remark 23.1) that the method which we present here for a nonlinearity of the form $\mathbf{v}(x,t)f(u)$ also yields the same results in the case of a nonlinearity of the form $F(x,t,u)$, see the recent work of CHAINAIS-HILLAIRET [1999].

The time implicit discretization adds the extra difficulties of proving the existence of the approximate solution (see Lemma 25.1) and proving a so called "strong time BV estimate" (see Lemma 26.3) in order to show that the error estimate for the implicit scheme may still be of order $h^{1/4}$ even if the time step k is of order $\sqrt{h}$, at least in particular cases.

We first describe in Section 24 finite volume schemes using a "general" mesh for the discretization of (23.1)–(23.2). In Sections 25 and 26 some estimates on the approximate solution given by the numerical schemes are shown and in Section 27 some entropy inequalities are proven. We then prove in Section 28 the convergence of convenient subsequences of sequences of approximate solutions towards an entropy process solution, by passing to the limit when the mesh size and the time step go to 0. A byproduct of this result is the existence of an entropy process solution to (23.1)–(23.2) (see Definition 28.1). The uniqueness of the entropy process solution to problem (23.1)–(23.2) is then proved; we can therefore conclude to the existence and uniqueness of the entropy weak solution and also to the L^p_{loc} convergence for any finite p of the approximate solution towards the entropy weak solution (Section 28). Using the existence of the entropy weak solution, an error estimate result is given in Section 29 (which also yields the convergence result). Therefore the main interest of this convergence result is precisely to prove the existence of the entropy weak solution to (23.1)–(23.2) without any regularity assumption on the initial data. Section 30 describes the notion of nonlinear weak-⋆ convergence, which is widely used in the proof of convergence of Section 28.

Section 31 is not related to the previous sections. It describes a finite volume approach which may be used to stabilize finite element schemes for the discretization of a hyperbolic equation (or system).

24. Meshes and schemes

Let us first define an admissible mesh of $\mathbb{R}^d$ as a generalization of the notion of admissible mesh of $\mathbb{R}$ as defined in Definition 20.1.

DEFINITION 24.1 (*Admissible meshes*). An admissible finite volume mesh of $\mathbb{R}^d$, with $d = 1, 2$ or 3 (for the discretization of problem (23.1)–(23.2)), denoted by $\mathcal{T}$, is given by a family of disjoint polygonal connected subsets of $\mathbb{R}^d$ such that $\mathbb{R}^d$ is the union of the closure of the elements of $\mathcal{T}$ (which are called control volumes in the following) and such that the common "interface" of any two control volumes is included in a hyperplane of $\mathbb{R}^d$ (this is not necessary but is introduced to simplify the formulation). Denoting by $h = \text{size}(\mathcal{T}) = \sup\{\text{diam}(K), K \in \mathcal{T}\}$, it is assumed that $h < +\infty$ and that, for some $\alpha > 0$,

$$\alpha h^d \leqslant \mathrm{m}(K), \qquad \mathrm{m}(\partial K) \leqslant \frac{1}{\alpha} h^{d-1}, \quad \forall K \in \mathcal{T}, \tag{24.1}$$

where $\mathrm{m}(K)$ denotes the d-dimensional Lebesgue measure of K, $\mathrm{m}(\partial K)$ denotes the $(d-1)$-dimensional Lebesgue measure of ∂K (∂K is the boundary of K) and $\mathcal{N}(K)$ denotes the set of neighbours of the control volume K; for $L \in \mathcal{N}(K)$, we denote by $K|L$ the common interface between K and L, and by $\boldsymbol{n}_{K,L}$ the unit normal vector to $K|L$ oriented from K to L. The set of all the interfaces is denoted by $\mathcal{E}$.

Note that, in this definition, the terminology is "mixed". For $d = 3$, "polygonal" stands for "polyhedral" and, for $d = 2$, "interface" stands for "edge". For $d = 1$ Definition 24.1 is equivalent to Definition 20.1.

In order to define the numerical flux, we consider functions $g \in C(\mathbb{R}^2, \mathbb{R})$ satisfying the following assumptions:

ASSUMPTION 24.1. Under Assumption 23.1 the function g, only depending on f, $\boldsymbol{v}$, U_m and U_M, satisfies

- g is locally Lipschitz continuous from $\mathbb{R}^2$ to $\mathbb{R}$,
- $g(s, s) = f(s)$, for all $s \in [U_m, U_M]$,
- $(a, b) \mapsto g(a, b)$, from $[U_m, U_M]^2$ to $\mathbb{R}$, is nondecreasing with respect to a and nonincreasing with respect to b.

Let us denote by g_1 and g_2 the Lipschitz constants of g on $[U_m, U_M]^2$ with respect to its two arguments.

The hypotheses on g are the same as those presented for monotone flux schemes in the one-dimensional case (see Definition 21.1); the function g allows the construction of a numerical flux, see Remark 24.2 below.

REMARK 24.1. In Assumption 24.1, the third item will ensure some stability properties of the schemes defined below. In particular, in the case of the "explicit scheme" (see (24.4)), it yields the monotonicity of the scheme under a CFL condition (namely, condition (24.3) with $\xi = 0$). The second item is essential since it ensures the consistency of the fluxes. All the examples of functions g given in Examples 21.1 satisfy these assumptions. We again give the important example of the "generalized 1D Godunov scheme" obtained with a one-dimensional Godunov scheme for each interface (see, e.g., for the explicit scheme, COCKBURN, COQUEL and LEFLOCH [1994], VILA [1994]),

$$g(a,b) = \begin{cases} \max\{f(s),\ b \leqslant s \leqslant a\} & \text{if } b \leqslant a, \\ \min\{f(s),\ a \leqslant s \leqslant b\} & \text{if } a \leqslant b, \end{cases}$$

and also the framework of some "flux splitting" schemes:

$$g(a,b) = f_1(a) + f_2(b),$$

with $f_1,\ f_2 \in C^1(\mathbb{R},\mathbb{R})$, $f = f_1 + f_2$, f_1 nondecreasing and f_2 nonincreasing (this framework is considerably more simple that the general framework, because it reduces the study to the particular case of two monotone nonlinearities).

Besides, it is possible to replace Assumption 24.1 on g by some slightly more general assumption, in order to handle, in particular, the case of some "Lax–Friedrichs type" schemes (see Remark 29.1 below).

In order to describe the numerical schemes considered here, let $\mathcal{T}$ be an admissible mesh in the sense of Definition 24.1 and $k > 0$ be the time step. The discrete unknowns are u_K^n, $n \in \mathbb{N}^\star$, $K \in \mathcal{T}$. The set $\{u_K^0,\ K \in \mathcal{T}\}$ is given by the initial condition,

$$u_K^0 = \frac{1}{\mathrm{m}(K)} \int_K u_0(x)\,\mathrm{d}x, \quad \forall K \in \mathcal{T}. \tag{24.2}$$

The equations satisfied by the discrete unknowns, u_K^n, $n \in \mathbb{N}^\star$, $K \in \mathcal{T}$, are obtained by discretizing Eq. (23.1). We now describe the explicit and implicit schemes.

24.1. Explicit schemes

We present here the "explicit scheme" associated to a function g satisfying Assumption 24.1. In this case, for stability reasons (see Lemmata 25.1 and 25.2), the time step $k \in \mathbb{R}_+^\star$ is chosen such that

$$k \leqslant (1-\xi)\frac{\alpha^2 h}{V(g_1+g_2)}, \tag{24.3}$$

where $\xi \in (0,1)$ is a given real value; recall that g_1 and g_2 are the Lipschitz constants of g with respect to the first and second variables on $[U_m, U_M]^2$ and that $U_m \leqslant u_0 \leqslant U_M$

a.e. and $|\boldsymbol{v}(x,t)| \leqslant V < +\infty$, for all $(x,t) \in \mathbb{R}^d \times \mathbb{R}_+$. Consider the following explicit numerical scheme:

$$\mathrm{m}(K)\frac{u_K^{n+1} - u_K^n}{k} + \sum_{L\in\mathcal{N}(K)} \big(v_{K,L}^n\, g\big(u_K^n, u_L^n\big) - v_{L,K}^n\, g\big(u_L^n, u_K^n\big)\big) = 0,$$
$$\forall K \in \mathcal{T},\ \forall n \in \mathbb{N}, \tag{24.4}$$

where

$$v_{K,L}^n = \frac{1}{k}\int_{nk}^{(n+1)k} \int_{K|L} \big(\boldsymbol{v}(x,t) \cdot \boldsymbol{n}_{K,L}\big)^+ \,\mathrm{d}\gamma(x)\,\mathrm{d}t$$

and

$$\begin{aligned} v_{L,K}^n &= \frac{1}{k}\int_{nk}^{(n+1)k} \int_{K|L} \big(\boldsymbol{v}(x,t) \cdot \boldsymbol{n}_{L,K}\big)^+ \,\mathrm{d}\gamma(x)\,\mathrm{d}t \\ &= \frac{1}{k}\int_{nk}^{(n+1)k} \int_{K|L} \big(\boldsymbol{v}(x,t) \cdot \boldsymbol{n}_{K,L}\big)^- \,\mathrm{d}\gamma(x)\,\mathrm{d}t. \end{aligned}$$

Recall that $a^+ = a\top 0$ and $a^- = -(a\bot 0)$ for all $a \in \mathbb{R}$ and that $\mathrm{d}\gamma$ is the integration symbol for the $(d-1)$-dimensional Lebesgue measure on the considered hyperplane.

REMARK 24.2 (*Numerical fluxes*). The numerical flux at the interface between the control volume K and the control volume $L \in \mathcal{N}(K)$ is then equal to $v_{K,L}^n\ g(u_K^n, u_L^n) - v_{L,K}^n\ g(u_L^n, u_K^n)$; this expression yields a monotone flux such as defined in Definition 21.1, given in the one-dimensional case. However, in the multidimensional case, the expression of the numerical flux depends on the considered interface; this was not so in the one-dimensional case for which the numerical flux is completely defined by the function g.

The approximate solution, denoted by $u_{\mathcal{T},k}$, is defined a.e. from $\mathbb{R}^d \times \mathbb{R}_+$ to $\mathbb{R}$ by

$$u_{\mathcal{T},k}(x,t) = u_K^n, \quad \text{if } x \in K,\ t \in \big[nk, (n+1)k\big),\ K \in \mathcal{T},\ n \in \mathbb{N}. \tag{24.5}$$

24.2. *Implicit schemes*

The use of implicit schemes is steadily increasing in industrial codes for reasons such as robustness and computational cost. Hence we consider in our analysis the following implicit numerical scheme (for which condition (24.3) is no longer needed) associated to a function g satisfying Assumption 24.1:

$$\mathrm{m}(K)\frac{u_K^{n+1} - u_K^n}{k} + \sum_{L\in\mathcal{N}(K)} \big(v_{K,L}^n\, g\big(u_K^{n+1}, u_L^{n+1}\big) - v_{L,K}^n\, g\big(u_L^{n+1}, u_K^{n+1}\big)\big) = 0,$$
$$\forall K \in \mathcal{T},\ \forall n \in \mathbb{N}, \tag{24.6}$$

where $\{u_K^0,\ K \in \mathcal{T}\}$ is still determined by (24.2). The implicit approximate solution $u_{\mathcal{T},k}$, is defined now a.e. from $\mathbb{R}^d \times \mathbb{R}_+$ to $\mathbb{R}$ by

$$u_{\mathcal{T},k}(x,t) = u_K^{n+1}, \quad \text{if } x \in K,\ t \in \big(nk, (n+1)k\big],\ K \in \mathcal{T},\ n \in \mathbb{N}. \tag{24.7}$$

24.3. Passing to the limit

We show in Section 28 the convergence of the approximate solutions $u_{\mathcal{T},k}$ (given by the numerical schemes above described) towards the unique entropy weak solution u to (23.1)–(23.2) in an adequate sense, when $\text{size}(\mathcal{T}) \to 0$ and $k \to 0$ (with, possibly, a stability condition). In order to describe the general line of thought leading to this convergence result, we shall simply consider the explicit scheme (i.e. (24.2), (24.4) and (24.5)) (the implicit scheme will also be fully investigated later).

First, in Section 25, by writing u_K^{n+1} as a convex combination of u_K^n and $(u_L^n)_{L\in\mathcal{N}(K)}$, the L^∞ stability is easily shown under the CFL condition (24.3) ($u_{\mathcal{T},k}$ is proved to be bounded in $L^\infty(\mathbb{R}^d \times \mathbb{R}_+^\star)$, independently of $\text{size}(\mathcal{T})$ and k).

By a classical argument, if any possible limit of a family of approximate solutions $u_{\mathcal{T},k}$ (where $\mathcal{T}$ is an admissible mesh in the sense of Definition 24.1 and k satisfies (24.3)) is the entropy weak solution to problem (23.1)–(23.2) then $u_{\mathcal{T},k}$ converges (in $L^\infty(\mathbb{R}^d \times \mathbb{R}_+^\star)$ for the weak-$\star$ topology, for instance), as $h = \text{size}(\mathcal{T}) \to 0$ (and k satisfies (24.3)), towards the unique entropy weak solution to problem (23.1)–(23.2). Unfortunately, the L^∞ estimate of Section 25 does not yield that any possible limit of a family of approximate is solution to problem (23.1)–(23.2), even in the linear case ($f(u) = u$) (see the proofs of convergence of Chapter V). The "BV stability" can be used (combined with the L^∞ stability) to show the convergence in the case of one space dimension (see Section 21.5) and in the case of Cartesian meshes in two or three space dimensions. Indeed, in the case of Cartesian meshes, assuming $u_0 \in BV(\mathbb{R}^d)$ and assuming (for simplicity) $\boldsymbol{v}$ to be constant (a generalization is possible for $\boldsymbol{v}$ regular enough), the following estimate holds, for all $T \geqslant k$:

$$k \sum_{n=0}^{N_{T,k}} \sum_{K|L\in\mathcal{E}} \text{m}(K|L)\big|u_K^n - u_L^n\big| \leqslant T|u_0|_{BV(\mathbb{R}^d)},$$

where $N_{T,k} \in \mathbb{N}$ is such that $(N_{T,k}+1)k \leqslant T < (N_{T,k}+2)k$, and the values u_K^n are given by (24.2) and (24.4). Such an estimate is wrong in the general case of admissible meshes in the sense of Definition 24.1, as it can be shown with easy counterexamples. It is, however, not necessary for the proof of convergence. A weaker inequality, which is called "weak BV" as in the one-dimensional case (see Lemma 21.3) will be shown in the multidimensional case for both explicit and implicit schemes (see Lemmata 25.2 and 26.1); the weak BV estimate yields the convergence of the scheme in the general case. As an illustration, consider the case $f' \geqslant 0$; using an upwind scheme, i.e.

$g(a,b) = f(a)$, the weak BV inequality (25.4), which is very close to that of the 1D case (Lemma 21.3), writes

$$\sum_{n=0}^{N_{T,k}} k \sum_{(K,L)\in\mathcal{E}_R^n} \left(v_{K,L}^n + v_{L,K}^n\right)\left|f\left(u_K^n\right) - f\left(u_L^n\right)\right| \leqslant \frac{C}{\sqrt{h}}, \tag{24.8}$$

where $\mathcal{E}_R^n = \{(K,L)\in\mathcal{T}^2, L\in\mathcal{N}(K), K|L\subset B(0,R)$ and $u_K^n > u_L^n\}$ and C only depends on $\boldsymbol{v}$, g, u_0, α, ξ, R and T (see Lemma 25.2).

We say that inequality (24.8) is "weak", but it is in fact "three times weak" for the following reasons:

(1) the inequality is of order $1/\sqrt{h}$, and not of order 1.

(2) In the left-hand side of (24.8), the quantity which is associated to the $K|L \in \mathcal{E}_R^n$ interface is zero if f is constant on the interval to which the values u_K^n and u_L^n belong; variations of the discrete unknowns in this interval are therefore not taken into account.

(3) The left-hand side of (24.8) involves terms $(v_{K,L}^n + v_{L,K}^n)$ which are not uniformly bounded from below by C m$(K|L)$ with some $C > 0$ only depending on the data (i.e. $\boldsymbol{v}$, u_0 and g) and not on $\mathcal{T}$ (note that, e.g., $v_{K,L}^n = v_{L,K}^n = 0$ if $\boldsymbol{v}\cdot\boldsymbol{n}_{K,L} = 0$).
For the convergence result (namely Theorem 28.2) the useful consequence of (24.8) is

$$h\sum_{n=0}^{N_{T,k}} k \sum_{(K,L)\in\mathcal{E}_R^n} \left(v_{K,L}^n + v_{L,K}^n\right)\left|f\left(u_K^n\right) - f\left(u_L^n\right)\right| \to 0 \quad \text{as } h\to 0,$$

as in the 1D case, see Theorem 21.1. For the error estimate in Theorem 29.1, the bound $C/\sqrt{h}$ in (24.8) is crucial. Note that a "twice weak BV" inequality in the sense (ii) and (iii), but of order 1 (that is C instead of $C/\sqrt{h}$ in the right-hand side of (24.8)), would yield a sharp error estimate, i.e. $C_e h^{1/2}$ instead of $C_e h^{1/4}$ in (29.1).

Note that, in order to obtain (24.8), $\xi > 0$ is crucial in the CFL condition (24.3).

Recall also that (24.8) together with the $L^\infty(\mathbb{R}^d \times \mathbb{R}_+^\star)$ bound does not yield any (strong) compactness property in $L^1_{\text{loc}}(\mathbb{R}^d \times \mathbb{R}_+)$ on a family of approximated solutions.

In the linear case (that is $f(s) = cs$ for all $s\in\mathbb{R}$, for some c in $\mathbb{R}$), the inequality (24.8) is used in the same manner as in the previous chapter; one proves that the approximate solution satisfies the weak formulation to (23.1)–(23.2) (which is equivalent to (23.3)) with an error which goes to 0 as $h\to 0$, under condition (24.3). We deduce from this the convergence of $u_{\mathcal{T},k}$ (as $h\to 0$ and under condition (24.3)) towards the unique weak solution of (23.1)–(23.2) in $L^\infty(\mathbb{R}^d \times \mathbb{R}_+^\star)$ for the weak-$\star$ topology. In fact, the convergence holds in $L^p_{\text{loc}}(\mathbb{R}^d \times \mathbb{R}_+)$ (strongly) for any $1\leqslant p<\infty$, thanks to the argument developed for the study of the nonlinear case.

The nonlinear case adds an extra difficulty, as in the 1D case; it will be handled in detail in the present chapter. This difficulty arises from the fact that, if $u_{\mathcal{T},k}$ converges to u (as $h\to 0$, under condition (24.3)) and $f(u_{\mathcal{T},k})$ to μ_f, in $L^\infty(\mathbb{R}^d \times \mathbb{R}_+^\star)$ for the weak-$\star$ topology, there remains to show that $\mu_f = f(u)$ and that u is the entropy weak solution to problem (23.1)–(23.2). The weak BV inequality (24.8) is used to show that, for any "entropy" function η, i.e. convex function of class C^1 from $\mathbb{R}$ to $\mathbb{R}$, with associated entropy flux ϕ, i.e. ϕ such that $\phi' = f'\eta'$, the following entropy inequality is satisfied:

$$\int_{\mathbb{R}_+}\int_{\mathbb{R}^d}\big(\mu_\eta(x,t)\varphi_t(x,t) + \mu_\phi(x,t)\boldsymbol{v}(x,t)\cdot\nabla\varphi(x,t)\big)\,\mathrm{d}x\,\mathrm{d}t$$
$$+\int_{\mathbb{R}^d}\eta\big(u_0(x)\big)\varphi(x,0)\,\mathrm{d}x \geqslant 0, \quad \forall\varphi\in C_c^\infty\big(\mathbb{R}^d\times\mathbb{R}_+,\mathbb{R}_+\big), \tag{24.9}$$

where μ_η (resp. μ_ϕ) is the limit of $\eta(u_{\mathcal{T},k})$ (resp. $\phi(u_{\mathcal{T},k})$) in $L^\infty(\mathbb{R}^d\times\mathbb{R}_+^\star)$ for the weak-$\star$ topology (the existence of these limits can indeed be assumed). From (24.9), it is shown that $u_{\mathcal{T},k}$ converges to u in $L^1_{\mathrm{loc}}(\mathbb{R}^d\times\mathbb{R}_+)$ (as $h\to 0$, k satisfying (24.3)), and that u is the entropy weak solution to problem (23.1)–(23.2). This last result uses a generalization of a result on measure valued solutions of DiPerna (see DIPERNA [1985], GALLOUËT and HERBIN [1994]), and is developed in Section 28.

25. Stability results for the explicit scheme

25.1. L^∞ stability

LEMMA 25.1. *Under Assumption* 23.1, *let $\mathcal{T}$ be an admissible mesh in the sense of Definition* 24.1 *and $k>0$, let $g\in C(\mathbb{R}^2,\mathbb{R})$ satisfy Assumption* 24.1 *and assume that* (24.3) *holds; let $u_{\mathcal{T},k}$ be given by* (24.5), (24.4), (24.2); *then,*

$$U_m \leqslant u_K^n \leqslant U_M, \quad \forall n\in\mathbb{N},\ \forall K\in\mathcal{T}, \tag{25.1}$$

and

$$\|u_{\mathcal{T},k}\|_{L^\infty(\mathbb{R}^d\times\mathbb{R}_+^\star)} \leqslant \|u_0\|_{L^\infty(\mathbb{R}^d)}. \tag{25.2}$$

PROOF. Note that (25.2) is a straightforward consequence of (25.1), which will be proved by induction. For $n=0$, since $U_m\leqslant u_0\leqslant U_M$ a.e., (25.1) follows from (24.2).

Let $n\in\mathbb{N}$, assume that $U_m\leqslant u_K^n\leqslant U_M$ for all $K\in\mathcal{T}$. Using the fact that $\operatorname{div}\boldsymbol{v}=0$, which yields $\sum_{L\in\mathcal{N}(K)}(v_{K,L}^n - v_{L,K}^n)=0$, we can rewrite (24.4) as

$$\mathrm{m}(K)\frac{u_K^{n+1}-u_K^n}{k} + \sum_{L\in\mathcal{N}(K)}\big(v_{K,L}^n\big(g(u_K^n,u_L^n) - f(u_K^n)\big)$$
$$- v_{L,K}^n\big(g(u_L^n,u_K^n) - f(u_K^n)\big)\big) = 0. \tag{25.3}$$

Set, for $u_K^n\neq u_L^n$,

$$\tau_{K,L}^n = v_{K,L}^n\frac{g(u_K^n,u_L^n)-f(u_K^n)}{u_K^n-u_L^n} - v_{L,K}^n\frac{g(u_L^n,u_K^n)-f(u_K^n)}{u_K^n-u_L^n},$$

and $\tau_{K,L}^n = 0$ if $u_K^n = u_L^n$.

Assumption 24.1 on g and Assumption 23.1 yields $0 \leqslant \tau^n_{K,L} \leqslant Vm(K|L)(g_1 + g_2)$. Using (25.3), we can write

$$u_K^{n+1} = \left(1 - \frac{k}{\mathrm{m}(K)} \sum_{L\in\mathcal{N}(K)} \tau^n_{K,L}\right) u^n_K + \frac{k}{\mathrm{m}(K)} \sum_{L\in\mathcal{N}(K)} \tau^n_{K,L} u^n_L,$$

which gives, under condition (24.3), $\inf_{L\in\mathcal{T}} u^n_L \leqslant u^{n+1}_K \leqslant \sup_{L\in\mathcal{T}} u^n_L$, for all $K \in \mathcal{T}$. This concludes the proof of (25.1), which, in turn, yields (25.2). □

REMARK 25.1. Note that the stability result (25.2) holds even if $\xi = 0$ in (24.3). However, we shall need $\xi > 0$ for the following "weak BV" inequality.

25.2. *A "weak BV" estimate*

In the following lemma, $B(0, R)$ denotes the ball of $\mathbb{R}^d$ of center 0 and radius R ($\mathbb{R}^d$ is always endowed with its usual scalar product).

LEMMA 25.2. *Under Assumption* 23.1, *let $\mathcal{T}$ be an admissible mesh in the sense of Definition* 24.1 *and $k > 0$. Let $g \in C(\mathbb{R}^2, \mathbb{R})$ satisfy Assumption* 24.1 *and assume that* (24.3) *holds. Let $u_{\mathcal{T},k}$ be given by* (24.5), (24.4), (24.2).

Let $T > 0$, $R > 0$, $N_{T,k} = \max\{n \in \mathbb{N}, n < T/k\}$, $\mathcal{T}_R = \{K \in \mathcal{T}, K \subset B(0, R)\}$ and $\mathcal{E}^n_R = \{(K, L) \in \mathcal{T}^2, L \in \mathcal{N}(K), K|L \subset B(0, R)$ and $u^n_K > u^n_L\}$.

Then there exists $C \in \mathbb{R}$, only depending on $\boldsymbol{v}$, g, u_0, α, ξ, R, T such that, for $h < R$ and $k < T$,

$$\begin{aligned}
&\sum_{n=0}^{N_{T,k}} k \sum_{(K,L)\in\mathcal{E}^n_R} \Big[v^n_{K,L} \Big(\max_{u^n_L \leqslant p \leqslant q \leqslant u^n_K} \big(g(q,p) - f(q)\big) \\
&\quad + \max_{u^n_L \leqslant p \leqslant q \leqslant u^n_K} \big(g(q,p) - f(p)\big)\Big) + v^n_{L,K} \Big(\max_{u^n_L \leqslant p \leqslant q \leqslant u^n_K} \big(f(q) - g(p,q)\big) \\
&\quad + \max_{u^n_L \leqslant p \leqslant q \leqslant u^n_K} \big(f(p) - g(p,q)\big)\Big)\Big] \leqslant \frac{C}{\sqrt{h}},
\end{aligned} \tag{25.4}$$

and

$$\sum_{n=0}^{N_{T,k}} \sum_{K\in\mathcal{T}_R} \mathrm{m}(K) \big| u^{n+1}_K - u^n_K \big| \leqslant \frac{C}{\sqrt{h}}. \tag{25.5}$$

PROOF. In this proof, we shall denote by C_i ($i \in \mathbb{N}$) various quantities only depending on $\boldsymbol{v}$, g, u_0, α, ξ, R, T.

Multiplying (25.3) by ku^n_K and summing the result over $K \in \mathcal{T}_R$, $n \in \{0, \ldots, N_{T,k}\}$ yields

$$B_1 + B_2 = 0, \tag{25.6}$$

with

$$B_1 = \sum_{n=0}^{N_{T,k}} \sum_{K \in \mathcal{T}_R} \mathrm{m}(K) u_K^n \left(u_K^{n+1} - u_K^n\right),$$

and

$$B_2 = \sum_{n=0}^{N_{T,k}} k \sum_{K \in \mathcal{T}_R} \sum_{L \in \mathcal{N}(K)} \left(v_{K,L}^n \left(g\left(u_K^n, u_L^n\right) - f\left(u_K^n\right)\right) u_K^n \right.$$
$$\left. - v_{L,K}^n \left(g\left(u_L^n, u_K^n\right) - f\left(u_K^n\right)\right) u_K^n\right).$$

Gathering the last two summations by edges in B_2 leads to the definition of B_3:

$$B_3 = \sum_{n=0}^{N_{T,k}} k \sum_{(K,L) \in \mathcal{E}_R^n} \left[v_{K,L}^n \left(u_K^n \left(g\left(u_K^n, u_L^n\right) - f\left(u_K^n\right)\right) - u_L^n \left(g\left(u_K^n, u_L^n\right) - f\left(u_L^n\right)\right)\right)\right.$$
$$\left. - v_{L,K}^n \left(u_K^n \left(g\left(u_L^n, u_K^n\right) - f\left(u_K^n\right)\right) - u_L^n \left(g\left(u_L^n, u_K^n\right) - f\left(u_L^n\right)\right)\right)\right].$$

The expression $|B_3 - B_2|$ can be reduced to a sum of terms each of which corresponds to the boundary of a control volume which is included in $B(0, R+h) \setminus B(0, R-h)$; since the measure of $B(0, R+h) \setminus B(0, R-h)$ is less than $C_2 h$, the number of such terms is, for n fixed, lower than $(C_2 h)/(\alpha h^d) = C_3 h^{1-d}$. Thanks to (25.2), using the fact that $\mathrm{m}(\partial K) \leqslant (1/\alpha) h^{d-1}$, that $|\boldsymbol{v}(x,t)| \leqslant V$, that g is bounded on $[U_m, U_M]^2$, and that $g(s,s) = f(s)$, one may show that each of the nonzero term in $|B_3 - B_2|$ is bounded by $C_1 h^{d-1}$. Furthermore, since $(N_{T,k}+1)k \leqslant 2k$, we deduce that

$$|B_3 - B_2| \leqslant C_4. \tag{25.7}$$

Denoting by Φ a primitive of the function $(\bullet) f'(\bullet)$, an integration by parts yields, for all $(a,b) \in \mathbb{R}^2$,

$$\Phi(b) - \Phi(a) = \int_a^b s f'(s)\,\mathrm{d}s = b\left(f(b) - g(a,b)\right) - a\left(f(a) - g(a,b)\right)$$
$$- \int_a^b \left(f(s) - g(a,b)\right)\mathrm{d}s. \tag{25.8}$$

Using (25.8), the term B_3 may be decomposed as

$$B_3 = B_4 - B_5,$$

where

$$B_4 = \sum_{n=0}^{N_{T,k}} k \sum_{(K,L)\in\mathcal{E}_R^n} \left(v_{K,L}^n \int_{u_K^n}^{u_L^n} \big(f(s) - g(u_K^n, u_L^n)\big)\,\mathrm{d}s \right.$$
$$\left. + v_{L,K}^n \int_{u_L^n}^{u_K^n} \big(f(s) - g(u_L^n, u_K^n)\big)\,\mathrm{d}s \right)$$

and

$$B_5 = \sum_{n=0}^{N_{T,k}} k \sum_{(K,L)\in\mathcal{E}_R^n} \big(v_{K,L}^n - v_{L,K}^n\big)\big(\Phi(u_K^n) - \Phi(u_L^n)\big).$$

The term B_5 is again reduced to a sum of terms corresponding to control volumes included in $B(0, R+h) \setminus B(0, R-h)$, thanks to $\operatorname{div} \boldsymbol{v} = 0$; therefore, as for (25.7), there exists $C_5 \in \mathbb{R}$ such that

$$B_5 \leqslant C_5.$$

Let us now turn to an estimate of B_4. To this purpose, let $a, b \in \mathbb{R}$, define $\mathcal{C}(a,b) = \{(p,q) \in [a\perp b, a\top b]^2;\ (q-p)(b-a) \geqslant 0\}$. Thanks to the monotonicity properties of g (and using the fact that $g(s,s) = f(s)$), the following inequality holds, for any $(p,q) \in \mathcal{C}(a,b)$:

$$\int_a^b \big(f(s) - g(a,b)\big)\,\mathrm{d}s \geqslant \int_p^q \big(f(s) - g(a,b)\big)\,\mathrm{d}s$$
$$\geqslant \int_p^q \big(f(s) - g(p,q)\big)\,\mathrm{d}s \geqslant 0. \tag{25.9}$$

The technical Lemma 18.5 (p. 854) can then be applied. It states that

$$\left|\int_p^q \big(\theta(s) - \theta(p)\big)\,\mathrm{d}s\right| \geqslant \frac{1}{2G}\big(\theta(q) - \theta(p)\big)^2, \quad \forall p, q \in \mathbb{R},$$

for all monotone, Lipschitz continuous function $\theta : \mathbb{R} \to \mathbb{R}$, with a Lipschitz constant $G > 0$.

From Lemma 18.5, we can notice that

$$\int_p^q \big(f(s) - g(p,q)\big)\,\mathrm{d}s \geqslant \int_p^q \big(g(p,s) - g(p,q)\big)\,\mathrm{d}s$$
$$\geqslant \frac{1}{2g_2}\big(f(p) - g(p,q)\big)^2, \tag{25.10}$$

and

$$\int_p^q \big(f(s) - g(p,q)\big)\,\mathrm{d}s \geqslant \int_p^q \big(g(s,q) - g(p,q)\big)\,\mathrm{d}s \geqslant \frac{1}{2g_1}\big(f(q) - g(p,q)\big)^2. \tag{25.11}$$

Multiplying (25.10) (resp. (25.11)) by $g_2/(g_1 + g_2)$ (resp. $g_1/(g_1 + g_2)$), taking the maximum for $(p,q) \in \mathcal{C}(a,b)$, and adding the two equations yields, with (25.9),

$$\int_a^b \big(f(s) - g(a,b)\big)\,\mathrm{d}s \geqslant \frac{1}{2(g_1 + g_2)}\Big(\max_{(p,q)\in\mathcal{C}(a,b)} \big(f(p) - g(p,q)\big)^2 + \max_{(p,q)\in\mathcal{C}(a,b)} \big(f(q) - g(p,q)\big)^2\Big). \tag{25.12}$$

We can then deduce, from (25.12):

$$\begin{aligned} B_4 \geqslant \frac{1}{2(g_1 + g_2)} \sum_{n=0}^{N_{T,k}} k \sum_{(K,L)\in\mathcal{E}_R^n} \Big[& v_{K,L}^n \Big(\max_{u_L^n \leqslant p \leqslant q \leqslant u_K^n} \big(g(q,p) - f(q)\big)^2 \\ & + \max_{u_L^n \leqslant p \leqslant q \leqslant u_K^n} \big(g(q,p) - f(p)\big)^2 \Big) \\ & + v_{L,K}^n \Big(\max_{u_L^n \leqslant p \leqslant q \leqslant u_K^n} \big(f(q) - g(p,q)\big)^2 \\ & + \max_{u_L^n \leqslant p \leqslant q \leqslant u_K^n} \big(f(p) - g(p,q)\big)^2 \Big) \Big]. \end{aligned} \tag{25.13}$$

This gives a bound on B_2, since (with $C_6 = C_4 + C_5$):

$$B_2 \geqslant B_4 - C_6. \tag{25.14}$$

Let us now turn to B_1. We have

$$\begin{aligned} B_1 = & -\tfrac{1}{2} \sum_{n=0}^{N_{T,k}} \sum_{K\in\mathcal{T}_R} \mathrm{m}(K)\big(u_K^{n+1} - u_K^n\big)^2 + \tfrac{1}{2} \sum_{K\in\mathcal{T}_R} \mathrm{m}(K)\big(u_K^{N_{T,k}+1}\big)^2 \\ & - \tfrac{1}{2} \sum_{K\in\mathcal{T}_R} \mathrm{m}(K)\big(u_K^0\big)^2. \end{aligned} \tag{25.15}$$

Using (25.3) and the Cauchy–Schwarz inequality yields the following inequality:

$$\big(u_K^{n+1} - u_K^n\big)^2 \leqslant \frac{k^2}{\mathrm{m}(K)^2} \sum_{L\in\mathcal{N}(K)} \big(v_{K,L}^n + v_{L,K}^n\big) \sum_{L\in\mathcal{N}(K)} \big[v_{K,L}^n \big(g\big(u_K^n, u_L^n\big)$$

$$- f\left(u_K^n\right)\right)^2 + v_{L,K}^n\left(g\left(u_L^n, u_K^n\right) - f\left(u_K^n\right)\right)^2\Big].$$

Then, using the CFL condition (24.3), Definition 24.1 and part (iv) of Assumption 23.1 gives

$$\mathrm{m}(K)\left(u_K^{n+1} - u_K^n\right)^2 \leqslant k\frac{1-\xi}{g_1+g_2}\sum_{L\in\mathcal{N}(K)}\Big[v_{K,L}^n\left(g\left(u_K^n, u_L^n\right) - f\left(u_K^n\right)\right)^2 + v_{L,K}^n\left(g\left(u_L^n, u_K^n\right) - f\left(u_K^n\right)\right)^2\Big]. \tag{25.16}$$

Summing Eq. (25.16) over $K \in \mathcal{T}_R$ and over $n = 0, \ldots, N_{T,k}$, and reordering the summation leads to

$$\begin{aligned}
&\tfrac{1}{2}\sum_{n=0}^{N_{T,k}}\sum_{K\in\mathcal{T}_R}\mathrm{m}(K)\left(u_K^{n+1} - u_K^n\right)^2 \\
&\quad\leqslant \frac{1-\xi}{2(g_1+g_2)}\sum_{n=0}^{N_{T,k}} k \sum_{(K,L)\in\mathcal{E}_R^n}\Big[v_{K,L}^n\Big(\left(g\left(u_K^n, u_L^n\right) - f\left(u_K^n\right)\right)^2 \\
&\qquad + \left(g\left(u_K^n, u_L^n\right) - f\left(u_L^n\right)\right)^2\Big) + v_{L,K}^n\Big(\left(f\left(u_K^n\right) - g\left(u_L^n, u_K^n\right)\right)^2 \\
&\qquad + \left(f\left(u_L^n\right) - g\left(u_L^n, u_K^n\right)\right)^2\Big)\Big] + C_7,
\end{aligned} \tag{25.17}$$

where C_7 accounts for the interfaces $K|L \subset B(0, R)$ such that $K \notin \mathcal{T}_R$ and/or $L \notin \mathcal{T}_R$ (these control volumes are included in $B(0, R+h) \setminus B(0, R-h)$).

Note that the right-hand side of (25.17) is bounded by $(1-\xi)B_4 + C_7$ (from (25.13)). Using (25.6), (25.14) and (25.15) gives

$$\begin{aligned}
&\frac{\xi}{2(g_1+g_2)}\sum_{n=0}^{N_{T,k}} k \sum_{(K,L)\in\mathcal{E}_R^n}\Big[v_{K,L}^n\Big(\max_{u_L^n\leqslant p\leqslant q\leqslant u_K^n}\left(g(q,p) - f(q)\right)^2 \\
&\qquad + \max_{u_L^n\leqslant p\leqslant q\leqslant u_K^n}\left(g(q,p) - f(p)\right)^2\Big) + v_{L,K}^n\Big(\max_{u_L^n\leqslant p\leqslant q\leqslant u_K^n}\left(f(q) - g(p,q)\right)^2 \\
&\qquad + \max_{u_L^n\leqslant p\leqslant q\leqslant u_K^n}\left(f(p) - g(p,q)\right)^2\Big)\Big] \\
&\quad\leqslant \tfrac{1}{2}\sum_{K\in\mathcal{T}_R}\mathrm{m}(K)\left(u_K^0\right)^2 + C_6 + C_7 = C_8.
\end{aligned} \tag{25.18}$$

Applying the Cauchy–Schwarz inequality to the left-hand side of (25.4) and using (25.18) yields

$$\sum_{n=0}^{N_{T,k}} k \sum_{(K,L)\in\mathcal{E}_R^n}\Big[v_{K,L}^n\Big(\max_{u_L^n\leqslant p\leqslant q\leqslant u_K^n}\left(g(q,p) - f(q)\right)$$

$$+ \max_{u_L^n \leqslant p \leqslant q \leqslant u_K^n} \big(g(q,p) - f(p)\big)\Big) + v_{L,K}^n \Big(\max_{u_L^n \leqslant p \leqslant q \leqslant u_K^n} \big(f(q) - g(p,q)\big)$$

$$+ \max_{u_L^n \leqslant p \leqslant q \leqslant u_K^n} \big(f(p) - g(p,q)\big)\Big)\Big]$$

$$\leqslant C_9 \Bigg(\sum_{n=0}^{N_{T,k}} k \sum_{(K,L)\in\mathcal{E}_R^n} \big(v_{K,L}^n + v_{L,K}^n\big)\Bigg)^{1/2}. \tag{25.19}$$

Noting that

$$\sum_{(K,L)\in\mathcal{E}_R^n} \big(v_{K,L}^n + v_{L,K}^n\big) \leqslant \sum_{K\in\mathcal{T}_{R+h}} V\mathrm{m}(\partial K) \leqslant V\frac{1}{\alpha}h^{d-1}\frac{\mathrm{m}(B(0,R+h))}{\alpha h^d} = \frac{C_{10}}{h}$$

and $(N_{T,k}+1)k \leqslant 2T$, one obtains (25.4) from (25.19).

Finally, since (25.3) yields

$$\mathrm{m}(K)\big|u_K^{n+1} - u_K^n\big| \leqslant k \sum_{L\in\mathcal{N}(K)} \big(v_{K,L}^n \big|g\big(u_K^n, u_L^n\big) - f\big(u_K^n\big)\big|$$
$$+ v_{L,K}^n \big|g\big(u_L^n, u_K^n\big) - f\big(u_K^n\big)\big|\big),$$

inequality (25.5) immediately follows from (25.4). This completes the proof of Lemma 25.2. □

26. Existence of the solution and stability results for the implicit scheme

This section is devoted to the time implicit scheme (given by (24.6) and (24.2)). We first prove the existence and uniqueness of the solution $\{u_K^n, n \in \mathbb{N}, K \in \mathcal{T}\}$ of (24.2), (24.6) and such that $u_K^n \in [U_m, U_M]$ for all $K \in \mathcal{T}$ and all $n \in \mathbb{N}$. Then, one gives a "weak space BV" inequality (this is equivalent to the inequality (25.4) for the explicit scheme) and a "(strong) time BV" estimate (estimate (26.14) below). This last estimate requires that $\boldsymbol{v}$ does not depend on t (and it leads to the term "k" in the right-hand side of (29.2) in Theorem 29.2). The error estimate, in the case where $\boldsymbol{v}$ depends on t, is given in Remark 29.2.

26.1. Existence, uniqueness and L^∞ stability

The following proposition gives an existence and uniqueness result of the solution to (24.2), (24.6). In this proposition, $\boldsymbol{v}$ may depend on t and one does not need to assume $u_0 \in BV(\mathbb{R}^d)$.

PROPOSITION 26.1. *Under Assumption 23.1, let $\mathcal{T}$ be an admissible mesh in the sense of Definition 24.1 and $k > 0$. Let $g \in C(\mathbb{R}^2, \mathbb{R})$ satisfy Assumption 24.1.*

Then there exists a unique solution $\{u_K^n,\ n \in \mathbb{N},\ K \in \mathcal{T}\} \subset [U_m, U_M]$ to (24.2), (24.6).

PROOF. One proves Proposition 26.1 by induction. Indeed, $\{u_K^0,\ K \in \mathcal{T}\}$ is uniquely defined by (24.2) and one has $u_K^0 \in [U_m, U_M]$, for all $K \in \mathcal{T}$, since $U_m \leqslant u_0 \leqslant U_M$ a.e. Assuming that, for some $n \in \mathbb{N}$, the set $\{u_K^n,\ K \in \mathcal{T}\}$ is given and that $u_K^n \in [U_m, U_M]$, for all $K \in \mathcal{T}$, the existence and uniqueness of $\{u_K^{n+1},\ K \in \mathcal{T}\}$, such that $u_K^{n+1} \in [U_m, U_M]$ for all $K \in \mathcal{T}$, solution of (24.6), must be shown.

Step 1 (*Uniqueness of* $\{u_K^{n+1},\ K \in \mathcal{T}\}$, *such that* $u_K^{n+1} \in [U_m, U_M]$ *for all* $K \in \mathcal{T}$, *solution of* (24.6)). Recall that $n \in \mathbb{N}$ and $\{u_K^n,\ K \in \mathcal{T}\}$ are given. Let us consider two solutions of (24.6), respectively denoted by $\{u_K,\ K \in \mathcal{T}\}$ and $\{w_K,\ K \in \mathcal{T}\}$; therefore, $\{u_K,\ K \in \mathcal{T}\}$ and $\{w_K,\ K \in \mathcal{T}\}$ satisfy $\{u_K,\ K \in \mathcal{T}\} \subset [U_m, U_M]$, $\{w_K,\ K \in \mathcal{T}\} \subset [U_m, U_M]$,

$$\mathrm{m}(K)\frac{u_K - u_K^n}{k} + \sum_{L\in\mathcal{N}(K)} \big(v_{K,L}^n\, g(u_K, u_L) - v_{L,K}^n\, g(u_L, u_K)\big) = 0, \qquad \forall K \in \mathcal{T}, \tag{26.1}$$

and

$$\mathrm{m}(K)\frac{w_K - u_K^n}{k} + \sum_{L\in\mathcal{N}(K)} \big(v_{K,L}^n\, g(w_K, w_L) - v_{L,K}^n\, g(w_L, w_K)\big) = 0, \qquad \forall K \in \mathcal{T}. \tag{26.2}$$

Then, subtracting (26.2) to (26.1), for all $K \in \mathcal{T}$,

$$\begin{aligned}
&\frac{\mathrm{m}(K)}{k}(u_K - w_K) + \sum_{L\in\mathcal{N}(K)} v_{K,L}^n\big(g(u_K, u_L) - g(w_K, u_L)\big)\\
&\quad + \sum_{L\in\mathcal{N}(K)} v_{K,L}^n\big(g(w_K, u_L) - g(w_K, w_L)\big)\\
&\quad - \sum_{L\in\mathcal{N}(K)} v_{L,K}^n\big(g(u_L, u_K) - g(w_L, u_K)\big)\\
&\quad - \sum_{L\in\mathcal{N}(K)} v_{L,K}^n\big(g(w_L, u_K) - g(w_L, w_K)\big) = 0
\end{aligned} \tag{26.3}$$

thanks to the monotonicity properties of g, (26.3) leads to

$$\begin{aligned}
&\frac{\mathrm{m}(K)}{k}|u_K - w_K| + \sum_{L\in\mathcal{N}(K)} v_{K,L}^n\big|g(u_K, u_L) - g(w_K, u_L)\big|\\
&\qquad + \sum_{L\in\mathcal{N}(K)} v_{L,K}^n\big|g(w_L, u_K) - g(w_L, w_K)\big|\\
&\quad \leqslant \sum_{L\in\mathcal{N}(K)} v_{K,L}^n\big|g(w_K, u_L) - g(w_K, w_L)\big|
\end{aligned}$$

$$+\sum_{L\in\mathcal{N}(K)} v^n_{L,K}\big|g(u_L,u_K)-g(w_L,u_K)\big|. \tag{26.4}$$

Let $\varphi:\mathbb{R}^d\mapsto\mathbb{R}^\star_+$ be defined by $\varphi(x)=\exp(-\gamma|x|)$, for some positive γ which will be specified later. For $K\in\mathcal{T}$, let φ_K be the mean value of φ on K. Since φ is integrable over $\mathbb{R}^d$ (and thanks to (24.1)), one has $\sum_{K\in\mathcal{T}}\varphi_K\leqslant(1/(\alpha h^d))\|\varphi\|_{L^1(\mathbb{R}^d)}<\infty$. Therefore the series

$$\sum_{K\in\mathcal{T}}\varphi_K\Big(\sum_{L\in\mathcal{N}(K)} v^n_{K,L}\big|g(w_K,u_L)-g(w_K,w_L)\big|\Big)\quad\text{and}$$

$$\sum_{K\in\mathcal{T}}\varphi_K\Big(\sum_{L\in\mathcal{N}(K)} v^n_{L,K}\big|g(u_L,u_K)-g(w_L,u_K)\big|\Big)$$

are convergent (thanks to (24.1) and the boundedness of $\boldsymbol{v}$ on $\mathbb{R}^d$ and g on $[U_m,U_M]^2$).

Multiplying (26.4) by φ_K and summing for $K\in\mathcal{T}$ yields five convergent series which can be reordered in order to give

$$\begin{aligned}&\sum_{K\in\mathcal{T}}\frac{\mathrm{m}(K)}{k}|u_K-w_K|\varphi_K\\&\quad\leqslant\sum_{K\in\mathcal{T}}\sum_{L\in\mathcal{N}(K)} v^n_{K,L}\big|g(u_K,u_L)-g(w_K,u_L)\big||\varphi_K-\varphi_L|\\&\qquad+\sum_{K\in\mathcal{T}}\sum_{L\in\mathcal{N}(K)} v^n_{L,K}\big|g(w_L,u_K)-g(w_L,w_K)\big||\varphi_K-\varphi_L|,\end{aligned}$$

from which one deduces

$$\sum_{K\in\mathcal{T}} a_K|u_K-w_K|\leqslant\sum_{K\in\mathcal{T}} b_K|u_K-w_K|, \tag{26.5}$$

with, for all $K\in\mathcal{T}$, $a_K=\frac{\mathrm{m}(K)}{k}\varphi_K$ and $b_K=\sum_{L\in\mathcal{N}(K)}(v^n_{K,L}g_1+v^n_{L,K}g_2)|\varphi_K-\varphi_L|$.

For $K\in\mathcal{T}$, let x_K be an arbitrary point of K. Then,

$$a_K\geqslant\frac{1}{k}\alpha h^d\inf\big\{\varphi(x),\ x\in B(x_K,h)\big\}$$

and

$$b_K\leqslant\frac{2V(g_1+g_2)}{\alpha}h^d\sup\big\{\big|\nabla\varphi(x)\big|,\ x\in B(x_K,2h)\big\}.$$

Therefore, taking $\gamma>0$ small enough in order to have

$$\inf\big\{\varphi(y),\ y\in B(x,h)\big\}>C\sup\big\{\big|\nabla\varphi(y)\big|,\ y\in B(x,2h)\big\},\quad\forall x\in\mathbb{R}^d, \tag{26.6}$$

with $C = (2kV(g_1 + g_2))/\alpha^2$, yields $a_K > b_K$ for all $K \in \mathcal{T}$. Hence (26.5) gives $u_K = w_K$, for all $K \in \mathcal{T}$.

A choice of $\gamma > 0$ verifying (26.6) is always possible. Indeed, since $|\nabla\varphi(z)| = \gamma\exp(-\gamma|z|)$, taking $\gamma > 0$ such that $\gamma\exp(3\gamma h) < 1/C$ is convenient.

This concludes Step 1.

Step 2 (*Existence of* $\{u_K^{n+1},\ K \in \mathcal{T}\}$, *such that* $u_K^{n+1} \in [U_m, U_M]$ *for all* $K \in \mathcal{T}$, *solution of* (24.6)). Recall that $n \in \mathbb{N}$ and $\{u_K^n,\ K \in \mathcal{T}\}$ are given. For $r \in \mathbb{N}^\star$, let $B_r = B(0,r) = \{x \in \mathbb{R}^d,\ |x| < r\}$ and $\mathcal{T}_r = \{K \in \mathcal{T},\ K \subset B_r\}$ (as in Lemma 25.2). Let us assume that r is large enough, say $r \geqslant r_0$, in order to have $\mathcal{T}_r \neq \emptyset$.

If $K \in \mathcal{T} \setminus \mathcal{T}_r$, set $u_K^{(r)} = u_K^n$. Let us first prove that there exists $\{u_K^{(r)},\ K \in \mathcal{T}_r\} \subset [U_m, U_M]$, solution to

$$\mathrm{m}(K)\frac{u_K^{(r)} - u_K^n}{k} + \sum_{L\in\mathcal{N}(K)} \left(v_{K,L}^n\, g\left(u_K^{(r)}, u_L^{(r)}\right) - v_{L,K}^n\, g\left(u_L^{(r)}, u_K^{(r)}\right)\right) = 0,$$
$$\forall K \in \mathcal{T}_r. \tag{26.7}$$

Then, we will prove that passing to the limit as $r \to \infty$ (up to a subsequence) leads to a solution $\{u_K^{n+1},\ K \in \mathcal{T}\}$ to (24.6) such that $u_K^{n+1} \in [U_m, U_M]$ for all $K \in \mathcal{T}$.

For a fixed $r \geqslant r_0$, in order to prove the existence of $\{u_K^{(r)},\ K \in \mathcal{T}_r\} \subset [U_m, U_M]$ solution to (26.7), a "topological degree" argument is used (see, e.g., DEIMLING [1980] for a presentation of the degree).

Let $U_r^n = \{u_K^n,\ K \in \mathcal{T}_r\}$ and assume that $U_r = \{u_K^{(r)},\ K \in \mathcal{T}_r\}$ is a solution of (26.7). The families U_r and U_r^n may be viewed as vectors of $\mathbb{R}^N$, with $N = \mathrm{card}(\mathcal{T}_r)$. Equation (26.7) gives

$$u_K^{(r)} + \frac{k}{\mathrm{m}(K)} \sum_{L\in\mathcal{N}(K)} \left(v_{K,L}^n\, g\left(u_K^{(r)}, u_L^{(r)}\right) - v_{L,K}^n\, g\left(u_L^{(r)}, u_K^{(r)}\right)\right) = u_K^n, \quad \forall K \in \mathcal{T}_r,$$

which can be written on the form

$$U_r - G_r(U_r) = U_r^n, \tag{26.8}$$

where G_r is a continuous map from $\mathbb{R}^N$ into $\mathbb{R}^N$.

One may assume that g is nondecreasing with respect to its first argument and nonincreasing with respect to its second argument on $\mathbb{R}^2$ (indeed, thanks to the monotonicity properties of g given by Assumption 24.1, it is sufficient to change, if necessary, g on $\mathbb{R}^2 \setminus [U_m, U_M]^2$, setting, e.g., $g(a,b) = g(U_m \top (U_M \bot a), U_m \top (U_M \bot b))$). Then, since $u_K^n \in [U_m, U_M]$, for all $K \in \mathcal{T}$, and $u_K^{(r)} = u_K^n \in [U_m, U_M]$, for all $K \in \mathcal{T} \setminus \mathcal{T}_r$, it is easy to show (using $\mathrm{div}(\boldsymbol{v}) = 0$) that if U_r satisfies (26.8), then one has $u_K^{(r)} \in [U_m, U_M]$, for all $K \in \mathcal{T}_r$. Therefore, if $\mathcal{C}_r$ is a ball of $\mathbb{R}^N$ of center 0 and of radius great enough, Eq. (26.8) has no solution on the boundary of $\mathcal{C}_r$, and one can define the topological degree of the application $\mathrm{Id} - G_r$ associated to the set $\mathcal{C}_r$ and to the point U_r^n, that is $\deg(\mathrm{Id} - G_r, \mathcal{C}_r, U_r^n)$. Furthermore, if $\lambda \in [0, 1]$, the same argument allows us to define

$\deg(\mathrm{Id} - \lambda G_r, \mathcal{C}_r, U_r^n)$. Then, the property of invariance of the degree by continuous transformation asserts that $\deg(\mathrm{Id} - \lambda G_r, \mathcal{C}_r, U_r^n)$ does not depend on $\lambda \in [0, 1]$. This gives

$$\deg(\mathrm{Id} - G_r, \mathcal{C}_r, U_r^n) = \deg(\mathrm{Id}, \mathcal{C}_r, U_r^n).$$

But, since $U_r^n \in \mathcal{C}_r$,

$$\deg(\mathrm{Id}, \mathcal{C}_r, U_r^n) = 1.$$

Hence

$$\deg(\mathrm{Id} - G_r, \mathcal{C}_r, U_r^n) \neq 0.$$

This proves that there exists a solution $U_r \in \mathcal{C}_r$ to (26.8). Recall also that we already proved that the components of U_r are necessarily in $[U_m, U_M]$.

In order to prove the existence of $\{u_K^{n+1},\ K \in \mathcal{T}\} \subset [U_m, U_M]$ solution to (24.6), let us pass to the limit as $r \to \infty$. For $r \geqslant r_0$, let $\{u_K^{(r)},\ K \in \mathcal{T}\}$ be a solution of (26.7) (given by the previous proof). Since $\{u_K^{(r)},\ r \in \mathbb{N}\}$ is included in $[U_m, U_M]$, for all $K \in \mathcal{T}$, one can find (using a "diagonal process") a sequence $(r_l)_{l \in \mathbb{N}}$, with $r_l \to \infty$, as $l \to \infty$, such that $(u_K^{r_l})_{l \in \mathbb{N}}$ converges (in $[U_m, U_M]$) for all $K \in \mathcal{T}$. One sets $u_K^{n+1} = \lim_{l \to \infty} u_K^{r_l}$. Passing to the limit in (26.7) (this is possible since for all $K \in \mathcal{T}$, this equation is satisfied for all $l \in \mathbb{N}$ large enough) shows that $\{u_K^{n+1},\ K \in \mathcal{T}\}$ is solution to (24.6).

Indeed, using the uniqueness of the solution of (24.6), one can show that $u_K^{(r)} \to u_K^{n+1}$, as $r \to \infty$, for all $K \in \mathcal{T}$.

This completes the proof of Proposition 26.1. □

26.2. *"Weak space BV" inequality*

One gives here an inequality similar to inequality (25.4) (proved for the explicit scheme). This inequality does not make use of $u_0 \in BV(\mathbb{R}^d)$ and $\boldsymbol{v}$ can depend on t. Inequality (25.5) also holds but is improved in Lemma 26.3 when $u_0 \in BV(\mathbb{R}^d)$ and $\boldsymbol{v}$ does not depend on t.

LEMMA 26.1. *Under Assumption* 23.1, *let $\mathcal{T}$ be an admissible mesh in the sense of Definition* 24.1 *and $k > 0$. Let $g \in C(\mathbb{R}^2, \mathbb{R})$ satisfy Assumption* 24.1 *and let $\{u_K^n, n \in \mathbb{N}, K \in \mathcal{T}\}$ be the solution of* (24.6), (24.2) *such that $u_K^{n+1} \in [U_m, U_M]$ for all $K \in \mathcal{T}$ and all $n \in \mathbb{N}$ (existence and uniqueness of such a solution is given by Proposition* 26.1).

Let $T > 0$, $R > 0$, $N_{T,k} = \max\{n \in \mathbb{N}, n < T/k\}$, $\mathcal{T}_R = \{K \in \mathcal{T}, K \subset B(0, R)\}$ and $\mathcal{E}_R^n = \{(K, L) \in \mathcal{T}^2, L \in \mathcal{N}(K), K|L \subset B(0, R)$ and $u_K^n > u_L^n\}$.

Then there exists $C_v \in \mathbb{R}$, only depending on $\boldsymbol{v}$, g, u_0, α, R, T such that, for $h < R$ and $k < T$,

$$\sum_{n=0}^{N_{T,k}} k \sum_{(K,L) \in \mathcal{E}_R^{n+1}} \Big[v_{K,L}^n \Big(\max_{u_L^{n+1} \leqslant p \leqslant q \leqslant u_K^{n+1}} \big(g(q, p) - f(q) \big)$$

$$+ \max_{u_L^{n+1} \leqslant p \leqslant q \leqslant u_K^{n+1}} \big(g(q,p) - f(p)\big)\Big)$$

$$+ v_{L,K}^n \Big(\max_{u_L^{n+1} \leqslant p \leqslant q \leqslant u_K^{n+1}} \big(f(q) - g(p,q)\big)$$

$$+ \max_{u_L^{n+1} \leqslant p \leqslant q \leqslant u_K^{n+1}} \big(f(p) - g(p,q)\big)\Big)\Big] \leqslant \frac{C_v}{\sqrt{h}}. \tag{26.9}$$

Furthermore, inequality (25.5) *holds.*

PROOF. We multiply (24.6) by ku_K^{n+1}, and sum the result over $K \in \mathcal{T}_R$ and $n \in \{0, \ldots, N_{T,k}\}$. We can then follow, step by step, the proof of Lemma 25.2 until Eq. (25.15) in which the first term of the right-hand side appears with the opposite sign. We can then directly conclude an inequality similar to (25.18), which is sufficient to conclude the proof of inequality (26.9). Inequality 25.5 follows easily from (26.9). □

26.3. "Time BV" estimate

This section gives a so called "strong time BV estimate" (estimate (26.14)). For this estimate, the fact that $u_0 \in BV(\mathbb{R}^d)$ and that $\boldsymbol{v}$ does not depend on t is required. Let us begin this section with a preliminary lemma on the space $BV(\mathbb{R}^d)$.

LEMMA 26.2. *Let $\mathcal{T}$ be an admissible mesh in the sense of Definition* 24.1 *and let $u \in BV(\mathbb{R}^d)$ (see Definition* 21.17). *For $K \in \mathcal{T}$, let u_K be the mean value of u over K. Then,*

$$\sum_{K|L \in \mathcal{E}} \mathrm{m}(K|L)|u_K - u_L| \leqslant \frac{C}{\alpha^4} |u|_{BV(\mathbb{R}^d)}, \tag{26.10}$$

where C only depends on the space dimension ($d = 1$, 2 or 3).

PROOF. Lemma 26.2 is proven in two steps. In the first step, it is proved that if (26.10) holds for all $u \in BV(\mathbb{R}^d) \cap C^1(\mathbb{R}^d, \mathbb{R})$ then (26.10) holds for all $u \in BV(\mathbb{R}^d)$. In Step 2, (26.10) is proved to hold for $u \in BV(\mathbb{R}^d) \cap C^1(\mathbb{R}^d, \mathbb{R})$.

Step 1 (*Passing from $BV(\mathbb{R}^d) \cap C^1(\mathbb{R}^d, \mathbb{R})$ to $BV(\mathbb{R}^d)$*). Recall that $BV(\mathbb{R}^d) \subset L^1_{\mathrm{loc}}(\mathbb{R}^d)$. Let $u \in BV(\mathbb{R}^d)$, let us regularize u by a sequence of mollifiers.

Let $\rho \in C_c^\infty(\mathbb{R}^d, \mathbb{R}_+)$ such that $\int_{\mathbb{R}^d} \rho(x)\,\mathrm{d}x = 1$. Define, for all $n \in \mathbb{N}^\star$, ρ_n by $\rho_n(x) = n^d \rho(nx)$ for all $x \in \mathbb{R}^d$ and $u_n = u \star \rho_n$, i.e.

$$u_n(x) = \int_{\mathbb{R}^d} u(y)\rho_n(x-y)\,\mathrm{d}y, \quad \forall x \in \mathbb{R}^d.$$

It is well known that $(u_n)_{n \in \mathbb{N}^\star}$ is included in $C^\infty(\mathbb{R}^d, \mathbb{R})$ and converges to u in $L^1_{\mathrm{loc}}(\mathbb{R}^d)$ as $n \to \infty$. Then, the mean value of u_n over K converges, as $n \to \infty$, to u_K, for all

$K \in \mathcal{T}$. Hence, if (26.10) holds with u_n instead of u (this will be proven in Step 2) and if $|u_n|_{BV(\mathbb{R}^d)} \leqslant |u|_{BV(\mathbb{R}^d)}$ for all $n \in \mathbb{N}^\star$, inequality (26.10) is proved by passing to the limit as $n \to \infty$.

In order to prove $|u_n|_{BV(\mathbb{R}^d)} \leqslant |u|_{BV(\mathbb{R}^d)}$ for all $n \in \mathbb{N}^\star$ (this will conclude Step 1), let $n \in \mathbb{N}^\star$ and $\varphi \in C_c^\infty(\mathbb{R}^d, \mathbb{R}^d)$ such that $|\varphi(x)| \leqslant 1$ for all $x \in \mathbb{R}^d$. A simple computation gives, using Fubini's theorem,

$$\begin{aligned}\int_{\mathbb{R}^d} u_n(x) \operatorname{div} \varphi(x)\, \mathrm{d}x &= \int_{\mathbb{R}^d} \left(\int_{\mathbb{R}^d} u(x-y) \operatorname{div} \varphi(x)\, \mathrm{d}x \right) \rho_n(y)\, \mathrm{d}y \\ &\leqslant |u|_{BV(\mathbb{R}^d)},\end{aligned} \tag{26.11}$$

since, setting $\psi_y = \varphi(y + \bullet) \in C_c^\infty(\mathbb{R}^d, \mathbb{R}^d)$ (for all $y \in \mathbb{R}^d$),

$$\int_{\mathbb{R}^d} u(x-y) \operatorname{div} \varphi(x)\, \mathrm{d}x = \int_{\mathbb{R}^d} u(z) \operatorname{div} \psi_y(z)\, \mathrm{d}z \leqslant |u|_{BV(\mathbb{R}^d)}, \quad \forall y \in \mathbb{R}^d,$$

and

$$\int_{\mathbb{R}^d} \rho_n(y)\, \mathrm{d}y = 1.$$

Then, taking in (26.11) the supremum over $\varphi \in C_c^\infty(\mathbb{R}^d, \mathbb{R}^d)$ such that $|\varphi(x)| \leqslant 1$ for all $x \in \mathbb{R}^d$ leads to $|u_n|_{BV(\mathbb{R}^d)} \leqslant |u|_{BV(\mathbb{R}^d)}$.

Step 2 (*Proving* (26.10) *if* $u \in BV(\mathbb{R}^d) \cap C^1(\mathbb{R}^d, \mathbb{R})$). Recall that $B(x, R)$ denotes the ball of $\mathbb{R}^d$ of center x and radius R. Since $u \in C^1(\mathbb{R}^d, \mathbb{R})$,

$$\int_{\mathbb{R}^d} u(x) \operatorname{div} \varphi(x)\, \mathrm{d}x = -\int_{\mathbb{R}^d} \nabla u(x) \bullet \varphi(x)\, \mathrm{d}x.$$

Then $|u|_{BV(\mathbb{R}^d)} = \|(|\nabla u|)\|_{L^1(\mathbb{R}^d)}$ and we will prove (26.10) with $\|(|\nabla u|)\|_{L^1(\mathbb{R}^d)}$ instead of $|u|_{BV(\mathbb{R}^d)}$.

Let $K|L \in \mathcal{E}$, then $K \in \mathcal{T}$, $L \in \mathcal{N}(K)$ and

$$u_K - u_L = \frac{1}{\mathrm{m}(K)\mathrm{m}(L)} \int_L \int_K \big(u(x) - u(y)\big)\, \mathrm{d}x\, \mathrm{d}y.$$

For all $x \in K$ and all $y \in L$,

$$u(x) - u(y) = \int_0^1 \nabla u\big(y + t(x-y)\big) \bullet (x-y)\, \mathrm{d}t.$$

Then,

$$\mathrm{m}(K)\mathrm{m}(L)|u_K - u_L| \leqslant \int_L \left(\int_K \int_0^1 \big|\nabla u\big(y + t(x-y)\big)\big| |x-y|\, \mathrm{d}t\, \mathrm{d}x \right) \mathrm{d}y$$

$$\leqslant \int_L \left(\int_0^1 \int_K \left|\nabla u\big(y + t(x-y)\big)\right| |x-y| \,\mathrm{d}x \,\mathrm{d}t \right) \mathrm{d}y.$$

Using $|x-y| \leqslant 2h$ and changing the variable x in $z = x - y$ (for all fixed $y \in L$ and $t \in (0,1)$) yields

$$\mathrm{m}(K)\mathrm{m}(L)|u_K - u_L| \leqslant 2h \int_L \left(\int_0^1 \int_{B(0,2h)} \left|\nabla u(y+tz)\right| \mathrm{d}z \,\mathrm{d}t \right) \mathrm{d}y,$$

which may also be written (using Fubini's theorem)

$$\mathrm{m}(K)\mathrm{m}(L)|u_K - u_L| \leqslant 2h \int_{B(0,2h)} \left(\int_0^1 \int_L \left|\nabla u(y+tz)\right| \mathrm{d}y \,\mathrm{d}t \right) \mathrm{d}z. \tag{26.12}$$

For all $K \in \mathcal{T}$, let x_K be an arbitrary point of K.

Then, changing the variable y in $\xi = y + tz$ (for all fixed $z \in L$ and $t \in (0,1)$) in (26.12),

$$\mathrm{m}(K)\mathrm{m}(L)|u_K - u_L| \leqslant 2h \int_{B(0,2h)} \left(\int_0^1 \int_{B(x_L,3h)} \left|\nabla u(\xi)\right| \mathrm{d}\xi \,\mathrm{d}t \right) \mathrm{d}z,$$

which yields, since $\mathcal{T}$ is an admissible mesh in the sense of Definition 24.1,

$$\mathrm{m}(K|L)|u_K - u_L| \leqslant \frac{2h^d}{\alpha^3 h^{2d}} \mathrm{m}\big(B(0,2h)\big) \int_{B(x_L,3h)} \left|\nabla u(\xi)\right| \mathrm{d}\xi.$$

Therefore there exists C_1, only depending on the space dimension, such that

$$\mathrm{m}(K|L)|u_K - u_L| \leqslant \frac{C_1}{\alpha^3} \int_{B(x_L,3h)} \left|\nabla u(\xi)\right| \mathrm{d}\xi, \quad \forall K|L \in \mathcal{E}. \tag{26.13}$$

Let us now remark that, if $M \in \mathcal{T}$ and $L \in \mathcal{T}$, $M \cap B(x_L, 3h) \neq \emptyset$ implies $L \subset B(x_M, 5h)$. Then, for a fixed $M \in \mathcal{T}$, the number of $L \in \mathcal{T}$ such that $M \cap B(x_L, 3h) \neq \emptyset$ is less or equal to $\mathrm{m}(B(0,5h))/(\alpha h^d)$ that is less or equal C_2/α where C_2 only depends on the space dimension.

Then, summing (26.13) over $K|L \in \mathcal{E}$ leads to

$$\sum_{K|L \in \mathcal{E}} \mathrm{m}(K|L)|u_K - u_L| \leqslant \frac{C_1 C_2}{\alpha^4} \sum_{M \in \mathcal{T}} \int_M \left|\nabla u(\xi)\right| \mathrm{d}\xi = \frac{C_1 C_2}{\alpha^4} \left\| \big(|\nabla u|\big) \right\|_{L^1(\mathbb{R}^d)},$$

i.e. (26.10) with $C = C_1 C_2$. $\square$

Note that, in Lemma 26.2 the estimate (26.10) depends on α. This dependency on α is not necessary in the one-dimensional case (see (19.6) in Remark 19.4) and for particular meshes in the two- and three-dimensional cases. Recall also that, except if $d = 1$, the space $BV(\mathbb{R}^d)$ is not included in $L^\infty(\mathbb{R}^d)$. In particular, it is then quite easy to prove that, contrary to the 1D case given in Remark 19.4, it is not possible, for $d = 2$ or 3, to replace, in (26.10), u_K by the mean value of u over an arbitrary ball (e.g.) included in K.

Let us now give the "strong time BV estimate".

LEMMA 26.3. *Under Assumption* 23.1, *let* $\mathcal{T}$ *be an admissible mesh in the sense of Definition* 24.1 *and* $k > 0$. *Let* $g \in C(\mathbb{R}^2, \mathbb{R})$ *satisfy Assumption* 24.1. *Assume that* $u_0 \in BV(\mathbb{R}^d)$ *and that* $\boldsymbol{v}$ *does not depend on* t.

Let $\{u_K^n, n \in \mathbb{N}, K \in \mathcal{T}\}$ *be the solution of* (24.6), (24.2) *such that* $u_K^n \in [U_m, U_M]$ *for all* $K \in \mathcal{T}$ *and all* $n \in \mathbb{N}$ (*existence and uniqueness of such a solution is given by Proposition* 26.1).

Then, there exists C_b, *only depending on* $\boldsymbol{v}$, g, u_0 *and* α *such that*

$$\sum_{K \in \mathcal{T}} \frac{\mathrm{m}(K)}{k} \left|u_K^{n+1} - u_K^n\right| \leqslant C_b, \quad \forall n \in \mathbb{N}. \tag{26.14}$$

PROOF. Since $\boldsymbol{v}$ does not depend on t, one denotes $v_{K,L} = v_{K,L}^n$, for all $K \in \mathcal{T}$ and all $L \in \mathcal{N}(K)$.

For $n \in \mathbb{N}$, let

$$A_n = \sum_{K \in \mathcal{T}} \mathrm{m}(K) \frac{|u_K^{n+1} - u_K^n|}{k}$$

and

$$B_n = \sum_{K \in \mathcal{T}} \left| \sum_{L \in \mathcal{N}(K)} \left[v_{K,L}\, g\left(u_K^n, u_L^n\right) - v_{L,K}\, g\left(u_L^n, u_K^n\right)\right]\right|.$$

Since $u_0 \in BV(\mathbb{R}^d)$ and $\operatorname{div} \boldsymbol{v} = 0$, there exists $C_b > 0$, only depending on $\boldsymbol{v}$, g, u_0 and α, such that $B_0 \leqslant C_b$. Indeed,

$$B_0 \leqslant \sum_{K \in \mathcal{T}} \sum_{L \in \mathcal{N}(K)} V(g_1 + g_2)\mathrm{m}(K|L)\left|u_K^0 - u_L^0\right|.$$

Thanks to Lemma 26.2, $B_0 \leqslant C_b$ with $C_b = 2V(g_1 + g_2)C(1/\alpha^4)|u_0|_{BV(\mathbb{R}^d)}$, where C only depends on the space dimension ($d = 1$, 2 or 3).

From (24.6), one deduces that $B_{n+1} \leqslant A_n$, for all $n \in \mathbb{N}$. In order to prove Lemma 26.3, there only remains to prove that $A_n \leqslant B_n$ for all $n \in \mathbb{N}$ (and to conclude by induction).

Let $n \in \mathbb{N}$, in order to prove that $A_n \leqslant B_n$, recall that the implicit scheme (24.6) writes

$$\mathrm{m}(K)\frac{u_K^{n+1}-u_K^n}{k}+\sum_{L\in\mathcal{N}(K)}\big(v_{K,L}\,g\big(u_K^{n+1},u_L^{n+1}\big)$$
$$-\,v_{L,K}\,g\big(u_L^{n+1},u_K^{n+1}\big)\big)=0. \tag{26.15}$$

From (26.15), one deduces, for all $K\in\mathcal{T}$,

$$\begin{aligned}
&\mathrm{m}(K)\frac{u_K^{n+1}-u_K^n}{k}+\sum_{L\in\mathcal{N}(K)}v_{K,L}\big(g\big(u_K^{n+1},u_L^{n+1}\big)-g\big(u_K^n,u_L^{n+1}\big)\big)\\
&\quad+\sum_{L\in\mathcal{N}(K)}v_{K,L}\big(g\big(u_K^n,u_L^{n+1}\big)-g\big(u_K^n,u_L^n\big)\big)\\
&\quad-\sum_{L\in\mathcal{N}(K)}v_{L,K}\big(g\big(u_L^{n+1},u_K^{n+1}\big)-g\big(u_L^n,u_K^{n+1}\big)\big)\\
&\quad-\sum_{L\in\mathcal{N}(K)}v_{L,K}\big(g\big(u_L^n,u_K^{n+1}\big)-g\big(u_L^n,u_K^n\big)\big)\\
&=-\sum_{L\in\mathcal{N}(K)}v_{K,L}\,g\big(u_K^n,u_L^n\big)+\sum_{L\in\mathcal{N}(K)}v_{L,K}\,g\big(u_L^n,u_K^n\big).
\end{aligned}$$

Using the monotonicity properties of g, one obtains for all $K\in\mathcal{T}$,

$$\begin{aligned}
&\mathrm{m}(K)\frac{|u_K^{n+1}-u_K^n|}{k}+\sum_{L\in\mathcal{N}(K)}v_{K,L}\big|g\big(u_K^{n+1},u_L^{n+1}\big)-g\big(u_K^n,u_L^{n+1}\big)\big|\\
&\quad+\sum_{L\in\mathcal{N}(K)}v_{L,K}\big|g\big(u_L^n,u_K^{n+1}\big)-g\big(u_L^n,u_K^n\big)\big|\\
&\leqslant\Big|-\sum_{L\in\mathcal{N}(K)}v_{K,L}\,g\big(u_K^n,u_L^n\big)+\sum_{L\in\mathcal{N}(K)}v_{L,K}\,g\big(u_L^n,u_K^n\big)\Big|\\
&\quad+\sum_{L\in\mathcal{N}(K)}v_{K,L}\big|g\big(u_K^n,u_L^{n+1}\big)-g\big(u_K^n,u_L^n\big)\big|\\
&\quad+\sum_{L\in\mathcal{N}(K)}v_{L,K}\big|g\big(u_L^{n+1},u_K^{n+1}\big)-g\big(u_L^n,u_K^{n+1}\big)\big|.
\end{aligned} \tag{26.16}$$

In order to deal with convergent series, let us proceed as in the proof of Proposition 26.1. For $0<\gamma<1$, let $\varphi_\gamma:\mathbb{R}^d\mapsto\mathbb{R}_+^\star$ be defined by $\varphi_\gamma(x)=\exp(-\gamma|x|)$.

For $K\in\mathcal{T}$, let $\varphi_{\gamma,K}$ be the mean value of φ_γ on K. As in Proposition 26.1, since φ_γ is integrable over $\mathbb{R}^d$, $\sum_{K\in\mathcal{T}}\varphi_{\gamma,K}<\infty$. Therefore, multiplying (26.16) by $\varphi_{\gamma,K}$ (for a fixed γ) and summing over $K\in\mathcal{T}$ yields six convergent series which can be reordered to give

$$\sum_{K\in\mathcal{T}} \mathrm{m}(K)\frac{|u_K^{n+1}-u_K^n|}{k}\varphi_{\gamma,K}$$
$$\leqslant \sum_{K\in\mathcal{T}}\Big|-\sum_{L\in\mathcal{N}(K)} v_{K,L}\, g\big(u_K^n,u_L^n\big)+\sum_{L\in\mathcal{N}(K)} v_{L,K}\, g\big(u_L^n,u_K^n\big)\Big|\varphi_{\gamma,K}$$
$$+\sum_{K\in\mathcal{T}}\sum_{L\in\mathcal{N}(K)} v_{K,L}\big|g\big(u_K^{n+1},u_L^{n+1}\big)-g\big(u_K^n,u_L^{n+1}\big)\big||\varphi_{\gamma,K}-\varphi_{\gamma,L}|$$
$$+\sum_{K\in\mathcal{T}}\sum_{L\in\mathcal{N}(K)} v_{L,K}\big|g\big(u_L^n,u_K^{n+1}\big)-g\big(u_L^n,u_K^n\big)\big||\varphi_{\gamma,K}-\varphi_{\gamma,L}|.$$

For $K\in\mathcal{T}$, let $x_K\in\overline{K}$ be such that $\varphi_{\gamma,K}=\varphi_\gamma(x_K)$. Let $K\in\mathcal{T}$ and $L\in\mathcal{N}(K)$. Then there exists $s\in(0,1)$ such that $\varphi_{\gamma,L}-\varphi_{\gamma,K}=\nabla\varphi_\gamma(x_K+s(x_L-x_K))\bullet(x_L-x_K)$. Using $|\nabla\varphi_\gamma(x)|=\gamma\exp(-\gamma|x|)$, this yields $|\varphi_{\gamma,L}-\varphi_{\gamma,K}|\leqslant 2h\gamma\exp(2h\gamma)\varphi_{\gamma,K}\leqslant 2h\gamma\exp(2h)\varphi_{\gamma,K}$.

Then, using Assumptions 23.1 and 24.1, there exists some a only depending on k, V, h, α, g_1 and g_2 such that

$$\sum_{K\in\mathcal{T}} \mathrm{m}(K)\frac{|u_K^{n+1}-u_K^n|}{k}\varphi_{\gamma,K}(1-\gamma a)$$
$$\leqslant \sum_{K\in\mathcal{T}}\Big|-\sum_{L\in\mathcal{N}(K)} v_{K,L}g\big(u_K^n,u_L^n\big)+\sum_{L\in\mathcal{N}(K)} v_{L,K}\, g\big(u_L^n,u_K^n\big)\Big|\varphi_{\gamma,K}\leqslant B_n.$$

Passing to the limit in the latter inequality as $\gamma\to 0$ yields $A_n\leqslant B_n$. This completes the proof of Lemma 26.3. □

27. Entropy inequalities for the approximate solution

In this section, an entropy estimate on the approximate solution is proved (Theorem 27.1), which will be used in the proofs of convergence and error estimate of the numerical scheme. In order to obtain this entropy estimate, some discrete entropy inequalities satisfied by the approximate solution are first derived.

27.1. Discrete entropy inequalities

In the case of the explicit scheme, the following lemma asserts that the scheme (24.4) satisfies a discrete entropy condition (this is classical in the study of 1D schemes, see, e.g., GODLEWSKI and RAVIART [1991, 1996]).

LEMMA 27.1. *Under Assumption* 23.1, *let $\mathcal{T}$ be an admissible mesh in the sense of Definition* 24.1 *and $k>0$. Let $g\in C(\mathbb{R}^2,\mathbb{R})$ satisfying Assumption* 24.1 *and assume that* (24.3) *holds.*

Let $u_{\mathcal{T},k}$ be given by (24.5), (24.4), (24.2); *then, for all $\kappa \in \mathbb{R}$, $K \in \mathcal{T}$ and $n \in \mathbb{N}$, the following inequality holds*:

$$\begin{aligned}
&\mathrm{m}(K)\frac{|u_K^{n+1}-\kappa| - |u_K^n - \kappa|}{k} \\
&\quad + \sum_{L\in\mathcal{N}(K)} \big[v_{K,L}^n\big(g\big(u_K^n\top\kappa, u_L^n\top\kappa\big) - g\big(u_K^n\bot\kappa, u_L^n\bot\kappa\big)\big) \\
&\quad - v_{L,K}^n\big(g\big(u_L^n\top\kappa, u_K^n\top\kappa\big) - g\big(u_L^n\bot\kappa, u_K^n\bot\kappa\big)\big)\big] \leqslant 0.
\end{aligned} \tag{27.1}$$

PROOF. From relation (24.4), we express u_K^{n+1} as a function of u_K^n and u_L^n, $L \in \mathcal{N}(K)$,

$$u_K^{n+1} = u_K^n + \frac{k}{\mathrm{m}(K)} \sum_{L\in\mathcal{N}(K)} \big(v_{L,K}^n\, g\big(u_L^n, u_K^n\big) - v_{K,L}^n\, g\big(u_K^n, u_L^n\big)\big).$$

The right-hand side is nondecreasing with respect to u_L^n, $L \in \mathcal{N}(K)$. It is also nondecreasing with respect to u_K^n, thanks to the Courant–Friedrichs–Levy condition (24.3), and the Lipschitz continuity of g.

Therefore, for all $\kappa \in \mathbb{R}$, using $\operatorname{div} \boldsymbol{v} = 0$, we have:

$$\begin{aligned}
u_K^{n+1}\top\kappa \leqslant u_K^n\top\kappa + \frac{k}{\mathrm{m}(K)} \sum_{L\in\mathcal{N}(K)} &\big[v_{L,K}^n\, g\big(u_L^n\top\kappa, u_K^n\top\kappa\big) \\
&- v_{K,L}^n\, g\big(u_K^n\top\kappa, u_L^n\top\kappa\big)\big]
\end{aligned} \tag{27.2}$$

and

$$\begin{aligned}
u_K^{n+1}\bot\kappa \geqslant u_K^n\bot\kappa + \frac{k}{\mathrm{m}(K)} \sum_{L\in\mathcal{N}(K)} &\big(v_{L,K}^n\, g\big(u_L^n\bot\kappa, u_K^n\bot\kappa\big) \\
&- v_{K,L}^n\, g\big(u_K^n\bot\kappa, u_L^n\bot\kappa\big)\big).
\end{aligned} \tag{27.3}$$

The difference between (27.2) and (27.3) leads directly to (27.1). Note that using $\operatorname{div} \boldsymbol{v} = 0$ leads to

$$\begin{aligned}
&\mathrm{m}(K)\frac{|u_K^{n+1}-\kappa| - |u_K^n - \kappa|}{k} \\
&\quad + \sum_{L\in\mathcal{N}(K)} \big[v_{K,L}^n\big(g\big(u_K^n\top\kappa, u_L^n\top\kappa\big) - f\big(u_K^n\top\kappa\big) \\
&\quad - g\big(u_K^n\bot\kappa, u_L^n\bot\kappa\big) + f\big(u_K^n\bot\kappa\big)\big) - v_{L,K}^n\big(g\big(u_L^n\top\kappa, u_K^n\top\kappa\big) - f\big(u_K^n\top\kappa\big) \\
&\quad - g\big(u_L^n\bot\kappa, u_K^n\bot\kappa\big) + f\big(u_K^n\bot\kappa\big)\big)\big] \leqslant 0. \quad \square
\end{aligned} \tag{27.4}$$

For the implicit scheme, one obtains the same kind of discrete entropy inequalities.

LEMMA 27.2. *Under Assumption* 23.1, *let* $\mathcal{T}$ *be an admissible mesh in the sense of Definition* 24.1 *and* $k > 0$. *Let* $g \in C(\mathbb{R}^2, \mathbb{R})$ *satisfying Assumption* 24.1.

Let $\{u_K^n,\ n \in \mathbb{N},\ K \in \mathcal{T}\} \subset [U_m, U_M]$ *be the solution of* (24.6), (24.2) (*the existence and uniqueness of such a solution is given by Proposition* 26.1). *Then, for all* $\kappa \in \mathbb{R}$, $K \in \mathcal{T}$ *and* $n \in \mathbb{N}$, *the following inequality holds*:

$$\begin{aligned}
&\mathrm{m}(K)\frac{|u_K^{n+1}-\kappa| - |u_K^n-\kappa|}{k}\\
&\quad + \sum_{L\in\mathcal{N}(K)} \big[v_{K,L}^n\big(g\big(u_K^{n+1}\top\kappa, u_L^{n+1}\top\kappa\big) - g\big(u_K^{n+1}\bot\kappa, u_L^{n+1}\bot\kappa\big)\big)\\
&\quad - v_{L,K}^n\big(g\big(u_L^{n+1}\top\kappa, u_K^{n+1}\top\kappa\big) - g\big(u_L^{n+1}\bot\kappa, u_K^{n+1}\bot\kappa\big)\big)\big] \leqslant 0.
\end{aligned}\tag{27.5}$$

PROOF. Let $\kappa \in \mathbb{R}$, $K \in \mathcal{T}$ and $n \in \mathbb{N}$. Equation (24.6) may be written as

$$u_K^{n+1} = u_K^n - \frac{k}{\mathrm{m}(K)} \sum_{L\in\mathcal{N}(K)} \big(v_{K,L}^n\, g\big(u_K^{n+1}, u_L^{n+1}\big) - v_{L,K}^n\, g\big(u_L^{n+1}, u_K^{n+1}\big)\big).$$

The right-hand side of this last equation is nondecreasing with respect to u_K^n and with respect to u_L^{n+1} for all $L \in \mathcal{N}(K)$. Thus,

$$\begin{aligned}
u_K^{n+1} \leqslant{}& u_K^n\top\kappa - \frac{k}{\mathrm{m}(K)} \sum_{L\in\mathcal{N}(K)} \big(v_{K,L}^n\, g\big(u_K^{n+1}, u_L^{n+1}\top\kappa\big)\\
&- v_{L,K}^n\, g\big(u_L^{n+1}\top\kappa, u_K^{n+1}\big)\big).
\end{aligned}$$

Writing $\kappa = \kappa - \frac{k}{\mathrm{m}(K)} \sum_{L\in\mathcal{N}(K)}(v_{K,L}^n\ g(\kappa,\kappa) - v_{L,K}^n\ g(\kappa,\kappa))$, one may remark that

$$\kappa \leqslant u_K^n\top\kappa - \frac{k}{\mathrm{m}(K)} \sum_{L\in\mathcal{N}(K)} \big(v_{K,L}^n\, g\big(\kappa, u_L^{n+1}\top\kappa\big) - v_{L,K}^n\, g\big(u_L^{n+1}\top\kappa, \kappa\big)\big).$$

Therefore, since $u_K^{n+1}\top\kappa = u_K^{n+1}$ or κ,

$$\begin{aligned}
u_K^{n+1}\top\kappa \leqslant{}& u_K^n\top\kappa - \frac{k}{\mathrm{m}(K)} \sum_{L\in\mathcal{N}(K)} \big(v_{K,L}^n\, g\big(u_K^{n+1}\top\kappa, u_L^{n+1}\top\kappa\big)\\
&- v_{L,K}^n\, g\big(u_L^{n+1}\top\kappa, u_K^{n+1}\top\kappa\big)\big).
\end{aligned}\tag{27.6}$$

A similar argument yields

$$\begin{aligned}
u_K^{n+1}\bot\kappa \geqslant{}& u_K^n\bot\kappa - \frac{k}{\mathrm{m}(K)} \sum_{L\in\mathcal{N}(K)} \big(v_{K,L}^n\, g\big(u_K^{n+1}\bot\kappa, u_L^{n+1}\bot\kappa\big)\\
&- v_{L,K}^n\, g\big(u_L^{n+1}\bot\kappa, u_K^{n+1}\bot\kappa\big)\big).
\end{aligned}\tag{27.7}$$

Hence, subtracting (27.7) to (27.6) gives (27.5). □

27.2. Continuous entropy estimates for the approximate solution

For $\Omega = \mathbb{R}^d$ or $\mathbb{R}^d \times \mathbb{R}_+$, we denote by $\mathcal{M}(\Omega)$ the set of positive measures on Ω, that is of σ-additive applications from the Borel σ-algebra of Ω in $\overline{\mathbb{R}}_+$. If $\mu \in \mathcal{M}(\Omega)$ and $\psi \in C_c(\Omega)$, one sets $\langle \mu, \psi \rangle = \int \psi \,\mathrm{d}\mu$.

The following theorems investigate the entropy inequalities which are satisfied by the approximate solutions $u_{\mathcal{T},k}$ in the case of the time explicit scheme (Theorem 27.1) and in the case of the time implicit scheme (Theorem 27.2).

THEOREM 27.1. *Under Assumption* 23.1, *let $\mathcal{T}$ be an admissible mesh in the sense of Definition* 24.1 *and $k > 0$. Let $g \in C(\mathbb{R}^2, \mathbb{R})$ satisfy Assumption* 24.1 *and assume that* (24.3) *holds.*

Let $u_{\mathcal{T},k}$ be given by (24.5), (24.4), (24.2); *then there exist $\mu_{\mathcal{T},k} \in \mathcal{M}(\mathbb{R}^d \times \mathbb{R}_+)$ and $\mu_{\mathcal{T}} \in \mathcal{M}(\mathbb{R}^d)$ such that*

$$\begin{aligned}
&\int_{\mathbb{R}_+} \int_{\mathbb{R}^d} \Big(\big|u_{\mathcal{T},k}(x,t) - \kappa\big| \varphi_t(x,t) \\
&\quad + \big(f\big(u_{\mathcal{T},k}(x,t) \top \kappa\big) - f\big(u_{\mathcal{T},k}(x,t) \bot \kappa\big)\big) \boldsymbol{v}(x,t) \cdot \nabla\varphi(x,t)\Big) \,\mathrm{d}x \,\mathrm{d}t \\
&\quad + \int_{\mathbb{R}^d} \big|u_0(x) - \kappa\big| \varphi(x,0) \,\mathrm{d}x \\
&\geqslant - \int_{\mathbb{R}^d \times \mathbb{R}_+} \big(\big|\varphi_t(x,t)\big| + \big|\nabla\varphi(x,t)\big|\big) \,\mathrm{d}\mu_{\mathcal{T},k}(x,t) - \int_{\mathbb{R}^d} \varphi(x,0) \,\mathrm{d}\mu_{\mathcal{T}}(x), \\
&\forall \kappa \in \mathbb{R},\ \forall \varphi \in C_c^\infty\big(\mathbb{R}^d \times \mathbb{R}_+, \mathbb{R}_+\big).
\end{aligned} \tag{27.8}$$

The measures $\mu_{\mathcal{T},k}$ and $\mu_{\mathcal{T}}$ verify the following properties:

(1) *For all $R > 0$ and $T > 0$, there exists C depending only on $\boldsymbol{v}$, g, u_0, α, ξ, R and T such that, for $h < R$ and $k < T$,*

$$\mu_{\mathcal{T},k}\big(B(0,R) \times [0,T]\big) \leqslant C\sqrt{h}. \tag{27.9}$$

(2) *The measure $\mu_{\mathcal{T}}$ is the measure of density $|u_0(\cdot) - u_{\mathcal{T},0}(\cdot)|$ with respect to the Lebesgue measure, where $u_{\mathcal{T},0}$ is defined by $u_{\mathcal{T},0}(x) = u_K^0$ for a.e. $x \in K$, for all $K \in \mathcal{T}$.*

If $u_0 \in BV(\mathbb{R}^d)$, then there exists D, only depending on u_0 and α, such that

$$\mu_{\mathcal{T}}\big(\mathbb{R}^d\big) \leqslant Dh. \tag{27.10}$$

REMARK 27.1. (1) Let u be the weak entropy solution to (23.1)–(23.2). Then (27.8) is satisfied with u instead of $u_{\mathcal{T},k}$ and $\mu_{\mathcal{T},k} = 0$ and $\mu_{\mathcal{T}} = 0$.

(2) Let $BV_{\mathrm{loc}}(\mathbb{R}^d)$ be the set of $v \in L^1_{\mathrm{loc}}(\mathbb{R}^d)$ such that the restriction of v to Ω belongs to $BV(\Omega)$ for all open bounded subset Ω of $\mathbb{R}^d$.

An easy adaptation of the following proof gives that if $u_0 \in BV_{\mathrm{loc}}(\mathbb{R}^d)$ instead of $BV(\mathbb{R}^d)$ (in the second item of Theorem 27.1) then, for all $R > 0$, there exists D, only depending on u_0, α and R, such that $\mu_{\mathcal{T}}(B(0, R)) \leqslant Dh$.

PROOF OF THEOREM 27.1. Let $\varphi \in C_c^\infty(\mathbb{R}^d \times \mathbb{R}_+, \mathbb{R}_+)$ and $\kappa \in \mathbb{R}$.

Multiplying (27.4) by $k\varphi_K^n = (1/\mathrm{m}(K)) \int_{nk}^{(n+1)k} \int_K \varphi(x,t)\,\mathrm{d}x\,\mathrm{d}t$ and summing the result for all $K \in \mathcal{T}$ and $n \in \mathbb{N}$ yields

$$T_1 + T_2 \leqslant 0,$$

with

$$T_1 = \sum_{n\in\mathbb{N}} \sum_{K\in\mathcal{T}} \frac{|u_K^{n+1} - \kappa| - |u_K^n - \kappa|}{k} \int_{nk}^{(n+1)k} \int_K \varphi(x,t)\,\mathrm{d}x\,\mathrm{d}t, \tag{27.11}$$

and

$$\begin{aligned} T_2 = k \sum_{n\in\mathbb{N}} \sum_{(K,L)\in\mathcal{E}_n} \Big[& v_{K,L}^n \varphi_K^n \big(g\big(u_K^n \top \kappa, u_L^n \top \kappa\big) - f\big(u_K^n \top \kappa\big) \\ & - g\big(u_K^n \bot \kappa, u_L^n \bot \kappa\big) + f\big(u_K^n \bot \kappa\big)\big) - v_{K,L}^n \varphi_L^n \big(g\big(u_K^n \top \kappa, u_L^n \top \kappa\big) \\ & - f\big(u_L^n \top \kappa\big) - g\big(u_K^n \bot \kappa, u_L^n \bot \kappa\big) + f\big(u_L^n \bot \kappa\big)\big) \\ & - v_{L,K}^n \varphi_K^n \big(g\big(u_L^n \top \kappa, u_K^n \top \kappa\big) - f\big(u_K^n \top \kappa\big) \\ & - g\big(u_L^n \bot \kappa, u_K^n \bot \kappa\big) + f\big(u_K^n \bot \kappa\big)\big) \\ & + v_{L,K}^n \varphi_L^n \big(g\big(u_L^n \top \kappa, u_K^n \top \kappa\big) - f\big(u_L^n \top \kappa\big) \\ & - g\big(u_L^n \bot \kappa, u_K^n \bot \kappa\big) + f\big(u_L^n \bot \kappa\big)\big)\Big], \end{aligned} \tag{27.12}$$

where $\mathcal{E}_n = \{(K, L) \in \mathcal{T}^2, u_K^n > u_L^n\}$.

One has to prove

$$\begin{aligned} T_{10} + T_{20} \leqslant & \int_{\mathbb{R}^d \times \mathbb{R}_+} \big(|\varphi_t(x,t)| + |\nabla\varphi(x,t)|\big)\,\mathrm{d}\mu_{\mathcal{T},k}(x,t) \\ & + \int_{\mathbb{R}^d} \varphi(x,0)\,\mathrm{d}\mu_{\mathcal{T}}(x), \end{aligned} \tag{27.13}$$

for some convenient measures $\mu_{\mathcal{T},k}$ and $\mu_{\mathcal{T}}$, and T_{10}, T_{20} defined as follows

$$T_{10} = -\int_{\mathbb{R}_+} \int_{\mathbb{R}^d} |u_{\mathcal{T},k}(x,t) - \kappa| \varphi_t(x,t)\,\mathrm{d}x\,\mathrm{d}t - \int_{\mathbb{R}^d} |u_0(x) - \kappa| \varphi(x,0)\,\mathrm{d}x,$$

$$T_{20} = -\int_{\mathbb{R}_+}\int_{\mathbb{R}^d}\Big(\big(f\big(u_{\mathcal{T},k}(x,t)\top\kappa\big) \\ - f\big(u_{\mathcal{T},k}(x,t)\bot\kappa\big)\big)\boldsymbol{v}(x,t)\cdot\nabla\varphi(x,t)\Big)\,\mathrm{d}x\,\mathrm{d}t. \qquad (27.14)$$

In order to prove (27.13), one compares T_1 and T_{10} (this will give $\mu_{\mathcal{T}}$, and a part of $\mu_{\mathcal{T},k}$) and one compares T_2 and T_{20} (this will give another part of $\mu_{\mathcal{T},k}$).

Inequality (25.5) (in the comparison of T_1 and T_{10}) and inequality (25.4) (in the comparison of T_2 and T_{20}) will be used in order to obtain (27.9).

Comparison of T_1 *and* T_{10}. Using the definition of $u_{\mathcal{T},k}$ and introducing the function $u_{\mathcal{T},0}$ (defined by $u_{\mathcal{T},0}(x) = u_K^0$, for a.e. $x \in K$, for all $K \in \mathcal{T}$) yields

$$T_{10} = \sum_{n\in\mathbb{N}}\sum_{K\in\mathcal{T}} \frac{|u_K^{n+1} - \kappa| - |u_K^n - \kappa|}{k}\int_{nk}^{(n+1)k}\int_K \varphi\big(x,(n+1)k\big)\,\mathrm{d}x\,\mathrm{d}t \\ + \int_{\mathbb{R}^d}\big(\big|u_{\mathcal{T},0}(x) - \kappa\big| - \big|u_0(x) - \kappa\big|\big)\varphi(x,0)\,\mathrm{d}x.$$

The function $|\cdot - \kappa|$ is Lipschitz continuous with a Lipschitz constant equal to 1, we then obtain

$$|T_1 - T_{10}| \leqslant \sum_{n\in\mathbb{N}}\sum_{K\in\mathcal{T}} \frac{|u_K^{n+1} - u_K^n|}{k}\int_{nk}^{(n+1)k}\int_K \big|\varphi\big(x,(n+1)k\big) - \varphi(x,t)\big|\,\mathrm{d}x\,\mathrm{d}t \\ + \int_{\mathbb{R}^d}\big|u_0(x) - u_{\mathcal{T},0}(x)\big|\varphi(x,0)\,\mathrm{d}x,$$

which leads to

$$|T_1 - T_{10}| \leqslant \sum_{n\in\mathbb{N}}\sum_{K\in\mathcal{T}} \big|u_K^{n+1} - u_K^n\big|\int_{nk}^{(n+1)k}\int_K \big|\varphi_t(x,t)\big|\,\mathrm{d}x\,\mathrm{d}t \\ + \int_{\mathbb{R}^d}\big|u_0(x) - u_{\mathcal{T},0}(x)\big|\varphi(x,0)\,\mathrm{d}x. \qquad (27.15)$$

Inequality (27.15) gives

$$|T_1 - T_{10}| \leqslant \int_{\mathbb{R}^d\times\mathbb{R}_+}\big|\varphi_t(x,t)\big|\,\mathrm{d}\nu_{\mathcal{T},k}(x,t) + \int_{\mathbb{R}^d}\varphi(x,0)\,\mathrm{d}\mu_{\mathcal{T}}(x), \qquad (27.16)$$

where the measures $\mu_{\mathcal{T}} \in \mathcal{M}(\mathbb{R}^d)$ and $\nu_{\mathcal{T},k} \in \mathcal{M}(\mathbb{R}^d \times \mathbb{R}_+)$ are defined, by their action on $C_c(\mathbb{R}^d)$ and $C_c(\mathbb{R}^d \times \mathbb{R}_+)$, as follows

$$\langle\mu_{\mathcal{T}},\psi\rangle = \int_{\mathbb{R}^d}\big|u_0(x) - u_{\mathcal{T},0}(x)\big|\psi(x)\,\mathrm{d}x, \quad \forall\psi \in C_c\big(\mathbb{R}^d\big),$$

$$\langle \nu_{\mathcal{T},k}, \psi \rangle = \sum_{n\in\mathbb{N}} \sum_{K\in\mathcal{T}} \left| u_K^{n+1} - u_K^n \right| \int_{nk}^{(n+1)k} \int_K \psi(x,t)\,\mathrm{d}x\,\mathrm{d}t,$$
$$\forall \psi \in C_c\left(\mathbb{R}^d \times \mathbb{R}_+\right).$$

The measures $\mu_{\mathcal{T}}$ and $\nu_{\mathcal{T},k}$ are absolutely continuous with respect to the Lebesgue measure. Indeed, one has $\mathrm{d}\mu_{\mathcal{T}}(x) = |u_0(x) - u_{\mathcal{T},0}(x)|\,\mathrm{d}x$ and $\mathrm{d}\nu_{\mathcal{T},k}(x,t) = \sum_{n\in\mathbb{N}} \sum_{K\in\mathcal{T}} |u_K^{n+1} - u_K^n| 1_{K\times[nk,(n+1)k)})\,\mathrm{d}x\,\mathrm{d}t$ (where 1_Ω denotes the characteristic function of Ω for any Borel subset Ω of $\mathbb{R}^{d+1}$).

If $u_0 \in BV(\mathbb{R}^d)$, the measure $\mu_{\mathcal{T}}$ verifies (27.10) with some D only depending on $|u_0|_{BV(\mathbb{R}^d)}$ and α (this is classical result which is given in Lemma 27.3 below for the sake of completeness).

The measure $\nu_{\mathcal{T},k}$ satisfies (27.9), with $\nu_{\mathcal{T},k}$ instead of $\mu_{\mathcal{T},k}$, thanks to (25.5) and condition (24.3). Indeed, for $R > 0$ and $T > 0$,

$$\nu_{\mathcal{T},k}\left(B(0,R)\times[0,T]\right) = \int_0^T \int_{B(0,R)} \sum_{n\in\mathbb{N}} \sum_{K\in\mathcal{T}} \left| u_K^{n+1} - u_K^n \right| 1_{K\times[nk,(n+1)k)}\,\mathrm{d}x\,\mathrm{d}t,$$

which yields, with $\mathcal{T}_{2R} = \{K \in \mathcal{T},\ K \subset B(0,2R)\}$ and $N_{\mathcal{T},k}k < T \leqslant (N_{\mathcal{T},k}+1)k$, $h < R$ and $k < T$,

$$\nu_{\mathcal{T},k}\left(B(0,R)\times[0,T]\right) \leqslant k \sum_{n=0}^{N_{\mathcal{T},k}} \sum_{K\in\mathcal{T}_{2R}} \mathrm{m}(K) \left| u_K^{n+1} - u_K^n \right| \leqslant \frac{kC_1}{\sqrt{h}},$$

where C_1 is given by Lemma 25.2 and only depends on $\boldsymbol{v}$, g, u_0, α, ξ, R, T. Finally, since the condition (24.3) gives $k \leqslant C_2 h$, where C_2 only depends on $\boldsymbol{v}$, g, u_0, α, ξ, the last inequality yields, for $h < R$ and $k < T$,

$$\nu_{\mathcal{T},k}\left(B(0,R)\times[0,T]\right) \leqslant C_3\sqrt{h}, \tag{27.17}$$

with $C_3 = C_1 C_2$.

Comparison of T_2 and T_{20}. Using $\operatorname{div} \boldsymbol{v} = 0$, and gathering (27.14) by interfaces, we get

$$T_{20} = -\sum_{n\in\mathbb{N}} \sum_{(K,L)\in\mathcal{E}_n} \Big[\left(\left(f\left(u_K^n \top \kappa\right) - f\left(u_K^n \bot \kappa\right) \right) - \left(f\left(u_L^n \top \kappa\right) - f\left(u_L^n \bot \kappa\right) \right) \right)$$
$$\times \int_{K|L} \int_{nk}^{(n+1)k} \left(\boldsymbol{v}(x,t) \bullet \boldsymbol{n}_{K,L} \varphi(x,t) \right) \mathrm{d}\gamma(x)\,\mathrm{d}t \Big]. \tag{27.18}$$

Define, for all $K \in \mathcal{T}$, all $L \in \mathcal{N}(K)$ and all $n \in \mathbb{N}$,

$$(v\varphi)_{K,L}^{n,+} = \frac{1}{k} \int_{nk}^{(n+1)k} \int_{K|L} \left(\boldsymbol{v}(x,t) \bullet \boldsymbol{n}_{K,L} \right)^+ \varphi(x,t)\,\mathrm{d}\gamma(x)\,\mathrm{d}t$$

and

$$(v\varphi)^{n,-}_{K,L} = \frac{1}{k}\int_{nk}^{(n+1)k}\int_{K|L}\big(\boldsymbol{v}(x,t)\bullet\boldsymbol{n}_{K,L}\big)^{-}\varphi(x,t)\,\mathrm{d}\gamma(x)\,\mathrm{d}t.$$

Note that $(v\varphi)^{n,+}_{K,L} = (v\varphi)^{n,-}_{L,K}$. Then, (27.18) gives

$$\begin{aligned}
T_{20} = k\sum_{n\in\mathbb{N}}\sum_{(K,L)\in\mathcal{E}_n}\Big[&(v\varphi)^{n,+}_{K,L}\big(g\big(u^n_K\top\kappa, u^n_L\top\kappa\big) - f\big(u^n_K\top\kappa\big)\\
&- g\big(u^n_K\bot\kappa, u^n_L\bot\kappa\big) + f\big(u^n_K\bot\kappa\big)\big)\\
&- (v\varphi)^{n,-}_{L,K}\big(g\big(u^n_K\top\kappa, u^n_L\top\kappa\big) - f\big(u^n_L\top\kappa\big)\\
&- g\big(u^n_K\bot\kappa, u^n_L\bot\kappa\big) + f\big(u^n_L\bot\kappa\big)\big)\\
&- (v\varphi)^{n,-}_{K,L}\big(g\big(u^n_L\top\kappa, u^n_K\top\kappa\big) - f\big(u^n_K\top\kappa\big)\\
&- g\big(u^n_L\bot\kappa, u^n_K\bot\kappa\big) + f\big(u^n_K\bot\kappa\big)\big)\\
&+ (v\varphi)^{n,+}_{L,K}\big(g\big(u^n_L\top\kappa, u^n_K\top\kappa\big) - f\big(u^n_L\top\kappa\big)\\
&- g\big(u^n_L\bot\kappa, u^n_K\bot\kappa\big) + f\big(u^n_L\bot\kappa\big)\big)\Big].
\end{aligned} \tag{27.19}$$

Let us introduce some terms related to the difference between φ on $K\in\mathcal{T}$ and $K|L\in\mathcal{E}$,

$$r^{n,+}_{K,L} = \big|v^n_{K,L}\varphi^n_K - (v\varphi)^{n,+}_{K,L}\big|$$

and

$$r^{n,-}_{K,L} = \big|v^n_{L,K}\varphi^n_K - (v\varphi)^{n,-}_{K,L}\big|.$$

Then, from (27.12) and (27.19),

$$\begin{aligned}
|T_2 - T_{20}| \leqslant \sum_{n\in\mathbb{N}} k\sum_{(K,L)\in\mathcal{E}_n}\Big[&r^{n,+}_{K,L}\big(g\big(u^n_K\top\kappa, u^n_L\top\kappa\big) - f\big(u^n_K\top\kappa\big)\\
&+ g\big(u^n_K\bot\kappa, u^n_L\bot\kappa\big) - f\big(u^n_K\bot\kappa\big)\big)\\
&+ r^{n,-}_{L,K}\big(g\big(u^n_K\top\kappa, u^n_L\top\kappa\big) - f\big(u^n_L\top\kappa\big)\\
&+ g\big(u^n_K\bot\kappa, u^n_L\bot\kappa\big) - f\big(u^n_L\bot\kappa\big)\big)\\
&+ r^{n,-}_{K,L}\big(f\big(u^n_K\top\kappa\big) - g\big(u^n_L\top\kappa, u^n_K\top\kappa\big) + f\big(u^n_K\bot\kappa\big)\\
&- g\big(u^n_L\bot\kappa, u^n_K\bot\kappa\big)\big)\\
&+ r^{n,+}_{L,K}\big(f\big(u^n_L\top\kappa\big) - g\big(u^n_L\top\kappa, u^n_K\top\kappa\big) + f\big(u^n_L\bot\kappa\big)\\
&- g\big(u^n_L\bot\kappa, u^n_K\bot\kappa\big)\big)\Big].
\end{aligned} \tag{27.20}$$

For all $(K, L) \in \mathcal{E}_n$, the following inequality holds:

$$0 \leqslant g\big(u_K^n \top \kappa, u_L^n \top \kappa\big) - f\big(u_K^n \top \kappa\big) \leqslant \max_{u_L^n \leqslant p \leqslant q \leqslant u_K^n} \big(g(q, p) - f(q)\big),$$

more precisely, one has $g(u_K^n \top \kappa, u_L^n \top \kappa) - f(u_K^n \top \kappa) = 0$, if $\kappa \geqslant u_K^n$, and one has $g(u_K^n \top \kappa, u_L^n \top \kappa) - f(u_K^n \top \kappa) = g(q, p) - f(q)$ with $p = \kappa$ and $q = u_K^n$ if $\kappa \in [u_L^n, u_K^n]$, and with $p = u_L^n$ and $q = u_K^n$ if $\kappa \leqslant u_L^n$.

In the same way, we can assert that

$$0 \leqslant g\big(u_K^n \perp \kappa, u_L^n \perp \kappa\big) - f\big(u_K^n \perp \kappa\big) \leqslant \max_{u_L^n \leqslant p \leqslant q \leqslant u_K^n} \big(g(q, p) - f(q)\big).$$

The same analysis can be applied to the six other terms of (27.20).

To conclude the estimate on $|T_2 - T_{20}|$, there remains to estimate the two quantities $r_{K,L}^{n,\pm}$. This will be done with convenient measures applied to $|\nabla\varphi|$ and $|\varphi_t|$. To estimate $r_{K,L}^{n,+}$, e.g., one remarks that

$$r_{K,L}^{n,+} \leqslant \frac{1}{k^2 \mathrm{m}(K)} \int_{nk}^{(n+1)k} \int_{nk}^{(n+1)k} \int_K \int_{K|L} \big|\varphi(x, t) - \varphi(y, s)\big| \times \big(\boldsymbol{v}(y, s) \cdot \boldsymbol{n}_{K,L}\big)^+ \,\mathrm{d}\gamma(y)\,\mathrm{d}x\,\mathrm{d}t\,\mathrm{d}s.$$

Hence

$$r_{K,L}^{n,+} \leqslant \frac{1}{k^2 \mathrm{m}(K)} \int_{nk}^{(n+1)k} \int_{nk}^{(n+1)k} \int_K \int_{K|L} \int_0^1 \big|\nabla\varphi\big(x + \theta(y - x), t + \theta(s - t)\big) \cdot (y - x) + \varphi_t\big(x + \theta(y - x), t + \theta(s - t)\big)(s - t)\big| \times \big(\boldsymbol{v}(y, s) \cdot \boldsymbol{n}_{K,L}\big)^+ \,\mathrm{d}\theta\,\mathrm{d}\gamma(y)\,\mathrm{d}x\,\mathrm{d}t\,\mathrm{d}s$$

which yields

$$r_{K,L}^{n,+} \leqslant \frac{1}{k^2 \mathrm{m}(K)} \int_{nk}^{(n+1)k} \int_{nk}^{(n+1)k} \int_K \int_{K|L} \int_0^1 \Big(h\big|\nabla\varphi\big(x + \theta(y - x), t + \theta(s - t)\big)\big| + k\big|\varphi_t\big(x + \theta(y - x), t + \theta(s - t)\big)\big|\Big) \times \big(\boldsymbol{v}(y, s) \cdot \boldsymbol{n}_{K,L}\big)^+ \,\mathrm{d}\theta\,\mathrm{d}\gamma(y)\,\mathrm{d}x\,\mathrm{d}t\,\mathrm{d}s.$$

This leads to the definition of a measure $\mu_{K,L}^{n,+}$, given by its action on $C_c(\mathbb{R}^d \times \mathbb{R}_+)$:

$$\langle \mu_{K,L}^{n,+}, \psi \rangle = \frac{2}{k^2 \mathrm{m}(K)} \int_{nk}^{(n+1)k} \int_{nk}^{(n+1)k} \int_K \int_{K|L} \int_0^1 \Big((h + k)\psi\big(x + \theta(y - x), t + \theta(s - t)\big)\Big)\big(\boldsymbol{v}(y, s) \cdot \boldsymbol{n}_{K,L}\big)^+ \,\mathrm{d}\theta\,\mathrm{d}\gamma(y)\,\mathrm{d}x\,\mathrm{d}t\,\mathrm{d}s,$$
$$\forall \psi \in C_c\big(\mathbb{R}^d \times \mathbb{R}_+\big),$$

in order to have $2r^{n,+}_{K,L} \leqslant \langle \mu^{n,+}_{K,L}, |\nabla\varphi| + |\varphi_t|\rangle$.

We define in the same way $\mu^{n,-}_{K,L}$, changing $(\boldsymbol{v}(y,s)\bullet \boldsymbol{n}_{K,L})^+$ in $(\boldsymbol{v}(y,s)\bullet \boldsymbol{n}_{K,L})^-$. We finally define the measure $\tilde{\nu}_{\mathcal{T},k}$ by

$$\begin{aligned}\langle \tilde{\nu}_{\mathcal{T},k}, \psi\rangle = \sum_{n\in\mathbb{N}} k \sum_{(K,L)\in\mathcal{E}_n} \Big[\Big(&\max_{u^n_L\leqslant p\leqslant q\leqslant u^n_K}\big(g(q,p) - f(q)\big)\Big)\langle \mu^{n,+}_{K,L}, \psi\rangle\\ &+ \Big(\max_{u^n_L\leqslant p\leqslant q\leqslant u^n_K}\big(g(q,p) - f(p)\big)\Big)\langle \mu^{n,-}_{L,K}, \psi\rangle\\ &+ \Big(\max_{u^n_L\leqslant p\leqslant q\leqslant u^n_K}\big(f(q) - g(p,q)\big)\Big)\langle \mu^{n,-}_{K,L}, \psi\rangle\\ &+ \Big(\max_{u^n_L\leqslant p\leqslant q\leqslant u^n_K}\big(f(p) - g(p,q)\big)\Big)\langle \mu^{n,+}_{L,K}, \psi\rangle\Big].\end{aligned} \tag{27.21}$$

Since $2r^{n,\pm}_{K,L} \leqslant \langle \mu^{n,\pm}_{K,L}, |\nabla\varphi| + |\varphi_t|\rangle$, (27.20) and (27.21) leads to $|T_2 - T_{20}| \leqslant \langle \tilde{\nu}_{\mathcal{T},k}, |\nabla\varphi| + |\varphi_t|\rangle$. Therefore, setting $\mu_{\mathcal{T},k} = \nu_{\mathcal{T},k} + \tilde{\nu}_{\mathcal{T},k}$, using (27.16) and $T_1 + T_2 \leqslant 0$,

$$T_{10} + T_{20} \leqslant \int_{\mathbb{R}^d\times\mathbb{R}_+} \big(|\varphi_t(x,t)| + |\nabla\varphi(x,t)|\big)\,\mathrm{d}\mu_{\mathcal{T},k}(x,t) + \int_{\mathbb{R}^d} \varphi(x,0)\,\mathrm{d}\mu_{\mathcal{T}}(x),$$

which is (27.13) and yields (27.8).

There remains to prove (27.9).

For all $K \in \mathcal{T}$, let x_K be an arbitrary point of K. For all $K \in \mathcal{T}$, all $K \in \mathcal{N}(K)$ and all $n \in \mathbb{N}$, the supports of the measures $\mu^{n,\pm}_{K,L}$ are included in the closed set $\bar{B}(x_K, h) \cap [nk, (n+1)k]$. Furthermore,

$$\mu^{n,+}_{K,L}\big(\mathbb{R}^d \times \mathbb{R}_+\big) \leqslant 2v^n_{K,L}(h+k) \quad \text{and} \quad \mu^{n,-}_{K,L}\big(\mathbb{R}^d \times \mathbb{R}_+\big) \leqslant 2v^n_{L,K}(h+k).$$

Then, for all $R > 0$ and $T > 0$, the definition of $\mu_{\mathcal{T},k}$ (i.e. $\mu_{\mathcal{T},k} = \nu_{\mathcal{T},k} + \tilde{\nu}_{\mathcal{T},k}$)) leads to

$$\begin{aligned}&\mu_{\mathcal{T},k}\big(B(0,R)\times[0,T]\big)\\ &\quad\leqslant C_3\sqrt{h} + 2(h+k)\sum_{n=0}^{N_{T,k}} k \sum_{(K,L)\in\mathcal{E}^n_{2R}} \Big[v^n_{K,L}\Big(\max_{u^n_L\leqslant p\leqslant q\leqslant u^n_K}\big(g(q,p) - f(q)\big)\\ &\qquad + \max_{u^n_L\leqslant p\leqslant q\leqslant u^n_K}\big(g(q,p) - f(p)\big)\Big)\\ &\qquad + v^n_{L,K}\Big(\max_{u^n_L\leqslant p\leqslant q\leqslant u^n_K}\big(f(q) - g(p,q)\big) + \max_{u^n_L\leqslant p\leqslant q\leqslant u^n_K}\big(f(p) - g(p,q)\big)\Big)\Big],\end{aligned}$$

for $h < R$ and $k < T$, where $C_3\sqrt{h}$ is the bound of $\nu_{\mathcal{T},k}(B(0,R)\times[0,T])$ given in (27.17). Therefore, thanks to Lemma 25.2,

$$\mu_{\mathcal{T},k}\big(B(0,R)\times[0,T]\big)\leqslant C_3\sqrt{h}+(1+C_2)h\frac{C_4}{\sqrt{h}}=C\sqrt{h},$$

where C only depends on $\boldsymbol{v}$, g, u_0, α, ξ, R and T. The proof of Theorem 27.1 is complete.

The following theorem investigates the case of the implicit scheme.

THEOREM 27.2. *Under Assumption* 23.1, *let $\mathcal{T}$ be an admissible mesh in the sense of Definition* 24.1 *and $k>0$. Let $g\in C(\mathbb{R}^2,\mathbb{R})$ satisfy Assumption* 24.1.

Let $\{u_K^n, n\in\mathbb{N}, K\in\mathcal{T}\}$, such that $u_K^n\in[U_m,U_M]$ for all $K\in\mathcal{T}$ and $n\in\mathbb{N}$, be the solution of (24.6), (24.2) (*existence and uniqueness of such a solution are given by Proposition* 26.1). *Let $u_{\mathcal{T},k}$ be given by* (24.5). *Assume that $\boldsymbol{v}$ does not depend on t and that $u_0\in BV(\mathbb{R}^d)$.*

Then, there exist $\mu_{\mathcal{T},k}\in\mathcal{M}(\mathbb{R}^d\times\mathbb{R}_+)$ and $\mu_{\mathcal{T}}\in\mathcal{M}(\mathbb{R}^d)$ such that

$$\begin{aligned}&\int_{\mathbb{R}_+}\int_{\mathbb{R}^d}\Big(\big|u_{\mathcal{T},k}(x,t)-\kappa\big|\varphi_t(x,t)\\&\qquad+\big(f\big(u_{\mathcal{T},k}(x,t)\top\kappa\big)-f\big(u_{\mathcal{T},k}(x,t)\bot\kappa\big)\big)\boldsymbol{v}(x,t)\cdot\nabla\varphi(x,t)\Big)\,\mathrm{d}x\,\mathrm{d}t\\&\qquad+\int_{\mathbb{R}^d}\big|u_0(x)-\kappa\big|\varphi(x,0)\,\mathrm{d}x\\&\qquad\geqslant-\int_{\mathbb{R}^d\times\mathbb{R}_+}\big(\big|\varphi_t(x,t)\big|+\big|\nabla\varphi(x,t)\big|\big)\,\mathrm{d}\mu_{\mathcal{T},k}(x,t)-\int_{\mathbb{R}^d}\varphi(x,0)\,\mathrm{d}\mu_{\mathcal{T}}(x),\\&\qquad\forall\kappa\in\mathbb{R},\ \forall\varphi\in C_c^\infty\big(\mathbb{R}^d\times\mathbb{R}_+,\mathbb{R}_+\big).\end{aligned}\tag{27.22}$$

The measures $\mu_{\mathcal{T},k}$ and $\mu_{\mathcal{T}}$ verify the following properties:

(1) *For all $R>0$ and $T>0$, there exists C, only depending on $\boldsymbol{v}$, g, u_0, α, R, T such that, for $h<R$ and $k<T$,*

$$\mu_{\mathcal{T},k}\big(B(0,R)\times[0,T]\big)\leqslant C\big(k+\sqrt{h}\big).\tag{27.23}$$

(2) *The measure $\mu_{\mathcal{T}}$ is the measure of density $|u_0(\cdot)-u_{\mathcal{T},0}(\cdot)|$ with respect to the Lebesgue measure and there exists D, only depending on u_0 and α, such that*

$$\mu_{\mathcal{T}}\big(\mathbb{R}^d\big)\leqslant Dh.\tag{27.24}$$

PROOF. Similarly to the proof of Theorem 27.1, we introduce a test function $\varphi\in C_c^\infty(\mathbb{R}^d\times\mathbb{R}_+,\mathbb{R}_+)$ and a real number $\kappa\in\mathbb{R}$. We multiply (27.5) by

$$(1/\mathrm{m}(K))\int_{nk}^{(n+1)k}\int_K\varphi(x,t)\,\mathrm{d}x\,\mathrm{d}t,$$

and sum the result for all $K \in \mathcal{T}$ and $n \in \mathbb{N}$. We then define T_1 and T_2 such that $T_1 + T_2 \leqslant 0$ using equations (27.11) and (27.12) in which we replace u_K^n by u_K^{n+1} and u_L^n by u_L^{n+1}. Therefore we get (27.16), where the measure $\nu_{\mathcal{T},k}$ is such that for all $T > 0$, there exists C_1 only depending on $\boldsymbol{v}$, g, u_0 α and T, such that, for $k < T$,

$$\nu_{\mathcal{T},k}\big(\mathbb{R}^d \times [0, T]\big) \leqslant C_1 k,$$

using Lemma 26.3, which is available if $\boldsymbol{v}$ does not depend on t (and for which one needs that $u_0 \in BV(\mathbb{R}^d)$).

The treatment of T_2 is very similar to that of Theorem 27.1, replacing u_K^n by u_K^{n+1} and u_L^n by u_L^{n+1}. But, since $\boldsymbol{v}$ does not depend on t, the bounds on $r_{K,L}^{n,\pm}$ are simpler. Indeed,

$$r_{K,L}^{n,\pm} \leqslant \frac{1}{k\mathrm{m}(K)} \int_{nk}^{(n+1)k} \int_K \int_{K|L} \big|\varphi(x,t) - \varphi(y,t)\big| \big(\boldsymbol{v}(y) \cdot \boldsymbol{n}_{K,L}\big)^{\pm} \,\mathrm{d}\gamma(y)\,\mathrm{d}x\,\mathrm{d}t.$$

Now $2r_{K,L}^{n,\pm} \leqslant \langle \mu_{K,L}^{n,\pm}, |\nabla\varphi|\rangle$ where $\mu_{K,L}^{n,\pm}$ is defined by

$$\begin{aligned}\langle \mu_{K,L}^{n,\pm} \psi\rangle = {} & \frac{2}{k\mathrm{m}(K)} \int_{nk}^{(n+1)k} \int_K \int_{K|L} \int_0^1 \big(h\,\psi\big(x + \theta(y-x), t\big)\big) \\ & \times \big(\boldsymbol{v}(y) \cdot \boldsymbol{n}_{K,L}\big)^{\pm} \,\mathrm{d}\theta\,\mathrm{d}\gamma(y)\,\mathrm{d}x\,\mathrm{d}t, \quad \forall \psi \in C_c\big(\mathbb{R}^d \times \mathbb{R}_+\big).\end{aligned}$$

With this definition of $\mu_{K,L}^{n,\pm}$, the bound on $\tilde{\nu}_{\mathcal{T},k}$ (defined by (27.21), replacing u_K^n by u_K^{n+1} and u_L^n by u_L^{n+1}) becomes, thanks to Lemma 26.1,

$$\tilde{\nu}_{\mathcal{T},k}\big(B(0,R) \times [0,T]\big) \leqslant C_2\sqrt{h},$$

for $h < R$ and $k < T$, where C_2 only depends on $\boldsymbol{v}$, g, u_0, α, R and T.

Hence, defining (as in Theorem 27.1) $\mu_{\mathcal{T},k} = \nu_{\mathcal{T},k} + \tilde{\nu}_{\mathcal{T},k}$, for all $R > 0$ and all $T > 0$ there exists C, only depending on $\boldsymbol{v}$, g, u_0, α, R, T such that, for $h < R$ and $k < T$,

$$\mu_{\mathcal{T},k}\big(B(0,R) \times [0,T]\big) \leqslant C\big(k + \sqrt{h}\big),$$

which is (27.23) and concludes the proof of Theorem 27.2. □

REMARK 27.2. In the case where $\boldsymbol{v}$ depends on t, Lemma 26.3 cannot be used. However, it is easy to show (the proof follows that of Theorem 27.1) that Theorem 27.2 is true if (27.23) is replaced by

$$\mu_{\mathcal{T},k}\big(B(0,R) \times [0,T]\big) \leqslant C\left(\frac{k}{\sqrt{h}} + \sqrt{h}\right), \tag{27.25}$$

which leads to the result given in Remark 29.2. The estimate (27.25) may be obtained without assuming that $u_0 \in BV(\mathbb{R}^d)$ (it is sufficient that $u_0 \in L^\infty(\mathbb{R}^d)$).

For the sake of completeness we now prove a lemma which gives the bound on the measure $\mu_{\mathcal{T}}$ in the two last theorems.

LEMMA 27.3. *Let $\mathcal{T}$ be an admissible mesh in the sense of Definition* 24.1 *and let $u \in BV(\mathbb{R}^d)$ (see Definition* 21.17*). For $K \in \mathcal{T}$, let u_K be the mean value of u over K. Define $u_{\mathcal{T}}$ by $u_{\mathcal{T}}(x) = u_K$ for a.e. $x \in K$, for all $K \in \mathcal{T}$. Then,*

$$\|u - u_{\mathcal{T}}\|_{L^1(\mathbb{R}^d)} \leqslant \frac{C}{\alpha^2} h |u|_{BV(\mathbb{R}^d)}, \tag{27.26}$$

where C only depends on the space dimension ($d = 1$, 2 or 3).

PROOF. The proof is very similar to that of Lemma 26.2 and we will mainly refer to the proof of Lemma 26.2.

First, remark that if (27.26) holds for all $u \in BV(\mathbb{R}^d) \cap C^1(\mathbb{R}^d, \mathbb{R})$ then (27.26) holds for all $u \in BV(\mathbb{R}^d)$. Indeed, let $u \in BV(\mathbb{R}^d)$, it is proven in Step 1 of the proof of Lemma 26.2 that there exists a sequence $(u_n)_{n \in \mathbb{N}} \subset C^\infty(\mathbb{R}^d, \mathbb{R})$ such that $u_n \to u$ in $L^1_{\mathrm{loc}}(\mathbb{R}^d)$, as $n \to \infty$, and $\|u_n\|_{BV(\mathbb{R}^d)} \leqslant \|u\|_{BV(\mathbb{R}^d)}$ for all $n \in \mathbb{N}$. One may also assume, up to a subsequence, that $u_n \to u$ a.e. on $\mathbb{R}^d$. Then, if (27.26) is true with u_n instead of u, passing to the limit in (27.26) (for u_n) as $n \to \infty$ leads to (27.26) (for u) thanks to Fatou's lemma.

Let us now prove (27.26) if $u \in BV(\mathbb{R}^d) \cap C^1(\mathbb{R}^d, \mathbb{R})$ (this concludes the proof of Lemma 27.3). Since $u \in C^1(\mathbb{R}^d, \mathbb{R})$,

$$|u|_{BV(\mathbb{R}^d)} = \big\| (|\nabla u|) \big\|_{L^1(\mathbb{R}^d)};$$

hence we shall prove (27.26) with $\|(|\nabla u|)\|_{L^1(\mathbb{R}^d)}$ instead of $|u|_{BV(\mathbb{R}^d)}$.

For $K \in \mathcal{T}$,

$$\int_K \big|u(x) - u_K\big| \,\mathrm{d}x \leqslant \frac{1}{\mathrm{m}(K)} \int_K \left(\int_K \big|u(x) - u(y)\big| \,\mathrm{d}x \right) \mathrm{d}y.$$

Then, following the lines of Step 2 of Lemma 26.2,

$$\int_K \big|u(x) - u_K\big| \,\mathrm{d}x \leqslant \frac{1}{\mathrm{m}(K)} h \int_{B(0,h)} \left(\int_0^1 \int_K \big|\nabla u(y + tz)\big| \,\mathrm{d}y \,\mathrm{d}t \right) \mathrm{d}z. \tag{27.27}$$

For all $K \in \mathcal{T}$, let x_K be an arbitrary point of K.

Then, changing the variable y in $\xi = y + tz$ (for all fixed $z \in K$ and $t \in (0, 1)$) in (27.27),

$$\int_K \big|u(x) - u_K\big| \,\mathrm{d}x \leqslant \frac{1}{\mathrm{m}(K)} h \int_{B(0,h)} \left(\int_0^1 \int_{B(x_K, 2h)} \big|\nabla u(\xi)\big| \,\mathrm{d}\xi \,\mathrm{d}t \right) \mathrm{d}z,$$

which yields, since $\mathcal{T}$ is an admissible mesh in the sense of Definition 24.1,

$$\int_K |u(x) - u_K| \, dx \leqslant \frac{1}{\alpha h^d} m(B(0,h))h \int_{B(x_K,2h)} |\nabla u(\xi)| \, d\xi.$$

Therefore there exists C_1, only depending on the space dimension, such that

$$\int_K |u(x) - u_K| \, dx \leqslant \frac{C_1}{\alpha} h \int_{B(x_K,2h)} |\nabla u(\xi)| \, d\xi, \quad \forall K \in \mathcal{T}. \tag{27.28}$$

As in Lemma 26.2, for a fixed $M \in \mathcal{T}$, the number of $K \in \mathcal{T}$ such that $M \cap B(x_K, 2h) \neq \emptyset$ is less or equal to $m(B(0,4h))/(\alpha h^d)$ that is less or equal to C_2/α where C_2 only depends on the space dimension.

Then, summing (27.28) over $K \in \mathcal{T}$ leads to

$$\sum_{K\in\mathcal{T}} \int_K |u(x) - u_K| \, dx \leqslant \frac{C_1 C_2}{\alpha^2} h \sum_{M\in\mathcal{T}} \int_M |\nabla u(\xi)| \, d\xi = \frac{C_1 C_2}{\alpha^2} h \| (|\nabla u|) \|_{L^1(\mathbb{R}^d)},$$

that is (27.26) with $C = C_1 C_2$. □

28. Convergence of the scheme

This section is devoted to the proof of the existence and uniqueness of the entropy weak solution and of the convergence of the approximate solution towards the entropy weak solution as the mesh size and time step tend to 0. This proof will be performed in two steps. We first prove in Section 28.1 the convergence of the approximate solution towards an entropy process solution which is defined in Definition 28.1 below (note that the convergence also yields the existence of an entropy process solution).

DEFINITION 28.1. A function μ is an entropy process solution to problem (23.1)–(23.2) if μ satisfies

$$\begin{cases} \mu \in L^\infty(\mathbb{R}^d \times \mathbb{R}_+^\star \times (0,1)), \\ \displaystyle\int_{\mathbb{R}^d} \int_0^{+\infty} \int_0^1 \big(\eta(\mu(x,t,\alpha))\varphi_t(x,t) + \Phi(\mu(x,t,\alpha)) \\ \quad \times \boldsymbol{v}(x,t) \cdot \nabla\varphi(x,t)\big) \, d\alpha \, dt \, dx \\ \quad + \displaystyle\int_{\mathbb{R}^d} \eta(u_0(x))\varphi(x,0) \, dx \geqslant 0, \\ \quad \text{for any } \varphi \in C_c^1(\mathbb{R}^d \times \mathbb{R}_+, \mathbb{R}_+), \\ \quad \text{for any convex function } \eta \in C^1(\mathbb{R},\mathbb{R}), \text{ and} \\ \quad \Phi \in C^1(\mathbb{R},\mathbb{R}) \text{ such that } \Phi' = f'\eta'. \end{cases} \tag{28.1}$$

REMARK 28.1. From an entropy weak solution u to problem (23.1)–(23.2), one may easily construct an entropy process solution to problem (23.1)–(23.2) by setting

$\mu(x,t,\alpha) = u(x,t)$ for a.e. $(x,t,\alpha) \in \mathbb{R}^d \times \mathbb{R}_+^\star \times (0,1)$. Reciprocally, if μ is an entropy process solution to problem (23.1)–(23.2) such that there exists $u \in L^\infty(\mathbb{R}^d \times \mathbb{R}_+^\star)$ such that $\mu(x,t,\alpha) = u(x,t)$, for a.e. $(x,t,\alpha) \in \mathbb{R}^d \times \mathbb{R}_+^\star \times (0,1)$, then u is an entropy weak solution to problem (23.1)–(23.2).

In Section 28.2, we show the uniqueness of the entropy process solution, which, thanks to Remark 28.1, also yields the existence and uniqueness of the entropy weak solution. This allows us to state and prove, in Section 28.3, the convergence of the approximate solution towards the entropy weak solution.

We now give a useful characterization of an entropy process solution in terms of Krushkov's entropies (as for the entropy weak solution).

PROPOSITION 28.1. *A function μ is an entropy process solution of problem* (23.1)–(23.2) *if and only if,*

$$\begin{cases} \mu \in L^\infty\big(\mathbb{R}^d \times \mathbb{R}_+^\star \times (0,1)\big), \\ \displaystyle\int_{\mathbb{R}^d}\int_0^{+\infty}\int_0^1 \Big(\big|\mu(x,t,\alpha)-\kappa\big|\varphi_t(x,t) + \Phi\big(\mu(x,t,\alpha),\kappa\big) \\ \quad \times \boldsymbol{v}(x,t)\cdot\nabla\varphi(x,t)\Big)\,\mathrm{d}\alpha\,\mathrm{d}t\,\mathrm{d}x \\ \quad + \displaystyle\int_{\mathbb{R}^d}\big|u_0(x)-\kappa\big|\varphi(x,0)\,\mathrm{d}x \geqslant 0, \\ \forall\kappa \in \mathbb{R},\ \forall\varphi \in C_c^1\big(\mathbb{R}^d \times \mathbb{R}_+, \mathbb{R}_+\big), \end{cases} \tag{28.2}$$

where we set $\Phi(a,b) = f(a\top b) - f(a\bot b)$, for all a, $b \in \mathbb{R}$.

PROOF. The proof of this result is similar to the case of classical entropy weak solutions. The characterization (28.2) can be obtained from (28.1), by using regularizations of the function $|\bullet - \kappa|$. Conversely, (28.1) may be obtained from (28.2) by approximating any convex function $\eta \in C^1(\mathbb{R},\mathbb{R})$ by functions of the form: $\eta_n(\bullet) = \sum_{i=1}^n \alpha_i^{(n)}|\bullet - \kappa_i^{(n)}|$, with $\alpha_i^{(n)} \geqslant 0$. □

28.1. Convergence towards an entropy process solution

Let $\alpha > 0$ and $0 < \xi < 1$. Let $(\mathcal{T}_m, k_m)_{m\in\mathbb{N}}$ be a sequence of admissible meshes in the sense of Definition 24.1 and time steps. Note that $\mathcal{T}_m$ is admissible with α independent of m. Assume that k_m satisfies (24.3), for $\mathcal{T} = \mathcal{T}_m$ and $k = k_m$, and that $\text{size}(\mathcal{T}_m) \to 0$ as $m \to \infty$.

By Lemma 25.1, the sequence $(u_{\mathcal{T}_m,k_m})_{m\in\mathbb{N}}$ of approximate solutions defined by the finite volume scheme (24.2) and (24.4), with $\mathcal{T} = \mathcal{T}_m$ and $k = k_m$, is bounded in $L^\infty(\mathbb{R}^d \times \mathbb{R}_+^\star)$; therefore, there exists $\mu \in L^\infty(\mathbb{R}^d \times \mathbb{R}_+^\star \times (0,1))$ such that $u_{\mathcal{T}_m,k_m}$ converges, as m tends to ∞, towards μ in the nonlinear weak-$\star$ sense (see Definition 30.1

and Proposition 30.1), i.e.:

$$\lim_{m\to\infty}\int_{\mathbb{R}^d}\int_{\mathbb{R}_+}\theta\big(u_{\mathcal{T}_m,k_m}(x,t)\big)\varphi(x,t)\,\mathrm{d}t\,\mathrm{d}x$$
$$=\int_{\mathbb{R}^d}\int_{\mathbb{R}_+}\int_0^1\theta\big(\mu(x,t,\alpha)\big)\varphi(x,t)\,\mathrm{d}\alpha\,\mathrm{d}t\,\mathrm{d}x,$$
$$\forall\varphi\in L^1\big(\mathbb{R}\times\mathbb{R}_+^\star\big),\ \forall\theta\in C(\mathbb{R},\mathbb{R}).\tag{28.3}$$

Taking for θ, in (28.3), the Krushkov entropies (namely $\theta=|\bullet-\kappa|$, for all $\kappa\in\mathbb{R}$) and the associated functions defining the entropy fluxes (namely $\theta=f(\bullet,\kappa)=f(\bullet\top\kappa)-f(\bullet\bot\kappa)$) and using Theorem 27.1 (that is passing to the limit, as $m\to\infty$, in (27.8) written with $u_{\mathcal{T},k}=u_{\mathcal{T}_m,k_m}$) yields that μ is an entropy process solution. Hence the following result holds:

PROPOSITION 28.2. *Under Assumptions* 23.1, *let* $\alpha>0$ *and* $0<\xi<1$. *Let* $(\mathcal{T}_m,k_m)_{m\in\mathbb{N}}$ *be a sequence of admissible meshes in the sense of Definition* 24.1 *and time steps. Note that* $\mathcal{T}_m$ *is admissible with* α *independent of* m. *Assume that* k_m *satisfy* (24.3), *for* $\mathcal{T}=\mathcal{T}_m$ *and* $k=k_m$, *and that* $\mathrm{size}(\mathcal{T}_m)\to 0$ *as* $m\to\infty$.

Then there exists a subsequence, still denoted by $(\mathcal{T}_m,k_m)_{m\in\mathbb{N}}$, *and a function* $\mu\in L^\infty(\mathbb{R}^d\times\mathbb{R}_+^\star\times(0,1))$ *such that*

(1) *the approximate solution defined by* (24.4), (24.2) *and* (24.5) *with* $\mathcal{T}=\mathcal{T}_m$ *and* $k=k_m$, *that is* $u_{\mathcal{T}_m,k_m}$, *converges towards* μ *in the nonlinear weak-$\star$ sense, i.e.* (28.3) *holds,*

(2) μ *is an entropy process solution of* (23.1)–(23.2).

REMARK 28.2. The same theorem can be proved for the implicit scheme without condition (24.3) (and thus without ξ).

REMARK 28.3. Note that a consequence of Proposition 28.2 is the existence of an entropy process solution to problem (23.1)–(23.2).

28.2. Uniqueness of the entropy process solution

In order to show the uniqueness of an entropy process solution, we shall use the characterization of an entropy process solution given in Proposition 28.1.

THEOREM 28.1. *Under Assumption* 23.1, *the entropy process solution* μ *of problem* (23.1), (23.2), *as defined in Definition* 28.1, *is unique. Moreover, there exists a function* $u\in L^\infty(\mathbb{R}^d\times\mathbb{R}_+^\star)$ *such that* $u(x,t)=\mu(x,t,\alpha)$, *for a.e.* $(x,t,\alpha)\in\mathbb{R}^d\times\mathbb{R}_+^\star\times(0,1)$. (*Hence, with Proposition* 28.2 *and Remark* 28.1, *there exists a unique entropy weak solution to problem* (23.1)–(23.2).)

PROOF. Let μ and ν be two entropy process solutions to problem (23.1)–(23.2). Then, one has $\mu \in L^\infty(\mathbb{R}^d \times \mathbb{R}_+^\star \times (0,1))$, $\nu \in L^\infty(\mathbb{R}^d \times \mathbb{R}_+^\star \times (0,1))$ and

$$\begin{aligned}&\int_{\mathbb{R}^d}\int_0^{+\infty}\int_0^1 \big(\big|\mu(x,t,\alpha)-\kappa\big|\varphi_t(x,t)\\&\quad+\big(f\big(\mu(x,t,\alpha)\top\kappa\big)-f\big(\mu(x,t,\alpha)\bot\kappa\big)\big)\boldsymbol{v}(x,t)\cdot\nabla\varphi(x,t)\big)\,\mathrm{d}\alpha\,\mathrm{d}t\,\mathrm{d}x\\&\quad+\int_{\mathbb{R}^d}\big|u_0(x)-\kappa\big|\varphi(x,0)\,\mathrm{d}x\geqslant 0,\\&\forall\kappa\in\mathbb{R},\ \forall\varphi\in C_c^1\big(\mathbb{R}^d\times\mathbb{R}_+,\mathbb{R}_+\big),\end{aligned}\tag{28.4}$$

$$\begin{aligned}&\int_{\mathbb{R}^d}\int_0^{+\infty}\int_0^1 \big(\big|\nu(y,s,\beta)-\kappa\big|\varphi_s(y,s)\\&\quad+\big(f\big(\nu(y,s,\beta)\top\kappa\big)-f\big(\nu(y,s,\beta)\bot\kappa\big)\big)\boldsymbol{v}(y,s)\cdot\nabla\varphi(y,s)\big)\,\mathrm{d}\beta\,\mathrm{d}s\,\mathrm{d}y\\&\quad+\int_{\mathbb{R}^d}\big|u_0(y)-\kappa\big|\varphi(y,0)\,\mathrm{d}y\geqslant 0,\\&\forall\kappa\in\mathbb{R},\ \forall\varphi\in C_c^1\big(\mathbb{R}^d\times\mathbb{R}_+,\mathbb{R}_+\big).\end{aligned}\tag{28.5}$$

The proof of Theorem 28.1 contains 2 steps. In Step 1, it is proven that

$$\begin{aligned}&\int_0^1\int_0^1\int_{\mathbb{R}_+}\int_{\mathbb{R}^d}\big[\big|\mu(x,t,\alpha)-\nu(x,t,\beta)\big|\psi_t(x,t)\\&\quad+\big(f\big(\mu(x,t,\alpha)\top\nu(x,t,\beta)\big)-f\big(\mu(x,t,\alpha)\bot\nu(x,t,\beta)\big)\big)\\&\quad\times\boldsymbol{v}(x,t)\cdot\nabla\psi(x,t)\big]\,\mathrm{d}x\,\mathrm{d}t\,\mathrm{d}\alpha\,\mathrm{d}\beta\geqslant 0,\\&\forall\psi\in C_c^1\big(\mathbb{R}^d\times\mathbb{R}_+,\mathbb{R}_+\big).\end{aligned}\tag{28.6}$$

In Step 2, it is proven that $\mu(x,t,\alpha)=\nu(x,t,\beta)$ for a.e. $(x,t,\alpha,\beta)\in\mathbb{R}^d\times\mathbb{R}_+^\star\times(0,1)\times(0,1)$. We then deduce that there exists $u\in L^\infty(\mathbb{R}^d\times\mathbb{R}_+^\star)$ such that $\mu(x,t,\alpha)=u(x,t)$ for a.e. $(x,t,\alpha)\in\mathbb{R}^d\times\mathbb{R}_+^\star\times(0,1)$ (therefore u is necessarily the unique entropy weak solution to (23.1)–(23.2)).

Step 1 (*Proof of relation* (28.6)). In order to prove relation (28.6), a sequence of mollifiers in $\mathbb{R}$ and $\mathbb{R}^d$ is introduced.

Let $\rho\in C_c^\infty(\mathbb{R}^d,\mathbb{R}_+)$ and $\bar\rho\in C_c^\infty(\mathbb{R},\mathbb{R}_+)$ be such that

$$\begin{aligned}&\big\{x\in\mathbb{R}^d;\ \rho(x)\neq 0\big\}\subset\big\{x\in\mathbb{R}^d;\ |x|\leqslant 1\big\},\\&\big\{x\in\mathbb{R};\ \bar\rho(x)\neq 0\big\}\subset[-1,0]\end{aligned}\tag{28.7}$$

and

$$\int_{\mathbb{R}^d}\rho(x)\,\mathrm{d}x=1,\qquad\int_{\mathbb{R}}\bar\rho(x)\,\mathrm{d}x=1.$$

For $n \in \mathbb{N}^\star$, define $\rho_n = n^d \rho(nx)$ for all $x \in \mathbb{R}^d$ and $\bar{\rho}_n = n\bar{\rho}(nx)$ for all $x \in \mathbb{R}$.

Let $\psi \in C_c^1(\mathbb{R}^d \times \mathbb{R}_+, \mathbb{R}_+)$. For $(y, s, \beta) \in \mathbb{R}^d \times \mathbb{R}_+ \times (0, 1)$, let us take, in (28.4), $\varphi(x, t) = \psi(x, t)\rho_n(x - y)\bar{\rho}_n(t - s)$ and $\kappa = \nu(y, s, \beta)$. Then, integrating the result over $\mathbb{R}^d \times \mathbb{R}_+ \times (0, 1)$ leads to

$$A_1 + A_2 + A_3 + A_4 + A_5 \geqslant 0, \tag{28.8}$$

where

$$\begin{aligned}
A_1 = {} & \int_0^1 \int_0^1 \int_0^\infty \int_{\mathbb{R}^d} \int_0^\infty \int_{\mathbb{R}^d} \Big[\big| \mu(x, t, \alpha) - \nu(y, s, \beta) \big| \psi_t(x, t) \rho_n(x - y) \\
& \times \bar{\rho}_n(t - s) \Big] \, dx \, dt \, dy \, ds \, d\alpha \, d\beta, \\
A_2 = {} & \int_0^1 \int_0^1 \int_0^\infty \int_{\mathbb{R}^d} \int_0^\infty \int_{\mathbb{R}^d} \Big[\big| \mu(x, t, \alpha) - \nu(y, s, \beta) \big| \psi(x, t) \rho_n(x - y) \\
& \times \bar{\rho}_n'(t - s) \Big] \, dx \, dt \, dy \, ds \, d\alpha \, d\beta, \\
A_3 = {} & \int_0^1 \int_0^1 \int_0^\infty \int_{\mathbb{R}^d} \int_0^\infty \int_{\mathbb{R}^d} \Big[\Big(f\big(\mu(x, t, \alpha) \top \nu(y, s, \beta)\big) \\
& - f\big(\mu(x, t, \alpha) \bot \nu(y, s, \beta)\big) \Big) \\
& \times \boldsymbol{v}(x, t) \cdot \nabla \psi(x, t) \rho_n(x - y) \bar{\rho}_n(t - s) \Big] \, dx \, dt \, dy \, ds \, d\alpha \, d\beta, \\
A_4 = {} & \int_0^1 \int_0^1 \int_0^\infty \int_{\mathbb{R}^d} \int_0^\infty \int_{\mathbb{R}^d} \Big[\Big(f\big(\mu(x, t, \alpha) \top \nu(y, s, \beta)\big) \\
& - f\big(\mu(x, t, \alpha) \bot \nu(y, s, \beta)\big) \Big) \\
& \times \boldsymbol{v}(x, t) \cdot \nabla \rho_n(x - y) \psi(x, t) \bar{\rho}_n(t - s) \Big] \, dx \, dt \, dy \, ds \, d\alpha \, d\beta
\end{aligned}$$

and

$$A_5 = \int_0^1 \int_{\mathbb{R}^d} \int_0^\infty \int_{\mathbb{R}^d} \big| u_0(x) - \nu(y, s, \beta) \big| \psi(x, 0) \rho_n(x - y) \bar{\rho}_n(-s) \, dy \, ds \, dx \, d\beta.$$

Passing to the limit in (28.8) as $n \to \infty$ (using (28.5) for the study of $A_2 + A_4$ and A_5) will give (28.6).

Let us first consider A_1 and A_3. Note that, using (28.7),

$$\int_{\mathbb{R}^d} \int_0^\infty \rho_n(x - y) \bar{\rho}_n(t - s) \, ds \, dy = 1, \quad \forall x \in \mathbb{R}^d, \ \forall t \in \mathbb{R}_+.$$

Then,

$$\left| A_1 - \int_0^1 \int_0^1 \int_{\mathbb{R}_+} \int_{\mathbb{R}^d} \Big[\big| \mu(x, t, \alpha) - \nu(x, t, \beta) \big| \psi_t(x, t) \Big] \, dx \, dt \, d\alpha \, d\beta \right|$$

$$\leqslant \int_0^1 \int_0^\infty \int_{\mathbb{R}^d} \int_0^\infty \int_{\mathbb{R}^d} \big[|\nu(x,t,\beta) - \nu(y,s,\beta)| |\psi_t(x,t)| \\ \times \rho_n(x-y)\bar{\rho}_n(t-s)\big] \,\mathrm{d}x\,\mathrm{d}t\,\mathrm{d}y\,\mathrm{d}s\,\mathrm{d}\beta \\ \leqslant \|\psi_t\|_{L^\infty(\mathbb{R}^d \times \mathbb{R}_+^\star)} \varepsilon(n,S),$$

with $S = \{(x,t) \in \mathbb{R}^d \times \mathbb{R}_+;\ \psi(x,t) \neq 0\}$ and

$$\varepsilon(n,S) = \sup\left\{ \big\| \nu - \nu(\bullet + \eta, \bullet + \tau, \bullet)\big\|_{L^1(S\times(0,1))};\ |\eta| \leqslant \frac{1}{n},\ 0 \leqslant \tau \leqslant \frac{1}{n} \right\}.$$

Since $\nu \in L^1_{\mathrm{loc}}(\mathbb{R}^d \times \mathbb{R}_+ \times [0,1])$ and S is bounded, one has $\varepsilon(n,S) \to 0$ as $n \to \infty$. Hence,

$$A_1 \to \int_0^1 \int_0^1 \int_{\mathbb{R}_+} \int_{\mathbb{R}^d} \big[|\mu(x,t,\alpha) - \nu(x,t,\beta)| \psi_t(x,t)\big] \,\mathrm{d}x\,\mathrm{d}t\,\mathrm{d}\alpha\,\mathrm{d}\beta, \\ \text{as } n \to \infty. \tag{28.9}$$

Similarly, let M be the Lipschitz constant of f on $[-D, D]$ where $D = \max\{\|\mu\|_\infty, \|\nu\|_\infty\}$, with $\|\bullet\|_\infty = \|\bullet\|_{L^\infty(\mathbb{R}^d \times \mathbb{R}_+^\star \times (0,1))}$,

$$\bigg| A_3 - \int_0^1 \int_0^1 \int_{\mathbb{R}_+} \int_{\mathbb{R}^d} \big(f\big(\mu(x,t,\alpha) \top \nu(x,t,\beta)\big) - f\big(\mu(x,t,\alpha) \bot \nu(x,t,\beta)\big)\big) \\ \times \boldsymbol{v}(x,t) \bullet \nabla\psi(x,t) \,\mathrm{d}x\,\mathrm{d}t\,\mathrm{d}\alpha\,\mathrm{d}\beta \bigg| \\ \leqslant 2MV \big\| \big(|\nabla\psi|\big)\big\|_{L^\infty(\mathbb{R}^d \times \mathbb{R}_+^\star)} \varepsilon(n,S),$$

which yields

$$A_3 \to \int_0^1 \int_0^1 \int_{\mathbb{R}_+} \int_{\mathbb{R}^d} \big(f\big(\mu(x,t,\alpha) \top \nu(x,t,\beta)\big) - f\big(\mu(x,t,\alpha) \bot \nu(x,t,\beta)\big)\big) \\ \times \boldsymbol{v}(x,t) \bullet \nabla\psi(x,t) \,\mathrm{d}x\,\mathrm{d}t\,\mathrm{d}\alpha\,\mathrm{d}\beta, \quad \text{as } n \to \infty. \tag{28.10}$$

Let us now consider $A_2 + A_4$.

For $(x,t,\alpha) \in \mathbb{R}^d \times \mathbb{R}_+ \times (0,1)$, let us take $\varphi(y,s) = \psi(x,t)\rho_n(x-y)\bar{\rho}_n(t-s)$ and $\kappa = \mu(x,t,\alpha)$ in (28.5). Integrating the result over $\mathbb{R}^d \times \mathbb{R}_+ \times (0,1)$ leads to

$$-A_2 - B_4 \geqslant 0, \tag{28.11}$$

with

$$A_4 - B_4 = \int_0^1 \int_0^1 \int_0^\infty \int_{\mathbb{R}^d} \int_0^\infty \int_{\mathbb{R}^d} \big[\big(f\big(\mu(x,t,\alpha) \top \nu(y,s,\beta)\big)$$

$$- f\big(\mu(x,t,\alpha)\bot\nu(y,s,\beta)\big)\big)$$
$$\times \big(\boldsymbol{v}(x,t) - \boldsymbol{v}(y,s)\big) \bullet \nabla\rho_n(x-y)\psi(x,t)\bar{\rho}_n(t-s)\big]\,\mathrm{d}x\,\mathrm{d}t\,\mathrm{d}y\,\mathrm{d}s\,\mathrm{d}\alpha\,\mathrm{d}\beta.$$

Note that $B_4 = A_4$ if $\boldsymbol{v}$ is constant (and one directly obtains (28.13) below). In the general case, in order to prove that $A_4 - B_4 \to 0$ as $n \to \infty$ (which then gives (28.13)), let us remark that, using $\operatorname{div} \boldsymbol{v} = 0$,

$$\begin{aligned}&\int_0^1\int_0^1\int_0^\infty\int_{\mathbb{R}^d}\int_0^\infty\int_{\mathbb{R}^d}\big[\big(f\big(\mu(x,t,\alpha)\top\nu(x,t,\beta)\big)\\&\quad - f\big(\mu(x,t,\alpha)\bot\nu(x,t,\beta)\big)\big)\big(\boldsymbol{v}(x,t)-\boldsymbol{v}(y,s)\big)\bullet\nabla\rho_n(x-y)\psi(x,t)\\&\quad\times\bar{\rho}_n(t-s)\big]\,\mathrm{d}x\,\mathrm{d}t\,\mathrm{d}y\,\mathrm{d}s\,\mathrm{d}\alpha\,\mathrm{d}\beta = 0.\end{aligned} \tag{28.12}$$

Indeed, the latter equality follows from an integration by parts for the variable $y \in \mathbb{R}^d$. Then, subtracting the left-hand side of (28.12) to $A_4 - B_4$ and using the regularity of $\boldsymbol{v}$, there exists C_1, only depending on M, $\boldsymbol{v}$ and ψ, such that $|A_4 - B_4| \leqslant C_1\varepsilon(n,S)$. This gives $A_4 - B_4 \to 0$ as $n \to \infty$ and, thanks to (28.11),

$$\limsup_{n\to\infty}(A_2 + A_4) \leqslant 0. \tag{28.13}$$

Finally, let us consider A_5.

For $x \in \mathbb{R}^d$, let us take $\varphi(y,s) = \psi(x,0)\rho_n(x-y)\int_s^\infty \bar{\rho}_n(-\tau)\,\mathrm{d}\tau$ and $\kappa = u_0(x)$ in (28.5). Integrating the resulting inequality with respect to $x \in \mathbb{R}^d$ gives

$$-A_5 + B_{5\mathrm{a}} + B_{5\mathrm{b}} \geqslant 0, \tag{28.14}$$

with

$$\begin{aligned}B_{5\mathrm{a}} &= -\int_0^1\int_0^\infty\int_{\mathbb{R}^d}\int_{\mathbb{R}^d}\int_s^\infty \big(f\big(\nu(y,s,\beta)\top u_0(x)\big) - f\big(\nu(y,s,\beta)\bot u_0(x)\big)\big)\\&\qquad\times \boldsymbol{v}(y,s)\bullet\nabla\rho_n(x-y)\psi(x,0)\bar{\rho}_n(-\tau)\,\mathrm{d}\tau\,\mathrm{d}y\,\mathrm{d}x\,\mathrm{d}s\,\mathrm{d}\beta,\\B_{5\mathrm{b}} &= \int_{\mathbb{R}^d}\int_{\mathbb{R}^d}\psi(x,0)\rho_n(x-y)\big|u_0(x)-u_0(y)\big|\,\mathrm{d}y\,\mathrm{d}x.\end{aligned}$$

Let $S_0 = \{x \in \mathbb{R}^d;\ \psi(x,0) \neq 0\}$ and

$$\varepsilon_0(n,S_0) = \sup\left\{\int_{S_0}\big|u_0(x) - u_0(x+\eta)\big|\,\mathrm{d}x;\ |\eta| \leqslant \frac{1}{n}\right\},$$

so that $B_{5\mathrm{b}} \leqslant \|\psi(\bullet,0)\|_{L^\infty(\mathbb{R}^d)}\varepsilon_0(n,S_0)$.

Since $u_0 \in L^1_{\mathrm{loc}}(\mathbb{R}^d)$ and since S_0 is bounded, one has $\varepsilon_0(n,S_0) \to 0$ as $n \to \infty$. Then, $B_{5\mathrm{b}} \to 0$ as $n \to \infty$.

Let us now prove that $B_{5a} \to 0$ as $n \to \infty$ (then, (28.14) will give (28.15) below). Note that $B_{5a} = -B_{5c} + (B_{5a} + B_{5c})$ with

$$B_{5c} = \int_0^1 \int_0^\infty \int_{\mathbb{R}^d} \int_{\mathbb{R}^d} \int_s^\infty \big(f\big(\nu(y,s,\beta) \top u_0(y)\big) - f\big(\nu(y,s,\beta) \bot u_0(y)\big)\big) \\ \times \boldsymbol{v}(y,s) \cdot \nabla \rho_n(x-y) \psi(x,0) \bar{\rho}_n(-\tau)\, d\tau\, dy\, dx\, ds\, d\beta.$$

Integrating by parts for the x variable yields

$$B_{5c} = \int_0^1 \int_0^\infty \int_{\mathbb{R}^d} \int_{\mathbb{R}^d} \int_s^\infty \big(f\big(\nu(y,s,\beta) \top u_0(y)\big) - f\big(\nu(y,s,\beta) \bot u_0(y)\big)\big) \\ \times \boldsymbol{v}(y,s) \cdot \nabla \psi(x,0) \rho_n(x-y) \bar{\rho}_n(-\tau)\, d\tau\, dy\, dx\, ds\, d\beta.$$

Noting that the integration with respect to s is reduced to $[0, 1/n]$, $B_{5c} \to 0$ as $n \to \infty$.

There remains to study $B_{5a} + B_{5c}$. Noting that $|f(a \top b) - f(a \top c)| \leqslant \overline{M}|b-c|$ and $|f(a \bot b) - f(a \bot c)| \leqslant \overline{M}|b-c|$ if $b, c \in [-\overline{D}, \overline{D}]$, where $\overline{D} = \|u_0\|_{L^\infty(\mathbb{R}^d)}$ and $\overline{M}$ is the Lipschitz constant to f on $[-\overline{D}, \overline{D}]$,

$$|B_{5a} + B_{5c}| \leqslant 2\overline{M}V \int_0^\infty \int_{\mathbb{R}^d} \int_{\mathbb{R}^d} \int_s^\infty \big|u_0(x) - u_0(y)\big| \big|\nabla \rho_n(x-y)\big| \\ \times \psi(x,0) \bar{\rho}_n(-\tau)\, d\tau\, dy\, dx\, ds,$$

which yields the existence of C_2, only depending on $\overline{M}$, V and ψ, such that

$$|B_{5a} + B_{5c}| \leqslant C_2 \int_0^{1/n} \int_{S_0} \int_{B(0,1/n)} \big|u_0(x) - u_0(x-z)\big| n^{d+1}\, dz\, dx\, ds.$$

Therefore, $|B_{5a} + B_{5c}| \leqslant C_3 \varepsilon_0(n, S_0)$, with some C_3 only depending on $\overline{M}$, V and ψ. Since $\varepsilon_0(n, S_0) \to 0$ as $n \to \infty$, one deduces $|B_{5a} + B_{5c}| \to 0$ as $n \to \infty$. Hence, $B_{5a} \to 0$ as $n \to \infty$ and (28.14) yields

$$\limsup_{n \to \infty} A_5 \leqslant 0. \tag{28.15}$$

It is now possible to conclude Step 1. Passing to the limit as $n \to \infty$ in (28.8) and using (28.9), (28.10), (28.13) and (28.15) yields (28.6).

Step 2 (*Proof of* $\mu = \nu$ *and conclusion*). Let $R > 0$ and $T > 0$. One sets $\omega = VM$ (recall that V is given in Assumption 23.1 and that M is given in Step 1).

Let $\varphi \in C_c^1(\mathbb{R}_+, [0,1])$ be a function such that $\varphi(r) = 1$ if $r \in [0, R + \omega T]$, $\varphi(r) = 0$ if $r \in [R + \omega T + 1, \infty)$ and $\varphi'(r) \leqslant 0$, for all $r \in \mathbb{R}_+$.

One takes, in (28.6), ψ defined by

$$\begin{cases} \psi(x,t) = \varphi\big(|x| + \omega t\big) \dfrac{T-t}{T}, & \text{for } x \in \mathbb{R}^d \text{ and } t \in [0,T], \\ \psi(x,t) = 0, & \text{for } x \in \mathbb{R}^d \text{ and } t \geqslant T. \end{cases}$$

The function ψ is not in $C_c^\infty(\mathbb{R}^d \times \mathbb{R}_+, \mathbb{R}_+)$, but, using a usual regularization technique, it may be proved that such a function can be considered in (28.6), in which case inequality (28.6) writes

$$\int_0^1\int_0^1\int_0^T\int_{\mathbb{R}^d}\biggl[\bigl|\mu(x,t,\alpha)-\nu(x,t,\beta)\bigr|\biggl(\frac{T-t}{T}\omega\varphi'\bigl(|x|+\omega t\bigr)-\frac{1}{T}\varphi\bigl(|x|+\omega t\bigr)\biggr)$$
$$+\bigl(f\bigl(\mu(x,t,\alpha)\top\nu(x,t,\beta)\bigr)-f\bigl(\mu(x,t,\alpha)\bot\nu(x,t,\beta)\bigr)\bigr)$$
$$\times\frac{T-t}{T}\varphi'\bigl(|x|+\omega t\bigr)\boldsymbol{v}(x,t)\bullet\frac{x}{|x|}\biggr]\,\mathrm{d}x\,\mathrm{d}t\,\mathrm{d}\alpha\,\mathrm{d}\beta\geqslant 0.$$

Since $\omega = VM$ and $\varphi' \leqslant 0$, one has $(f(a\top b) - f(a\bot b))\varphi'(|x| + \omega t)\boldsymbol{v}(x,t)\bullet(x/|x|) \leqslant |a-b|\omega(-\varphi'(|x|+\omega t))$, for a.e. $(x,t) \in \mathbb{R}^d \times \mathbb{R}_+^\star$ and all $a, b \in [-D, D]$ (D is defined in Step 1). Therefore, since $\varphi(|x| + \omega t) = 1$ if $(x,t) \in B(0,R) \times [0,T]$, the preceding inequality gives

$$\int_0^1\int_0^1\int_0^T\int_{B(0,R)}\bigl|\mu(x,t,\alpha)-\nu(x,t,\beta)\bigr|\,\mathrm{d}x\,\mathrm{d}t\,\mathrm{d}\alpha\,\mathrm{d}\beta\leqslant 0,$$

which yields, since R and T are arbitrary, $\mu(x,t,\alpha) = \nu(x,t,\beta)$ for a.e. $(x,t,\alpha,\beta) \in \mathbb{R}^d \times \mathbb{R}_+^\star \times (0,1) \times (0,1)$.

Let us now deduce also from this uniqueness result that there exists $u \in L^\infty(\mathbb{R}^d \times \mathbb{R}_+^\star)$ such that $\mu(x,t,\alpha) = u(x,t)$, for a.e. $(x,t,\alpha) \in \mathbb{R}^d \times \mathbb{R}_+^\star \times (0,1)$ (then it is easy to see, with Definition 28.1, that u is the entropy weak solution to problem (23.1)–(23.2)).

Indeed, it is possible to take, in the preceding proof, $\mu = \nu$ (recall that Proposition 28.2 gives the existence of an entropy process solution to problem (23.1)–(23.2), see Remark 28.3). This yields $\mu(x,t,\alpha) = \mu(x,t,\beta)$ for a.e. $(x,t,\alpha,\beta) \in \mathbb{R}^d \times \mathbb{R}_+^\star \times (0,1) \times (0,1)$. Then, for a.e. $(x,t) \in \mathbb{R}^d \times \mathbb{R}_+^\star$, one has

$$\mu(x,t,\alpha)=\mu(x,t,\beta)\quad\text{for a.e. }(\alpha,\beta)\in(0,1)\times(0,1)$$

and, for a.e. $\alpha \in (0,1)$,

$$\mu(x,t,\alpha)=\mu(x,t,\beta)\quad\text{for a.e. }\beta\in(0,1).$$

Thus, defining u from $\mathbb{R}^d \times \mathbb{R}_+^\star$ to $\mathbb{R}$ by

$$u(x,t)=\int_0^1\mu(x,t,\beta)\,\mathrm{d}\beta,$$

one obtains $\mu(x,t,\alpha) = u(x,t)$, for a.e. $(x,t,\alpha) \in \mathbb{R}^d \times \mathbb{R}_+^\star \times (0,1)$, and u is the entropy weak solution to problem (23.1)–(23.2). This completes the proof of Theorem 28.1. □

28.3. Convergence towards the entropy weak solution

We now know that there exists a unique entropy process solution to problem (23.1)–(23.2), which is identical to the entropy weak solution of problem (23.1)–(23.2); we may now prove the convergence of the approximate solution given by the finite volume scheme (24.4), (24.2) and (24.5) towards the entropy weak solution as the mesh size tends to 0.

THEOREM 28.2. *Under Assumptions* 23.1, *let* $\alpha \in \mathbb{R}_+^\star$ *and* $\xi \in (0,1)$ *be given. For an admissible mesh* $\mathcal{T}$ *in the sense of Definition* 24.1 *and for* $k > 0$ *satisfying* (24.3) (*note that* α *and* ξ *are fixed*), *let* $u_{\mathcal{T},k}$ *be the solution to* (24.4), (24.2) *and* (24.5).

Then, $u_{\mathcal{T},k} \to u$ *in* $L^p_{\mathrm{loc}}(\mathbb{R}^d \times \mathbb{R}_+)$ *for all* $p \in [1,\infty)$, *as* $h = \mathrm{size}(\mathcal{T}) \to 0$, *where* u *is the entropy weak solution to* (23.1)–(23.2).

PROOF. In order to prove that $u_{\mathcal{T},k} \to u$ (in $L^p_{\mathrm{loc}}(\mathbb{R}^d \times \mathbb{R}_+)$ for all $p \in [1,\infty)$, as $h = \mathrm{size}(\mathcal{T}) \to 0$), let us proceed by a classical way of contradiction which uses the uniqueness of the entropy process solution to problem (23.1)–(23.2). Assume that there exists $1 \leqslant p_0 < \infty$, $\varepsilon > 0$, $\bar{\omega}$ a compact subset of $\mathbb{R}^d$, $T > 0$ and a sequence $((\mathcal{T}_m, k_m))_{m\in\mathbb{N}}$ such that, for any $m \in \mathbb{N}$, $\mathcal{T}_m$ is an admissible mesh, k_m satisfies (24.3) (with $\mathcal{T} = \mathcal{T}_m$ and $k = k_m$, note that α and ξ are independent of m), $\mathrm{size}(\mathcal{T}_m) \to 0$ as $m \to \infty$ and

$$\int_0^T \int_{\bar{\omega}} |u_{\mathcal{T}_m,k_m} - u|^{p_0}\,\mathrm{d}x\,\mathrm{d}t \geqslant \varepsilon, \quad \forall m \in \mathbb{N}, \tag{28.16}$$

where $u_{\mathcal{T}_m,k_m}$ is the solution to (24.4), (24.2) and (24.5) with $\mathcal{T} = \mathcal{T}_m$ and $k = k_m$ and u is the entropy weak solution to (23.1)–(23.2).

Using Proposition 28.2, there exists a subsequence of the sequence $((\mathcal{T}_m, k_m))_{m\in\mathbb{N}}$, still denoted by $((\mathcal{T}_m, k_m))_{m\in\mathbb{N}}$, and a function $\mu \in L^\infty(\mathbb{R}^d \times \mathbb{R}_+^\star \times (0,1))$ such that

(1) $u_{\mathcal{T}_m,k_m} \to \mu$, as $m \to \infty$, in the nonlinear weak-$\star$ sense, i.e.:

$$\begin{aligned} &\lim_{m\to\infty} \int_0^\infty \int_{\mathbb{R}^d} \theta\big(u_{\mathcal{T}_m,k_m}(x,t)\big)\varphi(x,t)\,\mathrm{d}x\,\mathrm{d}t \\ &\quad = \int_0^1 \int_0^\infty \int_{\mathbb{R}^d} \theta\big(\mu(x,t,\alpha)\big)\varphi(x,t)\,\mathrm{d}x\,\mathrm{d}t\,\mathrm{d}\alpha, \\ &\quad \forall \varphi \in L^1\big(\mathbb{R}^d \times \mathbb{R}_+^\star\big),\ \forall \theta \in C(\mathbb{R},\mathbb{R}), \end{aligned} \tag{28.17}$$

(2) μ is an entropy process solution to (23.1)–(23.2).

By Theorem 28.1, one has $\mu(\bullet,\bullet,\alpha) = u$, for a.e. $\alpha \in [0,1]$ (and u is the entropy weak solution to (23.1)–(23.2)). Taking first $\theta(s) = s^2$ in (28.17) and then $\theta(s) = s$ and φu instead of φ in (28.17) one obtains:

$$\int_0^\infty \int_{\mathbb{R}^d} \big(u_{\mathcal{T}_m,k_m}(x,t) - u(x,t)\big)^2 \varphi(x,t)\,\mathrm{d}x\,\mathrm{d}t \to 0, \quad \text{as } m \to \infty, \tag{28.18}$$

for any function $\varphi \in L^1(\mathbb{R}^d \times (0, T))$. From (28.18), and thanks to the L^∞-bound on $(u_{\mathcal{T}_m,k_m})_{m\in\mathbb{N}}$, one deduces the convergence of $(u_{\mathcal{T}_m,k_m})_{m\in\mathbb{N}}$ towards u in $L^p_{\mathrm{loc}}(\mathbb{R}^d \times \mathbb{R}_+)$ for all $p \in [1, \infty)$, which is in contradiction with (28.16).

This completes the proof of our convergence theorem. □

REMARK 28.4. (1) Theorem 28.2 is also true with the implicit scheme instead of the explicit scheme (i.e. (24.6) and (24.7) instead of (24.4) and (24.5)) without the condition (24.3) (and thus without ξ).

(2) The following section improves this convergence result and gives an error estimate.

29. Error estimate

29.1. Statement of the results

This section is devoted to the proof of an error estimate of time explicit and time implicit finite volume approximations to the solution $u \in L^\infty(\mathbb{R}^d \times \mathbb{R}_+^\star)$ of problem (23.1)–(23.2). Assuming that $u_0 \in BV(\mathbb{R}^d)$, a "$h^{1/4}$" error estimate is shown for a large variety of finite volume monotone flux schemes such as those which were presented in Section 24.

Under Assumption 23.1, let $\mathcal{T}$ be an admissible mesh in the sense of Definition 24.1 and $k > 0$. Let $g \in C(\mathbb{R}^2, \mathbb{R})$ satisfying Assumption 24.1.

Let u be the entropy weak solution of (23.1)–(23.2) and let $u_{\mathcal{T},k}$ be the solution of the time explicit scheme (24.4), (24.2), (24.5), assuming that (24.3) holds, or $u_{\mathcal{T},k}$ be the solution of the time implicit scheme (24.6), (24.2), (24.7). Our aim is to give an error estimate between u and $u_{\mathcal{T},k}$.

In the case of the explicit scheme, one proves, in this section, the following theorem.

THEOREM 29.1. *Under Assumption* 23.1, *let $\mathcal{T}$ be an admissible mesh in the sense of Definition* 24.1 *and $k > 0$. Let $g \in C(\mathbb{R}^2, \mathbb{R})$ satisfy Assumption* 24.1 *and assume that condition* (24.3) *holds. Let u be the unique entropy weak solution of* (23.1)–(23.2) *and $u_{\mathcal{T},k}$ be given by* (24.5), (24.4), (24.2). *Assume $u_0 \in BV(\mathbb{R}^d)$. Then, for all $R > 0$ and all $T > 0$ there exists $C_e \in \mathbb{R}_+$, only depending on R, T, $\mathbf{v}$, g, u_0, α and ξ, such that the following inequality holds*:

$$\int_0^T \int_{B(0,R)} \left| u_{\mathcal{T},k}(x,t) - u(x,t) \right| \mathrm{d}x\, \mathrm{d}t \leqslant C_e h^{1/4}. \tag{29.1}$$

(Recall that $B(0, R) = \{x \in \mathbb{R}^d, |x| < R\}$.)

In Theorem 29.1, u_0 is assumed to belong to $BV(\mathbb{R}^d)$ (recall that $u_0 \in BV(\mathbb{R}^d)$ if $\sup\{\int u_0(x)\mathrm{div}\varphi(x)\,\mathrm{d}x,\ \varphi \in C_c^\infty(\mathbb{R}^d, \mathbb{R}^d);\ |\varphi(x)| \leqslant 1, \forall x \in \mathbb{R}^d\} < \infty$). This assumption allows us to obtain an $h^{1/4}$ estimate in (29.1). If $u_0 \notin BV(\mathbb{R}^d)$ (but u_0 still belongs to $L^\infty(\mathbb{R}^d)$), one can also give an error estimate which depends on the functions $\varepsilon(r, S)$ and $\varepsilon_0(r, S)$ defined in (29.16) and (29.23).

A slight improvement of Theorem 29.1 (and also Theorem 29.2 below) is possible. Using the fact that $u \in C(\mathbb{R}_+, L^1_{\mathrm{loc}}(\mathbb{R}^d))$ and thus $u(\bullet, t)$ is defined for all $t \in \mathbb{R}_+$, Theorem 29.1 remains true with

$$\int_{B(0,R)} \left|u_{\mathcal{T},k}(x,t) - u(x,t)\right| \mathrm{d}x \leqslant C_e h^{1/4}, \quad \forall t \in [0, T],$$

instead of (29.1). The proof of such a result may be handled with an adaptation of the proof a uniqueness of the entropy process solution given, e.g., in EYMARD, GALLOUËT and HERBIN [1995], see VILA [1994] and COCKBURN, COQUEL and LEFLOCH [1994] for some similar results.

In some cases, it is possible to obtain $h^{1/2}$, instead of $h^{1/4}$, in Theorem 29.1. This is the case, e.g., when the mesh $\mathcal{T}$ is composed of rectangles ($d = 2$) and when $\boldsymbol{v}$ does not depend on (x, t), since, in this case, one obtains a "BV estimate" on $u_{\mathcal{T},k}$. In this case, the right-hand sides of inequalities (25.4) and (25.5), proven above, are changed from $C/\sqrt{h}$ to C, so that the right-hand side of (27.9) becomes Ch instead of $C\sqrt{h}$, which in turn yields $C_e h^{1/2}$ in (29.1) instead of $C_e h^{1/4}$. It is, however, still an open problem to know whether it is possible to obtain an error estimate with $h^{1/2}$, instead of $h^{1/4}$, in Theorem 29.1 (under the hypotheses of Theorem 29.1), even in the case where $\boldsymbol{v}$ does not depend on (x, t) (see COCKBURN and GREMAUD [1996a] for an attempt in this direction).

REMARK 29.1. Theorem 29.1 (and also Theorem 29.2) remains true with some slightly more general assumption on g, instead of 24.1, in order to allow g to depend on $\mathcal{T}$ and k. Indeed, in (24.4), one can replace $g(u^n_K, u^n_L)$ (and $g(u^n_L, u^n_K)$) by $g_{K,L}(u^n_K, u^n_L, \mathcal{T}, k)$ (and $g_{L,K}(u^n_L, u^n_K, \mathcal{T}, k)$). Assume that, for all $K \in \mathcal{T}$ and all $L \in \mathcal{N}(K)$, the function $(a, b) \mapsto g_{K,L}(a, b, \mathcal{T}, k)$, from $[U_m, U_M]^2$ to $\mathbb{R}$, is nondecreasing with respect to a, nonincreasing with respect to b, Lipschitz continuous uniformly with respect to K and L and that $g_{K,L}(a, a, \mathcal{T}, k) = f(a)$ for all $a \in [U_m, U_M]$ (recall that $U_m \leqslant u_0 \leqslant U_M$ a.e. on $\mathbb{R}^d$). Then Theorem 29.1 remains true.

However, note that condition (24.3) and C_e in the estimate (29.1) of Theorem 29.1 depend on the Lipschitz constants of $g_{K,L}(\bullet, \bullet, \mathcal{T}, k)$ on $[U_m, U_M]^2$. An interesting form for $g_{K,L}$ is $g_{K,L}(a, b, \mathcal{T}, k) = c_{K,L}(\mathcal{T}, k) f(a) + (1 - c_{K,L}(\mathcal{T}, k))\ f(b) + D_{K,L}(\mathcal{T}, k)\ (a - b)$, with some $c_{K,L}(\mathcal{T}, k) \in [0, 1]$ and $D_{K,L}(\mathcal{T}, k) \geqslant 0$. In order to obtain the desired properties on $g_{K,L}$, it is sufficient to take $\max\{|f'(s)|,\ s \in [U_m, U_M]\} \leqslant D_{K,L}(\mathcal{T}, k) \leqslant D$ (for all K, L), with some $D \in \mathbb{R}$. The Lipschitz constants of $g_{K,L}$ on $[U_m, U_M]^2$ only depend on D, f, U_m and U_M.

For instance, a "Lax–Friedrichs type" scheme consists, roughly speaking, in taking $D_{K,L}(\mathcal{T}, k)$ of order "h/k". The desired properties on $g_{K,L}$ are satisfied, provided that $k/h \leqslant C$, with some C depending on $\max\{|f'(s)|,\ s \in [U_m, U_M]\}$. Note, however, that the condition $k/h \leqslant C$ is not sufficient to give a real "$h^{1/4}$" estimate, since the coefficient C_e in (29.1) depends on D. Taking, e.g., k of order "h^2" leads to an estimate "$C_e h^{1/4}$" which do not goes to 0 as h goes to 0 (indeed, it is known, in this case, that the approximate solution does not converge towards the entropy weak solution to (23.1)–(23.2)). One obtains a real "$h^{1/4}$" estimate, in the case of that "Lax–Friedrichs type" scheme, by taking $C_1 \leqslant (k/h) \leqslant C_2$. In order to avoid the condition $C_1 \leqslant (k/h)$ (note that $(k/h) \leqslant C_2$ is imposed by the Courant–Friedrichs–Levy condition (24.3), a possibility is to take $D_{K,L}(\mathcal{T}, k) = D = \max\{|f'(s)|,\ s \in [U_m, U_M]\}$ (this is related to the "modified Lax–Friedrichs" of Example 21.1 in the 1D case). Then D only depends on f and u_0 and, in the estimate "$C_e h^{1/4}$" of Theorem 29.1, C_e only depends on R, T, $\boldsymbol{v}$, f, u_0, α and ξ, which leads to a convergence result at rate "$h^{1/4}$" as $h \to 0$ (with fixed α and ξ).

In the case of the implicit scheme, one proves the following theorem.

THEOREM 29.2. *Under Assumption* 23.1, *let $\mathcal{T}$ be an admissible mesh in the sense of Definition* 24.1 *and $k > 0$. Let $g \in C(\mathbb{R}^2, \mathbb{R})$ satisfy Assumption* 24.1. *Let u be the unique entropy weak solution of* (23.1)–(23.2). *Assume that $u_0 \in BV(\mathbb{R}^d)$ and that $\boldsymbol{v}$ does not depend on t.*

Let $\{u_K^n,\ n \in \mathbb{N},\ K \in \mathcal{T}\}$ be the unique solution to (24.6) *and* (24.2) *such that $u_K^n \in [U_m, U_M]$ for all $K \in \mathcal{T}$ and $n \in \mathbb{N}$ (existence and uniqueness of such a solution is given by Proposition* 26.1). *Let $u_{\mathcal{T},k}$ be defined by* (24.7).

Then, for all $R > 0$ and $T > 0$, there exists C_e, only depending on R, T, $\boldsymbol{v}$, g, u_0 and α, such that the following inequality holds:

$$\int_0^T \int_{B(0,R)} \left| u_{\mathcal{T},k}(x,t) - u(x,t) \right| \mathrm{d}x\,\mathrm{d}t \leqslant C_e \left(k + h^{1/2}\right)^{1/2}. \tag{29.2}$$

REMARK 29.2. Note that, in Theorem 29.2, there is no restriction on k (this is usual for an implicit scheme), and one obtains an "$h^{1/4}$" error estimate for some "large" k, namely if $k \leqslant h^{1/2}$. In Theorem 29.2, if $\boldsymbol{v}$ depends on t and $u_0 \in L^\infty(\mathbb{R}^d)$ (but u_0 not necessarily in $BV(\mathbb{R}^d)$), one can also give an error estimate. Indeed one obtains

$$\int_0^T \int_{B(0,R)} \left| u_{\mathcal{T},k}(x,t) - u(x,t) \right| \mathrm{d}x\,\mathrm{d}t \leqslant C_e \left(\frac{k}{h^{1/2}} + h^{1/2} \right)^{1/2},$$

which yields an "$h^{1/4}$" error estimate if k is of order "h".

Theorem 29.1 (resp. Theorem 29.2) is an easy consequence of Theorem 27.1 (resp. 27.2) and of a quite general theorem of comparison between the entropy weak solution to (23.1)–(23.2) and an approximate solution. This theorem of comparison (Theorem 29.3) may be used in other frameworks (e.g., to compare the entropy weak solution to (23.1)–(23.2) and the approximate solution obtained with a parabolic regularization of (23.1)). It is stated and proved in Section 29.3 where the proofs of Theorems 29.1 and 29.2 are also given. First, in Section 29.2, two preliminary lemmata are given. Indeed, Lemma 29.2 is the crucial part of the two following sections.

29.2. Preliminary lemmata

Let us first give a classical lemma on the space BV.

LEMMA 29.1. *Let $u \in BV_{\mathrm{loc}}(\mathbb{R}^p)$, $p \in \mathbb{N}^\star$, i.e. $u \in L^1_{\mathrm{loc}}(\mathbb{R}^p)$ and the restriction of u to Ω belongs to $BV(\Omega)$ for all open bounded subset Ω of $\mathbb{R}^p$ (see Definition 21.17 for the definition of $BV(\Omega)$).*

Then, for all bounded subset Ω of $\mathbb{R}^p$ and for all $a > 0$,

$$\left\| u(\bullet + \eta) - u \right\|_{L^1(\Omega)} \leqslant |\eta| |u|_{BV(\Omega_a)}, \quad \forall \eta \in \mathbb{R}^p,\ |\eta| \leqslant a, \tag{29.3}$$

where $\Omega_a = \{x \in \mathbb{R}^p;\ d(x, \Omega) < a\}$ and $d(x, \Omega) = \inf\{|x - y|,\ y \in \Omega\}$ is the distance from x to Ω.

PROOF. Let Ω be a bounded subset of $\mathbb{R}^p$ and $\eta \in \mathbb{R}^p$. The following equality classically holds:

$$\begin{aligned}
&\left\| u(\bullet + \eta) - u \right\|_{L^1(\Omega)} \\
&\quad = \sup\left\{ \int_\Omega \big(u(x+\eta) - u(x)\big)\varphi(x)\,\mathrm{d}x,\ \varphi \in C_c^\infty(\Omega, \mathbb{R}),\ \left\|\varphi\right\|_{L^\infty(\Omega)} \leqslant 1 \right\}.
\end{aligned} \tag{29.4}$$

Let $\varphi \in C_c^\infty(\Omega, \mathbb{R})$ such that $\|\varphi\|_{L^\infty(\Omega)} \leqslant 1$.

Since $\varphi(x) = 0$ if $x \in \Omega_{|\eta|} \setminus \Omega$ (recall that $\Omega_{|\eta|} = \{x \in \mathbb{R}^p;\ d(x, \Omega) < \eta\}$),

$$\int_\Omega u(x)\varphi(x)\,\mathrm{d}x = \int_{\Omega_{|\eta|}} u(x)\varphi(x)\,\mathrm{d}x.$$

Similarly, using an obvious change of variables,

$$\int_\Omega u(x+\eta)\varphi(x)\,\mathrm{d}x = \int_{\Omega_{|\eta|}} u(x)\varphi(x-\eta)\,\mathrm{d}x.$$

Therefore,

$$\begin{aligned}
\int_\Omega \big(u(x+\eta) - u(x)\big)\varphi(x)\,\mathrm{d}x &= \int_{\Omega_{|\eta|}} u(x)\big(\varphi(x-\eta) - \varphi(x)\big)\,\mathrm{d}x \\
&= -\int_{\Omega_{|\eta|}} u(x)\left(\int_0^1 \nabla\varphi(x - s\eta) \bullet \eta\,\mathrm{d}s\right)\mathrm{d}x
\end{aligned}$$

and, with Fubini's theorem,

$$\int_\Omega \big(u(x+\eta) - u(x)\big)\varphi(x)\,\mathrm{d}x = \int_0^1 \left(\int_{\Omega_{|\eta|}} u(x)\nabla\varphi(x - s\eta) \bullet \eta\,\mathrm{d}x\right)\mathrm{d}s. \tag{29.5}$$

For all $s \in (0,1)$, Define $\psi_s \in C_c^\infty(\Omega_{|\eta|}, \mathbb{R}^p)$ by $\psi_s(x) = \varphi(x - s\eta)\eta$; since $\psi_s \in C_c^\infty(\Omega_{|\eta|}, \mathbb{R}^p)$ and $|\psi_s(x)| \leqslant |\eta|$ for all $x \in \mathbb{R}^p$, the definition of $|u|_{BV(\Omega_{|\eta|})}$ yields

$$\int_{\Omega_{|\eta|}} u(x)\nabla\varphi(x - s\eta) \bullet \eta\,\mathrm{d}x = \int_{\Omega_{|\eta|}} u(x)\operatorname{div}\psi_s(x)\,\mathrm{d}x \leqslant |\eta| |u|_{BV(\Omega_{|\eta|})}.$$

Then, (29.5) gives

$$\int_{\Omega} \big(u(x+\eta) - u(x)\big)\varphi(x)\,\mathrm{d}x \leqslant |\eta| |u|_{BV(\Omega_{|\eta|})}. \tag{29.6}$$

Taking in (29.6) the supremum over $\varphi \in C_c^{\infty}(\Omega, \mathbb{R})$ such that $\|\varphi\|_{L^{\infty}(\Omega)} \leqslant 1$ yields, thanks to (29.4),

$$\big\|u(\bullet + \eta) - u\big\|_{L^1(\Omega)} \leqslant |\eta| |u|_{BV(\Omega_{|\eta|})}, \quad \forall \eta \in \mathbb{R}^p,$$

and (29.3) follows, since $\Omega_{|\eta|} \subset \Omega_a$ if $|\eta| \leqslant a$. □

REMARK 29.3. Let us give an application of Lemma 29.1 which will be quite useful further on. Let $u \in BV_{\mathrm{loc}}(\mathbb{R}^p)$, $p \in \mathbb{N}^{\star}$. Let $\psi, \varphi \in C_c(\mathbb{R}^p, \mathbb{R}_+)$, $a > 0$ and $0 < \varepsilon < a$ such that $\int_{\mathbb{R}^p} \varphi(x)\,\mathrm{d}x = 1$ and $\varphi(x) = 0$ for all $x \in \mathbb{R}^p$, $|x| > \varepsilon$. Let $S = \{x \in \mathbb{R}^p, \psi(x) \neq 0\}$.

Then,

$$\int_{\mathbb{R}^p} \int_{\mathbb{R}^p} \big|u(x) - u(y)\big| \psi(x)\varphi(x-y)\,\mathrm{d}y\,\mathrm{d}x \leqslant \varepsilon \|\psi\|_{L^{\infty}(\mathbb{R}^p)} |u|_{BV(S_a)}, \tag{29.7}$$

where $S_a = \{x \in \mathbb{R}_p, d(x, S) < a\}$.

Indeed, Lemma 29.1 gives

$$\big\|u(\bullet + \eta) - u\big\|_{L^1(S)} \leqslant |\eta| |u|_{BV(S_a)}, \quad \forall \eta \in \mathbb{R}^p,\ |\eta| \leqslant a. \tag{29.8}$$

Using a change of variables in the left-hand side of (29.7),

$$\begin{aligned} &\int_{\mathbb{R}^p} \int_{\mathbb{R}^p} \big|u(x) - u(y)\big| \psi(x)\varphi(x-y)\,\mathrm{d}y\,\mathrm{d}x \\ &\quad \leqslant \|\psi\|_{L^{\infty}(\mathbb{R}^p)} \int_{B(0,\varepsilon)} \left(\int_S \big|u(x) - u(x-z)\big|\,\mathrm{d}x \right) \varphi(z)\,\mathrm{d}z. \end{aligned}$$

Then, (29.8) yields

$$\int_{\mathbb{R}^p} \int_{\mathbb{R}^p} \big|u(x) - u(y)\big| \psi(x)\varphi(x-y)\,\mathrm{d}y\,\mathrm{d}x \leqslant \varepsilon \|\psi\|_{L^{\infty}(\mathbb{R}^p)} |u|_{BV(S_a)} \int_{\mathbb{R}^p} \varphi(z)\,\mathrm{d}z,$$

which gives (29.7).

LEMMA 29.2. *Under Assumption* 23.1, *let* $u_0 \in BV(\mathbb{R}^d)$ *and* $\tilde{u} \in L^{\infty}(\mathbb{R}^d \times \mathbb{R}_+^{\star})$ *such that* $U_m \leqslant \tilde{u} \leqslant U_M$ *a.e. on* $\mathbb{R}^d \times \mathbb{R}_+^{\star}$. *Assume that there exist* $\mu \in \mathcal{M}(\mathbb{R}^d \times \mathbb{R}_+)$ *and*

$\mu_0 \in \mathcal{M}(\mathbb{R}^d)$ *such that*

$$\begin{aligned}
&\int_{\mathbb{R}_+}\int_{\mathbb{R}^d}\Big(\big|\tilde{u}(x,t)-\kappa\big|\varphi_t(x,t)\\
&\qquad + \big(f\big(\tilde{u}(x,t)\top\kappa\big)-f\big(\tilde{u}(x,t)\bot\kappa\big)\big)\boldsymbol{v}(x,t)\cdot\nabla\varphi(x,t)\Big)\,\mathrm{d}x\,\mathrm{d}t\\
&\qquad + \int_{\mathbb{R}^d}\big|u_0(x)-\kappa\big|\varphi(x,0)\,\mathrm{d}x\\
&\quad \geqslant -\int_{\mathbb{R}^d\times\mathbb{R}_+}\big(\big|\varphi_t(x,t)\big|+\big|\nabla\varphi(x,t)\big|\big)\,\mathrm{d}\mu(x,t)-\int_{\mathbb{R}^d}\big|\varphi(x,0)\big|\,\mathrm{d}\mu_0(x),\\
&\quad \forall\kappa\in\mathbb{R},\ \forall\varphi\in C_c^\infty\big(\mathbb{R}^d\times\mathbb{R}_+,\mathbb{R}_+\big).
\end{aligned} \tag{29.9}$$

Let u be the unique entropy weak solution of (23.1)–(23.2) (*i.e. $u \in L^\infty(\mathbb{R}^d \times \mathbb{R}_+^\star)$ is the unique solution to* (29.9) *with u instead of $\tilde{u}$ and $\mu = 0$, $\mu_0 = 0$*).

Then for all $\psi \in C_c^\infty(\mathbb{R}^d \times \mathbb{R}_+, \mathbb{R}_+)$ there exists C only depending on ψ (more precisely on $\|\psi\|_\infty$, $\|\psi_t\|_\infty$, $\|\nabla\psi\|_\infty$, and on the support of ψ), $\boldsymbol{v}$, f, and u_0, such that

$$\begin{aligned}
&\int_{\mathbb{R}_+}\int_{\mathbb{R}^d}\Big[\big|\tilde{u}(x,t)-u(x,t)\big|\psi_t(x,t)\\
&\qquad + \big(f\big(\tilde{u}(x,t)\top u(x,t)\big)-f\big(\tilde{u}(x,t)\bot u(x,t)\big)\big)\big(\boldsymbol{v}(x,t)\cdot\nabla\psi(x,t)\big)\Big]\,\mathrm{d}x\,\mathrm{d}t\\
&\quad \geqslant -C\big(\mu_0\big(\{\psi(\cdot,0)\neq 0\}\big)+\big(\mu\big(\{\psi\neq 0\}\big)\big)^{1/2}+\mu\big(\{\psi\neq 0\}\big)\big),
\end{aligned} \tag{29.10}$$

where $\{\psi \neq 0\} = \{(x,t) \in \mathbb{R}^d \times \mathbb{R}_+,\ \psi(x,t) \neq 0\}$ *and* $\{\psi(\cdot,0) \neq 0\} = \{x \in \mathbb{R}^d,\ \psi(x,0) \neq 0\}$. (*Note that* $\|\cdot\|_\infty = \|\cdot\|_{L^\infty(\mathbb{R}^d\times\mathbb{R}_+^\star)}$.)

PROOF. The proof of Lemma 29.2 is close to that of Step 1 in the proof of Theorem 28.1. Let us first define mollifiers in $\mathbb{R}$ and $\mathbb{R}^d$. For $p = 1$ and $p = d$, one defines $\rho_p \in C_c^\infty(\mathbb{R}^p, \mathbb{R})$ satisfying the following properties:

$$\mathrm{supp}(\rho_p) = \big\{x \in \mathbb{R}^p;\ \rho_p(x) \neq 0\big\} \subset \big\{x \in \mathbb{R}^p;\ |x| \leqslant 1\big\},$$

$$\rho_p(x) \geqslant 0, \quad \forall x \in \mathbb{R}^p,$$

$$\int_{\mathbb{R}^p} \rho_p(x)\,\mathrm{d}x = 1$$

and furthermore, for $p = 1$,

$$\rho_1(x) = 0, \quad \forall x \in \mathbb{R}_+. \tag{29.11}$$

For $r \in \mathbb{R}$, $r \geqslant 1$, one defines $\rho_{p,r}(x) = r^p \rho_p(rx)$, for all $x \in \mathbb{R}^p$.

Using the mollifiers $\rho_{p,r}$ will allow to choose convenient test functions in (29.9) (which are the inequations satisfied by $\tilde{u}$) and in the analogous inequalities satisfied by

u which are

$$\int_{\mathbb{R}_+}\int_{\mathbb{R}^d}\Big[\big|u(y,s)-\kappa\big|\varphi_s(y,s)+\big(f\big(u(y,s)\top\kappa\big)-f\big(u(y,s)\bot\kappa\big)\big)$$
$$\times\, \boldsymbol{v}(y,s)\cdot\nabla\varphi(y,s)\Big]\,\mathrm{d}y\,\mathrm{d}s+\int_{\mathbb{R}^d}\big|u_0(y)-\kappa\big|\varphi(y,0)\,\mathrm{d}y\geqslant 0,$$
$$\forall\kappa\in\mathbb{R},\ \forall\varphi\in C_c^\infty\big(\mathbb{R}^d\times\mathbb{R}_+,\mathbb{R}_+\big). \tag{29.12}$$

Indeed, the main tool is to take $\kappa=u(y,s)$ in (29.9), $\kappa=\tilde{u}(x,t)$ in (29.12) and to introduce mollifiers in order to have y close to x and s close to t.

Let $\psi\in C_c^\infty(\mathbb{R}^d\times\mathbb{R}_+,\mathbb{R}_+)$, and let $\varphi:(\mathbb{R}^d\times\mathbb{R}_+)^2\to\mathbb{R}_+$ be defined by:

$$\varphi(x,t,y,s)=\psi(x,t)\rho_{d,r}(x-y)\rho_{1,r}(t-s).$$

Note that, for any $(y,s)\in\mathbb{R}^d\times\mathbb{R}_+$, one has $\varphi(\bullet,\bullet,y,s)\in C_c^\infty(\mathbb{R}^d\times\mathbb{R}_+,\mathbb{R}_+)$ and, for any $(x,t)\in\mathbb{R}^d\times\mathbb{R}_+$, one has $\varphi(x,t,\bullet,\bullet)\in C_c^\infty(\mathbb{R}^d\times\mathbb{R}_+,\mathbb{R}_+)$. Let us take $\varphi(\bullet,\bullet,y,s)$ as test function φ in (29.9) and $\varphi(x,t,\bullet,\bullet)$ as test function φ in (29.12). We take, in (29.9), $\kappa=u(y,s)$ and we take, in (29.12), $\kappa=\tilde{u}(x,t)$. We then integrate (29.9) for $(y,s)\in\mathbb{R}^d\times\mathbb{R}_+$, and (29.12) for $(x,t)\in\mathbb{R}^d\times\mathbb{R}_+$. Adding the two inequations yields

$$E_{11}+E_{12}+E_{13}+E_{14}\geqslant -E_2, \tag{29.13}$$

where

$$E_{11}=\int_0^\infty\int_{\mathbb{R}^d}\int_0^\infty\int_{\mathbb{R}^d}\Big[\big|\tilde{u}(x,t)-u(y,s)\big|$$
$$\times\,\psi_t(x,t)\rho_{d,r}(x-y)\rho_{1,r}(t-s)\Big]\,\mathrm{d}x\,\mathrm{d}t\,\mathrm{d}y\,\mathrm{d}s,$$
$$E_{12}=\int_0^\infty\int_{\mathbb{R}^d}\int_0^\infty\int_{\mathbb{R}^d}\Big[\big(f\big(\tilde{u}(x,t)\top u(y,s)\big)-f\big(\tilde{u}(x,t)\bot u(y,s)\big)\big)$$
$$\times\,\boldsymbol{v}(x,t)\cdot\nabla\psi(x,t)\rho_{d,r}(x-y)\rho_{1,r}(t-s)\Big]\,\mathrm{d}x\,\mathrm{d}t\,\mathrm{d}y\,\mathrm{d}s,$$
$$E_{13}=-\int_0^\infty\int_{\mathbb{R}^d}\int_0^\infty\int_{\mathbb{R}^d}\big(f\big(\tilde{u}(x,t)\top u(y,s)\big)-f\big(\tilde{u}(x,t)\bot u(y,s)\big)\big)\psi(x,t)$$
$$\times\,\big(\boldsymbol{v}(y,s)-\boldsymbol{v}(x,t)\big)\cdot\nabla\rho_{d,r}(x-y)\rho_{1,r}(t-s)\,\mathrm{d}x\,\mathrm{d}t\,\mathrm{d}y\,\mathrm{d}s,$$
$$E_{14}=\int_{\mathbb{R}^d}\int_0^\infty\int_{\mathbb{R}^d}\big|u_0(x)-u(y,s)\big|\psi(x,0)\rho_{d,r}(x-y)\rho_{1,r}(-s)\,\mathrm{d}y\,\mathrm{d}s\,\mathrm{d}x$$

and

$$E_2=\int_0^\infty\int_{\mathbb{R}^d}\int_{\mathbb{R}^d\times\mathbb{R}_+}\Big(\big|\rho_{d,r}(x-y)\big(\psi_t(x,t)\rho_{1,r}(t-s)+\psi(x,t)\rho_{1,r}'(t-s)\big)\big|$$
$$+\big|\rho_{1,r}(t-s)\big(\nabla\psi(x,t)\rho_{d,r}(x-y)+\psi(x,t)\nabla\rho_{d,r}(x-y)\big)\big|\Big)$$

$$\times\, \mathrm{d}\mu(x,t)\,\mathrm{d}y\,\mathrm{d}s + \int_0^\infty \int_{\mathbb{R}^d}\int_{\mathbb{R}^d} \bigl|\psi(x,0)\rho_{d,r}(x-y)\rho_{1,r}(-s)\bigr|\,\mathrm{d}\mu_0(x)\,\mathrm{d}y\,\mathrm{d}s. \tag{29.14}$$

One may be surprised by the fact that the inequation (29.13) is obtained without using the initial condition which is satisfied by the entropy weak solution u of (23.1)–(23.2). Indeed, this initial condition appears only in the third term of the left-hand side of (29.12); since $\varphi(x,t,\bullet,0)=0$ for all $(x,t)\in\mathbb{R}^d\times\mathbb{R}_+$, the third term of the left hand side of (29.12) is zero when $\varphi(x,t,\bullet,\bullet)$ is chosen as a test function in (29.12). However, the fact that u satisfies the initial condition of (23.1)–(23.2) will be used later in order to get a bound on E_{14}.

Let us now study the five terms of (29.13). One sets $S=\{\psi\neq 0\}=\{(x,t)\in\mathbb{R}^d\times\mathbb{R}_+;\ \psi(x,t)\neq 0\}$ and $S_0=\{\psi(\bullet,0)\neq 0\}=\{x\in\mathbb{R}^d;\ \psi(x,0)\neq 0\}$. In the following, the notation C_i ($i\in\mathbb{N}$) will refer to various real quantities only depending on $\|\psi\|_\infty$, $\|\psi_t\|_\infty$, $\|\nabla\psi\|_\infty$, S, S_0, $\boldsymbol{v}$, f, and u_0.

Equality (29.14) leads to

$$E_2\leqslant (r+1)C_1\mu(S)+C_2\mu_0(S_0). \tag{29.15}$$

Let us handle the term E_{11}. For all $x\in\mathbb{R}^d$ and for all $t\in\mathbb{R}_+$, one has, using (29.11),

$$\int_{\mathbb{R}^d}\int_0^\infty \rho_{d,r}(x-y)\rho_{1,r}(t-s)\,\mathrm{d}s\,\mathrm{d}y=1.$$

Then,

$$\left|E_{11}-\int_{\mathbb{R}_+}\int_{\mathbb{R}^d}\bigl[\bigl|\tilde u(x,t)-u(x,t)\bigr|\psi_t(x,t)\bigr]\,\mathrm{d}x\,\mathrm{d}t\right| \leqslant \int_0^\infty\int_{\mathbb{R}^d}\int_0^\infty\int_{\mathbb{R}^d}\bigl[\bigl|u(x,t)-u(y,s)\bigr|\bigl|\psi_t(x,t)\bigr|\rho_{d,r}(x-y) \times \rho_{1,r}(t-s)\bigr]\,\mathrm{d}x\,\mathrm{d}t\,\mathrm{d}y\,\mathrm{d}s\leqslant \|\psi_t\|_\infty\varepsilon(r,S),$$

with

$$\varepsilon(r,S)=\sup\left\{\bigl\|u-u(\bullet+\eta,\bullet+\tau)\bigr\|_{L^1(S)},\ |\eta|\leqslant\frac{1}{r},\ 0\leqslant\tau\leqslant\frac{1}{r}\right\}. \tag{29.16}$$

Since $u_0\in BV(\mathbb{R}^d)$, the function u (entropy weak solution to (23.1)–(23.2)) belongs to $BV(\mathbb{R}^d\times(-T,T))$, for all $T>0$, setting, i.e., $u(.,t)=u_0$ for $t<0$ (see KRUSHKOV [1970] or CHAINAIS-HILLAIRET [1999] where this result is proven passing to the limit on numerical schemes).

Then, Lemma 29.1 gives, since $r\geqslant 1$ (taking $p=d+1$, $\Omega=S$ and $a=\sqrt{2}$ in Lemma 29.1),

$$\varepsilon(r,S)\leqslant\frac{C_3}{r}. \tag{29.17}$$

Hence,

$$\left| E_{11} - \int_{\mathbb{R}_+} \int_{\mathbb{R}^d} \left[\left| \tilde{u}(x,t) - u(x,t) \right| \psi_t(x,t) \right] \mathrm{d}x \, \mathrm{d}t \right| \leqslant \frac{C_4}{r}. \tag{29.18}$$

In the same way, using $|f(a \top b) - f(a \top c)| \leqslant M|b - c|$ and $|f(a \bot b) - f(a \bot c)| \leqslant M|b-c|$ for all $a, b, c \in [U_m, U_M]$ where M is the Lipschitz constant of f in $[U_m, U_M]$,

$$\begin{aligned} \Bigg| E_{12} - \int_{\mathbb{R}_+} \int_{\mathbb{R}^d} \big(f\big(\tilde{u}(x,t) \top u(x,t)\big) - f\big(\tilde{u}(x,t) \bot u(x,t)\big)\big) \\ \times \big(\boldsymbol{v}(x,t) \bullet \nabla \psi(x,t)\big) \, \mathrm{d}x \, \mathrm{d}t \Bigg| \leqslant C_5 \varepsilon(r,S) \leqslant \frac{C_6}{r}. \end{aligned} \tag{29.19}$$

Let us now turn to E_{13}. We compare this term with

$$\begin{aligned} E_{13\mathrm{b}} = - \int_0^\infty \int_{\mathbb{R}^d} \int_0^\infty \int_{\mathbb{R}^d} \big(f\big(\tilde{u}(x,t) \top u(x,t)\big) - f\big(\tilde{u}(x,t) \bot u(x,t)\big)\big) \psi(x,t) \\ \times \big(\boldsymbol{v}(y,s) - \boldsymbol{v}(x,t)\big) \bullet \nabla \rho_{d,r}(x-y) \rho_{1,r}(t-s) \, \mathrm{d}x \, \mathrm{d}t \, \mathrm{d}y \, \mathrm{d}s. \end{aligned}$$

Since $\operatorname{div}(\boldsymbol{v}(\bullet, s) - \boldsymbol{v}(x,t)) = 0$ (on $\mathbb{R}^d$) for all $x \in \mathbb{R}^d$, $t \in \mathbb{R}_+$ and $s \in \mathbb{R}_+$, one has $E_{13\mathrm{b}} = 0$. Therefore, subtracting $E_{13\mathrm{b}}$ from E_{13} yields

$$\begin{aligned} E_{13} \leqslant C_7 \int_0^\infty \int_{\mathbb{R}^d} \int_0^\infty \int_{\mathbb{R}^d} \left| u(x,t) - u(y,s) \right| \psi(x,t) \\ \times \left| \big(\boldsymbol{v}(y,s) - \boldsymbol{v}(x,t)\big) \bullet \nabla \rho_{d,r}(x-y) \right| \rho_{1,r}(t-s) \, \mathrm{d}x \, \mathrm{d}t \, \mathrm{d}y \, \mathrm{d}s. \end{aligned} \tag{29.20}$$

The right-hand side of (29.20) is then smaller than $C_8 \varepsilon(r,S)$, since $|(\boldsymbol{v}(y,s) - \boldsymbol{v}(x,t)) \bullet \nabla \rho_{d,r}(x-y)|$ is bounded by $C_9 r^d$ (noting that $|x-y| \leqslant 1/r$). Then, with (29.17), one has

$$E_{13} \leqslant \frac{C_{10}}{r}. \tag{29.21}$$

In order to estimate E_{14}, let us take in (29.12), for $x \in \mathbb{R}^d$ fixed, $\varphi = \varphi(x, \bullet, \bullet)$, with

$$\varphi(x,y,s) = \psi(x,0) \rho_{d,r}(x-y) \int_s^\infty \rho_{1,r}(-\tau) \, \mathrm{d}\tau,$$

and $\kappa = u_0(x)$. Note that $\varphi(x, \bullet, \bullet) \in C_c^\infty(\mathbb{R}^d \times \mathbb{R}_+, \mathbb{R}_+)$. We then integrate the resulting inequality with respect to $x \in \mathbb{R}^d$. We get

$$-E_{14} + E_{15} + E_{16} \geqslant 0,$$

with

$$E_{15} = -\int_0^\infty \int_{\mathbb{R}^d}\int_{\mathbb{R}^d}\int_s^\infty \big(f\big(u(y,s)\top u_0(x)\big) - f\big(u(y,s)\bot u_0(x)\big)\big)$$
$$\times\, \boldsymbol{v}(y,s)\bullet\big(\psi(x,0)\nabla\rho_{d,r}(x-y)\big)\rho_{1,r}(-\tau)\,\mathrm{d}\tau\,\mathrm{d}y\,\mathrm{d}x\,\mathrm{d}s,$$
$$E_{16} = \int_{\mathbb{R}^d}\int_{\mathbb{R}^d}\int_0^\infty \psi(x,0)\rho_{d,r}(x-y)\rho_{1,r}(-\tau)\big|u_0(x)-u_0(y)\big|\,\mathrm{d}\tau\,\mathrm{d}y\,\mathrm{d}x.$$

To bound E_{15}, one introduces $E_{15\mathrm{b}}$ defined as

$$E_{15\mathrm{b}} = \int_0^\infty \int_{\mathbb{R}^d}\int_{\mathbb{R}^d}\int_s^\infty \big(f\big(u(y,s)\top u_0(y)\big) - f\big(u(y,s)\bot u_0(y)\big)\big)$$
$$\times\, \big(\boldsymbol{v}(y,s)\bullet\nabla\rho_{d,r}(x-y)\big)\psi(x,0)\rho_{1,r}(-\tau)\,\mathrm{d}\tau\,\mathrm{d}y\,\mathrm{d}x\,\mathrm{d}s.$$

Integrating by parts for the x variable yields

$$E_{15\mathrm{b}} = -\int_0^\infty \int_{\mathbb{R}^d}\int_{\mathbb{R}^d}\int_s^\infty \big(f\big(u(y,s)\top u_0(y)\big) - f\big(u(y,s)\bot u_0(y)\big)\big)$$
$$\times\, \big(\boldsymbol{v}(y,s)\bullet\nabla\psi(x,0)\big)\rho_{d,r}(x-y)\rho_{1,r}(-\tau)\,\mathrm{d}\tau\,\mathrm{d}y\,\mathrm{d}x\,\mathrm{d}s.$$

Then, noting that the time support of this integration is reduced to $s\in[0,1/r]$, one has

$$E_{15\mathrm{b}} \leqslant \frac{C_{11}}{r}. \tag{29.22}$$

Furthermore, one has

$$|E_{15}+E_{15\mathrm{b}}| \leqslant C_{12}\int_0^\infty \int_{\mathbb{R}^d}\int_{\mathbb{R}^d}\int_s^\infty \big|u_0(x)-u_0(y)\big|\big|\boldsymbol{v}(y,s)\bullet\nabla\rho_{d,r}(x-y)\big|$$
$$\times\, \psi(x,0)\rho_{1,r}(-\tau)\,\mathrm{d}\tau\,\mathrm{d}y\,\mathrm{d}x\,\mathrm{d}s,$$

which is bounded by $C_{13}\varepsilon_0(r,S_0)$, since the time support of the integration is reduced to $s\in[0,1/r]$, where $\varepsilon_0(r,S_0)$ is defined by

$$\varepsilon_0(r,S_0) = \sup\left\{\int_{S_0}\big|u_0(x)-u_0(x+\eta)\big|\,\mathrm{d}x;\ |\eta|\leqslant\frac{1}{r}\right\}. \tag{29.23}$$

Since $u_0\in BV(\mathbb{R}^d)$, one has (thanks to Lemma 29.1) $\varepsilon_0(r,S_0)\leqslant C_{14}/r$ and therefore, with (29.22), $E_{15}\leqslant C_{15}/r$.

Since $u_0\in BV(\mathbb{R}^d)$, again thanks to Lemma 29.1, see Remark 29.3, the term E_{16} is also bounded by C_{16}/r.

Hence, since $E_{14}\leqslant E_{15}+E_{16}$,

$$E_{14} \leqslant \frac{C_{17}}{r}. \tag{29.24}$$

Using (29.13), (29.15), (29.18), (29.19),(29.21), (29.24), one obtains

$$\int_{\mathbb{R}_+}\int_{\mathbb{R}^d}\big[\big|\tilde{u}(x,t)-u(x,t)\big|\psi_t(x,t)$$
$$+\big(f\big(\tilde{u}(x,t)\top u(x,t)\big)-f\big(\tilde{u}(x,t)\bot u(x,t)\big)\big)\big(\boldsymbol{v}(x,t)\bullet\nabla\psi(x,t)\big)\big]\,\mathrm{d}x\,\mathrm{d}t$$
$$\geqslant -C_1(r+1)\mu(S)-C_2\mu_0(S_0)-\frac{C_{18}}{r},$$

which, taking $r=1/\sqrt{\mu(S)}$ if $0<\mu(S)\leqslant 1$ ($r\to\infty$ if $\mu(S)=0$ and $r=1$ if $\mu(S)>1$), gives (29.10).

This concludes the proof of Lemma 29.2. □

29.3. Proof of the error estimates

Let us now prove a quite general theorem of comparison between the entropy weak solution to (23.1)–(23.2) and an approximate solution, from which Theorems 29.1 and 29.2 will be deduced.

THEOREM 29.3. *Under Assumption 23.1, let $u_0\in BV(\mathbb{R}^d)$ and $\tilde{u}\in L^\infty(\mathbb{R}^d\times\mathbb{R}_+^\star)$ such that $U_m\leqslant\tilde{u}\leqslant U_M$ a.e. on $\mathbb{R}^d\times\mathbb{R}_+^\star$. Assume that there exist $\mu\in\mathcal{M}(\mathbb{R}^d\times\mathbb{R}_+)$ and $\mu_0\in\mathcal{M}(\mathbb{R}^d)$ such that (29.9) holds. Let u be the unique entropy weak solution of (23.1)–(23.2) (note that $u\in L^\infty(\mathbb{R}^d\times\mathbb{R}_+^\star)$ is solution to (29.9) with u instead of $\tilde{u}$ and $\mu=0$, $\mu_0=0$).*

Then, for all $R>0$ and all $T>0$ there exists C_e and $\overline{R}$, only depending on R, T, $\boldsymbol{v}$, f and u_0, such that the following inequality holds:

$$\int_0^T\int_{B(0,R)}\big|\tilde{u}(x,t)-u(x,t)\big|\,\mathrm{d}x\,\mathrm{d}t$$
$$\leqslant C_e\Big(\mu_0\big(B(0,\overline{R})\big)+\big[\mu\big(B(0,\overline{R})\times[0,T]\big)\big]^{1/2}+\mu\big(B(0,\overline{R})\times[0,T]\big)\Big).$$

Recall that $B(0,R)=\{x\in\mathbb{R}^d;\ |x|<R\}$.

PROOF. The proof of Theorem 29.3 is close to that of Step 2 in the proof of Theorem 28.1. It uses Lemma 29.2, the proof of which is given in Section 29.2 above.

Let $R>0$ and $T>0$. One sets $\omega=VM$, where V is given in Assumption 23.1 and M is the Lipschitz constant of f in $[U_m,U_M]$ (indeed, since $f\in C^1(\mathbb{R},\mathbb{R})$, one has $M=\sup\{|f'(s)|;\ s\in[U_m,U_M]\}$).

Let $\rho\in C_c^1(\mathbb{R}_+,[0,1])$ be a function such that $\rho(r)=1$ if $r\in[0,R+\omega T]$, $\rho(r)=0$ if $r\in[R+\omega T+1,\infty)$ and $\rho'(r)\leqslant 0$, for all $r\in\mathbb{R}_+$ (ρ only depends on R, T, $\boldsymbol{v}$, f and u_0).

One takes, in (29.10), ψ defined by

$$\psi(x,t)=\begin{cases}\rho\big(|x|+\omega t\big)\dfrac{T-t}{T}, & \text{for } x\in\mathbb{R}^d \text{ and } t\in[0,T],\\[1ex] 0, & \text{for } x\in\mathbb{R}^d \text{ and } t\geqslant T.\end{cases}$$

Note that $\rho(|x|+\omega t)=1$, if $(x,t)\in B(0,R)\times[0,T]$.

The function ψ is not in $C_c^\infty(\mathbb{R}^d\times\mathbb{R}_+,\mathbb{R}_+)$, but, using a usual regularization technique, it may be proved that such a function can be considered in (29.10), in which case inequality (29.10) writes, with $\overline{R}=R+\omega T+1$,

$$\begin{aligned}\int_0^T\int_{\mathbb{R}^d}\Bigg[&\big|\tilde{u}(x,t)-u(x,t)\big|\bigg(\frac{T-t}{T}\omega\rho'\big(|x|+\omega t\big)-\frac{1}{T}\rho\big(|x|+\omega t\big)\bigg)\\&+\big(f\big(\tilde{u}(x,t)\top u(x,t)\big)-f\big(\tilde{u}(x,t)\bot u(x,t)\big)\big)\\&\times\frac{T-t}{T}\rho'\big(|x|+\omega t\big)\bigg(\boldsymbol{v}(x,t)\cdot\frac{x}{|x|}\bigg)\Bigg]\,\mathrm{d}x\,\mathrm{d}t\\&\geqslant -C\big(\mu_0\big(B(0,\overline{R})\big)+\big(\mu\big(B(0,\overline{R})\times[0,T]\big)\big)^{1/2}+\mu\big(B(0,\overline{R})\times[0,T]\big)\big),\end{aligned}$$

where C only depends on R, T, $\boldsymbol{v}$, f and u_0.

Since $\omega=VM$ and $\rho'\leqslant 0$, one has

$$\begin{aligned}&\big(f\big(\tilde{u}(x,t)\top u(x,t)\big)-f\big(\tilde{u}(x,t)\bot u(x,t)\big)\big)\frac{T-t}{T}\rho'\big(|x|+\omega t\big)\bigg(\boldsymbol{v}(x,t)\cdot\frac{x}{|x|}\bigg)\\&\quad\leqslant\big|\tilde{u}(x,t)-u(x,t)\big|\frac{T-t}{T}\omega\big(-\rho'\big(|x|+\omega t\big)\big),\end{aligned}$$

and therefore, since $\rho(|x|+\omega t)=1$, if $(x,t)\in B(0,R)\times[0,T]$,

$$\begin{aligned}&\int_0^T\int_{B(0,R)}\big|\tilde{u}(x,t)-u(x,t)\big|\,\mathrm{d}x\,\mathrm{d}t\\&\quad\leqslant CT\big(\mu_0\big(B(0,\overline{R})\big)+\big(\mu\big(B(0,\overline{R})\times[0,T]\big)\big)^{1/2}+\mu\big(B(0,\overline{R})\times[0,T]\big)\big).\end{aligned}$$

This completes the proof of Theorem 29.3. □

Let us now conclude with the proofs of Theorems 29.1 (which gives an error estimate for the time explicit numerical scheme (24.4), (24.2)) and 24.3 (which gives an error estimate for the time implicit numerical scheme (24.6), (24.2)). There are easy consequences of Theorems 27.1 and 27.2 and of Theorem 29.3.

PROOF OF THEOREM 29.1. Under the assumptions of Theorem 29.1, let $\tilde{u}=u_{\mathcal{T},k}$. Thanks to the L^∞ estimate on $u_{\mathcal{T},k}$ (Lemma 25.1) and to Theorem 27.1, $\tilde{u}=u_{\mathcal{T},k}$ satisfies the hypotheses of Theorem 29.3 with $\mu=\mu_{\mathcal{T},k}$ and $\mu_0=\mu_{\mathcal{T}}$ (the measures $\mu_{\mathcal{T},k}$ and $\mu_{\mathcal{T}}$ are given in Theorem 27.1).

Let $R>0$ and $T>0$. Then, Theorem 29.3 gives the existence of C_1 and $\overline{R}$, only depending on R, T, $\boldsymbol{v}$, f and u_0, such that

$$\int_0^T\int_{B(0,R)}\big|u_{\mathcal{T},k}(x,t)-u(x,t)\big|\,\mathrm{d}x\,\mathrm{d}t$$

$$\leqslant C_1\big(\mu_{\mathcal{T}}\big(B(0,\overline{R})\big)+\big[\mu_{\mathcal{T},k}\big(B(0,\overline{R})\times[0,T]\big)\big]^{1/2}$$
$$+\,\mu_{\mathcal{T},k}\big(B(0,\overline{R})\times[0,T]\big)\big). \tag{29.25}$$

For h small enough, say $h\leqslant R_0$, one has $h<\overline{R}$ and $k<T$ (thanks to condition (24.3), note that R_0 only depends on R, T, $\boldsymbol{v}$, g, u_0, α and ξ).

Then, for $h<R_0$, Theorem 27.1 gives, with (29.25),

$$\int_0^T\int_{B(0,R)}\big|u_{\mathcal{T},k}(x,t)-u(x,t)\big|\,\mathrm{d}x\,\mathrm{d}t\leqslant C_1\big(Dh+\sqrt{C}h^{1/4}+C\sqrt{h}\big)\leqslant C_2h^{1/4},$$

where C_2 only depends on R, T, $\boldsymbol{v}$, g, u_0, α and ξ.

This gives the desired estimate (29.1) of Theorem 29.1 for $h<R_0$.

There remains the case $h\geqslant R_0$. This case is trivial since, for $h\geqslant R_0$,

$$\int_0^T\int_{B(0,R)}\big|u_{\mathcal{T},k}(x,t)-u(x,t)\big|\,\mathrm{d}x\,\mathrm{d}t$$
$$\leqslant 2\max\{-U_m,U_M\}\mathrm{m}\big(B(0,R)\times(0,T)\big)\leqslant C_3(R_0)^{1/4}\leqslant C_3h^{1/4},$$

for some C_3 only depending on R, T, $\boldsymbol{v}$, g, u_0, α and ξ.

This completes the proof of Theorem 29.1. □

PROOF OF THEOREM 29.2. The proof of Theorem 29.2 is very similar to that of Theorem 29.1 and we follow the proof of Theorem 29.1.

Under the assumptions of Theorem 29.2, using Theorem 27.2 instead of Theorem 27.1 gives that $\tilde{u}=u_{\mathcal{T},k}$ satisfies the hypotheses of Theorem 29.3 with $\mu=\mu_{\mathcal{T},k}$ and $\mu_0=\mu_{\mathcal{T}}$ (the measures $\mu_{\mathcal{T},k}$ and $\mu_{\mathcal{T}}$ are given in Theorem 27.2).

Let $R>0$ and $T>0$. Theorem 29.3 gives the existence of C_1 and $\overline{R}$, only depending on R, T, $\boldsymbol{v}$, f and u_0, such that (29.25) holds.

For $h<\overline{R}$ and $k<T$ Theorem 27.1 gives with (29.25),

$$\int_0^T\int_{B(0,R)}\big|u_{\mathcal{T},k}(x,t)-u(x,t)\big|\,\mathrm{d}x\,\mathrm{d}t$$
$$\leqslant C_1\big(Dh+\sqrt{C}\big(k+h^{1/2}\big)^{1/2}+C\big(k+h^{1/2}\big)\big)\leqslant C_2\big(k+h^{1/2}\big)^{1/2},$$

where C_2 only depends on R, T, $\boldsymbol{v}$, g, u_0, α.

This gives the desired estimate (29.2) of Theorem 29.2 for $h<\overline{R}$ and $k<T$.

There remains the cases $h\geqslant\overline{R}$ and $k\geqslant T$. These cases are trivial since

$$\int_0^T\int_{B(0,R)}\big|u_{\mathcal{T},k}(x,t)-u(x,t)\big|\,\mathrm{d}x\,\mathrm{d}t$$
$$\leqslant 2\max\{-U_m,U_M\}\mathrm{m}\big(B(0,R)\times(0,T)\big)\leqslant C_3\inf\big\{\overline{R}^{1/4},T^{1/2}\big\}$$

for some C_3 only depending on R, T, $\boldsymbol{v}$, g, u_0.

This completes the proof of Theorem 29.2. $\square$

29.4. Remarks and open problems

Theorem 29.1 gives an error estimate of order $h^{1/4}$ for the approximate solution of a nonlinear hyperbolic equation of the form $u_t + \operatorname{div}(\boldsymbol{v} f(u)) = 0$, with initial data in $L^\infty \cap BV$ by the explicit finite volume scheme (24.4) and (24.2), under a usual CFL condition $k \leqslant Ch$ (see (24.3)).

Note that, in fact, the same estimate holds if u_0 is only locally BV. More generally, if the initial data u_0 is only in L^∞, then one still obtains an error estimate in terms of the quantities

$$\varepsilon(r, S) = \sup\left\{\int_S \left|u(x,t) - u(x+\eta, t+\tau)\right| \mathrm{d}x\,\mathrm{d}t;\ |\eta| \leqslant \frac{1}{r},\ 0 \leqslant \tau \leqslant \frac{1}{r}\right\}$$

and

$$\varepsilon_0(r, S_0) = \sup\left\{\int_{S_0} \left|u_0(x) - u_0(x+\eta)\right| \mathrm{d}x;\ |\eta| \leqslant \frac{1}{r}\right\},$$

see (29.16) and (29.23). This is again an obvious consequence of Theorem 27.1 and Theorem 29.3.

We also considered the implicit schemes, which seem to be much more widely used in industrial codes in order to ensure their robustness. The implicit case required additional work in order

(i) to prove the existence of the solution to the finite volume scheme,

(ii) to obtain the "strong time BV" estimate (26.14) if $\boldsymbol{v}$ does not depend on t.

For $\boldsymbol{v}$ depending on t, Remark 29.2 yields an estimate of order $h^{1/4}$ if k behaves as h; however, in the case where $\boldsymbol{v}$ does not depend on t, then an estimate of order $h^{1/4}$ is obtained (in Theorem 29.2) for a behaviour of k as $\sqrt{h}$. Indeed, recent numerical experiments suggest that taking k of the order of $\sqrt{h}$ yields results of the same precision than taking k of the order of h, with an obvious reduction of the computational cost.

Note that the method described here may also be extended to higher-order schemes for the same equation, see CHAINAIS-HILLAIRET [1996]; other methods have been used for error estimates for higher-order schemes with a nonlinearity of the form $F(u)$, as in NOËLLE [1996]. However, it is still an open problem, to our knowledge, to improve the order of the error estimate in the case of higher-order schemes.

30. Nonlinear weak-⋆ convergence

The notion of nonlinear weak-⋆ convergence was used in Section 28.3. We give here the definition of this type of convergence and we prove that a bounded sequence of L^∞ converges, up to a subsequence, in the nonlinear weak-⋆ sense.

DEFINITION 30.1 (*Nonlinear weak-⋆ convergence*). Let Ω be an open subset of $\mathbb{R}^N$ ($N \geqslant 1$), $(u_n)_{n\in\mathbb{N}} \subset L^\infty(\Omega)$ and $u \in L^\infty(\Omega \times (0,1))$. The sequence $(u_n)_{n\in\mathbb{N}}$ converges towards u in the "nonlinear weak-⋆ sense" if

$$\int_\Omega g\big(u_n(x)\big)\varphi(x)\,\mathrm{d}x \to \int_0^1 \int_\Omega g\big(u(x,\alpha)\big)\varphi(x)\,\mathrm{d}x\,\mathrm{d}\alpha, \quad \text{as } n \to +\infty,$$
$$\forall \varphi \in L^1(\Omega),\ \forall g \in C(\mathbb{R},\mathbb{R}). \tag{30.1}$$

REMARK 30.1. Let Ω be an open subset of $\mathbb{R}^N$ ($N \geqslant 1$), $(u_n)_{n\in\mathbb{N}} \subset L^\infty(\Omega)$ and $u \in L^\infty(\Omega \times (0,1))$ such that $(u_n)_{n\in\mathbb{N}}$ converges towards u in the nonlinear weak-⋆ sense. Then, in particular, the sequence $(u_n)_{n\in\mathbb{N}}$ converges towards v in $L^\infty(\Omega)$, for the weak-⋆ topology, where v is defined by

$$v(x) = \int_0^1 u(x,\alpha)\,\mathrm{d}\alpha, \quad \text{for a.e. } x \in \Omega.$$

Therefore, the sequence $(u_n)_{n\in\mathbb{N}}$ is bounded in $L^\infty(\Omega)$ (thanks to the Banach–Steinhaus theorem). The following proposition gives that, up to a subsequence, a bounded sequence of $L^\infty(\Omega)$ converges in the nonlinear weak-⋆ sense.

PROPOSITION 30.1. *Let Ω be an open subset of $\mathbb{R}^N$ ($N \geqslant 1$) and $(u_n)_{n\in\mathbb{N}}$ be a bounded sequence of $L^\infty(\Omega)$. Then there exists a subsequence of $(u_n)_{n\in\mathbb{N}}$, which will still be denoted by $(u_n)_{n\in\mathbb{N}}$, and a function $u \in L^\infty(\Omega \times (0,1))$ such that the subsequence $(u_n)_{n\in\mathbb{N}}$ converges towards u in the nonlinear weak-⋆ sense.*

PROOF. This proposition is classical in the framework of "Young measures" and we only sketch the proof for the sake of completeness.

Let $(u_n)_{n\in\mathbb{N}}$ be a bounded sequence of $L^\infty(\Omega)$ and $r \geqslant 0$ such that $\|u_n\|_{L^\infty(\Omega)} \leqslant r$, $\forall n \in \mathbb{N}$.

Step 1 (*Diagonal process*). Thanks to the separability of the set of continuous functions defined from $[-r,r]$ into $\mathbb{R}$ (this set is endowed with the uniform norm) and the sequential weak-⋆ relative compactness of the bounded sets of $L^\infty(\Omega)$, there exists (using a diagonal process) a subsequence, which will still be denoted by $(u_n)_{n\in\mathbb{N}}$, such that, for any function $g \in C(\mathbb{R},\mathbb{R})$, the sequence $(g(u_n))_{n\in\mathbb{N}}$ converges in $L^\infty(\Omega)$ for the weak-⋆ topology towards a function $\mu_g \in L^\infty(\Omega)$.

Step 2 (*Young measure*). In this step, we prove the existence of a family $(m_x)_{x\in\Omega}$ such that

(1) for all $x \in \Omega$, m_x is a probability on $\mathbb{R}$ whose support is included in $[-r,+r]$ (i.e. m_x is a σ-additive application from the Borel σ-algebra of $\mathbb{R}$ in $\mathbb{R}_+$ such that $m_x(\mathbb{R}) = 1$ and $m_x(\mathbb{R} \setminus [-r,r]) = 0$),

(2) $\mu_g(x) = \int_\mathbb{R} g(s)\,\mathrm{d}m_x(s)$ for a.e. $x \in \Omega$ and for all $g \in C(\mathbb{R},\mathbb{R})$.

The family $m = (m_x)_{x\in\Omega}$ is called a "Young measure".

Let us first claim that it is possible to define $\mu_g \in L^\infty(\Omega)$ for $g \in C([-r,r],\mathbb{R})$ by setting $\mu_g = \mu_f$ where $f \in C(\mathbb{R},\mathbb{R})$ is such that $f = g$ on $[-r,r]$. Indeed, this definition is meaningful since if f and h are two elements of $C(\mathbb{R},\mathbb{R})$ such that $f = g$

on $[-r,r]$ then μ_f and μ_h are the same element of $L^\infty(\Omega)$ (i.e. $\mu_f = \mu_h$ a.e. on Ω) thanks to the fact that $-r \leqslant u_n \leqslant r$ a.e. on Ω and for all $n \in \mathbb{N}$.

For $x \in \Omega$, let

$$E_x = \left\{ g \in C\big([-r,r],\mathbb{R}\big);\ \lim_{h\to 0} \frac{1}{\mathrm{m}(B(0,h))} \int_{B(x,h)} \mu_g(z)\,\mathrm{d}z \text{ exists in } \mathbb{R} \right\},$$

where $B(x,h)$ is the ball of center x and radius h (note that $B(x,h) \subset \Omega$ for h small enough).

If $g \in E_x$, we set

$$\bar{\mu}_g(x) = \lim_{h\to 0} \frac{1}{\mathrm{m}(B(0,h))} \int_{B(x,h)} \mu_g(z)\,\mathrm{d}z.$$

Then, we define T_x from E_x in $\mathbb{R}$ by $T_x(g) = \bar{\mu}_g(x)$. It is easily seen that E_x is a vector space which contains the constant functions, that T_x is a linear application from E_x to $\mathbb{R}$ and that T_x is nonnegative (i.e. $g(s) \geqslant 0$ for all $s \in \mathbb{R}$ implies $T_x(g) \geqslant 0$). Hence, using a modified version of the Hahn–Banach theorem, one can prolong T_x into a linear nonnegative application $\overline{T}_x$ defined on the whole set $C([-r,r],\mathbb{R})$. By a classical Riesz theorem, there exists a (nonnegative) measure m_x on the Borel sets of $[-r,r]$ such that

$$\overline{T}_x(g) = \int_{-r}^{r} g(s)\,\mathrm{d}m_x(s), \quad \forall g \in C\big([-r,r],\mathbb{R}\big). \tag{30.2}$$

If $g(s) = 1$ for all $s \in [-r,r]$, the function g belongs to E_x and $\bar{\mu}_g(x) = 1$ (note that $\mu_g = 1$ a.e. on Ω). Hence, from (30.2), m_x is a probability over $[-r,r]$, and therefore a probability over $\mathbb{R}$ by prolonging it by 0 outside of $[-r,r]$. This gives the first item on the family $(m_x)_{x\in\Omega}$.

Let us prove now the second item on the family $(m_x)_{x\in\Omega}$. If $g \in C([-r,r],\mathbb{R})$ then $g \in E_x$ for a.e. $x \in \Omega$ and $\mu_g(x) = \bar{\mu}_g(x)$ for a.e. $x \in \Omega$ (this is a classical result, since $\mu_g \in L^1_{\mathrm{loc}}(\Omega)$, see RUDIN [1987]). Therefore, $\mu_g(x) = T_x(g) = \overline{T}_x(g)$ for a.e. $x \in \Omega$. Hence,

$$\mu_g(x) = \int_{-r}^{r} g(s)\,\mathrm{d}m_x(s) \quad \text{for a.e. } x \in \Omega,$$

for all $g \in C([-r,r],\mathbb{R})$ and therefore for all $g \in C(\mathbb{R},\mathbb{R})$. Finally, since the support of m_x is included in $[-r,r]$,

$$\mu_g(x) = \int_{\mathbb{R}} g(s)\,\mathrm{d}m_x(s) \quad \text{for a.e. } x \in \Omega,\ \forall g \in C(\mathbb{R},\mathbb{R}).$$

This completes Step 2.

Step 3 (*Construction of* u). It is well known that, if $\bar{m}$ is a probability on $\mathbb{R}$, one has

$$\int_{\mathbb{R}} g(s)\,\mathrm{d}\bar{m}(s) = \int_0^1 g\big(u(\alpha)\big)\,\mathrm{d}\alpha, \quad \forall g \in \mathcal{M}_b, \tag{30.3}$$

where $\mathcal{M}_b$ is the set of bounded measurable functions from $\mathbb{R}$ to $\mathbb{R}$ and with

$$u(\alpha) = \sup\{c \in \mathbb{R};\ \bar{m}\big((-\infty, c)\big) < \alpha\}, \quad \forall \alpha \in (0, 1).$$

Note that the function u is measurable, nondecreasing and left continuous. Furthermore, if the support of $\bar{m}$ is included in $[a, b]$ (for some $a, b \in \mathbb{R}$, $a < b$) then $u(\alpha) \in [a, b]$ for all $\alpha \in (0, 1)$ and (30.3) holds for all $g \in C(\mathbb{R}, \mathbb{R})$.

Applying this result to the measures m_x leads to the definition of u as

$$u(x, \alpha) = \sup\{c \in \mathbb{R};\ m_x\big((-\infty, c)\big) < \alpha\}, \quad \forall \alpha \in (0, 1),\ \forall x \in \Omega.$$

For all $x \in \Omega$, the function $u(x, \bullet)$ is measurable (from $(0, 1)$ to $\mathbb{R}$), nondecreasing, left continuous and takes its values in $[-r, r]$. Furthermore,

$$\mu_g(x) = \int_0^1 g\big(u(x, \alpha)\big)\,\mathrm{d}\alpha \quad \text{for a.e. } x \in \Omega,\ \forall g \in C(\mathbb{R}, \mathbb{R}).$$

Therefore,

$$\int_\Omega g\big(u_n(x)\big)\varphi(x)\,\mathrm{d}x \to \int_\Omega \left(\int_0^1 g\big(u(x, \alpha)\big)\,\mathrm{d}\alpha\right)\varphi(x)\,\mathrm{d}x, \quad \text{as } n \to \infty,$$
$$\forall \varphi \in L^1(\Omega),\ \forall g \in C(\mathbb{R}, \mathbb{R}).$$

In order to conclude the proof of Proposition 30.1, there remains to show that modifying u on a negligible set leads to a function (still denoted by u) measurable with respect to $(x, \alpha) \in \Omega \times (0, 1)$. Indeed, this measurability is needed in order to assert, e.g., applying Fubini's Theorem (see RUDIN [1987]), that

$$\int_\Omega \left(\int_0^1 g\big(u(x, \alpha)\big)\,\mathrm{d}\alpha\right)\varphi(x)\,\mathrm{d}x = \int_0^1 \left(\int_\Omega g\big(u(x, \alpha)\big)\varphi(x)\,\mathrm{d}x\right)\mathrm{d}\alpha,$$

for all $\varphi \in L^1(\Omega)$ and for all $g \in C(\mathbb{R}, \mathbb{R})$.

For all $g \in C(\mathbb{R}, \mathbb{R})$, one chooses for μ_g (which belongs to $L^\infty(\Omega)$) a bounded measurable function from Ω to $\mathbb{R}$.

Let us define $\mathcal{E} = \{g_{a,b};\ a, b \in \mathbb{Q},\ a < b\}$ where $g_{a,b} \in C(\mathbb{R}, \mathbb{R})$ is defined by

$$g_{a,b}(x) = \begin{cases} 1 & \text{if } x \leqslant a, \\ \dfrac{x - b}{a - b} & \text{if } a < x < b, \\ 0 & \text{if } x \geqslant b. \end{cases}$$

Since $\mathcal{E}$ is a countable subset of $C(\mathbb{R}, \mathbb{R})$, there exists a Borel subset A of Ω such that $\mathrm{m}(A) = 0$ and

$$\mu_g(x) = \int_\mathbb{R} g(s)\,\mathrm{d}m_x(s), \quad \forall x \in \Omega \setminus A,\ \forall g \in \mathcal{E}. \tag{30.4}$$

Define for all $\alpha \in (0, 1)$ $v(., \alpha)$ by

$$v(x, \alpha) = \begin{cases} 0 & \text{if } x \in A, \\ \sup\{c \in \mathbb{R},\ m_x((-\infty, c)) < \alpha\} & \text{if } x \in \Omega \setminus A, \end{cases}$$

so that $u = v$ on $(\Omega \setminus A) \times (0, 1)$ (and then $u = v$ a.e. on $\Omega \times (0, 1)$).

Let us now prove that v is measurable from $\Omega \times (0, 1)$ to $\mathbb{R}$ (this will conclude the proof of Proposition 30.1).

Since $v(x, \bullet)$ is left continuous on $(0, 1)$ for all $x \in \Omega$, proving that $v(\bullet, \alpha)$ is measurable (from Ω to $\mathbb{R}$) for all $\alpha \in (0, 1)$ leads to the measurability of v on $\Omega \times (0, 1)$ (this is also classical, see RUDIN [1987]).

There remains to show the measurability of $v(\bullet, \alpha)$ for all $\alpha \in (0, 1)$.

Let $\alpha \in (0, 1)$ (in the following, α is fixed). Let us set $w = v(\bullet, \alpha)$ and define, for $c \in \mathbb{R}$,

$$f_c(x) = m_x\big((-\infty, c)\big) - \alpha, \quad x \in \Omega \setminus A,$$

so that $v(x, \alpha) = w(x) = \sup\{c \in \mathbb{R},\ f_c(x) < 0\}$ for all $x \in \Omega \setminus A$.

Using (30.4) leads to

$$m_x\big((-\infty, c)\big) = \sup\big\{\mu_g(x),\ g \leqslant 1_{(-\infty,c)} \text{ and } g \in \mathcal{E}\big\}, \quad \forall x \in \Omega \setminus A.$$

Then, the function $f_c : \Omega \setminus A \to \mathbb{R}$ is measurable as the supremum of a countable set of measurable functions (recall that μ_g is measurable for all $g \in \mathcal{E}$).

In order to prove the measurability of w (from Ω to $\mathbb{R}$), it is sufficient to prove that $\{x \in \Omega \setminus A;\ w(x) \geqslant a\}$ is a Borel set, for all $a \in \mathbb{R}$ (recall that $w = 0$ on A).

Let $a \in \mathbb{R}$, since $f_c(x)$ is nondecreasing with respect to c, one has

$$\big\{x \in \Omega \setminus A;\ w(x) \geqslant a\big\} = \bigcap_{n>0} \big\{x \in \Omega \setminus A;\ f_{a-1/n}(x) < 0\big\}.$$

Then $\{x \in \Omega \setminus A;\ w(x) \geqslant a\}$ is measurable, thanks to the measurability of f_c for all $c \in \mathbb{R}$.

This concludes the proof of Proposition 30.1. □

REMARK 30.2. Let Ω be an open subset of $\mathbb{R}^N$ $(N \geqslant 1)$, $(u_n)_{n\in\mathbb{N}} \subset L^\infty(\Omega)$ and $u \in L^\infty(\Omega \times (0, 1))$ such that $(u_n)_{n\in\mathbb{N}}$ converges towards u in the nonlinear weak-$\star$ sense. Assume that u does not depend on α, i.e. there exists $v \in L^\infty(\Omega)$ such that $u(x, \alpha) = v(x)$ for a.e. $(x, \alpha) \in \Omega \times (0, 1)$. Then, it is easy to prove that $(u_n)_{n\in\mathbb{N}}$ converges towards u in $L^p(B)$ for all $1 \leqslant p < \infty$ and all bounded subset B of Ω. Indeed, let B be a bounded subset of Ω. Taking, in (30.1), $g(s) = s^2$ (for all $s \in \mathbb{R}$) and $\varphi = 1_B$ and also $g(s) = s$ (for all $s \in \mathbb{R}$) and $\varphi = 1_B v$ leads to

$$\int_B \big(u_n(x) - v(x)\big)^2 \,\mathrm{d}x \to 0, \quad \text{as } n \to \infty.$$

This proves that $(u_n)_{n\in\mathbb{N}}$ converges towards u in $L^2(B)$. The convergence of $(u_n)_{n\in\mathbb{N}}$ towards u in $L^p(B)$ for all $1 \leqslant p < \infty$ is then an easy consequence of the $L^\infty(\Omega)$ bound on $(u_n)_{n\in\mathbb{N}}$ (see Remark 30.1).

31. A stabilized finite element method

In this section, we shall try to compare the finite element method to the finite volume method for the discretization of a nonlinear hyperbolic equation. It is well known that the use of the finite element is not straightforward in the case of hyperbolic equations, since the lack of coerciveness of the operator yields a lack of stability of the finite element scheme. There are several techniques to stabilize these schemes, which are beyond the scope of this work. Here, as in SELMIN [1993], we are interested in viewing the finite element as a finite volume method, by writing it in a conservative form, and using a stabilization as in the third item of Example 21.1.

Let $F \in C^1(\mathbb{R}, \mathbb{R}^2)$, consider the following scalar conservation law:

$$u_t(x,t) + \operatorname{div}\bigl(F(u)\bigr)(x,t) = 0, \quad x \in \mathbb{R}^2,\ t \in \mathbb{R}_+, \tag{31.1}$$

with an initial condition. Let $\mathcal{T}$ be a triangular mesh of $\mathbb{R}^2$, well suited for the finite element method. Let $\mathcal{S}$ denote the set of nodes of this mesh, and let $(\phi_j)_{j\in\mathcal{S}}$ be the classical piecewise bilinear shape functions. Following the finite element principles, let us look for an approximation of u in the space spanned by the shape functions ϕ_j; hence, at time $t_n = nk$ (where k is the time step), we look for an approximate solution of the form

$$u(\bullet, t_n) = \sum_{j\in\mathcal{S}} u_j^n \phi_j;$$

then, multiplying (31.1) by ϕ_i, integrating over $\mathbb{R}^2$, approximating $F(\sum_{j\in\mathcal{S}} u_j^n \phi_j)$ by $\sum_{j\in\mathcal{S}} F(u_j^n)\phi_j$ and using the mass lumping technique on the mass matrix yields the following scheme (with the explicit Euler scheme for the time discretization):

$$\frac{u_i^{n+1} - u_i^n}{k} \int_{\mathbb{R}^2} \phi_i(x)\,\mathrm{d}x - \sum_{j\in\mathcal{S}} F\bigl(u_j^n\bigr) \cdot \int_{\mathbb{R}^2} \phi_j(x)\nabla\phi_i(x)\,\mathrm{d}x = 0,$$

which writes, noting that $\int \phi_j(x)\nabla\phi_i(x)\,\mathrm{d}x = -\int \phi_i(x)\nabla\phi_j(x)\,\mathrm{d}x$ and that $\sum_{j\in\mathcal{S}} \nabla\phi_j(x) = 0$,

$$\frac{u_i^{n+1} - u_i^n}{k} \int_{\mathbb{R}^2} \phi_i(x)\,\mathrm{d}x + \sum_{j\in\mathcal{S}} \bigl(F\bigl(u_i^n\bigr) + F\bigl(u_j^n\bigr)\bigr) \cdot \int_{\mathbb{R}^2} \phi_i(x)\nabla\phi_j(x)\,\mathrm{d}x = 0.$$

This last equality may also be written

$$\frac{u_i^{n+1} - u_i^n}{k} \int_{\mathbb{R}^2} \phi_i(x)\,\mathrm{d}x + \sum_{j \in \mathcal{S}} E_{i,j} = 0,$$

where

$$E_{i,j} = \tfrac{1}{2}\big(F(u_i^n) + F(u_j^n)\big) \cdot \int_{\mathbb{R}^2} \big(\phi_i(x)\nabla\phi_j(x) - \phi_j(x)\nabla\phi_i(x)\big)\,\mathrm{d}x.$$

Note that $E_{j,i} = -E_{i,j}$.

This is a centered and therefore unstable scheme. One way to stabilize it is to replace $E_{i,j}^n$ by

$$\tilde{E}_{i,j}^n = E_{i,j}^n + D_{i,j}\big(u_i^n - u_j^n\big),$$

where $D_{i,j} = D_{j,i}$ (in order for the scheme to remain "conservative") and $D_{i,j} \geqslant 0$ is chosen large enough so that $E_{i,j}^n$ is a nondecreasing function of u_i^n and a nonincreasing function of u_j^n, which ensure the stability of the scheme, under a so called CFL condition, and does not change the "consistency" (see (21.6) and Remark 29.1).

32. Moving meshes

For some evolution problems the use of time variable control volumes is advisable, e.g., when the domain of study changes with time. This is the case, e.g., for the simulation of a flow in a porous medium, when the porous medium is heterogeneous and its geometry changes with time. In this case, the mesh is required to move with the medium. The influence of the moving mesh on the finite volume formulation can be explained by considering the following simple transport equation:

$$\mu_t(x,t) + \operatorname{div}(u\boldsymbol{v})(x,t) = 0, \quad x \in \mathbb{R}^2,\ t \in \mathbb{R}_+, \tag{32.1}$$

where $\boldsymbol{v}$ depends on the unknown u (and possibly on other unknowns). Let k be the time step, and set $t_n = nk$, $n \in \mathbb{N}$. Let $\mathcal{T}(t)$ be the mesh at time t. Since the mesh moves, the elements of the mesh vary in time. For a fixed $n \in \mathbb{N}$, let $R(K,t)$ be the domain of $\mathbb{R}^2$ occupied by the element K ($K \in \mathcal{T}(t_n)$) at time t, $t \in [t_n, t_{n+1}]$, that is $R(K,t_n) = K$. Let $\boldsymbol{v}_s(x,t)$ be the velocity of the displacement of the mesh at point $x \in \mathbb{R}^2$ and for all $t \in [t_n, t_{n+1}]$ (note that $\boldsymbol{v}_s(x,t) \in \mathbb{R}^2$). Let u_K^n and u_K^{n+1} be the discrete unknowns associated to element K at times t_n and t_{n+1} (they can be considered as the approximations of the mean values of $u(\bullet, t_n)$ and $u(\bullet, t_{n+1})$ over $R(K,t_n)$ and $R(K,t_{n+1})$ respectively). The discretization of (32.1) must take into account the evolution of the mesh in time. In order to do so, let us first consider the following differential equation with initial condition:

$$\begin{aligned}\frac{\partial y}{\partial t}(x,t) &= -\boldsymbol{v}_s\big(y(x,t),t\big), \quad t\in[t_n,t_{n+1}],\\ y(x,t_n) &= x.\end{aligned} \tag{32.2}$$

Under suitable assumptions on $\boldsymbol{v}_s$ (assume, e.g., that $\boldsymbol{v}_s$ is continuous, Lipschitz continuous with respect to its first variable and that the Lipschitz constant is integrable with respect to its second variable), problem (32.2) has, for all $x\in\mathbb{R}^2$, a unique (global) solution. For $x\in\mathbb{R}^2$, define the function $y(x,\bullet)$ from $[t_n,t_{n+1}]$ to $\mathbb{R}^2$ as the solution of problem (32.2). Let $(\varphi_p)_{p\in\mathbb{N}}\subset C_c^1(\mathbb{R}^2,\mathbb{R}_+)$ such that $0\leqslant\varphi_p(x)\leqslant 1$ for $x\in\mathbb{R}^2$ and for all $p\in\mathbb{N}$, and such that $\varphi_p\to 1_K$ a.e. as $p\to+\infty$. Multiplying (32.1) by $\psi_p(x,t)=\varphi_p(y(x,t))$ and integrating over $\mathbb{R}^2$ yields

$$\begin{aligned}\int_{\mathbb{R}^2}\bigg(&\frac{\partial(u\psi_p)}{\partial t}(x,t)+u(x,t)\nabla\varphi_p\big(y(x,t)\big)\bullet\boldsymbol{v}_s\big(y(x,t),t\big)\\ &-(u\boldsymbol{v})(x,t)\bullet\nabla\psi_p(x,t)\bigg)\,\mathrm{d}x=0.\end{aligned} \tag{32.3}$$

Using the explicit Euler discretization in time on Eq. (32.3) and denoting by $u^n(x)$ a (regular) approximate value of $u(x,t_n)$ yields

$$\begin{aligned}&\int_{\mathbb{R}^2}\frac{1}{k}\big(u^{n+1}(x)\psi_p(x,t_{n+1})-u^n(x)\psi_p(x,t_n)\big)\,\mathrm{d}x\\ &\quad+\int_{\mathbb{R}^2}u^n(x)\big(\boldsymbol{v}_s(x,t_n)-\boldsymbol{v}(x,t_n)\big)\bullet\nabla\varphi_p(x)\,\mathrm{d}x=0,\end{aligned}$$

which also gives (noting that $\psi_p(x,t)=\varphi_p(y(x,t))$)

$$\begin{aligned}&\int_{\mathbb{R}^2}\frac{1}{k}\big(u^{n+1}(x)\varphi_p\big(y(x,t_{n+1})\big)-u^n(x)\varphi_p\big(y(x,t_n)\big)\big)\,\mathrm{d}x\\ &\quad-\int_{\mathbb{R}^2}\mathrm{div}\big(u^n(\boldsymbol{v}_s-\boldsymbol{v})\big)(x,t_n)\bullet\varphi_p(x)\,\mathrm{d}x=0.\end{aligned} \tag{32.4}$$

Letting p tend to infinity and noting that $1_K(y(x,t_n))=1_{R(K,t_n)}(x)$ and $1_K(y(x,t_{n+1}))=1_{R(K,t_{n+1})}(x)$, (32.4) becomes

$$\begin{aligned}&\frac{1}{k}\bigg(\int_{R(K,t_{n+1})}u^{n+1}(x)\,\mathrm{d}x-\int_{R(K,t_n)}u^n(x)\,\mathrm{d}x\bigg)\\ &\quad+\int_{R(K,t_n)}\mathrm{div}\big((\boldsymbol{v}-\boldsymbol{v}_s)u^n\big)(x,t_n)\,\mathrm{d}x=0,\end{aligned}$$

which can also be written

$$\frac{1}{k}\big(u_K^{n+1}\,\mathrm{m}\big(R(K,t_{n+1})\big) - u_K^n\,\mathrm{m}\big(R(K,t_n)\big)\big)$$
$$+\int_{\partial R(K,t_n)} (\boldsymbol{v}-\boldsymbol{v}_s)(x,t_n)\cdot\boldsymbol{n}_K(x,t_n)u^n(x)\,\mathrm{d}\gamma(x)=0,$$

where $u_K^n = [1/\mathrm{m}(R(K,t_n))]\int_{R(K,t_n)} u^n(x)\,\mathrm{d}x$ and $u_K^{n+1} = [1/\mathrm{m}(R(K,t_{n+1}))] \times \int_{R(K,t_{n+1})} u^{n+1}(x)\,\mathrm{d}x$. Recall that $\boldsymbol{n}_K$ denotes the normal to ∂K, outward to K. The complete discretization of the problem uses some additional equations (on $\boldsymbol{v}$, $\boldsymbol{v}_s,\ldots$).

REMARK 32.1. The above considerations concern a pure convection equation. In the case of a convection–diffusion equation, such a moving mesh may become nonadmissible in the sense of Definitions 9.1 or 10.1. It is an interesting open problem to understand what should be done in that case.

CHAPTER VII

Systems

In Chapters II–VI, the finite volume was successively investigated for the discretization of elliptic, parabolic, and hyperbolic equations. In most scientific models, however, systems of equations have to be discretized. These may be partial differential equations of the same type or of different types, and they may also be coupled to ordinary differential equations or algebraic equations.

The discretization of systems of elliptic equations by the finite volume method is straightforward, following the principles which were introduced in Chapters II and III. Examples of the performance of the finite volume method for systems of elliptic equations on rectangular meshes, with "unusual" source terms (in particular, with source terms located on the edges or interfaces of the mesh) may be found in, e.g., ANGOT [1989] (see also references therein), FIARD and HERBIN [1994] (where a comparison to a mixed finite element formulation is also performed). Parabolic systems are treated similarly as elliptic systems, with the addition of a convenient time discretization.

A huge literature is devoted to the discretization of hyperbolic systems of equations, in particular to systems related to the compressible Euler equations, using structured or unstructured meshes. We shall give only a short insight on this subject in Section 33, without any convergence result. Indeed, very few theoretical results of convergence of numerical schemes are known on this subject. We refer to GODLEWSKI and RAVIART [1996] and references therein for a more complete description of the numerical schemes for hyperbolic systems.

Finite volume methods are also well adapted to the discretization of systems of equations of different types (e.g., an elliptic or parabolic equation coupled with hyperbolic equations). Some examples are considered in Sections 34 and 35. The classical case of incompressible Navier–Stokes (for which, generally, staggered grids are used) and examples which arise in the simulation of a multiphase flow in a porous medium are described. The latter example also serves as an illustration of how to deal with algebraic equations and inequations.

33. Hyperbolic systems of equations

Let us consider a hyperbolic system consisting of m equations (with $m \geqslant 1$). The unknown of the system is a function $u = (u_1, \dots, u_m)^t$, from $\overline{\Omega} \times [0, T]$ to $\mathbb{R}^m$, where

Ω is an open set of $\mathbb{R}^d$ (i.e. $d \geqslant 1$ is the space dimension), and u is a solution of the following system:

$$\frac{\partial u_i}{\partial t}(x,t) + \sum_{j=1}^{d} \frac{\partial G_{i,j}}{\partial x_j}(x,t) = g_i\big(x,t,u(x,t)\big),$$
$$x = (x_1, \dots, x_d) \in \Omega,\ t \in (0,T),\ i = 1, \dots, m, \tag{33.1}$$

where

$$G_{i,j}(x,t) = F_{i,j}\big(x,t,u(x,t)\big),$$

and the functions $F_j = (F_{1,j}, \dots, F_{m,j})^t$ $(j = 1, \dots, d)$ and $g = (g_1, \dots, g_m)^t$ are given functions from $\overline{\Omega} \times [0,T] \times \mathbb{R}^m$ (indeed, generally, a part of $\mathbb{R}^m$, instead of $\mathbb{R}^m$) to $\mathbb{R}^m$. The function $F = (F_1, \dots, F_d)$ is assumed to satisfy the usual hyperbolicity condition, i.e., for any (unit) vector of $\mathbb{R}^d$, $\boldsymbol{n}$, the derivative of $F \bullet \boldsymbol{n}$ with respect to its third argument (which can be considered as an $m \times m$ matrix) has only real eigenvalues and is diagonalizable.

Note that in real applications, diffusion terms may also be present in the equations, we shall omit them here. In order to complete system (33.1), an initial condition for $t = 0$ and adequate boundary conditions for $x \in \partial\Omega$ must be specified.

In the first section (Section 33.1), we shall only briefly describe the general method of discretization by finite volume and some classical schemes. In the subsequent sections, some possible treatments of difficulties appearing in real simulations will be given.

33.1. Classical schemes

Let us first describe some classical finite volume schemes for the discretization of (33.1) with initial and boundary conditions, using the concepts and notations which were introduced in Chapter VI. Let $\mathcal{T}$ be an admissible mesh in the sense of Definition 24.1 and k be the time step, which is assumed to be constant (the generalization to a variable time step is easy). We recall that the interface, $K|L$, between any two elements K and L of $\mathcal{T}$ is assumed to be included in a hyperplane of $\mathbb{R}^d$. The discrete unknowns are the u_K^n, $K \in \mathcal{T}$, $n \in \{0, \dots, N_k + 1\}$, with $N_k \in \mathbb{N}$, $(N_k + 1)k = T$. For $K \in \mathcal{T}$, let $\mathcal{N}(K)$ be the set of its neighbours, i.e. the set of elements L of $\mathcal{T}$ such that the $(d-1)$ Lebesgue measure of $K|L$ is positive. For $L \in \mathcal{N}(K)$, let $\boldsymbol{n}_{K,L}$ be the unit normal vector to $K|L$ oriented from K to L. Let $t_n = nk$, for $n \in \{0, \dots, N_k + 1\}$.

A finite volume scheme writes

$$\mathrm{m}(K)\frac{u_K^{n+1} - u_K^n}{k} + \sum_{L \in \mathcal{N}(K)} \mathrm{m}(K|L) F_{K,L}^n = \mathrm{m}(K) g_K^n,$$
$$K \in \mathcal{T},\ n \in \{0, \dots, N_k\}, \tag{33.2}$$

where

(1) $\mathrm{m}(K)$ (resp. $\mathrm{m}(K|L)$) denotes the d (resp. $d-1$) Lebesgue measure of K (resp. $K|L$),

(2) the quantity g_K^n, which depends on u_K^n (or u_K^{n+1} or u_K^n and u_K^{n+1}), for $K \in \mathcal{T}$, is some "consistent" approximation of g on element K, between times t_n and t_{n+1} (we do not discuss this approximation here),

(3) the quantity $F_{K,L}^n$, which depends on the set of discrete unknowns u_M^n (or u_M^{n+1} or u_M^n and u_M^{n+1}) for $M \in \mathcal{T}$, is an approximation of $F \bullet \boldsymbol{n}_{K,L}$ on $K|L$ between times t_n and t_{n+1}.

In order to obtain a "good" scheme, this approximation of $F \bullet \boldsymbol{n}_{K,L}$ has to be consistent, conservative (that is $F_{K,L}^n = -F_{L,K}^n$) and must ensure some stability properties on the approximate solution given by the scheme (indeed, one also needs some consistency with respect to entropies, when entropies exist...). Except in the scalar case, it is not so easy to see what kind of stability properties is needed.... Indeed, in the scalar case, i.e. $m = 1$, taking $g = 0$ and $\Omega = \mathbb{R}^d$ (for simplicity), it is essentially sufficient to have an L^∞ estimate (that is a bound on u_K^n independent of K, n, and of the time and space discretizations) and a "touch" of "BV estimate" (see, e.g., Chapters V and VI and CHAINAIS-HILLAIRET [1996] for more precise assumptions). In the case $m > 1$, it is not generally possible to give stability properties from which a mathematical proof of convergence could be deduced. However, it is advisable to require some stability properties such as the positivity of some quantities depending on the unknowns; in the case of flows, the required stability may be the positivity of the density, energy, pressure...; the positivity of these quantities may be essential for the computation of $F(u)$ or for its hyperbolicity.

The computation of $F_{K,L}^n$ is often performed, at each "interface", by solving the following 1D (for the space variable) system (where, for simplicity, the possible dependency of F with respect to x and t is omitted):

$$\frac{\partial u}{\partial t}(z,t) + \frac{\partial f_{K,L}(u)}{\partial z}(z,t) = 0, \tag{33.3}$$

where $f_{K,L}(u)(z,t) = F \bullet \boldsymbol{n}_{K,L}(u(z,t))$, for all $z \in \mathbb{R}$ and $t \in (0,T)$, which gives consistency, conservativity (and, hopefully, stability) of the final scheme (i.e. (33.2)). To be more precise, in the case of lower order schemes, $F_{K,L}^n$ may be taken as: $F_{K,L}^n = F \bullet \boldsymbol{n}_{K,L}(w)$ where w is the solution for $z = 0$ of (33.3) with initial conditions $u(x,0) = u_K^n$ if $x < 0$ and $u(x,0) = u_L^n$ if $x > 0$. Note that the variable z lies in $\mathbb{R}$, so that the multidimensional problem has therefore been transformed (as in Chapter VI) into a succession of one-dimensional problems. Hence, in the following, we shall mainly keep to the case $d = 1$.

Let us describe two classical schemes, namely the Godunov scheme and the Roe scheme, in the case $d = 1$, $\Omega = \mathbb{R}$, $F(x,t,u) = F(u)$ and $g = 0$ (but $m \geqslant 1$), in which case system (33.1) becomes

$$\frac{\partial u}{\partial t}(x,t) + \frac{\partial F(u)}{\partial x}(x,t) = 0, \quad x \in \mathbb{R},\ t \in (0,T). \tag{33.4}$$

In order to complete this system, an initial condition must be specified, the discretization of which is standard.

Let $\mathcal{T}$ be an admissible mesh in the sense of Definition 20.1, i.e. $\mathcal{T} = (K_i)_{i\in\mathbb{Z}}$, with $K_i = (x_{i-1/2}, x_{i+1/2})$, with $x_{i-1/2} < x_{i+1/2}$, $i \in \mathbb{Z}$. One sets $h_i = x_{i+1/2} - x_{i-1/2}$, $i \in \mathbb{Z}$. The discrete unknowns are u_i^n, $i \in \mathbb{Z}$, $n \in \{0, \dots, N_k + 1\}$ and the scheme (33.2) then writes

$$h_i \frac{u_i^{n+1} - u_i^n}{k} + F_{i+1/2}^n - F_{i-1/2}^n = 0, \quad i \in \mathbb{Z},\ n \in \{0, \dots, N_k\}, \tag{33.5}$$

where $F_{i+1/2}^n$ is a consistent approximation of $F(u(x_{i+1/2}, t_n))$. This scheme is clearly conservative (in the sense defined above). Let us consider explicit schemes, so that $F_{i+1/2}^n$ is a function of u_j^n, $j \in \mathbb{Z}$. The principle of the Godunov scheme (GODUNOV [1976]) is to take $F_{i+1/2}^n = F(w)$ where w is the solution, for $x = 0$ (and any $t > 0$), of the following (Riemann) problem

$$\frac{\partial u}{\partial t}(x,t) + \frac{\partial F(u)}{\partial x}(x,t) = 0, \quad x \in \mathbb{R},\ t \in \mathbb{R}_+, \tag{33.6}$$

$$u(x,0) = \begin{cases} u_i^n, & \text{if } x < 0, \\ u_{i+1}^n, & \text{if } x > 0. \end{cases} \tag{33.7}$$

Then, w depends on u_i^n, u_{i+1}^n and F.

The time step is limited by the so called "CFL condition", which writes $k \leqslant Lh_i$, for all $i \in \mathbb{Z}$, where L is given by F and the initial condition. The quantity u_i^{n+1}, given by the Godunov scheme, see GODUNOV [1976], is, for all $i \in \mathbb{Z}$, the mean value on K_i of the exact solution at time k of (33.4) with the initial condition (at time $t = 0$) u_0 defined, a.e. on $\mathbb{R}$, by $u_0(x) = u_i^n$ if $x_{i-1/2} < x < x_{i+1/2}$.

The Godunov scheme is an efficient scheme (consistent, conservative, stable), sometimes too diffusive (especially if k is far from Lh_i defined above), but easy improvements are possible, such as the MUSCL technique, see below and Section 22. Its principal drawback is its difficult implementation for many problems, indeed the computation of $F(w)$ can be impossible or too expensive. For instance, this computation may need a nontrivial parametrization of the nonlinear waves. Note also that F is generally not given directly as a function of u (the components of u are called "conservative unknowns") but as a function of some "physical" unknowns (e.g., pressure, velocity, energy...), and the passage from u to these physical unknowns (or the converse) is often not so easy... it may be the consequence of expensive and implicit calculations, using, e.g., Newton's algorithm.

Due to this difficulty of implementation, some "Godunov type" schemes were developed (see HARTEN, LAX and VAN LEER [1983]). The idea is to take, for u_i^{n+1}, the mean value on K_i of an *approximate* solution at time k of (33.4) with the initial condition (at time $t = 0$), u_0, defined by $u_0(x) = u_i^n$, if $x_{i-1/2} < x < x_{i+1/2}$. In order for the scheme to be written under the conservative form (33.5), with a consistent approximation of the fluxes, this approximate solution must satisfy some consistency relation (another relation is needed for the consistency with entropies). One of the best known of this family of schemes is the Roe scheme (see ROE [1980, 1981]), where this approximate solution is computed by the solution of the following linearized Riemann problems:

$$\frac{\partial u(x,t)}{\partial t} + A\big(u_i^n, u_{i+1}^n\big)\frac{\partial u(x,t)}{\partial x} = 0, \quad x \in \mathbb{R},\ t \in \mathbb{R}_+, \tag{33.8}$$

$$u(x,0) = \begin{cases} u_i^n, & \text{if } x < 0, \\ u_{i+1}^n, & \text{if } x > 0, \end{cases} \tag{33.9}$$

where $A(\bullet,\bullet)$ is an $m \times m$ matrix, continuously depending on its two arguments, with only real eigenvalues, diagonalizable and satisfying the so called "Roe condition":

$$A(u,v)(u-v) = F(u) - F(v), \quad \forall u, v \in \mathbb{R}^m. \tag{33.10}$$

Thanks to (33.10), the Roe scheme can be written as (33.5) with

$$\begin{aligned} F_{i+1/2}^n &= F\big(u_i^n\big) + A^-\big(u_i^n, u_{i+1}^n\big)\big(u_i^n - u_{i+1}^n\big) \\ &\big(= F\big(u_{i+1}^n\big) + A^+\big(u_i^n, u_{i+1}^n\big)\big(u_i^n - u_{i+1}^n\big)\big), \end{aligned} \tag{33.11}$$

where $A^{\pm}$ are the classical nonnegative and nonpositive parts of the matrix A: let A be a matrix with only real eigenvalues, $(\lambda_p)_{p=1,\dots,m}$, and diagonalizable, let $(\varphi_p)_{p=1,\dots,m}$ be a basis of $\mathbb{R}^m$ associated to these eigenvalues. Then, the matrix A^+ is the matrix which has the same eigenvectors as A and has $(\max\{\lambda_p, 0\})_{p=1,\dots,m}$ as corresponding eigenvalues. The matrix A^- is $(-A)^+$.

Roe's scheme was proved to be an efficient scheme, often less expensive than Godunov's scheme, with, more or less the same limitation on the time step, the same diffusion effect and some lack of entropy consistency, which can be corrected. It has some properties of consistency and stability. Its main drawback is the difficulty of the computation of a matrix $A(u,v)$ satisfying (33.10). For instance, when it is possible to compute and diagonalize the derivative of F, $DF(u)$, one can take $A(u,v) = DF(u^\star)$, but the difficulty is to find $u^\star$ such that (33.10) holds (note that this condition is crucial in order to ensure conservativity of Roe's scheme). In some difficult cases, the Roe matrix is computed approximately by using a "limited expansion" with respect to some small parameter.

33.2. Rough schemes for complex hyperbolic systems

The aim of this section is to present some discretization techniques for "complex" hyperbolic systems. In many applications, the expressions of g and F which appear in (33.1) are rather "complex", and it is difficult or impossible to use classical schemes such as the 1D Godunov or Roe schemes or their standard extensions, for multidimensional problems, using 1D solvers on the interfaces of the mesh. This is the case of gas dynamics (Euler equations) with real gas, for which the state law (pressure as a function of density and internal energy) is tabulated or given by some complex analytical expressions. This is also the case when modelling multiphase flows in pipe-lines: the function

F is difficult to handle and highly depends on x and u, because, e.g., of changes of the geometry and slope of the pipe, of changes of the friction law or, more generally, of the varying nature of the flow. Most of the attempts given below were developed for this last situation. Other interesting cases of "complexity" are the treatment of boundary conditions (mathematical literature is rather scarce on this subject, see Section 33.4 for a first insight), and the way to handle the case where the eigenvalues (of the derivative of $F \cdot \boldsymbol{n}$ with respect to its third argument) are of very different magnitude, see Section 33.3. Another case of complexity is the treatment of nonconservative terms in the equations. One refers, e.g., to BRUN, HÉRARD, LEAL DE SOUSA and UHLMANN [1996] and references therein, for this important case.

Possible modifications of Godunov and Roe schemes (including "classical" improvements to avoid excessive artificial diffusion) are described now to handle "complex" systems. Because of the complexity of the models, the justification of the schemes presented here is rather numerical than mathematical. Many variations have also been developed, which are not presented here. Note that other approaches are also possible, see, e.g., GHIDAGLIA, KUMBARO and LE COQ [1996]. For simplicity, one considers the case $d = 1$, $\Omega = \mathbb{R}$, $F(x, t, u) = F(u)$ and $g = 0$ (but $m \geqslant 1$) described in Section 33.1, with the same notations. The Godunov and Roe schemes can both be written under the form (33.5) with $F^n_{i+1/2}$ computed as a function of u^n_i and u^n_{i+1}; both schemes are consistent (in the sense of Section 33.1, i.e. consistency of the "fluxes") since $F^n_{i+1/2} = F(u)$ if $u^n_i = u^n_{i+1} = u$.

Going further along this line of thought yields (among other possibilities, see below) the "VFRoe" scheme which is (33.5), i.e.:

$$h_i \frac{u_i^{n+1} - u_i^n}{k} + F^n_{i+1/2} - F^n_{i-1/2} = 0, \quad i \in \mathbb{Z},\ n \in \{0, \ldots, N_k\}, \tag{33.12}$$

with $F^n_{i+1/2} = F(w)$, where w is the solution of the linearized Riemann problem (33.8), (33.9), with $A(u^n_i, u^n_{i+1}) = DF(w^\star)$, i.e.:

$$\frac{\partial u(x,t)}{\partial t} + DF(w^\star)\frac{\partial u(x,t)}{\partial x} = 0, \quad x \in \mathbb{R},\ t \in \mathbb{R}_+, \tag{33.13}$$

$$u(x,0) = \begin{cases} u^n_i, & \text{if } x < 0, \\ u^n_{i+1}, & \text{if } x > 0, \end{cases} \tag{33.14}$$

where $w^\star$ is some value between u^n_i and u^n_{i+1} (e.g., $w^\star = (1/2)(u^n_i + u^n_{i+1})$). In this scheme, the Roe condition (33.10) is not required (note that it is naturally conservative, thanks to its finite volume origin). Hence, the VFRoe scheme appears to be a simplified version of the Godunov and Roe schemes. The study of the scalar case ($m = 1$) shows that, in order to have some stability, at least as much as in Roe's scheme, the choice of $w^\star$ is essential. In practice, the choice $w^\star = (1/2)(u^n_i + u^n_{i+1})$ is often adequate, at least for regular meshes.

REMARK 33.1. In Roe's scheme, the Roe condition (33.10) ensures conservativity . The VFRoe scheme is "naturally" conservative, and therefore no such condition is needed. Also note that the VFRoe scheme yields precise approximations of the shock velocities, without Roe's condition.

Numerical tests show the good behaviour of the VFRoe scheme. Its two main flaws are a lack of entropy consistency (as in Roe's scheme) and a large diffusion effect (as in the Godunov and Roe schemes). The first drawback can be corrected, as for Roe's scheme, with a nonparametric entropy correction inspired from HARTEN, HYMAN and LAX [1976] (see MASELLA, FAILLE, and GALLOUËT [1996]). The two drawbacks can be corrected with a classical MUSCL technique, which consists in replacing, in (33.9), u_i^n and u_{i+1}^n by $u_{i+1/2,-}^n$ and $u_{i+1/2,+}^n$, which depend on $\{u_j^n,\ j = i-1, i, i+1, i+2\}$ (see, e.g., Section 22 and GODLEWSKI and RAVIART [1996] or LEVEQUE [1990]). For stability reasons, the computation of the gradient of the unknown (cell by cell) and of the "limiters" is performed on some "physical" quantities (such as density, pressure, velocity for Euler equations) instead of u. The extension of the MUSCL technique to the case $d > 1$ is more or less straightforward.

This MUSCL technique improves the space accuracy (in the truncation error) and the numerical results are significantly better. However, stability is sometimes lost. Indeed, considering the linear scalar equation, one remarks that the scheme is antidiffusive when the limiters are not active, this might lead to a loss of stability. The time step must then be reduced (it is reduced by a factor 10 in severe situations...). In order to allow larger time steps, the time accuracy should be improved by using, e.g., an order 2 Runge–Kutta scheme (in the severe situations suggested above, the time step is then multiplied by a factor 4). Surprisingly, this improvement of time accuracy is used to gain stability rather than precision....

Several numerical experiments (see MASELLA, FAILLE, and GALLOUËT [1996]) were performed which prove the efficiency of the VFRoe scheme, such as the classical Sod tests (SOD [1978]). The shock velocities are exact, there are no oscillations.... For these tests, the treatment of the boundary conditions is straightforward. Throughout these experiments, the use of a MUSCL technique yields a significant improvement, while the use of a higher-order time scheme is not necessary. In one of the Sod tests, the entropy correction is needed.

A comparison between the VFRoe scheme and the Godunov scheme was performed by J.M. Hérard (personal communication) for the Euler equations on a Van Der Wals gas, for which a matrix satisfying (33.10) seems difficult to find. The numerical results are better with the VFRoe scheme, which is also much cheaper computationally. An improvment of the VFRoe scheme is possible, using, instead of (33.13)–(33.14), linearized Riemann problems associated to a nonconservative form of the initial system, namely system (33.4) or more generally System (33.1), for the computation of w (which gives the flux $F_{i+1/2}^n$ in (33.12) by the formula $F_{i+1/2}^n = F(w)$), see, e.g., BUFFARD, GALLOUËT and HÉRARD [1998] for a simple example.

In some more complex cases, the flux F may also highly, and not continuously, depend on the space variable x. In the space discretization, it is "natural" to set the discontinuities of F with respect to x on the boundaries of the mesh. The function F may change drastically from K_i to K_{i+1}. In this case, the implementation of the VFRoe scheme yields two additional difficulties:

(i) The matrix $A(u_i^n, u_{i+1}^n)$ in the linearized Riemann problem (33.8), (33.9) now depends on x:
$A(u_i^n, u_{i+1}^n) = D_u F(x, w^\star)$, where $w^\star$ is some value between u_i^n and u_{i+1}^n and $D_u F$ denotes the derivative of F with respect to its "u" argument.

ii) once the solution, w, of the linearized problem (33.8), (33.9), for $x = 0$ and any $t > 0$, is calculated, the choice $F_{i+1/2}^n = F(x, w)$ again depends on x.
The choice of $F_{i+1/2}^n$ (point (ii)) may be solved by remarking that, in Roe's scheme, $F_{i+1/2}^n$ may be written (thanks to (33.10)) as

$$F_{i+1/2}^n = \tfrac{1}{2}\big(F(u_i^n) + F(u_{i+1}^n)\big) + \tfrac{1}{2}A_{i+1/2}^n\big(u_i^n - u_{i+1}^n\big), \tag{33.15}$$

where $A_{i+1/2}^n = |A(u_i^n, u_{i+1}^n)|$, and $|A| = A^+ + A^-$.

Under this form, the second term of the right-hand side of (33.15) appears to be a stabilization term, which does not affect the consistency. Indeed, in the scalar case ($m = 1$), one has $A_{i+1/2}^n = |F(u_i^n) - F(u_{i+1}^n)|/|u_i^n - u_{i+1}^n|$, which easily yields the L^∞ stability of the scheme (but not the consistency with respect to the entropies). Moreover, the scheme is stable and consistent with respect to the entropies, under a Courant–Friedrichs–Levy (CFL) condition, if $F_{i+1/2}^n$ is nondecreasing with respect to u_i^n and nonincreasing with respect to u_{i+1}^n, which holds if $A_{i+1/2}^n \geqslant \sup\{|F'(s)|,\ s \in [u_i^n, u_{i+1}^n]$ or $[u_{i+1}^n, u_i^n]\}$. This remark suggests a slightly different version of the VFRoe scheme (closer to Roe's scheme), which is the scheme (33.12)–(33.14), taking

$$F_{i+1/2}^n = \tfrac{1}{2}\big(F(u_i^n) + F(u_{i+1}^n)\big) + \tfrac{1}{2}\big|DF(w^\star)\big|\big(u_i^n - u_{i+1}^n\big),$$

in (33.12), instead of $F_{i+1/2}^n = F(w)$. Note that it is also possible to take other convex combinations of $F(u_i^n)$ and $F(u_{i+1}^n)$ in the latter expression of $F_{i+1/2}^n$, without modifying the consistency of the scheme.

When F depends on x, the discontinuities of F being on the boundaries of the control volumes, the generalization of (33.15) is obvious, except for the choice of $A_{i+1/2}^n$. The quantity $F(u_i^n)$ is replaced by $F(x_i, u_i^n)$, where x_i is the center of K_i. Let us now turn to the choice of a convenient matrix $A_{i+1/2}^n$ for this modified VFRoe scheme, when F highly depends on x. A first possible choice is

$$A_{i+1/2}^n = \tfrac{1}{2}\big(\big|D_u F(x_i, u_i^n)\big| + \big|D_u F(x_{i+1}, u_{i+1}^n)\big|\big).$$

The following slightly different choice for $A_{i+1/2}^n$ seems, however, to give better numerical results (see FAILLE and HEINTZÉ [1999]). Let us define

$$A_i = D_u F(x_i, u_i^n), \quad \forall i \in \mathbb{Z}$$

(for the determination of $A_{i+1/2}^n$ the fixed index n is omitted). Let $(\lambda_p^{(i)})_{p=1,\dots,m}$ be the eigenvalues of A_i (with $\lambda_{p-1}^{(i)} \leqslant \lambda_p^{(i)}$, for all p) and $(\varphi_p^{(i)})_{p=1,\dots,m}$ a basis of $\mathbb{R}^m$ associated to these eigenvalues. Then, the matrix $A_{i+1/2}^{(-)}$ (resp. $A_{i+1/2}^{(+)}$) is the matrix which has the same eigenvectors as A_i (resp. A_{i+1}) and has $(\max\{|\lambda_p^{(i)}|, |\lambda_p^{(i+1)}|\})_{p=1,\dots,m}$ as corresponding eigenvalues. The choice of $A^n_{i+1/2}$ is

$$A^n_{i+1/2} = \frac{\lambda}{2}\left(A^{(-)}_{i+1/2} + A^{(+)}_{i+1/2}\right), \tag{33.16}$$

where λ is a parameter, the "normal" value of which is 1. Numerically, larger values of λ, say $\lambda = 2$ or $\lambda = 3$, are sometimes needed, in severe situations, to obtain enough stability. Too large values of λ yield too much artificial diffusion.

The new scheme is then (33.12)–(33.14), taking

$$F^n_{i+1/2} = \tfrac{1}{2}\left(F\left(x_i, u^n_i\right) + F\left(x_i, u^n_{i+1}\right)\right) + \tfrac{1}{2}A^n_{i+1/2}\left(u^n_i - u^n_{i+1}\right), \tag{33.17}$$

where $A_{i+1/2}t^n$ is defined by (33.16). It has, more or less, the same properties as the Roe and VFRoe schemes but allows the simulation of more complex systems. It needs a MUSCL technique to reduce diffusion effects and order 2 Runge–Kutta for stability. It was implemented for the simulation of multiphase flows in pipe lines (see FAILLE and HEINTZÉ [1999]). The other difficulties encountered in this case are the treatment of the boundary conditions and the different magnitude of the eigenvalues, which are discussed in the next sections.

33.3. Partial implicitation of explicit scheme

In the modelling of flows, where "propagation" phenomena and "convection" phenomena coexist, the Jacobian matrix of F often has eigenvalues of different magnitude, the "large" eigenvalues (large meaning "far from 0", positive or negative) corresponding to the propagation phenomena and "small" eigenvalues corresponding to the "convection" phenomena. Large and small eigenvalues may differ by a factor 10 or 100.

With the explicit schemes described in the previous sections, the time step is limited by the CFL condition corresponding to the large eigenvalues. Roughly speaking, with the notations of Section 33.1, this condition is (for all $i \in \mathbb{Z}$) $k \leqslant |\lambda|^{-1}h_i$, where λ is the largest eigenvalue. In some cases, this limitation can be unsatisfactory for two reasons. Firstly, the time step is too small and implies a prohibitive computational cost. Secondly, the discontinuities in the solutions, associated to the small eigenvalues, are not sharp because the time step is far from the CFL condition of the small eigenvalues (however, this can be somewhat corrected with a MUSCL method). This is in fact a major problem when the discontinuities associated to the small eigenvalues need to be computed precisely. It is the case of interest here.

A first method to avoid the time step limitation is to take a "fully implicit" version of the schemes developed in the previous sections, that is $F^n_{i+1/2}$ function of u^{n+1}_j, $j \in \mathbb{Z}$, instead of u^n_j, $j \in \mathbb{Z}$ (the terminology "fully implicit" is by opposition to "linearly implicit", see below and FERNANDEZ (1989]). However, in order to be competitive with explicit schemes, the fully implicit scheme is used with large time steps. In practice, this prohibits the use of a MUSCL technique in the computation of the solution at time t_{n+1} by, e.g., a Newton algorithm. This implicit scheme is therefore very diffusive and will smear discontinuities.

A second method consists in splitting the system into two systems, the first one is associated with the "small" eigenvalues, and the second one with the "large" eigenvalues (in the case of the Euler equations, this splitting may correspond to a "convection" system and a "propagation" system). At each time step, the first system is solved with an explicit scheme and the second one with an implicit scheme. Both use the same time step, which is limited by the CFL condition of the small eigenvalues. Using a MUSCL technique and an order 2 Runge–Kutta method for the first system yields sharp discontinuities associated to the small eigenvalues. This method is often satisfactory, but is difficult to handle in the case of severe boundary conditions, since the convenient boundary conditions for each system may be difficult to determine.

Another method, developed by E. Turkel (see TURKEL [1987]), in connexion with Roe's scheme, uses a change of variables in order to reduce the ratio between large and small eigenvalues.

Let us now describe a partially linearly implicit method ("turbo" scheme) which was successfully tested for multiphase flows in pipe lines (see FAILLE and HEINTZÉ [1999]) and other cases (see FERNANDEZ (1989]). For the sake of simplicity, the method is described for the last scheme of Section 33.2, i.e. the scheme defined by (33.12)–(33.14), where $F^n_{i+1/2}$ is defined by (33.17) and (33.16) (recall that F may depend on x).

Assume that $I \subset \{1, \ldots, m\}$ is the set of index of large eigenvalues (and does not depend on i). The aim here is to "implicit" the unknowns coresponding to the large eigenvalues only: let $\tilde{A}_i$, $\tilde{A}^{(-)}_{i+1/2}$ and $\tilde{A}^{(+)}_{i+1/2}$ be the matrix having the same eigenvectors as A_i, $A^{(-)}_{i+1/2}$ and $A^{(+)}_{i+1/2}$, with the same large eigenvalues (i.e. corresponding to $p \in I$) and 0 as small eigenvalues. Let

$$\tilde{A}^n_{i+1/2} = \frac{\lambda}{2}\left(\tilde{A}^{(-)}_{i+1/2} + \tilde{A}^{(+)}_{i+1/2}\right).$$

Then, the partially linearly implicit scheme is obtained by replacing $F^n_{i+1/2}$ in (33.5) by $\tilde{F}^n_{i+1/2}$ defined by

$$\begin{aligned}\tilde{F}^n_{i+1/2} = {} & F^n_{i+1/2} + \tfrac{1}{2}\left(\tilde{A}_i\left(u^{n+1}_i - u^n_i\right) + \tilde{A}_{i+1}\left(u^{n+1}_{i+1} - u^n_{i+1}\right)\right) \\ & + \tfrac{1}{2}\tilde{A}^n_{i+1/2}\left(u^{n+1}_i - u^n_i + u^n_{i+1} - u^{n+1}_{i+1}\right).\end{aligned}$$

In order to obtain sharp discontinuities corresponding to the small eigenvalues, a MUSCL technique is used for the computation of $F^n_{i+1/2}$. Then, again for stability reasons, it is preferable to add an order 2 Runge–Kutta method for the time discretization. Although it is not so easy to implement, the order 2 Runge–Kutta method is needed to enable the use of "large" time steps. The time step is, in severe situations, very close to that given by the usual CFL condition corresponding to the small eigenvalues, and can be considerably larger than that given by the large eigenvalues (see FAILLE and HEINTZÉ [1999] for several tests).

33.4. Boundary conditions

In many simulations of real situations, the treatment of the boundary conditions is not easy (in particular in the case of sign change of eigenvalues). We give here a classical possible mean (see, e.g., KUMBARO [1992] and DUBOIS and LEFLOCH [1988]) of handling boundary conditions (a more detailed description may be found in MASELLA [1997] for the case of multiphase flows in pipe lines).

Let us consider now the system (33.4) where "$x \in \mathbb{R}$" is replaced by "$x \in \Omega$" with $\Omega = (0, 1)$. In order for the system to be well-posed, an initial condition (for $t = 0$) and some convenient boundary conditions for $x = 0$ and $x = 1$ are needed; these boundary conditions will appear later in the discretization (we do not detail here the mathematical analysis of the problem of the adequacy of the boundary conditions, see, e.g., SERRE [1996] and references therein). Let us now explain the numerical treatment of the boundary condition at $x = 0$.

With the notations of Section 33.1, the space mesh is given by $\{K_i, i \in \{0, \dots, N_{\mathcal{T}}\}\}$, with $\sum_{i=1}^{N_{\mathcal{T}}} h_i = 1$. Using the finite volume scheme (33.5) with $i \in \{1, \dots, N_{\mathcal{T}}\}$ instead of $i \in \mathbb{Z}$ needs, for the computation of u_1^{n+1}, with $\{u_i^n, i \in \{1, \dots, N_{\mathcal{T}}\}\}$ given, a value for $F_{1/2}^n$ (which corresponds to the flux at point $x = 0$ and time $t = t_n$).

For the sake of simplicity, consider only the case of the Roe and VFRoe schemes. Then, the "interior fluxes", i.e. $F_{i+1/2}^n$ for $i \in \{1, \dots, N_{\mathcal{T}} - 1\}$, are determined by using matrices $A(u_i^n, u_{i+1}^n)$ ($i \in \{1, \dots, N_{\mathcal{T}} - 1\}$). In the case of the Roe scheme, $F_{i+1/2}^n$ is given by (33.11) or (33.15) and $A(\bullet, \bullet)$ satisfies the Roe condition (33.10). In the case of the VFRoe scheme, $F_{i+1/2}^n$ is given through the resolution of the linearized Riemann problem (33.8), (33.9) with, e.g., $A(u_i^n, u_{i+1}^n) = DF((1/2)(u_i^n + u_{i+1}^n))$. In order to compute $F_{1/2}^n$, a possibility is to take the same method as for the interior fluxes; this requires the determination of some u_0^n. In some cases (e.g., when all the eigenvalues of $D_u F(u)$ are nonnegative), the given boundary conditions at $x = 0$ are sufficient to determine the value u_0^n, or directly $F_{1/2}^n$, but this is not true in the general case.... In the general case, there are not enough given boundary conditions to determine u_0^n and missing equations need to be introduced. The idea is to use an iterative process. Since $A(u_0^n, u_1^n)$ is diagonalizable and has only real eigenvalues, let $\lambda_1, \dots, \lambda_m$ be the eigenvalues of $A(u_0^n, u_1^n)$ and $\varphi_1, \dots, \varphi_m$ a basis of $\mathbb{R}^m$ associated to these eigenvalues. Then the vectors u_0^n and u_1^n may be decomposed on this basis, this yields

$$u_0^n = \sum_{i=1}^{m} \alpha_{0,i} \varphi_i, \qquad u_1^n = \sum_{i=1}^{m} \alpha_{1,i} \varphi_i.$$

Assume that the number of negative eigenvalues of $A(u_0^n, u_1^n)$ does not depend on u_0^n (this is a simplifying assumption); let p be the number of negative eigenvalues and $m - p$ the number of positive eigenvalues of $A(u_0^n, u_1^n)$.

Then, the number of (scalar) given boundary conditions is (hopefully ...) $m - p$. Therefore, one takes, for u_0^n, the solution of the (nonlinear) system of m (scalar) unknowns, and m (scalar) equations. The m unknowns are the components of u_0^n and the

m equations are obtained with the $m - p$ boundary conditions and the p following equations:

$$\alpha_{0,i} = \alpha_{1,i}, \quad \text{if } \lambda_i < 0. \tag{33.18}$$

Note that the quantities $\alpha_{0,i}$ depend on $A(u_0^n, u_1^n)$; the resulting system is therefore nonlinear and may be solved with, e.g., a Newton algorithm.

Other possibilities around this method are possible. For instance, another possibility, perhaps more natural, consists in writing the $m-p$ boundary conditions on $u_{1/2}^n$ instead of u_0^n and to take (33.18) with the components of $u_{1/2}^n$ instead of those of u_0^n, where $u_{1/2}^n$ is the solution at $x = 0$ of (33.8), (33.9) with $i = 0$. With the VFRoe scheme, the flux at the boundary $x = 0$ is then $F_{1/2}^n = F(u_{1/2}^n)$. In the case of a linear system with linear boundary conditions and with the VFRoe scheme, this method gives the same flux $F_{1/2}^n$ as the preceding method, the value $u_{1/2}^n$ is completely determined although u_0^n is not completely determined.

In the case of the scheme described in the second part of Section 33.2, the following "simpler" possibility was implemented. For this scheme, $F_{i+1/2}^n$ is given, for $i \in \{1, \ldots, N_{\mathcal{T}} - 1\}$, by (33.15) with (33.16). Then, the idea is to take the same equation for the computation of $F_{1/2}^n$ but to compute u_0^n as above (i.e. with $m - p$ boundary conditions and (33.18)) with the choice $A(u_0^n, u_1^n) = D_u F(x_1, u_1^n)$.

This method of computation of the boundary fluxes gives good results but is not adapted to all cases (e.g., if p changes during the Newton iterations or if the number of boundary conditions is not equal to $m - p$ ...). Some particular methods, depending on the problems under consideration, have to be developped.

We now give an attempt for the justification of this treatment of the boundary conditions, at least for a linear system with linear boundary conditions.

Consider the system

$$\begin{aligned} &u_t(x,t) + u_x(x,t) = 0, \quad x \in (0,1),\ t \in \mathbb{R}_+, \\ &v_t(x,t) - v_x(x,t) = 0, \quad x \in (0,1),\ t \in \mathbb{R}_+, \end{aligned} \tag{33.19}$$

with the boundary conditions

$$\begin{aligned} &u(0,t) + \alpha v(0,t) = 0, \quad t \in \mathbb{R}_+, \\ &v(1,t) + \beta u(1,t) = 0, \quad t \in \mathbb{R}_+, \end{aligned} \tag{33.20}$$

and the initial conditions

$$\begin{aligned} &u(x,0) = u_0(x), \quad x \in (0,1), \\ &v(x,0) = v_0(x), \quad x \in (0,1), \end{aligned} \tag{33.21}$$

where $\alpha \in \mathbb{R}^\star$, $\beta \in \mathbb{R}^\star$, $u_0 \in L^\infty(\Omega)$ and $v_0 \in L^\infty(\Omega)$ are given. It is well known that problem (33.19)–(33.21) admits a unique weak solution (entropy conditions are not necessary to obtain uniqueness of the solution of this linear system).

A stable numerical scheme for the discretization of problem (33.19)–(33.21) will add some numerical diffusion terms. It seems quite natural to assume that this diffusion does not lead a coupling between the two equations of (33.19). Then, roughly speaking, the numerical scheme will consist in an approximation of the following parabolic system:

$$\begin{aligned} &u_t(x,t)+u_x(x,t)-\varepsilon u_{xx}(x,t)=0, \quad x\in(0,1),\ t\in\mathbb{R}_+,\\ &v_t(x,t)-v_x(x,t)-\eta v_{xx}(x,t)=0, \quad x\in(0,1),\ t\in\mathbb{R}_+, \end{aligned} \tag{33.22}$$

for some $\varepsilon>0$ and $\eta>0$ depending on the mesh (and time step) and $\varepsilon\to 0$, $\eta\to 0$ as the space and time steps tend to 0.

In order to be well posed, this parabolic system has to be completed with the initial conditions (33.21) and (for all $t>0$) four boundary conditions, i.e. two conditions at $x=0$ and two conditions at $x=1$. This is also the case for the numerical scheme which may be viewed as a discretization of (33.22). There are two boundary conditions given by (33.20). Hence two other boundary conditions must be found, one at $x=0$ and the other at $x=1$.

If these two additional conditions are, e.g., $v(0,t)=u(1,t)=0$, then the (unique) solution to (33.20)–(33.22) with these two additional conditions does not converge, as $\varepsilon\to 0$ and $\eta\to 0$, to the weak solution of (33.19)–(33.21). This negative result is also true for a large choice of other additional boundary conditions. However, if the additional boundary conditions are (wisely) chosen to be $v_x(0,t)=u_x(1,t)=0$, the solution to (33.20)–(33.22) with these two additional conditions converges to the weak solution of (33.19)–(33.21).

The numerical treatment of the boundary conditions described above may be viewed as a discretization of (33.20) and $v_x(0,t)=u_x(1,t)=0$; this remark gives a formal justification to such a choice.

33.5. *Staggered grids*

For some systems of equations it may be "natural" (in the sense that the discretization seems simpler) to associate different grids to different unknowns of the problem. To each unknown is associated an equation and this equation is integrated over the elements (which are the control volumes) of the corresponding mesh, and then discretized by using one discrete unknown per control volume (and time step, for evolution problems). This is the case, e.g., of the well known discretization of the incompressible Navier–Stokes equations with staggered grids, see PATANKAR [1980] and Section 34.2.

Let us now give an example in order to show that staggered grids should be avoided in the case of nonlinear hyperbolic systems since they may yield some kind of "instability". As an illustration, let us consider the following "academic" problem:

$$\begin{aligned} &u_t(x,t)+(vu)_x(x,t)=0, \quad x\in\mathbb{R},\ t\in\mathbb{R}_+,\\ &v_t(x,t)+(v^2)_x(x,t)=0, \quad x\in\mathbb{R},\ t\in\mathbb{R}_+,\\ &u(x,0)=u_0(x), \quad x\in\mathbb{R},\\ &v(x,0)=u_0(x), \quad x\in\mathbb{R}, \end{aligned} \tag{33.23}$$

where u_0 is a bounded function from $\mathbb{R}$ to $[0, 1]$. Taking $u = v$ equal to the weak entropy solution of the Bürgers equation (namely $u_t + (u^2)_x = 0$), with initial condition u_0, leads to a solution of problem (33.23). One would expect a numerical scheme to give an approximation of this solution. Note that the solution of the Bürgers equation, with initial condition u_0, also takes its values in $[0, 1]$, and hence, a "good" numerical scheme can be expected to give approximate solutions taking values in $[0, 1]$. Let us show that this property is not satisfied when using staggered grids.

Let k be the time step and h be the (uniform) space step. Let $x_i = ih$ and $x_{i+1/2} = (i + 1/2)h$, for $i \in \mathbb{Z}$. Define, for $i \in \mathbb{Z}$, $K_i = (x_{i-1/2}, x_{i+1/2})$ and $K_{i+1/2} = (x_i, x_{i+1})$.

The mesh associated to u is $\{K_i, i \in \mathbb{Z}\}$ and the mesh associated to v is $\{K_{i+1/2}, i \in \mathbb{Z}\}$. Using the principle of staggered grids, the discrete unknowns are u_i^n, $i \in \mathbb{Z}$, $n \in \mathbb{N}^\star$, and $v_{i+1/2}^n$, $i \in \mathbb{Z}$, $n \in \mathbb{N}^\star$. The discretization of the initial conditions is, e.g.,

$$
\begin{aligned}
u_i^0 &= \frac{1}{h}\int_{K_i} u_0(x)\,\mathrm{d}x, \quad i \in \mathbb{Z},\\
v_{i+1/2}^0 &= \frac{1}{h}\int_{K_{i+1/2}} u_0(x)\,\mathrm{d}x, \quad i \in \mathbb{Z}.
\end{aligned}
\tag{33.24}
$$

The second equation of (33.23) does not depend on u. It seems reasonable to discretize this equation with the Godunov scheme, which is here the upstream scheme, since u_0 is nonnegative. The discretization of the first equation of (33.23) with the principle of staggered grids is easy. Since $v_{i+1/2}^n$ is always nonnegative, we also take an upstream value for u at the extremities of the cell K_i. Then, with the explicit Euler scheme in time, the scheme becomes

$$
\begin{aligned}
&\frac{1}{k}\left(u_i^{n+1} - u_i^n\right) + \frac{1}{h}\left(v_{i+1/2}^n u_i^n - v_{i-1/2}^n u_{i-1}^n\right) = 0, \quad i \in \mathbb{Z},\ n \in \mathbb{N},\\
&\frac{1}{k}\left(v_{i+1/2}^{n+1} - v_{i+1/2}^n\right) + \frac{1}{h}\left(\left(v_{i+1/2}^n\right)^2 - \left(v_{i-1/2}^n\right)^2\right) = 0, \quad i \in \mathbb{Z},\ n \in \mathbb{N}.
\end{aligned}
\tag{33.25}
$$

It is easy to show that, whatever k and h, there exists u_0 (function from $\mathbb{R}$ to $[0, 1]$) such that $\sup\{u_i^1, i \in \mathbb{Z}\}$ is strictly larger than 1. In fact, it is possible to have, e.g., $\sup\{u_i^1, i \in \mathbb{Z}\} = 1 + k/(2h)$. In this sense the scheme (33.25) appears to be unstable. Note that the same phenomenon exists with the implicit Euler scheme instead of the explicit Euler scheme. Hence staggered grids do not seem to be the best choice for nonlinear hyperbolic systems.

34. Incompressible Navier–Stokes equations

The discretization of the stationary Navier–Stokes equations by the finite volume method is presented in this section. We first recall the classical discretization on cartesian staggered grids. We then study, in the linear case of the Stokes equations, a finite volume method on a staggered triangular grid, for which we show, in a particular case, the convergence of the method.

34.1. The continuous equation

Let us consider here the stationary Navier–Stokes equations:

$$
\begin{aligned}
&-\nu \Delta u^{(i)}(x) + \sum_{j=1}^{d} u^{(j)}(x) \frac{\partial u^{(i)}}{\partial x_j}(x) + \frac{\partial p}{\partial x_i}(x) = f^{(i)}(x), \quad x \in \Omega, \ \forall i = 1, \ldots, d, \\
&\sum_{i=1}^{d} \frac{\partial u^{(i)}}{\partial x_i}(x) = 0, \quad x \in \Omega,
\end{aligned}
\tag{34.1}
$$

with Dirichlet boundary condition

$$
u^{(i)}(x) = 0, \quad x \in \partial\Omega, \ \forall i = 1, \ldots, d, \tag{34.2}
$$

under the following assumption:

ASSUMPTION 34.1.
(i) Ω is an open bounded connected polygonal subset of $\mathbb{R}^d$, $d = 2, 3$,
(ii) $\nu > 0$,
(iii) $f^{(i)} \in L^2(\Omega)$, $\forall i = 1, \ldots, d$.

In the above equations, $u^{(i)}$ represents the ith component of the velocity of a fluid, ν the kinematic viscosity and p the pressure. The unknowns of the problem are $u^{(i)}$, $i \in \{1, \ldots, d\}$ and p. The number of unknown functions from Ω to $\mathbb{R}$ which are to be computed is therefore $d + 1$. Note that (34.1) yields $d + 1$ (scalar) equations.

We shall also consider the Stokes equations, which are obtained by neglecting the nonlinear convection term.

$$
\begin{aligned}
&-\nu \Delta u^{(i)}(x) + \frac{\partial p}{\partial x_i}(x) = f^{(i)}(x), \quad x \in \Omega, \ \forall i = 1, \ldots, d, \\
&\sum_{i=1}^{d} \frac{\partial u^{(i)}}{\partial x_i} = 0, \quad x \in \Omega.
\end{aligned}
\tag{34.3}
$$

There exist several convenient mathematical formulations of (34.1)–(34.2) and (34.3)–(34.2), see, e.g., TEMAM [1977]. Let us give one of them for the Stokes problem. Let

$$
V = \left\{ u = \left(u^{(1)}, \ldots, u^{(q)}\right)^t \in \left(H_0^1(\Omega)\right)^d, \ \sum_{i=1}^{d} \frac{\partial u^{(i)}}{\partial x_i} = 0 \right\}.
$$

Under Assumption 34.1, there exists a unique function u such that

$$
u \in V,
$$

$$\nu \sum_{i=1}^{d} \int_{\Omega} \nabla u^{(i)}(x) \cdot \nabla v^{(i)}(x)\,\mathrm{d}x = \sum_{i=1}^{d} \int_{\Omega} f^{(i)}(x) v^{(i)}(x)\,\mathrm{d}x,$$

$$\forall v = \left(v^{(1)}, \ldots, v^{(q)}\right)^t \in V. \tag{34.4}$$

Equation (34.4) yields the existence of $p \in L^2$ (unique if $\int_{\Omega} p(x)\,\mathrm{d}x = 0$) such that

$$-\nu \Delta u^{(i)} + \frac{\partial p}{\partial x_i} = f^{(i)} \quad \text{in } \mathcal{D}'(\Omega),\ \forall i \in \{1, \ldots, d\}. \tag{34.5}$$

In the following, we shall study finite volume schemes for the discretization of problem (34.1)–(34.2) and (34.3)–(34.2). Note that the Stokes equations may also be successfully discretized by the finite element method, see, e.g., GIRAULT and RAVIART [1986] and references therein.

34.2. *Structured staggered grids*

The discretization of the incompressible Navier–Stokes equations with staggered grids is classical (see PATANKAR [1980]): the idea is to associate different control volume grids to the different unknowns. In the two-dimensional case, the meshes consist in rectangles. Consider, e.g., the mesh, say $\mathcal{T}$, for the pressure p. Then, considering that the discrete unknowns are located at the centers of the elements of their associated mesh, the discrete unknowns for p are, of course, located at the centers of the element of $\mathcal{T}$. The meshes are staggered such that the discrete unknowns for the x-velocity are located at the centers of the edges of $\mathcal{T}$ parallel to the y-axis, and the discrete unknowns for the y-velocity are located at the centers of the edges of $\mathcal{T}$ parallel to the x-axis. The two equations of "momentum" are associated to the x- and y-velocity (and integrated over the control volumes of the considered mesh) and the "divergence free" equation is associated to the pressure (and integrated over the control volume of $\mathcal{T}$). Then the discretization of all the terms of the equations is straightforward, except for the convection terms (in the momentum equations) which, eventually, have to be discretized according to the Reynolds number (upstream or centered discretization...). The convergence analysis of this so-called "MAC" (Marker and Cell) is performed in NICOLAIDES [1992] in the linear case and NICOLAIDES and WU [1996] in the case of the Navier–Stokes equations.

34.3. *A finite volume scheme on unstructured staggered grids*

Let us now turn to the case of unstructured grids; the scheme we shall study uses the same control volumes for all the components of the velocity. The pressure unknowns are located at the vertices, and a Galerkin expansion is used for the approximation of the pressure. Note that other finite volume schemes have been proposed for the discretization of the Stokes and incompressible Navier–Stokes equations on unstructured grids (BOTTA and HEMPEL [1996]), but, to our knowledge, no proof of convergence has been given yet.

We again use the notion of admissible mesh, introduced in Definition 9.1, in the particular case of triangles, if $d=2$, or tetrahedra, if $d=3$. We limit the description below to the case $d=2$ and to the Stokes equations. Let Ω be an open bounded polygonal connected subset Ω of $\mathbb{R}^2$. Let $\mathcal{T}$ be a mesh of Ω consisting of triangles, satisfying the properties required for the finite element method (see, e.g., CIARLET [1978]), with acute angles only. Defining, for all $K \in \mathcal{T}$, the point x_K as the intersection of the orthogonal bisectors of the sides of the triangle K yields that $\mathcal{T}$ is an admissible mesh in the sense of Definition 9.1. Let $\mathcal{S}_\mathcal{T}$ be the set of vertices of $\mathcal{T}$. For $S \in \mathcal{S}_\mathcal{T}$, let ϕ_S be the shape function associated to S in the piecewise linear finite element method for the mesh $\mathcal{T}$. For all $K \in \mathcal{T}$, let $\mathcal{S}_K \subset \mathcal{S}_\mathcal{T}$ be the set of the vertices of K.

A possible finite volume scheme using a Galerkin expansion for the pressure is defined by the following equations, with the notations of Definition 9.1:

$$\nu \sum_{\sigma \in \mathcal{E}_K} F_{K,\sigma}^{(i)} + \sum_{S \in \mathcal{S}_K} p_S \int_K \frac{\partial \phi_S}{\partial x_i}(x)\,\mathrm{d}x = \mathrm{m}(K) f_K^{(i)},$$
$$\forall K \in \mathcal{T},\ \forall i = 1, \ldots, d, \tag{34.6}$$

$$F_{K,\sigma}^{(i)} = \begin{cases} \tau_\sigma (u_K^{(i)} - u_L^{(i)}), & \text{if } \sigma \in \mathcal{E}_{\mathrm{int}}, \sigma = K|L,\ i = 1, \ldots, d, \\ \tau_\sigma u_K^{(i)}, & \text{if } \sigma \in \mathcal{E}_{\mathrm{ext}} \cap \mathcal{E}_K,\ i = 1, \ldots, d, \end{cases} \tag{34.7}$$

$$\sum_{K \in \mathcal{T}} \sum_{i=1}^{d} u_K^{(i)} \int_K \frac{\partial \phi_S}{\partial x_i}(x)\,\mathrm{d}x = 0, \quad \forall S \in \mathcal{S}_\mathcal{T}, \tag{34.8}$$

$$\int_\Omega \sum_{S \in \mathcal{S}_\mathcal{T}} p_S \phi_S(x)\,\mathrm{d}x = 0, \tag{34.9}$$

$$f_K^{(i)} = \frac{1}{\mathrm{m}(K)} \int_K f(x)\,\mathrm{d}x, \quad \forall K \in \mathcal{T}. \tag{34.10}$$

The discrete unknowns of (34.6)–(34.10) are $u_K^{(i)}$, $K \in \mathcal{T}$, $i = 1, \ldots, d$, and p_S, $S \in \mathcal{S}_\mathcal{T}$.

The approximate solution is defined by

$$p_\mathcal{T} = \sum_{S \in \mathcal{S}_\mathcal{T}} p_S \phi_S, \tag{34.11}$$

$$u_\mathcal{T}^{(i)}(x) = u_K^{(i)}, \quad \text{a.e. } x \in K,\ \forall K \in \mathcal{T},\ \forall i = 1, \ldots, d. \tag{34.12}$$

The proof of the convergence of the scheme is not straightforward in the general case. We shall prove in the following proposition the convergence of the discrete velocities given by the finite volume scheme (34.6)–(34.10) in the simple case of a mesh consisting of equilateral triangles.

PROPOSITION 34.1. *Under Assumption* 34.1, *let* $\mathcal{T}$ *be a triangular finite element mesh of* Ω, *with acute angles only, and let, for all* $K \in \mathcal{T}$, x_K *be the intersection of the orthogonal bisectors of the sides of the triangle* K (*hence* $\mathcal{T}$ *is an admissible mesh in the sense of Definition* 9.1). *Then, there exists a unique solution to* (34.6)–(34.10), *denoted by* $\{u_K^{(i)}, K \in \mathcal{T}, i = 1, \dots, d\}$ *and* $\{p_S, S \in \mathcal{S}_\mathcal{T}\}$. *Furthermore, if the elements of* $\mathcal{T}$ *are equilateral triangles, then* $u_\mathcal{T} \to u$ *in* $(L^2(\Omega))^d$, *as* $\mathrm{size}(\mathcal{T}) \to 0$, *where* u *is the* (*unique*) *solution to* (34.4) *and* $u_\mathcal{T} = (u_\mathcal{T}^{(1)}, \dots, u_\mathcal{T}^{(d)})^d$ *is defined by* (34.12).

PROOF.

Step 1 (*Estimate on* $u_\mathcal{T}$). Let $\mathcal{T}$ be an admissible mesh, in the sense of Proposition 34.1, and $\{u_K^{(i)}, K \in \mathcal{T}, i = 1, \dots, d\}$, $\{p_S, S \in \mathcal{S}_\mathcal{T}\}$ be a solution of (34.6)–(34.8) with (34.10).

Multiplying Eqs. (34.6) by $u_K^{(i)}$, summing over $i = 1, \dots, d$ and $K \in \mathcal{T}$ and using (34.8) yields

$$\nu \sum_{i=1}^{d} \sum_{\sigma \in \mathcal{E}} \tau_\sigma \left(D_\sigma u^{(i)}\right)^2 = \sum_{i=1}^{d} \sum_{K \in \mathcal{T}} \mathrm{m}(K) u_K^{(i)} f_K^{(i)}, \tag{34.13}$$

with $D_\sigma u^{(i)} = |u_L^{(i)} - u_K^{(i)}|$ if $\sigma \in \mathcal{E}_{\mathrm{int}}$, $\sigma = K|L$, $i \in \{1, \dots, d\}$ and $D_\sigma u^{(i)} = |u_K^{(i)}|$ if $\sigma \in \mathcal{E}_{\mathrm{ext}} \cap \mathcal{E}_K$, $i \in \{1, \dots, d\}$.

In Step 2, the existence and the uniqueness of the solution of (34.6)–(34.10) will be essentially deduced from (34.13).

Using the discrete Poincaré inequality (9.13) in (34.13) gives an L^2 estimate and an estimate on the "discrete H_0^1 norm" on the component of the approximate velocities, as in Lemma 9.2, i.e.:

$$\|u_\mathcal{T}^{(i)}\|_{1,\mathcal{T}} \leqslant C, \qquad \|u_\mathcal{T}^{(i)}\|_{L^2(\Omega)} \leqslant C, \quad \forall i \in \{1, \dots, d\},$$

where C only depends on Ω, ν, and $f^{(i)}$, $i = 1, \dots, d$.

As in Theorem 9.1 (thanks to Lemma 9.3 and Theorem 14.2), this estimate gives the relative compactness in $(L^2(\Omega))^d$ of the set of approximate solutions $u_\mathcal{T}$, for $\mathcal{T}$ in the set of admissible meshes in the sense of Proposition 34.1. It also gives that if $u_{\mathcal{T}_n} \to u$ in $(L^2(\Omega))^d$, as $n \to \infty$, where $u_{\mathcal{T}_n}$ is the solution associated to the mesh $\mathcal{T}_n$, and $\mathrm{size}(\mathcal{T}_n) \to 0$ as $n \to \infty$, then $u \in (H_0^1(\Omega))^d$. This will be used in Step 3 in order to prove the convergence of $u_\mathcal{T}$ to the solution of (34.4).

Step 2 (*Existence and uniqueness of* $u_\mathcal{T}$ *and* $p_\mathcal{T}$). Let $\mathcal{T}$ be an admissible mesh, in the sense of Proposition 34.1. Replace, in the right hand side of (34.8), "0" by "g_S" with some $\{g_S, S \in \mathcal{S}_\mathcal{T}\} \subset \mathbb{R}$. Eliminating $F_{K,\sigma}^{(i)}$, the system (34.6)–(34.8) becomes a linear system with as many equations as unknowns. The sets of unknowns are $\{u_K^{(i)}, K \in \mathcal{T}, i = 1, \dots, d\}$ and $\{p_S, S \in \mathcal{S}_\mathcal{T}\}$. Ordering the equations and the unknowns yields a matrix, say A, defining this system.

Let us determine the kernel of A; let $f_K^{(i)} = 0$ and $g_S = 0$ for all $K \in \mathcal{T}$, all $S \in \mathcal{S}_\mathcal{T}$ and all $i \in \{1, \dots, d\}$. Then, (34.13) leads to $u_K^{(i)} = 0$ for all $K \in \mathcal{T}$ and all $i \in$

$\{1, \ldots, d\}$. Turning back to (34.6) yields that $p_{\mathcal{T}}$ (defined by (34.11)) is constant on K for all $K \in \mathcal{T}$. Therefore, since Ω is connected, $p_{\mathcal{T}}$ is constant on Ω. Hence, the dimension of the kernel of A is 1 and so is the codimension of the range of A. In order to determine the range of A, note that

$$\sum_{S \in \mathcal{S}_{\mathcal{T}}} \varphi_S(x) = 1, \quad \forall x \in \Omega.$$

Then, a necessary condition in order that the linear system (34.6)–(34.8) has a solution is

$$\sum_{S \in \mathcal{S}_{\mathcal{T}}} g_S = 0 \tag{34.14}$$

and, since the codimension of the range of A is 1, this condition is also sufficient. Therefore, under the condition (34.14), the linear system (34.6)–(34.8) has a solution, this solution is unique up to an additive constant for $p_{\mathcal{T}}$. In the particular case $g_S = 0$ for all $S \in \mathcal{S}_{\mathcal{T}}$, this yields that (34.6)–(34.10) has a unique solution.

Step 3 (*Convergence of* $u_{\mathcal{T}}$ *to* u). In this step the convergence of $u_{\mathcal{T}}$ towards u in $(L^2(\Omega))^d$ as $\operatorname{size}(\mathcal{T}) \to 0$ is shown for meshes consisting of equilateral triangles. Let $(\mathcal{T}_n)_{n \in \mathbb{N}}$ be a sequence of meshes (such as defined in Proposition 34.1) consisting of equilateral triangles and let $(u_{\mathcal{T}_n})_{n \in \mathbb{N}}$ be the associated solutions. Assume that $\operatorname{size}(\mathcal{T}_n) \to 0$ and $u_{\mathcal{T}_n} \to u$ in $(L^2(\Omega))^d$ as $n \to \infty$. Thanks to the compactness result of Step 1, proving that u is the solution of (34.4) is sufficient to conclude this step and to conclude Proposition 34.1.

By Step 1, $u \in (H_0^1(\Omega))^d$. It remains to show that $u \in V$ (which is the first part of (34.4)) and that u satisfies the second part of (34.4).

For the sake of simplicity of the notations, let us omit, from now on, the index n in $\mathcal{T}_n$ and let $h = \operatorname{size}(\mathcal{T})$. Note that x_K (which is the intersection of the orthogonal bisectors of the sides of the triangle K) is the center of gravity of K, for all $K \in \mathcal{T}$. Let $\varphi = (\varphi^{(1)}, \ldots, \varphi^{(d)})^t \in V$ and assume that the functions $\varphi^{(i)}$ are regular functions with compact support in Ω, say $\varphi^{(i)} \in C_c^\infty(\Omega)$ for all $i \in \{1, \ldots, d\}$. There exists $C > 0$ only depending on φ such that

$$\left| \varphi^{(i)}(x_K) - \frac{1}{\mathrm{m}(K)} \int_K \varphi^{(i)}(x)\, \mathrm{d}x \right| \leqslant Ch^2, \tag{34.15}$$

for all $K \in \mathcal{T}$ and $i = 1, \ldots, d$. Let us proceed as in the proof of convergence of the finite volume scheme for the Dirichlet problem (Theorem 9.1).

Assume that h is small enough so that $\varphi(x) = 0$ for all x such that $x \in K$, $K \in \mathcal{T}$ and $\mathcal{E}_K \cap \mathcal{E}_{\text{ext}} \neq \emptyset$.

Note that $(\partial \phi_S)/(\partial x_i)$ is constant in each $K \in \mathcal{T}$ and that

$$\sum_{i=1}^d \int_\Omega \frac{\partial \phi_S}{\partial x_i}(x) \varphi^{(i)}(x)\, \mathrm{d}x = -\int_\Omega \phi_S(x) \sum_{i=1}^d \frac{\partial \varphi^{(i)}}{\partial x_i}(x)\, \mathrm{d}x = 0.$$

Then,

$$\sum_{i=1}^{d}\sum_{K\in\mathcal{T}}\sum_{S\in\mathcal{S}_K} p_S\int_K \frac{\partial\phi_S}{\partial x_i}(x)\,\mathrm{d}x\frac{1}{\mathrm{m}(K)}\int_K \varphi^{(i)}(x)\,\mathrm{d}x=0.$$

Therefore, multiplying Eqs. (34.6) by $(1/\mathrm{m}(K))\int_K \varphi^{(i)}(x)\,\mathrm{d}x$, for each $i=1,\dots,d$, summing the results over $K\in\mathcal{T}$ and $i\in\{i\dots,d\}$ yields

$$\begin{aligned}\nu\sum_{i=1}^{d}\sum_{K|L\in\mathcal{E}_{\mathrm{int}}} &\tau_{K|L}\big(u_L^{(i)}-u_K^{(i)}\big)\bigg(\frac{1}{\mathrm{m}(L)}\int_L\varphi^{(i)}(x)\,\mathrm{d}x-\frac{1}{\mathrm{m}(K)}\int_K\varphi^{(i)}(x)\,\mathrm{d}x\bigg)\\ &=\sum_{i=1}^{d}\sum_{K\in\mathcal{T}} f_K^{(i)}\int_K\varphi^{(i)}(x)\,\mathrm{d}x.\end{aligned}\tag{34.16}$$

Passing to the limit in (34.16) as $n\to\infty$ and using (34.15) gives, in the same way as for the Dirichlet problem (see Theorem 9.1), that u satisfies the equation given in (34.4), at least for $v\in V\cap(C_c^\infty(\Omega))^d$. Then, since $V\cap(C_c^\infty(\Omega))^d$ is dense (for the $(H_0^1(\Omega))^d$-norm) in V (see, e.g., LIONS [1996] for a proof of this result), u satisfies the equation given in (34.4).

Since $u\in(H_0^1(\Omega))^d$, it remains to show that u is divergence free. Let $\varphi\in C_c^\infty(\Omega)$. Multiplying (34.8) by $\varphi(S)$, summing over $S\in\mathcal{S}_\mathcal{T}$ and noting that the function $\sum_{S\in\mathcal{S}_\mathcal{T}}\varphi(S)\phi_S$ converges to φ in $H^1(\Omega)$, one obtains that u is divergence free and then belongs to V. This completes the proof that u is the (unique) solution of (34.4) and concludes the proof of Proposition 34.1. □

35. Flows in porous media

35.1. *Two phase flow*

This section is devoted to the discretization of a system which may be viewed as an elliptic equation coupled to a hyperbolic equation. This system appears in the modelling of a two phase flow in a porous medium. Let Ω be an open bounded polygonal subset of $\mathbb{R}^d$, $d=2$ or 3, and let a and b be functions of class C^1 from $\mathbb{R}$ to $\mathbb{R}_+$. Assume that a is nondecreasing and b is nonincreasing. Let g and $\bar{u}$ be bounded functions from $\partial\Omega\times\mathbb{R}_+$ to $\mathbb{R}$, and u_0 be a bounded function from Ω to $\mathbb{R}$. Consider the following problem:

$$\begin{aligned} u_t(x,t)-\mathrm{div}\big(a(u)\nabla p\big)(x,t)&=0,\quad (x,t)\in\Omega\times\mathbb{R}_+,\\ (1-u)_t(x,t)-\mathrm{div}\big(b(u)\nabla p\big)(x,t)&=0,\quad (x,t)\in\Omega\times\mathbb{R}_+,\\ \nabla p(x,t)\cdot\boldsymbol{n}(x)&=g(x,t),\quad (x,t)\in\partial\Omega\times\mathbb{R}_+,\\ u(x,t)&=\bar{u}(x,t),\quad (x,t)\in\partial\Omega\times\mathbb{R}_+;\ g(x,t)\geqslant 0,\\ u(x,0)&=u_0(x),\quad x\in\Omega,\end{aligned}\tag{35.1}$$

where $\boldsymbol{n}$ is the normal to $\partial\Omega$, outward to Ω. The unknowns of this system are the functions p and u (from $\Omega \times \mathbb{R}_+$ to $\mathbb{R}$). Adding the two first equations of (35.1), this system may be viewed as an elliptic equation with respect to the unknown p, for a given u (note that there is no time derivative in this equation), with a Neumann condition, coupled to a hyperbolic equation with respect to the unknown u (for a given p). Note that, for the elliptic problem with the Neumann condition, the compatibility condition on g writes

$$\int_{\partial\Omega} M\big(u(x,t)\big)g(x,t)\,\mathrm{d}\gamma(x) = 0, \quad t \in \mathbb{R}_+,$$

where $M = a + b$. It is not known whether the system (35.1) has a solution, except in the simple case where the function M is a positive constant (which is, however, already an interesting case for real applications).

In order to discretize (35.1), let $\mathcal{T}$ be an admissible mesh of Ω in the sense of Definition 10.1 and $k > 0$ be the time step. The discrete unknowns are p_K^n and u_K^n for $K \in \mathcal{T}$ and $n \in \mathbb{N}^\star$. The discretization of the initial condition is

$$u_K^0 = \frac{1}{\mathrm{m}(K)} \int_K u_0(x)\,\mathrm{d}x, \quad K \in \mathcal{T}.$$

In order to take into account the boundary condition on u, define, with $t_n = nk$,

$$\bar{u}_K^n = \frac{1}{k\,\mathrm{m}(\partial K \cap \partial\Omega)} \int_{\partial K \cap \partial\Omega} \int_{t_n}^{t_{n+1}} \bar{u}(x,t)\,\mathrm{d}\gamma(x)\,\mathrm{d}t, \quad K \in \mathcal{T},\ n \in \mathbb{N}.$$

The scheme will use an "upstream choice" of $a(u)$ and $b(u)$ on each "interface" of the mesh, that is, for all $K \in \mathcal{T}$, $L \in \mathcal{N}(K)$,

$$\big(a(u)\big)_{K,L}^n = \begin{cases} a\big(u_K^n\big) & \text{if } p_K^{n+1} \geqslant p_L^{n+1}, \\ a\big(u_L^n\big) & \text{if } p_K^{n+1} < p_L^{n+1}, \end{cases}$$

$$\big(b(u)\big)_{K,L}^n = \begin{cases} b\big(u_K^n\big) & \text{if } p_K^{n+1} \geqslant p_L^{n+1}, \\ b\big(u_L^n\big) & \text{if } p_K^{n+1} < p_L^{n+1}. \end{cases}$$

The discrete equations are, for all $K \in \mathcal{T}$, $n \in \mathbb{N}$,

$$\begin{aligned} &\mathrm{m}(K)\frac{u_K^{n+1} - u_K^n}{k} - \sum_{L \in \mathcal{N}(K)} \tau_{K|L}\big(p_L^{n+1} - p_K^{n+1}\big)\big(a(u)\big)_{K,L}^n \\ &\quad - \frac{a(\bar{u}_K^n)}{k} \int_{\partial K \cap \partial\Omega} \int_{t_n}^{t_{n+1}} g^+(x,t)\,\mathrm{d}\gamma(x)\,\mathrm{d}t \\ &\quad + \frac{a(u_K^n)}{k} \int_{\partial K \cap \partial\Omega} \int_{t_n}^{t_{n+1}} g^-(x,t)\,\mathrm{d}\gamma(x)\,\mathrm{d}t = 0, \end{aligned}$$

$$
\begin{aligned}
&-\mathrm{m}(K)\frac{u_K^{n+1}-u_K^n}{k} - \sum_{L\in\mathcal{N}(K)} \tau_{K|L}\big(p_L^{n+1}-p_K^{n+1}\big)\big(b(u)\big)_{K,L}^n \\
&\quad - \frac{b(\bar{u}_K^n)}{k}\int_{\partial K\cap\partial\Omega}\int_{t_n}^{t_{n+1}} g^+(x,t)\,\mathrm{d}\gamma(x)\,\mathrm{d}t \\
&\quad + \frac{b(u_K^n)}{k}\int_{\partial K\cap\partial\Omega}\int_{t_n}^{t_{n+1}} g^-(x,t)\,\mathrm{d}\gamma(x)\,\mathrm{d}t = 0.
\end{aligned}
$$

Recall that $g^+(x,t) = \max\{g(x,t),0\}$, $g^- = (-g)^+$ and $\tau_{K|L} = \mathrm{m}(K|L)/d_{K|L}$ (see Definition 9.1). This finite volume scheme gives very good numerical results under a usual stability condition on the time step with respect to the space mesh. It can be generalized to more complicated systems (in particular, for the simulation of multiphase flows in porous medium such as the "black oil" case of reservoir engineering, see EYMARD [1992]). It is possible to prove the convergence of this scheme in the case where the function M is constant and the function g does not depend on t. In this case, the scheme may be written as a finite volume scheme for a stationary diffusion equation with respect to the unknown p (which does not depend on t) and an upstream finite volume scheme for a hyperbolic equation with respect to the unknown u. The proof of this convergence is given below (Theorem 35.1) under the assumptions that $a(u) = u$ and $b(u) = 1-u$ (see also VIGNAL [1996a]). Note that the elliptic equation with respect to the pressure may also be discretized with a finite element method, and coupled to the finite volume scheme for the hyperbolic equation. This coupling of finite elements and finite volumes was introduced in FORSYTH [1991], where it is called "CVFE" (Control Volume Finite Element), in SONIER and EYMARD [1993] and in EYMARD and GALLOUËT [1993], where the convergence of the finite element-finite volume scheme is shown under the same assumptions.

35.2. Compositional multiphase flow

Let us now turn to the study of a system of partial differential equations which arises in the simulation of a multiphase flow in a porous medium (the so called "Black Oil" case in petroleum engineering, see, e.g., EYMARD [1992]). This system consists in a parabolic equation coupled with hyperbolic equations and algebraic equations and inequations (these algebraic equations and inequations are given by an assumption of thermodynamical equilibrium). It may be written, for $x \in \Omega$ and $t \in \mathbb{R}_+$, as:

$$
\frac{\partial}{\partial t}\big(\rho_1(p)u\big)(x,t) - \mathrm{div}\big(f_1(u,v,c)\nabla p\big)(x,t) = 0, \tag{35.2}
$$

$$
\frac{\partial}{\partial t}\big(\rho_2(p,c)(1-u-v)(1-c)\big)(x,t) - \mathrm{div}\big(f_2(u,v,c)\nabla p\big)(x,t) = 0, \tag{35.3}
$$

$$
\frac{\partial}{\partial t}\big(\rho_2(p,c)(1-u-v)c + \rho_3(p)v\big)(x,t) - \mathrm{div}\big(f_3(u,v,c)\nabla p\big)(x,t) = 0, \tag{35.4}
$$

$$\big(v(x,t)=0 \text{ and } c(x,t)\leqslant f\big(p(x,t)\big)\big) \quad \text{or}$$
$$\big(c(x,t)=f\big(p(x,t)\big) \text{ and } v(x,t)\geqslant 0\big), \tag{35.5}$$

where Ω is a given open bounded polygonal subset of $\mathbb{R}^d$ ($d=2$ or 3), f_1, f_2, f_3 are given functions from $\mathbb{R}^3$ to $\mathbb{R}_+$, f, ρ_1, ρ_3 are given functions from $\mathbb{R}$ to $\mathbb{R}_+$ and ρ_2 is a given function from $\mathbb{R}^2$ to $\mathbb{R}_+$. The problem is completed by initial and boundary conditions which are omitted here. The unknowns of this problem are the functions u, v, c, p from $\Omega \times \mathbb{R}_+$ to $\mathbb{R}$.

In order to discretize this problem, let k be the time step (as usual, k may in fact be variable) and $\mathcal{T}$ be a cartesian mesh of Ω. Following the ideas (and notations) of the previous chapters, the discrete unknowns are u_K^n, v_K^n, c_K^n and p_K^n, for $K \in \mathcal{T}$ and $n \in \mathbb{N}^\star$ and it is quite easy to discretize (35.2)–(35.4) with a classical finite volume method. Note that the time discretization of the unknown p must generally be implicit while the time discretization of the unknowns u, v, c may be explicit or implicit. The explicit choice requires a usual restriction on the time step (linearly with respect to the space step). The only new problem is the discretization of (35.5), which is now described.

Let $n \in \mathbb{N}$. The discrete unknowns at time t_{n+1}, namely u_K^{n+1}, v_K^{n+1}, c_K^{n+1} and p_K^{n+1}, $K \in \mathcal{T}$, have to be computed from the discrete unknowns at time t_n, namely u_K^n, v_K^n, c_K^n and p_K^n, $K \in \mathcal{T}$. Even if the time discretization of (35.2)–(35.4) is explicit with respect to the unknowns u, v and c, the system of discrete equations (with unknowns u_K^{n+1}, v_K^{n+1}, c_K^{n+1} and p_K^{n+1}, $K \in \mathcal{T}$) is nonlinear, whatever the discretization of (35.5). It can be solved by, say, a Newton process. Let $l \in \mathbb{N}$ be the index of the "Newton iteration", and $u_K^{n+1,l}$, $v_K^{n+1,l}$, $c_K^{n+1,l}$ and $p_K^{n+1,l}$ ($K \in \mathcal{T}$) be the computed unknowns at iteration l. As usual, these unknowns are, for $l=0$, taken equal to u_K^n, v_K^n, c_K^n and p_K^n. In order to discretize (35.5), a "phase index" is introduced; it is denoted by i_K^n, for all $K \in \mathcal{T}$ and $n \in \mathbb{N}$ and it is defined by:

$$\text{if } i_K^n=0 \text{ then } v_K^n=0 \quad \big(\text{and } c_K^n \leqslant f\big(p_K^n\big)\big),$$
$$\text{if } i_K^n=1 \text{ then } c_K^n=f\big(p_K^n\big) \quad \big(\text{and } v_K^n \geqslant 0\big).$$

In the Newton process for the computation of the unknowns at time t_{n+1}, a "phase index", denoted by $i_K^{n+1,l}$ is also introduced, with $i_K^{n+1,0}=i_K^n$. This phase index is used in the computation of $u_K^{n+1,l+1}$, $v_K^{n+1,l+1}$, $c_K^{n+1,l+1}$, $p_K^{n+1,l+1}$ and $i_K^{n+1,l+1}$ ($K \in \mathcal{T}$), starting from $u_K^{n+1,l}$, $v_K^{n+1,l}$, $c_K^{n+1,l}$, $p_K^{n+1,l}$ and $i_K^{n+1,l}$. Setting $v_K^{n+1,l+1}=0$ if $i_K^{n+1,l}=0$, and $c_K^{n+1,l+1}=f(p_K^{n+1,l+1})$ if $i_K^{n+1,l}=1$, the computation of (intermediate) values of $u_K^{n+1,l+1}$, $v_K^{n+1,l+1}$, $c_K^{n+1,l+1}$, $p_K^{n+1,l+1}$ is possible with a "Newton iteration" on (35.2), (35.3), (35.4) (note that the number of unknowns is equal to the number of equations). Then, for each $K \in \mathcal{T}$, three cases are possible:

(1) if $c_K^{n+1,l+1} \leqslant f(p_K^{n+1,l+1})$ and $v_K^{n+1,l+1} \geqslant 0$, then set $i_K^{n+1,l+1}=i_K^{n+1,l}$,

(2) if $c_K^{n+1,l+1} > f(p_K^{n+1,l+1})$ (and necessarily $i_K^{n+1,l}=0$), then set $c_K^{n+1,l+1}=f(p_K^{n+1,l+1})$ and $i_K^{n+1,l+1}=1$,

(3) if $v_K^{n+1,l+1} < 0$ (and necessarily $i_K^{n+1,l} = 1$), then set $v_K^{n+1,l+1} = 0$ and $i_K^{n+1,l+1} = 0$.

This yields the final values of $u_K^{n+1,l+1}$, $v_K^{n+1,l+1}$, $c_K^{n+1,l+1}$, $p_K^{n+1,l+1}$ and $i_K^{n+1,l+1}$ $(K \in \mathcal{T})$.

When the "convergence" of the Newton process is achieved, say at iteration $l^\star$, the values of the unknowns at time t_{n+1} are found. They are taken equal to those indexed by $(n+1, l^\star)$ (for u, v, c, p, i). It can be proved, under convenient hypotheses on the function f (which are realistic in the applications), that there is no "oscillation" of the "phase index" during the Newton iterations performed from time t_n to time t_{n+1} (see EYMARD and GALLOUËT [1991]). This method, using the phase index, was also successfully adapted for the treatment of the obstacle problem and the Signorini problem, see HERBIN and MARCHAND [1997].

35.3. A simplified case

The aim of this section and of the following sections is the study of the convergence of two coupled finite volume schemes, for the system of equations $u_t - \operatorname{div}(u\nabla p) = 0$ and $\Delta p = 0$, defined on an open set Ω. A finite volume mesh $\mathcal{T}$ is used for the discretization in space, together with an explicit Euler time discretization. Similar results are in VIGNAL [1996a] and VIGNAL and VERDIÈRE [1998] where the case of different space meshes for the two equations is also studied.

We assume that the following assumption is satisfied.

ASSUMPTION 35.1. Let Ω be an open polygonal bounded connected subset of $\mathbb{R}^d$, $d = 2$ or 3, and $\partial\Omega$ its boundary. We denote by $\boldsymbol{n}$ the normal vector to $\partial\Omega$ outward to Ω.

Let $g \in L^2(\partial\Omega)$ be a function such that

$$\int_{\partial\Omega} g(x)\,\mathrm{d}\gamma(x) = 0,$$

and let $\partial\Omega^+ = \{x \in \partial\Omega,\ g(x) \geqslant 0\}$, $\Omega^+ = \Omega \cup \partial\Omega^+$ and $\partial\Omega^- = \{x \in \partial\Omega,\ g(x) \leqslant 0\}$. Let $u_0 \in L^\infty(\Omega)$ and $\bar{u} \in L^\infty(\partial\Omega^+ \times \mathbb{R}_+^\star)$ represent respectively the initial condition and the boundary condition for the unknown u.

The set

$$\mathcal{D}(\Omega^+ \times \mathbb{R}_+) = \{\varphi \in C_c^\infty(\mathbb{R}^d \times \mathbb{R}, \mathbb{R}),\ \varphi = 0 \text{ on } \partial\Omega^- \times \mathbb{R}_+\}$$

will be the set of test functions for Eq. (35.10) in the weak formulation of the problem, which is given below.

DEFINITION 35.1. A pair $(u, p) \in L^\infty(\Omega \times \mathbb{R}_+^\star) \times H^1(\Omega)$ (u is the saturation, p is the pressure) is a weak solution of

$$\begin{cases} \Delta p(x) = 0, & \forall x \in \Omega, \\ \nabla p(x) \cdot \boldsymbol{n}(x) = g(x), & \forall x \in \partial\Omega, \\ u_t(x,t) - \operatorname{div}(u \nabla p)(x,t) = 0, & \forall x \in \Omega,\ \forall t \in \mathbb{R}_+, \\ u(x,0) = u_0(x), & \forall x \in \Omega, \\ u(x,t) = \bar{u}(x,t), & \forall x \in \partial\Omega^+,\ \forall t \in \mathbb{R}_+, \end{cases} \tag{35.6}$$

if it verifies

$$p \in H^1(\Omega), \tag{35.7}$$

$$u \in L^\infty\big(\Omega \times \mathbb{R}_+^\star\big), \tag{35.8}$$

$$\int_\Omega \nabla p(x) \cdot \nabla X(x)\,\mathrm{d}x - \int_{\partial\Omega} X(x) g(x)\,\mathrm{d}\gamma(x) = 0, \quad \forall X \in H^1(\Omega). \tag{35.9}$$

and

$$\int_{\mathbb{R}_+} \int_\Omega u(x,t)(\varphi_t(x,t) - \nabla p(x) \cdot \nabla\varphi(x,t)\,\mathrm{d}x\,\mathrm{d}t + \int_\Omega u_0(x)\varphi(x,0)\,\mathrm{d}x$$
$$+ \int_{\mathbb{R}_+} \int_{\partial\Omega^+} \bar{u}(x,t)\varphi(x,t)g(x)\mathrm{d}\gamma(x)\,\mathrm{d}t = 0, \quad \forall \varphi \in \mathcal{D}\big(\Omega^+ \times \mathbb{R}_+\big). \tag{35.10}$$

Under Assumption 35.1, a classical result gives the existence of $p \in H^1(\Omega)$ and the uniqueness of ∇p where p is the solution of (35.7), (35.9), which is a variational formulation of the classical Neumann problem. Additional hypotheses on the function g are necessary to get the uniqueness of $u \in L^\infty(\mathbb{R}^d \times \mathbb{R}_+^\star)$ solution of (35.10). The existence of u results from the convergence of the scheme, but not its uniqueness, which could be obtained thanks to regularity properties of ∇p. We shall assume such regularity, which ensures the uniqueness of the function u and allows an error estimate between the finite volume scheme approximation of the pressure and the exact pressure. In fact, for the sake of simplicity, we assume (in Assumption 35.2 below) that $p \in C^2(\overline{\Omega})$. This is a rather "strong" assumption which can be weakened. However, a convergence result (such as in Theorem 35.1) with the only assumption $p \in H^1(\Omega)$ seems not easy to obtain. Note also that similar results of convergence (for the "pressure scheme" and for the "saturation scheme") are possible with an open bounded connected subset of $\mathbb{R}^d$ with a C^2 boundary (instead of an open bounded connected polygonal subset of $\mathbb{R}^d$) using Definition 18.4 of admissible meshes.

ASSUMPTION 35.2. The pressure p, weak solution in $H^1(\Omega)$ to (35.9), belongs to $C^2(\overline{\Omega})$.

REMARK 35.1. The solution (u, p) of (35.7)–(35.10) is also a weak solution of

$$(1-u)_t(x,t) - \operatorname{div}\big((1-u)\nabla p\big)(x,t) = 0.$$

REMARK 35.2. The finite volume scheme will ensure the conservation of each of the quantities u and $1-u$. It can be extended to more complex phenomena such as compressibility, thermodynamic equilibrium... (see Section 35.2).

REMARK 35.3. The proof which is given here can easily be extended to the case of the existence of a source term which writes

$$\begin{aligned}
&-\Delta p(x) = v(x), \quad x \in \Omega,\\
&\nabla p(x) \cdot \boldsymbol{n}(x) = g(x), \quad x \in \partial\Omega,\\
&u_t(x,t) - \operatorname{div}(u\nabla p)(x,t) + u(x,t)v^-(x) = s(x,t)v^+(x), \quad x \in \Omega,\ t \in \mathbb{R}_+,\\
&u(x,0) = u_0(x), \quad x \in \Omega,\\
&u(x,t) = \bar{u}(x,t), \quad x \in \partial\Omega^+,\ t \in \mathbb{R}_+,
\end{aligned}$$

where $v \in L^2(\Omega)$ with $\int_{\partial\Omega} g(x)\,\mathrm{d}\gamma(x) + \int_\Omega v(x)\,\mathrm{d}x = 0$ and $s \in L^\infty(\Omega \times \mathbb{R}_+^\star)$. All modifications which are connected to such terms will be stated in remarks.

35.4. The scheme for the simplified case

Let Ω be an open polygonal bounded connected subset of $\mathbb{R}^d$. Let $\mathcal{T}$ be an admissible mesh, in the sense of Definition 10.1, and let $h = \operatorname{size}(\mathcal{T})$. Assume furthermore that, for some $\alpha > 0$, $d_\sigma \geqslant \alpha h$ for all $\sigma \in \mathcal{E}_{\text{int}}$.

35.4.1. The pressure finite volume scheme

We first define the approximate pressure, using the finite volume scheme defined in Section 10 (i.e. (10.6)–(10.8)).

(i) The values G_K, for $K \in \mathcal{T}$, are defined by

$$\begin{aligned}
G_K &= \int_{\partial K \cap \partial\Omega} g(x)\,\mathrm{d}\gamma(x) \quad \text{if } \mathrm{m}(\partial K \cap \partial\Omega) \neq 0,\\
G_K &= 0, \quad \text{if } \mathrm{m}(\partial K \cap \partial\Omega) = 0.
\end{aligned} \tag{35.11}$$

(ii) The scheme is defined by

$$-\sum_{L \in \mathcal{N}(K)} \tau_{K|L}(p_L - p_K) = G_K, \quad \forall K \in \mathcal{T}, \tag{35.12}$$

and

$$\sum_{K \in \mathcal{T}} \mathrm{m}(K) p_K = 0. \tag{35.13}$$

We recall that, from Lemma 10.1, there exists a unique function $p_{\mathcal{T}} \in X(\mathcal{T})$ defined by $p_{\mathcal{T}}(x) = p_K$ for a.e. $x \in K$, for all $K \in \mathcal{T}$, where $(p_K)_{K \in \mathcal{T}}$ satisfy Eqs. (35.11)–(35.13). Then, using Theorem 10.1, there exist C_1 and C_2, only depending on p and Ω, such that

$$\|p_{\mathcal{T}} - p\|_{L^2(\Omega)} \leqslant C_1 h \tag{35.14}$$

and

$$\sum_{K|L \in \mathcal{E}_{\text{int}}} \mathrm{m}(K|L) d_{K|L} \left(\frac{p_L - p_K}{d_{K|L}} - \frac{1}{\mathrm{m}(K|L)} \int_{K|L} \nabla p(x) \cdot \boldsymbol{n}_{K,L} \, \mathrm{d}\gamma(x) \right)^2 \leqslant (C_2 h)^2. \tag{35.15}$$

Last but not least, using Lemma 10.6, there exists C_3, only depending on g and Ω, such that

$$\sum_{K|L \in \mathcal{E}_{\text{int}}} \tau_{K|L} (p_L - p_K)^2 \leqslant (C_3)^2. \tag{35.16}$$

The saturation finite volume scheme

Let us now turn to the finite volume discretization of the hyperbolic equation (35.10). In order to write the scheme, let us introduce the following notations: let

$$G_K^{(+)} = \int_{\partial K \cap \partial \Omega} g^+(x) \, \mathrm{d}\gamma(x) \quad \text{and} \quad G_K^{(-)} = \int_{\partial K \cap \partial \Omega} g^-(x) \, \mathrm{d}\gamma(x),$$

so that $G_K^{(+)} - G_K^{(-)} = G_K$. Let

$$G^{(+)} = \int_{\partial \Omega} g^+(x) \, \mathrm{d}\gamma(x) = \sum_{K \in \mathcal{T}} G_K^{(+)}$$

(note that $G^{(+)}$ does not depend on $\mathcal{T}$). The scheme (35.12) may also be written

$$\sum_{L \in \mathcal{N}(K)} \tau_{K|L} (p_L - p_K) + G_K^{(+)} - G_K^{(-)} = 0, \quad \forall K \in \mathcal{T}. \tag{35.17}$$

REMARK 35.4. In the case of the problem with source terms, the right-hand side of Eq. (35.12) is replaced by $G_K + V_K^{(+)} - V_K^{(-)}$ with

$$V_K^{(\pm)} = \int_K v^{\pm}(x) \, \mathrm{d}x.$$

Then, in Eq. (35.17) the quantities $G_K^{(\pm)}$ are replaced by $G_K^{(\pm)} + V_K^{(\pm)}$.

Let $\xi \in (0,1)$. Given an admissible mesh $\mathcal{T}$, the time step is defined by a real value $k > 0$ such that

$$k \leqslant \inf_{K \in \mathcal{T}} \frac{\mathrm{m}(K)\,(1-\xi)}{\sum_{L \in \mathcal{N}(K)} \tau_{K|L}(p_L - p_K)^+ + G_K^{(+)}}. \tag{35.18}$$

REMARK 35.5. Since the right-hand side of (35.18) has a strictly positive lower bound, it is always possible to find values $k > 0$ which satisfy (35.18). Roughly speaking, the condition (35.18) is a linear condition between the time step and the size of the mesh. Let us explain this point in more detail: in most practical cases, function g is regular enough so that $|p_L - p_K|/d_{K|L}$ is bounded by some C only depending on g and Ω. Assume furthermore that the mesh $\mathcal{T}$ is admissible in the sense of Definition 10.1 and that, for some $\alpha > 0$, $d_{K,\sigma} \geqslant \alpha h$, for all $\sigma \in \mathcal{E}_K$, $K \in \mathcal{T}$. Then the condition $k \leqslant Dh$, with $D = ((1-\xi)\alpha)/(d(C + \|g\|_{L^\infty(\partial\Omega)}))$, implies the condition (35.18). Note also that for all $g \in L^2(\partial\Omega)$ we already have a bound for $|p_{\mathcal{T}}|_{1,\mathcal{T}}$ (but this does not yield a bound on $|p_L - p_K|/d_{K|L}$). Finally, note that condition (35.18) is easy to implement in practise, since the values $\tau_{K|L}$ and p_K are available by the pressure scheme.

REMARK 35.6. In the problem with source terms, the condition (35.18) will be modified as follows:

$$k \leqslant \inf_{K \in \mathcal{T}} \frac{\mathrm{m}(K)\,(1-\xi)}{\sum_{L \in \mathcal{N}(K)} \tau_{K|L}(p_L - p_K)^+ + G_K^{(+)} + V_K^{(+)}}.$$

The initial condition is discretized by:

$$u_K^0 = \frac{1}{\mathrm{m}(K)} \int_K u_0(x)\,\mathrm{d}x, \quad \forall K \in \mathcal{T}. \tag{35.19}$$

We extend the definition of $\bar{u}$ by 0 on $\partial\Omega^- \times \mathbb{R}_+$, and we define $\bar{u}_K^n$, for $K \in \mathcal{T}$ and $n \in \mathbb{N}$, by

$$\begin{aligned} \bar{u}_K^n &= \frac{1}{k\,\mathrm{m}(\partial K \cap \partial\Omega)} \int_{nk}^{(n+1)k} \int_{\partial K \cap \partial\Omega} \bar{u}(x,t)\,\mathrm{d}\gamma(x)\,\mathrm{d}t, \quad \text{if } \mathrm{m}(\partial K \cap \partial\Omega) \neq 0, \\ \bar{u}_K^n &= 0, \quad \text{if } \mathrm{m}(\partial K \cap \partial\Omega) = 0. \end{aligned} \tag{35.20}$$

Hence the following function may be defined on $\partial\Omega \times \mathbb{R}_+$:

$$\bar{u}_{\mathcal{T},k}(x,t) = \bar{u}_K^n, \quad \forall x \in \partial K \cap \partial\Omega,\ \forall K \in \mathcal{T},\ \forall t \in [nk,(n+1)k),\ n \in \mathbb{N}.$$

The finite volume discretization of the hyperbolic equation (35.10) is then written as the following relation between u_K^{n+1} and all u_L^n, $L \in \mathcal{T}$.

$$\mathrm{m}(K)\big(u_K^{n+1} - u_K^n\big) - k\Bigg[\sum_{L \in \mathcal{N}(K)} \tau_{K|L} u_{K,L}^n (p_L - p_K) + \bar{u}_K^n G_K^{(+)} - u_K^n G_K^{(-)}\Bigg] = 0,$$

$$\forall K \in \mathcal{T},\ \forall n \in \mathbb{N}, \tag{35.21}$$

in which the upstream value $u^n_{K,L}$ is defined by

$$u^n_{K,L} = \begin{cases} u^n_K, & \text{if } p_K \geqslant p_L, \\ u^n_L, & \text{if } p_L > p_K. \end{cases} \tag{35.22}$$

The approximate solution, denoted by $u_{\mathcal{T},k}$, is defined a.e. from $\Omega \times \mathbb{R}_+ \to$ to $\mathbb{R}$ by

$$u_{\mathcal{T},k}(x,t) = u^n_K, \quad \forall x \in K,\ \forall K \in \mathcal{T},\ \forall t \in \big[nk, (n+1)k\big),\ \forall n \in \mathbb{N}. \tag{35.23}$$

REMARK 35.7. In the case of source terms, the following term is defined:

$$s^n_K = \frac{1}{\mathrm{m}(K)k} \int_{nk}^{(n+1)k} \int_K s(x,t)\,\mathrm{d}x\,\mathrm{d}t$$

and the term $k(s^n_K V^{(+)}_K - u^n_K V^{(-)}_K)$ is added to the right-hand side of (35.21).

35.5. Estimates on the approximate solution

Estimate in $L^\infty(\Omega \times \mathbb{R}^\star_+)$

LEMMA 35.1. *Under the Assumptions* 35.1 *and* 35.2, *let* $\mathcal{T}$ *be an admissible mesh in the sense of Definition* 10.1 *and* $k > 0$ *satisfying* (35.18). *Then, the function* $u_{\mathcal{T},k}$ *defined by* (35.11)–(35.13) *and* (35.19)–(35.23) *satisfies*

$$\|u_{\mathcal{T},k}\|_{L^\infty(\Omega\times\mathbb{R}^\star_+)} \leqslant \max\big\{\|u_0\|_{L^\infty(\Omega)}, \|\bar{u}\|_{L^\infty(\partial\Omega^+\times\mathbb{R}^\star_+)}\big\}. \tag{35.24}$$

PROOF. Relation (35.21) can be written as

$$\begin{aligned} u^{n+1}_K &= u^n_K\Bigg[1 - \frac{k}{\mathrm{m}(K)}\bigg(\sum_{L\in\mathcal{N}(K)} \tau_{K|L}(p_K - p_L)^- + G^{(-)}_K\bigg)\Bigg] \\ &\quad + \frac{k}{\mathrm{m}(K)}\bigg(\sum_{L\in\mathcal{N}(K)} \tau_{K|L} u^n_L (p_L - p_K)^+ + G^{(+)}_K \bar{u}^n_K\bigg). \end{aligned}$$

Using

$$\sum_{L\in\mathcal{N}(K)} \tau_{K|L}(p_L - p_K)^+ + G^{(+)}_K = \sum_{L\in\mathcal{N}(K)} \tau_{K|L}(p_K - p_L)^- + G^{(-)}_K,$$

and inequality (35.18), the term u^{n+1}_K may be expressed as a linear combination of terms u^n_L, $L \in \mathcal{T}$, and $\bar{u}^n_K$, with positive coefficients. Thanks to relation (35.17), the sum of these coefficients is equal to 1. The estimate (35.24) follows by an easy induction. □

REMARK 35.8. In the case of source terms, Lemma 35.1 remains true with the following estimate instead of (35.24):

$$\|u_{\mathcal{T},k}\|_{L^\infty(\Omega\times\mathbb{R}_+^\star)} \leqslant \max\big\{\|u_0\|_{L^\infty(\Omega)}, \|\bar{u}\|_{L^\infty(\partial\Omega^+\times\mathbb{R}_+^\star)}, \|s\|_{L^\infty(\Omega\times\mathbb{R}_+^\star)}\big\}.$$

Weak BV estimate

LEMMA 35.2. *Under Assumptions* 35.1 *and* 35.2, *let $\mathcal{T}$ be an admissible mesh in the sense of Definition* 10.1. *Let $h = \mathrm{size}(\mathcal{T})$ and $\alpha > 0$ be such that $d_\sigma \geqslant \alpha h$ for all $\sigma \in \mathcal{E}_{\mathrm{int}}$. Let $k > 0$ satisfying* (35.18). *Let $\{u_K^n,\ K \in \mathcal{T},\ n \in \mathbb{N}\}$ be the solution to* (35.19)–(35.22) *with $\{p_K,\ K \in \mathcal{T}\}$ given by* (35.11)–(35.13). *Let $T > k$ be a given real value, and let $N_{T,k}$ be the integer value such that $N_{T,k}k < T \leqslant (N_{T,k}+1)k$. Then there exists H, which only depends on T, Ω, u_0, $\bar{u}$, g, α and ξ, such that the following inequality holds*:

$$k\sum_{n=0}^{N_{T,k}}\sum_{K|L\in\mathcal{E}_{\mathrm{int}}} \tau_{K|L}|p_K - p_L|\big|u_K^n - u_L^n\big| + k\sum_{n=0}^{N_{T,k}}\sum_{K\in\mathcal{T}} G_K^{(+)}\big|u_K^n - \bar{u}_K^n\big| \leqslant \frac{H}{\sqrt{h}}. \tag{35.25}$$

PROOF. For $n \in \mathbb{N}$ and $K \in \mathcal{T}$, multiplying (35.21) by u_K^n yields

$$\mathrm{m}(K)\big(u_K^{n+1}u_K^n - u_K^n u_K^n\big) - k\Big(\sum_{L\in\mathcal{N}(K)} \tau_{K|L}u_{K,L}^n u_K^n(p_L - p_K) + \bar{u}_K^n u_K^n G_K^{(+)} - \big(u_K^n\big)^2 G_K^{(-)}\Big) = 0. \tag{35.26}$$

Writing $u_K^{n+1}u_K^n - u_K^n u_K^n = -\frac{1}{2}(u_K^{n+1} - u_K^n)^2 - \frac{1}{2}(u_K^n)^2 + \frac{1}{2}(u_K^{n+1})^2$ and summing (35.26) on $K \in \mathcal{T}$ and $n \in \{0, \dots, N_{T,k}\}$ gives

$$-\frac{1}{2}\sum_{n=0}^{N_{T,k}}\sum_{K\in\mathcal{T}} \mathrm{m}(K)\big(u_K^{n+1} - u_K^n\big)^2 + \frac{1}{2}\sum_{K\in\mathcal{T}} \mathrm{m}(K)\big(\big(u_K^{N_{T,k}+1}\big)^2 - \big(u_K^0\big)^2\big) - k\sum_{n=0}^{N_{T,k}}\sum_{K\in\mathcal{T}}\Big(\sum_{L\in\mathcal{N}(K)} \tau_{K|L}u_{K,L}^n u_K^n(p_L - p_K) + \bar{u}_K^n u_K^n G_K^{(+)} - \big(u_K^n\big)^2 G_K^{(-)}\Big) = 0. \tag{35.27}$$

Using (35.22) gives, for all $K \in \mathcal{T}$,

$$-\sum_{L\in\mathcal{N}(K)} \tau_{K|L}u_{K,L}^n u_K^n(p_L - p_K) = \sum_{L\in\mathcal{N}(K)} \tau_{K|L}\big(u_K^n\big)^2(p_K - p_L)^+ - \sum_{L\in\mathcal{N}(K)} \tau_{K|L}u_L^n u_K^n(p_L - p_K)^+.$$

Then,

$$-\sum_{K\in\mathcal{T}}\sum_{L\in\mathcal{N}(K)}\tau_{K|L}u^n_{K,L}u^n_K(p_L-p_K)$$
$$=\sum_{K\in\mathcal{T}}\sum_{L\in\mathcal{N}(K)}\tau_{K|L}\big(\big(u^n_K\big)^2-u^n_Lu^n_K\big)(p_K-p_L)^+.$$

Therefore, since $(u^n_K)^2-u^n_Ku^n_L=\frac{1}{2}(u^n_K-u^n_L)^2+\frac{1}{2}((u^n_K)^2-(u^n_L)^2)$,

$$-\sum_{K\in\mathcal{T}}\sum_{L\in\mathcal{N}(K)}\tau_{K|L}u^n_{K,L}u^n_K(p_L-p_K)$$
$$=\frac{1}{2}\sum_{K\in\mathcal{T}}\sum_{L\in\mathcal{N}(K)}\tau_{K|L}\big(u^n_K-u^n_L\big)^2(p_K-p_L)^+$$
$$+\frac{1}{2}\sum_{K\in\mathcal{T}}\sum_{L\in\mathcal{N}(K)}\tau_{K|L}\big(u^n_K\big)^2(p_K-p_L)^+$$
$$-\frac{1}{2}\sum_{K\in\mathcal{T}}\sum_{L\in\mathcal{N}(K)}\tau_{K|L}\big(u^n_L\big)^2(p_K-p_L)^+$$
$$=\frac{1}{2}\sum_{K\in\mathcal{T}}\sum_{L\in\mathcal{N}(K)}\tau_{K|L}\big(u^n_K-u^n_L\big)^2(p_K-p_L)^+$$
$$+\frac{1}{2}\sum_{K\in\mathcal{T}}\sum_{L\in\mathcal{N}(K)}\tau_{K|L}\big(u^n_K\big)^2(p_K-p_L)$$

and, using (35.17),

$$-\sum_{K\in\mathcal{T}}\sum_{L\in\mathcal{N}(K)}\tau_{K|L}u^n_{K,L}u^n_K(p_L-p_K)$$
$$=\frac{1}{2}\sum_{K\in\mathcal{T}}\sum_{L\in\mathcal{N}(K)}\tau_{K|L}\big(u^n_K-u^n_L\big)^2(p_K-p_L)^+$$
$$+\frac{1}{2}\sum_{K\in\mathcal{T}}G^{(+)}_K\big(u^n_K\big)^2-\frac{1}{2}\sum_{K\in\mathcal{T}}G^{(-)}_K\big(u^n_K\big)^2.$$

Hence

$$-k\sum_{n=0}^{N_{T,k}}\sum_{K\in\mathcal{T}}\bigg(\sum_{L\in\mathcal{N}(K)}\tau_{K|L}u^n_{K,L}u^n_K(p_L-p_K)+\bar{u}^n_Ku^n_KG^{(+)}_K-\big(u^n_K\big)^2G^{(-)}_K\bigg)$$
$$=\frac{1}{2}k\sum_{n=0}^{N_{T,k}}\bigg(\sum_{K|L\in\mathcal{E}_{\rm int}}\tau_{K|L}|p_K-p_L|\big(u^n_K-u^n_L\big)^2+\sum_{K\in\mathcal{T}}G^{(+)}_K\big(u^n_K-\bar{u}^n_K\big)^2\bigg)$$

$$- \tfrac{1}{2} k \sum_{n=0}^{N_{T,k}} \sum_{K \in \mathcal{T}} \left(G_K^{(+)} \left(\bar{u}_K^n \right)^2 - G_K^{(-)} \left(u_K^n \right)^2 \right). \tag{35.28}$$

Using (35.21), we get

$$\sum_{n=0}^{N_{T,k}} \sum_{K \in \mathcal{T}} \mathrm{m}(K) \left(u_K^{n+1} - u_K^n \right)^2$$
$$= \sum_{n=0}^{N_{T,k}} \sum_{K \in \mathcal{T}} \frac{k^2}{\mathrm{m}(K)} \left(\sum_{L \in \mathcal{N}(K)} \tau_{K|L} u_{K,L}^n (p_L - p_K) + \bar{u}_K^n G_K^{(+)} - u_K^n G_K^{(-)} \right)^2.$$

Then, for all $K \in \mathcal{T}$, using again (35.17) and the definition (35.22),

$$\sum_{n=0}^{N_{T,k}} \sum_{K \in \mathcal{T}} \mathrm{m}(K) \left(u_K^{n+1} - u_K^n \right)^2$$
$$= \sum_{n=0}^{N_{T,k}} \sum_{K \in \mathcal{T}} \frac{k^2}{\mathrm{m}(K)} \left(\sum_{L \in \mathcal{N}(K)} \tau_{K|L} \left(u_L^n - u_K^n \right) (p_L - p_K)^+ + G_K^{(+)} \left(\bar{u}_K^n - u_K^n \right) \right)^2.$$

The Cauchy–Schwarz inequality yields

$$\sum_{n=0}^{N_{T,k}} \sum_{K \in \mathcal{T}} \mathrm{m}(K) \left(u_K^{n+1} - u_K^n \right)^2$$
$$\leqslant \sum_{n=0}^{N_{T,k}} \sum_{K \in \mathcal{T}} \frac{k^2}{\mathrm{m}(K)} \left(\sum_{L \in \mathcal{N}(K)} \tau_{K|L} (p_L - p_K)^+ + G_K^{(+)} \right)$$
$$\times \left(\sum_{L \in \mathcal{N}(K)} \tau_{K|L} (p_L - p_K)^+ \left(u_L^n - u_K^n \right)^2 + G_K^{(+)} \left(\bar{u}_K^n - u_K^n \right)^2 \right).$$

Using the stability condition (35.18) and reordering the summations gives

$$\sum_{n=0}^{N_{T,k}} \sum_{K \in \mathcal{T}} \mathrm{m}(K) \left(u_K^{n+1} - u_K^n \right)^2$$
$$\leqslant \sum_{n=0}^{N_{T,k}} k (1 - \xi) \left(\sum_{K|L \in \mathcal{E}_{\mathrm{int}}} \tau_{K|L} |p_L - p_K| \left(u_L^n - u_K^n \right)^2 \right.$$
$$\left. + \sum_{K \in \mathcal{T}} G_K^{(+)} \left(\bar{u}_K^n - u_K^n \right)^2 \right). \tag{35.29}$$

Using (35.27), (35.28) and (35.29), we obtain

$$\sum_{K\in\mathcal{T}} \mathrm{m}(K)\Big(\big(u_K^{N_{T,k}+1}\big)^2-\big(u_K^0\big)^2\Big)$$
$$+\,\xi k\sum_{n=0}^{N_{T,k}}\Big(\sum_{K|L\in\mathcal{E}_{\mathrm{int}}}\tau_{K|L}|p_K-p_L|\big(u_K^n-u_L^n\big)^2+\sum_{K\in\mathcal{T}}G_K^{(+)}\big(u_K^n-\bar u_K^n\big)^2\Big)$$
$$-\,k\sum_{n=0}^{N_{T,k}}\sum_{K\in\mathcal{T}}\big(G_K^{(+)}\big(\bar u_K^n\big)^2-G_K^{(-)}\big(u_K^n\big)^2\big)\leqslant 0. \tag{35.30}$$

Then, setting $C_4=\mathrm{m}(\Omega)\|u_0\|^2_{L^\infty(\Omega)}+2TG^{(+)}\|\bar u\|^2_{L^\infty(\partial\Omega^+\times\mathbb{R}_+^\star)}$ which only depends on Ω, u_0, T, g and $\bar u$,

$$\sum_{K\in\mathcal{T}}\mathrm{m}(K)\big(u_K^{N_{T,k}+1}\big)^2+k\sum_{n=0}^{N_{T,k}}\sum_{K\in\mathcal{T}}G_K^{(-)}\big(u_K^n\big)^2\leqslant C_4$$

(this inequality will not be used in the sequel) and

$$k\sum_{n=0}^{N_{T,k}}\sum_{K|L\in\mathcal{E}_{\mathrm{int}}}\tau_{K|L}|p_K-p_L|\big(u_K^n-u_L^n\big)^2+k\sum_{n=0}^{N_{T,k}}\sum_{K\in\mathcal{T}}G_K^{(+)}\big(u_K^n-\bar u_K^n\big)^2$$
$$\leqslant\frac{C_4}{\xi}. \tag{35.31}$$

The Cauchy–Schwarz inequality yields

$$k\sum_{n=0}^{N_{T,k}}\sum_{K|L\in\mathcal{E}_{\mathrm{int}}}\tau_{K|L}|p_K-p_L|\big|u_K^n-u_L^n\big|+k\sum_{n=0}^{N_{T,k}}\sum_{K\in\mathcal{T}}G_K^{(+)}\big|u_K^n-\bar u_K^n\big|$$
$$\leqslant\Big(k\sum_{n=0}^{N_{T,k}}\sum_{K|L\in\mathcal{E}_{\mathrm{int}}}\tau_{K|L}|p_K-p_L|\big(u_K^n-u_L^n\big)^2$$
$$+\,k\sum_{n=0}^{N_{T,k}}\sum_{K\in\mathcal{T}}G_K^{(+)}\big(u_K^n-\bar u_K^n\big)^2\Big)^{1/2}$$
$$\times\Big(k\sum_{n=0}^{N_{T,k}}\Big(\sum_{K|L\in\mathcal{E}_{\mathrm{int}}}\tau_{K|L}|p_K-p_L|+\sum_{K\in\mathcal{T}}G_K^{(+)}\Big)\Big)^{1/2}. \tag{35.32}$$

The expression W, defined by $W = \sum_{K|L\in\mathcal{E}_{\text{int}}} \tau_{K|L}|p_K - p_L|$, verifies

$$W \leqslant \Big(\sum_{K|L\in\mathcal{E}_{\text{int}}} \tau_{K|L}\Big)^{1/2}\Big(\sum_{K|L\in\mathcal{E}_{\text{int}}} \tau_{K|L}(p_K - p_L)^2\Big)^{1/2}$$
$$\leqslant C_3\Big(\sum_{K|L\in\mathcal{E}_{\text{int}}} \tau_{K|L}\Big)^{1/2} \tag{35.33}$$

using (35.16). Recall that C_3 only depends on g and Ω.

Since

$$\sum_{K|L\in\mathcal{E}_{\text{int}}} \tau_{K|L} \leqslant \Big(\sum_{K|L\in\mathcal{E}_{\text{int}}} \mathrm{m}(K|L)d_{K|L}\Big)\frac{1}{\alpha^2h^2} \leqslant \frac{d\mathrm{m}(\Omega)}{\alpha^2h^2} \tag{35.34}$$

and

$$\sum_{K\in\mathcal{T}} G_K^{(+)} = \int_{\partial\Omega} g^+(x)\,\mathrm{d}\gamma(x),$$

we finally conclude that (35.25) holds. □

REMARK 35.9. In the case of source terms, one adds the term $k\sum_{n=0}^{N_{\mathcal{T},k}}\sum_{K\in\mathcal{T}} V_K^{(+)}|u_K^n - s_K^n|$ in the left-hand side of (35.25) (and H also depends on v and s).

35.6. Theorem of convergence

We already know, by the results of Section 10, that the pressure scheme converges. Let us now prove the convergence of the saturation scheme (35.21). Thanks to the estimate (35.24) in $L^\infty(\Omega\times\mathbb{R}_+^\star)$ (Lemma 35.1), for any sequence of meshes and time steps, such that the size of the mesh tends to 0, we can extract a subsequence such that the approximate saturation converges to a function u in $L^\infty(\Omega\times\mathbb{R}_+^\star)$ for the weak-$\star$ topology. We have to show that u is the (unique) solution of (35.8), (35.10) (the uniqueness of the solution is given by Assumption 35.2).

THEOREM 35.1. *Under Assumptions* 35.1 *and* 35.2, *let* $\xi\in(0,1)$ *and* $\alpha>0$ *be given. For an admissible mesh* $\mathcal{T}$, *in the sense of Definition* 10.1, *such that* $d_\sigma \geqslant \alpha\,\text{size}(\mathcal{T})$ *for all* $\sigma\in\mathcal{E}_{\text{int}}$ *and for a time step* $k>0$ *satisfying* (35.18), *let* $u_{\mathcal{T},k}$ *be defined by* (35.11)–(35.13) *and* (35.19)–(35.23). *Then* $u_{\mathcal{T},k}$ *converges to the solution* u *of* (35.8), (35.10) *in* $L^\infty(\Omega\times\mathbb{R}_+^\star)$ *for the weak-$\star$ topology, as* $\text{size}(\mathcal{T})\to 0$.

PROOF. In the case $g(x)=0$ for a.e. (for the $(d-1)$-dimensional Lebesgue measure) $x\in\partial\Omega$, the proof of Theorem 35.1 is easy. Indeed, $\nabla p(x)=0$ for a.e. $x\in\Omega$ and, for any mesh and time step, $p_K - p_L = 0$ for all $K, L\in\mathcal{T}$. Then, $u_K^n = u_K^0$ for all $K\in\mathcal{T}$ and all $n\in\mathbb{N}$. Therefore, it is easy to prove that the sequence $u_{\mathcal{T},k}$ converges, as

$\text{size}(\mathcal{T}) \to 0$ (for any k ...), to u, defined by $u(x,t) = u_0(x)$ for a.e. $(x,t) \in \Omega \times \mathbb{R}_+$; note that u is the unique solution to (35.8), (35.10).

Let us now assume that g is not the null function in $L^2(\partial\Omega)$.

Let $(\mathcal{T}_m, k_m)_{m\in\mathbb{N}}$ be a sequence of space meshes and time steps. For all $m \in \mathbb{N}$, assume that $\mathcal{T}_m$ is an admissible mesh in the sense of Definition 10.1, that $d_\sigma \geqslant \alpha\,\text{size}(\mathcal{T}_m)$ for all $\sigma \in \mathcal{E}_{\text{int}}$ and that $k_m > 0$ satisfies (35.18) (with $k = k_m$ and $\mathcal{T} = \mathcal{T}_m$). Assume also that $\text{size}(\mathcal{T}_m) \to 0$ as $m \to \infty$.

Let u_m be the function $u_{\mathcal{T},k}$ defined by (35.11)–(35.13) and (35.19)–(35.23), for $\mathcal{T} = \mathcal{T}_m$ and $k = k_m$. By Lemma 35.1, the sequence $(u_m)_{m\in\mathbb{N}}$ is bounded in $L^\infty(\Omega \times \mathbb{R}_+^\star)$. In order to prove that the sequence $(u_m)_{m\in\mathbb{N}}$ converges in $L^\infty(\Omega \times \mathbb{R}_+^\star)$ for the weak-$\star$ topology to the solution of (35.8), (35.10), using a classical contradiction argument, it is sufficient to prove that if $u_m \to u$ in $L^\infty(\Omega \times \mathbb{R}_+^\star)$ for the weak-$\star$ topology then the function u is a solution of (35.8), (35.10).

Let us proceed in two steps. In the first step, it is proved that $k_m \to 0$ as $m \to \infty$. Then, in the second step, it is proved that the function u is a solution of (35.8), (35.10).

From now on, the index "m" is omitted.

Step 1 (*Proof of $k \to 0$ as $m \to \infty$*). The proof that $k \to 0$ (as $m \to \infty$) uses (35.18) and the fact that $\text{size}(\mathcal{T}) \to 0$. Indeed, define

$$A_{\mathcal{T}} = \sum_{K|L \in \mathcal{E}_{\text{int}}} \text{m}(K|L)|p_K - p_L|,$$

and, for $\sigma \in \mathcal{E}_{\text{int}}$, define χ_σ from $\Omega \times \Omega$ to $\{0,1\}$ by

$$\chi_\sigma(x,y) = \begin{cases} 1, & \text{if } \sigma \cap [x,y] \neq \emptyset, \\ 0, & \text{if } \sigma \cap [x,y] = \emptyset. \end{cases}$$

Let $\eta \in \mathbb{R}^d \setminus \{0\}$ and $\bar{\omega} \subset \Omega$ be a compact set such that $d(\bar{\omega}, \Omega^c) \geqslant \eta$. Recall that $p_{\mathcal{T}}$ is defined by $p_{\mathcal{T}}(x) = p_K$ for a.e. $x \in K$ and all $K \in \mathcal{T}$. For a.e. $x \in \bar{\omega}$ one has

$$\left|p_{\mathcal{T}}(x+\eta) - p_{\mathcal{T}}(x)\right| \leqslant \sum_{\sigma = K|L \in \mathcal{E}_{\text{int}}} \chi_\sigma(x, x+\eta)|p_K - p_L|,$$

integrating this inequality over $\bar{\omega}$ yields, using $\int_{\bar{\omega}} \chi_\sigma(x, x+\eta)\,\text{d}x \leqslant |\eta|\text{m}(\sigma)$,

$$\left\|p_{\mathcal{T}}(\bullet + \eta) - p_{\mathcal{T}}\right\|_{L^1(\bar{\omega})} \leqslant |\eta| A_{\mathcal{T}}. \tag{35.35}$$

Assume $A_{\mathcal{T}} \to 0$ as $m \to \infty$. Then, since $p_{\mathcal{T}} \to p$ in $L^1(\Omega)$, one deduces from (35.35) that $\nabla p = 0$ a.e. on Ω which is impossible (since g is not the null function in $L^2(\partial\Omega)$). By the same way, it is also impossible that $A_{\mathcal{T}} \to 0$ for a subsequence. Then there exists $a > 0$ (only depending on the sequence $(p_{\mathcal{T}})_{m\in\mathbb{N}}$, recall that $p_{\mathcal{T}} = p_{\mathcal{T}_m}$ since we omit the index m) such that $A_{\mathcal{T}} \geqslant a$ for all $m \in \mathbb{N}$.

Therefore, since $A_{\mathcal{T}} = \sum_{K\in\mathcal{T}} \sum_{L\in\mathcal{N}(K)} \mathrm{m}(K|L)(p_L - p_K)^+ \geqslant a$, there exists $K \in \mathcal{T}$ such that

$$\sum_{L\in\mathcal{N}(K)} \mathrm{m}(K|L)(p_L - p_K)^+ \geqslant a\frac{\mathrm{m}(K)}{\mathrm{m}(\Omega)}.$$

Then, since $\tau_{K|L} = \mathrm{m}(K|L)/d_{K|L}$ and $d_{K|L} \leqslant 2h$,

$$\sum_{L\in\mathcal{N}(K)} \tau_{K|L}(p_L - p_K)^+ \geqslant a\frac{\mathrm{m}(K)}{2h\mathrm{m}(\Omega)},$$

which yields, using (35.18),

$$k \leqslant (1-\xi)\mathrm{m}(\Omega)\frac{2}{a}h.$$

Hence $k \to 0$ as $m \to \infty$ (since $h \to 0$ as $m \to \infty$). This concludes Step 1.

Step 2 (*Proof of u solution to* (35.10)). Let $\varphi \in \mathcal{D}(\Omega^+ \times \mathbb{R}_+)$. Let $T > 0$ such that, for all $t > T - 1$ and all $x \in \Omega$, $\varphi(x,t) = 0$. Let $m \in \mathbb{N}$ such that $h < 1$ and $k < 1$ (thanks to Step 1, this is true for m large enough). Recall that we denote $\mathcal{T} = \mathcal{T}_m$, $h = \mathrm{size}(\mathcal{T}_m)$ and $k = k_m$. Let $N_{T,k} \in \mathbb{N}$ be such that $N_{T,k}k < T \leqslant (N_{T,k}+1)k$. Multiplying Eq. (35.21) by $\varphi(x_K, nk)$ and summing the result on $K \in \mathcal{T}$ and $n \in \mathbb{N}$ yields

$$E_{1,m} + E_{2,m} = 0,$$

with

$$E_{1,m} = \sum_{n=0}^{N_{T,k}} \sum_{K\in\mathcal{T}} \mathrm{m}(K)\big(u_K^{n+1} - u_K^n\big)\varphi(x_K, nk)$$

and

$$E_{2,m} = -\sum_{n=0}^{N_{T,k}} k \sum_{K\in\mathcal{T}} \Bigg(\sum_{L\in\mathcal{N}(K)} \tau_{K|L} u_{K,L}^n (p_L - p_K) + G_K^{(+)} \bar{u}_K^n - G_K^{(-)} u_K^n \Bigg) \times \varphi(x_K, nk).$$

It is shown below that

$$\lim_{m\to\infty} E_{1,m} = T_1, \tag{35.36}$$

where

$$T_1 = -\int_{\mathbb{R}_+}\int_\Omega u(x,t)\varphi_t(x,t)\,\mathrm{d}x\,\mathrm{d}t - \int_\Omega u_0(x)\varphi(x,0)\,\mathrm{d}x,$$

and that

$$\lim_{m\to\infty} E_{2,m} = T_2, \tag{35.37}$$

where

$$T_2 = \int_{\mathbb{R}_+}\int_{\Omega} u(x,t)\nabla p(x)\cdot\nabla\varphi(x,t)\,\mathrm{d}x\,\mathrm{d}t - \int_{\mathbb{R}_+}\int_{\partial\Omega} \bar{u}(x,t)\varphi(x,t)g(x)\,\mathrm{d}\gamma(x)\,\mathrm{d}t.$$

Then, passing to the limit in $E_{1,m} + E_{2,m} = 0$ proves that u is the (unique) solution of (35.8), (35.10) and concludes the proof of Theorem 35.1.

Let us first prove (35.36). Writing $E_{1,m}$ in the following way:

$$E_{1,m} = \sum_{n=1}^{N_{T,k}}\sum_{K\in\mathcal{T}} \mathrm{m}(K)\frac{\varphi(x_K,(n-1)k) - \varphi(x_K,nk)}{k}u_K^n - \sum_{K\in\mathcal{T}} \mathrm{m}(K)u_K^0\varphi(x_K,0),$$

the assertion (35.36) is easily proved, in the same way as, e.g., in the proof of Theorem 18.1.

Let us prove now (35.37). To this purpose, we need auxiliary expressions, which make use of the convergence of the approximate pressure to the continuous one. Define $E_{3,m}$ and $E_{4,m}$ by

$$\begin{aligned} E_{3,m} &= \sum_{n=0}^{N_{T,k}} k \sum_{K|L\in\mathcal{E}_{\mathrm{int}}} \left(u_K^n - u_L^n\right)\frac{p_L - p_K}{d_{K|L}}\int_{K|L}\varphi(x,nk)\,\mathrm{d}\gamma(x) \\ &\quad + \sum_{n=0}^{N_{T,k}} k \sum_{K\in\mathcal{T}} \left(u_K^n - \bar{u}_K^n\right)\int_{\partial K\cap\partial\Omega} g(x)\varphi(x,nk)\,\mathrm{d}\gamma(x) \end{aligned}$$

and

$$\begin{aligned} E_{4,m} &= \sum_{n\in\mathbb{N}}\int_{nk}^{(n+1)k}\left(\int_{\Omega} u_{\mathcal{T},k}(x,t)\nabla p(x)\cdot\nabla\varphi(x,nk)\,\mathrm{d}x \right.\\ &\quad \left. - \int_{\partial\Omega} \bar{u}_{\mathcal{T},k}(x,t)\varphi(x,nk)g(x)\,\mathrm{d}\gamma(x)\right)\mathrm{d}t. \end{aligned}$$

We have $E_{4,m} \to T_2$ as $m\to\infty$ thanks to the convergence of $u_{\mathcal{T},k}$ to u in $L^\infty(\Omega\times\mathbb{R})$ for the weak-$\star$ topology and to the convergence of $\bar{u}_{\mathcal{T},k}$ to $\bar{u}$ in $L^\infty(\partial\Omega^+\times\mathbb{R}_+)$ for the weak-$\star$ topology (the latter convergence holds also in $L^p(\partial\Omega^+\times(0,S))$ for all $1\leqslant p<\infty$ and all $0<S<\infty$). Let us prove that $|E_{3,m} - E_{4,m}|\to 0$ as $m\to\infty$ (which gives $E_{3,m}\to T_2$ as $m\to\infty$).

Using the equation satisfied by p leads to

$$E_{4,m} = \sum_{n=0}^{N_{T,k}} k \sum_{K|L\in\mathcal{E}_{\text{int}}} (u_K^n - u_L^n) \int_{K|L} \varphi(x, nk) \nabla p(x) \cdot \boldsymbol{n}_{K,L} \, \mathrm{d}\gamma(x)$$

$$+ \sum_{n=0}^{N_{T,k}} k \sum_{K\in\mathcal{T}} (u_K^n - \bar{u}_K^n) \int_{\partial K\cap\partial\Omega} g(x)\varphi(x, nk) \, \mathrm{d}\gamma(x).$$

Therefore,

$$E_{3,m} - E_{4,m} = \sum_{n=0}^{N_{T,k}} k \sum_{K|L\in\mathcal{E}_{\text{int}}} (u_K^n - u_L^n) \times \int_{K|L} \left(\frac{p_L - p_K}{d_{K|L}} - \nabla p(x) \cdot \boldsymbol{n}_{K,L} \right) \varphi(x, nk) \, \mathrm{d}\gamma(x)$$

$$= \sum_{n=0}^{N_{T,k}} k \sum_{K\in\mathcal{T}} u_K^n \left(\sum_{L\in\mathcal{N}(K)} \int_{K|L} \left(\frac{p_L - p_K}{d_{K|L}} - \nabla p(x) \cdot \boldsymbol{n}_{K,L} \right) \times \varphi(x, nk) \, \mathrm{d}\gamma(x) \right).$$

Using the equation satisfied by the pressure in (35.6) and the pressure scheme (35.12) yields

$$E_{3,m} - E_{4,m} = \sum_{n=0}^{N_{T,k}} k \sum_{K\in\mathcal{T}} u_K^n \left(\sum_{L\in\mathcal{N}(K)} \int_{K|L} \left(\frac{p_L - p_K}{d_{K|L}} - \nabla p(x) \cdot \boldsymbol{n}_{K,L} \right) \times \big(\varphi(x, nk) - \varphi(x_K, nk)\big) \, \mathrm{d}\gamma(x) \right).$$

Thanks to the regularity of φ and p, there exists $C_5 > 0$, only depending on p, and C_6, only depending on φ, such that, for all $K|L \in \mathcal{E}_{\text{int}}$,

$$\left| \frac{p_L - p_K}{d_{K|L}} - \nabla p(x) \cdot \boldsymbol{n}_{K,L} \right| \leqslant \left| \frac{p_L - p_K}{d_{K|L}} - \frac{1}{\mathrm{m}(K|L)} \int_\sigma \nabla p(x) \cdot \boldsymbol{n}_{K,L} \, \mathrm{d}\gamma(x) \right| + C_5 h, \quad \forall x \in K|L,$$

and, for all $K \in \mathcal{T}$,

$$\big|\varphi(x, nk) - \varphi(x_K, nk)\big| \leqslant C_6 h, \quad \forall x \in \overline{K}, \ \forall n \in \mathbb{N}.$$

Thus,

$$
\begin{aligned}
&|E_{3,m} - E_{4,m}| \\
&\quad \leqslant \sum_{n=0}^{N_{T,k}} k \sum_{K\in\mathcal{T}} |u_K^n| \Bigg(\sum_{L\in\mathcal{N}(K)} \bigg| \tau_{K|L}(p_L - p_K) - \int_{K|L} \nabla p(x) \cdot \boldsymbol{n}_{K,L}\, \mathrm{d}\gamma(x) \bigg| \Bigg) C_6 h \\
&\qquad + \sum_{n=0}^{N_{T,k}} k \sum_{K\in\mathcal{T}} |u_K^n| \Bigg(\sum_{L\in\mathcal{N}(K)} \mathrm{m}(K|L) C_6 C_5 h^2 \Bigg),
\end{aligned}
$$

which leads to $|E_{3,m} - E_{4,m}| \to 0$ as $m \to \infty$, using (35.15), (35.34) and the Cauchy–Schwarz inequality.

In order to prove that $E_{2,m} \to T_2$ as $m \to \infty$ (which concludes the proof of Theorem 35.1), let us show that $|E_{2,m} - E_{3,m}| \to 0$ as $m \to \infty$.

We get, using (35.17) and (35.22)

$$
\begin{aligned}
E_{2,m} = &- \sum_{n=0}^{N_{T,k}} k \sum_{K|L\in\mathcal{E}_{\mathrm{int}}} \tau_{K|L} \big(u_L^n - u_K^n\big)(p_L - p_K)\varphi(x_K, nk) \\
&- \sum_{n=0}^{N_{T,k}} k \sum_{K\in\mathcal{T}} \big(\bar{u}_K^n - u_K^n\big) G_K^{(+)} \varphi(x_K, nk).
\end{aligned}
$$

This yields

$$
\begin{aligned}
E_{3,m} - E_{2,m} = &\sum_{n=0}^{N_{T,k}} k \sum_{K|L\in\mathcal{E}_{\mathrm{int}}} \tau_{K|L} \big(u_K^n - u_L^n\big)(p_L - p_K)\phi_{K,L}^n \\
&+ \sum_{n=0}^{N_{T,k}} k \sum_{K\in\mathcal{T}} \big(u_K^n - \bar{u}_K^n\big) G_K^{(+)} \phi_K^n,
\end{aligned}
\tag{35.38}
$$

where

$$
\phi_{K,L}^n = \frac{1}{\mathrm{m}(K|L)} \int_{K|L} \varphi(x, nk)\, \mathrm{d}\gamma(x) - \varphi(x_K, nk), \quad \forall K \in \mathcal{T},\ \forall L \in \mathcal{N}(K),
$$

and

$$
G_K^{(+)} \phi_K^n = \int_{\partial K\cap\partial\Omega} \varphi(x, nk) g(x)\, \mathrm{d}\gamma(x) - G_K^{(+)} \varphi(x_K, nk).
$$

We recall that, for all $x \in \partial\Omega$, $\varphi(x, nk)g^+(x) = \varphi(x, nk)g(x)$, by definition of $\mathcal{D}(\Omega^+ \times \mathbb{R}_+)$. Therefore, there exists C_7, which only depends on φ, such that $|\phi_{K,L}^n| \leqslant C_7 h$

and $G_K^{(+)}|\phi_K^n| \leqslant G_K^{(+)} C_7 h$, for all $K \in \mathcal{T}$, $L \in \mathcal{N}(K)$ and all $n \in \mathbb{N}$. Therefore, using Lemma 35.2, we get $|E_{3,m} - E_{2,m}| \leqslant C_7 h \frac{H}{\sqrt{h}}$ which yields $|E_{2,m} - E_{3,m}| \to 0$ and then $E_{2,m} \to T_2$ as $m \to \infty$. This concludes the proof of Theorem 35.1. □

REMARK 35.10. In the case of source terms, the convergence Theorem 35.1 still holds. There are some minor modifications in the proof. The definitions of $E_{2,m}$, $E_{3,m}$ and $E_{4,m}$ change. In the definition of $E_{2,m}$, the quantity $G_K^{(+)}\bar{u}_K^n - G_K^{(-)}u_K^n$ is replaced by $G_K^{(+)}\bar{u}_K^n - G_K^{(-)}u_K^n + V_K^{(+)}s_K^n - V_K^{(-)}u_K^n$. In the definition of $E_{3,m}$ one adds

$$\sum_{n=0}^{N_{T,k}} k \sum_{K \in \mathcal{T}} \left(u_K^n - s_K^n\right) \int_K v^+(x)\varphi(x, nk)\,\mathrm{d}x.$$

The quantity $E_{3,m} - E_{4,m}$ does not change and in order to prove $E_{3,m} - E_{2,m} \to 0$ it is sufficient to remark that there exists C_8, only depending on φ, such that

$$\left| \int_K \varphi(x, nk) v^+(x)\,\mathrm{d}x - V_K^{(+)}\varphi(x_K, nk) \right| \leqslant V_K^{(+)} C_8 h.$$

References

AMIEZ, G. and P.A. GREMAUD (1991), On a numerical approach to Stefan-like problems, *Numer. Math.* **59**, 71–89.

ANGERMANN, L. (1996), Finite volume schemes as non-conforming Petrov–Galerkin approximations of primal-dual mixed formulations, Report #181, Institut für Angewandte Mathematik, Universität Erlangen–Nürnberg.

ANGOT, P. (1989), Contribution à l'étude des transferts thermiques dans les systèmes complexes, application aux composants électroniques, Thesis, Université de Bordeaux 1.

AGOUZAL, A., J. BARANGER, J.-F. MAITRE and F. OUDIN (1995), Connection between finite volume and mixed finite element methods for a diffusion problem with non constant coefficients, with application to Convection Diffusion, *East-West J. Numer. Math.* **3** (4), 237–254.

ARBOGAST, T., M.F. WHEELER and I. YOTOV (1997), Mixed finite elements for elliptic problems with tensor coefficients as cell-centered finite differences, *SIAM J. Numer. Anal.* **34** (2), 828–852.

ATTHEY, D.R. (1974), A finite difference scheme for melting problems, *J. Inst. Math. Appl.* **13**, 353–366.

BANK R.E. and D.J. ROSE (1986), Error estimates for the box method, *SIAM J. Numer. Anal.* 777–790

BARANGER, J., J.-F. MAITRE and F. OUDIN (1996), Connection between finite volume and mixed finite element methods, *Modél. Math. Anal. Numér.* **30** (3–4), 444–465.

BARTH, T.J. (1994), Aspects of unstructured grids and finite volume solvers for the Euler and Navier–Stokes equations, Von Karmann Institute Lecture.

BAUGHMAN, L.A. and N.J. WALKINGTON (1993), Co-volume methods for degenerate parabolic problems, *Numer. Math.* **64**, 45–67.

BELMOUHOUB, R. (1996), Modélisation tridimensionnelle de la genèse des bassins sédimentaires, Thesis, Ecole Nationale Supérieure des Mines de Paris, 1996.

BERGER, A.E., H. BREZIS and J.C.W. ROGERS (1979), A numerical method for solving the problem $u_t - \Delta f(u) = 0$, *RAIRO Numer. Anal.* **13** (4), 297–312.

BERTSCH, M., R. KERSNER and L.A. PELETIER (1995), Positivity versus localization in degenerate diffusion equations, *Nonlinear Anal.* **9** (9), 987–1008.

BOTTA, N. and D. HEMPEL (1996), A finite volume projection method for the numerical solution of the incompressible Navier–Stokes equations on triangular grids, in: F. Benkhaldoun and R. Vilsmeier, eds., *Finite Volumes for Complex Applications, Problems and Perspectives* (Hermes, Paris) 355–363.

BREZIS, H. (1983), *Analyse Fonctionnelle: Théorie et Applications* (Masson, Paris).

BRUN, G., J.M. HÉRARD, L. LEAL DE SOUSA and M. UHLMANN (1996), Numerical modelling of turbulent compressible flows using second order models, in: F. Benkhaldoun and R. Vilsmeier, eds., *Finite Volumes for Complex Applications, Problems and Perspectives* (Hermes, Paris) 338–346.

BUFFARD, T., T. GALLOUËT and J.M. HÉRARD (1998), Un schéma simple pour les équations de Saint-Venant, *C. R. Acad. Sci. Paris Sér. I* **326**, 386–390.

CAI, Z. (1991), On the finite volume element method, *Numer. Math.* **58**, 713–735.

CAI, Z., J. MANDEL and S. MCCORMICK (1991), The finite volume element method for diffusion equations on general triangulations, *SIAM J. Numer. Anal.* **28** (2), 392–402.

CHAINAIS-HILLAIRET, C. (1996), First and second order schemes for a hyperbolic equation: Convergence and error estimate, in: F. Benkhaldoun and R. Vilsmeier, eds., *Finite Volumes for Complex Applications, Problems and Perspectives* (Hermes, Paris) 137–144.

CHAINAIS-HILLAIRET, C. (1999), Finite volume schemes for a nonlinear hyperbolic equation. Convergence towards the entropy solution and error estimate, *M2AN* **33** (1), 129–156.

CHAMPIER, S. and T. GALLOUËT (1992), Convergence d'un schéma décentré amont pour une équation hyperbolique linéaire sur un maillage triangulaire, *Modél. Math. Anal. Numér.* **26** (7), 835–853.

CHAMPIER, S., T. GALLOUËT and R. HERBIN (1993), Convergence of an upstream finite volume scheme on a triangular mesh for a nonlinear hyperbolic equation, *Numer. Math.* **66**, 139–157.

CHAVENT, G. and J. JAFFRÉ (1990), *Mathematical Models and Finite Element for Reservoir Simulation*, Stud. Math. Appl. (North-Holland, Amsterdam).

CHEVRIER, P. and H. GALLEY (1993), A Van Leer finite volume scheme for the Euler equations on unstructured meshes, *Modél. Math. Anal. Numér.* **27** (2), 183–201.

CIARLET, P.G. (1978), *The Finite Element Method for Elliptic Problems* (North-Holland, Amsterdam).

CIARLET, P.G. (1991), Basic error estimates for elliptic problems in: *Handbook of Numerical Analysis* **II** (North-Holland, Amsterdam) 17–352.

CIAVALDINI, J.F. (1975), Analyse numérique d'un problème de Stefan à deux phases par une méthode d'éléments finis, *SIAM J. Numer. Anal.* **12**, 464–488.

COCKBURN, B., F. COQUEL and P. LEFLOCH (1994), An error estimate for finite volume methods for multidimensional conservation laws, *Math. Comp.* **63** (207), 77–103.

COCKBURN, B., F. COQUEL and P. LEFLOCH (1995), Convergence of the finite volume method for multidimensional conservation laws, *SIAM J. Numer. Anal.* **32**, 687–705. s

COCKBURN, B. and P.A. GREMAUD (1996a), A priori error estimates for numerical methods for scalar conservation laws. I. The general approach, *Math. Comp.* **65**, 522–554.

COCKBURN, B. and P.A. GREMAUD (1996b), Error estimates for finite element methods for scalar conservation laws, *SIAM J. Numer. Anal.* **33** (2), 522–554.

CONWAY E. and J. SMOLLER (1966), Global solutions of the Cauchy problem for quasi-linear first-order equations in several space variables, *Comm. Pure Appl. Math.* **19**, 95–105.

COQUEL, F. and P. LEFLOCH (1996), An entropy satisfying MUSCL scheme for systems of conservation laws, *Numer. Math.* **74**, 1–33.

CORDES, C. and M. PUTTI (1998), Finite element approximation of the diffusion operator on tetrahedra, *SIAM J. Sci. Comput.* **19** (4), 1154–1168.

COUDIÈRE, Y., T. GALLOUËT and R. HERBIN (1998), Discrete Sobolev inequlities and L^p error estimates for approximate finite volume solutions of convection diffusion equations, submitted.

COUDIÈRE, Y., J.P. VILA and P. VILLEDIEU (1996), Convergence of a finite volume scheme for a diffusion problem, in: F. Benkhaldoun and R. Vilsmeier, eds., *Finite Volumes for Complex Applications, Problems and Perspectives* (Hermes, Paris) 161–168.

COUDIÈRE, Y., J.P. VILA and P. VILLEDIEU (1999), Convergence rate of a finite volume scheme for a two-dimensional convection diffusion problem, *M2AN* **33** (3), 493–516.

COURBET, B. and J.P. CROISILLE (1998), Finite-volume box-schemes on triangular meshes, *M2AN* **32** (5), 631–649.

CRANDALL, M.G. and A. MAJDA (1980), Monotone difference approximations for scalar conservation laws, *Math. Comp.* **34** (149), 1–21.

DAHLQUIST, G. and A. BJÖRCK (1974), *Numerical Methods*, Prentice-Hall Series in Automatic Computation.

DEIMLING, K. (1980), *Nonlinear Functional Analysis* (Springer, New York).

DIPERNA, R. (1985), Measure-valued solutions to conservation laws, *Arch. Rat. Mech. Anal.* **88**, 223–270.

DUBOIS, F. and P. LEFLOCH (1988), Boundary conditions for nonlinear hyperbolic systems of conservation laws, *J. Differential Equations* **71**, 93–122.

EYMARD, R.(1992), Application à la simulation de réservoir des méthodes volumes-éléments finis; problèmes de mise en oeuvre, Cours CEA-EDF-INRIA.

EYMARD, R. and T. GALLOUËT (1993), Convergence d'un schéma de type eléments finis – volumes finis pour un système couplé elliptique–hyperbolique, *Modél. Math. Anal. Numér.* **27** (7), 843–861.

EYMARD, R. and T. GALLOUËT (1991), Traitement des changements de phase dans la modélisation de gisements pétroliers, *J. Numér. de Besançon*.

EYMARD, R., T. GALLOUËT, M. GHILANI and R. HERBIN (1998), Error estimates for the approximate solutions of a nonlinear hyperbolic equation given by finite volume schemes, *IMA J. Numer. Anal.* **18**, 563–594.

EYMARD, R., T. GALLOUËT and R. HERBIN (1995), Existence and uniqueness of the entropy solution to a nonlinear hyperbolic equation, *Chinese Ann. Math. Ser. B* **16** (1), 1–14.

EYMARD, R., T. GALLOUËT and R. HERBIN (1999), Convergence of finite volume approximations to the solutions of semilinear convection diffusion reaction equations, *Numer. Math.* **82**, 91–116.

EYMARD, R. and M. GHILANI (1994), Convergence d'un schéma implicite de type éléments finis-volumes finis pour un système formé d'une équation elliptique et d'une équation hyperbolique, *C. R. Acad. Sci. Paris Ser. I* **319**, 1095–1100.

FAILLE, I. (1992a), Modélisation bidimensionnelle de la genèse et la migration des hydrocarbures dans un bassin sédimentaire, Thesis, Université de Grenoble.

FAILLE, I. (1992b), A control volume method to solve an elliptic equation on a two-dimensional irregular meshing, *Comput. Methods Appl. Mech. Engrg.* **100**, 275–290.

FAILLE, I. and E. HEINTZÉ (1999), A rough finite volume scheme for modeling two-phase flow in a pipeline, *Computers and Fluids* **28**, 213–241.

FEISTAUER, M., J. FELCMAN and M. LUKACOVA-MEDVIDOVA (1995), Combined finite element-finite volume solution of compressible flow, *J. Comput. Appl. Math.* **63**, 179–199.

FEISTAUER, M., J. FELCMAN and M. LUKACOVA-MEDVIDOVA (1997), On the convergence of a combined finite volume–finite element method for nonlinear convection–diffusion problems, *Numer. Methods Partial Differential Equations* **13**, 163–190.

FERNANDEZ, G. (1989), Simulation numérique d'écoulements réactifs à petit nombre de Mach, Thèse de Doctorat, Université de Nice.

FEZOUI, L., S. LANTERI, B. LARROUTUROU and C. OLIVIER (1989), Résolution numérique des équations de Navier–Stokes pour un Fluide Compressible en Maillage Triangulaire, INRIA Rept. 1033.

FIARD, J.M. (1994), Modélisation mathématique et simulation numérique des piles au gaz naturel à oxyde solide, thèse de Doctorat, Université de Chambéry.

FIARD, J.M. and R. HERBIN (1994), Comparison between finite volume finite element methods for the numerical simulation of an elliptic problem arising in electrochemical engineering, *Comput. Methods Appl. Mech. Engrg.* **115**, 315–338.

FORSYTH, P.A. (1989), A control volume finite element method for local mesh refinement, SPE 18415, 85–96.

FORSYTH, P.A. (1991), A control volume finite element approach to NAPL groundwater contamination, *SIAM J. Sci. Statist. Comput.* **12** (5), 1029–1057.

FORSYTH, P.A. and P.H. SAMMON (1988), Quadratic convergence for cell-centered grids, *Appl. Numer. Math.* **4**, 377–394.

GALLOUËT, T. (1996), Rough schemes for systems for complex hyperbolic systems, in: F. Benkhaldoun and R. Vilsmeier, eds., *Finite Volumes for Complex Applications, Problems and Perspectives* (Hermes, Paris) 10–21.

GALLOUËT, T. and R. HERBIN (1994), A uniqueness result for measure valued solutions of a nonlinear hyperbolic equations, *Int. Diff. Int. Equations* **6** (6), 1383–1394.

GALLOUËT, T., R. HERBIN and M.H. VIGNAL, Error estimate for the approximate finite volume solutions of convection diffusion equations with Dirichlet, Neumann or Fourier boundary conditions, to appear in *SIAM J. Numer. Anal.*

GIRAULT, V. and P.A. RAVIART (1986), *Finite Element Approximation of the Navier–Stokes Equations* (Springer-Verlag).

GHIDAGLIA, J.M., A. KUMBARO and G. LE COQ (1996), Une méthode "volumes finis" à flux caractéristiques pour la résolution numérique de lois de conservation, *C. R. Acad. Sci. Paris Sér. I* **332**, 981–988.

GODLEWSKI, E. and P.A. RAVIART (1991), *Hyperbolic Systems of Conservation Laws* (Ellipses).

GODLEWSKI, E. and P.A. RAVIART (1996), *Numerical Approximation of Hyperbolic Systems of Conservation Laws*, Appl. Math. Sci. 118 (Springer, New York).

GODUNOV, S. (1976), *Résolution Numérique des Problèmes Multidimensionnels de la Dynamique des Gaz* (Editions de Moscou).

GUEDDA, M., D. HILHORST and M.A. PELETIER (1997), Disappearing interfaces in nonlinear diffusion, *Adv. Math. Sci. Appl.* **7**, 695–710.

GUO, W. and M. STYNES (1997), An analysis of a cell-vertex finite volume method for a parabolic convection–diffusion problem, *Math. Comp.* **66** (217), 105–124.

HARTEN, A. (1983), On a class of high resolution total-variation-stable finite-difference schemes, *J. Comput. Phys.* **49**, 357–393.

HARTEN, A., P.D. LAX and B. VAN LEER (1983), On upstream differencing and Godunov-type schemes for hyperbolic conservations laws, *SIAM Rev.* **25**, 35–61.

HARTEN, A., J.M. HYMAN and P.D. LAX (1976), On finite difference approximations and entropy conditions, *Comm. Pure Appl. Math.* **29**, 297–322.

HEINRICH, B. (1986), *Finite Diference Methods on Irregular Networks*, I.S.N.M. 82 (Birkhauser).

HERBIN, R. (1995), An error estimate for a finite volume scheme for a diffusion–convection problem on a triangular mesh, *Numer. Methods Partial Differential Equations* **11**, 165–173.

HERBIN, R. (1996), Finite volume methods for diffusion convection equations on general meshes, in: F. Benkhaldoun and R. Vilsmeier, eds., *Finite Volumes for Complex Applications, Problems and Perspectives* (Hermes) 153–160.

HERBIN, R. and O. LABERGERIE (1997), Finite volume schemes for elliptic and elliptic-hyperbolic problems on triangular meshes, *Comput. Methods Appl. Mech. Engrg.* **147**, 85–103.

HERBIN, R. and E. MARCHAND (1997), Numerical approximation of a nonlinear problem with a Signorini condition, in: *Third IMACS International Symposium on Iterative Methods in Scientific Computation*, Wyoming, USA, July 97.

KAMENOMOSTSKAJA, S.L. (1995), On the Stefan problem, *Mat. Sb.* **53**, 489–514 (1961 in Russian).

KELLER, H.B. (1971), A new difference sheme for parabolic problems, in: B. Hubbard ed., *Numerical Solutions of Partial Differential Equations* **II** (Academic Press, New York) 327–350.

KRÖNER, D. (1997), *Numerical Schemes for Conservation Laws in Two Dimensions*, Wiley–Teubner Series Advances in Numerical Mathematics (Wiley, Ltd., Chichester; B.G. Teubner, Stuttgart).

KRÖNER, D. and M. ROKYTA (1994), Convergence of upwind finite volume schemes on unstructured grids for scalar conservation laws in two dimensions, *SIAM J. Numer. Anal.* **31** (2), 324–343.

KRÖNER, D., S. NOELLE and M. ROKYTA (1995), Convergence of higher order upwind finite volume schemes on unstructured grids for scalar conservation laws in several space dimensions, *Numer. Math.* **71**, 527–560.

KRUSHKOV, S.N. (1970), First Order quasilinear equations with several space variables, *Math. USSR Sb.* **10**, 217–243.

KUMBARO, A. (1992), Modélisation, analyse mathématique et numérique des modèles bi-fluides d'écoulement diphasique, Thesis, Université Paris XI Orsay.

KUZNETSOV (1976), Accuracy of some approximate methods for computing the weak solutions of a first-order quasi-linear equation, *USSR Comput. Math. and Math. Phys.* **16**, 105–119.

LADYŽENSKAJA, O.A., V.A. SOLONNIKOV and N.N. URAL'CEVA (1968), *Linear and Quasilinear Equations of Parabolic Type*, Transl. Math. Monographs 23.

LAZAROV, R.D. and I.D. MISHEV (1996), Finite volume methods for reaction diffusion problems, in: F. Benkhaldoun and R. Vilsmeier, eds., *Finite Volumes for Complex Applications, Problems and Perspectives* (Hermes, Paris) 233–240.

LAZAROV, R.D., I.D. MISHEV and P.S VASSILEVSKI (1996), Finite volume methods for convection–diffusion problems, *SIAM J. Numer. Anal.* **33**, 31–55.

LEVEQUE, R.J. (1990), *Numerical Methods for Conservation Laws* (Birkhäuser).

LIONS, P.L. (1996), *Mathematical Topics in Fluid Mechanics*; Vol. 1: *Incompressible Models*, Oxford Lecture Series in Math. and Its Appl. 3 (Oxford Univ. Press).

MACKENZIE, J.A. and K.W. MORTON (1992), Finite volume solutions of convection–diffusion test problems, *Math. Comp.* **60** (201), 189–220.

MANTEUFFEL, T. and A.B. WHITE (1986), The numerical solution of second order boundary value problem on non uniform meshes, *Math. Comp.* **47**, 511–536.

MASELLA, J.M. (1997), Quelques méthodes numériques pour les écoulements diphasiques bi-fluide en conduites petrolières, Thesis, Université Paris VI.

MASELLA, J.M., I. FAILLE and T. GALLOUËT (1996), On a rough Godunov scheme, accepted for publication in *Intl. J. Comput. Fluid Dynamics*.

MEIRMANOV, A.M. (1992), *The Stefan Problem* (Walter de Gruyter, New York).

MEYER, G.H. (1973), Multidimensional Stefan problems, *SIAM J. Numer. Anal.* **10**, 522–538.

MISHEV, I.D. (1998), Finite volume methods on Voronoï meshes, *Numer. Methods Partial Differential Equations* **14** (2), 193–212.

MORTON, K.W. (1996), *Numerical Solutions of Convection–Diffusion problems* (Chapman & Hall, London).

MORTON, K.W. and E. SÜLI (1991), Finite volume methods and their analysis, *IMA J. Numer. Anal.* **11**, 241–260.

MORTON, K.W., M. STYNES and E. SÜLI (1997), Analysis of a cell-vertex finite volume method for a convection–diffusion problems, *Math. Comp.* **66** (220), 1389–1406.

NEČAS, J. (1967), *Les Méthodes Directes en Théorie des Équations Elliptiques* (Masson, Paris).

NICOLAIDES, R.A. (1992), Analysis and convergence of the MAC scheme I. The linear problem, *SIAM J. Numer. Anal.* **29** (6), 1579–1591.

NICOLAIDES, R.A. and X. WU (1996), Analysis and convergence of the MAC scheme II. Navier–Stokes equations, *Math. Comp.* **65** (213), 29–44.

NOËLLE, S. (1996), A note on entropy inequalities and error estimates for higher-order accurate finite volume schemes on irregular grids, *Math. Comp.* **65**, 1155–1163.

ODEN, J.T. (1991), Finite elements: An introduction, in: *Handbook of Numerical Analysis* **II** (North-Holland, Amsterdam) 3–15.

OLEINIK, O.A. (1960), A method of solution of the general Stefan problem, *Soviet Math. Dokl.* **1**, 1350–1354.

OLEINIK, O.A. (1963), On discontinuous solutions of nonlinear differential equations, *Amer. Math. Soc. Transl. Ser. 2* **26**, 95–172.

OSHER, S. (1984), Riemann solvers, the entropy condition, and difference approximations, *SIAM J. Numer. Anal.* **21**, 217–235.

PATANKAR, S.V. (1980), *Numerical Heat Transfer and Fluid Flow*, Series in Computational Methods in Mechanics and Thermal Sciences, Minkowycz and Sparrow, eds. (McGraw-Hill).

PFERTZEL, A. (1987), Sur quelques schémas numériques pour la résolution des écoulements diphasiques en milieux poreux, Thesis, Université de Paris 6.

ROBERTS, J.E. and J.M. Thomas (1991), Mixed and hybrids methods, in: *Handbook of Numerical Analysis* **II** (North-Holland, Amsterdam) 523–640.

ROE, P.L. (1980), The use of Riemann problem in finite difference schemes, *Lectures Notes in Phys.* **141**, 354–359.

ROE, P.L. (1981), Approximate Riemann solvers, parameter vectors, and difference schemes, *J. Comput. Phys.* **43**, 357–372.

RUDIN, W. (1987), *Real and Complex Analysis* (McGraw-Hill).

SAMARSKI, A.A. (1965), On monotone difference schemes for elliptic and parabolic equations in the case of a nonselfadjoint elliptic operator, *Zh. Vychisl. Mat. i. Mat. Fiz.* **5**, 548–551 (Russian).

SAMARSKII, A.A. (1971), *Introduction to the Theory of Difference Schemes* (Nauka, Moscow) (Russian).

SAMARSKII, A.A., R.D. LAZAROV and V.L. MAKAROV (1987), *Difference Schemes for Differential Equations Having Generalized Solutions* (Vysshaya Shkola, Moscow) (Russian).

SANDERS, R. (1983), On the convergence of monotone finite difference schemes with variable spatial differencing, *Math. Comp.* **40** (161), 91–106.

SELMIN, V. (1993), The node-centered finite volume approach: Bridge between finite differences and finite elements, *Comput. Methods Appl. Mech. Engrg.* **102**, 107–138.

SERRE, D. (1996), *Systèmes de Lois de Conservation* (Diderot).

SHASHKOV, M. (1996), *Conservative Finite-Difference Methods on General Grids* (CRC Press, New York).

SOD, G.A. (1978), A survey of several finite difference methods for systems of nonlinear hyperbolic conservation laws, *J. Comput. Phys.* **27**, 1–31.

SONIER, F. and R. EYMARD (1993), Mathematical and numerical properties of control-volume finite-element scheme for reservoir simulation, Paper SPE 25267, 12th Symposium on Reservoir Simulation, New Orleans.

SÜLI, E. (1992), The accuracy of cell vertex finite volume methods on quadrilateral meshes, *Math. Comp.* **59** (200), 359–382.

SZEPESSY, A. (1989), An existence result for scalar conservation laws using measure valued solutions, *Comm. Partial Differential Equations* **14** (10), 1329–1350.

TEMAM, R. (1977), *Navier–Stokes Equations* (North-Holland, Amsterdam).

TICHONOV, A.N. and A.A. SAMARSKII (1962), Homogeneous difference schemes on nonuniform nets, *Zh. Vychisl. Mat. i. Mat. Fiz.* **2**, 812–832 (Russian).

THOMAS, J.M. (1977), Sur l'analyse numérique des méthodes d'éléments finis hybrides et mixtes, Thesis, Université Pierre et Marie Curie.

THOMÉE, V. (1991), Finite differences for linear parabolic equations, in: *Handbook of Numerical Analysis* **I** (North-Holland, Amsterdam) 5–196.

TURKEL, E. (1987), Preconditioned methods for solving the imcompressible and low speed compressible equations, *J. Comput. Phys.* **72**, 277–298.

VAN LEER, B. (1974), Towards the ultimate conservative difference scheme, II. Monotonicity and conservation combined in a second-order scheme, *J. Comput. Phys.* **14**, 361–370.

VAN LEER, B. (1977), Towards the ultimate conservative difference scheme, IV. A new approach to numerical convection, *J. Comput. Phys.* **23**, 276–299.

VAN LEER, B. (1979), Towards the ultimate conservative difference scheme, V, *J. Comput. Phys.* **32**, 101–136.

VANSELOW, R. (1996), Relations between FEM and FVM, in: F. Benkhaldoun and R. Vilsmeier, eds., *Finite Volumes for Complex Applications, Problems and Perspectives* (Hermes, Paris) 217–223.

VASSILESKI, P.S., S.I. PETROVA and R.D. LAZAROV (1992), Finite difference schemes on triangular cell-centered grids with local refinement, *SIAM J. Sci. Statist. Comput.* **13** (6), 1287–1313.

VIGNAL, M.H. (1996), Convergence of a finite volume scheme for a system of an elliptic equation and a hyperbolic equation, *Modél. Math. Anal. Numér.* **30** (7), 841–872.

VIGNAL, M.H. (1996), Convergence of finite volumes schemes for an elliptic hyperbolic system with boundary conditions, in: F. Benkhaldoun and R. Vilsmeier, eds., *Finite Volumes for Complex Applications, Problems and Perspectives* (Hermes, Paris) 145–152.

VIGNAL, M.H. and S. VERDIÈRE (1998), Numerical and theoretical study of a dual mesh method using finite volume schemes for two phase flow problems in porous media, *Numer. Math.* **80** (4), 601–639.

VILA, J.P. (1986), Sur la théorie et l'approximation numérique de problèmes hyperboliques non linéaires. Application aux équations de saint Venant et à la modélisation des avalanches de neige dense, Thesis, Université Paris VI.

VILA, J.P. (1994), Convergence and error estimate in finite volume schemes for general multidimensional conservation laws, I. Explicit monotone schemes, *Modél. Math. Anal. Numér.* **28** (3), 267–285.

VOL'PERT, A.I. (1967), The spaces *BV* and quasilinear equations, *Math. USSR Sb.* **2**, 225–267.

WEISER, A. and M.F. WHEELER (1988), On convergence of block-centered finite-differences for elliptic problems, *SIAM J. Numer. Anal.* **25**, 351–375.

Subject Index